Lecture Notes in Computer Science 4221

Lecture Notes in Computer Science 4221

Licheng Jiao Lipo Wang Xinbo Gao
Jing Liu Feng Wu (Eds.)

Advances in Natural Computation

Second International Conference, ICNC 2006
Xi'an, China, September 24-28, 2006
Proceedings, Part I

Volume Editors

Licheng Jiao
Xidian University, Xi'an 710071, China
E-mail: lchjiao@mail.xidian.edu.cn

Lipo Wang
Nanyang Technological University, Singapore
E-mail: elpwang@ntu.edu.sg

Xinbo Gao
Xidian University, Xi'an, 710071, China
E-mail: xbgao@mail.xidian.edu.cn

Jing Liu
Xidian University, Xi'ian 710071, China
E-mail: neouma@mail.xidian.edu.cn

Feng Wu
Microsoft Research Asia, 100080 Beijing, China
E-mail: fengwu@microsoft.com

Library of Congress Control Number: 20069322971

CR Subject Classification (1998): F.1, F.2, I.2, G.2, I.4, I.5, J.3, J.4

LNCS Sublibrary: SL 1 – Theoretical Computer Science and General Issues

ISSN 0302-9743
ISBN-10 3-540-45901-4 Springer Berlin Heidelberg New York
ISBN-13 978-3-540-45901-9 Springer Berlin Heidelberg New York

Springer is a part of Springer Science+Business Media

springer.com

Printed in Germany

Typesetting: Camera-ready by author, data conversion by Scientific Publishing Services, Chennai, India
Printed on acid-free paper SPIN: 11881070 06/3142 5 4 3 2 1 0

Preface

This book and its sister volumes, i.e., LNCS volumes 4221 and 4222, constitute the proceedings of the 2nd International Conference on Natural Computation (ICNC 2006), jointly held with the 3rd International Conference on Fuzzy Systems and Knowledge Discovery (FSKD 2006, LNAI volume 4223) on 24-28 September 2006 in Xi'an, Shaanxi, China. In its budding run, ICNC 2006 successfully attracted 1915 submissions from 35 countries/regions (the joint ICNC-FSKD 2006 event received 3189 submissions). After rigorous reviews, 254 high-quality papers, i.e., 168 long papers and 86 short papers, were included in the ICNC 2006 proceedings, representing an acceptance rate of 13.3%.

ICNC-FSKD 2006 featured the most up-to-date research results in computational algorithms inspired from nature, including biological, ecological, and physical systems. It is an exciting and emerging interdisciplinary area in which a wide range of techniques and methods are being studied for dealing with large, complex, and dynamic problems. The joint conferences also promoted cross-fertilization over these exciting and yet closely-related areas, which had a significant impact on the advancement of these important technologies. Specific areas included neural computation, quantum computation, evolutionary computation, DNA computation, fuzzy computation, granular computation, artificial life, etc., with innovative applications to knowledge discovery, finance, operations research, and more. In addition to the large number of submitted papers, we were blessed with the presence of six renowned keynote speakers.

On behalf of the Organizing Committee, we thank Xidian University for sponsorship, and the National Natural Science Foundation of China, the International Neural Network Society, the Asia-Pacific Neural Network Assembly, the IEEE Circuits and Systems Society, the IEEE Computational Intelligence Society, the IEEE Computational Intelligence Singapore Chapter, and the Chinese Association for Artificial Intelligence for technical co-sponsorship. We thank the members of the Organizing Committee, the Advisory Board, and the Program Committee for their hard work in the past 12 months. We wish to express our heartfelt appreciation to the keynote speakers, session chairs, reviewers, and student helpers. Our special thanks go to the publisher, Springer, for publishing the ICNC 2006 proceedings as two volumes of the Lecture Notes in Computer Science series (and the FSKD 2006 proceedings as one volume of the Lecture Notes in Artificial Intelligence series). Finally, we thank all the authors and participants for their great contributions that made this conference possible and all the hard work worthwhile.

September 2006

Lipo Wang
Licheng Jiao

Organization

ICNC 2006 was organized by Xidian University and technically co-sponsored by the National Natural Science Foundation of China, the International Neural Network Society, the Asia-Pacific Neural Network Assembly, the IEEE Circuits and Systems Society, the IEEE Computational Intelligence Society, the IEEE Computational Intelligence Singapore Chapter, and the Chinese Association for Artificial Intelligence.

Organizing Committee

Honorary Conference Chairs:	Shun-ichi Amari (RIKEN BSI, Japan)
	Xin Yao (University of Birmingham, UK)
General Co-chairs:	Lipo Wang (Nanyang Technological University, Singapore)
	Licheng Jiao (Xidian University, China)
Program Committee Chairs:	Xinbo Gao (Xidian University, China)
	Feng Wu (Microsoft Research Asia, China)
Local Arrangement Chairs:	Yuanyuan Zuo (Xidian University, China)
	Xiaowei Shi (Xidian University, China)
Proceedings Chair:	Jing Liu (Xidian University, China)
Publicity Chair:	Yuping Wang (Xidian University, China)
Sponsorship Chair:	Yongchang Jiao (Xidian University, China)
Secretaries:	Bin Lu (Xidian University, China)
	Tiantian Su (Xidian University, China)
Webmasters:	Yinfeng Li (Xidian University, China)
	Maoguo Gong (Xidian University, China)

Advisory Board

Zheng Bao	Xidian University, China
Zixing Cai	Central South University, China
Guoliang Chen	University of Science and Technology of China, China
Huowang Chen	National University of Defense Technology, China
David Corne	The University of Exeter, UK
Dipankar Dasgupta	University of Memphis, USA
Kalyanmoy Deb	Indian Institute of Technology Kanpur, India
Baoyan Duan	Xidian University, China
Kunihiko Fukushima	Tokyo University of Technology, Japan
Tom Gedeon	The Australian National University, Australia

Aike Guo	Chinese Academy of Science, China
Yao Hao	Xidian University, China
Zhenya He	Southeastern University, China
Fan Jin	Southwest Jiaotong University, China
Yaochu Jin	Honda Research Institute Europe, Germany
Janusz Kacprzyk	Polish Academy of Sciences, Poland
Lishan Kang	China University of Geosciences, China
Nikola Kasabov	Auckland University of Technology, New Zealand
John A. Keane	The University of Manchester, UK
Soo-Young Lee	KAIST, Korea
Yanda Li	Tsinghua University, China
Zhiyong Liu	National Natural Science Foundation of China, China
Erkki Oja	Helsinki University of Technology, Finland
Nikhil R. Pal	Indian Statistical Institute, India
Yunhe Pan	Zhe Jiang University, China
Jose Principe	University of Florida, USA
Witold Pedrycz	University of Alberta, Canada
Marc Schoenauer	University of Paris Sud, France
Zhongzhi Shi	Chinese Academy of Science, China
Harold Szu	Office of Naval Research, USA
Shiro Usui	RIKEN BSI, Japan
Shoujue Wang	Chinese Academy of Science, China
Xindong Wu	University of Vermont, USA
Lei Xu	Chinese University of Hong Kong, HK
Bo Zhang	Tsinghua University, China
Nanning Zheng	Xi'an Jiaotong University, China
Yixin Zhong	University of Posts & Telecommunications, China
Syozo Yasui	Kyushu Institute of Technology, Japan
Jacek M. Zurada	University of Louisville, USA

Program Committee

Shigeo Abe	Kobe University, Japan
Davide Anguita	University of Trento, Italy
Abdesselam Bouzerdoum	University of Wollongong, Australia
Laiwan Chan	The Chinese University of Hong Kong, HK
Li Chen	Northwest University, China
Guanrong Chen	City University of Hong Kong, HK
Shu-Heng Chen	National Chengchi University, Taiwan
Tianping Chen	Fudan University, China
YanQiu Chen	Fudan University, China
Vladimir Cherkassky	University of Minnesota, USA

Sung-Bae Cho	Yonsei University, Korea
Sungzoon Cho	Seoul National University, Korea
Tommy W.S. Chow	City University of Hong Kong, China
Vic Ciesielski	RMIT, Australia
Keshav Dahal	University of Bradford, UK
L.N. de Castro	Catholic University of Santos, Brazil
Emilio Del-Moral-Hernandez	University of Sao Paulo, Brazil
Andries Engelbrecht	University of Pretoria, South Africa
Tomoki Fukai	Tamagawa University, Japan
Lance Fung	Murdoch University, Australia
Takeshi Furuhashi	Nagoya University, Japan
Hiroshi Furutani	University of Miyazaki, Japan
John Q. Gan	The University of Essex, UK
Wen Gao	The Chinese Academy of Science, China
Peter Geczy	AIST, Japan
Zengguang Hou	University of Saskatchewan, Canada
Jiwu Huang	Sun Yat-Sen University, China
Masumi Ishikawa	Kyushu Institute of Technology, Japan
Yongchang Jiao	Xidian University, China
Robert John	De Montfort University, UK
Mohamed Kamel	University of Waterloo, Canada
Yoshiki Kashimori	University of Electro-Communications, Japan
Samuel Kaski	Helsinki University of Technology, Finland
Andy Keane	University of Southampton, UK
Graham Kendall	The University of Nottingham, UK
Jong-Hwan Kim	KAIST, Korea
JungWon Kim	University College London, UK
Natalio Krasnogor	University of Nottingham, UK
Vincent C.S. Lee	Monash University, Australia
Stan Z. Li	Chinese Academy of Science, China
Yangmin Li	University of Macau, Macau
Xiaofeng Liao	Chongqing University, China
Derong Liu	University of Illinois at Chicago, USA
Ding Liu	Xi'an University of Technology, China
Jing Liu	Xidian University, China
Ke Liu	National Natural Science Foundation of China, China
Baoliang Lu	Shanghai Jiao Tong University, China
Frederic Maire	Queensland University of Technology, Australia
Jacek Mandziuk	Warsaw University of Technology, Poland
Satoshi Matsuda	Nihon University, Japan
Masakazu Matsugu	Canon Research Center, Japan
Bob McKay	University of New South Wales, Australia
Ali A. Minai	University of Cincinnati, USA
Hiromi Miyajima	Kagoshima University, Japan
Hongwei Mo Harbin	Engineering University, China

Mark Neal	University of Wales, Aberystwyth, UK
Pedja Neskovic	Brown University, USA
Richard Neville	The University of Manchester, UK
Tohru Nitta	National Institute of Advanced Industrial Science and Technology, Japan
Yusuke Nojima	Osaka Prefecture University, Japan
Takashi Omori	Hokkaido University, Japan
Yew Soon Ong	Nanyang Technological University, Singapore
M. Palaniswami	The University of Melbourne, Australia
Andrew P. Paplinski	Monash University, Australia
Asim Roy	University of Arizona, USA
Bernhard Sendhoff	Honda Research Centre Europe, Germany
Leslie Smith	University of Stirling, UK
Andy Song	RMIT, Australia
Lambert Spaanenburg	Lund University, Sweden
Changyin Sun	Southeast University, China
Mingui Sun	University of Pittsburgh, USA
Johan Suykens	KULeuven, Belgium
Kay Chen Tan	National University of Singapore, Singapore
Jonathan Timmis	University of York, UK
Seow Kiam Tian	Nanyang Technological University, Singapore
Peter Tino	The University of Birmingham, UK
Kar-Ann Toh	Institute of Infocomm Research, Singapore
Yasuhiro Tsujimura	Nippon Institute of Technology, Japan
Ganesh Kumar Venayagamoorthy	University of Missouri-Rolla, USA
Ray Walshe	Dublin City University, Ireland
Lei Wang	Xi'an University of Technology, China
Xiaofan Wang	Shanghai Jiaotong University, China
Xufa Wang	University of Science and Technology of China, China
Yuping Wang	Xidian University, China
Sumio Watanabe	Tokyo Institute of Technology, Japan
Gang Wei	South China University of Technology, China
Stefan Wermter	University of Sunderland, UK
Kok Wai Wong	Murdoch University, Australia
Feng Wu	Microsoft Research Asia, China
Xihong Wu	Peking University, China
Zongben Xu	Xi'an Jiaotong University, China
Ron Yang	University of Exeter, UK
Li Yao	Beijing Normal University, China
Daniel Yeung	The Hong Kong Polytechnic University, HK
Ali M.S. Zalzala	Heriot-Watt University, UK
Hongbin Zha	Peking University, China
Liming Zhang	Fudan University, China
Qingfu Zhang	The University of Essex, UK

Wenxiu Zhang	Xi'an Jiaotong University, China
Yanning Zhang	Northwestern Polytechnical University, China
Yi Zhang	University of Electronic Science and Technology of China, China
Zhaotian Zhang	National Natural Science Foundation of China, China
Liang Zhao	University of Sao Paulo, Brazil
Mingsheng Zhao	Tsinghua University, China
Qiangfu Zhao	University of Aizu, Japan

Reviewers

A. Attila ISLIER
Abdesselam BOUZERDOUM
Adel AZAR
Ah-Kat TAN
Aifeng REN
Aifeng REN
Ailun LIU
Aimin HOU
Aizhen LIU
Ales KEPRT
Ali Bekir YILDIZ
Alparslan TURANBOY
Anan FANG
Andreas HERZOG
Andrew TEOH
Andrew P. PAPLINSKI
Andries ENGELBRECHT
Andy SONG
Anni CAI
Ariel GOMEZ
Arumugam S.
Ay-Hwa A. LIOU
Bang-Hua YANG
Baolong GUO
Bei-Ping HOU
Bekir CAKIR
Ben-Shun YI
Bharat BHASKER
Bin XU
Bin JIAO
Bin LI
Bing HAN
Binghai ZHOU
Bo FU
Bob MCKAY
Bohdan MACUKOW
Bo-Qin FENG
Brijesh VERMA
Caihong MU
Ce FAN
Changyin SUN
Changzhen HU
Chao DENG
Chaojian SHI
Chaowan YIN
Chen YONG
Cheng WANG
Chengxian XU
Cheng-Yuan CHANG
Cheol-Hong MOON
Chi XIE
Ching-Hung LEE
Chong FU
Chonghui GUO
Chong-Zhao HAN
Chor Min TAN
Chu WU
Chuang GUO
Chuanhan LIU
Chun JIN
Chun CHEN
Chung-Li TSENG
Chunshien LI
Cong-Kha PHAM
Cuiqin HOU
Cunchen GAO

Daehyeon CHO
Dat TRAN
Davide ANGUITA
De XU
Deqin YAN
Dewu WANG
Dexi ZHANG
Deyun CHEN
Diangang WANG
Dong LIU
Dong Hwa KIM
Dongbo ZHANG
Dongfeng HAN
Donghu NIE
Dong-Min WOO
Du-Yun BI
Emilio DEL-MORAL-HERNANDEZ
En-Min FENG
Ergun ERASLAN
Euntai KIM
Fajun ZHANG
Fang LIU
Fangshi WANG
Fan-Hua YU
Fei HAO
Fei GAO
Feng SHI
Feng XUE
Feng DING
Feng CHEN
Feng JIAO
Feng GAO
Fenlin LIU
Fu-Ming LI
Gabriel CIOBANU
Gang WANG
Gang CHEN
Gaofeng WANG
Gaoping WANG
Gary YEN
Gexiang ZHANG
Golayoglu Fatullayev AFET
Graham KENDALL
Guang REN
Guang TIAN
Guang LI
Guangming SHI
Guangqiang LI
Guang-Qiu HUANG
Guangrui WEN
Guang-Zhao CUI
Guanjun WANG
Guanlong CHEN
Guanzheng TAN
Gui-Cheng WANG
Guixi LIU
Guojun ZHANG
Guowei YANG
Guoyin WANG
Guo-Zheng LI
Gurvinder BAICHER
Gwi-Tae PARK
Hai TAO
Hai-Bin DUAN
Haifeng DU
Haiqi ZHENG
Haixian WANG
Haixiang GUO
Haiyan JIN
Hajime NOBUHARA
Hanjun JIN
Hao WANG
Haoran ZHANG
Haoyong CHEN
He JIANG
Hengqing TONG
Hiroshi FURUTANI
Hong JIN
Hong ZHANG
Hong LIU
Hong Jie YU
Hongan WANG
Hongbin DONG
Hongbing JI
Hongcai ZHANG
Honghua SHI
Hongsheng SU
Hongwei MO
Hongwei LI
Hongwei SI

Hongwei HUO
Hongxin ZHANG
Hongyu LI
Hongzhang JIN
Hua-An ZHAO
Huaxiang LU
Hua-Xiang WANG
Huayong LIU
Hui YIN
Hui LI
Hui WANG
Huizhong YANG
Hyun YOE
Hyun Chan CHO
Hyun-Cheol JEONG
Ihn-Han BAE
Ilhong SUH
In-Chan CHOI
I-Shyan HWANG
Ivan Nunes Da SILVA
Jae Hung YOO
Jae Yong SEO
Jae-Jeong HWANG
Jae-Wan LEE
Jea Soo KIM
Jia LIU
Jiafan ZHANG
Jian YU
Jian SHI
Jian CHENG
Jian XIAO
Jianbin SONG
Jiang CUI
Jiangang LU
Jianguo JIANG
Jianhua PENG
Jianjun WANG
Jianling WANG
Jian-Sheng QIAN
Jianwei YIN
Jianwu DANG
Jianyuan JIA
Jiating LUO
Jidong SUO
Jie LI
Jie HU
Jie WANG
Jie LI
Jih-Chang HSIEH
Jih-Fu TU
Jih-Gau JUANG
Jili TAO
Jil-Lin LIU
Jin YANG
Jinchao LI
Jinfeng YANG
Jing LIU
Jing-Min WANG
Jingwei LIU
Jingxin DONG
Jin-Ho KIM
Jinhui ZHANG
Jinling ZHANG
Jinping LI
Jintang YANG
Jin-Young KIM
Jiqing QIU
Jiquan SHEN
Ji-Song KOU
Jiu-Chao FENG
Jiulong ZHANG
Jiuying DENG
Jiyang DONG
Jiyi WANG
Johan SUYKENS
John Q GAN
Jong-Min KIM
Joong-Hwan BAEK
Jorge CASILLAS
Jose SEIXAS
Jr-Syu YANG
Ju Cheng YANG
Ju Han KIM
Juan LIU
Jumin ZHAO
Jun GAO
Jun YANG
Jun CAO
Jun JING
Jun MENG

Jun-An LU
Jung-Hyun YUN
Jungsik LEE
Junguo SUN
Junping ZHANG
Jun-Seok LIM
Jun-Wei LU
Junyi SHEN
Junying ZHANG
Kay Chen TAN
Kay-Soon LOW
Ke LU
Kefeng FAN
Kenneth REVETT
Keun-Sang PARK
Khamron SUNAT
Kok Wai WONG
Kwan Houng LEE
Kwang-Baek KIM
Kyung-Woo KANG
Laicheng CAO
Laiwan CHAN
Lambert SPAANENBURG
Lan GAO
Lance FUNG
Lean YU
Lei LIN
Lei WANG
Leichun WANG
Li MEIJUAN
Li WU
Li DAYONG
Li SUN
Li ZHANG
Liang GAO
Liang XIAO
Liang MING
Lian-Wei ZHAO
Lianxi WU
Liefeng BO
Lili ZHOU
Liming CHEN
Li-Ming WANG
Lin CONG
Lincheng SHEN
Ling WANG
Ling CHEN
Ling1 WANG
Liqing ZHANG
Liquan SHEN
Lixin ZHENG
Luo ZHONG
Lusheng ZHONG
Luyang GUAN
Manjaiah D H
Maoguo GONG
Maoguo GONG
Maoyuan ZHANG
Masahiko TOYONAGA
Masakazu MATSUGU
Masumi ISHIKAWA
Mehmet Zeki BILGIN
Mei TIAN
Meihong SHI
Meiyi LI
Mengxin LI
Michael MARGALIOT
Min LIU
Min FANG
Ming BAO
Ming LI
Ming LI
Ming CHEN
Mingbao LI
Mingguang WU
Minghui LI
Mingquan ZHOU
Moh Lim SIM
Mudar SAREM
Nagabhushan P.
Naigang CUI
Nak Yong KO
Naoko TAKAYAMA
Naoyuki KUBOTA
Ning CHEN
Otvio Noura TEIXEIRA
Pei-Chann CHANG
Peide LIU
Peixin YE
Peizhi WEN

Peng TIAN
Peter TINO
Phill Kyu RHEE
Ping JI
Pu WANG
Qi WANG
Qi LUO
Qiang LV
Qiang SUN
Qijuan CHEN
Qing GUO
Qing LI
Qinghe MING
Qingming YI
Qingqi PEI
Qiongshui WU
Qiyong GONG
Quan ZHANG
Renbiao WU
Renpu LI
Renren LIU
Richard EPSTEIN
Richard NEVILLE
Robo ZHANG
Roman NERUDA
Rong LUO
Rongfang BIE
Ronghua SHANG
Ronghua SHANG
Rubin WANG
Rui XU
Ruijun ZHU
Ruiming FANG
Ruixuan LI
Ruochen LIU
S.G. LEE
Sanyang LIU
Satoshi MATSUDA
Seok-Lyong LEE
Seong Whan KIM
Serdar KUCUK
Seunggwan LEE
Sezai TOKAT
Shan TAN
Shangmin LUAN
Shao-Ming FEI
Shao-Xiong WU
Shigeo ABE
Shiqiang ZHENG
Shuguang ZHAO
Shuiping GOU
Shui-Sen CHEN
Shui-Sheng ZHOU
Shunman WANG
Shunsheng GUO
Shutao LI
Shuyuan YANG
Soo-Hong PARK
Soon Cheol PARK
Sung-Bae CHO
Sungshin KIM
Sunjun LIU
Sunkook YOO
Tae Ho CHO
Tae-Chon AHN
Tai Hoon CHO
Takao TERANO
Takeshi FURUHASHI
Tan LIU
Tao SHEN
Tao WANG
Taoshen LI
Thi Ngoc Yen PHAM
Tianding CHEN
Tiantian SU
Tianyun CHEN
Tie-Jun ZHOU
Ting WU
Tong-Zhu FANG
Vianey Guadalupe CRUZ SANCHEZ
Vic CIESIELSKI
Wang LEI
Wanli MA
Wei ZOU
Wei WU
Wei LI
Wei FANG
Weida ZHOU
Wei-Hua LI
Weiqin YIN

Weiyou CAI
Wei-Yu YU
Wen ZHU
Wenbing XIAO
Wenbo XU
Wenchuan YANG
Wenhui LI
Wenping MA
Wenping MA
Wenqing ZHAO
Wen-Shyong TZOU
Wentao HUANG
Wentao HUANG
Wenxing ZHU
Wenxue HONG
Wenyu LIU
X.B. CAO
Xian-Chuan YU
Xianghui LIU
Xiangrong ZHANG
Xiangwei LAI
Xiaobing LIU
Xiaodong KONG
Xiaofeng SONG
Xiaoguang ZHANG
Xiaoguang LIU
Xiaohe LI
Xiaohua YANG
Xiaohua WANG
Xiaohua ZHANG
Xiaohui YUAN
Xiaohui YANG
Xiaojian SHAO
Xiao-Jie ZHAO
Xiaojun WU
Xiaoli LI
Xiaosi ZHAN
Xiaosuo LU
Xiaoyi FENG
Xiaoying PAN
Xiaoyuan WANG
Xin XU
Xin YUAN
Xinbo GAO
Xinchao ZHAO
Xingming SUN
Xinsheng YAO
Xinyu WANG
Xiu JIN
Xiu-Fen YU
Xiufeng WANG
Xiuhua GAO
Xiuli MA
Xiyang LIU
Xiyue HUANG
Xu YANG
Xu CHEN
Xuejun XU
Xueliang BI
Xuerong CHEN
Xuezhou XU
Xun WANG
Xuyan TU
Yan ZHANG
Yan LIANG
Yan ZHANG
Yang YAN
Yangmi LIM
Yangmin LI
Yangyang LI
Yangyang WU
Yanling WU
Yanning ZHANG
Yanning ZHANG
Yanpeng LIU
Yanping LV
Yanxia ZHANG
Yanxin ZHANG
Yan-Xin ZHANG
Yaoguo DANG
Yaping DAI
Yaw-Jen CHANG
Yeon-Pun CHANG
Yezheng LIU
Yidan SU
Yifeng NIU
Yimin YU
Ying GUO
Ying GAO
Ying TIAN

Yingfang FAN
Yingfeng QIU
Yinghong PENG
Yingying LIU
Yong ZHAO
Yong YANG
Yong FAN
Yong-Chang JIAO
Yonggui KAO
Yonghui JIANG
Yong-Kab KIM
Yongqiang ZHANG
Yongsheng DING
Yongsheng ZHAO
Yongzhong ZHAO
Yoshikii KASHIMORI
You-Feng LI
Youguo PI
You-Ren WANG
Yu GUO
Yu GAO
Yuan KANG
Yuehui CHEN
Yuehui CHEN
Yufeng LIAO
Yuheng SHA
Yukun BAO
Yulong LEI
Yumin LIU
Yumin TIAN
Yun-Chia LIANG
Yunjie ZHANG
Yuping WANG
Yurong ZENG
Yusuke NOJIMA
Yutao QI
Yutian LIU
Yuyao HE
Yu-Yen OU
Yuzhong CHEN
Zafer BINGUL
Zeng-Guang HOU
Zhang YANG
Zhanli LI
Zhao ZHAO
Zhaoyang ZHANG
Zhe-Ming LU
Zhen YANG
Zhenbing ZENG
Zhengxing CHENG
Zhengyou XIA
Zhi LIU
Zhidong ZHAO
Zhifeng HAO
Zhigang XU
Zhigeng FANG
Zhihui LI
Zhiqing MENG
Zhixiong LIU
Zhiyong ZHANG
Zhiyu ZHANG
Zhonghua LI
Zhurong WANG
Zi-Ang LV
Zixing CAI
Zong Woo GEEM
Zongmin LI
Zongying OU
Zoran BOJKOVIC

Table of Contents – Part I

Artificial Neural Networks

Natural Neural Systems and Cognitive Science

Neural Network Applications

Evolutionary Computation: Theory and Algorithms

Table of Contents – Part II

Other Topics in Natural Computation

Natural Computation Techniques Applications

Hardware

Cross-Disciplinary Topics

Hypersphere Support Vector Machines Based on Multiplicative Updates

Qing Wu, Sanyang Liu, and Leyou Zhang

Department of Applied Mathematics,
Xidian University, Xi'an, Shaanxi 710071, China
qwu@mail.xidian.edu.cn

Abstract. This paper proposes a novel hypersphere support vector machines based on multiplicative updates. This algorithm can obtain the boundary of hypersphere containing one class of samples by the description of the training samples from one class and uses this boundary to classify the test samples. Moreover, new multiplicative updates are derived to solve sum and box constrained quadratic programming. The experiments show the superiority of our new algorithm.

1 Introduction

In this paper, a hypersphere support vector machine is designed based on support vector domain description(SVDD) [5]. The classifier can be constructed immediately from the description without the use of the unknown training samples, which makes this method efficient.

Multiplicative updates [1] are just suited for nonnegative quadratic programming. In this paper, new multiplicative updates are derived for solving box and sum constrained quadratic programming in hypersphere support vector machines. The multiplicative updates converge monotonically to the solution of the maximum margin hyperplane. All the variables are updated in parallel. They provided an extremely straightforward way to implement SVMs.

2 Hypersphere SVMs (HSVMs)

let $\{(x_i, y_i)\}_{i=1}^n$ denote a training set of labeled samples with binary labels $y_i = \pm 1$, construct a hypersphere using the samples labeled $y_i = 1$ and do not consider those samples labeled $y_i = -1$. The sphere is characterized by center a and radius $R > 0$. Consider the minimization of the programming (P1):

$$\begin{array}{ll} \textit{minimise} & R^2 + C\sum_{i=1}^{l} \xi_i \\ \textit{subject to} & y_i(R^2 - (x_i - a)^T(x_i - a)) \geq -\xi_i, \\ & \xi_i \geq 0,\ i = 1, \cdots, l \end{array} \tag{1}$$

where $y_i = 1$, slack variables $\xi_i \geq 0$.

L. Jiao et al. (Eds.): ICNC 2006, Part I, LNCS 4221, pp. 1–4, 2006.

According to KKT conditions, the dual programming (P1) is (P2):

$$\begin{array}{ll} minimise & -\sum_{i=1}^{l}\alpha_i y_i k(x_i,x_i)+\sum_{i,j=1}^{l}\alpha_i\alpha_j y_i y_j k(x_i,x_j) \\ subject\ to & 0\le\alpha_i\le c \\ & \sum_{i=1}^{l} y_i\alpha_i=1,\ i=1,\cdots,l \end{array} \tag{2}$$

The kernel function is used to compute dot products, so one can design the following classifier:

$$f(x)=sgn(R^2-k(x,x)+2\sum_i\alpha_i y_i(x,x_i)-\sum_{i,j}\alpha_i\alpha_j y_i y_j k(x_i,x_j)) \tag{3}$$

where

$$R^2=\frac{1}{k}\sum_{\substack{\{k|y_k=1,\\ x_k\in\{sv\}\}}}(k(x_k,x_k)-2\sum_i\alpha_i y_i k(x_k,x_i)+\sum_{i,j}\alpha_i\alpha_j y_i y_j k(x_i,x_j)) \tag{4}$$

for any $x_k\in\{sv\}$, and k is the number of support vectors.

3 Extension of Multiplicative Updates for HSVMs

The multiplicative updates [1] are just suited for nonnegative quadratic programming. Therefore, the multiplicative updates can be extended to incorporate sum and box constraints beyond nonnegativity in this paper.

Consider the minimization of the quadratic programming (P3):

$$minimise\quad F(v)=\tfrac{1}{2}v^T Av+b^T v \tag{5}$$

$$subject\ to\quad 0\le v_i\le C \tag{6}$$

$$\sum_i v_i=1 \tag{7}$$

The programming (P3) is equivalent to the following programming (P4):

$$\begin{array}{ll} minimise & F(v,\lambda)=\tfrac{1}{2}v^T Av+b^T v+\lambda(\sum_i v_i-1) \\ subject\ to & 0\le v_i\le C \end{array} \tag{8}$$

where the lagrange multiplier $\lambda\ge 0$. Define the new variables:

$$\hat{v}_i=\begin{cases} v_i & if\ \frac{\partial F}{\partial v_i}\ge 0 \\ C-v_i & otherwise \end{cases} \tag{9}$$

Let $\hat{F}(\hat{v})=\frac{1}{2}\hat{v}^T\hat{A}\hat{v}+\hat{b}^T\hat{v}$, where the coefficients $\hat{A}_{ij}$ and $\hat{b}_i$ are chosen such that $\hat{F}(\hat{v})-F(v)$ is a constant that does not depend on v. We can ensure that all the elements of its gradient are nonnegative: $\frac{\partial\hat{F}}{\partial\hat{v}_i}\ge 0$.

Let $s_i=sgn(\frac{\partial F}{\partial v_i})$ denote the sign of $\frac{\partial F}{\partial v_i}$, with $sgn(0)$ equal to 1. Note that v_i and $\hat{v}_i$ are linearly related, then v_i as :

$$v_i=\hat{v}_i s_i+C\frac{(1-s_i)}{2}. \tag{10}$$

From $\sum_i v_i = 1$, we obtain $\sum_i s_i \hat{v}_i = \beta$, where $\beta = 1 - C\sum_i \frac{(1-s_i)}{2}$. So the programming (P4) is equivalent to minimizing the function :

$$\hat{F}(\hat{v}, \lambda) = \frac{1}{2}\hat{v}^T \hat{A}\hat{v} + \hat{b}^T \hat{v} + \lambda(\sum_i s_i \hat{v}_i - \beta). \tag{11}$$

Substituting (10) into $F(v) = \frac{1}{2}v^T Av + b^T v$ and extracting the coefficients of the terms $\hat{v}_i^T \hat{v}_j$ and $\hat{v}_i$, one can obtain:

$$\hat{A}_{ij} = s_i s_j A_{ij},\ \hat{b}_i = b_i s_i + \tfrac{C}{2}\sum_j s_i(1-s_j)A_{ij}\,.$$

Let

$$r(\lambda) = \frac{-(\hat{b}_i + s_i\lambda) + \sqrt{(\hat{b}_i + s_i\lambda)^2 + 4\hat{a}_i\hat{c}_i}}{2\hat{a}_i}, \tag{12}$$

define matrices $\hat{A}^{\pm}$ from $\hat{A}$ using the some construction as in [1] and define $\hat{a}_i = (\hat{A}^{+}\hat{v})_i$ and $\hat{c}_i = (\hat{A}^{-}\hat{v})_i$, then the multiplicative update $\hat{v}_i := \hat{v}_i r_i(\lambda)$ will decrease $\hat{F}(\hat{v})$ by deriving all the variables $\hat{v}_i$ toward zero. Note that by decreasing $\hat{F}(\hat{v})$, the update also decreases $F(v)$, since the two differ only by a constant. At each iteration, the unknown λ should be chosen so that the updated values of $\hat{v}_i$ satisfy the sum constraint in (7). That is done by choosing λ to satisfy:

$$\sum_i s_i \hat{v}_i r_i(\lambda) = \beta. \tag{13}$$

A simple iterative procedure exists to solve (12) for λ . In particular, we prove the following:

Theorem 1. *Equation (13) has a unique solution λ^*, that can be computed by iterating*

$$\lambda := \frac{1}{R}(\sum_i s_i \hat{v}_i r_i(\lambda) - \beta) + \lambda\ , \tag{14}$$

where R is any positive constant satisfying $|\sum_i s_i \hat{v}_i \frac{dr_i}{d\lambda}| < R$ for all λ.

Proof. It is easy to prove the fact that (13) has a unique solution and (14) converge monotonically.

4 Experimental Results

In order to compare our algorithm HSVM with the SMO [6] and SVM^{light} [7], we use the database from the MNIST database in [8]. Since the database is so large, we only have obtained two-class examples belonging to classes "7" and "9" respectively. In our experiment, 10-block cross validation is employed for parameter C, the Gaussion kernel with $\sigma = 0.2$ is employed. It is easy to see that our algorithm leads to high classification accuracy, though it is slightly less than that of SMO and SVM^{light}. The time consumed and the number of support vectors by our algorithm is much less than the others.

Table 1. Results comparison

Learning model	Training samples "7" + "9"	Testing samples	Time (s)	SVs	Accuracy %
SVM^{light}	65+64	1028+1009	136	56	95.7
SMO	65+64	1028+1009	112	58	95.4
HSVM	65+64	1028+1009	78	29	93.3

5 Conclusion

Due to the fact that training phase will occupy main time, the new algorithm proposed in this paper can greatly reduce computational cost and training time. The new multiplicative updates are straightforward to implement and have a rigorous guarantee of monotonic convergence.

Acknowledgements

We would like to thank the anonymous reviewers for their valuable comments and suggestions. This work is supported by the National Natural Science Foundation of China under contract No.60574075

References

1. F.sha,L.K.Saul, and D.D.Lee. Multiplicative updates for nonnegative quadratic programming in support vector machines. Neural and Information processing systems. Cambridge, MA: MIT press. **15**(2003)
2. V.Vapnik. Nature of satistical learning theory. Newyork: Springer-verlat. (2000)
3. Lu C. D.,Zhang T.Y.,Hu J.Y. Support vector Domain Classifier Based on Multiplicative Updates. Chinese Journal of Computers. **27**(2004) pages: 690–694
4. C.Gentile. A new approximate maximal margin classification algorithm. Journal of Machine Learning Research. **2**(2001) pages: 213–242
5. Tax.D.,Ypma.A and Duin.R.. Support vector domain description. Pattern Recognition letters. **20**(1999) pages: 1191–1199
6. Platt.J.C. Sequential minimal optimization: A fast algorithm for training support vector machines. Advance in kernel methods: Support vector machines. Cambridge, MA: MIT Press. (1998)
7. Joachims.T.. Making Large-scale Support Vector Machine Learning Practical. Advance in kernel methods: Support vector machines. Cambridge, MA: MIT Press. (1999) pages: 169–184
8. Blake.C.L.,Merz.C.J., UCI Repository of Machine Learning Databases. University of California, Irvine(1998) *http://www.ics.vci.edu/mlearn/MLRepostory.html*

The Study of Leave-One-Out Error-Based Classification Learning Algorithm for Generalization Performance*

Bin Zou[1], Jie Xu[1,2], and Luoqing Li[1]

[1] Faculty of Mathematics and Computer Science, Hubei University, Wuhan 430062, P.R. China

[2] College of Computer Science, Huazhong University of Science and Technology, Wuhan 430074, P.R. China

Abstract. This note mainly focuses on a theoretical analysis of the generalization ability of classification learning algorithm. The explicit bound is derived on the relative difference between the generalization error and leave-one-out error for classification learning algorithm under the condition of leave-one-out stability by using Markov's inequality, and then this bound is used to estimate the generalization error of classification learning algorithm. We compare the result in this paper with previous results in the end.

1 Introduction

A key issue in the design of efficient machine learning systems is the estimation of the accuracy of learning algorithms. Among the several approaches that have been proposed to this problem, one of the most prominent is based on the theory of uniform convergence of empirical quantities to their mean [3,11]. This theory provides ways to estimate the risk of a learning system based on empirical measurement of its accuracy and a measure of its complexity [1]. Bousquet and Elisseeff [2,7,9] explored a different approach which is based on sensitivity analysis. Sensitivity analysis aims at determining how much the variation of the input can influence the output of a system. It has been applied to many area such as statistics and mathematical programming.

In this paper, the objects of interest are learning algorithms. they take as input a learning set made of instance-label pairs and output a function that maps instances to the corresponding labels. The sensitivity in that case is thus related to changes of the outcome of the algorithm when the learning set is changed. There are two sources of randomness such algorithms have to cope with: the first one comes from the sampling mechanism used to generate the learning set and the second one is due to noise in the measurements. We mainly focus on the sampling randomness in the sequel and we are interested in how changes in the composition of the learning set influence the function produced by

* Supported in part by NSFC under grant 60403011.

L. Jiao et al. (Eds.): ICNC 2006, Part I, LNCS 4221, pp. 5–14, 2006.

the algorithm. The outcome of such an approach is a principled way of getting bound on the difference between empirical error and the generalization error.

There are some quantitative results relating the generalization error to the stability of learning algorithm with respect to changes in the training set. The first such results were obtained by Devroye, Rogers and Wagner [4] in the seventies. Rogers and Wagner [10] first showed that the variance of the leave-one-out error can be upper bounded by what Kearns and Ron [6] later called hypothesis stability. The quantity measures how much the function learned by the algorithm will change when one point in the training set is removed. Devroye [4] first applied concentration inequalities to bound the generalization error for learning algorithms. Bousquet and Elisseeff [2] derived exponential upper bounds on the generalization error based on the notation of stability. They considered both the leave-one-out error and the empirical error as possible estimates of the generalization error, they proved stability bounds for a large class of algorithms which includes the Support Vector Machines (SVM), both in the regression and in the classification cases.

To bound the accuracy of the predictions of learning algorithms, we introduce a new approach in this paper, that is, we first establish the bound on the relative difference between the generalization error and leave-one-out empirical error for classification learning algorithm under the condition of leave-one-out stability by using Markov's inequality, then derive the bound on the generalization error for classification learning algorithm by using the bound of relative difference.

The paper is organized as follows: In section 2 we introduce some notations. In section 3, we derive the bound on the relative difference between the generalization error and leave-one-out empirical error for classification learning algorithm under the condition of leave-one-out stability, and obtain the bound on the generalization error for classification learning algorithm. We compare this result with known results in section 4.

2 Preliminaries

We introduce some notations and do some preparations.

$\mathcal{X}$ and $\mathcal{Y} \subset \mathbf{R}$ being respectively an input and an output space, we consider a training set

$$S = \{z_1 = (x_1, y_1), z_2 = (x_2, y_2), \cdots, z_m = (x_m, y_m)\}$$

of size m in $\mathcal{Z} = \mathcal{X} \times \mathcal{Y}$ drawn i.i.d. from an unknown distribution $\mu(\mathcal{Z})$. A learning algorithm is a function $\mathcal{A}$ from $\mathcal{Z}^m$ into $\mathcal{F} \subset \mathcal{Y}^{\mathcal{X}}$ which maps a learning set S onto a function (or hypothesis) f_S from $\mathcal{X}$ to $\mathcal{Y}$. To avoid complex notation, we consider only deterministic algorithms. It is also assumed that the algorithm $\mathcal{A}$ is symmetric with respect to S, i.e., it does not depend on the order of the elements in the training set. Furthermore, we assume that all functions are measurable and all sets are countable which does not limit the interest of the results presented here.

Given a training set S of size m, we will build, for any $i = 1, 2, \cdots, m$, modified training sets as follows: By removing the i-th element

$$S^i = \{z_1, \cdots, z_{i-1}, z_{i+1}, \cdots, z_m\};$$

By replacing the i-th element

$$S^{i,u} = \{z_1, \cdots, z_{i-1}, u, z_{i+1}, \cdots, z_m\},$$

where the replacement example u is assumed to be drawn from $\mu(\mathcal{Z})$ and is independent from S.

To measure the accuracy of the predictions of learning algorithm, we will use a cost function $\mathcal{C} : \mathcal{Y} \times \mathcal{Y} \to \mathbf{R}^+$. The loss of the hypothesis f_S with respect to an example $z = (x, y)$ is then defined as

$$\mathcal{L}(f_S, z) = \mathcal{C}(f_S(x), y).$$

We will consider several measures of the performance of learning algorithm. The main quantity we are interested in is the risk (or the generalization error). This is a random variable depending on the training set S and it is defined as

$$\mathcal{R}(\mathcal{A}, S) = \mathrm{E}_{\mathcal{Z}}[\mathcal{L}(f_S, z)] = \int_{\mathcal{Z}} \mathcal{L}(f_S, z) d\mu(\mathcal{Z}). \tag{1}$$

Unfortunately, $\mathcal{R}(\mathcal{A}, S)$ cannot be computed directly since $\mu(\mathcal{Z})$ is unknown. So we have to estimate it from the available data S. We will consider estimators for this quantity.

The simplest estimator is the so-called empirical error defined as

$$\mathcal{R}_{emp}(\mathcal{A}, S) = \frac{1}{m} \sum_{i=1}^{m} \mathcal{L}(f_S, z_i).$$

Another classical estimator is the leave-one-out error defined as

$$\mathcal{R}_{loo}(\mathcal{A}, S) = \frac{1}{m} \sum_{i=1}^{m} \mathcal{L}(f_{S^i}, z_i).$$

Devroye and Wagner [5] first introduced the notation of leave-one-out stability.

Definition 1. *A learning algorithm $\mathcal{A}$ has leave-one-out stability β with respect to the loss function $\mathcal{L}$ if the following holds:*

$$\forall S \in \mathcal{Z}^m,\ \forall i \in \{1, 2, \cdots, m\}, \forall z \in \mathcal{Z},\ \ |\mathcal{L}(f_S, z) - \mathcal{L}(f_{S^i}, z)| \le \beta.$$

Bousquet and Elisseeff [2] introduced the notation of change-one stability.

Definition 2. *A learning algorithm $\mathcal{A}$ has change-one stability β with respect to the loss function $\mathcal{L}$ if the following holds:*

$$\forall S \in \mathcal{Z}^m,\ \forall i \in \{1, 2, \cdots, m\}, \forall u, z \in \mathcal{Z},\ \ |\mathcal{L}(f_S, z) - \mathcal{L}(f_{S^{i,u}}, z)| \le \beta.$$

Note that, in both of these definitions, we view β as a function of m, we are most interested in the case where $\beta = \lambda/m$ for a positive constant λ.

By these definitions, we can obtain easily

$$|\mathcal{L}(f_S, z) - \mathcal{L}(f_{S^{i,u}}, z)| \leq |\mathcal{L}(f_S, z) - \mathcal{L}(f_{S^i}, z)| + |\mathcal{L}(f_{S^{i,u}}, z) - \mathcal{L}(f_{S^i}, z)| \leq 2\beta,$$

which states that leave-one out stability implies change-one stability, the converse is not true [7]. This is reason that we consider leave-one-out empirical error and leave-one-out stability.

3 The Bound on the Relative Difference

The study we describe here intends to bound the relative difference between leave-one-out error and the generalization error for classification learning algorithm. For any $\varepsilon > 0$, and any p, $2 \geq p > 1$ our goal is to bound the term

$$\mathrm{P}\{\frac{\mathcal{R}(\mathcal{A}, S) - \mathcal{R}_{loo}(\mathcal{A}, S)}{\sqrt[p]{\mathcal{R}(\mathcal{A}, S)}} > \varepsilon\}, \tag{2}$$

which differs from what is usually studied in learning theory [2,7]

$$\mathrm{P}\{|\mathcal{R}(\mathcal{A}, S) - \mathcal{R}_{loo}(\mathcal{A}, \mathcal{S})| > \varepsilon\}. \tag{3}$$

As we know that the estimate in (3) fails to capture the phenomenon that for those $\mathcal{L}(f_S, \mathbf{z})$ and S for which $R(\mathcal{A}, S)$ is small, the deviation $R(\mathcal{A}, S) - R_{\mathrm{emp}}(\mathcal{A}, S)$ is also small with large probability. In order to improve the estimate in (3), so we bound term (2). Then we obtain the following theorem on the relative difference between leave-one-out error and the generalization error for classification learning algorithm.

Theorem 1. *Let $\mathcal{A}$ be a classification learning algorithm with leave-one-out stability β with respect to the loss function $\mathcal{L}$. Then for any $\varepsilon > 0$, and for all p, $2 \geq p > 1$, when $m > \varepsilon^{-\frac{p}{p-1}}$, the inequality*

$$\mathrm{P}\{\frac{\mathcal{R}(\mathcal{A}, S) - \mathcal{R}_{loo}(\mathcal{A}, S)}{\sqrt[p]{\mathcal{R}(\mathcal{A}, S)}} > \varepsilon\} \leq \frac{4[2 + \beta + (1 + \beta)m](2m)^{\frac{2}{p}}}{m\varepsilon^2(1 + m\varepsilon^{\frac{p}{p-1}})^{\frac{2}{p}}} \tag{4}$$

is valid, and when $p = 2$, for all δ, $0 < \delta \leq 1$, with probability at least $1 - \delta$, the inequality

$$\mathcal{R}(\mathcal{A}, S) \leq \mathcal{R}_{loo}(\mathcal{A}, S) + \frac{1}{2}\varepsilon^2(m)(1 + \sqrt{\frac{4\mathcal{R}_{loo}(\mathcal{A}, S)}{\varepsilon^2(m)}})$$

holds true, where

$$\varepsilon(m) = -\frac{1}{2m} + \frac{1}{2m}\sqrt{1 + \frac{32m[2 + \beta + (1 + \beta)m]}{\delta}}.$$

In order to prove Theorem 1, we first prove the following lemmas. We thus consider two events constructed from a random and independent sample of size $2m$:

$$Q_1 = \{z : \frac{\mathrm{P}(T_1) - \nu(\mathcal{Z}_1)}{\sqrt[p]{\mathrm{P}(T_1)}} > \varepsilon\},$$

$$Q_2 = \{z : \frac{\nu(\mathcal{Z}_2) - \nu(\mathcal{Z}_1)}{\sqrt[p]{\nu(\mathcal{Z}) + \frac{1}{2m}}} > \varepsilon\},$$

where $\mathrm{P}(T_1)$ is probability of event $T_1 = \{z : \mathcal{L}(f_S, z) = 1\}$

$$\mathrm{P}(T_1) = \int_{\mathcal{Z}} \mathcal{L}(f_S, z)\mathrm{d}\mu(\mathcal{Z}).$$

$\nu(\mathcal{Z}_1)$ is the frequency of event $T_2 = \{z : \mathcal{L}(f_{S^i}, z) = 1\}$ computed from the first half-sample $\mathcal{Z}_1 = z_1, z_2, \cdots, z_m$ of the sample $z_1, z_2, \cdots, z_m, z_{m+1}, \cdots, z_{2m}$

$$\nu(\mathcal{Z}_1) = \frac{1}{m}\sum_{i=1}^{m} \mathcal{L}(f_{S^i}, z_i),$$

and $\nu(\mathcal{Z}_2)$ is the frequency of event T_1 computed from the second half-sample $\mathcal{Z}_2 = z_{m+1}, z_{m+2}, \cdots, z_{2m}$

$$\nu(\mathcal{Z}_2) = \frac{1}{m}\sum_{i=1}^{m} \mathcal{L}(f_S, z_{i+m}).$$

Denote

$$\nu(\mathcal{Z}) = \frac{\nu(\mathcal{Z}_1) + \nu(\mathcal{Z}_2)}{2}.$$

Accordingly we shall prove the theorem as follows: first we show that for $m > \varepsilon^{-\frac{p}{p-1}}$, the inequality $\mathrm{P(Q_1)} < 4\mathrm{P(Q_2)}$ is valid, and then we bound $\mathrm{P(Q_2)}$. Thus we shall prove the following lemma.

Lemma 1. *For any classification learning algorithm $\mathcal{A}$, for any $\varepsilon > 0$ and all $m > \varepsilon^{-\frac{p}{p-1}}$, the inequality*

$$\mathrm{P}(Q_1) \leq 4\mathrm{P}(Q_2)$$

is valid.

Proof. Assume that event Q_1 has occurred, this means that

$$\mathrm{P}(T_1) - \nu(\mathcal{Z}_1) > \varepsilon \sqrt[p]{\mathrm{P}(T_1)}.$$

Since $\nu(\mathcal{Z}_1) \geq 0$, this implies that $\mathrm{P}(T_1) > \varepsilon^{\frac{p}{p-1}}$. Assume that

$$\nu(\mathcal{Z}_2) > \mathrm{P}(T_1).$$

Recall now that $m > \varepsilon^{-\frac{p}{p-1}}$, under these conditions, event Q_2 will definitely occur.

To show this we bound the quantity

$$\xi = \frac{\nu(\mathcal{Z}_2) - \nu(\mathcal{Z}_1)}{\sqrt[p]{\nu(\mathcal{Z}) + \frac{1}{2m}}} \tag{5}$$

under the conditions

$$\nu(\mathcal{Z}_1) < \mathrm{P}(T_1) - \varepsilon \sqrt[p]{\mathrm{P}(T_1)},$$
$$\mathrm{P}(T_1) > \varepsilon^{\frac{p}{p-1}},$$
$$\nu(\mathcal{Z}_2) > \mathrm{P}(T_1).$$

For this purpose we find the minimum of the function

$$\psi = \frac{x - y}{\sqrt[p]{x + y + c}}$$

in the domain $0 < a \leq x \leq 1,\ 0 < y \leq b,\ c > 0$. We have for $p > 1$

$$\frac{\partial \psi}{\partial x} = \frac{1}{p} \frac{(p-1)x + (p+1)y + pc}{(x+y+c)^{\frac{p+1}{p}}} > 0,$$

$$\frac{\partial \psi}{\partial y} = -\frac{1}{p} \frac{(p+1)x + (p-1)y + pc}{(x+y+c)^{\frac{p+1}{p}}} < 0.$$

Consequently ψ attains its minimum in the admissible domain at the boundary points $x = a$ and $y = b$. Therefore the quantity ξ is bounded from below. If in (5) one replaces $\nu(\mathcal{Z}_1)$ by $\mathrm{P}(T_1) - \varepsilon \sqrt[p]{\mathrm{P}(T_1)}$ and $\nu(\mathcal{Z}_2)$ by $\mathrm{P}(T_1)$. Thus

$$\xi \geq \frac{\varepsilon \sqrt[p]{2\mathrm{P}(T_1)}}{\sqrt[p]{2\mathrm{P}(T_1) - \varepsilon \sqrt[p]{\mathrm{P}(T_1)} + \frac{1}{m}}}.$$

Since $\mathrm{P}(T_1) > \varepsilon^{\frac{p}{p-1}}$, $m > \varepsilon^{-\frac{p}{p-1}}$, we have that

$$\xi > \frac{\varepsilon \sqrt[p]{2\mathrm{P}(T_1)}}{\sqrt[p]{2\mathrm{P}(T_1) - \varepsilon^{\frac{p}{p-1}} + \varepsilon^{\frac{p}{p-1}}}} = \varepsilon.$$

Thus, if Q_1 occurs and the conditions $\nu(\mathcal{Z}_2) > \mathrm{P}(T_1)$ is satisfied, then Q_2 occurs as well.

$$\mathrm{P}(Q_2) = \mathrm{P}\{\frac{\nu(\mathcal{Z}_2) - \nu(\mathcal{Z}_1)}{\sqrt[p]{\nu(\mathcal{Z}) + \frac{1}{2m}}} > \varepsilon\} = \int_{\mathcal{Z}^{2m}} \theta[\frac{\nu(\mathcal{Z}_2) - \nu(\mathcal{Z}_1)}{\sqrt[p]{\nu(\mathcal{Z}) + \frac{1}{2m}}} - \varepsilon] d\mu(\mathcal{Z}^{2m}).$$

Taking into account that the space $\mathcal{Z}^{2m}$ of samples of size $2m$ is the direct product of two subspaces $\mathcal{Z}_1(m)$ and $\mathcal{Z}_2(m)$, one can assert that the following inequalities

$$\mathrm{P}(Q_2) = \int_{\mathcal{Z}_1(m)} d\mu(\mathcal{Z}_1) \int_{\mathcal{Z}_2(m)} \theta[\frac{\nu(\mathcal{Z}_2) - \nu(\mathcal{Z}_1)}{\sqrt[p]{\nu(\mathcal{Z}) + \frac{1}{2m}}} - \varepsilon] d\mu(\mathcal{Z}_2)$$

$$\mathrm{P}(Q_2) > \int_{Q_1} d\mu(\mathcal{Z}_1) \int_{\mathcal{Z}_2(m)} \theta[\frac{\nu(\mathcal{Z}_2) - \nu(\mathcal{Z}_1)}{\sqrt[p]{\nu(\mathcal{Z}) + \frac{1}{2m}}} - \varepsilon] d\mu(\mathcal{Z}_2)$$

$$\mathrm{P}(Q_2) > \int_{Q_1} d\mu(\mathcal{Z}_1) \int_{\mathcal{Z}_2(m)} \theta[\nu(\mathcal{Z}_2) - \mathrm{P}(T_1)] d\mu(\mathcal{Z}_2) \tag{6}$$

is valid (Fubini's Theorem).

We now bound the inner integral on the right-hand side of inequality which we denote by I, then

$$I = \int_{\mathcal{Z}_2(m)} \theta[\nu(\mathcal{Z}_2) - \mathrm{P}(T_1)] d\mu(\mathcal{Z}_2) = \sum_{\frac{k}{m} \geq \mathrm{P}(T_1)} C_m^k [\mathrm{P}(T_1)]^k [1 - \mathrm{P}(T_1)]^{m-k}.$$

Since the samples are drawn i.i.d. from $\mu(\mathcal{Z})$, the last sum exceeds $\frac{1}{4}$. Returning to (6) we obtain that

$$\mathrm{P}(Q_2) > \frac{1}{4} \int_{Q_1} d\mu(\mathcal{Z}_1) = \frac{1}{4} \mathrm{P}(Q_1).$$

Lemma 2. *Let $\mathcal{A}$ be a classification learning algorithm with leave-one-out stability β with respect to the loss function $\mathcal{L}$. Then for any $\varepsilon > 0$,*

$$\mathrm{P}\{\frac{\nu(\mathcal{Z}_2) - \nu(\mathcal{Z}_1)}{\sqrt[p]{\nu(\mathcal{Z}) + \frac{1}{2m}}} > \varepsilon\} \leq \frac{[2 + \beta + (1+\beta)m](2m)^{\frac{2}{p}}}{m\varepsilon^2 (1 + m\varepsilon^{\frac{p}{p-1}})^{\frac{2}{p}}}.$$

Proof. Since

$$\mathrm{P}\{\frac{\nu(\mathcal{Z}_2) - \nu(\mathcal{Z}_1)}{\sqrt[p]{\nu(\mathcal{Z}) + \frac{1}{2m}}} > \varepsilon\} < \mathrm{P}\{\nu(\mathcal{Z}_2) - \nu(\mathcal{Z}_1) > \varepsilon \sqrt[p]{\frac{1}{2m} + \frac{1}{2}\varepsilon^{\frac{p}{p-1}}}\}.$$

$$\mathrm{E}_S[|\nu(\mathcal{Z}_2) - \nu(\mathcal{Z}_1)|^2] = \mathrm{E}_S[(\nu(\mathcal{Z}_2))^2] - 2\mathrm{E}_S[\nu(\mathcal{Z}_1)\nu(\mathcal{Z}_2)] + \mathrm{E}_S[(\nu(\mathcal{Z}_2))^2],$$

where

$$\mathrm{E}_S[(\nu(\mathcal{Z}_1))^2] = \mathrm{E}_S[(\frac{1}{m}\sum_{i=1}^m \mathcal{L}(f_{S^i}, z_i))^2].$$

It follows that

$$\mathrm{E}_S[(\nu(\mathcal{Z}_1))^2] \leq \frac{1}{m}\mathrm{E}_S[\frac{1}{m}\sum_{i=1}^m \mathcal{L}(f_{S^i}, z_i)] + \frac{1}{m^2}\sum_{i \neq j} \mathrm{E}_S[\mathcal{L}(f_{S^i}, z_i)\mathcal{L}(f_{S^j}, z_j)]$$

$$\mathrm{E}_S[(\nu(\mathcal{Z}_1))^2] \leq \frac{1}{m}\mathrm{E}_S[\mathcal{L}(f_{S^i}, z_i)] + \frac{m-1}{m}\mathrm{E}_S[\mathcal{L}(f_{S^i}, z_i)\mathcal{L}(f_{S^j}, z_j)].$$

Since

$$\mathrm{E}_S[\nu(\mathcal{Z}_1)\nu(\mathcal{Z}_2)] = \mathrm{E}_S[(\frac{1}{m}\sum_{i=1}^m \mathcal{L}(f_{S^i}, z_i))(\frac{1}{m}\sum_{i=1}^m \mathcal{L}(f_S, z_{i+m}))],$$

then it follows that

$$\mathrm{E}_S[\nu(\mathcal{Z}_1)\nu(\mathcal{Z}_2)] = \frac{1}{m^2}\sum_{i=1}^{m}\sum_{j=1}^{m}\mathrm{E}_S[\mathcal{L}(f_{S^i}, z_i)\mathcal{L}(f_S, z_{j+m})],$$

$$\mathrm{E}_S[(\nu(\mathcal{Z}_2))^2] \leq \frac{1}{m^2}\sum_{i=1}^{m}\mathrm{E}_S[\mathcal{L}(f_S, z_{i+m})^2] + \frac{1}{m^2}\sum_{i\neq j}\mathrm{E}_S[\mathcal{L}(f_S, z_{i+m})\mathcal{L}(f_S, z_{j+m})].$$

Therefore, we have

$$\mathrm{E}_S[|\nu(\mathcal{Z}_1) - \nu(\mathcal{Z}_2)|^2] \leq \frac{1}{m}\mathrm{E}_S[\mathcal{L}(f_{S^i}, z_i)] + A_1 + A_2,$$

where

$$A_1 = \frac{m-1}{m}\mathrm{E}_S[\mathcal{L}(f_{S^i}, z_i)\mathcal{L}(f_{S^j}, z_j)] - \frac{2}{m^2}\sum_{i=1}^{m}\sum_{j=1}^{m}\mathrm{E}_S[\mathcal{L}(f_{S^i}, z_i)\mathcal{L}(f_S, z_{j+m})],$$

$$A_2 = \frac{1}{m^2}\sum_{i=1}^{m}\mathrm{E}_S[\mathcal{L}(f_S, z_{i+m})^2] + \frac{1}{m^2}\sum_{i\neq j}\mathrm{E}_S[\mathcal{L}(f_S, z_{i+m})\mathcal{L}(f_S, z_{j+m})].$$

Thus we get

$$\mathrm{E}_S[|\nu(\mathcal{Z}_2) - \nu(\mathcal{Z}_1)|^2] \leq B_1 + B_2 + B_3 + B_4,$$

where

$$B_1 = \frac{1}{m}\mathrm{E}_S[\mathcal{L}(f_{S^i}, z_i)(1-\mathcal{L}(f_{S^j}, z_j))],\ B_2 = \frac{1}{m^2}\sum_{i=1}^{m}\mathrm{E}_S[|\mathcal{L}(f_S, z_{i+m})-\mathcal{L}(f_{S^i}, z_i)|]$$

$$B_3 = \frac{1}{m^2}\sum_{i=1}^{m}\sum_{j=1}^{m}\mathrm{E}_S[\mathcal{L}(f_{S^i}, z_i)|\mathcal{L}(f_{S^j}, z_j) - \mathcal{L}(f_S, z_{j+m})|],$$

$$B_4 = \frac{1}{m^2}\sum_{i\neq j}\mathrm{E}_S[|\mathcal{L}(f_S, z_{i+m}) - \mathcal{L}(f_{S^i}, z_i)|].$$

Using Definition 1, for all $j \in \{1, 2, \cdots, m\}$,

$$\mathrm{E}_S[|\mathcal{L}(f_{S^j}, z_j)-\mathcal{L}(f_S, z_{j+m})|] \leq \mathrm{E}_S[|\mathcal{L}(f_{S^j}, z_j)-\mathcal{L}(f_S, z_j)+\mathcal{L}(f_S, z_j)-\mathcal{L}(f_S, z_{j+m})|],$$

it follows that

$$\mathrm{E}_S[|\mathcal{L}(f_{S^j}, z_j) - \mathcal{L}(f_S, z_{j+m})|] \leq \beta + 1.$$

Thus we obtain

$$\mathrm{E}_S[|\nu(\mathcal{Z}_2) - \nu(\mathcal{Z}_1)|^2] \leq \frac{1}{m} + \frac{(1+\beta)(1+m)}{m}.$$

By Markov's inequality, we get

$$\mathrm{P}\{\nu(\mathcal{Z}_2) - \nu(\mathcal{Z}_1) > \varepsilon\sqrt[p]{\frac{1}{2m} + \frac{1}{2}\varepsilon^{\frac{p}{p-1}}}\} \leq \frac{[2+\beta+(1+\beta)m](2m)^{\frac{2}{p}}}{m\varepsilon^2(1+m\varepsilon^{\frac{p}{p-1}})^{\frac{2}{p}}}.$$

Proof of Theorem 1. We can get directly the first inequality of Theorem 1 By Lemma 1 and Lemma 2. Now we prove that the second inequality of Theorem 1 is also valid.

Let us rewrite the inequality (4) in the equivalent form. We equate the right-hand side of inequality (4) to a positive value δ $(0 < \delta \leq 1)$

$$\delta = \frac{4[2+\beta+(1+\beta)m](2m)^{\frac{2}{p}}}{m\varepsilon^2(1+m\varepsilon^{\frac{p}{p-1}})^{\frac{2}{p}}},$$

and when $p = 2$, we have

$$\delta = \frac{8[2+\beta+(1+\beta)m]}{\varepsilon^2(1+m\varepsilon^2)}. \tag{7}$$

We solve equation (7) with respect to ε^2, the solution

$$\varepsilon(m) = \varepsilon^2 = -\frac{1}{2m} + \frac{1}{2m}\sqrt{1 + \frac{32m[2+\beta+(1+\beta)m]}{\delta}}$$

is used to solve inequality

$$\frac{\mathcal{R}(\mathcal{A},S) - \mathcal{R}_{loo}(\mathcal{A},S)}{\sqrt{\mathcal{R}(\mathcal{A},S)}} \leq \varepsilon(m).$$

As a result we obtain that with probability at least $1-\delta$, the inequality

$$\mathcal{R}(\mathcal{A},S) \leq \mathcal{R}_{loo}(\mathcal{A},S) + \frac{1}{2}\varepsilon^2(m)\big(1 + \sqrt{\frac{4\mathcal{R}_{loo}(\mathcal{A},S)}{\varepsilon^2(m)}}\big) \tag{8}$$

is valid.

4 Conclusion

In this section we compare the result in this paper with previous results.

Kutin and Niyogi [7] bounded the term

$$\mathrm{P}\{|\mathcal{R}(\mathcal{A},S) - \mathcal{R}_{emp}(\mathcal{A},\mathcal{S})| > \varepsilon\} \tag{9}$$

by McDiarmid's inequality [8]. Bousquet and Elisseeff [2] bounded directly the term (3)(Theorem 11) under the condition of hypothesis stability by Markov's inequality. In contrast to the results (Theorem 11) in [2] (the case of term (9) in [7] is dealt with analgously), we can find that the bound on the generalization error (Theorem 1) in this paper has the type

$$\mathcal{R}(\mathcal{A},S) \leq \mathcal{R}_{loo}(\mathcal{A},S) + O(\frac{1}{\sqrt{m}}),$$

which is same as Theorem 11 in [2]. However, the term (3) fails to capture the phenomenon that if the generalization error is small, the deviation between generalization error and leave-one-out error is also small with large probability.

Therefore, the result (Theorem 1) in this paper is as good as that in [2], and the result in this paper improve the result (Theorem 11) in [2].

References

1. Alon, N., Ben-David, S., Cesa-Bianchi, N., and Haussler, D.: Scale-sensitive dimensions, uniform convergence, and learnability. Journal of the ACM. **44**(1997) 615-631
2. Bousquet, O. and Elisseeff, A.: Stability and generalization. Journal Machine Learning Research 2(2001) 499-526
3. Cucker, F., Smale, S.: On the mathematical foundations of learning. Bulletin of the American Mathematical Society **39**(2002) 1-49
4. Devroye, L., Wagner, T.: Distribution-free inequalities for the deleted and holdout error estimates. IEEE trans. inform. theory **25**(1979) 202-207
5. Devroye, L., Wagner, T.: Distribution-free performance bounds for potential function rules. IEEE Trans. Inform. Theory **25**(1979) 601-604
6. Kearns, M., Ron, D.: Algorithmic stability and sanity-check bounds for leave-one-out cross-validation. Neural Computation **11**(1999) 1427-1453
7. Kutin, S., Niyogi, P.: Almost-everywhere algorithmic stability and generalization error. IN Proceedings of Uncertainty in AI. Edmonton, Canda (2002)
8. McDiarmid, C.: On the method of bounded defferences. London Mathematical Lecture Note Series **141**(1989) 148-188
9. Mukherjee, S., Rifkin, R., Poggio, T.: Regression and classification with regularization. In Lectures Notes in Statistics **171**(2002) 107-124
10. Rogers, W., Wagner, T.: A finite sample distribution-free performance bound for local discrimination rules. Annals of Statistics **6**(1978) 506-514
11. Vapnik, V. N.: Statistical Learning Theroy. Wiley, NewYork (1998)

Gabor Feature Based Classification Using LDA/QZ Algorithm for Face Recognition

Weihong Deng, Jiani Hu, and Jun Guo

Beijing University of Posts and Telecommunications, 100876, Beijing, China
{cvpr_dwh, cughu}@126.com, junguo@bupt.edu.cn

Abstract. This paper proposes a LDA/QZ algorithm and its combination of Gabor Filter-based features for the face recognition. The LDA/QZ algorithm follows the common "PCA+LDA" framework, but it has two significant virtues compared with previous algorithms: 1) In PCA step, LDA/QZ transforms the feature space into complete PCA space, so that all discriminatory information is preserved, and 2) In LDA step, the QZ-decomposition is applied to solve the generalized eigenvalue problem, so that LDA can be performed stably even when within-class scatter matrix is singular. Moreover, the Gabor Filter-based Features and the new LDA/QZ algorithm are combined for face recognition. We also performed comparative experimental studies of several state-of-art dimension reduction algorithms and their combinations of Gabor feature for face recognition. The evaluation is based on six experiments involving various types of face images from ORL, FERET, and AR database and experimental results show the LDA/QZ algorithm is always the best or comparable to the best in term of recognition accuracy.

1 Introduction

Due to the military, commercial, and law enforcement application, there has been much interest in automatically recognizing faces in still and video images recently. Many methods have been proposed for face recognition within the last two decades [1]. Among these methods, the linear discriminant analysis (LDA) based methods are most widely used. The LDA, also called Fisher linear discriminant (FLD), defines a projection that makes the within-class scatter small and the between-scatter large. However, it is well-known that classical LDA requires that one of the scatter matrices is nonsingular. For the face recognition application, all scatter matrices can be singular since the feature dimension, in general, exceeds the number of sample size. This is known as the singularity, undersampled, or Small Sample Size(SSS) problem [4].

Many LDA extensions were proposed to overcome the singularity problem in face recognition, such as PCA+LDA [2][17][3], Regularized LDA [8], LDA/GSVD [20], LDA/QR [19]. Among these extensions, PCA+LDA, a two-stage method, received relatively more attention. By applying first principle component analysis (PCA) for dimensionality reduction and then LDA for discriminant analysis, Belhumire *et. al.* developed an approach called Fisherfaces. Using a similar

L. Jiao et al. (Eds.): ICNC 2006, Part I, LNCS 4221, pp. 15–24, 2006.

scheme, Swets and Weng [3] have pointed out the Eigenfaces [7] derived using PCA are only the most expressive features (MEF). The MEF are unrelated to actual recognition, and in order to derive the most discriminant features (MDF), one needs a subsequent LDA projection. More recently, Liu and Wechsler presented an enhanced Fisher discriminant model (EFM), which only selects the dominant PCs that capture most spectral energy for LDA. Jing et. al [21] proposed to select the PCs according to their "Fisher discriminability". Zheng et.al [16] developed a GA-Fisher method which uses genetic algorithm to select appropriate PCs for LDA. This type of approaches have two major limitations: 1) the optimal value of the reduced dimension for PCA is difficult to determine [19][20], and 2) the dimension reduction stage using PCA may potentially lose some useful information for discrimination [13].

This paper introduces a novel LDA/QZ algorithm to address the limitation of previous PCA+LDA approaches. Specifically, the LDA/QZ algorithm uses the QZ decomposition to circumvent non-singularity requirement of the LDA and thus can perform discriminant analysis in the complete PCA space. Thus, the LDA/QZ algorithm has two advantages compared with previous "PCA+LDA" methods: 1) no intermediate dimensionality selection procedure is required 2) no discriminatory information is lost theoretically [13]. Moreover, the Gabor Filter-based Features [9][10][11][12] and the new LDA/QZ algorithm are combined for face recognition. The feasibility of the LDA/QZ algorithm and its combination of Gabor feature have been successfully tested on six face recognition experiments using three data sets from the ORL, AR [6], and FERET [5] database. In addition, Comparative experimental studies of several state-of-art dimension reduction algorithms and their combination of Gabor feature for face recognition are given.

2 Linear Discriminant Analysis by QZ Decomposition (LDA/QZ)

2.1 PCA Plus LDA and Its Limitation

For simplicity of discussion, we will assume the training data vectors $a_1, \ldots, a_n$ form column of a matrix $A \in \mathbb{R}^{m \times n}$ and are grouped into k class as $A = [A_1, A_2, \ldots, A_k]$, where $A_i \in \mathbb{R}^{m \times n_i}$ and $\sum_{i=1}^{k} n_i = n$. Let N_i denotes the set of column indices that belong to class i. The class centroid $c^{(i)}$ is computed by taking the average of columns in the class i, i.e., $c^i = \frac{1}{n_i}\sum_{j \in N_i} a_j$, and the global centroid c is defined as $c = \frac{1}{n}\sum_{j=1}^{n} a_j$. Then the within-class and between-class, and mixture scatter matrices are defined [4] as

$$S_w = \frac{1}{n}\sum_{i=1}^{k}\sum_{j \in N_i}(a_j - c^{(i)})(a_j - c^{(i)})^T, \tag{1}$$

$$S_b = \frac{1}{n}\sum_{i=1}^{k} n_i(c^{(i)} - c)(c^{(i)} - c)^T, \text{ and} \tag{2}$$

$$S_m = \frac{1}{n}\sum_{i=1}^{n}(a_j - c)(a_j - c)^T, \tag{3}$$

respectively. LDA wants to find a linear transformation W, $W \in \mathbb{R}^{m \times l}$, with $l \ll m$ that makes the within-class scatter small and the between-class scatter large in the projected space. This simultaneous optimization can be approximated the criterion as follows:

$$J(W) = \arg\max_{W} \frac{|W^T S_b W|}{|W^T S_w W|}, \tag{4}$$

However, this criterion cannot be applied when the matrix S_w is singular, a situation that occurs when that data matrix $A \in \mathbb{R}^{m \times n}$ has $m > n$. Such is often the case of the face image database. This is referred to as the small sample size problem in face recognition. As mentioned in introduction, a common approach for dealing with the singularity of S_w uses PCA as a first stage, followed by the discriminant analysis stage. The goal of the first stage is to reduce the dimension of the data matrix enough so that the new S_w is nonsingular, and LDA can be performed. Specifically, the PCA factorizes S_m in to the following form:

$$S_m = U\Sigma U^T \text{ with } U = [\phi_1\phi_2\cdots\phi_m],\ \Sigma = \text{diag}\{\lambda_1, \lambda_2, \ldots, \lambda_m\} \tag{5}$$

where $U \in \mathbb{R}^{m \times m}$ is an orthogonal eigenvector matrix and $\Sigma \in \mathbb{R}^{m \times m}$ is a diagonal eigenvalue matrix with diagonal elements in decreasing order ($\lambda_1 \geq \lambda_2 \geq \cdots \geq \lambda_N$). An important property of PCA is its optimal singal reconstruction in the sense of minimum mean-square error when only a subset of principal components (eigenvectors) is used to represent the original singal. Following this property, the data dimension can be reduce from m to l as follows:

$$\tilde{A} = U_l A, \tag{6}$$

where $U_l = [\phi_1\phi_2\cdots\phi_l]$, $U_l \in \mathbb{R}^{m \times l}$, and $\tilde{A} \in \mathbb{R}^{l \times n}$. In the PCA space, the reduced within- and between- class matrices can be derived as $\tilde{S}_w = U_l^T S_w U_l$ and $\tilde{S}_b = U_l^T S_b U_l$ respectively. Then there are two common procedures to solve the optimization of $J(W)$ for the LDA stage: 1) computing the eigenvectors of matrix $\tilde{S}_w^{-1}\tilde{S}_b$, as in the cases of Fisherfaces [2], ILDA [21], and GA-Fisher [16]; 2) diagonalizing $\tilde{S}_w$ and $\tilde{S}_b$ simultaneously, as in the cases of EFM [17] and MDF [3].

The major limitation of the "PCA+LDA" approaches is that the optimal value of the reduced dimension (l) for PCA is difficult to determine. On one hand, a large value of l may make the $\tilde{S}_w$ ill-conditioned, and lead the LDA procedure to overfit. On the other hand, a small l may lose some useful information for discrimination.

2.2 QZ Algorithm for LDA

It is well known that LDA can be performed in the PCA space by solving following generalized eigenvalue problem:

$$\tilde{S}_b x = \lambda \tilde{S}_w x, \tag{7}$$

where $\tilde{S}_w$ and $\tilde{S}_b$ are the within- and between- class scatter matrices in the PCA space. To avoid losing discriminatory information in the PCA stage, LDA should be performed in the complete PCA space, with dimension of $l = rank(S_m)$ [13]. However, $\tilde{S}_w$ remains singular in this situation, and thus most of the algorithms which purport to solve the generalized eigenvalue problem are less than satisfactory. An exception is the QZ algorithm, which does not require matrix inversion and it unaffected by the condition of $\tilde{S}_w$ [15].

The QZ decomposition attempts to find orthogonal matrices Q and Z which simultaneously reduce $\tilde{S}_b$ and $\tilde{S}_w$ to the triangular form. The eigenvalues of the original problem can be determined by dividing the diagonal elements of the triangularized S_b by the corresponding diagonal elements of the triangularized S_w matrix. Although QZ-algorithm provide an stable solution to solve generalized eigenvalue problems, it has *not* been explicitly used for LDA under the singularity or ill-condition situation. The detailed description of the QZ algorithm can be found in [8] and a standard implementation of the QZ algorithm in Matlab 7.0 is used in the following experiments reported in this paper.

2.3 LDA/QZ Algorithm

Based on the above discussion, we propose an extension of classical LDA, namely, LDA/QZ. This algorithm follows the common "PCA+LDA" framework, but it has two significant virtues compared with previous algorithms: 1) In PCA step, LDA/QZ uses all positive principal components and transform the feature space into $\mathbb{R}^l$, where $l = rank(S_m)$, and 2) In LDA step, the QZ-decomposition is applied to solve the generalized eigenvalue problem, so that LDA can be performed even when S_w is singular. The detail algorithm is as follows.

LDA/QZ Algorithm
Input: Data matrix $A \in \mathbb{R}^{m \times n}$ with k class and an input vector $a \in \mathbb{R}^{m \times 1}$
Output: Optimal transformation matrix $G \in \mathbb{R}^{m \times (k-1)}$ and the $k-1$ dimensional representation y of a

---

PCA stage:

1. Compute the l eigenvectors $\{u_i\}_{i=1}^{l}$ of S_m with positive eigenvalues, where $l = rank(S_m)$.
2. $\tilde{A} \leftarrow U_l^T A$, where $U_l = [u_1, \cdots, u_l] \in \mathbb{R}^{m \times l}$, $\tilde{A} \in \mathbb{R}^{l \times n}$.

LDA stage:

3. Construct the matrices $\tilde{S}_w$ and $\tilde{S}_b$ based on $\tilde{A}$.
4. Solve equation $\tilde{S}_b w_i = \lambda_i \tilde{S}_w w_i$ using QZ algorithm, select $k-1$ eigenvectors $\{w_i\}_{i=1}^{k-1}$ with largest eigenvalues.
5. $G \leftarrow U_l W$, where $W = [w_1, \cdots, w_{k-1}]$.
6. $y \leftarrow G^T a$.

---

3 Combination of Gabor Feature and LDA/QZ for Face Recognition

3.1 Gabor Filter-Based Features

For almost three decades the use of features based on Gabor Filters has been promoted for their useful properties, which are mainly related the invariance to illumination, rotation, scale, and translation [9][10]. Gabor filters have succeeded in many application, from texture analysis to iris and face recognition. In this paper, the Gabor wavelet representation of an face image is the convolution of the image with a family of Gabor kernels (filters) defined as:

$$\psi(z) = \frac{\|k_{\mu,\nu}\|^2}{\sigma^2} e^{(-\|k_{\mu,\nu}\|^2 \|z\|^2 / 2\sigma^2)} [e^{ik_{\mu,\nu} z} - e^{-\sigma^2/2}], \tag{8}$$

where μ and ν define the orientation and scale of the Gabor kernels, $z = (x, y)$, $\|\cdot\|$ denotes the norm operator, and the wave vector $k_{\mu,\nu}$ is defined as $k_{\mu,\nu} = k_\nu e^{i\phi_u}$ where $k_\nu = k_{max}/f^\nu$ and $\phi_u = \pi\mu/8$. k_{max} is the maximum frequency, and f is the spacing factor between kernels in the frequency domain.

As in most cases of face recognition [11][12][17], we use Gabor wavelets of five different scales, $\nu \in \{0, \dots, 4\}$, eight orientations, $\mu \in \{0, \dots, 7\}$, and with following parameters: $\sigma = 2\pi$, $k_{max} = \pi/2$, and $f = \sqrt{2}$. The output of the 40 Gabor filters are first down-sampled by a factor $\rho = 64$, and then concatenated to an augmented feature vectors $\mathcal{X}^{(\rho)}$, with a dimensionality of 10, 240.

3.2 Discriminant Analysis and Classification

The combination method applies the LDA/QZ algorithm on the augmented Gabor feature vector $\mathcal{X}^{(\rho)}$. The discriminative Gabor feature vector with low dimensionality, $\mathcal{U}^{(\rho)}$, of the image is defined as follows:

$$\mathcal{U}^{(\rho)} = G^T \mathcal{X}^{(\rho)} \tag{9}$$

Let $\mathcal{M}_i^0$, $i = 1, 2, \dots, k$, be the prototype, the mean of training samples, for class ω_i in the feature space. The combination method applies the nearest neighbor (to the mean) rule for classification using similarity measure δ

$$\delta(\mathcal{U}^{(\rho)}, \mathcal{M}_i^0) = \min_j \delta(\mathcal{U}^{(\rho)}, \mathcal{M}_j^0) \rightarrow \mathcal{U}^{(\rho)} \in \omega_j. \tag{10}$$

The discriminant Gabor feature vector, $\mathcal{U}^{(\rho)}$, is classified as belonging to the class of the closet mean, $\mathcal{M}_j^0$), using the cosine similarity measure, δ_{cos}, which is defined as follows:

$$\delta_{cos}(\mathcal{X}, \mathcal{Y}) = \frac{-\mathcal{X}^T \mathcal{Y}}{\|\mathcal{X}\| \, \|\mathcal{Y}\|}. \tag{11}$$

4 Experimental Result

We assess the feasibility and performance of the LDA/QZ method and its combination with Gabor feature on the face recognition task, using the data sets from ORL, AR, FERET database respectively. Specifically, the ORL dataset contains the total 400 images corresponding to 40 subjects. The FERET subset involves 1400 images corresponding to 200 individuals (each individual has seven images marked with "ba", "bj", "bk", "be", "bf", "bd", and "bg"). The AR subset includes 1000 image of 100 individuals (each individual has ten images marked with "1", ..., "7", "11", "12", and "13"). For the FERET and AR datasets, the facial region of each original image was automatically cropped based on the location of the eyes. All images are resized to 128×128 pixels in the experiments. Fig.1 shows some cropped facial images used in our experiments and one can see that they vary in illumination, expression, pose, accessory and scale.

Two or three images of each subject are randomly chosen for training, while the remaining images are used for testing. Thus, using the three data sets, we can set up six experiments as show in Table 1. All the input feature vectors are normalized to unit length in the experiments. As the dimension of the image vector(16,384) and Gabor feature vector (10,240) both are much larger than the training sample size (n), the small sample size problem will occur in all the experiments. To make a fair comparison, all the output features derived by the tested algorithms are used for classification. Specifically, the tested algorithms and corresponding output dimension are PCA ($n-1$), Orthogonal Centroid (k), LDA/QZ ($k-1$), Fisherfaces ($min(k-1,l)$), EFM ($min(k-1,l)$), LDA/QR ($k-1$), LDA/GSVD ($k-1$). For all experiments, the classification is performed by the cosine similarity based nearest center classifier, as described in section 3.2, and the performance is evaluated in term of the recognition rate. In order to show significant results, we randomly obtain 10 different training and testing sets in each experiment, and run the algorithms 10 times. The average recognition rate and standard deviation are reported.

Table 1. The Six Experiments Used for Evaluation

Experiments	ORL-2	ORL-3	FERET-2	FERET-3	AR-2	AR-3
number of classes (k)	40	40	200	200	100	100
trainging sample size (n)	80	120	400	600	200	300
testing sample size	320	280	1000	800	800	700
feature dimension	16,384 (pixel) *or* 10,240 (Gabor)					

In first series of experiments, the LDA/QZ algorithm is compared with other two two-stage PCA+LDA methods in face recognition, namely the Fisherfaces method and the EFM method. The results in the six experiments are summarized in Fig. 1, where the x-axis shows the intermediate dimension after the PCA stage (l), and y-axis shows the mean recognition rate from ten runs. The Fisherfaces and EFM are tested under varying l, ranging from a low of 5 to $rank(S_w)$.

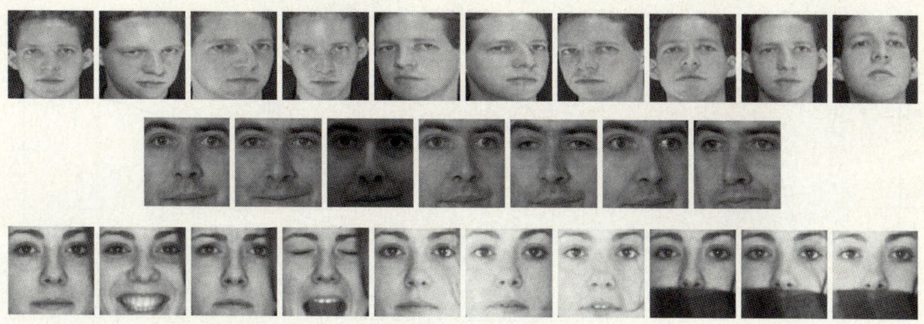

Fig. 1. Example (cropped) images used in our experiments. Top: ten image samples from one person in the ORL dataset. Middle: seven images of one person in the FERET dataset. Button: ten images of one individual in the AR dataset.

The main observations from these experiments are: 1) The Fisherfaces method generally shows better performance with increasing l. However, its best performance may lower than that of the EFM method, as it is in case of the FERET-2 and FERET-3 experiments. 2) The EFM method performs best with a middle value of l. Its performance will deteriorate with further increase of l, especially when the value of l gets close to $rank(S_w)$. 3) The overall results show that the LDA/QZ method is better than both the Fisherfaces and the EFM in term of recognition accuracy, even if they have a optimal choice of l. The superiority of the LDA/QZ is highlighted when only two training samples per class is available.

The next series of experiments compares the LDA/QZ algorithm with other State-of-Art dimension reduction algorithms, including PCA [7], Orthogonal Centroid (OC) [20], LDA/QR [19], LDA/GSVD [20], Improved LDA [21], CKFD [18]. Note that the accuracies of the ILDA and CKFD are cited from the original papers, which may use different preprocessing and classification procedures. The results on the six experiments are shown in Table 2. The main observations from these experiments are: 1) LDA based dimension reduction algorithms have much better performance than the PCA based algorithms (PCA and OC), when the data sets contain large within-class variations, as it is the case with the FERET and AR database. 2) LDA/QR algorithm has considerably lower accuracies than the LDA/GSVD when large within-class variations are presented. This may because the LDA/QR performs discriminant analysis in the subspace spanned by the class centers, which loses the discriminatory information in the null space of S_b. 3) The LDA/QZ performs best in the first five experiments, and slightly worse than LDA/GSVD in the sixth experiment. Since both LDA/QZ and LDA/GSVD perform discriminant analysis in the complete sample space, the better accuracies of the LDA/QZ may come from the numerical advantages of the QZ-decomposition over the generalized SVD.

The final series of experiments exploits the Gabor wavelet representation. We apply the dimension reduction algorithms to the Gabor feature vector $\mathcal{X}^{(\rho)}$, described in section 3.1. The results are listed in Table 3, and one can see that the accuracies improve by considerable margin for all the dimension reduction

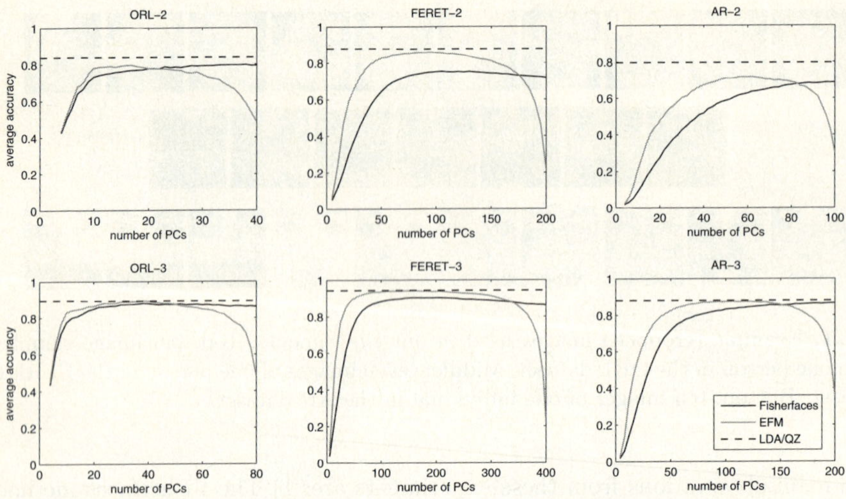

Fig. 2. Comparative Performance of the Fisherfaces, EFM, and LDA/QZ in the Six Experiments Based on Image Pixel Feature. The mean of accuracies (%) from ten runs are shown.

algorithms in all data sets. This finding qualifies the Gabor representation as a discriminating representation method. Besides the major observation mentioned above, we also find: 1) The performance enhancement of the LDA based algorithms is generally small than the PCA based algorithms. This may because the within-class variations are partially suppressed by the Gabor feature, and then the performance gap between PCA and LDA is narrowed. 2) The LDA/QZ has similar performance with the LDA/GSVD when using Gabor filter-based feature. This also may because the within-class variations become smaller when the face images are

5 Conclusion

In this paper, we propose a LDA/QZ algorithm and its combination of Gabor Filter-based features for the face recognition. The LDA/QZ algorithm follows the common "PCA+LDA" framework, but it has two significant virtues compared with previous algorithms: 1) In PCA step, LDA/QZ use all positive principal components and transform the feature space into complete PCA space, so that all discriminatory information is preserved, and 2) In LDA step, the QZ-decomposition is applied to solve the generalized eigenvalue problem, so that LDA can be performed stably even when S_w is singular. Moreover, the Gabor Filter-based Features and the new LDA/QZ algorithm are combined for face recognition. Comparative experimental studies of several state-of-art dimension reduction algorithms and their combinations of Gabor feature for face recognition are also performed, and the results show our LDA/QZ algorithm is always the best or comparable to best in term of recognition accuracy.

Table 2. Comparative Performance of the LDA/QZ and the state-of-art dimension reduction algorithms in the Six Experiments Based on Image Pixel Feature. The mean and standard deviation of accuracies (%) from ten runs are shown.

Method	ORL-2	ORL-3	FERET-2	FERET-3	AR-2	AR-3
PCA	79.88(3.40)	85.50(1.94)	36.75(1.30)	48.16(1.35)	20.25(1.57)	25.94(2.39)
OC	79.56(3.37)	85.39(1.75)	36.69(1.31)	48.15(1.29)	20.24(1.55)	25.93(2.40)
LDA/QR	76.88(2.52)	84.18(3.81)	66.91(1.14)	79.70(1.19)	69.03(2.25)	77.57(2.19)
LDA/GSVD	80.81(2.82)	88.04(2.20)	78.57(0.74)	89.99(0.62)	76.65(2.20)	**88.56**(1.35)
LDA/QZ	**83.78**(3.09)	**89.32(2.57)**	**87.69**(2.04)	**94.31**(1.18)	**79.66**(2.80)	88.10(3.24)
ILDA [21]	83.69(1.70)	87.71(1.31)	N/A	N/A	N/A	N/A
CKFD [18]	N/A	N/A	N/A	88.38(1.57)	N/A	N/A

Table 3. Comparative Performance of the LDA/QZ and other dimension reduction algorithms in the Six Experiments Based on Gabor Filter-based Feature. The mean and standard deviation of accuracies (%) from ten runs are shown.

Method	ORL-2	ORL-3	FERET-2	FERET-3	AR-2	AR-3
PCA	83.69(2.58)	90.14(2.08)	59.30(1.25)	71.02(1.45)	55.69(2.75)	66.46(2.44)
OC	83.31(2.54)	89.86(2.20)	59.28(1.37)	70.96(1.37)	55.26(2.70)	65.91(2.21)
LDA/QR	83.31(3.60)	90.07(2.56)	85.41(1.46)	91.19(1.22)	82.80(1.43)	87.46(0.94)
LDA/GSVD	**86.84**(2.59)	**93.32**(2.45)	90.64(0.77)	96.29(0.56)	**86.14**(1.42)	93.73(0.81)
LDA/QZ	82.47(4.53)	92.11(3.40)	**92.63**(2.04)	**97.04**(0.69)	86.00(2.57)	**94.04**(1.59)

Acknowledgements

This research was sponsored by NSFC (National Natural Science Foundation of China) under Grant No.60475007 and the Foundation of China Education Ministry for Century Spanning Talent.

References

1. Zhao W., Chellappa R., Rosenfeld A., and Phillips P.J.: Face Recognition: A Literature Survey. ACM Computing Surveys. (2003) 399–458
2. Belhumeour P.N., Hespanha J.P., and Kriegman D.J.: Eigenfaces versus Fisherfaces: Recognition Using Class Specific Linear Projection. IEEE Trans. Pattern Analysis and Machine Intelligence, **19** (1997) 711–720
3. Swets D.L., and Weng J.: Using discriminant eigenfeatures for image retrieval. IEEE Trans. Pattern Anal. Machine Intell. **18** (1996) 831–836
4. Fukunaga K.: Introduction to Statistical Pattern Recognition, second ed. Academic Press. (1990)

5. Phillips P.J., Wechsler H., and Rauss P.: The FERET Database and Evaluation Procedure for Face-Recognition Algorithms. Image and Vision Computing. **16** (1998) 295–306
6. Martinez, A.M., and Benavente, R.: The AR Face Database. http://rvl1.ecn. purdue.edu/ aleix/aleix_face_DB.html. (2003)
7. Turk,M., and Pentland, A.: Eigenfaces for Recognition. J. Cognitive Neuroscience. **13** (1991) 71–86
8. Friedman, J.H.: Regularized Discriminant Analysis. J. Am. Statistical Assoc. **84** (1989) 165-175
9. Daugman, J.G.: Uncertainty Relation for Resolution in Space, Spatial Frequency, and Orientation Optimized by Two-Dimensional Cortical Filters. J. Optical Soc. Am.. **2** (1985) 1160–1169 .
10. Daugman, J.G: Two-Dimensional Spectral Analysis of Cortical Receptive Field Profiles. Vision Research. **20** (1980) 847-856
11. Deng, W., Hu, J., and Guo, J.: Gabor-Eigen-Whiten-Cosine: A Robust Scheme for Face Recognition. Lecture Notes in Computer Science. **3723** (2005) 336-349.
12. Liu, C.: Capitalize on Dimensionality Increasing Techniques for Improving Face Recognition Grand Challenge Performance. IEEE Trans. Pattern Anal. Machine Intell. **28** (2006) 725–737
13. Yang, J., and Yang, J.Y.: Why Can LDA Be Performed in PCA Transformed Space? Pattern Recognition. **36** (2003) 563–566.
14. Moler, C.B., and Stewart, G.W.: An Algorithm for Generalized Matrix Eigenvalue Problems. SIAM J. Num. Anal. **10** (1973) 241–256
15. Kaufman, L.: Some Thoughts on the QZ Algorithm for Solving the Generalized Eigenvalue Problem. ACM Trans. Math. Soft. **3**, (1977) 65–75
16. Zheng, W.S., Lai, J.H., and Yuen, P.C.: GA-Fisher: A New LDA-Based Face Recognition Algorithm With Selection of Principal Components. IEEE Trans. Syst., Man, Cybern. B. **35** (2005) 1065–1078
17. Liu, C., and Wechsler, H.: Gabor Feature Based Classification Using the Enhanced Fisher Linear Discriminant Model for Face Recognition. IEEE Trans. Image Processing. **11** (2002) 467–476
18. Yang, J., Alejandro, F.F., Yang, J., Zhang, D., and Jin, Z.: KPCA Plus LDA: A Complete Kernel Fisher Discriminant Framework for Feature Extraction and Recognition. IEEE Trans. Pattern Anal. Machine Intell. **27** (2005) 230-244
19. Ye, J., and Li, Q.: A Two-Stage Linear Discriminant Analysis via QR-Decomposition. IEEE Trans. Pattern Anal. Machine Intell. **27** (2005) 929–941
20. Howland, P., and Park, H.: Generalizing Discriminant Analysis Using the Generalized Singular Value Decomposition. IEEE Trans. Pattern Anal. Machine Intell. **26** (2004) 995–1006
21. Jing, X.Y., Zhang, D., and Tang, Y.Y.: An Improved LDA Approach. IEEE Trans. Syst., Man, Cybern. B. **34** (2004) 1942–1951

Breast Cancer Detection Using Hierarchical B-Spline Networks

Yuehui Chen, Mingjun Liu, and Bo Yang

School of Information Science and Engineering
Jinan University, Jinan 250022, P.R. China
yhchen@ujn.edu.cn

Abstract. In this paper, an optimized hierarchical B-spline network was employed to detect the breast cancel. For evolving a hierarchical B-spline network model, a tree-structure based evolutionary algorithm and the Particle Swarm Optimization (PSO) are used to find an optimal detection model. The performance of proposed method was then compared with Flexible Neural Tree (FNT), Neural Network (NN), and Wavelet Neural Network (WNN) by using the same breast cancer data set. Simulation results show that the obtained hierarchical B-spline network model has a fewer number of variables with reduced number of input features and with the high detection accuracy.

1 Introduction

Breast cancer is the most common cancer in women in many countries. Most breast cancers are detected as a lump/mass on the breast, or through self-examination or mammography [1]. Various artificial intelligence techniques have been used to improve the diagnoses procedures and to aid the physician's efforts [2][3]. In our previous studies, the performance of Flexible Neural Tree (FNT) [5], Neural Network (NN), Wavelet Neural Network (WNN) and an ensemble method to detect breast-cancer have been evaluated [6].

Hierarchical B-spline networks consist of multiple B-spline networks assembled in different level or cascade architecture. In this paper, an automatic method for constructing hierarchical B-spline network is proposed. Based on a pre-defined instruction/operator set, the hierarchical B-spline networks can be created and evolved. The hierarchical B-spline network allows input variables selection. In our previous studies, in order to optimize Flexible Neural Tree (FNT) the hierarchical structure of FNT was evolved using Probabilistic Incremental Program Evolution algorithm (PIPE) [4][5] and Ant Programming [7] with specific instructions. In this research, the hierarchical structure is evolved using the Extended Compact Genetic Programming (ECGP). The fine tuning of the parameters encoded in the structure is accomplished using Particle Swarm Optimization (PSO). The novelty of this paper is in the usage of hierarchical B-spline model for selecting the important input variables and for breast cancel detection.

L. Jiao et al. (Eds.): ICNC 2006, Part I, LNCS 4221, pp. 25–28, 2006.

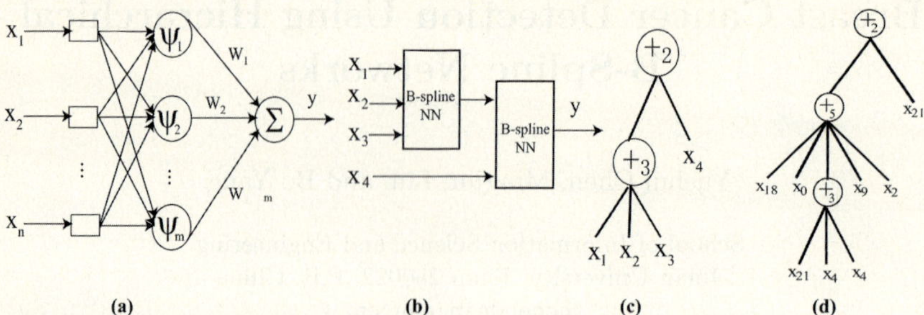

Fig. 1. A B-spline network (a), a hierarchical B-spline network (b), a tree-structural representation of the B-spline network (c), and the optimized hierarchical B-spline network for breast cancel detection

2 The Hierarchical B-Spline Network

Encode and Calculation. A function set F and terminal instruction set T used for generating a hierarchical B-spline network model are described as $S = F \bigcup T = \{+_2, +_3, \ldots, +_N\} \bigcup \{x_1, \ldots, x_n\}$, where $+_i (i = 2, 3, \ldots, N)$ denote non-leaf nodes' instructions and taking i arguments. $x_1, x_2, \ldots, x_n$ are leaf nodes' instructions. The output of a non-leaf node is calculated as a B-spline neural network model (see Fig.1 (a)(b)(c)). From this point of view, the instruction $+_i$ is also called a basis function operator with i inputs.

In this research, the translation and dilation of order 3 B-spline function is used as basis function,

$$N_3(a, b, x) = \begin{cases} \frac{9}{8} + \frac{3}{2}(\frac{x-b}{a}) + \frac{1}{2}(\frac{x-b}{a})^2, & x \in [-\frac{3}{2}a + b, -\frac{1}{2}a + b) \\ \frac{3}{4} - (\frac{x-b}{a})^2, & x \in [-\frac{1}{2}a + b, \frac{1}{2}a + b) \\ \frac{9}{8} - \frac{3}{2}(\frac{x-b}{a}) + \frac{1}{2}(\frac{x-b}{a})^2, & x \in [\frac{1}{2}a + b, \frac{3}{2}a + b] \\ 0, & otherwise \end{cases} \quad (1)$$

and the number of B-spline basis functions used in hidden layer of the network is same with the number of inputs, that is, $m = n$.

In the creation process of hierarchical B-spline network tree, if a nonterminal instruction, i.e., $+_i (i = 2, 3, 4, \ldots, N)$ is selected, i real values are randomly generated and used for representing the connection strength between the node $+_i$ and its children. In addition, $2 \times n^2$ adjustable parameters a_i and b_i are randomly created as B-spline basis function parameters. Finding an optimal or near-optimal hierarchical B-spline network structure is formulated as a product of evolution. In our previously studies, the Genetic Programming (GP), Probabilistic Incremental Program Evolution (PIPE) have been explored for structure optimization of the FNT. In this paper, the ECGP is employed to find an optimal or near-optimal structure of hierarchical B-spline networks.

Optimization. The ECGP [8] is employed to find an optimal or near-optimal HRBF structure. ECGP is a direct extension of ECGA to the tree representation which is based on the PIPE prototype tree. In ECGA, Marginal Product Models

(MPMs) are used to model the interaction among genes, represented as random variables, given a population of Genetic Algorithm individuals. MPMs are represented as measures of marginal distributions on partitions of random variables. ECGP is based on the PIPE prototype tree, and thus each node in the prototype tree is a random variable. ECGP decomposes or partitions the prototype tree into sub-trees, and the MPM factorises the joint probability of all nodes of the prototype tree, to a product of marginal distributions on a partition of its sub-trees. A greedy search heuristic is used to find an optimal MPM mode under the framework of minimum encoding inference. ECGP can represent the probability distribution for more than one node at a time. Thus, it extends PIPE in that the interactions among multiple nodes are considered. The parameters embedded in the hierarchical B-spline network is optimized by PSO algorithm.

3 Simulations

The Wisconsin breast cancer data set has 30 attributes and 569 instances of which 357 are of benign and 212 are of malignant type. The data set is randomly divided into a training data set and a test data set. The first 285 data is used for training and the remaining 284 data is used for testing the performance of the different models. Binary classification is adopted in this research.

All the models were trained and tested with the same set of data. The instruction sets used to create an optimal hierarchical B-spline network classifier is $S = F \bigcup T = \{+_2, \ldots, +_5\} \bigcup \{x_0, x_1, \ldots, x_{29}\}$. Where $x_i (i = 0, 1, \ldots, 29)$ denotes the 30 input features. The optimal hierarchical B-spline network for breast cancel detection problem is shown in Figure 1(d). The classification results for testing data set are shown in Table 1. For comparison purpose, the detection performances of the FNT, NN and WNN are also shown in Table 1 (for details, see [6]). The impor-

Table 1. Comparative results of the FNT, NN, WNN [6] and the proposed hierarchical B-spline network classification methods for the detection of breast cancer

Cancer	FNT(%)[6]	NN(%)[6]	WNN(%)[6]	H-Bspline(%)
Benign	93.31	94.01	94.37	96.83
Malignant	93.45	95.42	92.96	96.83

Table 2. Comparison of false positive rate (fp) and true positive rate (tp) for FNT, NN, WNN [6] and hierarchical B-spline network

Cancer	FNT[6]		NN[6]		WNN[6]		H-Bspline	
diagnosis	fp(%)	tp(%)	fp(%)	tp(%)	fp(%)	tp(%)	fp(%)	tp(%)
Benign	3.88	91.71	4.85	93.37	6.8	98.34	2.91	96.69
Malignant	2.76	86.41	4.97	96.12	9.4	97.09	3.31	97.09

tant features for constructing the hierarchical B-spline models are x_0, x_2, x_3, x_7, x_9, x_{18}, x_{21}. It should be noted that the obtained hierarchical B-spline network classifier has smaller size and reduced features and with high accuracy in breast cancel detection. Receiver Operating Characteristics (ROC) analysis of the FNT, NN, WNN and the hierarchical B-spline network model is shown in Table 2.

4 Conclusion

In this paper, we presented an optimized hierarchical B-spline network for the detection of breast cancel and compared the results with some advanced artificial intelligence techniques, i.e., FNT, NN and WNN. As depicted in Table 1, the preliminary results are very encouraging. The best accuracy was offered by the hierarchical B-spline network method followed by the wavelet neural network for detecting benign types and PSO trained neural network for detecting the malignant type of cancer. ROC analysis (Table 2) illustrates that wavelet neural network has the highest false positive rate and the H-Bspline and FNT models have the lowest false positive rates for detecting benign and malignant cancer, respectively.

Acknowledgement

This research was partially supported the Natural Science Foundation of China under contract number 60573065, and The Provincial Science and Technology Development Program of Shandong under contract number SDSP2004-0720-03.

References

1. DeSilva, C.J.S. et al., Artificial Neural networks and Breast Cancer Prognosis, The Australian Computer Journal, 26, pp. 78-81, 1994.
2. David B. Fogel , Eugene C. Wasson , Edward M. Boughton and Vincent W. Porto, A step toward computer-assisted mammography using evolutionary programming and neural networks, Cancer Letters, Volume 119, Issue 1, pp. 93-97, 1997.
3. Barbara S. Hulka and Patricia G. Moorman, Breast Cancer: Hormones and Other Risk Factors, Maturitas, Volume 38, Issue 1, pp. 103-113, 2001.
4. Chen, Y., Yang, B., Dong, J., Nonlinear systems modelling via optimal design of neural trees, International Journal of Neural ystems, **14**, pp. 125-138, 2004.
5. Chen, Y., Yang, B., Dong, J., Abraham A., Time-series forcasting using flexible neural tree model, Information Science, Vol.174, Issues 3/4, pp.219-235, 2005.
6. Chen, Y., Abraham, A., Yang, B., Hybrid Neurocomputing for Breast Cancer Detection, The Fourth International Workshop on Soft Computing as Transdisciplinary Science and Technology (WSTST'05), pp.884-892, Springer, 2005.
7. Chen, Y., Yang, B. and Dong, J., Evolving Flexible Neural Networks using Ant Programming and PSO algorithm, International Symposium on Neural Networks (ISNN'04), LNCS 3173, pp. 211-216, 2004.
8. K. Sastry and D. E. Goldberg. "Probabilistic model building and competent genetic programming", In R. L. Riolo and B. Worzel, editors, Genetic Programming Theory and Practise, chapter 13, pp. 205-220. Kluwer, 2003.
9. Merz J., and Murphy, P.M., UCI repository of machine learning databases, http://www.ics.uci.edu/- learn/MLRepository.html, 1996.

Ensemble-Based Discriminant Manifold Learning for Face Recognition

Junping Zhang[1,2], Li He[1], and Zhi-Hua Zhou[3]

[1] Shanghai Key Laboratory of Intelligent Information Processing
Department of Computer Science and Engineering
Fudan University, Shanghai 200433, China
jpzhang@fudan.edu.cn, demonstrate@163.com

[2] The Key Laboratory of Complex Systems and Intelligence Science
Institute of Automation, Chinese Academy of Sciences, Beijing, 100080, China

[3] National Laboratory for Novel Software Technology
Nanjing University, Nanjing 210093, China
zhouzh@nju.edu.cn

Abstract. The locally linear embedding (LLE) algorithm can be used to discover a low-dimensional subspace from face manifolds. However, it does not mean that a good accuracy can be obtained when classifiers work under the subspace. Based on the proposed ULLELDA (Unified LLE and linear discriminant analysis) algorithm, an ensemble version of the ULLELDA (En-ULLELDA) is proposed by perturbing the neighbor factors of the LLE algorithm. Here many component learners are generated, each of which produces a single face subspace through some neighborhood parameter of the ULLELDA algorithm and is trained by a classifier. The classification results of these component learners are then combined through majority voting to produce the final prediction. Experiments on several face databases show the promising of the En-ULLELDA algorithm.

1 Introduction

During the past decades, much research on face recognition has been done by computer scientists [1]. Many assumptions are generated for building a high-accuracy recognition system. One assumption is that a facial image can be regarded as single or multiple arrays of pixel values. The other one is that face images are empirically assumed to points lying on manifolds which are embedded in the high-dimensional observation space [2,3]. As a result, many manifold learning approaches are proposed for discovering the intrinsic face dimensions, such as expression and pose, and so on [4,5]. However, when the manifold learning-based approaches are employed for discriminant learning, the accuracy of classifiers easily suffer. A possible reason is that the distances among the smallest d eigenvalues or eigenvectors which are obtained by the LLE algorithm are so small and close [6] that the face subspace is ill-posed and instability. Furthermore, the choose of neighbor factors influences the accuracy of classifiers.

L. Jiao et al. (Eds.): ICNC 2006, Part I, LNCS 4221, pp. 29–38, 2006.

To overcome the mentioned two disadvantages, we propose an ensemble ULLELDA (En-ULLEDA) algorithm. First, the previous proposed ULLELDA algorithm with a single-value neighborhood parameter is performed for obtaining a single subspace [7]. Second, a set of neighborhood parameters are employed for generating a collection of subspaces, each of which is used for discriminant learning with a specific classifier. Finally, the classification is performed by majority voting. Experiments on several face databases show the promising of the En-ULLELDA algorithm when compared with the ULLELDA algorithm.

The rest of the paper is organized as follows. In Section 2 we propose the En-ULLELDA algorithm. In Section 3 the experimental results are reported. Finally, In section 4 we conclude the paper with some discusses.

2 Ensemble of ULLELDA (En-ULLELDA)

For better understanding the proposed En-ULLELDA algorithm, we first give a brief introduction on the ULLELDA algorithm. The objective of the previous proposed ULLELDA algorithm is to refine the classification ability of manifold learning. The basic procedure of the algorithm is demonstrated as follows:

Step 1. Approximate manifold around sample $\boldsymbol{x}_i$ with a linear hyperplane passing through its neighbors $\{\boldsymbol{x}_j, j \in \mathcal{N}(\boldsymbol{x}_i)\}$. Therefore, define

$$\psi(\boldsymbol{W}) = \sum_{i=1}^{N} \|\boldsymbol{x}_i - \sum_{j \in \mathcal{N}(\boldsymbol{x}_i)} \boldsymbol{W}_{i,j} \boldsymbol{x}_j\|^2 \tag{1}$$

Subject to constraint $\sum_{j \in \mathcal{N}(\boldsymbol{x}_i)} \boldsymbol{W}_{i,j} = 1$, and $\boldsymbol{W}_{i,j} = 0, j \notin \mathcal{N}(x_i)$, all the weights $\boldsymbol{W}$ in Eq. 1 are calculated in the least square sense [4]. Where $\mathcal{N}(\boldsymbol{x}_i)$ denotes the index set of neighbor samples of x_i.

Step 2. The LLE algorithm assumes that the weight relationship of sample and its neighborhood samples is invariant when data are mapped into a low-dimensional subspace. Let $\boldsymbol{Y} = (\boldsymbol{y}_1, \ldots, \boldsymbol{y}_N)$. Therefore, define

$$\varphi(\boldsymbol{Y}) = \sum_{i=1}^{N} \|\boldsymbol{y}_i - \sum_{j \in \mathcal{N}(\boldsymbol{x}_i)} \boldsymbol{W}_{i,j} \boldsymbol{y}_j\|^2 \tag{2}$$

where $\boldsymbol{W}$ can minimize Eq. 1 and satisfies the constraints, and

$$\sum_i \boldsymbol{y}_i = 0 \tag{3}$$

$$\frac{1}{N} \sum_i \boldsymbol{y}_i \boldsymbol{y}_i^\top = \boldsymbol{I} \tag{4}$$

The optimal solution of $\boldsymbol{Y}^*$ in Eq. 2 can be transformed into computing the smallest or bottom $d+1$ eigenvectors of matrix $(\boldsymbol{I} - \boldsymbol{W})^T(\boldsymbol{I} - \boldsymbol{W})$ except that the zero eigenvalue needs to be discarded.

Step 3. The d-dimensional data $\boldsymbol{Y}$ are further mapped into a p-dimensional discriminant subspace through the LDA algorithm (linear discriminant analysis) [8]. Therefore, we have

$$\boldsymbol{Z} = \boldsymbol{DY} \tag{5}$$

Where $\boldsymbol{Z} = (z_i \in \mathbb{R}^p, i = 1, \cdots, N)$, and the size of matrix $\boldsymbol{D}$ is p rows with d columns which is calculated based on the LDA algorithm [8].

Step 4. To project out-of-the-samples into the discriminant subspace, the mapping idea of the LLE algorithm is employed [4]. A main difference between the LLE and the ULLELDA algorithms is that in the latter one, the out-of-the-samples are directly projected into a discriminant subspace without the computation of the LLE algorithm.

Step 5. A specific classifier is employed in the subspace obtained by the ULLELDA algorithm for discriminant learning.

The previous experiments on several face databases show that the combination of the proposed ULLELDA algorithm and some specific classifier has better accuracy than the combination of the traditional PCA (principal component analysis) algorithm and classifier [7]. However, two potential instabilities influence the performance of the ULLELDA algorithm:

1. Spectral decomposition in the LLE algorithm may generate different intrinsic low-dimensional subspaces even if the neighborhood parameters are fixed. The reason is that the differences among the principal eigenvalues obtained by the LLE algorithm are so small and close that the sequences among the principal eigenvalues are easily alternated.
2. The neighbor parameter in the LLE algorithm plays a trade-off role between global measure and local one. If the size is very large, the LLE algorithm is approximately equivalent to the classical linear dimensionality reduction approach. And if it is very small, the LLE algorithm cannot achieve an effective dimensionality reduction.

While some refinements on how to select a suitable neighborhood size had been proposed [9] for unsupervised manifold learning, it is still difficult for supervised learning to choose an optimal parameter because data are noisy and the mentioned two instabilities are dependent each other. As a result, choosing the neighbor parameter for supervised manifold learning still depends on user's experience. These problems motivate us to further improve the proposed ULLELDA algorithm through ensemble learning.

Krogh and Vedelsby [10] have derived a famous equation $\boldsymbol{E} = \overline{\boldsymbol{E}} - \overline{\boldsymbol{A}}$ in the case of regression, where $\boldsymbol{E}$ is the generalization error of an ensemble, while $\overline{\boldsymbol{E}}$ and $\overline{\boldsymbol{A}}$ are the average generalization error and average ambiguity of the component learners, respectively. The ambiguity was defined as the variance of the component predictions around the ensemble prediction, which measures the disagreement among the component learners [10]. This equation discloses that the more accurate and the more diverse the component learners are, the better the ensemble is. However, measuring diversity is not straightforward because there is no generally accepted formal definition, and so it remains a trick at present to generate accurate but diverse component learners. Several known tricks include perturbing the training data, perturbing input attributes and learning parameters.

As for the ULLELDA algorithm, it is clear that when neighbor factor is perturbed, a set of different discriminant subspaces which result in diversity will be generated. When a test sample is classified under the subspaces, the accuracy may be different from one subspace to the other one. Therefore, we can use majority voting to model ensemble ULLELDA algorithm (En-ULLELDA). Let ULLELDA_K be the ULLELDA algorithm with some neighbor factor K (Here K denotes the number of $\mathcal{N}(\cdot)$), and base classifier be BC, then the classification criterion of the En-ULLELDA algorithm is written as follows:

$$C(\boldsymbol{v}_i) = \arg\max_l \{K|(\text{ULLELDA}_K(v_i), \text{BC}) = l\} \tag{6}$$

Where $C(\boldsymbol{v}_i)$ denotes the label of test sample $\boldsymbol{v}_i$, and $(\text{ULLELDA}_K(\boldsymbol{v}_i), \text{BC})$ denotes a component learner with neighbor factor K and some specific base classifier. It is noticeable that a base classifier is explicit contained by the proposed En-ULLELDA framework.

A pseudo-code of the En-ULLELDA algorithm is illustrated as follows:

Table 1. The En-ULLELDA algorithm

Input:
  Data set X; Learner L; Neighborhood Parameters Set $K = \{k_1, k_2, \cdots, k_m\}$,
  Stepsize= $k_i - k_j, i - j = 1$
  Reduced Dimension d and Reduced Discriminant Dimension d', Trials T
Procedure:
1. Normalization of X
2. for $t = 1$ to T {
3.   Let $Enerror$ be an empty set
4.   for $k = k_1$: Stepsize: k_m
5.   {
6.   Generate training set $X_{training}$ and test set X_{test} with random partition
7.   Based on LLE, k and d,
    Calculate the corresponding one $Y_{training}$ of training set $X_{training}$
8.   Based on LDA and d',
    Calculate the corresponding one $Z_{training}$ of $Y_{training}$
9.   Project test set X_{test} into the subspace of $Z_{training}$ with ULLELDA,
    and obtain the corresponding one Z_{test}
10.   Compute error rate $e(Z_{test}, t, k, L)$ of Z_{test} based on $Z_{training}$ and learner L
11.   Storage classification labels $Lab(test, k, t, L)$ of test samples.
12.   }
13. $Enerror = \{Enerror; \text{MajorityVoting}(Lab(test, K, t, L))\}$
14. }
15. Output:
  The average error and standard deviation of $Enerror, t = 1, \cdots, T$

3 Experiments

To test the proposed En-ULLELDA algorithm, three face databases (the ORL database (40 subjects and 10 images per person) [11], the UMIST database (575 multi-view facial images with 20 subjects) [12], and the Yale database (15 individuals and 11 images per person) [13]) are used. By combining the ORL database and the UMIST database with the Yale database, a large database Henceforth the OUY face database) including 75 individuals and 1140 images is built. In the paper, the intensity of each pixel is characterized by one dimension, and the size of an image is $112 * 92$ pixels which form a 10304 dimensional vector. All the dimensions are standardized to the range $[0, 1]$. The training samples and test samples are randomly separated without overlapping. All the reported results are the average of 100 repetitions.

For overcoming the curse of dimensionality, the reduced dimensions of data based on the LLE algorithm are set to be 150. For the 2nd mapping (LDA-based reduction) of the ULLELDA algorithm, the reduced dimension is generally no more than $L - 1$ (Where L means the number of classes). Otherwise eigenvalues and eigenvectors will appear complex numbers. Actually, we only keep the real part of complex values when the 2th reduced dimensions are higher than $L - 1$. When the ULLELDA algorithm is used, the neighbor parameter need to be predefined. With broad experiments, let the neighbor size K be 40 for ORL, UMIST, and be 15 for OUY databases.

Finally, four base classifiers (1-nearest neighbor algorithm (NN) and nearest feature line (NFL) algorithm [14], the nearest mean algorithm (M) and the Nearest-Manifold-based (NM) algorithm) are employed to test classification performance based on the face subspace(s). Here the NFL algorithm denotes that classification is achieved by searching the nearest projection distance from sample to line segments of each class. And the NM algorithm is to calculate the minimum projection distance from each unknown sample to hyperplanes of different classes where each hyperplane is made up of three prototypes of the same class [15].

A set of comparative experiments between the En-ULLELDA algorithm and the ULLELDA algorithm are performed on three face databases. It is noticeable that due to the fact that the number of face images each class is different in the OUY database, we divide the database into training /test set based on the ratio of the number of training samples to the number of samples of the same class. And the ratio is equal to 0.4 when we investigate the influence of neighbor factor K. The reported results are shown in Fig. 1 to Fig. 3. In these figures, the ranges of neighbor factors are shown in the horizontal axis of Fig. 1 to Fig. 3. The vertical axis denotes the ratio of the error rate of the ULLELDA algorithm against that of the En-ULLELDA algorithm. Also, in the title of subplots on the OUY database, the abbreviation “ROUY” denotes that the training samples is sampled based on the mentioned ratio. If a value in vertical direction is greater than 1, it means that the En-ULLELDA algorithm has better recognition performance than the ULLELDA with single neighbor parameter. Furthermore, the experimental results obtained by the En-ULLELDA algorithm with base classifiers are

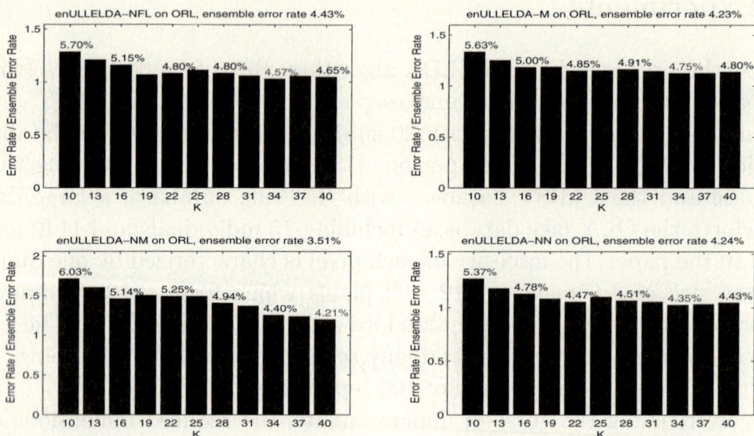

Fig. 1. The En-ULLELDA and The ULLELDA algorithms on The ORL Database

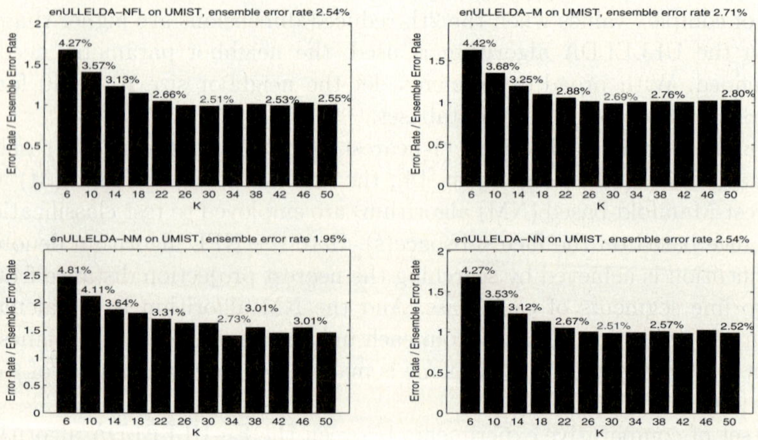

Fig. 2. The En-ULLELDA and The ULLELDA algorithms on The UMIST Database

shown in the titles of subplots. By analyzing the experimental results, it is not difficult to see that as K varies, the error rates have remarkable difference. For example, when $K = 10$, the error rate based on the ULLELDA+NM is 6.32%; and when $K = 40$, the error rate is 4.21% in the ORL database. Due to the fact that in real applications, it is almost always impossible to prior know which K value is the best, therefore it is not easy for the ULLELDA algorithm to get its lowest error rate. From the figures we can also see that the lowest error rates are obtained by the proposed En-ULLELDA algorithm. So, the En-ULLELDA algorithm is a better choice. In summary, the results argue that 1) the selection of K is a crucial factor to the performance of face recognition. 2) The accuracy

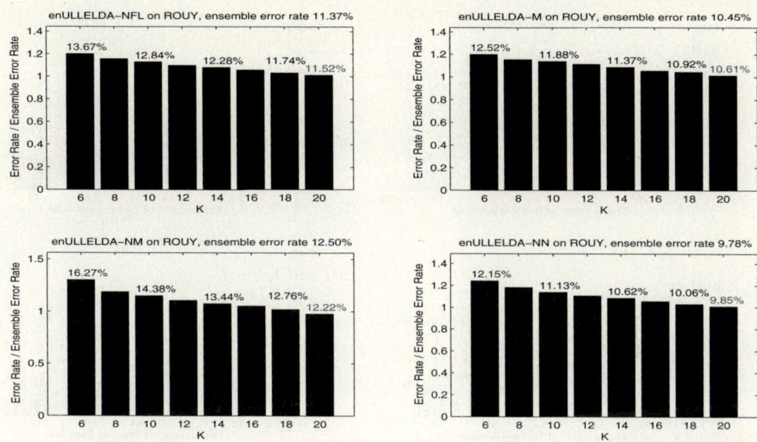

Fig. 3. The En-ULLELDA and The ULLELDA algorithms on The OUY face Database

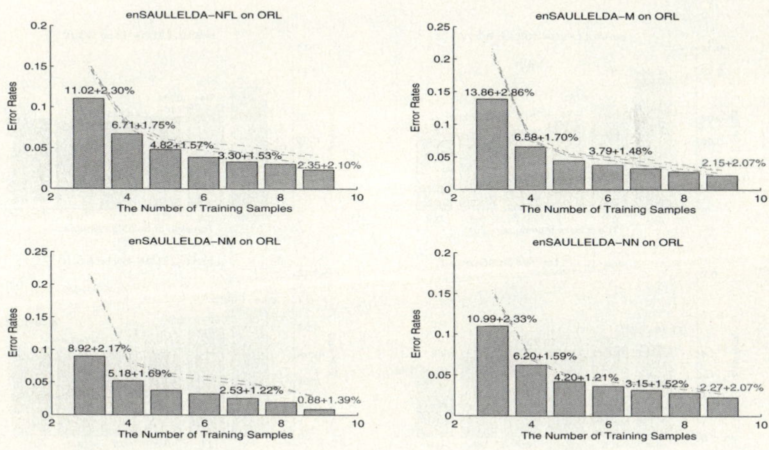

Fig. 4. The Influence of Training Samples on The ORL Database

of the En-ULLELDA algorithm is superior to that of the ULLELDA algorithm in all of the mentioned face databases.

For testing the generalization performance of the En-ULLELDA algorithm, the influence of the number of training samples is also studied. The results are displayed as in Fig. 4 through Fig. 6. It is worth noting that for better visualization, we only draw some main results achieved by the ULLELDA algorithm. In these figures, all the dashed lines represent the error rates of the ULLELDA algorithm with some specific neighbor factors. Each factor is in the range of $[10, 40]$. Also, each bar is error rate of the En-ULLELDA algorithm with fixed number of training samples. Meanwhile, in the top of each bar, "$x + x\%$" denotes the

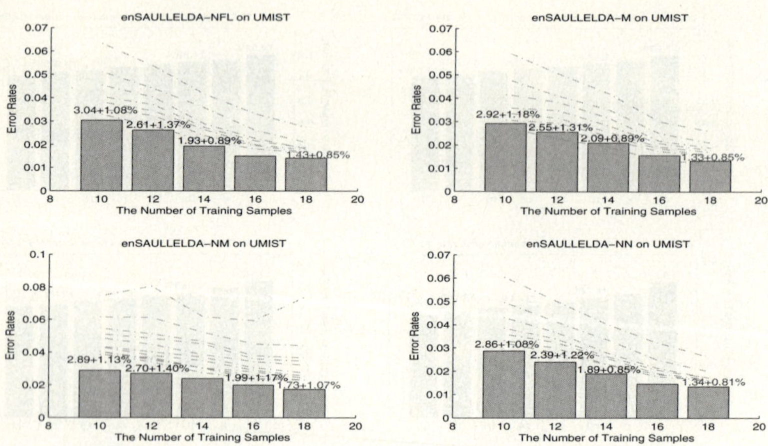

Fig. 5. The Influence of Training Samples on The UMIST Database

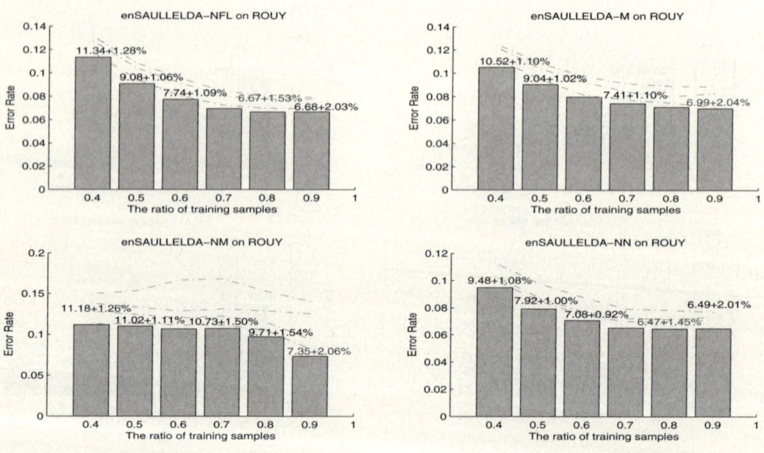

Fig. 6. The Influence of The Training Samples on The OUY Database

error rate plus standard deviation of the En-ULLELDA algorithm. From experimental results it can be seen that the En-ULLELDA with base learners obtains better recognition performance than the ULLELDA algorithm with single neighbor factor. For example, when the En-ULLEDA with NM algorithm is used for ORL face database, the lowest error rate and standard deviation are 0.88% and 1.39%, respectively (where training samples= 9). From these figures it can be seen that compared with the combination of the ULLELDA algorithm and three base classifiers, the En-ULLELDA algorithm has better recognition ability. It show again that the En-ULLELDA algorithm is a refinement of the ULLELDA algorithm in enhancing the recognition ability of base classifiers.

4 Conclusions

In this paper, we study ensemble-based discriminant manifold learning in face subspaces and propose the En-ULLELDA algorithm to improve the recognition ability of base classifiers and reduce the influences of choosing neighbor factors and small eigenvalues.

By perturbing the neighbor factor of the LLE algorithm and introducing majority voting, a group of discriminant subspaces with base classifiers are integrated for classifying face images. Experiments show that the classification performance of the En-ULLELDA algorithm is better than that of the combination of the ULLELDA algorithm and base classifier.

In the future, we will consider the combination of ULLELDA with other ensemble learning methods, such as boosting algorithms and bagging algorithms, to further enhance the separability of the common discriminant subspace and decrease the computational complexity of the proposed En-ULLELDA algorithm.

Acknowledgements

This work is sponsored by NSF of China (60505002), the Fok Ying Tung Education Foundation (91067) and the Excellent Young Teachers Program of MOE of China.

References

1. Samal, A., Iyengar, P. A.: Automatic Recognition and Analysis of Human Faces and Facial Expressions: A Survey. Pattern Recognition. 25 (1992) 65-77
2. Nayar, S. K., Nene, S. A., Murase, H.: Subspace Methods for Robot Vision. Technical Report CUCS-06-95, Columbia University, New York (1995)
3. Seung H. S., Daniel, D. L.: The Manifold Ways of Perception. Science 290 (2000) 2268-2269
4. Roweis S. T., Lawrance, K. S.: Nonlinear Dimensionality Reduction by Locally Linear Embedding. Science 290 (2000) 2323-2326
5. Tenenbaum, J. B., Silva, de, Langford, J. C.: A Global Geometric Framework for Nonlinear Dimensionality Reduction. Science 290 (2000)2319-2323
6. Sha, F., Saul, L. K.: Analysis and Extension of Spectral Methods for Nonlinear Dimensionality Reduction. In: Proceedings of the 22nd International Conference on Machine Learning, Bonn, Germany, 15 (2005) 721–728
7. Zhang, J., Shen, H., Zhou, Z. –H.: Unified Locally Linear Embedding and Linear Discriminant Analysis Algorithm (ULLELDA) for Face Recognition. In: Stan Z. Li, Jianhuang Lai, Tieniu Tan, Guocan Feng, Yunhong Wang (eds.): Advances in Biometric Personal Authentication. Lecture Notes in Computer Science, Vol. 3338. Springer-Verlag, Berlin Heidelberg New York (2004) 209-307
8. Swets D. L., Weng, J.: Using Discriminant Eigenfeatures for Image Retrieval. IEEE Trans. on PAMI. 18 (1996) 831-836
9. Kouropteva, O., Okun, O., PietikÄinen, M.: Selection of the Optimal Parameter Value for the Locally Linear Embedding Algorithm. In: Wang L, Halgamuge SK, Yao X, (eds.): Computational Intelligence for the E-Age, Singapore, (2002) 359-363

10. Zhou Z.–H., Yu, Y.: Ensembling local learners through multimodal perturbation. IEEE Trans. on SMCB, 35(2005) 725-735
11. Samaria, F. S.: Face Recognition Using Hidden Markov Models. PhD thesis, University of Cambridge, (1994)
12. Daniel, B., Graham, Allinson, N. M.: em Characterizing virtual Eigensignatures for General Purpose Face Recognition. In: H. Wechsler, P. J. Phillips, V. Bruce, F. Fogelman-Soulie and T. S. Huang (eds.): Face Recognition: From Theory to Applications. NATO ASI Series F, Computer and Systems Sciences, 163 (1998) 446-456
13. Bellhumer, P. N., Hespanha, J., Kriegman, D.: Eigenfaces vs. fisherfaces: Recognition using class specific linear projection. IEEE Trans. on PAMI. 17(1997) 711–720
14. Li, S. Z., Chan K. L., Wang, C. L.: Performance Evaluation of the Nearest Feature Line Method in Image Classification and Retrieval. IEEE Trans. on PAMI, 22 (2000) 1335-1339
15. Zhang, J., Li, S. Z., Wang, J.: Nearest Manifold Approach for Face Recognition. The 6th IEEE International Conference on Automatic Face and Gesture Recognition (2004) 223-228

Perceptual Learning Inspired Model Selection Method of Neural Networks

Ziang Lv, Siwei Luo,
Yunhui Liu, and Yu Zheng

School of Computer and Information Technology, Beijing Jiaotong University,
Beijing 100044, China
lvziang@tsinghua.org.cn

Abstract. Perceptual learning is the improvement in performance on a variety of simple sensory tasks. Current neural network models mostly concerned with bottom-up processes, and do not incorporate top-down information. Model selection is the crux of learning. To obtain good model we must make balance between the goodness of fit and the complexity of the model. Inspired by perceptual learning, we studied on the model selection of neuro-manifold, use the geometrical method. We propose that the Gauss-Kronecker curvature of the statistical manifold is the natural measurement of the nonlinearity of the manifold. This approach provides a clear intuitive understanding of the model complexity.

1 Introduction

Recent study of psychology and psychophysics experimental give adequate evidences that even the adult primary sensory cortices have a fair amount of plasticity to perform changes of information processing as a result of training. Sensory perception is a learned trait [1]. The strategies we use to perceive the world are constantly modified by experience. That means we learning to perceive while perceiving to learn. According to Gibson's definition [2], "Any relatively permanent change in the perception of a stimulus array following practice or experience with this array will be considered perceptual learning." Perceptual learning is a new and exciting aspect of learning and one that has seen tremendous progress during the last decade. Perceptual learning differs from other forms of learning in that it can be highly specific, that is, transferring little to very similar tasks, and hence at least partly involves structural and functional changes in primary sensory cortices. Perceptual learning can be quantitatively assessed using three approaches [1]: psychophysical measurement of behavior, physiological recording of living cortical neurons and computer modeling of well-defined neuronal networks.

In machine learning, our goal is to gain rules from observed data and use these rules to predict the future data. This procedure includes two aspects: modeling and model selection. The goal of model selection is to improve generalization capability, that is, to choose the model have best prediction. We desire a model that is complex

L. Jiao et al. (Eds.): ICNC 2006, Part I, LNCS 4221, pp. 39–42, 2006.

enough to describe the data sample accurately but without over-fitting and thus losing generalization capability.

Inspired by perceptual learning, we propose a neural network model selection method that uses top-down information as the complexity measure of a model. The top-down information we used here is the geometrical property of the model, that is, the curvature of the manifold of the model.

2 Neural Network Models of Perceptual Learning

Current neural network models, mostly concerned with bottom-up processes, such as finding optimal parameters for a given set of data (which correspond to the stimuli of experiments), do not incorporate top-down information, such as preselecting features or internal knowledge. New experimental results, however, show that attention and other higher cortical processes play an important role in perceptual learning issues. One important topic is the question of how specific the improvement achieved through learning is, and how much it can generalize. [3] The Gestalt's prior rule and Marr's constraint can also be viewed as top-down information.

Models of perceptual learning can be broadly divided into two classes: feedforward versus feedback. The best-known model of feedforward type is that conceived by Poggio on visual hyperacuity. The main appeal of the model is that it can perform the specific of perceptual learning, use very limited number of neurons and learned quickly. However, it has some drawbacks that it needs a teaching signal. One of the feedback network model is Adini's contrast discrimination model. An attractive feature of this model is that it does not require a teaching signal, but it cannot account for task-specificity. Combined model and reinforcement learning algorithm also proposed by researchers. In general, perceptual learning is the real brain's challenges to models: perceptual learning is highly specific, it requires repetition but not feedback, and it involves top-down influences.

This idea was proposed by many learning theories. Regularization theory [4] demonstrated that the not "self-evident" method of minimizing the regularized functional does work. Moreover, Vapnik's statistical learning theory [5] uses the Structural Risk Minimization (SRM) inductive principle to substitute the Empirical Risk Minimization (ERM) inductive principal. The Minimum Description Length (MDL) principle [6] originated in algebra coding theory states that the best model for describing a set of data is the one that permits the greatest compression of the data description. Many researcher proposed the comprise method to interpret and realize this occam' razor. These methods are similar at fitting term but are quite different at complexity term. Akaike Information Criterion (AIC) [7] only use the number of parameters as the term of complexity; Bayesian Information Criterion (BIC) [8] include the sample size in its complexity term; Rianen's Stochastic Complexity (SC) based on MDL [9,10], Most of these theories fail to meet the crucial requirement of being invariant under reparameterization of the model, the condition that any meaningfully interpretable method of model selection must satisfy.

3 Top-Down Information Used in Model Selection

In differential geometric perspective, a statistical model is a set of probability distributions described by a parameter set [11]. This parameter set forms a manifold which is called model manifold. It is the submanifold of the manifold formed by all the probability distributions. The probability distributions describing true information process can also form a submanifold but it is generally not the same with the model manifold. Thus to approximate the true information process with statistical model can be changed into research the geometrical relation (shape and relative position) of the two submanifolds in their enveloping manifold. A Neural Networks (NN) including modifiable parameters (connection weights and thresholds) $\theta=(\theta_1 ... \theta_n)$ can be described by a statistical model, thus the set of all the possible neural networks realized by changing θ in the parameter space Θ forms a n-dimensional manifold S called the neuro-manifold, where θ plays the role of a coordinate system of S [11]. S is a flat dual Riemannian manifold and the submanifold of manifold P formed by all probability distributions. Each point in S represents a network depicted by a probability distribution $p(x;\theta)$ or $p(y \mid x;\theta)$.

A manifold usually are not flat, it can be analogy with the curved surface. When we define the Riemannian metric on the manifold, we can get the unique Riemannian curvature tensor, and can use the curvature to describe the property of this manifold.

If we denote the volume element of n-dimension manifold with ds, then the r th integral of mean curvature and Gauss-Kronecker curvature [12]:

$$M_r(S)=\begin{pmatrix} n-1 \\ r \end{pmatrix}^{-1} \int_s \{k_{i1}, k_{i1}, ..., k_{ir}\} ds \tag{1}$$

$$M_{n-1}=K=\prod_{i=1}^{n-1} k_i \tag{2}$$

where $\{k_{i1}, k_{i1}, ..., k_{ir}\}$ denotes the r th elementary symmetric function of the principal curvatures.

Because that the hyperspherical image of S is U_{n-1}, which maps a point P of S to the end point of the unit vector through the origin parallel to the normal vector to S at P, Now we can see the geometric meaning of the K is a limitation, which equal the volume element of hyperspherical image of the manifold divided by the volume element of the manifold. In another word, the local curvature of the manifold is the ratio that it compared with the unit hypersphere On a n-dimension manifold S with Riemann metric, we can define the volume element as follows: [13]

$$ds=\prod_{i=1}^{n} d\theta^i \sqrt{\det I(\theta)} \tag{3}$$

This natural property of Gauss-Kronecker is very remarkable. It said that the curvature is the intrinsic property. Gauss first find this theorem in 2-dimentionan, that is Gauss's Remarkable Theorem, the most important context of differential geometry, and then Riemann gives the n-dimension form. We can see the Gauss-Kronecker curvature not only describe the local property of the manifold, but also describe the global property of the manifold. Finally, we give the Gauss-Kronecker Curvature Criteria (GCIC) as:

$$GCIC = -\ln f(x\,|\,\hat{\theta}) + \frac{p}{2}\ln\left(\frac{n}{2\pi}\right) + \ln\int K d\theta \tag{4}$$

where $\hat{\theta}$ is the Maximum Likelihood Estimators (MLE) of the parameters of the model, p is the number of the parameters of the model, n is the size of the sample set, K is the Gauss-Kronecker curvature.

Acknowledgement

The research is supported by the National Nature Science Foundation of China (60373029), the research Fund for Doctoral Program of Higher Education of China (20050004001) and the Scientific Foundation of Beijing Jiaotong University (2005RC004).

References

1. Misha Tsodyks and Charles Gilbert: Neural Networks and Perceptual Learning. Nature, Vol.431 (2004)775-781
2. Gibson, E. J., Perceptual Learning, Annu. Rev. Psychol., Vol. 14(1963)29-56
3. Manfred Fahle: Perceptual Learning: Specificity versus generalization. Current Opinion in Neurobiology. Vol.15, No. 14(2005)154-160
4. Tikhonov, A. N., solution of ill-posed problems, W.H. Winston, Washington DC (1977)
5. Vapnik,V.N.: The Nature of Statistical Learning Theory. Springer-Verlag (1999)
6. Rissanen, J.: Modeling by shortest data description. Automatica, No. 14(1978)465-471
7. Akaike, H.: Information theory and an extension of the maximum likelihood principle. In Petrox, B.N., Caski, F. (eds): Second International Symposium on Information Theory. Budapest. (1973) 267-281
8. Schwarz, G.: Estimation the dimension of a model. Annals of Statistics, Vol. 7, No. 2 (1978) 461-464
9. Rissanen, J.: Stochastic Complexity and Modeling. Annals of Statistics, Vol. 14, No. 3(1986) 1080-1100
10. Rissanen, J.: Fisher Information and Stochastic Complexity. IEEE Transaction on Information Theory, Vol. 42, No. 1 (1996) 40-47
11. Amari, S.: Methods of Information Geometry. Oxford University Press. (2000)
12. Santalo, L.A.: Integral Geometry and Geometric Probability. Addison-Wesley (1979)
13. Spivak, M.: A Comprehensive Introduction to Differential Geometry. Vol. 2 of 5. (1979)

Improving Nearest Neighbor Rule with a Simple Adaptive Distance Measure

Jigang Wang, Predrag Neskovic, and Leon N Cooper

Institute for Brain and Neural Systems,
Department of Physics,
Brown University, Providence RI 02912, USA
jigang@brown.edu, pedja@brown.edu, Leon_Cooper@brown.edu

Abstract. The k-nearest neighbor rule is one of the simplest and most attractive pattern classification algorithms. However, it faces serious challenges when patterns of different classes overlap in some regions in the feature space. In the past, many researchers developed various adaptive or discriminant metrics to improve its performance. In this paper, we demonstrate that an extremely simple adaptive distance measure significantly improves the performance of the k-nearest neighbor rule.

1 Introduction

The k-nearest neighbor (k-NN) rule is one of the oldest and simplest pattern classification algorithms [1]. Given a set of n labeled examples $D_n = \{(\boldsymbol{X}_1, Y_1), \ldots, (\boldsymbol{X}_n, Y_n)\}$ with inputs $\boldsymbol{X}_i \in \mathbb{R}^d$ and class labels $Y_i \in \{\omega_1, \ldots, \omega_M\}$, the k-NN rule classifies an unseen pattern $\boldsymbol{X}$ to the class that appears most often among its k nearest neighbors. To identify the nearest neighbors of a query pattern, a distance function has to be defined to measure the similarity between two patterns. In the absence of prior knowledge, the Euclidean and Manhattan distance functions have conventionally been used as similarity measures for computational convenience.

In practice, the success of the k-NN rule depends crucially on how well the constant *a posteriori* probability assumption is met and it is desirable to choose a suitable metric so that the majority of the k nearest neighbors of a query pattern are from the desired class. In the past, many methods have been developed for locally adapting the metric so that neighborhoods of approximately constant *a posteriori* probabilities can be produced (c.f. [2,3,4,5]). In this paper, we show that, an extremely simple adaptive distance measure, which basically normalizes the ordinary Euclidean or Manhattan distance from a query pattern to each training example by the shortest distance between the corresponding training example to training examples of a different class, significantly improves the generalization performance of the k-NN rule.

The remainder of the paper is organized as follows. In Section 2, we describe the locally adaptive distance measure and the resulting adaptive k-NN rule using the adaptive distance measure. In Section 3, we present experimental results of

L. Jiao et al. (Eds.): ICNC 2006, Part I, LNCS 4221, pp. 43–46, 2006.

Therefore, the theory and algorithms of reinforcement learning have attracted lots of research interests in recent years [1][2].

In reinforcement learning, value function estimation for MDPs is one of the central problems since the policies of MDPs can be easily obtained based on corresponding value functions. However, due to the delayed reward problem, the value function estimation techniques in RL are more difficult than function regression in supervised learning. One major advance in solving this problem is the temporal difference (TD) learning method originally proposed by R. Sutton [3], which is to update the value functions based on the differences between two temporally successive predictions rather than the traditional errors between the real value and predicted one. And it has been proved that TD learning is more efficient than supervised learning in multi-step prediction problems [3].

Until now, lots of work has been done on tabular temporal difference learning algorithms which represent the value functions or policies of MDPs in discrete tables. But in many applications, the state spaces of the underlying MDPs are large or continuous, which makes the existing tabular algorithms be impractical due to the curse of dimensionality. To solve this problem, TD learning algorithms using various linear function approximators have been widely studied in the literature [4][5][6] and some theoretical results were also established for the convergence of linear TD(λ) algorithms [6][7]. Nevertheless, the approximation ability of linear approximators is limited and the performance of linear TD learning is greatly influenced by the selection of linear basis functions. In addition, conventional gradient-based TD(λ) learning algorithms with nonlinear function approximators, such as neural networks, were shown to be unstable in some cases [6][8]. Therefore, how to implement TD learning with nonlinear function approximators becomes an important problem both in RL theory and applications.

In recent years, kernel methods [9] were popularly studied in supervised learning to realize nonlinear versions of previous linear classifiers or regression methods, e.g., support vector machines (SVMs). However, there are relatively few results on kernel methods in RL. In [10], a kernel-based RL approach was suggested by constructing implicitly an approximate model using kernels with smoothness assumptions, but the computational costs may increase fast with the number of training samples and as indicated in [11], the learning efficiency of the kernel-based RL method in [10] is worse than RL methods that approximating value functions directly. In [12], a kernel-based least-squares TD (LSTD) learning algorithm was presented for value function approximation of Markov chains, but the dimension of the kernel LSTD solution is equal to the number of observation data, which will hinder the algorithm to be applied in larger problems. In this paper, we will extend the previous results in [12] by presenting a sparse kernel-based LSTD learning algorithm for reinforcement learning. The sparsity of the kernel LSTD solution is guaranteed by a kernel sparsification method called approximately linear dependence (ALD) analysis, which was initially used in supervised learning tasks based on the kernel recursive least-squares methods [13] and later in reinforcement learning with Gaussian process models (GPTD) [14]. The sparse kernel LSTD(λ) algorithm presented in this paper is different from the GPTD algorithm [14] in three aspects. The first aspect is that eligibility traces are designed in the sparse kernel LSTD method, which may contribute to improve prediction performance further. For the GPTD method in [14], it was pointed out by the authors in [14] that it is necessary to study the eligibility traces in kernel-based TD learning. The second aspect is that in GPTD, it is assumed that measurements are corrupted by additive independent identical distribution (IID) Gaussian noise, and in consequence GPTD requires an additional parameter estimating the measurement noise variance which is not required in our approach. Moreover, the derivation of the sparse kernel LSTD(λ) algorithm is simpler and the framework of kernel LSTD(λ) is within that of linear TD(λ) algorithms so that the convergence theory of linear TD algorithms [6] can be easily extended to the sparse kernel LSTD(λ) algorithm.

This paper is organized as follows. In Section 2, a brief introduction on temporal-difference methods in RL is given. In Section 3, the sparse kernel LSTD(λ) algorithm is presented. In Section 4, learning prediction experiments on a Markov chain are conducted to illustrate the effectiveness of the proposed method. Some conclusions and remarks on future work are given in Section 5.

2 Temporal-Difference Methods in RL

A Markov decision process is denoted as a tuple $\{S, A, R, P\}$, where S is the state space, A is the action space, P is the state transition probability and R is the reward function. The policy of an MDP is defined as a function $\pi : S \rightarrow \Pr(A)$, where $\Pr(A)$ is a probability distribution in the action space.

The state value function of a stationary policy π is defined as follows:

$$V^{\pi}(s) = E_{\pi}[\sum_{t=0}^{\infty} \gamma^t r_t | s_0 = s] \tag{1}$$

For a stationary action policy π, the value function satisfies the Bellman equation:

$$V^{\pi}(s_t) = R(s_t, s_{t+1}) + \gamma E^{\pi}[V(s_{t+1})] \tag{2}$$

where $R(s_t, s_{t+1})$ is the expected reward received after the state transition from s_t to s_{t+1}.

Based on the above model definition, the learning prediction problem can be formally described as follows: Given observation data from state transitions of an MDP $\{(x_i, r_i, x_{i+1}), i=1,2,\ldots t\}$, and assuming the action policy π is stationary, the goal of learning prediction or policy evaluation is to compute the value functions without knowing the state transition probability and the reward function as *a priori*.

As an efficient approach to the above learning prediction problem, tabular TD(λ) algorithms have been well studied in the literature. However, in most applications, function approximators need to be used for generalization in large and continuous state spaces. Until now, there have been many research works on linear TD(λ) learning algorithms, where the value functions were represented as a weighted sum of linear basis functions. Let $W=[w_1, w_2,\ldots,w_n]$ denote the weight vector and $[\phi_1(x), \phi_2(x),\ldots,\phi_n(x)]$ denote the linear basis functions, then the value functions in linear TD(λ) can be approximated as

$$\tilde{V}_t(x_t) = \phi^T(x_t)W = \sum_i w_i\phi_i(x_t) \tag{3}$$

In TD(λ), there are two basic mechanisms which are the temporal difference and the eligibility trace, respectively. Temporal differences are defined as the differences between two temporally successive estimations and have the following form

$$\delta_t = r_t + \gamma\phi^T(x_{t+1})W - \phi^T(x_t)W \tag{4}$$

where x_{t+1} is the successive state of x_t, and r_t is the reward received after the state transition from x_t to x_{t+1}.

To realize incremental or online learning, eligibility traces are defined for each state in equation (5), where $0 \le \lambda \le 1$ is the parameter for eligibility traces.

$$\vec{z}_t(s_i) = \gamma\lambda\,\vec{z}_{t-1}(s_i) + \phi(s_i) \tag{5}$$

The online linear TD(λ) update rule with eligibility traces is

$$\tilde{V}_{t+1}(s_i) = \tilde{V}_t(s_i) + \alpha_t\delta_t\vec{z}_t(s_i) \tag{6}$$

where $\vec{z}_0(s)=0$ for all s.

In [6], the convergence of linear TD(λ) algorithms was analyzed and it was proved that under certain assumptions, linear TD(λ) converges and the limit of convergence W^* satisfies the following equation:

$$E_0[A(X_t)]W^* - E_0[b(X_t)] = 0 \tag{7}$$

where $X_t = (x_t, x_{t+1}, z_{t+1})$ (t=1,2,…) form a Markov process, $E_0[\cdot]$ stands for the expectation with respect to the unique invariant distribution of $\{X_t\}$, and $A(X_t)$ and $b(X_t)$ are defined as

$$A(X_t) = \vec{z}_t(\phi^T(x_t) - \gamma\phi^T(x_{t+1})) \tag{8}$$

$$b(X_t) = \vec{z}_t r_t \tag{9}$$

3 Sparse Kernel-Based LS-TD Learning

Although the convergence theory of linear TD(λ) learning algorithms has been proved, the approximation ability of linear function approximators is greatly influenced by the selection of linear basis functions. To realize stable nonlinear TD learning algorithms, in our previous work [12], a kernel-based LS-TD(λ) learning method was proposed but it suffered from high computational costs since the dimension of kernel matrix is equal to the number of observation states. In the following, we will present the sparse kernel-based LSTD(λ) algorithm by integrating the ALD-based kernel sparsification method [13] to the kernel-based LS-TD(λ) algorithm.

By using the average value of observations as the estimation of expectation $E_0[\cdot]$, the least-squares TD regression equation (7) can be expressed as follows:

$$\sum_{i=1}^{T}[\vec{z}(s_i)(\phi^T(s_i)-\gamma\phi^T(s_{i+1})]W=\sum_{i=1}^{T}\vec{z}(s_i)r_i \tag{10}$$

Based on the idea of kernel methods, a high-dimensional nonlinear feature mapping can be constructed by specifying a Mercer kernel function $k(x_1, x_2)$ in a reproducing kernel Hilbert space (RKHS). In the following, the nonlinear feature mapping based on the kernel function $k(.,.)$ is also denoted by $\phi(s)$ and according to the Mercer Theorem [9], the inner product of two feature vectors is computed by

$$k(x_i,x_j)=\phi^T(x_i)\phi(x_j) \tag{11}$$

Due to the properties of RKHS [9], the weight vector W can be represented as the weighted sum of the state feature vectors:

$$W=\Phi_N\alpha=\sum_{t=1}^{N}\phi(s(x_t))\alpha_i \tag{12}$$

where $x_i\,(i=1,2,...,N)$ are the observed states and $\alpha=[\alpha_1,\alpha_2,...,\alpha_N]^T$ are the corresponding coefficients, and the matrix notation of the feature vectors is denoted as

$$\Phi_N=(\phi(s_1),\phi(s_2),...,\phi(s_N)) \tag{13}$$

For a state sequence $x_i\,(i=1,2,...,N)$, let the corresponding kernel matrix K be denoted as $K=(k_{ij})_{N\times N}$, where $k_{ij}=k(x_i, x_j)$.

By substituting (11), (12) and (13) into (10), we can get

$$Z_N H_N \Phi_N^T \Phi_N \alpha = Z_N H_N K\alpha = Z_N R_N \tag{14}$$

where Z_N, H_N and R_N are defined as

$$Z_N=[\vec{z}_1 \quad \phi(x_2)+\gamma\lambda\vec{z}_1 \quad ... \quad \phi(x_{N-1})+\gamma\lambda\vec{z}_{N-2}] \tag{15}$$

$$R_N=[r_1,r_2,...,r_{N-1}]^T \tag{16}$$

$$H_N=\begin{bmatrix}1 & \beta_1\gamma & & & \\ & 1 & \beta_2\gamma & & \\ & & \cdots & & \\ & & 1 & \beta_{N-1}\gamma\end{bmatrix}_{(N-1)\times N} \tag{17}$$

In (17), the values of β_i (i=1,2,…,N-1) are determined by the following rule: when state x_i is not an absorbing state, β_i is equal to -1, otherwise, β_i is set to zero.

By multiplying the two sides of (14) with Φ_N^T, we can get

$$\Phi_N^T Z_N H_N K\alpha = \Phi_N^T Z_N R_N \tag{18}$$

Let $\overline{Z}_N = \Phi_N^T Z_N$, then we have

$$\overline{Z}_N = [\overline{z}_1, \overline{z}_2, ..., \overline{z}_{N-1}] = [k_{1N}, k_{2N} + \gamma\lambda\overline{z}_1, ..., k_{(N-1)N} + \gamma\lambda\overline{z}_{N-2}] \tag{19}$$

$$k_{iN} = [k(x_1, x_i), k(x_2, x_i), ..., k(x_N, x_i)]^T \tag{20}$$

$$\overline{Z}_N H_N K\alpha = \overline{Z}_N R_N \tag{21}$$

In (19), for any absorbing state x_i ($1<i<N-1$), since it is necessary to clear the previous eligibility traces, we set $\overline{z}_{i+1} = k_{(i+1)N}$. In addition, since the matrix in the left side of equation (18) may not be invertible, in the following, we will use the techniques of generalized inverse matrix in [16] to obtain the least-squares solution with minimum norm. Let A^+ denote the generalized inverse of matrix A. Then, the kernel-based least-squares solution to the TD learning problem is as follows:

$$\alpha = (H_N K)^+ \overline{Z}_N^+ \overline{Z}_N R_N \tag{22}$$

As discussed before, the dimension of the kernel-based LS-TD solution is equal to the number of state transition samples, which may increase fast and cause huge computational costs in complex problems. To make the above algorithm be practical, one key problem is to decrease the dimension of kernel matrix K as well as the dimension of α. In the following, we will use the approximately linear dependence (ALD) analysis method [13] for the sparsification of kernel matrix K.

The main idea of ALD-based kernel sparsification is to find approximately linearly independent feature or data dictionary $D_N = \{\phi(x_1), \phi(x_2), ..., \phi(x_n)\}$ from the observation state sequence $\{x_1, x_2, \ldots, x_N\}$, where the corresponding kernel function is $k(x_i, x_j) = <\phi(x_i), \phi(x_j)>$. Then, any feature vector $\phi(x_t)$ can be approximately represented by the feature vectors $\Phi_n = \{\phi(x_1), \phi(x_2), ..., \phi(x_n)\}$ as

$$\phi(x_t) = \Phi_n^T \vec{a}_t = \sum_{j=1}^{n} a_t(j)\phi(x_j) \tag{23}$$

The sparsification procedure using ALD analysis mainly includes two steps. The first step is to compute the following optimization solutions

$$\delta_t = \min_{\vec{a}} \left\| \sum_j \vec{a}_j \phi(x_j) - \phi(x_t) \right\|^2 \tag{24}$$

which is equivalent to

$$\delta_t = \min_{\vec{a}}\{\sum_{i,j} \vec{a}_i\vec{a}_j\langle\phi(x_i),\phi(x_j)\rangle - 2\sum_i \vec{a}_i\langle\phi(x_i),\phi(x_t)\rangle + \langle\phi(x_t),\phi(x_t)\rangle\} \quad (25)$$

Using the kernel trick in (11), we can obtain the approximation error as

$$\delta_t = \min_{\vec{a}_t}\{\vec{a}_t^T K_n \vec{a}_t - 2\vec{a}_t^T k_n(x_t) + k_{tt}\} \quad (26)$$

where $[K_n]_{i,j}=k(x_i,x_j)$, x_i $(i=1,2,\ldots,n)$ are the elements in the dictionary, n is the dimension of the data dictionary, which is usually much smaller than the original data number, $k_n(x_t)=[k(x_1, x_t), k(x_2, x_t),\ldots,k(x_n, x_t)]^T$, $\vec{a}_t=[a_1, a_2,\ldots,a_n]^T$ and $k_{tt}=k(x_t, x_t)$.

As discussed in [13], the optimal solution to (26) is:

$$\vec{a}_t = K_n^{-1}k_n(x_t) \quad (27)$$

$$\delta_t = k_{tt} - k_n^T(x_t)\vec{a}_t \quad (28)$$

The second step of the ALD-based sparsification is to update the data dictionary by comparing δ_t with a predefined threshold μ. If $\delta_t<\mu$, the dictionary is unchanged, otherwise, x_t is added to the dictionary, i.e., $D_t=D_{t-1}\cup x_t$.

Let $n(t)$ denote the dimension of the sparsified dictionary D_t for state sequence $\{x_1, x_2,\ldots, x_t\}$, and $\vec{a}_i^{'}=[\vec{a}_i,0,\ldots,0]^T_{n(t)\times 1}$ $(1\le i\le n(t))$ be the expanded vector of $\vec{a}_i$. For the final dictionary D_N, define a coefficient matrix

$$A_N = [\vec{a}_1^{'},\vec{a}_2^{'},\ldots,\vec{a}_N^{'}]_{n(N)\times N} \quad (29)$$

Denote $\hat{K}$ as the kernel matrix obtained from the dictionary D_N, which has the dimension of $n(N)$. Then it is easy to verified that [13]

$$K = A_N^T\hat{K}A_N \quad (30)$$

Let the transformed solution vector based on the sparsified dictionary D_N be

$$\hat{\alpha} = A_N\alpha \quad (31)$$

Let $\hat{Z}_N = \Phi^T_{n(N)}Z_N$, then we have $\hat{Z}_N = [\hat{z}_1,\hat{z}_2,\ldots,\hat{z}_N]$, which has similar form as (19) but with the dimension of $n(N)\times N$, where $n(N)<<N$.

Substituting (30) and (31) to (21), we can get

$$\hat{Z}_N H_N A_N \hat{K}\hat{\alpha} = \hat{Z}_N R_N \quad (32)$$

Define

$$C_N = (H_N A_N \hat{K})^T(H_N A_N \hat{K}),\ D_N = (\hat{Z}_N)^T(\hat{Z}_N) \quad (33)$$

Then, the LS-TD solution using the sparsified kernel matrix is

$$\hat{\alpha} = C_N^{+}(H_N A_N \hat{K})^T (D_N)^{+} D_N R_N \tag{34}$$

Since the dimensions of $\hat{\alpha}$, C_N , D_N and $\hat{K}$ are both equal to $n(N)$, which will usually be much smaller than the original number N of observation data samples, the computational complexity of the algorithm, which is mainly due to the generalized inversion of C_N, as well as the memory cost of kernel matrix K can be greatly reduced without scarifying much in approximation precision. Moreover, according to the theoretical analysis in statistical learning theory [15], better generalization ability may also be obtained for learning machines with lower model dimension and complexity.

After the sparsification procedure, a data dictionary D_N with reduced number of feature vectors will be obtained and the approximated state value function can be represented as:

$$\tilde{V}(x) = \sum_{j=1}^{n(N)} \hat{\alpha}_j k(x_j, x) \tag{35}$$

4 Learning Prediction Experiments

In this section, the proposed sparse kernel LS-TD(λ) learning algorithm will be applied to the learning prediction problem of a Markov reward process, which is a stochastic 13-state Markov chain with an absorbing state. The problem is called the HopWorld problem and was firstly introduced in [4] and later in [12], the linear value functions were transformed to nonlinear ones to test the performance of Kernel LS-TD algorithms without sparsification. In our experiments, we used the same nonlinear problem to compare the performance between sparse kernel LSTD(λ) and the original kernel LSTD(λ) as well as previous linear TD(λ) algorithms.

In the HopWorld problem, the states of the Markov chain, from left to right, are numbered from 12 to 0. And state 12, which is the left-end state in the Markov chain, is the initial state for each trajectory and state 0 is the absorbing state at the right-end. Each non-absorbing state i ($i \geq 2$) has two possible state transitions: one is to go to the next state i-1, and the other is to go to state i-2. Both state transitions have a transition probability of 0.5. For state 1 and 0, the only state transition is to go to the absorbing state 0 with probability 1. The true value function for state i ($0 \leq i \leq 12$) is defined as $V(i) = \sin(i/12.0)/(i/12.0 + 0.1)$, which is only used to compute the prediction errors. From the above nonlinear value functions, it is easily to compute the reward function for each state transition which is used in the simulation. However, in the experiments, the state transition probabilities as well as the true value functions are assumed to be unknown for TD algorithms. The task of TD learning is to estimate the true value function only by observing state transition data and the corresponding rewards.

In the experiments, the sparse kernel-based LS-TD(λ) algorithm and previous linear TD(λ) algorithms, i.e., LS-TD(λ) and TD(λ), as well as the kernel LS-TD(λ) without kernel sparsification [12], were all implemented for the learning prediction task. In the kernel LS-TD(λ) algorithm, a radius basis function (RBF) kernel was selected and its width parameter was set to 0.02 in all the experiments. A threshold parameter δ=0.001

was selected for the ALD-based kernel sparsification procedure. The LS-TD(λ) algorithm used 12 RBFs as linear basis functions and the centers of the RBFs were equally distributed in the interval of [0, 12]. The width of the linear RBF functions was also set to 0.02. In the experiments, we also tested other parameter settings for the RBF width and similar results were obtained. The linear TD(λ) algorithm used the same basis functions as LS-TD(λ) and the learning rate was manually selected and optimized as 0.01. Different values of parameter λ were tested for the above four TD learning algorithms, where the number of data samples from the Markov chain was 100 for all the algorithms. The performance of different TD learning methods was evaluated by the root-mean-squares (RMS) errors of value function prediction.

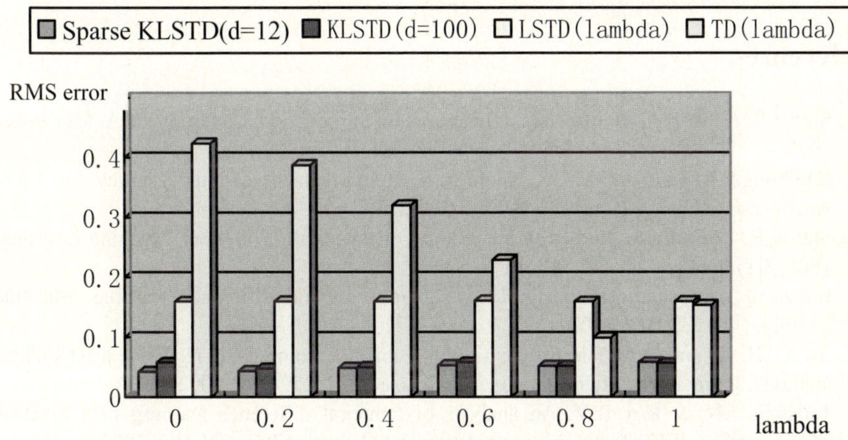

Fig. 1. Performance comparisons among different TD algorithms

The above Fig.1 shows the experimental results, where the value of parameter λ varies from 0 to 1 and the RMS errors of learning prediction are compared. It is illustrated that the proposed sparse kernel LS-TD(λ) algorithm can realize similar prediction precision as previous kernel LS-TD(λ) with much lower dimension of kernel matrix and solution vector. In the experiments, the kernel dimension *d* for the sparse kernel LSTD(λ) algorithm is equal to 12 while the original KLSTD in [12] has a kernel dimension of 100. The prediction errors of conventional LS-TD(λ) and TD(λ) are much larger than the kernel-based LS-TD(λ) algorithms, which demonstrates that the kernel methods in TD learning can greatly improve the approximation ability of conventional linear TD learning methods.

5 Conclusion and Future Work

Kernel methods in reinforcement learning, especially in temporal difference learning, are very promising to improve the generalization ability of RL algorithms in large and nonlinear spaces since by utilizing Mercer kernels, nonlinear feature mappings can be implicitly constructed and linear forms of computation can be implemented in the feature space. However, compared to the kernel methods in supervised learning, there have been relatively few works on kernel-based reinforcement learning. In this paper, a sparse kernel-based least-squares TD learning algorithm is proposed, where an ALD-based kernel sparsification approach is integrated with a kernel-based least-squares solution to the TD learning prediction problem. Experimental results in a typical Markov chain with nonlinear value functions show that the sparse kernel LSTD(λ) algorithm can obtain high approximation precision with low feature dimension, which is superior to the previous kernel LSTD(λ) algorithm without sparsification [12], and it also has much better nonlinear approximation ability than previous linear TD(λ) algorithms. Future work may include the extension of the proposed method in learning control problems of MDPs.

References

1. Sutton R. & Barto A. Reinforcement Learning, an Introduction. Cambridge MA, MIT Press, 1998.
2. Kaelbling L.P., Littman M.L., & Moore A.W. Reinforcement learning: a survey. Journal of Artificial Intelligence Research, 1996, 4, 237-285.
3. Sutton R. Learning to predict by the method of temporal differences. Machine Learning, 1988, 3(1):9-44.
4. Boyan, J. A. Technical update: Least-squares temporal difference learning. Machine Learning, 49: 233-246, 2002.
5. Xu X., H. He and D.W. Hu. Efficient reinforcement learning using recursive least-squares methods. Journal of Artificial Intelligence Research, 16:259–292, 2002
6. Tsitsiklis J.N. & Roy B.V. An analysis of temporal difference learning with function approximation. IEEE Transactions on Automatic Control. 42(5), 674-690, 1997.
7. Vladislav Tadić. On the Convergence of Temporal-Difference Learning with Linear Function Approximation. Machine Learning, 42 (3): 241-267, 2001
8. Baird L C. Residual algorithms: reinforcement learning with function approximation. In: Proc. of the 12th International Conference on Machine Learning (ICML95), Tahoe City, California, USA, 1995. 30-37
9. Schölkopf B. and A. Smola, Learning With Kernels. Cambridge, MA: MIT Press, 2002.
10. Ormoneit D. and S. Sen, Kernel-based reinforcement learning, Machine Learning, vol.49, no.2–3, pp.161–178, 2002.
11. Lagoudakis M. G., R. Parr, Least-squares policy iteration, Journal of Machine Learning Research, vol.4 , pp.1107-1149, 2003
12. Xu X., et al. Kernel Least-squares Temporal difference learning. International Journal of Information Technology, 2005, Vol.11, no.9, pp:54-63.
13. Engel Y., S. Mannor, and R. Meir, The kernel recursive least-squares algorithm, IEEE Transactions on Signal Processing, vol. 52, no. 8, pp.2275-2285, 2004.
14. Engel Y., S. Mannor and R. Meir, Bayes meets bellman: the Gaussian Process approach to temporal difference learning. In: Proc. of the 20th International Conference on Machine Learning, ICML-03 , 2003.
15. Vapnik V., Statistical Learning Theory. NewYork:Wiley Interscience,1998.
16. Nashed M. Z., ed., Generalized Inverses and Applications, Academic Press, New York, 1976.

Independent Component Analysis Based Blind Deconvolution of Spectroscopic Data

Jinghe Yuan[1], Shengjiang Chang[2], Ziqiang Hu[1], and Yanxin Zhang[2]

[1] Institute of Science and Technology for Opto-electron Information, Yantai University, Yantai 264005, China
[2] Institute of Modern Optics, Nankai University, Key Laboratory of Opto-electronics Information Technical Science, EMC, Tianjin 300071 China
jacobyuan@yahoo.com.cn

Abstract. The spectroscopic data recorded by dispersion spectrophotometer are usually degraded by the response function of the instrument. To improve the resolving power, double or triple cascade spectrophotometer, and narrow slits have been employed, but the total flux of the radiation available decreases accordingly, resulting in a lower signal-to-noise ratio (SNR) and a longer measure time. However, the spectral resolution can be improved by mathematically removing the effect of the instrument response function. An independent component analysis based algorithm is proposed to blindly deconvolve the measured spectroscopic data. The true spectrum and the instrument response function are estimated simultaneously. In the preprocessing stage, the noise can be reduced in some degree. Experiments on some real measured spectroscopic data demonstrate the feasibility of this method.

1 Introduction

The spectroscopic data recorded by dispersion spectrophotometer are usually degraded by the response function of the instrument. The major factors degrading the resolving power are the slit width and the diffractive limit of the dispersive grating. To improve the resolving power, double or triple cascade spectrophotometer, and narrow slits have been employed, but the total flux of the radiation available decreases accordingly, resulting in a lower signal-to-noise ratio (SNR) and a longer measure time. Hence, there is a trade-off between the resolving power, the SNR and the measure time. The impulse response of electronic circuits also degrades the spectrum. Nevertheless, the degraded spectrum can be significantly improved by mathematically removing the effect of the instrument response function, which is named deconvolution. The most popular method for resolving this problem is Wiener filtering [1]. However, in most practical applications, the response function of the instrument is unknown, even if can it be measured approximately, it will change along with the time. So the blind deconvolution is a more significant and challenging task because the true spectroscopic data and the instrument response function must be estimated simultaneously from the measured data. Homomorphic filtering method developed by Yasuhiro Senga etc. was applied to

L. Jiao et al. (Eds.): ICNC 2006, Part I, LNCS 4221, pp. 57–65, 2006.

the infrared molecular absorption spectra successfully [2]. But this method needs to calculate the logarithm of the Fourier spectrum, which is time-costly, and is limited to the cases where the slit function is triangle. The Jansson algorithm [3] starting from an initial guess of the instrument response function, estimates the true spectrum iteratively. But the guess of the convolution kernel is often inaccurate. Yuan et al. [4, 5] proposed a high-order statistical algorithm of blind deconvolution for spectroscopic data. But the length of the deconvolution kernel influences the performance of the deconvolution dramatically.

Based on the independent component analysis (ICA) algorithm [6], an artificial neural networks method is proposed for spectroscopic data blind deconvolution. The true spectrum and the instrument response function are estimated simultaneously. Experiments on artificial and real measured spectroscopic data demonstrate the feasibility of this method.

The rest of this paper is organized as follows. Section 2 introduces the ICA based blind deconvolution method of spectroscopic data. Section 3 presents the deconvolution results of some real measured Raman spectroscopic data. Finally, some discussions are drawn in section 4.

2 ICA Based Blind Deconvolution

For both coherent and incoherent irradiation, the spectroscopic data measured by spectrophotometer can usually be mathematically modeled as a convolution of the true spectrum and the instrument response function (called convolution kernel) [3], i.e.

$$\mathbf{x}=\mathbf{f}*\mathbf{w} \tag{1}$$

where $*$ denotes the convolution operation, and $\mathbf{w}=\left(w(1),w(2),\cdots w(N)\right)^T$ is the true spectrum, $\mathbf{x}=\left(x(1),x(2),\cdots x(N)\right)^T$ is the measured spectroscopic data, $\mathbf{f}=\left(f_0,f_1,f_2,\cdots\right)^T$ is the convolution kernel.

In principle, the true spectrum can be restored by the inverse operation of the convolution,

$$\mathbf{y}=\mathbf{g}*\mathbf{x} \tag{2}$$

where $\mathbf{y}=\left(y(1),y(2),\cdots y(N)\right)^T$ is the deconvolved data, and $\mathbf{g}=\left(g_1,g_2,\cdots g_Q\right)^T$ is the response function of the deconvolution operation (called deconvolution kernel) with length Q. For the infinite long convolution kernel $\mathbf{f}$, to achieve an ideal deconvolution result, the deconvolution kernel $\mathbf{g}$ should be infinite long too, but this is unfeasible in practice. It can be proven that a sufficiently long $\mathbf{g}$ can restore the true spectrum adequately [7].

The matrix format of equation (2) is

$$\mathbf{y}=\mathbf{X}\mathbf{g} \tag{3}$$

where $\mathbf{X}$ is a $N\times Q$ Teoplitz matrix whose elements are

$$\mathbf{X}_{i,j}=x_{i-j},\quad 1\le i\le N,\quad 1\le j\le Q \tag{4}$$

It is very like the ICA. Where the matrix $\mathbf{g}$ is a column of the mixing matrix $\mathbf{G}$ and $\mathbf{y}$ is the corresponding independent component(IC). So the equation can be resolved by artificial neural networks (ANN) based ICA algorithm, such as fastICA developed by Dr. Aapo Hyvärinen [6]. Generally, we can obtain Q ICs corresponding to each $\mathbf{g}$. However, only that one with maximum correlation with $\mathbf{x}$ is the deconvolved result required.

However, before applying an ICA algorithm on the data, some preprocessing will make the problem simpler and better conditioned.

The most basic preprocessing is to center $\mathbf{x}$, i.e. subtract its mean $m_{\mathbf{x}}=E\{\mathbf{x}\}$ so as to make $\mathbf{x}$ a zero-mean vector. This preprocessing is made solely to simplify the ICA algorithms. It does not mean that the mean could not be estimated. After estimating the mixing matrix $\mathbf{G}$ with centered data, we can complete the estimation by adding the mean vector of $\mathbf{y}$ back to the centered estimates of $\mathbf{y}$. The mean vector of $\mathbf{y}$ is given by $\mathbf{G}^{-1}m_{\mathbf{x}}$.

Another useful preprocessing strategy in ICA is to whiten the observed variables. This means that before the application of the ICA algorithm (and after centering), we transform the observed vector $\mathbf{x}$ linearly so that we obtain a new vector $\tilde{\mathbf{x}}$ which is white, i.e. its components are uncorrelated and their variances equal unity. In other words, the covariance matrix of $\tilde{\mathbf{x}}$ equals the identity matrix,

$$E\left\{\tilde{\mathbf{x}}\tilde{\mathbf{x}}^T\right\}=I \tag{5}$$

The whitening transformation is always possible. One popular method for whitening is the principle component analysis (PCA). It may also be quite useful to reduce the dimension of the data at the same time as we do the whitening. This has often the effect of reducing noise. Moreover, dimension reduction prevents over-learning, which can sometimes be observed in ICA.

Before applying the method above-proposed to the spectroscopic data, we have to select the length of the deconvolution kernel Q properly. In principle, the selection of the length is not rigorous. However, too small Q can not to descript the convolution kernel and induce some pseudo spectral lines. Too large Q may result in inaccurate estimates of the empirical central moments.

3 Experiments

In this section, three spectra are used to demonstrate the feasibility of this method.

An 832 length Raman spectroscopic data of (D+)-Glucopyranose from 149-980cm^{-1} with the spectral step 1.0cm^{-1} is displayed in figure 1 (A). The FWHM of peak 441cm^{-1} and 542cm^{-1} is 10cm^{-1} and 13cm^{-1} respectively. Figure 1(B) is its deconvolved data. Here Q=15. In figure 1(B), the peak at 406cm^{-1} is split into two peaks at 396cm^{-1} and 407cm^{-1} respectively, and the peak at 542cm^{-1} is split into two peaks at 542cm^{-1} and 559cm^{-1} respectively. The FWHM of peak 441 and 542cm^{-1} becomes 7 and 9cm^{-1} respectively. Figure 1(C) is the estimated convolution kernel with FWHM 9cm^{-1}.

Figure 2(A) is a 475 length Raman spectroscopic data of Benzene from -613.7-4800cm^{-1} with the spectral step 10.41cm^{-1}. The FWHM of peak 3069.5cm^{-1} is 93.7cm^{-1}(9 steps). Figure 2(B) is its deconvolved data. Here, Q=15 . The two-peak excitation line is distinctly split into -104.5 cm^{-1} and 0cm^{-1} respectively. The peak 992cm^{-1} is split into two peaks at 958.8 and 992.0cm^{-1} respectively. The FWHM of peak 3069.5 cm^{-1} has become 41.6cm^{-1}(4 steps). Figure 2(C) is the estimated convolution kernel which FWHM is about 140cm^{-1}.

Figure 3(A) is a 902 length Raman spectroscopic data of D-Glucuronic from 700-1601cm^{-1} with the spectral step 1cm^{-1}. The FWHM of peak 884cm^{-1} is 17cm^{-1}. Figure 3(B) is its deconvolved data. Here, Q=15 . The peak 884cm^{-1} is distinctly split into three peaks at 870, 884 and 892cm^{-1} respectively. The FWHM of peak 884cm^{-1} become 5cm^{-1}. The peak 1452cm^{-1} is split into two peaks at 1452 and 1462cm^{-1} respectively. Figure 3(C) is the estimated convolution kernel which FWHM is about 6cm^{-1}.

4 Discussions

Based on the ICA method, a blind deconvolution algorithm for spectroscopic data is proposed. The true spectrum and the response function of instruments are estimated simultaneously. The spectral distortion is improved considerably. In the preprocessing stage, the noise is reduced by reducing the dimension of the data.

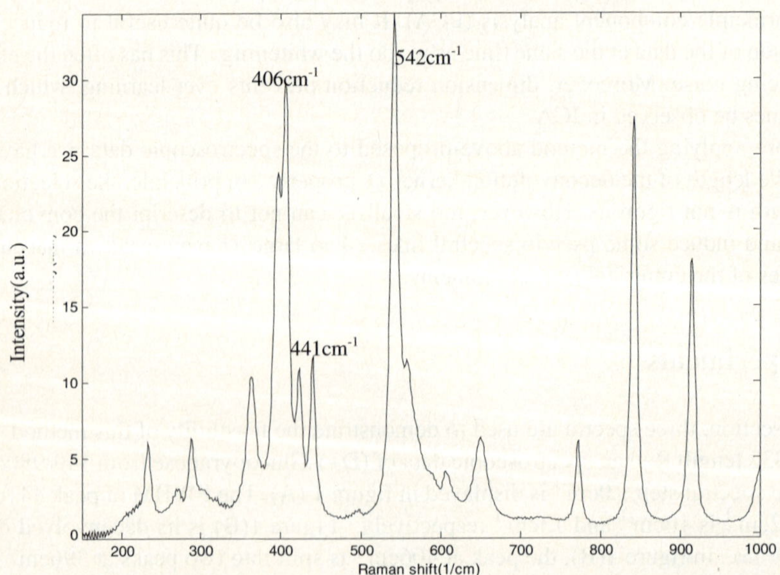

Fig. 1(A). 832 length Raman spectroscopic data of (D+)-Glucopyranose from 149-980cm^{-1}

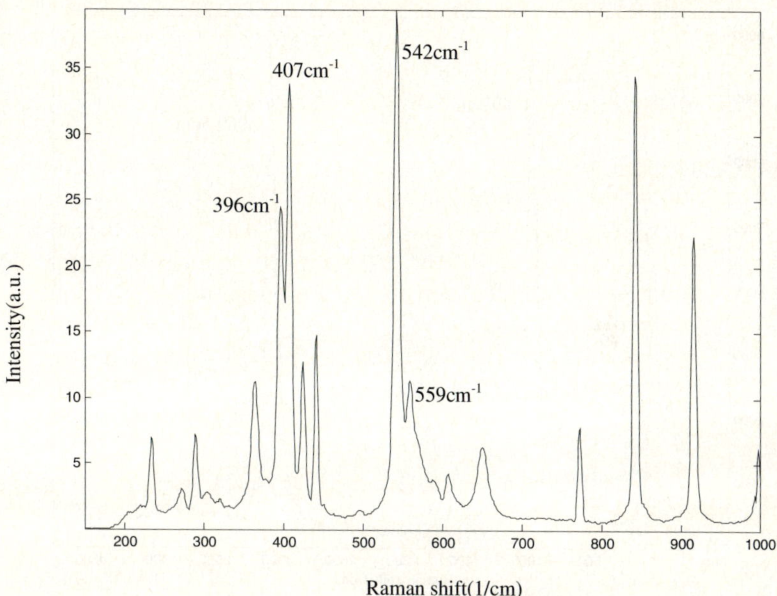

Fig. 1(B). The deconvolved data of (A)

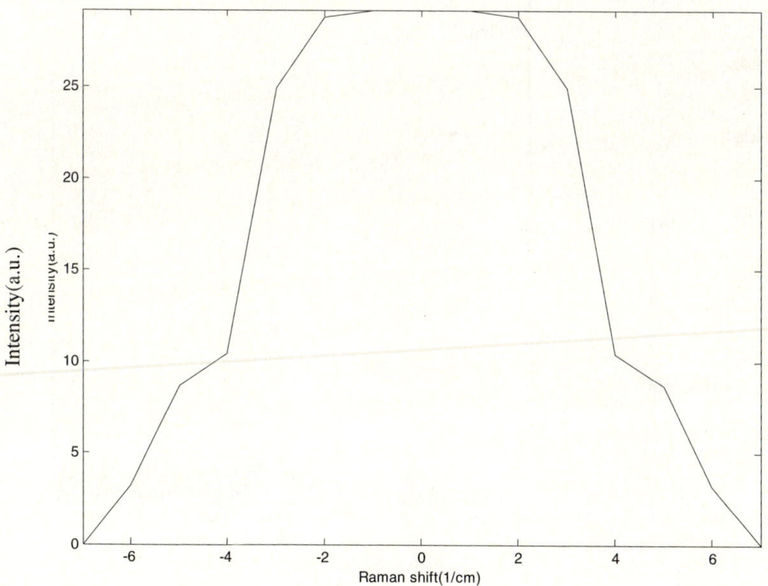

Fig. 1(C). The estimated convolution kernel

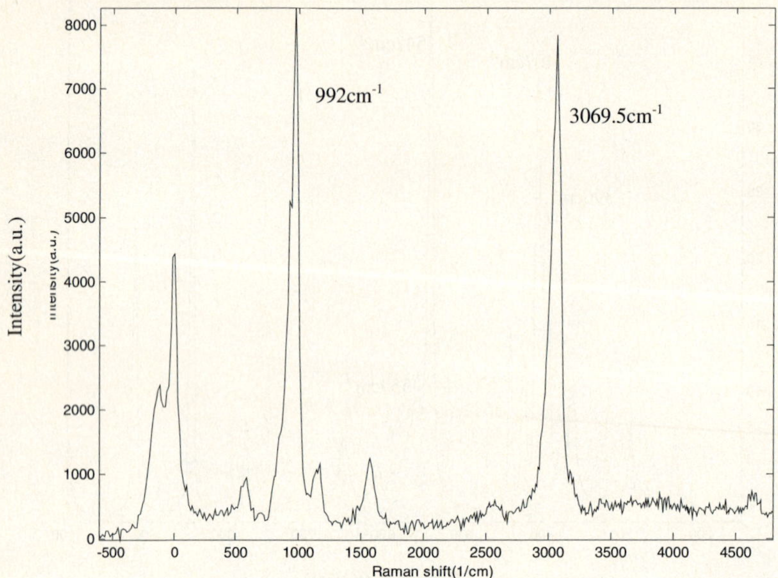

Fig. 2(A). 475 length Raman spectroscopic data of Benzene from -613.7-4800cm-1

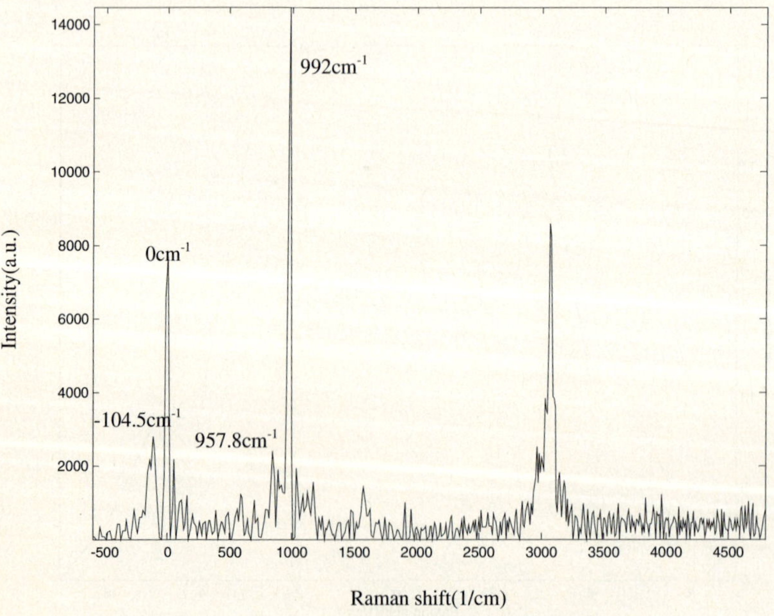

Fig. 2(B). The deconvolved data of (A)

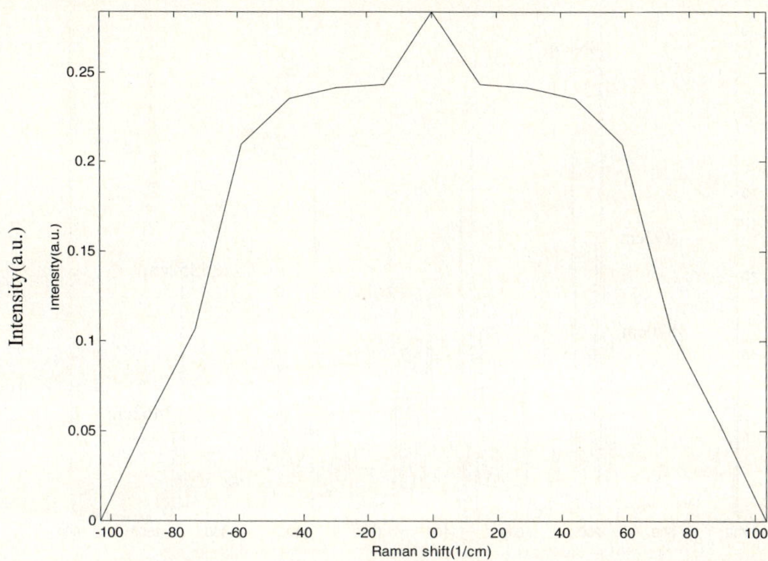

Fig. 2(C). The estimated convolution kernel

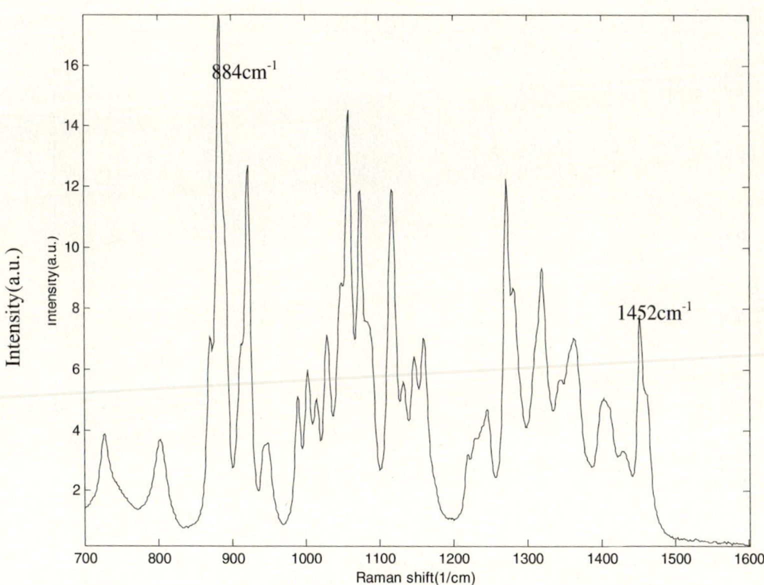

Fig. 3(A). 902 length Raman spectroscopic data of D-Glucuronic from 700-1601cm-1

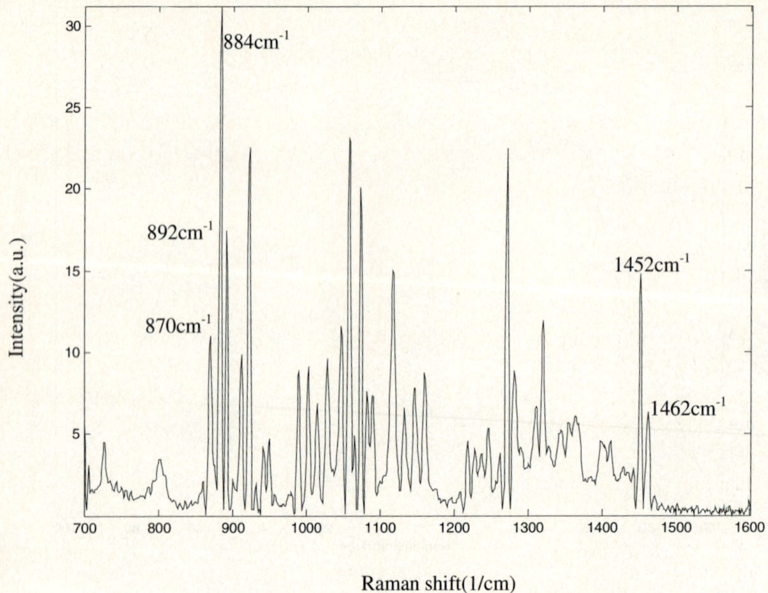

Fig. 3(B). The deconvolved data of (A)

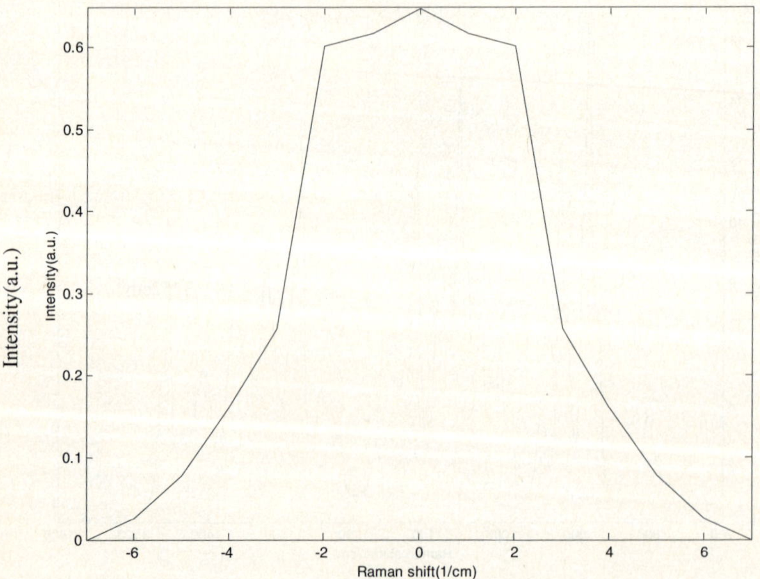

Fig. 3(C). The estimated convolution kernel

In principle, to achieve an ideal deconvolution, an infinite long deconvolution kernel is necessary. But in practice, this is impossible. It can be proven that if the deconvolution kernel length Q is long enough, a sufficient deconvolution can be obtained. However, too long Q will induce an inaccurate estimate of the empirical statistics. So a longer spectroscopic data series is profitable. By experiments, Q being larger than the FWHM of the narrowest spectral line and smaller than ten percent of the data length is a reasonable selection.

References

1. C. W. Helstrom: Image Restoration by the Method of Least Squares. J. Opt. Soc. Am. 57 (1967) 297-303.
2. Y. Senga, K. Minami, S. Kawata, and S. Minami: Estimation of Spectral Slit Width and Blind Deconvolution of Spectroscopic Data by Homomorphic Filtering. Appl. Opt. 23 (1984) 1601-1608.
3. P. A. Jansson (eds.): Deconvolution of Image and Spectra. 2nd edn. Academic Press. Inc. (1997) 9.
4. J. Yuan, Z. Hu: High-order Cumulant Based Blind Deconvolution of Raman Spectra. Appl. Opt. 44 (2005) 7595-7601.
5. J. Yuan, Z. Hu, G. Wang, Z. Xu: Constrained High-order Statistical Blind Deconvolution of Spectral Data. Chinese Optics Letters, 3 (2005) 552-555.
6. Aapo Hyvärinen and Erkki Oja: Independent Component Analysis: Algorithms and Applications. Neural Networks, 13 (2000) 411-430.
7. O. Shalvi and E. Weinstein: Universal methods for blind deconvolution. In: S. Haykin (eds.): Blind deconvolution. Englewood Cliffs, NJ: Ptr prentice-hall. (1994) 122.

Parameterized Semi-supervised Classification Based on Support Vector for Multi-relational Data

Ling Ping[1,2] and Zhou Chun-Guang[1]

[1] College of Computer Science, Jilin University, Key Laboratory of Symbol Computation and Knowledge Engineering of the Ministry of Education, Changchun 130012, China

[2] School of Computer Science, Xuzhou Normal University, Xuzhou, 221116, China

lingicehan@yahoo.com.cn

Abstract. A Parameterized Semi-supervised Classification algorithm based on Support Vector (PSCSV) for multi-relational data is presented in this paper. PSCSV produces class contours with support vectors, and further extracts center information of classes. Data is labeled according to its affinity to class centers. A novel Kernel function encoded in PSCSV is defined for multi-relational version and parameterized by supervisory information. Another point is the self learning of penalty parameter and Kernel scale parameter in the support-vector-based procedures, which eliminates the need to search parameter spaces. Experiments on real datasets demonstrate performance and efficiency of PSCSV.

1 Introduction

Multi-Relational Data Mining (MRDM) [1] promotes mining tasks from traditional single-table case to Multi-Relational (MR) environment. It aims to look for patterns that involve some connected tables. A straightforward solution to MRDM is to integrate all involved tables into a comprehensive table, so that traditional mining algorithms can be performed on it. This, however, would cause loss of structured information, and what makes it worse, it is unwieldy to operate this table. Naturally, MRDM requires mining from multi tables directly.

Kernel function is often used to tackle mining tasks of MRDM due to its fine quality that it implicitly defines the nonlinear map from original space to feature space. A suitably designed Kernel creates the linear version of the problem that would otherwise be nonlinear version in input space while not introducing extra computation efforts. This makes Kernel and Kernel-based methods become hot research area of MRDM. For example, Support Vector Clustering (SVC) [2] and Support Vector Machine (SVM) [3, 4] exhibit impressive performance in concerned application areas.

In MR environment, the description of objects may cover many tables, so the definite label information might be unavailable. Often some supervisory information is provided in the form of pairwise similarity [5]. With this weak label information or side-information, classification issue is equivalent to semi-supervised clustering where the decision model is learned from all data. This paper proposes a Parameterized Semi-supervised Classification algorithm based on Support Vector technique (PSCSV) to

L. Jiao et al. (Eds.): ICNC 2006, Part I, LNCS 4221, pp. 66 – 75, 2006.

address semi-supervised classification of MR data. The Kernel frame encoded in PSCSV is tailored to MR version and it is parameterized by learning from side-information. PSCSV is equipped with the tuning strategies of penalty parameter and Kernel scale parameter. This benefits PSCSV without suffering from searching parameter spaces that is necessary in traditional SV-based algorithms.

2 Related Work

2.1 Support Vector Clustering

Let $x_i \in \Re^n$, $\Re^n$ be the input space, Φ be the nonlinear map from $\Re^n$ to the feature space. a is the center of sphere, R is the radius of sphere. The objective function is:

$$\min_{R,\xi} \; R^2 + C\sum_i \xi_i \tag{1}$$

$$\text{s.t.} \;\; \| \Phi(x_i) - a \|^2 \le R^2 + \xi_i \;,\; \xi_i \ge 0 .$$

Here C is penalty parameter to tradeoff radius and slack variable ξ_i. Transfer it to the Lagrangian function, then to the Wolfe dual, with Kernel trick, leading to:

$$\max_{\gamma} \; \sum_i \gamma_i K(x_i, x_i) - \sum_{i,j} \gamma_i \gamma_j K(x_i, x_j) \tag{2}$$

$$\text{s.t.} \;\; \sum_i \gamma_i = 1 ,\; 0 \le \gamma_i \le C .$$

Gaussian Kernel $k(x_i, x_j) = \exp(-q \| x_i - x_j \|^2)$ is adopted into (2). Following (2), those points with $\xi_i = 0$ and $0 < \gamma_i < C$ are mapped to the surface of sphere, referred as non-bounded Support Vector (nbSV). They describe cluster contours. Those points with $\xi_i > 0$, and $\gamma_i = C$ are located outside the hyper sphere, and are called bounded Support Vector (bSV). Cluster assignment is done based on the deduction of an adjacent matrix of point pairs.

2.2 Support Vector Classification

For t samples: $(x_1, y_1), (x_2, y_2)......(x_t, y_t) \in X \times Y$, where $X = \Re^n$ and $Y = \{1, -1\}$, the optimal classification interface is determined by:

$$g(x) = \sum_i \alpha_i y_i K(x_i, x) - b \tag{3}$$

The orientation vector α and offset vector b are obtained by optimizing:

$$\max_{\alpha} \ \sum_i \alpha_i - \frac{1}{2}\sum_i \sum_j \alpha_i \alpha_j y_i y_j K(x_i, x_j) \tag{4}$$

$$\text{s.t. } 0 \le \alpha_i \le C, \ \sum_i \alpha_i y_i = 0 .$$

Similarly, nbSV refers to points with $0 < \alpha_i < C$, and bSV is points with $\alpha_i = C$.

3 PSCSV Algorithm

3.1 Algorithm Steps

The idea of PSCSV is that those SVs (nbSVs and bSVs) generated by SVC are utilized to describe data contours firstly. Then an iterative classification procedure to separate set of SVs from set of inner-boundary points is appended so as to obtain the interfaces ever-closing to class centers, from which center information will be extracted. The final label is assigned based on affinity criteria. Main steps are as follows.

1) **SVC** produces {*nbSV*} and {*bSV*};
2) *A* = {*data*} - {*nbSV*} - {*bSV*}; *B* = {*nbSV*} + {*bSV*};
3) While (*iteration condition*)
4)   **SVM** (*A, B*);
5)   *A* = *A* + *B* - *new*{*nbSV*} - *new*{*bSV*};
6)   *B* = *new*{*nbSV*} + *new*{*bSV*};
7) End
8) Aggregate centers according to *Affinity* (*Cen_i*, *Cen_j*);
9) Clusters' assignment according to *Affinity (x, Cen_j)*;

Line1 executes SVC process. Instead of only using nbSVs to describe boundary lines conventionally, here nbSVs and bSVs are united together to give a broad description about the decision bands, which are wider than the decision lines. When the boundary bands are engaged in the following iterative classification, they reach class central zones in less runs than the boundary lines do.

The extraction of cluster center information is from Line2 to Line7. Based on these contour bands gained, SVM is performed between the set of points located inside boundaries, *A*, and the set of SVs, *B*. Update *A* as the new inner contents *A* = *new*{*x* | *x is within clusters*}, and *B* as the new decision bands *B* = *new*{*nbSV*}+*new*{*bSV*}. This makes dividing interfaces generated in each classification move closer and closer to central zones. The interfaces decrease their sizes gradually and are quite expected to converge to class centers. This idea, approximation step by step to detect class centers using SVM classifier, is inspired by literature [6]. But in [6], *B* set is initialized as the set of random points and updates two sets as: *A* = {*x* | *x*∈*clusters*} + {*bSV*}, *B* = {*nbSV*}. These are not effective settings. Moreover [6] performs SVM at the fixed runs, lack of consideration of "over-classification", that is, *A* or *B* maybe becomes empty during one time iteration. We modify this procedure and give a heuristics for *iterative condition* as following:

(a) Initialize set B as $B = \{nbSV, bSV \mid obtained\ from\ SVC\}$.

(b) Update $A = A + B$ - $new\{nbSV\}$ - $new\{bSV\}$ and $B = new\{nbSV\} + new\{bSV\}$.

(c) Set $i \le \frac{\sqrt{MinSize}}{coef}$ as the upper bound of runs of SVM iterations, where *Minsize* is the size estimate of the minimum cluster. *Minsize* is evaluated by following steps: i) Sort rows of Kernel matrix in a descending order. ii) Find $Kgap(i) = \max_j\{k(i,j) - k(i,j+1)\}$. iii) $MinSize = \min\{Kgap(i)\}$. Intuitively, the maximum distance between neighboring affinities reveals the edge of dense distributing area around x_i. The corresponding $Kgap(i)$ is the number of points that are closely related with x_i, and this number formulates the natural size estimate of x_i neighborhood. *Minsize* is the rough estimate of the smallest cluster size. So $\sqrt{MinSize}$ has some reason to represent the smallest cluster's radius provided that cluster is of circle shape. *coef* is to indicate the width of decisive bands. It is set as 2 in this paper.

Generally, it is believed that some number of SVs together have the power to completely describe one cluster center. So Line 8 computes the affinity between two cluster centers and aggregates them. Assuming two centers be $Cen\_i=\{si_1,si_2,\ldots si_n\}$ and $Cen\_i=\{sj_1,sj_2,\ldots sj_m\}$, where si_t and sj_t are SVs, their affinity is:

$$Affinity(Cen\_i, Cen\_j) = \sum_{u=1}^{n}\sum_{v=1}^{m}\frac{k(si_u, sj_v)}{m \cdot n} \tag{5}$$

In Line 9, a similarity definition (6) is used to measure the distance between point and class centers to assign labels:

$$Affinity(x, Cen\_i) = \sum_{u=1}^{n}\frac{k(x, si_u)}{n} \tag{6}$$

3.2 Parameterization of Scale *q*

In algorithms based on Gaussian Kernel, width q plays an important role that controls the scale of affinity measurement. Conventionally, algorithms have to be conducted several runs to find a good setting. In MR environment, distribution density varies from table to table. If specifying one single scale, the proper affinities can be formulated on only a few tables because one scale works well under several densities. So q value should be investigated on each involved table. It costs much extremely to explore these scale spaces. We design an auto parameterized approach by learning a fine setting from data's neighborhood context. That is, q is developed as:

$$q = 1/\sigma^2, \quad \text{where } \sigma = average\{\| x - x_r \|\}. \tag{7}$$

Here, x_r is the rth nearest point of x. In other words, x_r is the rth point in the ascending list of distance from x to other points. Intuitively, for the same r, if $\| x- x_r \| < \| y- y_r \|$, it means the dispersion around x is denser than that around y. r may be viewed as the size estimate of natural class containing x_r. $\| x- x_r \|$ reveals something about density of x's class. Taking the average of density information from all neighborhoods, it will present a general distance scale that can be employed as the reference in affinity measurement. Here, r is probed by steps: a) Sort rows of distance matrix d_{ij} in an ascending order. b) Find $r = \max_j\{dis(i,j) - dis(i,j-1)\}$.

3.3 Parameterization of Penalty *C*

Firstly, we analyze the role of *C* in SVM version, and this insight can be promoted to SVC version similarly. The introduction of slack variable ξ_i gives rise to the appearance of penalty parameter *C*. It acts as the balancer to maximize the margin and minimize the error. Traditionally, *C* is selected using crossing validation without taking data distribution and their individual demands on penalty term into consideration. Much consumption and difficulty in grasping suitable range weaken algorithm performance. This paper proposes an approach to parameterize *C* by integrating diverse C_i values that are expected by data points individually.

Collecting factors that can suggest the location of x_i and assuming neighborhood size is expressed by the number of points in neighborhood, we find that, the size of point neighborhood, the distribution density of inner and outer class make influence on point location. Assuming the amount of data is *N*, and the Kernel matrix has been sorted in a descending order, the definition of C_i is:

$$C_i = \frac{1 - Kgap(i) / N}{aveIn(i) / aveAll(i) - 1} \tag{8}$$

where $aveIn(i) = (\Sigma_{j=1}^{Kgap(i)} k(i,j)) / Kgap(i)$, $aveAll(i) = (\Sigma_{j=1}^{N} k(i,j)) / N$.

Kgap(*i*) acts as the size estimate of neighborhood. *aveIn*(*i*) is the average affinity of x_i neighborhood. *aveAll*(*i*) is the average affinity of x_i to all other points. Both of them tell density information. C_i reflects the personal demand to penalty term of x_i. So the global *C* is defined as the average of all individual C_i:

$$C = average\{C_i\} \tag{9}$$

There is no upper bound for *C* in SVM, so values formulated by (11) can be used correctly. But in SVC algorithm, where *C* is bounded in the range of $0 < C \le 1$, some modification is needed to scale C_i in the specified range. We rewrite C_i as:

$$C_i = \exp\left(-\frac{aveIn(i) / aveAll(i) - 1}{1 - Kgap(i) / N}\right) \tag{10}$$

4 Kernel Definition

Some basic concepts are covered firstly. **Key**: It is the attribute that exclusively identifies a unique record. **Foreign Key (*FK*)**: It is the attribute that references some *Key* of another table. **Referenced Key (*RK*)**: The attribute that corresponds to some *FK* of another table is considered as ***RK*** in the original table. *FK* is not necessarily *Key* but *RK* is bound to be *Key*. *FK* and *RK* are **Association Key (*AK*)**. **Main Table:** It is the table that is mainly investigated, where the class of the instance is stated. **Extending Table:** It is the table associated with Main table directly or indirectly. They provide extending descriptions of Main table object. The corresponding instance of Main table object to another record through association key is called as **Extending Instance**. **Main Attribute:** It is the attribute on which data mining tasks are focused. **Common Attribute:** The attribute fields in Main table are Common Attributes.

4.1 Kernel Frame for MR Data

Some literatures have proposed Kernel definition of structured data based on convolution thoughts or summing thoughts [7]. Our Kernel is guided by the weighted summing idea. In MR environment, starting from Main table, extending tables in association frame provide the direct or indirect description of objects. For a pair of objects, we think a local affinity is formulated in each extending table. The global affinity is accumulated in a weighted fashion. So the global affinity is developed by collecting all elementary affinities together in the combination as:

$$K(x,y)=\Sigma_{i=0}^{m}\ \mu_i k_i(x,y) \qquad \text{s.t. } \Sigma_{i=0}^{m}\ \mu_i=1,\ \mu_i\geq 0. \tag{11}$$

Here, weight coefficients tradeoff the contributions of extending tables to the global affinity. They will be parameterized by side-information in Section 4.3. k_0 is to compute the affinity on Main table. k_i $(i=1...m)$ is to compute local affinity on *ith* extending table. The local affinity is obtained by the computation of elementary Kernel on the corresponding extending instances in each extending table. Thus the computation of global Kernel is reduced to the computation of elementary Kernel on single table. This meets the requirement of MRDM to mine from multi sources directly. Usually, Gaussian Kernel serves as elementary Kernel. Its positive definite property makes (11) hold this property too. Note that the scale of elementary Gaussian Kernel is tuned in each table.

4.2 Elementary Kernel Computation

Given x and y are Main table objects, their local affinity in extending table is addressed in two cases. If the current *i*th table is extended by the map from *FK* to *RK*, which means the current table contains *RK*, x and y correspond to the unique extending instances respectively x^+ and y^+. Then the local affinity between them is the affinity between their extending instances:

$$k_i(x,y)=k(x_i^+,y_i^+) \tag{12}$$

Here x_i^+ is the extending instance of x in *ith* extending table. But as for the map from *RK* to *FK*, which means the current *j*th table contains *FK*, object x may correspond to a set of extending instances. Then the local affinity between x and y in this direction is decided by the affinity of their extending instance sets. That is:

$$k_j(x,y)=k_{set}(\{x_j^+\},\{y_j^+\}) \tag{13}$$

Where $k_{set}(S1,S2)=\dfrac{\Sigma_{x1\in S1,x2\in S2}\ k(x1,x2)}{|S1\times S2|}$.

4.3 Learning Weight Coefficients

Given the side-information in form of $S=\{(x,y)\mid similar(x,y)\}$, we learn weighted coefficients of formula (11) from S by optimizing below objection function:

$$\min \| \eta - K \|_F \tag{14}$$

where $\| M \|_F = \Sigma_{i,j} M_{ij}^2$, and K is the designed global Kernel. η is the expected affinity matrix on S. Its entry is defined as:

$$\eta_{ij} = \frac{1}{2} \cdot \delta_{ij} + \frac{1}{2} \exp(-q \| dis_{ij} \|^2) \quad \text{where } \delta_{ij} = \begin{cases} 1 & (x_i, x_j) \in S^+ \\ 0 & (x_i, x_j) \notin S^+ \end{cases}. \tag{15}$$

In the above definition, S^+ is the reflective and symmetric closure of S. dis_{ij} is the average of distances in m extending tables between object x_i and x_j. Width q in (15) is the average of all widths from m extending tables. This affinity fulfills the property that the affinity of similar points should larger than the dissimilar points. Adopt global Kernel into (14), and it turns to a quadratic optimization problem:

$$\min \| \eta - K \|_F = \min \Sigma_{i,j}^t (\eta_{ij} - V_{ij}^T \mu)^2 = \min(\mu^T A \mu + B \mu) \tag{16}$$

Here, $V_{ij}^T = (k_1(x_i, x_j), k_2(x_i, x_j), ..., k_m(x_i, x_j))$, which is local affinities of one pair of objects on all extending tables. And $A = \Sigma_{i,j}^t V_{ij} V_{ij}^T$, $B = -2\Sigma_{i,j}^t \eta_{ij} V_{ij}^T$.

5 Experiment Results

5.1 Tests on Points

Firstly pure PSVSC algorithm is performed to cluster plane points without side information, with intension to verify its correctness visually (with 2G P4 CPU PC, 256M memory, WinXP, MatLab7.0).There are 3 Gaussian distributions with means and covariance [(1, 1), 1], [(5.8, 5.8), 1.7], [(11, 6), 2]. Fig. 1 shows data contours after SVC with the adaptive width, where "+" denotes SVs. Fig. 2 and Fig.3 illustrate the results after 1 run and 2 runs of SVM iterations, where the moving of "+" reveals the process that surfaces converge to class centers.

Then, to examine the quality of tuning approaches, with the definite label information, the classical SVM is performed on real datasets: Winc and Breast Cancer [8]. In this single-table data environment, a single Gaussian Kernel with the tuned q is used. We classify data in even-sampling cases and uneven-sampling cases, to access the performance parameterized C mechanism. To Wine dataset with 3 classes, we

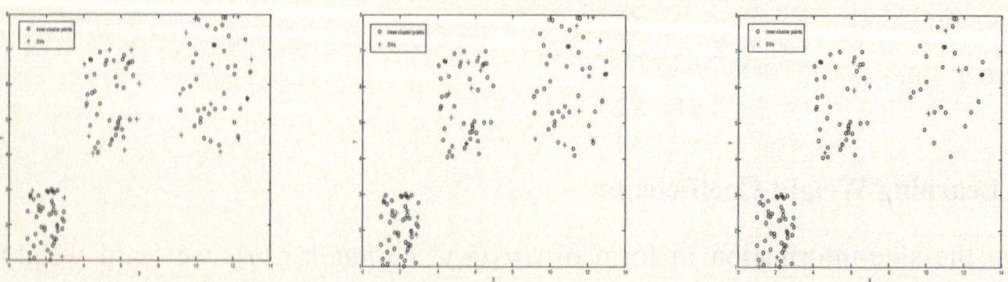

Fig. 1. Result after SVC **Fig. 2.** Result after 1 run SVM **Fig. 3.** Result after 2 runs SVM

perform one-VS-one strategy to generate multi decision functions, and label data according to voting approach. In Table 1, ES means the even-sampling manner, US means the uneven-sampling way, and numbers in '{}' refer to the number of samples from each class. *pC-SVM* refers to the SVM using parameterized *C*, and *C-SVM* refers to traditional SVM with *C* being set as 1.8.

Table 1. Clustering Accuracy Comparison on real dataset. (%) (*N* is the size of dataset, *D* is the dimension, and *NC* is the number of classes. 20 runs of algorithm.)

Wine ($N = 178$, $D = 13$, $NC = 3$)	ES {10, 10, 10}	ES {25, 25, 25}	US {8, 20, 15}	US {5, 25, 15}
C-SVM	96.17	96.56	94.31	94.15
pC-SVM	95.97	96.33	96.02	95.61
BC ($N = 683$, $D = 9$, $NC = 2$)	ES {40, 40}	ES {80, 80}	US {30, 60}	US {10, 30}
C-SVM	97.82	99.01	97.44	97.01
pC-SVM	97.74	98.52	98.51	97.68

From Table 1, it is easy to see *pC-SVM* is competitive with *C-SVM* in the even-sampling cases. It produces better performance that traditional SVM in uneven-sampling cases due to its adaptive learning method of penalty coefficients.

5.2 Tests on MR Data

Now, PSVSC is applied on relational problems: Musk [9] and Mutagenesis [10]. Musk has two versions Musk1 and Musk2. They record 476 and 6598 conformations for musk molecules and non-musk molecules. We fix the data with some normalization and develop its relation frame as shown in Fig. 4. Mutagenesis dataset records 230 aromatic and heteroaromatic nitro compounds. They are divided into two groups based on the mutagenicity: the active and the inactive. Its relation schema is shown in Fig. 5. Since there are two types of Mutagenesis molecules: regression-friendly and regression-unfriendly, we sample from two subsets and the whole dataset respectively.

To access quality of the designed Kernel, we employ it into classical SVM to classify Musk1 and Mutagenesis. In Table 2, the results are compared with other classifiers: TILDE [11], SVM-MM and SVM-MI [12]. There, *S*1 refers to the regression-friendly subset and *S*2 is the unfriendly one. Our SVM procedure achieves better results, showing the designed Kernel can capture relational features effectively. Then, PSCSV is performed. We also perform another PSCSV version: PSCSV-*q*. PSCSV-*q* finds *q* width for each extending table by searching in their parameter spaces. From Table 3, both of two algorithms can produce fine results. PSCSV is competitive with PSCSV-*q*, but we prefer PSCSV because of its ease of parameterization.

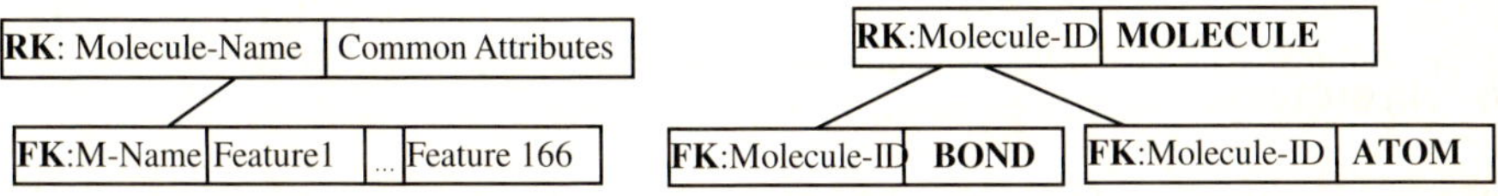

Fig. 4. Relation Schema of Musk

Fig. 5. Relation Schema of Mutagenesis

where $\Delta e(l) = e(l+1) - e(l) = \left(\frac{\partial e(l)}{\partial w}\right)^{\mathrm{T}} \Delta w$, and

$$\Delta w = -\eta \frac{\partial E}{\partial e(l)} \frac{\partial e(l)}{\partial w} = -\eta e(l) \frac{\partial e(l)}{\partial w} \tag{9}$$

where $w = (w_0, w_1, w_2, \cdots, w_{N-1})^{\mathrm{T}}$, so

$$\Delta e(l) = -\eta \left(\frac{\partial e(l)}{\partial w}\right)^{\mathrm{T}} e(l) \frac{\partial e(l)}{\partial w} = -\eta e(l) \left|\frac{\partial e(l)}{\partial w}\right|^2 \tag{10}$$

Substituting (10) into (8) yields

$$\Delta V(l) = = e^2(l) \left|\frac{\partial e(l)}{\partial w}\right|^2 \left[-\eta + \frac{1}{2}\eta^2 \left|\frac{\partial e(l)}{\partial w}\right|^2\right] \tag{11}$$

From (11) and the convergent condition of NN, it only satisfies

$$-\eta + \frac{1}{2}\eta^2 \left|\frac{\partial e(l)}{\partial w}\right|^2 < 0 \tag{12}$$

Using $\eta > 0$ and $\frac{\partial e(l)}{\partial w_n} = \frac{\partial e(l)}{\partial w_{n\mathrm{R}}} + \mathrm{j}\frac{\partial e(l)}{\partial w_{n\mathrm{I}}} = \frac{1}{e(l)}\left[H(\mathrm{e}^{\mathrm{j}\Omega_l}) - H_{\mathrm{d}}(\mathrm{e}^{\mathrm{j}\Omega_l})\right]\mathrm{e}^{\mathrm{j}\Omega_l n}$ $n = 0, 2, \cdots, N-1$, so

$$\left|\frac{\partial e(l)}{\partial w}\right|^2 = \frac{1}{e^2(l)} e^2(l) \sum_{n=0}^{N-1} \left|\mathrm{e}^{\mathrm{j}n\Omega_l}\right| = N \tag{13}$$

From (13), the convergent condition is

$$0 < \eta < \frac{2}{N} \tag{14}$$

In fact, in the case of that different case N (odd or even) and $h(n)$ ($h(n) = h(N-n-1)$ or $h(n) = -h(N-n-1)$), the Eq.(1) can be equivalent to four types of classical linear phase FIR filters. For the limited space of the paper, we don't prove it here. That is to say, the CFNN model has universal performance theoretically, it can implement designing of complex filter with arbitrary amplitude frequency and phase frequency characteristic.

4 Design Examples

Example 1. Design a band-pass FIR filter whose ideal frequency response and ideal amplitude-frequency response are respectively

$$H_{\mathrm{d}}(\mathrm{e}^{\mathrm{j}\Omega}) = \begin{cases} \mathrm{e}^{-\mathrm{j}T_{\mathrm{d}}\Omega} & \Omega \in [0.25\pi, \pi] \\ 0 & \Omega \in [0, 0.25\pi) \cup (\pi, 2\pi] \end{cases} \qquad \left|H_{\mathrm{d}}(\mathrm{e}^{\mathrm{j}\Omega})\right| = \begin{cases} 1 & \Omega \in [0.4\pi, 1.2\pi] \\ 0 & \Omega \in [0, 0.4\pi) \cup (1.2\pi, 2\pi] \end{cases}$$

The following designs are required: (1) Type-1 (cosine base) FIR filter with N=81, where $T_{\mathrm{d}} = \frac{N-1}{2}$; (2) Type-4 (sine base) FIR with N= 80, where $T_{\mathrm{d}} = \frac{\pi}{2\Omega} + \frac{N-1}{2}$.

For the two networks, their free parameters are N, and type-1 FIR filter can be designed by cosine NN, and type-4 FIR filter designed by sine NN. In addition, in order that there are no excess impulse and no fluctuation in the range of pass-band and stop-band, we sample two points 0.78 and 0.25 respectively in each transition band. The iterative step η is 0.03, the global error is set to $E = 1.5\times10^{-10}$.

Design the two types of FIR filters by using CFNN. And $\eta = 0.01$, $E=1.5\times10^{-10}$. Amplitude-frequency characteristics of three types of trained FIR filter are showed in Fig. 2. From the Fig.2, we can see that the pass-band and stop-band characteristics of CFNN is better than those of cosine base and sine base.

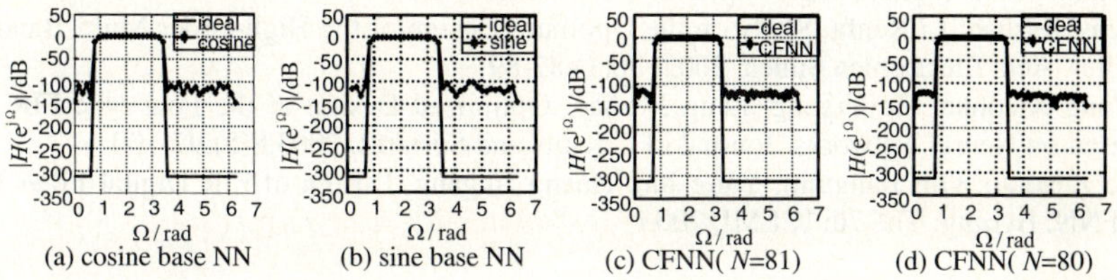

Fig. 2. Amplitude-frequency characteristics of NN models

Example 2. Design a 200-order (i.e. N=200)nonlinear two pass-band FIR filter. The ideal frequency response of this filter is

$$H_d(e^{j\Omega}) = \begin{cases} e^{-j12.5\Omega} & \Omega\in[0.2\pi,0.8\pi] \\ e^{-j27\Omega} & \Omega\in[1.2\pi,1.6\pi] \\ 0 & \text{others} \end{cases}, \text{ where } \Omega\in[0,2\pi].$$

Under the condition of N=200, $\eta = 0.008$, $E = 2\times10^{-10}$.

After the network training, the amplitude-frequency characteristics is showed in Fig.3. Pass-band and stop-band performance indexesare respectively δ_{p1} = 0.0278 dB, δ_{p2}=0.0332dB, δ_{s1}=113.38dB, δ_{s2}=109.50dB, δ_{s3}=115.52dB.

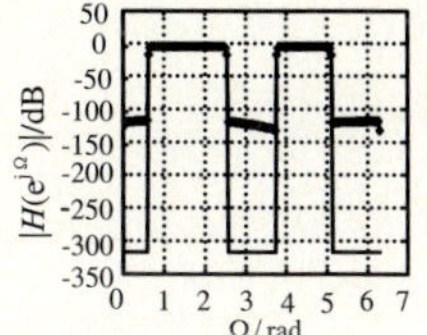

Fig. 3. $|H(e^{j\Omega})|$by CFNN

These simulations are completed by Matlab codes in Dell computer which CPU is 1.7GHz and memory is 1GB. For the example 1, the training time is 0.1250s for sine case and cosine case while 0.4690s for CFFN case. For the example 2, its training time is 1.4060s. The two examples above show that CFNN model is not only used to design linear FIR digital filter, but also to design high-order multi-band-pass nonlinear one, which verifies the universality of CFNN model.

5 Conclusion

The paper proposes a CFNN model of filter optimum design based on the complex frequency response of FIR digital filter directly, proves its convergence, and draws the conclusion that CFNN has universality, finally gives two examples. The simulation result shows that it's not only CFNN model can design linear phase and high-order multi-band-pass nonlinear FIR digital filter, but also its pass-band and stop-band characteristics is better than that of cosine base and sine base, which verifies the universality of CFNN model in the aspect of designing filter.

References

1. Ding Yumei, Gao Xiquan. Digital Signal Processing. Xi'an: Xidian University Press, 2001:195-197
2. Wang Yan, Liao Xiaofeng, Yu Juebang. Complex Neural Network on Designing FIR Digital Filter. Signal Processing, 1999, 15(3):193-198
3. Zeng Jiezhao, Li Renfa. Study on the Optimum Design of the High-Order Multi-Band-Pass Filter. Acta Electronica Sinica, 2002, 30(1):87-89
4. Wang Xiaohua, He Yigang, Zeng Jiezhao. Optimized Design of the Four-type FIR Filter based on Neural Networks. Journal of Circuits and Systems, 2003,8(5):97-100
5. He Zhaohui, Qin Jiangmin, Yang Jun, Huang Jingang. Design of FIR Digital Filter Using CFNN. Beijing: The 7th ICEMI, 2005

SAPSO Neural Network for Inspection of Non-development Hatching Eggs

Yu Zhi-hong, Wang Chun-guang, and Feng Jun-qing

College of Mechanical and Electrical Engineering,
Inner Mongolia Agricultural University, Huhhot 010018, China
yzhyq@sohu.com

Abstract. Detection fertility and development in hatchery eggs could increase efficiency in commercial hatcheries. A new algorithm named simulated annealing particle swarm optimization algorithm (SAPSO) is proposed, and it is used to optimize topology structure of multi-layer feedback forward neural network for classification of hatching eggs. Trained and tested by a great deal of samples, a reasonable neural network model is obtained. Its performance is measured in terms of two parameters: short computing time and accuracy in the classification process.

1 Introduction

Recognition of hatching eggs could benefit the hatchery by culling infertile or problem ones, reducing crowding of incubator space, producing better airflow, possibly reducing utility costs, and preventing contamination. Manual inspection lacks of objectivity and accuracy and it is time-consuming. Developing an automated system capable of detecting eggs would be a major benefit to the poultry industry, which suffer from visual stress and tiredness, perform the inspection for improving quality control and productivity. Previous studies[1-5] related to the poultry industry deal with hatching eggs for classification purposes. But those studies couldn't satisfy the request of on-line detection at accuracy and speed. Therefore, present work deals with the automatic visual inspection infertile or non-developing hatching eggs during early incubation to perform a more accurate classification.

2 Materials and Method

2.1 Feature Parameters Extraction

Hatching day 3 to 7 eggs are obtained from a commercial hatchery. RGB images of eggs are obtained by computer vision system and converted into gray images. In order to avoid time-consuming computation and statistics histogram of the eggs inner pixels accurately, the background influence of the pixels must be done away with upon images. After pre-processing, gray values of histograms on hatching day 3 to 7 eggs are between 0-60, a number of pixels in those cells are used as inputs of a neural network.

L. Jiao et al. (Eds.): ICNC 2006, Part I, LNCS 4221, pp. 85 – 88, 2006.

The numbers of inputs are reduced to 30 by combining two adjacent cells in each histogram for an effective neural network structure. A neural network with fewer inputs is desired because it would require less computer resources.

2.2 Simulated Annealing Particle Swarm Optimization (SAPSO)

BP (Back Propagation) network has the advantage of parallel processing and extrapolation capabilities. It needs huge amount of training data and has drawbacks of over-training and may end up at local minimum [6]. Particle Swarm Optimization (PSO) [7-9] is a kind of global optimization evolution algorithm. It can get better results in a faster, cheaper way. Therefore PSO is used to train reasonable BP neural network structure for detecting fertility of hatching eggs. The PSO system is initialized with a population of random solutions and searches for optima by updating generations. The potential solutions, called particles update their velocities and positions with following equation.

$$v_{id}(t)=x(v_{id}(t-1)+w_1(p_{id}-x_{id}(t-1))+w_2(p_{gd}-x_{id}(t-1))) \quad (1)$$

$$x_{id}(t)=x_{id}(t-1)+v_{id}(t) \quad (2)$$

Where $x=2/\left|2-w-\sqrt{w^2-4w}\right|$ and w=w1+w2, w>4.0

v_{id} is the particle velocity, x_{id} is the current particle (solution), $w_1$and w_2 are learning factors that are positive constants, p_{id} is the best solution of each particle it has achieved so far , p_{gd} is the global best .

Sometimes basic PSO tends to converge to the local best solution not the global best solution quickly, SAPSO is put forward combined basic PSO with Simulated Annealing (SA) [10] to overcome this limitation. It is described as below:

1. Randomly initialize particle velocities v_{id} and positions x_{id} on n dimensions in the real number problem space, set iteration counter t=1;
2. Calculate fitness function $f(x_{id})$, average fitness f_{avg}, maximum fitness f_{max}, save p_{gd};
3. Resort particles according to their fitness, select $m=n\times p_t$ particles of bigger fitness, $p_t=k_1\times f_{avg}/f_{max}$, get the new population x_{jd} (j = 1 , 2 , …,m ; m< n) ;
4. Set the start temperature T and speed of annealing β. For each particle x_{jd}, produce a new solution y_{jd} randomly in its neighborhood respectively, if $f(y_{jd})>f(x_{jd})$, then accept y_{jd} ,else if $\exp((f(y_{jd})-f_{avg})/T)>rand(0,1)$,then accept y_{jd} either, the temperature is decremented, T←βT, else hold T. Repeated until a frozen state is achieved at T=0.
5. Compare each particle's fitness value with its p_{id}, If the current value is better, then set p_{id} equal to the current value, and update p_{gd} ;
6. While a criterion is met or t equals the maximum iteration counter, then output the current solution ,else t:t+1 ; Loop to step 2.

2.3 Training BP Neural Network by SAPSO

Each particle corresponds to the complete augmented weight vector that connects SAPSO-trained BP neural network's input-to-hidden and hidden-to-output layers. A

3-layer neural network is implemented, using sigmoid activation functions in the hidden and output layer. All sample data are normalized. The network weights are initialized in the range of (-1, 1). The initial network structure is 30-8-3. The training process is considered with 10, 12 ,14 ,16 ,18and 20 hidden units respectively. Mean square error (MSE) of neural network output values is used to generate goal function, set fitness function f= 1/ MSE , the little error, the better performance in correspondence with the particle. Parameters of SAPSO are set as follow: Population size n=20;w_1=w_2=2.05; v_{max}=4.0, v_{max} was set to be the lower and upper bounds of the allowed variable ranges; maximum iteration counter t=1000; k_1=0.20; T=10000; β=0.8; goal function error ε=0.01.

3 Results and Discussion

In order to reduce the sensitivity of neural network, the network training samples selected are representative and comprehensive. Pre-classified by farm experts, 60 alive eggs, 60 non-alive eggs and 30 infertile eggs on every hatching day 3 to 7 are selected as the network training samples. A training set of 2/3 patterns is constructed randomly. The rest is used as a testing set to evaluate the accuracy of neural network. A 30-12-3 network topology structure is obtained. The valid weight values are achieved by training network, and saved into weight files.

Table 1. Algorithm MSE of iteration counter t= N

Algorithm	MSE of iteration counter t= N		
	N=50	N=100	N=200
BP neural network (BPNN)	0.02641	0.01079	0.00226
PSO neural network (PSONN)	0.63217	0.02476	0.00998
SA neural network (SANN)	0.88763	0.03128	0.01023
SAPSO neural network (SAPSONN)	0.01065	0.00087	8.9015e-4

Table 1 shows the MSE of different iteration counters of BPNN, PSONN , SANN and SAPSONN. While error is 0.01, their average convergence rate is 60.61%, 41.83% , 38.45%, 89.14%,their average convergence iteration is 67.6080 ,213.70543, 256.89741, 56.47262 after running 50 times respectively. So convergence process of SAPSONN has the most obvious advantage. SAPSO algorithm has a stronger local and global searching performance at all evolution periods, all particles converge to the best solution quickly. At the same time, SA operation allows particles accept worse solution, it can revise particles, expand new solution space continuously, prevent from the individual degeneration, accelerate the convergence of algorithm. The rest of the patterns form testing set to evaluate the accuracy of neural network. The result shows that the imaging system appears capable of detecting hatching eggs. Accuracy rates for alive eggs are: day 3, 86.7%; day 4, 90.0%; day 5, 91.4%; day 6, 95.0%; day 7, 98.2%.

4 Conclusions

An effective method of automatic detecting fertility of hatching egg using neural network system is put forward. Structure of multi-layer feedback forward neural network is optimized by SAPSO algorithm, The training phase is accomplished by means of a set of examples and an adequate fitness function. It improves the precision of former classification algorithms[1-6] with similar computational cost. It proved by plenty of experiment that the neural network system for fertility of hatching eggs detection has a high accuracy and efficiency, the algorithm is robust and reliable.

Further study from now on is to investigate and improve the representation of the data, to use Principal Component Analysis to select the most significant feature variables, to consider comprehensively the fitness function, population size, characteristic parameters influence upon algorithm, to improve the algorithm performance according to three criteria: accuracy of the result, performance efficiency and robusticity of the algorithm for raising accurate rate and speeds of detection.

References

1. Liu, Y., A. Ouyang, C. Zhong, and Y. Ting: Ultraviolet and visible transmittance techniques for detection of quality in poultry eggs. Proc. SPIE (2004) 5587:47-52
2. Schouenberg, K. O. P.: Method and device for detecting undesired matter in eggs. US Patent (2003) 6,504,603
3. Chalker, II, B. A.: Methods and apparatus for non-invasively identifying conditions of eggs via multi-wavelength spectral comparison. US Patent (2003) 6,535,277
4. Smith, D.P., Mauldin, J.M., Lawrence, K.C., Park, B., Heitschmidt, G.R. : Detection Of Early Changes In Fertile Eggs During Incubation Using A Hyperspectral Imaging System. Poultry Science (2004) 83(SUPPL.1):75
5. Smith, D.P., Mauldin, J.M., Lawrence, K.C., Park, B., Heitschmidt, G.W. : Detection Of Fertility And Early Development Of Hatching Eggs With Hyperspectral Imaging. Proceedings 17th European Symposium On Quality Of Poultry Meat (2005)139–144
6. Martin T.Hagan. Neural network design. Machine industry publisher (2002)
7. Zeng Jian-chao , Jie Qian ,Cui Zhi-hua. Particle swarm optimization algorithm. Science publisher (2004)
8. Eberhart, R. C. , Shi, Y. : Particle swarm optimization: developments, applications and resources. Proceedings of IEEE Congress on Evolutionary Computation. IEEE service center, Piscataway, NJ., Seoul, Korea.(2001a)81–86
9. Eberhart, R. C. ,Shi,Y. : Tracking and optimizing dynamic systems with particle swarms. Proceedings of IEEE Congress on Evolutionary Computation. IEEE service center, Piscataway, NJ., Seoul, Korea.(2001b) 94–97
10. http://www.cs.sandia.gov/opt/survey/sa.html

Research on Stereographic Projection and It's Application on Feed Forward Neural Network*

Zhenya Zhang[1,3], Hongmei Cheng[2], and Xufa Wang[4]

[1] Computer and Information Engineering Department of Anhui Institute of Architecture & Industry (AIAI), 230022, Hefei, China
zhenyazhang@ustc.edu
[2] Management Engineering Department of AIAI, 230022, Hefei, China
[3] MOE-Microsoft Key Laboratory of Multimedia Computing and Communication, University of Science and Technology of China (USTC), 230027, Hefei, China
[4]Computer Science Department of USTC, 230027, Hefei, China

Abstract. Feed forward neural network for classification instantly requires that the modular length of input vector is 1. On the other hand, Stereographic projection can map a point in n dimensional real space into the surface of unit sphere in (n+1) dimensional real space. Because the modular length of any point in the unit sphere of (n+1) dimensional real surface is 1 and stereographic projection is a bijective mapping, Stereographic projection can be treated as an implementation for the normalization of vector in n dimensional real space. Experimental results shown that feed forward neural network can classify data instantly and accurately if stereographic projection is used to normalized input vector for feed forward network.

1 Motivation and Method

Feed forward neural network can be used to classify documents instantly [1~3]. When feed forward neural network is prepared for document classification, it is necessary to represent each document as vector. TF vector is used widely [2~6]. If the modular length of each represented vector 1, feed forward network can classify documents more instantly. To normalize the input vector for TextCC [3], which is a kind of feed forward neural network for corner classification instantly with cosine similarity as criteria, stereographic projection is introduced in this letter.

Let $S=\{X=(x_1,\cdots,x_n,x_{n+1})\,|\,(x_1,\cdots,x_n,x_{n+1})\in R^{n+1}, i=1\cdots n+1, \sum_{i=1}^{n+1} x_i^2=1\}$. S is the surface of the unit sphere in n+1 dimensional real space. $(0,0,\cdots,0)\in R^{n+1}$ is the center of the n+1 dimensional unit sphere. $(0,\cdots,0,1)\in S$ is the polar of S.

* This paper was supported in part by "the Science Research Fund of MOE-Microsoft Key Laboratory of Multimedia Computing and Communication (Grant No.05071807)" and post-doc's research fund of Anhui Institute of Architecture&Industry.

L. Jiao et al. (Eds.): ICNC 2006, Part I, LNCS 4221, pp. 89 – 92, 2006.

Let $X=(x_1,\cdots,x_n)\in R^n$. The point $X^{'}=(x_1,\cdots,x_n,0)$ is the extension of X in (n+1) dimensional real space.

If $X^{'}=(x_1,\cdots,x_n,0)$ be the extension of $X=(x_1,\cdots,x_n)\in R^n$, S be the unit sphere in (n+1) dimensional real space and $P=(0,\cdots,0,1)\in R^{n+1}$ be the polar of S, the line PX' will intersect with S at $Y=(y_1,\cdots y_{n+1})$. Y is the stereographic projection of X in S.

Because the modular length of the stereographic projection for each vector is 1 and stereographic projection is a bijective mapping, stereographic projection can be treated as one mapping for normalization.

If $X=(x_1,\cdots,x_n)\in R^n$ and $Y=(y_1,\cdots y_{n+1})$ be the stereographic projection of X in the surface of n+1 dimension unit sphere, $y_i=\dfrac{2x_i}{1+\sum_{i=1}^{n}x_i^2}, i=1\cdots n$ and

$$y_{n+1}=\frac{\sum_{i=1}^{n}x_i^2-1}{1+\sum_{i=1}^{n}x_i^2}.$$

Assume the stereographic projection of $X_1=(x_{11},\cdots,x_{1n})\in R^n$ is $Y_1=(k_1x_{11},\cdots,k_1x_{1n},1-k_1)\in R^{n+1}$ and the stereographic projection of $X_2=(x_{21},\cdots,x_{2n})\in R^n$ is $Y_2=(k_2x_{21},\cdots,k_2x_{2n},1-k_2)\in R^{n+1}$ here $k_1=\dfrac{2}{1+\sum_{i=1}^{n}x_{1i}^2}$ and

$k_2=\dfrac{2}{1+\sum_{i=1}^{n}x_{2i}^2}$. If we denote the distance from X_1 to X_2 as d(X_1,X_2) and the angle from Y_1 to Y_2 is θ. $d^2(X_1,X_2)=\dfrac{2-2\cos\theta}{k_1k_2}$. With these results, it is easy to know that $d^2(X_1,X_2)\le r^2$ equal to $\cos\theta\ge 1-\dfrac{k_1k_2r^2}{2}$ here $r\in R^+$, $d(X_1,X_2)$ is the distance from X_1 to X_2 and θ is the angle from Y_1 to Y_2.

If TextCC is used to classify data in n dimensional real space with input vector normalized as stereographic projection, the criteria for classification is Euclidean distance in n dimensional space. Because the classification criteria of TextCC is cosine similarity, there are some modification on the weight matrix between input and hidden layer of TextCC.

Let the weight matrix of hidden layer of TextCC is W and $W_i=(w_{i1},\cdots,w_{in}), i=1\cdots H$ be all sample vectors for TextCC training. Let $W_i^s=(k_iw_{i1},\cdots,k_iw_{in},1-k_i)$ be stereographic projection of W_i, here

$k_i = \dfrac{2}{1+\sum_{j=1}^{n} w_{ij}^2}, i = 1 \cdots H$, W, the weight matrix of hidden layer of TextCC can be constructed according to equation (1).

$$W = \begin{pmatrix} k_1 w_{11} & k_2 w_{21} & \cdots & k_H w_{H1} \\ \vdots & \vdots & \cdots & \vdots \\ k_1 w_{1n} & k_2 w_{2n} & \cdots & k_H w_{Hn} \\ 1-k_1 & 1-k_2 & \cdots & 1-k_H \\ 0 & 0 & \cdots & 0 \end{pmatrix} \tag{1}$$

Let $P = (p_1, \cdots, p_n) \in R^n$ be a vector. According to the definition of stereographic projection, the stereographic projection of P is $P^S = (k_p p_1, \cdots, k_p p_n, 1-k_p)$ with $k_p = \dfrac{2}{1+\sum_{j=1}^{n} p_j^2}$. It would be the representation of P. Because the construction of W is not completed according to equation (1), W would be constructed completely while a probe vector is presented. When $(P^S, 1)$ is presented, W can be constructed completely according to equation (2).

To construct W completely, when a probe vector is given, elements in the last row of W are values according to equation (2).

$$W = \begin{pmatrix} k_1 w_{11} & k_2 w_{21} & \cdots & k_H w_{H1} \\ \vdots & \vdots & & \vdots \\ k_1 w_{1n} & k_2 w_{2n} & \cdots & k_H w_{Hn} \\ 1-k_1 & 1-k_2 & \cdots & 1-k_H \\ \dfrac{k_1 k_p r^2}{2} - 1 + \varepsilon & \dfrac{k_2 k_p r^2}{2} - 1 + \varepsilon & \cdots & \dfrac{k_H k_p r^2}{2} - 1 + \varepsilon \end{pmatrix} \tag{2}$$

In equation (2), $k_i = \dfrac{2}{1+\sum_{j=1}^{n} w_{ij}^2}, i = 1 \cdots H$, $k_p = \dfrac{2}{1+\sum_{j=1}^{n} p_j^2}$ and ε is a positive infinitesimal. In equation (2), r is parameters named as generalized radius.

2 Experimental Results and Conclusion

Performances of TextCC based on stereographic projection are tested with random spatial data in plan (2 dimensional real space) firstly. Coordinates of 1000 points in plan are generated. 500 points are in each regions, $[0,0.5]\times[0.5,1]$ and $[0.5,1]\times[0,0.5]$. The distribution of those points in plan is shown at fig1. While TextCC is classifying those points into two regions, data for the relation between generalized radius and precision, and data for the relation between sample step and the maximum precision are collected. All those 1000 random points will be normalized according to stereographic projection. The relation between generalized radius and the classification precision of TextCC is shown at fig2. In fig2, when TextCC is trained, samples are regular sampled from those 1000 random points. The sample step for fig2 is 4 (sample ration=0.2). It is obvious that classification precision of TextCC based on stereographic projection is high with appropriate generalized radius.

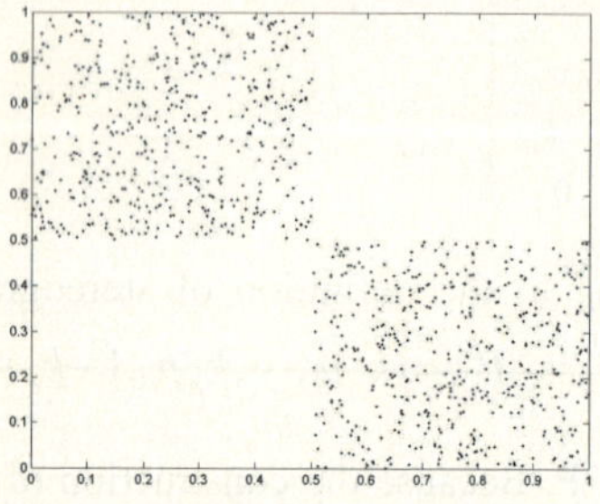

Fig. 1. 1000 random points in $[0,0.5]\times[0.5,1]$ and $[0.5,1]\times[0,0.5]$

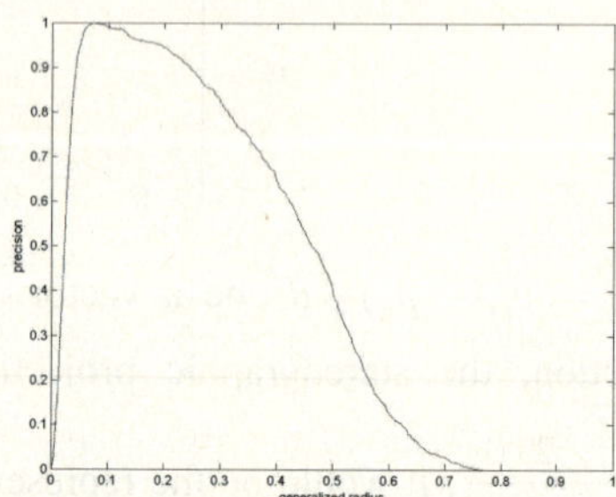

Fig. 2. relation between generalized radius and classification precision

Because stereographic projection is one-one onto mapping, information implicated in data is not losing when data is normalized by stereographic projection and can be rediscovered with mathematical tools.

The normalization and classification of human attention is focused by our research in chance/sign discovery. Stereographic projection would give us new powerful mathematical tool for our research and the combination of TextCC and stereographic projection will be an important technology for the implementation of demo system on chance/sign discover.

References

1. M. IKONOMAKIS, S. KOTSIANTIS, V. TAMPAKAS, Text Classification Using Machine Learning Techniques, WSEAS TRANSACTIONS on COMPUTERS, Issue 8, Volume 4, August 2005, p966-974
2. Shu, B., Kak, S., A neural network-based intelligent meta search engine, Information Sciences, 1999,120(1), p1-11.
3. Zhenya Zhang, Shuguang Zhang, Xufa Wang, Enhong Chen, Hongmei Cheng, TextCC: New Feed Forward Neural Network for Classifying Documents Instantly, Lecture Notes in Computer Science, Volume 3497, Jan 2005, pp232-237
4. S. Brin and L. Page, Anatomy of a large scale hypertextual web search engine, Proc. of the Seventh International World Wide Web Conference, Amsterdam, 1998, p107-117.
5. Venkat N. Gudivada, Vijay V. Raghavan, William I. Grosky, Information retrieval on the world wide web, IEEE Internet Computing, 1997, 1(5), p59-68
6. Arwar,Item-Based collaborative filtering recommendation algorithms, Proceedings of the 10th International World Wide Web Conference (WWW10), 2001, p 285-295.

Fuzzy CMAC with Online Learning Ability and Its Application

Shixia Lv, Gang Wang, Zhanhui Yuan, and Jihua Yang

Mechanical Engineering School, Tianjin University
92 weijin Road, Nankai Division ,Tianjin City, China
lsx_robot@126.com

Abstract. The binary behavior of activation function in receptive field of conventional cerebellar model articulation controller (CMAC) affects the continuity of the network output. In addition, the original learning scheme of CMAC may corrupt the previous learning data. A control scheme, which parallely combines the fuzzy CMAC (FCMAC) and PID, is proposed in the paper. The weights are updated according to the credits which are assigned to the hypercubers according to their learning histories and fuzzy membership degrees. The FCMAC is powerful in control time-varying processes due to the online learning ability of the FCMAC. Experimental results of temperature control have shown that the FCMAC with online learning ability can accurately follow the control trajectory and reduce the tracking errors.

1 Introduction

FCMAC is a modification of CMAC by introducing fuzzy logic operations into neurons of CMAC. The FCMAC overcomes the binary behavior in receptive field of CMAC and can learn both the function and its derivatives.

One drawback of the CMAC learning algorithm is that the correcting amounts of errors are equally distributed into all addressed hypercubers. This learning scheme may corrupt the previous learning results. Sun et al. [2] proposed a learning scheme in which credits is assigned to each hypercuber. The weights are updated according to their credits. The learning effect is improved greatly.

A control scheme which combines FCMAC and PID is proposed in this paper. The credit, which considers both the fuzzy membership degrees and learning history, is taken into account in the learning procedure. An application example of the controller is given.

2 The Learning Scheme of Conventional CMAC

In the learning phase of the conventional CMAC, the stored weights are changed by the following equation

$$w_j(i)=w_j(i-1)+\frac{\alpha}{m}(y_d-y_a)S_j \tag{1}$$

L. Jiao et al. (Eds.): ICNC 2006, Part I, LNCS 4221, pp. 93 – 96, 2006.

where y_a and y_d are actual output and desired output of the network, respectively. $S_j(x)$ is a binary indicating whether the jth neuron is activated. $w_j(i)$ is the weight stored into the stored space. α is the learning rate. m is generalization parameter.

From equation (1), the correcting amounts of errors between desired outputs and actual outputs are equally distributed into all addressed hypercubers. The stored data may be corrupted by new training. If the samples are presented for several cycles, this situation may be smoothed out. But in real-time application, a sample is presented only once. Repeated training may not be accepted.

3 FCMAC and Credit Assigned Learning Scheme

In the conventional CMAC, a neuron in the receptive field is activated or not activated at all. This binary behavior of the activation function may result in the discontinuities in the output of the network.

The FCMAC employs Gaussian curve as activation function. The active degree of a neuron in receptive field takes the value in the range [0, 1]. In the fuzzy theory, this value is considered as membership degree. The architecture of FCMAC is shown in Fig.1.

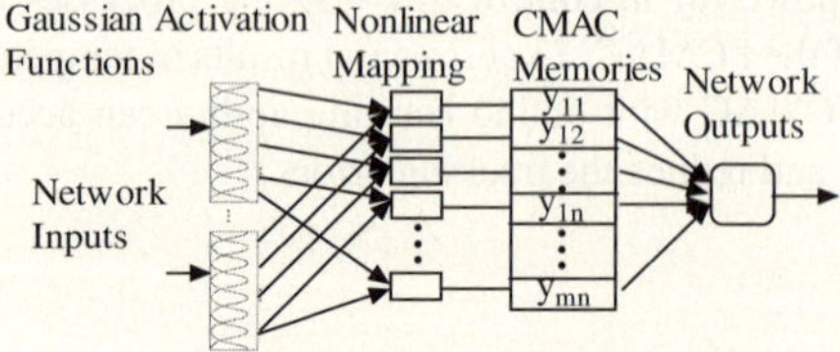

Fig. 1. The architecture of fuzzy CMAC

In the working phase, a neuron in the internal layer is viewed as a fuzzy rule. The active degree of the jth neuron for an input x is

$$S_j(x)=\prod_{i=1}^{n} S_{ji}(x_i) \tag{2}$$

where x_i is the ith element of the input vector. S_{ji} is the active degree of the element x_i belonging to the jth fuzzy set. n is the dimension of input vector.

The output of the network is a defuzzific action procedure.

$$y=\sum_{j=1}^{N} (w_j S_j(x)/\sum_{j=1}^{N} S_j(x)) \tag{3}$$

In order to overcome the disadvantage of conventional CMAC learning model, the credit for FCMAC learning is defined as

$$C_j=S_j(x)(f(j)+1)^{-1}/\sum_{l=1}^{m} (S_l(x)(f(l)+1)^{-1}) \qquad \text{subject to} \quad \sum_{j=1}^{m} c_j=1 \tag{4}$$

where f(j) is the learning times of the jth hypercuber. m is the number of addressed hypercuber for a state.

The credit is proportional to the active degree and the inverse of learning times of the addressed hypercuber. The hypercubers with bigger active degrees and less learned times would be responsible for the errors. The learning times must include the current one to prevent dividing by zero. The FCMAC learning algorithm can be rewritten as

$$w_j(i)=w_j(i-1)+\alpha C_j(y_d-y_a) \quad (5)$$

With this modification, the errors can be appropriately distributed into the addressed hypercubers according to the creditability of hypercubers.

4 The Structure of Control System Based on FCMAC

Fig.2 shows the structure of control system which combines the FCMAC and PID. The overall control signal applied to the plant is the sum of the FCMAC output and PID controller output. In the early stage of the control process, the control is dominated by PID controller. As the FCMAC network gradually learns the inverse dynamics of the plant, the control is shifted from PID controller to FCMAC network. The PID controller can not be removed since possible variation of external loads and other outer conditions may result in unstable phenomenon.

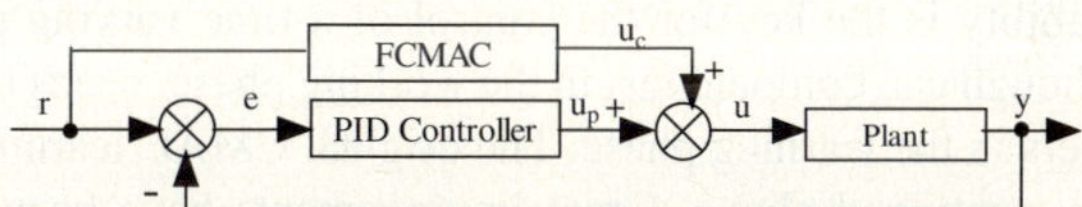

Fig. 2. The structure of control system based on FCMAC

The structure is similar to the Albus' approach[1] in which a conventional CMAC is used and the learning scheme is original one. In this structure, however, FCMAC is used and the credit assigned learning scheme is employed.

5 Application of FCMAC to Temperature Control

We consider the control of hot wind drier depicted in Fig.3. The hot air at 80 ^{0}C is supplied by the heat exchanger, the fan velocity and state of electricity controlled valve control the flux entering the drier. A temperature sensor is placed at the center of the drier. The products which are to be dried are moving slowly on the opposite direction of the hot wind. The controller should keep the temperature to a give reference value for ensuring satisfactory drying effect.

The control scheme of FCMAC is employed to control the plant. The online learning ability is capable of following the variation of the plant and achieving satisfactory control performance. For demonstration of the performance improvements, the conventional PID controller and proposed hybrid control scheme based on FCMAC are tested, respectively.

Experimental results of traditional PID control and FCMAC control scheme are shown in Fig.4a and Fig.4b. In the figure, the dash line shows the reference and solid line is the measured output. The results reveal that both PID and FCMAC control scheme can effectively follow the reference trajectory. But the FCMAC controller is more effective than PID.

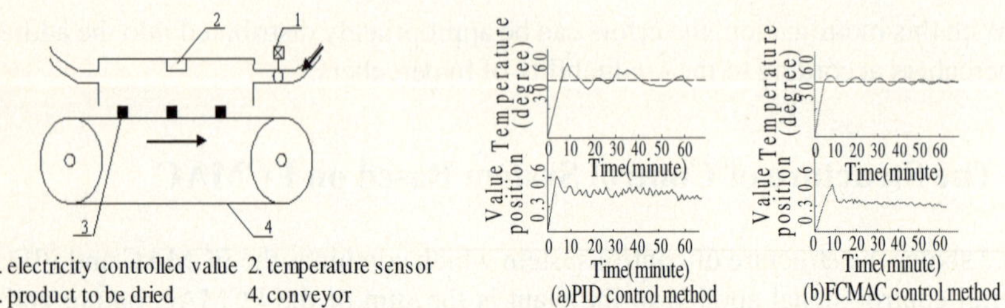

Fig. 3. The structure of hot wind drier **Fig. 4.** Experimental results of hot air drier

6 Conclusion

Online learning ability is the key for the control of a time varying process. The BP neural network, though can compute fast in the working phase, needs update all neuron weights in all layers in the learning phase. The original CMAC learning algorithm has critical drawback mentioned above. Great improvements have been made by Sun's proposition. In this paper, fuzzy membership degrees are taken into accounts of the credits for FCMAC. The application example of the FCMAC controller demonstrated the effectiveness of the scheme.

References

[1] Albus J. S.: A new approach to manipulator control: the cerebellar model articulation controller, Trans. ASME J. Dyn. Sys. Meas.Control, vol.97, no.9, (1975)220-227.

[2] Sun S. F., Tao T. and Huang T. H.:Credit assigned CMAC and its application to online learning robust controller,IEEE Trans. Syst. Man Cybern., vol.33, no.2, (2003)202-213.

[3] Lane S.H., handelman D.A., and Gelfand J.J.:Theory and development of higher-order CMAC neural networks, IEEE Control Syst., vol.12, (1992) 23-30.

[4] Almeida P.E.M., and Simoes M.G.: Parametric CMAC networks: fundamentals and applications of the a fast convergences neural structure, IEEE Trans. Industrial Applications, vol.39, no.5, (2003) 1551-1557.

[5] Miller W.T., Glanz F.H., and Kraft L.G.: Real-time dynamic control of a industrial manipulator using a neural network based learning controller, IEEE Trans. Robotics Automat., vol.6, no.1,(1990)1-9.

Multiresolution Neural Networks Based on Immune Particle Swarm Algorithm

Li Ying[1,2] and Deng Zhidong[1,⋆]

[1] Department of Computer Science and Technology State Key Laboratory of Intelligent Technology and Systems, Tsinghua University, Beijing 100084
{liyingqh, michael}@tsinghua.edu.cn
[2] College of Computer Science and Technology Jilin University, Changchun 130021, P.R. China

Abstract. Inspired by the theory of multiresolution analysis (MRA) of wavelets and artificial neural networks, a multiresolution neural network (MRNN) for approximating arbitrary nonlinear functions is proposed in this paper. MRNN consists of a scaling function neural network (SNN) and a set of sub-wavelet neural networks, in which each sub-neural network can capture the specific approximation behavior (global and local) at different resolution of the approximated function. The structure of MRNN has explicit physical meaning, which indeed embodies the spirit of multiresolution analysis of wavelets. A hierarchical construction algorithm is designed to gradually approximate unknown complex nonlinear relationship between input data and output data from coarse resolution to fine resolution. Furthermore, A new algorithm based on immune particle swarm optimization (IPSO) is proposed to train MRNN. To illustrate the effectiveness of our proposed MRNN, experiments are carried out with different kinds of wavelets from orthonormal wavelets to prewavelets. Simulation results show that MRNN provides accurate approximation and good generalization.

1 Introduction

Artificial neural networks (ANN) and wavelet theory become popular tools in various applications such as engineering problems, pattern recognition and nonlinear system control. Incorporating ANN with wavelet theory, wavelet neural networks (WNN) was first proposed by Zhang and Benveniste [1] to approximate nonlinear functions. WNNs are feedforward neural networks with one hidden layer, where wavelets were introduced as activation functions of the hidden neurons instead of the usual sigmoid functions. As a result of the excellent properties of wavelet theory and the adaptive learning ability of ANN, WNN can make an remarkable improvement for some complex nonlinear system control and identification. Consequently, WNN was received considerable attention [2,3,4,5,6,12].

⋆ Correspondence author. This work was supported by the National Science Foundation of China under Grant No. 60321002.

L. Jiao et al. (Eds.): ICNC 2006, Part I, LNCS 4221, pp. 97–106, 2006.

However, in general the hidden layer of WNN consists of only wavelet nodes and the common used wavelets are Mexican hat function. Multiresolution analysis, the main characteristic of wavelet theory, has still not been efficiently taken into account. In addition, the number of the hidden layer nodes and efficient learning algorithms of WNN are still challenging problems. The most used learning algorithm is backpropagation based on gradient descent,quasi-Newton, Levenburg-Marquardt and conjugate gradient techniques, which easily leads to a local minimum especially for multimodal function optimization problem partially attributed to the lack of a stochastic component in the training procedure. Particle swarm optimization (PSO) in [8,9] is a stochastic population-based optimization technique. The simplicity of implementation and weak dependence on the optimized model of PSO make it a popular tool for a wide range of optimization problems.

In this paper, a multiresolution neural network (MRNN) is proposed, which consists of a scaling function neural network (SNN) and a set of sub-wavelet neural networks (sub-WNN), in which each sub-neural network can capture the specific approximation behavior (global and local information) at different resolution of the approximated function. Such sub-WNN have a specific division of work. The adjusted parameters of MRNN include linear connective weights and nonlinear parameters: dilation and translation, which have explicit physical meaning, i.e., resolution and domain of signification (defined by the training set). The structure of MRNN is designed according to the theory of multiresolution analysis, which plays a significant role in wavelet analysis and approximation of a given function. To determine the structure of MRNN, we give a hierarchical construction algorithm to gradually approximate the unknown function from global to local behavior. In addition, a new learning algorithm referred as immune particle swarm optimization (IPSO) is proposed to train MRNN. In this approach, the introduction of immune behavior contributes to increase the diversity of particles and improve the speed and quality of convergence. For validating the effectiveness of our proposed MRNN, we take different kinds of wavelets from orthonormal wavelets to prewavelets. Experiments were carried out on different types of functions such as discontinuous, continuous and infinitely differentiable functions. Simulation results show that MRNN provides accurate approximation and good generalization. In addition, our proposed learning algorithm based on IPSO can efficiently reduce the probability of trapping in the local minimum.

This paper is organized as follows. The MRNN is introduced in section 2. In section 3, the hierarchical construction algorithm is described,and the training algorithm based on IPSO is presented to update the unknown parameters of MRNN. Finally, simulation examples are utilized to illustrate the good performance of the method in section 4.

2 Multiresolution Neural Networks

The main characteristics of wavelets are their excellent properties of time-frequency localization and multi-resolution. The wavelet transform can capture high frequency (local) behavior and can focus on any detail of the observed object through modulating the scale parameters, while the scale function captures the low frequency (global behavior). In this sense wavelets can be referred to as a mathematical microscope. Firstly, we give a simplified review of wavelets and multiresolution analysis of $L^2(\mathbb{R})$.

A series sequence $V_j, j \in \mathbb{Z}$ of closed subspaces in $L^2(\mathbb{R})$ is called a multiresolution analysis (MRA) if the following holds:
(1) $\cdots \subset V_j \subset V_{j+1} \cdots$
(2) $f(\cdot) \in V_j$ if and only if $f(2\cdot) \in V_{j+1}$
(3) $\bigcap_{j\in\mathbb{Z}} V_j = \{0\}$, $\overline{\bigcap_{j\in\mathbb{Z}} V_j} = L^2(\mathbb{R})$
(4) There exists a function $\phi \in V_0$, such that its integer shift $\{\phi(\cdot - k), k \in \mathbb{Z}\}$ form a Riesz basis of V_0.

It is noted that ϕ is a scaling function. Let $\psi(x)$ be the associated wavelet function. For $j \in \mathbb{Z}$, $\{\phi_{j,k}(x) = 2^{j/2}\phi(2^j x - k), k \in \mathbb{Z}\}$ forms a Riesz basis of V_j, $\{\psi_{j,k}(x) = 2^{j/2}\psi(2^j x - k), k \in \mathbb{Z}\}$ forms a Riesz basis of W_j, where W_j is the orthogonal complement of V_j in V_{j+1}, i.e. $V_{j+1} = V_j \oplus W_j$, where V_j and W_j are referred to as scaling space and detail space (wavelet space) respectively. Obviously,$L^2(\mathbb{R}) = V_J \bigcup_{j=J}^{\infty} \oplus W_j$, and $L^2(\mathbb{R}) = \bigcup_{j=-\infty}^{\infty} \oplus W_j$.

For any function $f(x) \in L^2(\mathbb{R})$ could be written

$$f(x) = \sum_{k\in\mathbb{Z}} c_{J,k}\phi_{J,k}(x) + \sum_{j\geq J}^{\infty} \sum_{k\in\mathbb{Z}} d_{j,k}\psi_{j,k}(x)$$

as a combination of global information and local information at different resolution levels. On the other hand, $f(x)$ could also be represented as

$$f(x) = \sum_{j\in\mathbb{Z}} \sum_{k\in\mathbb{Z}} d_{j,k}\psi_{j,k}(x)$$

as a linear combination of wavelets at different resolutions levels, i.e. the combination of all local information at different resolution levels. The above equations indeed embody the concept of multiresolution analysis and imply the ideal of hierarchical and successive approximation, which is the motivation of our work.

Inspired by the theory of multiresolution analysis, we present a new wavelet neural networks model, called multiresolution neural networks (MRNN). The MRNN consists of a scaling function neural network (SNN) and a set of sub-wavelet neural networks (WNN). In SNN, scaling function at a specific resolution are introduced as active functions, while the active functions of each sub-WNN correspond to wavelet functions at different resolution. The unknown parameters of MRNN include linear connective weights and nonlinear parameters: dilation and translation parameters. The architecture of the proposed MRNN is shown in Fig 1. The formulation of MRNN is given as follows:

$$\hat{y}(x) = \sum_{i=0}^{C} \hat{y}^i(x)$$
$$\hat{y}^0(x) = \sum_{j=1}^{N^0} w_{0,j}\phi(a_0 x - b_{0,j}), y^i(x) = \sum_{i=1}^{C} \sum_{j=1}^{N^i} w_{i,j}\psi(a_i x - b_{i,j})$$

Here $\hat{y}^0(x)$ represents the output of SNN and $\hat{y}^i(x)$ is the output of the i-th sub-wavelet neural network,i.e. the output of WNN-i. Note that C is the number of sub-wavelet neural networks, i.e. the number of different resolution level of wavelets. N^0 denotes the number of neurons of SNN, and $N^i, 1 \le i \le C$, is the number of WNN-i. The number of whole neurons of NRNN is $N = \sum_{i=0}^{C} N^i$. $a_i, b_{i,j}$ and $w_{i,j} \in \mathbb{R}, 1 \le i \le N^i, 0 \le i \le C$ are dilation parameters, transition parameters and linear output connective weights, which need to be adjusted to fit the known data optimally. Dilation parameters have explicit meaning, i.e., resolution, which satisfies $a_i > 0$ and $a_0 \le a_1 < a_2 \cdots < a_C$. The number of unknown parameters to be adjusted sums up to $C + 2N$. The other parameters including C and $N^i, i = 1, \cdots, C$, are called structure parameters. In general, such structure parameters must be settled in advance. But in this paper, due to the hierarchical learning ability of our proposed MRNN, structure parameters can be partially determined in the training process.

For SNN: $\{\phi(a_0 x - b_{0,j}), j = 1, 2, \cdots, N^0\}$,are taken as active functions, which capture the global behavior of the approximated function at the a_0 coarse resolution level.

For WNN-i: $\{\psi(a_i x - b_{i,j}), j = 1, 2, \cdots, N^i\}$,are taken as active functions of the i-th sub-wavelet neural networks, which capture the local behavior of approximated function at the a_i finer resolution level. And with an increase of i, the WNN-i focus on the more local behavior of the unknown function. The adaption of scaling function can rapidly implement a global approximation and probably obtain good performance with a relative small model size.

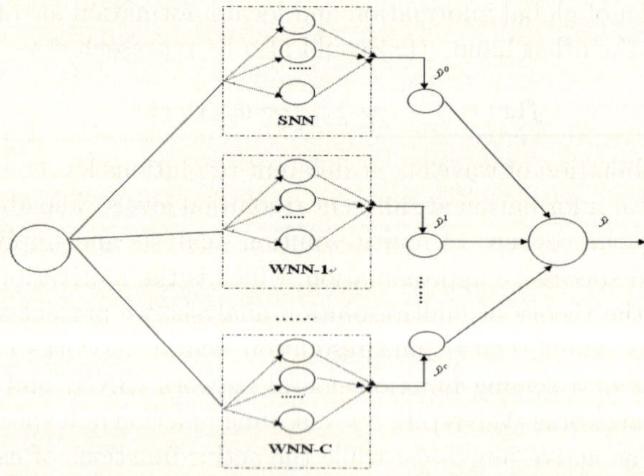

Fig. 1. Architecture of Multiresolution Neural Networks: a scaling function neural netwok and C sub-wavelet neural networks (WNN-1,$\cdots$,WNN-C)

The most commonly used wavelets are continuous wavelets, Daubecies wavelets and spline wavelets and so on. For the general WNN, the most frequently adopted active functions are continuous wavelets such as Morlet, Mexican Hat and Gaussian derivatives wavelets, which do not exist scaling functions. Daubechies compactly supported orthonormal wavelets are provided by scaling and wavelets filters, while they have no analytic formula. In this paper, the cascade algorithm i.e. the repeated single-level inverse wavelet transform, is employed to approximate the corresponding scale function and wavelet function, which are adopted as active functions in our proposed MRNN. Considering that such functions are the approximated form, an learning algorithm based on gradient descent seems unreasonable. A training algorithm not demanding gradient information will be described in the next. In addition,pre-wavelets are also considered as candidates for active functions of our MRNN. Compared with orthogonal wavelets, pre-wavelets have more freedom degrees and potential advantages. Let $N_m(x)$ be the spline function of m-th order supported on $[0, m]$, then a m-th order pre-wavelet corresponding to $N_m(x)$ is defined as : $\psi_m(x) = \frac{1}{2^{m-1}} \sum_{n=0}^{3m-2} (-1)^n [\sum_{j=0}^{m} \binom{m}{j} N_{2m}(n-j+1)] N_m(2x-n)$.

Given a training set $S : S = \{(x_1, y_1), \cdots, (x_L, y_L)\}, \quad x_i, y_i \in \mathbb{R}$ where L is the total number of training samples and (x_i, y_i) is input-output pair. Denote the input sample vector $\mathbf{X} = [\mathbf{x_1}, \cdots, \mathbf{x_L}]^\mathbf{t} \in \mathbb{R}^\mathbf{L}$ and output sample vector $\mathbf{Y} = [\mathbf{y_1}, \cdots, \mathbf{y_L}]^\mathbf{t} \in \mathbb{R}^\mathbf{L}$, where 't' represents the transpose.

For a training set S, there must exit an unknown function $f : \mathbb{R} \to \mathbb{R}$, s.t.

$$y_i = f(x_i), 1 \le i \le L$$

The network deals with finding a function $\tilde{f}$ that can approximate the unknown function f based on the training set S. The next problems are how to determine the structure parameters and train MRNN based on the given training set S so that the error between the output of MRNN and **Y** is minimal.

3 Hierarchical Training Algorithm Based on Immune Particle Swarm Optimization

3.1 Hierarchical Training Strategy

According to the multiresolution property of our proposed neural networks, we employ a hierarchical learning strategy at the start of SNN, which is the coarsest resolution. The concrete steps are given in the following:

(1). Determine the scaling function and the associated wavelet and give an expected error index. Let $[A, B]$ and $[C, D]$ be the main range of scaling function and mother function respectivley. Note that, the main range refer to the support for the compactly supported scaling function and wavelet , while for a non compactly supported functions, due to the rapid vanishing of wavelets, roughly take a range where the value of non compactly supported functions is usually bigger. Let $I_1, I_2]$ be the significant domain of the input vectors.

(2). First train SNN-the coarsest resolution level. The dilation parameter a_0 of SNN is random in a reasonable range. For covering the significant domain of the input vectors, SNN may demand more neurons (translation parameters), because the support of active functions at resolution a_0 becomes smaller when a relative larger a_0 is taken. In detail, the translation parameters belong to the rang of $[-B + a_0 I_1, -A + a_0 I_2]$. According to the width of $[-B + a_0 I_1, -A +$

$a_0I_2]$, a ratio is designed to determine the number of neurons N^0, then we randomly chosen the translation parameters in $[-B + a_0I_1, -A + a_0I_2]$. The linear connective weights are initialized with small random values, taken from a normal distribution with zero mean. Now we compute the current error for such initialization method. Repeating the above process several times, we chose a good trade-off between the error index, the size of SNN and resolution. After we determine the structure parameters, our proposed new training algorithm based on immune particle swarm optimization is used to update the dilation, translation parameters and linear weights until the current error reduces very slowly or obtains a certain admissible error range.

(3). After training SNN, a sub-WNN (WNN-1) is added to approximate the difference between the expected output vector and the estimated output of SNN. Define $D^1 = \{d_j^1, j = 1, \cdots, L\}$ as the expected output of WNN-1, where $d_j^1 = Y_j - \hat{y}^0(x_j)$. Similar to the above, the resolution of WNN-1 a_1 needs to be randomly chosen in a relative small reasonable range larger than a_0. Then the translation parameters are initialized from the range $[-D + a_1I_1, -C + a_1I_2]$, in which the number of neurons of WNN-1 is also determined using a width of $[-D+a_1I_1, -C+a_1I_2]$ and the comparison of several trial results for reference. Then the unknown parameters of WNN-1 are trained based on the current input data X and expected output data D^1 using the IPSO algorithm.

(4). While the approximation error of WNN-1 dose not yet reach the expected error, repeat step (3) to gradually add WNN-i+1 until the expected error is reached. Note that the expected output vector of WNN-i+1 is the difference between the expected output vector WNN-i and the estimated output vector of WNN-i. Here, denote WNN-0 as SNN. In addition, the dilation parameter of WNN-i+1 needs to be larger than that of WNN-i.

From the above hierarchical training strategy, the approximation of an unknown function is gradually performed from coarse resolution to fine resolution, then from fine resolution to finer resolution. This process gradually focuses on different details at different resolutions of the approximated function. Each sub-neural network of MRNN has a definite division of work. Additionally through this strategy, the complexity of the training problem of MRNN is greatly reduced, MRNN can be rapidly and effectively constructed.

3.2 Training Algorithm Based on Immune Particle Swarm Optimization

The particle swarm optimization method mimics the animal social behaviors such as schools of fish, flocks of birds etc and mimics the way they find food sources. PSO has gained much attention and wide applications in different optimization problems. PSO has the ability to escape from local minima traps due to its stochastic nature. Assume that a particle swarm contains M particles. Denote Z_i and V_i as the position and velocity of the i-th particle.In fact the position of each particle is a potential solution. Define f_i as the fitness value of the i-th particle, which is used to evaluate the goodness of a position. P_i represents the best position of the i-th particle so far. P_g is the best position of all the particles in a particle swarm so far. The velocity and position of each particle are updated based on the following equations:

$$\begin{aligned} V_i(k+1) &= wV_i(k) + c_1r_1(P_i - x) + c_2r_2(P_g - Z_i) \\ Z_i(k+1) &= Z_i(k) + Z_i(k+1), \quad i = 1, 2, \cdots, M \end{aligned} \tag{1}$$

where w is inertial weight, c_1 and c_2 are constriction factors, r_1 and r_2 are random numbers in $[0, 1]$, and k represents the iteration step. The empirical investigations shows the importance of decreasing the value of w from a higher valve (usually 0.9 to 0.4) during the searching process.

In optimization processing of PSO, premature convergence would often take place especially for the case of an uneven solution population distribution. In this paper, we propose two kinds of mechanism to improve the the diversity of particles and reduce the premature convergence.

One mechanism is the combination of an immune system and PSO. Biological research on the immune system indicates that the immune principle can provide inspiration for improvement of the performance of evolution computations [11]. Immune behavior is beneficial for retaining diversity, which can effectively avoid premature convergence and improve the search speed and quality of the evolution algorithm based on populations. Inspired by the diversity of biological immune systems, we introduce a diversity retaining mechanism based on concentration and immune memory for PSO.

On the other hand, we find another efficient mechanism for diversity, called a forgetfulness mechanism. The basic ideal is that we impose each particle and the swarm a slight amnesia when the particle swarm trends to premature convergence. This mechanism gives the particles a new chance based on existing better conditions. Firstly, we define a merit τ to reflect the convergent status of the particle swarm as follows:

$$\tau = \frac{\sigma_f^2}{max_{1 \le j \le M}\{(f_j - \bar{f})^2\}}, \quad \sigma_f^2 = \frac{1}{M}\sum_{i=1}^{M}(f_i - \bar{f})^2, \quad \bar{f} = \frac{1}{M}\sum_{i=1}^{M} f_i \tag{2}$$

Here M is the size of the population. $\bar{f}$ is the average fitness of all particles, and σ_f^2 is the covariance of fitness. For a given small threshold, if τ is less than this threshold and the expected solution have not been reached, then we think that this particle swarm tends to premature convergence. Under this condition, we impose each particle and the swarm a slight amnesia. Each particle forgets its historical best position and considers the current position as its best position. Similarly, the swarm does not remember its historical global best position and choose the best position from the current positions of all particles. The implementation of our algorithm can be categorized into the following steps:

Step-1. Define the solution space and a fitness function and determine c_1, c_2, the size of particles - M and the size of the immune memory library - U.

Step-2. Initialize randomly the position and velocity of each particle.

Step-3. Proceed PSO under a given iteration steps (100 steps), and store the best position P_g in the immune memory library at each iteration. Then evaluate the convergence status using (2). If the convergence status is good, continue *Step-3* until the expected error is reached. Otherwise, go to *Step-4*.

Step-4. Introduce the forgetfulness mechanism: the particle swarm and each particle all forget their historical best position.

Step-5. Build a candidate particle set including 2M particles, where M particles are obtained in terms of the update formulation (1), while the remaining M particles are generated randomly.

Step-6. Compute each candidate particle concentration d_i and selection probability s_i, which are defined as : $d_i = (\sum_{j=1}^{2M} |f_i - f_j|)^{-1}, s_i = d_i / \sum_{j=1}^{2M} d_j), \quad i = 1, 2, \cdots, 2M$.

Step-7. Undate the particle swarm: chose M particles with large selection probabilities from the candidate set and replace them with lower fitness among the chosen M particles by the particles in the immune memory library, then turn to *Step-3*.

4 Numerical Experiments

In this section, for evaluating the performance of the network, we take the following merits defined in [1] as a criterion to compare the performance of various methods.

$$J = \sqrt{(\sum_{l=1}^{N} y_l - \hat{y}_l)/(\sum_{l=1}^{N} y_l - \bar{y})}, \bar{y} = \frac{1}{N} \sum_{l=1}^{N} y_l$$

where y_l is the desired output and $\hat{y}_l$ is the estimated output from the constructed neural networks.

Experiments are carried out on different types of functions such as discontinuous , continuous and infinitely differentiable in order to to demonstrate the validity of the presented MRNN. We respectively selected spline wavelets with order 3 and Daubechies wavelet with support width 2N-1 and vanishing moment N. The approximated functions from $f_1(x)$ to $f_2(x)$ are given in the following:

$$f_1(x) = \begin{cases} -2.186x - 12.864, -10 \le x < -2 \\ 4.246x, -2 \le x < 0 \\ 10e^{-0.05x-0.5} sin[(0.03x + 0.7)x], 0 \le x \le 10 \end{cases}$$

$$f_2(x) = \begin{cases} 0, & -10 \le x < -7.5 \\ 0.2x + 1.5, & -7.5 \le x < -5 \\ 0.5, & -5 \le x < -2.5 \\ 0.2x + 1, & -2.5 \le x < 0 \\ -0.2x + 1, & 0 \le x < 2.5 \\ 0.5, & 2.5 \le x < 5 \\ -0.2x + 1.5, & 5 \le x < 7.5 \\ 0, & 7.5 \le x \le 10 \end{cases} \qquad f_2(x) = \begin{cases} 0, & -10 \le x < -7.5 \\ 1/3, & -7.5 \le x < -5 \\ 2/3, & -5 \le x < -2.5 \\ 1, & -2.5 \le x < 2.5 \\ 1/3, & 2.5 \le x < 5 \\ 0, & 7.5 \le x \le 10 \end{cases}$$

$$f_4(x) = sin(6\pi/20(x + 10)), \quad -10 \le x \le 10$$

For each approximated function, we sampled 200 points distributed uniformly over $[-10, 10]$ as training data to construct and train the associated MRNN. The test data was obtained through uniformly sampled data of 200 points. Each simulation is computed on uniformly sampled test data of 200 points. In the following, we show the maximal performance index and minimal performance index among nine simulations to evaluate the performance of our MRNN. The simulation results of the approximated functions can be seen in Table I. From Fig 2 and Fig 3 we show the good approximation performance of our MRNN for $f_1(x), f_2(x), f_3(x)$ and $f_4(x)$. Additionally, in order to evaluate the performance of our MRNN, for $f_i(x)$, we make a comparison between our MRNN and other works [1,12]. The Mexican Hat function are taken as the wavelet function in [1,12], whereas we take the B-spline wavelet with order 4 as the active function. The comparised results are given in Table II. It should be noticed that the performance of MRNN shown in Table II is the average of the nine simulations.

Table 1. Simulation results of our MRNN

Example	parameters number	Maximal performance index	Minimal performance index
$f_1(x)$	23	0.040297	0.021052
$f_2(x)$	13	0.00010712	8.6695e-005
$f_3(x)$	22	0.060172	2.1689e-005
$f_4(x)$	17	0.023313	0.033459

Table 2. Comparison of our MRNN with others work for $f_1(x)$

Method	Number of unknown parameters	Performance index
MRNN-prewavelets	23	0.03259
WNN [1]	22	0.05057
WNN [12]	23	0.0480

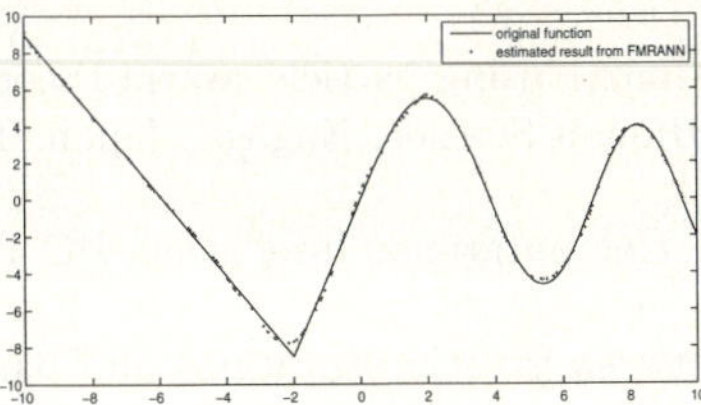

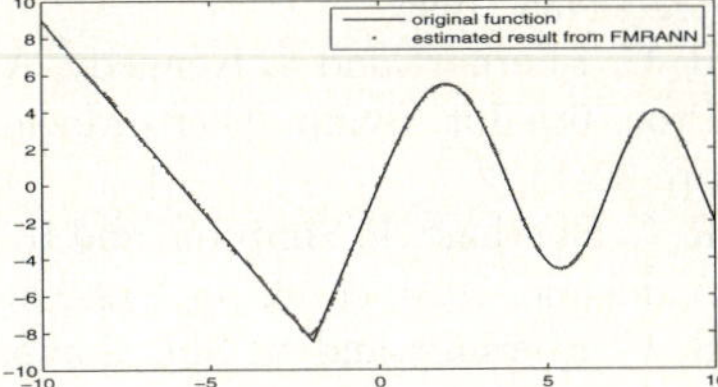

Fig. 2. For $f_1(x)$: at left is a comparison between the original function (solid line) and the estimated result (dotted point) from MRNN with the maximal performance index 0.040297 during the nine simulations, at the right is the comparison with the minimal performance index 0.021052

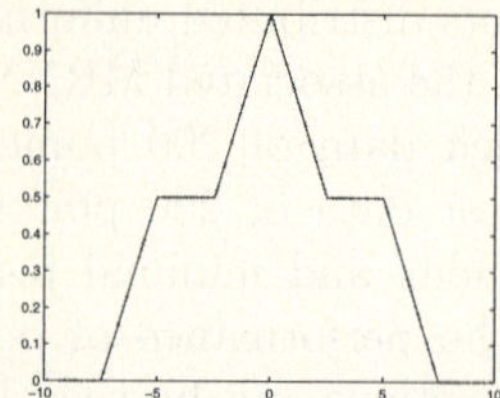

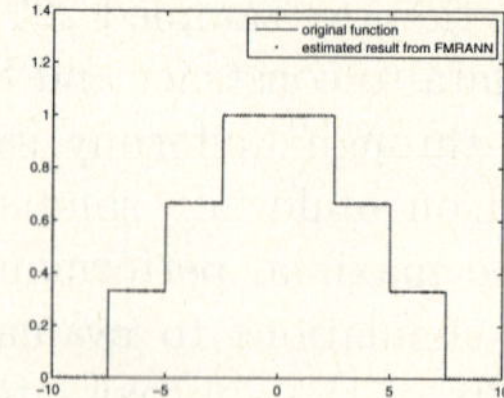

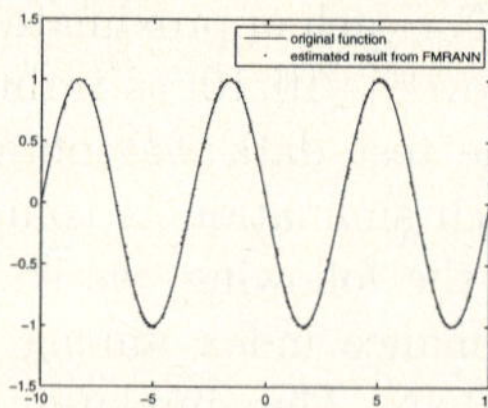

Fig. 3. the comparison between the original function (solid line) and the estimated result (dotted point) from MRNN: at the left is the comparison of $f_2(x)$ with the minimal performance index 8.6695e-005 during the nine simulations, at the middle is the comparison of $f_3(x)$ with the maximal performance index 0.06060172, at the right is the comparison of $f_4(x)$ with the maximal performance index 0.023313

References

1. Q. H. Zhang and A. Benveniste, Wavelet networks, IEEE Trans. Neural Netw., vol. 3, no. 6, pp. 889-898, Nov. 1992.
2. Y. C. Pati and P. S. Krishnaprasad, Analysis and synthesis of feedforward neural networks using discrete affine wavelet transformation, IEEE Trans. Neural Networks, vol. 4, pp. 73-85, 1993.
3. J. Zhang, G. G. Walter, and W. N. W. Lee, Wavelet neural networks for function learning, IEEE Trans. Signal Processing, vol. 43, pp. 1485-1497, June 1995.
4. Q. Zhang, Using wavelet networks in nonparametric estimation, IEEE Trans. Neural Networks, vol. 8, pp. 227-236, Mar. 1997. Beijing, P.R. China, June 1999, pp. 451-456.
5. F. Alonge, F. Dippolito, S. Mantione, and F. M. Raimondi, A new method for optimal synthesis of wavelet-based neural networks suitable for identification purposes, in Proc. 14th IFAC, Beijing, P.R. China, June 1999, pp. 445-450.
6. X. Li, Z. Wang, L. Xu, and J. Liu, Combined construction of wavelet neural networks for nonlinear system modeling, in Proc. 14th IFAC, Beijing, P.R. China, June 1999, pp. 451-456.
7. J. Chen and D. D. Bruns, WaveARX neural network development for system identification using a systematic design synthesis, Ind. Eng. Chem. Res., vol. 34, pp. 4420-4435, 1995.
8. R. C. Eberhart and J. Kennedy, A new optimizer using particle swarm theory, in Proc. 6th Int. Symp. Micro Machine and Human Science, Nagoya, Japan, 1995, pp. 39-43.
9. R. C. Eberhart, P. Simpson, and R. Dobbins, Computational Intelligence PC Tools: Academic, 1996, ch. 6, pp. 212-226.
10. R. C. Eberhart and Y. Shi, Comparison between genetic algorithms and particle swarm optimization, Evolutionary Prograinniing VII: Proc. 71h Ann. Conf. on Evolutionary Prograriirnirig Conf., San Diego, CA. Berlin: Springer-Verlag 1998.
11. Dipankar Dusgupta, Artificial Immnue Systems and their a pplications, Springer-Verlag, Berlin Heidelberg, 1999.
12. J. Chen and D. D. Bruns, WaveARX neural network development for system identification using a systematic design synthesis, Ind. Eng. Chem. Res., vol. 34, pp. 4420-4435, 1995.

Multicategory Classification Based on the Hypercube Self-Organizing Mapping (SOM) Scheme

Lan Du, Junying Zhang, and Zheng Bao

National Lab. of Radar Signal Processing, Xidian University,
Xi'an, Shaanxi 710071, China
dulan@mail.xidian.edu.cn

Abstract. A new multicalss recognition strategy is proposed in this paper, where the self-organizing mapping (SOM) scheme with a hypercube mapped space is used to represent each category in a binary string format and a binary classifier is assigned to each bit in the string. Our strategy outperforms the existing approaches in the prior knowledge requirement, the number of binary classifiers, computation complexity, storage requirement, decision boundary complexity and recognition rate.

1 Introduction

In a multicategory classification problem, if each adjacent class pair can be robustly separated with a binary classifier, all of the K classes can be consequentially separated by the fusion of all binary classifiers. The result is that a binary code representation for each category, which is of a string format, is obtained, and each bit in the string corresponds to a binary classifier. In this approach, the optimal hypercube mapped space, which can preserve topological structure of the multiclass dataset and separation information among separate classes as largely as possible, is the key for getting simple decision boundaries and a small number of classifiers. The former, implemented with a self-organizing mapping (SOM) scheme [1] due to its topological order preservation property, and the latter, implemented with the hypercube structured mapped space in the SOM and our proposed circulation training algorithm (CiTA) for training the SOM due to its good separation structure, are used simultaneously and their tradeoff is made for solving a multicategory classification problem with smaller number of binary classifiers and each classifier corresponding to a simpler decision boundary.

2 Hypercube SOM Scheme Based Multicategory Classification

2.1 Ideal Cases

Following assumptions on training data are the ideal case for the start of our discussions: a) compactness and exclusiveness of the classes in the training data, i.e., samples belonging to each class form a compact cluster, and each compact cluster includes samples belonging to only this class; b) exclusive center-vertex mapping, i.e., the K class centers can be exclusively mapped to K vertices of a hypercube of the

L. Jiao et al. (Eds.): ICNC 2006, Part I, LNCS 4221, pp. 107 – 110, 2006.

smallest dimensions by the hypercube SOM, of which separate class centers can be mapped to separate vertices.

If the above two assumptions hold simultaneously for the training data, at the convergence of the learning process of the SOM trained with the centers of classes in the data, an exclusive mapping from the centers of classes in the input space to the vertices of the M -dimensional hypercube ($M=\lceil \log_2 K \rceil$) in the mapped space is obtained. We denote each vertex in the hypercube as its coordinate which is a string of M -bit length, $(z_1 z_2 \cdots z_j \cdots z_M)$ with $z_j \in \{0,1\}$ for $j=1,2,\cdots,M$.

Notice that the hamming distance between the vertices coded with $(z_1 z_2 \cdots z_{j-1} 0 z_{j+1} \cdots z_M)$ (i.e. $z_j=0$) and with $(z_1 z_2 \cdots z_{j-1} 1 z_{j+1} \cdots z_M)$ (i.e. $z_j=1$) is one, which is the smallest hamming distance for different vertices in the hypercube, no matter what values of z_1 , z_2 ,…, z_{j-1} , z_{j+1} ,…, z_M are. Therefore, from the topological order preservation of the hypercube SOM mapping from input space to hypercube vertices, the class samples belonging to the class code of $(z_1 z_2 \cdots z_{j-1} 0 z_{j+1} \cdots z_M)$, denoted by X_{j0} , and the class samples belonging to the class code of $(z_1 z_2 \cdots z_{j-1} 1 z_{j+1} \cdots z_M)$, denoted by X_{j1} , has the smallest topological distance in the input space, meaning that they are adjacent in the input space, and should be separated by a decision boundary. Therefore, the j th binary classifier should be trained with the training data $\Omega_j = \{(x_i, z_j^{(i)}) \mid z_j^{(i)} = 0 \text{ for } x_i \in X_{j0};\ z_j^{(i)} = 1 \text{ for } x_i \in X_{j1}\}$ for $j=1,2,\cdots,M$. This means that each bit of the class code corresponds to a binary classifier, and altogether M binary classifiers, each trained with all the N training samples, are needed for the K -class recognition problem. We take an example of $K=6$ and $M=\lceil \log_2 K \rceil = 3$ for showing the mechanism of the hypercube SOM scheme for solving the K -class recognition problem.

Two test calibrations are studied for multicategory classification based on hypercube SOM scheme. In the first test calibration, referred to as the sign test calibration, a test sample x is inputted to the j th binary classifier to get its corresponding belonging $z_j \in \{0,1\}$ for $j=1,2,\cdots,M$. $(z_1 z_2 \cdots z_j \cdots z_M)$ is then obtained when x is inputted to all the M binary classifiers. The class label y of the test sample x is directly the one which corresponds to the class code $(z_1 z_2 \cdots z_j \cdots z_M)$ in the list form of the correspondences between class labels and class codes. Another test calibration is what we call the nearest neighbor test calibration. The test sample x is inputted to the j th binary classifier for getting its corresponding discriminatory function value $z_j \in [0,1]$ for $j=1,2,\cdots,M$. Then we let the test sample x belong to the class label of its closest non-vacant vertex in the hypercube.

2.2 Some Problems in Realistic Cases

The problems which embarrass its realistic applications and need to be tackled includes: a) the training data are compact and exclusive but could not mapped onto the vertices of the smallest-scale hypercube SOM; b) the training data are not compact or exclusive.

For problem b), due to the clustering distribution of each class, class center is meaningless, but cluster center is meaningful. However, the clustering structure of the data is unknown in prior, which includes how many clusters there is for each class of data, and how samples belonging to that class of data is clustered in the input space. Any clustering algorithm which can find both the solutions can be employed. In our experiments mentioned in this paper, we select the rival penalized competitive learning (RPCL) algorithm [2] as the clustering algorithm. Generally speaking, cluster centers of different classes in the data may not always be mapped onto exclusive vertices of the hypercube. Therefore, we should pursue two directions for modifying the above method used in the ideal cases to be suitable to more realistic applications: a) to enlarge the scale of the hypercube, such that it becomes smallest without overlapped vertices in the hypercube SOM; b) to force the cluster centers, which are mapped to overlapped vertices, to be moved to their most adjacent vacant/same-class vertices. Due the unawareness of the clustering structure of the data in prior, we suggest to have a tradeoff for getting simple decision boundaries and a small number of classifiers. Hence, an exclusively mapped class ratio (EMCR) is introduced, which is defined as the ratio between the number of the exclusively mapped classes to that of the all classes, to be compared with a predetermined threshold for determining whether the hypercube mapped space should be enlarged or overlapped cluster centers should be moved. The whole procedure referred to as CiTA for training SOM is as follows.

Parameter: EMCR threshold Th.

Input:

an initial hypercube structure as the mapped space; a set of cluster centers, and their belongings to classes.

Output:

a hypercube SOM of a suitable size.

learning process:

Step 1: learn the map with cluster centers of all the classes with conventional SOM learning algorithm; calculate the EMCR from the obtained map;

Step 2: repeat this step until the EMCR equals one:

If the EMCR $>Th$, prohibit the update of synaptic weight vectors relating to exclusive vertices, set the obtained weight vectors relating to other vertices to be their initial ones and keep the mapped space unchanged; continue to train the map with only cluster centers which are mapped onto mixed vertices with conventional SOM learning algorithm; calculate the EMCR from the obtained map;

Otherwise, enlarge the hypercube, and go to Step 1.

3 Experimental Results

The digit dataset consisting of handwritten numerals ("0"~"9") was extracted from a collection of Dutch utility maps. Details of this dataset are available in the literature [3]. In our experiments we always used the same subset of 1000 Karhunen-Loève feature samples for testing and various subsets of the remaining 1000 patterns for training. After the RPCL, there are 2, 2, 3, 2, 3, 3, 3, 3, 2 and 2 cluster centers for the classes "0, 1, 2, 3, 4, 5, 6, 7, 8 and 9" respectively. Nevertheless, by our $2\times2\times2\times2$ structured hypercube SOM based multicategory classification strategy, the cluster centers of each class are mapped onto one vertex of the hypercube in the class coding phase

Table 1. The recognition results of the OAA based SVM strategy and the hypercube SOM based strategy using binary SVMs for a set of handwritten digits

OAA based SVM strategy			
number of classifiers		10	
kernel		gaussian	poly
kernel parameter		9	3
number of support vectors		3136	1550
recognition rates (%)	sign	86.40	91.90
	maximum	97.70	97.40
hypercube SOM based strategy using binary SVMs			
hypercube		2×2×2×2	
number of classifiers		4	
kernel		gaussian	poly
kernel parameter		9	3
number of support vectors		1476	929
recognition rates (%)	sign	97.00	96.40
	nearest neighbor	98.10	97.50

with EMCR $Th = 80\%$, and there are 6 vacant vertices. As shown in Table 1, our hypercube SOM based strategy for multicategory classification not only reduces the number of binary classifiers and the storage requirement, but also improves the recognition rate to some extent.

4 Conclusion

In this paper, an effective K-class recognition strategy using a least number of binary classifiers with simplest decision boundaries is proposed. Recognition experiments show that the proposed method outperforms OAA based SVM classifier in the number of binary classifiers, computation complexity, storage requirement, decision boundary complexity and recognition rate.

References

1. R. O. Duda, P. E. Hart and D. G. Stork, Pattern Classification (Second Edition), New York: John Wiley and Sons..
2. L. Xu, A. Krzyżak and E. Oja, "Rival Penalized Competitive Learning for Clustering Analysis, RBF Net, and Curve Detection", IEEE Transaction on N.N., Vol. 4, pp. 636-648.
3. A. K. Jain, R. P. W. Duin and J. Mao, Statistical pattern recognition: A review. IEEE Trans. on P.A.M.I., Vol.22(1), 4-37

Increased Storage Capacity in Hopfield Networks by Small-World Topology

Karsten Kube, Andreas Herzog, and Bernd Michaelis

Institute of Electronics, Signal Processing and Communications,
Otto-von-Guericke University Magdeburg,
P.O. Box 4120, 39114 Magdeburg, Germany
kkube@iesk.et.uni-magdeburg.de

Abstract. We found via numerical simulations, that connectivity structure in sparsely connected hopfield networks for random bit patterns affects the storage capacity. Not only the number of local connections is important, but also, and in contrast, the recently found small-world-topology will increase the quality of recalled patterns. Here, we propose and investigate the impact from network network architecture to pattern storage capacity capabilities.

1 Introduction

One of the most widely used artifical neural network for modeling memory in the brain is the Hopfield network [1,2], a symmetrical recurrent network that converges to point attractors which represent individual memories. In the case of the hippocampus, the area CA3 is in focus of interest, since there a high interconnectivity can be found, where ist topology-related behavior are intensively investigated in the 'small word' network model [3,4]. The 'small word' network topologies are considered as biologically plausible [5] and considered as a fundamental structural principle in biological networks.

2 Implementation

Hopfield networks [1] consist of units which may take possible states $S_i(i = 1, .., N)$, where N is the number of units in the net. For our network, we define the possible unit states as $S_i = -1; 1$. If a connection between the two units i and j is established then the element c_{ij} of the connection matrix C is 1 and 0 otherwise. Since all connections are considered as bi-directional C is symmetric. No self connections are allowed ($c_{ii} = 0$). The weight $w_{ij}(i, j = 1, 2, ..., N)$ is the strength with which unit j affects unit i. The network store maximal M patterns of N bits, $\zeta_i^\mu = \pm 1$ $(i = 1, 2, ..., N)$ $(\mu = 1, 2, ..., M)$. Pattern learning is simply done by using the Hebb rule: $w_{ij} = \frac{1}{N} \sum_{\mu=1}^{P} (\zeta_i^\mu \cdot \zeta_j^\mu) \Leftrightarrow c_{ij} = 1$. The states are calculated by: $S_i = sgn(\sum_{j=1}^{N} w_{ij} \cdot S_j)$. For the pattern recall, the states s_i of a sequentially randomly picked unit u_i are updated, until $10 * N$ iterations with no changed state s_i are proceeded.

L. Jiao et al. (Eds.): ICNC 2006, Part I, LNCS 4221, pp. 111–114, 2006.

2.1 Mapping Small-World Topology onto the Hopfield Model

Watts and Strogatz examined structurally simple, circular networks, in a sliding transition of regular, locally coupled networks too purely randomly coupled networks adjusting only one free parameter ρ. In the mature cerebral cortex, the densely connected local neuron networks contain an additional small fraction of long-range connections [7], which effectively reduces the synaptic path lengths between distant cell assemblies [3]. This architectural design, reminiscent of the mathematically defined 'small-world' networks [3], keeps the synaptic path lengths short and maintains fundamental functions in growing brains without excessive wiring [7].

For the hopfield connectivity, at first, the neurons are only connected with the k nearest neurons in the direct neighborhood using algorithm [6]: I) Begin with a nearest-neighbor coupled network consisting of N nodes arranged in a ring, where each node i is adjacent to its neighbor nodes, $i = 1, 2, ..., k/2$, with k being even. II) Randomly rewire each edge of the network with probability ρ; varying ρ in such a way that the transition between order ($\rho = 0$) and randomness $\rho = 1$ can be closely monitored.

By applying C, the connectivity matrix, elements c_{ij} means that a connection between node i and node j is present, while the C's diagonal is zero, since no self-connections, the locality of neighborhood can be observed as one values around the diagonal of C. By the rewiring procedure, connections are deleted from that zone and randomly appear elsewhere in the matrix, see Fig 1.

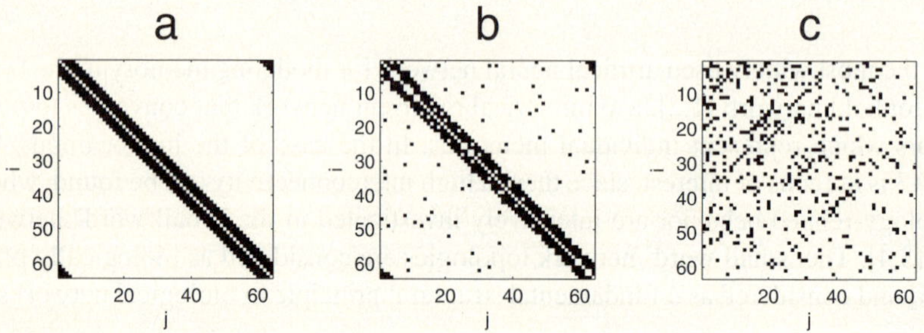

Fig. 1. Parameter ρ of the "Small World" networks in connectivity matrices. Three connectivity matrices, where black represents the existence of a connection between two units; left 0% random connectivity, center 10% random connectivity, and right 100% random connectivity. All matrices are derived from the 100% matrix. The average connectivity in all three matrices is 10%. No self-connections.

2.2 Error Recoverage and Recall Quality

Implementation of the algorithm showed generally good ability to reproduce the desired output vectors. Errors were generally infrequent but tended to cluster in particular output vectors. The majority of patterns were learned perfectly or very well, but those patterns that were not well learned tended to show many errors. The error matrix $E = (e_1...e_n)$ is achieved by the actual recalled output matrix $O = (o_1...o_n)$ and the original target pattern matrix $P = (p_1...p_n)$. $E = |O - P|$. The recall quality q

of all patterns is calculated by first measuring the Hamming distance (the number of bits that are wrong) h_i of each pattern i: $h_i = \sum_{j=1}^{n} e_{ij}$ and then the average number $\overline{h}$ of correct recognized bits are normalized by the number of bits per pattern (m): $q = (m - \overline{h})/m$.

3 Impact on Network Capacity

Performing many simulations, it appeared that network recall quality was related not only to the total number of connections $k \cdot N$, but also to the ratio of long-distance-connections, ρ (Fig 3). Obviously, the storage capacity decreases with more number of learned patterns (Fig 2a).

The structure of local connectivity has an important impact. With rising the number of neighbor nodes of each node k, the local connection areas become larger, and the quality q of recognition decreases. due to the fact of information overloading in sparse coupling, as depicted in Fig 2b. Comparing local with global and small world methods we also have to consider the different effects of scaling the network, see Fig 2c. On global and small world methods the minimal network distance between neurons do not depend on placement area but only on number of neurons [8]. To overcome these side

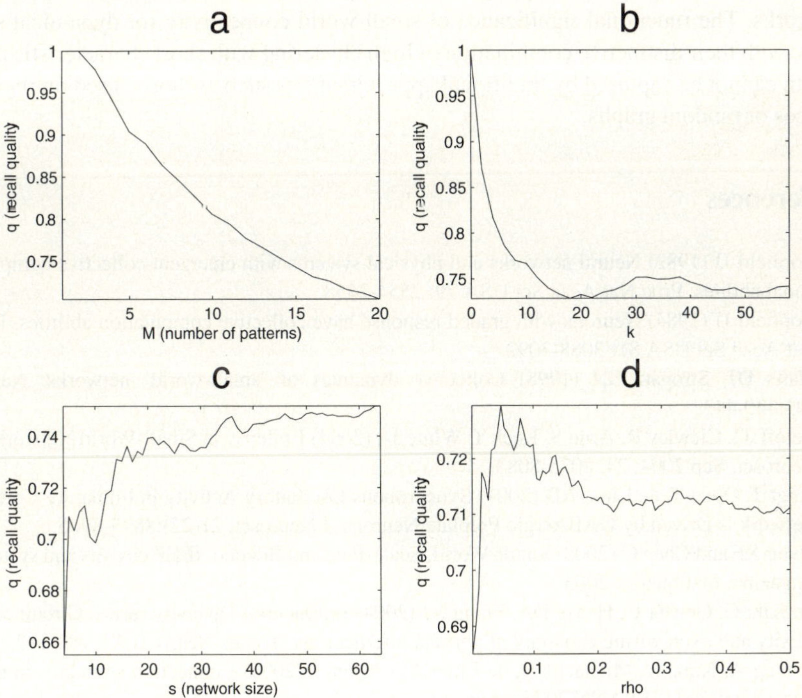

Fig. 2. Recall quality q in "small world" networks. a)number of patterns M, b) coupling locality k, c) network size s, d) ratio ρ of long-distance connections in network topology.

effects, both the number of neurons N and the number of neighborhood connections k are scaled simultaneously. In this way, the impact of network size is very small (Fig 2c). Fig 2d shows the impact of the ratio ρ of long-distance connections, the storage capacity is maximized at a value of $\rho \approx 0.1$. Interestingly, this is the close to the parameter range of oscillation in small world networks, so there $\rho \approx 0.1$ seems to be a ratio of long-distance-connections in networks which has an universal optimal meaning, here only a small number of connections exist, which have to be optimized. For a pattern recall, strong feedback loops are essential, which can be generated only locally. An update between distant neurons in a small number of iterations is achieved by a small number of long-distance-connections. So if the ratio of local connections is too small, primarily pattern-recall driving feedback loops are missed.

4 Conclusion

We have shown that in attractor networks with limited number of connections, 'small world' connectivity leads to a larger storage capacity than regular local or unspecific global connectivity. minimizes wiring (and thus volume) while maximizing functional abilities. Taken together, it is obvious that a 'small world' connectivity is very advantageous. Given the physical and computational benefits, it seems very plausible that this is a generic design principle [6] which also applies to other types of optimized networks. The functional significance of small-world connectivity for dynamical systems, with their distinctive combination of high clustering with short characteristic path length cannot be captured by traditional approximations such as those based on regular lattices or random graphs.

References

1. Hopfield JJ (1982) Neural networks and physical systems with emergent collective computational abilites. Proc Nat Acac Sci USA 79: 2554-2558.
2. Hopfield JJ (1984) Neurons with graded response have collective computation abilities. Proc Nat Acad Sci USA 81: 3088-3092.
3. Watts DJ, Strogatz SH (1998) Collective dynamics of 'small-world' networks. Nature 393:440-442.
4. Netoff TI, Clewley R, Arno S, Keck T, White JA (2004) Epilepsy in Small-World Networks J. Neurosci. Sep 2004, 24, 8075-8083
5. Voigt T, Opitz T, de Lima AD (2001) Synchronous Oscillatory Activity in Immature Cortical Network is Driven by GABAergic Preplate Neurons. J Neurosci, 21(22):8895-8905.
6. Wang XF and Chen G (2003) Small-World, Scale-Free and Beyond, IEEE circuits and systems magazine, first quarter 2003
7. Buzsaki G, Geisler C, Henze DA, Wang XJ (2004) Interneuron Diversity series: Circuit complexity and axon wiring economy of cortical interneurons. Trends Neurosci 27: 186-193.
8. Herzog A, Kube K, Michaelis B, de Lima AD, Voigt T (2006) Connection strategies in neocortical networks, ESANN 2006 (in press)

Associative Memory with Small World Connectivity Built on Watts-Strogatz Model*

Xu Zhi[1,2], Gao Jun[1,3], Shao Jing[1,3], and Zhou Yajin[1,2]

[1] Center for Biomimetic Sensing and Control Research, Institute of Intelligent Machines, Chinese Academy of Sciences, Anhui Hefei 230031, China
xuzhi@mail.ustc.edu.cn
gaojun@iim.ac.cn

[2] Department of Automation University of Science and Technology of China, Anhui Hefei 230026, China

[3] Lab. of Image Information Processing, School of Computer & Information, Hefei University of Technology, Anhui Hefei 230009, China

Abstract. Most models of neural associative memory have used networks with full connectivity. However, this seems unrealistic from a neuroanatomical perspective and a VLSI implementation viewpoint. In this study, we built a new associative memory network based on the Watts-Strogatz model. The results indicate that this new network can recall the memorized patterns even with only a small fraction of total connections and is more sufficient than other networks with sparse topologies, such as randomly connected network and regularly network.

1 Introduction

In 1982 Professor John Hopfield proposed a fully connected artificial neural network, introduced the idea of an energy function to formulate a way of understanding the computation performed by this neural network, and showed that a combinatorial optimization problem (TSP) can be solved by this network [1]. From then on many results on their dynamics, capacity, applications, learning rules, etc., have been established [2].But most researches are on the canonical Hopfield model and this fully connected network is not a realistic structure of real biological associative memories in the brain, which have sparse connectivity.

Based on the evidence of biology (in the cortex of the mouse each neuron is connected to about 0.1% of the other neurons [3]), some researchers proposed a Hopfield network with randomly diluted connectivity [5, 6]. Furthermore some purely local connected sparse Hopfield networks have been established, such as modular model [7] and Cellular Neural Network [14].

* Project supported by National Natural Science Foundation of China (No. 60375011, 60575028) and Program for New Century Excellent Talents in University(NCET-04-0560).

L. Jiao et al. (Eds.): ICNC 2006, Part I, LNCS 4221, pp. 115 – 122, 2006.

It is also known that in cortex of the brain the neural networks have a small world characteristic [8, 9]: a high degree of node clustering and short minimum path lengths between nodes, which motivated the investigation undertaken here. Based on other people's previous work [10, 11], we studied a network built upon the Watts-Strogatz model for small world and found some different but interesting results.

In this paper, Sections 2 and 3 describe the background and the network model. After the simulations and results are explained in Section 4, the discussion is given in Section 5.

2 Background and Motivation

The theoretical storage limitation M_{max} of fully connected Hopfield networks [15] is

$$M_{\max} = \frac{N}{2\log_e N} \tag{1}$$

where N is the number of neurons in the network. It is clear that there is a approximate linear relationship between the storage limitation and N, that is M_{max} increases with N. But due to the full connectivity of Hopfield network, when N arrives at some certain value, it is hard for current technology to implement the artificial neural network in VLSI. So it is necessary to build a new topology to replace this fully connected network.

Many real networks have been shown to have a small world architecture, including the internet, metabolic networks, human acquaintance networks and neural networks [8]. Theoretical work [4] has now shown the detailed relationship between the characteristics of such networks.

The notion of small world networks was first introduced in a social experiment by Stanley Milgram [12]. The experiments probed the distribution of path lengths in an acquaintance network by asking participants to pass a letter to one of their first-name acquaintances in an attempt to get it to an assigned target individual. Milgram demonstrated that, despite the high amount of clustering in social networks (meaning that two acquaintances are likely to have other common acquaintances), any two individuals could be "linked" through a surprisingly small number of others. This idea is commonly referred to as "6 degrees of separation", implying that the average "distance" between any two people in the world is about 6.

The simple small world model of Watts and Strogatz [8] consists of a regular N-node ring lattice. Each node is connected to its $\alpha N/2$ nearest neighbors on either side, where αN is typically small compared to N. The mean path length (the minimum number of nodes on a path) between any pair of random nodes is therefore high. A fraction p of these connections are then re-wired to other randomly selected nodes. For low p, this creates a network with primarily local connectivity and a few randomly placed long-range connections termed "short cut" (see Fig. 1). They showed that at surprisingly low values of p, the mean path length in the network dropped dramatically, resulting in a small world regime: short path lengths between any pair of neurons, but also show a cliquish behavior, with locally clustered connections.

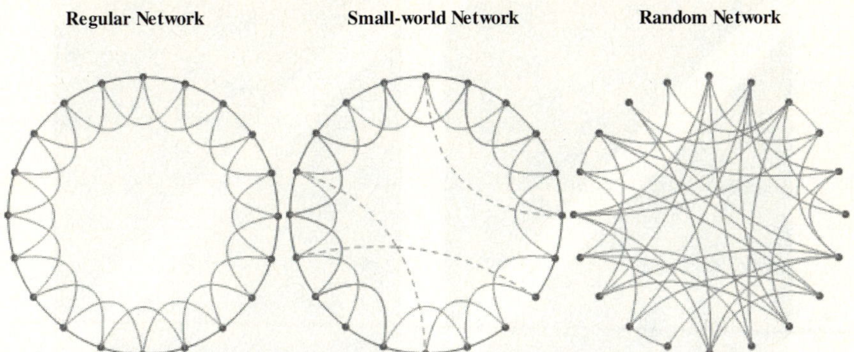

Fig. 1. A ring lattice, with regular connectivity on the left, small world connectivity on the middle and random connectivity on the left

So the motivation for introducing the small world model to associative memory is very apparent. From the neurobiological viewpoint, it has been shown that the nervous systems of the worm C. elegans and hippocampus show small world properties [8]. From the implementation perspective, the fully connected artificial neural network is not easily implementable in silicon chips using current VLSI technology. And theoretical work has shown systems with small-world coupling could enhance signal-propagation speed, computational power, and synchronizability [8].

3 Network Model

In order to investigate the effect of connectivity profiles on the properties of associative memories, we create networks of N neurons which assume binary states (either +1 or –1). Initially, the basic network topology is a one-dimensional ring lattice on which we place the neurons with periodic boundary conditions. Each neuron feeds its output to its $k = \alpha N$ nearest neighbors, where α is the overall connectivity of the network. Then, by keeping the overall connectivity α constant, we create increasingly random graphs through the procedure of Watts and Strogatz [8]. Each neuron is visited, and with probability p, each edge is rewired to a randomly chosen neuron in the network. But self connections and repeated connections are not allowed.

This introduces long-range or "short-cut" connections. The network changes from regular connectivity at p=0 to random connectivity at p=1, with small world connectivity in the middle. The constraint of exactly k efferent connections per neuron reduces the probability of isolated vertices, and, if $k >> \ln N$, the graph will remain connected [13].

The different neural interconnectivities we define are specified by an undirected graph G on the nodes[N]×[N] where a connection from neuron i to neuron j exists if $\{i,j\} \in G$. The connectivity structure of the network is expressed by the connectivity matrix with white pixel representing two neurons connected and black pixel representing two neurons disconnected (see Fig. 2).

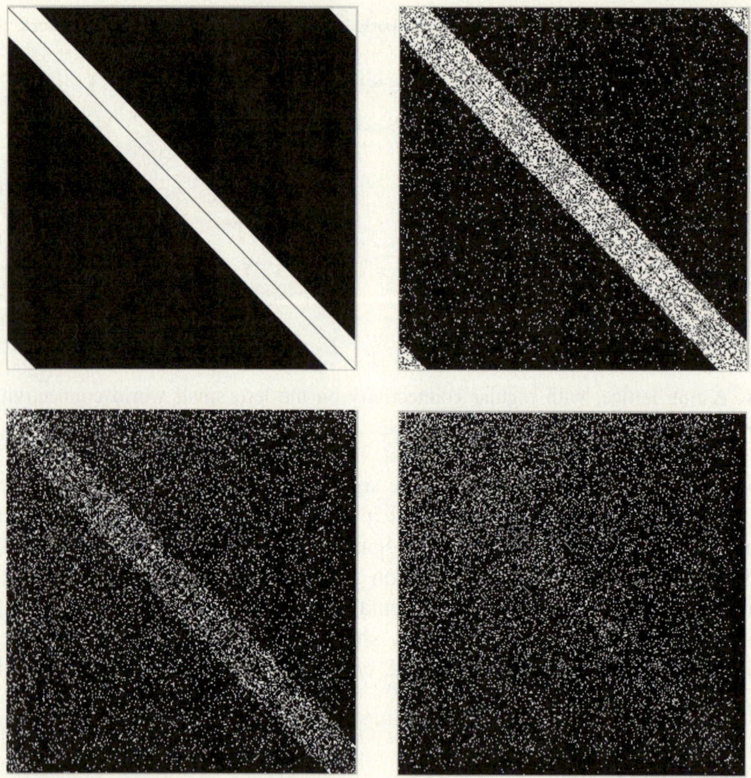

Fig. 2. The connectivity structure of the network. Upper left rewiring probability p =0,upper right p =0.25 ,lower left p =0.75,lower right p =1.0.

The network is trained with M randomly generated unbiased patterns ζ of length N and the connection strength W_{ij} between node i and node j is computed by the Hebbian learning rule:

$$w_{ij} = \frac{1}{N}\sum_{\mu=1}^{M} \zeta_i^{\mu} \zeta_j^{\mu} c_{ij} \tag{2}$$

Where

$$c_{ij} = \begin{cases} 0 & if\,\{i,j\} \notin G \\ 1 & if\,\{i,j\} \in G \end{cases} \tag{3}$$

The state of each neuron in the system is updated in discrete time by the following rule:

$$S_i(t+1) = \mathrm{sgn}\left(\sum_{j \neq i}^{N} w_{ij} S_j(t)\right) \tag{4}$$

where sgn() is the sign function and $S_j(t)$ is the output activation of neuron j at time t. Updating the activations is done asynchronously by randomly picking a single neuron and evaluating whether its current state has to change or not. This guarantees the convergence of the network into a fixed point attractor (either the correct pattern or spurious attractors) after a sufficient number of iterations.

4 Simulations

We summarize some key parameters in Table 1.

Table 1. Parameters of the Model

N	The number of neurons in the network
k	The number of connections each neuron has
p	rewiring probability
α	The loading per connection (k/N)
M	The number of randomly generated unbiased patterns required to memorize

The network was simulated on N=1000 neurons ,M= 25 and α =0.15. The ratio (M/N) was chosen largely below the theoretical storage limitation of about 0.138N for fully connected Hopfield networks [15] in order to avoid any catastrophic forgetting of memories. In fact, it was shown that, when a Hopfield network is trained on a proportion of memories above this criterion value, the entire set of memories is not learned and the memory model becomes useless.

The performance and dynamics of the network are investigated for p=0 (only short-range connections), intermediate values of p (the small-world regime), and p=1 (random connections).

To test the networks, we probed them with test patterns containing two different kind of corrupted bits in order to evaluate the generalization capacities of the networks and to roughly check the sizes of the basins of attractors.

To mark the quality of recall, we computed the mean overlap between the expected attractor (the stored pattern) and the actual attractor reached by the network. A high mean overlap indicates a good recall and/or an accurate generalization. On the other hand, a low overlap is a sign of an inability to store patterns and/or a poor capacity to generalize (this might indicate the small size of the basins of attractors).

In case 1, a continuous fraction (25%) of the bits of an initial pattern are inverted on each run. So the test patterns are presented with a continuous portion of errors (25%). In case 2, a fraction of the randomly chosen bits (25%) of an initial pattern are inverted on each run. So the test patterns are presented with a portion (25%) of random errors. Figure 3 and Figure 4 show the performance of the network as a function of rewiring probability in two cases, respectively. All the results reported here are averaged on 1000 runs performed on each type of simulation.

Based on the above results, it is clear that, when the test patterns are with continuous fraction of errors and networks are primarily local connectivity, that is p is very small, the network could not correct the errors very well even after many iterations. Compared

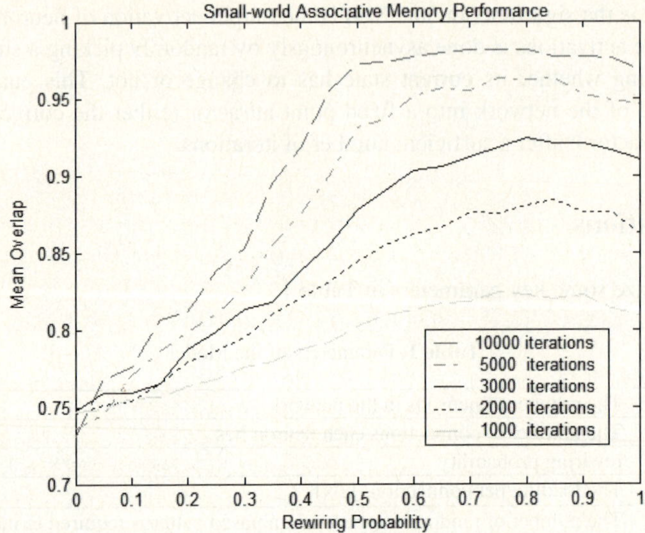

Fig. 3. The performance of the network as a function of rewiring probability

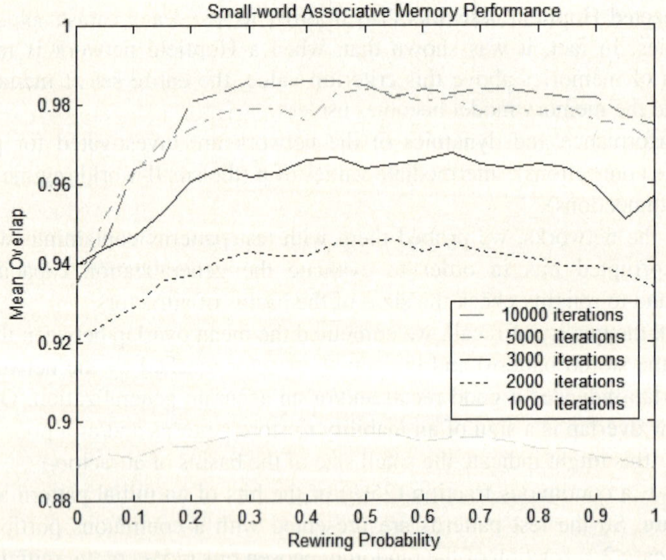

Fig. 4. The performance of the network as a function of rewiring probability

with case 1, the performance of the network in case 2 is improved to a certain extent, when the test patterns are with random errors and networks are primarily local connectivity.

And the performance is improving dramatically with the increase of p in both cases. What is different from the conclusion on [10] is that the network does not always perform better along with the increase of p. At some certain value of p, the network achieves the best performance. The best performances in two cases are obtained when p is about 0.8 and about 0.4, respectively, rather than 1.0. So the network with small world connectivity can achieve the best performance.

5 Discussion

The fully connected models of associative memory, such as the standard Hopfield network, are difficult to realize in VLSI implementations. For the sake of reducing the complexity of connectivity and the small world characteristic discovered in the neural system [8, 9], we studied a neural network built upon the Watts-Strogatz model for small world.

The results presented here show that the neural network with purely local connectivity primarily failed to recall the test patterns with a continuous domain of errors. The regular network is not able to propagate the correcting signals to the center of the error region, even after many iterations. But with the introducing of small world characteristic to the network, the small-world architecture helps break up these domains by injecting signals from distant low-error domains. The addition of a small percentage of long range connections "short cut" greatly improves the network's signal propagation ability, and, hence, the ability to recall the test patterns to the expected attractor.

Whatever the test patterns are, with random errors or with continuous errors, the performance of the network is not getting better while the topology of the network is becoming more random. In other words, the network with small world connectivity could achieve the best performance. Compared with common randomly sparse Hopfield network, there is a further benefit that in the network with small world connectivity the wiring complexity is much reduced due to the predominantly local connections

In further work we will explore other types of complex network connectivity models, such as scale free networks [16]. It is expected to find the topology of connectivity that maximizes performance while minimizing wiring cost.

References

1. J.J. Hopfield, Neural networks and physical systems with emergent collective computational abilities, Proc. Natl. Acad. Sci. USA 79 (1982) 2445-2458.
2. D. J. Amit, Modeling brain function: the world of attractor neural networks, (Cambridge University Press, Cambridge,1989).
3. Braitenberg, V. and A. Schüz, Cortex: Statistics and Geometry of Neuronal Connectivity. 1998,Berlin: Springer-Verlag.
4. Newman, M.E.J., Models of the Small World. Journal of Statistical Physics, 2000. 101(3/4): p.819-841.
5. S. Venkatesh, Robustness in neural computation: Random graphs and sparsity, IEEE Trans. Inf.Theory 38 (1992) 1114-1119.
6. J. KomloH s, Effect of connectivity in an associative memory model, J. Comput. System Sciences 47(1993) 350-373.

7. N. Levy, E. Ruppin, Associative memory in a multi-modular network, Neural Comput. 11 (1999)1717-1737.
8. Watts, D. and S. Strogatz, Collective Dynamics of 'small-world' networks. Nature, 1998. 393: p.440-442.
9. Shefi, O., et al., Morphological characterization of in vitro neuronal networks. Physical Review E, 2002. 66(021905).
10. Bohland, J. W. and A. A. Minai (2001). "Efficient associative memory using small-world architecture. " Neurocomputing 38-40: 489-496.
11. Davey, N., B. Christianson and R. Adams (2004). "High Capacity Associative Memories and Small World Networks." In Proc. of IEEE IJCNN'04 25-29 July, Budapest, Hungary.
12. S. Milgram, The small-world problem, Psychol. Today 2 (1967) 60}67.
13. B. Bollobás B, Random Graphs, Academic Press, New York, 1985.
14. L O Chua, L Yang. Cellular neural network: applications. IEEE Transactions on Circuits and Systems 35:1273-1290,1988.
15. R.J. McEliece, E.C. Posner, E.R. Rodemich, S.S. Venkatesh, The capacity of the Hopfield associative memory, IEEE Trans. on Information Theory 33 (1987) 461-482.
16. Stauffer, D., et al., Efficient Hopfield pattern recognition on a scale-free neural network. European Physical Journal B, 2003. 32(3): p. 395-399.

A Hopfiled Neural Network Based on Penalty Function with Objective Parameters

Zhiqing Meng, Gengui Zhou, and Yihua Zhu

College of Business and Administration
Zhejiang University of Technology, Zhejiang 310032, China

Abstract. This paper introduces a new Hopfiled neural network for nonlinear constrained optimization problem based on penalty function with objective parameters. The energy function for the neural network with its neural dynamics is defined, which differs from some known Hopfiled neural networks. The system of the neural networks is stable, and its equilibrium point of the neural dynamics corresponds to a solution for the nonlinear constrained optimization problem under some condition. Based on the relationship between the equilibrium points and the energy function, an algorithm is developed for computing an equilibrium point of the system or an optimal solution to its optimization problem. One example is given to show the efficiency of the algorithm.

1 Introduction

Hopfield neural networks (HNNs) are important tools to solve optimization problems[1-6]. Studies on systems of nonlinear dynamics mainly deal with the stability of the system. It is a challenge to define energy functions and to make them stable. Usually, penalty functions are used to define energy functions [5-6]. In order to make an HNN stable, energy functions are usually needed to be two-order differential. To solve optimization problems, this paper introduces an algorithm that a two-order differential penalty function with objective parameters is constructed and this function is then used as an energy function of an HNN in order to keep the HNN stable. It is shown that an equilibrium point of the HNN can be obtained by finding a solution of the penalty optimization problem under some mild condition. Finally, an algorithm is presented to find out an approximate solution to the corresponding optimization problem.

2 A Nonlinear Neural Network

This paper studies a nonlinear neural network for the following constrained optimal problem.

(P) minimize $f(\mathbf{x})$

Subject to $g_i(\mathbf{x}) \le 0, i = 1, 2, ..., m,$

where f, g_i:$\mathrm{R}^n \to \mathrm{R}^1$, $i \in I = \{1, 2, ..., m\}$ are two-order differential. Let $X = \{\mathbf{x} | g_i(\mathbf{x}) \le 0, i = 1, 2, ..., m\}$.

L. Jiao et al. (Eds.): ICNC 2006, Part I, LNCS 4221, pp. 123–126, 2006.

2.1 The Construction of a New Neural Network

Assuming that **b**: $R^n \times R^1 \rightarrow R^n$, **b** is a vector, we define a dynamical differentiable system:

$$\frac{\mathrm{d}\mathbf{x}}{\mathrm{d}t} = B\mathbf{b}(\mathbf{x}(t), t), \tag{1}$$

where, $\mathbf{x}(t) = (x_1(t), x_2(t), \ldots, x_n(t))^T$ is a state vector and $B = \mathrm{diag}(\beta_1, \beta_2, ..., \beta_n)$ with β_i, $i = 1, 2, \cdots, n$.

Definition 2.1.1. If a point $\mathbf{x}^* \in X$, and satisfies:

$$\mathbf{b}(\mathbf{x}^*, t) = 0, \forall t,$$

then $\mathbf{x}^*$ is called an equilibrium point of dynamics system (1).

Define the nonlinear penalty functions for (P) with objective parameters:

$$H(\mathbf{x}, M) = e^{\rho(f(\mathbf{x})-M)^2} + \alpha \sum_{i=1}^{m} \max\{g_i(\mathbf{x}), 0\}^{2.1}, \tag{2}$$

where ρ and α are given positive numbers and $M \in R^1$ is a parameter. $H(\mathbf{x}, M)$ is twice differentiable at every $x \in X$. The energy function $E(\mathbf{x})$ is defined as: $E(\mathbf{x}) = H(\mathbf{x}, M)$, which is twice differential. And the corresponding dynamical differentiable system is:

$$\frac{\mathrm{d}\mathbf{x}}{\mathrm{d}t} = -B\nabla E(\mathbf{x}). \tag{3}$$

The new neural network of size n is a fully connected network with n continuous valued units. Let ω_{ij} be the weight of the connection from neuron i to neuron j. Since $E(\mathbf{x})$ are twice continuous differentiable, we can define the connection coefficients as follows:

$$\omega_{ij} = \frac{\partial E^2(\mathbf{x})}{\partial x_1 \partial x_2}, \qquad i, j = 1, 2, \cdots, n. \tag{4}$$

Now, let's describe the structure of our neural network. We define $\mathbf{x} = (x_1, x_2, \cdots, x_n)$ as the input vector of the neural network, $\mathbf{y} = (y_1, y_2, \cdots, y_n)$ as the output vector, and $V(t) = (v_1(t), v_2(t), \cdots, v_n(t))$ as the state vector of neurons. Here, $v_i(t)$ is the state of neuron i at the time t. And this new neural network is a type of Hopfield-like neural network.

2.2 Analysis of Stability

Consider the unconstrained problem:

(P(M)): $\min H(\mathbf{x}, M)$ s.t. $\mathbf{x} \in R^n$.

For the stability analysis of (2), we have some theorems as follows.

Theorem 2.2.1. Let $\mathbf{x}^*$ be an equilibrium point of dynamics system (3) under the parameter M. If $\mathbf{x} \neq 0$ and $E(\mathbf{x}) \neq 0$, then $\mathbf{x}^*$ is the stable point of dynamics system (3). Additionally, if weight coefficient matrix $(\omega_{ij})_{n\times n}$ is positive semi-definite, then $\mathbf{x}^*$ is a locally optimal solution to the problem (P(M)).

Theorem 2.2.2. If $\mathbf{x}^*$ is an optimal solution to (P) and $M = f(\mathbf{x}^*)$, then, for a given $\rho > 0$ and $\alpha > 0$, $\mathbf{x}^*$ is also an optimal solution to (P(M)) with $H(\mathbf{x}^*, M) = 0$.

Theorem 2.2.3. Suppose that $\mathbf{x}^*$ is an optimal solution to (P). For some $\rho > 0$,$\alpha > 0$, M and $\mathbf{x}_M^*$, if $\mathbf{x}_M^*$ is an optimal solution to (P(M)) and a feasible solution to (P), $H(\mathbf{x}_M^*, M) \neq 0$, and $M \leq f(\mathbf{x}^*)$, then $\mathbf{x}_M^*$ is an optimal solution to (P).

Theorem 2.2.4. Suppose that the feasible set X is connected and compact and that, for some $\rho > 0$ and $\alpha > 0$, and M, $\mathbf{x}_M^*$ is an optimal solution to $(P(M))$. Then

(i) $M_* \leq M \leq M^*$ if $H(\mathbf{x}_M^*, M) = 0$.

(ii) $M < M_*$ if $H(\mathbf{x}_M^*, M) \neq 0$ and $M \leq M^*$. Furthermore, $\mathbf{x}_M^*$ is an optimal solution to (P) if $\mathbf{x}_M^*$ is a feasible solution to (P).

Theorem 2.2.5. If $\mathbf{x}^*$ is an optimal solution to the problem (P (M)), then $\mathbf{x}^*$ is an equilibrium point of dynamics system (3) under the parameter M.

Theorems 2.2.1 and 2.2.5 show that an equilibrium point of the dynamic system yields an optimal solution to the optimization problem (P(M)). Theorem 2.2.3 and 2.2.4 mean that an optimal solution to (P(M)) is also an optimal solution to (P) for some M. Therefore, we may obtain an optimal solution to (P) by finding an solution to (P(M)) or an equilibrium point of the dynamic system (3).

3 Algorithm and Numerical Examples

We propose the following Algorithm I, by which and Theorem 2.2.4, we can obtain an approximate optimal solution to (P), and an equilibrium point of the dynamic system (3) of the neural network.

Algorithm I

Step 1: Choose $\rho > 0$,$\alpha > 0$, $\epsilon \geq 0$, $\mathbf{x}^0 \in X$, and a_1 satisfying $a_1 < \min_{\mathbf{x} \in X} f(\mathbf{x})$. Let $k = 1$, $b_1 = f_0(\mathbf{x}^0)$, and $M_1 = \frac{a_1+b_1}{2}$. Go to Step 2.

Step 2: Solve $\min_{\mathbf{x} \in R^n} H(\mathbf{x}, M_k)$. Let $\mathbf{x}^k$ be an optimal solution to it.

Step 3: If $\mathbf{x}^k$ is not feasible to (P), let $b_{k+1} = b_k$, $a_{k+1} = M_k$, $M_{k+1} = \frac{a_{k+1}+b_{k+1}}{2}$, and go to Step 5. Otherwise, $\mathbf{x}^k \in X$, and go to Step 4.

Step 4: If $H(\mathbf{x}^k, M_k) = 0$, let $a_{k+1} = a_k$, $b_{k+1} = M_k$, $M_{k+1} = \frac{a_{k+1}+b_{k+1}}{2}$, and go to Step 5. Otherwise, $H(\mathbf{x}^k, M_k) > 0$, stop. $\mathbf{x}^k$ is a solution to (P).

Step 5: If $|b_{k+1} - a_{k+1}| \leq \epsilon$ and $g_i(\mathbf{x}^k) \leq \epsilon, i = 1, 2, \cdots, m$, stop. $\mathbf{x}^k$ is an ϵ-solution to (P). Otherwise, let $k = k + 1$ and go to Step 2.

Remark 3.1. A vector $\mathbf{x} \in X$ is ϵ-feasible or ϵ-solution if $g_i(\mathbf{x}) \leq \epsilon, \forall i \in I$. It is easy to prove the Algorithm I that is convergent under some conditions.

Example 3.1. Consider the Rosen-Suzki problem{[7]}:

$$(RSP) \min f(\mathbf{x}) = x_1^2 + x_2^2 + 2x_3^2 + x_4^2 - 5x_1 - 5x_2 - 21x_3 + 7x_4$$
$$\text{s.t. } g_1(\mathbf{x}) = 2x_1^2 + x_2^2 + x_3^2 + 2x_1 + x_2 + x_4 - 5 \leq 0$$

$$g_2(\mathbf{x}) = x_1^2 + x_2^2 + x_3^2 + x_4^2 + x_1 - x_2 + x_3 - x_4 - 8 \leq 0$$
$$g_3(\mathbf{x}) = x_1^2 + 2x_2^2 + x_3^2 + 2x_4^2 - x_1 - x_4 - 10 \leq 0.$$

Let $\rho = 0.0001, \alpha = 10\ \mathbf{x}^0 = (0,0,0,0) \in X,\ a_1 = -200,\ b_1 = 0,\ M_1 = -100.$ The numerical results are given in Table 3.1, where $\mathbf{x}^k$ is the solution to (RSP) in the kth iteration and

$$e(\mathbf{x}) = \max\{g_1(\mathbf{x}), 0\} + \max\{g_2(\mathbf{x}), 0\} + \max\{g_3(\mathbf{x}), 0\}$$

is the constraint error that is used to determine if a solution is $\epsilon-$ feasible to (RSP) or not. It is clear that a solution $\mathbf{x}$ is $\epsilon-$ feasible to (RSP) when $e(\mathbf{x}) < \epsilon = 10^{-10}$. The algorithm yields an approximate solution $\mathbf{x}^2 = (0.172460, 0.837638, 2.004541, -0.969747) \in X$ to (RSP) with $f(\mathbf{x}^2) = -44.225919$, while, in [7], the solution is $\mathbf{x}' = (0, 1, 2, -1)$ with $f(\mathbf{x}') = -44$, which were thought as an optimal solution. In fact, it is easy to check that $\mathbf{x}^2$ is feasible.

Table 3.1. Results of (RSP)

k	$g_1(\mathbf{x}^k)$	$g_2(\mathbf{x}^k)$	$g_3(\mathbf{x}^k)$	$e(\mathbf{x}^k)$	$\mathbf{x}^k$	$f(\mathbf{x}^k)$	M_k
1	0.001241	0.002499	-1.898479	0.003741	(0.173584,0.838770,2.005145,-0.969105)	-44.239505	-100.000000
2	-0.007882	-0.000918	-1.900437	0.000000	(0.172460,0.837638,2.004541,-0.969747)	-44.225919	-50.000000

Acknowledgements

This research work was partially supported by grant No. Z105185 from Zhejiang Provincial Nature Science Foundation.

References

1. Hopfield,J.J., Tank, D.W.: Neural Computation of Decision in Optimization Problems. Biological Cybernetics. **58** (1985) 67-70
2. Joya,G., Atencia,M.A., Sandoval,F.: Hopfield Neural Networks for Optimizatiom: Study of the Different Dynamics. Neurocomputing. **43** (2002) 219-237
3. Chen,Y.H., Fang,S.C.: Solving Convex Programming Problems with Equality Constraints by Neural Networks. Computers Math. Applic. **36** (1998) 41-68
4. Staoshi M.: Optimal Hopfield Network for Combinatorial Optimization with Linear Cost Function, IEEE Tans. On Neural Networks. **9** (1998) 1319-1329
5. Meng,Z.Q., Dang,C.Y., Zhou G.,Zhu Y., Jiang M.: A New Neural Network for Nonlinear Constrained Optimization Problems, Vol.3173, Lecture Notes in Computer Science, (2004) 406-411
6. Meng, Z.Q., Dang, C.Y.:A Hopfiled Neural Network for Nonlinear Constrained Optimization Problems based on Penalty Function, Vol.3496, Lecture Notes in Computer Science,(2005) 712-717
7. Lasserre,J.B.: A Globally Convergent Algorithm for Exact Penalty Functions, European Journal of Opterational Research,**7** (1981) 389-395

Study on Discharge Patterns of Hindmarsh-Rose Neurons Under Slow Wave Current Stimulation

Yueping Peng, Zhong Jian, and Jue Wang

Key Laboratory of Biomedical Information Engineering of Education Ministry,
Xi'an Jiaotong University, 710049, Xi'an, China
Percy001@163.com, Jz68720@263.net,
Juewang@mail.xjtu.edu.cn

Abstract. The Hindmarsh-Rose neuron under different initial discharge patterns is stimulated by the half wave sine current and the ramp current; and the discharge pattern of the neuron is discussed by analyzing its membrane potential's interspike interval(ISI) distribution. Under the ramp current stimulation, the neuron's discharge pattern gradually changes into dynamic period 1 discharge pattern whose ISI drops off with the ramp's amplitude increasing; and slow adaptation current gradually increases according to the linear function with the ramp's amplitude increasing, which reflects the linear cumulation of intramembranous calcium ion. Under the half wave sine current stimulation, the current frequency affects greatly the neuron's discharge patterns; and under the fixedness of the current's amplitude, the neuron presents the integral multiple period discharge pattern, the periodic parabolic bursting pattern and the chaos discharge pattern under different frequency current Stimulation. This investigation shows the mechanism of the frequency and the amplitude of the slow wave current stimulating the neuron, and the neuron's discharge patterns can be adjusted and controlled by the slow wave current. This result is of far reaching importance to study synchronization and encode of many neurons or neural network, and provides the theoretic basis for studying the mechanism of some nervous diseases such as epilepsy by the slow wave of EEG.

1 Introduction

In recent years, slow wave of EEG is regarded widely, which may be the agent that some neural diseases form or burst. During epileptic seizure prophase, slow wave energy has the trend of augmenting remarkably, and too large slow wave may be the main cause that epilepsy seizures form and change[2, 3]. Bursting of Hodgkin-Huxley neuron model under the synapse slow current stimulation was studied by Zhu Jingling, et al, and they concluded that enough large synapse slow response can cause the bursting or over excitement of neuron[4, 7]. The neuron membrane potential under different noises was discussed by applying different frequency sine current to stimulate Hodgkin-Huxley neuron model, and the conclusion that neurons have the resonance characteristic was drawn[6]. In addition, the effect of the neuron under noise or chaos stimulation was also discussed[5, 8, 9, 10]. However, it is done little how slow wave current

L. Jiao et al. (Eds.): ICNC 2006, Part I, LNCS 4221, pp. 127 – 134, 2006.

affects neuron discharge patterns. In this study, we take one Hindmarsh-Rose neuron(HR neuron)[1] as the object, and make the HR neuron different initial discharge patterns by setting the value of the parameter *r*, and apply the ramp current and the different frequency and amplitude half wave sine current to stimulate the HR neuron, and discuss how the HR neuron discharge patterns change by calculating and analyzing the neuron membrane potential's interspike interval (ISI).

2 The HR Neuron Model and Its Discharge Patterns

The HR neuron has many time scale dynamics action, and its equation is set of three dimension ordinary nondimensional differential equations[1],

$$\dot{x} = y - ax^3 + bx^2 - z + I \tag{1}$$

$$\dot{y} = c - dx^2 - y \tag{2}$$

$$\dot{z} = r[s(x - X) - z] \tag{3}$$

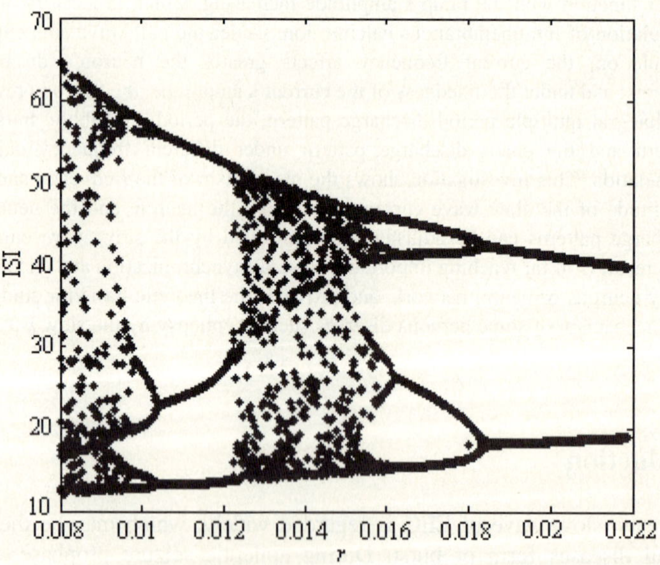

Fig. 1. Bifurcation figure of the HR neuron under the parameter *r* changing from 0.008 to 0.022

The HR neuron has three time variables: the membrane potential variable *x* which has the quick depolarization ability, the quick recovery variable *y* and slow adaptation current variable *z*. *I* is input stimulation current; *a*, *b*, *c*, *d*, *s*, *r* and *X* are parameters. The parameter *r* is related to the membrane penetration of calcium ion, and other parameters have no specific physical meaning. The equations are nondimensional, and at numerical calculation, value of parameters is as follows: a=1.0, b=3.0, c=1.0, d=5.0, s=4.0, X=-1.56, I=3.0, and you can make the neuron different discharge patterns by controlling the parameter r variation. Fig.1 is the ISI bifurcation figure of the neuron, where parameters except r are set to the above values, and the initial state of the neuron is (1.0, 0.2, 0.2), and parameter r is 0.008~0.022. From fig.1, the discharge pattern of the neuron begins from the chaos state(r is about 0.008~0.009), and evolves period 6 discharge pattern(r is near 0.01), and via the adverse period doubling bifurcation passes period 3(r is about 0.0105~0.012) and enters the chaos state(r is about 0.0125~0.015) again, and at last via the adverse period doubling bifurcation passes period 4(r is about 0.016~0.018) and comes into period 2(r is about 0.0185 ~0.022).

3 Discharge Patterns of the HR Neuron Under the Slow Wave Current Stimulation

Total stimulation current I(t) includes two parts: the bias current I_{BIAS} and the input stimulation current I_S(t). Where I_{BIAS} is 3, the neuron initial discharge pattern can be controlled by the value of parameter r. The ramp current and the half wave sine current are taken as I_S(t) and begin to stimulate the neuron after I_{BIAS} has been stimulating the neuron for some time, which are the two typical slow wave current and better simulate the slow postsynaptic potential of nervous system, especially these slow postsynaptic potentials arise slowly and their time courses are several seconds or several minutes. In addition, the characteristics of the ramp current are that its amplitude is on the advance and its moment energy is buildingup, and it is convenient to adjust the amplitude increasing speed by setting its slope value. For low frequency sine current, it is convenient to adjust the frequency and the amplitude, especially to discuss the problems related to frequency response.

3.1 Discharge Patterns of the HR Neuron Under the Ramp Current Stimulation

The values of the parameters except r of the neuron model are above given values, and the slope value of the ramp current is 0.001, which stimulates the neuron model after I_{BIAS} has been working for 500.

Discharge Patterns Under the Parameter *r* Being Different Values. When the parameter r is respectively 0.01, 0.013 and 0.02, the discharge pattern of the neuron is respectively period 6, the chaos and period 2. As the ramp current begins to stimulate the neuron, with the ramp's amplitude increasing, the neuron's discharge pattern gradually approaches each other from the initial state and finally changes into dynamic period 1 discharge pattern whose ISI drops off, and its dynamical trend approximates to an exponent function. Moreover, the slow adaptation current gradually increases with the ramp's amplitude increasing, whose dynamical trend approximates to a linear function. Fig. 2 shows the discharge process of the neuron under the ramp current stimulation, where the parameter r is 0.013 and the initial discharge state of the neuron is the chaos.

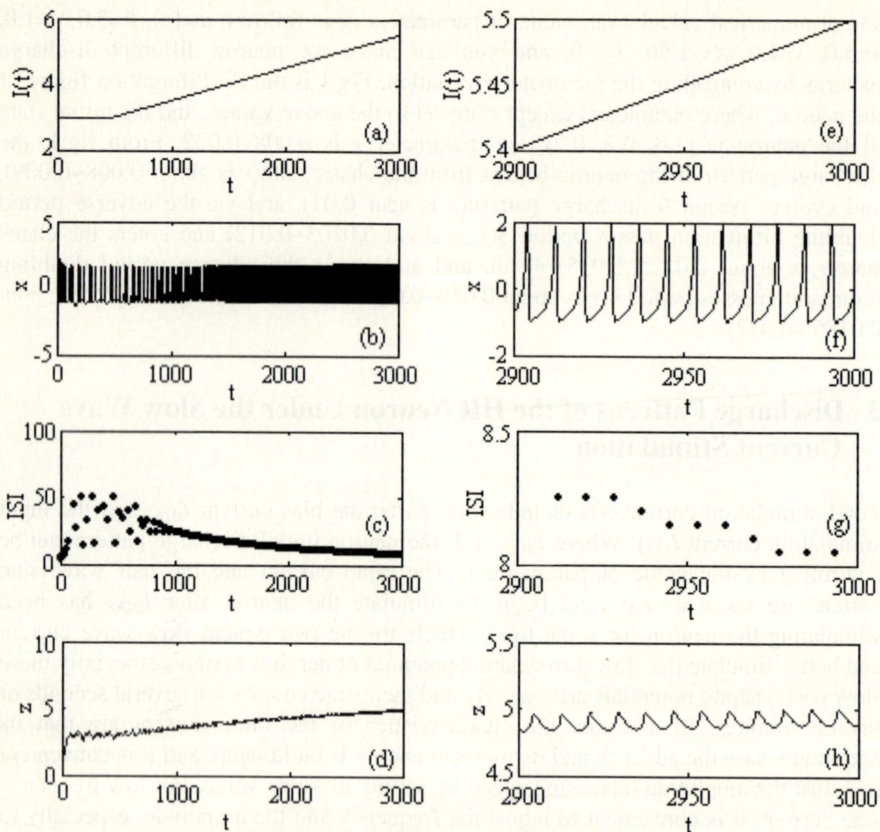

Fig. 2. The discharge process of the neuron, where *r* is 0.013 and the initial discharge state of the neuron is the chaos, and the slope value of the ramp is 0.001. (a) and (e). The changing chart of the stimulation current with time, and the ramp current begins at 500. (b) and (f). The changing chart of the membrane potential *x* with time. (c) and (g). The changing chart of the membrane potential's ISI with time. (d) and (h). The changing chart of the slow adaptation current *z* with time. From fig. (c), the ISI distribution begins from the chaos state, and gradually approaches each other and evolves into dynamic period 1 discharge pattern with the ramp amplitude increasing. From fig. (d), the slow adaptation current *z* gradually increases with the ramp's amplitude increasing, whose dynamical trend approximates to a linear function.

Discharge Patterns under the Value of the Parameter *r* Changing Continuously. When the value of the parameter *r* changes from 0.008 to 0.022 according to the step 0.0002, the initial discharge patterns of the neuron are presented in Fig. 1. As the ramp current begins to stimulate the neuron, its membrane potential's ISI distribution is showed in Fig.3.

From Fig.3, whatever the parameter *r* is, namely whatever the initial discharge state of the neuron is, when the ramp current changes and reaches a certain value(about 0.5), namely the stimulation current amplitude is 3.5, the dynamical decreasing trend of the neuron membrane potential's ISI is the same, namely approximating to an exponent function. And, whatever the parameter *r* is, the slow adaptation current gradually increases with the ramp's amplitude increasing and its dynamical trend approximates to a linear function.

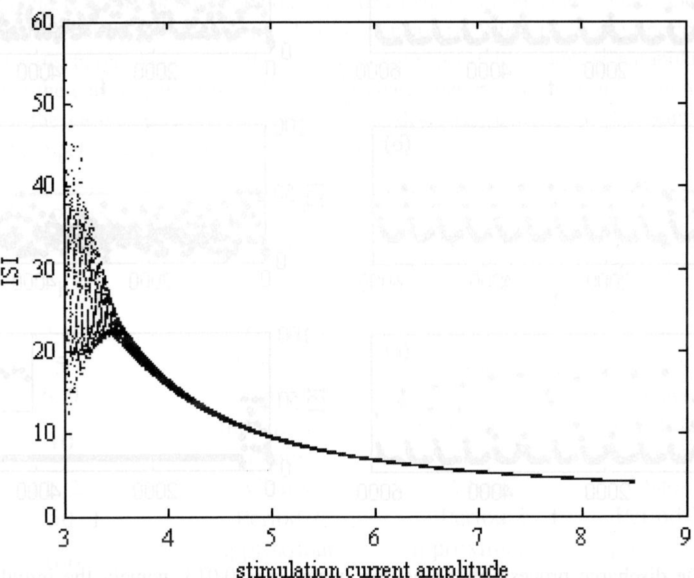

Fig. 3. The changing chart of the neuron's membrane potential's ISI with the stimulation current amplitude, where *r* changes from 0.008 to 0.022 according to the step 0.0002, and the slope value of ramp is 0.001

3.2 Discharge Patterns of the HR Neuron Under the Half Wave Sine Current Stimulation

The values of the parameters except *r* of the neuron model are above given values, and the half wave sine current begins to stimulate the neuron model after I_{BIAS} has been working for 500.

Discharge Patterns under the Stable Period and Variable Amplitude of the Stimulation Current. The value of the stimulation current period is 500 and 8, and its amplitude is 0.5, 1 and 3 under each period value. The parameter *r* of the neuron is 0.01, 0.013 and 0.02, and its discharge pattern is respectively period 6, the chaos and period 2.

As the current whose period is 500 begin to stimulate the neuron, the membrane potential's ISI distributions of the neuron all approximate to the periodic quadratic parabola bursting pattern under the three amplitudes and the three initial discharge patterns. As the current whose period is 8 begin to stimulate the neuron, the discharge patterns of the neuron are all the chaos under the amplitude of the current being 0.5 and 1, and when the amplitude of the current is 3, the discharge pattern is period 5 under the parameter *r* being 0.01 and is period 2 under the parameter *r* being 0.013 and 0.02. Fig.4 represents the discharge process of the neuron, where the parameter *r* is 0.013.

As the half wave sine current stimulates the neuron, its amplitude and its frequency (or period) affect greatly the discharge pattern of the neuron, especially the frequency. Under the stable frequency, the discharge patterns are getting simpler with the amplitude increasing, which is the same as the result under the ramp current stimulation. Under the stable amplitude, the discharge patterns are plentiful with the frequency changing, which is the periodic approximate quadratic parabola bursting, integral multiple periodic discharge and the chaos, and which is nothing to do with the initial discharge pattern, and the stimulation current frequency of whose chaos area is 0.0025~1. So the discharge pattern of the neuron can be adjusted by adjusting the stimulation current period (or frequency).

This investigation shows the mechanism of the frequency and the amplitude of slow wave current stimulating the neuron. As the stimulation frequency changes, the discharge pattern of the neuron also changes with it, which has plentiful discharge actions. As the stimulation amplitude changes, the membrane potential's ISI series of the neuron are getting smaller and smaller, and the discharge pattern are getting simpler. All these changes have nothing to do with the initial discharge pattern of the neuron. So the neuron's discharge patterns can be adjusted and controlled by the frequency and the amplitude of the slow wave current. This result is of far reaching importance to study synchronization and encode of many neurons or neural network, and provides the theoretic basis for studying the mechanism of some nervous diseases such as epilepsy by the slow wave of EEG.

References

1. Hindmarsh J L, Rose R M: A Model of the Nerve Impulse Using Two First-order Differential Equation. Nature, Vol. 296. (1982) 162-165
2. Zhu Junling, et al: The Statuses and Problems of Engineering Methods Being Applied to Epileptic Researches. Biomedical Engineering Foreign Medical Sciences, Vol. 27. (2004) 344-347
3. Zhu Junling, et al: The Roles of Different Components of EEGs for Seizure Prediction Wavelet Energy Evaluation. ACTA Biophysica Sinica, Vol. 19. (2003) 73-77
4. Zhu Jingling, et al: Bursting of Neurons under Slow Wave Stimulation. ACTA Biophysica Sinica. (2001) 632–636
5. Yu Hongjie, et al: Chaotic Control of the Hindmarsh-Rose Neuron Model. ACTA Biophysica Sinica, Vol. 21. (2005) 295-301
6. Wen Zhihong, et al: Noise-induced Changes of Dynamic Characteristics of Neurons. J Fourth Mil Med University, Vol. 25. (2004) 948-949
7. Zhu Jingling, et al: Discharge Patterns of Neurons under Sinusoidal Current Stimulation. Space Medicine & Medical Engineering, Vol. 15. (2002) 108-111
8. Wen Zhihong, et al: Dynamic Characteristics of the Fitzhugh-Nagumo Neuron Model Induced by Quasi-monochromatic Noise. Chinese Journal of Clinical Rehabilitation, Vol. 8. (2004) 1266-1267
9. M. La Rosa, et al: Slow Regularization Through Chaotic Oscillation Transfer in an Unidirectional Chain of Hindmarsh–Rose Models. Physics Letters, Vol. 266. (2000) 88–93
10. Daihai He, et al: Noise-induced Synchronization in Realistic Models. Physical Review, Vol. 67. (2003)

Proximal SVM Ensemble Based on Feature Selection

Xiaoyan Tao[1,2], Hongbing Ji[3], and Zhiqiang Ma[4]

[1] School of Electronic Engineering, Xidian University, Xi'an, China
[2] Telecommunication Engineering Institute,
Air Force Engineering University, Xi'an, China
taoxiaoyan@lab202.xidian.edu.cn
[3] School of Electronic Engineering, Xidian University, Xi'an, China
hbji@xidian.edu.cn
[4] Telecommunication Engineering Institute,
Air Force Engineering University, Xi'an, China
mzq123@sina.com

Abstract. Ensemble is a very popular learning method. Among most of the existing approaches, Bagging is commonly used. However, Bagging is not very effective on the stable learners. Proximal SVM, a variant of SVM, is a stable learner, so Bagging does not work well for PSVM. For this, two new feature selection based PSVM ensemble methods are proposed, i.e. BRFS and BR. Through perturbing both the training set and the input features, component learners with high accuracy as well as high diversity can be obtained. The experimental results on four datasets from UCI demonstrate that the new approaches perform over a single PSVM and Bagging.

1 Introduction

As a variant of SVM, Proximal SVM (PSVM) [1] improves the computation efficiency with the classification performance unchanged. In recent years, combining machines instead of using a single one for increasing accuracy is a hot topic [2]. And to improve the generalization performance, Kim etc. [3] construct the SVM ensembles by Bagging and AdaBoost respectively. In this paper, we introduce the PSVM ensemble based on Bagging, and the feature set perturbation is added to generate accurate but diverse component PSVM. Meanwhile, in order to demonstrate their superiority, two algorithms named BRFS and BR for PSVM ensemble are proposed.

This paper is organized as follows. In Section 2, we discuss the principle of PSVM ensemble first, followed by the introduction of Relief feature selection algorithm. And PSVM ensemble based on feature selection is developed at last. Section 3 gives the experimental results and discussions. Finally, conclusive remarks are given in Section 4.

2 PSVM Ensemble Based on Feature Selection

Current ensemble learning algorithms include two classes, namely AdaBoost and Bagging, where Bagging is the most representative one. But it could hardly work on stable learners such as PSVM. A possible way to improve the performance of the

L. Jiao et al. (Eds.): ICNC 2006, Part I, LNCS 4221, pp. 135 – 138, 2006.

PSVM ensemble may be the employment of other kinds of perturbations [4]. In this paper, the combination of the perturbation on the training set and that on the input features is utilized in building ensemble of PSVM. Besides the random selection scheme, Relief technique is also used. The key idea of Relief is to assign attributes weights according to how well their values distinguish among samples that are near each other. The formula is as follows:

$$W_p^{i+1} = W_p^i - \mathrm{diff}(p,x,H)/m + \mathrm{diff}(p,x,M)/m \tag{1}$$

Here W is the weight value, H is its nearest hit and M is its nearest miss. $\mathrm{diff}(p, I_1, I_2)$ calculates the difference between sample I_1 and I_2 on the *p-th* attribute. And the difference function is (2). Fig. 1 summarizes the experimental procedure of the new methods.

$$\mathrm{diff}(p,x,x') = \begin{cases} \dfrac{|value(p,x) - value(p,x')|}{\max(p) - \min(p)} & x_p \neq x_p' \\ 0 & x_p = x_p' \end{cases} \tag{2}$$

```
Input: training set S, PSVM classifier, trials T, test set
  Process:
     for i:=1to T {
          Si =bootstrap sample from S
          // calculate the feature weights from the whole subset Si
             Wi =Ran-subspace(Si)   // for BRFS
           Wi =Relief(Si)          // for BR
        Train a PSVM classifier using the selected features in Si
                    }
     Give the final results via the majority voting
  Output: the final decisions of the test samples
```

Fig. 1. Procedure to BRFS and BR

3 Experiment Results and Discussion

To evaluate the performance of the proposed methods, four data sets are selected for binary classification from UCI machine learning repository. In all experiments described below we always use 30% as the training set while leaving the rest for testing and in the Relief algorithm m equals the number of the samples in the training set.

To measure the classification error, we performed 10 runs on each data set and the average error rates on the test set are listed in Table 1, where the values following "±" are standard deviations. Here the ν value is fixed to 50 and ensemble size is 20. The results confirm that BR is the best ensemble algorithm while the classification result of Bagging is a little worse than that of the single PSVM, which mainly lies in that Bagging could be more effective on the unstable learners.

Table 1. Comparison on error rates with ν =50 and ensemble size set to 20(%)

Data Set	Single	Bagging	BRFS	BR
Ionosphere	16.67 ± 0.46	17.10 ± 0.43	14.15 ± 0.64	13.76 ± 0.58
Breast-cancer-w	3.86 ± 0.24	4.29 ± 0.27	3.84 ± 0.21	3.73 ± 0.20
Heart	21.43 ± 1.07	21.43 ± 1.10	21.53 ± 0.97	18.27 ± 0.99
Credit-a	14.57 ± 1.04	15.77 ± 0.65	13.85 ± 0.57	13.85 ± 0.60

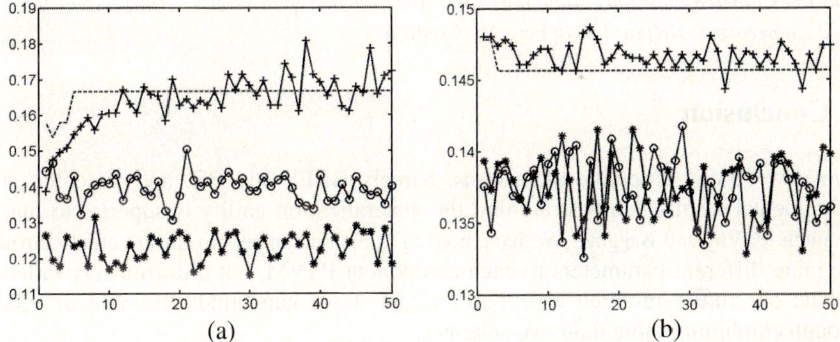

Fig. 2. Generalization Error with respect to ν *(dashed lines: Single PSVM; solid lines labeled with plus: Bagging; solid lines labeled with circles: BRFS; solid lines labeled with crosses: BR)* (a) Ionosphere, (b) Credit-a

In addition, two data sets, Ionosphere and Credit-a, are used to test the relationships between generalization error and parameter ν. Fig. 2 summarizes the results obtained by varying ν from 1 to 50 with the ensemble size fixed to 20. In this figure, the error rate of the single PSVM remains substantially unchanged independently of parameter ν. On Ionosphere, the error rate of Bagging is lower than that of the single method for small ν values, while fluctuates as ν increases. Both new methods, as expected, achieve the smaller test error with BR having the best result. On Credit-a, BRFS and BR get the similar error rate while they show better performance with respect to single PSVM and Bagging. Thus, it can be concluded that PSVM ensemble based on the feature selection can get a good classification result although the parameter is changed. In Fig. 3, we study the impact of the ensemble size on the results when the ν value is fixed to 50. Bagging shows the similar error rate with single PSVM for Credit-a data set and Bagging obtains the worst performance for Ionosphere. Fig. 3 also demonstrates that the error rate of BRFS or BR is lower than that of the single and bagged PSVM when the ensemble size is increased.

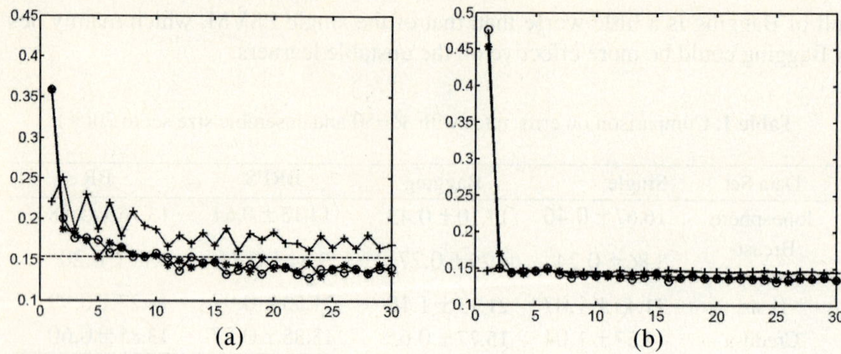

Fig. 3. Generalization Error with respect to the ensemble size *(dashed lines: Single PSVM; solid lines labeled with plus: Bagging; solid lines labeled with circles: BRFS; solid lines labeled with crosses: BR)* (a) Ionosphere, (b) Credit-a

4 Conclusion

Two new PSVM ensemble algorithms, namely BRFS and BR, are presented. The experimental results demonstrate that the generalization ability is superior to that of the single PSVM and Bagging. We have tried to inject randomness to linear kernels through assigning different parameters to each component PSVM, but unfortunately failed. It will be our future research aspect to explore if we can build ensemble of PSVM through combining more than two schemes.

References

1. Fung, G. and Mangasarian, O.L.: Proximal Support Vector Machine Classifiers. In: Provost, F. and Srikant, R. (Eds.): Proceedings of the Knowledge Discovery and Data Mining, (2001) 77-86
2. Breiman, L.: Bagging Predictors. Machine Learning, 24 (1996) 123-140
3. Kim, H.C, Pang, S.N, Je, H.M.: Pattern Classification using Support Vector Machine Ensemble. Proceedings of the 16th International Conference on Pattern Recognition, (2002) 160-163
4. Kononenko, I.: Estimating Attributes: Analysis and Extensions of Relief. In: Raedt, De. L and Bergadano, F. (Eds.): Proceedings of the 7th European Conference on Machine Learning, (1994) 171-182

Exact Semismooth Newton SVM*

Zhou Shui-Sheng[1], Liu Hong-Wei[1], Cui Jiang-Tao[2], and Zhou Li-Hua[2]

[1] School of Science, Xidian University, Xi'an 710071, P.R. China
{sszhou, hwliu}@mail.xidian.edu.cn

[2] School of Computer, Xidian University, Xi'an 710071, P.R. China
{cuijt, zhlh}@mti.xidian.edu.cn

Abstract. The Support vector machines can be posed as quadratic program problems in a variety of ways.This paper investigates a formulation using the two-norm for the misclassification error and appending a bias norm to objective function that leads to a positive definite quadratic program only with the nonnegative constraint under a duality construction. An unconstrained convex program problem, which minimizes a differentiable convex piecewise quadratic function, is proposed as the Lagrangian dual of the quadratic program. Then an exact semismooth Newton support vector machine (ESNSVM) is obtained to solve the program speedily. Some numerical experiments demonstrate that our algorithm is very efficient comparing with the similar algorithms such as LSVM.

1 Introduction

Based on the Vanpnik and Chervonkis' structural risk minimization principle, Support Vector Machine (SVM) is proposed as computationally powerful tools for supervised learning[1].

Note matrix $A = [x_1, x_2, \cdots, x_m]^T$, $y = [y_1, y_2, \cdots, y_m]^T$ and $D = diag(y)$, where $x_i \in R^n$ are training samples and y_i=1 or -1 are the target labels. The support vector machine for classification achieves the optimal classification function by solving the following quadric program with parameter $\nu > 0$:

$$\min_{u \in R^m} q(u) = \frac{1}{2} u^T DK(A,A)Du - e^T u \quad s.t. \;\; y^T u = 0, \; 0 \leq u \leq \nu. \tag{1}$$

where $K(x, y)$ is kernel function, $K(A, A) = (K(x_i, x_j))_{m \times n}$ is kernel matrix.

Currently the decomposition method is one of the major methods to solve SVM. The representative training algorithms is SMO[2,3].

Mangasarian etc in [4,5] use the two-norm for the misclassification error and appending a bias norm to objective function that leads to a positive definite quadratic program only with the nonnegative constraint under a duality construction:

$$\min_{0 \leq u \in R^m} q(u) = \frac{1}{2} u^T Qu - e^T u \tag{2}$$

* This work was supported in part by the National Natural Science Foundation of China under Grant No.60572150 and the Scientific Research Foundation of Weinan Normal Institute under Grant No.06YKS021.

L. Jiao et al. (Eds.): ICNC 2006, Part I, LNCS 4221, pp. 139–142, 2006.

where $Q = I/\nu + HH^T$ and $H = D[A \ \ -e]$ for linear problem, $Q = I/\nu + D[K(A,A) + e^T e]D$ for nonlinear problem. Then many type perturbation reformation of SVMs are proposed, such as Smooth Support Vector Machine(SSVM) [4], Lagrangian Support Vector Machines (LSVM)[5] etc.

In [5], Mangasarian and Musicant write the KKT optimality conditions of program (2) in the equivalent form $Qu - e = [(Qu - e) - \gamma u]_+$ and a very simple iterative scheme $u^{i+1} = Q^{-1}[e - ((Qu^i - e) - \gamma u^i)_+]$, which constitutes LSVM algorithm, is proposed. The linear convergence of LSVM is proved under some conditions.

All the algorithms above are solved a dual program with high dimension([2,3,5]) or solved a low dimension program approximately with smooth penalization skill [4], then the efficiency or the precision of the algorithm is limited.

In our paper [6], the Lagrangian dual technique is used to convert the Mangasarian's perturbation quadric program (2) to an unconstrained convex program with low dimension equivalently. The resulting unconstrained program (Eq. (3) in following) minimizes a differentiable convex piecewise quadratic function and a conjugate gradient support vector machine (CGSVM) is proposed to solve it. In this paper, following by the results in [6] and the definitions of semismooth in [7], we introduce an Exact Semismooth Newton Support Vector Machine (ESNSVM) to train SVM within finite number iterations, and the efficiency and the precision are improved also.

2 Exact Semismooth Newton SVM

Let $H = D[A \ \ -e]$ for linear problem and $H = D[G \ \ -e]$ for nonlinear problem, where $G \in R^{m \times l}(l \le m)$ satisfied $K(A,A) = GG^T$. In [6], we prove that the program (2) is equivalent to the following unconstrained minimization problem:

$$\min_x \frac{1}{2} x^T x + \frac{1}{2\nu}(e - Hx)^T (e - Hx)_+ \tag{3}$$

where $x \in R^{n+1}$ for linear problem or $x \in R^{l+1}$ for nonlinear problem, and satisfied $u = \nu(e - Hx)_+$.

The objective function of the program (3) is a continuous differentiable piecewise quadratic function and its gradient is $G(x) = x - \nu H^T(e - Hx)_+$. We can prove that $G(x)$ is strongly semismooth, and every $B \subset \partial G(x)$ is positive definite matrix, where $\partial G(x)$ is Clarke generalized Jacobian of $G(x)$. The unique minima solution of (3) is just corresponding to the solution of the system of the equations $G(x) = 0$, which is non-differentiable and semismooth piecewise linear equations. A method called Exact Semismooth Newton Support Vector Machine (ESNSVM) is proposed to solve it.

Algorithm 3.1 **ESNSVM**

0) (Initialization) For input data H and ν, let $x_0 \in R^{n+1}$ or R^{l+1}, $\varepsilon > 0$ be given. Calculate $g_0 = G(x_0)$. Set $k = 0$.

1) (Termination)If $\| g_k \| < \varepsilon$, end the algorithm.

2) (Direction generation) Otherwise, select $B_k \in \partial G(x_k)$, and calculate d_k by solving the Newton systems $B_k d = -g_k$.

3) (Exact line search) Choose t_k satisfies $t_k = argmin_{t>0} f(x_k + td_k)$.

4) (Update) Let $x_{x+1} = x_k + t_k d_k$, $g_{k+1} = G(x_{k+1})$ and $k := k + 1$. Go to 1).

Because the objective function is strict convex and the optimal solution is unique, the start point is not important. In this work, we set $x_0 = e$. The exact line search procedure in step 3) is similar with that proposed in [6]. The following convergence result can be gotten by the Theorem 2.4 in paper [7].

Theorem 1. *ESNSVM has a Q-quadratic rate of convergence when the Armijo line search technique is used in step 3).*

Furthermore when using the exact line search technique (see [6] in detail), a stronger convergence result is obtained. We know that Newton method with exact line search can find the minimum of the strict convex quadratic function in one iteration. Our objective function is a strict convex quadratic function in a polyhedron that enclosed by some hyperplanes $1 - H_i x = 0$. Simplified Theorem 3 of [7], we have:

Theorem 2. *The convergence of the ESNSVM is finite, i.e. for some $\bar{k}$, one will have $x_k = x^*$ for all $k > \bar{k}$.*

3 Numerical Implementation and Comparisons

In this section we compare the running times and classification correctness between ESNSVM and LSVM in linear case and nonlinear case. The source code of LSVM is obtained from the author's web site[8], and the program of ESNSVM is written by pure MATLAB language.

The first results in Table 1 are designed to compare the training time, iterations, and the training correctness between ESNSVM and LSVM for linear classification problem. Six datasets is available from the UCI Machine Learning Repository[9]. For LSVM, we use an optimality tolerance of 10^{-4} to determine when to terminate, and for ESNSVM the stopping criterion is $\varepsilon = 10^{-8}$. The results show that the training accuracies are almost the same but the training time of new method is fast than the training time of LSVM and the iterations of ESNSVM are very few.

The second experiment is designed to demonstrate the effectiveness of ESNSVM in solving nonlinear classification problems through the use of kernel functions. One highly nonlinearly separable but simple example is the checkerboard dataset[10]. The Gaussian kernel function $K(x, y) = \exp(-0.0011 \parallel x - y \parallel^2)$ is used. The first trial of the training set contains 1000 points obtained from [10] and the same as [5]. Total training time using ESNSVM is 6.41 seconds, and test set accuracy is 98.28% on test set. Only 81 iterations need. In [5] total time using LSVM was 2.85 hours with a 97.0% accuracy on the same test set. The second trial of the training set contains the 3000-point training set randomly sampled from the checkerboard. Total time using ESNSVM is 16.48 seconds with

Table 1. ESNSVM compared with LSVM([5]) on six UCI datasets in linear classification

dataset	Algorithm	Time(sec)	# SVs	Iterations	Correctness
Liver disorder(345x6, ν=1)	LSVM	0. 0203	332	112	70.15%
	ESNSVM	**0.0031**	332	**4**	70.15%
Pima diabetes(768x8, ν=1)	LSVM	0.0351	708	117	78.39%
	ESNSVM	**0.0079**	708	**4**	78.39%
Tic tac toe(958x9, ν=1)	LSVM	0.0297	941	90	98.33%
	ESNSVM	**0.0062**	941	**3**	98.33%
Ionosphere(351x34, ν=4)	LSVM	0.0313	152	118	93.73%
	ESNSVM	**0.0125**	**151**	**6**	93.73%
Adult(30162x14, ν=1)	LSVM	3.0205	20926	318	83.78%
	ESNSVM	**0.5108**	**20200**	**7**	83.78%
Connect-4(61108x42, ν=1)	LSVM	7.7452	60767	107	73.00%
	ESNSVM	**1.0843**	60767	**4**	73.00%

a 99.57% accuracy on test set and only 102 iterations need, but LSVM cannot fulfill the experiment on my computer because the memory is overflowing.

4 Conclusion

We propose an ESNSVM in this work. It only needs to minimize a low-dimension differentiable unconstrained convex piecewise quadratic function. It has many advantages, such as only solving a low-dimension problem, speedy convergence rate, less training time, ability to solve very massive problems etc. Those are illustrated by experiments in section 3.

References

1. V. N. Vapnik: The Nature of Statistical Learning Theory, NY: Springer-Verlag, 2000.
2. J. C. Platt: Fast Training of Support Vector Machines using Sequential Minimal Optimization. In B. Scholkopf et al.(ed.), Advances in Kernel Method-Support Vector Learning, Cam-bridge, MIT Press,1999, 185-208.
3. S. Keerthi, S. Shevade, C. Bhattacharyya et al: Improvements to Platt's SMO Algorithm for SVM Classifier Design. Neural Computation, 2001,**13**: 637-649.
4. Yuh-Jye Lee, O. L. Mangasarian: SSVM: A smooth support vector machine. Computational Optimization and Applications, 2001,**20**(1):5-22.
5. O. L. Mangasarian, D.R. Musicant: Lagrangian Support Vector Machines. Journal of Machine Learning Research **1**, March 2001, 161-177.
6. Zhou Shui-sheng, Zhou Li-hua: Conjugate Gradients support vector machine. Pattern Recognition and Artificial Intelligence, 2006,**19**(2):129-136.
7. J. Sun: On piecewise quadratic Newton and trust region problems. Mathematical programming, 1997, **76**,451-467.
8. D. R. Musicant, O. L. Managsarian: LSVM: Lagrangian Support Vector Machine. 2000, http://www.cs.wisc.edu/dmi/svm/.
9. P. M. Murphy, D. W. Aha: UCI repository of machine learning databases, 1992. www.ics.uci.edu/~mlearn/MLRepository.html.
10. T. K. Ho, E. M. Kleinberg: Checkerboard dataset, 1996. http://www.cs.wisc.edu/math-prog/mpml.html

General Kernel Optimization Model Based on Kernel Fisher Criterion

Bo Chen, Hongwei Liu, and Zheng Bao

National Lab of Radar Signal Processing, Xidian University
Xi'an, Shaanxi, 710071, P.R. China
bchen@mail.xidian.edu.cn

Abstract. In this paper a general kernel optimization model based on kernel Fisher criterion (GKOM) is presented. Via a data-dependent kernel function and maximizing the kernel Fisher criterion, the combination coefficients of different kernels can be learned adaptive to the input data. Finally positive empirical results on benchmark datasets are reported.

1 Introduction

The kernel K is the dot product of x in the high dimension feature space, F, where $\Phi(\cdot)$ is a mapping function

$$K(x, y) = (\Phi(x) \cdot \Phi(y)) \tag{1}$$

Xiong et al. [1] proposed an alternate method for optimizing the kernel function by maximizing a class separability criterion in the empirical feature space using the data-dependent model [2]. However the method is based on the single kernel, which limits the performance of kernel optimization. In this paper we extend the method to give a general kernel optimization model based on kernel Fisher.

2 General Kernel Optimization Model Based on Kernel Fisher

In [2] a data-dependent kernel is employed as the objective kernel to be optimized

$$k(x, y) = q(x)q(y)k_0(x, y), \tag{2}$$

where $k_0(x, y)$, the basic kernel, is an ordinary kernel such as a Gaussian kernel, and $q(\cdot)$ is a factor function of the form

$$q(x) = \alpha_0 + \sum_{i=1}^{n} \alpha_i k_1(x, a_i), \tag{3}$$

where $k_1(x, a_i) = \exp(-\|x - a_i\|^2 / \sigma_1^2)$, $\{a_i\}$, called the "empirical cores," and α_i's are the combination coefficients which need normalizing.

Then according to Theorem1 in [2], Fisher criterion in the kernel-induced feature space can be represented as the following

L. Jiao et al. (Eds.): ICNC 2006, Part I, LNCS 4221, pp. 143 – 146, 2006.

$$J(\boldsymbol{\alpha}) = \frac{tr(S_b)}{tr(S_w)} = \frac{1_m^T B 1_m}{1_m^T W 1_m} = \frac{\mathbf{q}^T B_0 \mathbf{q}}{\mathbf{q}^T W_0 \mathbf{q}}, \tag{4}$$

where 1_m is the vector of ones of the length m, $\boldsymbol{\alpha} = (\alpha_0, \alpha_1, ..., \alpha_n)^T$ and $\mathbf{q} = (q(x_1), q(x_2), ..., q(x_m))^T$, B and W are respectively between-class and within-class kernel scatter matrix.

So an updating equation for maximizing the class separability J through the standard gradient approach is given as the following

$$\boldsymbol{\alpha}_{(n+1)} = \boldsymbol{\alpha}_{(n)} + \eta\left(\frac{K_1^T B_0 K_1}{\mathbf{q}^T W_0 \mathbf{q}} - J(\boldsymbol{\alpha}_{(n)}) \frac{K_1^T W_0 K_1}{\mathbf{q}^T W_0 \mathbf{q}}\right)\boldsymbol{\alpha}_{(n)}, \tag{5}$$

where $\eta(t) = \eta_0 (1 - t/T)$, in which η_0 is the initial learning rate, T denotes iteration number, and t represents the current iteration number. The detail algorithm can be found in [2].

It is evident that the method in [2] is effective to improve the performance of the kernel machines. Also the experiment results in [2] prove its validity. However the kernel optimization procedure is under a given kernel model, so we have to optimize the kernel based on the given embedding space. The optimization capability will be limited. Below we generalize the kernel optimization method based on single kernel (KOS) to propose a general kernel optimization model based on kernel Fisher.

Since the sum of multiple kernels is still a kernel, (2) can be represented as

$$K = \sum_{i=1}^{L} Q_i K_0^{(i)} Q_i, \tag{6}$$

where L is the number of selected kernels, $K_0^{(i)}$ is the i-th basic kernel, Q_i is the factor matrix corresponding to $K_0^{(i)}$. Therefore B and W are modified as

$$B^{\text{model}} = \sum_{i=1}^{L} B_i, \quad W^{\text{model}} = \sum_{i=1}^{L} W_i, \tag{7}$$

According to (4), the general kernel quality function, J^{model} can be written as

$$J^{\text{model}} = (1_m^T B^{\text{model}} 1_m)/(1_m^T W^{\text{model}} 1_m) = \sum_{i=1}^{L} q_i^T B_0^{(i)} q_i \Big/ \sum_{i=1}^{L} q_i^T W_0^{(i)} q_i, \tag{8}$$

where $B_0^{(i)}$ and $W_0^{(i)}$, the between-class and the within-class kernel scatter matrices, correspond to the basic kernel $K_0^{(i)}$. $\alpha^{(i)}$ is the combination coefficient vector corresponding to $K_0^{(i)}$. Therefore (8) can be derived as

$$J^{\text{model}} = (\mathbf{q}^T B_0^{\text{model}} \mathbf{q})/(\mathbf{q}^T W_0^{\text{model}} \mathbf{q}) \tag{9}$$

where $B_0^{\text{model}} = diag(B_0^{(1)}, B_0^{(2)}, ..., B_0^{(L)})$, $W_0^{\text{model}} = diag(W_0^{(1)}, W_0^{(2)}, ..., W_0^{(L)})$, $\mathbf{q} = \mathbf{q}(\alpha^{\text{model}}) = [q_1 \ q_2 \ \cdots \ q_L] = K_1^{\text{model}} \alpha^{\text{model}}$. K_1^{model} is a $Lm \times L(n+1)$ matrix, α^{model} is a vector of length $L(n+1)$.

Apparently the form of (9) is the same as that of the right side of (4), so through (5) our result can also be given by the following

$$\alpha_{(n+1)}^{\text{model}} = \alpha_{(n)}^{\text{model}} + \eta\left(\frac{(K_1^{\text{model}})^T B_0^{\text{model}} K_1^{\text{model}}}{\mathbf{q}(\alpha_{(n)}^{\text{model}})^T W_0^{\text{model}} \mathbf{q}(\alpha_{(n)}^{\text{model}})} - J^{\text{model}}(\alpha_{(n)}^{\text{model}}) \frac{(K_1^{\text{model}})^T W_0^{\text{model}} K_1^{\text{model}}}{\mathbf{q}(\alpha_{(n)}^{\text{model}})^T W_0^{\text{model}} \mathbf{q}(\alpha_{(n)}^{\text{model}})}\right)\alpha_{(n)}^{\text{model}}. \tag{10}$$

From (9) we can find that the KOS method just is a special case of our model (when $L=1$). Therefore, the proposed kernel optimization model based on kernel Fisher criterion is of $O(NL^2n^2)$ computational complexity where N denotes the prespecified iteration number and n stands for the data size.

3 Experimental Results on Benchmark Data

Our method has been also tested on the other four benchmark data, namely, the Ionosphere, Pima Indians Diabetes, Liver disorder, Wisconsin Breast Cancer (Available from http://www.ics.uci.edu/mlearn). Kernel optimization methods were applied to KPCA. Linear SVM classifier was utilized to evaluate the classification performances. We used Gaussian kernel $K_1(\mathrm{x},\mathrm{y})=\exp\left(-\|x-y\|^2/\sigma\right)$, polynomial kernel $K_2(\mathrm{x},\mathrm{y})=((x^T\cdot y)+1)^p$ and linear kernel $K_3(\mathrm{x},\mathrm{y})=x^T\cdot y$ as initial basic kernel. And all kernels were normalized. Firstly kernel parameters for the three kernels of KPCA without kernel optimization were respectively selected by 10-fold CV. Then the chosen kernels were applied as the basis ones in (2). σ_1's in (3) were also selected using 10-fold CV. 20 local centers were selected as the empirical core set $\{a_i\}$. η_0 was set to 0.08 and T was set to 400. Meanwhile the procedure of determining the parameters of KOS was the same as GKOM.

Experimental results were summarized in Table 1. Evidently GKOM can further improve the classification performance and at least as the same as the KOS method. The combination coefficients of three kernels were also been illustrated in Fig. 1. We find that the combination coefficients of GKOM are dependent on the classification performance of the corresponding kernel in KOS. The better the kernels can work after the optimization of KOS, the larger the combination coefficients of GKOM are. Apparently GKOM can automatically combine three fixed parameter kernels.

Table 1. The comparison of recognition rates of different methods in different experiments. K_1, K_2 and K_3 respectively correspond to Gaussian, polynomial and linear kernels.

Data sets		K_1	K_2	K_3
	KPCA	88.96%	90.45%	88.58%
BCW	KPCA with KOS	88.96%	96.94%	97.1%
	KPCA with GKOM		97.33%	
	KPCA	73.72%	64.15%	63.48%
Pima	KPCA with KOS	73.72%	66.21%	64.63%
	KPCA with GKOM		74.10%	
	KPCA	71.19%	69.47%	66.19%
Liver	KPCA with KOS	74.67%	73.36%	73.47%
	KPCA with GKOM		75.17%	
	KPCA	93.11%	93.11%	89.73%
Ionosphere	KPCA with KOS	93.11%	93.55%	89.38%
	KPCA with GKOM		93.55%	

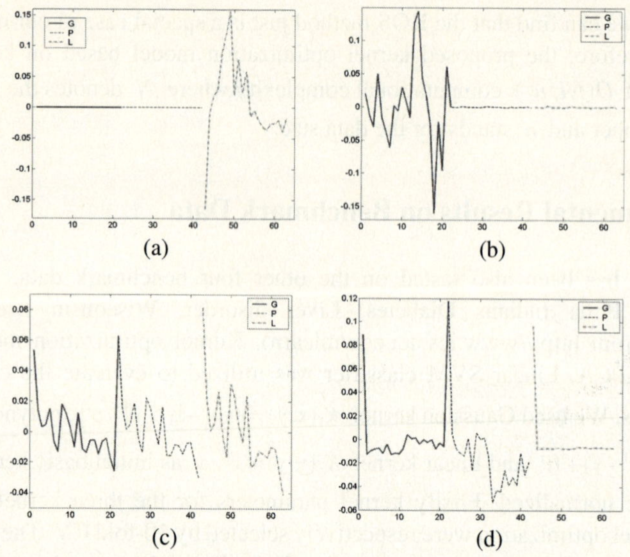

Fig. 1. The combination coefficients corresponding to four data sets. (a) BCW ; (b) Pima; (c) Liver; (d) Ionosphere.

4 Conclusion

In this paper a general kernel optimization model based on kernel Fisher criterion (GKOM) is proposed, that makes use of a data-dependent kernel structure. The KOS by [2] just is the special case of GKOM when $L = 1$. The proposed method maximizes the kernel Fisher criterion to automatically determine the combination coefficients of different kernels, which includes selecting better kernels and combining several kernels with similar classification performance. The learned kernel possesses better flexibility and can be more adaptive to the input data. The positive experimental results prove the power of the proposed method.

Acknowledgement

This work is partially supported by the NSFC under grant 60302009.

References

1. Xiong, H. L., Swamy, M. N. S., Ahmad, M. O.: Optimizing the kernel in the empirical feature space. IEEE Trans. Neural Networks, Vol. 16, No. 2, (2005) 460-474.
2. Amari S. and Wu, S.: Improving support vector machine classifiers by modifying kernel functions. Neural Networks, Vol. 12, No. 6, (1999) 783–789.

A Novel Multiple Support Vector Machines Architecture for Chaotic Time Series Prediction

Jian-sheng Qian, Jian Cheng, and Yi-nan Guo

School of Information and Electrical Engineering, China University of Mining and Technology, 221008, Xu Zhou, China
chjpaper@126.com

Abstract. Inspired by the so-called "divide-and-conquer" principle that is often used to attack a complex problem by dividing it into simpler problems, a two-stage multiple support vector machines (SVMs) architecture is proposed to improve its prediction accuracy and generalization performance for chaotic time series prediction. Fuzzy C-means (FCM) clustering algorithm is adopted in the first stage to partition the input dataset into several subnets. Then, in the second stage, multiple SVMs that best fit partitioned subsets are constructed by Gaussian radial basis function kernel and the optimal free parameters of SVMs. All the models are evaluated by Mackey-Glass chaotic time series and used for coal mine gas concentration in the experiment. The simulation shows that the multiple SVMs achieve significant improvement in the generalization performance in comparison with the single SVM model. In addition, the multiple SVMs also converges faster and uses fewer support vectors.

1 Introduction

Successful time series prediction is a major goal in many areas of research, however, most practical time series are of nonlinear and chaotic nature that makes conventional, linear prediction methods inapplicable. Although the neural networks (NN) is developed in chaotic time series prediction, some inherent drawbacks, e.g., the multiple local minima problem, the choice of the number of hidden units and the danger of over fitting, etc., would make it difficult to put the NN into some practice.

Recently, support vector machine (SVM) has been proposed as a novel technique in time series prediction [1]. SVM is a new approach of pattern recognition established on the unique theory of the structural risk minimization principle to estimate a function by minimizing an upper bound of the generalization error via the kernel functions and the sparsity of the solution [2]. SVM usually achieves higher generalization performance than traditional NN that implement the empirical risk minimization principle in solving many machine learning problems. Another key characteristic of SVM is that training SVM is equivalent to solving a linearly constrained quadratic programming problem so that the solution of SVM is always unique and globally optimal. One disadvantage of SVM is that the training time scales somewhere between quadratic and cubic with respect to the number of training samples [3]. So a large amount of computation time will be involved when SVM is applied for solving

L. Jiao et al. (Eds.): ICNC 2006, Part I, LNCS 4221, pp. 147 – 156, 2006.

large-size problems. Furthermore, using a single model to learn the data is somewhat mismatch as there are different noise levels in different input regions.

Inspired by the so-called "divide-and-conquer" principle that is often used to attack a complex problem by dividing it into simpler problems [4], a potential solution to the above problems is proposed by using a multiple SVMs architecture based on fuzzy c-means clustering (FCM) algorithm. The feasibility of applying the multiple SVMs in Mackey-Glass chaotic time series is first examined by comparing it with the single SVM, then it is used to forecast the coal mine gas concentration. The experimental results show that the multiple SVMs based on FCM outperforms the single SVM model in generalization performance and consumed CPU time for chaotic time series prediction.

This paper is structured as follows: Section 2 provides a brief introduction to SVM in regression approximation. Section 3 presents FCM clustering algorithm. Structure and algorithm of the multiple SVMs are given in Section 4. Section 5 presents the results and discussions on the experimental validation. Finally, some concluding remarks are drawn in Section 6.

2 SVM for Regression Estimation

Consider a given training set $\{x_i, y_i : i = 1,\cdots,l\}$ ($x_i \in X \subseteq R^n$, $y_i \in Y \subseteq R$, l is the total number of training samples) randomly and independently generated from an unknown function, SVM approximates the function using the following form:

$$f(x) = w \cdot \phi(x) + b\,. \tag{1}$$

Where $\phi(x)$ represents the high-dimensional feature space which is nonlinearly mapped from the input space x, so nonlinear function estimation in input space becomes linear function estimation in feature space. By Structure Risk Minimum principle, the coefficients w and b are estimated by minimizing the regularized risk function:

$$\text{minimize } \frac{1}{2}\|w\|^2 + C \cdot R_{emp}\,. \tag{2}$$

Where, the first term $\|w\|^2$ is called the regularized term, the second term $R_{emp} = (1/l)\sum_{i=1}^{l} L(y_i, f(x_i))$ is the empirical risk, and C is a regularization parameter. $L(y_i, f(x))$ is the loss function as following:

$$L(y_i, f(x_i)) = \begin{cases} |y_i - f(x_i)| - \varepsilon, & |y_i - f(x_i)| \geq \varepsilon \\ 0, & otherwise \end{cases}\,. \tag{3}$$

Different SVM can be constructed by selecting different ε-insensitive loss function. The optimization goal of standard SVM is formulated as

$$\text{minimize } \frac{1}{2}\|w\|^2 + C \cdot \sum_{i=1}^{l} (\xi_i + \xi_i^*) . \tag{4}$$

subject to

$$\begin{cases} y_i - w \cdot \phi(x_i) - b \le \varepsilon + \xi_i \\ w \cdot \phi(x_i) + b - y_i \le \varepsilon + \xi_i^* \\ \xi_i^*, \xi_i \ge 0, \quad i = 1, \cdots, l \end{cases} \tag{5}$$

Introducing Lagrange multipliers:

$$\begin{aligned} L(w, b, \xi_i, \xi_i^*, a, a^*, \beta, \beta^*) &= \frac{1}{2} w \cdot w + C \sum_{i=1}^{l} (\xi_i + \xi_i^*) \\ &- \sum_{i=1}^{l} a_i ((w \cdot \phi(x_i)) + b - y_i + \varepsilon + \xi_i) \\ &- \sum_{i=1}^{l} a_i^* (y_i - (w \cdot \phi(x_i)) - b + \varepsilon + \xi_i^*) - \sum_{i=1}^{l} (\beta_i \xi_i + \beta_i^* \xi_i^*) \end{aligned} \tag{6}$$

In Eq.(6), a_i and a_i^* are Lagrange multipliers, they satisfy $a_i, a_i^* \ge 0$ and $\beta_i, \beta_i^* \ge 0, i = 1, \cdots, l$, and they are obtained by maximizing the dual function of Eq.(4), which has the following explicit form:

$$\begin{aligned} W(a, a^*) &= \sum_{i=1}^{l} y_i (a_i - a_i^*) - \varepsilon \sum_{i=1}^{l} (a_i + a_i^*) \\ &- \frac{1}{2} \sum_{i=1}^{l} \sum_{j=1}^{l} (a_i - a_i^*)(a_j - a_j^*) K(x_i, x_j) \end{aligned} \tag{7}$$

with the following constraints:

$$\sum_{i=1}^{l} (a_i - a_i^*) = 0 , \; 0 \le a_i, a_i^* \le C , \; i = 1, \cdots, l . \tag{8}$$

Finally, the decision function (1) has the following explicit form:

$$f(x) = \sum_{i=1}^{l} (a_i - a_i^*) K(x_i, x) + b . \tag{9}$$

$K(x_i, x_j)$ is defined as the kernel function. The value of the kernel is equal to the inner product of two vectors x_i and x_j in the feature space $\phi(x_i)$ and $\phi(x_j)$, that is, $K(x_i, x_j) = \phi(x_i) \cdot \phi(x_j)$. The elegance of using the kernel function is that one can deal with feature spaces of arbitrary dimensionality without having to compute the map $\phi(x)$ explicitly. Any function that satisfies Mercer's condition [3] can be used as the kernel function. Common examples of kernel function are the polynomial kernel

$K(x_i, x_j) = (x_i x_j + 1)^d$ and the Gaussian radial basis function (RBF) kernel $K(x_i, x_j) = \exp(-(x_i - x_j)^2 / \sigma^2)$, where d and σ^2 are the kernel parameters.

3 Fuzzy C-Means Clustering Algorithm

Fuzzy C-means (FCM) clustering algorithm is one of the most important and popular fuzzy clustering algorithm. Its main idea is to obtain a partition that minimizes the within-cluster scatter or maximizes the between-cluster scatter. The FCM was proposed first by Dunn in 1973, and generalized by Bezdek [5]. Input data set $X = \{x_1, x_2, \cdots, x_l\}$ is classified into c clusters. Each sample data x_j includes n features, i.e. $x_j = \{x_{j1}, x_{j2}, \cdots, x_{jn}\}$, where x_j in the set X (X is a n-dimensional space). Because these features all can have different units in general, each of the features has to be normalized to unified scale before classification.

Objective function approach is adopted for classifying l data points to c clusters. In this approach, each cluster is considered as one hyper spherical shape with hypothetical geometric cluster center. The main aim of the objective function is to minimize the Euclidian distance between each data point within the cluster, and maximize the Euclidian distance between other cluster centers. The classification matrix $U = \{u_{ij}\}$ is a fuzzy matrix, also u_{ij} should satisfy the following condition:

$$\sum_{i=1}^{c} u_{ij} = 1, \quad \forall j,\ u_{ij} \in [0, \ 1],\ j = 1,2,\cdots,l,\ i = 1,2,\cdots,c \tag{10}$$

where, u_{ij} is the degree of membership of x_j in the i-th cluster. The object function is defined as following:

$$J_m(U,Z) = \sum_{i=1}^{c} J_i = \sum_{i=1}^{c}\sum_{j=1}^{l} u_{ij}^m d_{ij}^2 \tag{11}$$

where, $Z = \{z_1, z_2, \cdots, z_c\}$ is the cluster centers, $d_{ij} = \|z_i - x_j\|$ is the distance between x_j and the i-th cluster center z_i, u_{ij} is the membership function value of j-th sample data belongs to the i-th cluster, $m \in [1, \ \infty)$ is the weighted exponent on each fuzzy membership. They are defined as following:

$$z_i^{(k)} = \left(\sum_{j=1}^{l} u_{ij}^m x_j\right) \Big/ \left(\sum_{j=1}^{l} u_{ij}^m\right) \tag{12}$$

$$u_{ij} = 1 \Big/ \left(\sum_{k=1}^{c} (d_{ij}/d_{kj})^{2/(m-1)}\right) \tag{13}$$

If $d_{ij} = 0$, then $u_{ij} = 1$, and $u_{kj} = 0$, ($k = i$).

Because we do not have a clear idea on how many cluster should be for a given data set. The initial clusters are chosen randomly. The algorithm, described in [5], can be used to minimize the objective function shown in formula (11).

4 The Multiple Support Vector Machines

4.1 Constructing the Multiple SVMs

The basic idea underlying the multiple SVMs is to use FCM to partition the whole input space into several subsets and to use SVM for solving these partitioned subsets. Fig. 1 shows how the multiple SVMs is built.

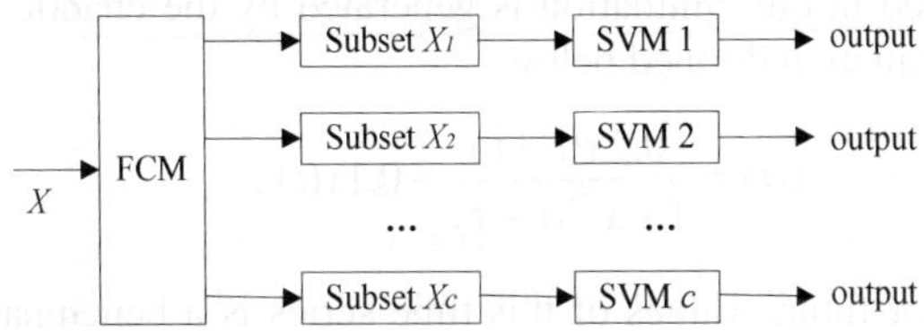

Fig. 1. A generalized two-stage multiple SVMs architecture

As illustrated in Fig.1, for a large input data set X, FCM plays a "divide" role to cluster X into c subsets, and then each SVM plays a "conquer" role to tackle each partitioned subset respectively. As a result, the input data set is partitioned into c non-overlapping subsets X_1, X_2, …, X_c, where $X_i \bigcap X_j = \phi$ and $\bigcup_{i=1}^{c} X_i = X$ ($i, j = 1,2,\cdots,c$, $i \neq j$, ϕ denotes the null set). The aforementioned procedures are applied in the input space until the number of training data points in the following partitioned subsets is appropriate to construct and train the SVM respectively.

4.2 The Multiple SVMs Learning Algorithm

A learning algorithm for the proposed architecture is outlined as follows:

(1) The training, validating and the checking data set are chosen in the chaotic time series for the input of the multiple SVMs. The input data is smoothed and normalized before simulation.
(2) Then the training data set is clustered into c subsets (X_1, X_2, …, X_c) with corresponding clustering center z_i ($i = 1,2,\cdots,c$) by using FCM through several trials.
(3) The individual SVM is built and trained according to the subset respectively.
(4) Choose the most adequate SVM that produces the smallest error on the subset respectively.

For an unknown data point X_{new} in checking, it is first classified into one of the partitioned subsets by following algorithm:

Calculate the distance between X_{new} and the i-th subset clustering center z_i:

$$d_i = \left\| X_{new} - z_i \right\|, \; i = 1,2,\cdots,c . \tag{14}$$

The results of the single SVM and the multiple SVMs models are given in Table 2 (S-SVM and M-SVM represent the single SVM and the multiple SVMs respectively). Comparing the results of the multiple SVMs with the single SVM model, it can be observed that the multiple SVMs achieve a much smaller *RMSE* than the single SVM model. In addition, the consumed CPU time and the number of converged support vectors are less for in the multiple SVMs than the sum of all single SVM model.

Table 2. The converged *RMSE* of the model in Mackey-Glass chaotic time series

RMSE	Model						
	S-SVM	SVM 1	SVM 2	SVM 3	SVM 4	SVM 5	M-SVMs
Training	0.0418	0.0178	0.0151	0.0264	0.0342	0.0172	0.0232
Checking	0.0404	0.0173	0.0130	0.0269	0.0295	0.0172	0.0167

5.2 Coal Mine Gas Concentration Prediction

The gas is one of most important factors that endanger the produce in coal mines. It has very highly social and economic benefits to strengthen the forecast and control over the coal mine gas concentration. From the coal mine, 1500 samples are collected from online sensor underground after eliminating abnormal data and normalizing all samples. The first 1000 samples are used for training the model, while the remaining 500 samples used for verifying the identified model only.

Through several trials, the input data are constructed by the values $D=6$ and $\Delta=P=2$ in the simulation, then the result of FCM and the parameters of every SVM are obtained. Where the number of the clusters c is 10 (the number of the subsets in input space) and the optimal values of ε, σ^2 (the Gaussian RBF kernel function) and C are, respectively, fixed at 0.03, 0.01 and 10.

Fig.4 illustrate the actual (the solid line) and the predicted values (the thick solid line and the dotted line represent the output of the S-SVM and M-SVMs respectively) in checking set. Obviously, the multiple SVMs predict more closely to the actual values than the single SVM in most of the checking data set. The results of the multiple SVMs and the single SVM model are given in Table 3. The table shows that the multiple SVMs achieves better performance than the single SVM. The converged *RMSE* in the multiple SVMs is smaller than the single SVM by using the Gaussian RBF kernel function.

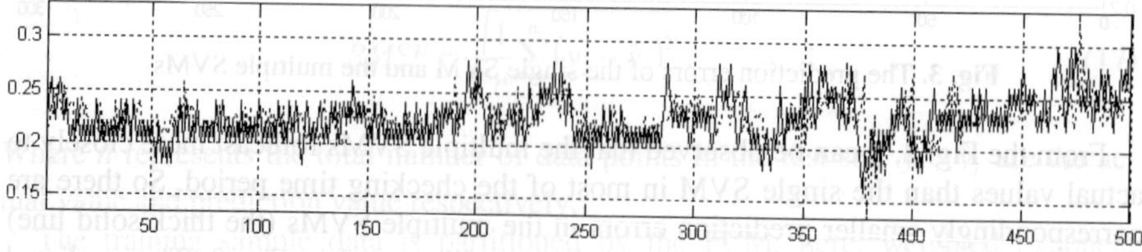

Fig. 4. The predicted and actual value in coal mine gas concentration

Table 3. The converged *RMSE* of the models in coal mine gas concentration prediction

RMSE	Models	
	The Single SVM	The Multiple SVM
Training	0.0437	0.0217
Checking	0.0540	0.0223

6 Conclusions

The multiple SVMs model is developed by compounding SVMs with FCM in a two-stage architecture based on the principle of "divide-and-conquer", In the first stage, FCM is used to is used to cluster a given input data set into several individual subsets. Then, at the second stage, the corresponding SVM is used to produce the output. There are several advantages in the multiple SVMs. First, it achieves high prediction performance because each input subset is learned by the SVM separately. Due to the number of training data points getting smaller in each SVM, the convergence speed of SVM is largely increased. Second, the multiple SVMs converges to fewer support vectors. Thus, the solution can be represented more sparsely and simply. The multiple SVMs model has been evaluated by the checking data set to use for prediction of Mackey-Glass chaotic time series and coal mine gas concentration. Its superiority is demonstrated by comparing it with the single SVM model. All the simulation results shows that the multiple SVMs model is more effective and efficient in predicting chaotic time series than the single SVM model.

Although this paper shows the effectiveness of the multiple SVMs model, there are more issues need to be investigated. Firstly, how to ascertain the number of the subsets in input space affects deeply the performance of the whole model. The other methods to partition the input space should be investigated in future work. Secondly, how to determine the optimal parameters of SVM is an important issue needs to be research. Finally, in this paper only the Gaussian RBF kernel function is investigated. Future work needs to explore more useful kernel functions for improving the performance of the multiple SVMs.

Acknowledgements

This research is supported by National Natural Science Foundation of China under grant 70533050, Postdoctoral Science Foundation of China under grant 2005037225, Postdoctoral Science Foundation of Jiangsu Province under grant [2004]300 and Young Science Foundation of CUMT under grant OC4465.

References

1. L. J. Cao, F. E. H. Tay, Support Vector Machine With Adaptive Parameters in Financial Time Series Forecasting, IEEE Transactions on Neural Networks, Vol. 14, No. 6, (2003) 1506-1518
2. V. N. Vapnik, An Overview of Statistical Learning Theory, IEEE Transactions Neural Networks, Vol.10, No.5, (1999)988-999

3. N. Cristianini, J. S. Taylor, An Introduction to Support Vector Machines: and Other Kernel-based Learning Methods, Cambridge University Press, New York (2000)
4. R. L. Milidiu, R. J. Machado, R. P. Rentera, Time-series Forecasting Through Wavelets Transformation and a Mixture of Expert Models, Neurocomputing, Vol. 20 (1999) 145-146
5. J C. Bezdek, Pattern Recognition with Fuzzy Objective Function Algorithms. New York: Plenum Press (1981)
6. M. C. Mackey, L. Glass. Oscillation and Chaos in Physiological Control System, Science, Vol.197, (1977)287-289
7. T. D. Sanger. A Tree-structured Adaptive Network for Function Approximation in High-dimensional Spaces, IEEE Transaction on Neural Networks, vol. 2, No. 2 (1991)285-293
8. J.-S.R Jang, ANFIS: Adaptive-Network-based Fuzzy Inference System, IEEE Transactions on System, Man and Cybernetics, Vol. 23, No. 3 (1993) 665-685
9. W. Wan, K. Hirasawa, J. Hu, Relation between Weight Initialization of Neural networks and Pruning Algorithms Case Study on Mackey-Glass Time Series, Proceedings of the International Joint Conference on Neural Networks, (2001)1750-1755
10. K. J. Kim, Financial Time Series Forecasting Using Support Vector Machines, Neurocomputing, Vol. 55 (2003)307-319

Robust LS-SVM Regression Using Fuzzy C-Means Clustering

Jooyong Shim[1], Changha Hwang[2,*] and Sungkyun Nau[3]

[1] Department of Applied Statistics, Catholic University of Daegu, Kyungbuk 702-701, South Korea
jyshim@cu.ac.kr

[2] Division of Information and Computer Science, Dankook University, Yongsan Seoul, 140-714, South Korea
chwang@dankook.ac.kr

[3] Division of Information and Computer Science, Dankook University, Yongsan Seoul, 140-714, South Korea
nuhsam@nate.com

Abstract. The least squares support vector machine(LS-SVM) is a widely applicable and useful machine learning technique for classification and regression. The solution of LS-SVM is easily obtained from the linear Karush-Kuhn-Tucker conditions instead of a quadratic programming problem of SVM. However, LS-SVM is less robust due to the assumption of the errors and the use of a squared loss function. In this paper we propose a robust LS-SVM regression method which imposes the robustness on the estimation of LS-SVM regression by assigning weight to each data point, which represents the membership degree to cluster. In the numerical studies, the robust LS-SVM regression is compared with the ordinary LS-SVM regression.

1 Introduction

The support vector machine(SVM) introduced by Vapnik[9] for classification and function estimation is an important methodology in the area of neural networks and nonlinear modeling. The solution of SVM is characterized by convex optimization problem with determination of several tuning parameters. Typically, a convex quadratic programming(QP) is solved in the dual space in order to determine SVM model. The formulation of the optimization problem in the primal space associated with QP problem involves the inequality constraints. The introductions and overviews of recent developments of SVM regression can be found in Vapnik[9], Kecman[4], and Wang[10]. The least squares support vector machine(LS-SVM), a modified version of SVM in a least squares sense, has been proposed for classification and regression by Suykens and Vanderwalle[8]. In LS-SVM the solution is obtained from the linear Karush-Kuhn-Tucker conditions instead of a quadratic programming(QP) problem. A drawback of LS-SVM is

* Corresponding Author.

L. Jiao et al. (Eds.): ICNC 2006, Part I, LNCS 4221, pp. 157–166, 2006.

that the use of a squared loss function without regularization might lead to less robust estimates with respect to outliers on the data or when the underlying assumption of a Normal distribution for errors is not realistic.

In this paper we propose the robust version of LS-SVM by assigning a weight to each data point, which represents fuzzy membership degree of the cluster whose prototype is the curve expressed by regression function. This paper is organized as follows. In Section 2 and 3 we give brief reviews of LS-SVM and fuzzy c-means(FCM) clustering, respectively. In Section 4 we define the objective functions of the robust LS-SVM and the weight which represent the membership degree. And we give an iterative scheme to solve minimization problem. In Section 5 we illustrate the new algorithm with examples to compare its performance with LS–SVM. In Section 6 we give conclusions.

2 LS-SVM

Given a training data set of n points $\{\boldsymbol{x}_i, y_i\}_{i=1}^n$ with each input $\boldsymbol{x}_i \in \boldsymbol{R}^d$ and output $y_i \in \boldsymbol{R}$, we consider the following optimization problem in primal weight space:

$$L(\boldsymbol{w}, b, \boldsymbol{e}) = \frac{1}{2}\boldsymbol{w}'\boldsymbol{w} + \frac{\gamma}{2}\sum_{i=1}^{n} e_i^2 \tag{1}$$

subject to equality constraints

$$y_i = \boldsymbol{w}'\boldsymbol{\Phi}(\boldsymbol{x}_i) + b + e_i, \; i = 1, \cdots, n \tag{2}$$

with $\boldsymbol{\Phi}(\cdot) : \boldsymbol{R}^d \rightarrow \boldsymbol{R}^{d_f}$ a function which maps the input space into a higher dimensional(possibly infinite dimensional) feature space, weight vector $\boldsymbol{w} \in \boldsymbol{R}^{d_f}$ in primal weight space, error variables $e_i \in \boldsymbol{R}$ and bias term b. To find minimizers of the objective function, we can construct the Lagrangian function as follows,

$$L(\boldsymbol{w}, b, \boldsymbol{e}; \boldsymbol{\alpha}) = \frac{1}{2}\boldsymbol{w}'\boldsymbol{w} + \frac{\gamma}{2}\sum_{i=1}^{n} e_i^2 - \sum_{i=1}^{n} \alpha_i(\boldsymbol{w}'\boldsymbol{\Phi}(\boldsymbol{x}_i) + b + e_i - y_i) \tag{3}$$

where α_i's are the Lagrange multipliers. We obtain the equations for optimal solutions by partial differentiating the Lagrangian function over $\{\boldsymbol{w}, b, \boldsymbol{e}\}$, which lead to the solution of a linear equation

$$\begin{bmatrix} 0 & \boldsymbol{1}_n' \\ \boldsymbol{1}_n & \boldsymbol{K} + \gamma^{-1}\boldsymbol{I}_n \end{bmatrix} \begin{bmatrix} b \\ \boldsymbol{\alpha} \end{bmatrix} = \begin{bmatrix} 0 \\ \boldsymbol{y} \end{bmatrix}. \tag{4}$$

Here $\boldsymbol{K} = \{K_{kl}\}$ with $K_{kl} = \boldsymbol{\Phi}(\boldsymbol{x}_k)'\boldsymbol{\Phi}(\boldsymbol{x}_l), \; k, l = 1, \cdots, n$. From application of the Mercer condition we can choose a kernel $K(\cdot, \cdot)$ such that

$$K(\boldsymbol{x}_k, \boldsymbol{x}_l) = \boldsymbol{\Phi}(\boldsymbol{x}_k)'\boldsymbol{\Phi}(\boldsymbol{x}_l), \; k, l = 1, \cdots, n. \tag{5}$$

By solving the linear system, Lagrange multipliers $\alpha_i,\ i = 1, \cdots, n$ and bias b can be obtained. With these estimates we can get the optimal regression function $f(\boldsymbol{x})$ for given $\boldsymbol{x}$ as follows

$$f(\boldsymbol{x}) = \sum_{i=1}^{n} \alpha_i K(\boldsymbol{x}_i, \boldsymbol{x}) + b. \tag{6}$$

We focus on the choice of an Gaussian kernel $K(\boldsymbol{x}_k, \boldsymbol{x}_l) = \exp\left(-\frac{\|\boldsymbol{x}_k - \boldsymbol{x}_l\|^2}{\sigma^2}\right)$ for the sequel.

When we use LS-SVM approach for regression, we need to determine an optimal choice of the kernel parameter σ and the regularization parameter γ. There could be several parameter selection methods such as cross validation type methods, bootstraping and Bayesian learning methods. In this paper we use generalized cross validation(GCV) method. Let $\boldsymbol{\lambda} = (\sigma, \gamma)'$. The GCV score is constructed as follows:

$$GCV(\boldsymbol{\lambda}) = \frac{1}{n} \frac{\sum_{i=1}^{n} (y_i - f(\boldsymbol{x}_i; \boldsymbol{\lambda}))^2}{(1 - n^{-1} tr[\boldsymbol{S}(\boldsymbol{\lambda})])^2}, \tag{7}$$

where $\boldsymbol{S}(\boldsymbol{\lambda}) = \boldsymbol{K}\left(\boldsymbol{Z}^{-1} - \boldsymbol{Z}^{-1}\frac{\boldsymbol{J}_n}{c}\boldsymbol{Z}^{-1}\right) + \frac{\boldsymbol{J}_n}{c}\boldsymbol{Z}^{-1}$. Here $c = \boldsymbol{1}_n'\left(\boldsymbol{K} + \frac{1}{\gamma}\boldsymbol{I}_n\right)^{-1}\boldsymbol{1}_n$, $\boldsymbol{Z} = \left(\boldsymbol{K} + \frac{1}{\gamma}\boldsymbol{I}_n\right)^{-1}$ and $\boldsymbol{J}_n$ is a square matrix with all elements equal to 1. See for details De Brabanter et al.[2].

3 Fuzzy C-Means Clustering

The FCM algorithm is one of the most widely used fuzzy clustering algorithms. This technique was developed by Dunn[3] and improved by Bezdek[1]. The FCM algorithm attempts to partition a finite data set into c fuzzy clusters with respect to some given criterion.

Let $\boldsymbol{L} = \{L_1, \cdots, L_c\} \subset \boldsymbol{R}^{d_f+1}$ be an c-tuple of prototypes characterizing each cluster. Partitioning the data $\{\boldsymbol{z}_i\}_{i=1}^n = \{(y_i, \boldsymbol{\Phi}(\boldsymbol{x}_i)')'\}_{i=1}^n$ into c fuzzy clusters is performed by minimizing the objective function

$$J_c = \sum_{k=1}^{c} \sum_{i=1}^{n} U_{ik}^m d^2(\boldsymbol{z}_i, L_k) \tag{8}$$

where $m > 1$ is weighting exponent determining the fuzzyness degree, $U_{ik} \in \boldsymbol{U}$ represents the membership degree of the data point $\boldsymbol{z}_i$ to cluster A_k where $A = \{A_1, \cdots, A_c\}$ is the fuzzy partition, and $d(\boldsymbol{z}_i, L_k)$ is the distance between the data point $\boldsymbol{z}_i$ and cluster A_k, which is usually defined by the Euclidean distance, $d^2(\boldsymbol{z}_i, L_k) = \|\boldsymbol{z}_i - L_k\|^2$.

The optimal fuzzy partition is obtained by the following iterative algorithm where the objective function is successively minimized with respect to $\boldsymbol{U}$ and $\boldsymbol{L}$.

step 0) Given data set $\{z_i\}_{i=1}^n = \{(y_i, \Phi(x_i)')'\}_{i=1}^n$, specify $m > 1$ and choose a termination threshold $\epsilon > 0$. Initialize the partition matrix $U^{(0)}$.

step 1) Calculate $L_k = L_k^{(r)}$ which minimize the objective function J_c,

$$L_k = \frac{\sum_{i=1}^n U_{ik}^m z_i}{\sum_{i=1}^n U_{ik}^m}, \quad k = 1, \cdots, c$$

step 2) Update $U^{(r)}$ to $U^{(r+1)}$ from

$$U_{ik} = \frac{1}{\sum_{j=1}^c \left(\frac{d^2(z_i, L_k)}{d^2(z_i, L_j)}\right)^{\frac{1}{m-1}}}, \quad i = 1, \cdots, n, \quad k = 1, \cdots, c.$$

step 3) Stop iteration if $\|U^{(r)} - U^{(r+1)}\| < \epsilon$, otherwise set $r = r+1$ and return step 1).

Each component of L_k is a weighted average of corresponding components of all data points. Components with high degree of membership in cluster k contribute significantly to weighted average(the center of cluster k) and components with lower degree of membership can be interpreted as being far from the center.

4 Robust LS-SVM Using FCM

In this section we illustrate the whole procedure of the robust LS-SVM using FCM. This procedure consists of two stages. The first stage is to determine a fuzzy partition of separating the training data set into the set of normal data points and the set of outliers. The second stage is to derive the robust version of LS-SVM by using the membership degrees of the fuzzy partition as weights.

First, we illustrate the first stage of the robust LS-SVM using FCM. To apply a FCM method to the regression, we choose the curve as the prototype of cluster, a fuzzy partition as $\{A, A^c\}$ and the fuzziness degree $m = 2$. The set A is characterized by curve prototype $L(u, v)$, where v is the center of the cluster and u is the normal vector. The normal vector u is a $(d_f + 1) \times 1$ vector defined by using the regression coefficients as $u = (1, -w')'$.

Following Sarbu[7], we set the dissimilarity between the hypothetical prototype of the complementary fuzzy set A^c and the data point z_i equal to $\frac{\theta}{1-\theta}$ with $\theta \in [0, 1]$. It is explained in Sarbu[7] that θ represents the membership degree of the point farthest(the largest outlier) from the curve expressed as regression function, and its optimal value can be determined according to the quality coefficients. In this paper we use a normalized quality coefficient NQC_1 to determine θ and the other hyperparameters in the next process of updating the membership degrees of the fuzzy partition.

As in Sarbu[7], we define the objective function as follows:

$$J(U, L; \theta) = \sum_{i=1}^n U_i^2 d^2(z_i, L) + \sum_{i=1}^n (1 - U_i)^2 \frac{\theta}{1-\theta}, \tag{9}$$

where $U_i \in [0,1]$ represents the membership degree of the data point $\boldsymbol{z}_i$ to the cluster A and $d(\boldsymbol{z}_i, L)$ is the distance between the data point $\boldsymbol{z}_i$ and the cluster A. Then the prototype $L(\boldsymbol{u}, \boldsymbol{v})$ minimizing the objective function $J(\boldsymbol{U}, L; \theta)$ is given by

$$\boldsymbol{v} = \begin{pmatrix} y^v \\ \Phi^v \end{pmatrix} = \frac{\sum_{i=1}^n U_i^2 \boldsymbol{z}_i}{\sum_{i=1}^n U_i^2} = [\boldsymbol{z}_1, \cdots, \boldsymbol{z}_n]\boldsymbol{\nu}, \tag{10}$$

where

$$\boldsymbol{\nu} = \left(\frac{U_1^2}{\sum_{i=1}^n U_i^2}, \cdots, \frac{U_n^2}{\sum_{i=1}^n U_i^2} \right)'$$

and

$$U_i = \frac{\frac{\theta}{1-\theta}}{\frac{\theta}{1-\theta} + d^2(\boldsymbol{z}_i, L)}, \; i = 1, \cdots, n.$$

Let an $n \times n$ matrix $\boldsymbol{K}_c$ be defined by

$$\boldsymbol{K}_c(i,j) = (\Phi(\boldsymbol{x}_i) - \Phi^v)'(\Phi(\boldsymbol{x}_j) - \Phi^v).$$

Then it can be rewritten as

$$\boldsymbol{K}_c = \{(\Phi(\boldsymbol{x}_i) - \Phi^v)'(\Phi(\boldsymbol{x}_j) - \Phi^v)\} = \boldsymbol{K} - \boldsymbol{1}_n \boldsymbol{\nu}' \boldsymbol{K} - \boldsymbol{K}\boldsymbol{\nu}\boldsymbol{1}_n' + \boldsymbol{1}_n \boldsymbol{\nu}' \boldsymbol{K} \boldsymbol{\nu} \boldsymbol{1}_n'.$$

Thus, the distance from the point $\boldsymbol{z}_i$ to the curve (passing through $\boldsymbol{v}$ and having direction $\boldsymbol{u}$) is given by

$$d(\boldsymbol{z}_i, L) = \frac{|y_i - \boldsymbol{K}_c(i,:)\boldsymbol{\alpha} - b|}{\sqrt{1 + \boldsymbol{\alpha}' \boldsymbol{K}_c \boldsymbol{\alpha}}}, \; i = 1, \cdots, n, \tag{11}$$

where $\boldsymbol{K}_c(i,:)$ is the i-th row of $\boldsymbol{K}_c$.

We now need to illustrate how to get the vector of the Lagrange multipliers $\boldsymbol{\alpha}$ and the bias b in the equation (11). Let the curve passing through $\boldsymbol{v} = \begin{pmatrix} y^v \\ \Phi^v \end{pmatrix}$ be

$$y = \boldsymbol{w}'\Phi(\boldsymbol{x}) + b.$$

Then the estimates of $\boldsymbol{\alpha}$ and b can be obtained from the following optimization problem:

$$\min \frac{1}{2}\boldsymbol{w}'\boldsymbol{w} + \frac{\gamma}{2}\sum_{i=1}^n e_i^2 \tag{12}$$

subject to equality constraints

$$\begin{aligned} y_i - \boldsymbol{w}'\Phi(\boldsymbol{x}_i) - b &= e_i, \; i = 1, \cdots, n \\ y^v - \boldsymbol{w}'\Phi^v - b &= 0, \end{aligned} \tag{13}$$

where $y^v = \boldsymbol{\nu}'\boldsymbol{y}$ and $\Phi^v = [\Phi(\boldsymbol{x}_1), \cdots, \Phi(\boldsymbol{x}_n)]\,\boldsymbol{\nu}$.

It is noted that the optimization problem (12) becomes to the following optimization problem:

$$\min \frac{1}{2}\boldsymbol{w}'\boldsymbol{w} + \frac{\gamma}{2}\sum_{i=1}^{n} e_i^2 \tag{14}$$

subject to equality constraints

$$y_i - y^v - \boldsymbol{w}'(\Phi(\boldsymbol{x}_i) - \Phi^v) = e_i, \; i = 1, \cdots, n. \tag{15}$$

Thus we have

$$\boldsymbol{\alpha} = (\boldsymbol{K}_c + \gamma^{-1}\boldsymbol{I}_n)^{-1}(\boldsymbol{y} - y^v\boldsymbol{1}_n) \tag{16}$$

and

$$b = y^v - \boldsymbol{\nu}'\boldsymbol{K}_c\boldsymbol{\alpha}. \tag{17}$$

We can summarize how to get the optimal fuzzy partition vector $\boldsymbol{U}$ in what follows:

step 0) Given data set $\{\boldsymbol{z}_i\}_{i=1}^n = \{(y_i, \Phi(\boldsymbol{x}_i)')'\}_{i=1}^n$, specify $m > 1$ and choose a termination threshold $\epsilon > 0$. Initialize the partition vector $\boldsymbol{U}^{(0)}$.

step 1) Calculate the weight vector of the cluster center $\boldsymbol{\nu}$ with $\boldsymbol{U} = \boldsymbol{U}^{(r)}$, $\boldsymbol{\nu} = \left(\frac{U_1^2}{\sum_{i=1}^n U_i^2}, \cdots, \frac{U_n^2}{\sum_{i=1}^n U_i^2}\right)'$ and the distance from the data point $\boldsymbol{z}_i$ to the cluster A with $L = L^{(r)}$, $d(\boldsymbol{z}_i, L) = \frac{|y_i - \boldsymbol{K}_c(i,:)\boldsymbol{\alpha} - b|}{\sqrt{1+\boldsymbol{\alpha}'\boldsymbol{K}_c\boldsymbol{\alpha}}}$, $i = 1, \cdots, n$. Here, $\boldsymbol{\alpha} = (\boldsymbol{K}_c + \gamma^{-1}\boldsymbol{I}_n)^{-1}(\boldsymbol{y} - y^v\boldsymbol{1}_n)$ and $b = y^v - \boldsymbol{\nu}'\boldsymbol{K}_c\boldsymbol{\alpha}$.

step 2) Update $\boldsymbol{U}^{(r)}$ to $\boldsymbol{U}^{(r+1)}$ from $U_i = \frac{\frac{\theta}{1-\theta}}{\frac{\theta}{1-\theta} + d^2(\boldsymbol{z}_i, L)}$, $i = 1, \cdots, n$.

step 3) Stop iteration if $\|\boldsymbol{U}^{(r)} - \boldsymbol{U}^{(r+1)}\| < \epsilon$, otherwise set $r = r+1$ and return step 1).

We now illustrate the way of determining θ, the kernel and the regularization parameters in the process of updating U_i's using the normalized quality coefficient NQC_1, which is defined as

$$QC_1 = \sqrt{\sum_{i=1}^{n}\left(\frac{r_i}{\max_{k=1}^{n}|r_k|}\right)^2} \tag{18}$$

$$NQC_1 = \frac{QC_1 - 1}{\sqrt{n} - 1}, \tag{19}$$

where r_i is residual. The smaller the NQC_1, the better fit of model. Therefore, we choose the parameters which minimize NQC_1 in updating U_i's.

Once we have the partition vector $\boldsymbol{U}$, we can obtain a robust estimate

$$f^*(\boldsymbol{x}) = \sum_{i=1}^{n} \alpha_i^* K(\boldsymbol{x}_i, \boldsymbol{x}) + b^* \tag{20}$$

by solving the following weighted LS-SVM problem:

$$\begin{bmatrix} 0 & \mathbf{1}_n' \\ \mathbf{1}_n & \boldsymbol{K} + \boldsymbol{V}_\gamma \end{bmatrix} \begin{bmatrix} b^* \\ \boldsymbol{\alpha}^* \end{bmatrix} = \begin{bmatrix} 0 \\ \boldsymbol{y} \end{bmatrix}, \tag{21}$$

where the diagonal matrix $\boldsymbol{V}_\gamma$ is given by

$$\boldsymbol{V}_\gamma = diag\{\frac{1}{\gamma U_1}, \cdots, \frac{1}{\gamma U_n}\}. \tag{22}$$

Here the unknown variables for this weighted LS-SVM problem are denoted by the * symbol.

For the robust LS-SVM we should also determine the kernel and the regularization parameters. These can be determined by using the GCV score for weighted LS-SVM, which is constructed as follows:

$$GCV(\boldsymbol{\lambda}) = \frac{1}{n} \frac{\sum_{i=1}^{n} (y_i - f^*(\boldsymbol{x}_i; \boldsymbol{\lambda})^2}{(1 - (1/\sum_{k=1}^{n} U_k) tr[\boldsymbol{S}^*(\boldsymbol{\lambda})])^2}, \tag{23}$$

where $\boldsymbol{S}^*(\boldsymbol{\lambda}) = \boldsymbol{K}\left(\boldsymbol{Z}^{*-1} - \boldsymbol{Z}^{*-1}\frac{\boldsymbol{J}_n}{c}\boldsymbol{Z}^{*-1}\right) + \frac{\boldsymbol{J}_n}{c}\boldsymbol{Z}^{*-1}$ with $\boldsymbol{Z}^* = \boldsymbol{K} + \boldsymbol{V}_\gamma$. See for details De Brabanter et al.[2].

5 Numerical Studies

In this section we illustrate the performance of the proposed algorithm by comparing with ordinary LS-SVM algorithm through two simulated examples. The Gaussian kernel function is employed for the nonlinear regression, which is $K(\boldsymbol{x}_k, \boldsymbol{x}_l) = \exp\left(-\frac{\|\boldsymbol{x}_k - \boldsymbol{x}_l\|^2}{\sigma^2}\right)$.

For the first example, the data set without outliers is generated by

$$y_i = 1 + 2\sin(2\pi x_i) + \epsilon_i, \ \ i = 1, \cdots, 40 \tag{24}$$

where the noise ϵ_i represents a real number which is normally distributed with mean 0 and variance 0.04. For data set with outliers, 2 of 40 data points are moved away from their original locations as outliers. For data set without outliers, the parameters (γ, σ^2) of the ordinary LS-SVM were chosen as (7500, 0.5) by $GCV(\boldsymbol{\lambda})$ of equation (7). The parameters $(\theta, \gamma, \sigma^2)$ of the first stage of the robust LS-SVM were chosen as (0.002, 70, 0.5) by NQC_1 given in equation (19). The parameters (γ, σ^2) of the second stage of the robust LS-SVM were chosen

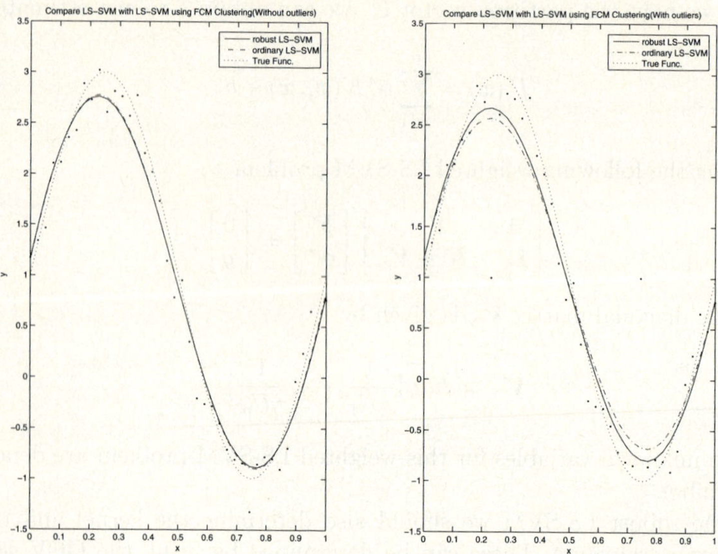

Fig. 1. Scatter plots of data and estimated regression functions for the first example

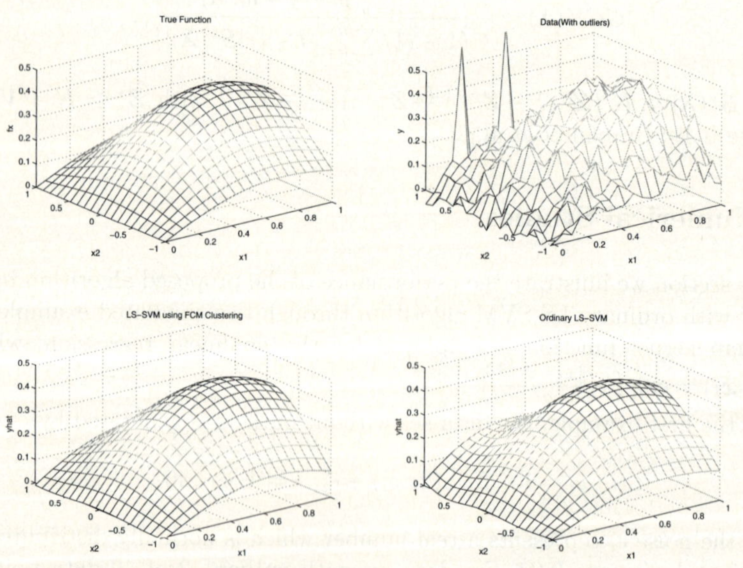

Fig. 2. Meshed plots of data and estimated regression functions for the second example

as (300, 0.3) by $GCV(\boldsymbol{\lambda})$ of equation (23). On the other hand, for data set with outliers the parameters (γ, σ^2) of the ordinary LS-SVM were chosen as (100, 0.4) by $GCV(\boldsymbol{\lambda})$ of equation (7). The parameters $(\theta, \gamma, \sigma^2)$ of the first stage of the robust LS-SVM were chosen as (0.002, 100, 0.4) by NQC_1 of equation (19). The parameters (γ, σ^2) of the second stage of the robust LS-SVM were chosen as (70, 0.5) by $GCV(\boldsymbol{\lambda})$ of equation (23).

Fig. 1 shows the scatter plot of outputs versus input values of 40 data. In Fig. 1, the left subplot shows the results for data set without outliers, whereas the right one for data set with outliers. The dotted, dash-dotted and solid curves represent the true function, the estimates of the ordinary and the robust LS-SVMs, respectively. From the left subplot we can see both regression estimates give almost the same results. From the right one we observe that the robust LS-SVM estimator is less influenced by outliers and thus slightly performs better than the ordinary LS-SVM.

For the second example, the data set without outliers is generated by

$$y_{ik} = x_{1i}e^{(-x_{1i}^2 - x_{2k}^2)} + \epsilon_{ik}, \; i = 1, \cdots, 20, \; k = 1, \cdots, 20 \tag{25}$$

where the noise ϵ_{ik} represents a real number which is normally distributed with mean 0 and variance 0.001. For data set with outliers, 2 of 400 data points are moved away from their original locations as outliers. Unlike Fig.1, Fig. 2 shows the results only for data set with outliers. The parameters (γ, σ^2) of the ordinary LS-SVM were chosen as (20, 0.9) by $GCV(\boldsymbol{\lambda})$ of equation (7). On the other hand, the parameters $(\theta, \gamma, \sigma^2)$ of the first stage of the robust LS-SVM were chosen as (0.002, 50, 3) by NQC_1. The parameters (γ, σ^2) of the second stage of the robust LS-SVM were chosen as (50, 2) by $GCV(\boldsymbol{\lambda})$ of equation (23). From two subplots in the bottom we can observe the robust LS-SVM(left) is less influenced by outliers and thus slightly performs better than the ordinary LS-SVM(right).

6 Conclusions

In this paper, we have proposed a new robust LS-SVM algorithm by combining ordinary LS-SVM and FCM method which is used to reduce the influence of outliers on estimating regression function. The experimental results show that the proposed robust LS-SVM derives the satisfying solution to estimating regression functions and captures well the characteristics of both data sets with and without outliers. Our robust LS-SVM performs better than the ordinary LS-SVM for the data set with outliers. Hence, our robust LS-SVMS appears to be useful in estimating regression functions. The functional characteristics are obtained through the selection of the related parameters. These parameters have been tuned using cross-validation.

Acknowledgement

This work was supported by grant No. R01-2006-000-10226-0 from the Basic Research Program of the Korea Science & Engineering Foundation.

References

1. Bezdek, J. C. : Pattern recognition with fuzzy objective function algoritms. Plenum Press, New York (1981)
2. De Brabanter, J., Pelckmans, K., Suykens, J. A. K., Vandewalle, J. and De Moor, B. : Robust cross-validation score functions with application to weighted least squares support vector machine function estimation. Tech. Report, Department of Electrical Engineering, ESAT-SISTA (2003)
3. Dunn, J. C. : A fuzzy relative of the ISODATA process and its use in detecting compact well-separated clusters. Journal of Cybernetics **3** (1973) 32–57
4. Kecman, V. : Learning and soft computing, support vector machines, neural networks and fuzzy logic models. The MIT Press, Cambridge, MA (2001)
5. Mercer, J. : Functions of positive and negative type and their connection with theory of integral equations. Philosophical Transactions of Royal Society **A** (1909) 415–446
6. Smola, A . and Schölkopf, B. : On a kernel-based method for pattern recognition, regression, approximation and operator inversion. Algorithmica **22** (1998) 211–231
7. Sarbu, C. : Use of fuzzy regression for calibration in thin-layer chromatography/densitometry. Journal of AOAC International **83** (2000) 1463–1467
8. Suykens, J. A. K. and Vanderwalle, J. : Least square support vector machine classifier. Neural Processing Letters **9** (1999) 293–300
9. Vapnik, V. N. : The nature of statistical learning theory. Springer, New York (1995)
10. Wang, L.(Ed.) : Support vector machines: theory and application. Springer, Berlin Heidelberg New York (2005)

Support Vector Regression Based on Unconstrained Convex Quadratic Programming

Weida Zhou[1], Li Zhang[1], Licheng Jiao[1], and Jin Pan[2]

[1] Institute of Intelligence Information Processing, Xidian University, Xi'an 710071, China
{wdzhou, zhangli, lichjiao}@mail.xidian.edu.cn
[2] Xi'an Communications Institute, Xi'an 710106, China
panjin_163@163.com

Abstract. Support vector regression (SVR) based on unconstrained convex quadratic programming is proposed, in which Gaussian loss function is adopted. Compared with standard SVR, this method has a fast training speed and can be generalized into the complex-valued field directly. Experimental results confirm the feasibility and the validity of our method.

1 Introduction

Support vector machines have been successfully applied to regression estimation problem of the real-valued functions [1-4]. The procedure of a standard SVR is to solve a constrained convex quadratic programming (QP). So the complexity of its computation is relatively large, especially for large-scale data. In addition signals or functions in practice are not only real but also complex, such as communication signals, radar signals and so on. Standard SVMs can not be applied to complex-valued functions directly because of the linear constraints of the SVMs programming. The details about this problem were discussed in document [5].

In this paper support vector regression based on unconstrained convex quadratic programming is proposed, in which Gaussian loss function is adopted. In principle, if the condition number of the Hessian matrix is good, we can use the pseudo inverse matrix method to solve this convex QP problem. Otherwise if the Hessian matrix is close to ill condition or the scale of it is very large, we can adopt steepest gradient descent method, Newton method, quasi-Newton method, conjugate gradient descent method and others. Our method has a faster training speed than standard SVR does in the real-valued flied. More important, our method can be generalized into the complex-valued field directly. Experimental results confirm the feasibility and the validity of our method.

2 SVR Base on Unconstrained Convex QP

Optimization problem of the standard SVMs is a convex quadratic programming with linear constraints where the Hessian matrix is $2l \times 2l$. We can transform it into a convex QP without constraints. In doing so, the computation complexity will be decreased largely without a loss in performance. It is important that the unconstrained QP can be applied to the complex-valued field.

L. Jiao et al. (Eds.): ICNC 2006, Part I, LNCS 4221, pp. 167 – 174, 2006.

2.1 SVR Base on Unconstrained Convex QP in the Real-Valued Field

Given a set of training data $\{(\mathbf{x}_i, y_i) \mid \mathbf{x}_i \in \mathbb{R}^d, y_i \in \mathbb{R}, i = 1,\cdots,l\}$. In SVMs, if Gaussian loss function $c(\xi) = \frac{1}{2}\xi^2$ is adopted, a regression estimation problem can be changed to the following optimization problem

$$\min \quad \frac{1}{2}\|\mathbf{w}\|^2 + \frac{C}{2}\sum_{i=1}^{l}(\xi_i^2 + (\xi_i^*)^2) \tag{1}$$

subject to $\quad \mathbf{w}\cdot\mathbf{x}_i + b - y_i \le \xi_i$, $y_i - \mathbf{w}\cdot\mathbf{x}_i - b \le \xi_i^*$, $\xi_i^{(*)} \ge 0$, $i = 1,\cdots,l$

Here let

$$\xi_i = (f(\mathbf{x}_i) - y)^2 \text{ and } \xi_i^* = (y - f(\mathbf{x}_i))^2 \tag{2}$$

Then we have $\xi = \xi^*$. Substituting (2) into (1), we can have

$$\min \quad \frac{1}{2}\|\mathbf{w}\|^2 + C\sum_{i=1}^{l}(y_i - \mathbf{w}\cdot\mathbf{x}_i - b)^2 \tag{3}$$

which is an unconstrained convex QP. Let $\beta = \left[\mathbf{w}^T, b\right]^T \in \mathbb{R}^{d+1}$, then (3) can be rewritten as

$$\min \quad \frac{1}{2}\beta^T\mathbf{Q}\beta + \mathbf{d}^T\beta \tag{4}$$

where

$$\mathbf{Q} = \begin{bmatrix} 2C\sum_{i=1}^{l}\mathbf{x}_i\mathbf{x}_i^T + \mathbf{I}, & 2C\sum_{i=1}^{l}\mathbf{x}_i \\ 2C\sum_{i=1}^{l}\mathbf{x}_i^T, & 2Cl \end{bmatrix} \in \mathbb{R}^{(d+1)\times(d+1)} \tag{5}$$

and

$$\mathbf{d} = \left[-2C\sum_{i=1}^{l} y_i\mathbf{x}_i^T, \quad -2C\sum_{i=1}^{l} y_i\right]^T \in \mathbb{R}^{d+1} \tag{6}$$

where the matrix $\mathbf{I}$ is the $d\times d$ unit matrix and l is the number of the training patterns. Obviously, the Programming (4) is a convex QP, or the Hessian matrix $\mathbf{Q}$ is a positive semi-definite and symmetric matrix. The estimation function can be expressed as

$$f(\mathbf{x}) = \mathbf{w}\cdot\mathbf{x} + b$$

For nonlinear regression estimation problems, we introduce kernel functions. Let the nonlinear regression estimation has the form

$$f(\mathbf{x}) = \sum_{i=1}^{l}\eta_i K(\mathbf{x}, \mathbf{x}_i) + b \tag{7}$$

Finally we can obtain the following unconstrained QP

$$\min \quad \frac{1}{2}\beta^T \mathbf{Q}\beta + \mathbf{d}^T\beta \tag{8}$$

where $\beta = [\eta_1 \quad \cdots \quad \eta_l \quad b]^T \in \mathbb{R}^{l+1}$,

$$\mathbf{Q} = \begin{bmatrix} 2C\mathbf{K}\mathbf{K}^T + \mathbf{K} & 2C\sum_{i=1}^{l}\mathbf{K}_{\cdot,i} \\ 2C\sum_{i=1}^{l}\mathbf{K}_{i,\cdot} & 2Cl \end{bmatrix} \in \mathbb{R}^{(l+1)\times(l+1)} \tag{9}$$

and

$$\mathbf{d} = \left[-2C\sum_{i=1}^{l} y_i\mathbf{K}_{i,\cdot} \quad -2C\sum_{i=1}^{l} y_i\right]^T \in \mathbb{R}^{(l+1)} \tag{10}$$

where $\mathbf{K}$ denotes the kernel matrix, $\mathbf{K}_{i,k} = K(\mathbf{x}_i, \mathbf{x}_k)$ denotes the i th row and the k th column of kernel matrix. Here the QP (8) is a convex QP, too.

As stated above, we have obtained SVR based on unconstrained convex quadratic programming with Gaussian loss function in the real-valued field. In what follows, we will generalize it in the complex-valued field. In the following, we only consider nonlinear SVR. If the kernel function $K(\mathbf{x}_i, \mathbf{x}_k) = \mathbf{x}_i^T\mathbf{x}_k$, then nonlinear SVR becomes linear one.

2.2 SVR Based on Unconstrained QP in the Complex-Valued Field

Given a set of training patterns $X = \{(\mathbf{x}_i, y_i) \mid \mathbf{x}_i \in \mathcal{X} \subset \mathbb{C}^d, y_i \in \mathbb{C}, i = 1, \cdots, l\}$. Assume that there exists complex Mercer kernels $K(\mathbf{x}, \mathbf{y}) = (\Phi(\mathbf{x}) \cdot \Phi(\mathbf{y}))$ and the corresponding feature mapping $\Phi : \mathbf{x} \in \mathcal{X} \mapsto \Phi(\mathbf{x}) \in \mathcal{F}$. Then the optimization problem becomes

$$\min \quad \frac{1}{2}\mathbf{w}^H\mathbf{w} + C\sum_{i=1}^{l}\left[y_i - (\mathbf{w}\cdot\Phi(\mathbf{x}_i)) - b\right]^H\left[y_i - (\mathbf{w}\cdot\Phi(\mathbf{x}_i)) - b\right] \tag{11}$$

where $\mathbf{w} = \sum_{i=1}^{l}\eta_i\Phi(\mathbf{x}_i) \in \mathcal{F}$ denotes the weight vector, $\eta_i, b \in \mathbb{C}$ and superscript ^{H} denotes the conjugate transpose of a matrix or a vector. Let complex-valued column vector be $\beta = [\eta_1 \quad \cdots \quad \eta_l \quad b]^H \in \mathbb{C}^{l+1}$. Then Eq. (21) can be rewritten as

$$\frac{1}{2}\left(\beta^H\mathbf{Q}\beta + \mathbf{d}^H\beta + \beta^H\mathbf{d}\right) \tag{12}$$

where

$$\mathbf{Q} = \begin{bmatrix} 2C\mathbf{K}\mathbf{K}^H + \mathbf{K} & 2C\sum_{i=1}^{l}\left(\mathbf{K}^H\right)_{\cdot,i} \\ 2C\sum_{i=1}^{l}\mathbf{K}_{i,\cdot} & 2Cl \end{bmatrix} \in \mathbb{C}^{(l+1)\times(l+1)} \tag{13}$$

and

$$\mathbf{d} = \left[-2C\sum_{i=1}^{l} y_i\left(\mathbf{K}^H\right)_{\cdot,i}, \quad -2C\sum_{i=1}^{l} y_i\right]^H \in \mathbb{C}^{l+1} \tag{14}$$

Then QP (12) is a convex QP, or the Hessian matrix $\mathbf{Q}$ is a positive semi-definite and symmetrical conjugate matrix. QP (12) is a quadratic function with complex-valued variables, which indicates that we can use the derivation of (12) to obtain its extreme points. The derivation of a complex-valued variable can be divided into two groups: the derivation of the real part and the imaginary part of the complex-valued variable. Then we can obtain a complex-valued gradient. Here, a complex-valued gradient is defined by

$$\nabla_{\mathbf{z}} g(\mathbf{z}) = \left(\frac{\partial}{\partial \mathbf{z}_1} + j \frac{\partial}{\partial \mathbf{z}_2} \right) g(\mathbf{z}) \tag{15}$$

where $g(\mathbf{z})$ is a scalar function and $\mathbf{z} = \mathbf{z}_1 + j\mathbf{z}_2 \in \mathbb{C}$. Therefore the complex-valued gradient of Eq. (12) is

$$\mathbf{Q}\beta + \mathbf{d} \tag{16}$$

Let Eq. (16) be equal to zero. Hence we can obtain the optimal solution of the QP (12) has the form

$$\beta = -Q^{\dagger} d$$

where $\mathbf{Q}^{\dagger}$ denotes the generalized inverse matrix of $\mathbf{Q}$. For large-scale data problems, we can adopt steepest descent method, Newton method, conjugate gradient descent and other optimization methods to solve (12) [6].

3 Simulation

In order to validate the performance of our method, we took regression estimation on one-dimensional real-valued sinc function, two-dimensional real-valued sinc function and one-dimensional complex-valued sinc function. These experiments were implemented in Matlab 5.3. We adopted Gaussian kernel function in our experiments. For SVR based on unconstrained convex QP, we used three optimization methods: pseudo inverse method, conjugate gradient method and QP method.

The performance "Flops" shows the number of computing of floating point numbers, which can measure the computation complexity of a learning algorithm. "Approximation error" is defined by the mean square deviation between estimated and actual values y on the test set, $\sqrt{\left(\sum_{i=1}^{l}(y_i - f_i)^2\right)/l}$.

3.1 Regression on One-Dimensional Real-Valued Sinc Function

Let the set of training data $\{(x_k, y_k) \mid x_k \in \mathbb{R}, y_k \in \mathbb{R}, k = 1, \cdots, l\}$ be defined by one-dimensional real-valued function on the interval $[-10,10]$, y_k is corrupted by noise with normal distribution, $y = \sin(x)/x + n$, where $n \sim N(0, \sigma^2)$. We sampled 201 pointes uniformly on the interval $[-10,10]$. 100 of them were chosen uniformly for the training and the others for testing. The parameters of both our methods and standard SVR are the same ones: $C = 1$, $\varepsilon = 0$, when the variance of the noise $\sigma = 0.1$, $p = 1.6$ and when the variance of the noise $\sigma = 0.5$, $p = 2.1$. We took 100 runs totally. The average results on 100 runs are shown in Table 1. SVR based on unconstrained convex QP is comparable to standard SVR on the estimation performance. But the computation complexity of our method is smaller than that of standard SVR. In Table 1, it looks unreasonable that standard SVR has smaller "Flops" but longer "training time" than generalized inverse method. We

Table 1. Results of regression on the one-dimensional real-valued sinc function

Noise	Methods		Training time (s)	Flops	Approximation error
$\sigma = 0.1$	Standard QP SVR		7.14	2,406,848	0.03171
	SVR based on unconstrained convex QP	Generalized inverse matrix	0.621	13,842,472	0.03173
		Conjugate gradient descent	0.29	922,274	0.03175
		QP	0.19	482,379	0.03173
$\sigma = 0.5$	Standard QP SVR		6.96	2406,848	0.1369
	SVR based on unconstrained convex QP	Generalized inverse matrix	0.64	13,842,472	0.1377
		Conjugate gradient descent	0.42	1,645,923	0.1372
		QP	0.20	482,379	0.1377

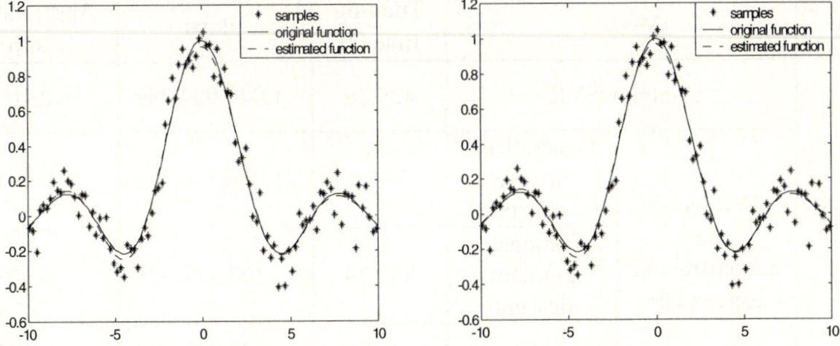

Fig. 1. The one-dimensional real-valued sinc function (real lines) and its approximation (dotted lines) with $\sigma = 0.1$, (a) Standard SVR, (b) SVR based on unconstrained convex QP

think that it does not need to be explained and compiled for the function $\text{pinv}(\cdot)$ in Matlab5.3 and it needs to do for $\text{qp}(\cdot)$. So the training time of standard QP SVR is longer. Figure 1 is shown the approximation results with $\sigma = 0.1$, in which the asterisks are corrupted training patterns, the real lines are the original function and the dotted lines the approximation. The result in Figure 1(a) was obtained by standard SVR and that in Figure 1(b) by our method. Figure 2 is shown the approximation results with $\sigma = 0.5$.

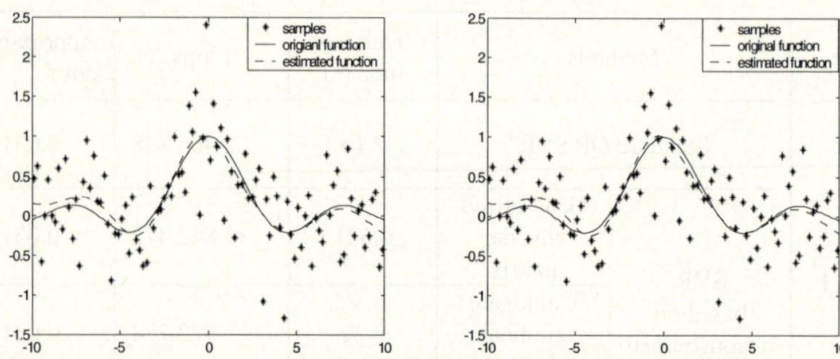

Fig. 2. The one-dimensional real-valued sinc function (real lines) and its approximation (dotted lines) with $\sigma = 0.5$, (a) Standard SVR, (b) SVR based on unconstrained convex QP

3.2 Regression on the One-Dimensional Complex-Valued Sinc Function

Let the set of training data $\{(x_k, y_k) \mid x_k \in \mathbb{C}, y_k \in \mathbb{C}, k = 1, \cdots, l\}$ be defined by the one-dimensional complex-valued sinc function on the interval $[-10,10] + j[-10,10]$, y_k is corrupted by noise with normal distribution

$$y_k = \frac{\sin(|x_k|)}{x_k} + n_k$$

Table 2. Results of regression on the one-dimensional complex-valued sinc function

Noise	Method		Training time (s)	Flops	Approximation error
$\sigma = 0.1$	Complex SVR[1]		429.38	1,049,934,799	0.1031
	SVR based on unconstrained convex QP	Generalized inverse matrix	35.11	1,050,316,175	0.1010
		Conjugate gradient descent	38.139	123,187,938	0.1033
		QP	60.59	24,643,863	0.1016

[1] This method was presented in [5].

where $n = (nr + j\,ni)$, $nr \sim N(0,\sigma^2)$ 和 $ni \sim N(0,\sigma^2)$. We sampled uniformly 14×14 training patterns and 67×67 testing ones on the interval $[-10,10]+j[-10,10]$. Let $\sigma = 0.1$. The parameters of both our methods and standard

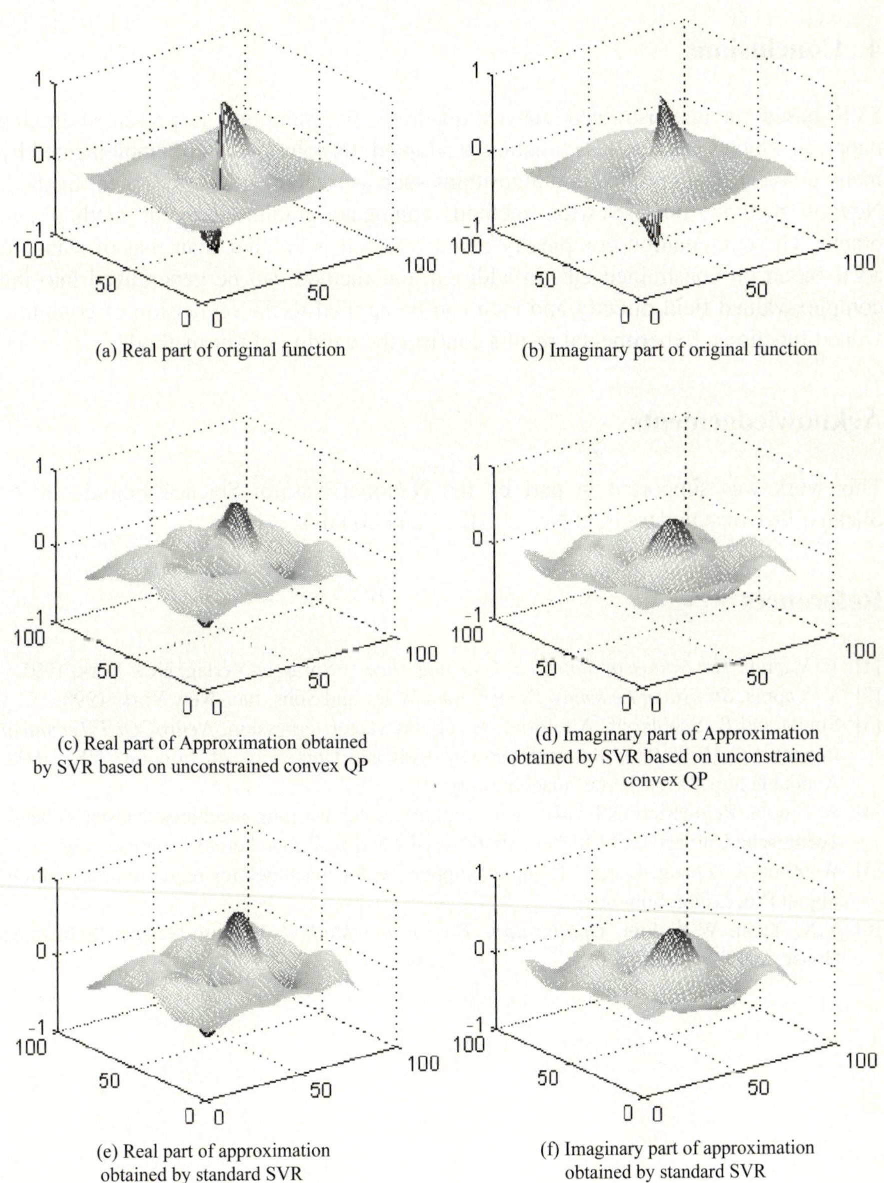

(a) Real part of original function

(b) Imaginary part of original function

(c) Real part of Approximation obtained by SVR based on unconstrained convex QP

(d) Imaginary part of Approximation obtained by SVR based on unconstrained convex QP

(e) Real part of approximation obtained by standard SVR

(f) Imaginary part of approximation obtained by standard SVR

Fig. 3. The one-dimensional complex-valued sinc function and its approximation with $\sigma = 0.1$

SVR are the same ones: $p = 2.0$, $C = 1$, $\varepsilon = 0$. We performed 100 runs. The average results on 100 runs are shown in Table 2. Figure 3(a) and (b) are shown the real part and the imaginary part of the one-dimensional complex-valued sinc function. Figure 3(c) and (d) are shown the approximation to the original function obtained by our method. Figure 3(e) and (f) are shown the approximation obtained by standard SVR.

4 Conclusions

SVR based on unconstrained convex quadratic programming is presented in this paper, in which Gaussian loss function is adopted. Its solution can be implemented by many conventional optimization algorithms such as steepest gradient descent method, Newton method, quasi-Newton method, conjugate gradient descent method and others. The computation complexity of our method is smaller than that of standard SVR based on constrained QP. In addition, our method can be generalized into the complex-valued field directly and then can be applied to the regression of complex-valued functions. Experimental results confirm the validity of our method.

Acknowledgements

This work was supported in part by the National Natural Science Foundation of Shaanxi Province under Grant Nos.2004F11 and 2005F32.

References

[1] V. Vapnik. *The Nature of Statistical Learning Theory*. Springer-Verlag, New York, 1995.

[2] V. Vapnik. *Statistical Learning Theory*. John Wiley and Sons, Inc., New York, 1998.

[3] Smola and B. Schölkopf, A tutorial on support vector regression, *NeuroCOLT Technical Report NC-TR-98-030*, Royal Holloway College, University of London, UK, 1998. Available http://www.kernel-machines.org/

[4] A. Smola. Regression estimation with support vector learning machines. Master's thesis, Technische Universität München, 1996. Available http://www.kernel-machines.org/

[5] W. Zhou, L. Zhang, L. Jiao. Complex support vector machines for regression estimation. Signal Processing. Submitted..

[6] Y.X. Yuan, W.Y. Sun. *Optimization Theory and Method*. Beijing: Science Publishing House, 1999.

Base Vector Selection for Support Vector Machine

Qing Li

Institute of Intelligent Information Processing, Xidian university, 224#, Xi'an, 710071, PR
kingdomyangfan@hotmail.com

Abstract. SVM has been receiving increasing interest in areas ranging from its original application in pattern recognition to other applications such as regression estimation due to its remarkable generalization performance. However, it also contains some defects such as storage problem (in training process) and sparsity problem. In this paper, a new method is proposed to pre-select the base vectors from the original data according to vector correlation principle, which could greatly reduce the scale of the optimization problem and improve the sparsity of the solution. The method could capture the structure of the data space by approximating a basis of the subspace of the data; therefore, the statistical information of the training samples is preserved. In the paper, the process of mathematical deduction is given in details and results of simulations on artificial data and practical data have been done to validate the performance of base vector selection (BVS) algorithm. The experimental results show the combination of such algorithm with SVM can make great progress while can't sacrifice the SVM's performance.

1 Introduction

The support vector machine (SVM) is a new and promising classification and regression technique proposed by Vapnik and his group at AT&T Bell Laboratories. The theory of SVM is based on the idea of structural risk minimization (SRM) [1,2]. It has been shown to provide higher performance than traditional learning machines [3,4] and has been introduced as powerful tools for solving both pattern recognition and regression estimation problems. SVM uses a device called kernel mapping to map the data in input space to a high-dimensional feature space in which the problem becomes linear. The decision function obtained by SVM is related not only to the number of SVs and their weights but also to the priori chosen kernel that is called support vector kernel.It has been successfully applied in many areas such as the face detection, the hand-writing digital character recognition, the data mining, etc.

However, there still exist some drawbacks in SVM. First, from the implementation points of view, training a SVM is equivalent to solving a linearly constrained Quadratic Programming (QP) problem in a number of variables equal to the number of data points. Second, training a SVM needs to handle with a very large and fully dense kernel matrix [5]. For large-scale problems, perhaps it cannot be saved in the main memory at all. Thus, traditional optimization algorithms like Newton or Quasi-Newton cannot be directly used. Currently major methods are the decomposition methods which solves a sequence of smaller size problems so the memory difficulty is avoided. However, the

L. Jiao et al. (Eds.): ICNC 2006, Part I, LNCS 4221, pp. 175 – 184, 2006.

decomposition method still suffers from slow convergence for huge problems. Third, SVM is a sparse algorithm in theory, but the sparsity of the solution is not as good as what we expect, so many problems are presented such as the reduction of the time when test a new samples.

Vapnik shows that the capacity of generalization depends on the geometrical characteristics of the training data and not on their dimensionality. Therefore, if these characteristics are well chosen, the expected generalization error could be small even if the feature space has a huge dimensionality [6,7].The former work on solving the storage of the kernel matrix mostly concentrates on the decomposition methods and only a few of works have been done to improve the sparsity of the SVM. In this paper, we proposed a new insight to pre-select a base subset of the training samples according to vector correlation principle. Base vectors samples are far fewer than the original data in many circumstances, so training SVM by these samples could greatly reduce the scale of the optimization problem and improve the sparsity of the support vector solution.

Given a nonlinear mapping function ϕ, it maps the training data $(\mathbf{x}_i, y_i)_{1\le i\le l}$ in input space into a high-dimensional feature space $(\phi(\mathbf{x}_i), y_i)_{1\le i\le l}$. According to the linear theory, these feature vector $\{\phi(\mathbf{x}_1), \phi(\mathbf{x}_2), ..., \phi(\mathbf{x}_l)\}$ maybe not linear independent. Suppose $\{\phi(\tilde{\mathbf{x}}_1), \phi(\tilde{\mathbf{x}}_2), ..., \phi(\tilde{\mathbf{x}}_M)\}$ (often M is far less than l) are the base vectors of the $\{\phi(\mathbf{x}_1), \phi(\mathbf{x}_2), ..., \phi(\mathbf{x}_l)\}$, then any vector $\phi(\mathbf{x}_i)$ can be expressed in the linear form of the base vector $\sum \beta_j \bullet \phi(\tilde{\mathbf{x}}_j)$ .Therefore, training on the original data will be equal to training on the base vector samples as well as we exactly know the corresponding coefficients matrix $\boldsymbol{\beta} = [\vec{\beta}_1, \vec{\beta}_2, ..., \vec{\beta}_l]^T$, because

$$\begin{bmatrix} \phi(\mathbf{x}_1) \\ \phi(\mathbf{x}_2) \\ \vdots \\ \phi(\mathbf{x}_l) \end{bmatrix} = \begin{bmatrix} \beta_{11} & \beta_{12} & \cdots & \beta_{1M} \\ \beta_{21} & \beta_{22} & \cdots & \beta_{2M} \\ \vdots & \vdots & \ddots & \vdots \\ \beta_{l1} & \beta_{l2} & \cdots & \beta_{lM} \end{bmatrix} \begin{bmatrix} \phi(\tilde{\mathbf{x}}_1) \\ \phi(\tilde{\mathbf{x}}_2) \\ \dots \\ \phi(\tilde{\mathbf{x}}_M) \end{bmatrix} \tag{1}$$

where $[\phi(\mathbf{x}_1), \phi(\mathbf{x}_2), ..., \phi(\mathbf{x}_l)]^T$ is the actual training samples in feature space.

In this paper, we aim to pre-select the base vectors $X_B = \{x_{B_1}, x_{B_2}, ..., x_{B_M}\}$ $(1 \le B_i \le M)$ according to the vector correlation principle and the Mercer kernel function, and then training SVM with the base vectors to reduce the scale of the optimization problem and improve the sparsity of the solution.

2 Support Vector Machine

SVM seeks to minimize an upper bound of the generalization error based on the structural risk minimization (SRM) principle. It uses SV kernel to map the data in input space to a high-dimension feature space in which we can solve the problem in linear form.

2.1 SVM for Regression Estimation

Given a set of data points $(\mathbf{x}_i, y_i)_{1\le i\le l}$ ($\mathbf{x}_i \in R^N$, $y \in R$, R^N is the input space and l is the total number of training samples) randomly generated from an unknown function, SVM approximates the function using the following form:

$$y = f(\mathbf{x},\boldsymbol{\omega}) = \boldsymbol{\omega}^T \phi(\mathbf{x}) + b \tag{2}$$

where ϕ is a nonlinear mapping function, and it maps the input vector $\mathbf{x}$ into the high-dimensional feature space $\phi(\mathbf{x})$. In order to obtain a small risk when estimate (2), SVM using the following regularized risk functional:

$$\min\left(\frac{1}{2}\|\boldsymbol{\omega}\|^2 + \frac{C}{l}\sum_{i=1}^{l} L_\varepsilon(y_i, f(\mathbf{x}_i,\boldsymbol{\omega}))\right) \tag{3}$$

The second term of (3) is the empirical error measured by the ε-insensitive loss function defined as:

$$L_\varepsilon(y, f(\mathbf{x},\boldsymbol{\omega})) = \begin{cases} 0, & \text{if } |y - f(\mathbf{x},\boldsymbol{\omega})| < \varepsilon \\ |y - f(\mathbf{x},\boldsymbol{\omega})| - \varepsilon, & \text{otherwise.} \end{cases} \tag{4}$$

In (3), $C > 0$ is a constant and $\varepsilon > 0$ is called tube size which defines the width of the ε-insensitive zone of the cost function [1,2].

By introducing Lagrange multiplier techniques, the minimization of (2) leads to the following dual optimization problem.

$$\begin{aligned} \max\ W\left(\alpha_i^{(*)}\right) &= \sum_{i=1}^{l} y_i(\alpha_i - \alpha_i^*) - \varepsilon\sum_{i=1}^{l}(\alpha_i + \alpha_i^*) \\ &- \frac{1}{2}\sum_{i=1}^{l}\sum_{j=1}^{l}(\alpha_i - \alpha_i^*)(\alpha_j - \alpha_j^*)K(\mathbf{x}_i,\mathbf{x}_j) \end{aligned} \tag{5}$$

with the following constraints:

$$\sum_{i=1}^{l}(\alpha_i - \alpha_i^*) = 0, \qquad 0 \le \alpha_i^{(*)} \le C,\ i = 1,\ldots,l. \tag{6}$$

$K(\mathbf{x},\mathbf{x}_i)$ is defined as the SV kernel function. The value of the kernel is equal to the inner product of two vectors $\mathbf{x}_i$ and $\mathbf{x}_j$ in the feature space $\phi(\mathbf{x}_i)$ and $\phi(\mathbf{x}_j)$, that is, $K(\mathbf{x},\mathbf{x}_i) = \langle \Phi(\mathbf{x}_i), \Phi(\mathbf{x}_j) \rangle$. Any function that satisfies Mercer's condition can be used as the kernel function. Common examples of the SV kernel function are the polynomial kernel $K(\mathbf{x},\mathbf{x}_j) = (\langle \mathbf{x},\mathbf{x}_j \rangle + 1)^d$ and the Gaussian kernel $K(\mathbf{x},\mathbf{x}_j) = \exp(-(1/\rho^2)\|\mathbf{x}-\mathbf{x}_j\|^2)$, where d and ρ are the kernel parameters. Finally, the resulting regression estimation is

$$f(\mathbf{x}) = \sum_{i=1}^{l}(\alpha_i - \alpha_i^*)K(\mathbf{x},\mathbf{x}_i) + b \tag{7}$$

2.2 SVM for Pattern Recognition

SVM for pattern recognition is similar to it for regression, which is to solve a constrained quadratic optimization problem as well. The only difference between them is the expression of the optimization problem and the decision function. Suppose a binary classification and let $(\mathbf{x}_i, y_i) \in R^N \times \{\pm 1\}, i = 1,...,l$ be a set of training samples. Kernel mapping can map the training samples in input space into a high-dimension feature space in which the mapped training samples become linearly separable with introduction of the slack variance ξ. Finding the classification hyperplane by SVM can be cast as a quadratic optimization problem:

$$\begin{aligned} &\min \Psi(\omega,\xi) = \frac{1}{2}\|\omega\|^2 + C\bullet(\sum_{i=1}^{l}\xi_i) \\ &\text{s.t. } y_i[(\omega\bullet\mathbf{x}_i) - b] \geq 1 - \xi_i, \\ &\xi_i \geq 0, \ i = 1,...,l \end{aligned} \tag{8}$$

The final decision function becomes

$$f(\mathbf{x}) = \text{sgn}(\sum\nolimits_{i=1}^{l} \alpha_i y_i K(\mathbf{x},\mathbf{x}_i) + b) \tag{9}$$

3 Base Vector Selection

Let $(\mathbf{x}_i, y_i)_{1\leq i\leq l}$ ($\mathbf{x}_i \in R^N$, $y \in R$) be the training samples. A nonlinear mapping function ϕ maps the input space of the training data into a feature Hilbert space H:

$$\begin{aligned} \Phi &: R^N \rightarrow H \\ \mathbf{x} &\rightarrow \phi(\mathbf{x}) \end{aligned} \tag{10}$$

Therefore, the mapping training set in feature space are $(\phi(\mathbf{x}_i), y_i)_{1\leq i\leq l}$ ($\phi(\mathbf{x}_i) \in H$, $y \in R$), which lies in a subspace H_s of the H with the dimension up to l. In practice, the dimension of this subspace is far lower than l and equal to the number of its base vector. As we have shown in the introduction, training on the original samples will be equal to training on the base vector samples as well as we exactly know the corresponding coefficients matrix. So we propose a preprocessing method to select base vectors of the original data and use these base vectors as the training samples to train SVM so that to reduce the scale of the optimization problem.

In the following, the method of select base vector will be introduced. For notation simplification: for each $\mathbf{x}_i$ the mapping is noted $\phi(\mathbf{x}_i) = \phi_i$ for $1 \leq i \leq l$ and the selected base vectors are noted by $\mathbf{x}_{B_j}$ and $\phi(\mathbf{x}_{B_j}) = \phi_{B_j}$ for $1 \leq j \leq M$ (M is the number of the base vectors). For a given base vectors $X_B = \{x_{B_1}, x_{B_2}, ..., x_{B_M}\}$, the mapping of any vector x_i can be expressed as a linear combination of X_s with the form

$$\hat{\phi}_i = \boldsymbol{\beta}_i^T \Phi_B \tag{11}$$

where $\Phi_B = (\phi_{B_1}, ..., \phi_{B_M})^T$ is the matrix of the mapping base vectors and $\boldsymbol{\beta}_i = (\beta_{i1}, ..., \beta_{iM})^T$ is the corresponding coefficient vector.

Given $(\mathbf{x}_i, y_i)_{1 \le i \le l}$, the goal is to find the base vectors $X_B = \{x_{B_1}, x_{B_2}, ..., x_{B_M}\}$ such that for any mapping ϕ_i, the estimated mapping $\hat{\phi}_i$ is as close as possible to ϕ_i. For this purpose, we minimize the following form to select the base vectors:

$$\min_{X_B} \left(\sum_{\mathbf{x}_i \in X} \left(\left\| \phi_i - \hat{\phi}_i \right\|^2 \right) \right) \tag{12}$$

Let $\rho_i = \left\| \phi_i - \hat{\phi}_i \right\|^2$, we can get:

$$\begin{aligned} \rho_i &= \left\| \phi_i - \boldsymbol{\beta}_i^T \Phi_B \right\|^2 = (\phi_i - \boldsymbol{\beta}_i^T \Phi_B)^T (\phi_i - \boldsymbol{\beta}_i^T \Phi_B) \\ &= \phi_i^T \phi_i - 2\Phi_B^T \phi_i \boldsymbol{\beta}_i + \boldsymbol{\beta}_i^T \Phi_B^T \Phi_B \boldsymbol{\beta}_i \end{aligned} \tag{13}$$

Putting the derivative of ρ_i to zero gives the coefficient vector $\boldsymbol{\beta}_i$:

$$\begin{aligned} &\frac{\partial \rho_i}{\partial \boldsymbol{\beta}_i} = 2(\Phi_B^T \Phi_B)\boldsymbol{\beta}_i - 2\Phi_B^T \phi_i, \\ &\frac{\partial \rho_i}{\partial \boldsymbol{\beta}_i} = 0 \Rightarrow \boldsymbol{\beta}_i = (\Phi_B^T \Phi_B)^{-1} \Phi_B^T \phi_i \end{aligned} \tag{14}$$

$(\Phi_B^T \Phi_B)^{-1}$ exists if the mapping base vectors are linear independent. Substituting (13) and (14) to (12), we get

$$\min_{X_B} \left(\sum_{\mathbf{x}_i \in X} \left(\phi_i^T \phi_i - \phi_i^T \Phi_B (\Phi_B^T \Phi_B)^{-1} \Phi_B^T \phi_i \right) \right) \tag{15}$$

By Mercer's theorem, we can replace the inner product between feature space vectors by a positive defined kernel function over pairs of vectors in input space. In other word, we use the substitution $\phi(\mathbf{x})^T \phi(\mathbf{y}) = \langle \phi(\mathbf{x}), \phi(\mathbf{y}) \rangle = k(\mathbf{x}, \mathbf{y})$ to (15) and get

$$\min_{X_B} \left(\sum_{\mathbf{x}_i \in X} \left(k_{ii} - \vec{K}_{Bi}^T K_{BB}^{-1} \vec{K}_{Bi} \right) \right) \tag{16}$$

We defined the base set fitness F_B and the arbitrary vector fitness F_{B_i} corresponding to a given base vector set X_B by

$$F_B = \frac{1}{l} \sum_{\mathbf{x}_i \in X} F_{B_i} \tag{17}$$

where

$$F_{B_i} = k_{ii} - \vec{K}_{Bi}^T K_{BB}^{-1} \vec{K}_{Bi} \tag{18}$$

Now, (15) is equivalent to:

$$\min_{X_B}(F_B) \tag{19}$$

The process of base vector selection is a greedy iterative algorithm. When selecting the first base vector, we look for the samples that gives the minimum F_B. In each iteration, (17) is used to estimate the performance of the current base set and (18) is used to select the next best candidate base vector. If the one has the maximal fitness F_{B_i} for the current base set, we select it as the next base vector. It is because the collinearity between such vector and the current base set is the worst (or in other word, such vector could hardly be expressed as the linear combination of the current base vectors). A pre-defined base set fitness and the expected maximal number of the base vector could be used to stop the iterative process. When the current base set fitness reaches the pre-defined fitness (which means that X_B is a good approximation of the basis for the original dataset in *H*) or the number of the base vectors reaches the expected value, the algorithm stops. Another important stop criterion must be noted that the algorithm should be stopped if the matrix of $K_{BB} = (\Phi_B^T \Phi_B)$ is not invertible anymore, which means the current base set X_B is the real basis in feature space *H*.

4 Validation of Effectiveness

In the paper, we compare the BVS SVM with the standard SVM and the kernel both for SVM and base vector selection adopt RBF kernel with the form $K(x, x_i) = \exp(-\|x - x_i\|^2 / 2p)$. We adopt the parameters' notation of BVS as follows: maxN—maximum of the base vectors; minFit—BVS stopping criterion (predefined accuracy). In the test of regression, we adopt the approximation error as $e_{ss} = \sqrt{(\sum_{i=1}^{l} (y_i - f_i)^2)/l}$, and the loss function of the SVM adopts ε-insensitive loss. For avoiding the weak problem, each experiment has been performed 30 independent runs, and all experiments were carried on a Pentium IV 2.6Ghz with 512 MB RAM using Matlab 7.01 compiler.

4.1 Regression Experiments

4.1.1 Approximation of Single-Variable Function

In this experiment, we approximate the 1-dimension function, $y = \sin(x)/x$, which is also named Sinc function. We uniformly sample 1200 points over the domain $[-10,10]$, 400 of which are taken as the training examples and others as testing examples.

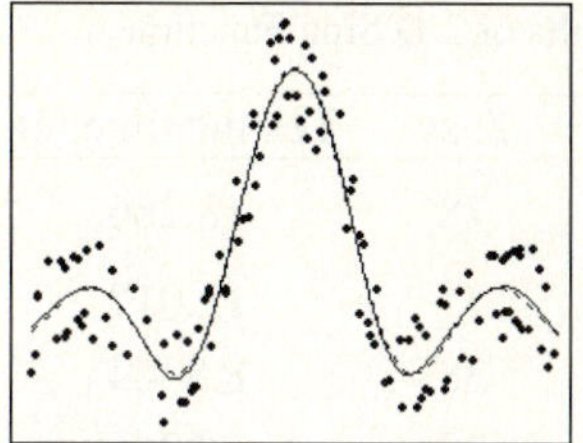
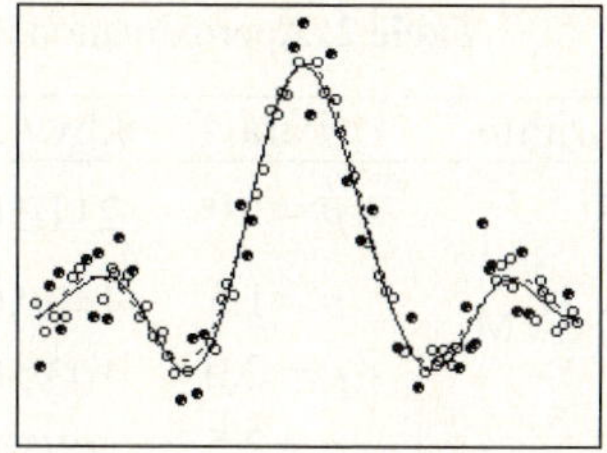

Fig. 1. Regression results by SVM (left with 124SV/400total[1]) and BVS SVM (right with 38SV/100BSV/400 total). The solid line represents approximated curve and dotted line represents original function while '•' representing SVs and 'o' representing base vectors (right).

Table 1. Approximation Results of 1-D Sinc Function

Algorithm	Para.[2]	#.bv/ #.tr[3]	#. sv	Test time (s)	Error
BVS SVM	$p = 0.13$	100/400	37	4.6832	0.033803
	$p = 0.16$	97/400	32	4.2031	0.035811
	$p = 0.19$	82/400	29	3.6094	0.037692
	$p = 0.22$	71/400	25	3.0938	0.39899
Standard SVM	None	None/400	124	17.281	0.031271

Before the training, each training point is added Gaussian noise with 0 mean and standard deviation 0.1. The kernel both for SVM and base vector selection adopt RBF kernel. For SVM, C=1, $\varepsilon = 0.1$ and RBF kernel parameter sigma=1.79; for BVS, maxN=100, minFit=0.05. Table 1 lists the approximation errors for BVS SVM and standard SVM algorithm. We also give the approximation drawings in Fig 1 with kernel parameter $\rho = 0.16$ for BVS.

4.1.2 Approximation of Two-Variable Function

This experiment is to approximate a two-dimension Sinc function with the form $y = \sin\left(\pi\sqrt{x_1^2 + x_2^2}\right)/\left(\pi\sqrt{x_1^2 + x_2^2}\right)$ over the domain $[-10,10]\times[-10,10]$. We uniformly sample 2500 points, 500 of which are taken as the training examples and others as testing examples.

Before the training, each training point is added Gaussian noise with 0 zero mean and standard deviation 0.1. The kernel both for SVM and base vector selection adopt RBF kernel. For SVM, C=1, $\varepsilon = 0.1$ and RBF kernel parameter sigma=0.49; for BVS, maxN=250, minFit=0.01. Table 2 lists the approximation errors for BVS SVM and standard SVM algorithm.

---

[1] 124 SV/ 400 total means 400 training samples in which contains 124 support vectors.

[2] In this paper, 'Para.' means the selected Kernel parameter for BVS.

[3] We adopt note '#' to present 'the number of' and 'bv', 'tr', 'sv' means the base vectors, training samples, support vectors, respectively.

Table 2. Approximation Results of 2-D Sinc Function

Algorithm	Para.	#.bv / #.tr	#. sv	Testing time (s)	Error
BVS SVM	$p=1.0$	211/500	78	18.286	0.052812
	$p=1.5$	140/500	51	12.017	0.053624
	$p=2.0$	103/500	38	8.8344	0.056971
	$p=2.5$	78/500	28	6.6266	0.057776
Standard SVM	None	None/500	177	44.984	0.051849

4.2 Experiments of Pattern Recognition

4.2.1 Two Spirals' Problem

Learning to tell two spirals apart is important both for purely academic reasons and for industrial application [8]. In the research of pattern recognition, it is a well-known problem for its difficulty. The parametric equation of the two spirals can be presented as follows:

$$\begin{aligned} \text{spiral-1:} \quad & x_1 = (k_1\theta + e_1)\cos(\theta) \\ & y_1 = (k_1\theta + e_1)\sin(\theta) \\ \text{spiral-2:} \quad & x_2 = (k_2\theta + e_2)\cos(\theta) \\ & y_2 = (k_2\theta + e_2)\sin(\theta) \end{aligned} \tag{20}$$

where k_1, k_2, e_1 and e_2 are parameters. In our experiment, we choose $k_1 = k_2 = 4$, $e_1 = 1$, $e_2 = 10$. We uniformly generate 2000 samples, and randomly choose 800 of them as training data, others as test data.

Table 3. Recognition Results of Two Spirals

Algorithm	Para.	#.bv / #.tr	#. sv	Test time (s)	Accuracy
BVS SVM	$p=1.15$	192/800	97	11.126	100%
	$p=1.71$	131/800	55	7.1406	100%
	$p=2.25$	103/800	41	5.7656	100%
	$p=2.81$	84/800	37	4.5313	100%
Standard SVM	None	None/800	597	42.969	100%

Before the training, we add noise to data—randomly choosing 80 training samples , changing its class attributes. The kernel both for SVM and base vector selection adopt RBF kernel. For SVM, C=10 and RBF kernel parameter sigma=8; for BVS, maxN=200, minFit=0.01. Table 3 lists the classification result by BVS SVM and standard SVM algorithm, respectively.

4.2.2 Artificial Data Experiment

We generate artificial dataset with the parametric equation of the data as $\begin{cases} x = \rho \bullet \sin\varphi \bullet \cos\theta \\ y = \rho \bullet \sin\varphi \bullet \sin\theta \quad \theta \in U[0, 2\pi] \quad \varphi \in U[0, \pi] \\ z = \rho \bullet \cos\varphi \end{cases}$. Parameter ρ of the first class is of continuous uniform distribution $U[0,50]$, and ρ in the second class is of $U[50,100]$. We randomly generate 8000 samples, and randomly choose 2000 samples as training data, others as test data.

Before the training, we add noise to data—randomly choosing 200 training samples, changing its class attributes. The kernel both for SVM and base vector selection adopt RBF kernel. For SVM, C=10 and RBF kernel parameter sigma=8; for BVS, maxN=300, minFit=0.01. Table 4 lists the classification result by BVS SVM and standard SVM algorithm., respectively.

Table 4. Recognition Results of Artificial Dataset

Algorithm	Para.	#.bv / #.tr	#. sv	Test time (s)	Accuracy
BVS SVM	$p=1.01$	300/2000	62	78.781	90.98%
	$p=1.25$	204/2000	49	53.75	90.92%
	$p=1.51$	147/2000	38	38.875	90.87%
	$p=1.75$	109/2000	30	28.961	90.01%
Standard SVM	None	None/2000	465	522.19	91.08%

4.2.3 Ionosphere Dataset Classification

We did experiment on a well-known dataset, ionosphere dataset coming from the UCI Benchmark Repository[4].

Ionosphere dataset is a binary problem with 34 continuous characteristic attributes and one class attribute. The dataset contains 351 instances and we randomly choose 250 instances as training samples, others as testing samples. The kernel both for SVM and base vector selection adopt RBF kernel. For SVM, C=4 and RBF kernel parameter sigma=1.3; for BVS, maxN=150, minFit=0.01. Table 5 lists the classification result by BVS SVM and standard SVM algorithm., respectively.

Table 5. Recognition Results of Ionosphere Dataset

Algorithm	Para.	#.bv / #.tr	#. sv	Test time (s)	Accuracy
BVS SVM	$p=2.01$	150/250	126	0.71979	95.24%
	$p=2.56$	144/250	123	0.69688	94.82%
	$p=3.11$	124/250	118	0.59375	94.81%
	$p=3.67$	110/250	109	0.52187	94.18%
	$p=3.94$	105/250	104	0.50104	93.56%
	$p=4.22$	99/250	99	0.46667	90.26%
Standard SVM	None	None/250	135	1.2344	95.37%

5 Concluding Remarks

SVM uses a device called kernel mapping to map the data in input space to a high-dimensional feature space in which the problem becomes linear. It has made a

[4] URL: http://www.ics.uci.edu/mlearn.

great progress on the fields of machine learning. However, there still exist some drawbacks in SVM such as the scale of the optimization problem and sparsity problem, etc. In this paper, a method of base vectors selection is introduced to improve these disadvantages of SVM. The method could capture the structure of the data by approximating a basis of the subspace of the data; therefore, the statistical information of the training samples is preserved. From the method, three advantages will be obtained: 1. the scale of the optimization problem is greatly reduced; 2. the solution of the SVM becomes much sparser (due to the less support vectors); 3. shorten the time when testing a new samples. We have tested many experiments on regression and classification problems. The data cited in the paper include both artificial data and practical data. The experimental results show that combination of such algorithm with SVM can make great progress while can't sacrifice the performance of the SVM. Therefore, it is valid and feasible to implement such algorithm for embedded applications.

References

1. Cortes C., Vapnik V. Support vector network. Mach. Learn. 20 (1995) 273-297.
2. Vapnik V. An overview of statistical learning theory. IEEE Trans. Neural Network, vol.10, no.5, pp.988-999, 1999.
3. Burges C. J. C. A tutorial on support vector machines for pattern recognition. Data Mining Knowledge Discovery. 2 (2), 1998, pp.121-167.
4. Scholkopf, B., Smola, A. Learning with Kernels. MIT Press, 1999.
5. Platt J. C. Fast training of support vector machines using sequential minimal optimization. In Advances in Kernel Methods-Support Vector Learning, A. J. Smola, C. J. C. Burges, and B. Schölkopf, Eds. Cambridge, MA: MIT Press, 1998.
6. V. Vapnik. Three remarks on support vector machine. In: S. A. Solla, T. K. Leen, K. R. Müller (Eds.), Advances in Neural Comput. 10 (1998), pp.1299-1319.
7. Baudat G., Anouar F. Generalized Discriminant Analysis Using a Kernel Approach. Neural Computation 12(10): 2385-2404 (2000).
8. Lang K.J., Witbrock M.J. Learning to tell two spirals apart. In Proc. 1989 Connectionist Models Summer School, 1989, pp.52-61.

How to Stop the Evolutionary Process in Evolving Neural Network Ensembles

Yong Liu

School of Computer Science
China University of Geosciences
Wuhan, 430074, P.R. China
yliu@u-aizu.ac.jp

Abstract. In practice, two criteria have often been used to stop the evolutionary process in evolving neural network (NN) ensembles. One criterion is to stop the evolution when the maxial generation is reached. The other criterion is to stop the evolution when the evolved NN ensemble, i.e., the whole population, is satisfactory according to a certain evaluation. This paper points out that NN ensembles evolved from these two criteria might not be robust by having different performance. In order to make the evolved NN ensemble more stable, an alternative solution is to combine a number of evolved NN ensembles. Experimental analyses based on n-fold cross-validation have been given to explain why the evolved NN ensembles could be very different and how such difference could disappear or be reduced in the combination.

1 Introduction

Evolutionary neural network (NN) learning is a population based learning in which evolutionary algorithms are used to to perform various tasks, such as connection weight training, architecture design, learning rule adaptation, input feature selection, connection weight initialization, rule extraction from NNs, etc. One distinct feature of evolutionary NNs is their adaptability to a dynamic environment. In other words, evolutionary NNs can adapt to an environment as well as changes in the environment. The two forms of adaptation, i.e., evolution and learning in evolutionary NNs, make their adaptation to a dynamic environment much more effective and efficient. In a broader sense, evolutionary NNs can be regarded as a general framework for adaptive systems, i.e., systems that can change their architectures at the end of the evolutionary process.

Although evolutionary NNs have their own distinct features, one common problem faced by both evolutionary NNs and other NN learning algorithms is when the learning should be stopped. In NN learning, if the learning would be stopped too early, the learned NNs might be underfitting. That is, the learned NNs have not learned well. In contrast, if the learning would be stopped too late, the learned NNs might likely be overfitting. In this case, the learned NNs could even have learned the noise contained in the training data.

L. Jiao et al. (Eds.): ICNC 2006, Part I, LNCS 4221, pp. 185–194, 2006.

Commonly, the stopping criteria used in either evolutionary algorithms or NN learning algorithms have been directly used to stop the evolutionary process in evolutionary NNs. Little research has been done on whether such stopping criteria match what evolutionary NNs require. Evolutionary NNs include evolving NNs and evolving NN ensembles in which evolving NNs is at the level of individual evolution while evolving NN ensembles are at the level of population evolution. Since there are two levels of evolution in evolutionary NNs, the stopping criteria suited to one level might not be good to the other level.

In evolving NNs, it is practical that the NN with the minimum error on a training data set or a validation data set is chosen to be the final learning system. However, learning is different from optimization in practice because the learned system should have best generalization, which is different from minimizing an error function on the training set or the validation set. On one hand, with the help of the evolution, the error on the training set or the validation set for an evolved NN might be even smaller than that of an NN with learning only. On the other hand, The NN with the minimum error on a training set or a validation set may not have best generalization unless there is an equivalence between generalization and the error on the training data or the validation data. Unfortunately, measuring generalization quantitatively and accurately is almost impossible in practice [12] although there are many theories and criteria on generalization, such as the minimum description length [10], Akaike information criteria [1], and minimum message length [11]. In practice, these criteria are often used to define better error functions in the hope that minimizing the functions will maximize generalization. While these functions often lead to better generalization of learned systems, there is no guarantee.

Since the maximum fitness or the minimum error may not be equivalent to best generalization in evolutionary NN learning, the best NN with the maximum fitness in a population may not be the desired one. Other NNs in the population may contain some useful information that will help to improve generalization of the learned systems. It is thus beneficial to make use of the whole population rather than any single individual. A population always contains at least as much information as any single individual. Hence, combining different NNs in the population to form an NN ensemble is expected to produce better results. There have been some very successful experiments which show that evolutionary learning can be used to evolve NN ensembles [6,7,8,9].

The difference between evolving NNs and evolving NN ensembles is that the solution of evolving NNs is an evolved NN while the solution of evolving NN ensemble is a population of NNs. The evolution cycle of evolving NNs stops when a satisfactory NN is found or the maximum number of generations has been reached. The evaluation of the satisfactory NN is commonly based on its error on a training data set or a validation data set. The experiments had shown that such an evolved neural network might overfit on the training set or the validation set. Anyhow, it is a stopping criterion that is used to stop the evolution.

In the practice of evolving NN ensembles, it is common that each individual rather the whole population is evaluated. During the evolution, each individual

NN would be better and better while the population with the better individual NNs might not better since not individual but the whole evolved population would be the final solution. It suggests that the evolved population in the last generation might be worse so that it is not wise to choose the final population as a solution. However, it is hard to know in which evolution stage the evolved population should be chosen to be the solution without being evaluated directly. Meanwhile, if the whole population would be evaluated directly, it would face the same problem of overfit on the training set or the validation set happened in evolving NNs. Such evolution behavior have been clearly shown in this paper through experimental studies. In order to deal with the dilemma of having the population been evaluated or not, an alternative solution is not to use one evolved population but to combine a number of evolved populations. This solution allows a stopping criterion to be setup for evolving NN ensembles while avoiding the problem of overfitting on the training set or the validation set.

The rest of this paper is organized as follows: Section 2 reviews the related work about stopping criteria used in single NN learning. Section 3 describes the evolutionary NN ensembles based on minimizing mutual information. Section 4 presents experimental analyses to explain why the performance of the evolved NN ensembles could be rather unstable. Finally, Section 5 concludes with a summary of the paper.

2 Stopping Criteria in Single Neural Network Learning

The basic principle adopted by single NN learning is Ockham's razor: prefer the simplest hypothesis consist with the data. That is, of all the NNs which will fit the training data, find the simplest. Since these NNs fit the training data, all of them have small biases. However, their variance is quite different. The simplest NN is chosen because of its smallest variance. Many learning methods have been developed for single NN learning.

2.1 Constructive Methods

Roughly speaking, a constructive method starts with a minimal NN (i.e., an NN with a minimal number of hidden layers, nodes, and weights) and adds new layers, nodes, and weights if necessary during training. One major problem with constructive methods is how to determine when to stop the addition of new layers, nodes, and weights. If the addition process is stopped too early, the constructed NN may be too small. This will cause the constructed NN to have a large bias. On the other hand, If the addition process is stopped too late, the constructed NN may be too complex. This will make the NN have large variance.

2.2 Pruning Methods

A pruning method does the opposite of the constructive methods. Starting with a larger than needed NN, the pruning method deletes unnecessary layers, nodes, and weights during training. However, pruning methods still face the similar problem in constructive methods. That is, one has to determine when to stop the deletion of unnecessary layers, nodes, and weights. Early stop of the deletion may lead to overly large NNs with large variance. If too many layers, nodes, and weight are pruned, large bias may be introduced in the pruned NN.

2.3 Combination of Addition and Deletion

Some learning algorithms combine node/connection addition and node/connection deletion in the same learning process. There are two phases in the learning process including addition phase and reduction phase. Note that the problem faced by constructive methods and pruning methods still need to be addressed by those learning algorithms which combine addition and deletion. For example, there is still a need to determine when to stop addition or deletion.

2.4 Complexity Regularization

In most of the constructive methods and pruning methods, networks need to be retrained after network architectures are altered. One way to train and select the networks simultaneously is through the use of complexity regularization. The basic idea of complexity regularization is to add a complexity term to the error function that discourages the learning algorithm from seeking solutions which are too complex.

In complexity regularization methods, there is a balance, controlled by the regularization parameter λ, between the error term and the complexity term. The difficulty in using such methods lies in the selection of suitable parameter λ, which often involves tedious trial-and-error experiments. Large parameter λ may introduce large bias, while small parameter λ may cause large variance.

3 Evolutionary Neural Network Ensembles

In single NN learning, the NN learning problem is often formulated as an optimization problem, i.e., minimizing certain criteria, e.g., minimum error, fastest learning, lowest complexity, etc., about architectures. Learning algorithms are used as optimization algorithms to minimize an error function. Despite the different error functions used, these learning algorithms reduce a learning problem to the same kind of optimization problem.

Learning is different from optimization because the learned system is expected to have best generalization, which is different from minimizing an error function. The NN with the minimum error does not necessarily have the best generalization unless there is an equivalence between generalization and the error function. Regardless of the error functions used, single network methods are still used as optimization algorithms. They just optimize different error functions. The nature of the problem is unchanged.

While there is little to be done in single NN learning, there are opportunities in ensemble learning. NN ensembles adopt the divide-and-conquer strategy. Instead of using a single network to solve a task, an NN ensemble combines a set of NNs which learn to subdivide the task and thereby solve it more efficiently and elegantly.

In this section, an evolutionary NN ensemble based on evolutionary learning and negative correlation learning [4,5] is used to explain how the evolutionary process should be stopped.

3.1 Evolving Neural Network Ensembles

The idea of the evolutionary NN ensemble is to regard the population of NNs as an ensemble, and the evolutionary process with fitness sharing as the design of NN ensembles. Fitness sharing refers to a class of speciation techniques in evolutionary algorithms. The fitness used in this paper is based on the mutual information between one individual and the rest of population. The mutual information between two variables, output F_i of network i and output F_j of network j, is given by

$$I(F_i; F_j) = H(F_i) + H(F_j) - H(F_i, F_j) \tag{1}$$

where $H(F_i)$ is the entropy of F_i, $H(F_j)$ is the entropy of F_j, and $H(F_i, F_j)$ is the joint differential entropy of F_i and F_j. The equation shows that joint differential entropy can only have high entropy if the mutual information between two variables is low, while each variable has high individual entropy. That is, the lower mutual information two variables have, the more different they are. By minimizing the mutual information between variables extracted by two NNs, two NNs are forced to convey different information about some features of their input.

The major steps of evolutionary NN ensembles are given as follows [6]:

1. Generate an initial population of M NNs, and set $k = 1$. The number of hidden nodes for each NN, n_h, is specified by the user. The random initial weights are distributed uniformly inside a small range.
2. Train each NN in the initial population on the training set for a certain number of epochs using negative correlation learning. The number of epochs, n_e, is specified by the user.
3. Calculate the fitness of M NNs in the population on the validation set.
4. Randomly choose a group of n_b NNs as parents to create n_b offspring NNs by Gaussian mutation.
5. Replace the worst n_b NNs in the current population with the n_b offspring NNs, and train the new population using negative correlation learning for n_e epochs.
6. Stop the evolution process if the maximum number of generations has been reached. Otherwise, $k = k + 1$ and go to Step 2.

There are two levels of adaptation in evolutionary NN ensembles: negative correlation learning at the individual level and evolutionary learning based on evolutionary programming (EP) [3] at the population level. In evolutionary NN ensembles, an evolutionary algorithm based on evolutionary programming [3] has been used to search for a population of diverse individual NNs that solve a problem together. The fitness evaluation in evolutionary NN ensembles is carried out by fitness sharing based on the minimization of mutual information. In order to create a population of NNs that are as uncorrelated as possible, the mutual information between each individual NN and the rest of population should be minimized. The fitness f_i of individual network i in the population can therefore be evaluated by the mutual information:

$$f_i = \frac{1}{\sum_{j \neq i} I(F_i, F_j)} \tag{2}$$

Minimization of mutual information has the similar motivations as fitness sharing. Both of them try to generate individuals that are different from others, though overlaps are allowed.

3.2 Negative Correlation Learning

We consider estimating y by forming an NN ensemble whose output is a simple averaging of outputs F_i of a set of NNs. Given the training data set $D = \{(\mathbf{x}(1), y(1)), \cdots, (\mathbf{x}(N), y(N))\}$, all the individual networks in the ensemble are trained on the same training data set D

$$F(n) = \frac{1}{M} \Sigma_{i=1}^{M} F_i(n) \tag{3}$$

where $F_i(n)$ is the output of individual network i on the nth training pattern $\mathbf{x}(n)$, $F(n)$ is the output of the NN ensemble on the nth training pattern, and M is the number of individual networks in the NN ensemble.

The idea of negative correlation learning [4,5] is to introduce a correlation penalty term into the error function of each individual network so that the mutual information among the ensemble can be minimized. The error function E_i for individual i on the training data set $D = \{(\mathbf{x}(1), y(1)), \cdots, (\mathbf{x}(N), y(N))\}$ in negative correlation learning is defined by

$$E_i = \frac{1}{N} \Sigma_{n=1}^{N} E_i(n) = \frac{1}{N} \Sigma_{n=1}^{N} \left[\frac{1}{2} (F_i(n) - y(n))^2 + \lambda p_i(n) \right] \tag{4}$$

where N is the number of training patterns, $E_i(n)$ is the value of the error function of network i at presentation of the nth training pattern, and $y(n)$ is the desired output of the nth training pattern. The first term in the right side of Eq.(4) is the mean-squared error of individual network i. The second term p_i is a correlation penalty function. The purpose of minimizing p_i is to negatively correlate each individual's error with errors for the rest of the ensemble. The parameter λ is used to adjust the strength of the penalty.

The penalty function p_i has the form

$$p_i(n) = -\frac{1}{2} (F_i(n) - F(n))^2 \tag{5}$$

$$e_{1s}\dot{e}_{1s} = e_{1s}[\Delta h_{f1}(x) + b_1(x_1)(x_2 - x_2^d) - \tilde{f}_1(x_1) - k_{1d}e_{1s}$$
$$+[-\tilde{b}_1(x_1) + \Delta h_{b1}(x)]\frac{-\hat{f}_1(x_1) + \dot{z}_1 - \dot{e}_{1s} - k_{1d}e_{1s} + \dot{x}_1^d}{\hat{b}_1(x_1)} + b_1(x_1)u_{2c}]$$

If choose the adjusting law of the weight value of LPNN as:

$$\dot{\hat{W}}_{f1} = e_{1s}\eta_{f1}B_{f1}(x_1) \tag{6}$$

$$\dot{\hat{W}}_{b1} = e_{1s}\eta_{g1}B_{b1}(x_1)\frac{-\hat{f}_1(x_1) + \dot{z}_1 - \dot{e}_{1s} - k_{1d}e_{1s} + \dot{x}_1^d}{\hat{b}_1(x_1)} \tag{7}$$

Then the following equation can be satisfied:

$$\begin{aligned} e_{1s}\dot{e}_{1s} + \tilde{W}_{f1}^T\eta_{f1}^{-1}\tilde{W}_{f1} + \tilde{W}_{b1}^T\eta_{b1}^{-1}\tilde{W}_{b1} &= e_{1s}[\Delta h_{f1}(x) + b_1(x_1)(x_2 - x_2^d) \\ -k_{1d}e_{1s} + \Delta h_{b1}(x)\frac{-\hat{f}_1(x_1) + \dot{z}_1 - \dot{e}_{1s} - k_{1d}e_{1s} + \dot{x}_1^d}{\hat{b}_1(x_1)} &+ b_1(x_1)u_{2c}] \end{aligned} \tag{8}$$

Then we discuss how to design u_{2c} to make the left of the equation (8) less than 0. But considering the introduce of u_{2c}, we redesign the u_{2c} to estimate the affect of $-\tilde{b}_1(x_1)e_{1s}u_{2c}$ and $e_{1s}u_{2c}\Delta h_{b1}(x)$. Design u_{2c} satisfies the following equation:

$$u_{2c} = -\frac{sign(e_{1s})\hat{\xi}_{f1}}{\hat{b}_1(x_1)} - \frac{sign(e_{1s})\hat{\xi}_{b1}}{\hat{b}_1(x_1)}\left|\frac{-\hat{f}_1(x_1) + \dot{z}_1 - \dot{e}_{1s} - k_{1d}e_{1s} + \dot{x}_1^d}{\hat{b}_1(x_1)} + u_{2c}\right|$$

And by choosing a proper k_{1d} the solution of the equation always exist two solutions, and we choose the small one as:

$$u_{2c} = \min(u_{2c1}, u_{2c2})$$

Design u_{2c} as above paragraphs, we get: Obviously, the bad affect of the error of estimation of unknown bound can be eliminate by constructing the tuning laws of $\hat{\xi}_{f1}$ and $\hat{\xi}_{b1}$ as follows:

$$\dot{\hat{\xi}}_{f1} = \rho_{f1}|e_{1s}|$$

$$\dot{\hat{\xi}}_{b1} = \rho_{b1}|e_{1s}|\left|\frac{-\hat{f}_1(x_1) + \dot{z}_1 - \dot{e}_{1s} - k_{1d}e_{1s} + \dot{x}_1^d}{\hat{b}_1(x_1)} + u_{2c}\right|$$

Define $z_2 = x_2 - x_2^d$, leave $\hat{b}_1(x_1)e_{1s}z_2$ to be disposed in the following text, and change the tuning law of $\hat{\xi}_{f1}$ and $\hat{\xi}_{b1}$ as:

$$\dot{\hat{\xi}}_{b1} = \rho_{b1}|e_{1s}|\left|\frac{-\hat{f}_1(x_1)+\dot{z}_1-\dot{e}_{1s}-k_{1d}e_{1s}+\dot{x}_1^d}{\hat{b}_1(x_1)}+u_{2c}+z_2\right|$$

$$\dot{\hat{W}}_{b1} = e_{1s}\eta_{g1}B_{b1}(x_1)\left(\frac{-\hat{f}_1(x_1)+\dot{z}_1-\dot{e}_{1s}-k_{1d}e_{1s}+\dot{x}_1^d}{\hat{b}_1(x_1)}+u_{2c}+z_2\right)$$

It is easy to get that:

$$e_{1s}\dot{e}_{1s}+\tilde{\xi}_{f1}\rho_{f1}^{-1}\dot{\tilde{\xi}}_{f1}+\tilde{\xi}_{b1}\rho_{b1}^{-1}\dot{\tilde{\xi}}_{b1}+\tilde{W}_{f1}^T\eta_{f1}^{-1}\dot{\tilde{W}}_{f1}+\tilde{W}_{b1}^T\eta_{b1}^{-1}\dot{\tilde{W}}_{b1} < -k_{1d}e_{1s}^2+\hat{b}_1(x_1)e_{1s}z_2$$

Then:

$$V_1 = e_{1s}e_{1s}+\tilde{\xi}_{f1}\rho_{f1}^{-1}\tilde{\xi}_{f1}+\tilde{\xi}_{b1}\rho_{b1}^{-1}\tilde{\xi}_{b1}+\tilde{W}_{f1}^T\eta_{f1}^{-1}\tilde{W}_{f1}+\tilde{W}_{b1}^T\eta_{b1}^{-1}\tilde{W}_{b1} > 0 \tag{9}$$

To improve the performance of converging, introduce the initial value of weight to the tuning law of the weight of LPNN as follows:

$$\dot{\hat{W}}_{f1} = e_{1s}\eta_{f1}B_{f1}(x_1)-\eta_{f1}\sigma_{f1}(\hat{W}_{f1}-W_{f1}^0)$$

$$\dot{\hat{W}}_{b1} = e_{1s}\eta_{b1}B_{b1}(x_1)\left(\frac{-\hat{f}_1(x_1)+\dot{z}_1-\dot{e}_{1s}-k_{1d}e_{1s}+\dot{x}_1^d}{\hat{b}_1(x_1)}+u_{2c}+z_2\right)-\eta_{b1}\sigma_{b1}(\hat{W}_{b1}-W_{b1}^0)$$

$$\dot{\hat{\xi}}_{b1} = \rho_{b1}|e_{1s}|\left|\frac{-\hat{f}_1(x_1)+\dot{z}_1-\dot{e}_{1s}-k_{1d}e_{1s}+\dot{x}_1^d}{\hat{b}_1(x_1)}+u_{2c}+z_2\right|-\rho_{b1}\sigma_{eb1}(\hat{\xi}_{b1}-\xi_{b1}^0)$$

$$\dot{\hat{\xi}}_{f1} = \rho_{f1}|e_{1s}|-\rho_{f1}\sigma_{ef1}(\hat{\xi}_{f1}-\xi_{f1}^0)$$

It is easy to get :

$$-\tilde{\xi}_{f1}\rho_{f1}^{-1}\rho_{f1}\sigma_{ef1}(\hat{\xi}_{f1}-\xi_{f1}^0) < -\frac{\sigma_{ef1}}{2}\left|\tilde{\xi}_{f1}\right|^2+\frac{\sigma_{ef1}}{2}\left|\xi_{f1}^*-\xi_{f1}^0\right|^2$$

Define d_1 and k_1 as follows:

$$d_1 = \min\left(k_{1d},\frac{\sigma_{ef1}}{2},\frac{\sigma_{eb1}}{2},\frac{\sigma_{f1}}{2},\frac{\sigma_{b1}}{2}\right)$$

$$k_1 = \frac{\sigma_{ef1}}{2}\left|\xi_{f1}^*-\xi_{f1}^0\right|^2+\frac{\sigma_{eb1}}{2}\left|\xi_{b1}^*-\xi_{b1}^0\right|^2+\frac{\sigma_{f1}}{2}\left|W_{f1}^*-W_{f1}^0\right|^2+\frac{\sigma_{b1}}{2}\left|W_{b1}^*-W_{b1}^0\right|^2$$

Then we get:

$$\dot{V}_1 < -d_1 V_1 + k_1 + \hat{b}_1(x_1) e_{1s} z_2 \tag{10}$$

Consider the second subsystem:

$$\dot{z}_2 = f_2(x_1, x_2) + b_2(x_1, x_2) x_3 - \dot{x}_2^d \tag{11}$$

Adopt the same means described in the above paragraphs, we design x_3^d as:

$$x_3^d = \frac{-\hat{f}_2(x_1, x_2) + \dot{x}_2^d + \dot{z}_2 - \dot{e}_{2s} - k_{2d} e_{2s} + u_{3b}}{\hat{b}_2(x_1,, x_2)} + u_{3c} \tag{12}$$

Where u_{3b} is used to eliminate the affect of $\hat{b}_1(x_1)(x_2 - x_2^d)$, and it is defined as:

$$u_{3b} = -\hat{b}_1(x_1) z_2 e_{1s} / e_{2s} \tag{13}$$

Also, we design u_{3c} as:

$$u_{3c} = -\frac{sign(e_{2s})\hat{\xi}_{f2}}{\hat{b}_2(x_1, x_2)} - \frac{sign(e_{2s})\hat{\xi}_{b2}}{\hat{b}_2(x_1, x_2)} \left| \frac{-\hat{f}_2(x_1, x_2) + \dot{x}_2^d + \dot{z}_2 - \dot{e}_{2s} - k_{2d} e_{2s} + u_{3b}}{\hat{b}_2(x_1, x_2)} + u_{3c} \right| \tag{14}$$

Choose the proper k_{2d} to let the equation always exists two solutions. Choose the small one as:

$$u_{3c} = \min(u_{3c1}, u_{3c2}) \tag{15}$$

And choose the tuning law of the weight of LPNN as:

$$\dot{\hat{W}}_{b2} = e_{2s} \eta_{b2} B_{b2}(x_1, x_2) \left(\frac{-\hat{f}_2(x_1, x_2) + \dot{x}_2^d + \dot{z}_2 - \dot{e}_{2s} - k_{2d} e_{2s} + u_{3b}}{\hat{b}_2(x_1, x_2)} + u_{3c} + z_3 \right) - \eta_{b2} \sigma_{b2} (\hat{W}_{b2} - W_{b2}^0)$$

$$\dot{\hat{W}}_{f2} = e_{2s} \eta_{f2} B_{f2}(x_1, x_2) - \eta_{f2} \sigma_{f2} (\hat{W}_{f2} - W_{f2}^0)$$

$$\dot{\hat{\xi}}_{b2} = \rho_{b2} |e_{2s}| \left| \frac{-\hat{f}_2(x_1, x_2) + \dot{x}_2^d + \dot{z}_2 - \dot{e}_{2s} - k_{2d} e_{2s} + u_{3b}}{\hat{b}_2(x_1, x_2)} + u_{3c} + z_3 \right| - \rho_{b2} \sigma_{eb2} (\hat{\xi}_{b2} - \xi_{b2}^0)$$

$$\dot{\hat{\xi}}_{f2} = \rho_{f2} |e_{2s}| - \rho_{f2} \sigma_{ef2} (\hat{\xi}_{f2} - \xi_{f2}^0)$$

Choose Lyapunov function as:

$$V_2 = e_{2s} e_{2s} + \tilde{\xi}_{f2} \rho_{f2}^{-1} \tilde{\xi}_{f2} + \tilde{\xi}_{b2} \rho_{b2}^{-1} \tilde{\xi}_{b2} + \tilde{W}_{f2}^T \eta_{f2}^{-1} \tilde{W}_{f2} + \tilde{W}_{b2}^T \eta_{b2}^{-1} \tilde{W}_{b2} > 0 \tag{16}$$

Also, we get

$$\dot{V}_1+\dot{V}_2<-d_1V_1-d_2V_2+k_1+k_2+\hat{b}_2(x_1,x_2)e_{2s}z_3 \tag{17}$$

Adopt the same ways and means, consider the n th subsystem:

$$\dot{z}_n=f_n(x)+b_n(x)u-\dot{x}_n^d \tag{18}$$

We design u as:

$$u=\frac{-\hat{f}_n(x)+\dot{x}_n^d+\dot{z}_n-\dot{e}_{ns}-k_{nd}e_{ns}+u_{n+1b}}{\hat{b}_n(x)}+u_{n+1c} \tag{19}$$

$$u_{n+1b}=-\hat{b}_{n-1}(x)z_ne_{n-1s}/e_{ns} \tag{20}$$

And design u_{n+1c} as:

$$u_{n+1c}=\min(u_{n+1c1},u_{n+1c2}) \tag{21}$$

Choose the tuning law of the weights of LPNN as:

$$\dot{\hat{W}}_{bn}=e_{ns}\eta_{bn}B_{bn}(x)(\frac{-\hat{f}_n(x)+\dot{x}_n^d+\dot{z}_n-\dot{e}_{ns}-k_{nd}e_{ns}+u_{n+1b}}{\hat{b}_n(x)}+u_{n+1c})-\eta_{bn}\sigma_{bn}(\hat{W}_{bn}-W_{bn}^0)$$

$$\dot{\hat{W}}_{fn}=e_{ns}\eta_{fn}B_{fn}(x)-\eta_{fn}\sigma_{fn}(\hat{W}_{fn}-W_{fn}^0)$$

$$\dot{\hat{\xi}}_{bn}=\rho_{bn}|e_{ns}|\left|\frac{-\hat{f}_n(x)+\dot{x}_n^d+\dot{z}_n-\dot{e}_{ns}-k_{nd}e_{ns}+u_{n+1b}}{\hat{b}_n(x)}+u_{n+1c}\right|-\rho_{bn}\sigma_{ebn}(\hat{\xi}_{bn}-\xi_{bn}^0)$$

$$\dot{\hat{\xi}}_{fn}=\rho_{fn}|e_{ns}|-\rho_{fn}\sigma_{efn}(\hat{\xi}_{fn}-\xi_{fn}^0) \tag{21}$$

Define d and k as:

$$d=\min(d_i),i=1,\cdots,n\,,k=\sum_{i=1}^{n}k_i\,,V=\sum_{i=1}^{n}V_i \tag{22}$$

According to the Barbalat lemma, we get:

$$\dot{V}<-dV+k\,,\quad 0\le V(t)\le\frac{k}{d}+(V(0)-\frac{k}{d})e^{-dt}\,,\ \lim_{t\to\infty}V(t)=0 \tag{23}$$

When $t\to\infty$, the error of estimation of the weight of LNPP $\tilde{W}_{bi}\to 0$ and $\tilde{W}_{fi}\to 0$, and the error of estimation of the unknown bound

$\tilde{\xi}_{bi} \to 0$ and $\tilde{\xi}_{fi} \to 0$, also get $e_{is} \to 0, \hat{W}_{bi} \to W_{bi}^{*}, \hat{W}_{fi} \to W_{fi}^{*}$.And eventually we get $z_i \to 0$.

4 Simulation Study

The Duffing chaotic system is choose to do the simulation. And the Duffing chaotic system can be defined as:

$$\ddot{x} + a\dot{x} + bx + cx^3 = d\sin(\omega t)$$

Where d and ω are known, and a, b, c are unknown, by introducing u, the unknown Duffing chaotic system can be written as:

$$\ddot{x} = -a\dot{x} - bx - cx^3 d\sin(\omega t) + u$$

According the discuss of our paper, the system can be transformed to "standard block control type "as follows.

$$\dot{x}_1 = f_1(x_1) + b_1(x_1)x_2$$

$$\dot{x}_2 = f_2(x_1, x_2) + b_2(x_1, x_2)u_m$$

Because $d\sin(\omega t)$ is known, so we redefine the control variable u_m as:

$u_m = d\sin(\omega t) + u$, define $f_1(x_1)=b_1(x_1)=b_2(x_1,x_2)=1$, $f_2(x_1,x_2)=-ax_2-bx_1-cx_1^3$.

So the method discuss in our paper can be applied to the system which makes $x_1 \to x_1^d$. When choose the parameter as $\omega = 1.8$, $a = 0.05$, $b = 0.2$, $c = 2$, $d = 7$ the output of system is chaotic as figure1.and choose control parameter as $k_{1d} = 5$, $k_{2d} = 7.5$, $e_{1s} = 30z_1 + 11\dot{z}_1 + \ddot{z}_1$, $e_{2s} = 9z_2 + 6\dot{z}_2 + \ddot{z}_2$, $x_1^d = 1$, the result of simulation is fine(show by figure2).

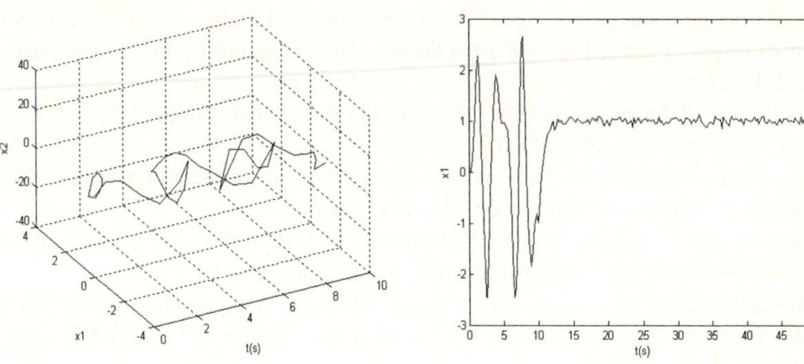

Fig. 1. The chaotic output of system **Fig. 2.** The trace of $x_1^d = 1$

5 Conclusion

A new robust controller based on linear-paremeter-neural-networks is designed for a class of nonlinear unkonwn chaotic systems which could be turned to "standard block control type" by using backstepping method. It was proved by constructing Lyapunov function step by step that all signals of the system are bounded and exponentially converge to the neighborhood of the origin globally and the weights of neural network converge to the optimal weights eventually. The assumption for unknown control function is reduced which stand for the innovation of our method compared with the traditional method. Also the unknown control function needn't to be positive or negative strictly in our paper. This assumption in the other papers is so strict that it couldn't be satisfied by many practical systems. So our method can be applied to a more extensive nonlinear systems. At last, take the unknown Duffing chaotic system for example, simulation study is given to demonstrate that the proposed method is effective. Though it is a sample system, but it shows that the method of our paper can be applied to the chaotic systems with complex mechanism.

References

1. Ungar, L. H., Powell, B. A., and Kamens, S. N. Adaptive networks for fault diagnosis and process control. Computers & Chemical Engineering. Vol. 14 (1990) 561-572
2. F.C. Chen, C.C. Liu. Adaptively controlling nonlinear continuous-time systems using multiplayer neural networks. IEEE Transactions on Automatic Control, Vol. 39 (1994) 1306-1310
3. Polycarpou, M. M., Ioannou, P. A. A robust adaptive nonlinear control design. Automatica, Vol. 32. No. 3 (1996) 423-427
4. Chiman Kwan and F. L. Lewis. Robust backstepping control of nonlinear systems using neural networks. IEEE Transactions on systems, man and cybernetics,Vol. 30. No. 6 (2000) 753-766
5. B. S. Kim, and A. J. Calise, "Nonlinear Flight Control Using Neural Networks," Proc. of rhe AZAA Guidance, Navigation and Coy2trol Conference, Scottsdale, AZ, 1994, pp. 930-940.
6. J. Leitner, Helicopter Nonlinear Control Using Adaptive Feedback Linearization, Ph.D. Thesis, Georgia Institute of Technology, Atlanta, CA, 1995.
7. M. B. McFarland, and A. J. Calise, "Neural Networks for Stable Adaptive Control of Air-to-Air Missiles," Proc. Of the AIM Guidance, Navigation and Control Conference, Baltimore, MD, 1995.
8. D. Serakos, "Nonlinear Controllers for a Tail-Fin Controlled Missile," Proc. IEEE Southeastcon, Charlotte, NC, 1993. 2000
9. Z. Qu, "Robust Control of Nonlinear Uncertain Systems Under Generalized Matching Conditions," Automatica, Vol. 29, pp. 985-998, 1993.
10. Marino, R., and P. Tomei, "Robust Stabilization of Feedback Linearizable Time-varying Uncertain Nonlinear Systems," Automatica, Vol. 29, pp. 181-189, 1993.
11. I. Kanellakopoulos, P. V. Kokotovic, and R. Marino, "An Extended Direct Scheme for Robust Adaptive Nonlinear Control," Automatica, Vol. 27, pp. 247-255, 1991.
Automation (ICCA'04)

Applications of Granger Causality Model to Connectivity Network Based on fMRI Time Series

Xiao-Tong Wen[2], Xiao-Jie Zhao[1,2], Li Yao[1,2], and Xia Wu[2]

[1] School of Information Science and Technology,
Beijing Normal University, Beijing, China, 100088
[2] State Key Laboratory of Cognitive Neuroscience and Learning,
Beijing Normal University, Beijing, China, 100088
zhaoxj86@hotmail.com

Abstract. The connectivity network with direction of brain is a significant work to reveal interaction and coordination between different brain areas. Because Granger causality model can explore causal relationship between time series, the direction of the network can be specified when the model is applied to connectivity network of brain. Although the model has been used in EEG time sires more and more, it was seldom used in fMRI time series because of lower time resolution of fMRI time series. In this paper, we introduced a pre-processing method to fMRI time series in order to alleviate the magnetic disturbance, and then expand the time series to fit the requirement of time-variant algorism. We applied recursive least square (RLS) algorithm to estimate time-variant parameters of Granger model, and introduced a time-variant index to describe the directional connectivity network in a typical finger tapping fMRI experiment. The results showed there were strong directional connectivity between the activated motor areas and gave a possibility to explain them.

1 Introduction

Granger causality model introduced by Granger in 1969 is an important approach to explore the dynamic causal relationships between two time series [1]. With its application to more and more research field, some researchers began to apply the model to time series from neurophysiologic and neuroimaging data [2][3]. The causal relationship based on brain data, such as EEG, fMRI, can help us to understand how the different brain regions coordinate and interact dynamically, and reveal the influence one brain region exerts over another. This kind of causal relationship can also be called effective connectivity network compared with correlation or synchronization network. However, many researchers studied the effective connectivity network mainly by EEG or MEG data because of their high time resolution. Moller introduced generalized recursive least square causality model to investigated Stroop effective based on EEG data, and Ding developed adaptive vector autoregressive causality model to investigate the connectivity network on cortex of monkey [4][5]. In recent years, several works were aimed the effective connectivity using fMRI data due to its inherit advantages for localization [6][7]. Although fMRI

L. Jiao et al. (Eds.): ICNC 2006, Part I, LNCS 4221, pp. 205 – 213, 2006.

2.3 Determination of VAR Model Order

The order of an M-dimensional VAR model in equation (6) can be determined by minimizing Akaike Information Criterion (AIC) defined as:

$$AIC(p) = N\ln[\det(\Sigma(p))] + 2pM^2. \tag{8}$$

Here N is the number of the time points. We applied the RLS algorithm to M-dimensional signal and obtained the time-variant matrix series $\{\Sigma_p(n)\}$ for each p. After the fluctuation of the $\Sigma(n)$ converged to a relatively low and stable range, $\hat{\Sigma}(p)$, the estimation of $\Sigma(p)$ was calculated by averaging the $\Sigma(n)$s along the time. Usually, for a signal with several hundred sampling points, the data of 100 time points was enough.

2.4 Statistical Test

To explore the significance of the time-variant Granger causality, we generated surrogate data using amplitude adjusted Fourier transform (AAFT) method which made the surrogate data remain the original frequency characteristic [11]. Firstly, the method standardized the signal and matched the signal to a random Gaussian progress. Secondly, it transformed the signal to frequency domain using FFT, then maintained its amplitude spectrum and randomized its phase spectrum according a uniform distribution on $[0, 2\pi]$. Thirdly, by using IFFT, the surrogate signal is obtained. To test Granger causality of Y to X, we generated 100 surrogate data X_i to calculate the $F_{Y\to Xi}(n)$, *i*=1, 2, 3, …, 100. For each time point t, we sorted the $F_{Y\to Xi}(t)$ s in a descendent order, picked out the 95th value, and obtained a threshold sequence $F_{thre}(n)$. We defined a significance index $I_{Y\to X}(n)$ as:

$$I_{Y\to X}(n) = \begin{cases} F_{Y\to X}(n) - F_{thre}(n), & (F_{Y\to X}(n) \ge F_{thre}(n)) \\ 0, & else \end{cases}. \tag{9}$$

Where $I_{Y\to X}(n)>0$ denoted there was significant Granger causality from Y to X.

3 Experiment and Results

3.1 Experiment

In this experiment, subjects did the finger-opposite movement according to the direction from the experimenter. The experiment paradigm was illustrated in Fig. 1. The scan sequences were arranged into 3 blocks and there were 10 scans in each block.

Each block would last 30 seconds. The total scan time was 210 seconds. The fMRI data was acquired in a 2.0 T GE/Elscint Prestige whole-body MRI scanner.

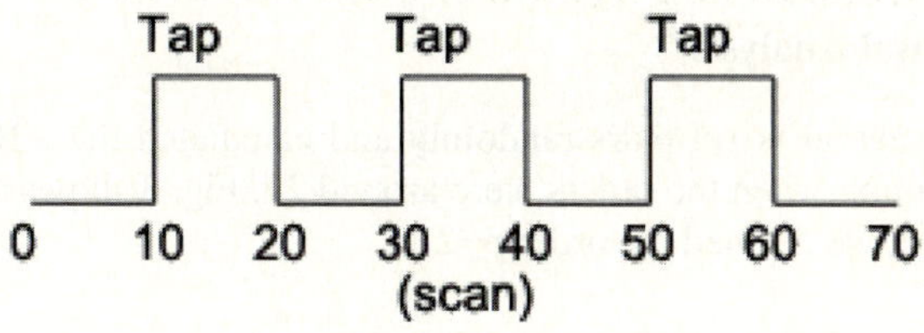

Fig. 1. The arrangement of block design in the experiment

3.2 Data Pre-processing and fMRI Time Series

70 images were obtained along 210 seconds and the last 65 images were selected. We use the famous software package SPM2 (Wellcome Department of Cognitive Neurology, London, UK, **http://www.fil.ion.ucl.ac.uk/spm/**) to find the activated areas in order to locate the region of interest (ROI) (Fig. 2).

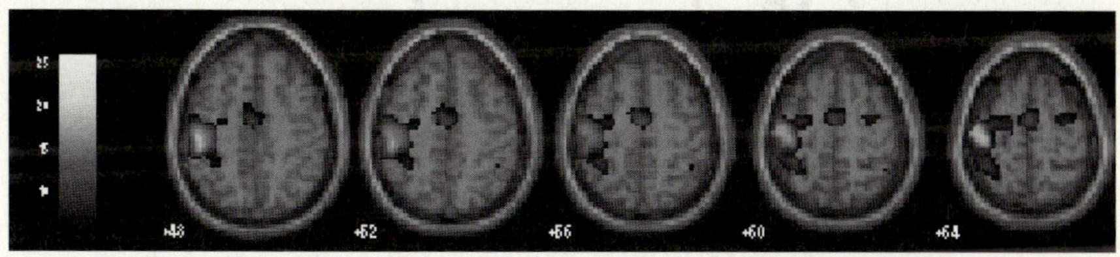

Fig. 2. The activated areas obtained by SPM2

After processed fMRI data using SPM2, we chose a voxel in the ROI and extracted the corresponding values of different scans. But these values couldn't describe the time variance clearly because of the influence of changing magnetic strength from one scan to another. Therefore, for value $v_0(n)$ in the *n*th scan, we normalized the value by dividing it with the averaged global value $\overline{v}(n)$ of this scan and obtained $v(n)$. Another problem for fMRI time series is lack of enough data points because of low temporal sampling rate. The RLS algorithm requires series long enough to make the recursive progress converge to a relatively stable state. To overcome this conflict, we interpolated $\{v(n)\}$ by means of lowpass interpolation algorithm [12]. The algorithm expands the input vector to the correct length by inserting zeros between the original data values. It designs a special symmetric FIR filter that allows the data to pass through and replaces the interpolated zeros with estimated values so that the mean-square errors between the interpolated points and their ideal values are minimized, then applies the filter to the input vector to produce the interpolated output vector. The length of the interpolating filter is $2lr+1$, the factor r denotes that the output vector is r times longer than the original vector. Factor l should be less than 10 and we set l=4 as common. In our work, we set r=10 and obtained an interpolated time series with 650 time points.

3.3 Granger Causal Analysis

Firstly, we chose several voxel pairs randomly and calculated the AICs. Most AICs attained their minimum when the orders were around 21. Fig. 3 shows a typical one of them. Consequently, we defined the order p=21.

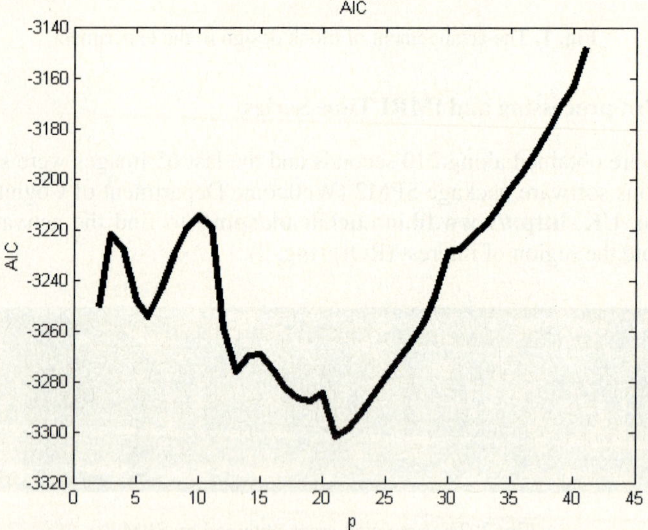

Fig. 3. The AIC curve for two voxels according to equation (8)

Then, we specified three separated spatial voxels A, B and C from different brain regions. Their details are listed in Table 1.

Table 1. The selected voxels

Voxel	Cortex region	State
A	Supplementary motor area	Activated
B	Left motor cortex	Activated
C	Right temporal cortex	Non-activated

Using their interpolated time series, we calculated the time-variant significant indices, $I_{B\to A}(n), I_{A\to B}(n)$, $I_{C\to A}(n)$ and $I_{A\to C}(n)$. To facilitate the analysis of the time-variant characteristic of the changing indices, the results were smoothed by a sliding 21 points wide average-window (Fig. 4 and Fig. 5).

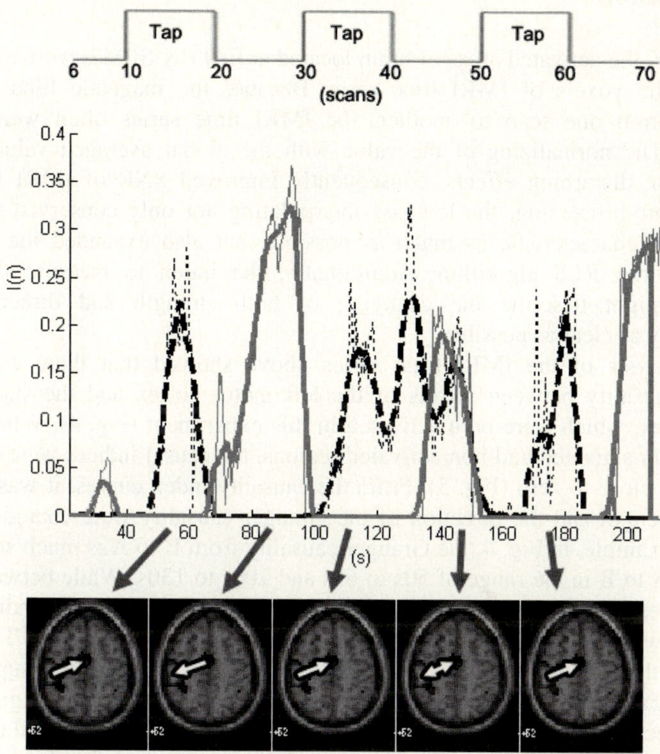

Fig. 4. Time-variant Granger causality between A and B both from activated areas with mapping to the same slice according to time sequence. The bold solid line represented Granger causality from A to B, and the bold dash line represented that from B to A. The direction of the arrow in brain slice referred causal relation between two voxels.

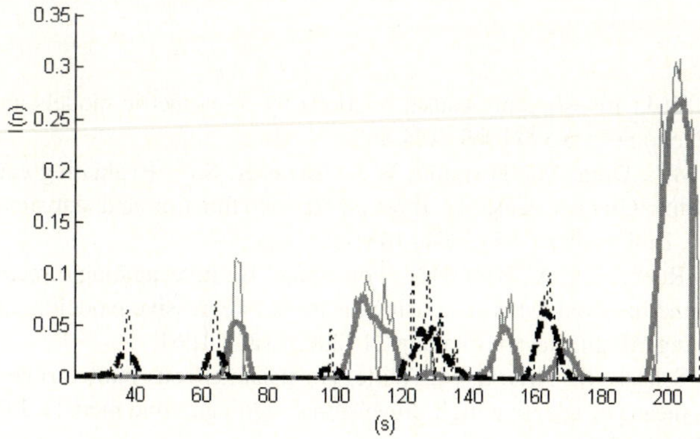

Fig. 5. Time-variant Granger causality between A and C from activated and non-activated areas separately

4 Discussion

In our work, the activated areas of brain located at first (by SPM2) were necessary for choosing the voxels of fMRI time serie. Because the magnetic filed were often changing from one scan to another, the fMRI time series often were disturbed seriously. The normalizing of the value with the global averaged value helped to alleviate the disturbing effects, consequently improved SNR of fMRI time series. After the pre-processing, the lowpass interpolating not only conserved the signal's time-variant characteristic as much as possible, but also expanded the series long enough to run RLS algorithm. Additionally, the index to measure the Granger causality might describe the changing of both strength and direction of the connectivity as clear as possible.

The analysis of the fMRI time series above showed that there exited strong Granger causality between voxels of the left motor cortex and the supplementary motor cortex, which were both activated in this experiment (Fig. 4). While, between the voxels in activated and non-activated regions, the causal indices were really weak and almost close to zero (Fig. 5). From the causality index curves, it was also found that the strength and the direction of the Granger causality were changing with the time. For example, in Fig. 4, the Granger causality from B to A is much stronger than that from A to B in the range of 50s to 65s and 100s to 130s. While between 65s and 100s, the result was inversed, and in the range of 130s to 150s, the indices of both directions were almost the same and remained a relatively high level. These results illustrated that the causality between the left motor cortex and the supplementary motor cortex was not unchangeable and related to the experiment design. The work in-depth needs the more particular experiment design and the anatomical explanation. And furthermore, it indicated that under the obvious strong relationships between the activated cortexes, there was undergoing connectivity with directions.

Acknowledgment. This work is supported by National Natural Science Foundation of China (60472016), National Natural Science Foundation of Beijing (4061004) and Technology and Science Project of Ministry of Education of China (105008).

References

1. Granger, C.W.J.: Investigating causal relations by econometric models and cross-spectral methods. Econometrica 37 (1969) 424-438
2. Kaminski, M.J., Ding, M., Truccolo, W.A., Bressler, S.L.: Evaluating causal relations in neural systems: Granger causality, directed transfer function and statistical assessment of significance. Biol. Cybern. 85 (2001) 145-157
3. Goebel, R., Roebroeck, A., Kim, D.S., Formisano, E.: Investigating directed cortical interactions in time-resolved fMRI data using vector autoregressive modeling and Granger causality mapping. Magn. Reson. Imaging 21 (2003) 1251-1261
4. Moller, E., Schack, B., Arnold, M., Witte, H.: Instantaneous multivariate EEG coherence analysis by means of adaptive high-dimensional autoregressive models. J. Neurosci. Methods 105 (2001) 143-58

5. Ding, M., Bressler, S.L., Yang, W., Liang, H.: Short-window spectral analysis of cortical event-related potentials by adaptive multivariate autoregressive modelling: data preprocess-ing, model validation, and variability assessment. Biol. Cybern. 83 (2000) 35-45
6. van de Ven VG, Formisano E, Prvulovic D, et al. Functional Connectivity as Revealed by Spatial Independent Component Analysis of fMRI Measurements Dur-ing Rest. Hum. Brain Mapp. 22 (2004) 165-178.
7. Astolfi, L., Cincotti, F., Mattia, D., etl.: Estimation of the effective and functional human cortical connectivity with structural quation modeling and directed transfer function applied to high-resolution EEG. Magn. Reson. Imaging 22 (2004) 1457-1470
8. Geweke, J.: Measurement of linear dependence and feedback between multiple time series. J. Amer. Statist. Assoc. 77 (1982) 304-324
9. Brockwell, P.J., Davis, R.A.: Time Series: Theory and Methods. New York: Springer, (1987) 417-420
10. Haykin, S.: Adaptive Filter Theory. Prentice Hall (1986) 381-407
11. Theiler, J., Eubank, Longtin, S.J., Galdrikian, B., Farmer, J.D.: Testing for nonlinearity in time series: The method of surrogate data. Phys. D 58 (1992) 77-94
12. Wiley, J.: Programs for Digital Signal Processing. IEEE Press, New York (1979)

A Spiking Neuron Model of Theta Phase Precession

Enhua Shen, Rubin Wang, Zhikang Zhang, and Jianhua Peng

Institute for Brain Information Processing and Cognitive Neurodynamics,
School of Information Science and Engineering,
East China University of Science and Technology,
130 Meilong Rd., Shanghai 200237, China
sheh@ecust.edu.cn

Abstract. Theta phase precession is an interesting phenomenon in hippocampus and may enhance learning and memory. Based on Harris KD et al. and Magee JC's electrophysiology experiments, a biology plausible spiking neuron model for theta phase precession was proposed. The model is both simple enough for constructing large scale network and realistic enough to match the biology context. The numerical results of our model were shown in this paper. The model can capture the main attributes of experimental result. The results of a simple neuron network were also showed in the paper, and were compared with single neuron result. The influence of network connections on theta phase precession was discussed. The relationship of phase shift with place shift in experiment was well repeated in our model. Such a model can mimic the biological phenomenon of theta phase precession, and preserve the main physiology factors underline theta phase precession.

1 Introduction

O'Keefe and Recce first discovered in rat hippocampus, the phase (defined by hippocampus local theta rhythm) at which a pyramidal cell fires advances as the rat passes through the place field, which they called 'theta phase precession' [1]. A more detailed experiment showed that this phenomenon is robust across hippocampus neural populations [2], and an important discovery of them is that the neurons with neighboring place fields show nearly fixed phase difference, and phase differences increase with place fields shift. This attribute of theta phase precession will cause the place sequence the rat passed to be represented by a neuron population firing pattern lasting for 0.1 second and repeating in each theta cycle. Such a pattern will enhance the memory of place sequence a lot [3].

Some theoretical models of theta phase precession have been proposed, e.g. asymmetric connection weights in the network [4], [5], mechanisms in neural oscillators needn't asymmetric connections [1], [3], [6]. While these models are hard to be evaluated by experiment. Harris KD et al. (2002) [9] and Magee JC (2001) [8] discovered by electrophysiology experiments that a large exciting input to a hippocampus pyramidal cell can overcome sinusoidal inhibitory input at early phase and then let the cell fire. Thus, the larger the exciting input is, the earlier the firing phase of pyramidal cell will be, then theta phase precession occurs. According to [9], the

L. Jiao et al. (Eds.): ICNC 2006, Part I, LNCS 4221, pp. 214 – 223, 2006.

place related exciting input first increasing then decreasing, which is in agree with [4]. If it's the case, the firing phase will firstly advance and then lag back. While the rat experiments show phase precession is almost monotonic. To "cut off" the lag back part, Harris KD et al. assumed that adaptation of a neuron will stop it's firing before the exciting input decreasing [9]. So there will only be phase advance. They checked their theory in a detailed neuron model and the model did show theta phase precession phenomenon. But according to [9], increasing exciting input will make the neuron stop firing later, which is different from rat experiments. To limit the firing of neurons in the 1st half of theta cycle, Kamondi A et al. (1998) proposed a kind of slow potassium current will limit the firing of neurons in the 1st half of theta cycle [10], which compensate Harris KD et al's model.

To capture the major attribute of experimental theta phase precession phenomenon, we constructed a model with a basic leakage integrate-and-fire neuron, adaptation mechanism and slow potassium current, and try to mimic the firing pattern in real hippocampus neuron in rats. Models constructed with H-H equations, such as [9]'s model and [10]'s model, are computing demanding, so they may not be suitable for large scale simulation of a network.

2 Model

The equation determines the membrane potential of a neuron is:

$$C_m \frac{dV_m}{dt} = -I_{ion} - I_L + I_{input}. \tag{1}$$

Where C_m means membrane capacity; V_m is membrane potential; I_{ion} is the sum of all kinds of ion current, in this model, only two ion currents are considered – slow potassium current and Ca^{2+} related potassium current (for adaptation), which will be introduced later. I_L is leakage current; I_{input} is the sum of inputs from outside.

During the process of integration, if the membrane potential V_m reaches to a threshold V_{th}, the neuron is seen as fire. We just record the time of threshold crossing as firing time t^f. After an absolute refractory period T_{abf} of 2 ms, the membrane potential V_m is reset to resting potential V_{rest}.

$$t_f = t, \quad if\ V_m(t) = V_{th}\ \&\ \frac{dV_m}{dt} > 0. \qquad \lim_{t \to t_f + T_{abr} + 0} V_m(t) = V_{rest}, T_{abr} = 2ms. \tag{2}$$

For the leakage current I_L, it can be calculated by following:

$$I_L = g_L \cdot (V_m - V_{rest}). \tag{3}$$

Where g_L is leakage conductance of membrane, it's a constant.

If we don't include any ion current I_{ion} in our model, under our input scheme, the model's output is like Fig. 1b, which is far from the real neuron's firing pattern (Fig. 1a). Ion currents will make the output more like the experiment result.

First, adaptation. The mechanism of adaptation is come from [11], through minor modification. The ion current underline adaptation is AHP potassium current I_{AHP}.

$$I_{AHP} = \bar{g}_{K(AHP)} \cdot m([Ca^{2+}], t) \cdot (V_m - V_K). \tag{4}$$

Here $\bar{g}_{K(AHP)}$ means the maximum conductance of AHP potassium channel. m, often called gating variable, is a proportion of open AHP potassium channels. During the simulation process, it varies with calcium concentration $[Ca^{2+}]$ and time. V_K is the balance potential of potassium ion. The gating variable m is determined by the following equation:

$$\frac{dm}{dt} = \min(\varphi \cdot [Ca^{2+}], 0.01) \cdot (1-m) - 0.001 \cdot m. \tag{5}$$

This equation comes from [11]. Here φ is a suitable selected constant for matching model in [11]. This equation is H-H like equation. By now we haven't found a simpler way to substitute this equation, so we currently reserve this H-H like equation.

In our neuron model, we use the method proposed by Liu YH & Wang XJ, 2001 [12] to simulate calcium concentration changing. When the neuron fires a spike, calcium concentration will be added a constant small amount Δc. In other time, calcium concentration decays to zero with a time constant τ_{Ca}.

$$\frac{d[Ca^{2+}]}{dt} = -\frac{[Ca^{2+}]}{\tau_{Ca}}; [Ca^{2+}] \to [Ca^{2+}] + \Delta c, \text{ when the neuron fires a pulse.} \tag{6}$$

Eq. (4)~(6) fully describe the adaptation mechanism.

Second, slow potassium current. We try to use a simpler way to capture the main behavior of slow potassium current in [10]. The conductance of inhibitory ion channel is a good index of inhibition extent. Slow potassium current is inhibitory. So in our model, we add an inhibitory input in proportion to conductance of slow potassium channel to simulate the effect of slow potassium current.

In our model, if a neuron fires a single pulse at time t^f, the strength of inhibition will change like this way: I_{Ks}=(t- t^f)2*exp(-(t- t^f)/τ_{Ks}), t≥t^f. Where I_{Ks} means the strength of inhibition, t^f means firing time and τ_{Ks} is time constant of inhibition decay. If there're more than one pulses, the inhibition changes caused by each pulse are integrable. So we can treat this process as a linear system with an impulse response h(t)= c*t^2/2 * exp(-t/τ_{Ks}), t≥0. The Laplacian transform of h(t), L{h(t)}=c/(s+ t/τ_{Ks})^3. The time course of inhibition strength changing can be got from the convolution of h(t) and pulse train.

$$I_{Ks}(t) = \int_0^\infty S(\tau) \cdot h(t-\tau) d\tau, \quad S(t) = \sum_i \delta(t^f(i)), \quad h(t) = c \cdot \frac{q \cdot t^2}{2\tau_{Ks}^2} \cdot e^{(\frac{-t}{\tau_{Ks}})}, \quad t \geq 0 \tag{7}$$

Where $I_{Ks}(t)$ is the time course of inhibition strength changing, S(t) is spike train fired by the neuron, $t^f(i)$ is the firing time of the ith spike, c is a constant to correct the amplitude of h(t).

Our model of single hippocampus neuron is thoroughly described by Eq. (1)~(7). In a neural network, the interaction between spiking neurons is implemented through post synaptic currents. In our research of a simple network, we haven't considered inhibitory neuron so far. The post synaptic current can be well fitted by follow [13]:

$$\alpha(t) = \frac{q}{\tau_s - \tau_f} \cdot [e^{\frac{-t}{\tau_s}} - e^{\frac{-t}{\tau_f}}], \quad t \geq 0. \tag{8}$$

Where α(t) is the waveform of post synaptic current, q is a constant to adjust the amplitude, τ_s and τ_f are parameters to determine the shape of the waveform. The synaptic current given by neuron i to neuron j can be got from the convolution of α(t) and pulse train of neuron i, and multiplies the connection weight from neuron i to neuron j.

$$I_{\mathrm{EPSC}(i \to j)}(t) = w_{ij} \cdot \int_0^\infty S_i(\tau - t_d) \cdot \alpha(t-\tau) d\tau. \tag{9}$$

$I_{EPSC(i\rightarrow j)}(t)$ is the synaptic current given by neuron i to neuron j, $S_i(t)$ is spike train fired by the neuron i, t_d is transmission delay of synapse, w_{ij} is the connection weight from neuron i to neuron j. In a network scheme, we just add the synaptic currents given by all neurons to neuron j to the input to neuron j.

$$I_{input(j)} = I_{outstimulus} + \sum_i I_{EPSC(i\rightarrow j)} . \tag{10}$$

$I_{input}(j)$ is the input to jth neuron, $I_{outstimulus}$ is the input we give to the neuron and will not be affected by network dynamics, the last term of Eq. (10) represents the coupling from the network.

Table 1. Parameters chosen in simulation

Parameter	Value	Eq.No.	Parameter	Value	Eq.No.
C_m	1 nF	1	V_{rest}	-70 mV	2
V_{th}	-54 mV	2	g_L	50 nS	3
$\bar{g}_{K(AHP)}$	4 μS	4	V_K	-80 mV	4
ϕ	2000	5	Δc	0.2 μM	6
τ_{Ca}	500msec	6	τ_{Ks}	30 msec	7
c	-1300	7	q	10^{-12}coulomb	8
τ_s	35msec	8	τ_f	4 msec	8
t_d	2 msec	9			

3 Results

The model was programmed through Simulink of MatLab. We uses fixed step ode5 method in simulating. The time step is 0.0002 sec.

We stimulated our single neuron model with an input show in Fig. 2a. According to [9], pyramid neuron in hippocampus will receive a place specified ramp-like excitatory input and a sinusoidal inhibitory input in theta frequency, which simulates the local field potential. The sinusoidal inhibitory input is: -0.2+0.2*sin(10*π*t), in nA. Its frequency is 5 Hz, in theta range (bottom line Fig. 2a). We replace the exciting ramp in [9] with a simpler one, show in the top line of Fig. 2a. The input has a basic level component with a strength of 0.7 nA, and adds a triangle peak centered at 2 sec with a width of 2 sec. The peak strength of exciting ramp is 1.5 nA. In our model we suppose the rat runs in a fixed speed. So the coordinate of place is in proportion to time and plot our model results verses time is equivalent to plot verses place coordinate. We selected the strength of input according to [8]. Besides the above we add white noise to the input, the strength of the noise (standard deviation) is 0.01 nA. The outside input is the sum of sinusoidal inhibitory input, the ramp and the noise.

The firing patterns of our model, subjected to input shown in Fig. 2a, are showed in Fig. 2b and c. The horizontal axes are time, but they're equivalent to place. These results are come from 2 trials. Due to noises, they are different trail by trail, but we can clearly see that these points seem to distribute along a line with a negative slope, which capture the major attribute of theta phase precession. Similar to experiment results in

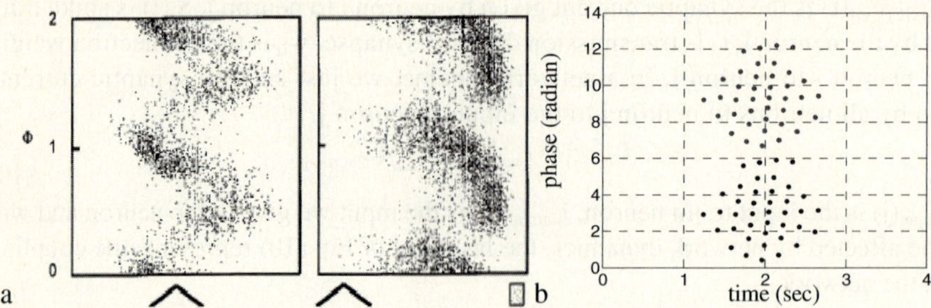

Fig. 1. a) Examples of experiment result chosen from Fig. 7 in [2]. Each point in the figure represents a spike recorded. The vertical axis is phase of the spike, in unit of one theta period. The horizontal axis is the place where the spike appears. b) The firing pattern of an integrate and fire neuron without any ion channels. The vertical axis is phase, but in a unit of radian. The horizontal axis is time. As explained in next paragraph, which is in proportion to place.

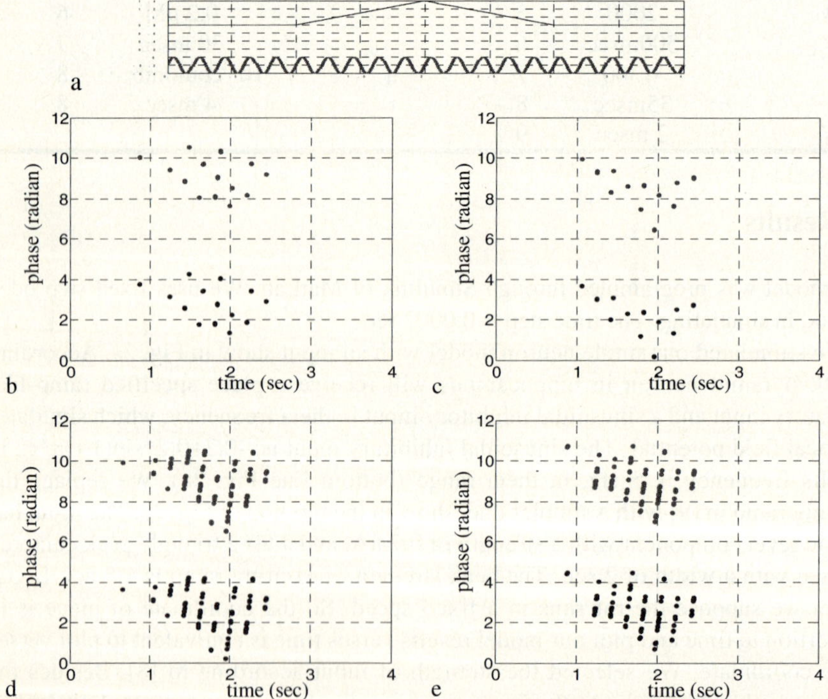

Fig. 2. a) An illustration of input to our model. The input lasts for 4 seconds. The peak is 1.5 nA, the level part is 0.7 nA. The amplitude of oscillatory inhibition is 0.2 nA. Other details can be found in the text. b, c) Results from 2 single trails of simulation of our model. The axes are same as Fig. 1b. d, e) Data overlapped from 6 single trails.

Fig. 1a, we overlapped data from several trails (6 trails) together, show in Fig. 2d and e. The overlapped results then show a relatively fixed pattern, and like the experiment results in Fig. 1a. In Fig. 2d and e, for spikes before 2 sec, which is the peak of our

asymmetric network weights may need to be combined to explain results in [2]. We will evaluate this idea in further research work.

Acknowledgement

This work was supported by the National Natural Science Foundation of China (30270339).

References

1. O'Keefe, J., Recce, M.L.: Phase Relationship between Hippocampal Place Units and the EEG Theta Rhythm. Hippocampus 3 (1993) 317–330
2. Skaggs, W.E., McNaughton, B.L., Wilson, M.A., Barnes, C.A.: Theta Phase Precession in Hippocampal Neuronal Populations and the Compression of Temporal Sequences. Hippocampus 6 (1996) 149–172
3. Yamaguchi, Y.: A Theory of Hippocampal Memory Based on Theta Phase Precession. Biological Cybernetics 89 (2003) 1-9
4. Tsodyks, M.V., Skaggs, W.E., Sejnowski, T.J., McNaughton, B.L.: Population Dynamics and Theta Rhythm Phase Precession of Hippocampal Place Cell Firing: a Spiking Neuron Model. Hippocampus 6 (1996) 271–280
5. Jensen, O., Lisman, J.E.: Hippocampal CA3 Region Predicts Memory Sequences: Accounting for the Phase Precession of Place Cells. Learning Memory 3 (1996) 279–287
6. Bose, A., Booth, V., Recce, M.: A Temporal Mechanism for Generating the Phase Precession of Hippocampal Place Cells. J. Comp. Neurosci. 9 (2000) 5–30
7. Mehta, M.R., Lee, A.K., Wilson, M.A.: Role of Experience and Oscillations in Transforming a Rate Code into a Temporal Code. Nature 417 (2002) 741–746
8. Magee, J.C.: Dendritic Mechanisms of Phase Precession in Hippocampal CA1 Pyramidal Neurons. J. Neurophysiol. 86 (2001) 528-532
9. Harris, K.D., Henze, D.A., Hirase, H., Leinekugel, X., Dragoi, G., Czurko, A., Buzsaki, G.: Spike Train Dynamics Predicts Theta Related Phase Precession in Hippocampal Pyramidal Cells. Nature 417 (2002) 738–741
10. Kamondi, A., Acsady, L., Wang, X.J., Buzsaki, G.: Theta Oscillations in Soma and Dendrites of Hippocampal Pyramidal Cells in vivo: Activity-Dependent Phase-Precession of Action Potentials. Hippocampus 8 (1998) 244–261
11. Traub, R.D., Wong, R.K.S., Miles, R., Michelson, H.: A Model of a CA3 Hippocampal Pyramidal Neuron Incorporating Voltage-Clamp Data on Intrinsic Conductances. J. Neurophysiol., 66 (1991) 635-650.
12. Liu, Y.H. Wang, X.J.: Spike-Frequency Adaptation of a Generalized Leaky Integrate-and-Fire Model Neuron. J. Comp. Neurosci. 10 (2001) 25–45
13. Gerstner, W. Kistler, W.M.: Spiking Neuron Models. Cambridge University Press, Cambridge (2002) Chapter 2
14. Yamaguchi, Y., Aota, Y., McNaughton, B.L., Lipa, P.: Bimodality of Theta Phase Precession in Hippocampal Place Cells in Freely Running Rats. J Neurophysiol 87 (2002) 2629–2642

Suprathreshold Stochastic Resonance in Single Neuron Using Sinusoidal Wave Sequence

Jun Liu, Jian Wu, and Zhengguo Lou

Department of Biomedical Engineering, Key Laboratory of Biomedical Engineering of Ministry of Education of China, Zhejiang University, 310027 Hangzhou, P.R. China
junliu@mail.bme.zju.cn

Abstract. This paper discussed suprathreshold stochastic resonance (SSR) in a HH model forced the periodic input. Temporal sequences of neuronal action potentials were transformed into the sinusoidal wave sequences and the sequences were analyzed by signal-to-noise ratio (SNR) in order to investigate the spectrum characteristic of a Hodgkin-Huxley (HH) neuron. The transformation reflected the pure information of interspike interval and provided a new tool to observe the suprathreshold behavior induced by periodic input. In contrast with the previous SSR in network devices, SSR of single neuron is observed.

1 Introduction

Stochastic resonance (SR) [1] is a counter intuitive nonlinear phenomenon, which has been proven that noise can enhance weak signal transduction in sensory neurons via SR by many experimental and theoretical studies [2-3] In these studies, temporal sequences of neuronal action potentials were analyzed by computer for signal-to-noise ratio (SNR) [2], Shannon information rate [4] and stimulus-response coherence [5]. Among these measurement index, SNR is generally calculated from the ratio of the height of the peak with the same frequency of periodic input to its nearby background in the power spectrum density (PSD), thus SR has usually been loosely defined as occurring when an increase in input noise leads to an increase in output SNR in a nonlinear system driven by a periodic force. An advantage of using SNR measurement is to analyze the frequency property of neuronal model when SR occurs [6]. Beside thisadvantage, there are some physiological bases to sustain rationality of SNR measurement. For example, the auditory, visual, and tactile sensory systems, at least, can be usefully modeled as spectral frequency analyzers [7].

In this paper, we studied numerically nonlinear responses of the periodically forced Hodgkin-Huxley neuron that is biologically realistic. Because of using the sinusoidal wave sequence, the SSR was observed in single neuron. SSR is a phenomenon that stochastic resonance can occur when input signal exceed the threshold and it has been initially defined in an array of threshold devices by maximum information transition rate [8]. In contrast with the previous SSR, which rely on collective dynamical behavior of many neurons, SSR of single neuron has the different mechanism that noises

L. Jiao et al. (Eds.): ICNC 2006, Part I, LNCS 4221, pp. 224 – 227, 2006.

contribute to overcome the barrier from refractory period during which the neuron cannot spike again.

2 Model Description and Response Analysis

A HH neuronal model is a useful paradigm that accounts naturally for both the spiking behavior and refractory properties of real neurons [9], which is described by four nonlinear coupled equations: one for the membrane potential V and the other three for the gating variables: m, n, and h. $I_1 \sin(2\pi f_s t)$ is a periodic signal with I_1 and f_s being the amplitude and the frequency of the signal respectively. I_0 is a constant stimulus and is regarded as the simplest modulation to the neuron. $\xi(t)$ is the Gaussian white noise ,satisfying $<\xi(t)> = 0$, $<\xi(t_1)\, \xi(t_2)> = 2D\delta(t_1\text{-}t_2)$, D is intensity of noises.

3 Suprathreshold Stochastic Resonance Evaluated by Sinusoidal Wave Sequences

According to the study of Yu [6], it can be known that the responses of the input signal with different frequency depend on the match between the input signal and the intrinsic characteristic of neuron. Especially, there is the supratheshold case that spikes cannot reflect the input signal correctly when the frequency of input signal is higher than that of intrinsic firing. Besides increasing the amplitude of input signal, noises are considered to enhance the synchronization between spike and input signal.

3.1 Sinusoidal Wave Sequence and SNR

Due to periodic signal inputted to neuron, SNR naturally acts as the measurement index to analyze the spectrum property. Calculating the SNR, previous methods firstly simplify the spike sequences into the standard rectangle pulse sequences (as shown the middle plot in Fig.1), and then obtain the power spectrum density (PSD) the FFT of the pulse sequences. It can be found some harmonious waves are introduced into the spectrum besides the component of basic frequency and the highest power is not the component of the basic frequency that is approximately the firing frequency. This may

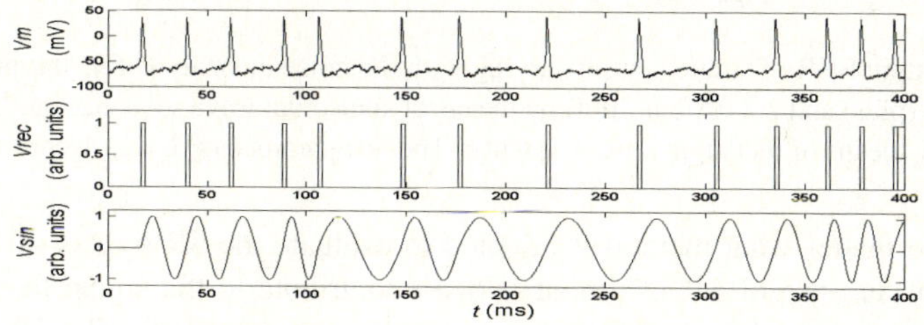

Fig. 1. The spike sequence (upper plot), the corresponding rectangle pulse sequence (middle plot) and the sinusoidal wave sequence (bottom plot)

mislead us to understand the encoding information of spikes. In view of the defect of rectangle pulse sequence, we introduce the sinusoidal wave sequence to replace the spike sequence, as shown in the bottom plot of Fig.1. Each interspike interval firstly is transformed into the period of sinusoidal wave, and then these sinusoidal waves form the sequence according to the place of interspike. It can be proved that the spectrum power concentrate nearby the firing frequency, thus this method reflect the real response of neuron.

After the PSDs of sinusoidal wave sequence are obtain, SNRs are defined as 10log10(G/*B*) with G and *B* representing the height of the signal peak and the mean amplitude of background noise at the input signal frequency in the power spectrum respectively.

3.2 Suprathreshold Stochastic Resonance

Here we investigate how noises influence a HH neuron when the amplitude of sprathreshold input signal change between 1.4μA/cm^2 and 2.5μA/cm^2. As shown in the left plot of Fig.2, three values of I_1 (1.5, 2.0 and 2.3 μA/cm^2) are chosen to simulate the process. It is noted that the sinusoidal wave sequence is used in simulation. We can find that the typical characteristic of SR occurs for three curves, i.e., the output SNR first raises up to a maximum at a certain noise intensity *D* and then drops as *D* increases. Due to the occurrence of SR at the case of suprathreshold input signal, we call it SSR. This is an interesting result that was observed in previous paper by use of the measurement index SNR. The reason is that different transformation methods are used. The right plot of Fig.2 shows that SR does not occur by use of rectangle pulse sequence when I_1=1.5 μA/cm^2 and the rest parameter is the same with left plot.

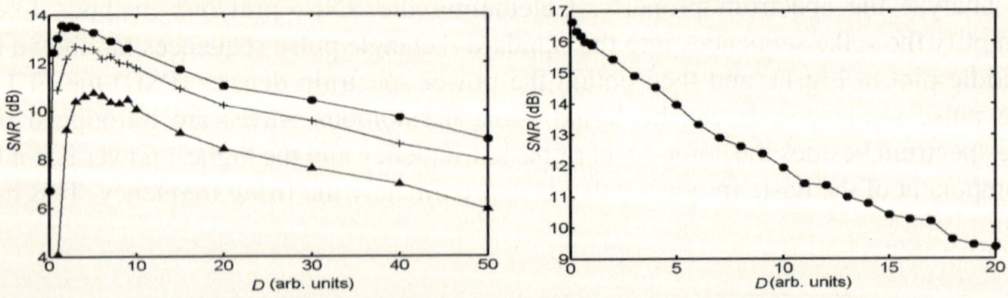

Fig. 2. The output SNR of single neuron varying with the noise intensity *D* for: left plot, I_1=1.5 (triangle), 2.0 (plus) and 2.3 μA/cm^2 (dot) by means of sinusoidal wave sequence, and right plot, 1.5μA/cm^2, by means of rectangle pulse sequence .The rest parameters: I_0=1μA/cm^2, *f*=70Hz.

In fact, no matter what method is adopted to evaluate the SR by use of SNR, the essence of occurrence of SR is similar. Noises contribute to the signal to overcome the barrier that may be the real threshold or the refractory period of spike. Here we use the new method to transform the spike sequence so that SSR can be observed.

4 Conclusion

This paper focuses on the influence of periodic input and noise on a HH neuronal model. Periodic input is conveniently used to study the frequency property of HH model by use of spectrum analysis. We firstly observed the response of a HH model forced by periodic signal, and found the suprathreshold input in a certain range cannot induce the optimal firing response. As compared with the threshold behavior of neuron, a "soft" threshold appears at the case of suprathreshold input. In order to analyze this behavior, the sinusoidal wave sequence is introduced to transform the interspike interval instead of using rectangle pulse sequence. The former removes the disturbance of harmonious wave, gives the concentrated power of spectrum and describe the global information. By use of the new method of sequence transformation, SSR is observed in single HH neuronal model. In contrast with the previous SSR, which rely on collective dynamical behavior of many neurons, SSR of single neuron has the different mechanism that noises contribute to overcome the barrier from refractory period during which the neuron cannot spike again.

Acknowledgment

This work was supported by the National Natural Science Foundation of China (projects No. 30470470) and the Project of Zhejiang Province Education Department of China (20050907).

References

1. Benzi, R., Sutera, A., Vulpiani, A.: The Mechanism of Stochastic Resonance. J. Phys. A: Math. Gen. 14 (1981) 453~457
2. Douglass, J.K., Wilkens, L., Pantazelou, E., Moss, F.: Noise Enhancement of Information Transfer in Crayfish Mechanoreceptors by Stochastic Resonance. Nature 365 (1993) 337~340
3. Gammaitoni, L., Hänggi, P., Jung, P., Marchesoni, F.: Stochastic Resonance. Rev. Mod. Phys. 70 (1998) 223~287
4. Heneghan, C., Chow, C. C., Collins, J. J.: Information Measures Quanifying Aperiodic Stochastic Resonance. Phys. Rev. E. 54 (1996) 2228-2231
5. Collins, J. J., Chow, C. C., Imhoff, T. T.: Aperiodic Stochastic Resonance in Excitable Systems. Phys. Rev. E. 52 (1995) R3321–R3324
6. Yu, Y. G., Liu, F., Wang. W.: Frequency Sensitivity in Hodgkin-Huxley Systems. Biological Cybernetics. 84 (2001) 227-235
7. Coren, S., Ward, L. M., Enns, J.: Sensation and perception, 5th edn. Harcourt, San Diego (1999)
8. Stocks, N. G.: Suprathreshold Stochastic Ressonanc in Multilevel Threshold Systems. Physical Review Letters. 84 (2000) 2310-2313
9. Hodgkin, A., Huxley, A.: A Quantitative Description of Membrane Current and its Application to Conduction and Excitation in Nerve. J. Physiol. 117 (1952) 500-544

Phase Coding on the Large-Scaled Neuronal Population Subjected to Stimulation

Rubin Wang[1,2], Xianfa Jiao[2,3], and Jianhua Peng[1]

[1] Institute for Brain Information Processing and Cognitive Neurodynamics, School of Information Science and Engineering, East China University of Science and Technology, 130 Meilong Road, Shanghai 200237, China

[2] College of Information Science and Technology, Donghua University, Shanghai 200051, China

[3] School of Science, Hefei University of Technology, He fei 230009, China

Abstract. A stochastic nonlinear model of neuronal activity in a neuronal population is proposed in this paper, where the combined dynamics of phase and amplitude is taken into account. An average number density is introduced to describe collective behavior of neuronal population, and a firing density of neurons in the neuronal population is referred to be neural coding. The numerical simulations show that with a weaker stimulation, the response of the neuronal population to stimulation grows up gradually, the coupling configuration among neurons dominates the evolution of the average number density, and new neural coding emerges. Whereas, with a stronger stimulation, the neuronal population responds to the stimulation rapidly, the stimulation dominates the evolution of the average number density, and changes the coupling configuration in the neuronal population.

1 Introduction

The brain is always in the state of receiving neural information flow. Various sensory inputs are processed in the cerebral cortex quickly, so that one can adjust consciousness and behavior to adapt oneself to the change in the environment. It has suggested that synchronous oscillation of neural activity in the central nervous system plays a vital role in information processing in the brain [1,2].In order to investigate the collective behavior of neurons and the candidate mechanisms of neural information processing, a great deal of theoretical work on synchronization in a neuronal population has been investigated [3-6].

Neuronal impulse aroused by external stimulation can form new synapse and change the density of the postsynaptic membrane due to the coupling of synapse, thus the information is stored for a long time. In other words, the long-term memory occurs [7]. Weng suggested that the stimulation's intensity play a prominent role in transformation from short-term memory to long-term memory [8]. Han suggested that the stimulation's duration is an important index in nervous foundation of memory [9]. Therefore, in the present paper, we assume the duration of stimulation coincides with

L. Jiao et al. (Eds.): ICNC 2006, Part I, LNCS 4221, pp. 228 – 236, 2006.

the coupling of neurons in numerical simulation. Based on our previous research [10-18], this paper will investigate the effect of external stimulation on the neural coding.

2 Derivation of Stochastic Model Equation

Letting the amplitudes and the phases of N oscillators are r_j, ψ_j (j = 1,2,…,N). The phase and the amplitude dynamics obey the following evolution equations

$$\dot{\psi}_j = \Omega + \frac{1}{N}\sum_{k=1}^{N} M(\psi_j - \psi_k, r_j, r_k) + S(\psi_j, r_j) + F_{j_1}(t). \tag{1}$$

$$\dot{r}_j = g(r_j) + F_{j_2}(t) \qquad (j = 1, \ldots, N). \tag{2}$$

Where all oscillators have the same eigenfrequency Ω, and the term $M(\psi_j - \psi_k, r_j, r_k)$ models interaction between the kth and the jth oscillator, $S(\psi_j, r_j)$ is considered to be the response of jth oscillator to external stimulation. $g(r_j)$ is a nonlinear function of the amplitude . For the sake of simplicity, the random force, $F_{ji}(t)$ (i=1,2) is modeled by Gaussian white noise, which is delta-correlated with zero mean value [12],

$$< F_{j1}(t) > = < F_{j2}(t) > = 0, \quad < F_{j1}(t) F_{j2}(t') > = 0,$$

$$< F_{j1}(t) F_{j1}(t') > = Q_1 \delta(t - t'), \quad < F_{j2}(t) F_{j2}(t') > = Q_2 \delta(t - t'),$$

where Q_1, Q_2 denote the intensities of the noise F_{j1}, F_{j2} .

According to Eqs.(1) and (2), We get the Fokker-Planck equation of probability density f

$$\frac{\partial f}{\partial t} = \frac{Q_1}{2}\sum_{j=1}^{N}\frac{\partial^2 f}{\partial \psi_j^2} + \frac{Q_2}{2}\sum_{j=1}^{N}\frac{\partial^2 f}{\partial r_j^2} - \sum_{j=1}^{N}\frac{\partial}{\partial \psi_j}\left[\frac{1}{N}\sum_{k=1}^{N}\Gamma(\psi_j, \psi_k, r_j, r_k) f\right]$$

$$-\sum_{j=1}^{N}\frac{\partial}{\partial r_j}[g(r_j) f]. \tag{3}$$

Where $f = f(\psi_1, \psi_2, \ldots, \psi_N, r_1, r_2, \ldots, r_N, t)$, f is the probability density of the oscillators' phases and amplitudes. For the brevity, letting

$$\Gamma(\psi_j, \psi_k, r_j, r_k) = \Omega + M(\psi_j - \psi_k, r_j, r_k) + S(\psi_j, r_j).$$

In order to describe collective behavior of population of neuronal oscillator, we introduce the number density of the neuronal oscillators with the same phase ψ and the same amplitude R as follows:

$$\tilde{n}(\psi, R) = \frac{1}{N}\sum_{k=1}^{N}\delta(\psi - \psi_k)\delta(R - r_k).$$

Taking into account the stochastic aspect of nervous system, the average number density is given by

$$n(\psi,R,t)=\int_0^{2\pi}\ldots\int_0^{2\pi}d\psi_l\int_0^{\infty}\ldots\int_0^{\infty}\frac{1}{N}\sum_{k=1}^{N}\delta(\psi-\psi_k)\delta(R-r_k)fdr_l. \quad (4)$$

Taking the partial derivative of $n(\psi,R,t)$ with respect to t and performing integrations by parts, we obtain

$$\frac{\partial n}{\partial t}=\frac{Q_1}{2}\frac{\partial^2 n}{\partial\psi^2}+\frac{Q_2}{2}\frac{\partial^2 n}{\partial R^2}-\frac{\partial}{\partial\psi}(n\Gamma(\psi,R,t))-\frac{\partial(g(R)n)}{\partial R} \quad (5)$$

Where

$$\Gamma(\psi,R,t)=\Omega+S(\psi,R)+\int_0^{2\pi}d\psi'\int_0^{R_0}M(\psi-\psi',R,R')n(\psi',R',t)dR' \quad (6)$$

$M(\psi-\psi',\mathrm{R},\mathrm{R}')$ can be expanded in Fourier modes, since it is a 2π -periodic function. For numerical simulation, the mutual interaction term, the nonlinear function of amplitude and the stimulation term is given by

$$M(\psi_j-\psi_k,r_j,r_k)=-\sum_{m=1}^{4}r_j^m r_k^m(K_m\sin m(\psi_j-\psi_k)+C_m\cos m(\psi_j-\psi_k)) \quad (7)$$

$$g(r_j)=r_j-r_j^3 \quad (8)$$

$$S(\psi_j,r_j)=\sum_{m=1}^{4}I_m r_j^m\cos(m\psi_j) \quad (9)$$

K_m denotes mth order coupling coefficient between neurons, I_m denotes mth order stimulation. The first term of the average number density in Fourier expansion is given by

$$\hat{n}(0,R,t)=B_1 e^{\frac{2}{Q_2}\int_0^2 g(x)dx}\int_0^R e^{-\frac{2}{Q_2}\int_0^2 g(r)dr}dx. \quad (10)$$

where B_1 is determined by the normalization condition.

3 Numerical Simulation

In order to investigate the response of the neuronal population to stimuli with different intensities, we choose the coupling with one mode, namely, $K_1=1$, and other coupling parameters are all considered to be zero. We describe the evolution of the average number density with respect to t by means of equation (5) as in figure 1. In moderate stimulation, the response of the neuronal population to stimulation grow up gradually

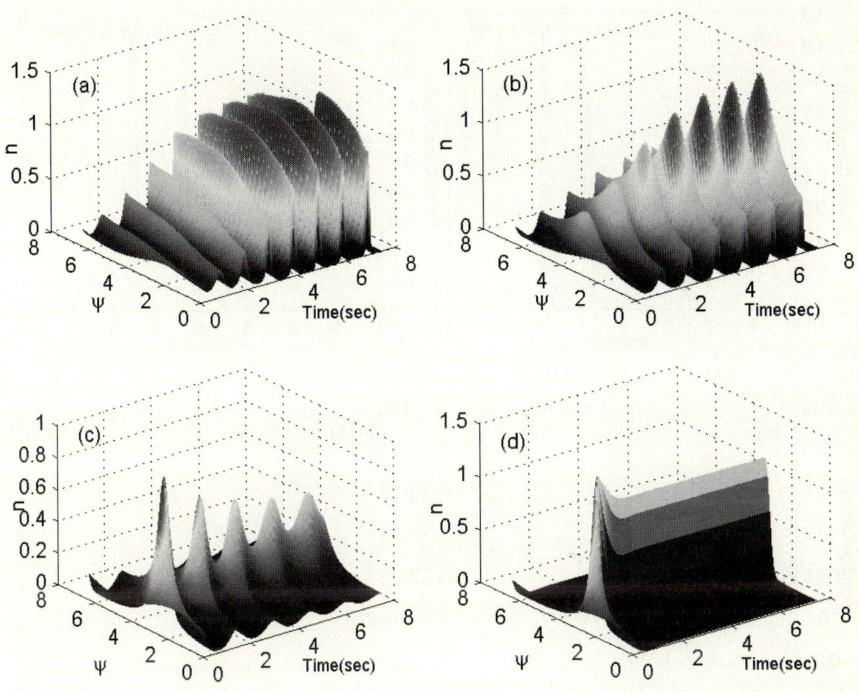

Fig. 1. Average number density's evolution with time. Stimulation parameters: a) $I_1 = 0.5$, b) $I_1 = 3$, c) $I_1 = 5$, d) $I_1 = 7$.

(Fig.1a and b); whereas in the stronger stimulation, the activity of the neuronal population responds to the stimulation rapidly (Fig.1c and d).

Experimental evidences have suggested that with a stronger stimulation, glucocorticoid releases continuously, or glucocorticoids reacts for a long time, the hippocampal capability reduces, the dendrite of the hippocampal CA3 atrophies, abundant pyramidal cells attenuate and fall off, the effect of dentate gyrus granular neuron is inhibited, so that the neuronal plasticity of the hippocampus and some cognitive functions depending on the hippocampus are injured. Figure 2 display the firing pattern of the neuronal population under four different intensities of stimulation. With a moderate stimulation, the neuronal firing density changes, stochastically, with increasing time, and the wave of firing density goes into stationary state (Fig.2a and b). Hence, new neural coding emerges. With a stronger stimulation, the amplitude of the firing density drops (Fig.2c), this shows that a stronger stimulation depresses the synchronization of neuronal firing. What is more, with a strong stimulation, the firing density approaches a constant (Fig.2d), this shows that the neuronal population is desynchronized, and then new neural coding cannot emerge. This behavior is consistent with the experimental evidences.

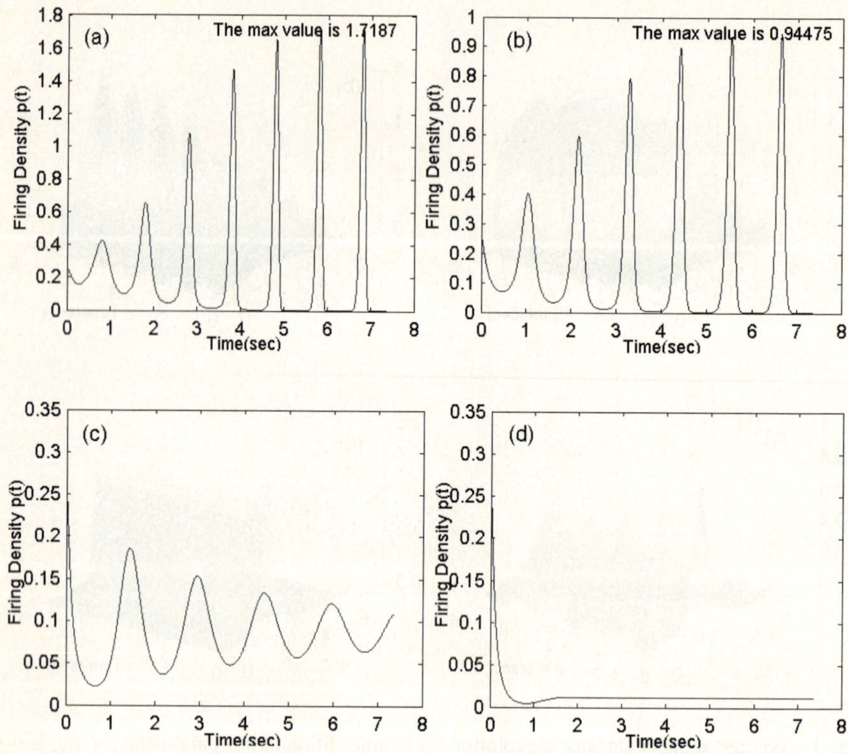

Fig. 2. The firing density of neurons on limit circle $p(t) = n(0,1,t)$ is plotted over time. Stimulation parameters: a) $I_1 = 0.5$, b) $I_1 = 3$, c) $I_1 = 5$, d) $I_1 = 7$.

In order to investigate the response of the neuronal population to the stimulation with different order harmonics, we choose the coupling configuration with one mode $K_2 = 1$, the combined stimulation of different modalities, composed of a first order harmonic and a second order harmonic. Fig.3 displays the evolution of the average number density with respect to phase on limit circle, with increasing time, under no stimulation, the stimulation with a second order harmonic, or the combined stimulation. The evolutions of the average number density differ remarkably. This shows that under the combined modalities of stimulation, these stimuli interact with each other.

In comparison with Fig.4a and Fig.4b, we found that the peak of the latter is lower than the former; this shows that the stimulation with only second order harmonic changes the coupling parameter K_2. In steady states, the periods of Fig.4a and Fig.4b are 0.5, but the period of Fig.4c is 1, and the peak of the latter is lower than the former. This indicates that the combined stimulation with the first order and the second order harmonic changes not only the coupling intensity among but also the coupling

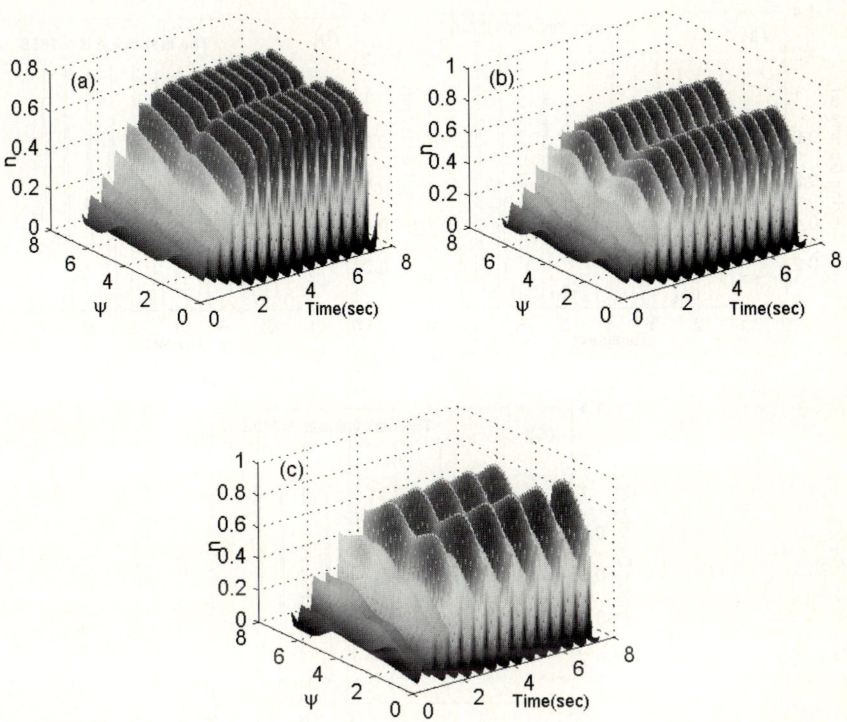

Fig. 3. Average number density's evolution with time on limit circle. Parameter $\Omega = 2\pi$, $Q_1 = 0.4$, $Q_2 = 0.2$, $K_2 = 1$, $K_1 = K_3 = K_4 = 0$, initial condition $n(\psi, R, 0) = \hat{n}(0, R, 0)(1 + 0.05\sin 2\psi)$ (a) $I_1 = I_2 = 0$, (b) $I_1 = 0, I_2 = 0.5$, (c) $I_1 = I_2 = 0.5$.

configuration in the neuronal population, here the new coupling configuration is two-mode configuration, namely $K_1 \neq 0, K_2 \neq 0$ (Fig.4c).

In order to investigate the effect of stimulation on the coupling configuration, we choose the coupling configuration with two modalities $K_1 = K_2 = 1$, and the stimulation with a first order harmonic. With a weaker stimulation, the coupling configuration dominates the evolution of the average number density (Fig.5a), and the firing pattern is also dominated by the coupling configuration, viz. the firing pattern is two synchronous clusters (Fig.6a). However, with a stronger stimulation, the stimulation dominates the evolution of the average number density (Fig.5b), and the firing pattern is one synchronous cluster (Fig.6b), in other words, a stronger stimulation changes the coupling configuration in the neuronal population. This indicates that, in nervous system, with a weaker stimulation, the coupling configuration can dominate neural coding, whereas with a stronger stimulation, the stimulation can dominate neural coding.

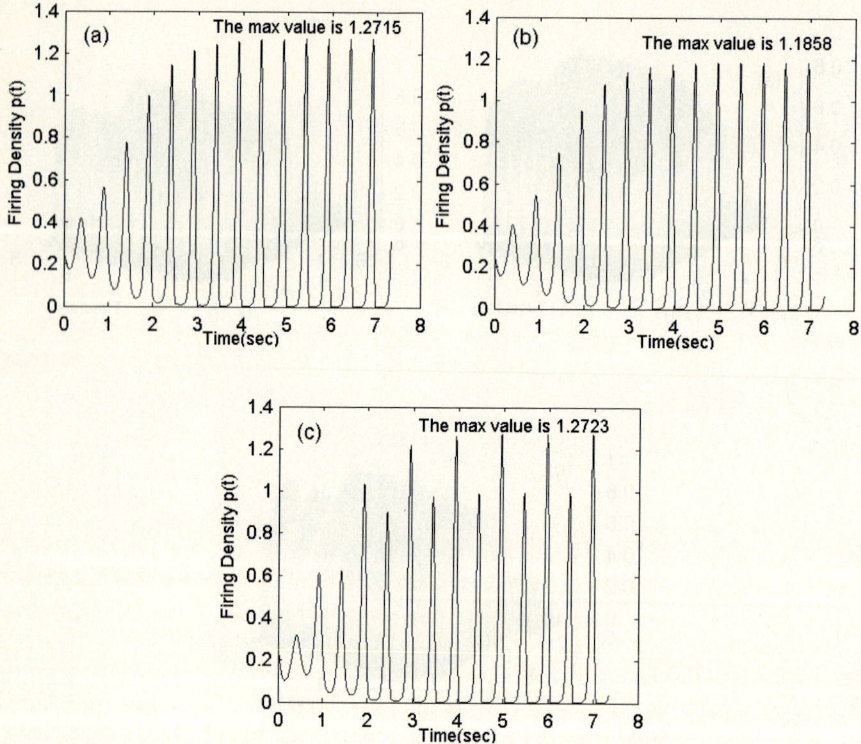

Fig. 4. The firing density of neurons on the limit circle $p(t) = n(0,1,t)$ is plotted over time. Parameters and initial condition are the same as in Fig.3.

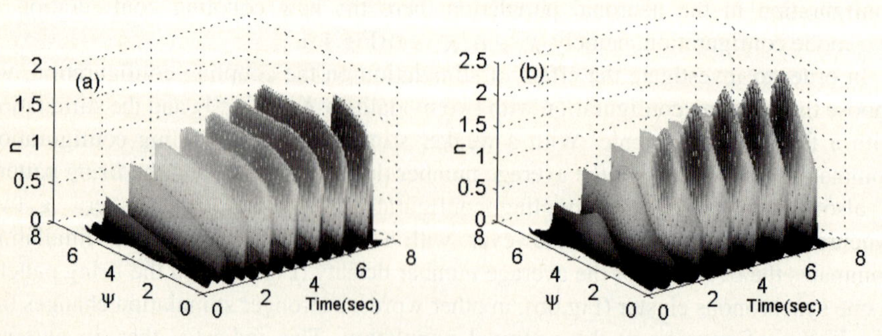

Fig. 5. Average number density's evolution with time on limit circle. Parameters are $K_1 = K_2 = 1, K_3 = K_4 = 0,$, $Q_1 = 0.4, Q_2 = 0.2, \Omega = 2\pi$. (a) $I_1 = 0.5$, (b) $I_1 = 3$.

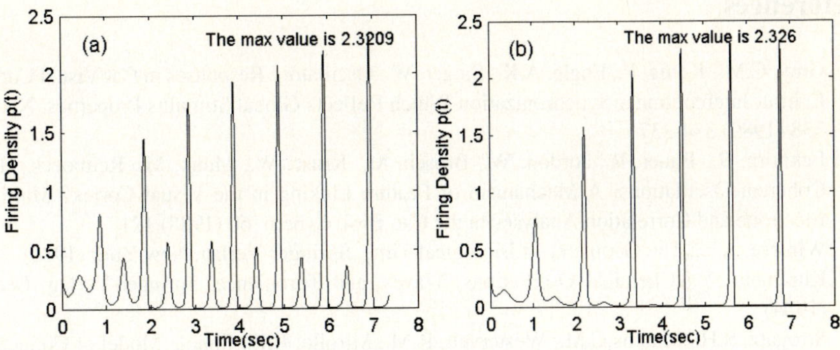

Fig. 6. The firing density of neurons on limit circle $p(t) = n(0,1,t)$ is plotted over time. Parameters and initial condition are the same as in Fig.5.

4 Conclusion

In the present work, taking into account the variation of amplitude of a neuronal oscillator, we have derived the stochastic nonlinear evolution model of neuronal activity in a neuronal population. In particular, the combined dynamics of phase and amplitude has been considered. The firing density of neurons in a neuronal population is considered as neural coding. In the presence of external stimulation, stimulation intensity, stimulation pattern, or coupling strength between neurons can affect the neural coding in the neuronal population. The numerical simulations have indicated that with a weaker stimulation, the response of the neuronal population to stimulation grows up gradually, the coupling configuration in the neuronal population dominates the evolution of the average number density, and new neural coding emerges. Whereas, with a stronger stimulation, the neuronal population responds to the stimulation rapidly, the stimulation dominates the evolution of the average number density, and changes the coupling configuration in the neuronal population. Numerical stimulations have also shown that with a stronger stimulation, or higher harmonic stimulation, the stimulation can alter the coupling configuration in a neuronal population. The theoretical model proposed in the present work is also applicable to many neuronal populations with combined dynamics of phase and amplitude. The proposed model will be very promising in neuronal information processing.

Acknowledgment

This work was supported by the National Natural Science Foundation of China. (30270339).

References

1. Gray, C.M., Konig, P., Engle, A.K., Singer, W.: Oscillatory Responses in Cat Visual Cortex Exhibit Intercolumnar Synchronization Which Reflects Global Stimulus Properties. Nature 338 (1989) 334-337
2. Eckhorn, R., Bauer, R., Jordon, W., Brosch, M., Kruse, W., Munk, M., Reitboeck, H.J.: Coherent Oscillations: A Mechanism of Feature Linking in the Visual Cortex? Multiple Electrode and Correlation Analyses in the Cat. Biol. Cybern. 60 (1988) 121
3. Winfree, A.T.: The Geometry of Biological Time. Springer-Verlag, New York (1980)
4. Kuramoto, Y.: Chemical Oscillations, Waves, and Turbulence. Springer-Verlag, Berlin (1984)
5. Strogatz, S.H., Marcus,C.M., Westervelt, R.M., Mirollo, R.E.: Simple Model of Collective Transport with Phase Slippage. Phys. Rev. Lett. 61 (1988) 2380-2383
6. Tass, P.A.: Phase resetting in Medicine and Biology. Spring-Verlag, Berlin (1999)
7. Colicos, M.A., Collins, B.E., Sailor, M.J., Goda, Y.: Remodeling of Synaptic Actin Induced by Photoconductive Stimulation. Cell, 107 (2001) 605 -616
8. Gazzaniga, M.S., Lvry, R.B., Mangum, G.R.: Cognitive Neuroscience. The Biology of the Mind. Second edition. W.W. Norton & Company, New York London (2002)
9. Han, T.Z., Wu, F.M.: Neurobiology of Learning and Memory. Beijing Medical University Publisher, Beijing (1998)
10. Wang, R., Yu, W.: A Stochastic Nonlinear Evolution Model and Dynamic Neural Coding on Spontaneous Behavior of Large-Scale Neuronal Population. Advances in Natural Computation Part 1 (2005) 490-497
11. Wang, R., Zhang, Z.: Nonlinear Stochastic Models of Neurons Activities. Neurocomputing 51C (2003) 401-411
12. Wang, R., Hayashi, H., Zhang, Z.: A Stochastic Nonlinear Evolution Model of Neuronal Activity with Random Amplitude. Proceedings of 9th International Conference on Neural information Processing 5 (2002) 2497-2502
13. Wang, R., Chen, H.: A Dynamic Evolution Model for the Set of Populations of Neurons. Int. J. Nonlinear Sci. Numer..Simul. 4 (2003) 203-208
14. Jiao, X., Wang, R.: Nonlinear Dynamic Model and Neural Coding of Neuronal Network with the Variable Coupling Strength in the Presence of External Stimuli. Applied Physics Letters 87 (2005) 083901-3
15. Wang, R., Jiao, X.: Stochastic Model and Neural Coding of Large-Scale Neuronal Population with Variable Coupling Strength. Neurocomputing 69 (2006) 778-785
16. Wang, R., Jiao, X., Yu, W.: Some Advance in Nonlinear Stochastic Evolution Models for Phase Resetting Dynamics on Populations of Neuronal Oscillators. Int. J. Nonlinear Sci. Numer..Simul. 4 (2003) 435-446
17. Wang, R., Hayashi, H.: An Exploration of Dynamics of the Moving Mechanism of the Growth Cone. Molecules 8 (2003) 127-138.
18. Jiao, X., Wang, R.: Nonlinear Stochastic Evolution Model of Variable Coupled Neuronal Oscillator Population in the Presence of External Stimuli. Control and Decision 20 (2005) 897-900

Coherent Sources Mapping by K-Means Cluster and Correlation Coefficient

Ling Li[1], Chunguang Li[2], Yongxiu Lai[1], Guoling Shi[1], and Dezhong Yao[1]

[1] School of Life Science and Technology,
University of Electronic Science and Technology of China,
610054 Chengdu, China
{liling, cgli, laiyx, dyao}@uestc.edu.cn
[2] Center of Neuro Informatics,
University of Electronic Science and Technology of China,
610054 Chengdu, China

Abstract. A new equivalent representation of the neural electric activities in the brain is presented here, which images the time courses of neural sources within several seconds by the methods of k-means cluster and correlation coefficient on the scalp EEG waveforms. The simulation results demonstrate the validity of this method: The correlation coefficients between time courses of simulation sources and those of estimated sources are higher than 0.974 for random noise with $NSR \leq 0.2$. The distances between the normalized locations of simulation sources and those of estimated sources are shorter than 0.108 for random noise with $NSR \leq 0.2$. The proposed approach has also been applied to a human study related to spatial selective attention. The results of real VEPs data show that three correlative sources can be located and time courses of those can describe the attention cognitive process correctly.

1 Introduction

Electroencephalography (EEG) and evoked potential (EP) are important tools for exploring the function of the nervous system. Because scalp EEG suffers from limited spatial resolution and distortions existing in the head volume conductor, many approaches have been developed to estimate activities of neural sources in recent years. The popular approaches include cortical surface potential mapping (CPM), scalp laplacian mapping (LM), low resolution electromagnetic tomography (LORETA) and focal underdetermined system solver (FOCUSS) [1] [2] [3] [4]. Because the measurements do not contain enough information about the generators, this gives rise to what is known as the non-uniqueness of the inverse solution [5]. K-means begins with a user-specified number of clusters. Cases are allocated into clusters, then reallocated on successive iterations until the within-cluster sums of squares are minimized [6]. In the present study, our main focus is the fast equivalent representation of the neural electric activities in the brain by the methods of k-means cluster and correlation coefficient. The contents include introduction of methods, forward simulation for neural electric sources in a 4-concentric-sphere head model, and the real visual evoked data tests.

L. Jiao et al. (Eds.): ICNC 2006, Part I, LNCS 4221, pp. 237 – 240, 2006.

2 Methods

K-means cluster aims to partition N inputs $x_1, x_2 \dots x_N$ into k clusters [7]. An input x_t is assigned into the jth cluster if indicator function $I(j \mid x_t) = 1$. It is the following:

$$I(j \mid x_t) = \begin{cases} 1 & if \quad j = \arg\min_{1 \le r \le k} \|x_t - m_r\|^2 ; \\ 0 & others. \end{cases} \tag{1}$$

Here, $m_1, m_2 \dots m_k$ are called seed points or units that can be learned in an adaptive way. Update the winning seed point m_w by

$$m_w^{new} = m_w^{old} + \eta(x_t - m_w^{old}), \tag{2}$$

where η is a small positive learning rate [8].

In the present study, a more quantitative way to determine the cluster number is to look at the average silhouette values (*ASVs*) for those cases. It is defined as:

$$s(i) = \frac{\min[b(i,k)] - a(i)}{\max[a(i), \min[b(i,k)]]}, \tag{3}$$

where $a(i)$ is the average distance from the ith point to the other points in its cluster, and $b(i,k)$ is the average distance from the ith point to points in another cluster k. When $s(i)$ is close to one, this implies that the object is well classified. Then the correlation coefficients (*CCs*) between N inputs $x_1, x_2 \dots x_N$ and k clusters centroid locations are calculated respectively in order to get k correlation topographies.

3 Computer Simulation

The head is modeled as a concentric 4-sphere model. The conductivities and radii were taken from Cuffin and Cohen [9]. Suppose that there are four radial dipoles (s_1, s_2, s_3, s_4), the locations of which in Cartesian coordinates (x, y, z) are (-0.4975, 0.7107, 0.4975) × 7.0356 cm, (0.4975, -0.7107, 0.4975) × 7.0356 cm, (0.4975, 0.7107, 0.4975) × 7.0356 cm and (-0.4975, -0.7107, 0.4975) × 7.0356 cm respectively. The forward scalp EEG map is shown in the first figure of Fig 1 [10]. Time courses of them are produced by the following two coupled Rössler oscillators' equations:

$$\begin{aligned} \dot{x}_{1,2} &= -\omega_{1,2} y_{1,2} - z_{1,2} + R(x_{2,1} - x_{1,2}), \\ \dot{y}_{1,2} &= \omega_{1,2} x_{1,2} + 0.15 y_{1,2}, \\ \dot{z}_{1,2} &= 0.2 + z_{1,2}(x_{1,2} - 10). \end{aligned} \tag{4}$$

In our study, natural frequencies of the two oscillators are 10+0.024 *Hz* and 10-0.024 Hz and coupling strength (R) are 0.011 and 0.035 respectively. Suppose that coupling strength is 0.011 and 0.035 between s_1 and s_2, s_3 and s_4, respectively. The mutual correlation coefficients (*CCs*) are 0.0977, -0.0360, 0.0486, 0.1305, -0.1476, 0.4892 between s_1 and s_2, s_1 and s_3, s_1 and s_4, s_2 and s_3, s_2 and s_4, s_3 and s_4 respectively.

Maximal *ASVs* is 0.6654 when k is five. So there are five clusters (c_1, c_2, c_3, c_4, c_5) and five cluster centroid vectors. Figure 1 (2) ~ (5) illustrates the correlation topographies. Five cluster centers in every cluster are (x, y, z) = (-0.4463, 0.6790, 0.4082), (0.5051, -0.6493, 0.4451), (0.4617, 0.6793, 0.4325), (-0.5051, -0.6493, 0.4451) and (0.0196, -0.1153, -0.0471) respectively. The variance of c_5 is close to zero, so it can be discarded. *CCs* are -0.2209, -0.0198, -0.1426, -0.0707, -0.2193, 0.2315 between c_1 and c_2, c_1 and c_3, c_1 and c_4, c_2 and c_3, c_2 and c_4, c_3 and c_4, respectively. *CCs* (*Distances*) between s_1~s_4 and c_1~c_4 are larger (smaller) than 0.979 (0.108). Furthermore, *CCs* (*Distances*) between s_1~s_4 and c_1~c_4 are larger (smaller) than 0.974 (0.108) for random noise with $NSR \le 0.2$.

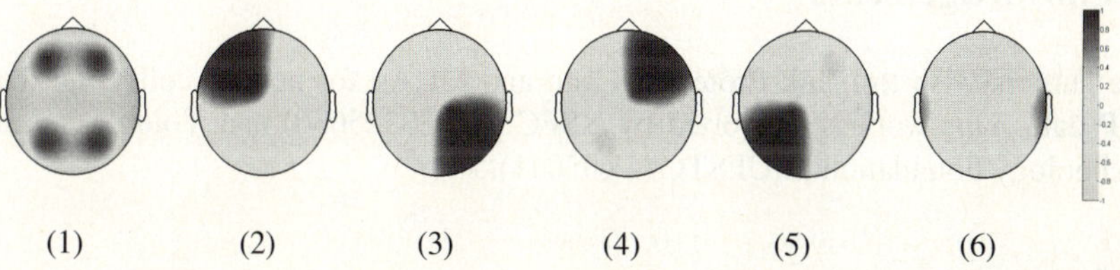

Fig. 1. The correlation topographies (k=5). (1) scalp potentials of four assumed dipoles; (2) ~ (6) the correlation topographies between scalp potentials and five cluster centroid vectors.

4 Real VEPs Data

Grand average VEPs over 10 subjects recorded from 119 scalp sites in a spatial selective attention task related to small circular checkerboard stimuli location in the right field and attention to the right visual field were analyzed. Maximal *ASVs* is 0.6378 when k is three. Imaging results are shown in Fig 2.

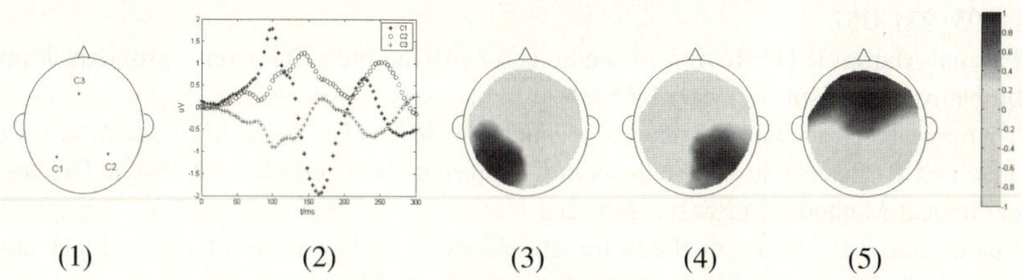

Fig. 2. The results of VEP data. (1) The normalized locations of three cluster centers, which are (-0.4652, -0.50078, 0.72994), (0.39371, -0.45218, 0.80033) and (-0.10252, 0.42576, 0.89901) respectively. (2) Three cluster centroid vectors, (3) ~ (5) the correlation topographies (k=3).

CCs are -0.2117, -0.7085 and -0.5397 between c_1 and c_2, c_1 and c_3, c_2 and c_3. There are three sources located at left, right occipital and frontal in spatial selective attention experiment after stimulus onset to 300 ms, which are according with the previous knowledge in attention process [11].

5 Discussion

A new equivalent representation of the neural electric activities in the brain is presented here, which images the time courses of neural sources in several seconds by the methods of k-means cluster and correlation coefficient on the scalp EEG waveforms. The results of simulation and real VEPs data demonstrate that the time course of sources can be calculated fast, the normalized locations of which can be determined easily, and the major relationships between sources can be differentiated correctly.

Acknowledgements

The authors wish to thank Professor Chen and Mr Ao for help in collecting the real ERP data. This work is supported by NSFC (No 30525030) and Youth Science and Technology Foundation of UESTC (JX05041).

References

1. Dezhong Yao.: Source potential mapping: a new modality to image neural electric activities. Phys. Med. Biol. 46 (2001) 3177–3189
2. Nunez P L, Silibertein R B, Cdush P J, Wijesinghe R S, Westdrop A F and Srinivasan R.: A theoretical and experimential study of high resolution EEG based on surface laplacian and cortical imaging Electroencephalogr.Clin. Neurophysiol. 90 (1994) 40–57
3. Pascual-Marqui R D, Michel C M and Lehmann D.: Low resolution electromagnetic tomography: a new method for localizing electrical activity in the brain. Int J Psychophysiol. 18 (1994) 49–65
4. Gorodnitsky I F, George J S and Rao B D.: Neuromagnetic source imaging with FOCUS: a recursive weighted minimum norm algorithm. Electroenceph & clin Neurophysiol. 95 (1995) 231–251
5. Pascual-Marqui R D.: Review of methods for solving the EEG inverse problem. Journal of bioeletromagnetism. 1 (1999) 1 75–86
6. Theodore P. Beauchaine, Robert J. Beauchaine III.: A Comparison of Maximum Covariance and *K*-Means Cluster Analysis in Classifying Cases Into Known Taxon Groups. Psychological Methods. 7 (2002) 2 245–261
7. MacQueen, J.B.: Some methods for classification and analysis of multivariate observations. In Proceedings of 5th Berkeley Symposium on Mathematical Statistics and Probability. University of California Press, Berkeley, CA. (1967) 281–297
8. Yiu-Ming Cheung.: k-Means: A new generalized k-means clustering algorithm. Pattern Recognition Letters. 24 (2003) 2883–2893
9. Cuffin B N and Cohen D: Magnetic fields of a dipole in spherical volume conductor shapes. IEEE Trans. Biomed. Eng. 24 (1997) 372–381
10. Dezhong Yao.: High-resolution EEG mapping: an equivalent charge-layer approach. Phys. Med. Biol. 48 (2003) 1997–2011
11. Banich, M. T.: neuropsychology: the neural base of mental function. Houghton Mifflin, New York (1997) 243

Measuring Usability: Use HMM Emotion Method and Parameter Optimize

Lai Xiangwei, Bai Yun, and Qiu Yuhui

Faculty of Computer and Information Science,
Southwest University,
Chongqing, China 400715
{lxw, baiyun, yhqiu}@swu.edu.cn

Abstract. Usability evaluation increasingly used in the software development, especially for evaluation of user interaction designs. Software's usability comprises the aspects effectiveness, efficiency and satisfaction. The user emotional is the best reflection of user's satisfaction degree. This paper presents the definition of a user satisfaction metric method based on the psychology emotion theories. A Hidden Makov model (HMM) was used to describe the experience-emotion mapping relationship. An artificial neural network was used to affirm the parameter of the emotional-satisfaction mapping. Moreover, we get an effective reflection either.

1 Introduction

Among the factors affecting the usability of the software, people are always on the tone of "User-centered design" [1]. To perfectly implementing the design of software's usability required Human-Computer interaction, User Interface Design, Human Factors, Human Ergonomics, and other relative knowledge [2]. The usability definition of ISO 9126[3][4][5] shows: Usability is the capability of the software to be understood, learned, used and liked by the user, when used under specified conditions (ISO 9126-1,2001).Usability comprises the aspects effectiveness, efficiency, and satisfaction. Satisfaction is the users' comfort with and positive attitudes towards the use of the system. Two techniques that can be carried out at reasonable costs to evaluate the usability product quality are heuristic evaluation and checklist. These techniques have the disadvantage that the real stakeholder, i.e. the user, often is not involved. The ease or comfort during usage is mainly determined by characteristics of the software product itself, such as the user interface. Within this type of scope, usability is part of the product quality characteristics.

Further more, users' self-reporting is the most simple and applicable method. But compensate on its afterwards and the poor contents of the reports given by users which is resulted from users' misunderstanding of the goals and means of the experiments. The development of the biology characteristics recognition technology makes the detection and justification of the users' biology characteristics to be an applicable method used in usability testing. It is researched by MIT [6], Carnegie Mellon University USA, etc, and has effective results.

L. Jiao et al. (Eds.): ICNC 2006, Part I, LNCS 4221, pp. 241 – 250, 2006.

Several studies have shown that in addition to functionality and reliability, usability is a very important success factor (Nielsen, 1993) (MultiSpace, 1997) [7]. So it's difficult to metric software's usability characters for the deferent people always give different appraisement to same software in deferent conditions. The traditional method is difficult to shows the user's experience during they use a software [8]. In a broader scope, usability is determined by using the product in its (operational) environment. The type of users, the tasks to be carried out and the physical and social aspects that can be related to the usage of the software products are taken into account. Hence, the user's affective is the best exhibition of software's usability. This paper is focused on the detail affective method, which is used to metric the software's usability.

2 Monitor and Collection the User's Experience Activity

User was asked to operate software in a special testing environment [8]. And they must report their planning and feeling during their operation process. Many technical have been used to inspect user's experience activities. A screen record software was used to catch all the user's operations and a tape was used to catch the user's speech. Some guide should be given to the participants when they falling into an inextricable conditions.

Some parameters which perhaps affect the system's usability was been record during the testing. Such as:

- the total time used to finish the mission
- the mistake ratio during the man-machine conversation
- the time resume from a mistake
- the number of mistake
- the total order/function number which user used
- the problem number which is solved by the help system
- the ratio of the favorite function
- the stagnation time during the man-machine conversation
- the number of the user's attention was allured by the unreal mission

Etc.

3 The Experience-Emotion Mapping Method Use HMM

During the user's experience process, they will have the directly feeling to the software's usability characters. And how to metric user's feeling is the most important problem we must face. So an experience-emotion mapping model is founded to describe the internal relation between the user's experience and their emotion.

3.1 The Psychology Background of Experience-Emotion Mapping

To specify the user's emotional states, we use Mehrabian's PAD emotion space model [9] [10]. The PAD Emotional State Model consists of three nearly independent dimensions that are used to describe and measure emotional states (or feeling, affective conditions): pleasure and displeasure, arousal and nonarousal, dominance and submissiveness. Pleasure-displeasure distinguishes the positive-negative affective quality of emotional states, arousal-nonarousal refers to a combination of physical activity and mental alertness, and dominance- submissiveness is defined in terms of control versus lack of control.

The following sample ratings illustrate definitions of various emotion terms when the scores on each PAD scale range from -1 to +1: angry (-.51, .59, .25), sleepy (.20, -.70, -.44), and bored (-.65, -.62, -.33). Thus, according to the ratings given for "angry," it is a highly unpleasant, highly aroused, and moderately dominant emotional state.

The psychology research shows that there is a direct connection between the outside stimulation and the person's susceptibility [11]. These connections defined as the follow laws.

Weber's Law: The radio between the Just Noticeable Difference (JND) and the intension of the stimulation is constant:

$$\Delta I / I = K_m \tag{1}$$

This law comes from the research of the eyes. And people found it is also useful to explain all the stimulation impressibility in the whole biology. This law means that the stimulation that can affect JND is increasing with the increase of the standard stimulation. It also used to explain why the user could not feel the increase of software's usability if the software's usability had reached a better level. It means the more maturity a software is and the more difficultly to amelioration its usability level.

Fechner's Law: It is a direct ratio relation between the sense intensity and the stimulation intensity.

$$\mu = K_m \log I + C \tag{2}$$

K_m and C is constant, μ is the sense intensity, I is stimulation intensity

To constant stimulation intensity, the law can be changed as:

$$\mu = K_m \log(1 + I) \tag{3}$$

It is easy to see:

- When $I \to 0$, $\mu \approx K_m \times I$. It means people always is sensitive to the important stimulation fountainhead which they attach important to. Then they can catch the stimulation movement by rule and line and got more information they wanted.
- When $I \to \infty$, $\mu \approx K_m \times \log I$, It means it's a direct ratio relation between the sense intensity and the logarithm of stimulation intensity. People always can't detect the stimulation movement if they don't attach importance to the outside stimulation fountainhead.

Hence, moderate explanation can make participants pay more attention to the software's usability characters. And it can help us to get more authentic data. A HMM model has been founded to specification the experience-emotional mapping in software usability testing.

3.2 The HMM Model Used in Experience-Emotion Mapping

The user's experience can be looked as a typical random process due to the subjectivity and unbending of the users' operation. And the arm of the operation process is self-satisfied. So users' self-report has inevitable contact to their emotional experience. A discrete HMM [12] can be used to descript these connection between users' operation experience and their emotional experience [13].

For the task-oriented characteristic of the users' operation, we can generally consider the users' affective is influence by the recently experience. Hence, the users' experience can be defined as a random process chain:

$$S(\xi_n = j \mid \xi_{n-1} = i_{n-1}, \xi_{n-2} = i_{n-2}, \ldots, \xi_0 = i_0) = S(\xi_n = j \mid \xi_{n-1} = i_{n-1}) \tag{4}$$

Where the symbol ξ_n figure to the user's experience state on the time n.

It easy to see:

Property 1. The state corresponding to the "finish" state must be a return station.

Property 2. All states except the "finish" state can be considered as temporary states.

Property 3. Every experience serial of each user is an orderliness Markov chain.

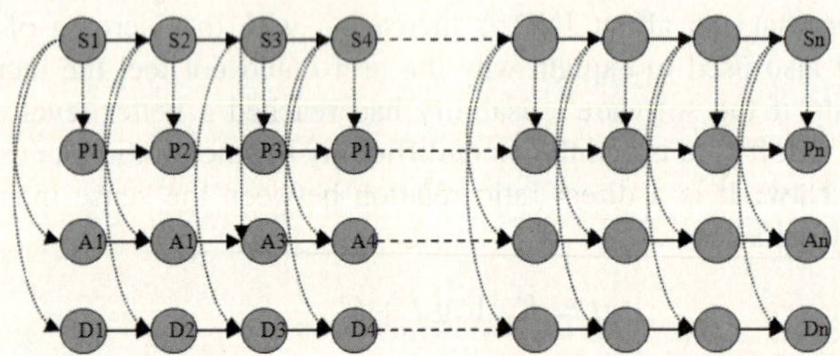

Fig. 1. The experience-emotion mapping HMM model

The HMM model can be defined as:

$$\begin{cases} \Delta PAD_{ij} = p(\omega_j(t) \mid \omega_i(t-1)) \\ S_{jk} = p(v_k(t) \mid \omega_j(t-1)) \end{cases} \tag{5}$$

ΔP, ΔA, ΔD is the transfer probability of the PAD model's three parameters. S_{jk} is the transfer probability of the observation chain. The symbol V is an affect gene vector.

After ensuring the parameters of the Markov transient matrix using the datum generating from the experiments, these models have a strong capability to describe the user's experience conditions. The main information includes:

3.2.1 User's First-Arrive Probability

The user's first-arrive probability can be define as:

$$a_{ij}^{(n)} = P(\xi_n = j, \xi \neq j, k = 1,2,..n-1 \mid \xi_0 = i) \qquad (n \geq 1) \tag{6}$$

The first-arrive probability can be used to judge the user's operation station. We can consider the user fall in some trouble if the fact operation step number is more than the first-arrive probability from the start station i to the object station j. The timely and relevant clews can effective help user to finish the aim. It also avoids time waste and emotional impact.

3.2.2 Return Station Visit Probability

In the Markov theory, if the return station i's first arrive time is T_0, the interval of the next return is T_1, T_2, $\ldots T_n$, it will know that $\{T_n : n \geq 1\}$ is independence distribution. A lingering update process can be create as $N_t^{(i)}$, T_i's distribution is d-distribution($d \geq 1$). When $d \geq 2$, d is the cycle of station i.

The average revisit probability S_a can get by the return station limit law. If $|S_t - S_a| \leq 2e$ (S_t is the operation step after the lately revisit) we can consider the user is in the correct operation process. Otherwise we can consider user is falling in some difficult in currently operation, some supervise should be afforded to the currently operation aim.

3.2.3 User Experience Revisits Probability

Some state's average visit time also can be used to estimate user's puzzled.

The less visit time to some return station always arose by two reasons:

- Operate favorably with higher efficiency(software design in reason with better usability)
- User is falling in big trouble. They had to back the station in order to attempt other possible roads after run into a blank wall.

The follow function can be use to judge which state the user is in.

$$f(n) = 1 \Leftrightarrow \exists l \bullet p_{ij}(n,n) = \sum_k p_{ik}(n,l) p_{kj}(l,n) \tag{7}$$

The function means we can consider user is in a normal revisits process if the users revisit the other primary stations got in function (7). Otherwise, it means user is falling in trouble. Although the method has not integrality in theory (it's possible to revisit the station without pass the primary stations). But the examination shows that the method has better partition ability in fact.

3.3 The Training of HMM

We rebuild the Baum-Welch algorithm [12] to adapt to the user's PAD emotion model. Every user experience's direction must be affirmed first. The normal method is the expert evaluation (just like the table 1).

Table 1. The initialize vector

S	User experience	P	A	D
1	Successfully operate	1	1	1
2	Unsuccessfully operate	-1	-1	-1
3	Help message	1	C	1
4	Error message	-1	-1	C
5	A broken operate	-1	C	-1
6	Backtrack operate	C	0	0
7	Stagnation	-1	-1	-1
8	Miss target	0	C	-1
9	Serially input	C	-1	0

Table 2. The affect parameter got from HMM

S	User experience	P	A	D
1	Successfully operate	0.134	0.064	0.059
2	Unsuccessfully operate	-0.073	-0.082	-0.012
3	Help message	-0.008	0.028	0.041
4	Error message	-0.081	-0.043	0.019
5	A broken operate	-0.031	0.009	-0.023
6	Backtrack operate	-0.004	0.012	-0.014
7	Stagnation	-0.025	-0.021	-0.016
8	Miss target	-0.005	-0.013	-0.025
9	Serially input	-0.051	-0.031	-0.012

Where the value “C” means chaos, “1” means a positive stimulation and “-1” means negative stimulation.

An affect vector table can be gotten from the translate table of the each experience state (Table 2). The vector shows the affect vector’s director and degree. We can rebuild the user’s PAD parameter real-time by the vector.

3.4 The Attenuation of User Emotion

With the experience of user’s operation to software, users will more and more familiar it. Then user’s feedback to the software’s usability stimulation will step down due to the sensory adaptation theory. The finally value will trend to a valve *C*. The attenuation processes obey the rule of follow:

Property 4. The attenuation of user emotion: user’s emotion will attenuate with the its persistence, and it obey the rule of:

$$\mu = \mu_0 \exp(-k, T) \tag{8}$$

Where μ is the new appraise intensity, μ_0 is the initialize appraise intensity, *k* is the attenuation coefficient and *T* is the duration.

3.5 The Accumulation of User Emotion

To any people, emotion can be looked as a sequential and correlative process, the passed station will affect the certainly station. We can use the follow method to specification this characters.

Property 5. The accumulation of user emotion: the emotional appraise intensity in time t is the summation of the emotional appraise intensity in time *t-1* and the new emotional affect by the new outside stimulations:

$$\mu(t) = \delta\mu(t-1) + v_k(t) \tag{9}$$

Where δ is emotional appraise intensity attenuation coefficient, and $v_k(t)$ is the emotional vector affect by the new outside stimulations.

Due to the function (9), the next time emotional intensity can be described as:

$$\begin{aligned} &\mu(t) = \delta\mu(t-1) + v_k(t) \\ &\Rightarrow \mu(t) = \mu_0 \exp(-k,T) + v_k(t) \\ &\Rightarrow \mu(t) = \mu_0 \exp(-k,T) + V_s \end{aligned} \tag{10}$$

Where V_s is the affect coefficient of every user experience in the table 2.

4 The Mapping Method from User Emotion to Software Usability

The best easily method the mapping the user emotion to software usability is a function:

$$M_u = \alpha P + \beta A + \gamma D \tag{11}$$

And M_u is the metric of the software's usability, P, A, D is the user's emotion parameters. α, β, γ are the right of the emotion parameters. The most important thing is to found out a better value of the three rights. Mehrabian and Blum (1995) used Mehrabian's (1978) PAD scales to study the physical attractiveness of a person as a function of emotional reactions elicited by him or her. The following regression equation, written for standardized variables and .05-level effects, described physical attraction to another as a function of the P A D reactions of the rater [14].

Physical Attraction = .40P + .16A -.24D.

A BP neural network [15] was used to finish this job.

User is ordered to finish the software usability checklist, which is designed by Purdue University [16]. This checklist mainly oriented the software's physical usability characters. It is combined by 100 problems. The main parameters are:

A, each problem's weightiness: L= {1, 2, 3}
B, each problem's grade: G= {1, 2, 3, 4, 5, 6, 7}

And the software's usability appraise coefficient can be describe by the follow expressions:

$$f = \sum_{n=1}^{100} L_n (G_n - 3.5) / (nL_{\max} (G_{\max} - 3.5)) \tag{12}$$

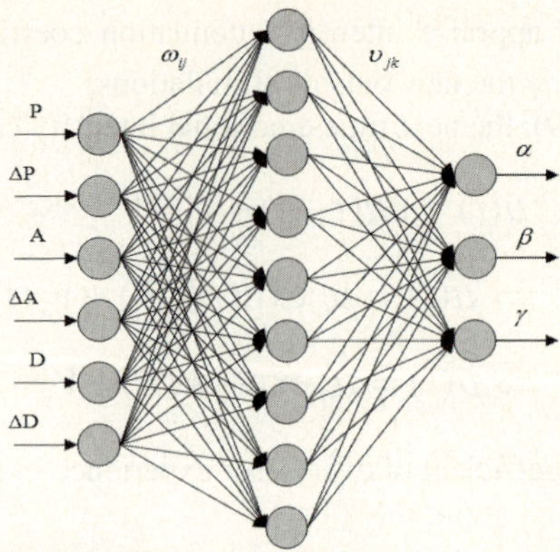

Fig. 2. The image of BP artificial neural network

A basic back propagation-training algorithm was used to training the neural network. And the training arm is a conic from zero to the final usability evaluation result got from the user self-report check list. It just likes some PID controller training algorithm. The weights connecting neurons in the hidden layer to those in the output layer are adjusted according to:

$$w_{jm}^{new} = w_{jm}^{old} + \eta \sum_{p=1}^{n} (O_j \partial_m)_p + \beta [\Delta w_{jm}^{(previous)}] \tag{13}$$

Where w_{jm} is the weight connecting neuron j in hidden layer to neuron m in the output layer, n is the number training set and $\Delta w_{jm}^{(previous)}$ is the previous weight change. The weight connecting neurons in the input layer to those in the hidden layer are adjusted as:

$$w_{ij}^{new} = w_{ij}^{old} + \eta \sum_{p=1}^{n} (O_i \partial_j)_p + \beta [\Delta w_{ij}^{(previous)}] \tag{14}$$

Where w_{ij} is the weight connecting neuron i in the input layer to j in the hidden layer.

After 200 times training, the finally result got by the BP neural network is:

$\alpha = 0.49$, $\beta = 0.20$, $\gamma = -0.31$

So we use the function 15 to compute the final usability metric result.

$$M_u = .49P + .20A - .31D \tag{15}$$

This expressions can be used to mapping the user emotional model to a software usability appraise coefficient.

5 Application and Experimental Result

To show the effectiveness and the usability of the metric method for composite software system we present two softwares [17]. The software we use in the experimental are some OA systems. One of them is a mature system and with little problems. User can finish their work easier and quickly. And the other is a new system. There are many problems in it. The software's interface is uncomfortable. User will always meet some troubles.

There are 24 participants were chosen to use the software. They all have no experience with the two software. The participants are assigned to 3 groups. The test times to the 3 group are 15 min, 30min and 60min. Three methods were used to metric the software's usability.

- DE: Directly user evaluation (the evaluate value is in the zone [-1, 1])
- CL: Check-list (use Purdue self-report checklist)
- EM: Emotional monitor

The participants are requested to use finish some works use the software. The operation process are all been recorded by different device. And they all required filling the Purdue self-report checklist. The follow tables can show the result.

Table 3. The experimental result

Software	A			B		
Method	DE	TM	EM	DE	TM	EM
Group1(15)	0.725	0.468	0.654	-0.823	0.173	-0.329
Group2(30)	0.604	0.623	0.724	-0.687	0.159	-0.358
Group3(60)	0.416	0.642	0.708	-0.739	0.258	-0.423

Some peculiarity can be easily to see in the experimental:

- The DE method result has biggish warp for the participants have not a unify criterion to evaluate the software.
- The checklist method can get better result if there is enough time to the participants.
- Emotional method has better divisive ability than the traditional method especially with a short test time limited.

6 Conclusions

It is important to point out that the research is one of the attempts to keep the software's quality. The main parameters we use is based the really examinations. They are probably not fit all the kinds of software due to the limited of the examination software is not bestrew some special software types. However, we think this process can be applied both in small and big organizations. It has their particular excellence such as quickly; low cost; better effect to less examination data and so on.

We plan to continue working in order to achieve the three goals:

- Defining and improving more detail model to fit the different connotation of the software's usability characters.

– Identify the usability contributions for the software internal and external elements.
– Defining a software develop process model to improve its usability level and fit the user operation affective.

We also intend to use some soft computing method in the SE process to improve the software develop quality and cost.

References

1. Bevan, N., Quality and usability: a new framework, in: E. van Veenendaal and J. McMullan(eds.), Achieving Software Product Quality, Tutein Nolthenius, 's Hertogenbosch, The Netherlands (1997)
2. Bos, R. and E.P.W.M. van Veenendaal, For quality of Multimedia systems: The MultiSpace approach (in Dutch), in: Information Management, (1998)
3. ISO/IEC FCD 9126-1, Information technology - Software product quality - Part 1 : Quality model, International Organization of Standardization (2001)
4. ISO 9421-10, Ergonomic Requirements for office work with visual display terminals (VDT's)-Part 10 : Dialogue principles, International Organization of Standardization (1994)
5. ISO 9241-11, Ergonomic Requirements for office work with visual display terminals (VDT's)-Part 11: Guidance on usability, International Organization of Standardization(1995)
6. Rosalind W.Picard and Jonathan Klein, Computers that Recognize and Respond to User Emotion: Theoretical and Practical Implications, MIT Media Lab Tech Report 538 (2002)
7. Kirakowski, J., The Software Usability Measurement Inventory: Background and Usage, in: Usability Evaluation in Industry, Taylor and Francis
8. Bass, L. J, John, B. E. & Kates, J. Achieving usability through software architecture. Carnegie Mellon University/Software Engineering Institute Technical Report No. CMU/SEI-2001-TR-005. (2001)
9. Andrew Ortony, Gerald L. Clore, and Allan Collins. The Cognitive Structure of Emotions. Cambridge University Press, Cambridge, MA,(1988)
10. Albert Mehrabian. Pleasure-arousal-dominance: A general framework for describing and measuring individual differences in temperament. Current Psychology: Developmental, Learning, Personality, Social, (1996) 14:261–292
11. Pierre Philippot, Robert S. Feldman, The Regulation of Emotion, Lawrence Erlbaum Associates, Publishers. (2004)
12. Timo Koski, Hidden Markov Models for Bioinformatics, Kluwer Academic Publishers,
13. H. & P. Johnson, "Task knowledge structures: psychological basis andintegration into system design", in Acta Psychologica 78, (1991), p. 3-26
14. Wilson, C., Rosenbaum, S. Categories of Return on Investment and Their Practical Implications. Cost Justifying Usability. An Update for the Internet Age. Randolph Bias & Deborah Mayhew Eds. Elsevier, USA, (2005)
15. M.Ananda Rao, J.Srinivas, Neural Networks-- Algorithms and Applications, Alpha Science International Ltd. (2003)
16. Benedikte S. Als, Janne J. Jensen, Mikael B. Skov, Comparison of Think-Aloud and Constructive Interaction in Usability Testing with Children, IDC, June 8-10, 2005, Boulder, Colorado, USA. (2005)
17. Lai Xiangwei, Yang juan, Qiu Yuhui, Zhang Weiqun, A HMM Based Adaptation Model Used in Software Usability Monitoring, ICYCS (2005)

Affective Computing Model Based on Emotional Psychology

Yang Guoliang, Wang Zhiliang, Wang Guojiang, and Chen Fengjun

School of Information and Engineering,
University of Science and Technology Beijing, 100083, China
ygliang30@126.com

Abstract. According to the basic emotions theory, the paper presents personality,mood, and emotion space. The mapping relationship among personality , mood and emotion is built. The equations for updating the affective and mood states are induced and a generic computing model for personality ,mood and emotion simulation for virtual human is constructed. The simulation results demonstrate that the affective model can better simulate the dynamic process of emotion and mood change under various environment stimulus. The model provides a valid method to the emotional robot modelling and affective decision system.

1 Introduction

Recently, some evidences indicated that the emotion plays an important role in judgment, perception, learn and many other cognition functions. Appropriate balance of emotion is essential to intelligence and makes human having batter creativity and flexibility in the aspects of solving problem. Similarly, if we want the computer to have the true intelligence, to adapt the environment in which human's living and to communicate with the human naturally, then the computer needs to have the capability of understanding and expressing emotion. Putting the intelligent ability of recognizing, understanding ,and expressing emotion to the computer is gained increasing attention in academia and industry. A remarkable work in the area is the research of affective computing [1], the conception of affective computing was proposed by Picard and the contents of research have made introduced systemically in the monograph "Affective Computing". Another noteworthy research is the artificial psychology[2]. The artificial psychology is computational realization of human's mental activity by using the method of information science. The emotion modeling is a very important part both in affective computing and in artificial psychology.

Based on the basic theory of emotion, The paper builds up personality space, mood space and emotion space respectively, describes the relation of among personality, mood and emotion, educes the affective computing model according to the space of personality, mood and emotion.

L. Jiao et al. (Eds.): ICNC 2006, Part I, LNCS 4221, pp. 251–260, 2006.

2 Basic Definitions

The affective computing model involves to the relations among personality, mood, emotion and the time responses of them under the influence of exterior incentives, so we introduce the correlative conception and the quantification processing method firstly.

2.1 The Inductive Variable of Emotion

Any occurrence of emotion is caused by exterior incitement, the common incentive signals mainly have facial expression, voice, gesture, posture and various physiology signal and so on. The human brain extracts emotion factor from all these incentive signal as the input signal of affective interaction which drives humanity's mood and emotion having the corresponding change. So the emotion information (i.e. the inductive variable A of emotion) by the model of cognition and appraisement of emotion is obtained.

According to [1], the paper supposes the emotion similar to the response of striking bell. It is expressed by the relation:

$$y = a_1 t^{b_1} \quad \text{if } 0 \leq \text{t} \leq \text{t}_1 \qquad \text{and} \qquad y = a_2 e^{-b_2(t-t_1)} \quad \text{if } \text{t} \geq \text{t}_1$$

Where y represents the intensity of exterior incentive signal, the parameters a_1 and b_1 control the response scope,b_2 controls the response attenuation velocity. we define the emotional inductive variable $A = (y_1, y_2, y_3, y_4, y_5)$,$y_i (i = 1, 2, 3, 4, 5)$ as the emotional information variable of 5 kind of basic emotion separately, i.e, anger, disgust, fear,happy and sad.

2.2 Personality Space

The emotion psychology defines the personality as a person's entire mental appearance, namely the whole mental characteristic which forms under a certain social environment and has a certain tendency and stability. The personality is regarded as the result effected by environment in a great extent, but some evidences indicate that personality has been decided from birth in a great extent. In this paper, personality is considered as constantly invariable and is initialized at $t = 0$.

The Five-Factor Model (FFM) is a personality model with extensively application [3]. Its five factors are openness, conscientiousness, extraversion, agreeableness and neuroticism. In this paper, we adopt the five factor models as personality model, the personality vector can be defined as a five dimensions vector,$P = [p_1, p_{2,3}, p_4, p_5]^T, \forall i \in [1, 5] : p_i \in [0, 1]$.So personality space can be described as a 5D hypercube, personality of somebody corresponds to a certain dot in the hypercube.

2.3 Mood Space

Mood is a kind of affective states which are longer lasting and weaker, affects human's entire mental activity. Mood has the characteristic of pervasion, it isn't particular experience to a certain event, but is by the similar mood condition experience, treats and experiences all things with the same emotion state. For example, one will feel happy when he gets ahead in job or study, and the experience can affect the other activities with satisfactory and joyful emotion state in a long time. Compared with personality, mood is variable. Usually it is divided into three kinds of states, i.e. good, bad and normal, for example, the bad mood caused by angry has an high incentive level.

Concerning the dimension of mood, it hasn't unified definition in psychology. Mehrabian [4] adopt a 3D characteristic to describes mood, i.e. pleasure (P), arousal(A) and dominance(D). The 3D characteristics are independent mutually and constitute 3D mood space (PAD space). According to PAD, this paper defines the variable of 3D mood space M as: $M = [m_P, m_A, m_D]^T, -1 \leq m_P, m_A, m_D \leq 1$

Where $M = [0, 0, 0]^T$ corresponds to the mood state of calmness.

Opposition to emotion space, the change of mood is slower and more stable. But comparison with personality, the mood is time variable.

2.4 Emotion Space

Emotion is the experience of human's attitude to objective things and the reflection whether one's needs is satisfied. It includes two layers, i.e., emotion process and emotion personality. For their classifications, there are some different classifications. The OCC model [5] divides emotion into 22 kinds, as well as Ekman divides emotion into six kinds [6], namely,happiness, anger, disgust, fear, sadness and surprise. This paper construct a five-dimension emotion space and the emotion variable is: $E = [e_{Anger}, e_{Disgust}, e_{Fear}, e_{Happiness}, e_{Sadness}]^T, \forall e_{[\cdot]} \in [0, 1]$

Compared with mood, the change of emotion is more rapid and fluctuating when it is stimulated by the exterior environment.

2.5 Decay of Mood

Under the situation that the exterior stimulus disappear, the mood will decay gradually and its intensity will be weaken .Supposed the ideal mood state is $\hat{M}$,the mood state at time t is $M(t)$,then change velocity of current mood state is in proportion to the difference between current time mood state and ideal mood state. The mood decay process can be described as follows:

$$edM(t)/dt = \alpha[\hat{M} - M(t)]. \tag{1}$$

Where α represents mood decay coefficient, its value is decided by the personality: $\alpha = \sum_i w_i p_i$

Research indicated that the mood of extravert is easier to reach stable state than that of introvert and the decay velocity of mood of extravert is also faster than that of introvert. Similarly, the decayed velocity of mood of agreeable person is faster than that of disagreeable person.

We can find by the equation (1), if the current mood state could not arrive at the ideal mood state (the three dimensions values of mood are lower than those of ideal mood state), then the change velocity of mood is positive, causes the current mood state shifting to the ideal mood state. If the current mood state surpasses the ideal mood state (the dimension values of current mood state are higher than those of ideal mood state), then the change velocity of mood is negative, causes the current mood state to recede to the ideal mood state.

2.6 Decay of Emotion

Similar to mood, the emotion will also decay gradually and it will trend to stability under the situation that the exterior stimulus disappear. Follow above-mentioned process, we can define the decay process of emotion as follows:

$$dE(t)/dt = \alpha[\hat{E} - E(t)] \tag{2}$$

Where $\hat{E}$ represents ideal emotion state, β is emotion decay coefficient and relate to personality.

In consideration of the change velocity of emotion is more rapid than that of mood, it has $\alpha < \beta$ for the person having the same personality generally.

3 Interrelation of Personality, Mood and Emotion

3.1 Transformation of Between Personality Space and Mood Space

Mood is a 3D space and each dimension value of mood is in the interval [-1,1].Each dimension value takes positive or negative respectively. So we can divide mood space into 8 subspaces, its description as shown in table 1[5].

Table 1. Description of each subspace of PAD space

+P+A+D	Exuberant	-P-A-D	Bored
+P+A-D	Dependent	-P-A+D	Disdainful
+P-A+D	Relaxed	-P+A-D	Anxious
+P-A-D	Docile	-P+A+D	Hostile

Personality space is a 5D space. We build up the transformation relations between personality and mood [4]:

$Pleasure = 0.21 \cdot Extraversion + 0.59 \cdot Agreeableness + 0.19 \cdot Neuroticism$

$Arousal = 0.15 \cdot Openness + 0.30 \cdot Agreeableness - 0.57 \cdot Neuroticism$

$Dominance = 0.25 \cdot Openness + 0.17 \cdot Conscientiousness + 0.60 \cdot Extraversion - 0.32 \cdot Agreeableness$

namely

$$M = K * P \tag{3}$$

where $K = \begin{pmatrix} 0 & 0 & 0.21 & 0.59 & 0.19 \\ 0.15 & 0 & 0 & 0.30 & -0.57 \\ 0.25 & 0.17 & 0.60 & -0.32 & 0 \end{pmatrix}$ is the transfer matrix between personality and mood.

3.2 Mapping Relation Between Moods Spaces and Emotion Space

The change of emotion not only is related to personality and exterior incentive, but also is influenced by the current mood state. Picard[1]establishes linear mapping between mood space and emotion space under one dimensional mood space. This paper uses the three dimensional mood space and the quantificational relations between three dimensional mood PAD space and emotion as shown in Table 2[8]. Table 2 only showes the coordinates of particular intensity of emotion

Table 2. Relations between PAD space and emotion space

Emotion	P	A	D	Sub-space of Mood
Anger	-0.51	0.59	0.25	-P+A+D
Disgust	-0.4	0.2	0.1	-P+A+D
Fear	-0.64	0.6	-0.43	-P+A-D
Happy	0.4	0.2	0.15	+P+A+D
Sad	-0.4	-0.2	-0.5	-P-A-D

in the mood PAD space, but the change intensity of mood is in the interval [0,1] and is a continuous quantity. To find the corresponding points of the continuous quantity in mood space, this paper first defines the mapping base of emotion in mood space:

$$PAD = [PAD_{Anger}, PAD_{Disgust}, PAD_{Fear}, PAD_{Happiness}, PAD_{Sadness}]$$
$$= \begin{pmatrix} -0.51 & -0.4 & -0.64 & 0.4 & -0.4 \\ 0.59 & 0.2 & 0.6 & 0.2 & -0.2 \\ 0.25 & 0.1 & -0.43 & 0.15 & -0.5 \end{pmatrix}$$

Defined the mapping relations between emotion variable and mood variable are:

$$E = f(M, PAD) = \frac{D}{\sum_{i=1}^{5} d_i} \qquad (4)$$

Where $D = [d_1, d_2, d_3, d_4, d_5]$,$d_i = [(M - PAD_i)^T(M - PAD_i)]^{\frac{1}{2}}$, correspond to five kinds of emotions, i.e.,anger, disgust, fear, happiness and sadness respectively.

4 Decisions-Making Model of Emotion

The external stimulus are translated into the inductive variable of emotion by perceptive appraisal. The inductive variable of emotion affects mood space and emotion space respectively, at the same time state transfer of emotion and mood will occur under the influence of personality. Finally expression synthesis is regarded as the output of the model.

4.1 Renewal Equation of Mood

The change of mood state is decided by the exterior stimulus and personality, at the same time the mood state will decay continuously along with the change of time. The renewal equation of mood should manifest the influence of the three kinds of factors. Supposed the mood state is $M(t)$ at time t , then the renewal equation of mood can be described as:

$$M_t = M_{t-1} + K * P_t + g(P_t, M_{t-1}) + h(A_t) - T_M \tag{5}$$

where M_{t-1} is the variable of mood state at $t-1$, $K * P_t$ is influence component of personality to mood, $g(P_t, M_{t-1})$ is decay component of mood which restrains the transfer of mood states,$h(A_t, P)$ represents change component of mood under the influence of inductive variable and T_M is threshold of change of mood, mood states can transfer only that mood variable is bigger than this threshold.

We can obtains the equation (6) by discreting the equation (1):

$$g(P_t, M_{t-1}) = M'_t - M'_{t-1} = -\frac{\alpha(M_{t-1} - \hat{M})}{1+\alpha} \tag{6}$$

The driving results of exterior stimulus with different attribute are different to mood and the functional relation between them is still unknown in psychology. For the purpose of simplification, this paper supposes the relation between inductive variable of emotion and state variable of mood is linear, then:

$$h(A_t, P) = \lambda_M * PAD * A_t \tag{7}$$

Whereλ_M represents the inductive factor of exterior stimulus to mood, it is related to personality.

4.2 Renewal Equation of Emotion

The change of emotion state is decided by the exterior stimulus, mood and personality, moreover the emotion state will decay continuously along with the change of time. The renewal equation of emotion can be described as:

$$E_t = E_{t-1} + f(M_t, PAD) + \Psi(P_t, E_{t-1}) + \Phi(A_t, P_t) - T_E \tag{8}$$

where E_{t-1} is the variable of emotion state at $t-1$, $f(M_t, PAD)$ is influence component of the variable of mood state to variable of emotion state, $\Psi(P_t, E_{t-1})$is decay component of emotion which restrains the transfer of emotion states,$\Phi(A_t, P_t)$ represents influence component of inductive variable of mood to emotion and T_E represents threshold of change of emotion, emotion states can transfer only that emotion variable is bigger than this threshold.

We can obtains the equation (9) by discreting the equation (2):

$$\Psi(P_t, E_{t-1}) = E'_t - E'_{t-1} = -\frac{\beta(E_{t-1} - \hat{E})}{1+\beta} \tag{9}$$

and

$$\Phi(A_t, P_t) = \lambda_E A_t \tag{10}$$

Where λ_E represents the inductive factor of emotion, it is related to personality.

4.3 Facial Expressions Synthesis

Emotion state can be obtained through the emotion decision-making model. At the same time the emotion state will be carried on the emotion expression. The emotion expression provides the possibility of emotional interaction and communication. This paper uses facial expression synthesis as the way of emotion expression. We use the keyframe interpolation technique[9] to synthesize the facial expression. The method can be described as follows:

$$V = \alpha V_1 + (1 - \alpha)V_2 \quad (0 \leq \alpha \leq 1) \tag{11}$$

Where V_1,V_2 are the keyframe in animation sequence, V is an synthesis frame by interpolation, α is the coefficient of inner-frame interpolation. The animation between V_1 and V_2 can be obtained by the keyframe interpolation method.

5 Experiment Results and Analysis

We carry on simulation experiment with the computational model of emotion proposed in this paper. The purpose of simulation is validate whether the method proposed in this paper conforms to the change rule of human's emotion, namely under the influence of exterior stimulation signal, the emotion state and the

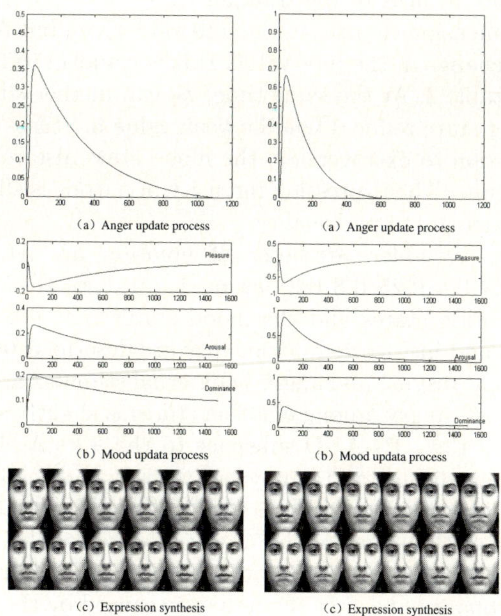

(a) Anger update process (a) Anger update process

(b) Mood updata process (b) Mood updata process

(c) Expression synthesis (c) Expression synthesis

Fig. 1. An overview of the anger state, mood and expression updata process with different personality(left:introverted;right: extroversive)

mood state of human should occur to transfer, and the fluctuant scope and velocity of emotion and mood should have a difference with different personality. At the same time, because human has a certain regulation ability of emotion, the state of emotion and mood should tend to the tranquil state gradually under the situation that the exterior stimulus disappearance.

We simulate the change process of emotion and mood states under the influence of exterior incentives by adjusting the stimulus signal parameters. Supposed emotion and mood are $(0,0,0,0,0)^T$ and $(0,0,0)^T$ respectively in the tranquil state . The attribute of given stimulus is unitary, for example, only angry will have the state transfer if the attribute of stimulus is angry.

Adjusting inductive variable parameter of emotion to $a1 = [0.5,0,0,0,0]$,$b1 = [0.8,0,0,0,0]$,$b2 = [1,0,0,0,0]$, personality $P = [0.5;0.3;0.2;0.2;0.1]$,the attribute of stimulation signal is anger and the exhibition of personality is quite introverted. The fluctuant process of emotion state and mood state and the synthetical facial expression are shown in the left of Fig.1.

Maintaining the inductive variable parameter of emotion invariable, we adjust personality $P = [0.4;0.3;0.8;0.2;0.1]$. At this time the personality is extrovert. The change process of anger state, mood state and the synthetical expression are shown in the right of Fig.1.

We can conclude by Fig. 1 that the angry state achieves the highest intensity in short time under the condition of the attribute of stimulus being angry. The intensity of anger state decays gradually with the exterior stimulus disappearing gradually and finally returns to the tranquil state.

The Pleasure dimension changes in the interval $[-1,0]$, but Arousal and Dominance dimension change in the interval $[0,1]$.It is conform to the corresponding relations given in table 2. At the same time, we can further educe that the fluctuant time of anger state reduces but the peak value increases when personality vary from introversion to extroversion, the mood state also restores to tranquil state in a shorter time. The expression intensity of output is slightly high on the condition of the extroverted personality.

When inductive variable parameter of emotion are $a1 = [0.3,0.4,0.2,0.5,0.15]$, $b1 = [0.4,0.5,0.25,0.8,0.2]$, $b2 = [1,1,1,1,1]$, the change process of the compound emotion states and the mood states that are caused by extroverted personality and introverted personality are described by Fig.2. Since the information of happy, disgust and angry in external stimulus are more, the intensity of happy, disgust, angry among emotion states is slightly strong. The mood states transfer from the - P+A+D subspace to the +P+A+D subspace,which corresponds to anger, disgust and happy respectively and is conform to Table 2. At the same time, the duration of mood and the emotion states of extroverted personality is shorter than that of the introverted personality ,it suggests that emotion and mood of extroverted personality are more stabilization than those of introverted personality and that the influence caused by the external stimulation of emotion and mood of extroverted personality is more weak than that of introverted personality. It is conform to the rule of human's emotion basically.

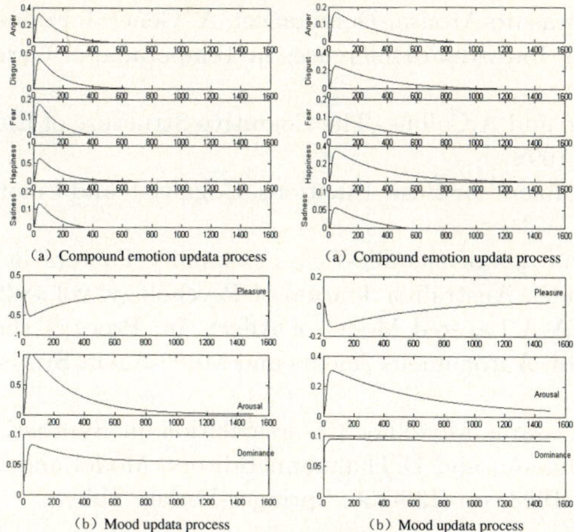

Fig. 2. An overview of the anger state, mood and expression updata process with different personality(left:introverted;right: extroversive)

6 Conclusion

Based on basic theory of emotion psychology, the paper presents the basic definitions of personality space, mood space, emotion decay, mood decay and quantification process of them. We build up the transformation relation between personality and mood space, and the mapping relations between personality and mood space. We propose an emotion decision-making model, which can reasonably represent the relations among external stimulation, personality, mood and emotion.

Through the Matlab simulation, the change process of emotion and mood under the influence of external stimulus can be displayed. The computational model of emotion conforms to the change rule of human's emotion and provides a new method for the automatic production of machine emotion.

Acknowledgments

This work was supported by the National Natural Science Foundation of China (No. 60573059), the Beijing Key Laboratory of Modern Information Science and Network Technology (No. TDXX0503) and Key Foundation of USTB.

References

1. Picard R W. Affective Computiong[M]. MIT Press, London, England,1997
2. Zhiliang Wang Lun Xie, Artificial Psychology-an Attainable Scientific Research on the Human Brain. IPMM'99, Honolulu, USA, July 10-15,1999
3. J.Wiggins.The Five-Factor Model of Personality: Theoretical Perspective. The Guilford Press, New York,1996

4. Mehrabian A. Pleasure-Arousal-Dominance: A General Framework for Describing and Measuring Individual Differences in Temperament. Current Psychology,vol 14,1996:261-292
5. A.Ortony,G.Clore, and A.Collins. The Cognitive Structure of Emotions. Cambridge University Press, 1998
6. Ekman.P. An Argument for Basic Emotions. Cognition and Emotion 6,3-4,1992:169-200
7. Mehrabian A. Analysis of the Big-Five-Personality Factors in Term of the PAD Temperament Model. Australian Journal of Psychology, vol.48,2,1996:86-92
8. Gebhard,P. ALMA-A Layered Model of Affect. In: Proc. of the 4th International Joint Conference on Autonomous Agents and Multi-Agent Systems, AAMAS'05:29-36
9. M. Cohen, D. Massara, Modeling Co-articulation in SyntheticVisual Speech. In N.Magnenat-Thalmann, and D.Thalmann Editors, Model and Technique in Computer Animation, 1993, pp. 139-156, Springer-Verlag, Tokyo

Locating Salient Edges for CBIR Based on Visual Attention Model

Feng Songhe and Xu De

Dept. of Computer Science & Technology, Beijing Jiaotong Univ.,
Beijing, China 100044
songhe_feng@163.com

Abstract. Visual attention model was usually used for salient region detection. However, little work has been employed to use the model for salient edge extraction. Since edge information is also important element to represent the semantic content of an image, in this paper, attention model is extended for salient edges detection. In our approach, an improved saliency map computing algorithm is employed first. Then, based on the saliency map, a novel and efficient salient edges detection method is introduced. Moreover, the concept of salient edge histogram descriptors (SEHDs) is proposed for image similarity comparison. Experiments show that the proposed algorithm works well.

1 Introduction

Content-based image retrieval (CBIR) has been widely investigated in the past few years [1] where the so called semantic gap still exists. Extracting salient objects from images is a promising solution which is useful for high-level visual processing such as image semantic understanding and retrieval. Visual attention model provides just the mechanism to locate the visual salient part of the image.

There exist some computational models of visual attention which are mainly used for salient region extraction while edge information is often ignored. [2] indicate that without edge information, the saliency map is only a blur map which can only provide the location of the attention, but not exact the scope of the region.

Existing salient edge detection algorithms are mainly based on graph theory to model the Gestalt laws, proximity, closure and continuity. However, all these algorithms aim to determine the global salient closed contour in the image for object identification, so the computational costs are very high. Since our goal here is image semantic retrieval, in this paper, an efficient and effective attention guided salient edge detection algorithm is proposed and the concept of salient edge histogram descriptors (SEHDs) is employed for CBIR.

The remainder of this paper is organized as follows. The details of salient edges detection algorithm is proposed and the Salient Edge Histogram Descriptors (SEHDs) is constructed in Section 2. Experimental results are reported in Section 3 and we give the conclusion in Section 4.

L. Jiao et al. (Eds.): ICNC 2006, Part I, LNCS 4221, pp. 261 – 264, 2006.

2 Salient Edges Detection Based on Visual Attention Model

Visual attention is one of the most important functions of the human vision system. Since the motivation of this paper is image semantic retrieval based on the salient edge information, saliency map construction is the basis of our work. In this paper, we adopt our previous method [6] to construct the saliency map. Compared to previous works, our method is more effective while used in CBIR. (See [6] for details)

2.1 Salient Edges Detection

Edge information is a basic feature of images. It contains contour information of a valuable object in the image and can be used to represent image contents, recognize the objects and further for object-based image retrieval [5]. Although some edge detectors such as Canny detector can filter part of an image background, not all extracted edges are beneficial to describe image content which may be composed of a large number of short lines or curves.

Some graph theory based methods have been proposed to extract salient boundaries from noisy images. [4] used the shortest-path algorithm to connect fragments to form salient closed boundaries. [3] proposed the ratio cut algorithm which formulate the salient boundary detection problem into a problem for finding an optimal cycle in an undirected graph. However, since gap filling and smoothing are very time-consuming, these algorithms may not be suitable for image retrieval.

Visual attention provides a mechanism to locate the visual salient part of the image. Since the human visual system is sensitive with the salient part in the image, certainly it would also be sensitive with the edges around the saliency map. So here we extend the saliency map for salient edge detection.

Initially we use the standard Canny edge detector to obtain the edge map from a given input image. Let $E=\{e_1, e_2, \cdots e_n\}$ be the set of all edges in the edge map. Since not all the edges in the edge map are useful for image retrieval especially the cluttered background edges, we define the edge saliency below:

$$S(e_i)=\lambda_1 \cdot L(e_i)+\lambda_2 \cdot VA(e_i) \quad where\ i=1\cdots n \tag{1}$$

Here, $L(e_i)$ is the length of edge e_i, $VA(e_i)$ denotes the average saliency value of edge e_i based on the saliency map. λ_1 and λ_2 denote two weights of $L(e_i)$ and $VA(e_i)$. Here, we set λ_1 and λ_2 equal to 0.3 and 0.7, respectively. $VA(e_i)$ is defined as follows. Different from directly compute the saliency value of each pixel on the edge, here we consider the 3×3 neighborhood of each pixel on the edge. Let $p_n^i \in e_i, n=1\cdots L(e_i)$ be the set of all pixels on edge e_i.

$$VA(e_i) = \sum_{n=1}^{L(e_i)} \sum_{x \in W_{p_n^i}} s_x \Big/ L(e_i) \tag{2}$$

Where $W_{p_n^i}$ denotes the 3×3 window centered at pixel p_n^i, s_x is the saliency value of pixel x.

After all the saliency values of edges are calculated, an empirical threshold $T_E = \max(S(e_i))/4$ is adopted here. So the final salient edge set is defined as

$$SE = \{ e_i | S(e_i) > T_E, i = 1 \cdots n \} \tag{3}$$

2.2 Salient Edge Histogram Descriptors Extraction for Image Semantic Retrieval

After salient edges are extracted, image features are extracted to represent the image content and for similarity measure. Here, the concept of salient edge histogram descriptors is proposed for CBIR. Since edges play an important role for image perception, they are frequently used as a shape describing feature descriptor in image retrieval. The edge histogram descriptor (EHD) is an example, which is one of three normative texture descriptors proposed for MPEG-7. It captures the spatial distribution of edges in an image and has been proven to be useful for image retrieval, especially for natural images with non-uniform textures and clip art images. Five types of edges, namely vertical, horizontal, 45° diagonal, 135° diagonal, and non-directional edges have been utilized to represent the edge orientation. The distance between SEHDs are denoted as

$$dis_E(A,B) = \sqrt{\sum_{i=1}^{5} \left(EH_A^i - EH_B^i \right)^2} \tag{4}$$

where EH_A^i denotes one orientation of edge histogram.

3 Experimental Results

We implement the proposed algorithm using MATLAB 6.5 on a PC with 3.0G Pentium IV CPU and 1G memory. To evaluate the performance of the proposed algorithm, we choose about 2000 images of 20 categories from the Corel database as our test image database.

We use the standard measures, precision and recall, in different forms to evaluate the results. The precision is the fraction of the returned images that is indeed relevant for the query while the recall is the fraction of relevant images that is returned by the query.

We test the proposed method using different query images and retrieval numbers. For comparison, we also use the features of global EHD for image retrieval. And the average precision and recall rates are seen in Fig.2.

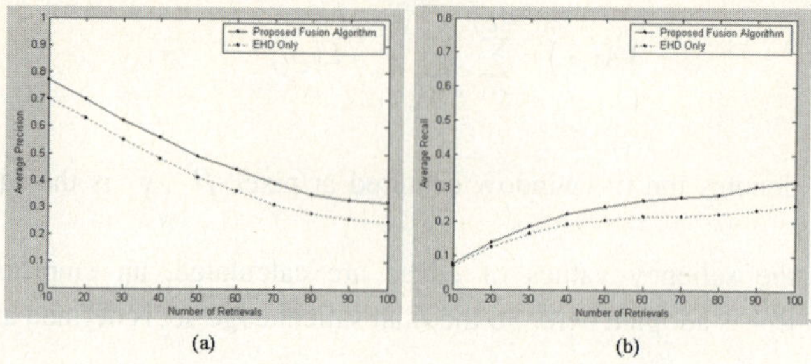

Fig. 1. (a) and (b) denote the comparison of average precision and recall rates respectively

From the experiment we can conclude that, since the salient edges information is considered, the retrieval performance of the proposed algorithm is better than the method which only consider global EHDs.

4 Conclusions and Future Work

This paper presents a novel image retrieval algorithm which uses salient edges information. The visual attention model is extended for salient edge extraction. After salient edges are extracted, the concept of salient edge histogram descriptors is proposed for image retrieval. The advantage lies in that the extracted features can represent the salient part of an image and characterize the human perception well. Experimental results have proved the proposed algorithm.

References

1. Datta, R., Li, J., Wang J.Z.: Content-Based Image Retrieval - Approaches and Trends of the New Age. ACM MIR (2005)
2. Hu, Y., Xie, X., Ma, W.Y., Rajan, D., Chia, L.T.: Salient object extraction combining visual attention and edge information. Technical Report (2004)
3. Wang, S., Kubota, T., Siskind, J.M., Wang, J.: Salient Closed Boundary Extraction with Ratio Contour. IEEE Trans on Pattern Analysis and Machine Intelligence. 27(4) (2005) 546-561
4. Elder, J., Zucker, S.: Computing contour closure. In European Conference on Computer Vision. (1996) 399–412
5. Zhou, X.S., Huang, T.S.: Edge-Based Structural Features for Content-Based Image Retrieval. Pattern Recognition Letters. 22(5) (2001) 457-468
6. Feng, S.H., Xu, D.: A Novel Region-Based Image Retrieval Algorithm Using Selective Visual Attention Model. Lecture Notes in Computer Science. 3708 (2005) 235-242

"What" and "Where" Information Based Attention Guidance Model

Mei Tian, Siwei Luo, Lingzhi Liao, and Lianwei Zhao

School of Computer and Information Technology,
Beijing Jiaotong University, Beijing 100044, China
tmlily@126.com

Abstract. Visual system can be defined as consisting of two pathways. The classic definition labeled a "what" pathway to process object information and a "where" pathway to process spatial information. In this paper, we propose a novel attention guidance model based on "what" and "where" information. Context-centered "where" information is used to control top-down attention, and guide bottom-up attention which is driven by "what" information. The procedure of top-down attention can be divided into two stages: pre-attention and focus attention. In the stage of pre-attention, "where" information can be used to provide prior knowledge of presence or absence of objects which decides whether search operation is followed. By integrating the result of focus attention with "what" information, attention is directed to the region that is most likely to contain the object and series of salient regions are detected. Results of experiment on natural images demonstrate its effectiveness.

1 Introduction

While navigating in an environment, visual system can be divided into two cortical visual subsystems—"what" and "where" pathways [1]. The information transferred through "what" and "where" pathways are about what the objects in external world are and where they are. This theory of visual system has the same view with Marr's [2]. Accordingly, by integrating the theory of two visual pathways with attention mechanism, we define new "what" and "where" information to drive bottom-up and top-down attention.

Since enormous progress has been made in recent years' research towards understanding the basic principles of information processing in visual cortex, bottom-up attention has drawn great interest, and some attention guidance models have been build up [3-5]. But the detection procedures of these models are irrelevant to objects, so the performance is not high when the object of interest is not the most salient object in a blurring image. To solve this difficulty, Sun presented an object-based visual attention model [6]. Among the applications of object-based attention, the model of automatic traffic sign detection [7] is more representative. However, all of these models use characters that belong to objects themselves and ignore high level information. Considering effects of high level information, some researchers introduced top-down

L. Jiao et al. (Eds.): ICNC 2006, Part I, LNCS 4221, pp. 265 – 274, 2006.

attention into their models [8-12]. But the "high level information" used in these top-down attention is only about threshold, weight, model and so on. They're not real high level information. Current attention models confront two main obstacles. First, these models will be less effective when image is degraded that information of object itself is not sufficient for reliable detection. Second, for different appearances and locations of the same object, they don't use priors well and needs exhaustive exploration on a large space of an image.

Distinct from previous work, in our model, bottom-up and top-down attention are combined by using our new definition of "what" and "where" information. The procedure of top-down attention can be divided into two stages: pre-attention and focus attention. These two stages are different from those defined in bottom-up attention mechanism [13]. Our model can solve the obstacles in traditional attention models.

In this paper, section 2 details the architecture of our model, which consists of 5 main components. Experiments are carried out to demonstrate efficacy of the model and discussion on future work is given in the last section.

2 Model

The architecture of our attention model is as shown in Fig.1. A set of Gabor filters receive input image and extract 16 feature maps. They are used to form "where" and "what" information. Then the attention model is divided into two parallel pathways. By using different coding algorithms, "where" and "what" information are represented by Ve and Vl, respectively. To achieve top-down attention control, we use the conditional probability function $p(O|Ve)$. Considering category property s and location property $l=(x,y)$ of the object O, we apply the Bayes rule successively

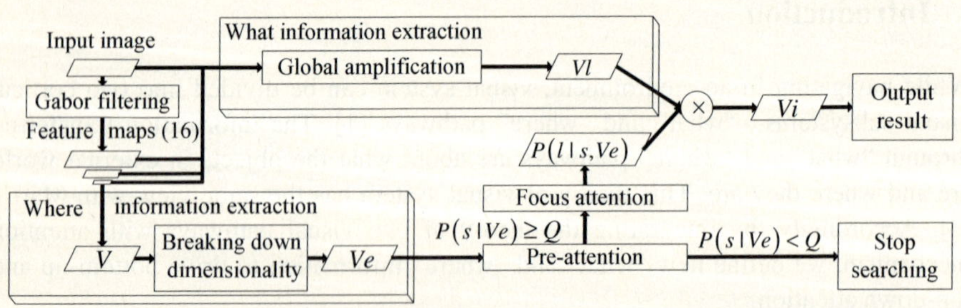

Fig. 1. Architecture of our model

$$p(O|Ve) = p(l|s,Ve)\,p(s|Ve) \tag{1}$$

where $p(s|Ve)$ and $p(l|s,Ve)$ correspond to pre-attention and focus attention, respectively. In the stage of pre-attention, the value of $p(s|Ve)$ is estimated and is used to decide whether search operation is followed. Q is threshold. The estimated $p(l|s,Ve)$ is used to guide bottom-up attention and direct attention to the focus attention region that is most likely to contain the object. The output result can be draw from integrated information Vi.

The following subsections detail the components of our model.

2.1 "Where" Information Extraction

Contextual information plays a major role in facilitating object detection by recent years' research. When subjects are asked to search for a specific object, that object is noticed faster when it is consistent with the scene [14]. Accordingly, we define an entire context-correlative feature as "where" information. In this paper, Gabor filter is chosen for its ability to simulate the early stages of the visual pathway. For each filter, the output response of image $I(x, y)$ is defined as

$$v(x, y) = I(x, y) * \psi(x - x_0, y - y_0) \tag{2}$$

where $\psi(x, y)$ is Gabor filter defined by $\psi(x, y) = e^{-\left(\alpha^2 x'^2 + \beta^2 y'^2\right)} e^{j2\pi f_0 x'}$, $x' = x\cos\varphi + y\sin\varphi$, $y' = -x\sin\varphi + y\cos\varphi$, α is sharpness of the Gaussian major axis, f_0 is frequency of the sinusoid, φ is rotation of the Gaussian and sinusoid and β is sharpness of the Gaussian minor axis. In this definition, (x_0, y_0) is the center of receptive field and $*$ represents convolution.

The outputs of all filters can be described as $\left\{v_{i,j}(x, y), i, j = 1, 2, 3, 4\right\}$, where the variable i and j are indexes of orientation and frequency, respectively. Each output serves as a sub-block and a high dimensional code V of "where" information is defined by $V_{i,j} = v_{i,j}(x, y)$. The indexes of orientation and frequency in the new code V increase from left to right and top to down.

For high-order statistics of code V, independent component analysis [15] is applied to reduce the dimensionality. Each sub-block $V_{i,j}$ serves as a sample. The rows of each sample are concatenated to a vector, top row first. By using ICA, different samples have the same linear basis function and the different coefficient vectors. So the coefficient vectors can be used to represent the samples. To the training set $\left\{V_{i,j}^{(k)}, k = 1, 2, \ldots, 16\right\}$, we use the fast fixed-point algorithm [15] to compute the feature coefficients $S_{i,j}^{(k)}$ defined by $S_{i,j}^{(k)} = \langle W_{i,j}, V_{i,j}^{(k)} \rangle$, where $W_{i,j}$ is the transformation matrix. The initial image is represented by a set of feature coefficients $S_{i,j}$ now.

Then we define the linear basis functions as a linear subspace and define the output response as the projective distances to subspace. All responses correspond to different orientation and frequency constitute a matrix $Ve = \left\{ve_{i,j}\right\}$, and each $ve_{i,j}$ is

$$ve_{i.j} = \sqrt{\sum_{n=1}^{d} \left(a_{i,j}^{(n)}\right)^2} \tag{3}$$

where $a_{i,j}$ is a scalar and represents the n th eigenvalue of $S_{i,j}$ and d is the number of the eigenvalue. Because all regions of the initial image contribute to the response matrix Ve and Ve doesn't contain any independent information belonging to the object, we define the response matrix Ve as the "where" information of the initial image.

2.2 "What" Information Extraction

We first define the input image $I(x,y)$ as the17th feature map. And the 17 feature maps are normalized to the same range. Then, for each feature map, we compute the global amplification factor which is defined by $(M-\bar{m})^2$ [16] and use it as weight, where M and $\bar{m}$ are its global maximum and the average of all other local maxima, respectively. The final salient map defined by Vl is the weighted sum of all feature maps and is defined as "what" information in our model.

2.3 Pre-attention

In the stage of pre-attention, the likelihood function $p(s \mid Ve)$ can be represented as

$$p(s \mid Ve) = \frac{p(Ve \mid s)\,p(s)}{p(Ve \mid s)\,p(s) + p(Ve \mid \neg s)\,p(\neg s)} \tag{4}$$

The training set consists of a large number of images that contain the object s, so we approximate the prior probability by $p(s) = p(\neg s) = 1/2$. The training data are "where" information $VE = \{Ve_1, Ve_2, \cdots, Ve_t, \cdots, Ve_N\}$, where N is the number of training pictures. The likelihood function $p(Ve \mid s)$ can be represented as

$$p(Ve \mid s) = \prod_{t=1}^{N} p(Ve_t \mid s) \tag{5}$$

We introduce contextual category information $C = \{C_i\}_{i=1,K}$ into our model

$$p(Ve_t \mid s) = \sum_{i=1}^{K} \omega_i\, p(Ve_t \mid C_i, s),\ \sum_{i=1}^{K} \omega_i = 1 \tag{6}$$

where ω_i represents the prior probability of the ith contextual category. The generalized Gaussian mixture model is then used to select the most appropriate mixture model to simulate the $p(Ve_t \mid C_i, s)$ [17]

$$p(Ve_t \mid \mu_i, \sigma_i, \varsigma) = \frac{\omega(\varsigma)}{\sigma_i} \exp\left[-c(\varsigma)\left|\frac{Ve_t - \mu_i}{\sigma_i}\right|^{2/(1+\varsigma)}\right] \tag{7}$$

where $c(\varsigma) = \left[\frac{\Gamma\left[\frac{3}{2}(1+\varsigma)\right]}{\Gamma\left[\frac{1}{2}(1+\varsigma)\right]}\right]^{1/(1+\varsigma)}$, $\omega(\varsigma) = \frac{\Gamma\left[\frac{3}{2}(1+\varsigma)\right]^{1/2}}{(1+\varsigma)\Gamma\left[\frac{1}{2}(1+\varsigma)\right]^{3/2}}$, $\Gamma(q) = \int_0^{\infty} e^{-t} t^{q-1} dt$. μ_i is the mean while σ_i^2 is the deviation, respectively. The parameter ς controls the deviation of distribution from normality. We use some models as candidates in which the parameter ς are -0.25, 0 (the normal distribution), 0.495 (Tanh distribution) and 1 (Laplace distribution), respectively. By using IGMSC model selection criterion [18], the normal distribution is selected to form the Gaussian mixture model.

Thus, the distribution of each data can be simulated by a Gaussian mixture model. The parameter for object s is $\Theta = \{\theta_i, \omega_i\}_{i=1,K}$, and the i th parameter of Gaussian distribution is $\theta_i = (\mu_i, \sigma_i)$

$$p(Ve_t \mid s) = p(Ve_t \mid \Theta) = \sum_{i=1}^{K} \omega_i p(Ve_t \mid C_i, \theta_i) \tag{8}$$

$$p(Ve_t \mid C_i, \theta_i) = \frac{1}{\sqrt{2\pi}\sigma_i} \exp\left[-\frac{(Ve_t - \mu_i)^2}{2\sigma_i^2}\right] \tag{9}$$

Supposing $p(Ve \mid \Theta)$ is differentiable to Θ, we define $L(\Theta \mid Ve)$ as

$$L(\Theta \mid Ve) = \log p(Ve \mid \Theta) = \sum_{t=1}^{N} \log p(Ve_t \mid \Theta) = \sum_{t=1}^{N} \log\left[\sum_{i=1}^{K} \omega_i p(Ve_t \mid C_i, \theta_i)\right] \tag{10}$$

The parameter Θ for object s is estimated by EM algorithm [19].

Based on Bayes rule, given the newest parameters $\hat{\theta}_i = (\hat{\mu}_i, \hat{\sigma}_i)$ and $\hat{\omega}_i$, the E-step computes the posterior probabilities

$$p\left(C_i \mid Ve_t, \hat{\theta}_i\right) = \frac{\hat{\omega}_i p\left(Ve_t \mid C_i, \hat{\theta}_i\right)}{\sum_{i=1}^{K} \hat{\omega}_i p\left(Ve_t \mid C_i, \hat{\theta}_i\right)} \tag{11}$$

Then, the current parameter $\hat{\Theta}$ and "where" information Ve are used to compute the expectation of complete-data $L(\Theta \mid Ve, C)$ [19]

$$\Lambda\left(\Theta, \hat{\Theta}\right) = \mathrm{E}\left[\log p(Ve, C \mid \Theta) \mid Ve, \hat{\Theta}\right] = \int_C \log p(Ve, C \mid \Theta)\, p\left(C \mid Ve, \hat{\Theta}\right) dC \tag{12}$$

The M-step selects the value of Θ which can maximize the $\Lambda\left(\Theta, \hat{\Theta}\right)$

$$\hat{\omega}_i^{new} = \frac{1}{N} \sum_{t=1}^{N} p(C_i \mid Ve_t, \hat{\theta}_i) \tag{13}$$

$$\hat{\mu}_i^{new} = \frac{1}{\sum_{t=1}^{N} p(C_i \mid Ve_t, \hat{\theta}_i)} \sum_{t=1}^{N} p(C_i \mid Ve_t, \hat{\theta}_i) Ve_t \tag{14}$$

$$\hat{\sigma}_i^{2new} = \frac{1}{\sum_{t=1}^{N} p(C_i \mid Ve_t, \hat{\theta}_i)} \sum_{t=1}^{N} p(C_i \mid Ve_t, \hat{\theta}_i)(Ve_t - \hat{\mu}_i^{new})\, (Ve_t - \hat{\mu}_i^{new})^T \tag{15}$$

These two steps are repeated until Θ converges at a stable value. The same algorithm holds for the likelihood function $p(Ve \mid \neg s)$. And the final result of pre-attention is obtained by computing equation (4).

2.4 Focus Attention

In the stage of focus attention, likelihood function $p(l \mid s,Ve)$ can be represented as

$$p(l \mid s,Ve)=\frac{p(l,Ve \mid s)}{p(Ve \mid s)} \tag{16}$$

$p(l,Ve \mid s)$ is simulated by a Gaussian mixture model too. The training data are "where" information $VE=\{Ve_1,Ve_2,\cdots,Ve_t,\cdots,Ve_N\}$ of the pictures and location information $L=\{l_1,l_2,\cdots,l_t,\cdots,l_N\}$ of the object s. $p(l,Ve \mid s)$ is simulated by K Gaussian clusters. And each cluster is decomposed into two Gaussian functions which correspond to "where" information ($\theta_i=(\mu_i,\sigma_i)$) and location information ($\delta_i=(\mu_i',\sigma_i')$). Thus, the parameter of the model is $\Theta=\{\theta_i,\delta_i,\omega_i\}_{i=1,\ldots,K}$. The distribution of each data can be represented as

$$p(l_t,Ve_t \mid s)=p(l_t,Ve_t \mid \Theta)=\sum_{i=1}^{K}\omega_i p(Ve_t \mid C_i,\theta_i)\,p(l_t \mid \delta_i) \tag{17}$$

The training process for parameter Θ is similar to the process in pre-attention.

2.5 Integration of Top-Down and Bottom-Up Attention

In the stage of focus attention, we compute $p(l \mid s,Ve)$ for each location. In order to find the focus attention region, we observe the distribution of $p(l \mid s,Ve)$ in testing image, and find that "where" information provides strong prior knowledge for the elevation coordinate y, but it does not allow the x coordinate. We define $l_{x_0,y_0}=\sum_{l_{x,y}} l_{x,y}\,p(l_{x,y} \mid s,Ve)$ as the center of focus attention region. The width of the region can be defined as the width of the image. The height is defined as

$$\Delta_+=\frac{\sum_{x=0}^{255}p(l_{x,y} \mid s,Ve)}{\sum_{x=0}^{255}p(l_{x,y+1} \mid s,Ve)},\quad y=y+1 \tag{18}$$

where Δ_+ is iteratively computed, and the initial value of y is y_0. Once $\Delta_+>10$, the iteration is stopped and the value of y_+ is set to the current value of y. The similar scheme holds for Δ_-, and the value of y_- is gained when iteration is stopped. At last, the height of the focus attention region is defined as $h=y_+ + y_-$.

The integrated information Vi is defined as the multiplication of Vl and $p(l \mid s,Ve)$. Supposing the focus attention region can be split into small $n\times n$ sub-blocks and the number of blocks is b, we use $4b$ random blocks to overlay the initial region to guarantee the reconstruct error is minimal. The integrated information Vi of each block is regarded as a sample and is transformed into a vector. Then, we compute the difference between each sample's Vi and all other samples' Vi, and use the square sum of these differences as saliency of corresponding sample.

3 Experiments

To test the model, we run it on 400 images (256×256), half of which contain cars while the other half contain people. These images come from the Database of Cars and Faces in Context. The contextual categories are indoors and outdoors, so the parameter K of Gaussian mixture model is set to 2. Fig.2 shows the distribution of $P(s\,|\,Ve)$ when looking for cars and people in 8 images. $P(s_{cars}\,|\,Ve)$ and $P(s_{people}\,|\,Ve)$ represent the probabilities of presence of cars and people, respectively. If $P(s\,|\,Ve)\approx 1$ or $P(s\,|\,Ve)\approx 0$, we can reliably determine the presence or absence of the object without scanning the whole image. We can see that cars are missing in the sixth image in Fig.2, but $P(s_{cars}\,|\,Ve)\sim 1$. It shows that attention is only driven by "where" information in this stage and is uncorrelated with "what" information which belongs to the object.

Note that almost all values of $P(s\,|\,Ve)$ approximate to 0 or 1 in Fig.2. We chose 50 images at random from database to compute $P(s_{people}\,|\,Ve)$. There are 29 and 17 images belong to the sets defined by $P(s_{people}\,|\,Ve)>0.95$ and $P(s_{people}\,|\,Ve)<0.05$ respectively. Fig.3 is about examples belonging to the two sets. It shows that the pre-attention can provide the reliable priors of presence or absence of objects for most of the images. If $P(s\,|\,Ve)<0.05$, the search operation is stopped for knowing the object can't be in the scene. This operation can save computing resources, and enhance object detection performance. We set $Q=0.05$ to guarantee reliable object detection.

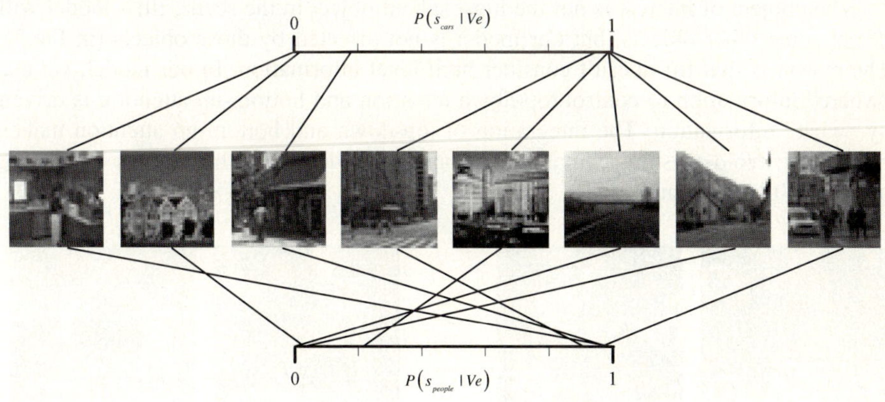

Fig. 2. $P(s\,|\,Ve)$ of object categories s_{cars} and s_{people}

(a) $P\left(s_{people} \mid Ve\right) > 0.95$ (b) $P\left(s_{people} \mid Ve\right) < 0.05$

Fig. 3. Examples belonging to the sets defined by $P\left(s_{people} \mid Ve\right) > 0.95$ and $P\left(s_{people} \mid Ve\right) < 0.05$

Fig.4 shows the experiment result when looking for cars in an image. The parameter n is set to 16. In the black regions in Fig.4(b), the value of $p(l \mid s, Ve)$ approximate to 0 and "what" information is suppressed. So there is no need to direct attention to these black regions. The rest regions in Fig.4(b) belong to the focus attention region where the probability of presence of cars is very high. In descending order of saliency, five selected regions are shown in Fig.4(c) and labeled from 1 to 5 (regions overlap each other are discarded).

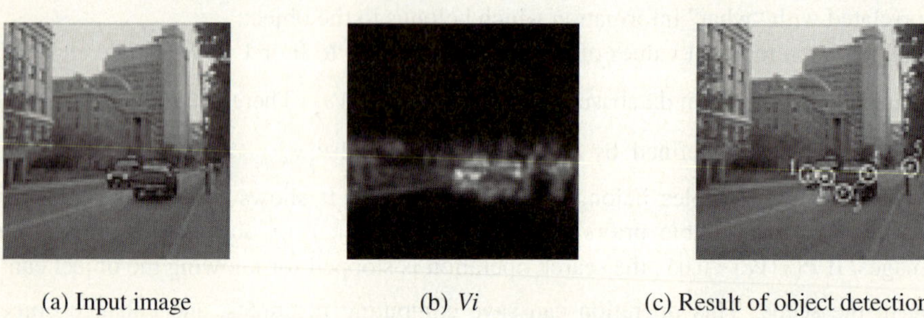

(a) Input image (b) *Vi* (c) Result of object detection

Fig. 4. Experimental results

When object of interest is not the most salient object in the scene, Itti's model will detect some other objects, but our model is not affected by those objects (in Fig.5). The reason is that Itti doesn't consider high level information. In our model, we use "where" information to control top-down attention and bottom-up attention is driven by "what" information. The integration of top-down and bottom-up attention makes our model avoid wasting computing resources on detecting the regions where the object can't be present.

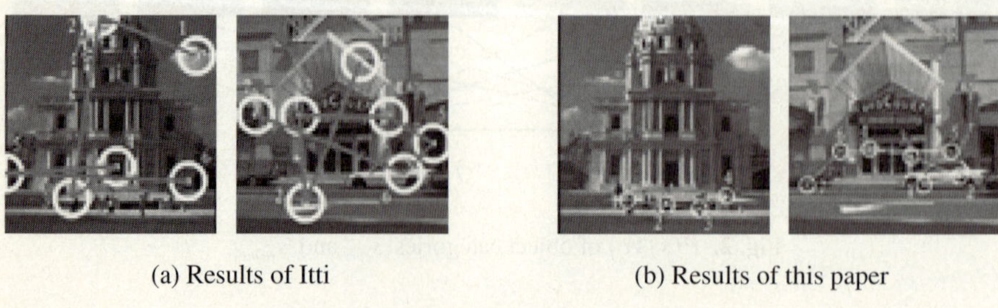

(a) Results of Itti (b) Results of this paper

Fig. 5. Experimental results of Itti and our model

4 Conclusion

In this paper we implement a new "what" and "where" information based attention guidance model. Experiments on 400 natural images demonstrate its effectiveness.

Contrasting with current models, our model has two advantages. First, we define "where" information as high level information. It contains the information of the entire scene and can provide reliable prior knowledge for bottom-up attention. Second, the procedure of top-down attention is divided into two stages. The entire detecting process can be stopped only after pre-attention. And the result of focus attention can be integrated with bottom-up attention.

The model has several limitations. It doesn't consider "what" information based perceptual grouping. But grouping is a common phenomenon in biology vision. Therefore, the next study is to use perceptual grouping as a bridge to connect "what" and "where" information. Additionally, according to resources share in the process of visual attention, the category information of context can be used to predict new class of context.

Acknowledgements

The authors would like to thank Itti and his iLab for providing source codes which were used for comparison in this paper.

This research is supported by the National Natural Science Foundation of China under Grant No. 60373029 and the National Research Foundation for the Doctoral Program of Higher Education of China under Grant No. 20050004001.

References

1. Creem, S.H. and Proffitt, D.R.: Defing the Cortical Cisual Systems: "What", "Where", and "How". Acta Psychologica, Vol.107 (2001) 43-68
2. Marr, D.: Vision: a Computational Investigation into the Human Representation and Processing of Visual Information. San Francisco: Freeman, W. H. (1982)
3. Itti, L., Gold, C., Koch, C.: Visual Attention and Target Detection in Cluttered Natural Scenes. Optical Engineering, Vol.40 (2001) 1784-1793
4. Zhang, P. and Wang, R.S.: Detecting Salient Regions Based on Location Shift and Extent Trace. Journal of Software, Vol.15 (2004) 891-898
5. Frintrop, S. and Rome, E.: Simulating Visual Attention for Object Recognition. Proceedings of the Workshop on Early Cognitive Vision (2004)
6. Sun, Y. and Fisher, R.: Object-Based Visual Attention for Computer Vision. Artificial Intelligence, Vol.146 (2003) 77-123
7. Ouerhani, N.: Visual Attention: from Bio-Inspired Modeling to Real-Time Implementation. Switzerland: Institute of Micro technology (2003)
8. Long, F.H. and Zheng, N.N.: A Visual Computing Model Based on Attention Mechanism. Journal of Image and Graphics, Vol.3 (1998) 592-595
9. Rybak, I.A., Gusakova, V.I., Golovan, A.V., Podladchikova, L.N., Shevtsova, N.A.: A Model of Attention-Guided Visual Perception and Recognition. Vision Research, Vol.38 (1998) 2387-2400

10. Salah, A.A., Alpaydin, E., Akarun, L.: A Selective Attention-Based Method for Visual Pattern Recognition with Application to Handwritten Digit Recognition and Face Recognition. IEEE Trans. on Pattern Analysis and Machine Intelligence, Vol.24 (2002) 420-425
11. Itti, L.: Models of Bottom-Up and Top-Down Visual Attention. Pasadena: California Institute of Technology (2000)
12. Navalpakkam, V. and Itti, L.: A Goal Oriented Attention Guidance Model. Lecture Notes in Computer Science, Vol.2525 (2002) 453-461
13. Itti, L and Koch, C.: Computational Modeling of Visual Attention. Nature Reviews Neuroscience, Vol.2 (2001) 194-230
14. Henderson, J.M.: Human Gaze Control during Real-World Scene Perception. Trends in Cognitive Sciences, Vol.7 (2003) 498-504
15. Hyvärinen, A.: Fast and Robust Fixed-Point Algorithms for Independent Component Analysis. IEEE Transactions on Neural Network, Vol.10 (1999) 626-634
16. Itti, L. and Koch, C.: Feature Combination Strategies for Saliency-Based Visual Attention Systems. Journal of Electronic Imaging, Vol.10 (2001) 161-169
17. Lee, T.W. and Lewicki, M.S.: The Generalized Gaussian Mixture Model Using ICA. International Workshop on Independent Component Analysis (ICA'00), (2000) 239-244,
18. Liu, Y.H., Luo, S.W., Li, A.J., Yu, H.B.: A New Model Selection Criterion Based on Information Geometry. 7th Intern. Conf. on Signal Processing (ICSP'04), Vol.2 (2004) 1562-1565
19. Bilmes, J.A.: A Gentle Tutorial of the EM Algorithm and Its Application to Parameter Estimation for Gaussian Mixture and Hidden Markov Models. Technical Report, ICSI-TR-97-021, Berkeley: University of Berkeley (1998)

Emotion Social Interaction for Virtual Characters

Zhen Liu

Faculty of Information Science and Technology,
Ningbo University, Ningbo 315211
liuzhen@nbu.edu.cn

Abstract. Nonverbal emotion social interaction is a direct communication manner for virtual characters. A believable virtual character has not only emotion, but also social norm, which include status, friendliness, time of interaction, interaction radius, and priority degree of interaction. In this paper, some new concepts on social norm are presented, a method of emotion social interaction is set up. A virtual character in the demo system has the ability of emotion social interaction with other virtual characters by the method.

1 Introduction

Social interaction of virtual characters is an interesting research subject in virtual environment. A believable virtual character has the ability of social interaction to other virtual characters with verbal and nonverbal manner. In fact, nonverbal interaction is even more important than verbal interaction, particularly in respect of social interaction [1][[4]. There are a lot of researches on nonverbal behavior expression of virtual characters, but little on how interaction triggers and ends when two virtual characters en-counter with social status. For example, Cassell et al. developed behavior expression toolkit for virtual characters with XML-based language [5], Pelachaud created a subtle method of facial expression for virtual characters [6], Chi et al. built a system called EMOTE to express natural posture movements of a virtual character [7]. Liu et al. presented an emotion model for virtual characters [8], a virtual character has a built-in cognitive model that include perception and emotion module. On the basis of these researches, this paper improve the result in previous work, the goal of the paper is to explore a model of nonverbal emotion social interaction.

2 Social Norm of Virtual Characters

In this section, some new definitions about social interaction process of virtual characters are described.

Definition 1. For a virtual character, a status is a social degree or position. In general, a virtual character may own many status, let $\boldsymbol{ST}(CA)$ is a status set for virtual character CA, $\boldsymbol{ST}(CA)$={st_1,..., st_N}, st_i is a status (such as mother or son), $i \in [0, N]$, N is the number of $\boldsymbol{ST}(CA)$.

L. Jiao et al. (Eds.): ICNC 2006, Part I, LNCS 4221, pp. 275–278, 2006.

Status plays an important role in a social interaction. For example, in a virtual office, there are two kinds of social status altogether, namely the manager and staff member. The manager's status is higher than the status of the staff member. In general, a person will control emotion expression by one's status.

Definition 2. For two virtual characters CA_1 and CA_2, let $FD(CA_1/CA_2)$ is friendliness degree from CA_1 to CA_2. If $FD(CA_1/CA_2)$=1, CA_2 is a friend of CA_1; If $FD(CA_1/CA_2)$=-1, CA_2 is a enemy of CA_1; If $FD(CA_1/CA_2)$=0, CA_2 is a stranger of CA_1; If $FD(CA_1/CA_2)$=2, CA_2 is a lover of CA_1; If $FD(CA_1/CA_2)$=3, CA_2 is a mother or father of CA_1; If $FD(CA_1/CA_2)$=4, CA_1 is a mother or father of CA_2.

A virtual character judges others with friendliness degree. In general, a virtual character will not interact with a stranger unless in some exceptive conditions (calling help in danger etc.).

Definition 3. For two virtual characters CA_1 and CA_2, let $ET_1(CA_1/CA_2)$ is default-ending time of interaction from CA_1 to CA_2, let $ET_2(CA_2/CA_1)$ is default-ending time of interaction from CA_2 to CA_1, and ET is the time from beginning to ending in interaction.

In general, if ET>min (ET_1 (CA_1/CA_2), ET_2 (CA_2/CA_1), the interaction will end.

Definition 4. For two virtual characters CA_1 and CA_2, let IR (CA_1/CA_2) is interaction radius of CA_1 to CA_2, let DS (CA_1/CA_2) is distance from CA_1 to CA_2. In general, if $DS(CA_1/CA_2)>IR(CA_1/ CA_2)$, CA_1 will not make interaction to CA_2; if $DS(CA_1/CA_2)\leq IR$ (CA_1/CA_2), CA_1 may make interaction to CA_2.

In default condition, when two agents encounter together, interaction radius is critical distance of interaction triggering.

Definition 5. For two virtual characters CA_1 and CA_2, let PN (CA_1, CA_2) is priority degree of social interaction between CA_1 and CA_2. If PN (CA_1, CA_2)=0, CA_1 first interact with CA_2, CA_1 is initiator; If PN (CA_1, CA_2)=1, CA_2 first interact with CA_1, CA_2 is initiator; If PN (CA_1, CA_2)=2, CA_1 and CA_2 interact each other at the same time.

In general, a virtual character acts different status with interaction to others. For instance, there are three virtual characters CA_1, CA_2 and CA_3, CA_2 is mother of CA_1, CA_3 is a student of CA_1, when CA_1 meets CA_2 or CA_3, CA_1 usually first interacts with CA_2, CA_3 usually first interacts with CA_1, and PN (CA_1, CA_2)=0, PN (CA_1, CA_3)=1.

Definition 6. For two virtual characters CA_1 and CA_2, let $\boldsymbol{INS}(CA_1 \leftarrow CA_2)$ is an interaction signal set from CA_2 to CA_1, $\boldsymbol{INS}$ $(CA_1, CA_2)=\{ins_1,\ldots,ins_M\}$, ins_j is a nonverbal interaction signal (such as "calling help pose"), $j \in [0, M]$, M is the number of $\boldsymbol{INS}$ $(CA_1 \leftarrow CA_2)$.

In a virtual environment, when two virtual characters begin to interact each other, we can suppose each of them is able to know interaction signal, in a practical demo system, interaction signals are sent to memory module by social norm module.

Definition 7. For a virtual characters CA, let $IRU(CA)$ is an interaction rule for virtual character CA, IRU control the manner of interaction, IRU include some production rulers.

A high-level algorithm can illustrate how to construct IRU, the details is related to context, there are two virtual characters CA_1 and CA_2 , the algorithm procedure of emotion social interaction is as follows step:

Step 1: The procedure read memory to get current emotion state [8] and social norms (status, all friendliness degree, default-ending time of interaction, all interaction radius, and all priority degree of social interaction), go to **Step 2** .
Step 2: The procedure judge whether a character can interact with others according to the current emotion state and social norm. If a character can interact with others, go to **Step 3** ; else go to **Step 4** .
Step 3: Interaction signals are transferred between two characters. If one of interaction signal is "ending interaction", then go to **Step 4** .
Step 4: Ending interaction.

3 A Demo System of Interaction

A demo system of emotion social interaction is realized on PC. Tom and Billy are two virtual characters in the demo system. Billy is an enemy of Tom, he thieved Tom's house two years ago. When Tom meets Billy, Tom will be angry with Billy, the social norm for Tom is recorded in a script file as follows:

Status (Tom):=enemy of Billy,worker;
Friendliness degree (to Billy)=-1;
Friendliness degree (to others)=0;
Default-ending time of interaction (to Billy)=2 minutes;
Default-ending time of interaction (to others)=0.1 minutes;
Interaction radius (to Billy)= 10 meter;
Interaction radius (to others)= 5 meter;
Priority degree of social interaction(to Billy)=0;
Priority degree of social interaction(to others)=1;
Interaction signal set=(angry, calling help, happy,...);
Emotion Interaction rules of sending information to others
If Friendliness degree=-1 then Emotion to other = angry
Else Emotion to other =Null; //no any emotion to others
End
Emotion Interaction rules of receiving information from others
If Emotion from enemy = sad then Emotion to enemy =happy
Else Emotion to enemy =angry
End
End

4 Conclusion

In a certain virtual environment, multi virtual characters interact with emotions and construct a virtual society. This paper presents some new concepts of social norm for virtual characters. A social norm includes status information, interaction signals and interaction rules. All these new concepts are illustrated by an example, and the corresponding demo system is realized on PC. Simulation of social interaction for virtual characters is a very difficult subject. This paper only gives a primary outline for social interaction model in this paper. In fact, social interaction is related to many factors, such as different culture, emotion, personality, motivation etc., the next work is to find a model that integrate all these factors together in future. .

Acknowledgements

The work described in this paper was co-supported by science and technology project of Zhejiang Province Science Department (grant no: 2006C33046), forepart professional research of ministry of science and technology of the People's Republic of China (grant no: 2005cca04400), University Research Project of Zhejiang Province Education Department (grant no: 20051731), and K.C.Wong Magna Fund in Ningbo University.

References

1. Tu, X., Terzopoulos, D.: Artificial fishes: Physics, locomotion, perception, behavior, In Proceedings of SIGGRAPH'94 (1994) 43-50
2. Funge, J., Tu, X., Terzopoulos, D.: Cognitive Modeling: Knowledge, Reasoning and Planning for Intelligent Agents, In Proceedings of SIGGRAPH'99 (1999) 29-38
3. Badler, N., Phillips, C., Webber, B.: Simulating Humans: Computer Graphics Animation and Control, New York: Oxford University Press (1993) 154-159
4. Thalmann,N.M., Thalmann,D.: Artifical Life and Virtual Reality, chichester: John Wiley Press(1994)1-10
5. Cassell, J., Vilhjalmsson, H.H., Bickmore, T.: BEAT: the behavior expression Animation toolkit. In: ACM SIGGRAPH. (2001) 477-486.
6. Pelachaud, C., Poggi, I.: Subtleties of facial expressions in embodied agents. Journal of Visualization and Computer Animation. **13** (2002) 287-300
7. Chi, D., Costa, M., Zhao, L., Badler, N.: The EMOTE model for effort and shape. In Proceedings of SIGGRAPH'00 (2000) 173-182
8. Liu, Z., Pan, Z.G.: An Emotion Model of 3D Virtual characters In Intelligent Virtual Environment, In First International conference on affective computing and intelligent in-teraction, ACII2005, LNCS 3784, Beijing, China, (2005) 629-636

Biologically Inspired Bayes Learning and Its Dependence on the Distribution of the Receptive Fields

Liang Wu, Predrag Neskovic, and Leon N Cooper

Institute for Brain and Neural Systems and Department of Physics
Brown University, Providence, RI 02912, USA

Abstract. In this work we explore the dependence of the Bayesian Integrate And Shift (BIAS) learning algorithm on various parameters associated with designing the retina-like distribution of the receptive fields. The parameters that we consider are: the rate of increase of the sizes of the receptive fields, the overlap among the receptive fields, the size of the central receptive field, and the number of directions along which the centers of the receptive fields are placed. We show that the learning algorithm is very robust to changes in parameter values and that the recognition rates are higher when using a retina-like distribution of receptive fields compared to uniform distributions.

1 Introduction

Although we perceive the world around us as a continuous flow of images of high resolution, our visual system processes information in a very discrete fashion and without a uniformly high resolution. Each neuron processes only information that is coming from a localized region, called a receptive field (RF), and our visual system somehow knows how to seamlessly connect the image patches and integrate information contained in them. Moreover, the distribution of the RFs in our retina, as in many other animals with foveal vision, is organized in a very specific way. The sizes of the RFs that cover the area around the fixation point are small whereas the sizes of the RFs that cover peripheral regions increase with the distance of those regions from the fixation point [6]. It is estimated that 33% of all the retina's ganglion cells receive input from cones occupying the fovea, which itself accounts for only 2% of the retina's total area [10]. The density of the cones decreases with the distance from the fovea, and as a consequence the acuity of our visual system also decreases with eccentricity. However, the dependence of the RF's size on its location with respect to the fixation point is not only a property of the ganglion cells in the retina but also extends to cortical neurons. Therefore, the features that we perceive with our peripheral vision are fuzzy and we are also not able to accurately determine their location.

We know what the consequences of the retina-like distribution of the RFs are but we still don't know what the benefits are from the learning point of view and why the nature chose such a distribution. It is certainly not necessary in

L. Jiao et al. (Eds.): ICNC 2006, Part I, LNCS 4221, pp. 279–288, 2006.

order to be able to recognize objects since some animals do not have a foveal vision and in some animals that do have it the organization of the RFs is quite different compared to human foveal vision. The pigeon's retina has two foveas whereas the ganglion cells of the rabbit are distributed mostly along the horizontal direction [4]. In contrast to uniform distribution of the RFs, the foveal distribution is much more complex and depends on more parameters that are difficult to estimate. For example, we do not yet know precisely how the number of cones that converge on a ganglion cell changes across the retina and the extensive physiological measurements required to answer that question have yet to be performed [10]. The question: "what are the optimal parameters of the foveal distribution of the RFs" therefore still remains open and is likely to depend on the function of the visual system and visual priorities.

In this work we consider the Bayesian Integrate And Shift (BIAS) [8] model for learning object categories and investigate how sensitive it is to the changes of the parameters that determine the distribution of the RFs. The specific parameters that we consider are: the rate of increase of the sizes of the RFs, the overlap between the RFs, the size of the central RF, and the number of directions along which the centers of the RFs are placed. We show that the learning algorithm is very robust to changes in parameter values and that for some parameters the range of acceptable values is quite large. Furthermore, we show that the recognition rates, when the system uses a retina-like distribution of the RFs, are much higher compared to those when the system uses various uniform distributions. We therefore demonstrate that for our model there is a clear advantage for using a retina-like distribution of the RFs.

The paper is organized as follows. In section 2 we give an overview of the BIAS model for learning new object categories. In section 3 we discuss implementation details. In section 4 we test the sensitivity of the model to changes of various parameters that determine the arrangement of the RFs. In section 5 we summarize the main results and outline the plans for the future work.

2 The Model

The input to our system is processed through an array of feature detectors whose RFs form a grid. The RFs completely cover the input image and their sizes and overlaps increase with their distance from the central location. We call a configuration consisting of the outputs of feature detectors associated with a specific fixation point a *view*. That means that there can be as many views for a given object as there are points within the object and that number is very large. In order to reduce the number of views, we will assume that some views are sufficiently similar to each other so that they can be clustered into the same view. A region that consists of points that constitute the same view we call a view region.

Notations. With symbol H we denote a random variable with values $H = (n, i)$ where n goes through all possible object classes and i goes through all possible views within the object. Instead of (n, i), we use the symbol H_i^n to denote the ith view of an object of the nth (object) class. The background class, by definition, has only one view. With variable $\boldsymbol{y}$ we measure the distances of the centers of the RFs from the fixation point. The symbol D_k^r denotes a random variable that takes values from a feature detector that is positioned within the RF centered at $\boldsymbol{y}_k$ from the central location, and is selective to the feature of the rth (feature) class, $D_k^r = d^r(\boldsymbol{y}_k)$. The symbol A_t denotes the outputs of all the feature detectors for a given fixation point $\boldsymbol{x}_t$ at time t.

What we want to calculate is how spatial information, coming from different feature detectors, influences our hypothesis, $p(H_i^n|A_t)$. In order to gain a better insight into dependence of these influences, we will start by including the evidence coming from one feature detector and then increase the number of feature detectors and fixation locations.

Combining information within a fixation. Let us now assume that for a given fixation point $\boldsymbol{x}_0$, the feature of the rth class is detected with confidence $d^r(\boldsymbol{y}_k)$ within the RF centered at $\boldsymbol{y}_k$. The influence of this information on our hypothesis, H_i^n, can be calculated using the Bayesian rule as

$$p(H_i^n|d^r(\boldsymbol{y}_k), \boldsymbol{x}_0) = \frac{p(d^r(\boldsymbol{y}_k)|H_i^n, \boldsymbol{x}_0)p(H_i^n|\boldsymbol{x}_0)}{p(d^r(\boldsymbol{y}_k)|\boldsymbol{x}_0)}, \tag{1}$$

where the normalization term indicates how likely it is that the same output of the feature detector can be obtained (or "generated") under any hypothesis, $p(d^r(\boldsymbol{y}_k)|\boldsymbol{x}_0) = \sum_{n,i} p(d^r(\boldsymbol{y}_k)|H_i^n, \boldsymbol{x}_0)p(H_i^n|\boldsymbol{x}_0)$.

We will now assume that a feature detector with RF centered around $\boldsymbol{y}_q$ and selective to the feature of the pth class outputs the value $d^p(\boldsymbol{y}_q)$. The influence of this new evidence on the hypothesis can be written as

$$p(H_i^n|d^p(\boldsymbol{y}_q), d^r(\boldsymbol{y}_k), \boldsymbol{x}_0) = \frac{p(d^p(\boldsymbol{y}_q)|d^r(\boldsymbol{y}_k), H_i^n, \boldsymbol{x}_0)p(H_i^n|d^r(\boldsymbol{y}_k), \boldsymbol{x}_0)}{p(d^p(\boldsymbol{y}_q)|d^r(\boldsymbol{y}_k), \boldsymbol{x}_0)}. \tag{2}$$

The main question is how to calculate the likelihood $p(d^p(\boldsymbol{y}_q)|d^r(\boldsymbol{y}_k), H_i^n, \boldsymbol{x}_0)$? In principle, if the pattern does not represent any object but just a random background image the outputs of the feature detectors $d^p(\boldsymbol{y}_q)$ and $d^r(\boldsymbol{y}_k)$ are independent of each other. If, on the other hand, the pattern represents a specific object, say an object of the nth class, then the local regions of the pattern within the detectors RFs, and therefore the features that capture the properties of those regions, are not independent from each other, $p(d^p(\boldsymbol{y}_q)|d^r(\boldsymbol{y}_k), H^n, \boldsymbol{x}_0) \neq p(d^p(\boldsymbol{y}_q)|H^n, \boldsymbol{x}_0)$. However, once we introduce a hypothesis of a specific view, the features become much less dependent on one another. This is because the hypothesis H_i^n is much more restrictive and at the same time more informative than the hypothesis about only the object class, H^n. Given the hypothesis H^n, each feature depends both on the locations of other features and the confidences with which they are detected (outputs of feature detectors). The hypothesis H_i^n significantly reduces the dependence on the locations of other features since it provides information about the exact location of each feature *within* the object up to the uncertainty given by the size of the feature's RF.

The likelihood term, under the independence assumption, can therefore be written as $p(d^p(\boldsymbol{y}_q)|d^r(\boldsymbol{y}_k), H_i^n, \boldsymbol{x}_0) = p(d^p(\boldsymbol{y}_q)|H_i^n, \boldsymbol{x}_0)$. Note that this property is very important from a computational point of view and allows for a very fast training procedure. The dependence of the hypothesis on the collection of outputs of feature detectors A_0 can be written as

$$p(H_i^n|, A_0, \boldsymbol{x}_0) = \frac{\prod_{rk \in A} p(d^r(\boldsymbol{y}_k)|H_i^n, \boldsymbol{x}_0) p(H_i^n|\boldsymbol{x}_0)}{\sum_{n,i} \prod_{rk \in A} p(d^r(\boldsymbol{y}_k)|H_i^n, \boldsymbol{x}_0) p(H_i^n|\boldsymbol{x}_0)} \tag{3}$$

where r, k goes over all possible feature detector outputs contained in the set A_0 and n, i goes over all possible hypotheses.

3 Implementation

In this section we briefly discuss implementation details related to modeling the likelihoods, feature extraction and design of the RFs. We explain the choice of the parameters that we use in constructing Gabor filters and define parameters associated with the organization of the RFs. Detailed dependence of the system on the RFs parameters is given in section 4.

3.1 Modeling Likelihoods

We model the likelihoods in Eq. (3) using Gaussian distributions. The probability that the output of the feature detector representing the feature of the rth class and positioned within the RF centered at $\boldsymbol{y}_k$ has a value $d^r(\boldsymbol{y}_k)$, given a specific hypothesis and the location of the fixation point, is calculated as

$$p(d^r(\boldsymbol{y}_k)|H_i^n, \boldsymbol{x}_t) = \frac{1}{\sigma_k^r \sqrt{2\pi}} exp \frac{-(\mu_k^r - d^r(\boldsymbol{y}_k))^2}{2(\sigma_k^r)^2} \tag{4}$$

This notation for the mean and the variance assumes a particular hypothesis so we omitted some indices, $\sigma_k^r = \sigma_k^r(n, i)$. The values for the mean and variance are calculated in the batch mode but only a small number of instances are used for training so the memory requirement is minimal.

3.2 Feature Extraction

In this work we extract features using a collection of Gabor filters where a Gabor function that we use is described with the following equation

$$\psi_{f_0,\theta,\sigma}(x, y) = \frac{e^{-\frac{1}{8\sigma^2}(4(x cos\theta + y sin\theta)^2 + (y cos\theta - x sin\theta)^2)}}{\sqrt{2\pi}\sigma} \cdot$$

$$sin(2\pi f_0(x cos\theta + y sin\theta)). \tag{5}$$

The inspiration for selecting these features comes from the fact that simple cells in the visual cortex can be modeled by Gabor functions as shown by Marcelja [7] and Daugman [1,2].

One way to constrain the values of the free parameters in Eq. (5) is to use information from neurophysiological data on simple cells as suggested by Lee [5]. More specifically, the relation between the spatial frequency and the bandwidth can be derived to be: $2\pi f_o\sigma = 2\sqrt{ln2}(2^\phi + 1)/(2^\phi - 1)$ (see [5] for more detail). Since the spatial frequency bandwidths of the simple and complex cells have been found to range from 0.5 to 2.5 octaves, clustering around 1.2 octaves, we set ϕ to 1.5 octaves. The orientations and bandwidths of the filters are set to: $\theta = \{0, \pi/4, \pi/2, 3\pi/4\}$ and $\sigma = \{2, 4, 6, 8\}$.

3.3 Receptive Fields

We use two different methods for arranging RFs. In the first method, which we call the *directional distribution*, we arrange the RFs along different directions while in the second method, which we call the *circular distribution*, we place the RFs along concentric rings [9,13,12,3].

a) *Directional distribution.* In this implementation, we use four parameters to control the distribution of the RFs: the rate of increase of the sizes of the RFs, the overlap between the adjacent RFs, the size of the central RF, and the number of directions along which the centers of the RFs are placed. Each RF has a square form and the RFs are arranged is such a way as to completely cover the input image. Therefore, in order to satisfy a complete coverage requirement, not all the values of the parameters can be used. The overlap between two RFs is defined as a percentage of the area of the smaller of the two RFs that is being covered. For example, if the $ovr = 50\%$, that means that the larger RF covers 50% of the area of the smaller receptive field.

The main shortcoming of the directional distribution is that it does not provide a sufficiently dense packing of the RFs, especially for regions that are further away from the central field. This is shown in Figure 1 (left) where, for illustrative purpose, we use the circular RFs. In order to prevent the gaps between the RFs, and completely cover the visual field, the rate of increase of the sizes of the RFs has to be sufficiently high.

b) *Circular distribution.* Another solution is to arrange the centers of the RFs using a hexagonal packing, as shown in Figure 1 (right). Within each ring we use a fixed number of RFs, which we set to 10. The radius of a RF, $r(n)$, whose center is on the nth ring, is calculated as $r(n) = B \cdot r(n-1)$, where B is the enlarge parameter. The angle between neighboring RFs from the same ring is called the *characteristic angle*, θ_o, and is calculated as $\theta_o = 2\pi/F$, where F is the number of RFs per ring. In this arrangement, the position of each RF is fully determined by the ring number and the angle, θ, with respect to a chosen direction. For example, the angle of the mth RF is calculated as $\theta_m = m \cdot \theta_o$.

A hexagonal packing is obtained by shifting the angles of all the RFs in all even rings by half the characteristic angle. This disposition of the centers of the

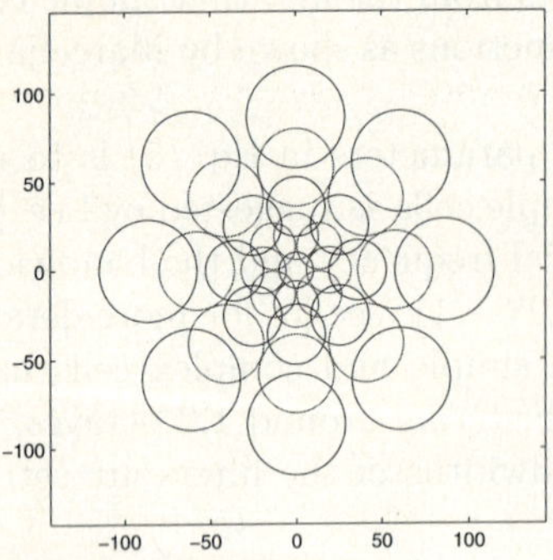

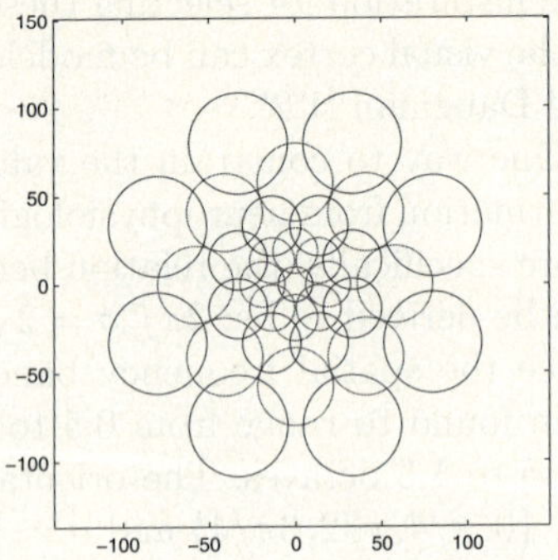

Fig. 1. Comparison between the directional and circular arrangements of the RFs. Left: RFs are arranged along 8 directions. Right: RFs are arranged along 4 concentric rings, each ring containing 8 RFs. The circular arrangement provides denser packing of the RFs, especially for the regions that are further away from the center.

RFs is also known as triangular tessellation. The radius of an nth ring, $R(n)$, is calculated using the following equation:

$$R(n) = R(n-1) + r(n) + r(n-1)(1 - 2 * ovr).$$

3.4 Feature Detectors

With each RF we associate 16 feature detectors where each feature detector signals the presence of a feature (i.e. a Gabor filter of specific orientation and size) to which it is selective no matter where the feature is within its receptive field. One way to implement this functionality is to use a max operator. The processing is done in the following way. On each region of the image, covered by a specific RF, we apply a collection of 16 Gabor filters (4 orientations and 4 sizes) and obtain 16 maps. Each map is then supplied to a corresponding feature detector and the feature detector then finds a maximum over all possible locations. As a result, each feature detector finds the strongest feature (to which it is selective) within its RF but does not provide any information about the location of that feature.

3.5 The Training Procedure

The training is done in a supervised way. We constructed an interactive environment that allows the user to mark a section of an object and label it as a fixation region associated with a specific view. Every point within this region can serve as the view center. Once the user marks a specific region, the system samples the points within it and calculates the mean and variance for each feature detector. Since the number of training examples is small the training is very fast.

Note that during the training procedure the input to the system is the whole image and the system learns to discriminate between an object and the background. It is important to stress that the system does not learn parts of the object, but the whole object from the perspective of the specific fixation point.

4 Results

We tested the performance of our system on four object categories (faces, cars, airplanes and motorcycles) using the Caltech database (www.vision.caltech.edu). For illustrative purposes, we chose a face category to present some of the properties of our system in more detail.

The system was first trained on background images in order to learn the "background" hypothesis. We used 20 random images and within each image the system made fixations at 100 random locations. The system was then trained on specific views of specific objects. For example, in training the system to learn the face from the perspective of the right eye, the user marks with the cursor the region around the right eye and the system then makes fixations within this region in order to learn it. During the testing phase, the system makes random fixations and for each fixation point we calculated the probability that the configuration of the outputs of feature detectors represents a face from the perspective of the right eye. To make sure that among the random fixations are also positive examples, each testing image is divided into the view region(s) (in this case the right eye region) and the rest of the image represents the "background" class. Therefore, positive examples consisted of random fixations within the region of the right eye and negative examples consisted of random fixations outside the region of the right eye. The system was tested on people that were not used for training. We used 200 positive examples and 1000 negative examples for testing.

As a measure of performance we use the error rate at equilibrium point (as in [11]) which means that the threshold is set so that the miss rate is equal to the false positive rate. We chose this measure over the Receiver Operator

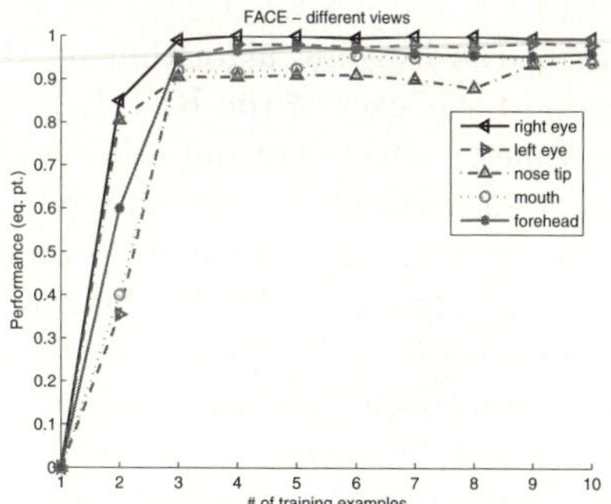

Fig. 2. Left: Performance comparison for different views using optimal values of RFs parameters. Right: View regions as selected by the teacher.

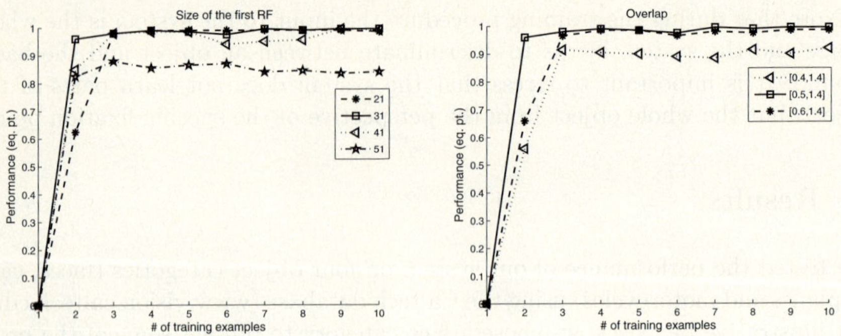

Fig. 3. Left: The sensitivity of the learning algorithm to variations of the size of the central RF. Right: The dependence of the learning algorithm on the overlap between neighboring RFs.

Characteristic since it provides more compact representation of the results in the sense that much more information can be represented in one graph.

In Figure2 (left) we show the performance of our system using the directional distribution and the best RFs parameters. The size of the first receptive field was set to 31, the overlap parameter was set to 0.5, the increase size parameter was set to 1.4, and the number of directions was set to 8. The highest accuracy rates are obtained for the right eye view. As an illustrative example, we will use this view as a baseline for future comparisons. Therefore, in all the graphs that follow we present the performance of the system using the right eye view.

The sensitivity of the learning algorithm to variations of the size of the central RF is illustrated in Figure 3 (left). It is clear that the system is robust to changes over the large range of values, approximately from 20 to 40 pixels. The value of the parameter that is much larger than 50 and much smaller than 20 gives noticeably lower recognition rates. The dependence of the learning algorithm on the size of the overlap region is illustrated in Figure 3 (right). The value of the increase size parameter was kept constant and set to 1.4. As one can see, the performance is not affected by the changes of the overlap parameter that are within the range 0.5 to 0.6.

In Figure 4 (left) we test the dependence of the learning algorithm on the parameter that determines the rate of increase of the sizes of the RFs. It is interesting to observe that the value of this parameter affects not only the recognition accuracy but also the learning speed. For the values that are between 1.4 and 1.6 the system can still achieve high accuracy, although with more training examples, but for the values below 1.3 the accuracy can not be further improved even if using more training examples. The right graph in Figure 4 shows that when implementing the directional distribution approach, the larger number of direction does not necessarily improve performance. The number of directions smaller than 6 gives much worse results due to the fact that the RFs can not completely cover the input without significant distortions of other parameters.

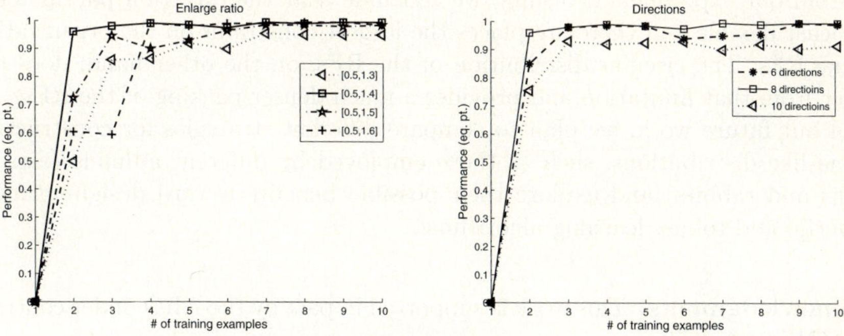

Fig. 4. Left: The dependence of the learning algorithm on the rate of increase of the sizes of the RFs. Right: The dependence of the learning algorithm on the number of directions along which the centers of the RFs are placed.

In order to verify whether high recognition rates can be achieved using a uniform distribution of the RFs, we tested the system using the grid of the RFs of the same size, and repeated the experiment choosing different RF sizes. In Table 1 we illustrate the performance of the system when trained on a face using the tip of the nose as a fixation region - the "nose view". The performance is much worse compared to retina-like distribution and, as expected, the system has difficulties learning the view using small RFs. Using large RFs recognition improves but only to the point and then decreases again.

Table 1. Performance Using Uniform Distribution of the RFs

RF Size	10	20	30	40	50	60	70	80	90	100	110
Performance (%)	60.0	60.0	57.5	59.5	60.8	70	69	75.7	78.5	73.5	73.5

5 Summary

In this work we explore the dependence of the Bayesian Integrate And Shift (BIAS) [8] learning algorithm on various parameters associated with designing the retina-like distribution of the RFs. The parameters that we consider are: the rate of increase of the sizes of the RFs, the overlap between the RFs, the size of the central RF, and the number of directions along which the centers of the RFs are placed. We demonstrate that the learning algorithm is very robust to changes in parameter values and that for some parameters the range of acceptable values is quite large. Furthermore, we show that the recognition rates, when the system uses a retina-like distribution of the RFs, are much higher compared to those when the system uses various uniform distributions.

From our experimental results, we conclude that the direction parameter is the least flexible and therefore places the largest constraint on the organization of the RFs. The circular distribution of the RFs, on the other hand, does not suffer from that limitation and provides a much denser packing of the RFs.

In our future work, we plan to compare different strategies for constructing retina-like distributions, such as those employed by different animals (e.g. pigeons and rabbits) and explore their possible benefits toward designing more powerful and robust learning algorithms.

Acknowledgments. This work is supported in part by the ARO under contract W911NF-04-1-0357.

References

1. J. G. Daugman. Two-dimensional spectral analysis of cortical receptive field profile. *Vision Research*, 20:847–856, 1980.
2. J. G. Daugman. Uncertainty relation for resolution in space, spatial frequency, and orientation optimized by two-dimensional visual cortical filters. *J. Optical Soc. Am.*, 2(7):1,160–1,169, 1985.
3. H. M. Gomes. Model learning in iconic vision. Ph. D. Thesis, University of Edinburgh, Scotland, 2002.
4. M. F. Land and D.-E. Nilsson. *Animal Eyes.* Oxford University Press, 2002.
5. T. S. Lee. Image representation using 2d gabor wavelets. *PAMI*, 18(10):1–13, 1996.
6. T. Lindeberg and L. Florack. Foveal scale-space and the linear increase of receptive field size as a function of eccentricity. Tech. Report, Royal Institute of Technology, S-100 44 Stockholm, Sweden, 1994.
7. S. Marcelja. Mathematical description of the responses of simple cortical cells. *J. Optical Soc. Am.*, 70:1,297–1,300, 1980.
8. P. Neskovic, L. Wu, and L. Cooper. Learning by integrating information within and across fixations. In *Proc. ICANN*, 2006.
9. E. Schwartz. Spatial mapping in primate sensory projection: analytic structure and relevance to perception. *Biological Cybernetics*, 25:181–194, 1977.
10. R. Sekuler and R. Blake. *Perception.* McGraw-Hill Companies, Inc., 2002.
11. T. Serre, L. Wolf, and T. Poggio. Object recognition with features inspired by visual cortex. In *Proc. CVPR*, 2005.
12. F. Smeraldi and J. Bigun. Retinal vision applied to facial features detection and face authentication. *Pattern Recognition Letters*, 23:463–475, 2002.
13. S. Wilson. On the retino-cortical mapping. *International Journal on Man-Machine Studies*, 18:361–389, 1983.

Using PCA-Based Neural Network Committee Model for Early Warning of Bank Failure

Sung Woo Shin[1] and Suleyman Biljin Kilic[2,*]

[1] School of Business Administration, Sungkyunkwan University
110-745, Seoul, Korea
shinswoo@skku.edu, shinswoo@kaist.ac.kr

[2] Faculty of Economic and Administrative Science, Cukurova University
01330 Balcali, Adana, Turkey
sbilgin@cu.edu.tr

Abstract. As the Basel-II Accord is deemed to be an international standard to require essential capital ratios for all commercial banks, early warning of bank failure becomes critical more than ever. In this study, we propose the use of combining multiple neural network models based on transformed input variables to predict bank failure in the early stage. Experimental results show that: 1) PCA-based feature transformation technique effectively promotes an early warning capability of neural network models by reducing type-I error rate; and 2) the committee of multiple neural networks can significantly improve the predictability of a single neural network model when PCA-based transformed features are employed, especially in the long-term forecasting by showing comparable predictability of raw features models in short-term period.

1 Introduction

The study for predicting corporate failure has a long history over four decades. From the seminal work by Beaver [4], many efforts have been devoted to develop accurate prediction models. Traditionally, conventional statistical models including multivariate discriminant analysis [1], logit [8] and probit [11] have been employed for this domain. On the other hand, as an alternative prediction model, the artificial neural network [9] was introduced [7], [10]. Recent literature review (eg., [2]) reveals that the use of a neural network model is a promising direction in diverse forms of corporate failures including commercial banks.

In this study, we extend previous work by suggesting a multiple neural network committee model focusing on promoting the early warning capability by employing a principal component analysis (PCA)-based feature transformation technique. The rationale of employing PCA lies in that reduced features by PCA can effectively represent different characteristics of financial ratios in a parsimonious and non-redundant fashion, showing improved early warning predictability for bank failure phenomenon.

[*] Corresponding author.

L. Jiao et al. (Eds.): ICNC 2006, Part I, LNCS 4221, pp. 289 – 292, 2006.

2 PCA-Based Neural Network Committee Model

Recently, researchers have begun experimenting with general algorithms for improving classification performance by combining multiple classifiers as a form of committee machine [6], [3].

The committee combines the outputs of multiple prediction models with the expectation that it can show enhanced predictability by aggregating partially independent predictions of individual members. For neural network case such as this study, different members are expected to converge to different local minima and thus generate uncorrelated errors. A summary of the Bagging algorithm [5] used in this study is presented in Figure 1.

```
PCA-based Neural Network Bagging(D, m, T)
{

//INPUT:   D – a training set
//       m – a number of training samples
//       T – a number of trials

//OUTPUT: a committee decision, containing T modules using majority voting rule
//Given: (x1, y1),...(xm, ym) where xi ∈ Principal components of X, yi = {-1, +1}

t = 1; T=10;
While (t ≤ T)
{
Set Dt = bootstrap sample from D;    // random sampling with replacement
Train single neural network model ANNt using Dt traning set
t = t+1;
}
Return (the most often predicted label y)
}
```

Fig. 1. Algorithm for PCA-based neural network committee predictions

3 Experimental Investigation

In this study we employed publicly available bank failure data set that can be reached at the Banks Association of Turkey (www.tbb.org.tr/english/default.htm). The data set consists of total 57 privately owned commercial banks where 21 banks failed during the period of 1997- 2003, and each bank is represented by five sets of 49 financial ratios for five years prior to failure. Consequently, the resulting data set is composed of 285 bank-years observations, 105 of which are labeled as failed.

Figure 2 contrasts the effect of PCA analysis in discriminating two groups of banks in year-1 period. Figure 2(a) displays complex class boundaries produced by most significant three raw variables while three factors in Figure 2(b) fairly well distinguish solvent banks from bankrupt ones. The Wilks' lambda was 0.622 and 0.476 for raw features and factor variables, respectively. This result clearly indicates

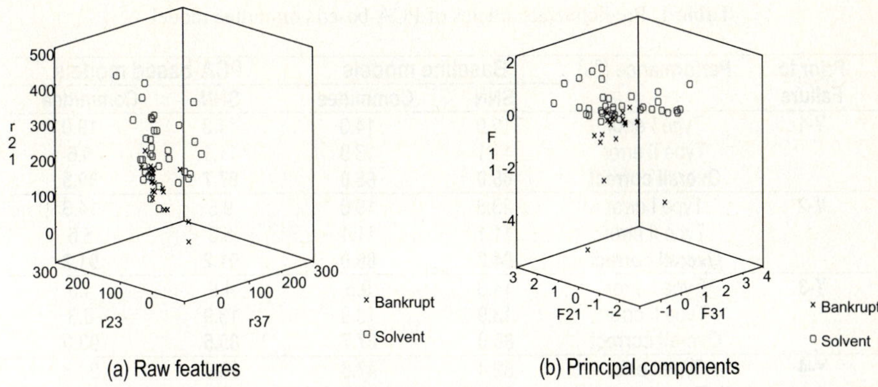

Fig. 2. Effect of PCA transformation on discriminating two groups of banks

that PCA preprocessor effectively makes the class boundary more linearly separable by transforming raw feature space, and thus prediction model can easily discriminate two groups of banks even for the nonlinear models such as a neural network model.

In Table 1, 10-fold cross-validated results are summarized over five year periods. Firstly, for the baseline models using raw features, both best single neural network and committee model show almost same performances: the average performances over five year period were slightly below than 80% (77.56% and 78.96%, respectively). From one to three years ahead forecasting (ie., year-1, -2, -3), both models show around 85% accuracies, but in four and five years ahead forecasting the performances are dramatically deteriorated to around 65%. Regarding the committee model, the performance is slightly superior to that of a single model over all five years period, mainly due to the reduction in costly type-I error rate 33.32% down to 27.6%.

On the other hand, for the PCA-based transformed features models both single and committee models commonly show notably different prediction behavior against the raw features models. Specifically, for the relatively short-term forecasting (year-1, -2, and -3) both models commonly slightly improve predictive accuracies of raw features models by showing near 90% accuracies. Surprisingly, however, for the long-term (year-4 and -5) forecasting they show approximately 80% accuracies. This translates into approximately 5% (from 85% to 90%) and 20% (from 65% to 80%) improvement rates from the raw features models, largely due to the remarkable reduction rate in type-I error (approximately 60%). Note that the predictability of committee model using PCA-based features in the early stage of year-4 and -5 periods is comparable with that of the raw features models in the short-term periods (85.6% versus 81.6%).

In summary, based on the first series of experiments using raw features models, consistent error patterns between single and committee models induce that one can infer that current set of financial ratios has a limited predictive power to maximally three years ahead in predicting bank failure. However, the latter experiments using PCA-based models show that long-term forecasting models can be sufficiently competitive with the short-term forecasting models when a highly accurate model is combined with properly preprocessed feature representation.

Table 1. Predictive accuracies of PCA-based committee model

Prior to Failure	Performance (%)	Baseline models		PCA-based models	
		SNN	Committee	SNN	Committee
Y-1	Type I error	19.0	14.3	14.3	19.0
	Type II error	11.1	13.9	11.1	5.6
	Overall correct	**86.0**	**86.0**	**87.7**	**89.5**
Y-2	Type I error	23.8	19.0	9.5	14.3
	Type II error	11.1	11.1	8.3	5.6
	Overall correct	**84.2**	**86.0**	**91.2**	**91.2**
Y-3	Type I error	14.3	9.5	4.8	4.8
	Type II error	13.9	13.9	13.9	8.3
	Overall correct	**86.0**	**87.7**	**89.5**	**93.0**
Y-4	Type I error	52.4	47.6	28.6	23.8
	Type II error	22.2	22.2	19.4	16.7
	Overall correct	**66.7**	**68.4**	**77.2**	**80.7**
Y-5	Type I error	57.1	47.6	19.0	14.3
	Type II error	22.2	25.0	27.8	19.4
	Overall correct	**64.9**	**66.7**	**75.4**	**82.5**

Note: SNN denotes the best single neural network model.

References

1. Altman E.I. 1968. Financial Ratios, Discriminant analysis and the prediction of corporate bankruptcy. *Journal of Finance* 589-609.
2. Atiya, A. 2001. Bankruptcy prediction for credit risk using neural networks: A survey and new results, *IEEE Trans. Neural Networks*, **12**(4) 929-935.
3. Battiti, R., A. M. Colla. 1994. Democracy in neural nets: voting schemes for classification. *Neural Networks* **7**(4) 691-707.
4. Beaver W. 1968. Market prices, financial ratios, and the prediction of failure, *Journal of Accounting Research* 170-192.
5. Breiman. 1996. Bagging predictors. *Machine Learning* **24**(2) 123-140.
6. Hansen, L. K., P. Salamon. 1990. Neural network ensembles. *IEEE Trans. Systems, Man, and Cybernetics* **12**(10) 993-1001.
7. Odom, M. D. and R., Sharda. 1990. A Neural Network Model for Bankruptcy Prediction, *International Joint Conference on Neural Networks*, **2** 163-168.
8. Ohlson, J. A. Financial ratios and the probabilistic prediction of bankruptcy. Journal of Accounting Research 1980; **18** (1): 109 - 131.
9. Rumelhart, D. E., G.E. Hinton, R.J. Williams. 1986. Learning internal representations by error propagation. in *Parallel Data Processing*, I, Chapter 8, the M.I.T. Press, Cambridge, MA, 318-362.
10. Tam, K., M. Kiang. 1992. Managerial application of neural networks: The case of bank failure predictions. *Management Science* **38**(7) 926-947.
11. Zmijewski, M. E. 1984. Methodological Issues Related to the Estimation of Financial Distress Prediction Models, *Journal of Accounting Research* **22** 59-82.

Theoretical Derivation of Minimum Mean Square Error of RBF Based Equalizer*

Jungsik Lee[1] and Ravi Sankar[2]

[1] School of Electronic and Information Engineering, Kunsan National University
Kunsan, Chonbuk 573-701, Korea
leejs@kunsan.ac.kr
[2] Department of Electrical Engineering, University of South Florida
Tampa, FL. 33620, U.S.A.
Sankar@eng.usf.edu
http://icons.eng.usf.edu

Abstract. In this paper, the minimum mean square error (MSE) convergence of the RBF equalizer is evaluated and compared with the linear equalizer based on the theoretical minimum MSE. The basic idea of comparing these two equalizers comes from the fact that the relationship between the hidden and output layers in the RBF equalizer is also linear. As extensive studies of this research, various channel models are selected, which include linearly separable channel, slightly distorted channel, and severely distorted channel models. The theoretical minimum MSE for both RBF and linear equalizers were computed, compared and the sensitivity of minimum MSE due to RBF center spreads was analyzed.

1 Introduction

For more than a decade, there has been much attention given to applying neural networks to the digital communication areas, including channel equalization problem [1]. Multi-layer perceptrons (MLP) equalizer is able to equalize non-minimum phase channels without the introduction of any time delay; and it is less susceptible than a linear equalizer to the effects of high levels of additive noise [2],[3]. However, the network architecture and training algorithm of the MLP equalizer is much more complex than the linear equalizer. Also, the radial basis functions (RBF) network has received a great deal of attention by many researchers because of its structural simplicity and more efficient learning [4-6]. In [7], Chen et al. applied RBF network to channel equalization problem to get the optimal Bayesian solution. Since then, many researchers applied RBF network to equalization problem [8-12].

Although many of studies, mentioned above, claims that RBF based equalizers are superior to conventional linear equalizer due to both RBF network's structural linearity (or simplicity) and efficient training, none of them tried to compare those two from the theoretical minimum mean square (MSE) error point of view. The basic idea of

* This work was financially supported by the Kunsan National University's Long-term Overseas Research Program for Faculty Member in the year 2004.

L. Jiao et al. (Eds.): ICNC 2006, Part I, LNCS 4221, pp. 293 – 302, 2006.

comparing these two equalizers comes from the fact that once input domain of RBF network is transformed to another domain through radial basis functions, the relationship between the hidden and output layers in the RBF equalizer is also linear. In this paper, the theoretical MSE for various channels for both RBF and the linear equalizers were evaluated and compared. Also, the sensitivity of minimum MSE due to RBF center spreads was analyzed.

2 Background of RBF Equalizer

Dispersion in the digital channel may be represented by the transfer function

$$H(z) = h_0 + h_1 z^{-1} + \ldots + h_d z^{-d} + \ldots + h_p z^{-p} \tag{1}$$

where p is the channel order. The channel is, in general, not minimum phase, so the dominant response of the channel is h_d, where d is the channel delay. The channel output at the kth time may be written

$$r_k = h_0 a_k + h_1 a_{k-1} + \ldots h_d a_{k-d} \ldots + h_p a_{k-p} + n_k = \hat{r}_k + n_k \tag{2}$$

where the vector $\mathbf{a}_k = [a_k, a_{k-1}, \ldots, a_{k-p}]$ is a length $p+1$ sequence of the transmitted data which affect the kth decision, and n_k is additive zero-mean white Gaussian noise (AWGN), assumed to be independent from one decision to the next. For a conventional equalizer, the input is the sequence of channel outputs, or a vector.

$$\mathbf{r}_k = [r_k, r_{k-1}, \ldots, r_{k-q}] \tag{3}$$

Where q is the RBF equalizer order (the number of tap delay elements in RBF equalizer). A radial basis function equalizer is a three-layer neural network structure as shown in Fig. 1.

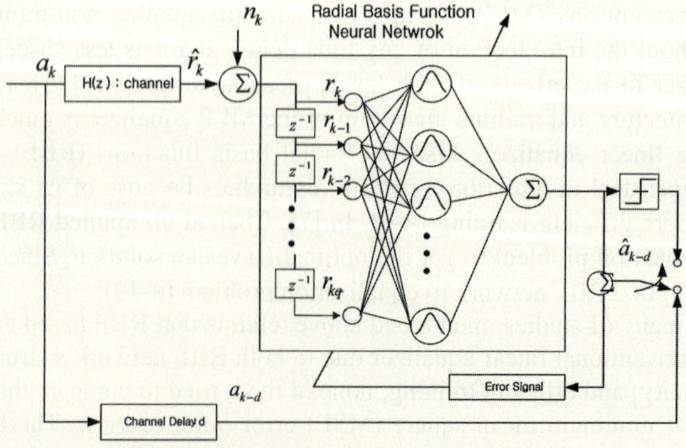

Fig. 1. The block diagram of RBF equalizer

The output layer forms a weighted sum of the outputs from the internal hidden layer. Each nose in the hidden layer is a RBF center vector with dimension $q+1$. At each node, the computed response depends on the proximity of the received input vector to the position vector defining the node. For each node, or RBF center, in the hidden layer the generated response is strongest when the input vector is closest to that center. The following equation is the output of RBF equalizer.

$$F = \sum_{i=1}^{M} w_i \exp\left(\|\mathbf{r}_k - \mathbf{g}_i\|^2 \Big/ 2\sigma^2 \right) \tag{4}$$

Where M is the number of RBF centers, the w_i are the output layer weights, the $\mathbf{g}_i$ are the Gaussian basis function center vectors, and the σ^2 are the RBF center spread parameters. Clustering techniques [4] are commonly used to determine the desired RBF centers from the number of noisy centers. Generally, the number of centers in the RBF equalizer increases as the channel order and equalizer order increase. To get around this problem, Lee et al. [12] provided a technique for selecting proper number of centers that are used for training by considering the knowledge of channel delay.

3 Derivation of MSE Errors of RBF Equalizer

The main objective of this paper is to derive the theoretical minimum mean square error of RBF based equalizer and compare it with that of the linear equalizer (finite transversal filter). The basic idea of comparing these two equalizers comes from the fact that the relationship between the hidden and output layers in the RBF equalizer is also linear. To do this, first the general theory of minimum MSE for linear equalizer was reviewed [13],[14]. The theoretical minimum mean square error of linear filter is as follows

$$\xi_{\min} = E\left[a_{k-d}^{2}\right] - \alpha^{*T}\Phi^{-1}\alpha \tag{5}$$

Where Φ and α are autocorrelation and cross-correlation matrix. Then, the theoretical minimum MSE for RBF based equalizer was derived using equations (5) and finally compared with linear equalizer. As shown in [7], the maximum number of Gaussian centers in the RBF equalizer is

$$M = 2^{p+q+1} \tag{6}$$

Accordingly, M numbers of Gaussian basis function outputs are generated from the hidden layer in the RBF equalizer systems

$$\left.\begin{aligned} F_i^k &= \exp\left[-\|\mathbf{r}_k - \mathbf{g}_i\|^2 \Big/ 2\sigma^2 \right] \\ \mathbf{g}_i &= \left[g_{i,0}, g_{i,1}, \ldots, g_{i,q}\right]^T, i = 1, 2, \ldots, M \end{aligned}\right\} \tag{7}$$

where F_i^k, and $\mathbf{g}_i$ denote the Gaussian basis function output in the ith hidden unit of the RBF equalizer and the desired Gaussian basis function vector. Then the autocorrelation matrix based on Gaussian basis function output is represented as

$$\Phi = \begin{bmatrix} \phi_{1,1} & \phi_{1,2} & \dots & \phi_{1,M} \\ \phi_{2,1} & \phi_{2,2} & \dots & \phi_{2,M} \\ . & . & . & . \\ . & . & . & . \\ \phi_{M,1} & \phi_{M,2} & \dots & \phi_{M,M} \end{bmatrix}_{M \times M} \tag{8}$$

where

$$\phi_{i,j} = E\left[F_i^k . F_j^k\right] = E\left[e^{-\left(\|\mathbf{r}_k - \mathbf{g}_i\|^2 + \|\mathbf{r}_k - \mathbf{g}_j\|^2\right)/2\sigma^2}\right], i, j = 1, 2, ..., M \tag{9}$$

The cross-correlation matrix between the desired symbol and the basis function outputs is

$$\alpha = E\begin{bmatrix} a_{k-d}.e^{-\frac{\|\mathbf{r}_k - \mathbf{g}_1\|^2}{2\sigma^2}} \\ a_{k-d}.e^{-\frac{\|\mathbf{r}_k - \mathbf{g}_2\|^2}{2\sigma^2}} \\ . \\ . \\ a_{k-d}.e^{-\frac{\|\mathbf{r}_k - \mathbf{g}_M\|^2}{2\sigma^2}} \end{bmatrix}_{M \times 1} \tag{10}$$

4 Simulation Studies by Theoretical Approach

The following channel model is selected as an example for calculating and comparing the $\xi_{\min}$ for both RBF and linear equalizer [7]

$$H(z) = 0.5 + z^{-1} \tag{11}$$

The equalizer order, q, is assumed to be 1. The transmitted sequences are assumed to be independent and equi-probable binary symbols, designated as a_k, that are either -1 or +1. Then the maximum number of RBF centers, M is equal to 8 ($M = 2^{p+q+1}$). To determine $\mathbf{g}_i = \left[g_{i,0}, g_{i,1}\right]^T$, the noise-free channel output, $\hat{r}_k$, and $\hat{r}_{k-1}$ are provided in Table 1. From (3), the $\mathbf{r}_k$ is represented as

$$\begin{aligned} \mathbf{r}_k &= \left[r_k, r_{k-1}\right]^T \\ &= \left[0.5a_k + a_{k-1} + n_k, 0.5a_{k-1} + a_{k-2} + n_{k-1}\right]^T \end{aligned} \tag{12}$$

Table 1. Input and desired channel states

RBF Center	a_k	a_{k-1}	a_{k-2}	$\hat{r}_k$	$\hat{r}_{k-1}$
$\mathbf{g}_1$	-1	-1	-1	-1.5	-1.5
$\mathbf{g}_2$	-1	-1	1	-1.5	0.5
$\mathbf{g}_3$	-1	1	-1	0.5	-0.5
$\mathbf{g}_4$	-1	1	1	0.5	1.5
$\mathbf{g}_5$	1	-1	-1	-0.5	-1.5
$\mathbf{g}_6$	1	-1	1	-0.5	0.5
$\mathbf{g}_7$	1	1	-1	1.5	-0.5
$\mathbf{g}_8$	1	1	1	1.5	1.5

The estimated auto correlation matrix as follows (refer to [10])

$$\Phi = \begin{bmatrix} 0.044 & 0 & 0 & 0 & 0.003 & 0 & 0 & 0 \\ 0 & 0.044 & 0 & 0 & 0 & 0.003 & 0 & 0 \\ 0 & 0 & 0.044 & 0 & 0 & 0 & 0.003 & 0 \\ 0 & 0 & 0 & 0.044 & 0 & 0 & 0 & 0.003 \\ 0.003 & 0 & 0 & 0 & 0.044 & 0 & 0 & 0 \\ 0 & 0.003 & 0 & 0 & 0 & 0.044 & 0 & 0 \\ 0 & 0 & 0.003 & 0 & 0 & 0 & 0.044 & 0 \\ 0 & 0 & 0 & 0.003 & 0 & 0 & 0 & 0.044 \end{bmatrix}_{8\times 8} \tag{13}$$

The estimated cross-correlation matrix is

$$\left.\begin{aligned} \alpha_{d=1} &= [-0.067, -0.067, 0.067, 0.067, -0.067, -0.067, 0.067, 0.067]^{\mathrm{T}} \\ \alpha_{d=0} &= [-0.055, -0.055, -0.055, -0.055, 0.055, 0.055, 0.055, 0.055]^{\mathrm{T}} \end{aligned}\right\} \tag{14}$$

Based on the equations (5), (13), and (14), the minimum MSE values are calculated as

$$\left.\begin{aligned} \xi_{\min}^{d=1} &= E\left[|a_{k-1}|^2\right] - \hat{\alpha}_{d=1}{}^T \hat{\Phi}^{-1} \hat{\alpha}_{d=1} = 0.22 \\ \xi_{\min}^{d=0} &= E\left[|a_k|^2\right] - \hat{\alpha}_{d=0}{}^T \hat{\Phi}^{-1} \hat{\alpha}_{d=0} = 0.35 \end{aligned}\right\} \tag{15}$$

For the purpose of comparison, the theoretical minimum MSE for the linear equalizer is described as follows. By considering equations (12), the following results come out

$$\left.\begin{aligned} &\Phi = E\left[\mathbf{r}_k.\mathbf{r}_k^{\ T}\right] = \begin{bmatrix} 1.25+\sigma_n^{\ 2} & 0.5 \\ 0.5 & 1.25+\sigma_n^{\ 2} \end{bmatrix} \\ &\alpha_{d=1} = E\left[a_{k-1}.\begin{pmatrix} \mathbf{r}_k \\ \mathbf{r}_{k-1} \end{pmatrix}\right] = \begin{bmatrix} 1 \\ 0.5 \end{bmatrix}, \alpha_{d=0} = E\left[a_k.\begin{pmatrix} \mathbf{r}_k \\ \mathbf{r}_{k-1} \end{pmatrix}\right] = \begin{bmatrix} 0.5 \\ 0 \end{bmatrix} \end{aligned}\right\} \quad (16)$$

where $\sigma_n^{\ 2} = E\left[n_k.n_k\right]$. By substituting $\sigma_n^{\ 2} = 0.1$ into equation (16), the result is

$$\left.\begin{aligned} \xi_{\min}^{d=1} &= E\left[\left|a_{k-1}\right|^2\right] - \alpha_{d=1}^{\ T}\Phi^{-1}\alpha_{d=1} = 0.2448 \\ \xi_{\min}^{d=0} &= E\left[\left|a_k\right|^2\right] - \alpha_{d=0}^{\ T}\Phi^{-1}\alpha_{d=0} = 0.7854 \end{aligned}\right\} \quad (17)$$

From the equation (15) and (17), we can see that the minimum MSE of the RBF equalizer is a little less than that of the linear equalizer when a_{k-1} is used as the desired symbol. In addition, the difference becomes more distinct when channel delay is not introduced. Fig. 2 shows the comparison of MSE convergence of both linear and RBF equalizers, which was obtained by stochastic gradient LMS learning. As shown in 2(a), the minimum MSE of the RBF equalizer with the introduction of proper channel delay is a little lower than the linear equalizer. For the case of not introducing channel delay, the minimum MSE of the RBF equalizer is much lower than the linear equalizer, as shown in Fig. 2(b).

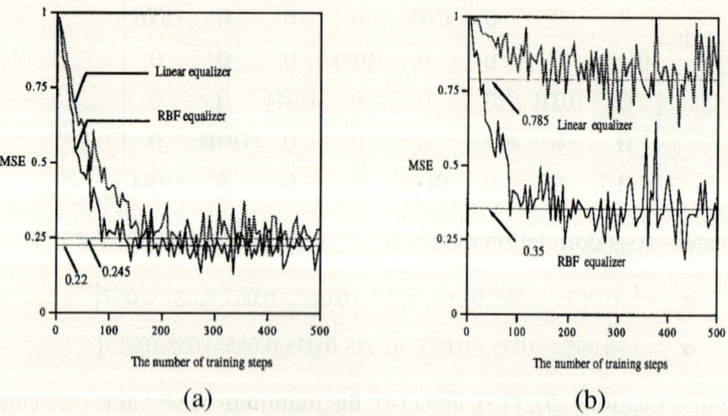

Fig. 2. Comparison of MSE Convergence: $H(z) = 0.5 + z^{-1}$, $\sigma_n^{\ 2} = 0.1$ (a) Desired symbol, $a_{k-d} = a_{k-1}$ (b) Desired symbol, $a_{k-d} = a_k$ (no channel delay)

Thus far, what has been described above are results with training noise variance $\sigma_n^{\ 2} = 0.1$. Table 2 shows how different noise variances affect the minimum MSE. Other channel impulse responses were selected for simulation as an extension to this

Table 2. Comparison of Minimum MSE: $H(z) = 0.5 + 1.0z^{-1}$

Noise Variance $\left(\sigma_n^{\ 2}\right)$	Minimum MSE			
	$\xi_{\min}^{d=1}$		$\xi_{\min}^{d=0}$	
	RBF	Linear	RBF	Linear
0.1	0.22	0.245	0.35	0.785
0.2	0.27	0.291	0.59	0.804
0.3	0.327	0.332	0.732	0.82

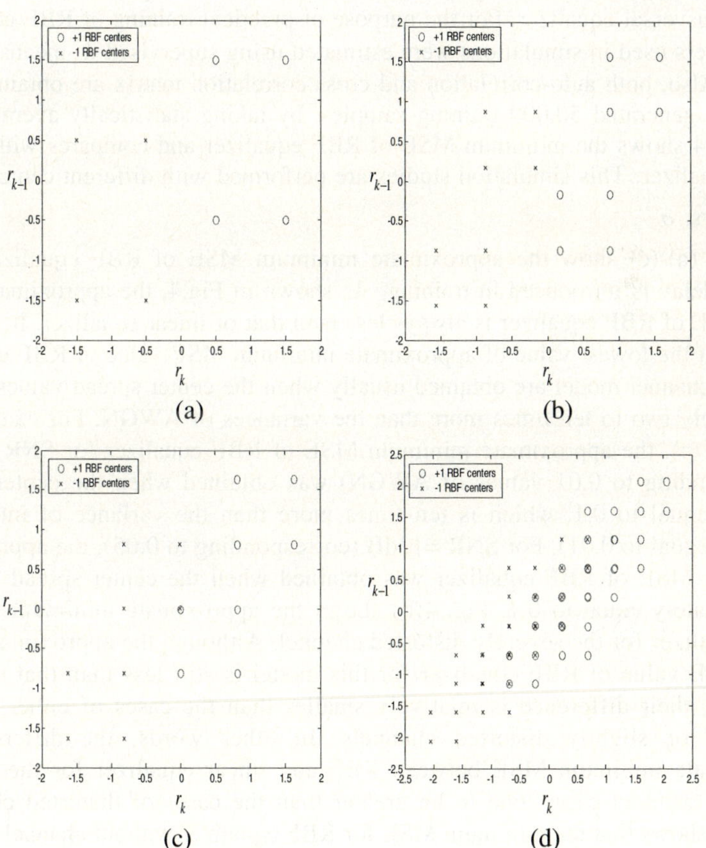

Fig. 3. Comparison of Noise-Free Center Distribution (a) $\mathbf{h} = [0.5\ \ 1.0]$ (b) $\mathbf{h} = [0.3482\ \ 0.8704\ \ 0.3482]$ (c) $\mathbf{h} = [0.407\ \ 0.815\ \ 0.407]$ (d) $\mathbf{h} = [0.227\ \ 0.460\ \ 0.688\ \ 0.460\ \ 0.688]$

research study, which are the linearly separable, slightly distorted, and severely distorted channel models [10],[15].

As mentioned in [3], distorted channel models contain deep nulls in their frequency spectrum which result in high MSE, and as a consequence high bit error rate probability. Fig. 3 shows the center distribution of four different kinds of linearly dispersive channel models. For the purpose of graphical illustration, it is assumed that the equalizer order, q , is one.

Fig. 3(a) and (b) show the distribution of noise free-RBF center, $\hat{\mathbf{r}}_k$, for linearly separable channels, (c) for slightly distorted, and (d) for severely distorted channel model. Through this research, the minimum MSE of RBF equalizer for the channel models (b), (c), and (d) in Fig. 3. are investigated and compared with that of the linear transversal equalizer. For the purpose of practical training of RBF equalizer, RBF centers used in simulations were estimated using supervised k -means clustering [7]. Also, both auto-correlation and cross correlation matrix are obtained from randomly generated 50,000 training samples, by taking statistically averaged values. Fig. 4 shows the minimum MSE of RBF equalizer and compares with that of linear equalizer. This simulation studies are performed with different center spread parameters, σ^2 .

Fig. 4 (a)-(c) show the approximate minimum MSE of RBF equalizer when channel delay is introduced in training. As shown in Fig.4, the approximate minimum MSE of RBF equalizer is always less than that of linear equalizer. It was also found that the lowest value of approximate minimum MSE value of RBF equalizer for each channel model are obtained usually when the center spread values are approximately two to ten times more than the variances of AWGN. For example, in the Fig.4 (c), the approximate minimum MSE of RBF equalizer for SNR = 20dB (corresponding to 0.01 values of AWGN) was obtained when the center spread value is equal to 0.1, which is ten times more than the variance of introduced AWGN (equal to 0.01). For SNR =13dB (corresponding to 0.05), the approximate minimum MSE of RBF equalizer was obtained when the center spread value is approximately equal to 0.1. Fig. 4(b) shows the approximate minimum MSE of RBF equalizer for the severely distorted channel. Although the approximate minimum MSE value of RBF equalizer for this model is still less than that of linear equalizer, their difference is relatively smaller than the cases of either linearly separable or slightly distorted channels. In other words, the difference of approximate minimum MSE between RBF and linear equalizer for the linearly separable channel cases tend to be greater than the cases of distorted channels. Fig. 4(d) shows that the minimum MSE for RBF equalizer without channel delay is almost the same as that of RBF equalizer with channel delay. Also, it was found that the difference of approximate minimum MSE between RBF and linear equalizer when not considering proper channel delay is much greater over all range of SNR than when considering channel delay. This properties prove that the RBF based equalizer recover the transmitted symbols successfully without introducing proper channel delay, because the lower value of MSE usually leads to the good bit error rate performance.

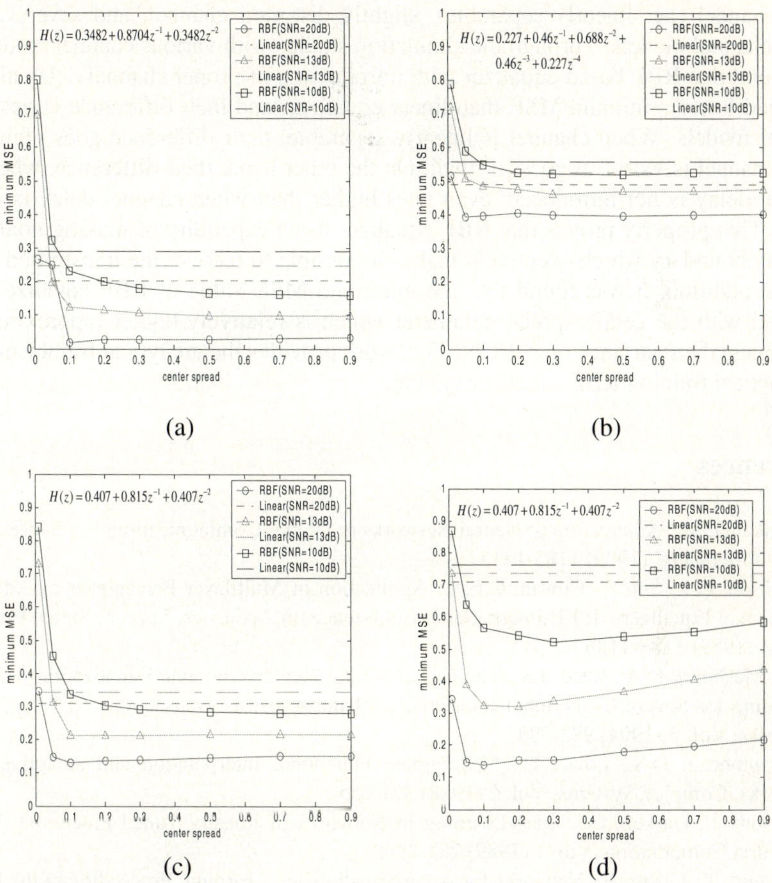

Fig. 4. Comparison of minimum MSE of RBF and Linear Equalizers (a) Desired symbol, $a_{k-d} = a_{k-1}$ (b) Desired symbol, $a_{k-d} = a_{k-2}$ (c) Desired symbol, $a_{k-d} = a_{k-1}$ (d) Desired symbol, $a_{k-d} = a_k$ (no channel delay)

5 Conclusion

Traditional adaptive algorithm for channel equalizers is based on the criterion of minimizing the mean square error between the desired filter output and actual filter output. In this paper, the theoretical minimum MSE of a RBF equalizer was evaluated and compared with that of a linear equalizer. The procedure of computing theoretical minimum MSE of RBF equalizer was derived using the same concepts of finding minimum MSE for linear equalizer, based on the fact that the relationship between the hidden and output layers in the RBF equalizer is also linear. For the purpose of theoretically exact minimum MSE for both RBF and linear transversal equalizer, a linear time dispersive channel whose order is one is selected. As extensive studies of this research, various channel models are selected for simulation which include fairly good channel (i.e. linearly separable), slightly distorted channel, and severely distorted channel models. Through the simulation studies with various channel models, it was found that RBF based equalizer with introduction of proper channel delay always produced lower minimum MSE than linear equalizer, and their difference varies with channel models. When channel is linearly separable, their difference goes high, and when channel is worse, it goes to small. On the other hand, their difference, when the channel delay is not introduced, even goes higher than when channel delay is introduced. This property proves that RBF equalizer has a capability of making nonlinear decision boundary which eventually makes it possible to recover the transmitted symbols. In addition, it was found that the minimum MSE value of RBF equalizer was obtained with the center spread parameter which is relatively higher (approximately 2-10 times) than variance of AWGN. This work provides the analytical framework for the practical training RBF equalizer system.

References

1. Ibnkahla, M.: Application of Neural Networks to Digital Communications – a Survey. Signal Processing, (1999) 1185-1215
2. Gibson, G.J., Siu, S., Cowan, C.F.N.: Application of Multilayer Perceptrons as Adaptive Channel Equalizers. IEEE International Conference on Acoustics, Speech, Signal Processing, (1989) 1183-1186
3. Al-Mashouq, K.A., Reed, I.S.: The Use of Neural Nets to Combine Equalization with Decoding for Severe Intersymbol Interference Channels. IEEE Transactions on Neural Networks, Vol. 5 (1994) 982-988
4. Broomhead, D.S., Lowe, D.: Multivariate Functional Interpolation and Adaptive Networks. Complex Systems, Vol. 2 (1988) 321-355
5. Moody J., Darken, C.J.: Fast Learning in Networks of Locally Tuned Processing Units. Neural Computation, Vol. 1 (1989) 281-294
6. Poggio, T., Girosi, F.: Networks for Approximation and Learning. Proceeding of the IEEE, Vol. 78 (1990) 1481-1497
7. Chen, S., Mulgrew, B., Grant, P.M.: A Clustering Technique for Digital Communication Channel Equalization using Radial Basis Function Networks. IEEE Transactions on Neural Networks, Vol. 4 (1993) 570-579
8. Mulgrew, B.: Applying Radial Basis Functions. IEEE Signal Processing Magazine, (1996) 50-65
9. Patra, S.K., Mulgrew, B.: Computational Aspects of Adaptive Radial Basis Function Equalizer Design. IEEE International Symposium on Circuits and Systems, (1997) 521-524
10. Lee, J.: A Radial Basis Function Equalizer with Reduced number of Centers. Ph. D. Dissertation, Florida Institute of Technology (1996)
11. Gan, Q., Subramanian, R., Sundararajan, N., Saratchandran, P.: Design for Centers of RBF Neural Networks for Fast Time-Varying Channel Equalization. Electronics Letters, Vol. 32 (1996) 2333-2334
12. Lee, J., Beach, C.D., Tepedelenlioglu, N.: A Practical Radial Basis Function Equalizer. IEEE Transactions on Neural Networks, Vol. 10 (1999) 450-455
13. Lee, E., Messerschmitt, D.: Digital Communication. Springer (1993)
14. Haykin, S.: Adaptive Filter Theory. PrenticeHall (1996)
15. Proakis, G.: Digital Communication. Mcgraw-Hill (1995)

A Hybrid Unscented Kalman Filter and Support Vector Machine Model in Option Price Forecasting

Shian-Chang Huang[1,⋆] and Tung-Kuang Wu[2]

[1] Dept. of Business Administration, National Changhua University of Education
No.2, Shi-Da Road, Changhua 500, Taiwan
Tel. no: 886-953092968, Fax no: 886-4-7211162
shhuang@cc.ncue.edu.tw

[2] Dept. of Information Management, National Changhua University of Education
No.2, Shi-Da Road, Changhua 500, Taiwan
tkwu@mail.tkwu.net

Abstract. This study develops a hybrid model that combines unscented Kalman filters (UKFs) and support vector machines (SVMs) to implement an online option price predictor. In the hybrid model, the UKF is used to infer latent variables and make a prediction based on the Black-Scholes formula, while the SVM is employed to capture the nonlinear residuals between the actual option prices and the UKF predictions. Taking option data traded in Taiwan Futures Exchange, this study examined the forecasting accuracy of the proposed model, and found that the new hybrid model is superior to pure SVM models or hybrid neural network models in terms of three types of options. This model can also help investors for reducing their risk in online trading.

1 Introduction

Because of the high risk in trading options, accurate online forecasts of option prices are important for an investor to control and hedge his risk. Traditionally, the literature of online forecasting focuses on using Kalman filters and extended Kalman filters (EKF) as prediction methodologies. However, the Kalman filter model requires linearity and Gaussian conditions of the underlying processes, and the EKF model uses a first order linear approximation to simplify the processes, which is a sub-optimal solution. These approximations are limiting because many financial time series are nonlinear. For example, option prices are nonlinear functions of the underlying stock price process.

Therefore, the study uses an unscented Kalman filter (UKF [6,13]) instead to improve the forecasting performance. Unlike the EKF, the UKF does not approximate the nonlinear process and observation models. The UKF uses a minimal set of sample points to capture the true mean and covariance of the nonlinear

⋆ Corresponding author.

L. Jiao et al. (Eds.): ICNC 2006, Part I, LNCS 4221, pp. 303–312, 2006.

process. When this set of sample points propagate through the nonlinear system, the posterior mean and covariance can be accurately estimated, with errors only introduced in the 3rd and higher orders. But there are still some nonlinear residuals which cannot be captured by the UKF, thus one needs a hybrid model to reduce the prediction errors further.

In recent years, neural networks (NN) are an emerging and challenging computational technology and they offer a new avenue to explore the dynamics of a variety of financial applications. Recent works on time series predictions using neural networks include Han et al. [5] and Wang et al. [15]. Han et al. [5] used a recurrent predictor neural network to forecast chaotic time series, while Wang et al. [15] used wavelet packet neural networks to predict time series. Neural networks are good at nonlinear inputs and outputs relationship modelling, and the wavelet packet can capture important time-scale features of a time series. [5] and [15] tend to give good performance on time series forecasting.

Recently, the support vector machine (SVM) method (Vapnik [12], Cristianini, N. and J. Shawe-Taylor [4], Schoelkopf, Burges, and Smola [11], Kecman [7], Wang [14]), another form of neural networks, has been gaining popularity and has been regarded as the state-of-the-art technique for regression and classification applications. We refer to Wang [14] for theory and typical applications of SVM. It is believed that the formulation of SVM embodies the structural risk minimization principle, thus combining excellent generalization properties with a sparse model representation. SVM is based on some beautifully simple ideas and provide a clear intuition of what learning from examples is all about. More importantly they are showing high performance in practical applications. Support vector regression (SVR), as described by Vapnik [12], exploit the idea of mapping input data into a high dimensional reproducing kernel Hilbert space (RKHS) where a linear regression is performed. The advantages of SVR are: the presence of a global minimum solution resulting from the minimization of a convex programming problem; relatively fast training speed; and sparseness in solution representation.

Different forecasting models can complement each other in capturing different patterns appearing in one data set, and both theatrical and empirical studies have concluded that a combination of forecast outperforms individual forecasting models ([1,8,9]). Typical examples of combining forecasting are Terui and Dijk [10] and Zhang [16]. Terui and Dijk [10] presented a linear and nonlinear time series model for forecasting the US monthly employment rate and production indices. Their results demonstrated that the combined forecasts outperformed the individual forecasts. Recently, Zhang [16] combined the ARIMA and feedforward neural networks models in forecasting typical time series data, and showed the superiority of the combined forecasting model. Furthermore, Castillo and Melin [3] developed hybrid intelligent systems for time series prediction based on neural networks, fuzzy logic, and fractal theory.

This study presents a new hybrid model that combines UKFs and SVMs to forecast the option prices. In the hybrid approach, the UKF serves as a processor to make predictions based on the Black-Scholes (BS, [2]) formula. The residuals between the actual prices and the Black-Scholes model are fed into the SVM in the hybrid model, and the SVM is conducted to further reduce the prediction errors. Empirical results of this study demonstrated that the new hybrid model outperforms another hybrid model based on UKFs and neural networks (UKF+NN). In terms of in-the-money, at-the-money, and out-the-money options, the empirical results revealed that the UKF or pure SVM models cannot capture all of the patterns in the option data, and the performance of UKF+NN is also not good. Only the new hybrid model can significantly reduce the forecasting errors.

The remainder of the paper is organized as follows. Section 2 describes the option price modelling and forecasting methods, including the BS model, the UKF and SVM models. Section 3 describes the data used in the study, and displays the empirical results with discuss on some empirical findings. Conclusions are given in Section 4.

2 Option Price Modelling and Forecasting

Black and Scholes [2] established the price of an european call option through a well known formula, which is the solution to a second order partial differential equation. This closed analytical solution conferred elegance to the proposed formulation and multiplied in extensive and complementary studies. Black and Scholes [2] assumed that the underlying stock price follows a geometric Brownian motion with constant volatility,

$$\frac{dS}{S} = \mu dt + \sigma dW_t, \tag{1}$$

where μ is the expected return and W_t is the Brownian motion. In their paper, Black and Scholes derived the following partial differential equation which governs prices of a call option or a put option,

$$\frac{\partial f}{\partial t} + rS\frac{\partial f}{\partial S} + \frac{1}{2}\sigma^2 S^2 \frac{\partial^2 f}{\partial S^2} = rf, \tag{2}$$

where r is the riskless interest rate. The solutions for the call and put option prices are thus

$$C = SN(d_1) - Ke^{-r\tau}N(d_2), \tag{3}$$

$$P = -SN(-d_1) + Ke^{-r\tau}N(-d_2), \tag{4}$$

where C is the call option price, P the put option price, K the strike price, and $\tau = T - t$ the maturity. Parameters d_1 and d_2 are as follows:

$$d_1 = \frac{\ln(S/K) + (r + \sigma^2/2)\tau}{\sigma\sqrt{\tau}}, \tag{5}$$

$$d_2 = d_1 - \sigma\sqrt{\tau}. \tag{6}$$

The above equations can be represented as a state-space form. r, σ can be viewed as the hidden states, and C, P the output observations, while τ, K are treated as the parameters or the input signals. Consequently, an EKF or UKF can be employed to infer the latent states and makes predictions. The unscented Kalman filter is a straightforward application of the scale unscented transformation (SUT) to recursively minimize the mean-squared estimation errors. We refer to [6, 13] for details.

The support vector machines (SVMs) were proposed by Vapnik [12]. Based on the structured risk minimization (SRM) principle, SVMs seek to minimize an upper bound of the generalization error instead of the empirical error as in other neural networks. Additionally, the SVMs models generate the regress function by applying a set of high dimensional linear functions. The SVM regression function is formulated as follows:

$$y = w\phi(x) + b, \tag{7}$$

where $\phi(x)$ is called the feature, which is nonlinear mapped from the input space x to the future space. The coefficients w and b are estimated by minimizing

$$R(C) = C\frac{1}{N}\sum_{i=1}^{N} L_{\varepsilon}(d_i, y_i) + \frac{1}{2}||w||^2, \tag{8}$$

where

$$L_{\varepsilon}(d, y) = \begin{cases} |d-y| - \varepsilon & |d-y| \geq \varepsilon, \\ 0 & \text{others,} \end{cases}, \tag{9}$$

where both C and ε are prescribed parameters. The first term $L_{\varepsilon}(d, y)$ is called the ε-intensive loss function. The d_i is the actual option price in the ith period. This function indicates that errors below ε are not penalized. The term $\frac{C}{N}\sum_{i=1}^{N} L_{\varepsilon}(d_i, y_i)$ is the empirical error. The second term, $\frac{1}{2}||w||^2$, measures the smoothness of the function. C evaluates the trade-off between the empirical risk and the smoothness of the model. Introducing the positive slack variables ξ and ξ^*, which represent the distance from the actual values to the corresponding boundary values of ε-tube. Equation (8) is transformed to the following constrained formation:

$$\min_{w,b,\xi,\xi^*} R(w, \xi, \xi^*) = \frac{1}{2}w^T w + C\left(\sum_{i=1}^{N}(\xi_i + \xi_i^*).\right) \tag{10}$$

Subject to

$$w\phi(x_i) + b_i - d_i \leq \varepsilon + \xi_i^*, \tag{11}$$

$$d_i - w\phi(x_i) - b_i \leq \varepsilon + \xi_i, \tag{12}$$

$$\xi_i, \xi_i^* \geq 0. \tag{13}$$

After taking the Lagrangian and conditions for optimality, one obtain the following solution:

$$y = f(x, \alpha, \alpha^*) = \sum_{i=1}^{N}(\alpha_i - \alpha_i^*)K(x, x_i) + b, \tag{14}$$

where $K(x, x_i)$ is called the kernel function. α_i, α_i^* are nonzero Lagrangian multipliers, and b follows from the complementarity Karush-Kuhn-Tucker (KKT) conditions.

The value of the kernel is equal to the inner product of two vectors x_i and x_j in the feature space, such that $K(x_i, x_j) = \phi(x_i)\phi(x_j)$. Any function that satisfying Mercer's condition (Vapnik [12]) can be used as the Kernel function. The Gaussian kernel function

$$K(x_i, x_j) = \exp\left(-\frac{||x_i - x_j||^2}{2\sigma^2}\right) \tag{15}$$

is specified in this study, because Gaussian kernels tend to give good performance under general smoothness assumptions.

2.1 Hybrid Approaches

The true dynamics of option prices are not easily modelled. Therefore, a hybrid strategy combining both Black-Scholes model and the support vector regression is a good alternative for forecasting option prices. Compared with general classes of non-parametric regression methods such as global parametric methods (i.e. linear regression), adaptive computation methods (i.e. projection pursuit regression), and neural networks, support vector regression are among the most accurate and efficient regression techniques. In the hybrid model, both the UKF and the support vector machine have different capabilities to capture data characteristics in different domains. Thus, the overall forecasting performance can be significantly improved.

The hybrid prediction model of call option price C_t can be represented as follows:

$$\widetilde{C}_t = \widetilde{BS}_t + \widetilde{f}_t, \tag{16}$$

where $\widetilde{BS}_t$ is the prediction based on Black-Scholes, $\widetilde{f}_t$ the nonlinear SVM prediction, and $\widetilde{C}_t$ the overall prediction. Let δ_t represent the residual between the actual option price and the BS model at time t, namely,

$$\delta_t = C_t - \widetilde{BS}_t. \tag{17}$$

The residuals are fed to the SVM, which further reduces the prediction errors, that is,

$$\delta_t = \widetilde{f}(S_t/K, T - t, \delta_{t-1}) + \varepsilon_t, \tag{18}$$

where ε_t is the final residual.

3 Experimental Results and Analysis

The data used in this research are the option prices on the Taiwan composite stock index traded in Taiwan Futures Exchange (TWIFEX). The transaction data of call and put option prices from 16 September 2004 to 14 June 2005 with expiration on 15 June 2005 were studied. Only traded prices were used. Bid and ask prices were not considered in this study. Totally, there are 184 observations.

This study chooses three types of options, namely, out-of-money, at-the-money, and in-the-money options, to test the forecasting performance. In each type of options, this study selects those with largest trading volumes in its group. Data with $K = 5800$ represents the in-the-money options in the sample period; data with $K = 6000$ approximates the at-the-money options, while data with $K = 6200$ represents the out-of-money options in the sample period.

The UKF was trained in a sequential manner, while the SVM was trained in a batch manner. The last eighty samples of every data set were used as the test set, and one hundred data points before the prediction day were used as the training set for the SVM. In this study, only one-step-ahead forecasting is considered. One-step-ahead forecasting can prevent problems associated with cumulative errors from the previous period for out-of-sample forecasting. This study adopted RMSE (root-mean-squared error) as the performance index, which is shown below:

$$RMSE = \left(\frac{1}{N} \sum_{t=1}^{N} (d_t - \widehat{d}_t)^2 \right)^{1/2} , \tag{19}$$

where N is the number of forecasting periods, d_i is the normalized actual option price at period t, namely, $d_i = \frac{C_i}{K}$ or $\frac{P_i}{K}$, where C_i is the actual call price and P_i the actual put price. $\widehat{d_t}$ is the forecasting price at period t.

Table 1 displayed the forecasting performance of the UKF model, the pure SVM model, and two hybrid models for call option prices with different Ks. Table 2 provided the performance of these models on predicting put option prices. Figures 1-4 plotted the predictions of the UKF and the hybrid model. For example, Figure 1 plotted the normalized actual prices, and the predicted prices of the UKF on the option of K=5800. Figure 2 displayed the SVM predictions on the residuals of Figure 1. Figures 3 and 4 plotted similar results on the option of K=6200.

Table 1. Forecasting Performance of Every Model on the Three Call Options

	$K = 6200$	$K = 6000$	$K = 5800$
UKF Predictions	0.00239	0.00341	0.00375
Pure SVM	0.00280	0.00420	0.00370
UKF+NN Predictions	0.00159	0.00252	0.00217
UKF+SVM Predictions	0.00120	0.00193	0.00159

The results in Table 1 and 2 indicated that the performance of the UKF were better than the pure SVM model, namely, the UKF is superior to capture important time series characteristics of the option prices. For the two hybrid models, the UKF model served as a processor to handle the BS model predictions. Then, the residuals between the actual prices and the BS model predictions were

Table 2. Forecasting Performance of Every Model on the Three Put Options

	$K = 6200$	$K = 6000$	$K = 5800$
UKF Predictions	0.00418	0.00339	0.00262
Pure SVM	0.00500	0.00490	0.00370
UKF+NN Predictions	0.00317	0.00234	0.00169
UKF+SVM Predictions	0.00304	0.00160	0.00095

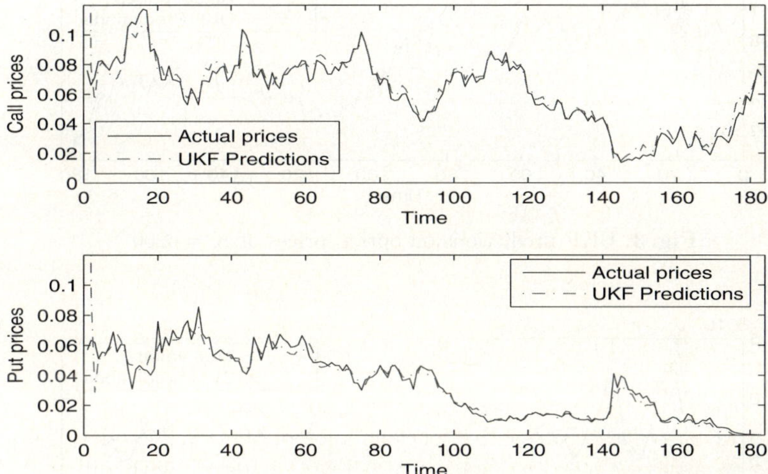

Fig. 1. UKF predictions on option prices of $K = 5800$

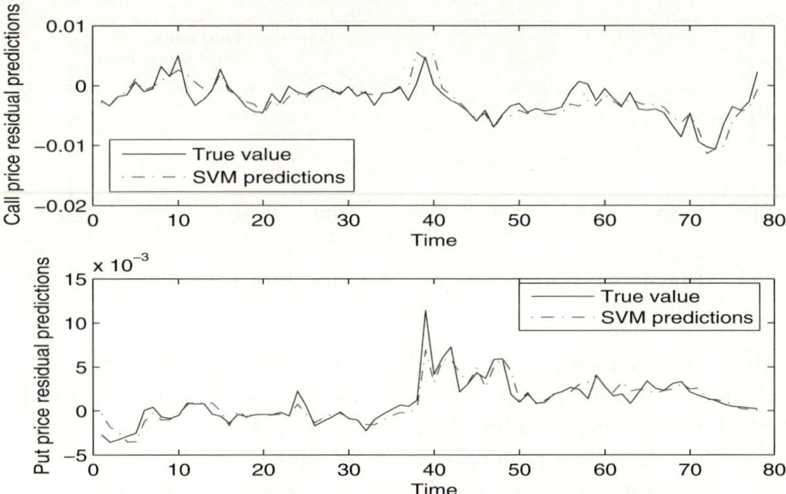

Fig. 2. SVM predictions on the residuals of $K = 5800$

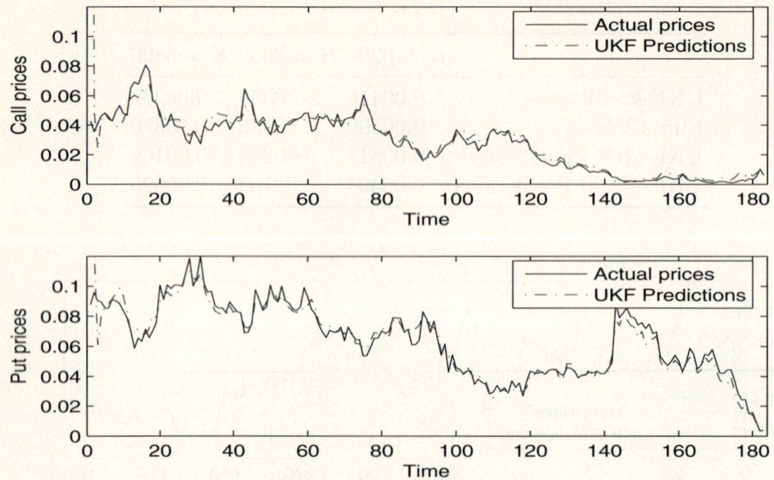

Fig. 3. UKF predictions on option prices of $K = 6200$

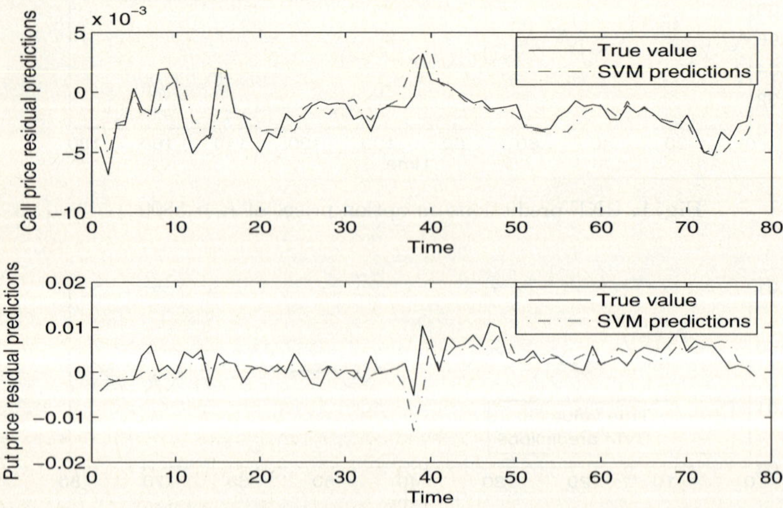

Fig. 4. SVM predictions on the residuals of $K = 6200$

fed into the neural network or the SVM in these hybrid models. NN or SVM were conducted to further reduce the prediction errors. This study trained the NN and SVM by the data including $S_t/K, T-t, \delta_{t-1}$. From Tables 1 and 2, one can compare the forecasting performance of different models.

The empirical results indicated that these hybrid model outperformed the UKF or pure SVM models in terms of all three options, and the hybrid UKF+SVM model outperformed the hybrid UKF+NN model. These results revealed that none of the three models (UKF, pure SVM, UKF+NN) can capture all of the nonlinear patterns in the option data. But the combination of UKF+SVM is better than UKF+NN, this is owing to that SVM models minimize the structural risk which prevents the problem of overfitting usually encountered in neural networks. From Figures 2 and 4, one can observe that the BS model residuals are still quite nonlinear. Comparing the forecasts of NNs and SVMs on these nonlinear residuals, SVMs outperformed NNs. Therefore, UKF+SVM is the best choice to capture nonlinear dynamics of option prices.

4 Conclusions

This study proposes to hybrid UKFs and SVMs as an option price predictor. The UKF serves as a state estimator and a predictor based on the Black-Schole formula, and the nonlinear price characteristics which cannot be captured by the UKF is captured by the SVM. Compared with another hybrid model based on UKF+NN, the proposed hybrid model can capture the maximum amount of dynamics in the option prices.

Due to the high risk in trading options, an accurate online forecast of option prices is important for an investor to control and hedge his risk. Options are nonlinear financial instruments. Their price dynamics are quite nonlinear. Linear prediction tools such as the Kalman filter is not a good choice for online forecasting. This study found that the new hybrid model was superior to traditional hybrid models. Compared with UKF+NN, the new hybrid model can significantly reduced the root-mean-square errors by nearly one half.

The powerful framework established in this study can be applied to high-frequency financial data, and other online forecasting problems. A challenging future task is the Bayesian selection of optimal weighting between the UKF and the SVM by data, and to train the SVM by a Bayesian method to adjust its internal parameters.

Acknowledgment

This work was supported in part by the National Science Foundation of Taiwan.

References

1. Clemen, R.: Combining forecasts: a review and annotated bibliography with discussion. *International Journal of Forecasting*, 5 (1989), 559-608.
2. Black, F. and Scholes, M. S.: The pricing of options and corporate liabilities, *Journal of Political Economy*, 81 (1973), 637-654.
3. Castillo, O., Melin, P.: Hybrid intelligent systems for time series prediction using neural networks, fuzzy logic, and fractal theory. IEEE Trans. Neural Networks, 13 (2002) 1395-1408.
4. Cristianini, N. and Shawe-Taylor, J.: An Introduction to Support Vector Machines, Cambridge University Press (2000).

5. Han, M., Xi, J., Xu, S., Yin, F.-L.: Prediction of chaotic time series based on the recurrent predictor neural network. IEEE Trans. Signal Processing, 52 (2004) 3409-3416.
6. Julier, S. J. and Uhlmann, J. K.: A New Extension of the Kalman Filter to Nonlinear Systems. In The Proceedings of AeroSense: The 11th International Symposium on Aerospace/Defense Sensing,Simulation and Controls, Multi Sensor Fusion, Tracking and Resource Management II, SPIE (1997).
7. Kecman, V.: Learning and Soft Computing, Support Vector machines, Neural Networks and Fuzzy Logic Models, The MIT Press, Cambridge, MA (2001).
8. Lawerence, M. J., Edmundson, R. H., and O'Connor, M. J.: The accuracy of combining judgemental and stastical forecasts, *Management Science*, 32 (1986), 1521-1532.
9. Makridakis, S.: Why combining works?, *International Journal of Forecasting*, 5 (1989), 601-603.
10. Terui, N. and Dijk, H. K.: Combined forecasts from linear and nonlinear time series models. *International Journal of Forecasting*, 18 (2002), 421-438.
11. Schoelkopf, B., Burges, C. J. C., and Smola, A. J.: Advances in kernel methods - support vector learning, MIT Press, Cambridge, MA (1999).
12. Vapnik, V. N.: The Nature of Statistical Learning Theory, New York, Springer-Verlag (1995).
13. Wan, E. A., and R. van der Merwe: The Unscented Kalman Filter for Nonlinear Estimation. In Proceedings of Symposium 2000 on Adaptive Systems for Signal Processing, Communication and Control(AS-SPCC), IEEE Press (2000).
14. Wang, L.P. (Ed.): Support Vector Machines: Theory and Application. Springer, Berlin Heidelberg New York (2005).
15. Wang, L.P., Teo, K.K., Lin, Z.: Predicting time series using wavelet packet neural networks. Proc. IJCNN 2001, 1593-1597.
16. Zhang, G. P.: Times series forecasting using a hybrid ARIMA and neural network model. *Neurocomputing*, 50 (2003), 159-175.

Empirical Study of Financial Affairs Early Warning Model on Companies Based on Artificial Neural Network

Tian Bo and Qin Zheng

School of Management, Xi'an Jiaotong University, Xi'an 710049, China
btian@stu.xjtu.edu.cn, youngtb@sina.com

Abstract. This paper attempts to develop an intelligent financial distress forecasting pattern using Artificial Neural Networks (ANNs) by taking advantage of the ANNs for recognizing complex patterns in data and universal functional approximation capability. Using STzhujiang and Non-STshenzhenye stocks as the study samples, the objective is to make ANN model as financial affairs early warning research tool through building an intelligent and individual financial distress forecasting patterns. The model built for individual industries would be even more predictive than general models built with multi-industry samples. Results show that ANNs are valuable tools for modeling and forecasting ST and Non-ST companies whether they are being in financial distress. The simulation result shown that ANNs models can be applied in financial affairs early warning system. The companies can build their own financial distress forecasting patterns based on their own running surroundings using proposed ANNs models.

1 Introduction

Prediction of financial distress has long been an important topic and has been studied extensively in the accounting and finance literature of foreign countries, and has formulated an effective financial distress forecasting pattern. We haven't set maturity financial distress forecasting pattern based on our own running surroundings, and only revealed certain financial forecasting information in corporation financial reports [1-6]. Financial forecasting data are only simple expression in forecasting operation for historical and currently working states. When being in real financial distress, data of financial forecasting information in corporation financial reports cannot accurately reflect the potential running in the future. Because of restriction of data collection, the present research only focused on corporations of certain trade in our country. When evaluate the risky operation, we have to use a universal model of certain trade. The universal model does not take into account microcosmic surrounding factors. When applied in forecasting the certain corporation financial condition, the result is of doubtful veracity.

This paper developed an intelligent financial distress forecasting system using ANNs models. The data were from a single corporation (the corporation belongs to real estate trade). The following of the paper consists of six sections: Structure of proposed financial distress analysis models is discussed in the second section. Index reduction using the method of Factor Analysis is done in the third section. The index

L. Jiao et al. (Eds.): ICNC 2006, Part I, LNCS 4221, pp. 313 – 320, 2006.

reduction is used as the ANNs inputs. The ANNs models and its learning rules are given in the fourth section. The method of bankruptcy prediction with ANNs is investigated in the fifth section. Simulation results are performed in the sixth section. At last section, some conclusions are drawn.

2 Structure of Proposed Financial Distress Analysis Models

There are some difference between inner research and foreign in financial analysis for technology gap. Our financial analysis has always utilized financial ratios computed from balance sheet and income statement data to measure the continuing strength of companies. The methods are simple lacking predictive models, which combine the use of these tradition financial ratios with statistical techniques. The advanced techniques including econometrics, statistics, information science, and computer science are not being used. It is necessary for us to combine the advanced techniques into a dynamic method system. Our financial analysis has always used dozens of financial ratios as inputs of predication systems, some of which are unique to the industry, but others were proved to be failure of prediction. While it was felt that biases might exist in the model. Several approaches were proposed for solving binary classification problems. The approaches can be categorized as linear and non-linear discriminant analyses approaches. The linear approaches use a line or a plane to separate two groups in a binary classification problem. Among the popular linear approaches are a statistical discriminant analysis model (Fisher's discriminant analysis), and the non-parametric linear discriminant analysis (LDA) model such as genetic algorithm (GA) based linear discriminant model. Non-linear approaches used for discriminant analysis fall into two categories, and in our country non-linear approaches not been used widely [7,8]. The judgement scope of financial distress is fixed, and lacks self organization and learning ability. The purpose of this paper is to present the initial results on this topic.

ANNs models for financial distress analysis were presented in this paper. Data were collected from two real estate corporations. Index come from the outputs of factors analysis approach were used as inputs of ANNs. The neural network model was then used to apply in financial affairs early warning system. In this paper, data were all from historic balance sheet and income statement data, so the model was individual industries model, not generic models built with multi-industry samples. This allowed analysis of the future performance of the operation project of its own.

It is often not an easy task to build ANNs for time-series analysis and forecasting because of the large number of factors related to the model selection process. In order to identify and explore the various components of critical factors associated with bankruptcy in the STzhujiang and Non-STshenzhenye corporations, we used factor analysis approach. The methodology was designed to identify the critical factors that affect the corporation on time. We employed three-phased approach to study the critical factors that affect the corporation. The outputs (critical factors) were fed into ANNs.

We focused on the feed-forward multi-layer perceptrons (MLPs) as predication tools [9]. It is the most popular type of ANNs used for forecasting purposes. For a univariate time-series forecasting problem, the inputs of the network are the past lagged observations and the outputs are the predicted values after training. Each input pattern is composed of a moving window of fixed length along the series.

The proposed method differs from prior studies in the following respects. Firstly, factor analysis approach was applied to refine critical factors. Secondly, ANNs were used to predict financial distress. Finally, using the proposed method, a corporation can select suitable samples, and develop an intelligent financial distress forecasting pattern based on its own operation condition. Following, each component of the proposed model is discussed in detail.

3 Selecting Predictor Variables by Factor Analysis Approach

The factors that lead corporations to failure vary. Many economists attribute the phenomena to more interests, recession-squeezed profits and heavy debt burden. Furthermore, industry-specific characteristics, such as government regulation and the nature of operation, can contribute to a firm's financial distress. The study employed a relatively large number of financial index, according to the Ministry of Finance promulgated "Regulations on Evaluating the Efficiency and Achievement of Government Capital". We adopted 18 financial indexes to the financial distress forecasting pattern. These indexes along with their definitions are shown as Table1.

Table 1. Selected 18 financial indexes

Index	Definition
X_1	Income Before Extraordinary Items / Average Total Assets
X_2	Income Before Extraordinary Items / Average Total Current Assets
X_3	Income Before Extraordinary Items / Net Accounts Receivable
X_4	Income Before Extraordinary Items /Net Fixed Assets
X_5	Earnings Before Interest & Taxes/Average Net Assets
X_6	Earnings Before Interest /Average Assets
X_7	Earnings Before Interest & Taxes/Total Assets
X_8	Earnings Before Interest & Taxes/(Net Fixed Assets+ Current Assets)
X_9	Sell Returns/(Net Fixed Assets +Current Assets)
X_{10}	Earnings Before Interest & Taxes/Total Costing
X_{11}	Total Liabilities/Total Assets
X_{12}	Total Liabilities/Shareholder's Equity
X_{13}	Earnings Before Interest & Taxes/ Interest
X_{14}	Cash Flow From Operations/Current Liabilities
X_{15}	Current Assets/ Current Liabilities
X_{16}	(Current Assets- stock-in-trade)/ Current Liabilities
X_{17}	(Income Before Extraordinary Items this year- Income Before Extraor dinary Items last year)/ Income Before Extraordinary Items last year
X_{18}	(Total Assets This year- Total Assets last year)/ Total Assets last year

To identify the significant factors in an individual corporation, the critical factor variables were adapted from the 18 variables by factors analysis approach (FAA). The primary objective was to obtain coherent and meaningful critical factors from variables (sub-factors). By examining the correlation between the variables, FAA can arrange critical variables into specific groups and assist in understanding the latent critical factor structure. Then let the reduction factors as inputs of the network, which make the network simpler and computational speed faster, and increase the accurate. Experiment in the sixth section also shown that FAA is a suitable approach.

4 ANNs and Learning Algorithm

ANNs are flexible, nonparametric modeling tools. They can perform any complex function mapping with arbitrarily desired accuracy [9]. ANNs in this paper are composed of three layers of computing elements called nodes. Each node receives an input signal from other layer nodes or external inputs and then after processing the signals locally through a transfer function, it outputs a transformed signal to other nodes or final result. The multi-layer perceptron (MLP) is used in this paper, all nodes and layers of which are arranged in a feed-forward manner. The first or the lowest layer is the input layer where external information is received. The last or the highest layer is the output layer where the network produces the model solution. In between, there is one hidden layer which is critical for ANNS to identify the complex patterns in the data. Three-layer MLP is a commonly used ANNs structure for two-group classification problems like the bankruptcy prediction. Each node in the neural network has some finite"fan-in" of connections represented by weight values from the previous nodes and "fan-out" of connections to the next nodes. Associated with the fan-in of a node is an integration function f which combines information, activation, and evidence from other nodes and provides the net input for this node.

$$I_{net\ input}=f\text{ (inputs to this node associated link weights)} \tag{1}$$

Each node also outputs an activation value as a function of its net input

$$O_{node\ output}=a\text{ (net input)} \tag{2}$$

where a denotes the activation function. The active function between the first two layers is Sigmoid function

$$O_i^1=(1+\exp(I_i^1))^{-1} \tag{3}$$

where I_i^1 are the net inputs between input nodes and hidden nodes, $i=1,2\cdots 5$, and the net input for these node are

$$I_i^1=w_{i1}x_1+w_{i2}x_2 \tag{4}$$

where x_1,x_2 are the inputs of the network., and w_{i1},w_{i2} are the weights between input nodes and hidden nodes to be adjusted by learning.

The active function between hidden layer and output layer is identification function

$$O_j^2=I_j^2 \tag{5}$$

where I_j^2 are the net inputs between hidden nodes and output nodes, $j=1,2\cdots 4$, and the net input for these node are

$$I_j^2 = \sum_{k=1}^{5} v_{jk} O_k^1 \tag{6}$$

where v_{jk} are the weights between hidden nodes and output nodes. Back-propagation algorithm is used for supervised training of the proposed neural network model to minimize the discrete form of error function (7), which is defined as

$$E(W) = \sum_{j=1}^{4} (D_j - O_j(W))^2 \tag{7}$$

where D_j denotes the jth output of the neural network, and O_j denotes the j th desired output. A forward process is used to calculate the output function values of all nodes in the network layer by layer using initializing the connection weights value between each two layers. Then, starting from the output layer, a backward pass is used to update the connection weights value sequentially. Following is the next iteration and update the weight until the stop criteria are satisfied. Gradient optimization algorithm containing momentum term [9] is used to training the neural network. The general parameter update rule is given by

$$\Delta W(k+1) = \alpha(-\frac{\partial E}{\partial W}) + \beta \Delta W(k) \tag{8}$$

where W denotes the adjustable parameters in the network, α denotes the learning rate, β is a momentum term to damp possibility of oscillations. The weights between hidden nodes and output nodes v_{jk} and the weights between input nodes and hidden nodes w_{i1}, w_{i2} are to be adjusted by learning (8).

The aim of training is to minimize the differences between the ANNs output values and the known target values for all training patterns. After the parameters of the ANNs are acquired by training they can be used to predict future.

5 Bankruptcy Prediction with ANNs

Currently there are no systematic principles to guide the design of a neural network model for a particular classification problem although heuristic methods such as the pruning algorithm [10]. The polynomial time algorithm [11], and the network information technique [12] have been proposed. Since many factors such as hidden layers, hidden nodes, data normalization and training methodology can affect the performance of neural network, the best network architecture is typically chosen through experiments. In this sense, neural network design is more an art than a science.

In our bankruptcy predictor model, the networks have two input nodes in the first layer corresponding to matching several predictor variables. The number of hidden nodes is not easy to determine a priori. Although there are several rules of thumb suggested for determining the number of hidden nodes, such as using $n/2$, n, n+1 and $2n$+1 where n is the number of input nodes, none of them works well for all situations. Determining the appropriate number of hidden nodes usually involves lengthy experimentation since this parameter is problem and/or data dependent. In this paper, the number of hidden nodes is $2n+1$. A neural network learns by updating its weights in accordance with a learning rule as (8) that is used to train it, which provides a method of updating the weights so that the errors are minimized[13,14,15]. During the learning state, examples are presented to the network in input-output pairs. For each example, the network computes the predicted outputs based on the inputs. The difference between the computed outputs and the desired output by the example is computed and back-propagated. This means that the weights are modified for subsequent predictions based on the observed errors.

We make the financial states four levels. They are: security, at sake, danger and imminent danger expressed as $(1,0,0,0)^T$, $(0,1,0,0)^T$, $(0,0,1,0)^T$ and $(0,0,0,1)^T$ respectively.

6 Experiment and Discussion

6.1 Data Collection and Reduction Factors by FAA

The data were collected from the CSMAR financial database covering the period 1993-2003[16]. Two kinds of firms were drawn, one of which is identified as financially distressed companies and another identified as financially viable company, named the "distressed kind" and the "viable kind", respectively. The distressed group was culled from the Industrial Research File which only lists firms that have ceased operations. We collected two firms in real estate industry. One is Shen Zhen Zhen Xing Corporation A stock, and another is Han Nan ZHu Jiang Corporation ST stock [16,17,18]. The data was divided into two sets: a training set and a test set. The training set involved 1993-2003 financial ratios, the 2004 financial ratios as the test set. The 1993-2003 financial ratios of STzhujiang and Non-STshenzhenye stocks were produced by computing based on table 1 using data of [16,17,18]. The financial state of each year of both companies can gained in [17,18]. To identify the critical factor we applied factor analysis approach, and the process of factor analysis was made with SPSS software. The results show as figure 1 and figure 2. From figure 1 and 2, we found 2 factors out of 18 factors are critical to each sample. The two factors reflect almost 100% information, and the two factors are unrelated to each other. STzhujiang and Non-STshenzhenye stocks factor load data can be calculated by SPSS software.

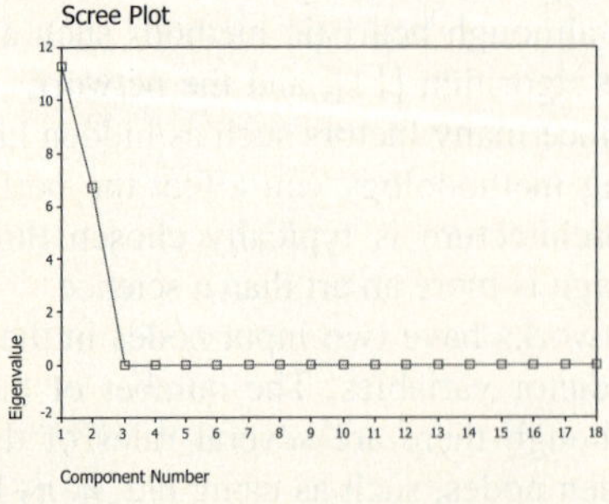

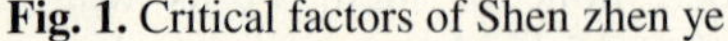

Fig. 1. Critical factors of Shen zhen ye

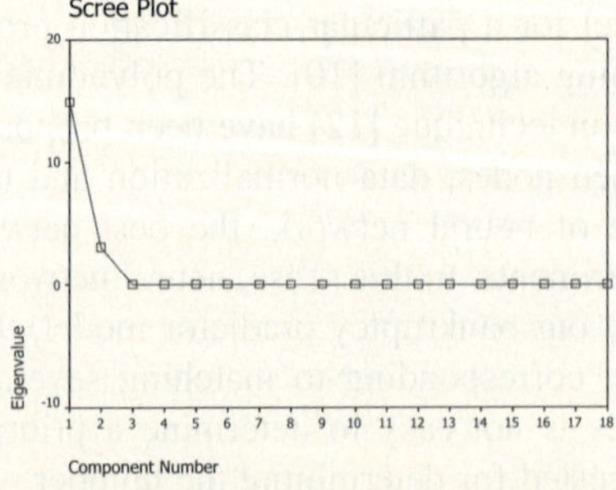

Fig. 2. Critical factors of Zhu jiang

6.2 Learning Process of ANNs

According to China Stock Network and financial statement, the financial states of STzhujiang and Non-STshenzhenye corporations were represented with 0 and 1, which stood for bankrupt or non-bankruptcy, respectively. Since the output values were not discrete, any value less than (or equal) 0.5 was considered to be 0 and above that was considered to be 1.

In this paper, the ANNs have 2 inputs, 5 hidden nodes and 4 outputs. The two critical financial factors of 1993-2003 database and corresponding financial states of two corporations were used as training set by using Matlab software. The training result of the network is consistent with the reality financial report forms. It turned out that the ANNs can be applied to simulate financial states of the two firms. It is feasible and practicability to apply the ANNs in forecasting financial distress.

6.3 Distress Forecasting Using ANNs

Using the trained network, we made 2004 financial ratios of STzhujiang and Non-ST shenzhenye as inputs of ANNs. The outputs of the ANNs are shown as table 2. From table 2, we can conclude that the operation state of Non-STshenzhenye is danger, and that of STzhujiang is danger too. So the two firms should take measures to make the loss minimum.

Table 2. The result of distress forecasting about two corporations

year	The result of Shen zhen ye	The result of Zhu jiang
2004	(-0.007,0.0007, 0.9729, 0.0033)	(-0.0030,0.0076,0.9797,0.0036)

7 Conclusion

Financial distress prediction is a class of important problem. A better understanding of the cause will have tremendous financial and managerial consequences. ANNs are promising alternative tools that should be given much consideration when solving real problems like bankruptcy prediction. This paper has applied a new approach, namely ANNs, to forecast financial distress in the ST and Non-ST industry. The neural network can both tolerate noise and missing data and learn by changing network connections with the environment. ANNs have the ability of generalizing from specific instances, and establishing complex relationships among variables. STzhujiang and Non-STshenzhenye stocks as selected study sample, intelligent financial affairs early warning patterns for individual financial distress forecasting were proposed based on ANNs. The model built for individual industries would be even more predictive than generic models built with multi-industry samples. Results show that ANNs are valuable tools for modeling and forecasting ST and Non-ST companies which whether they are being in financial distress. The main work of this paper is as following. The FAA was used to reduce the number of financial factor for a given firm. ANNs pattern were established to predict financial distress. And methodology for individual financial distress forecasting was proposed using individual industries sample instead of using models established by trade samples as usual.

The experiment simulation shows that the ANNs can be applied in financial affairs early warning system, and company can build individual financial distress forecasting patterns based on their own running surroundings.

References

1. Altman E.L.: Financial Ratios Discriminate Analysis and the Prediction of Corporate Bankruptcy. Journal of Finance 23 (3) (1968) 589-609
2. Altman E.L.: Accounting Implications of Failure Prediction Models. Journal of Accounting Auditing and Finance (1982) 4-19
3. Pendharkar PC, Rodger J.A.: An Empirical Study of Impact of Crossover Operators on the Performance of Non-binary Genetic Algorithm Based Neural Approaches for Classification. Computers & Operations Research 31 (2004) 481–498
4. Edmister R.: An Empirical Test of Financial Ratio Analysis for Small Business Failure Prediction. Journal of Finance and Quantitative Analysis 7 (1972) 1477-1493
5. Johnson C.: Ratio Analysis and the Prediction of Firm Failure. Journal of Finance 25 (1970) 1166-1168
6. Jones F.L.: Current Techniques in Bankruptcy Prediction. Journal of Accounting Literature 6 (1987) 131-164
7. Bhattacharyya S., Pendharkar P.C.: Inductive Evolutionary and Neural Techniques for Discrimination. Decision Sciences 29(4) (1998) 871–899
8. Nanda S, Pendharkar P.C.: Development and Comparison of Analytical Techniques for Predicting Insolvency Risk. International Journal of Intelligent Systems in Accounting, Finance and Management 10 (2001) 155–168
9. Pendharkar P.C.: A Computational Study on the Performance of ANNs under Changing Structural Design and Data Distributions. European Journal of Operational Research 138 (2002) 155–177
10. Hornik K.: Approximation Capabilities of Multilayer Feedforward Networks. Neural Networks 4 (1991) 251-257
11. Hornik K.: Some New Results on Neural Network Approximation. Neural Networks 6 (1993) 1069-1072
12. Nanda S, Pendharkar P.C.: Development and Comparison of Analytical Techniques for Predicting Insolvency Risk. International Journal of Intelligent Systems in Accounting, Finance and Management 10 (2001) 155–168
13. Reed R.: Pruning Algorithm -A survey. IEEE Transactions on Neural Networks 4 (5) (1993) 740-747
14. Roy, Kim L.S., Mukhopadhyay S.: A Polynomial Time Algorithm for the Construction and Training of a Class of Multilayer Perceptrons. Neural Networks 6 (1993) 535-545
15. Wang Z., Massimo C.D., Tham M.T., Morris A.J.: A Procedure for Determining the Topology of Multilayer Feed-forward Neural Networks. Neural Networks 7 (1994) 291-300
16. CSMAR database<<The Corporation Financial Statement Database of China (2004)>>, Shenzhen GuoTaiAn Information & Technology Corporation, Xiang Gang University Financial Institute
17. ShangHai Stock Exchange Financial Statement Database (2004)
18. ShenZhen stock exchange financial statement database (2004)

Rolling Bearings Fault Diagnosis Based on Adaptive Gaussian Chirplet Spectrogram and Independent Component Analysis*

Haibin Yu[1], Qianjin Guo[1,2], Jingtao Hu[1], and Aidong Xu[1]

[1] Shenyang Inst. of Automation, Chinese Academy of Sciences, Liaoning 110016, China
guoqianjin@sia.cn

[2] Graduate School of the Chinese Academy of Sciences, Beijing 100039, China
yhb@sia.cn

Abstract. Condition monitoring of rolling element bearings through the use of vibration analysis is an established technique for detecting early stages of component degradation. The location dependent characteristic defect frequencies make it possible to detect the presence of a defect and to diagnose on what part of the bearing the defect is. The difficulty of localized defect detection lies in the fact that the energy of the signature of a defective bearing is spread across a wide frequency band and hence can be easily buried by noise. To solve this problem, the adaptive Gaussian chirplet distribution for an integrated time-frequency signature extraction of the machine vibration is developed; the method offers the advantage of good localization of the vibration signal energy in the time-frequency domain. Independent component analysis (ICA) is used for the redundancy reduction and feature extraction in the time-frequency domain, and the self-organizing map (SOM) was employed to identify the faults of the rolling element bearings. Experimental results show that the proposed method is very effective.

1 Introduction

Rolling element bearings find many uses in today's machinery. They can be found in motors, slow-speed rollers, gas turbines, pumps, and many other machines. Some of the reasons rolling element bearings are used are: low starting friction, low operating friction, ability to support loads at low (even zero) speed, lower sensitivity to lubrication, and the ability to support both radial and axial loads in the same bearing. Compared to fluid film bearings which generally have a long life, rolling element bearings have a limited fatigue life due to the repeated stresses involved in their normal use. So it is necessary to detect changes in vibration signals caused by faulty components and to judge the conditions of the rolling element bearings. Traditional analysis has generally relied upon spectrum analysis based on FFT or STFT[1-2]. Due to the disadvantages of the FFT, it is required to find supplementary methods for vibration signal analysis. Hitherto, time-frequency analysis is the most popular method for the analysis

* Foundation item: Project supported by the Shenyang High-Tech. R&D Program, China (Grant No. 1053084-2-02).

L. Jiao et al. (Eds.): ICNC 2006, Part I, LNCS 4221, pp. 321 – 330, 2006.

of non-stationary vibration signals, such as the Gabor transform [3], and the bilinear time-frequency representation [4].

In this study, the adaptive Gaussian chirplet spectrogram is used for an integrated time-frequency signature extraction of the rolling bearings vibration. And then Independent component analysis (ICA) is used for feature extraction, which transforms a high-dimensional input vector into a low-dimensional one whose components are independent.

ICA [5-7] is a novel statistical technique that aims at finding linear projections of the data that maximize their mutual independence. Its main applications are in feature extraction [8], and blind source separation (BSS) [9], with special emphasis to physiological data analysis [10], and audio signal processing [11]. Independent component analysis (ICA) is a technique that stems out from the Blind Source Separation, and has been widely used in fields like voice and medical signal analysis and image processing. Unlike principal component analysis that decorrelates multivariate random vectors, ICA steps much further to project the vector to the space where the items of the vector are mutually statistically independent. Based on the features extracted from the time-frequency moments using ICA method, the machine fault diagnoses are to be classified through the SOM network [12].

2 Adaptive Chirplet-Based Signal Decomposition

It is well understood that the chirp is one of the most important functions in nature, particularly, the so-called Gaussian chirplets [13]

$$g_k(t) = \sqrt[4]{\frac{\alpha_k}{\pi}} \exp\left\{ -\frac{\alpha_k}{2}(t-t_k)^2 + j.\left(\omega_k + \frac{\beta_k}{2}(t-t_k) \right)(t-t_k) \right\} \qquad (1)$$

$$t_k, \omega_k, \beta_k \in R, \qquad \alpha_k \in R^+$$

where the parameter (t_k, ω_k) determines the time and frequency center of the linear chirp function. The variance α_k controls the width of the chirp function. Compared with the Gaussian function used for the Gabor expansion [14], the Gaussian chirplet in [15] has more freedom and thereby can better match the signal under consideration. Note that the Wigner–Ville distribution of the Gaussian chirplet has a form

$$WVD_g(t,\omega) = 2\exp\left\{ -\left[\alpha_k (t-t_k)^2 + \alpha_k^{-1} \times \left(\omega - \omega_k - \beta_k (t-t_k) \right)^2 \right] \right\} \qquad (2)$$

which can be thought of as a joint time–frequency energy density function, describing the signal's energy distribution in the joint time-frequency domain. Since the Gaussian chirplet-type function is the only set of functions, whose Wigner-Ville distribution is non-negative, the Gaussian chirplet plays a unique role in the area of time-frequency analysis.

In this section, the adaptive signal decomposition scheme independently developed in [16-18] will be briefly reviewed. That is, for a given signal $s(t)$, first select a function $g_0(t)$ from a set of predefined atoms such that the distance between $s(t)$ and its orthogonal projection on $g_0(t)$ is minimum in the sense of

$$\min_{g_0} \|s_1(t)\|^2 = \min_{g_0} \|s_0(t) - \langle s_0(t), g_0(t)\rangle g_0(t)\|^2 \tag{3}$$

where $s_0(t)=s(t)$. The elementary function set $\{g_k(t)\}$ composed with chirplet function in (1), can sufficiently span the $L^2(R)$ space. It is easy to verify that (3) is equivalent to finding $g_0(t)$ that is most similar to $s_0(t)$, i.e.,

$$\max_{g_0} |\langle s_0(t), g_0(t)\rangle|^2 \tag{4}$$

Upon selected $g_0(t)$, we can compute $s_1(t)$, which is the distance of $s_0(t)$ and its orthogonal projection on $g_0(t)$, that is:

$$s_1(t) = s_0(t) - \langle s_0(t), g_0(t)\rangle g_0(t) = s_0(t) - A_0 g_0(t) \tag{5}$$

$s_1(t)$ can also be called as the residual component after the first decomposition. Repeating this processing, we can obtain:

$$s(t) = \sum_{k=0}^{K} A_k g_k(t) + s_{K+1}(t) = \sum_{k=0}^{K} v_k(t) + s_{K+1}(t) \tag{6}$$

Consequently, with the elementary function's joint time-frequency energy density in (2), its adaptive spectrogram (AD) is defined as:

$$AD_s(t,\omega) = \sum_{k=0}^{\infty} |A_k|^2 \, WVD_{g_k}(t,\omega) \tag{7}$$

One of the most challenging tasks of the adaptive decomposition and adaptive spectrogram is to efficiently compute the set of optimal elementary functions $g_k(t)$. If the elementary function is selected to be the chirplet as in (1), the maximization of (4) is a multi-dimensional and non-linear optimization problem.

The chirplet-based fast algorithm [19] is the analytical solution of chirplet-based adaptive signal decomposition. Its basic idea is to convert the optimization process to a traditional curve-fitting problem. By applying different testing points, we can obtain a group of equations. The solution of the set of equations is the parameters of the desired function $g_k(t)$. Consequently, the multidimensional optimization problem becomes the solution of a group of simultaneous equations in four variables: a classical curve-fitting problem. The detailed derivation can be found in [19].

3 Independent Component Analysis (ICA)

3.1 ICA Model

ICA is closely related to the blind source separation (BBS) problem, where the goal is to separate mutually independent but otherwise unknown source signals from their observed mixtures without knowing the mixing process[7]. ICA can be regarded as an extension of principle component analysis (PCA). ICA decorrelates higher-order statistics from the training signals, while PCA decorrelates up to second-order statistics only.

In the simplest form of ICA(see figure 1), the problem is to separate *M* statistically independent inputs which have been mixed linearly in *N* output channels without further knowledge about their distributions or dynamics. The relation between *x*(*t*) and *s*(*t*) can be modeled as

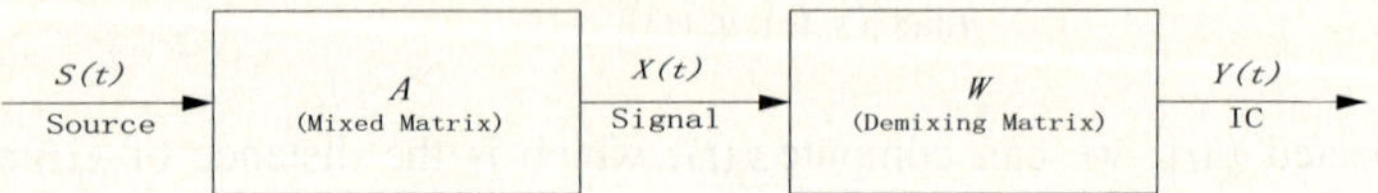

Fig. 1. The concept of ICA

$$x(t) = As(t) \tag{8}$$

where $x(t)=[x_1(t),x_2(t),...,x_N(t)]^T$ is an *m*-dimensional observed vector, $s(t)=[s_1(t),s_2(t),...,s_M(t)]^T$ is an unknown *n*-dimensional source vector containing the source signals $s_1(t)$, $s_2(t)$,…, $s_M(t)$, which are assumed to be statistically independent. Usually, the number of the sources is assumed to be smaller than or equal to the number of the observations, i.e. $n \le m$. *A* is an unknown $m \times n$ mixing matrix, which is written as

$$A = \begin{bmatrix} a_{11} & a_{12} & \cdots & a_{1n} \\ a_{21} & a_{22} & \cdots & a_{2n} \\ \cdots & \cdots & & \cdots \\ a_{m1} & a_{m2} & \cdots & a_{mn} \end{bmatrix} = [a_1| \ a_2| \ \dots \ a_n|] \tag{9}$$

where $a_i|, i=1,2,...,n$denotes the *i*th column of *A*. So Eq.(8) can be rewritten as

$$x(t) = As(t) \rightarrow \begin{bmatrix} x_1(t) \\ x_2(t) \\ \vdots \\ x_m(t) \end{bmatrix} = [a_1| \ a_2| \ \dots \ a_n|] \begin{bmatrix} s_1(t) \\ s_2(t) \\ \vdots \\ s_m(t) \end{bmatrix} \tag{10}$$

where, the columns of *A* represent features, and $s_i(t)$ is the coefficient of the *i*th feature in an observed data vector *x*(*t*). The aim of ICA is to find a linear transform denoted by *W*, which can be treated as an approximate pseudo-inverse of *A*, such that

$$y(t) = Wx(t) \tag{11}$$

where $y(t)=[y_1(t),y_2(t),...,y_n(t)]^T$ is the approximate estimate of the source signals, *W* is also called the ICA estimated basis, i.e. separating matrix. So an *m*-dimensional observed vector is transferred to an *n*-dimensional vector by the ICA transform. Owing to $m \ge n$, this implicates that the ICA transform primarily compresses the dimensions of the observation to some extent.

3.2 FastICA[20][21][22]

FastICA is a well known and studied ICA algorithm. The standard implementation uses an approximation of negentropy(which is less outlier prone than kurtosis) as non-Gaussianity measure (objective function), and a quasi-Newton method for optimization[21].To estimate a source y_i, the weight vector w(a row of W) is selected such that the negentropy of $w^T x$ is maximized. This follows from the additive central limit theorem, that an additive mixture of non-Gaussian distributions moves closer to Gaussian; thus, the w that creates the least Gaussian distribution must produce an independent source. The approximate negentropy of one source estimate is given by

$$J_G(w,x) = \left[E\{G(w^T x)\} - E\{G(v)\} \right]^2 \tag{12}$$

where G is a non-polynomial function, and v is a Gaussian variable of zero mean and unit variance. Proper selection of G is critical to the outlier robustness of the algorithm. The B-robustness of outliers is important[21]

$$G(w^T x) = -\frac{1}{a_2}\exp(-a_2 (w^T x)^2 / 2) \tag{13}$$

With $a_2 \approx 1$. Using a quasi-Newton method to find the optima of $E\{G(w^T x)\}$,and thus the maxima of the approximate negentropy, the FastICA algorithm as following.

1. Choose an initial weight vector.
2. $w^+ = E\{xG'(w^T x)\} - E\{G''(w^T x)\}w$
3. Let $w = w^+ / \|w^+\|$
4. If not converged, go back to Step 2.

As presented, this algorithm has estimated only one independent source. To estimate additional sources, methods based on orthogonalization of the weight vectors w of W are necessary. Source estimates maybe found one by one, or in parallel. The parallel method reduces the carry through of errors in estimates. A complete derivation of the FastICA algorithm is presented in [21,22].

4 Fault Detection Using ICA Based SOM

After the ICA processing, the conventional minimum Euclidean distance method is replaced by an SOM neural network to classify the fault diagnosis in our method. Furthermore, in order to improve the performance of the system, the original defect signals are preprocessed by lowpass filter and space sampling. The framework of ICA based SOM method is shown as Figure 2. The detailed algorithm is shown as follows:

1. The raw time signal is obtained by the accelerometer from the machinery fault simulator. Then the spectrogram gray images of the data are extracted through the adaptive Gaussian chirplet distribution.
2. Preprocess all the the adaptive Gaussian chirplet spectrogram images by mean filtering and space sampling. The computational load is reduced a lot, thus the speed of the algorithms is improved greatly.

3. Select some of the spectrogram images as the training samples. In order to enhance the robustness of the system, the mean of the selected images is used as the input.
4. Extract the features of the spectrogram images using FastICA method. Then a subspace is structured with these features.
5. Project all the images to the subspace and then the spectrogram images are replaced by the projection coefficients.
6. Train and recognize the projection coefficient by an SOM neural network.

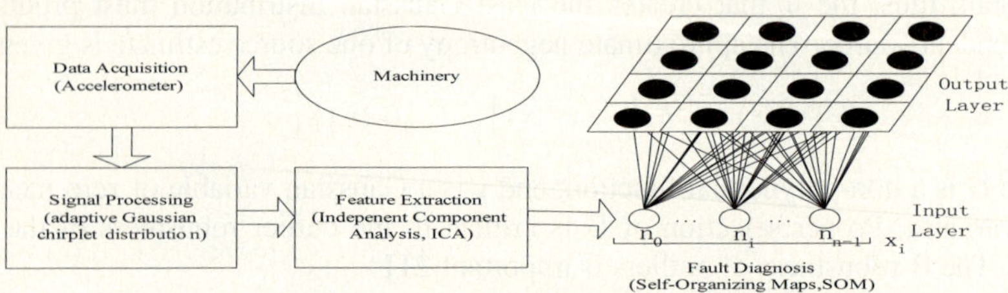

Fig. 2. Architecture of ICA-SOM fault diagnosis system

5 Experimental Case

Rolling bearings are installed in many kinds of machinery. A lot of problems of those machines may be caused by rolling bearings. Generally, local defects occur on outer-race, inner-race or rollers of bearings. When the rollers pass through the defect, an impulse may appear. According to the period of the impulse, we can judge the location of the defect using the spectrum analysis or time-frequency distribution method.

5.1 Signal Processing

The vibration signals of defective bearing is shown in Figures 3(a),Figure 4(a) and Figure 5(a), respectively. In order to assess the clarity in different bearings defects, the spectrum analysis is shown in Figures 3-5(a) and the time-frequency analysis is shown in Figures 3-5(b) for bearing with outer race defect, bearing with inner race defect and bearing with roller defect, respectively. The details analyses of outer race defect, inner race defect and roller defect are maken as follows, respectively.

5.1.1 Defect on the Outer Ring

In the case of a localised defect on the outer ring,successive impulses are produced when a ball carrying a portion of the radial load contacts with a local defect. The magnitude of the impulses depends on various parameters such as amplitude of the radial load,size of the defect,material and velocity. Figure 3 shows a set of the outer ring defect data, which were sampled at a speed of 1.6 kHz. The rotary speed is 1780 r.p.m. Compared with the spectrum(Figure 3(a)), the adaptive Gaussian chirplet distribution (Figure 3(b)) shows the components clearly, and the approximate time for the appearance of components is also shown. The comparison result indicates that the time-frequency method can reflect the inherent features of the outer ring defect signal and can provide information for the outer ring defect analysis and more useful features for fault diagnostics.

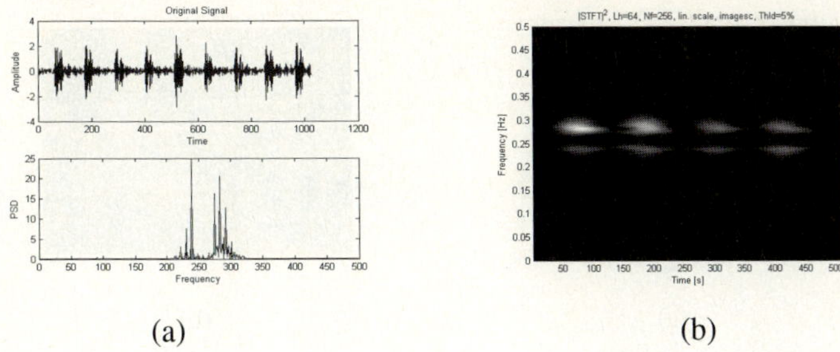

(a) (b)

Fig. 3. (a) vibrational signal and Spectrum of routrace (b) adaptive Gaussian chirplet distribution of outrace

5.1.2 Defect on the Inner Ring

Depending on the working conditions,a local defect may appear on the inner ring of a rolling element bearing. The rotational speed of the inner ring is greater than the cage speed and an impulse occurs when the defect on the inner ring strikes a ball. The inner ring defect produces periodic impulses if the defect is in the load zone. But,the magnitude of the impulse is different at every contact due to the changing ball load. In Fig. 4(a), a signal measured on a bearing with an inner race fault is shown. The measurement was made on a laboratory fault simulator. A ball bearing with an inner race fault was used. The rotation speed was 33 Hz and the sampling frequency 30 kHz. In Fig. 4(a, b), the spectrum and the adaptive Gaussian chirplet distribution of the signal is illustrated, respectively. This shows that the chirplet distribution analysis is more sensitive to impulsive oscillations, than the tradditional frequency spectrum.

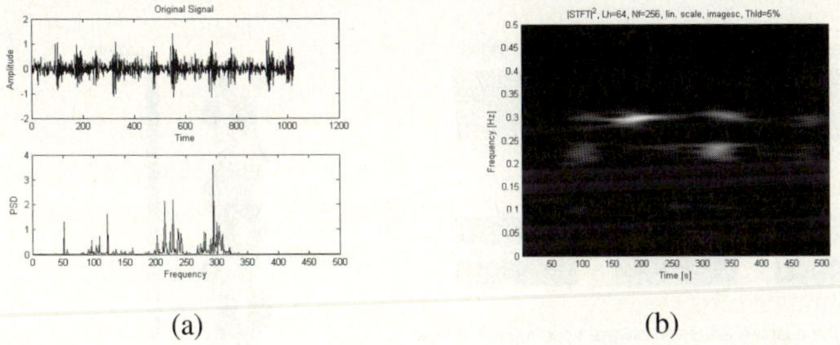

(a) (b)

Fig. 4. (a) vibrational signal and Spectrum of innerrace (b) adaptive Gaussian chirplet distribution of innerrace

5.1.3 Defect on the Rolling Element

The rolling element defect is another type of rolling element bearing failure. The rolling elements rotate with the cage about the bearing axis and spin about their own axis simultaneously. Therefore,a defect strikes both the inner ring and the outer ring. Compared with the spectrum, the adaptive Gaussian chirplet distribution (Figure 5) shows all components clearly, and the approximate time for the appearance of components is also shown. Through the adaptive Gaussian chirplet distribution, we can see the energy density is represented by different shades of the colour.

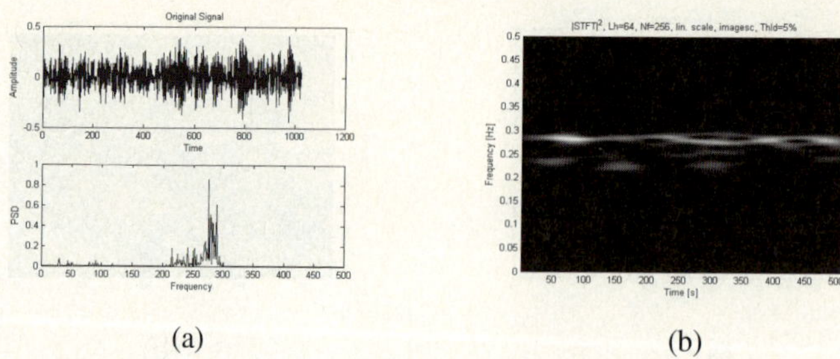

Fig. 5. (a) vibrational signal and Spectrum of rolling element (b) adaptive Gaussian chirplet distribution of rolling element

5.2 Feature Exteraction

There are 3 different defects in rolling bearings and each defect consists of 20 different defect samples. The length of one group of the sampling data is 512 points. Total 60 spectrogram views are acquired form these different defect samples. The first ten views for each type of defects are selected for training, and the rest for testing. The size of the adaptive Gaussian chirplet spectrogram images is 512×512. Some adaptive Gaussian chirplet spectrogram images in bearings are shown as Figure 6. In order to enhance the robustness of the system, a lowpass filter is used firstly. To decrease the computational load, the dimension of the adaptive Gaussian chirplet spectrogram images is reduced to a quarter of original by sampling secondly. Then the size of the adaptive Gaussian chirplet spectrogram images changes to 128×128. Figure 6 shows some basis of the adaptive Gaussian chirplet spectrogram images by FastICA method.

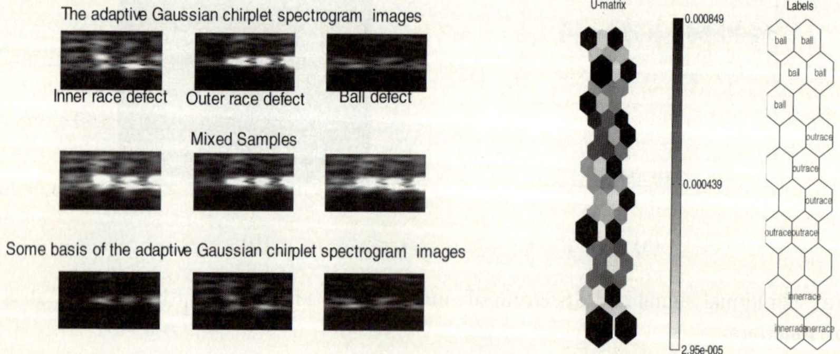

Fig. 6. Some adaptive Gaussian chirplet spectrogram and ICA basis images in bearings

Fig. 7. The U matrix describing the distances between adjacent SOM nodes and SOM tested with labeled data set

The input spectrogram image is projected into these ICA subspace. The projection is simply an inner product between the input spectrogram and each projection bases.

Thus the bearing defect can be identified by recognizing the projection coefficient using an SOM neural network.

5.3 Feature Classification

Based on the features extracted from the spectrogram moments using ICA method, the rolling bearings fault diagnoses were to be classified through the SOM network. The SOM was trained using the batch-training algorithm for each data set collected. In training the SOM, the selected 30 examples from the time-frequency moments using ICA method were used to train a SOM.

The unified distance matrix (u-matrix) is a simple and effective tool to show the possible cluster structure on SOM grid visualization. It shows the distances between neighbouring units using a gray scale representation on the map grid. The observations in the testing data set were labelled "innerrace", "outerrace" or "ball", corresponding to the rolling bearings fault components. The best matching units for these data were looked up on the SOM, and the program was instructed to place the labels of the test data on the corresponding best matching neurons. As shown in Figure 7, the three letters in the figure correspond to the according faults and the blanks denote the neurons that are not excited for all the input vectors. The map not only could figure out the fault modes but also indicate the similar samples. Because it indicates that the locations of two neurons in the output map will be close to each other if their fault symptoms are similar. It is an important topology-mapped characteristic of the SOM neural network. The monitoring system obtained a near 100% success rate in distinguishing among all test fault diagnosis.

6 Conclusions

The paper proposed a kind of rolling bearing fault diagnosis method based on the combination of the adaptive Gaussian chirplet distribution and independent component analysis. The adaptive Gaussian chirplet distribution has been applied to this bearing fault diagnosis system due to their advantages in the representation of signals in both the time-frequency domains. Based on the features extracted from the time-frequency moments using ICA method, the bearing fault diagnoses were to be classified through the SOM network. The experimental test results show that the new system can be effectively used in the diagnosis of various motor bearing faults through appropriate measurement and interpretation of motor bearing vibration signals.

References

1. Yu,H.B.,Guo,Q.J, Xu,A.D.: A self-constructing compensatory fuzzy wavelet network and Its applications. Springer LNAI.3613(2005)743-755.
2. Guo,Q.J, Yu,H.B,Xu,A,D.: Hybrid PSO based wavelet neural networks for intelligent fault diagnosis. Springer LNCS.3498(2005)521-530.
3. Bastiaans,M.J.:Gabor'expansion of a signal into Gaussian elementary signals. Proceedings of the IEEE.68(1980)538-539.
4. Cohen ,L.: Time-frequency distribution—a review. Proc. IEEE. vol.77, (1989)941–981.

5. Comon,P.: Independent component analysis—A new concept?. Signal Processing.36 (1994) 287–314.
6. Hyvärinen,A.: Survey on independent component analysis. Neura Computing Surveys. 2 (1999)94–128.
7. Jutten ,C.,Herault,J.: Blind separation of sources, part I: An adaptive algorithm based on neuromimetic architecture. Signal Processing.24 (1991)1–10.
8. Hurri,J., Hyvärinen,A.,et al.:Image feature extraction using independent component analysis. IEEE Nordic Signal Processing Symp, NORSIG'96. (1996)475–478.
9. Amari,S., Cichocki,A.:Adaptive blind signal processing—Neural network approaches. Proc. IEEE. 86(1998)2026–2048.
10. Vigário,R.:Extraction of ocular artifacts from EEG using independent component analysis, Electroenceph. Clin. Neurophysiol.. 103 (1997)395–404.
11. Torkkola,K.:Blind separation for audio signals: Are we there yet?, the Int.Workshop on Independent Component Analysis and Blind Separation of Signals (ICA'99), Aussois, France. (1999)239–244.
12. Kohonen,T.: Self-Organizing Maps, 3rd ed. Berlin, Germany:Springer, (2001)138–140.
13. Mann, Haykin,S.S.:The chirplet transform: Physical considerations, IEEE Trans.Signal Processing, vol. 43, (1995)2745–2461.
14. Gabor,D.:Theory of communication, in *J. Inst. Elect. Eng.*, vol. 93,(1946)429–457.
15. Mann,S. , Haykin,S.:'Chirplets' and 'warblets': Novel timefrequency methods, Electron. Lett.. 28 (1992)114–116.
16. Mallat ,S., Zhang,Z.: Matching pursuit with time-frequency dictionaries. IEEE Trans. Signal Processing. 41(1993)3397–3415.
17. Qian,S., Chen,D.:Signal representation in adaptive Gaussian functions and adaptive spectrogram, in Proc. Twenty-Seventh Annu. Conf. Inform. Sci. Syst.. (1993) 59–65.
18. Qian,S.,Chen, D.:Signal representation using adaptive normalized Gaussian functions, Signal Process..36(1)(1994)1–11.
19. Yin,Q., Qian,S., Feng,A.:A fast refinement for adaptive Gaussian chirplet decomposition, IEEE Trans. Signal Processing. 50(6)(2002)1298-1306.
20. Gadhok,N., Kinsner,W.:Estimating outlier impact on. FastICA using fuzzy inference, Conference of the North American Fuzzy Information Processing Society–NAFIPS. (2004) 832-837.
21. Hyvärinen,A.,Karhunen, A. J.:Oja,E.:Independent Component Analysis. New York, NY:John Wiley & Sons.2001.
22. Hyvärinen, A.,Oja, E.:A fast fixed-point algorithm for independent component analysis. Neural Comput. 9(7)(1997)1483–1492.

T-Test Model for Context Aware Classifier

Mi Young Nam, Battulga Bayarsaikhan, Suman Sedai, and Phill Kyu Rhee

Dept. of Computer Science & Engineering, Inha University
253, Yong-Hyun Dong, Incheon, Korea
{rera, battulga, suman}@im.inha.ac.kr, pkrhee@inha.ac.kr

Abstract. This paper proposes a t-test decision model for context aware classifier combination scheme based on the cascade of classifier selection and fusion. In the proposed scheme, system working environment is learned and the environmental context is identified. Best selection is applied to the environment context where one classifier strongly dominates the other. In the remaining context, fusion of multiple classifiers is applied. The decision of best selection or fusion is made using t-test decision model. Fusion methods namely Cosine based identify and Euclidian identify. In the proposed scheme, we are modeling for t-test based combination system. A group of classifiers are assigned to each environmental context in prior. Then the decision of fusion of more than one classifiers or selecting best classifier is made using proposed t-test decision model.

1 Introduction

Context can be various configurations, computing resource availability, dynamic task requirement, application condition, environmental condition, etc. [1]. Identified context describes a trigger an associated system action (combined classifier output) using a context stored in the context knowledge base over a period of time and/or the variation of a set of context data of the system over a period of time [1]. Context data is defined as any observable and relevant attributes, and its interaction with other entities and/or surrounding environment at an instance of time [2]. Classifier combination can be thought of as classifier selection and classifier fusion. During classifier selection, proper classifier that is most likely to produce accurate output for a local area of feature space is selected. Whereas In classifier fusion, individual classifiers are activated in parallel and group decision is made to combine the output of the classifiers. In real world analogy, classifier selection can be thought as choosing experts who are specialized in a particular field and classifier fusion can be thought as combining the decision of multiple experts to make the final decision.

Classifier selection methods are divided into static classifier selection and dynamic classifier selection methods [3, 4, 5].

The proposed method primarily aims at robust object recognition under uneven environments by cascading of selection and fusion of different classifiers using context awareness. The method makes the decision of whether to use multiple classifiers fusion or to use best single classifier using the proposed t-test decision model. The proposed method adopts the strategy of context knowledge accumulation so that classifier combination scheme is able to adapt itself under changing environments based on

L. Jiao et al. (Eds.): ICNC 2006, Part I, LNCS 4221, pp. 331 – 336, 2006.

experience. Context knowledge base consists of the knowledge of classifiers and there combination scheme for each environmental context. Once the context knowledge base is constructed; the system can react to changing environments at run-time. In session 2, we discuss general classifier fusion and proposed classifier and then session 3, we described experimental results of similarity measure.

2 Classification Selection Using T-Test Coefficient

In this section, we discuss about the model of context-aware classifier combination scheme with the capability of adaptation. The implementation issues of the proposed adaptive framework employing the context-awareness and the genetic algorithm will be discussed in the next section.

2.1 Exhaustive Classifier Combination

There is no general approach or theory for efficient classifier combination, yet [1]. Classifier combination can be thought as the generation of candidate classifiers and decision aggregation of candidate classifiers. For the simplicity of explanation, we assume that a classifier combination consists of four stages: preprocessing, feature representation, class decision, and aggregation stages. Each stage consists of several competitive and complementary components. An example of classifier combination scheme is given in Fig. 1.

Classifier combination scheme combines classifier components and produce the combined classifier system to achieve high performance. The model of classifier combination can be described as follows Fig 1. Fig. 1 shows that our proposed context based classification and classifier fusion scheme.

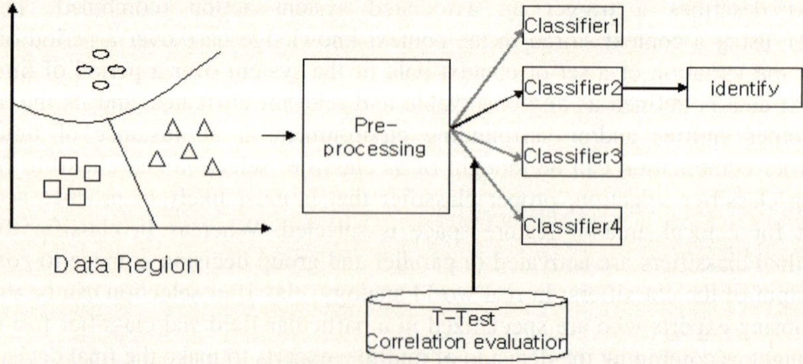

Fig. 1. Adaptive classifier combination scheme

In figure2, we decision classifier selection and fusion by using by t-test coefficient because face images have different weighted features. So we proposed t-test method and clustering method for face images including varying brightness. The selected most effective classifiers under an identified context are combined to produce a final output of the scheme using a traditional aggregation method.

In training step, we decide t-test coefficient and selected classifier and then in testing step, we would applied the classifier for face recognition. We assume that context data can be modeled as (clustered into) several contexts (context categories) in association with distinguishable application actions. The proposed scheme operates in two modes: the learning and action modes. The context learning is performed by an unsupervised learning method in the learning mode, and the context identification is implemented by a normal classification method in the action mode.

Our system consists of training step and testing step. In training step, we include face image clustering and classifier selecting each cluster and in testing step it includes data region classification and apples adaptive classifier.

2.2 Proposed Classifier Combination

In this session, classifier combination method following:

Classes: $w_1, w_2, \ldots, w_c$

Soft label Recognition Result of $D_A : t_{A1}, t_{A2}, \ldots, t_{Ac}$ D_A:

Product [8]: $\max(t_{Ai} \bullet t_{Bi})$

MV [8]: $\max(t_{Ai}, t_{Bi})$

Average [8]: $\max(\frac{t_{Ai} + t_{Bi}}{2})$

DT [2]: Second layer classifier.

D_{AB} is classifier after fusion.

Let e_{AB} is misclassification number of D_F .

a) We wants that $e_F < \min(e_A, e_B)$

b) In most case $e_F > \min(e_A, e_B)$, $e_F < \max(e_A, e_B)$

If $P_A > P_B$ => $P_A > P_F > P_B$

P_A is recognition rate for a single experiment using N number of images. P_A is ....not real recognition rate. So real recognition rate lies somewhere between $P_A + S_A$ and P_A - S_A by 95%. So our target is to find a classifier such that recognition rate is greater than P_A - S_A.

Classifier A: P_A - S_A is lower limit where sa is ... and classifier B is lowest

Classifier A is has highest recognition rate ie $P_A > P_B$.

Error rate of classifier F, we have to check whether to apply fusion of choose best one by the following condition C $\Rightarrow \Delta = |\hat{P_A} - \hat{P_B}| < \frac{3S_A}{2} - \frac{S_B}{2} \approx S_A$.

That gives the condition of better fusion, in order to increase the reliability by increasing lowest recognition rate of best classifier (P_A-S_A.) to P_F-S_F.

Kunchiva have applied a paired t-test to find whether the best classifier is different enough from rest in decision space [3].Here we try to estimate the fusion condition to achieve the reliability of the fusion system without much loss in accuracy. We want to increase the lower limit of the recognition rate of the fusion system than that of best classifier, hence achieving higher reliability. Classifier recognition rate is normally distributed with mean $\hat{P}_A$ and standard deviation δ_A and the distribution is denoted by $P_A \sim N(\hat{P}_A, \delta_A)$. Similarly for classifier distribution is $P_B \sim N(\hat{P}_B, \delta_B)$. Then distribution of recognition rate after fusion is $P_F \sim N(\hat{P}_F, \delta_F)$. If $\hat{P}_F = a\hat{P}_A + b\hat{P}_B$ such that a+b=1.

In the figure SA=2δA, $S_B=2\delta_B$ and $S_F= 2\delta F$ such that (P_A-S_A), (P_B-S_B) and (P_F-S_F) denotes the lower limit of the 95% confidence interval in respective distribution.

Here we are interested to find the gap $\Delta = \hat{P}_A - \hat{P}_B$ between the best classifier and second best classifiers recognition rate.

And assuming that standard deviation of P_A and P_B similar, i.e. $P_A \approx P_B$

$$\Delta < 0.6 S_A \tag{1}$$

S_A can be calculated from N number of observations of recognition rate as follows.

$$S_A = t_{(0.05, N-1)} \sqrt{\frac{\hat{P}_A(1-\hat{P}_A)}{N}} \tag{2}$$

3 Experiment on Face Recognition

Single image is used for registration of each people in training set containing 1196 images of FERET fbset. We used as the intermediate test images to perform t-test test to train classifier combiner. Face images are divided several region, we called context. We are comparison with six and nine clusters. We used Euclidian distance and Mahalanobis distance for classifier identify. Context modeling is implemented by K-means [6, 7]. Histogram equalization (HE) is used for preprocessing components and Gabor wavelet is used as feature representation [6]. Gabor wavelet shows desirable characteristics in orientation selectivity and special locality. Three classifiers namely Gabor 13, Gabor28 and Gabor32 are used as candidate classifier for each region. These classifiers are trained using enrollment images. Knowledge of classifier combination scheme for each context is again learned by t-test decision model.

It shows that t-test decides one of best selection or fusion for each context for reliable fusion. Again the result shows that, we get reliable fusion, while the fusion result approaches the best selection. In the case t-test decided for fusion see context model label 0, 1, 2, 3, 4 and 5, recognition rate for all the fusion strategies gives almost similar result. Since, it was stated that fusion method is not a primary factor when the ensemble contains diverse set of classifiers [1]. This gives insight that our classifier selection method has chosen optimal set of classifiers for each context.

Table 1 and 2 show the result of six identify measure experiments along with the distance of each combination strategy for 6 clusters. In Table 2, we used feature extraction method, Gabor 28. Table 3 and 4 show a recognition rate of proposed t-test

Table 1. Shows cosine distance comparison result over six experiments conducted on classifiers using t-test coefficient 0.95

95% CI Cosine	Cluster 0	Cluster 1	Cluster 2	Cluster 3	Cluster 4	Cluster 5
Cluster 0	0.029541	0.031991	0.031203	0.031991	0.031203	0.031203
Cluster 1	0.033771	0.033771	0.031694	0.035701	0.029438	0.033771
Cluster 2	0.038253	0.038253	0.033976	0.038253	0.038253	0.040178
Cluster 3	0.037049	0.037049	0.037049	0.032899	0.037049	0.038918
Cluster 4	0.039625	0.041108	0.043865	0.038062	0.036408	0.042519
Cluster 5	0.041319	0.043007	0.042177	0.041319	0.043007	0.041319

Table 2. Shows cosine distance comparison result over six experiments conducted on classifiers using t-test coefficient 0.95

95% CI Euclidian	Cluster 0	Cluster 1	Cluster 2	Cluster 3	Cluster 4	Cluster 5
Cluster 0	0.037489	0.039258	0.037489	0.03623	0.038683	0.031991
Cluster 1	0.039207	0.043784	0.046485	0.043784	0.047749	0.037508
Cluster 2	0.05233	0.055872	0.048314	0.054738	0.058013	0.048314
Cluster 3	0.040676	0.048192	0.042336	0.035051	0.042336	0.040676
Cluster 4	0.036408	0.043865	0.038062	0.039625	0.036408	0.046386
Cluster 5	0.048169	0.051285	0.048169	0.043812	0.048826	0.038563

Table 3. Classifier Selection using Propose t-test (measure – cosine distance)

Cosine Distance	Gabor13	Gabor28	Gabor32
Cluster 0	0.938596	0.938596	0.938596
Cluster 1	0.955414	0.968153	0.955414
Cluster 2	0.954198	0.931298	0.938931
Cluster 3	0.939189	0.918919	0.918919
Cluster 4	0.961783	0.929936	0.974522
Cluster 5	0.898618	0.898618	0.894009

method, t-test coefficient is 0.95. In experimental result, we know that our proposed method is better the other approach.

Each classifiers output with highest performance, proposed t-test classifier selection is highest another approaches [3, 4].

Table 4. Classifier Selection using Propose t-test (measure – cosine distance)

Euclidian Distance	Gabor13	Gabor28	Gabor32
Cluster 0	0.907895	0.921053	0.934211
Cluster 1	0.929936	0.942675	0.968153
Cluster 2	0.900763	0.900763	0.923664
Cluster 3	0.945946	0.918919	0.959459
Cluster 4	0.936306	0.904459	0.942675
Cluster 5	0.907834	0.894009	0.917051

4 Conclusion

The classifier fusion method tend to increase reliability of recognition rate of the system, by compromising on the minor decrease if accuracy. We trained and examined the system for three context models, in each model the number of context clusters were 6 and 9. Since no single classifier is best for all environmental contexts we choose the best classifiers for each context. The best classifier selection used proposed t-test methods, and face recognition performance is highest another algorithm. Face classifier's identify measure is important, because similarity measure is classifier selecting main cause. Cosine distance is highest performance.

References

1. S. Yau, Y. Wang, and F. Karim "Developing Situation-Awareness in Middleware for Ubicomp Environments," *Proc. 26th Int'l Computer Software and Applications Conference(COMPSAC 2002)*. 233-238..
2. S. Yau, F. Karim, Y. Wang, B. Wang, and S. Gupta, Reconfigurable Context-Sensitive Middleware for Pervasive Computing, IEEE Pervasive Computing, Vol.1, No.3, (2002) 33-40.
3. Ludmila I. Kuncheva, Switching Between Selection and Fusion in Combining Classifiers: An Experiment, IEEE Transactions on Systems, Man, and Cybernetics - part B: cybernetics, Vol.32, Vo.2, (2002) 146-156.
4. Ludmila I. Kuncheva, ,A Theoretical Study on Six Classifier Fusion Strategies IEEE S on PAMI, Vol. 24, No. 2, (2002).
5. Y.S. Huang and C.Y. Suen, "A Method of Combining Multiple Classifiers—A Neural Network Approach," Proc. 12th Int'l Conf. Pattern Recognition.
6. Y. Nam and P.K. Rhee, An Efficient Face Recognition for Variant Illumination Condition, ISPACS2005, Vol.1, (2004) 111-115.
7. M.Y. Nam and P.K. Rhee, A Novel Image Preprocessing by Evolvable Neural Network, LNAI3214, Vol.3, (2004) 843-854.
8. Kuncheva L.I. and L.C. Jain, Designing classifier fusion systems by genetic algorithms, IEEE Transactions on Evolutionary Computation, Vol.4, No.4, (2000) 327-336.

Face Recognition Using Probabilistic Two-Dimensional Principal Component Analysis and Its Mixture Model*

Haixian Wang[1] and Zilan Hu[2]

[1] Research Center for Learning Science, Southeast University, Nanjing, Jiangsu 210096, P.R. China
hxwang@seu.edu.cn

[2] School of Mathematics and Physics, Anhui University of Technology, Maanshan, Anhui 243002, P.R. China
hu_107@hotmail.com

Abstract. In this paper, by supposing a parametric Gaussian distribution over the *image space* (spanned by the row vectors of 2D image matrices) and a spherical Gaussian noise model for the image, we endow the two-dimensional principal component analysis (2DPCA) with a probabilistic framework called probabilistic 2DPCA (P2DPCA), which is robust to noise. Further, by using the probabilistic perspective of P2DPCA, we extend P2DPCA to a mixture of local P2DPCA models (MP2DPCA). MP2DPCA offers us a method of being able to model faces in unconstrained (complex) environment with possibly large variation. The model parameters could be fitted on the basis of maximum likelihood (ML) estimation via the expectation maximization (EM) algorithm. The experimental recognition results on UMIST face database confirm the effectivity of the proposed methods.

1 Introduction

The importance of research on face recognition (FR) is driven by both its wide range of potential applications and scientific challenges. And the appearance-based paradigm using the principal component analysis (PCA) [1], also known as Karhunen-Loéve transformation, to extract features has exhibited great advantage, producing the well-known *eigenfaces* method [2]. Now, the eigenfaces method has become one of the most successful approaches in FR (to see Ref. [3], for an early survey of FR). However, it could be noted that eigenfaces approach is vector-oriented; and thus it is difficult to handle the (possibly singular) large covariance matrix because of the high dimension and relatively small training sample size. Recently, to attack the problem, Yang et al. [4] proposed the two-dimensional principal component analysis (2DPCA), which was directly based

* This work was partly funded by National Natural Science Foundation of China (Grant No. 10571001, 60503023, and 60375010), and partly by Jiangsu Natural Science Foundation (Grant No. BK2005407) and Program for New Century Excellent Talents in University (Grant No. NCET-05-0467).

L. Jiao et al. (Eds.): ICNC 2006, Part I, LNCS 4221, pp. 337–340, 2006.

on 2D image matrices rather than 1D vectors. Following the introduction of 2DPCA immediately, Wang et al. [5] showed the equivalence of 2DPCA to line-based PCA. And the generalization of 2DPCA to bilateral and kernel-based versions were also presented [6]. The 2DPCA (-based) method is appealing, since the 2D spatial information of the image is well preserved and the computational complexity is significantly reduced compared with the PCA. Moreover, 2DPCA could naturally and effectively avoid the *small sample size* (SSS) problem.

Like PCA, one flaw of 2DPCA model, however, is the absence of an associated probability density; that is, the process of implementing 2DPCA is *distribution-free* — a mathematical method with no underlying statistical model. The proposal of probabilistic PCA (PPCA) [7], which defines PCA from the probabilistic perspective, aims to remedy the disadvantage of PCA. In this paper, by supposing a parametric Gaussian distribution over the *image space* (spanned by the row vectors of 2D image matrices) and a spherical Gaussian noise model for the image, we endow the 2DPCA with a probabilistic framework, which we will refer to as probabilistic 2DPCA (P2DPCA). Such a probabilistic formulation enables the application of Bayesian methods. And on the basis of maximum likelihood (ML) estimation, the estimates of model parameters could be implemented via the expectation maximization (EM) algorithm [8]. Besides, the probability model could fit the noise in the data whereas 2DPCA is not robust to independent noise. Further, whereas P2DPCA only defines a global projection of the sample vectors in the image space, we extend P2DPCA to a mixture of local P2DPCA models (MP2DPCA). We adopt P2DPCA and MP2DPCA as mechanisms for extracting facial information followed by discrimination. The experimental results demonstrate their superiority over 2DPCA.

2 Probabilistic Model for 2DPCA

Let $\mathbf{X}_1, \ldots, \mathbf{X}_n$ be a set of observed $s \times t$ image samples. For the sake of simplicity, we made the assumption that the image samples are already mean centered (which is easy to achieve). And we denote the kth row vector of the ith (mean-centered) image sample $\mathbf{X}_i$ as $\mathbf{X}_i^k$ for $k = 1, \ldots, s, i = 1, \ldots, n$; namely, $\mathbf{X}_i^{\mathsf{T}} = (\mathbf{X}_i^1, \ldots, \mathbf{X}_i^s)$. Then, we have the following

Proposition 1. *Regarding $\boldsymbol{X}_i^k$, $k = 1, \ldots, s, i = 1, \ldots, n$, as t-dimensional observations of size sn, and then applying the model of PPCA to these samples:*

$$\boldsymbol{X}_i^k = \Lambda F + \varepsilon, \; for \; k = 1, \ldots, s, i = 1, \ldots, n, \tag{1}$$

in which Λ is a $t \times q$ factor loading matrix, *F is q-dimensional latent (unobservable) variables (also known as* common factors*), and ε is t-dimensional* specific factors. *This generative model assumes that the latent variables are independently and identically distributed (i.i.d.) as Gaussian with zero-mean and identity covariance, the zero mean specific factors are also i.i.d. as Gaussian with covariance σ^2, and F is independent with ε and their joint distribution is Gaussian. Then one has that the columns of the ML estimator of Λ span the principal subspace as in the 2DPCA.*

From the probabilistic perspective, it is natural to adopt the posterior mean $E[F|\mathbf{X}^k]$ (an estimation of factor scores) as the reduced-dimensionality transformation for $\mathbf{X}^k$, which is the kth row vector of an image $\mathbf{X}$. Specifically, in the model of P2DPCA, the posterior mean becomes $\beta(\mathbf{X}^k) = (\hat{\Lambda}^\mathsf{T}\hat{\Lambda}+\hat{\sigma}^2 I_q)^{-1}\hat{\Lambda}^\mathsf{T}\mathbf{X}^k$, $k = 1, \ldots, s$. So, for an observed image $\mathbf{X}$, the reduced-dimensionality representation, that is an $s \times q$ feature matrix, is given by $Z = (\beta(\mathbf{X}^1), \ldots, \beta(\mathbf{X}^s))^\mathsf{T} = \mathbf{X}\hat{\Lambda}(\hat{\Lambda}^\mathsf{T}\hat{\Lambda} + \hat{\sigma}^2 I_q)^{-1}$. As a result, the reconstructed image of $\mathbf{X}$ could be $\hat{\mathbf{X}} = (\hat{\Lambda}\beta(\mathbf{X}^1), \ldots, \hat{\Lambda}\beta(\mathbf{X}^s))^\mathsf{T} + \bar{\mathbf{X}} = Z\hat{\Lambda}^\mathsf{T} + \bar{\mathbf{X}}$. It can be seen that when $\hat{\sigma}^2 \to 0$, the rows of Z represent projections onto the latent space and the conventional 2DPCA is recovered.

The P2DPCA (and P2DPCA) is, essentially, a linear model for data representation in a low dimensional subspace. It may be insufficient to model data with large variation caused by, for example, pose, expression and lighting. An alternative choice is to model the complex manifold with a mixture of local linear sub-models from the probabilistic formulation of P2DPCA. For $j = 1, \ldots, s; i = 1, \ldots, n$, we suppose that the t-dimensional samples $\mathbf{X}_i^j$ are generated independently from a mixture of g underlying populations with unknown proportion $\pi_1, \ldots, \pi_g$

$$\mathbf{X}_i^j = M_k + \Lambda_k F_k + \varepsilon_k, \text{ with probablilty } \pi_k \; (k = 1, \ldots, g), \tag{2}$$

where M_k are t-dimensional mean vectors, and π_k are mixing proportions with $\pi_k > 0$ and $\sum_{k=1}^{g} = 1$. All the model parameters could be estimated by using the EM algorithm, which is omitted here for limited space. We use the most appropriate local P2DPCA for a given sample in terms of the fitted posterior probability.

3 Experiments

In this section, we compare the recognition performances of 2DPCA, P2DPCA and MP2DPCA on the UMIST face image database [9]. The three methods are used for extracting features of facial images from the training samples, respectively, and then a *nearest-neighbor classifier* is used to find the most-similar face from the training samples for a queried face. The measure of distance between two feature matrices is defined as the sum of s Euclidean distances between each corresponding row of the two matrices. We adopt two strategies to evaluate the performances of the three methods. In the first series of experiment the *interpolation* performance is tested by training on a subset of the available views and testing on the intermediate views. The second series of experiments test on the *extrapolation* performance by training on a range of views and testing on novel views outside the training range. We select 10 images per person for training and another 10 images per person for testing. The experiment is run 10 times, and the average recognition rates obtained are recorded, as shown in Table 1. From Table 1, it could be seen that the three methods achieve high recognition rate in the case of interpolation. While in the case of extrapolation, the MP2DPCA method obtains the best recognition rate among the three systems. The reason may be that, when there is large pose variation, different pose may appear more separated by using the local subspace. And MP2DPCA is well suited to model such complex facial images. In this experiment, 2DPCA is a benchmark for evaluation, since for FR domain 2DPCA works better than many other methods, including Eigenfaces, Fisherfaces, ICA, and kernel Eigenfaces, in terms of recognition rate [4]. In this experiment, the best interpolation and extrapolation recognitions of PCA are 87.5%, 56.5% respectively.

Table 1. Interpolation and extrapolation performances of 2DPCA, P2DPCA and MP2DPCA with varying q on the UMIST database (%), where in MP2DPCA g takes the values $2, 3, 4, 5$ respectively

	interpolation						extrapolation					
q	2DPCA	P2DPCA	MP2DPCA				2DPCA	P2DPCA	MP2DPCA			
2	98.5	98.5	99.0	98.0	99.0	96.0	**73.0**	**73.0**	76.0	76.0	**79.0**	79.0
4	**100**	**100**	**100**	100	99.0	100	68.0	68.0	76.0	70.0	73.0	70.5
6	99.5	99.5	100	100	100	99.5	60.5	62.5	66.5	65.5	61.5	70.0
8	99.5	100	100	100	100	100	55.5	59.0	62.0	61.5	64.5	65.0

References

1. Jolliffe, I.T.: Principal Component Analysis. Springer-Verlag, New York (1986)
2. Turk, M., Pentland, A.: Eigenfaces for recognition. J. Cogn. Neurosci. **3** (1991) 71–86
3. Zhao, W., Chellappa, R., Rosenfeld, A., Phillips, P.J.: Face recognition: a literature survey. Technical Report CAR-TR-948. Vision Technologies Lab, Sarnoff Corporation, Princeton (2000)
4. Yang, J., Zhang, D., Frangi, A.F., Yang, J.y.: Two-dimensional PCA: a new approach to appearance-based face representation and recognition. IEEE Trans. Patt. Anal. Mach. Intell. **26** (2004) 1–7
5. Wang, L., Wang, X., Zhang, X., Feng, J.: The equivalence of two-dimensional PCA to line-based PCA. Pattern Recognition Lett. **26** (2005) 57–60
6. Kong, H., Wang, L., Teoh, E.K., Li, X., Wang, J.G., Venkateswarlu, R.: Generalized 2D principal component analysis for face image representation and recognition. Neural Networks **18** (2005) 585–594
7. Tipping, M.E., Bishop, C.M.: Mixtures of probabilistic principal component analysers. Neural Comput. **11** (1999) 443–482
8. McLachlan, G.J., Krishnan, T.: The EM Algorithm and Extensions. Wiley, New York (1997)
9. Graham, D.B., Allinson, N.M.: Characterizing virtual eigensignatures for general purpose face recognition. In: Wechsler, H., Phillips, P.J., Bruce, V., Fogelman-Soulie, F., Huang, T.S. (eds.): Face Recognition: From Theory to Applications, NATO ASI Series F, Computer and Systems Sciences, Vol. 163. (1998) 446–456

A Hybrid Bayesian Optimal Classifier Based on Neuro-fuzzy Logic

Hongsheng Su[1], Qunzhan Li[1], and Jianwu Dang[2]

[1] School of Electrical Engineering, Southwest Jiaotong University
Chengdu 610031, China
shsen@163.com

[2] School of Information and Electrical Engineering, Lanzhou Jiaotong University
Lanzhou 730070, China
dangjw@mail.lzjtu.cn

Abstract. Based on neural networks and fuzzy set theory, a hybrid Bayesian optimal classifier is proposed in the paper. It can implement fuzzy operation, and generate learning behaviour. The model firstly applies fuzzy membership function of the observed information to establish the posterior probabilities of original assumptions in Bayesian classification space, the classified results of all input information then are worked out. Across the calculation, the positive and reverse instances of all observed information are fully considered. The best classification result is acquired by incorporating with all possible classification results. The whole classifier adopts a hybrid four-layer forward neural network to implement. Fuzzy operations of input information are performed using fuzzy logic neurons. The investigation indicates that the proposed method expands the application scope and classification precision of Bayesian optimal classifier, and is an ideal patter classifier. In the end, an experiment in transformer insulation fault diagnosis shows the effectiveness of the method.

1 Introduction

Bayesian optimal classifier makes the likelihood of a new instance to be correctly classified up to maximum by incorporating with the posterior probabilities of all assumptions under the same assumption space and the observed data as well as the prior probabilities [1]. Bayesian optimal classifier can achieve the best classification result from the given data set, but the count quantities of the algorithm are very large. This mainly is because it need calculate the posterior probability of each hypothesis and incorporate with them to predict the likely classification of a new instance. An alternative suboptimal algorithm is called Gibbs algorithm [2], the expected error rate of the algorithm is twice than one of Bayesian optimal classifier at most under certain conditions [3]. However, due to the fuzziness of the observed data and the uncertainties of the prior probabilities of the assumptions, it is difficult to apply Bayesian optimal classifier to statistically infer, e.g., fault sources and fault symptoms information in transformer fault diagnosis [4]. To tackle the problem, fuzzy set is applied to dispose fuzzy symptoms information [5], [6], [7], but the classification errors generated by these algorithms are larger than one of Bayesian optimal classifier, and they can't implement automatic computation. Considering it, in this paper a hybrid Bayesian

L. Jiao et al. (Eds.): ICNC 2006, Part I, LNCS 4221, pp. 341 – 350, 2006.

optimal classifier based on neuro-fuzzy logic is proposed. It can perform automatic inference and dispose fuzzy data, and also, the count cost of the algorithm is for that reduced. Moreover, the positive and reverse instances of all observed information are considered, simultaneously. Thus, the classification precision of the algorithm is improved and the complexity of the algorithm is reduced, greatly.

The following sections would include fuzzy set foundation; Bayesian optimal classifier; hybrid Bayesian optimal classifier; neural networks model of Bayesian optimal classifier; practical example and conclusions, etc

2 Neuro-fuzzy Logic

2.1 Fuzzy Set Foundation

Fuzzy set is proposed by L. Zadeh in 1965[8], which can interpret and apply fuzzy information to infer. In fuzzy set theory, fuzzy membership function is used to describe the fuzziness of fuzzy information, and fuzzy algorithms are used to perform fuzzy operation.

Set V is an objects space, $\forall x\in V$, $A\subseteq V$. To study whether x belongs to A or not, a characteristic function $u_A(x)$ is defined to x, thus x, together with $u_A(x)$, constitutes a sequence couple $[x,\ u_A(x)]$. Fuzzy subset A in V may for that be defined as $A=\{x,u_A(x)|x\in V\}$, $u_A(x)$ is defined as fuzzy membership function of x to A, and $u_A(x)\in[0,1]$.

Let A an B respectively represent two fuzzy subsets in V, $u_A(x)$ and $u_B(x)$ respectively express their fuzzy membership functions, thus the basic fuzzy algorithms may be described as follows.

$$u_{A\cup B}(x)=\max[u_A(x),\ u_B(x)] = u_A(x) \vee u_A(x),\forall x \in V. \tag{1}$$

$$u_{A\cap B}(x)=\min[u_A(x),\ u_B(x)]= u_A(x) \wedge u_A(x),\forall x \in V. \tag{2}$$

$$u_{\bar{A}}(x)=1- u_A(x),\forall x \in V\ . \tag{3}$$

2.2 Logic Neuron

Logic neuron is proposed by W. Pedrycz in 1993[9], there are two types of neuron models, one is aggregative logic neuron (ALN), and another is referential logic-based neuron (RLN). As ALN can be conveniently used to describe control rules, we therefore mostly present it here. ALN has two types of models, one is OR model, another is AND model, they generate diverse logic operations, respectively.

2.2.1 OR Neuron

OR neuron refers to that each input signal as well as relevant weighted coefficient implements logical multiplicative operation, all the results then are performed a locally additive operation. Its mathematical model is expressed by

$$y=\mathrm{OR}(\boldsymbol{X};\ \boldsymbol{W}). \tag{4}$$

where y is the output of OR neuron, $\boldsymbol{X}$ is the input of OR neuron, $\boldsymbol{X}=\{x_1,x_2,\ldots,x_n\}$, $\boldsymbol{W}$ is the connection weight vector, $\boldsymbol{W}=\{\omega_1,\omega_2,\ldots,\omega_n\}$, $\omega_i\in[0,1]$, i=0,1,…,n. Hence, equation (**4**) may be described by y=OR[x_1 and ω_1, x_2 and ω_2, … , x_n and ω_n], or

$$y = \underset{i=1}{\overset{n}{\vee}}[x_i \wedge \omega_i]. \tag{5}$$

2.2.2 AND Neuron

AND neuron refers to that each input signal as well as relevant weighted coefficient implements logically additive operation, all the results then are performed a local multiplicative operation. Its mathematical model is expressed by

$$y=\text{AND}(\boldsymbol{X};\ \boldsymbol{W}). \tag{6}$$

where y is the output of AND neuron, $\boldsymbol{X}$ is the input of AND neuron, $\boldsymbol{X}$ = $\{x_1,x_2,\ldots,x_n\}$, $\boldsymbol{W}$ is the weight vector, $\boldsymbol{W}=\{\omega_1,\omega_2,\ldots,\omega_n\}$, $\omega_i\in[0,1]$, i=0,1,…,n. Hence, equation (**6**) may be described by y=AND[x_1 or ω_1, x_2 or ω_2, … , x_n], or

$$y = \underset{i=1}{\overset{n}{\wedge}}[x_i \vee \omega_i]. \tag{7}$$

Also, Logical processor (LP) composed of two types of logical neurons may realize more complex functions. LP adopts Delta law as its learning algorithm [1].

3 Bayesian Optimal Classifier

3.1 Bayesian Law

Set $P(h)$ is prior probability of the hypothesis h, $h\in H$, H is the hypothesis space, $P(D)$ denotes the prior probability of the observed data D, $P(D/h)$ specifies the likelihood of D when h occurrence, conversely, $P(h/D)$ specifies the likelihood of h occurrence while D is observed. $P(h/D)$ is the posterior probability of h, it reflects the influence of the observed data D to h. Thus, Bayesian law may be described as follows.

$$P(h/D) = \frac{P(D \mid h)P(h)}{P(D)}. \tag{8}$$

Since D is a constant independently of h, then

$$P(h/D) \propto P(D \mid h)P(h). \tag{9}$$

Thus, while a new instance D occurs, the most possible classification $h\in H$ on it is called maximum posteriori (MAP) hypothesis. h_{MAP} can be called as MAP hypothesis only when the following formula holds.

$$h_{\text{MAP}} \leftarrow \underset{h\in H}{\operatorname{argmax}}\, P(h/D). \tag{10}$$

3.2 Bayesian Optimal Classifier

In the former sections, we discuss which one is its most possible assumption while the observed data *D* is given. Actually, another more interesting problem with more close relations with it is which one is the most possible classification while *D* occurs. For the latter, we may simply use MAP hypothesis to get possible classification of a new instance, that is,

$$c_{\mathrm{MAP}} = \arg\max_{c\in C} P(C / h_{\mathrm{MAP}}) \,. \tag{11}$$

In (11), *C* is the classification space of the new instance; *c* is the possible classification of the new instance, $c\in C$; c_{MAP} is the most possible classification. However, we still have better algorithm, i.e., Bayesian optimal classifier.

In general, the most likely classification of the new instance may be gained by incorporating with the predictions of all assumptions. Let $P(c_j/D)$ express the probability that the new instance *h* is correctly classified, then

$$P\ (c_j \mid D) = \sum_{h_i\in H} P(c_j \mid h_i) P(h_i \mid D) \tag{12}$$

Then, the optimal classification of the new instance is c_j because it makes $P(c_j/D)$ up to the maximum, namely,

$$\arg\max_{c_j\in C} \sum_{h_i\in H} P(c_j \mid h_i) P(h_i \mid D) \cdot \tag{13}$$

The classification system generated by formula (13) is called as Bayesian optimal classifier. Under similar instances such as prior probability and hypothesis space, no any other methods can do better than it. According to (9), Bayesian optimal classifier may be described as

$$\arg\max_{c_j\in C} \sum_{h_i\in H} P(c_j \mid h_i) P(D \mid h_i) P(h_i) \,. \tag{14}$$

4 Fuzzy Bayesian Optimal Classifier

In (14), hypothesis *P*(*h*) as well as its support degree to *D* is usually fuzzy and indeterminate, the exact descriptions to them often are extremely difficult. To deal with the flaw, fuzzy set is applied to acquire fuzzy knowledge, that is, fuzzy membership function $u_h(D)$ is used to replace $P(h_i|D)$ in (13), that is,

$$\arg\max_{c_j\in C} \sum_{h_i\in H} P(c_j \mid h_i) u_{hi}(D) \,. \tag{15}$$

where $u_h(D)$ may be understood as under the observed information *D*, it can be interpreted as the probability of the known information type *h*. Clearly, it is fully consistent with $P(h_i|D)$ in numerical value.

5 Improved Bayesian Optimal Classifier

In Bayesian optimal classifier, only the positive factor of a new instance is considered, while the reverse one isn't considered. Hence, Bayesian optimal classifier still can be improved, further.

Set two sets ***A*** and ***B*** are associated by a number u, and defined as follows.

$$u=a+bi \tag{16}$$

where a expresses the consistency of the two sets to uniform solution, and b expresses the conflict to uniform solution, $a,b \in [0,1]$, and they must be probabilistically complete, that is , $a+b=1$, $i\equiv$-1. After the positive and reverse sets are weighted by (α, β), We get

$$u=\alpha a+\beta bi \tag{17}$$

where $\alpha+\beta=1$.

Set w_{ij} is the connecting strength from new instance i to hypothesis j, then the possible classifications of the new instance is

$$d_j=\alpha\sum_{i=1}^{n} w_{ij}a_i+\beta\sum_{i=1}^{n} w_{ij}b_i \tag{18}$$

Compared with (15), a_i and b_i may be respectively understood as a probability whether the new instance belongs to hypothesis h_i and not. Thus Bayesian optimal classifier searches for the maximum d_i of new instance in the classification space D, and described as

$$\underset{d_i\in D}{\arg\max}\, d_i \tag{19}$$

6 Neural Networks Model of Bayesian Optimal Classifier

According to (18) and (19), a Bayesian optimal classifier is proposed based on neuro-fuzzy logic. It not only can dispose fuzzy information, but also implement parallel diagnosis like neural networks. Hence, with the aid of neuro-fuzzy logic, Bayesian optimal classifier extends its capabilities. Moreover, it stores fielded knowledge in the connecting weights of the networks, and learns and updates them on and on. This reduces the count cost of Bayesian optimal classifier, dramatically. The structure of the model is shown in Fig.1.

In Fig.1, the upper part of the network realizes the class results of all positive instances, and the lower part of the network realizes the class results of all reverse instances, the two parts are incorporated with by weighted average in output-layer. Learning algorithms of the network comprise the two parts, one is Delta law for LP,

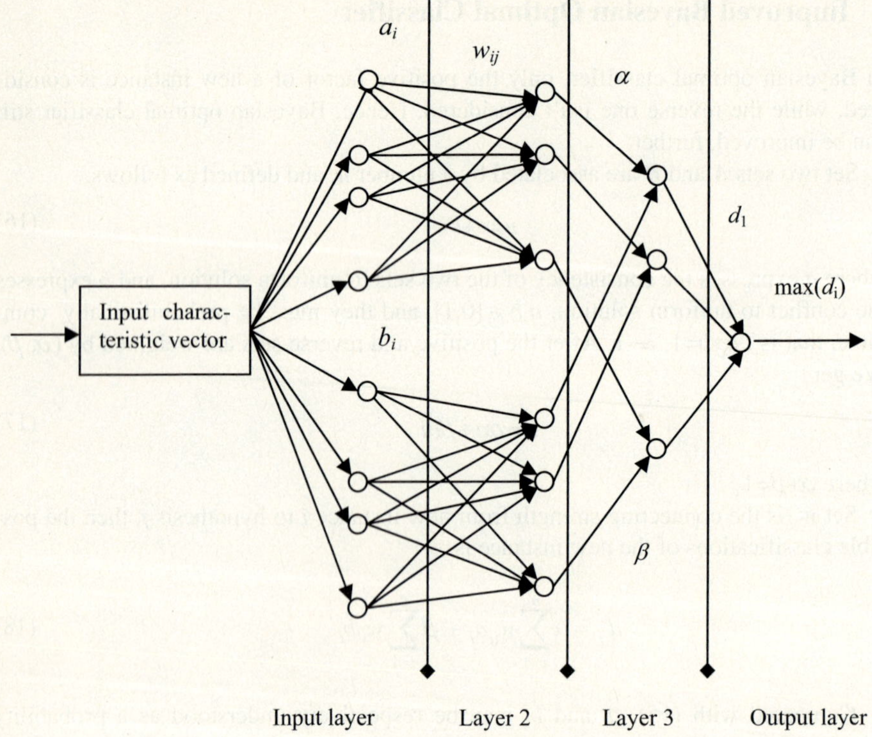

Fig. 1. Bayesian optimal classifier based on neuro-fuzzy logic

and another is increment learning algorithm of fuzzy neural networks [10]. Since all knowledge is stored in connecting weight values, they may be for that dynamic regulated.

7 Example

There are 296 data of DGA in knowledge base of one transformer. The fault symptoms and related fuzzy membership functions are established in Table 1.

According to the positive or reverse membership function, *a* and *b* in (18) may be woked out. Thus, for fuzzy fault symptoms, we may work out its positive and reverse influence coefficients. Fuzzy membership function adopts ascending or falling semi-trapezoid distribution, and respectively expressed as $u^{\uparrow}(a,b,x)$ and $u^{\downarrow}(a,b,x)$ as described below.

$$u^{\uparrow}(a,b,x)=\begin{cases}0 & x\le a\\ (x-a)/(b-a) & a<x\le b\\ 1 & x>b\end{cases};\quad u^{\downarrow}(a,b,x)=\begin{cases}1 & x\le a\\ (b-x)/(b-a) & a<x\le b\\ 0 & x>b\end{cases}$$

Table 1. Symptom types and fuzzy membership function

Fault symptom type		Positive membership function(M^+).	Reverse membership Function(M^-)
m_1: three-ratio-code based heat fault characteristics	C_2H_2/C_2H_4 ($A1$)	$u_\downarrow(0.08,0.12,x)$	$u^\uparrow(0.08,0.12,x)$
	CH_4/H_2 ($B1$)	$u^\uparrow(0.8,1.2,x)$	$u_\downarrow(0.8,1.2,x)$
	φ_{H2} ($\times10^{-6}$) ($C1$)	$u^\uparrow(120,180,x)$	$u_\downarrow(120,180,x)$
	φ_{C2H2} ($\times10^{-6}$) ($D1$)	$u^\uparrow(4,6,x)$	$u_\downarrow(4,6,x)$
	φ_{C1+C2} ($\times10^{-6}$) ($E1$)	$u^\uparrow(120,180,x)$	$u_\downarrow(120,180,x)$
	generation gas rate ($F1$)/(ml/h) — open type	$u^\uparrow(0.2,0.3,x)$	$u_\downarrow(0.2,0.3,x)$
	generation gas rate ($F1$)/(ml/h) — close type	$u^\uparrow(0.4,0.6,x)$	$u_\downarrow(0.4,0.6,x)$
	$A1\cap B1\cap(C1\cup D1\cup E1\cup F1)$		
m_2/(mg/L): water capacity in transformer oil	110kV downwards	$u^\uparrow(28,42,x)$	$u_\downarrow(28,42,x)$
	110kV upwards	$u^\uparrow(20,30,x)$	$u_\downarrow(20,30,x)$
m_3: earth current.		$u^\uparrow(0.196,0.144,x)$	$u_\downarrow(0.196,0.144,x)$
m_4: three-phase unbalanced coefficient.	1.6MVA downwards	$u^\uparrow(0.032,0.048,x)$	$u_\downarrow(0.032,0.048,x)$
	1.6MVA upwards	$u^\uparrow(0.016,0.024,x)$	$u_\downarrow(0.016,0.024,x)$
m_5(pC):local discharge capability		$u^\uparrow(300,900,x)$	$u_\downarrow(300,900,x)$
m_6 : three-ratio-code based discharge fault characteristics.	C_2H_2/C_2H_4 ($A2$)	$u^\uparrow(0.08,0.12,x)$	$u_\downarrow(0.08,0.12,x)$
	CH_4/CH_2 ($B2$)	$u_\downarrow(0.8,1.2,x)$	$u^\uparrow(0.8,1.2,x)$
	φ_{H2} ($\times10^{-6}$) ($C2$)	$u^\uparrow(120,180,x)$	$u_\downarrow(120,180,x)$
	φ_{C2H2} ($\times10^{-6}$) ($D2$)	$u^\uparrow(4,6,x)$	$u_\downarrow(4,6,x)$
	φ_{C1+C2} ($\times10^{-6}$) ($E2$)	$u^\uparrow(120,180,x)$	$u_\downarrow(120,180,x)$
	generation gas rate($F2$)/ ml/h — open type	$u^\uparrow(0.2,0.3,x)$	$u_\downarrow(0.2,0.3,x)$
	generation gas rate($F2$)/ ml/h — close type	$u^\uparrow(0.4,0.6,x)$	$u_\downarrow(0.4,0.6,x)$
	$A2\cap B2\cap(C2\cup D2\cup E2\cup F2)$		
m_7: absolute value of the warp of winding transform -ation ratio.	rated tapping	$u^\uparrow(0.004,0.006,x)$	$u_\downarrow(0.004,0.006,x)$
m_8(CO/CO_2)	A <0.09	$u_\downarrow(0.072,0.108,x)$	$u^\uparrow(0.072,0.108,x)$
	B >0.33	$u^\uparrow(0.264,0.396,x)$	$u_\downarrow(0.264,0.396,x)$
	$A\cup B$		
m_9	winding absorptance	$u_\downarrow(1.04,1.56,x)$	$u^\uparrow(1.04,1.56,x)$
	winding polarization index	$u_\downarrow(1.2,1.8,x)$	$u^\uparrow(1.2,1.8,x)$

Based on expert experience and fielded knowledge, connecting weights w_{ij} from symptom i to fault source j can be confirmed as shown in Table 2.

Clearly, the system has nine types of fault symptoms and fault sources. They respectively act as the inputs and outputs of neural computation model. Fuzzy membership

Table 2. Fault type and connection strength

Symptom type	Fault type	w_{ij}
m_1 m_3 m_5 m_6	*d*1: multi-point earth or local short in iron core	0.82 0.90 0.19 0.30
m_1 m_5 m_6	*d*2: leak magnetism heating or overheat	0.71 0.35 0.29
m_1 m_2 m_8	*d*3: insulation aging	0.22 0.27 0.82
m_2 m_9	*d*4: insulation damp	0.72 0.75
m_1 m_4 m_6	*d*5: tapping switch or down-lead fault	0.67 0.87 0.23
m_5 m_6	*d*6: suspend discharge	0.90 0.86
m_1 m_5 m_6 m_7 m_8	*d*7: winding distortion and circle short	0.15 0.75 0.68 0.80 0.72
m_5 m_6 m_7 m_8	*d*8: circle short and insulation damage	0.90 0.52 0.80 0.68
m_2 m_5 m_6 m_8	*d*9: encloser discharge	0.42 0.90 0.88 0.76

functions in Table 1 act as the base functions of fuzzy neurons. For *m*1 and *m*6 in Table 1, we adopt LP to emulate. Connection strength ***W*** may be used as prior weight values from input layer to class space. The two parameters α and β in output layer may be regulated in process of learning. Thus we can use the model in Fig.1 to realize fault diagnosis. The process is described as follows.

1) Applying LP to realize *m*1 or *m*6.

To attain fuzzy subjection degree of *m*1 in Table 1, fuzzy operation is needed, that is, $A1 \cap B1 \cap (C1 \cup D1 \cup E1 \cup F1)$. Below we divide two steps to complete it.

Step 1: Firstly, using OR to realize $C1 \cup D1 \cup E1 \cup F1$. Hence, order x_i in (5) equals, $C1, D1, E1, F1$, respectively, ω_i equals one, then

$$H = \overset{4}{\underset{i=1}{\vee}} [x_i \wedge 1] = \overset{4}{\underset{i=1}{\vee}} x_i = C1 \cup D1 \cup E1 \cup F1$$

Step 2: Secondly, applying AND to perform $A1 \cap B1 \cap H$, Let x_i in (7) equals A1,B1, H, respectively, ω_i =0, then

$$Z = \mathop{\wedge}_{i=1}^{3} [x_i \vee 0] = \mathop{\wedge}_{i=1}^{3} x_i = A1 \cap B1 \cap H = A1 \cap B1 \cap (C1 \cup D1 \cup E1 \cup F1)$$

Likewise, we also may emulate $m6$ using the same way.

2) The weight vector $\boldsymbol{W}$.

Fig.1 is a fully connecting network structure. The fielded knowledge in Table 1 and Table 2 may act as the initial weight values of the network. If prior information is fully correct, it would be unnecessary to train the network. If prior information isn't fully correct, then it would be necessary to train it by backward propagation (BP) algorithm to update fielded knowledge.

3) The weights of positive and reverse instances.

In general, the weight of positive inference is larger than the one of reverse inference. Hence, we may order α=0.85, β=0.15. They also may be regulated in process of learning.

A diagnosis example then is given out as follows.

The data of the DGA of one transformer are measured as $\varphi(H_2)$ =70.4×10^{-6}, $\varphi(CH_4)$=69.5×10^{-6}, $\varphi(C_2H_6)$=28.9×10^{-6}, $\varphi(C_2H_2)$=10.4×10^{-6}, $\varphi(C_2H_4)$=241.2×10^{-6}, $\varphi(CO)$=704×10^{-6}, $\varphi(CO_2)$=3350×10^{-6}, the unbalanced coefficient of the winding is 1.9%, earth current of iron core is 0.1A. The ratio of various characteristic gases are calculated as $\varphi(C_2H_2)/\varphi(C_2H_4)$=0.043, $\varphi(CH_4)/\varphi(H_2)$=0.99, $\varphi(C_2H_4)/\varphi(C_2H_6)$=8.35, $\varphi(CO)/\varphi(CO_2)$=0.21, three-ratio-code is 002. However, the code doesn't exist in the three-ratio-code table, and so the fault type is difficult to identify. But through fuzzy operation, we get M^{+}=0.475/m_1+0.083/m_3+0.375/m_4, M^{-}=0.525/m_6+0.917/m_3+ 0.625/m_4 +1/m_8. Hence, assume that prior knowledge in Table 1 and Table 2 is correct, then according to computation model in Fig.1, set α=0.85, β=0.15, the likelihood of each fault occurrence for that is calculated as d_1=0.249, d_2=0.275, d_3=0.064, d_4=0, d_5=0.495, d_6=-0.033, d_7=-0.055, d_8=-0.159, d_9=-0.1833. According to (19), since d_5=max{d_i, i=1,2,…,9}, the most likely diagnosis result is d_5, that is, tapping switch or down-lead fault. Field practical inspection proves the correctness of the diagnosis result. This result is full similar to one in [7], but here the applied approach is quite different.

8 Conclusions

1) By incorporating with all posterior probabilities of all assumptions, Bayesian optimal classifier makes the likelihood that a new instance is correctly classified up to maximum under similar conditions.

2) Due to fuzzy characteristic of the observed information, Bayesian optimal classifier is difficulty to dispose it. Hence, in the paper fuzzy subjection degree is applied to depict the information, which deals with the "bottle neck" in fuzzy knowledge acquisition.

3) The hybrid Bayesian optimal classifier based on neuro-fuzzy logic supports parallel automatic calculation, which effectively reduces the complexity of the algorithm, and improves the generalized ability of classification.

4) With the deeply study to transformer and the advancement of test means, fault symptom types and related learning factors require to be improved, further, which consequently facilitates the advancement of classification learning quality of Bayesian optimal learner.

References

1. Mitchhell Tom M. : Machine Learning. 1st edn. McGraw-Hill Companies, Inc., Columbus (1997)
2. Opper, M., and Haussler, D.: Generalization Behavior of Bayesian Optimal Prediction Algorithm for Learning a Perception. Physical Review Letters, 66(1991) 2677-2681
3. Haussler, D., Kearns, M., Schapire, R. E.: Bounds on The Sample Complexity of Bayesian Learning, Using Information Theory and the VC Dimension, Machine Learning, 14(1994) 79-83.
4. Yang, L., Shang, Y., Zhou, Y. F., Yan, Z.: Probability Reasoning and Fuzzy Technique Applied for Identifying Power Transformer Fault. Proceedings of the CSEE, 7(2000)19-23.
5. Zhang, J. W., Zhao, D. G., Dong, L. W.: An Expert System for Transformer Fault Diagnosis Based on Fuzzy Mathematics. High Voltage Engineering, 4(1998)6-8.
6. Chen, W. H., Liu, C. W., Tsai, M. S.: On-line Fault Diagnosis of Distribution Substations Using Hybrid Cause-Effect Network and Fuzzy Rule-based Method. IEEE Transactions on Power Delivery, 15(2000)710-717.
7. Su,H. S., Li, Q. Z.: Transformer Insulation Fault Diagnosis Method Based on Rough Set and Fuzzy Set and Evidence Theory. In: the 6th World Congress on Intelligent Control and Automation, Dailan(2006)
8. Zadeh, L.: Fuzzy Set. Information and Control, 8(1965)610-614.
9. Wang, Y. N.: Intelligent Information Processing. 1st edn. High Education Press, BJ(2003).
10. Zeng, H.: Intelligent Calculation. 1st edn. Chongqing University Press, Chongqing(2004)

Face Detection Using Kernel PCA and Imbalanced SVM

Yi-Hung Liu, Yen-Ting Chen, and Shey-Shin Lu

Visual-Servoing Control Lab., Department of Mechanical Engineering,
Chung Yuan Christian University, Chung-Li, 32023, Taiwan, China
lyh@cycu.edu.tw

Abstract. The task of face detection can be accomplished by performing a sequence of binary classification: face/nonface classification, in an image. Support vector machine (SVM) has shown to be successful in this task due to its excellent generalization ability. However, we find that the performance of SVM is actually limited in such a task due to the imbalanced face/nonface data structure: the face training images outnumbered by the nonface images in general, which causes the class-boundary-skew (CBS) problem. The CBS problem would greatly increase the false negatives, and result in an unsatisfactory face detection rate. This paper proposes the imbalanced SVM (ISVM), a variant of SVM, to deal with this problem. To enhance the detection rate and speed, the kernel principal component analysis (KPCA) is used for the representation and reduction of input dimensionality. Experimental results carried out on CYCU multiview face database show that the proposed system (KPCA+ISVM) outperforms SVM. Also, results indicate that without using KPCA as the feature extractor, ISVM is also superior to SVM in terms of multiview face detection rate.

1 Introduction

Face detection plays a key role in a fully face recognition system, which has been applied to many face recognition based systems such as visual surveillance and access control. Within the last two decades, there many face detection systems have been proposed with different approaches, and one of the most promising methods is the support vector machine (SVM) based face detection method [1-6].

The learning strategy of SVM is based on the principle of structural risk minimization [7-8], which makes SVM has better generalization ability than other traditional learning machines that are based on the learning principle of empirical risk minimization [9]. SVM has been recently successfully applied to the face detection [1-6]. The main idea behind those methods is to treat the face detection as the problem of SVM-based face/nonface classification. However, using SVM to accomplish the task of face/nonface classification suffers from a critical problem due to the very imbalanced training face/nonface dataset, called class-boundary-skew (CBS) problem, which will drop the classification performance. This paper aims to solve this problem for SVM-based face detection by introducing the imbalanced SVM (ISVM).

Aside from the face/nonface classifier, we also use the kernel principal component analysis (KPCA) [18] as the face/nonface feature extractor. KPCA has shown to be more effective than PCA in terms of representation due to the kernel trick used [18-19]. The proposed face detection system is composed of two kernel learning

L. Jiao et al. (Eds.): ICNC 2006, Part I, LNCS 4221, pp. 351 – 360, 2006.

machines: the former one is the unsupervised KPCA, and the latter is the supervised ISVM. KPCA is responsible for good representation and reduction of input dimensionality. ISVM is responsible for robust face/nonface classification. Accordingly, the proposed face detection system is a cascaded architecture of kernel learning machines.

This paper is organized as follows. Section 2 states the problem of applying SVM to face detection in details, and the solutions. The basic theory of KPCA and the reformulation of 1- and 2-norms ISVM are given in Section 3. Experimental results and discussions are given in Section 4. Conclusions are drawn in Section 5.

2 Problem Descriptions and Solutions

Before the problem description, a brief review of SVM is given as follows. Given the training set $S=\{x_i, y_i\}, i=1,...,L$, where $x_i \in R^n$ is the training vector, and y_i is its class label being either +1 or -1, let the weight vector and the bias of the separating hyperplane be w and b, the objective of SVM is to maximize the margin of separation and minimize the errors, formulated as the constrained optimization problem:

$$\textit{Minimize} \quad \frac{1}{2}\|w\|^2 + C\sum_{i=1}^{L}\xi_i \tag{1}$$

$$\begin{aligned} \textit{Subject to} \quad & y_i\left(w^T\Phi(x_i)+b\right)-1+\xi_i \geq 0, \quad \forall i \\ & \xi_i \geq 0, \quad \forall i \end{aligned} \tag{2}$$

where $\Phi(x): R^n \rightarrow R^m, m>n$, is a nonlinear mapping which maps the data into higher dimensional feature space from the original input space, ξ_i are slack variables representing the error measures of data. The error weight C is a free parameter; it measures the size of the penalties assigned to the errors.

Due to the property of very rich nonface patterns, the number of nonface patterns is in general much larger than that of face patterns when we prepare the training data for the SVM training. This leads to a very imbalanced training dataset. In SVM the error penalty C for each of the two classes are the same. This will make the learned optimal separating hyperplane (OSH) move toward to the smaller class, the face (positive) class, which will further result in numerous false negative errors. The "false negative error" means that face patterns are classified as nonface patterns. We call this phenomenon "class-boundary-skew (CBS) problem". Due to this problem, the success of using SVM in face detection is limited.

In face detection, the false negative error is the most critical because missing any test patterns that possibly belong to the face makes a detection system unreliable. If we can reduce the false negative errors, the SVM-based face detection system can be much more reliable and practicable. Therefore, how to solve the CBS problem when applying SVM to face detection becomes a very critical studying issue.

Several works have been proposed to study the CBS problem due to the very imbalanced training datasets [10-15], which can be divided into three categories. The methods of [13-15] use different sampling techniques to the training data before data enter the classifier. Secondly, The different error cost (DEC) algorithm of [10],[12] is embedded into the formulation of SVM such that the task of the skew phenomenon can be corrected. This method does not change the information of the data structure beforehand. The third is the SDC method [11], which combines the SMOTE [15] and the different error cost algorithm [10]. For face detection, since every face data stands for one particular face information, we intend not to use any pre-sampling techniques like those fall into the first category that may change the data structure. Therefore, the DEC algorithm proposed by Veropoulos et al. [10] is adopted in this paper to deal with the CBS problem due to the imbalanced face/nonface dataset. By introducing DEC algorithm to SVM, the imbalanced SVM (ISVM), a variant of SVM is proposed.

The feature extraction is also crucial to the face detection. Face patterns suffer from the problem of variations from facial expressions, facial details, viewpoints, and illuminations, which generally makes the face distribution non-Gaussian and nonlinearity [17]. So how to provide an effective representation method also dominates the detection accuracy. KPCA [18] is based on the principle that since PCA in *F* can be formulated in terms of the dot products in *F*, this same formulation can also be performed using kernel functions without explicitly performing in *F*. KPCA has been recently shown to have better performance than PCA in the face recognition application [19]. Therefore, KPCA is adopted as the feature extractor in this work.

3 Face Detection Method

3.1 Feature Extraction Via KPCA

The idea of KPCA is to map the input data x into the feature space F via a nonlinear mapping function Φ and then perform PCA in F. Suppose that the data are centered to have a zero mean distribution in F, i.e., $\sum_{i=1}^{M}\Phi(x_i)=0$ (the centering method of data in F can be found in [18]), where M is the number of input data, KPCA aims to diagonalize the estimate of the covariance matrix of the mapped data $\Phi(x_i)$:

$$\Gamma=\frac{1}{M}\sum_{i=1}^{M}\Phi(x_i)\Phi^T(x_i) \tag{3}$$

This is equivalent to solving the eigenvalue problem: $\lambda v=\Gamma v$, and finding eigenvectors $v\in F$ associated with nonzero eigenvalues λ. By substituting (3) into the eigenvalue problem we have

$$\Gamma v=\frac{1}{M}\sum_{i=1}^{M}(\Phi(x_i)\cdot v)\Phi(x_i) \tag{4}$$

Since all solutions v with $\lambda\neq 0$ lie within the span of $\Phi(x_i), i=1,..,M$, there must exist the expansion coefficients $a_i, i=1,...,M$ such that

$$v = \sum_{i=1}^{M} a_i \Phi(x_i) \tag{5}$$

Then the following equations are considered:

$$\lambda(\Phi(x_i) \cdot v) = (\Phi(x_i) \cdot \Gamma v) \quad \forall i = 1,\dots,M \tag{6}$$

By substituting (3) and (5) into (6) and defining a $M \times M$ kernel matrix *K*: $K_{ij} \equiv k(x_i, x_j) = (\Phi(x_i) \cdot \Phi(x_j))$, where $k(x_i, x_j)$ is the kernel function, then solving the eigenvalue problem $\lambda v = \Gamma v$ is equivalent to solving the following:

$$M\lambda a = Ka \tag{7}$$

for nonzero eigenvalues λ_l and eigenvectors $a^l = (a_1^l, \dots, a_M^l)^T$ subject to the normalization condition $\lambda_l (a^l \cdot a^l) = 1$.

For achieving the goal of reduction of input dimensionality in face detection while extracting face/nonface features, we choose the first *n* eigenvectors as the projection axes, associated with the first *n* largest nonzero eigenvalues such that $n << M$

3.2 Face Detection Via Imbalanced SVM

In SVM, the error penalty for each of the two classes are the same. However, DEC algorithm proposes that the cost of misclassifying a point from the small class should be heavier than the cost for errors on the large class. In face detection, the nonface patterns always much outnumber the face patterns. Therefore, the face class is a smaller class compared with the nonface class. The basic idea is to introduce different error weights C^+ and C^- for the positive and the negative class with $C^+ > C^-$, which result in a bias for larger Lagrange multipliersof the face class. This induces a decision boundary which is more distant from the smaller class than from the other.

Assume that a training set is given as $S = \{x_i, y_i\}_{i=1}^{l}$. Let $I_+ = \{i \mid y_i = +1\}$ and $I_- = \{i \mid y_i = -1\}$, then the optimization problem of ISVM is formulated as follows:

$$\begin{aligned} &\text{Minimize} \quad \frac{1}{2}\|w\|^2 + C^+ \sum_{i \in I_+} \xi_i^k + C^- \sum_{i \in I_-} \xi_i^k \\ &\text{subject to} \quad y_i\left(w^T \phi(x_i) + b\right) - 1 + \xi_i \qquad i = 1,\dots,l \\ &\qquad\qquad\quad \xi_i \geq 0 \end{aligned} \tag{8}$$

where ϕ is a nonlinear mapping.

1-Norm Soft Margin. For *k*=1 the primal Lagrangian is given by:

$$\begin{aligned} L_p(w,b,\xi,\alpha,\beta) = \frac{1}{2}\|w\|^2 + C^+ \sum_{i \in I_+} \xi_i + C^- \sum_{i \in I_-} \xi_i \; - \\ \sum_{i=1}^{l} \alpha_i \left(y_i\left(w^T \phi(x_i) + b\right) - 1 + \xi_i\right) - \sum_{i=1}^{l} \beta_i \xi_i \end{aligned} \tag{9}$$

By taking partial differential to the Lagrangian with respect to the variables, and by introducing the kernel function $k(x,y)$, the dual problem becomes

$$L_D(\alpha)=\sum_{i=1}^{l}\alpha_i-\frac{1}{2}\sum_{i=1}^{l}\sum_{j=1}^{l}\alpha_i\alpha_j y_i y_j k(x_i,x_j) \tag{10}$$

subject to the constraints:

$$\begin{aligned}&0\le\alpha_i\le C^+ \quad for \quad y_i=+1\\&0\le\alpha_i\le C^- \quad for \quad y_i=-1\\&\sum_{i=1}^{l}\alpha_i y_i=0\end{aligned} \tag{11}$$

2-Norm Soft Margin. For k=2 we get the Lagrangian:

$$\begin{aligned}L_p(w,b,\xi,\alpha,\beta)=\frac{1}{2}\|w\|^2+(C^+/2)\sum_{i\in I_+}\xi_i^2+(C^-/2)\sum_{i\in I_-}\xi_i^2-\\\sum_{i=1}^{l}\alpha_i\left(y_i\left(w^T\phi(x_i)+b\right)-1+\xi_i\right)-\sum_{i=1}^{l}\beta_i\xi_i\end{aligned} \tag{12}$$

This results in a dual formulation:

$$L_D(\alpha)=\sum_{i=1}^{l}\alpha_i-\frac{1}{2}\sum_{i=1}^{l}\sum_{j=1}^{l}\alpha_i\alpha_j y_i y_j(k(x_i,x_j)+\Pi_{[i\in I_+]}\frac{1}{C^+}\delta_{ij}+\Pi_{[i\in I_-]}\frac{1}{C^-}\delta_{ij}) \tag{13}$$

where $\Pi_{[\cdot]}$ is the indicator function. This can be viewed as a change in the *Gram matrix G*. Add $1/C^+$ to the elements of the diagonal of G corresponding to examples of positive class and $1/C^-$ to those corresponding to examples of the negative class:

$$G_{ii}'=\begin{cases}k(x_i,x_i)+\dfrac{1}{C^+} & for \quad y_i=+1\\ k(x_i,x_i)+\dfrac{1}{C^-} & for \quad y_i=-1\end{cases} \tag{14}$$

The Kuhn-Tucker (KT) complementary conditions are necessary for the optimality:

$$\alpha_i[y_i(w^T\phi(x_i)+b)-1+\xi_i]=0\,,\ i=1,2,...,l \tag{15}$$

$$\begin{aligned}&\left(C^+-\alpha_i\right)\cdot\xi_i=0 \quad for \quad y_i=+1\\&\left(C^--\alpha_i\right)\cdot\xi_i=0 \quad for \quad y_i=-1\end{aligned} \tag{16}$$

There are two types of α_i. If $0<\alpha_i\le C^+$ or $0<\alpha_i\le C^-$, the corresponding data points are called support vectors (SVs). The solution for the weight vector is given by

$$w_o = \sum_{i=1}^{N_S} \alpha_i y_i \phi(x_i) \tag{17}$$

where N_s is the number of SVs. In the case of $0 < \alpha_i < C^+$ or $0 < \alpha_i < C^-$, we have $\xi_i = 0$ according to the KT conditions. Hence, one can determine the optimal bias b_o by taking any data point in the training set by, for which we have $0 < \alpha_i < C^+$ or $0 < \alpha_i < C^-$ and therefore $\xi_i = 0$, and using data point in the KT condition. Once the optimal pair (w_o, b_o) is determined, for an unseen data x the decision function that classifies the face class and the nonface class is obtained by:

$$D(x) = sign\left(\sum_{i=1}^{N_S} \alpha_i y_i k(x_i, x) + b_o \right) \tag{18}$$

With this decision function, we can easily construct the OSH for ISVM for those data points x satisfying $D(x) = 0$. From the formulation, we know two things. First, by assigning proper values to the penalty weights such that $C^+ > C^-$, OSH will move toward the negative class (nonface class) more such that the false negative errors can be decreased, and thereby the CBS problem that occurs in SVM can be solved. As a result, the face detection rate can be enhanced. Second, in the training step we only need the training points as entries in the Gram matrix and in the test phase the new points only appear in an inner product with few support vectors in the training set.

4 Experimental Results

The face images used for training and testing are collected from the CYCU multiview face database [16],[20]. The CYCU multiview face database contains 3150 face images out of 30 subjects and involves variations of facial expression, viewpoint and sex. Each image is a 112×92 24-bit pixel matrix and each subject has 105 face images in total. Fig. 1 shows the total 30 subjects in this database. All images contain face contours and hairs. Fig. 2(a) shows the collected 21 images containing 21 different viewpoints of one subject. For achieving better detection rate, every image is cropped to a new image containing only the face region. Also, in order to reduce the effect due to illumination changes, each cropped image is transferred to a gray-level image, and then to enhance the contrast by using the histogram equalization, as shown in Fig. 2(b).

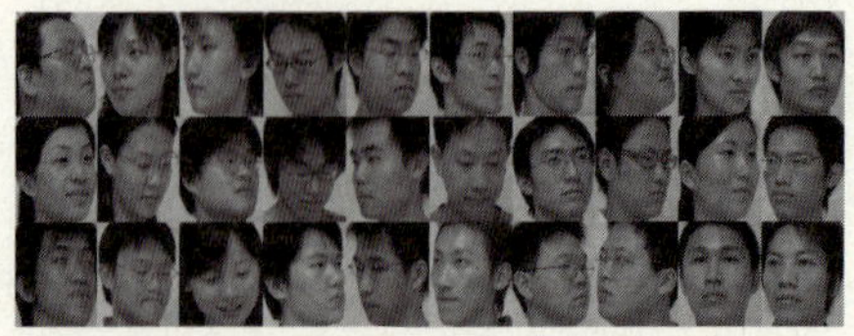

Fig. 1. 30 subjects in CYCU multiview face database

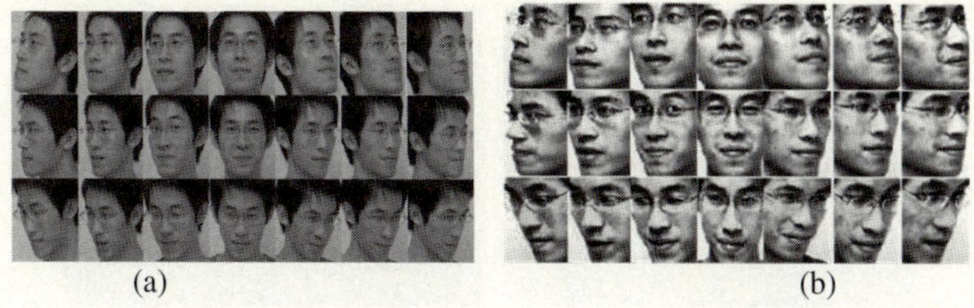

(a) (b)

Fig. 2. (a) 21 images of one subject in CYCU multiview face database, (b) cropped images after histogram equalization

In the experiments, we collect 630 face images from the CYCU face database by random. The face training set consists of 315 face images randomly selected from the 630 face images, and the remaining form the face test set. There is no overlap between the two sets. In addition, we collect 1575 nonface images from the real pictures on websites. 1260 images are randomly chosen from the 1575 nonface images for the nonface training set, and the remaining 315 images form the nonface test set. Therefore, the training set contains 315+1260=1575 images, and the test set contains 630 images. Also, the imbalanced ratio of nonface class to the face class in the training set is 1260/315=4. To reduce the computational effort, every image is normalized to a 28×23 pixel image before this image enters the system KPCA+ISVM.

Since KPCA and ISVM are kernel learning machines, the kernel function needs to be determined first. The radial-basis function (RBF) is adopted as the kernel in this study:

$$k(z_i, z_j) = \phi(z_i) \cdot \phi(z_j) = \exp\left(-0.5 \left\| z_i - z_j \right\|^2 / \sigma^2 \right) \tag{19}$$

where the width σ^2 is specified *a priori* by the user. The training phase is composed of two stages: KPCA training and ISVM training. First, we use the training set to find the optimal value of σ^2 and the optimal number of eigenvectors of KPCA. Following the parameter searching procedure suggested by [17], the optimal kernel parameter σ^2, and the optimal number of eigenvectors of KPCA are found to be 2 and 200, respectively. Therefore, the input dimensionality is reduced from 644 to 200.

Next, we train the ISVM with the kernel eigenimages. For the simplicity, we only use 1-Norm soft margin based ISVM as the classifier. The optimal parameters of $\sigma^2 = 64$ and $(C^+, C^-) = (50,1)$ are found for ISVM. Before the testing, we compare the OSH learned by SVM and the one learned by ISVM in the KPCA-based subspace, in which the patterns are the projections in the two eigenvectors associated with the largest two nonzero eigenvalues among the 200 ones, as shown in Fig. 3.

From Fig. 3, we can observe that the number of false negatives classified by SVM is much larger than that classified by ISVM. Contrarily, by using ISVM, the problem is solved since the OSH is moved toward the nonface class, as we can observe from Fig. 3(b). As a consequence, most face images will not be missed though the false positives will be increased. Nevertheless, the false positive errors can be further reduced by using other post-processing methods such as the skin-colored detection. It is noted that the overlaps in Fig. 3(a) and Fig. 3(b) are not small, because we only plot these patterns in two of the total 200 eigenvectors. To test the performance of KPCA+ISVM, the test set is fed into the system. The results are listed in Table 1.

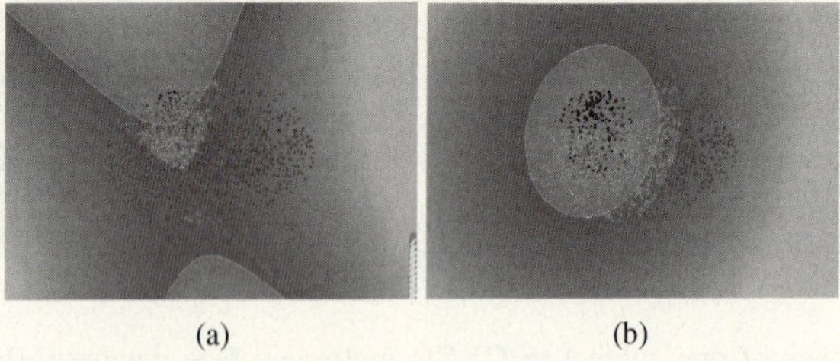

(a) (b)

Fig. 3. 315 face patterns are plotted with blue color, and 1260 nonface patterns are plotted with red color. The OSHs are learned by (a) SVM, and (b) ISVM, respectively.

Table 1. Comparison of face/nonface classification results among different systems

Systems	False negatives	False positives	Error rates (%)
SVM	111	0	17.62
ISVM	2	48	7.94
KPCA+ISVM	2	26	4.44

Form Table 1, we can see that by using SVM, the number of false negatives is very large (111 in total) though the number of false positive is equal to zero. This causes a SVM-based face detection system unreliable since too many face images are classified as the nonface images, while this critical problem is solved by ISVM by which the number of false negatives is substantially reduced from 111 to 2. However, compared with SVM, ISVM induces more false positives (48 in total) because the learned OSH is closer to the nonface class. By cascading KPCA to ISVM, the number of false positives is further reduced from 48 to 26 while keeping the same number of false negatives. In terms of overall performance, the system KPCA+ISVM achieves the lowest error rate (4.44%), compared with SVM (17.92%) and ISVM (7.94%).

Next, we employ the proposed system KPCA+ISVM as the face detector to detect faces from colored pictures. In the first-time scanning a scanning window of 28×23 is used to scan the whole picture from top-left to the bottom-right with a shifting interval of five pixels. The scanning window for the second-time scanning is enlarged with 1.2 times the height and the width of the scanning window in the first-time scanning, and so on. The region in a scanning window is transformed to gray-level image,

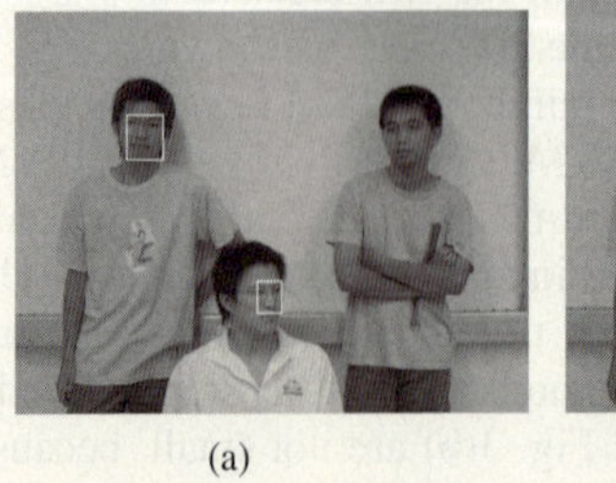

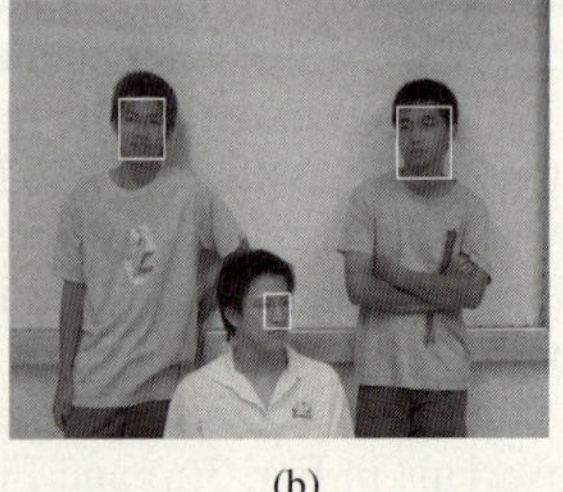

(a) (b)

Fig. 4. Face detection results by (a) SVM, and (b) KPCA+ISVM in pure background

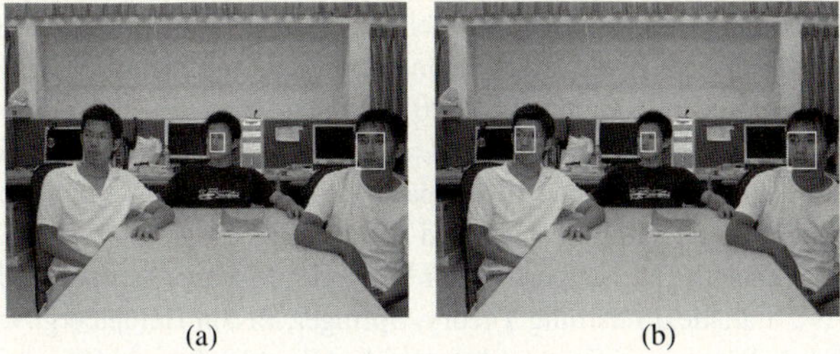

(a) (b)

Fig. 5. Detection results by (a) SVM, and (b) KPCA+ISVM in more complex background

through the histogram equalization, resized to a 28×23 pixel image, and then fed into the system for the face/nonface classification. If the region is classified as the face class, a squared bounding box equaling to the size of the current scanning window is used to fix this region. Two results are shown in Fig. 4 and Fig. 5 respectively. From these results, we can see that the proposed system KPCA+ISVM is able to detect all the faces in pictures, while SVM lost some of them.

5 Conclusions

This paper proposes a face detection method based on the cascaded architecture of kernel learning machines: KPCA + ISVM. We first point out the class-boundary-skew problem resulted from the imbalanced face/nonface training set, and suggest the solution based on different error cost (DEC) algorithm. Then, by introducing DEC algorithm to SVM, we reformulate the SVM to ISVM for 1- and 2-Norms soft margins, respectively. Results show that the proposed system KPCA+ISVM not only substantially reduces the number of false negatives, which is crucial to the face detection, but also achieves much lower error rate of face/nonface classification compared with SVM and ISVM. The effectiveness of KPCA+ISVM for solving CBS problem of SVM in face detection and achieving high face detection accuracy is demonstrated. The future work will be focused on combing more powerful SVM-based approaches [21-23] with the proposed methods to achieve better face detection performance.

Acknowledgements. This work was supported by National Science Council (NSC), Taiwan, R.O.C., under Grant No. 93-2212-E-033-014.

References

1. Osuna E., Freund R., and Girosit F.: Training Support Vector Machines: an Application to Face Detection. Proc. CVPR, (1997) 130-136
2. Buciu, L., Kotropoulos, C., and Pitas, I.: Combining Support Vector Machines for Accurate Face detection. Proc. Intl. Conf. Image Processing. **1** (2001) 1054-1057
3. Romdhani, S., Torr, P., Schölkopf, B. and Blake, A.: Computationally Efficient Face Detection. Proc. Eighth IEEE Intl. Conf. Computer Vision. **2** (2001) 695-700
4. Terrillon, T. J., Shirazi, M. N., Sadek, M., Fukamachi, H., and Akamatsu, S.: Invariant Face Detection with Support Vector Machines. Proc. Int. Conf. Pattern Recognition. **4** (2000) 210-217

5. Li, Yongmin, Gong, Shaogang, and Liddell, H.: Support Vector Regression and Classification Based Multi-view Face Detection and Recognition. Proc. 4th IEEE Intl. Conf. Automatic Face and Gesture Recognition. (2000) 300-305
6. Li, Yongmin, Gong, Shaogang, Sherrah, J. and Liddell, H.: Multi-view Face Detection using Support Vector Machines and Eigenspace Modeling. Proc. 4th Intl. Conf. Knowledge-Based Intelligent Engineering Systems and Allied Technologies. **1** (2000) 241-244
7. Corts, C. and Vapnik, V. N.: Support Vector Networks. Machine Learning. **20** (1995) 273-297
8. Vapnik, V. N.: Statistical Learning Theory, Springer, Berlin Heidelberg New York (1998)
9. Burges, J. C.: A Tutorial on Support Vector Machines for Pattern Recognition. Data Mining and Knowledge Discovery. **2** (1998) 121-167
10. Veropoulos, K., Campbell, C., and Cristianini, N.: Controlling the Sensitivity of Support Vector Machines. Proc. Int. Joint Conf. Artificial Intelligence (IJCAI99). (1999)
11. Akbani, R., Kwek, S., and Japkowicz, N.: Applying Support Vector Machines to Imbalanced Datasets. Proc. 15th European Conf. Machine Learning (ECML). (2004) 39-50
12. Wu, G., and Cheng, E.: Class-boundary alignment for imbalanced dataset learning. ICML 2003 Workshop on Learning from Imbalanced Data Sets II. (2003)
13. Japkowicz, N.: The Class Imbalance Problem: Significance and Strategies. Proc. 2000 Int. Conf. Artificial Intelligence: Special Track on Inductive Learning. (2000) 111-117
14. Ling, C., and Li, C.: Data Mining for Direct Marketing Problems and Solutions. Proc. 4th Int. Conf. Knowledge Discovery and Data Mining. (1998)
15. Chawla, N., Bowyer, K., and Kegelmeyer, W.: SMOTE: Synthetic Minority Oversampling Technique. Journal of Artificial Intelligence Research. **16** (2002) 321-357
16. Liu, Y. H., and Chen, Y. T.: Face Recognition Using Total Margin-Based Adaptive Fuzzy Support Vector Machines. IEEE Trans. Neural Networks. **17** (2006) In press
17. Lu, J. Plataniotis, K. N., and Venetsanopoulos, A. N.: Face Recognition Using Kernel Direct Discriminant Analysis Algorithms. IEEE Trans. Neural Networks. **14** (2003)
18. Schölkopf, B., Smola, A., and Müller, K. R.: Nonlinear Component Analysis as a Kernel Eigenvalue Problem. Neural Computation. **10** (1998) 1299-1319
19. Kim, K. I., Jung, K., and Kim, H. J.: Face Recognition Using Kernel Principal Component Analysis. IEEE Signal Processing Letters. **9** (2002) 40-42
20. CYCU face database. URL: http://vsclab.me.cycu.edu.tw/~face/face_index.html
21. Kecman, V.: Learning and Soft Computing, Support Vector machines, Neural Networks and Fuzzy Logic Models, The MIT Press, Cambridge, MA (2001)
22. Wang, L. P. (Ed.): Support Vector Machines: Theory and Application. Springer, Berlin Heidelberg New York (2005)
23. Tan, Y. P., Yap, K. H., Wang, L. P. (Eds.): Intelligent Multimedia Processing with Soft Computing. Springer, Berlin Heidelberg New York (2004)

Neural Networks Based Structural Model Updating Methodology Using Spatially Incomplete Accelerations

Bin Xu

College of Civil Engineering, Hunan University, Changsha, Hunan, 410082, P.R. China
binxu@hnu.cn

Abstract. Because it is difficult to obtain structural dynamic measurements of the whole structure in reality, it is critical to develop structural model updating methodologies using spatially incomplete dynamic response measurements. A general structural model updating methodology by the direct use of free vibration acceleration time histories without any eigenvalue extraction process that is required in many inverse analysis algorithms is proposed. An acceleration-based neural network(ANN) and a parametric evaluation neural network(PENN) are constructed to update the inter-storey stiffness and damping coefficients of the object structure using an evaluation index called root mean square of prediction difference vector(RMSPDV). The performance of the proposed methodology using spatially complete and incomplete acceleration measurements is examined by numerical simulations with a multi-degree-of-freedom(MDOF) shear structure involving all stiffness and damping coefficient values unknown. Numerical simulation results show that the proposed methodology is robust and may be a practical method for structural model updating and damage detection when structural dynamic responses measurements are incomplete.

1 Introduction

Many existing civil infrastructures are now deteriorating due to aging, misuse, lacking proper maintenance, and, in some cases, overstressing as a result of increasing load levels and changing environments. Developing efficient model updating algorithms is crucial for structural performance evaluation, crisis management and control system design. Structural model updating and identification methods can be classified under various categories. Comprehensive literature reviews on structural identification and model updating within the context of civil engineering can be found in references[1-3]. The most widely used global model updating and identification methodologies are based on eigen-values and/or mode shapes extracted from dynamic vibration measurements. Unfortunately, the identified mode shapes from dynamic measurements are so noise-contaminated that low to intermediate levels of local damage can not be identified correctly[4]. Moreover, the sensitivity of frequencies to changes in structural parameters is an obstacle in the application of the eigen-values based methodologies in practice. Error is incurred to obtain velocity and displacement signals from acceleration measurements by integration and acceleration measurements of infrastructures instrumented with monitoring system can be normally measured using accelerometers. Therefore, developing a model updating strategy, which is not based

L. Jiao et al. (Eds.): ICNC 2006, Part I, LNCS 4221, pp. 361 – 370, 2006.

on eigen-values and/or modes, using acceleration measurements is preferred over velocity and displacement signals.

With the ability to approximate any arbitrary continuous function and mapping, soft computing algorithms involving neural networks have recently drawn considerable attention for the modeling and identification of linear or nonlinear engineering structures[5-8]. Although several neural-network-based structural model updating strategies are available for qualitative evaluation of damages that may have taken place in a structure, it was not until recently that a quantitative way of detecting damage with neural networks has been proposed[9-13]. As described in reference[3], the novel and unique methodology for structural model updating using structural dynamic displacement and velocity measurements with the support of two neural networks proposed by Xu et al. is absolutely different from the existing traditional methodologies[12]. Unlike any conventional system model updating technique that involves the inverse analysis with an optimization process, those strategies proposed by Xu et al. and Wu et al. with the direct use of dynamic responses can give the model updating results in a substantially faster way and thus provides a viable tool for the near real-time model updating of structural parameters for a SHM system[10-13].

This study is aimed at the development and performance evaluation of a soft structural model updating methodology using spatially complete and incomplete acceleration measurements. The performance of the methodology is examined by numerical simulations when complete and incomplete acceleration measurements are available.

2 Structural Model Updating Using Acceleration Measurements

From vibration measurements from a long-term performance monitoring system instrumented with a structure under unknown/assumed stationary zero-mean Gaussian white noise ambient excitations, structural free vibration responses can be extracted using the Random Decrement(RD) technique. Therefore, developing a free vibration-based model updating strategy is critical for civil infrastructures.

To facilitate the structural model updating process, a reference structure and a number of associated structures that have the same overall dimension and topology as the object structure are created, and an ANN and a PENN are established and trained to update the physical parameters of the object structure. The basic three-step procedure for structural model updating methodology is shown in Fig. 1. In this study, acceleration measurements of object structure are assumed to be known and employed to update its model.

2.1 Construction of ANN for Acceleration Forecasting of Reference Structure

The free vibration of a linear structure system with n degrees of freedom (DOFs) under a certain initial condition can be characterized by the following equation,

$$M\ddot{x}+C\dot{x}+Kx=0\,,\quad x_{t=0}=x_0,\ddot{x}_{t=0}=0,\dot{x}_{t=0}=0 \tag{1}$$

where $\boldsymbol{M}$, $\boldsymbol{C}$ and $\boldsymbol{K}$ = the mass, damping, and stiffness matrices of the structure, $\ddot{x}$, $\dot{x}$ and x = the acceleration, velocity, and displacement vectors, x_0 is initial displacement condition for free vibration.

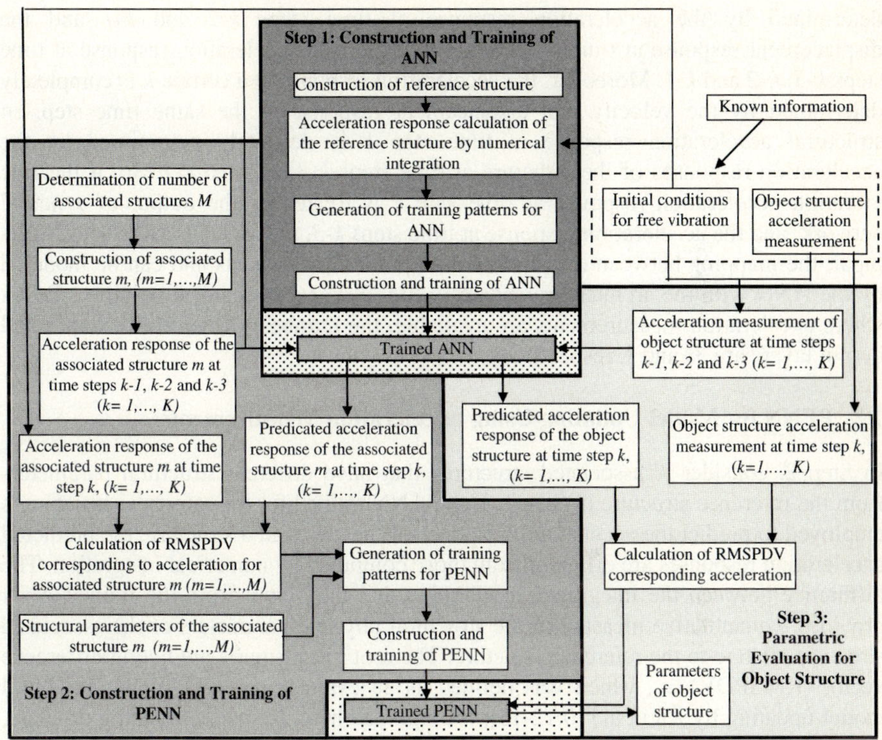

Fig. 1. Structural model updating methodology based on neural networks using acceleration measurements

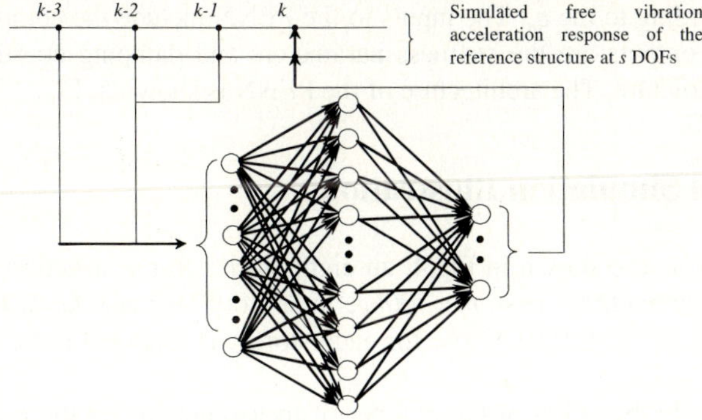

Fig. 2. Architecture of ANN

According to the description in reference[12,13], structural free vibration displacement and velocity response at time step k is uniquely and completely determined by them at time step k-1. The velocity response at time step k-1 is determined by the acceleration response at time steps *k-2* and *k-1* and the displacement response at time *k-1* is determined by the acceleration response at time steps *k-3*, *k-2* and *k-1*. Moreover, the acceleration response at a certain *k* is completely determined by the velocity and displacement response at the same time step, so structural acceleration response at time step *k* is definitely determined by the acceleration responses of the structure at time steps *k-3, k-2* and *k-1*. So, if the free vibration acceleration response at time step *k* is treated as the output of a neural network, and the acceleration response at time steps *k-3*, *k-2* and *k-1* are selected as its input, the mapping between the input and output uniquely exists and can be modeled by the ANN with the architecture shown in Fig. 2. For a structure involving *s* DOFs where acceleration measurements are available, the numbers of neurons in input and output layers are *3s* and *s*, respectively.

2.2 PENN for Model Updating Using Acceleration Measurements

In Step 2, consider *N* associated structures that have different structural parameters from the reference structure in Step 1. If the ANN trained for the reference structure is employed to predict the acceleration response of the associated structure, the predicted acceleration responses are different from those computed by numerical integration. The difference between the integrated acceleration and the output from the trained ANN provides a quantitative measure of the structural physical parameters of the associated structure relative to the reference structure. The root mean square prediction difference vector (RMSPDV), e_n, which was defined as an evaluation index for the structural model updating by Xu et al.[10,12-13], is employed here. In model updating problem, the structural mass distribution is usually treated as known. Therefore, the e_n is completely determined by the stiffness and damping parameters of the associated structure. The PENN is constructed and trained to describe the mapping between the RMSPDV and structural parameters of the corresponding associated structure. When the PENN is trained successfully, structural parameters of the object structure can be determined according to the e_n. The inputs to the PENN include the components of the RMSPDV. The outputs are the stiffness parameters and damping coefficients of the corresponding structure. The architecture of the PENN is shown in Fig. 3.

3 Numerical Simulation Illustration

For the object structure shown in Fig.4, an eight stories frame structure with the following known parameters, mass $m_i=300kg$, inter-story stiffness $k_i=6.0*10^5 N/m$ and damping coefficient $c_i=4.0*10^3 N*s/m$ for each storey, is assumed to be the reference structure.

In order to study the effect of the number of measurements on the performance of the proposed methodology, spatially complete acceleration measurement (Case 1) and spatially incomplete acceleration measurements (Cases 2) are considered. DOFs where acceleration measurements are available are listed in Table 1. Let the number of neurons of the hidden layer of the ANN to be two times it in the input layer. Therefore, the architecture of the ANN in case 1 and 2 is 24-48-8 and 9-18-3, respectively.

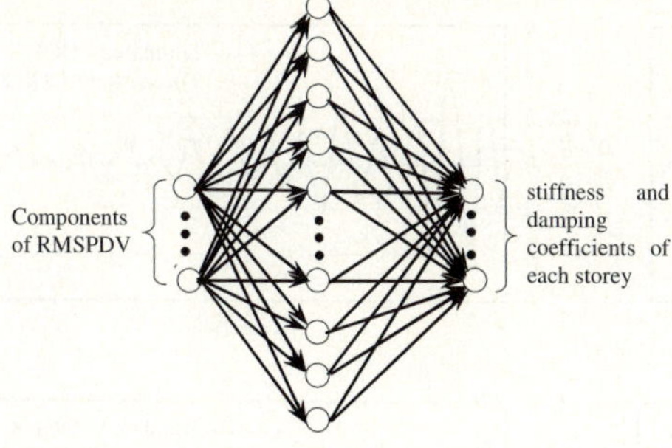

Fig. 3. Architecture of the PENN for structural parametric model updating

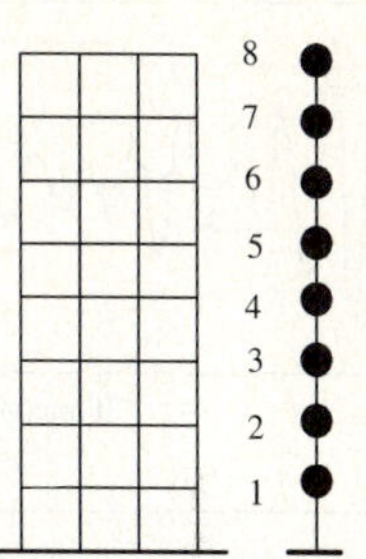

Fig. 4. A shear frame structure with 8 DOFs

Table 1. DOFs where acceleration measurements are available

Cases	DOFs with available acceleration measurements
1	1, 2, 3, 4, 5, 6, 7, 8
2	2, 5, 8

3.1 ANN for Acceleration Response Forecasting of Reference Structure

The data sets for the ANN training are constructed from the numerical integration results of the free vibration acceleration time histories of the reference structure under a certain initial displacement condition. Without loss of generality, the initial displacement condition is shown in Table 2. Velocity and acceleration initial condition are set to be 0.

Table 2. Initial displacement condition for free vibration

DOFs	1	2	3	4	5	6	7	8
Disp.(mm)	0.772	1.251	0.686	-0.331	-1.113	2.073	3.907	3.160

The numerical integration is carried out with integration time step of 0.002s. The training data sets are performed with the data taken at the intervals of the sampling period of 0.01s. Therefore, 197 patterns taken from the first 2s of acceleration response of the structure can be constructed. Based on the error back-propagation algorithm, one ANN is trained for each case. At the beginning of training, a linear normalization pre-conditioning for the training data sets is carried out and the weights are initialized with small random values.

Figs. 5 and 6 give the comparison between the acceleration responses determined from the numerical integration and those forecast by the trained ANN using spatially complete and incomplete measurements in cases 1 and 2, respectively. Figure 7 shows the forecast error in Cases 1 and 2 compared with the acceleration response corresponding to DOFs 2, 5, and 8. It can be found that the forecast error can reach a small value in both cases compared with the structural acceleration response.

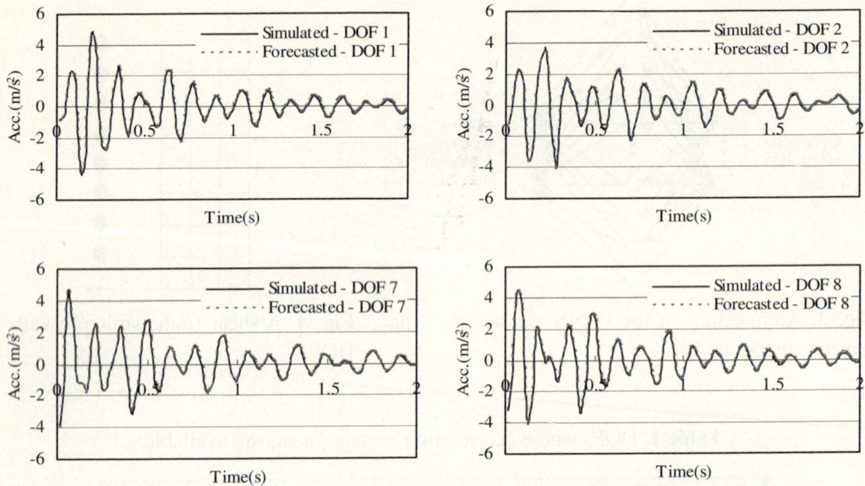

Fig. 5. Comparison between the acceleration responses determined from the numerical integration and those forecast by the trained ANN in Case 1 corresponding to DOFs 1, 2, 7 and 8

Table 3 shows the standard deviation of the forecast error. It can be seen that the forecast acceleration responses meet with the numerical integration results very well even incomplete dynamic responses are employed. This means that the trained ANN is a nonparametric modeling for the reference structure in the form of forecasting its dynamic responses using complete and incomplete dynamic measurements.

3.2 Structural Model Updating by PENN

For each case, a PENN is trained independently based on the corresponding trained ANN. For the purpose of training of the PENN, some associated substructures with different structural parameters within the interested space are assumed, and the RMSPDVs are calculated with the help of the corresponding trained ANN. The stiffness and damping coefficients of the associated structures and the corresponding RMSPDVs are used to train the PENN.

Suppose the interstory stiffness and damping coefficients corresponding to each DOF of the whole object structure are within $\pm 30\%$ of the values of the reference structure. 120 associated structures are constructed by randomly selecting the combination of the interstory stiffness and damping coefficients. Then, 120 pairs of training data composed of the structural parameters and corresponding RMSPDV are determined and employed to train the PENN.

After the PENN has been successfully trained, it will be applied in Step 3 into the object structure to update the structural stiffness and damping coefficients with RMSPDV determined from the corresponding ANN and the acceleration measurements of the object structure. The architecture of the three-layer PENN for case 1 and

2 is 8-50-16 and 3-50-16, respectively. The neurons in the input layer represent the components of the RMSPDV of each case. For the purpose of comparison, the number of neurons in the hidden layer is identical in both cases. The neurons in the output layer of each PENN for case 1 or 2 represent the stiffness and damping coefficients of the whole object structure. Here, the stiffness and damping coefficients updating results corresponding to each DOF of an object structure in Cases 1 and 2 are shown in Figure 8.

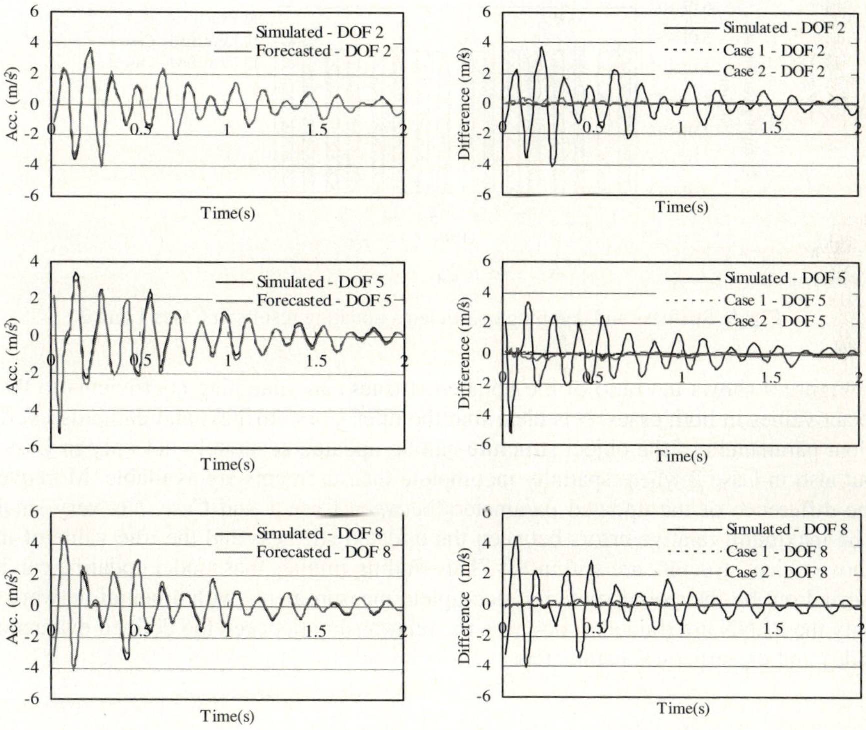

Fig. 6. Comparison between the acceleration responses determined from the numerical integration and those forecast in Case 2

Fig. 7. Forecast error in Cases 1 and 2 compared with the acceleration response determined by integration

Table 3. Standard deviation of the forecast error of accelerations

	Standard deviation (m/s^2)		
	DOFs		
Cases	2	5	8
1	0.0454	0.0538	0.0381
2	0.0894	0.1223	0.0993

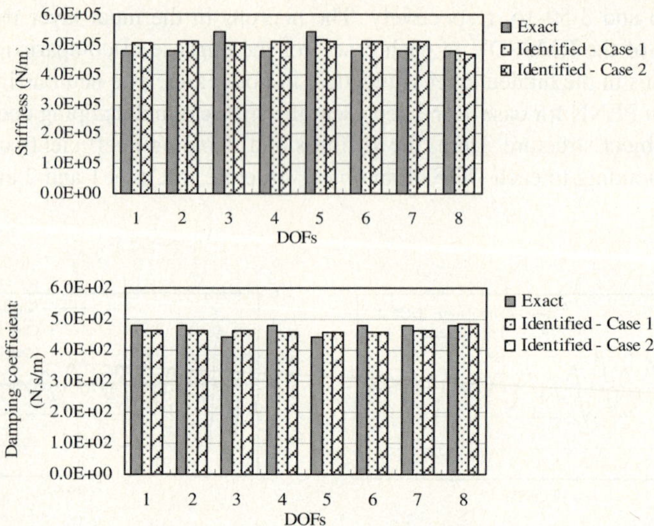

Fig. 8. Stiffness and damping coefficients updating results in Cases 1 and 2

Figure 9 shows the ratio of the updated stiffness and damping coefficients to their exact values in both cases. It is clear that the inter-story stiffness and damping coefficient parameters of the object structure can be updated accurately not only in Case 1 but also in Case 2 where spatially incomplete measurements are available. Moreover, the difference of the updated parameters between Case 1 and Case 2 is very small. The maximum relative errors between the updated stiffness and the true value of the above object structure are within 8%. This finding implies that model updating can be carried out by the use of spatially incomplete measurements with a neural network if only the PENN are trained to describe the relationship between the defined evaluation index and the structural parameters.

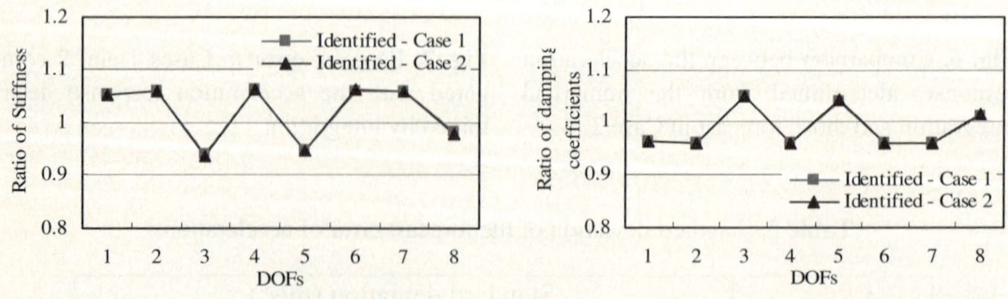

Fig. 9. Ratio of identified stiffness and damping coefficients to their exact values

In order to investigate the statistic performance of the model updating methodology in the interested scope, the parameter updating results for a number of object structures with different structural parameters are introduced. Let the stiffness and damping coefficients of DOFs 2-7 have 10% and 20% decrease and it of DOF 8 have 20% and 30% decrease respectively, damping coefficients have the same degree of increase simultaneously, totally 384 combinations are constructed. Each combination is treated as an object structure. The averages of the ratios of the stiffness and damping coefficients to their exact values corresponding to each DOF in Cases 1 and 2 is shown in Table 4. It is clear that the average ration of each parameter is very close to 1. The proposed algorithm can give a reliable model updating results even in Case 2 where only partial dynamic response measurements are available.

Table 4. Average of ratio of identified parameters to exact values of 384 object structures

		DOF							
Case	Parameters	1	2	3	4	5	6	7	8
1	Stiffness	0.996	1.000	0.997	1.003	1.009	1.002	1.000	0.994
	Damping	1.008	1.004	1.007	1.003	0.998	1.003	1.004	1.016
2	Stiffness	0.994	1.001	0.996	1.003	1.008	1.003	1.000	0.989
	Damping	1.009	1.004	1.007	1.002	0.999	1.003	1.004	1.018

4 Conclusions

A neural networks based structural model updating methodology with the direct use of free vibration acceleration measurements is proposed. The performance of the proposed methodology for spatially incomplete measurements is investigated. Basing on the numerical simulations, the following conclusions can be made:

1. The rationality of the proposed methodology is explained and the theory basis for the construction of acceleration-based neural network(ANN) and parametric evaluation neural network(PENN) are described.
2. ANN is a robust way to identify structures in a nonparametric format though spatially complete acceleration measurements are unavailable.
3. Based on the trained ANN, a PENN can be trained corresponding to the available dynamic measurements. Structural inter-storey stiffness and damping coefficient parameters of the object structure are updated with enough accuracy by the direct use of acceleration measurements. The difference of the updating accuracy between the complete and incomplete measurements is very small as long as the PENN can be trained to describe the relationship between the evaluation index and the corresponding structural parameters. This finding is very important for real structural model updating because the dynamic response measurements of existing structures are usually limited and spatially incomplete.

The methodology does not require any time-consuming eigen-values or model extraction form measurement and uses is free vibration acceleration measurements directly, it can be a practical tool for on-line and near real-time identification of civil engineering structures instrumented with health monitoring system.

Acknowledgments

The author gratefully acknowledges the financial support from the "Lotus (Furong) Scholar Program" provided by Hunan provincial government and the corresponding Grant-in-aid for scientific research from Hunan University, P.R. China. Partial support from the "985 project" of the Center for Integrated Protection Research of Engineering Structures (CIPRES) at Hunan University is also appreciated.

References

1. Doebling, S.W., Farrar, C.R., Prime, M.B.: A Summary Review of Vibration-based Damage Model Updating Methods, Shock and Vibration Digest, 30(2) (1998) 91-105
2. Wu, Z.S., Xu, B., Harada, T.: Review on Structural Health Monitoring for Infrastructure, Journal of Applied Mechanics, JSCE, 6 (2003) 1043-1054
3. Li, H., Ding H.: Progress in Model Updating for Structural Dynamics, Advances in Mechanics, 35(2) (2005) 170-180
4. Alampalli, S., Fu G., Dillon E.W.: Signal versus Noise in Damage Detection by Experimental Modal Analysis, Journal of Structural Engineering, ASCE, 123(2) (1997) 237-245
5. Masri, S.F., Smyth, A.W., Chassiakos, A.G., Caughey, T.K., Hunter, N.F.: Application of Neural Networks for Detection of Changes in Nonlinear Systems, Journal of Engineering Mechanics, ASCE, 126(7) (2000) 666-676
6. Smyth, A.W., Pei, J.S., Masri, S.F.: System Model Updating of the Vincent Thomas Suspension Bridge Using Earthquake Inputs, Earthquake Engineering & Structural Dynamics, 32 (2003) 339-367
7. Chang, C.C., Zhou L.: Neural Network Emulation of Inverse Dynamics for a Magnetorheological Damper, Journal of Structural Engineering, ASCE, 128(2) (2002) 231-239
8. Worden, K.: Structural Fault Detection Using a Novelty Measure, Journal of Sound and Vibration, 201(1) (1997) 85-101
9. Yun, C.B., Bahng, E.Y.: Substructural Model Updating Using Neural Networks, Computers and Structures, 77 (2000) 41-52
10. Xu, B.: Time Domain Substructural Post-earthquake Damage Detection Methodology with Neural Networks, Lecture Note of Computer Science 3611 (2005) 520-529
11. Wu, Z.S., Xu, B., Yokoyama, K.: Decentralized Parametric Damage Based on Neural Networks, Computer-Aided Civil and Infrastructure Engineering, 17 (2002) 175-184
12. Xu, B., Wu, Z.S. Chen, G., Yokoyama, K.: Direct Model Updating of Structural Parameters from Dynamic Responses with Neural Networks, Engineering Applications of Artificial Intelligence, 17(8) (2004) 931-943
13. Xu, B., Wu, Z.S., Yokoyama, K., Harada, T., Chen, G.: A Soft Post-earthquake Damage Identification Methodology Using Vibration Time Series, Smart Materials and Structures, 14(3) (2005) s116-s124

Appearance-Based Gait Recognition Using Independent Component Analysis*

Jimin Liang, Yan Chen, Haihong Hu, and Heng Zhao

School of Electronic Engineering, Xidian University
Xi'an, Shaanxi 710071, China
{jimleung, chenyan}@mail.xidian.edu.cn

Abstract. For human identification at distance (HID) applications, gait characteristics are hard to conceal and has the inherent merits such as non-contact and unobtrusive. In this paper, a novel appearance-based method for automatic gait recognition is proposed using independent component analysis (ICA). Principal component analysis (PCA) is performed on image sequences of all persons to get the uncorrelated PC coefficients. Then, ICA is performed on the PC coefficients to obtain the more independent IC coefficients. The IC coefficients from the same person are averaged and the mean coefficients are used to represent individual gait characteristics. For improving computational efficiency, a fast and robust method named InfoMax algorithm is used for calculating independent components. Gait recognition performance of the proposed method was evaluated by using CMU MoBo dataset and USF Challenge gait dataset. Experiment results show the efficiency and advantages of the proposed method.

1 Introduction

Gait recognition tries to identify a person by the manner he walks. It has been attracting more and more interests from biometric recognition researchers. This is driven by the increasing demand for automated human identification systems in security sensitive occasions. Compared with other kinds of biometrics (such as face, iris, and fingerprint), gait has the merit of non-contact, unobtrusive and can be used for human recognition at a distance when other biometrics are obscured. What's more, gait characteristics are hard to disguise and conceal, especially on occasions of committing crime.

Different human have different body structures and walking manners. Psychological researches [1] have also showed that gait signatures provide reliable clue for human identification. In recent years, various approaches have been proposed for gait recognition. These approaches can be divided into two categories: model-based [2, 3] and appearance-based [4, 5]. Model-based approaches focus on setting up a model and perform recognition by model matching. Appearance-based approaches perform recognition using some spatial and/or temporal features extracted from the gait image sequences.

* Project supported by NSF of China (60402038) and NSF of Shannxi (2004f39).

L. Jiao et al. (Eds.): ICNC 2006, Part I, LNCS 4221, pp. 371 – 380, 2006.

In this paper, we introduce a new gait recognition method using independent component analysis (ICA). Our method is motivated by the successful applications of ICA in face recognition [6]. In [6], Bartlett proposed two architectures for performing ICA on images, namely ICA Architecture I and II. Architecture I found spatially local basis images for the faces. Architecture II produced a factorial face code. The ICA based face recognition method shows good classification performance. As for gait recognition, there are not many reports on ICA-based methods. Lu and Zhang [7, 8] constructed one dimensional (1D) signals from the human silhouette and performed ICA on these signals for gait recognition. However, Lu and Zhang evaluated their methods using small size datasets (6 subjects and 20 subjects respectively) and got limited results.

Our proposed method is appearance-based. Firstly, we perform principal component analysis (PCA) directly on image sequences of all persons to get the uncorrelated PC coefficients. Then, ICA is performed on the PC coefficients to obtain the more independent IC coefficients. The IC coefficients from the same person are averaged and the mean coefficients are used to represent individual gait characteristics. ICA Architecture II is used in our method because of its better classification performance than Architecture I. We evaluated the proposed method using CMU MoBo dataset (25 subjects) and USF Gait Challenge dataset (71 subjects) respectively.

Section 2 introduces the principle of ICA Architecture II, and Section 3 outlines our new method. The experiment data and experiment results are shown in section 4. The last section is conclusion and future work.

2 Image Sequence Representation Using ICA

Independent component analysis (ICA) has close relationship with principal component analysis (PCA). PCA is a popular statistical method for image representation. The goal of PCA is to find a set of basis images as coordinates axes on which an image can be projected, and the projection coefficients are uncorrelated. The individual image can be represented as a linear combination of the basis images weighted by the uncorrelated PC coefficients. However, PCA can only separate first-order and second-order linear dependencies between pixels and high-order dependencies maybe still exist in the joint distribution of PC coefficients. Statistically, some important information may exist in the high-order relationships among the image pixels. It seems reasonable to expect that better basis images may be found by methods which are sensitive to high-order statistics. ICA, a generalization of PCA, is one such method. ICA can minimize both second-order and high-order dependencies of the coefficients and can obtain the more independent IC coefficients. In some applications, ICA has better performance than PCA.

Motivated by the ICA based image representation, we apply ICA to dynamic image sequence representation for gait recognition. We perform principal component analysis (PCA) directly on image sequences of all persons to get the uncorrelated PC coefficients. Then, ICA is performed on the PC coefficients to obtain the more independent IC coefficients. This follows the ICA Architecture II proposed by Bartlett in [6]. The difference between two architectures proposed in [6] is that

Architecture I tries to find independent basis images, while Architecture II tries to find independent projection coefficients. ICA Architecture II is used in our method because of its better classification performance than Architecture I.

The diagram of ICA Architecture II is depicted in Fig.1 [9], where each column of S is a set of unknown independent coefficients which linearly combine the unknown basis image sequences in columns of A to form the input gait image sequences in columns of X. The ICA algorithm learns the separating matrix W, which is used to recover S in C, i.e. C is the estimation of S. Each row of the separating matrix W is a basis image sequence, i.e. W is the inverse matrix of A.

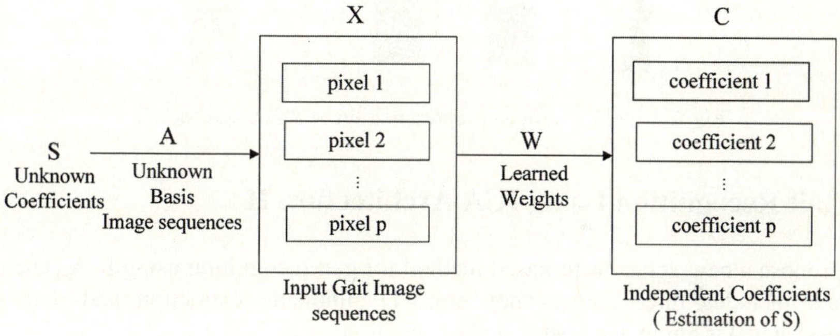

Fig. 1. ICA Architecture II: finding statistically independent coefficients

The relationships depicted in Fig.1 can be expressed by equations as follows.

$$X = AS, \tag{1}$$

$$C = WX = WAS, \tag{2}$$

$$W = A^{-1}. \tag{3}$$

ICA attempts to make the elements in each column of C as independent as possible.

In this paper, ICA is performed on the PC coefficients instead of directly on the input image sequences to reduce the dimensionality and computational complexity [6]. In this way, C can be estimated by

$$C = WP^T, \tag{4}$$

where $P = (p_1, p_2, \cdots, p_i, \cdots p_n)$ in which p_i is a set of PC coefficients corresponding to a gait sequence in one column of X.

For gait image sequence representation, a gait sequence can be expressed by a linear combination of a set of basis image sequences weighted by a set of independent IC coefficients. Their relationship is depicted in Fig.2, where x_i is a gait image sequence, $u_1, u_2, \cdots, u_n$ are the independent IC coefficients corresponding to x_i, $a_1, a_2, \cdots, a_n$ are the basis image sequences in columns of A.

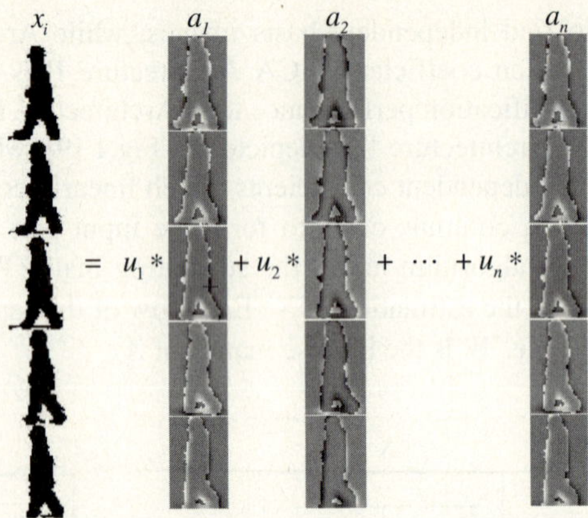

Fig. 2. The IC coefficient representation of a gait sequence x_i

3 Gait Recognition Using ICA Architecture II

We propose a new appearance-based method for gait recognition using ICA. The new method includes three steps. They are: (1) silhouette extraction and data pre-processing, (2) training, (3) testing and recognition.

3.1 Silhouette Extraction and Pre-processing

In order to reduce the sensitivity to variations of clothes color and texture, the proposed method uses the binary silhouette for gait recognition. Human body silhouette extraction is achieved by simple background subtraction. The extracted silhouettes may have holes, the boundary of the silhouette may be discontinuous, and static background pixels can be mistakenly included (such as the shadows). However, as shown in section 4, promising recognition results are obtained in our experiment.

In order to reduce the computational complexity, all images are cropped and scaled to 44 pixels wide by 64 pixels high.

3.2 Training

Two periods from each subject's image sequence are chosen as the training data. To make the training data of the same size, only the first 28 frames of each period are used no matter how many frames a period has. All the 28 frames are concatenated into a long vector. Thus, each period can be denoted by a long vector, which is referred as periodic vector in the following text.

All the periodic vectors are put into one matrix X . If the gait dataset has N subjects, and each subject has 2 gait periods used for training, the matrix X will have 2N columns, given by

$$X = \{x_{1,1}, x_{1,2}, x_{2,1}, x_{2,2}, \cdots, x_{N,1}, x_{N,2}\}, \tag{5}$$

where $x_{i,j}$ ($i = 1 \cdots N$, $j = 1,2$) represents each periodic vector, with index i ranging over all subjects and j ranging over all periodic vectors for subject i.

Firstly, PCA is performed on X. Certain number of eigenvectors correspond to the largest number of eigenvalues are chosen as the PC eigenspace. Each column of X is projected into the PC eigenspace to get the PC coefficients P. The PCA operation largely reduces the dimension of the data. Then, ICA is performed on the PC coefficients P (as in equation (4)) and the more independent IC coefficients C are obtained, which can be denoted as

$$C = \{c_{1,1}, c_{1,2}, c_{2,1}, c_{2,2}, \cdots, c_{N,1}, c_{N,2}\}, \tag{6}$$

where $c_{i,j}$ ($i = 1 \cdots N$, $j = 1,2$) is a set of independent IC coefficients corresponding to each $x_{i,j}$, i and j have the same meaning as in $x_{i,j}$.

Secondly, the coefficient vectors from the same subject are averaged. The result is denoted by

$$C_{average} = \{ \frac{(c_{1,1} + c_{1,2})}{2}, \frac{(c_{2,1} + c_{2,2})}{2}, \cdots, \frac{(c_{N,1} + c_{N,2})}{2} \}. \tag{7}$$

The averaging step can reduce the randomicity of the data. It has been confirmed by experiment results that the averaging step can increase the recognition rate.

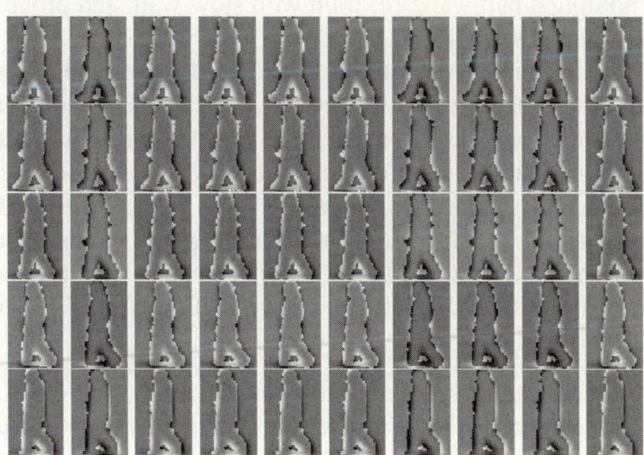

Fig. 3. The first 10 IC axes (columns of $A = W^{-1}$) obtained in Architecture II; it only shows the first 5 images of each IC axis

From the IC coefficients, the IC axes which from the IC eigenspace can be inversely calculated[1]. The first 10 IC axes are shown in Fig.3 and it only shows the first 5 images of each IC axis.

---

[1] In this paper, ICA is performed on the PC coefficients, we didn't use the IC axes directly, but we can calculate the IC axes from the IC coefficients inversely.

For improving computational efficiency, a fast and robust Information-Maximization based algorithm named InfoMax for calculating independent components is used. The InfoMax algorithm [10] was derived from the principle of optimal information transfer in neurons with sigmoidal transfer functions [11].

3.3 Testing and Recognition

Two periods from the subject's probe image sequences are chosen as the probe data. The probe data are projected onto the PC eigenspace obtained in section 3.2 and then ICA is performed to compute the probe IC coefficients. The nearest neighbor algorithm based on Mahalanobis distance is used for recognition.

4 Numerical Experiment

4.1 Experiment Data

We use two datasets to evaluate the proposed method. The first dataset is CMU MoBo gait dataset [12]. The other dataset is USF Gait Challenge dataset [13].

The CMU MoBo dataset contains 25 subjects walking on a treadmill under four different conditions: slow walk, fast walk, holding a ball walk and inclined walk. Seven cameras are mounted at different angles. In this experiment, we only use the profile data. We carried out 6 experiments on CMU MoBo dataset: (a) train with slow walk and probe with slow walk; (b) train with fast walk and probe with fast walk; (c) train with incline walk and probe with incline walk; (d) train with holding a ball walk and probe with holding a ball walk; (e) train with slow walk and probe with holding a ball walk; (f) train with fast walk and probe with slow walk.

There are 25 subjects in experiment a, b, c and f , 24 subjects in experiment d and e, this because the CMU MoBo dataset only provide 24 subjects in holding a ball gait data.

The USF Gait Challenge dataset is made up of outdoor gait sequences of 71 subjects walking in an elliptical path in front of the cameras. Different sequences of the same person are collected to explore variations in gait recognition performance with variations in three factors: (1) difference in ground surfaces (Concrete and Grass); (2) difference in view angles (Right and Left); (3) difference in shoe types (A and B). We conduct 7 experiments on USF dataset. The experiment details are shown in Table 1.

Table 1. USF Gait Challenge dataset: the common training set (G, A, R consisting 71 subjects) with 7 probe sets .The numbers in the brackets are the number of subjects in each probe set.

Experiment label	Probe	Difference
A	G,A,L(71)	View
B	G,B,R(41)	Shoe
C	G,B,L(41)	Shoe, View
D	C,A,R(70)	Surface
E	C,B,R(44)	Surface, Shoe
F	C,A,L(70)	Surface, View
G	C,B,L(44)	Surface, Shoe , View

4.2 Experiment Results and Analysis

The experiment results are shown in Table.2 for the CMU MoBo dataset and Fig.6 for the USF Gait Challenge dataset respectively.

Table 2. Experiment results for CMU MoBo dataset in terms of the recognition rate P at ranks 1, 2 and 5

Train vs. Probe (profile)	Method in [5]			Method in [15]			Our method		
	P (%) at rank			P (%) at rank			P (%) at rank		
	1	2	5	1	2	5	1	2	5
Slow vs. Slow	95.8	95.8	100	100	100	100	100	100	100
Fast vs. Fast	95.8	95.8	100	96.0	100	100	100	100	100
Incline vs. Incline		\		95.8	100	100	100	100	100
Ball vs. Ball	95.4	100	100	100	100	100	100	100	100
Slow vs. Ball		\		52.2	60.9	69.6	79.2	87.5	91.7
Fast vs. Slow	75.0	83.3	87.5		\		64.0	88.0	96.0

Table 2 shows that our proposed method performed quite well for the CMU MoBo dataset when the condition of the training data and the probe data are the same. But it doesn't work so well when train with slow walk data while probe with holding a ball data (experiment e) and train with fast walk data while train with slow walk data (experiment f), this is mainly because of the variations of the profile appearance when differences in walking action and walking speed. For experiment Slow vs. Ball, our method still performs better than that in [15]. For experiment Fast vs. Slow, although our recognition rate at rank=1 is lower than that in [5], our proposed method have much faster convergent speed and only at rank=2, our recognition rate exceeds that in [5].

Fig.4.(a) shows the cumulative match characteristic curves of the baseline recognition algorithm in [15] for the USF Gait Challenge dataset. Fig. 6 (b) shows the recognition results of our method. It can be seen clearly that our method has better recognition rate and convergent speed than that in [15]. Especially for experiment B, our method's recognition rate can reach 100% at rank 4. For experiment A and C, their recognition rate can exceed 90% at rank 4. This mainly owe to the reason that the probe data are on the same grass (G) road as the training data. The recognition rates of D and E are not as high as A, B and C, but higher than that of F and G. This is due to that D and E have the same view angle (Right) as the training data does, but F and G are not. From the analysis above, we can get the conclusion that the road condition plays more important role than the view angle does in gait recognition, and the shoe type nearly has no affect on gait recognition.

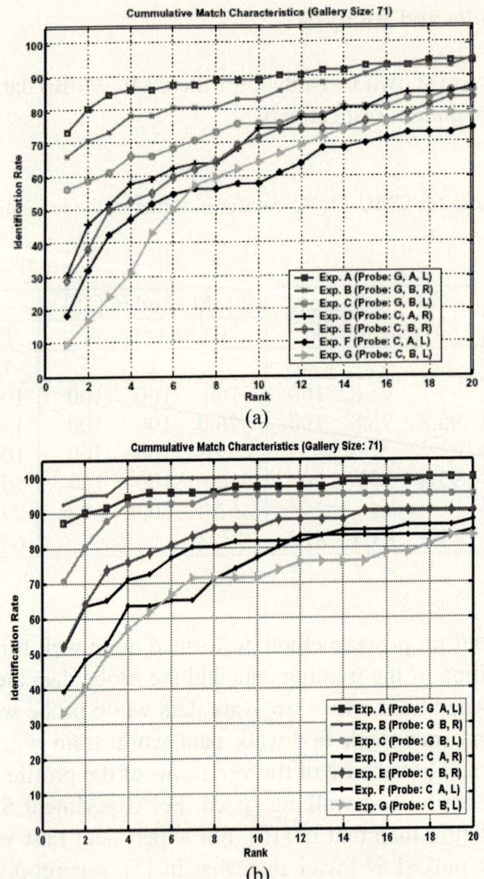

Fig. 4. (a): The CMC curves of the baseline algorithm experiment results [15]. Each curve corresponds to a probe set in Table 1 **(b)**: The CMC curves of our experiment results. Each curve corresponds to a probe set in Table 1.

The relationship of recognition rate and the number of PC axes is analyzed[2]. The results are shown in Fig.5. It can be seen that the main tendency of the recognition curves are increasing with the increasing number of PC axes. But the highest points do not always appear at the point when the percent of the image variance reach 100% (140 PC axes). This mainly due to the fact that there may exist some noise in the last several PC axes which can influence the recognition capability. So we only choose the number of the PC axes which account for about 95% of the variance in gait images (120 to130 PC axes) in the preceding experiments.

---

[2] In this paper, ICA is performed on the PC coefficients, we didn't use the IC axes directly. So we only consider the relationship between recognition rate and the number of the PC axes.

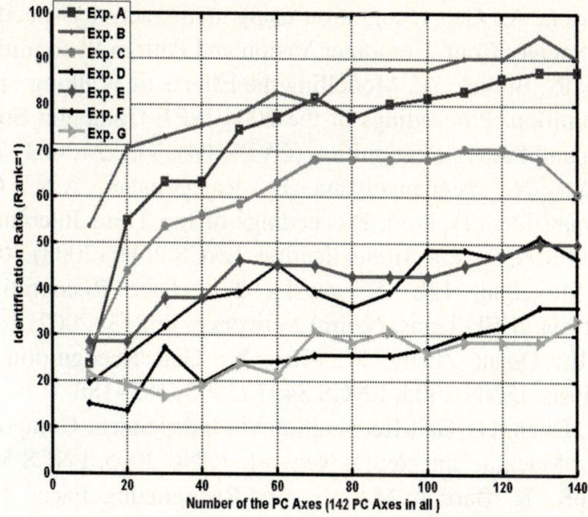

Fig. 5. Recognition rates (rank=1) for ICA Architecture II as a function of the number of subspace dimensions. Each curve corresponds to a probe set in Table 1. Recognition rates were measured for PC eigenspace dimensionalities starting at 10 and increasing by 20 dimensions up to a total of 140.

5 Conclusion and Future Work

We propose a new appearance-based method for gait recognition using independent component analysis. ICA is performed on PC coefficients to obtain more independent IC coefficients. Then, IC coefficients from the same person are averaged. Classification is performed on the averaged IC coefficients by nearest neighbor matching using Mahalanobis distance.

The merits of the proposed method are: 1) it is easy to understand and implement; 2) it can tolerant noisy image data; 3) it is insensitive to clothing color and texture; 4) it is robust under different walking conditions.

Our method is evaluated with two datasets (having 25 and 71 person respectively) with different view angles (left and right), different background conditions (indoors and outdoors), different road surfaces (grass and concrete), different shoe types (A and B), different walking actions (slow and ball) and different walking speed (fast and slow). Generally speaking, our proposed method produces promising results and show that our method is robust, efficient and advantageous in recognition.

However, there still are some problems to be solved, such as the computational complexity and the influence of the walking speed on gait recognition. It also worthies further research on the comparative study about ICA Architecture I and II.

References

1. Stevenage, S.V., Nixon, M.S., Vince, K.: Visual Analysis of Gait as a Cue to Identity. Applied Cognitive Psychology (1999) 13:513–526
2. BenAbdelkader, C., Culter, R., and Davis, L.: Stride and cadence as a biometric in automatic person identification and verification. Proc. Int. Conf. Automatic Face and Gesture Recognition, Washington, DC (2002) 372–376

3. Bobick, A., Johnson, A.: Gait recognition using static, activity-specific parameters. Proc. IEEE Computer Society Conf. Computer Vision and Pattern Recognition (2001)
4. Tanawongsuwan, R., Bobick, A.: Modelling the Effects of Walking Speed on Appearance-based Gait Recognition. Proceedings of the 2004 IEEE Computer Society Conference on Computer Vision and Pattern Recognition (CVPR'04), Vol. 2 (2004) 783-790
5. Kale, A., Cuntoor, N., Yegnanarayana, B., Rajagopalan, A.N., Chellappa, R.: Gait Analysis for Human Identification. Proceedings of the Third International conference on Audio and Video Based Person Authentication, LNCS 2688 (2003) 706–714
6. Bartlett, M.S., Movellan, J.R., Sejnowski, T.J.: Face Recognition by Independent Component Analysis. IEEE Trans. Neural Network*s*, Vol.13 (2002) 1450–1464
7. Lu, J., Zhang, E., Duan, Zhang, Z., Xue, Y.: Gait Recognition Using Independent Component Analysis. ISNN 2005, LNCS 3497 (2005) 183–188
8. Zhang, E., Lu, J., Duan, G.: Gait Recognition via Independent Component Analysis Based on Support Vector Machine and Neural Network. ICNC 2005, LNCS 3610 (2005) 640-649.
9. Draper B.A, BaeK, K, Bartlett M.S., et al.: Recognizing Faces with PCA and ICA. Computer Vision and Image Understanding 91(2003) 115 – 137
10. Bell, A.J., Sejnowski, T.J.: An Information-maximization Approach to Blind Separation and Blind Deconvolution. Neural Comput, Vol. 7, no. 6 (1995) 1129–1159
11. Laughlin, S.: A Simple Coding Procedure Enhances a Neuron's Information Capacity. Z. Naturforsch., Vol. 36 (1981) 910–912
12. http://hid.ri.cmu.edu
13. http://marathon.csee.usf.edu/GaitBaseline/
14. Zhang, R., Vogler, C., Metaxas, D.: Human Gait Recignition. IEEE Computer Society Conference on Computer Vision and Pattern Recognition Workshops (CVPRW'04) (2004) 1063-6919
15. Phillips, P.J., Sarkar, S., Robledo, I., et al.: Baseline Results for the Challenge Problem of Human ID Using Gait Analysis. Proceedings of the Fifth IEEE International Conference on Automatic Face and Gesture Recognition (FGR.02) (2002) 0-7695-1602-5/02

Combining Apriori Algorithm and Constraint-Based Genetic Algorithm for Tree Induction for Aircraft Electronic Ballasts Troubleshooting

Chaochang Chiu[1], Pei-Lun Hsu[2], and Nan-Hsing Chiu[1]

[1] Department of Information Management, Yuan Ze University, Taiwan, R.O.C.
imchiu@saturn.yzu.edu.tw

[2] Department of Electronic Engineering, Ching Yun University, Taiwan, R.O.C.
hsupl@mail.cyu.edu.tw

Abstract. Reliable and effective maintenance support is vital to the airline operations and flight safety. This research proposes the hybrid of apriori algorithm and constraint-based genetic algorithm (ACBGA) approach to discover a classification tree for electronic ballasts troubleshooting. Compared with a simple GA (SGA) and the Apriori algorithms with GA (AGA), the ACBGA achieves higher classification accuracy for electronic ballast data.

1 Introduction

The electronic ballast plays an important role in providing proper lights for passengers and flight crews during a flight. Unstable cabin lighting, such as flash and ON/OFF problems, would degrade the flight quality. The maintenance records of electronic ballasts generally contain information about the number of defective units found, the procedures taken, and the inspection or repair status. In this paper we propose a novel approach that combines the Apriori algorithm (Agrawal et al., 1993) and constraint-based genetic algorithm (CBGA) (Chiu and Hsu, 2005) to discover classification trees. Apriori algorithm, one of the common seen association rule algorithms, is used for attributes selection (Agrawal et al., 1993); therefore those related input attributes can be determined before proceeding the GA's evolution for producing classification rules (Carvalho & Freitas, 2004). The constraint-based reasoning is used to push constraints along with data insights into the rule set construction. This research applied tree search and forward checking (Haralick & Elliott, 1980; Brailsford et al, 1999) techniques to reduce the search space from possible gene values that can not meet predefined constraints during the evolution process. This approach provides a chromosome-filtering mechanism prior to generating and evaluating a chromosome. Thus insignificant or irreverent rules can be precluded in advance via the constraint network. A prototype troubleshooting system based on ACBGA approach was developed using the aircraft maintenance records of electronic ballasts.

2 The Aircraft Electronic Ballasts

The aircraft electronic ballasts used to drive fluorescent lamps can be mounted on a panel such as the light deflector of a fluorescent lamp fixture. There are two control

L. Jiao et al. (Eds.): ICNC 2006, Part I, LNCS 4221, pp. 381 – 384, 2006.

lines connecting with the ballast set and control panel for ON/OFF as well as BRIGHT/DIM modes among which DIM mode is used at night when the cabin personnel attempts to decrease the level of ambient light. To diagnose which component is of malfunction, the mechanics usually measure the alternating current in BRIGHT mode, DIM mode when the electronic ballast turns on or off. In addition, the check of light stability and the illumination status is also important to aid the maintenance decision. Each maintenance record is summarized in Table 1. Each category in the outcome attribute represents a different set of replacement parts. For instance, category C_1 denotes the replacement parts of a transformer and a capacitor. Category C_2 denotes the replaced parts of an integrated circuit, a transistor and a fuse.

Table 1. The Record Description

Input Attributes	Data Type	Range
Alternating Current on Bright Mode When Electronic Ballast Turns On	Categorical	15 intervals of (amp)
Alternating Current on DIM Mode When Electronic Ballast Turns On	Categorical	11 intervals of (amp)
Alternating Current on Bright Mode When Electronic Ballast Turns Off	Categorical	16 intervals of (amp)
Alternating Current on DIM Mode When Electronic Ballast Turns Off	Categorical	16 intervals of (amp)
Is Light Unstable When Electronic Ballast Turns On	Categorical	0 and 1
Is It Not Illuminated When Electronic Ballast Turns On	Categorical	0 and 1
Outcome Attribute		
Replacement Parts	Categorical	$C_1, C_2, \ldots, C_{10}$

3 The Hybrid of Association Rule Algorithms and Constraint-Based Genetic Algorithms (ACBGA)

The proposed ACBGA approach consists of three modules. According to Fig. 1, these modules are: Rule Association; GA Initialization; Fitness Evaluation; Chromo some Screening; and GA Operation. By executing apriori algorithm the Rule Association module produces association rules. In this research, the data items of categorical attributes are used to construct the association rules. The association rule here is an implication of the form X→Y where X is the conjunction of conditions, and Y is the classification result. The rule X→Y has to satisfy user-defined minimum support and minimum confidence levels. The formal form of a tree node is expressed by the following expression:

A and B → C;

where A denotes the antecedent part of the association rule; B is the conjunction of inequality functions in which the continuous attributes, relational operators and splitting values are determined by the GA; and C is the classification result directly obtained from the association rule. The association rules for each classification result are used not only to construct the candidate classification trees for GA initialization, but also filter the insignificant or irreverent attributes in rules.

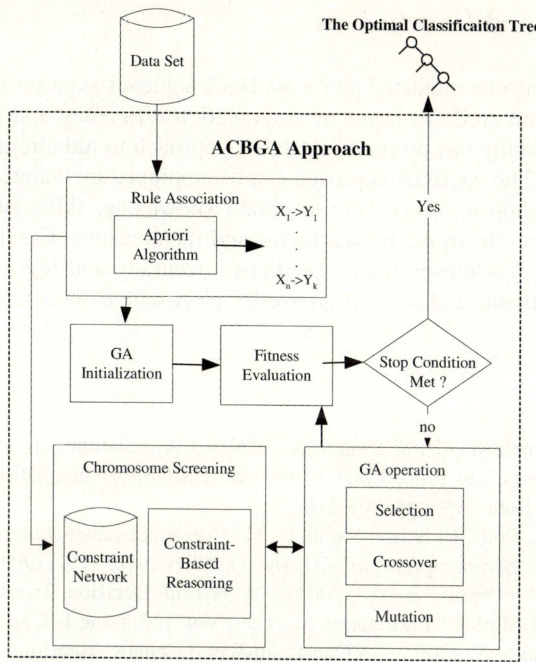

Fig. 1. The Conceptual Diagram of ACBGA

4 The Experiments and Results

The evaluation of the classification tree generated by each ACBGA was based on a five-fold cross validation. That is, each training stage used 4/5 of the entire data records; with the rest 1/5 data records used for the testing stage. Two hundred and fifty electric ballast maintenance records of Boeing 747-400 from the accessory shop were used to construct the trouble-shooting model. Both the training and testing performance are summarized in Table 2. These results are also based on SupportDiscount (=10%) and confidence value (=100). Based on this produced classification tree, the accuracy rates for the training and testing stages are able to reach 84.5% and 84%, respectively.

Table 2. The Summarized Learning Performance for SGA, AGA, and ACBGA (based on 5-fold average)

Gen.	100		200		300		400		500	
	Train	Test	Train	Test	Train	Test	Train	Test	Train	Test
SGA	58.30%	56.40%	67.60%	63.20%	68.80%	64.00%	71.60%	68.00%	72.40%	68.00%
AGA	76.90%	76.40%	78.60%	75.20%	79.30%	74.80%	79.30%	74.80%	80.00%	74.80%
ACBGA	83.30%	79.60%	84.10%	79.60%	84.50%	80.40%	85.00%	80.80%	85.10%	80.40%

5 Discussion and Conclusions

According to the results indicated above ACBGA achieves superior learning performance than SGA and AGA in terms of computation efficiency and accuracy for this application. This study has proposed ACBGA approach to aid aircraft electronic ballast maintenance. The ACBGA approach can be employed by maintenance mechanics in the accessory shop to assist them to obtain knowledge, skills and experience required for effective electronic ballast repair and maintenance. Comparing with SGA and AGA, ACBGA achieves higher predictive accuracy and less computation time required in constructing a classification tree for electronic ballasts troubleshooting.

References

1. Agrawal, R., Imielinski, T., & Swami, A., "Mining Association between Sets of Items in Massive Databases," In Proceedings of the ACM SIGMOD International Conference on Management of Data, 1993, pp. 207-216.
2. Brailsford, S. C., Potts, C. N. and Smith, B. M., Constraint satisfaction problem: Algorithm and applications, *European Journal of Operational Research,* 119, (1999) 557-581.
3. Carvalho, D. R., Freitas, A. A. (2004). "A Hybrid Decision Tree/Genetic Algorithm Method for Data Mining," Information Sciences, Vol. 163, Issue 1-3, pp. 13-35.
4. Chiu, C. and Hsu, P. L. (2005). "A Constraint-Based Genetic Algorithm Approach for Mining Classification Rules," *IEEE Trans. On Systems, Man & Cybernetics, Part C,* Vol.35, No.2, May 2005, pp. 205-220.
5. Haralick, R. and Elliott, G., Increasing Tree Search Efficiency for Constraint Satisfaction Problems, *Artificial Intelligence,* 14, (1980) 263-313.

Container Image Recognition Using ART2-Based Self-organizing Supervised Learning Algorithm

Kwang-Baek Kim[1], Sungshin Kim[2], and Young-Ju Kim[3]

[1] Dept. of Computer Engineering, Silla University, Korea
gbkim@silla.ac.kr

[2] School of Electrical Engineering, Pusan National University, Korea
sskim@pusan.ac.kr

[3] Dept. of Computer Engineering, Silla University, Korea
yjkim@silla.ac.kr

Abstract. This paper proposed an automatic recognition system of shipping container identifiers using fuzzy-based noise removal method and ART2-based self-organizing supervised learning algorithm. Generally, identifiers of a shipping container have a feature that the color of characters is black or white. Considering such a feature, in a container image, all areas excepting areas with black or white colors are regarded as noises, and areas of identifiers and noises are discriminated by using a fuzzy-based noise detection method. Noise areas are replaced with a mean pixel value of the whole image and areas of identifiers are extracted by applying the edge detection by Sobel masking operation and the vertical and horizontal block extraction in turn to the noise-removed image. Extracted areas are binarized by using the iteration binarization algorithm, and individual identifiers are extracted by applying 8-directional contour tacking method. This paper proposed an ART2-based self-organizing supervised learning algorithm for the identifier recognition, which creates nodes of the hidden layer by applying ART2 between the input and the hidden layers and improves the performance of learning by applying generalized delta learning and Delta-bar-Delta algorithm between the hidden and the output layers. Experiments using many images of shipping containers showed that the proposed identifier extraction method and the ART2-based self-organizing supervised learning algorithm are more improved compared with the methods previously proposed.

1 Introduction

Identifiers of shipping containers are given in accordance with the terms of ISO standard, which consist of 4 code groups such as shipping company codes, container serial codes, check digit codes and container type codes [1][2]. And, only the first 11's identifier characters are prescribed in the ISO standard and shipping containers are able to be discriminated by automatically recognizing the first 11's characters. But, other features such as the foreground and background colors, the font type and the size of container identifiers, etc., vary from one container to another since the ISO standard doesn't prescribes other features except code type [2][3]. Since identifiers are printed on the surface of containers, shapes of identifiers are often impaired by the environmental factors during the transportation by sea. The damage to a container

L. Jiao et al. (Eds.): ICNC 2006, Part I, LNCS 4221, pp. 385 – 394, 2006.

surface may lead to a distortion of shapes of identifier characters in a container image. So, the variations in the feature of container identifiers and noises make it quite difficult the extraction and recognition of identifiers using simple information like color values [4].

Generally, container identifiers have another feature that the color of characters is black or white. Considering such a feature, in a container image, all areas excepting areas with black or white colors are regarded as noises, and areas of identifiers and noises are discriminated by using a fuzzy-based noise detection method. Noise areas are replaced with a mean pixel value of the whole image area, and areas of identifiers are extracted and binarized by applying the edge detection by Sobel masking operation and the vertical and horizontal block extraction to the conversed image one by one. In the extracted areas, the color of identifiers is converted to black and one of background to white, and individual identifiers are extracted by using 8-directional contour tacking algorithm. This paper proposed an ART2-based self-organizing supervised learning algorithm for the identifier recognition, which creates nodes of the hidden layer by applying ART2 between the input layer and the hidden one and improves performance of learning by applying generalized delta learning and the Delta-bar-Delta algorithm [5]. Experiments using many images of shipping containers showed that the proposed identifier extraction method and the ART2-based supervised learning algorithm is more improved compared with the methods proposed previously.

2 Proposed Container Identifier Recognition Method

2.1 Extraction of Container Identifier Areas

Due to the rugged surface shape of containers and noises vertically appeared by an external light, a failure may occur in the extraction of container identifier areas from a container image. To refine the failure problem, this paper proposed a novel method for extraction of identifier areas based on a fuzzy-based noise detection method.

This paper detects edges of identifiers by applying Sobel masking operation to a grayscale image of the original image and extracts areas of identifiers using information on edges. Sobel masking operation is sensitive to noises so that it detects noises by an external light as edges. To remove an effect of noises in the edge detection, first, this paper detects noise pixels by using a fuzzy method and replaces the pixels with a mean gray value. Next, Applying Sobel masking to the noise-removed image, areas of container identifiers are separated from background areas.

2.1.1 Fuzzy-Based Noise Detection

To remove noises by an external light, this paper convert an container image to a grayscale one and apply the membership function like Fig. 1 to each pixel of the grayscale image, deciding whether the pixel is a noise or not. In Fig. 1, C and E are categories being likely to belong to an area of identifiers, and D is the category being likely to be a noise. Eq. (1) shows the expression for the membership function of Fig. 1. The criterion to distinguish pixels of noise and non-noise using the degree of membership in this paper is given in Table 1.

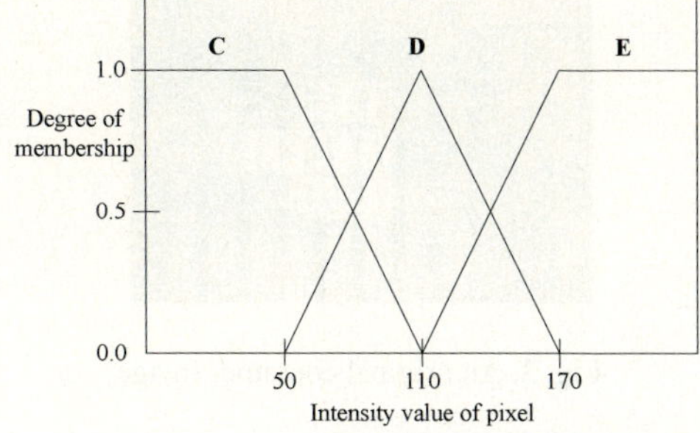

Fig. 1. Membership function(G) for gray-level pixels

$$
\begin{aligned}
&if\ (G \prec 50)\ or\ (G \succ 170)\ then\ u(G) = 0 \\
&else\ if\ (G \succ 50)\ or\ (G \le 110)\ then\ u(G) = \frac{G-50}{110-50} \\
&else\ if\ (G \succ 110)\ or\ (G \le 170)\ then\ u(G) = \frac{110-G}{170-110}
\end{aligned}
\tag{1}
$$

Table 1. Criterion to distinguish pixels of noise and non-noise

pixel of non-noise	$u(\mathrm{G}) < 0.42$
pixel of noise	$u(\mathrm{G}) \geq 0.42$

To observe the effectiveness of the fuzzy-based noise detection, results of edge detection by Sobel masking were compared between the original image and the noise-removed image by the proposed method. Fig. 2 is the original container image, and Fig. 3 is the output image generated by applying only Sobel masking to a grayscale image of Fig. 2. Fig. 4 is results of edge detection obtained by applying the fuzzy-based noise removal and Sobel masking to Fig. 2. First, the fuzzy-based noise detection method is applied to a grayscale image of the original image and pixels detected as noises are replaced with a mean gray value. Next, edges of container identifiers are detected by applying Sobel masking to the noise-removed image. As shown in Fig. 3, noise removal by the proposed fuzzy method generates more efficient results in the extraction of areas of identifiers.

2.2 Binarization of Container Identifier Areas

Currently, the iterative binarization algorithm is mainly used in the preprocessing of pattern recognition.

The iterative binarization algorithm, first, roughly determines an initial threshold, divides an input image to two pixel groups using the threshold, calculates a mean value for each pixel group, and sets the arithmetic mean of two mean values to a new

Fig. 2. An original container image

Fig. 3. Result of edge detection by only Sobel masking

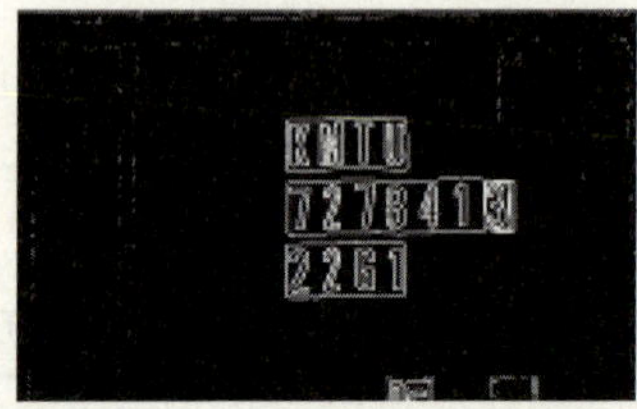

Fig. 4. Result of edge detection by fuzzy-based noise-removal and Sobel masking

threshold. And, the algorithm repeats the above processing until there is no variation of threshold value and sets the last value to the threshold value for binarization operation. In the case of a noise-removed container image, since the difference of intensity between the background and the identifiers is great, the iterative algorithm is able to provide a good threshold value. The binarization process of identifier areas using the iterative algorithm is as follow:

Step 1. Set an initial threshold value T^0.

Step 2. Classify all pixels using the threshold value T^{t-1} given in the previous iteration, and calculate the mean value of background pixels, u_B^t and the mean value of identifier pixels, u_O^t using Eq. (2).

$$u_B^t = \frac{\sum f(i,j)}{N_B}, \quad u_O^t = \frac{\sum f(i,j)}{N_O} \tag{2}$$

where, N_B and N_O are the numbers of background pixels and identifier pixels, respectively.

Step 3. Set the arithmetic mean of u_B^t and u_O^t to new threshold value T^t like Eq. (3).

$$T^t = \frac{\left(u_B^t + u_O^t\right)}{2} \tag{3}$$

Step 4. Terminate the iteration if $T^t = T^{t-1}$. Otherwise, repeat from Step.2 to Step. 4.

2.3 Extraction of Individual Identifiers

This paper extracts individual identifiers by applying 8-directional contour tracking method [6] to binarized areas of container identifiers. In the extraction process, the extraction of individual identifiers is successful in the case that the background color is a general color except white one like Fig. 5, and on the other hand, the extraction is failed in the case with white background color as shown in Fig. 6.

In the binarization process, background pixels of a bright intensity are converted to black and identifier pixels of a dark intensity are converted to white. And, the contour tracking method detects edges of an area with black color, so that it can not detect edges of identifiers from target areas with white background. This paper, for identifier areas with white background, reverses a result of binarization process. That is,

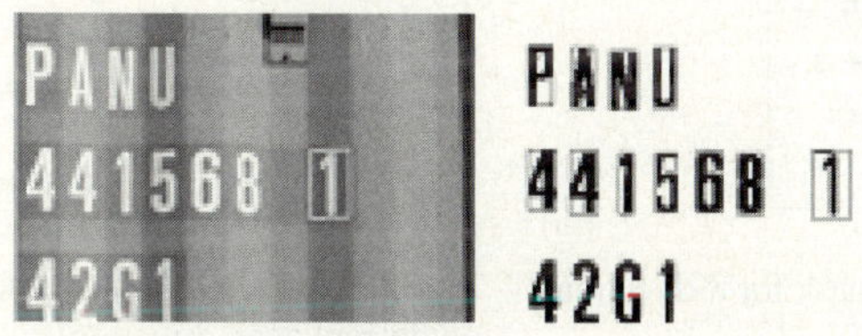

Fig. 5. Identifier area with a general color and successful results of edge extraction

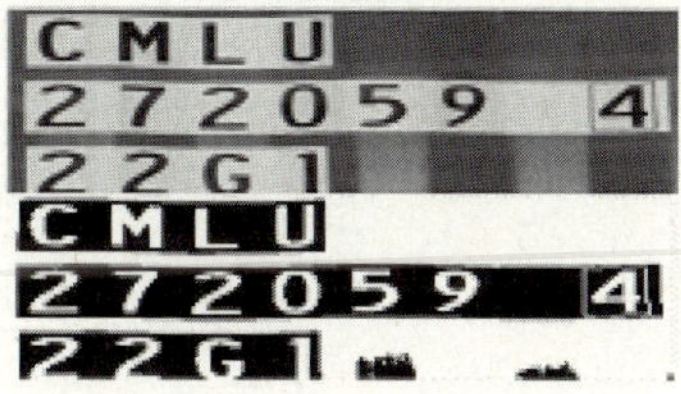

Fig. 6. Identifier area with white color and failed results of edge extraction

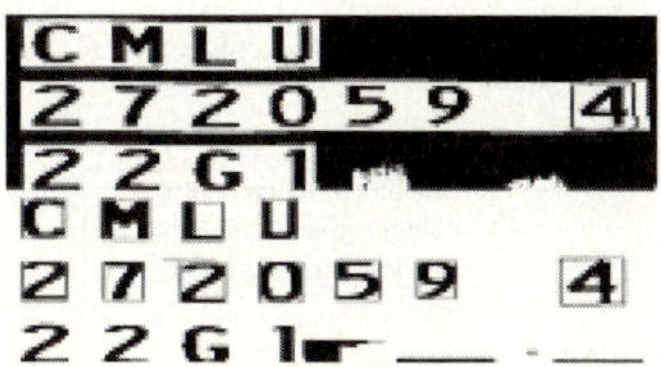

Fig. 7. Reversed binarized area of Fig. 6 and successful result of edge detection

background pixels are converted to white and identifier pixels to black. Fig. 7 shows that the pixel reversal lead to a success of edge detection in an identifier area with white background presented in Fig. 6.

For extracting individual identifiers from binarized areas of identifiers, 8-directional contour tracking method overlays a 3x3 tracking mask like Fig. 8 on a pixel in the target area so that corresponding the center pixel of mask to the target pixel and decides the next scan direction by using the algorithm shown in Fig. 10. If the target pixel is a black one, the method includes the target pixel among contour pixels of a identifier and creates a new tracking mask for next scan direction by setting the priority of 1 to the vertical neighbor pixel to the current scan direction and priorities from 2 to 8 to neighbors in the clockwise direction as shown in Fig. 9. Otherwise, the method selects a target pixel based on priorities in the current tracking mask and continues the contour tracking.

Fig. 5 and Fig. 7 shows extraction results of individual identifiers by using 8-directional contour tracking method.

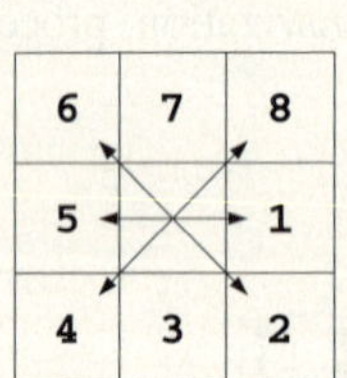

Fig. 8. 3x3 basic tracking mask for 8-directional contour tracking

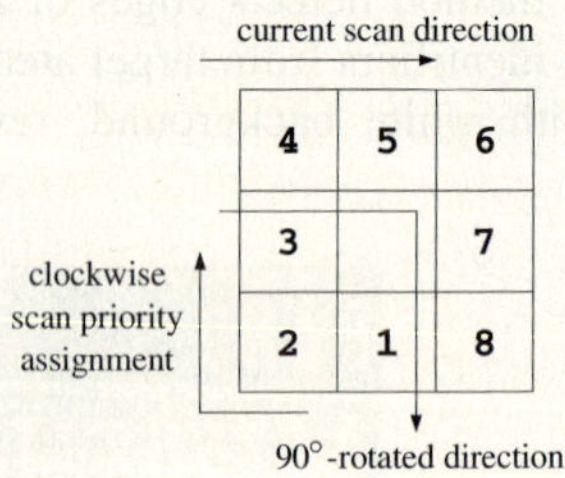

Fig. 9. Creation of a next tracking mask based on current scan direction

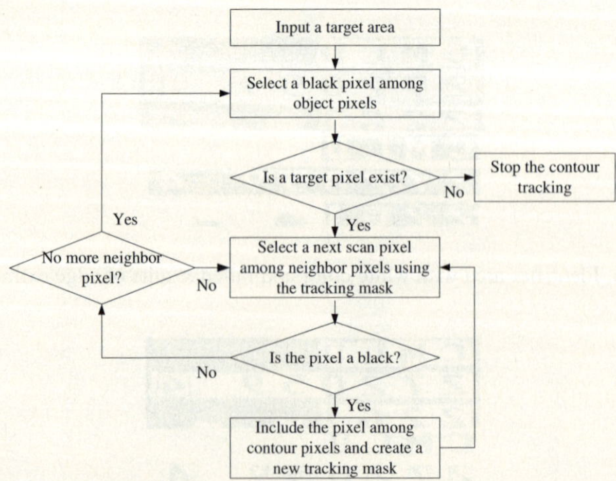

Fig. 10. 8-directional contour tracking algorithm

2.4 Recognition of Container Identifiers Using ART2-Based Self-organizing Supervised Leaning Algorithm

This paper proposed an ART2-based self-organizing supervised learning algorithm for the recognition of container identifiers. First, a new leaning structure is applied between the input and the middle layers, which applies ART2 algorithm between the two layers, select a node with maximum output value as a winner node, and transmits the selected node to the middle layer. Next, generalized Delta learning algorithm and Delta-bar-Delta algorithm are applied in the learning between the middle and the output layers, improving the performance of learning. The proposed learning algorithm is summarized as follows:

1. The connection structure between the input and the middle layers is like ART2 algorithm and the output layer of ART2 becomes the middle layer of the proposed learning algorithm.
2. Nodes of the middle layer mean individual classes. Therefore, while the proposed algorithm has a fully-connected structure on the whole, it takes the winner node method that compares target vectors and output vectors and back-propagates a representative class and the connection weight.
3. The proposed algorithm performs the supervised learning by applying generalized Delta learning as the learning structure between the middle and the output layers.
4. The proposed algorithm improves the performance of learning by applying Delta-bar-Delta algorithm to generalized Delta learning for the dynamical adjustment of a learning rate. When defining the case that the difference between the target vector and the output vector is less than 0.1 as an accuracy and the opposite case as an inaccuracy, Delta-bar-Delta algorithm is applied restrictively in the case that the number of accuracies is greater than or equal to inaccuracies with respect to total patterns. This prevents no progress or an oscillation of learning keeping almost constant level of error by early premature situation incurred by competition in the learning process.

The detailed description of ART2-based self-organizing supervised learning algorithm is like Fig. 11.

3 Performance Evaluation

The proposed algorithm was implemented by using Microsoft Visual C++ 6.0 on the IBM-compatible Pentium-IV PC for performance evaluation. 79's container images with size of 640x480 were used in the experiments for extraction and recognition of container identifiers. The implemented result screen of identifier extraction and recognition is like Fig. 12.

In the extraction of identifier areas, the previously proposed method fails to extract in images containing noises vertically appearing by an external light and the rugged surface shape of containers. On the other hand, the proposed extraction method detects and removes noises by using a fuzzy method, improving the success rate of extraction compared with the previously proposed. The comparison of the success rate of identifier area extraction between the proposed in this paper and the previously proposed is like Table 2.

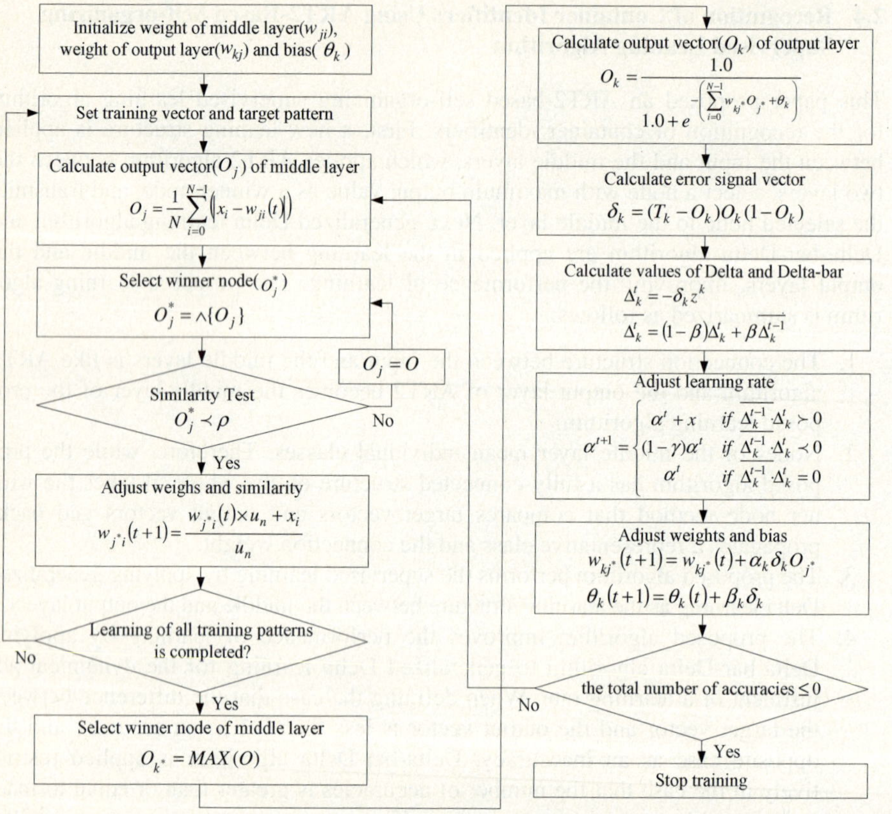

Initialize weight of middle layer(w_{ji}), weight of output layer(w_{kj}) and bias(θ_k)

Set training vector and target pattern

Calculate output vector(O_j) of middle layer

$$O_j = \frac{1}{N}\sum_{i=0}^{N-1}\left(\left|x_i - w_{ji}(t)\right|\right)$$

Select winner node(O_j^*)

$$O_j^* = \wedge\{O_j\}$$

Similarity Test

$$O_j^* \prec \rho$$

No → $O_j = O$

Yes

Adjust weighs and similarity

$$w_{j^*i}(t+1) = \frac{w_{j^*i}(t)\times u_n + x_i}{u_n}$$

Learning of all training patterns is completed?

No

Yes

Select winner node of middle layer

$$O_{k^*} = MAX(O)$$

Calculate output vector(O_k) of output layer

$$O_k = \frac{1.0}{1.0 + e^{-\left(\sum_{i=0}^{N-1} w_{kj^*} O_{j^*} + \theta_k\right)}}$$

Calculate error signal vector

$$\delta_k = (T_k - O_k)O_k(1 - O_k)$$

Calculate values of Delta and Delta-bar

$$\Delta_k^t = -\delta_k z^k$$

$$\bar{\Delta}_k^t = (1-\beta)\Delta_k^t + \beta\bar{\Delta}_k^{t-1}$$

Adjust learning rate

$$\alpha^{t+1} = \begin{cases} \alpha^t + x & \text{if } \bar{\Delta}_k^{t-1}\cdot\Delta_k^t \succ 0 \\ (1-\gamma)\alpha^t & \text{if } \bar{\Delta}_k^{t-1}\cdot\Delta_k^t \prec 0 \\ \alpha^t & \text{if } \bar{\Delta}_k^{t-1}\cdot\Delta_k^t = 0 \end{cases}$$

Adjust weights and bias

$$w_{kj^*}(t+1) = w_{kj^*}(t) + \alpha_k\delta_k O_{j^*}$$

$$\theta_k(t+1) = \theta_k(t) + \beta_k\delta_k$$

the total number of accuracies ≤ 0

No

Yes

Stop training

Fig. 11. ART2-based self-organizing supervised learning algorithm

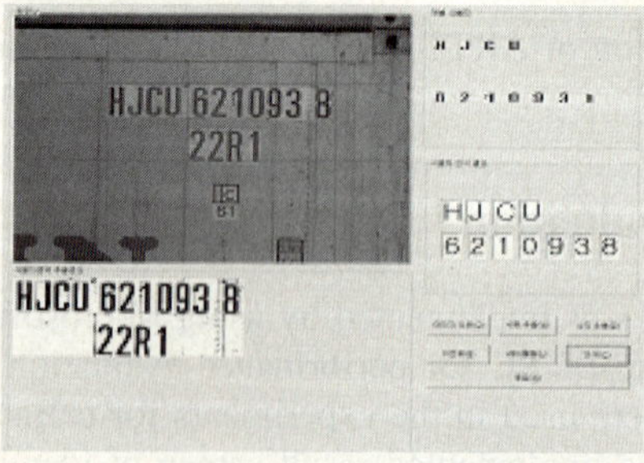

Fig. 12. Result screen of identifier extraction and recognition

Table 2. Comparison of the success rate of identifier area extraction

	Previously-proposed method	Proposed method in this paper
Success rate	55/79(69.6%)	72/79(91.1%)

For the experiment of identifier recognition, applying 8-directional contour tracking method to 72's identifier areas extracted by the proposed extraction algorithm, 284's alphabetic characters and 500's numeric characters were extracted. This paper performed the recognition experiments with the FCM-based RBF network and the proposed ART2-based self-organizing supervised learning algorithm using extracted identifier characters and compared the recognition performance in Table 3.

Table 3. Evaluation of recognition performance

	FCM-based RBF network		ART2-base self-organizing supervised learning algorithm	
	# of Epoch	# of success of recognition	# of Epoch	# of success of recognition
Alphabetic characters (284)	236	240 (84.5%)	221	280 (98.5%)
Numeric Characters (500)	161	422 (84.4%)	151	487 (97.4%)

In the experiment of identifier recognition, the learning rate and the momentum were set to 0.4 and 0.3 for the two recognition algorithms, respectively. And, for ART2 algorithm generating nodes of the middle layer in the proposed algorithm, vigilance variables of two character types were set to 0.4.

When comparing the number of nodes of the middle layer between the two algorithms, the proposed algorithm creates more nodes than FCM-based RBF network, but via the comparison of the number of Epochs, it is known that the number of iteration of learning in the proposed algorithm is less than FCM-based RBF network. That is, the proposed algorithm improves the performance of learning. Also, comparing the success rate of recognition, it is able to be known that the proposed algorithm improves the performance of recognition compared with FCM-based RBF network. Failures of recognition in the proposed algorithm were incurred by the damage of shapes of individual identifiers in original images and the information loss of identifiers in the binarzation process.

4 Conclusions

This paper proposed an automatic recognition system of shipping container identifiers using fuzzy-based noise removal method and ART2-based self-organizing supervised learning algorithm. In this paper, after detecting and removing noises from an original image by using a fuzzy method, areas of identifiers are extracted. In detail, the performance of identifier area extraction is improved by removing noises incurring errors using a fuzzy method based on the feature that the color of container identifiers is white or black on the whole. And, individual identifiers are extracted by applying 8-directional contour tracking method to extracted areas of identifiers. This paper proposed an ART2-based self-organizing supervised learning algorithm and applied to the recognition of individual identifiers. Experiments using 79's container images showed that 72's areas of identifiers and 784's individual identifiers were extracted successfully and 767's identifiers among the extracted were recognized by the proposed recognition algorithm. Failures of recognition in the proposed algorithm were incurred by the damage of shapes of individual identifiers in original images and the information loss of identifiers in the binarzation process.

A Future work is the development of fuzzy association algorithm that may recover damaged identifiers to improve the performance of extraction and recognition of individual identifiers.

References

1. ISO-6346, Freight Containers-Coding -Identification and Marking, (1995)
2. Kim, K. B.: Recognition of Identifiers from Shipping Container Images using Fuzzy Binarization and Neural Network with Enhanced Learning Algorithm. Applied Computational Intelligence. World Scientific, (2004) 215-221
3. Nam, M. Y., Lim, E. K., Heo, N. S., Kim, K. B.: A Study on Character Recognition of Container Image using Brightness Variation and Canny Edge. Proceedings of Korea Multimedia Society, 4(1) (2001) 111-115
4. Kim, N. B.: Character Segmentation from Shipping Container Image using Morphological Operation. Journal of Korea Multimedia Society, 2(4) (1999) 390-399
5. Vogl, T. P., Mangis, J. K., Zigler, A. K., Zink, W. T., Alkon, D. L.: Accelerating the convergence of the backpropagation method. Biological Cybernetics., 59 (1998) 256-264
6. Chen, Y. S., Hsu, W. H.: A systematic approach for designing 2-Subcycle and pseudo 1-Subcycle parallel thinning algorithms. Pattern Recognition, 22(3) (1989) 267-282

Fingerprint Classification by SPCNN and Combined LVQ Networks

Luping Ji, Zhang Yi, and Xiaorong Pu

Computational Intelligence Laboratory
School of Computer Science and Engineering
University of Electronic Science and Technology of China
Chengdu 610054, P.R. China
jlp0813@163.com

Abstract. This paper proposes a novel fingerprint classification method. It uses an SPCNN (Simplified Pulse Coupled Neural Network) to estimate directional image of fingerprint, and quantizes them to obtain fingerprint vector. Then, a fully trained LVQ (Learning Vector Quantization) neural network is used as classifier for the fingerprint vector to determine the corresponding fingerprint classification. Experiments show this proposed method is robust and has high classification accuracy.

1 Introduction

Fingerprints have been used in personal identification for about 100 years. Referenced on Henry system [1], fingerprints can be classified into six classes: whorl (W), right loop (R), left loop (L), arch (A), tented arch (TA) and twin loops (TL). From viewpoint of statistics, L: 33.8%, R: 31.7%, W and TL: 27.9%, A and TA: 6.6%. So, fingerprint classification can largely increase the matching precision, shorten the matching time, and improve the performance of automatic fingerprint identification system, so it is very important.

Many algorithms have been proposed for fingerprint classification. Sum up, some are based on syntactic method, some algorithms are based on singularities, and some are based on neural networks. Thereinto, those neural approaches are mostly based on multi-layer perceptrons or Kohcnen self-organizing networks [5]-[7]. In this paper, we propose a new fingerprint classification method using a SPCNN and a combined LVQ neural networks. A SPCNN is used to estimate direction image, then a combined LVQ neural network is used to classify fingerprint vector to obtain the corresponding fingerprint classification.

2 Fingerprint Image Quantization

The main purpose to estimate direction image of fingerprint is reducing the dimension of fingerprint data for the neural classifier, see [3]. In this paper, we define four directions: $d_1 = 0°$, $d_2 = 45°$, $d_3 = 90°$ and $d_4 = 135°$, and they respectively are corresponding to the four direction column vectors, v_i, where

L. Jiao et al. (Eds.): ICNC 2006, Part I, LNCS 4221, pp. 395–398, 2006.

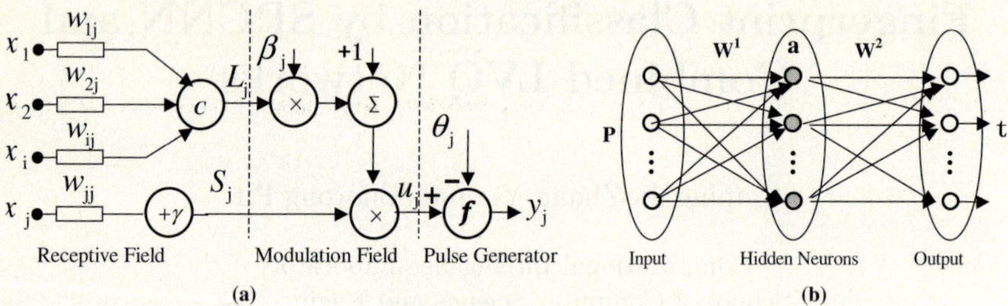

Fig. 1. SPCNN Neuron Model and LVQ Neural Network Structure. (a) SPCNN Neuron Model. (b) Combined LVQ Neural Network Structure.

$i \in \{1,2,3,4\}$ and $v_i \in \{-1,1\}^{2\times1}$, $v_1 = (1,-1)^T$, $v_2 = (1,1)^T$, $v_3 = (-1,1)^T$, and $v_4 = (-1,-1)^T$.

The proposed SPCNN neuron model [2] is shown in Fig. 1 (a). For the jth neuron in network, it follows:

$$\begin{cases} u_j = (w_{jj}x_j + \gamma)(\beta_j \sum\limits_{\substack{i \in N(j) \\ i \neq j}} w_{ij}x_i + 1)\ , \\ y_j = f(u_j - \theta_j). \qquad\qquad j = 1,2,3,...,n \end{cases} \tag{1}$$

where n is the number of neurons, w_{ij} is linking weight between two neurons, x_i is neighbor input, γ is modulation constant for neuron, $N(j)$ is neighbor field determined by coupled matrix $C \in \{0,1\}^{3\times3}$, θ_j is a threshold for jth neuron, y_j is a pulse output (if fires), and $f(.)$ is a pulse generation function defined as: if $\varphi > 0$ then $f(\varphi) = 1$ (fire), otherwise $f(\varphi) = 0$ (no fire).

In this paper, it is one-to-one corresponding between pixel and neuron, we set w_{ij} to 1, let x_i be the binary value of pixel, $\gamma = 1$, $\beta = 1$, $N(j)$ is 3×3 neighborhoods, and θ_j is experimentally determined. The four used coupled feature matrixes are respectively defined as follows:

$$C_1 = \begin{pmatrix} 0\,0\,0 \\ 1\,1\,1 \\ 0\,0\,0 \end{pmatrix}, C_2 = \begin{pmatrix} 0\,0\,1 \\ 0\,1\,0 \\ 1\,0\,0 \end{pmatrix}, C_3 = \begin{pmatrix} 0\,1\,0 \\ 0\,1\,0 \\ 0\,1\,0 \end{pmatrix}, C_4 = \begin{pmatrix} 1\,0\,0 \\ 0\,1\,0 \\ 0\,0\,1 \end{pmatrix}.$$

Moreover, define $y^i(k) = \sum^{j \in B_k} y_j$, where i is index of coupled feature matrix C_i, where C_i is corresponding to direction vector v_i, $i = 1,2,3,4$, and j is index of pixel in B_k, $j = 1,2,...,64$. The given image, IMG, firstly is divide into 8×8 non-overlapped sub-images, B_k, where $k = 1,2,...,(m \times n)/64$, and m, n respectively are the width and height of image, $\bigcup B_k$=IMG. Then using the SPCNN to compute maximal $y^i(k)$ of each B_k. Note, to the maximal $y^i(k)$, the i just is index of true direction column vector v_i. Use v_i to encode each B_k, and all the true direction vectors of sub-images are clockwise linked together to generate the full fingerprint vector $F \in \{-1,1\}^{2\times a}$, where a is the number of sub-images. It will be the input vector of designed LVQ Neural Classifier.

3 Combined LVQ Neural Classifier

To respectively denote each of the six fingerprint classes, we define output vector $\mathbf{t}_i$ $(i = 1, 2, ..., 6) \in \{0, 1\}^{6\times 1}$, in which no more than one element is 1, and different $\mathbf{t}$ indicates different classification, but if each element in output vector $\mathbf{t}$ is 0, it indicates unknown classification.

The simple architecture of combined LVQ network is shown in Fig. 1 (b). To any hidden neuron j, the output $\mathbf{n}_j^1$ is defined as:

$$\mathbf{n}_j^1 = b_j - \|W_j^1 - P\|$$

So, the output of first layer can be expressed as a vector $\mathbf{a}^1$, and it follows:

$$\mathbf{a}^1 = compet(\mathbf{n}^1)$$

where *Compet* is a competitive function, $\|\cdot\|$ indicates the direct Euclidean distance, b_j is a bias, W_j^1 is the jth weight vector of first layer, and P is the input fingerprint vector F. Moreover, the output vector $\mathbf{t}$ is defined as:

$$\mathbf{t} = W^2\mathbf{a}^1$$

where W^2 is weight matrix of the second layer, and $\mathbf{a}^1$ is the output vector of the first layer. In this network, the winner neuron indicates a subclass, and there may be some different neurons that make up a class. To the W^2, the columns represent subclasses, and rows represent classes, and with a single 1 in each column, so the row in which the 1 occurs indicates which class the appropriate subclass belongs to.

In this paper, we take a combined learning rules for training the LVQ network [8]. To a series of training samples, $\{P_1, t_1\}, \{P_2, t_2\}, ..., \{P_Q, t_Q\}$, we use the Kohonen rule and "conscience" rule to improve the competitive learning of network as follows.

$$W_{i^*}^1(k) = W_{i^*}^1(k-1) + \alpha(k)s(k)[P(k) - W_{i^*}^1(k)(k-1)] \tag{2}$$

where k is update times, $k = 0, 1, 2, ..., n$, i^* is the winner neuron, $\alpha(k) \in (0, 1)$ is learning rate updated with k, $s(k)$ is followed as

$$s(k) = \begin{cases} +1, \text{if } t_{i^*}' = t_{i^*} = 1 \text{ (correct classification)}; \\ -1, \text{otherwise (error classification)}. \end{cases}$$

As the learning rate α of winner, it is updated as

$$\alpha_{i^*}(k) = \frac{\alpha_{i^*}(k-1)}{1 + s(k)\alpha_{i^*}(k-1)}$$

Furthermore, based on "conscience" rule, the bias b will be updated as:

$$b_i(k) = \begin{cases} 0.95b_i(k-1), & \text{if } i \neq i^*; \\ b_i(k-1) - 0.2, \text{if } i = i^*. \end{cases}$$

In our algorithm, let $N_{input} = L_F$, where N_{input} indicates the number of neurons in input layer, and L_F is length of the input fingerprint vector F. The hidden layer consists of six neurons, and the number of neurons in output layer is also six. From the output vector $\mathbf{t}_i$ of designed LVQ classifier, the given fingerprint can be classified into one of six classifications it belongs to.

4 Experiment and Conclusions

We have tested the proposed algorithm using approximate 600 binary fingerprint images randomly acquired from different persons, and each image size is 256 × 256. Each fingerprint classification contains approximately 100 images.

The experiments for five classifications (W, R, L, A, TL) and six (W, R, L, A, TA, TL) classifications are compared with other methods, and table 1 shows the synthetical comparison data. It indicates the performance of proposed algorithm is a little better than others, especially, to the five classifications.

Table 1. Comparison of Different Classification Methods

Researchers	Class	Accurateness	Reject	Approaches
Jain at al, 1999 [5]	5	91.3%	0.0	Gabor filters+K-nearest neighbor
Yuan, et al, 2003 [6]	5	90.0%	1.8%	SVM+RNN
Ugur, et al, 1996	5	94.0%	0.0	Modified SOM
		94.0%	0.0	PCA+MSOM
Sankar K, et al, 1996	5	96.9%	0.5%	MLP+ Fuzzy geometric
Candela, et al, 1995 [4]	5	94.9%	0.0	PCASYS(PNN+Pseudoridge)
This paper	6	88.5%	0.4%	SPCNN+CLVQ
	5	96.4 (average)	0.2%	

The experiment comparison shows the proposed algorithm for binary fingerprint classification is very good, and it shows better synthetical classifying performance than other fingerprint classification methods.

References

1. Henry, E. R.: Classification and Uses of Finger Prints. London: Routledge, (1900)
2. Johnson, J. L., Padgett, M. L.: PCNN Models and Applications. IEEE Transactions On Neural Networks, **10**(1999) 480-498
3. Khaled, A. N.: On learning to estimate the block directional image of a fingerprint using a hierarchical neural network. Neural Networks, **16**(2003) 133-144
4. Candela, G.T., et al.: PCASYSA Pattern-Level Classification Automation System for Fingerprints. NIST Technical Report NISTIR 5647, Aug. 1995
5. Jain, A. K., Prabhakar, S., Hong, L.: A Multichannel Approach to Fingerprint Classification. IEEE Transactions on Pattern Analysis and Machine Intelligence, **21**(1999) 348-359
6. Yuan, Y., Gian, L., etc.: Combining flat and structured representations for fingerprint classification with recursive neural networks and support vector machines. Pattern Recognition, **36**(2003) 397-406
7. Khaled, A. N.: Fingerprints classification using artificial neural networks: a combined structural and statistical approach. Neural Networks, **14**(2001) 1293-1395
8. Mohhamad, T., Vakil, B., Nikola, P.: Premature clustering phenomenon and new trainning algorithms for LVQ. Pattern Recognition, **36**(2003) 1901-1912

Gait Recognition Using Hidden Markov Model*

Changhong Chen, Jimin Liang, Heng Zhao, and Haihong Hu

School of Electronic Engineering, Xidian University
Xi'an, Shaanxi 710071, China
{chhchen, jimleung}@mail.xidian.edu.cn

Abstract. Gait-based human identification is a challenging problem and has gained significant attention. In this paper, a new gait recognition algorithm using Hidden Markov Model (HMM) is proposed. The input binary silhouette images are preprocessed by morphological operations to fill the holes and remove noise regions. The width vector of the outer contour is used as the image feature. A set of initial exemplars is constructed from the feature vectors of a gait cycle. The similarity between the feature vector and the exemplar is measured by the inner product distance. A HMM is trained iteratively using Viterbi algorithm and Baum-Welch algorithm and then used for recognition. The proposed method reduces image feature from the two-dimensional space to a one-dimensional vector in order to best fit the characteristics of one-dimensional HMM. The statistical nature of the HMM makes it robust to gait representation and recognition. The performance of the proposed HMM-based method is evaluated using the CMU MoBo database.

1 Introduction

Gait recognition is a subfield of biometric based human identification. Compared with other kind of human biometrics, such as face, iris and fingerprints, gait has the advantages of hard to conceal, less obtrusive and non-contact. These offer the possibility to identify people at a distance, without any interaction or co-operation from the subject.

Gait recognition has received a lot attention recent years and many methods have been proposed. These methods can be roughly divided into two major categories, namely model-based methods and appearance-based methods. The model-based methods [1-4] aim to model human body and incorporate knowledge of the shape and dynamics of human gait into extraction process. They usually perform model matching in each frame in order to measure the parameters such as trajectories, limb lengths, and angular speeds. These methods are easy to be understood, however, they tend to be complex and need high computational cost. The appearance-based methods are more popularly used recently, which usually take some statistical theories to characterize the whole motion pattern by a compact representation regardless of the underlying structure. Principal component analysis (PCA) are used to determine the relative role of temporal and spatial information in discriminating walking from running[5] and in [6], [7], it was used to get eigengaits using image self-similarity

* Project supported by NSF of China (60402038) and NSF of Shaanxi (2004f39).

L. Jiao et al. (Eds.): ICNC 2006, Part I, LNCS 4221, pp. 399 – 407, 2006.

plots. Murase and Sakai [8] presented a template matching method based on eigenspace representation to distinguish different gaits. Independent component analysis (ICA) is used in [9] to get the independent components of each 2D binary silhouette.

Hidden Markov Model (HMM) is suitable for gait recognition because of its statistical feature and it can reflect the temporal state-transition nature of gait. HMM has been applied to human identification in [10-13]. The methods in these literatures differ mainly in the image feature selection. He and Debrunner used the Hu moments of the motion segmentation in each frame as the image feature [10]. But their method was evaluated using image sequences of three individuals only. In [11], the outlines of person are represented by P-style Fourier Description. In [12], the binarized background-subtracted image is adopted directly as the feature vector to train HMM. Kale, et al., used the width of the outer contour of the silhouette as the image feature and trained a continuous HMM using several frame to exemplar distance (FED) vector sequences [13].

In this paper, we propose a new HMM-based approach for gait representation and recognition. Similar to the method in [13], we also use the width vector of the outer contour as the image feature. But we use the width vectors to train the HMM directly, instead of using the FED vectors. The structural aspect of the person is represented by a set of exemplars corresponding to different states. A probability distribution is defined similar to that in [12] for the observations based on the exemplars and the inner product distance is used to measure the distance between the feature vector and the exemplar. Unlike [12], we define the parameter in the equation of the probability distribution as a function, which makes the probability distribution adaptively. Also, we train HMMs respectively using individual cycles and average the corresponding parameters of the resulted models to represent the people.

This paper is structured as follows. Section 2 introduces the image preprocessing and feature extraction method. Section 3 describes the proposed HMM based gait recognition method in details. In section 4, the proposed method is evaluated using the CMU MoBo database [14] and its performance is compared with that in [13] and [15]. Section 5 concludes the paper.

2 Gait Feature Vector Extraction

The background-subtracted silhouette image usually contains many holes, noise structure and shadows. One such image is shown in fig. 1(a). The recognition performance may be degraded if the original silhouette is used directly. In this paper, some mathematical morphological operations are used to fill the holes, remove noise regions and extract the outer contour of the people. The image is cropped to remove the shadows and the result is depicted in fig.1 (b). A one-dimensional signal, which is defined as the outer contour width of each image row, can be generated (fig.1 (c)). To further reduce influence of the remaining noise regions, the width signal is filtered using some specific rules. The smoothed width signal (fig.1 (d)) is then used as the image feature. The one-dimensional feature vector fits the one-dimensional HMM well and has the advantages of noise insensitive and human location independent, thus we do not need to align the outer contour.

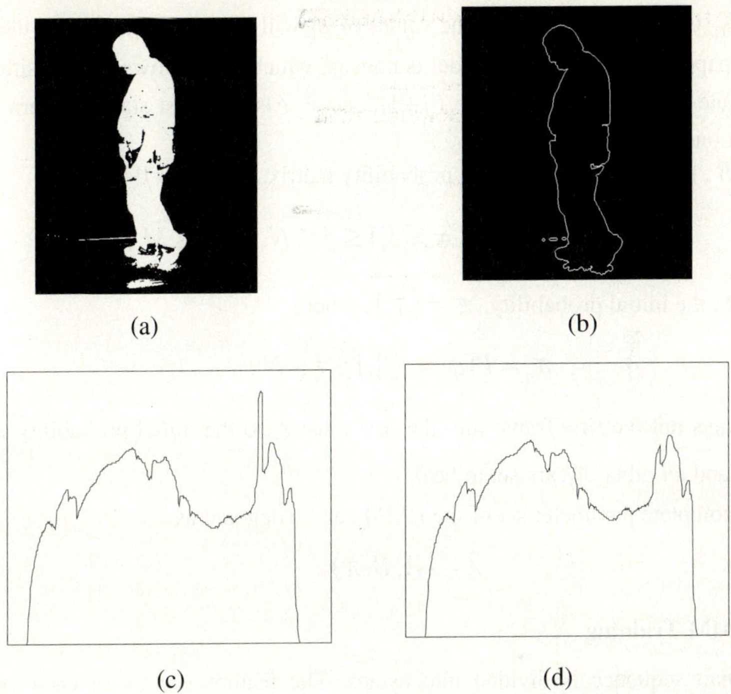

Fig. 1. Use morphological operations to extract contour (b) from silhouette image (a), and the feature vector (d) can be obtained after filtering the width signal (c) of the contour

3 HMM-Based Gait Recognition

We proposed a new HMM-based gait recognition method. An HMM is characterized by the following parameters.

(1) N, the number of states in the model. How to choose N is important and this is a classical problem of choosing the appropriate dimensionality of a model that will fit a given set of observations. For CMU MoBo database, $N = 5$ is suggested [13]. The HMM states are denoted as $S = \{S_1, S_2, \cdots, S_N\}$.

(2) M, the number of distinct observation symbols per state. For gait recognition, every frame's feature vector is treated as an observation symbol. The number M depends on the number of frames per cycle, the number of states in the model and how to divide one cycle into clusters. The frames in a gait cycle are a consecutive transition along with time. We divide each cycle into N clusters of approximately the same size of M. The observation symbols for one HMM state are denoted as $V = \{v_1, v_2, \cdots, v_M\}$.

(3) A, the transition probability matrix. $A = \{a_{ij}\}$, and a_{ij} is defined as

$$a_{ij} = P[q_{t+1} = S_j \mid q_t = S_i], 1 \le i, j \le N, \tag{1}$$

where q_t is the state at time t. The values of a_{ij} will determine the type of the HMM. In this paper, the left-to-right model is chosen, which only allows the transition from the jth state to either the jth or the $(j+1)^{th}$ state. Also the last state can turn back to the first one.

(4) B, the observation symbol probability matrix. $B=\{b_j(k)\}$, where

$$b_j(k)=P[v_k \text{ at } t|q_t=S_j],1\le j\le N,1\le k\le M. \tag{2}$$

(5) π, the initial probability. $\pi=\{\pi_i\}$, where

$$\pi_i=P[q_1=S_i],1\le i\le N. \tag{3}$$

We always put the first frame into the first cluster, so the initial probability π_1 is set to be 1 and all other π_i are set to be 0.

The complete parameter set of the HMM can be denoted as

$$\lambda=(A,B,\pi). \tag{4}$$

3.1 HMM Training

Every gait sequence is divided into cycles. The feature vectors of each cycle are further divided into clusters with about the same size. Each cluster center is treated as an exemplar. An exemplar is defined as

$$e_n=\frac{1}{N_n}\sum_{f_t\in C_n} f_t, \tag{5}$$

where f_t is the feature vector of the tth frame, C_n represents the nth cluster, N_n is the number of the frames in the nth cluster. The exemplar set is denoted as $E=\{e_1,e_2,\cdots,e_N\}$.

The initial exemplars of slow walk (a), fast walk (b) and walking with a ball (c) of the same person are shown in fig.2. The similarities between (a) and (b) show the effectiveness of the feature extraction method, however, greater changes i.e. (c) cause the similarities decline.

For each feature vector f in a cycle, its distance from an exemplar e is measured by the inner product (IP) as defined in the following equation.

$$D(f,e)=1-\frac{f^Te}{\sqrt{f^Tfe^Te}}. \tag{6}$$

The transition probability matrix A is initialized using trial and error method. The initial observation symbol probability matrix $B=\{b_j(k)\}$ is defined as

$$b_n(f_t)=0.01\delta_n e^{-\delta_n\times D(f_t,e_n)}, \tag{7}$$

$$\delta_n = \frac{N_n}{\sum_{f_t \in e_n} D(f_t, e_n)}. \tag{8}$$

The parameter δ in equation (8) can reflect how much a feature vector belongs to the cluster C_n.

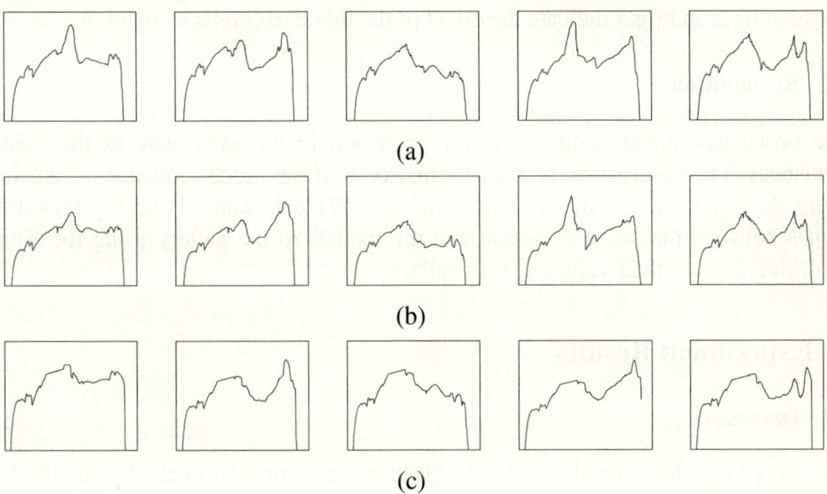

Fig. 2. The initial exemplars of a person, (a) slow walk, (b) fast walk and (c) walking with a ball

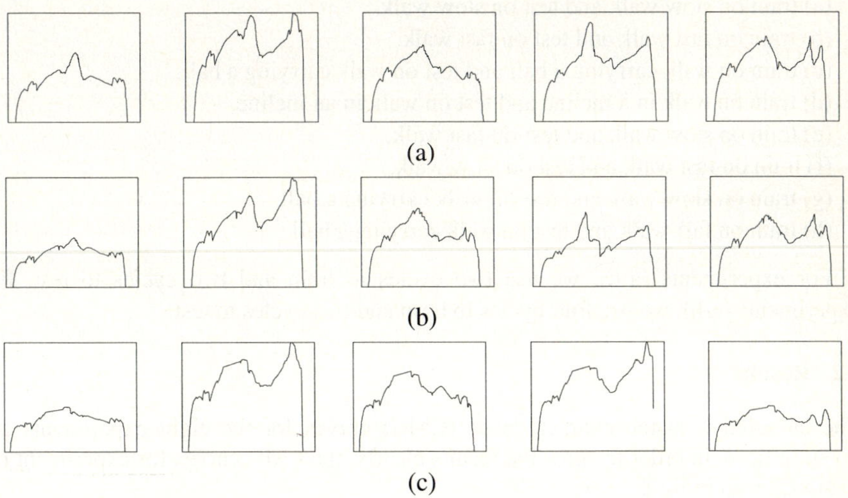

Fig. 3. The exemplars estimated after training. (a) slow walk, (b) fast walk and (c) walking with a ball.

The HMM is trained iteratively. Viterbi decoding is performed on the cycle to obtain the most probable path $Q=\{q_1^{(i)},q_2^{(i)},\cdots,q_T^{(i)}\}$, where $q_t^{(i)}$ is the state at time t after i^{th} iteration. The new exemplars $E^{(i)}$ can be obtained from the most probable path using equation (5) and $B^{(i)}$ can be calculated using equations (6)-(8). Next $A^{(i)}$ and $\pi^{(i)}$ are updated using the Baum-Welch algorithm. The training results consist of the exemplars and the HMM model parameters. It usually takes a few iterations to obtain a better estimate. The exemplars after training are shown in fig.3. The similarities between fig.2 and fig.3 indicate the effect of the initial exemplar estimation.

3.2 Recognition

The probe gait image sequence is pre-processed in the same way as the training sequences. The observation probability matrix B of the probe sequence is calculated using the exemplars of the training sequence. We calculate all the log probability values between the probe sequence and the models in the gallery using the Viterbi algorithm and use their values as the similarity scores.

4 Experiment Results

4.1 Database

We use CMU MoBo database [14] to evaluate our proposed method. This database has 25 people walking at a fast pace (2.82 miles/hr), slow pace (2.06 miles/hr), carrying a ball and walking in an incline. Fronto-parallel sequences are adopted and the image size is 640×480. Following experiments are done on this database:

(a) train on slow walk and test on slow walk.
(b) train on fast walk and test on fast walk.
(c) train on walk carrying a ball and test on walk carrying a ball.
(d) train on walk in a incline and test on walk in an incline.
(e) train on slow walk and test on fast walk.
(f) train on fast walk and test on slow walk.
(g) train on slow walk and test on walk carrying a ball.
(h) train on fast walk and test on walk carrying a ball.

For experiments (a-d), we use two cycles to train and two cycles to test. For experiments (e-h), we use four cycles to train and two cycles to test.

4.2 Results

The cumulative match characteristics (CMC) curves for the eight experiments are given in fig.4. In order to show the results clearly, the CMC curves for experiment (a-h) are showed in fig.4.

For experiments (a-d), the training and the testing data set are of the same motion style. All of experiments (a), (b) and (d) hit 100% and (c) matches more than 95% at the top match. For experiments (e-f), the training and the testing data set are of the different motion styles. Experiments (e) and (f) also achieve very high recognition rate. The CMC of experiment (g) goes beyond 90% at rank three. The CMC of experiment (h) goes beyond 90% at rank six. The results show that the proposed method is robust to speed. But the drastic changes may cause the recognition performance drop slightly.

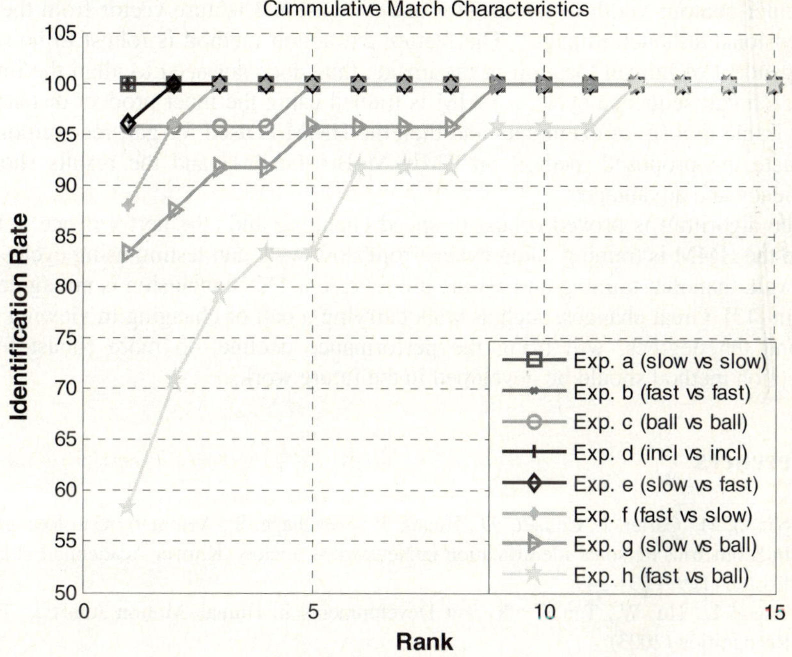

Fig. 4. The CMC curves for Experiments (a-h)

Table 1. Performances comparison with other methods

Train vs. Probe	Maryland [13]			Rutgers [15]			Our method		
	P (%) at rank			P (%) at rank			P (%) at rank		
	1	5	10	1	5	10	1	5	10
Slow vs. Slow	72.0	96.0	100	100	100	100	100	100	100
Fast vs. Fast	68.0	92.0	96.0	96.0	100	100	100	100	100
Ball vs. Ball	91.7	100	100	100	100	100	95.8	100	100
Incline vs. Incline	---	---	---	95.8	100	100	100	100	100
Slow vs. Fast	32.0	72.0	88.0	---	---	---	96.0	100	100
Fast vs. Slow	56.0	80.0	88.0	---	---	---	88.0	100	100
Slow vs. Ball	---	---	---	52.2	69.6	91.3	83.3	95.8	100
Fast vs. Ball	---	---	---	---	---	---	58.3	83.3	95.8

We compare our results with that of Maryland [13] and Rutgers [15]. The details are shown in Table.1, where P is the identification rate and P values at rank 1, 5 and 10 are listed. It can be seen that our results are much better than that of [13] and [15].

5 Conclusion

We propose a HMM-based framework to represent and recognize human gait. We use the outer contour width to construct a one-dimensional feature vector from the two-dimensional silhouette image. The feature extraction method is robust to noise and independent of human location in the image, thus does not need to align the images. For each gait sequence cycle, a HMM is trained using the inner product distances of each frame to a set of exemplars and then the HMM is used for gait recognition. We evaluate the proposed method on CMU MoBo database and the results show its efficiency and advantages.

The algorithm is proved robust to speed changing and the performance is better when the HMM is training using cycles from slow walk and testing using cycles from fast walk than that training and testing are reversed. This conclusion is not agree with that in [13]. Great changes, such as walk carrying a ball or changing in viewing angle beyond ten degrees, will bring the performance decline, so more robust feature extraction method should be developed in the future work.

References

1. Nixon, M., Carter, J., Cunado, D., Huang, P., Stevenage, S.: Automatic Gait Recognition. In Biometrics Personal Identification in Networked Society, Kluwer Academic Publishers, Boston (1999) 231-249
2. Wang, L., Hu, W., Tan, T.: Recent Developments in Human Motion Analysis. Pattern Recognition (2003)
3. Collins, R.T., Gross, R., Shi, J.: Silhouette-based Human Identification from Body Shape and Gait. Proceedings of the Fifth IEEE International Conference on Automatic Face and Gesture Recognition (2002)
4. Wagg, M., Nixon, M.: On Automated Model-Based Extraction and Analysis of Gait. Proceedings of the Sixth IEEE International Conference on Automatic Face and Gesture Recognition (2004)
5. Das, S., Lazarewicz M, Finkel, L.: Principal Component Analysis of Temporal and Spatial Information for Human Gait Recognition. Proceedings of the 26th Annua International Conference of the IEEE Engineering in medicine and Biology Society (2004)
6. BenAbdelkader, C., Cutler, R., Davis, L.: Motion-based Recognition of People in Eigengait Space. Proceedings of the Fifth IEEE International Conference on Automatic Face and Gesture Recognition (2002)
7. BenAbdelkader, C., Culter, R., Nanda, H., Davis, L.: EigenGait: Motion-based Recognition of People Using Image Self-similarity, Proceedings of the Third International Conference on Audio- and Video-Based Biometric Person Authentication (2001) 284–294
8. Murase, H., Sakai, R.: Moving Object Recognition in Eigenspace Representation. Gait analysis and lip reading, Pattern Recognition (1996)
9. Lu, J., Zhang, E., Duan, Zhang, Z., Xue, Y.: Gait Recognition Using Independent Component Analysis. ISNN 2005, LNCS 3497 (2005) 183–188

10. He, Q., Debrunner, C.: Individual Recognition from Periodic Activity Using Hidden Markov Models, Proceedings of IEEE Workshop on Human Motion (2000)
11. Iwamoto, K., Sonobe, K., Komatsu, N.: A Gait Recognition Method using HMM. SICE Annual Conference in Fukui, Japan (2003)
12. Sundaresan, A., RoyChowdhury, A., Chellappa, R.: A Hidden Markov Model Based Framework for Recognition of Humans from Gait Sequences. Proceedings of IEEE International Conference on Image Processing (2003)
13. Kale, A., Cuntoor, N., Chellappa, R.: A Framework for Activity Specific Human Identification, Proceedings of IEEE Acoustics, Speech, and Signal Processing (2002)
14. Gross, R., Shi, J.: The Cmu Motion of Body (mobo) Database. Technical report, Robotics Institute (2001)
15. Zhang, R., Vogler, C., Metaxas, D.: Human Gait Recognition. Proceedings of the IEEE Computer Society Conference on Computer Vision and Pattern Recognition Workshops (2004)

Neurocontroller Via Adaptive Learning Rates for Stable Path Tracking of Mobile Robots

Sung Jin Yoo[1], Jin Bae Park[1], and Yoon Ho Choi[2]

[1] Yonsei University, Seodaemun-gu, Seoul, 120-749, Korea
{niceguy1201, jbpark}@yonsei.ac.kr
[2] Kyonggi University, Kyonggi-Do, Suwon 443-760, Korea
yhchoi@kyonggi.ac.kr

Abstract. In this paper, we present a neurocontroller via adaptive learning rates (ALRs) for stable path tracking of mobile robots. The self recurrent wavelet neural networks (SRWNNs) are employed as two neurocontrollers for the control of the mobile robot. Since the SRWNN combines the advantages such as the multi-resolution of the wavelet neural network and the information storage of the recurrent neural network, it can easily cope with the unexpected change of the system. Specially, the ALR algorithm in the gradient-descent method is extended for the multi-input multi-output system and is applied to train the parameters of the SRWNN controllers. The ALRs are derived from the discrete Lyapunov stability theorem, which are used to guarantee the stable path tracking of mobile robots. Finally, through computer simulations, we demonstrate the effectiveness and stability of the proposed controller.

1 Introduction

In recent years, neural networks (NNs) have been used as a good tool to control an autonomous mobile robot [1, 2] because no mathematical models are needed and they can easily be applied to the nonlinear and linear systems. But, NNs have some drawbacks such as slow convergence, complex structure. To solve these defects, recently wavelet neural network (WNN), which absorbs the advantages of multi-resolution of wavelets and learning of NN, has been proposed to guarantee the fast convergence and has been used for the identification and control of the nonlinear systems [3, 5, 4, 6]. However, the WNN does not require prior knowledge about the plant to be controlled due to its feedforward structure. Therefore, the WNN cannot adapt easily under the circumstances, such as the operation environment of mobile robots, to change frequently the operating conditions and parameters of dynamics. To overcome these problems, we apply the self recurrent wavelet neural network (SRWNN) [7], which is a modified structure of the WNN proposed in [5], as the controller for the stable path tracking of mobile robots.

The back-propagation (BP) or the gradient-descent (GD) method has been used for training the weights of NNs for many years [5,8,7,9,10,6]. The learning rates in the GD method are sensitive factors for guaranteeing the convergence of

L. Jiao et al. (Eds.): ICNC 2006, Part I, LNCS 4221, pp. 408–417, 2006.

the NNs. Since the existing GD method using the fixed learning rate (FLR) has the problem that the optimal learning rates cannot easily be found, the adaptive learning rates (ALRs), which can adapt rapidly the change of the plant, have been researched. The ALRs have been derived from the Lyapunov stability theorem and applied to the various networks such as the diagonal recurrent neural network [9], the recurrent fuzzy neural network [10], and the WNN [6]. But these works only applied the ALRs for the single-input single-output (SISO) systems. If the ALRs are applied to the multi-input multi-output (MIMO) practical systems, they must be induced considering the relation between the input and output of the plant. Specially, like the kinematic model of the mobile robot, in case that the number of inputs and outputs of the plant are different, it is not easy to derive the ALR theorems. Accordingly, in this paper, we extend these ideas to the adaptive control for the stable path tracking of the mobile robot which has two inputs and three outputs. Two SRWNNs trained by ALRs are used as each kinematic controller in our control scheme for generating two control inputs: the translational and rotational displacement of the mobile robot. The ALRs are derived in the sense of discrete Lyapunov stability analysis, which are used to guarantee the convergence of the SRWNN controllers (SRWNNCs) in the proposed control system.

2 Preliminaries

2.1 The Kinematic Model of Mobile Robot

The mobile robot model used in this paper has two opposed drive wheels, mounted on the left and right sides of the mobile robot, and a caster. In this model, the location of the mobile robot is represented by three states: the coordinates (x_c, y_c) of the midpoint between the two driving wheels and the orientation angle θ [11]. The kinematic model of the mobile robot in a global coordinate frame can then be expressed as follows:

$$\begin{bmatrix} x_c(n+1) \\ y_c(n+1) \\ \theta(n+1) \end{bmatrix} = \begin{bmatrix} x_c(n) \\ y_c(n) \\ \theta(n) \end{bmatrix} + \begin{bmatrix} \delta d(n)cos(\theta(n) + \frac{\delta\theta(n)}{2}) \\ \delta d(n)sin(\theta(n) + \frac{\delta\theta(n)}{2}) \\ \delta\theta(n) \end{bmatrix}, \tag{1}$$

where, $\delta d = \frac{d_R + d_L}{2}$ and $\delta\theta = \frac{d_R - d_L}{b}$ are used as control inputs. Here, d_R and d_L denote the distances, traveled by the right and the left wheel, respectively. Also, b is the distance between the wheels.

2.2 Self Recurrent Wavelet Neural Network

The SRWNN structure has N_i inputs, one output, and $N_i \times N_w$ mother wavelets [7]. The SRWNN consists of four layers. *The layer 1* is an input layer. This layer accepts the input variables and transmits the accepted variables to the next layer directly. *The layer 2* is a mother wavelet layer. Each node of this layer has a mother wavelet and a self-feedback loop. In this paper, we select the first derivative of a Gaussian function, $\phi(x) = -x\exp(-\frac{1}{2}x^2)$ as a mother wavelet function. A wavelet ϕ_{jk} of each node is derived from its mother wavelet ϕ as follows: $\phi_{jk}(z_{jk}) = \phi(\frac{u_{jk}-m_{jk}}{d_{jk}})$, with $z_{jk} = \frac{u_{jk}-m_{jk}}{d_{jk}}$, where, m_{jk} and d_{jk} are the translation factor and the dilation factor of the wavelets, respectively. The subscript jk indicates the k-th input term of the j-th wavelet. In addition, the inputs of this layer for discrete time n can be denoted by $u_{jk}(n) = x_k(n)+\phi_{jk}(n-1)\cdot\alpha_{jk}$, where, α_{jk} denotes the weight of the self-feedback loop. The input of this layer contains the memory term $\phi_{jk}(n-1)$, which can store the past information of the network. That is, the current dynamics of system is conserved for the next sample step. Thus, even if the SRWNN has less mother wavelets than the WNN, the SRWNN can attract nicely the system with complex dynamics. Here, α_{jk} is a factor to represent the rate of information storage. These aspects are the apparent dissimilar point between the WNN and the SRWNN. And also, the SRWNN is a generalization system of the WNN because the SRWNN structure is the same as the WNN structure when $\alpha_{jk} = 0$. *The layer 3* is a product layer. The nodes in this layer are given by the product of the mother wavelets as follows: $\Phi_j(\mathrm{x}) = \prod_{k=1}^{N_i} \phi(z_{jk})$. *The layer 4* is an output layer. The node output is a linear combination of consequences obtained from the output of the layer 3. In addition, the output node accepts directly input values from the input layer. Therefore, the SRWNN output $y(n)$ is composed by self-recurrent wavelets and parameters as follows:

$$y(n) = \sum_{j=1}^{N_w} w_j \Phi_j(\mathrm{x}) + \sum_{k=1}^{N_i} a_k x_k, \tag{2}$$

where, w_j is the connection weight between product nodes and output nodes, and a_k is the connection weight between the input nodes and the output node. By using the direct term, the SRWNN has a number of advantages such as a direct linear feedthrough network, including initialization of network parameters based on process knowledge and enhanced extrapolation outside of examples of the learning data sets [12]. The weighting vector W of SRWNN is represented by $W = [a_k \quad m_{jk} \quad d_{jk} \quad \alpha_{jk} \quad w_j]^T$, where, the initial values of tuning parameters a_k, m_{jk}, d_{jk}, and w_j are given randomly in the range of [-1 1], but $d_{jk} > 0$. And also, the initial values of α_{jk} are given by 0. That is, there are no feedback units initially.

3 Neurocontroller for the Mobile Robot

3.1 SRWNN Controller

In this subsection, we design the SRWNN based direct adaptive control system for path tracking of the mobile robot. Since the kinematics of the mobile robot given in (1) consists of two inputs and three outputs, two SRWNNCs must be used for generating each control input δd and $\delta\theta$. The overall controller architecture is shown in Fig. 1. In this figure, the SRWNNC1 and SRWNNC2 denote two SRWNNCs for controlling the control input δd and $\delta\theta$, respectively. In order to consider the accurate position of the mobile robot in global coordinate frame, the sum of the squared past errors $\sqrt{e_x^2(n-1)+e_y^2(n-1)}$ is used as the input of the controllers. The past control signal $\delta d(n-1)$ and the sum of the squared past errors $\sqrt{e_x^2(n-1)+e_y^2(n-1)}$ are fed into the SRWNNC1 so that the current control input $\delta d(n)$ is generated. And also, $\delta\theta(n-1)$, $\sqrt{e_x^2(n-1)+e_y^2(n-1)}$ and $e_\theta(n-1)$ are used as the input of the SRWNNC2 for generating the current control signal $\delta\theta(n)$. Accordingly, two cost functions must be defined to select the optimal control signals.

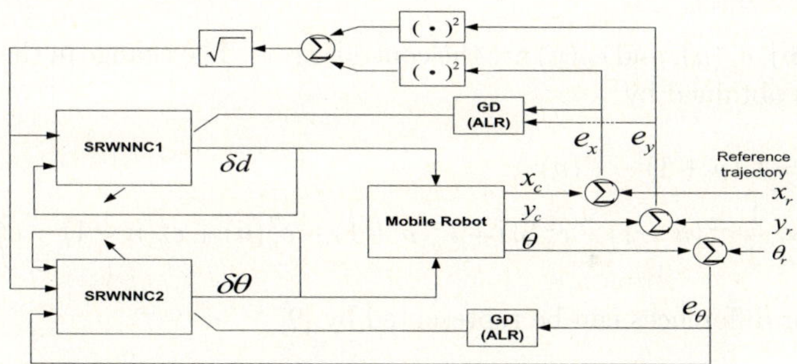

Fig. 1. The proposed control structure for mobile robot

3.2 Training Algorithm

Let us define two cost functions as follows: $J_1(n) = \frac{1}{2}e_x^2(n) + \frac{1}{2}e_y^2(n)$, $J_2(n) = \frac{1}{2}e_\theta^2(n)$, where, $e_x(n) = x_r(n) - x_c(n)$, $e_y(n) = y_r(n) - y_c(n)$, and $e_\theta(n) = \theta_r(n) - \theta(n)$. Here, $x_r(n)$, $y_r(n)$, and $\theta_r(n)$ denote the current states of the mobile robot for the reference trajectory.

By using the GD method, the weight values of SRWNNC1 and SRWNNC2 are adjusted so that cost functions are minimized after a given number of training cycles. The GD method for each cost functions may be defined as

$$W_\zeta(n+1) = W_\zeta(n) + \Delta W_\zeta(n) = W_\zeta(n) + \bar{\eta}_\zeta\left(-\frac{\partial J_\zeta(n)}{\partial W_\zeta(n)}\right), \tag{3}$$

where $\zeta = 1, 2$. W_ζ denote a weighting vector of the ζ-th SRWNNC. $\bar{\eta}_\zeta = diag[\eta_\zeta^a,\ \eta_\zeta^m,\ \eta_\zeta^d,\ \eta_\zeta^\alpha,\ \eta_\zeta^w]$ is a learning rate matrix for weights of the ζ-th SRWNNC. The gradient of cost functions J_1 and J_2 with respect to weighting vectors W_1 and W_2 of the controllers, respectively, are

$$\frac{\partial J_1(n)}{\partial W_1(n)} = -\left[e_x(n)\frac{\partial x_c(n)}{\partial u_1(n)} + e_y(n)\frac{\partial y_c(n)}{\partial u_1(n)}\right]\frac{\partial u_1(n)}{\partial W_1(n)}, \tag{4}$$

$$\frac{\partial J_2(n)}{\partial W_2(n)} = -e_\theta(n)\frac{\partial \theta(n)}{\partial u_2(n)}\frac{\partial u_2(n)}{\partial W_2(n)}, \tag{5}$$

where, $u_1(n) = \delta d(n)$ and $u_2(n) = \delta\theta(n)$. $\frac{\partial x_c(n)}{\partial u_1(n)}$, $\frac{\partial y_c(n)}{\partial u_1(n)}$, and $\frac{\partial \theta(n)}{\partial u_2(n)}$ denote the system sensitivity. It can be computed from (1). And also, the components of the Jacobian of the control input $u_\zeta(n)$ with respect to the ζ-th weighting vector W_ζ are computed by (2) [7].

4 Convergence Analysis Via ALRs

Let us define a discrete Lyapunov function as

$$V(n) = \frac{1}{2}[e_x^2(n) + e_y^2(n) + e_\theta^2(n)], \tag{6}$$

where, $e_x(n)$, $e_y(n)$, and $e_\theta(n)$ are the control errors. The change in the Lyapunov function is obtained by

$$\begin{aligned}\Delta V(n) &= V(n+1) - V(n)\\ &= \frac{1}{2}[e_x^2(n+1) - e_x^2(n) + e_y^2(n+1) - e_y^2(n) + e_\theta^2(n+1) - e_\theta^2(n)].\end{aligned} \tag{7}$$

Three error differences can be represented by [9]

$$\Delta e_x(n) \approx \left[\frac{\partial e_x(n)}{\partial W_1^i(n)}\right]^T \Delta W_1^i(n), \tag{8}$$

$$\Delta e_y(n) \approx \left[\frac{\partial e_y(n)}{\partial W_1^i(n)}\right]^T \Delta W_1^i(n), \tag{9}$$

$$\Delta e_\theta(n) \approx \left[\frac{\partial e_\theta(n)}{\partial W_2^i(n)}\right]^T \Delta W_2^i(n), \tag{10}$$

where, $W_1^i(n)$ and $W_2^i(n)$ are the arbitrary component of the weighting vectors $W_1(n)$ and $W_2(n)$, respectively. And the corresponding changes of them are denoted by $\Delta W_1^i(n)$ and $\Delta W_2^i(n)$. Using (3)–(5), ΔW_1 and ΔW_2 are obtained by

$$\Delta W_1^i(n) = \eta_1^i \left[e_x(n)\frac{\partial x_c(n)}{\partial u_1(n)} + e_y(n)\frac{\partial y_c(n)}{\partial u_1(n)}\right]\frac{\partial u_1(n)}{\partial W_1^i(n)}, \tag{11}$$

$$\Delta W_2^i(n) = \eta_2^i e_\theta(n)\frac{\partial \theta(n)}{\partial u_2(n)}\frac{\partial u_2(n)}{\partial W_2^i(n)}, \tag{12}$$

where η_1^i and η_2^i are arbitrary diagonal elements of the learning rate matrices $\bar{\eta}_1$ and $\bar{\eta}_2$ corresponding to the weight component $W_1^i(n)$ and $W_2^i(n)$, respectively.

Theorem 1. *Let* $\bar{\eta}_\zeta = [\eta_\zeta^1 \;\; \eta_\zeta^2 \;\; \eta_\zeta^3 \;\; \eta_\zeta^4 \;\; \eta_\zeta^5] = [\eta_\zeta^a \;\; \eta_\zeta^m \;\; \eta_\zeta^d \;\; \eta_\zeta^\alpha \;\; \eta_\zeta^w]$ *be the learning rates for the weights of the SRWNNC1 and SRWNNC2, respectively and define* $\mathbf{C}_{\zeta,max}$ *as*

$$\begin{aligned} \boldsymbol{C}_{\zeta,max} &= [C^1_{\zeta,max} \quad C^2_{\zeta,max} \quad C^3_{\zeta,max} \quad C^4_{\zeta,max} \quad C^5_{\zeta,max}]^T \\ &= \left[max_n \left\| \frac{\partial u_\zeta(n)}{\partial a_\zeta(n)} \right\| \quad max_n \left\| \frac{\partial u_\zeta(n)}{\partial m_\zeta(n)} \right\| \right. \\ &\quad \left. max_n \left\| \frac{\partial u_\zeta(n)}{\partial d_\zeta(n)} \right\| \quad max_n \left\| \frac{\partial u_\zeta(n)}{\partial \alpha_\zeta(n)} \right\| \quad max_n \left\| \frac{\partial u_\zeta(n)}{\partial w_\zeta(n)} \right\| \right]^T, \end{aligned}$$

where $\|\cdot\|$ *represents the Euclidean norm, and* $\zeta = 1, 2$. *Then, the asymptotic convergence of the SRWNNC1 and SRWNNC2 for mobile robots is guaranteed if* η_1^i *and* η_2^i *are chosen to satisfy*

$$0 < \eta_1^i < 2/((S_x^2 + S_y^2)(C^i_{1,max})^2), \quad 0 < \eta_2^i < 2/(S_\theta C^i_{2,max})^2, \tag{13}$$

where $i = 1, ...5$, $S_x = \frac{\partial x_c(n)}{\partial u_1(n)}$, $S_y = \frac{\partial y_c(n)}{\partial u_1(n)}$, *and* $S_\theta = \frac{\partial \theta(n)}{\partial u_2(n)}$.

Proof. See Appendix A.

Corollary 1. *From conditions of Theorem 1, the learning rates which guarantee the maximum convergence are* $\eta_1^{i,M} = 1/((S_x^2 + S_y^2)(C^i_{1,max})^2)$, $\eta_2^{i,M} = 1/(S_\theta C^i_{2,max})^2$.

Theorem 2. *Let* η_1^a *and* η_2^a *be the learning rates for the input direct weights of the SRWNNC1 and the SRWNNC2 respectively. The asymptotic convergence of the SRWNNC1 and the SRWNNC2 for mobile robots is guaranteed if the learning rates* η_1^a *and* η_2^a *satisfy:*

$$0 < \eta_1^a < 2/((S_x^2 + S_y^2)N_{1i}|x_{1,max}|^2), \quad 0 < \eta_2^a < 2/(S_\theta^2 N_{2i}|x_{2,max}|^2), \tag{14}$$

where, N_{1i} *and* N_{2i} *denote the input number of the SRWNNC1 and the SRWNNC2, respectively.* $|x_{1,max}|$ *and* $|x_{2,max}|$ *are the maximum of the absolute value of each controller's input, respectively.*

Theorem 3. *Let* $\eta^m_{1,2}$, $\eta^d_{1,2}$ *and* $\eta^\alpha_{1,2}$ *be the learning rates of the translation, dilation and self-feedback weights for the SRWNNC1 and the SRWNNC2, respectively. The asymptotic convergence is guaranteed if the learning rates satisfy:*

$$0 < \eta_1^m, \eta_1^\alpha < \frac{2}{(S_x^2 + S_y^2)N_{1w}N_{1i}} \left[\frac{1}{|w_{1,max}| \left(\frac{2exp(-0.5)}{|d_{1,min}|} \right)} \right]^2, \tag{15}$$

$$0 < \eta_2^m, \eta_2^\alpha < \frac{2}{S_\theta^2 N_{2w}N_{2i}} \left[\frac{1}{|w_{2,max}| \left(\frac{2exp(-0.5)}{|d_{2,min}|} \right)} \right]^2, \tag{16}$$

$$0 < \eta_1^d < \frac{2}{(S_x^2 + S_y^2)N_{1w}N_{1i}} \left[\frac{1}{|w_{1,max}| \left(\frac{2exp(0.5)}{|d_{1,min}|} \right)} \right]^2, \tag{17}$$

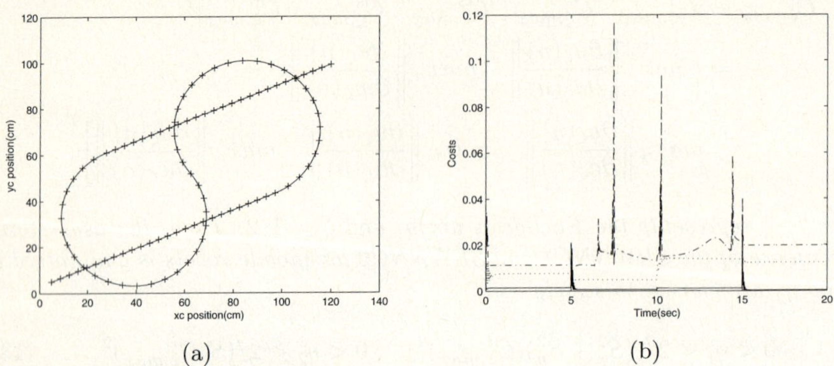

(a) (b)

Fig. 2. (a) Tracking result of SRWNNCs using ALRs for the mixed line (Reference trajectory (solid line) and the actual output (+ line)) (b) Comparison of Tracking costs for the mixed line. (SRWNNCs using ALRs (solid line), SRWNNCs using FLRs (dotted line), and WNNCs using FLRs (dash-dotted line)).

$$0 < \eta_2^d < \frac{2}{S_\theta^2 N_{2w} N_{2i}} \left[\frac{1}{|w_{2,max}| \left(\frac{2exp(0.5)}{|d_{2,min}|} \right)} \right]^2 , \tag{18}$$

where N_{1w} *and* N_{2w} *are the number of nodes in the product layer of the SRWNNC1 and the SRWNNC2, respectively.*

Theorem 4. *Let* η_1^w *and* η_2^w *be the learning rates for the weight* w_1 *of the SRWNNC1 and the weight* w_2 *of the SRWNNC2, respectively. Then, the asymptotic convergence is guaranteed if the learning rates satisfy:* $0 < \eta_1^w < 2/((S_x^2 + S_y^2)N_{1w})$, $0 < \eta_2^w < 2/S_\theta^2 N_{2w}$.

The proofs of Corollary 1 and Theorems 2-4 are omitted due to the lack of space.

Remark 1. From Corollary 1, the maximum learning rates of the SRWNNC1 and SRWNNC2 are chosen as the middle values in the range of the learning rates induced from Theorems 2, 3, and 4.

Remark 2. The GD method using maximum ALRs in Remark 1 is used to train the weights of the SRWNNCs. That is, the ALRs for learning of the SRWNNCs are updated to adapt the variation of the control environment on-line. From (8) –(12), note that the learning rates depend on the control errors e_x, e_y, and e_θ. So, the ALRs affect directly the convergence of all signals of the closed-loop system.

5 Simulation Results

To visualize the validity of the proposed SRWNNCs, we present simulation results for the stable path tracking of the mobile robot. In order to examine the

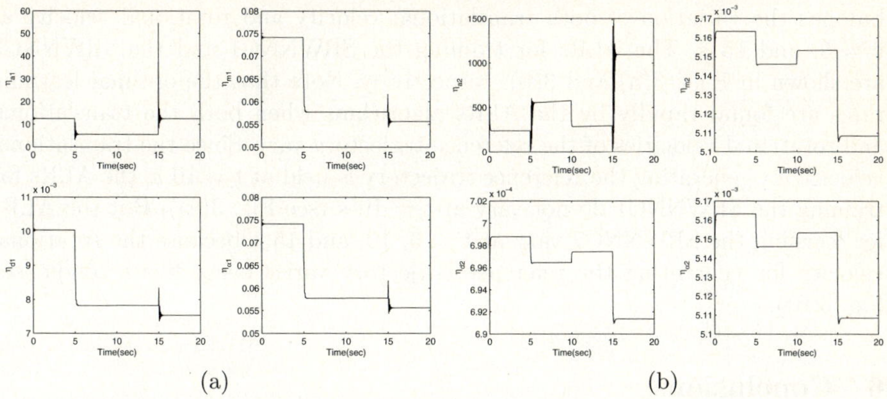

Fig. 3. The ALRs for the SRWNNCs (a) SRWNNC1 (b) SRWNNC2

tracking performances, we consider the complex trajectory combined with the straight line and the curved line. In addition, we compare the performance of the SRWNNCs using the ALRs with that of the SRWNNCs and WNN controllers (WNNCs) using the FLRs, respectively. The ALRs defined in Remark 1 are used for training the SRWNNC1 and SRWNNC2. To evaluate the performance of our controller, we define the cost and the mean cost as follows [4]:

$$\mathrm{Cost}(n) = e_x^2(n) + e_y^2(n) + e_\theta^2(n), \tag{19}$$

$$\text{Mean cost} = \frac{1}{T}\sum_{n=1}^{T}[e_x^2(n) + e_y^2(n) + e_\theta^2(n)], \tag{20}$$

where, T is the total number of samples. The reference trajectory is generated by the following control inputs:

$$\begin{aligned}
u_1 &= 20\ cm/s, \quad u_2 = 0\ rad/s \quad (0 \le t < 5),\\
u_1 &= 30\ cm/s, \quad u_2 = 1\ rad/s \quad (5 \le t < 10),\\
u_1 &= 30\ cm/s, \quad u_2 = -1\ rad/s \quad (10 \le t < 15),\\
u_1 &= 20\ cm/s, \quad u_2 = 0\ rad/s \quad (15 \le t \le 20).
\end{aligned}$$

The departure posture vector is $(5, 5, \pi/8)$ in this simulation and this trajectory has the variation of both the translational and rotational velocities. Fig. 2(a) displays the tracking control results of the SRWNN system using the ALRs, respectively. The simulation result demonstrates the good tracking capability of the suggested control method. The tracking costs of the SRWNNCs using the ALRs and the FLRs are compared with the WNNCs using the FLRs in Fig. 2(b). This Figure reveals that the SRWNNCs system using ALRs has the smallest control cost and fastest convergence among three cases. Also, the tracking cost of the SRWNNCs using the ALRs increases suddenly at t = 5, and 15 s because the reference path has only the variation of rotational velocity at t = 10 s

but has the variation of both translational velocity and rotational velocity at t = 5, and 15 s. The ALRs for training the SRWNNC1 and the SRWNNC2 are shown in Figs. 3(a) and 3(b), respectively. Note that the optimal learning rates are found rapidly by the ALRs algorithms when both the translational and rotational velocities of the reference trajectory vary. Since the translational velocity for generating the reference trajectory is held at t = 10 s, the ALRs for training the SRWNNC1 do not vary at t = 10 s (see Fig. 3(a)). But the ALRs for learning the SRWNNC2 vary at t = 5, 10, and 15 s because the rotational velocity for generating the reference trajectory varies every five seconds (see Fig. 3(b)).

6 Conclusions

In this paper, we have proposed a neurocontroller via adaptive learning rates for the stable path tracking of mobile robots. In the control scheme, two SRWNNCs have been designed for generating the control inputs. Since the SRWNN has the ability for storing the past information of the network, it can adapt rapidly to the dynamic environment of mobile robots. Using the discrete Lyapunov theorem, the stability for the whole control scheme has been carried out and the ALRs, which are suitable for the mobile robot, have been also established for the stable path tracking control of the mobile robot. Simulation results have shown that the proposed control system has an on-line adapting ability for controlling the mobile robot although the desired trajectories are complex and the SRWNNCs have very simple structures. And also, note that the ALRs have faster adaptability and convergence in the complex trajectory than the FLRs.

Acknowledgment

This work was supported by the Korea Research Foundation Grant funded by the Korean Government (MOEHRD) (KRF-2005-041-D00277).

References

1. Fierro R., and Lewis F.L.: Control of a nonholonomic mobile robot using neural networks. IEEE Trans. Neural Networks 9 (1998) 389-400
2. Hu T., Yang S.X., Wang F., and Mittal G.S.: A neural network controller for a nonholonomic mobile robot with unknown robot parameters. In Proc. of the IEEE Int. Conf. Robotics and Automation (2002) 3540-3545
3. Zhang Q., and Benveniste A.: Wavelet networks. IEEE Trans. Neural Networks 3 (1992) 889-898.
4. Sousa C.D., Hemerly E.M., and Galvão R.K.H.: Adaptive control for mobile robot using wavelet networks. IEEE Trans. Systems, Man, and Cybernetics 32 (2002) 493-504
5. Oussar Y., Rivals I., Personnaz L., and Dreyfus G.: Training wavelet networks for nonlinear dynamic input-output modeling. Neurocomputing 20 (1998) 173-188

6. Wai R.J., and Chang J.M.: Intelligent control of induction servo motor drive via wavelet neural network. Electric Power Systems Research 61 (2002) 67-76
7. Yoo S.J., Choi Y.H., and Park J.B.: Stable predictive control of chaotic systems using self-recurrent wavelet neural network. Int. Jour. Control, Automation and Systems 3 (2005) 43-55
8. Kim K.B., Park J.B., Choi Y.H., and Chen G.: Control of chaotic nonlinear systems using radial basis network approximatiors. Information Science 130 (2000) 165-183
9. Ku C.C., and Lee K.Y.: Diagonal recurrent neural networks for dynamics systems control. IEEE Trans. Neural Networks 6 (1995) 144-156
10. Lee C.H., and Teng C.C.: Identification and control of dynamic systems using recurrent fuzzy neural network. IEEE Trans. Fuzzy Systems 8 (2000) 349-366
11. Wang C.M.: Location estimation and uncertainty analysis for mobile robots. In Proc. of the IEEE Int. Conf. Robotics and Automation (1988) 1230-1235
12. Haesloop D., and Holt B.R.: A neural network structure for system identification. In Proc. of the American Control Conference (1990) 2460-2465

A The Proof of Theorem 1

From (6), $V(n) > 0$. Using (7)–(12), the change in the Lyapunov function is

$$\begin{aligned}
\Delta V(n) &= \Delta e_x(n)\left[e_x(n) + \frac{1}{2}\Delta e_x(n)\right] + \Delta e_y(n)\left[e_y(n) + \frac{1}{2}\Delta e_y(n)\right] \\
&\quad + \Delta e_\theta(n)\left[e_\theta(n) + \frac{1}{2}\Delta e_\theta(n)\right] \\
&= -(e_x(n)S_x + e_y(n)S_y)^2 \\
&\quad \cdot \left[\eta_1^i \left\|\frac{\partial u_1(n)}{\partial W_1^i(n)}\right\|^2 \left(1 - \frac{1}{2}(S_x^2 + S_y^2)\eta_1^i \left\|\frac{\partial u_1(n)}{\partial W_1^i(n)}\right\|^2\right)\right] \\
&\quad - e_\theta^2(n)S_\theta^2 \left[\eta_2^i \left\|\frac{\partial u_2(n)}{\partial W_2^i(n)}\right\|^2 \left(1 - \frac{1}{2}\eta_2^i S_\theta^2 \left\|\frac{\partial u_2(n)}{\partial W_2^i(n)}\right\|^2\right)\right] \\
&= -(e_x(n)S_x + e_y(n)S_y)^2\rho - e_\theta^2(n)\gamma,
\end{aligned}$$

where

$$\begin{aligned}
\rho &= \eta_1^i \left\|\frac{\partial u_1(n)}{\partial W_1^i(n)}\right\|^2 \left(1 - \frac{1}{2}(S_x^2 + S_y^2)\eta_1^i \left\|\frac{\partial u_1(n)}{\partial W_1^i(n)}\right\|^2\right) \\
\gamma &= \eta_2^i \left\|\frac{\partial u_2(n)}{\partial W_2^i(n)}\right\|^2 \left(1 - \frac{1}{2}\eta_2^i S_\theta^2 \left\|\frac{\partial u_2(n)}{\partial W_2^i(n)}\right\|^2\right).
\end{aligned}$$

If $\rho > 0$ and $\gamma > 0$ are satisfied, $\Delta V(n) < 0$. Thus, the asymptotic convergence of the proposed control system are guaranteed. Here, we obtain (13). This completes the proof of the theorem.

Neuro-PID Position Controller Design for Permanent Magnet Synchronous Motor

Mehmet Zeki Bilgin and Bekir Çakir

Kocaeli University, Department of Electrical Engineering, Vinsan Kampusu,
41300, Kocaeli, Turkey
{bilgin, bcakir}@kou.edu.tr
http://www.kou.edu.tr

Abstract. A new speed control strategy is presented for high performance control of a Permanent Magnet Synchronous Motor (PMSM). A self-tuning Neuro-PID controller is developed for speed control. The PID gains are tuned automatically by the neural network in an on-line way. In recent years, the researches on the control of electrical machines based on ANN are increased. ANN's, developed controller in this work, offer inherent advantages over conventional PID controller for PMSM , namely: Reduction of the effects of motor parameter variations, improvement of controller time response and improvement of drive robustness. The PMSM drive system was simulated by using MATLAB 5.0/Simulink software package. The performance of the proposed method is compared with the conventional PID methods. At the result, the control based on self-tuning Neuro-PID control has better performance than the conventional PID controller.

1 Introduction

Recent advances in power semiconductor devices, microprocessor, converter design technology and control theory have enabled ac servo drives to satisfy the high performance requirements in many industrial application. [1],[2]. Current-regulated Permanent Magnet Synchronous Motors (PMSM) are used in many applications that require rapid torque response and high-performance operation such as robotics, vehicle propulsion, heat pumps, actuators, computer numerically machine tools and ship propulsion.[3],[4].

The control performance of the PMSM servo drive is still influenced by uncertainties, which usually are composed of unpredictable plant parameter variations, external load disturbance, and unmodeled and nonlinear dynamics of the plant. In the past decade, many modern control theories, such as nonlinear control [5], variable structure system control[6], adaptive control [7] ,optimal control[8],and the robust control[9] have been developed for the PMSM drive to deal with uncertainties. In the application of such techniques, development of mathematical models is a prior necessity. However, such mathematical modelling which is largely based on the assumption of linearization of system might not reflect the true physical properties of the system.

L. Jiao et al. (Eds.): ICNC 2006, Part I, LNCS 4221, pp. 418 – 426, 2006.

The complex mathematical models which do reflect precise input-output physically relation of the system can be build, but the sensivity of parameters should be low in order to make the control system useful. And , if some changes in the plant occur, the model must be re-build and the new control law must be determined. Therefore, these control theory is difficult to apply for real world problem.

In recent years, intelligent control in general has been quite readily acceptable for real control applications. In the last decade, the back propagation algorithm has been recognized as an efficient tool to realize the learning mechanism after its discovery. After, the researches on the control of electrical machines based on ANN are increased [10]-[14].

The purpose of this paper is to develop a self tuning Neuro-PID position control drive system for PMSM. The analysis, design and simulation of the purposed controller are described. Good and robust control performance , both in command tracking and the load regulating characteristics of the rotor position, is achieved.

2 System Description and Machine Model

Fig. 1 shows the total configuration of a field-oriented Permanent Magnet Synchronous Motor drive system investigated in this work.

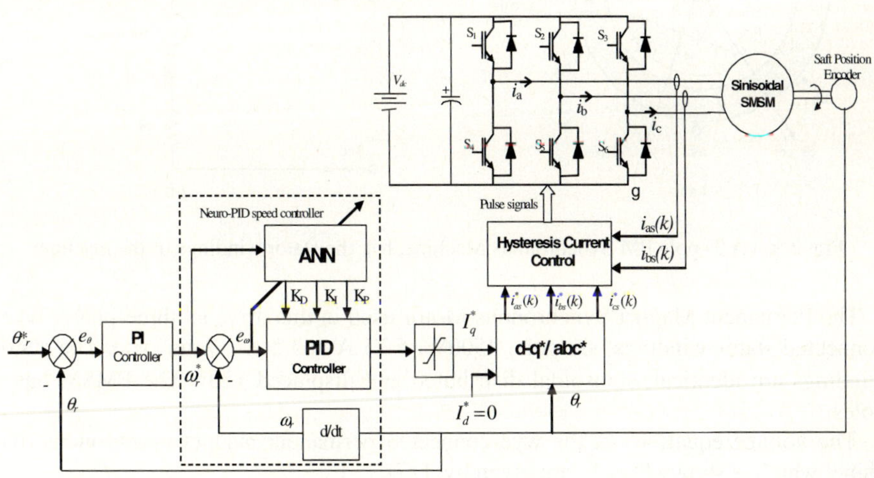

Fig. 1. The block diagram of proposed PMSM motor drive system

The system consists of a PMSM and load, a hysteresis current controlled voltage source inverters (VSI), a field-orientation mechanism and a coordinate translator, a position control loop and a self-tuning Neuro-PID speed controller.

The torque in PMSM's is usually controlled by controlling armature current. In high-performance drives, pulse-width modulated (PWM) inverters are used to provide effective current control. Various techniques and control algorithm of current control of PWM inverters have been studied and reported in the literature [15]-[16]. In one of these control schemes called hysteresis on-off current control, the motor currents are compared to their reference currents and the switching instants for the inverter power switches are determined using hysteresis control strategy. In this work, the hysteresis control strategy is used for current control of PWM inverter and the reference currents are produced by self-tuning PID-neuro controller.

2.1 The Model of Permanent Magnet Synchronous Motor

The machine model is developed for the simulation work. At the developed machine model, magnetic saturation is neglected, all parameters of the motor are not assumed to be constant and dependent operation condition. All harmonic torques resulting from supply harmonics and operation temperature are neglected. The inverter is assumed to be ideal and the machine has damper windings. The machine model is shown Fig 2.

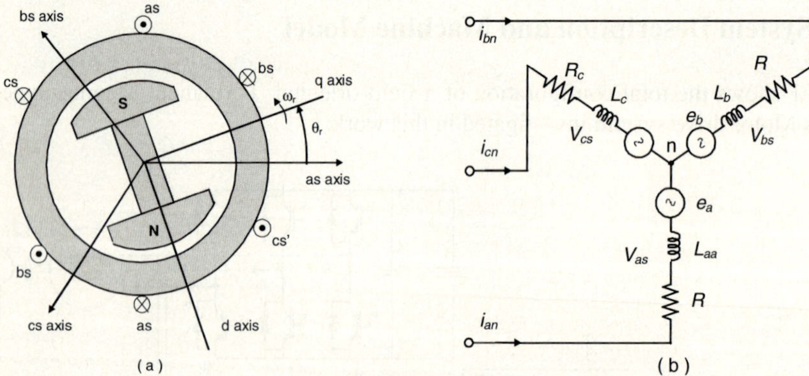

Fig. 2. a-) A 2- pole PM Synchronous Machine, b-) the stator windings of the machine

The Permanent Magnet Synchronous Motor used in this drive is three-phase, wye-connected stator windings, six pole, 2500W, 5.75 A and 5000 r/min type. The stator windings are identical, sinusoidal distributed and displaced 120°. The PMSM has 6 poles.

The voltage equations of the wye-connected permanent magnet synchronous machine, which is shown Fig. 2 , are given by, [17],

$$\begin{aligned} V_{as} &= r_s.i_{as} + L_{ss}\frac{di_{as}}{dt} + \omega_r\lambda_m Cos(\theta_r) \\ V_{bs} &= r_s.i_{bs} + L_{ss}\frac{di_{bs}}{dt} + \omega_r\lambda_m Cos\left(\theta_r - \frac{2\pi}{3}\right) \\ V_{cs} &= r_s.i_{cs} + L_{ss}\frac{di_{cs}}{dt} + \omega_r\lambda_m Cos\left(\theta_r + \frac{2\pi}{3}\right) \end{aligned} \tag{1}$$

where r_s , L_{ss}, θ_r , ω_r, and λ_m denote the stator resistance, stator self inductance, the position of the rotor, angular shaft speed and the flux linkage due to permanent magnet, respectively. In this paper, the voltage equation of the PMSM is established using reference frame theory to express the variables in the rotor reference frame. This transform is;

$$f^{r}{}_{qdo} = K^{r}_{s} f_{abcs} \tag{2}$$

where f may represent voltage ,current or flux linkage. K^{r}_{s} , which is a matrix, are as follows,

$$K^{r}_{s} = \begin{bmatrix} Cos(\theta_r) & Cos\left(\theta_r - \frac{2\pi}{3}\right) & Cos\left(\theta_r + \frac{2\pi}{3}\right) \\ Sin(\theta_r) & Sin\left(\theta_r - \frac{2\pi}{3}\right) & Sin\left(\theta_r + \frac{2\pi}{3}\right) \\ \frac{1}{2} & \frac{1}{2} & \frac{1}{2} \end{bmatrix} \tag{3}$$

The voltage equations of PMSM are transformed from abc variable to qdo variable using Eq.(3). A circuit model of a PMSM, which is used predicting its transient behaviour , can be obtained using either of two equivalent circuits representation of PMSM. The equivalent qdo circuits shown in Fig. 3. for PMSM with damper cage windings.[16]

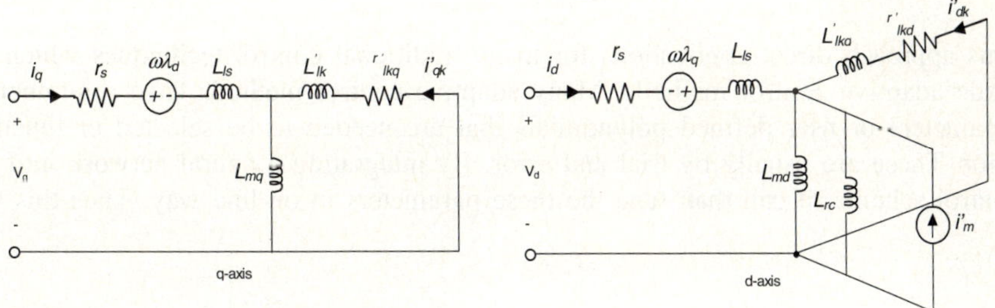

Fig. 3. Equivalent q and d axis circuits of a PMSM

The qd voltage equation of the PMSM are,

$$\begin{aligned} V_q &= r_s i_q + \frac{d\lambda_q}{dt} + \lambda_d \frac{d\theta_r}{dt} \\ V_d &= r_s i_d + \frac{d\lambda_d}{dt} - \lambda_q \frac{d\theta_r}{dt} \\ 0 &= r'_{kd} i'_{kd} + \frac{d\lambda'_{kd}}{dt} \\ 0 &= r'_{kd} i'_{kd} + \frac{d\lambda'_{kd}}{dt} \end{aligned} \tag{4}$$

The *qd* flux linkage equation is;

$$
\begin{aligned}
\lambda_q &= L_q i_q + L_{mq} i'_{kq} \\
\lambda_d &= L_d i_d + L_{md} i'_{kd} + \lambda'_m \\
\lambda'_{kq} &= L_{mq} i_q + L'_{kqkq} i'_{kq} \\
\lambda'_{kd} &= L_{md} i_d + L'_{kdkd} i'_{kd} + L_{md} i'_m
\end{aligned}
\tag{5}
$$

The instantaneous electromagnetic torque of motor is;

$$T_e = \left(\frac{3}{2}\right)\left(\frac{P}{2}\right)\left(L_d - L_q\right) i_d i_q + \frac{3}{2}\frac{P}{2}\left(L_{md} i'_{kd} i_q - L_{mq} i'_{kq} i_d\right) + \frac{3}{2}\frac{P}{2} L_{md} i'_m i_q \tag{6}$$

The relation of moment and motor speed is ;

$$T_e = J\left(\frac{2}{P}\right)\cdot\frac{d\omega_r}{dt} + B_m\left(\frac{2}{P}\right)\cdot\omega_r + T_L \tag{7}$$

where J, T_L, B_m and P denote the moment of inertia, external load torque, viscous friction coefficient of the rotating parts and poles of machine, respectively.

3 Design of the Self -tuning PID-Neuro Controller

This approach direct applications for many traditional control techniques which include adaptive control methods. Many adaptive control methods have a number of parameters or user defined polynomials that are needed to be selected or tuned in prior. These are usually by trial and error. By integrating a neural network into the control scheme, it can than tune the these parameters in on-line way. Thus this self

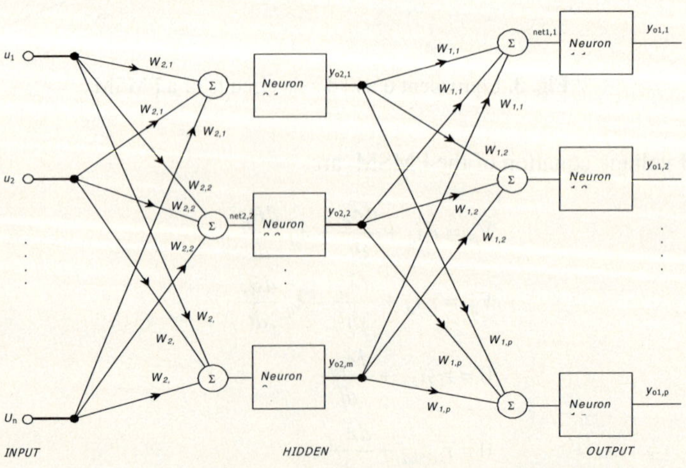

Fig. 4. Topology of multi input ,multi output 2 layer feed-forward ANN

tuning neuro-control strategy has possible application in many traditional control approaches. In this paper, the neural network is used to tune the parameters of the PID controller. The neural network is minimized error function by adjusting the PID gain, such as, K_P, K_I, K_D. There are many artificial neural network architectures that have been proposed.[17]. One of these architecture is the feed-forward neural network (FFNN). A typical multi input, multi output two layer FNN structure is illustrated in Fig 4. [18].

The input and output signal equations of the hidden and output layer can be written from Fig.4. as:

$$\left.\begin{aligned} net_{2,j} &= \sum_{i=1}^{n} W_{2,ji} \cdot U_i \\ y_{o_{2,1}} &= f_i\left(y_{net_{2,j}}\right) \end{aligned}\right\} \quad and \quad \left.\begin{aligned} net_{1,k} &= \sum_{j=1}^{m} W_{1,kj} \cdot y_{out_{2,j}} \\ y_{o_{1,k}} &= f_k\left(y_{net_{1,k}}\right) \end{aligned}\right\} \tag{8}$$

where j=1,2,m and k=1,2,..p. Using a three layer neural network, find the suitable PID gains. Thus, the outputs of neural network are K_P, K_I and K_D which are denoted $y_{o1,1}$, $y_{o1,2}$, $y_{o1,3}$ respectively. Based on the steepest descent method, for the output layer,

$$\Delta W_{1,1x}(t+1) = -\eta \frac{\partial E}{\partial W_{1,1x}} + \alpha \Delta W_{1,1x}(t) \tag{9}$$

and for the hidden layer,

$$\Delta W_{2,x1}(t+1) = -\eta \frac{\partial E}{\partial W_{2,x1}} + \alpha \Delta W_{2,x1}(t) \tag{10}$$

$$\frac{\partial E}{\partial W_{1,kx}} = \delta_{1,k} \cdot \frac{\partial net_{1,k}}{\partial W_{1,kx}} \tag{11}$$

$$\delta_{1,k} = e \cdot y_{o1,k}\left(1 - y_{o1,k}\right) \frac{\partial y}{\partial u} \cdot \frac{\partial u}{\partial y_{o1,k}} \qquad (k = 1,2,3) \tag{12}$$

$$\frac{\partial u}{\partial y_{o1,k}} = \begin{cases} e(t) - e(t-1) & k = 1 \\ e(t) & k = 2 \\ e(t) - 2.e(t-1) + e(t-2) & k = 3 \end{cases} \tag{13}$$

For the hidden layer,

$$\delta_{2,x} = \sum_{k} \delta_{1,k} \cdot W_{1,kx} \cdot y_{o2,x} \cdot (1 - y_{o2,x}) \tag{14}$$

Where, $\partial y / \partial u$ is system (PMSM) Jacobean and it has been estimated. The self-tuning PID type neuro-control strategy is employed from Eq.(8) to Eq.(14).

4 Simulation of the Drive System

The dynamic performance of the PM synchronous motor drive is evaluated by using computer simulation. The control system is shown in Fig. 5.

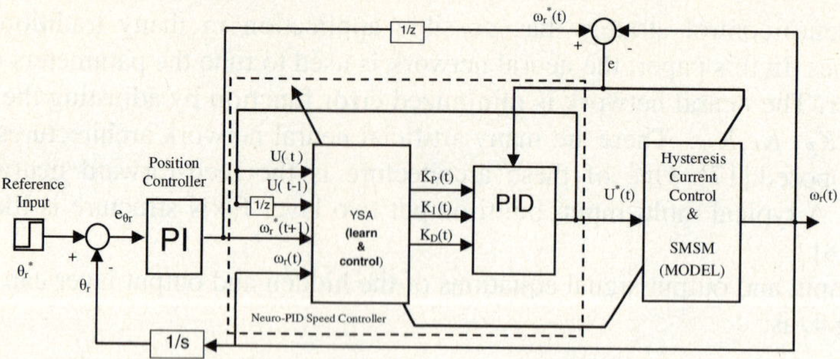

Fig. 5. The block diagram of the self tuning neuro-PID control system

The MATLAB/Simulink software package was used to analyze the position controller. The simulation was run three different times for each specified reference trajectory. The first run was to demonstrate position response with nominal machine parameters. PID and Neuro-PID position controller responses are shown in Fig. 6. The load torque is increased to 5 Nm at 0.3s.

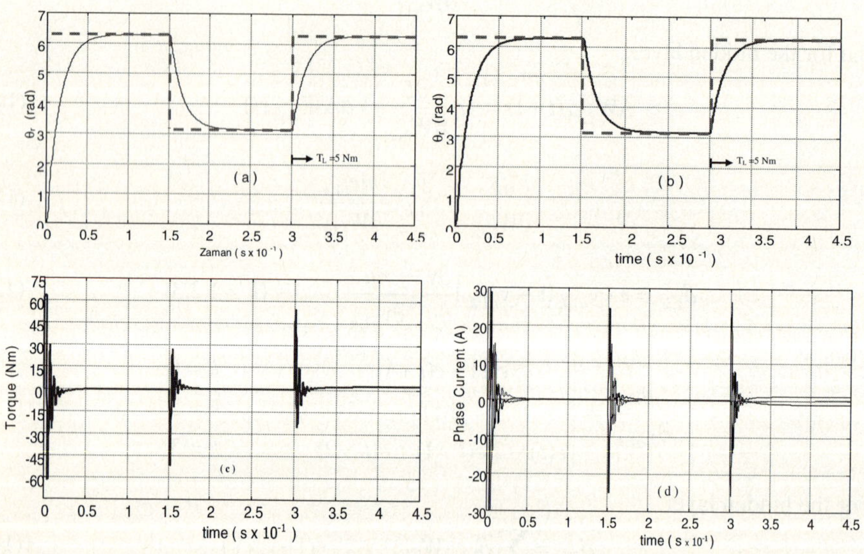

Fig. 6. The simulation results with nominal parameters. a–position response with only PID controller, b- position response with neuro-PID controller, c-electrical torque with Neuro-PID controller d- phase current of PMSM with Neuro-PID controller.

The electrical time constant is increased to 1.5 times the nominal value for PID and neuro-PID controllers. The simulation results is shown in Fig. 7.

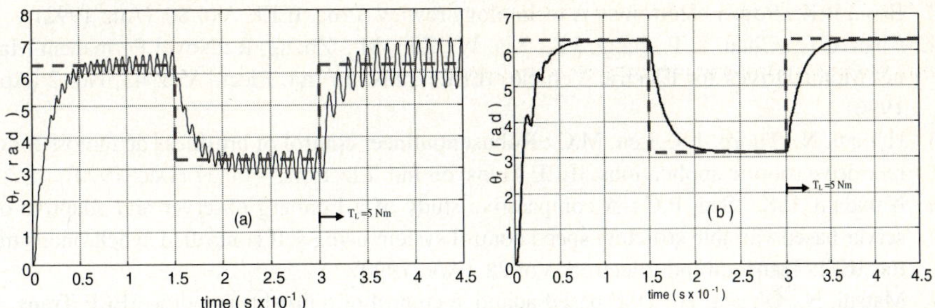

Fig. 7. Simulation results with τ_e variation, a- PID, b- proposed Neuro-PID control

Fig 8. shows the responses of controllers under mechanical time constant(τ_m)variation. τ_m changed to 4 times the nominal value.

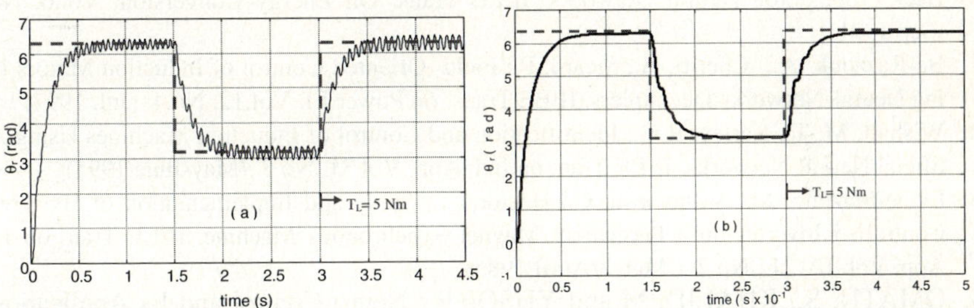

Fig. 8. Simulation results with τ_m variation, a- with PID, b- with Neuro-PID controller

5 Conclusion

A robust Permanent Magnet Synchronous Motor drive system, which is based on self-tuning PID neuro-control structure , has been presented in this paper. A PI position controller was described first. Next self tuning Neuro-PID speed controller is designed with three layer and three output neural network. The Neural Network is tuned PID gains on-line. Then , the self-tuning Neuro-PID controller was on line designed to match the time domain reference tracking specification under the parameter variation.

In this work, the mathematical basis of the proposed controller was derived and the drive system simulated. The simulation result showed that this self-tuning Neuro-PID control strategy effectively achieved the desired dynamic performance of PMSM.

References

1. Bose, B.K.: Technology trends in microcomputer control of electrical machines. IEEE Trans. on Ind. Electron., Vol. 35 (Feb. 1988)
2. Sen, P.C.: Electric motor drives and control-Past, present, and future. IEEE Trans. on Ind. Electron.,Vol.37 (Dec 1990)

3. Bose, B.K.: Power electronics-A technology rewiev. Proc. IEEE, Vol.80 (Aug 1992)
4. Chan, C.C., Chau, K.T., Jiang, J.Z., Xia, W., Zhu, M., Zhang, R.: Novel Permanent Magnet Motor Drives for Electric Vehicle. IEEE Trans. on Ind. Elect., Vol. 43, No. 2 (April 1996)
5. Hemati, N., Thorp, J.S., Leu, M.C.: Robust nonlineer control of brushless dc motors for direct-drive robotic applications. IEEE Trans. on Ind. Electron., Vol. 37 (Dec. 1990)
6. Namdam, P.K., Sen, P.C.: A comparative study of a Lunberg observer and adaptive observer based variable structure speed control system using self controlled synchronous motor. IEEE Trans. on Ind. Electron.,Vol.28 (Apr. 1990)
7. Matsui, N., Ohashi, H.: DSP based adaptive control of a brushless motor. IEEE Trans. on Ind. Applicat.,Vol.28 (Mar./Apr 1992)
8. Cerruto, E., Consili, A., Raciti, A., Testa, A.: A robust adaptive controller for PM motor driver in robotic application. IEEE Trans. on Power Electron.,Vol.10 (Jan.1995)
9. Fukuda, T., Shibata, T.: Theory and Application of Neural Networks for Industrial Control System. IEEE Trans. on Ind. Elec. Vol.39, No.6 (Dec. 1992)
10. Weerasooriya,S., El-Sharkawi, M.A.: Identification and Control of a DC Motor using Back-Propagation Neural Networks. IEEE Trans. On Energy Conversion, Vol.6, No.4 (Sep 1991)
11. Ba-Razzouk, A., Cheriti, A., Sicard, P.: Field Oriented Control of Induction Motors Using Neural-Networks Decouplers. IEEE Trans. on Power El. Vol.12. No.4 (Jul. 1997)
12. Wishart, M. T., Karley, R.G.: Identification and Control of Induction Machines Using Artificial Neural Networks. IEEE Tran. on Ind. App. Vol.31, No.3 (May/June 1995)
13. Lajoe-Mazenc, M., Villanueva, C., Hector, J.: Study and Implementation of Hysteresis Controlled Inverter on a Permanent Magnet Synchronous Machine. IEEE Tran. on Ind. App. Vol. IA-21 , No.2 (March/April 1985)
14. OMATU, S., KHALID, M and YUSOF,R.: Neuro-Control and Its Applications. Springer-Verlag Berlin Heidelberg, New York (1996)
15. Le-Huy, H., Dessaint, L.A.: An Adaptive Current Scheme for PWM Synchronous Motor Drive: Analysis and Simulation. IEEE Tran. on Power Elect.,Vol.4,No.4 (October 1989)
16. Ong, C.M..: Dynamic Simulation of Electric Machinery, Prentice Hall (1998)
17. Narenda, K. S.: Identification and Control of Dynamical System Using Neural networks. IEEE Trans. On Energy Conversion, Vol.6, No.4 (Sep 1991)
18. Abulafya, N.: Neural Networks for System Identification and Control. Ph.D. Thesis, Imperial College of Science, Technology and Medicine, University of London (September 1995)

Robust Stability of Nonlinear Neural-Network Modeled Systems

Jong-Bae Lee, Chang-Woo Park, and Ha-Gyeong Sung

Intelligent mechatronics center, Korea Electronics Technology Institute
401-402 B/D 193, Yakdae-dong, Wonmi-gu
{drcwpark, jblee}@keti.re.kr

Abstract. In this paper, a robust stability analysis method for feedback linearization using neural networks is presented. The robust regulation problem of nonlinear system with external disturbance is considered. The feedforward neural networks with one hidden layer are used to approximate the uncertain nonlinear system. The approximation errors are treated as the structured uncertainties with the known bounds. For these external disturbance and structured uncertainties, stability robustness of the closed system is analyzed in both input-output sense and Lyapunov sense.

1 Introduction

A class of nonlinear plant can be transformed into the linear system model using the feedback linearization method. Since the transformed linear system model can be easily controlled by well-known and powerful linear control theory, feedback linearization is one of the important approaches in nonlinear control theory[1][2]. However, feedback linearization does not guarantee robustness in the face of parameter uncertainty or disturbances. Up to now, various research results have been published on robust or adaptive feedback linearization methods [3][4]. Recently, some studies have also been reported on feedback linearization using neural networks [5]-[8]. Feedback linearization using neural networks is a feedback linearization method which uses neural networks as a nonlinear system model. Neural networks have excellent capability in nonlinear system approximation and are particularly suitable for the complex and uncertain system [9][10]. In [7], the adaptive feedback linearization using neural networks was introduced. From a practical point of view, the robust approach is more suitable. In our paper, we deal with the robust stability analysis for feedback linearization using neural networks. We consider the robust regulation problem of nonlinear system with the external disturbance. The unknown nonlinear system is approximated by two neural networks. To analyze robust stability, we assume that the approximation errors are represented by the structured uncertainties with the known norm bounds. For these external disturbance and structured uncertainties, the robust stability of the closed system is analyzed in input-output sense and Lyapunov sense by applying multivariable circle criterion and the relationship between input-output stability and Lyapunov stability [11]-[13].

L. Jiao et al. (Eds.): ICNC 2006, Part I, LNCS 4221, pp. 427 – 436, 2006.

2 Problem Formulation

Consider the regulation problem of the following n-th order nonlinear system.

$$x^{(n)} = f(\mathbf{x}) + g(\mathbf{x})u + d \tag{1}$$

where $f(\mathbf{x})$ and $g(\mathbf{x})$ are unknown (uncertain) but bounded continuous nonlinear functions and d denotes the external disturbance which is unknown but bounded. The external disturbances are due to system load, external noise, etc. Let $\mathbf{x} = [x, \dot{x}, \cdots, x^{(n-1)}]^T \in R^n$ be the state vector of the system which is assumed to be available. Also, it is assumed that the system (1) has a well-defined equilibrium point, which without loss of generality is chosen to be $\mathbf{x} = \mathbf{0}$ (i.e., $f(\mathbf{0}) = 0$ and $g(\mathbf{0}) = 0$. The unknown nonlinear system (1) can be approximated by the feedforward neural networks architecture. In (2), $f(\mathbf{x})$ and $g(\mathbf{x})$ are approximated by two neural networks $\hat{f}(\mathbf{x})$ and $\hat{g}(\mathbf{x})$, respectively (see Fig. 1 and 2).

Fig. 1. The feed forward neural network model $\hat{f}(\mathbf{x})$

Fig. 2. The feed forward neural networks model $\hat{g}(\mathbf{x})$

$$x^{(n)} = \hat{f}(\mathbf{x}) + \hat{g}(\mathbf{x})u + d \tag{2}$$

$$\hat{f}(\mathbf{x}) = \sum_{i=1}^{L} \sigma(\mathbf{a}_\mathbf{i}^{\mathbf{f}T} \cdot \mathbf{x})\, c_i \ ,$$

$$\hat{g}(\mathbf{x}) = \sum_{i=1}^{M} \sigma(\mathbf{a}_\mathbf{i}^{\mathbf{g}T} \cdot \mathbf{x})\, b_i$$

where, L and M are the number of neurons in the hidden layer. Also, $\mathbf{a}_\mathbf{i}^\mathbf{f} \in R^{1\times n}$, $\mathbf{a}_\mathbf{i}^\mathbf{g} \in R^{1\times n}$, $b_i \in R$ and $c_i \in R$ are the weights for a particular node and $\sigma(\bullet)$ is the activation function. In this paper, we use the following hyperbolic tangent function (3).

$$\sigma(x) = tanh(x) = \frac{(1-\exp(-2x))}{(1+\exp(-2x))} \tag{3}$$

The derivative of the activation function is

$$\frac{d\sigma(x)}{dx} = 1 - \sigma^2(x) \tag{4}$$

For robust stability analysis, we assume that the nonlinear model (2) has the norm-bounded approximation errors in the weights b_i and c_i. From the above assumption, $f(\mathbf{x})$ and $g(\mathbf{x})$ of the original system (1) can be represented by the following equations (5).

$$x^{(n)} = f(\mathbf{x}) + g(\mathbf{x})u + d \tag{5}$$

$$f(\mathbf{x}) = \sum_{i=1}^{L} \sigma(\mathbf{a}_\mathbf{i}^{\mathbf{f}T} \cdot \mathbf{x})\ (c_i + \Delta c_i(t))$$

$$g(\mathbf{x}) = \sum_{i=1}^{M} \sigma(\mathbf{a}_\mathbf{i}^{\mathbf{g}T} \cdot \mathbf{x})\ (b_i + \Delta b_i(t))$$

where, $\Delta b_i(t), \Delta c_i(t) \in R$ denote the norm-bounded time-varying approximation errors or modeling uncertainties which are bounded by the following inequalities (6).

$$\left|\Delta b_i(t)\right| \le \Delta b_i\ , \quad \left|\Delta c_i(t)\right| \le \Delta c_i \quad \text{for all } i \tag{6}$$

To stabilize the nonlinear system (6), we have proposed the following feedback linearization regulator (7) based on neural networks approximation.

$$u = \frac{\hat{\mathbf{a}}^T \cdot \mathbf{x} - \hat{f}(\mathbf{x})}{\hat{g}(\mathbf{x})} = \frac{\hat{\mathbf{a}}^T \cdot \mathbf{x} - \sum_{i=1}^{L} \sigma(\mathbf{a}_\mathbf{i}^{\mathbf{f}T} \cdot \mathbf{x})\ c_i}{\sum_{i=1}^{M} \sigma(\mathbf{a}_\mathbf{i}^{\mathbf{g}T} \cdot \mathbf{x})\ b_i} \tag{7}$$

where we use the same $\mathbf{a}_\mathbf{i}^\mathbf{f}$, $\mathbf{a}_\mathbf{i}^\mathbf{g}$, b_i, c_i with the nonlinear model (2) for all i. In the feedback linearization regulator (7), $\hat{\mathbf{a}} \in R^{1 \times n}$ is the linear state feedback gain vector which provides the desired linear dynamics. If there is no uncertainty in (5) ($\Delta b_i(t) = 0$, $\Delta c_i(t) = 0$ for all i, $d = 0$), the above feedback linearization regulator (7) can cancel the nonlinearity of (5) and achieve perfect linearization (8).

$$x^{(n)} = \hat{\mathbf{a}}^T \cdot \mathbf{x} \tag{8}$$

In practical application, perfect linearization can not be achieved. By substituting (7) into (5), the imperfectly linearized system can be written as (9).

$$\begin{aligned} x^{(n)} &= \sum_{i=1}^{L} \sigma(\mathbf{a}_\mathbf{i}^{\mathbf{f}T} \cdot \mathbf{x})\ (c_i + \Delta c_i(t)) \\ &+ \frac{\sum_{i=1}^{M} \sigma(\mathbf{a}_\mathbf{i}^{\mathbf{g}T} \cdot \mathbf{x})\ (b_i + \Delta b_i(t))}{\sum_{i=1}^{M} \sigma(\mathbf{a}_\mathbf{i}^{\mathbf{g}T} \cdot \mathbf{x})\ b_i} \{\hat{\mathbf{a}}^T \cdot \mathbf{x} - \sum_{i=1}^{L} \sigma(\mathbf{a}_\mathbf{i}^{\mathbf{f}T} \cdot \mathbf{x})\ c_i\} + d \end{aligned} \tag{9}$$

In order to analyze the robust stability of (9) we should represent (9) as the connection between the linear part and the sector-bounded nonlinear part. To do this, we define

$$\phi(\mathbf{a}_\mathbf{i}^{\mathbf{f}T} \cdot \mathbf{x}) = \frac{\sigma(\mathbf{a}_\mathbf{i}^{\mathbf{f}T} \cdot \mathbf{x})}{\mathbf{a}_\mathbf{i}^{\mathbf{f}T} \cdot \mathbf{x}} \tag{10}$$

From the definition of $\phi(\mathbf{a}_\mathbf{i}^{\mathbf{f}T} \cdot \mathbf{x})$, the closed system (9) can be represented as (11).

$$\begin{aligned} x^{(n)} &= \sum_{i=1}^{L} \phi(\mathbf{a}_\mathbf{i}^{\mathbf{f}T} \cdot \mathbf{x})\,(c_i + \Delta c_i(t))(\mathbf{a}_\mathbf{i}^{\mathbf{f}T} \cdot \mathbf{x}) \\ &\quad + \frac{\sum_{i=1}^{M} \sigma(\mathbf{a}_\mathbf{i}^{\mathbf{g}T} \cdot \mathbf{x})\,(b_i + \Delta b_i(t))}{\sum_{i=1}^{M} \sigma(\mathbf{a}_\mathbf{i}^{\mathbf{g}T} \cdot \mathbf{x})\, b_i} \{\hat{\mathbf{a}}^T \cdot \mathbf{x} - \sum_{i=1}^{L} \phi(\mathbf{a}_\mathbf{i}^{\mathbf{f}T} \cdot \mathbf{x})\, c_i (\mathbf{a}_\mathbf{i}^T \cdot \mathbf{x})\} + d \\ &= \mathbf{a}_L^T \cdot \mathbf{x} + (\hat{\mathbf{a}} - \mathbf{a}_L)^T \cdot \mathbf{x} + \sum_{i=1}^{L} \phi(\mathbf{a}_\mathbf{i}^{\mathbf{f}T} \cdot \mathbf{x})\, \Delta c_i(t)\, (\mathbf{a}_\mathbf{i}^{\mathbf{f}T} \cdot \mathbf{x}) \\ &\quad + \frac{\sum_{i=1}^{M} \sigma(\mathbf{a}_\mathbf{i}^{\mathbf{g}T} \cdot \mathbf{x})\, \Delta b_i(t)}{\sum_{i=1}^{M} \sigma(\mathbf{a}_\mathbf{i}^{\mathbf{g}T} \cdot \mathbf{x})\, b_i} \{\hat{\mathbf{a}}^T \cdot \mathbf{x} - \sum_{i=1}^{L} \phi(\mathbf{a}_\mathbf{i}^{\mathbf{f}T} \cdot \mathbf{x})\, c_i (\mathbf{a}_\mathbf{i}^{\mathbf{f}T} \cdot \mathbf{x})\} + d \\ &= \mathbf{a}_L^T \cdot \mathbf{x} + \mathbf{a}_N(t)^T \cdot \mathbf{x} + d \end{aligned} \tag{11}$$

where,

$$\begin{aligned} \mathbf{a}_N(t)^T &= (\hat{\mathbf{a}} - \mathbf{a}_L)^T + \sum_{i=1}^{L} \phi(\mathbf{a}_\mathbf{i}^{\mathbf{f}T} \cdot \mathbf{x})\, \Delta c_i(t)\, \mathbf{a}_\mathbf{i}^{\mathbf{f}T} \\ &\quad + \frac{\sum_{i=1}^{M} \sigma(\mathbf{a}_\mathbf{i}^{\mathbf{g}T} \cdot \mathbf{x})\, \Delta b_i(t)}{\sum_{i=1}^{M} \sigma(\mathbf{a}_\mathbf{i}^{\mathbf{g}T} \cdot \mathbf{x})\, b_i} (\hat{\mathbf{a}}^T - \sum_{i=1}^{L} \phi(\mathbf{a}_\mathbf{i}^{\mathbf{f}T} \cdot \mathbf{x})\, c_i\, \mathbf{a}_\mathbf{i}^{\mathbf{f}T}) \end{aligned}$$

3 Robust Stability Analysis for Feedback Linearization Using Neural Networks

To analyze the robust stability of (11), consider two different cases, i) $d \neq 0$ ii) $d = 0$. In case of i) $d \neq 0$, the input-output stability should be guaranteed so as to bound the norm of the state vector $\mathbf{x}$ (output) with respect to the norm-bounded disturbance d (input). In our analysis, well-known multivariable circle criterion is used to analyze the input-output robust stability of (11) [11]-[13]. Since

$a_{Nj}(t)$ in (11) is bounded by the maximum and the minimum obtained in Appendix A for all j and t, it can be treated as time-varying sector bounded nonlinearity. Therefore, multivariable circle criterion can be applied to analyze L_2 stability of (11). To apply multivariable circle criterion, the closed system (11) should be transformed into the basic configuration of multivariable circle criterion as in Fig. 3 and Appendix B. In this basic configuration, the transfer function matrix **G(s)** can be computed from (26) in Appendix B. Applying multivariable circle criterion to the transformed basic configuration, we have proposed a sufficient condition for L_2 stability of feedback linearization using neural networks in Theorem1. In case of ii) $d=0$, Lyapunov stability of the equilibrium $\mathbf{x}=0$ of (11) is required with respect to the initial state $\mathbf{x_0}$. Lyapunov stability can be related to the input-output stability as in Theorem 2. Therefore, we can derive Lyapunov stability condition for $d=0$ using Theorem 1 and the relationship between the input-output stability and Lyapunov stability. In Theorem 3, Lyapunov stability condition for $d=0$ is proposed. The following analysis procedure can be commonly applied to analyze both L_2 stability and Lyapunov stability.

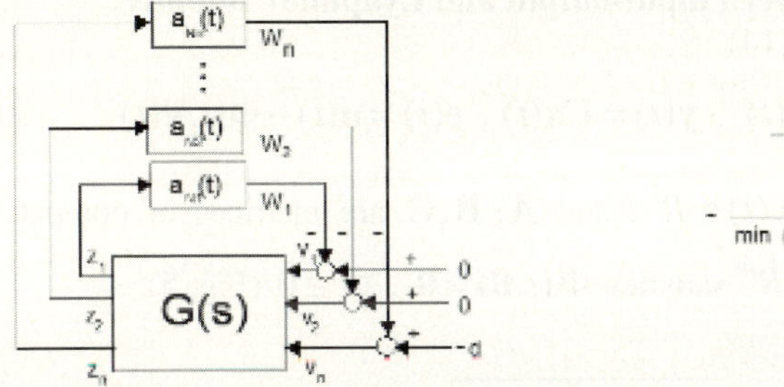

Fig. 3. Basic configuration for multivariable circle criterion

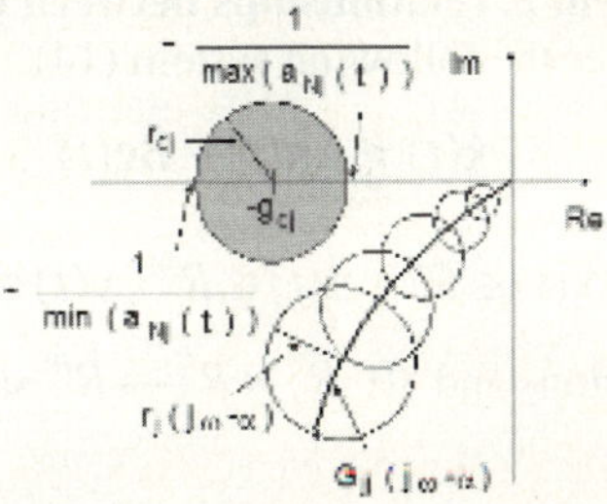

Fig. 4. Graphical analysis of multivariable circle criterion

Procedure for robust stability analysis

Step 1. Select the linear stable reference $\mathbf{a}_L$ so as to satisfy the basic assumption of Theorem 1 and compute the transfer function matrix **G(s)** from (26) in Appendix B.

Step 2. From (18) and (19) in Appendix A, find the maximum and minimum sector bounds of $a_{Nj}(t)$ for all j.

Step 3. Plot Gershigorin band and the sector disk for all j using the transfer function matrix **G(s)** and the sector bounds.

Step 4. Check if the sufficient condition of Theorem 1 or Theorem 3 is met.

Theorem 1. L_2 stability condition for feedback linearization using neural networks.

basic assumption : $\max\{a_{Nj}(t)\} \geq \min\{a_{Nj}(t)\} \geq 0, \quad \forall j$

L_2 stability of the overall system is guaranteed if

$$\left|G_{jj}(jw)+g_{cj}\right|-r_j(jw)>r_{cj}\ ,\ \forall j$$

or none of Gershigorin bands enter and encircle the disc centered at $-g_{cj}$ with radius r_{cj} (Fig. 4). where $r_j(jw)=\sum_{k=1,k\neq j}^{n}\left|G_{jk}(jw)\right|$ or $\sum_{k=1,k\neq j}^{n}\left|G_{kj}(jw)\right|$

$$g_{cj}=0.5\left(\frac{1}{\min(a_{Nj}(t))}+\frac{1}{\max(a_{Nj}(t))}\right) \tag{12}$$

$$r_{cj}=0.5\left(\frac{1}{\min(a_{Nj}(t))}-\frac{1}{\max(a_{Nj}(t))}\right) \tag{13}$$

Proof of the above theorem is almost same as the proofs of references [11]-[13] with slight modification and hence omitted here.

Theorem 2. relationships between input-output and Lyapunov stability. Consider the following system (14).

$$\dot{\mathbf{x}}(t)=\mathbf{A}\mathbf{x}(t)+\mathbf{B}\mathbf{e}(t)\ ,\ \mathbf{y}(t)=\mathbf{C}\mathbf{x}(t)\ ,\ \mathbf{e}(t)=\mathbf{u}(t)-\Phi[t,\mathbf{y}(t)] \tag{14}$$

where $\mathbf{x}(t)\in R^n$, $\mathbf{u}(t)\in R^m$, $y(t)\in R^l$, and $\mathbf{A},\mathbf{B},\mathbf{C}$ are matrices of compatible dimensions and $\Phi:R_+\times R^l\rightarrow R^m$ satisfies $\Phi(t,\mathbf{0})=\mathbf{0}$, $\forall t\geq 0$ (Fig. 5).

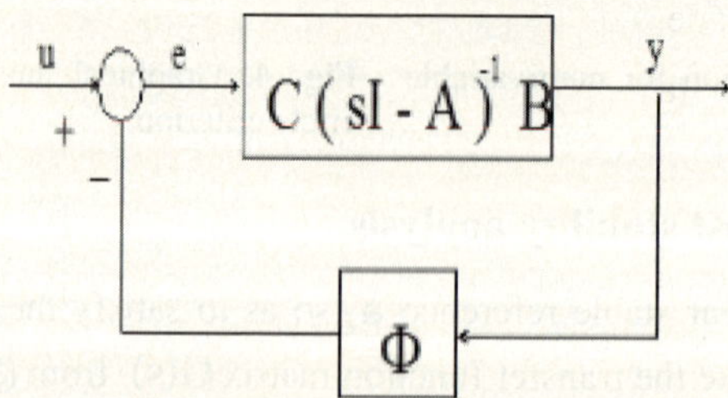

Fig. 5. Relationship between input-output and Lyapunov stability

if the following three conditions i), ii) and iii) are satisfied, then $\mathbf{x}=\mathbf{0}$ is a globally attractive equilibrium of the unforced system ($\mathbf{u}(t)=\mathbf{0}$).

i) Φ is globally Lipschitz continuous; i.e., there exists a finite constant μ such that

$\left\|\Phi(t,\mathbf{y}_1)-\Phi(t,\mathbf{y}_2)\right\|\leq\mu\left\|\mathbf{y}_1-\mathbf{y}_2\right\|$, $\forall t\geq 0$, $\forall \mathbf{y}_1,\mathbf{y}_2\in R^l$

ii) the pair $(\mathbf{A}\ ,\ \mathbf{B})$ is controllable, and the pair $(\mathbf{C}\ ,\ \mathbf{A})$ is observable.

iii) the forced system is L_2 stable.

Proof of this theorem can be found in reference [19] (Theorem(46)).

Theorem 3. Lyapunov stability condition for feedback linearization using neural networks.
$\mathbf{x}=\mathbf{0}$ is a globally attractive equilibrium of the unforced system of (11) (i.e. $d=0$)
if $\left|G_{jj}(jw)+g_{cj}\right|-r_j(jw)>r_{cj}$, $\forall j$
or none of Gershigorin bands enter and encircle the disc centered at $-g_{cj}$ with radius r_{cj} (Fig. 4). where $r_j(jw)=\sum_{k=1,k\neq j}^{n}\left|G_{jk}(jw)\right|$ or $\sum_{k=1,k\neq j}^{n}\left|G_{kj}(jw)\right|$

$$g_{cj}=0.5(\frac{1}{\min(a_{Nj}(t))}+\frac{1}{\max(a_{Nj}(t))}),\ r_{cj}=0.5(\frac{1}{\min(a_{Nj}(t))}-\frac{1}{\max(a_{Nj}(t))})$$

proof : To prove Theorem 3, first, we express the system (11) in the form of (14) as in (15).

$$\dot{\mathbf{x}}(t)=\mathbf{A}\mathbf{x}(t)+\mathbf{B}\mathbf{e}(t)\ ,\ \mathbf{y}(t)=\mathbf{C}\mathbf{x}(t)\ ,\ \mathbf{e}(t)=\mathbf{u}(t)-\Phi[t,\mathbf{y}(t)] \tag{15}$$

where $\mathbf{A},\mathbf{B},\mathbf{C}$ are $\mathbf{A}_L,\mathbf{B}_L,\mathbf{C}_L$ in Appendix B, respectively.

$$\text{and } u=\begin{bmatrix}0\\0\\0\\\cdots\\-d\end{bmatrix},\ \Phi=\mathbf{A_N}(t)=\begin{bmatrix}a_{N1}(t)&0&0&\cdots&0\\0&a_{N1}(t)&0&\cdots&0\\0&0&a_{N1}(t)&\cdots&0\\&&&\vdots&0\\0&0&0&\cdots&a_{Nn}(t)\end{bmatrix}$$

Then, for (15), we examine three sufficient conditions of Theorem 2.
a) Since $a_{Nj}(t)$ is bounded for all j and t, we can assume that $\left\|\mathbf{A_N}(t)\right\|\le\mu$ for all t, where μ is a finite constant. With this assumption and the property of the induced matrix norm, the following inequality holds for all $t\ge 0$ and for all $\mathbf{y}_1,\mathbf{y}_2$.

$$\left\|\mathbf{A_N}(t)\mathbf{y}_1-\mathbf{A_N}(t)\mathbf{y}_2\right\|=\left\|\mathbf{A_N}(t)(\mathbf{y}_1-\mathbf{y}_2)\right\|\le\left\|\mathbf{A_N}(t)\right\|\left\|\mathbf{y}_1-\mathbf{y}_2\right\|\le\mu\left\|\mathbf{y}_1-\mathbf{y}_2\right\|$$

Therefore, (15) is globally Lipschitz continuous.
b) The controllability and observability test shows that the pair $(\mathbf{A}\ ,\ \mathbf{B})$ is controllable and the pair $(\mathbf{C}\ ,\ \mathbf{A})$ is observable, independent of $\mathbf{a_L}$.
c) If the sufficient condition of Theorem 1 is met, the forced system (i.e. $d\neq 0$) is L_2 stable.
a) and b) show that (15) always satisfies the sufficient conditions i) and ii) of Theorem 2. Therefore, according to c), if the sufficient condition of Theorem 1 is met, $\mathbf{x}=\mathbf{0}$ is a globally attractive equilibrium of the unforced system of (11) (i.e. $d=0$).

4 Conclusion

In this paper, we have presented the robust stability analysis for feedback linearization using neural networks. Both input-output stability and Lyapunov stability can be analyzed by the proposed analysis method. In our work, feedback linearization regulator is designed based on the consideration of both the nominal neural networks model and some characterization of the model uncertainties and external disturbance. Also, we have proposed the connection between the control scheme based on neural networks and modern control theory. Using this connection, we have applied multivariable circle criterion to analyze the robust stability.

References

1. J.E. Slotine, W. Li, Applied nonlinear control, Englewood Cliffs, NJ: Prentice-Hall, 1991.
2. A. Isidori, Nonlinear control systems, Springer-Verlag, Berlin, 1989.
3. S. Sastry, M. Bodson, Adaptive Control: Stability, Convergence, and Robustness, Prentice-Hall, 1989.
4. R. Marino, P. Tomei, Nonlinear Control Design, Prentice-Hall, 1995.
5. K.S. Narendra, K. Paethasarathy, "Identification and control of dynamical systems using neural networks," IEEE Trans. Neural Networks, vol. 1, no. 1, pp. 1-27, 1990.
6. K.S. Narendra, K. Paethasarathy, "Gradient methods for the optimization of dynamical systems containing neural networks," IEEE Trans. Neural Networks, vol. 2, no. 2, pp. 252-262, 1991.
7. F.C. Chen, H.K. Khalil, "Adaptive control of a class of nonlinear discrete-time systems using neural networks," IEEE Trans. Automat. Contr., vol. 40, no. 5, pp. 791-801, 1995.
8. E.P. Teixeira, E.B. Faria, A. Breunig, "The use of feedforward neural networks to cancel nonlinearities of dynamic systems," in Proc. ICNN'97, Houston, vol. 2, pp. 767-772, 1997.
9. S. Chen, S.A. Billings, "Neural networks for nonlinear dynamic system modelling and identification," International journal of control, vol. 56, no. 2, pp.319-346, 1992.
10. J.A.K. Suykens, J.P.L Vandewalle, B.D.R. De Moor, "Artificial neural networks for modelling and control of non-linear systems," Boston, Kluwer Academic Publishers, 1995.
11. M. Safonov, M. Athans, "A multiloop generalization of the circle criterion for stability margin analysis," IEEE Trans. Automat. Contr., vol. 26, no. 2, pp. 415-421, 1981.
12. H.H. Rosenbrock, Multivariable circle criterion in recent Mathematical Development in Control, D.J. Bell, Ed. New York, Academic, 1973.
13. P.A. Cook, Modified multivariable circle theorems in recent Mathematical Development in Control, D.J. Bell, Ed. New York, Academic, 1973.
14. T. Parisini, R. Zoppoli, "Neural networks for feedback feedforward nonlinear control systems," IEEE Trans. Neural Networks, vol. 5, no. 3, pp. 436-449, 1994.
15. W.H. Schiffmann, H.W. Geffers, "Adaptive control of dynamic systems by back propagation networks," Neural Networks, vol. 6, pp. 517-524, 1993.
16. A. Meyer-Base, "Perturbation analysis of a class of neural networks" in Proc. ICNN'97, Houston, vol. 2, pp. 825-828, 1997.
17. P. Werbos, "Back propagation through time : What it does and how to do it," Proc. of the IEEE, 78 (10), pp. 1150-1560, 1990.
18. W.T. Miller, R.S. Sutton, P.J. Werbos, "Neural networks for control, Cambridge, MA, MIT Press, 1990.
19. M. Vidyasagar, Nonlinear system analysis, Englewood Cliffs, NJ: Prentice-Hall, 1993.

Appendix A: A Computing Method for the Maximum and Minimum Sector Bounds

$$-1 \le \sigma(\mathbf{a}_\mathbf{i}^{\mathbf{f}T} \cdot \mathbf{x}) \le 1$$

basic properties : $$-1 \le \sigma(\mathbf{a}_\mathbf{i}^{\mathbf{g}T} \cdot \mathbf{x}) \le 1$$

$$0 \le \phi(\mathbf{a}_\mathbf{i}^{\mathbf{f}T} \cdot \mathbf{x}) \le 1$$

From the definitions of σ and ϕ, the above basic properties for the function bounds can be easily verified. The maximum and minimum sector bounds of $a_{Nj}(t)$ can be computed from (16) and (17).

$$\max_t(a_{Nj}(t)) = (\hat{a}_j - a_{Lj}) + \max_t \left\{ \sum_{i=1}^{L} \phi(\mathbf{a}_\mathbf{i}^{\mathbf{f}T} \cdot \mathbf{x})\, \Delta c_i(t)\, a_{ij}^f \right\} + \max_t \left\{ \frac{\sum_{i=1}^{M} \sigma(\mathbf{a}_\mathbf{i}^{\mathbf{g}T} \cdot \mathbf{x})\, \Delta b_i(t)}{\sum_{i=1}^{M} \sigma(\mathbf{a}_\mathbf{i}^{\mathbf{g}T} \cdot \mathbf{x})\, b_i} e_j \right\} \tag{16}$$

$$\min_t(a_{Nj}(t)) = (\hat{a}_j - a_{Lj}) + \min_t \left\{ \sum_{i=1}^{L} \phi(\mathbf{a}_\mathbf{i}^{\mathbf{f}T} \cdot \mathbf{x})\, \Delta c_i(t)\, a_{ij}^f \right\} + \min_t \left\{ \frac{\sum_{i=1}^{M} \sigma(\mathbf{a}_\mathbf{i}^{\mathbf{g}T} \cdot \mathbf{x})\, \Delta b_i(t)}{\sum_{i=1}^{M} \sigma(\mathbf{a}_\mathbf{i}^{\mathbf{g}T} \cdot \mathbf{x})\, b_i} e_j \right\} \tag{17}$$

where, $e_j = \hat{a}_j - \sum_{i=1}^{L} \phi(\mathbf{a}_\mathbf{i}^{\mathbf{f}T} \cdot \mathbf{x})\, c_i\, a_{ij}^f$. The second terms in the right sides of (16) and (17) can be obtained using the following property.

$$-\sum_{i=1}^{L} \Delta c_i(t) \left| a_{ij}^f \right| \le \sum_{i=1}^{L} \phi(\mathbf{a}_\mathbf{i}^{\mathbf{f}T} \cdot \mathbf{x})\, \Delta c_i(t)\, a_{ij}^f \le \sum_{i=1}^{L} \Delta c_i(t) \left| a_{ij}^f \right| \tag{18}$$

The third terms in the right sides of (16) and (17) can be obtained from (19).

$$-\bar{b}\,\bar{e}_j \le \frac{\sum_{i=1}^{M} \sigma(\mathbf{a}_\mathbf{i}^{\mathbf{g}T} \cdot \mathbf{x})\, \Delta b_i(t)}{\sum_{i=1}^{M} \sigma(\mathbf{a}_\mathbf{i}^{\mathbf{g}T} \cdot \mathbf{x})\, b_i} e_j \le \bar{b}\,\bar{e}_j \tag{19}$$

$\overline{b}$ in (19) can be computed by (20).

$$\overline{b} = \max_i \frac{\Delta b_i}{|b_i|} \tag{20}$$

$\overline{e}_j$ in (19) can be computed by (21).

$$\overline{e}_j = \max\left(\left| \hat{a}_j - \sum_{i=1}^{L} (1-s_i) c_i a_{ij}^f \right|, \quad \left| \hat{a}_j - \sum_{i=1}^{L} s_i c_i a_{ij}^f \right| \right) \tag{21}$$

where $s_i = \begin{cases} 1 & if \ c_i a_{ij}^f \geq 0 \\ 0 & if \ c_i a_{ij}^f < 0 \end{cases}$, for all i.

Appendix B: A Computing Method for G(s)

To compute **G(s)** from (11), we divide (11) into the linear and the nonlinear part as

$$x^{(n)} - \mathbf{a}_L^T \cdot \mathbf{x} = \mathbf{a}_N(t)^T \cdot \mathbf{x} + d \tag{22}$$

Using the following state-space representation of (23), we can compute **G(s)**.

$$\begin{aligned} \dot{\mathbf{x}} &= \mathbf{A_L x} + \mathbf{B_L v} \\ \mathbf{z} &= \mathbf{C_L x} \\ \mathbf{v} &= \mathbf{-w} + \mathbf{d} \\ \mathbf{w} &= \mathbf{A_N}(t)\mathbf{z} \end{aligned} \tag{23}$$

where,

$$\mathbf{A_L} = \begin{bmatrix} 0 & 1 & 0 & \cdots & 0 \\ 0 & 0 & 1 & \cdots & 0 \\ 0 & 0 & 0 & \cdots & 0 \\ & & & \vdots & \\ a_{L1} & a_{L2} & a_{L3} & \cdots & a_{Ln} \end{bmatrix}, \quad \mathbf{B_L} = \begin{bmatrix} 0 & 0 & 0 & \cdots & 0 \\ 0 & 0 & 0 & \cdots & 0 \\ 0 & 0 & 0 & \cdots & 0 \\ & & & \vdots & \\ -1 & -1 & -1 & \cdots & -1 \end{bmatrix}$$

$$\mathbf{C_L} = \begin{bmatrix} 1 & 0 & 0 & \cdots & 0 \\ 0 & 1 & 0 & \cdots & 0 \\ 0 & 0 & 1 & \cdots & 0 \\ & & & \vdots & \\ 0 & 0 & 0 & \cdots & 1 \end{bmatrix}$$

$$\mathbf{A_N}(t) = \begin{bmatrix} a_{N1}(t) & 0 & 0 & \cdots & 0 \\ 0 & a_{N2}(t) & 0 & \cdots & 0 \\ 0 & 0 & a_{N3}(t) & \cdots & 0 \\ & & & \vdots & 0 \\ 0 & 0 & 0 & \cdots & a_{Nn}(t) \end{bmatrix} \quad and \quad \mathbf{d} = \begin{bmatrix} 0 \\ 0 \\ 0 \\ \vdots \\ -d \end{bmatrix}$$

$$\mathbf{G(s)} = \mathbf{C_L}(s\mathbf{I} - \mathbf{A_L})^{-1}\mathbf{B_L} \tag{24}$$

Effects of Using Different Neural Network Structures and Cost Functions in Locomotion Control

Jih-Gau Juang

Department of Communications and Guidance Engineering
National Taiwan Ocean University, Keelung 20224, Taiwan
jgjuang@mail.ntou.edu.tw

Abstract. Effects of using different neural network structures and cost functions in locomotion control are investigated. Simulations focus on refinement and a thorough understanding of an artificial intelligent learning scheme. This scheme uses a neural network controller with backpropagation through time learning rule. Through learning, the controller can generate locomotion trajectory along a pre-defined path. Different issues regarding the scheme have been examined. They include the effects of using different numbers of hidden units, the effects of using only angle parameters in the cost function, and the effects of including an energy criterion in the cost function.

1 Introduction

Artificial intelligence (AI) techniques have been applied to improve the accuracy of nonlinear system modeling and the performance of system dynamics for years. The use of AI methods to design and implement automatic control systems has been broadly referred to as "intelligent control". Today several AI methods are often used in control system design. Neural networks, fuzzy logic, and genetic algorithms are among them. Neural networks have been shown to possess good approximation capabilities for a wide range of nonlinear functions [1-4] and have been used in robot locomotion control [5-11] successfully.

Although much progress has occurred in recent researches of robot locomotion, the results are often not optimal if some special goals are desirable, such as crossing over a specific clearance, having a desired step length, and walking at a certain speed. To overcome these problems, a locomotion control technique using recurrent averaging learning scheme has been proposed [12-13]. This scheme can generate and refine the robot gait by using partial or inaccurate initial conditions. It takes the average of initial states and final states after a cycle of training and sets this value as the new initial and final states for next training cycle. Through learning, the robot can develop skills to walk along a pre-defined path with specified step length, walking speed, and crossing clearance.

In this paper, the effects of using different neural network structures and cost functions in locomotion control are investigated. Simulations focus on refinement and a thorough understanding of the previous proposed learning scheme. This scheme uses a neural network controller with backpropagation through time learning rule. Many algorithms have been developed for the training of neural network [14-15].

L. Jiao et al. (Eds.): ICNC 2006, Part I, LNCS 4221, pp. 437 – 446, 2006.

Backpropagation learning is the most popular one. In some applications—such as trajectory tracking or speech recognition—the result at time t will be more accurate if we can account for what we saw at earlier times. Since the locomotion control in this study is modeled as a trajectory-tracking problem. This means that the output of the controller network is affected by variables at earlier times. Thus, a modified backpropagation learning scheme called "backpropagation through time" [16] is required in the training process. Through learning, the controller can generate locomotion trajectory along a pre-defined path. Different issues regarding the scheme have been examined. They include the effects of using a different number of hidden units, the effects of using only position parameters in the cost function, and the effects of including an energy criterion in the cost function. By including an energy criterion in the cost function, energy minimization can be achieved.

2 Biped Model and Reference Trajectory

The walking machine BLR-G1 robot [17] is used as the simulation model. This robot consists of five links, a body, two lower legs and two upper legs, with two hip joints and two knee joints as shown in Fig. 1. It has no feet (no ankle joint). A steel pipe at the tip of each leg is used to maintain the lateral balance. Thus the motion of the robot is limited to only the sagittal plane. Since there is no foot, no ankle torque can be generated; the biped locomotion can be only indirectly controlled by using the effect of gravity (inverted pendulum). The ground condition is assumed to be rigid and non-slip. The biped always keeps only one leg in contact with the ground. The contact is assumed to be a single point. The dynamic equations of motion for the model are given in [17] as follows:

$$\mathbf{A}(\underline{\theta})\ddot{\underline{\theta}} + \mathbf{B}(\underline{\theta})h(\dot{\underline{\theta}}) + \mathbf{C}g(\underline{\theta}) = \mathbf{D}\boldsymbol{T}\,, \tag{1}$$

where

$$\boldsymbol{A}(\underline{\theta}) = \{q_{ij}\cos(\theta_i - \theta_j) + p_{ij}\},\quad \boldsymbol{B}(\underline{\theta}) = \{q_{ij}\sin(\theta_i - \theta_j)\},\quad \boldsymbol{C} = \mathrm{diag}\{-h_i\},\qquad \boldsymbol{D} = \begin{bmatrix} 1 & 0 & 0 & 0 \\ -1 & 1 & 0 & 0 \\ 0 & -1 & 1 & 0 \\ 0 & 0 & -1 & 1 \\ 0 & 0 & 0 & -1 \end{bmatrix},$$

$$\underline{\theta} = [\theta_1, \theta_2, \theta_3, \theta_4, \theta_5]^{\mathrm{T}},$$

$$\boldsymbol{T} = [\tau_1, \tau_2, \tau_3, \tau_4]^{\mathrm{T}},$$

q_{ij}, p_{ij} and h_i are constants derived by using Lagrange's equation of motion, τ_i is the torque at i^{th} joint, θ_i and $\dot{\theta}_i$ are the angle and angular velocity of link i.

A cycloidal profile [18] is used as the reference trajectory of the hip and ankle joints of the swinging leg for learning. This profile is used because it shows a similar pattern to a human's ankle trajectory in normal walking and it describes a simple function which can be easily changed for different walking patterns. From [18], one can generate different walking patterns based on the desired environment or specification. Fig. 2 shows a reference trajectory which is used in this study. The step length is 40cm, maximum crossing height of the swinging leg's ankle is 11cm, and the total sampling number for one gait is 25. It is assumed that the robot body is always upright.

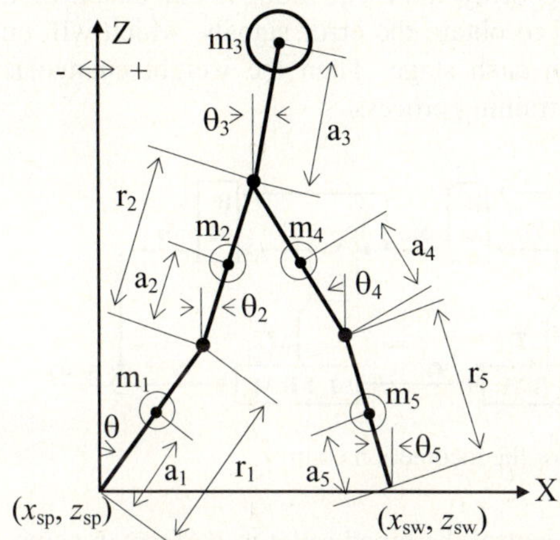

Fig. 1. The 5-link biped model: m_i is the mass of link i; r_i is the length of link i; a_i is the distance between the center of mass of link i and the lower joint of link i

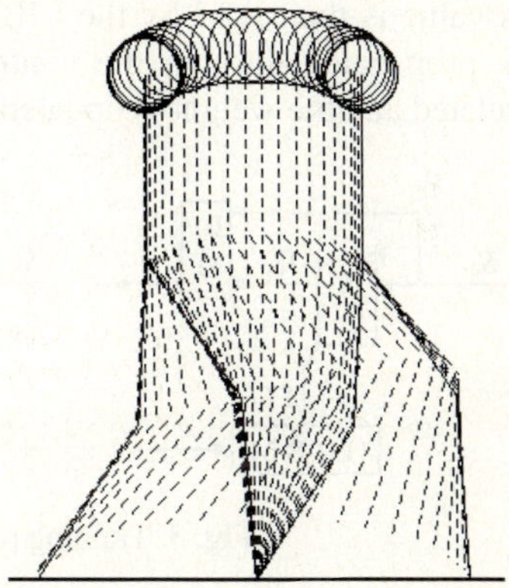

Fig. 2. Reference trajectories on level ground with *step_length* = 40cm, and *crossing_height* = 11cm

3 Locomotion Control

Since biped locomotion is periodic, robotic walking control can be considered as a recurrent trajectory-tracking problem. To investigate the biped locomotion, one can focus on gait synthesis for a single walking step. If we can obtain the first walking step properly, we will be able to reduce the difficulties of robotic walking control. The biped locomotion control will then be simplified to a periodic trajectory-tracking problem. Thus, gait synthesis of the first walking step becomes an important issue in locomotion control. To generate optimal locomotion trajectory that follows a pre-defined path requires not only the reference pattern as the target data but also the initial conditions, such as link positions or velocities. However, the adequate initial positions or velocities are not always available. These constraints for the initial conditions have limited the controller's capability of optimization using many proposed algorithms. To overcome this problem, a "recurrent averaging learning" scheme has been developed in previous studies [12-13]. This technique can use partial or inaccurate initial conditions to generate optimal trajectory and still follow the reference pattern.

The training process is illustrated in Fig. 3. The neural network controller (NC) is a three-layered feedforward neural network, and it provides the control signals in each stage of one gait. The linearized inverse biped model (LIBM) provides error signals which are used to train the controller. The biped model (BM) is given in Section 2. The learning strategy is the recurrent averaging learning. Since the training includes walking stages in the past, the weight calculations of the basic backpropagation are not capable of solving the time delay problem which involves past calculations. Thus, a modified method is needed in this case. This method is called backpropagation through time. In training, the difference between the desired state of the training data and the calculated output state from the robot dynamic equations is calculated first. This value is then used by the LIBM [13] to obtain the error signals, which will be back propagated through the controller in each stage. Then the weight change is calculated and the weight is updated in the training process.

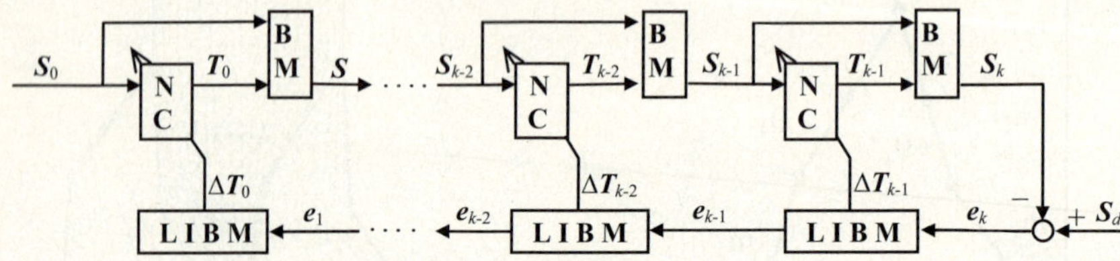

Fig. 3. Training process for the locomotion control

In this study, the presented method can control the biped robot to perform dynamic walking with ground impact reaction [19-21]. A robot gait is divided into several stages. The training patterns are obtained from the reference trajectory that defines the desired step length, crossing clearance, and speed. The controller learns to drive the biped from an initial state $\boldsymbol{S_0}$ to a desired state $\boldsymbol{S_d}$ in k time stages, where

$$\boldsymbol{S}_k = (\underline{\theta}^{k^{\mathrm{T}}}, \underline{\dot{\theta}}^{k^{\mathrm{T}}})^{\mathrm{T}},$$

$$\underline{\theta}^k = (\theta_1^k, \theta_2^k, \theta_3^k, \theta_4^k, \theta_5^k)^{\mathrm{T}},$$

$$\underline{\dot{\theta}}^k = (\dot{\theta}_1^k, \dot{\theta}_2^k, \dot{\theta}_3^k, \dot{\theta}_4^k, \dot{\theta}_5^k)^{\mathrm{T}}. \tag{2}$$

The objective of the learning process is to find a set of net weights that minimizes the cost function E, where E is the mean square error of the final state $\boldsymbol{S}_k$ and desired state $\boldsymbol{S}_d$.

$$E = \frac{1}{2}(\|\mathbf{e_k}\|^2),$$

$$\boldsymbol{e}_k = \boldsymbol{S}_d - \boldsymbol{S}_k. \tag{3}$$

The goal in this study is to generate a walking gait for which the final angle $\underline{\theta}^k$ approaches the mirror image of the initial angle $\underline{\theta}^0$, and the angular velocity after ground impact action of its final angular velocity $\underline{\dot{\theta}}^k$ approaches the mirror image of the initial angular velocity $\underline{\dot{\theta}}^0$. The meaning of "mirror image" is shown as follows

$$\underline{\theta}^k = (\theta_1^k, \theta_2^k, \theta_3^k, \theta_4^k, \theta_5^k)^{\mathrm{T}} = (\theta_5^0, \theta_4^0, \theta_3^0, \theta_2^0, \theta_1^0)^{\mathrm{T}},$$

$$Ip\{\, \underline{\dot{\theta}}^k : (\dot{\theta}_1^k, \dot{\theta}_2^k, \dot{\theta}_3^k, \dot{\theta}_4^k, \dot{\theta}_5^k)^{\mathrm{T}} \,\} = (\dot{\theta}_5^0, \dot{\theta}_4^0, \dot{\theta}_3^0, \dot{\theta}_2^0, \dot{\theta}_1^0)^{\mathrm{T}}, \tag{4}$$

where $Ip\{\phi\}$ is the function of ground impact action from [19]. This means the initial angle and angular velocity conditions of the second step are the same as the initial conditions of the first step. Once we reach this step, a conventional angle and angular velocity feedback control technique can be used for further walking control.

4 Simulations Based on Different Criteria

The following simulations focus on refinement and a thorough understanding of the proposed scheme. They include the effects of using a different number of hidden units, the effects of using only angle parameters in the cost function, and the effects of including an energy criterion in the cost function.

4.1 Effects of Using a Different Number of Hidden Units in the Neural Network Controller

It has been proved that any smooth function can be represented by a three-layer neural network with enough hidden neurons [22-29]. The second layer of the neural network controller is the hidden layer. In this layer, there must be enough units to cover the workspace required by a given problem. With too many hidden units, "over-fitting" could occur. That means the output errors for the training data decrease but the errors for the untrained data increase. Murata [30] presented a Network Information Criterion for selecting the optimal number of hidden units based on a given training set. Kung [31] used an Algebraic Projection Analysis to predict the number of hidden units. The number of units is problem dependent.

The number of hidden units used in the neural network controller will affect not only the learning time but also the controller performance. Here, we investigate the trade-off between the training time and the controller performance by using a different number of hidden units. The numbers of hidden units used are: 10, 30, 50, and 70. In order to make a comparison on the effect of a different number of hidden units, it is assumed that they all have the same stage number and the same desired pattern. Fig. 2 is used as the reference pattern in training. The corresponding cost functions vs. learning time are shown in Fig. 4. The controller with 30 hidden units has the least error while the one with 70 units has the worst performance. This might be because the number of training patterns being used is 25. Based on [30], with 25 training patterns, 30 units should be optimal, while 70 units are too many.

4.2 Effects of Using Only Position Parameters in the Cost Function

The cost function (or objective function) used in the previous subsection contains both angle and angular velocity information. Minimizing this cost function makes the total error of angle and angular velocity decline, but there is no guarantee that the angle error reaches a minimum. In this subsection, the effects of using only angle parameters in the cost function are investigated. Since the controller with 30 hidden units has the best performance, we will use this configuration in the following simulations. Fig. 5 shows the trained result. Fig. 6 shows the total cost function **J**, the cost function **JS** and the cost function **JP**, where **JS** is the angular velocity error component of **J,** and **JP** is the angle error. With the use of only the angle error in the training, the final angle error can be reduced by 16% (compared to the one with 30 hidden units in subsection 4.1), but the total cost function increases about 25%. This is because the angular velocity was not trained, and its error increases about 40%. This becomes a trade-off problem. If the accuracy of the angle is more important in a control problem, this approach would be a good choice.

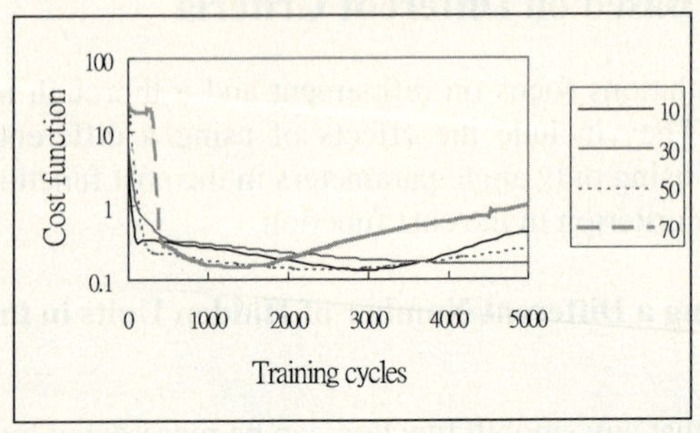

Fig. 4. Cost functions of different numbers of hidden units

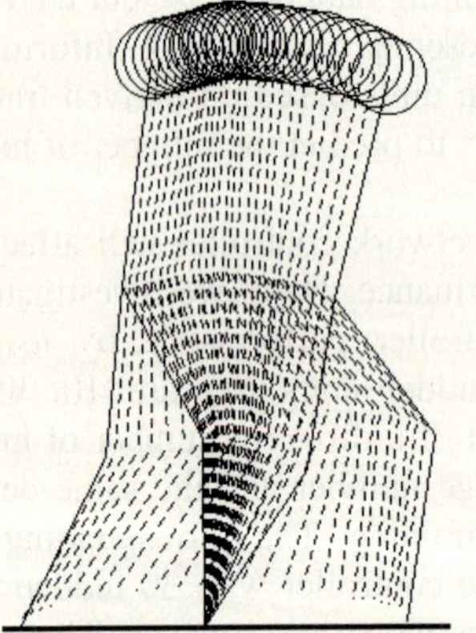

Fig. 5. Simulation result of using only angle error in the cost function

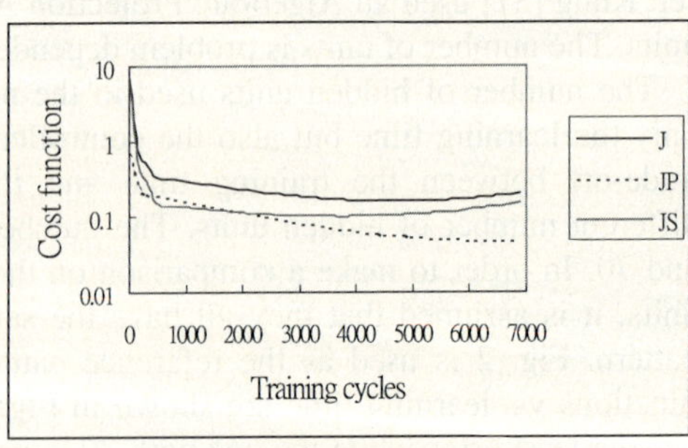

Fig. 6. Cost functions J, JP and JS

4.3 Effects of Including Energy Criterion in the Cost Function

The above-described controller was trained to minimize only the final state error. One can also train it to minimize the total control energy in addition to the final state. In this subsection, the effects of including an energy criterion in the cost function are investigated. Since this study is done by simulation, real electrical energy input to the robot cannot be obtained. Assume all the potential energy and the kinetic energy can be transferred to electrical energy. The total energy converted by the biped robot will be proportional to the summation of the potential energy and the kinetic energy of each link. Thus, minimizing the energy cost can be considered as minimizing the summation of the potential energy and the kinetic energy of each link. To simplify the kinetic energy, the translation energy of each link is not included in this study. One can use the following cost function to minimize the total energy:

$$E = \frac{1}{2}(S_d - S_k)^2 + \sum_{i=1}^{5}\left[\frac{1}{2}\mathrm{I}_i\dot{\theta}_i^2 + \mathrm{m}_i g \mathrm{h}_i\right]^2, \tag{5}$$

where S_d and S_k are given in Equation (3), I_i is the moment of inertia about the center of mass of link i, m_i is the mass of link i, and h_i is the height of the center of mass of link i. Total energy cost can be minimized by adding the following terms to the original error as in Equation (4).

For $\dot{\theta}_i$, the modified error is:

$$-\mathrm{I}_i\dot{\theta}_i \text{, for } i\text{=1 to 5.} \tag{6}$$

For θ_1, θ_2, θ_3, θ_4 and θ_5, the modified errors are shown in Equations (7), (8), (9), (10) and (11), respectively:

$$\mathrm{m}_1 g \mathrm{a}_1 \sin\theta_1 . \tag{7}$$

$$\mathrm{m}_2 g \mathrm{a}_2 \sin\theta_2 . \tag{8}$$

$$\mathrm{m}_3 g \mathrm{a}_3 \sin\theta_3 . \tag{9}$$

$$\mathrm{m}_4 g (\mathrm{a}_4 - \mathrm{r}_4)\sin\theta_4 . \tag{10}$$

$$\mathrm{m}_5 g (\mathrm{a}_5 - \mathrm{r}_5)\sin\theta_5 . \tag{11}$$

The parameters a_i and r_i are defined in Fig. 1. The parameter g is the gravity. The modified errors are then back propagated through the controller to update the controller's weights. These changes make sense, since adding the above terms to the original errors in the controller at each stage causes the controller to learn to make the second term in Equation (5) smaller.

Fig. 2 is used as the reference pattern in training. At first, 30 hidden units are used in the neural network controller. The simulation result is shown in Fig. 7. This walking pattern is very similar to Fig. 8 which is without energy criterion in the cost function. With an energy criterion in the cost function, the total energy is reduced by about 10%. The cost functions are shown in Fig. 9 and Fig. 10, where J is the total cost function, JS is the cost function with only angular velocity parameter, JP is the cost function with only angle parameter, and JE is the cost function with energy criterion. Based on different configuration of the controller with the same desired reference pattern, the total energy costs are shown in Table 1. Including energy in the cost function, the controller generates the gait that requires less control energy.

5 Conclusion

In this paper, we have briefly reviewed how the backpropagation through time learning algorithm with the recurrent averaging learning scheme can be applied to locomotion control of a biped robot. Given a desired trajectory, the neural network controller can generate control sequences and drive the robot along this pre-specified trajectory. The complex inverse dynamic computations in determining the error in joint force can be eliminated through the use of the linearized inverse biped model. The effects of using different numbers of hidden units in the neural network controller are investigated. The number of hidden units closest to the number of training patterns has the best performance. If only angle parameters are used in the cost function, the trained gait has a smaller angle error. Minimal energy performance has also been investigated. By including energy in the cost function, the learning scheme can generate a near-minimal energy gait. Simulation results show that the proposed scheme can train the robot by following a desired step length, crossing clearance, and walking speed. With different criterion in the cost function, special goals such as angle accuracy and energy cost can be achieved also.

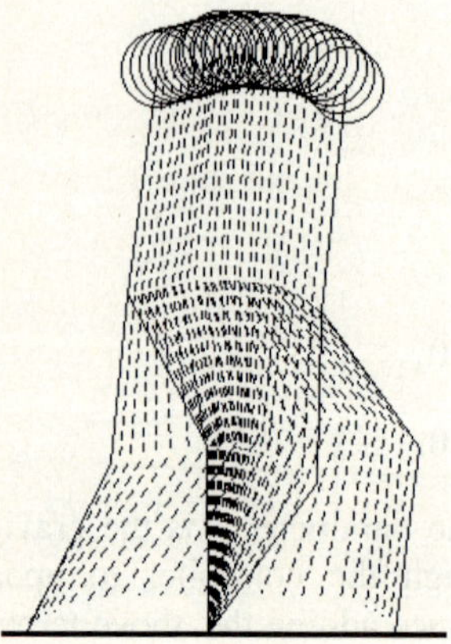

Fig. 7. With energy criterion in cost function

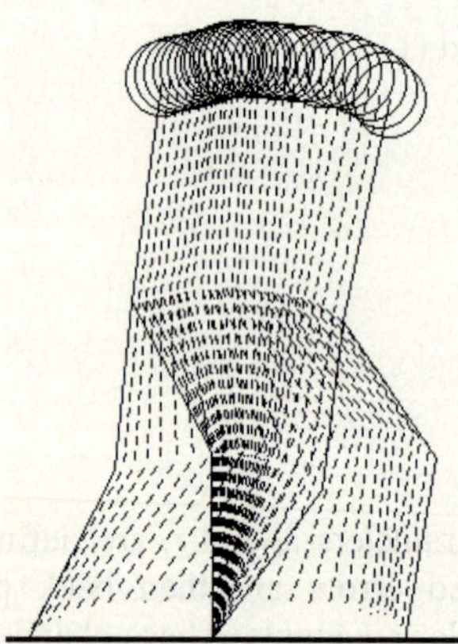

Fig. 8. Without energy criterion in cost function

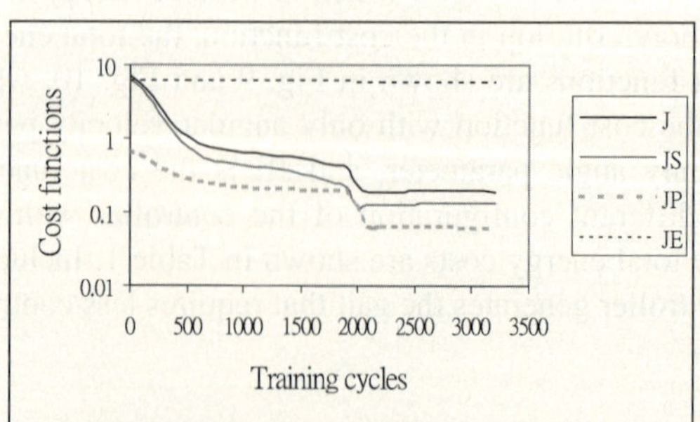

Fig. 9. Cost functions with energy criterion

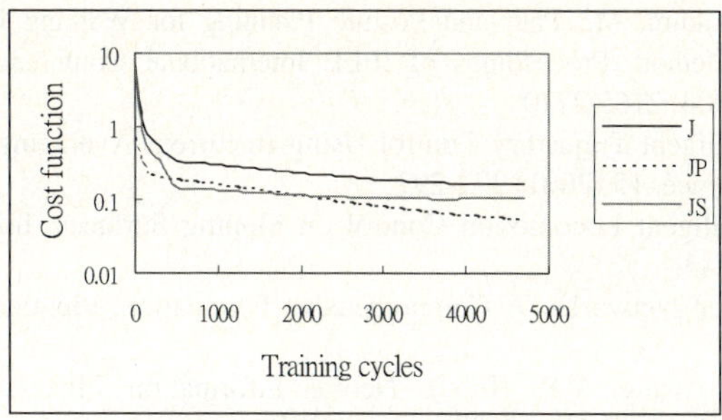

Fig. 10. Original cost functions without energy criterion

Table 1. Total energy cost based on different training configuration

Configuration	Total energy cost
10 hidden units	648 joule
30 hidden units	644 joule
50 hidden units	642 joule
70 hidden units	649 joule
energy training	598 joule

References

1. Cybenko, G.: Approximation by Superpositions of a Sigmoidal Function. Mathematics of Control, Signals, and Systems. 2 (1987) 303-314
2. Girosi, F., Poggio, T.: Networks and The Best Approximation Property. Biological Cybernetics. 63 (1990) 169-179
3. Park, J., Sanberg, I.W.: Universal Approximation Using Radial Basis Function Networks. Neur. Comp. 3 (1991) 246-257
4. Wang, Z, Tham, M.T, Morris, A.J.: Multilayer Neural Networks: Approximated Canonical Decomposition of Nonlinearity. International Journal of Control. 56 (1992) 655-672
5. Zhao, M., Liu, L., Wang, J., Chen, K.: Gaits Compensation Algorithm for Humanoid Robot Walking Control. Proc. of CLAWAR. (2002) 171-176
6. Fujimoto, Y., Kawamura, A.: Simulation of an Autonomous Biped Walking Robot Including Environmental Force Interaction. IEEE Robotics and Automation Magazine. 5 (1998) 33-42
7. Collins, S.H., Wisse, M., Ruina, A.: A Three-Dimensional Passive-Dynamic Walking Robot with Two Legs and Knees. International Journal of Robotics Research. 20 (2001) 607-615
8. Miller, W.T., Kun, A.L.: Dynamic Balance of a Biped Walking Robot. In: Omidvar, O., Smagt, P. (eds.): Neural Systems for Robotics. Academic Press (1997) 17-35
9. Park, S.H., Kim, D.S., Lee, Y.J.: Discontinuous Spinning Gait of a Quadruped Walking Robot with Waist-joint. Proc. of IEEE International Conference on Intelligent Robots and Systems. (2005) 2744-2749
10. Noh, K.K., Kim, J.G., Huh, U.Y.: Stability Experiment of a Biped Walking Robot with Inverted Pendulum. 3 (2004) 2475-2479

11. Igarashi, H., Kakikura, M.: Path and Posture Planning for Walking Robots by Artificial Potential Field Method. Proceedings of IEEE International Conference on Robotics and Automation. 3 (2004) 2165-2170
12. Juang, J.G.: Intelligent Trajectory Control Using Recurrent Averaging Learning. Applied Artificial Intelligence. 15 (2001) 277-297
13. Juang, J.G.: Intelligent Locomotion Control on Sloping Surfaces. Information Sciences. 147 (2002) 229-243
14. Haykin, S.: Neural Networks: A Comprehensive Foundation. Prentice-Hall, New Jersey, 2nd ed. (1999)
15. Rajapakse, J.C., Wang, L.P. (Eds.): Neural Information Processing: Research and Development. Springer, Berlin (2004)
16. Werbos, P.: Backpropagation Through Time: What It Does and How to Do It. Proceedings of the IEEE. 78 (1990) 1550-1560
17. Furusho, J, Masubuchi, M.: Control of a Dynamical Biped Locomotion System for Steady Walking. Journal of Dynamic Systems, Measurement, and Control. 108 (1986) 111-118
18. Kurematsu, Y., Kitamura, S, Kondo, Y.: Trajectory Planning and Control of a Biped Locomotive Robot -Simulation and Experiment-. In: Jamshidi, M. (ed.): Robotics and Manufacturing. Recent Trends in Research, Education and Applications. ASME Press (1988) 65-72
19. Zheng, Y.F., Hemami, H.: Mathematical Modeling of a Robot Collision with its Environment. Journal of Robotic Systems. 2 (1985) 289-307
20. Raibert, M., Tzafestas, S., Tzafestas, C.: Comparative Simulation Study of Three Control Techniques Applied to a Biped Robot. Proc. IEEE Int. Conf. on Systems, Man and Cybernetics. 1 (1993) 494-502
21. Pars, L.A.: A Treatise on Analytical Dynamics, Chapter 14. Heinemann Educational Books LTD, London (1965)
22. Hornik, K.: Some New Results on Neural Network Approximation. Neural Networks. 6 (1993) 1069-1072
23. Funahashi, K.: On the Approximate Realization of Continuous Mapping by Neural Networks. Neural Networks. 2 (1989) 183-192
24. Hornik, K., Stinchcombe, M., White, H.: Universal Approximation of an Unknown Mapping and Its Derivatives Using Multilayer Feedforward Networks. Neural Networks. 3 (1990) 551-560
25. Hornik, K., Stinchcombe, M., White, H.: Multilayer Feedforward Networks are Universal Approximators. Neural Networks. 2 (1989) 359-366
26. Hornik, K.: Approximation Capabilities of Multilayer Perceptrons. Neural Networks. 4 (1991) 251-257
27. Kreinovich, V.: Abitrary Nonlinearity is Sufficient to Represent all Functions by Neural Networks: A Theorem. Neural Networks. 4 (1991) 381-384
28. Ito, Y.: Approximation of Function on a Compact Set by Finite Sums of a Sigmoid Function without Scaling. Neural Networks. 4 (1991) 385-394
29. Ito, Y.: Approximation of Continuous Functions on R^d by Linear Combinations of Shifted Rotations of a Sigmoid Function with and without Scaling. Neural Networks. 5 (1992) 105-115
30. Murata, N., Yoshizawa, S., Amari, S.I.: Network Information Criterion-Determining the Number of Hidden Units for an Artificial Neural Network Model. IEEE Trans. Neural Networks. 5 (1994) 865-872
31. Kung, S.Y., Hwang, J.N.: Algebraic Projection Analysis for Optimal Hidden Units Size and Learning Rates in Back-Propagation Learning. Proc. IEEE Int. Conf. Neural Networks. 1 (1988) 363-370

Humanoid Robot Behavior Learning Based on ART Neural Network and Cross-Modality Learning

Lizhong Gu and Jianbo Su

Department of Automation & Research Center of Intelligent Robotics
Shanghai Jiaotong University, Shanghai, 200240, China
{gulizhong, jbsu}@sjtu.edu.cn

Abstract. This paper presents a novel robot behavior learning method based on Adaptive Resonance Theory (ART) neural network and cross-modality learning. We introduce the concept of classification learning and propose a new representation of observed behavior. Compared with previous robot behavior learning methods, this method has the property of learning a new behavior while at the same time preserving prior learned behaviors. Moreover, visual information and audio information are integrated to form a unified percept of the observed behavior, which facilitates robot behavior learning. We implement this learning method on a humanoid robot head for behavior learning and experimental results demonstrate the effectiveness of this method.

1 Introduction

It is critical to endow robot with learning ability to acquire human-like behaviors from human demonstration [1][2]. Hidden Markov Models (HMM) had been used in robot skill acquisition from human demonstration [3]. However, it could not deal with situations not encountered in the training phase. Recurrent neural network model with parametric bias (RNNPB) had been developed to realize imitative interaction between robot and human [4]. Although the robot could respond to new movement patterns, the prior learned movement patterns would be forgotten when the robot was to learn new patterns. Moreover existing methods used only visual information for robot behavior learning from human demonstration.

Actually the demonstrated behavior is a sequence of actions and human movement has subtle variations when repeated. The robot should learn general attributes of these action sequences not specific attributes of each action sequence. Therefore we introduce the concept of classification learning and consider robot behavior learning as a classification process. We notice that ART neural network is a powerful tool for incremental categorization learning and does not lose the memory of prior learned behaviors when learning new behaviors [5]. We propose a novel robot behavior learning method based on ART neural network and first investigate the representation of observed behaviors. Moreover, the cross-modality learning is also utilized to facilitate robot behavior learning where information from visual modality and audio modality are integrated to percept the observed behavior [6].

L. Jiao et al. (Eds.): ICNC 2006, Part I, LNCS 4221, pp. 447 – 450, 2006.

2 Description of the Robot Behavior Learning Method

We construct a robot behavior learning system based on ART neural network and cross-modality learning (Fig. 1). This system mainly consists of feature coding, behavior classifier, word set, behavior memory and action generator.

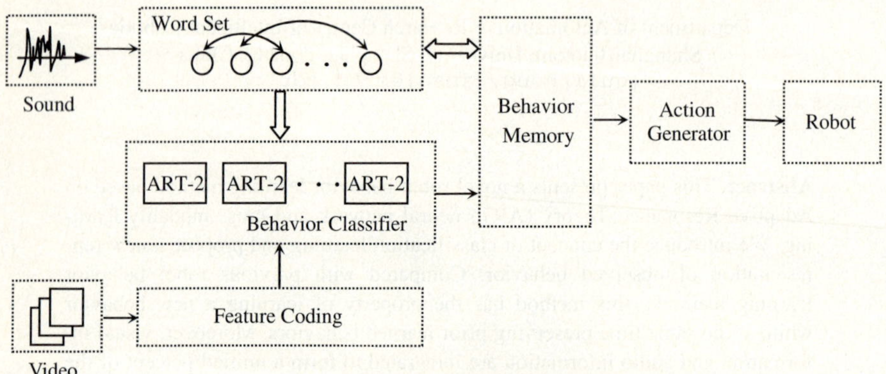

Fig. 1. Overview of the robot behavior learning system

Feature coding is to represent the observed human behavior. When a certain behavior is demonstrated, robot acquires an image sequence of actions. The observed action at a certain instant is spatially decomposed into several joint positions in the image plane. Then the corresponding joint position in time-sequential images is temporally combined to form a joint trajectory pattern. The number of joint trajectory patterns is equal to that of moving joints in action sequences. Then these joint trajectory patterns are fed to corresponding ART neural networks for behavior classification.

Behavior classifier is composed of several ART-2 neural networks, which recognizes corresponding joint trajectory pattern. Since observed joint positions in image plane are analog values, we adopt ART-2 neural network for its capability of processing analog input patterns [7]. Recognized human behaviors are represented by the long-term weights of these ART-2 neural networks and top-down weights are then stored into the behavior memory. Every ART-2 neural network has a vigilance parameter ρ , which is used to control the number and size of generated clusters. These vigilance parameters in ART-2 neural networks are adjusted together and set to be the same since a certain behavior is represented by several joint trajectory patterns.

Word set is a group of self-organized neurons and stores the word heard from audio modality. There are connections between the word set and behavior memory and the recognized behavior is labeled by the word uttered by the demonstrator. Sometimes different words are linked to each other for they are related to the same behavior. The connection weights between the word set and behavior memory will be adjusted to enhance the connection weights between them when voice command and demonstrated behavior are given to the robot simultaneously. Action generator is to map the learned joint trajectory into the robot's internal actuators representation.

3 Experimental Results

The first experiment tested the learning adaptability of the proposed method. During this experiment a man sit in front of the humanoid robot and demonstrated the following behaviors such as nod, head shaking, smooth pursuit, vergence, gaze control and nod. Fig. 2 shows activated neurons in each of ART-2 neural network when the robot was learning new behaviors. When nod behavior was given to the robot for the first time, a certain neuron in the NN1 was activated. After the robot had learned other behaviors, nod behavior was given to the robot again. The same neuron in NN1 was activated. As for the smooth pursuit behavior and the vergence behavior, the individual eye joint trajectory pattern of the two different behaviors might have similarity to some certain extent. However, the combinations of joint trajectory pattern of the two different behaviors were not the same so that different neurons in the neural networks were activated. This verified that the new learned behavior would not damage the memory of prior learned behaviors.

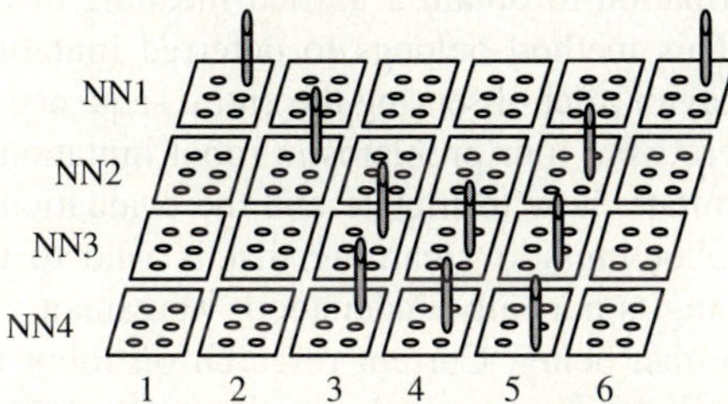

Fig. 2. Illustration of activated neurons in each of ART-2 neural network when the robot learns new behaviors. NN1 is an ART-2 neural network classifying neck tilt joint trajectory pattern. NN2, NN3 and NN4 are for neck pan joint, left eye pan joint, right eye pan joint, respectively. (1-nod behavior, 2-head shaking behavior, 3-smooth pursuit behavior, 4-vergence behavior, 5-gaze control behavior, 6-nod behavior).

The second experiment examined the effect of audio modality on the recognition rate of learned behaviors. The demonstrated behaviors were nod, head shaking, smooth pursuit, vergence and gaze control. Five persons performed each of the five behaviors ten times. Three persons' demonstrated behaviors were used for robot learning. The remains were used for testing the recognition rate of learned behaviors. We did this experiment under two different conditions, which audio information was integrated or not. We found that recognition rate of learned behaviors were increased when we utilized the audio information during the learning phase (Table 1).

Table 1. Recognition rate of learned behaviors with and without integrating audio information

Behavior	Recognition rate without audio modality (%)	Recognition rate with audio modality (%)
Nod	90 (18/20)	100 (20/20)
Head shaking	95 (19/20)	100 (20/20)
Smooth pursuit	70 (14/20)	100 (20/20)
Vergence	75 (15/20)	100 (20/20)
Gaze control	60 (12/20)	80 (16/20)

4 Discussions and Conclusions

This work presents a new approach for robot behavior learning from human demonstration, using ART neural network and cross-modality learning. Inspired from human learning mechanism, we introduce the concept of classification learning and consider robot behavior learning as a classification process. We propose a novel representation of observed behavior, which carries out spatial decomposition and temporal combination for moving joints in time-sequential images. Compared with previous robot behavior learning methods, this method has the advantage of responding to a new behavior while at the same time preserving prior learned behaviors. Cross-modality learning is adopted to facilitate learning convergence through adjusting the vigilance parameter ρ of behavior classifier. This adaptive vigilance parameter mechanism overcomes the problem of misclassification and coarse clustering of original ART neural networks for its fixed vigilance parameter.

The proposed method is different from supervising learning since robot itself integrates multi-modality information to obtain a unified meaning of observed behavior without a teacher signal. This method belongs to deferred imitation and robot will reproduce the observed behavior after observing the entire sequence of actions.

As mentioned in [8], there were four problems in robot imitation learning, such as when to imitate, what to imitate, how to imitate and the evaluation of imitation. We added another aspect of robot imitation learning that is who to imitate. Since the mechanism of human learning is not fully understood, we cannot expect the robot to autonomously learn like human being. Current research on robot imitation learning mainly focused on how to imitate. This work also explores the method of how to learn by imitation and contributes to the research on human movement recognition, multi-modal human-robot interaction and imitation learning.

References

1. Bakker, P., Kuniyoshi, Y.: Robot See, Robot Do: An Overview of Robot Imitation. AISB96 Workshop on Learning in Robots and Animals. (1996) 3-11
2. Amit, R., Matari , M.: Learning Movement Sequences from Demonstration. Proceedings of the 2nd International Conference on Development and Learning. (2002) 203-208
3. Hovland, G.E., Sikka, P., McCarragher, B,J.: Skill Acquisition from Human Demonstration Using a Hidden Markov Model. IEEE International Conference on Robotics and Automation. 3(1996) 2706-2711
4. Ito, M., Tani, J.: On-line Imitative Interaction with a Humanoid Robot Using a Dynamic Neural Network Model of a Mirror System. Adaptive Behavior. 12(2) (2004) 93-115
5. Carpenter, G.A., Grossberg, S.: The ART of Adaptive Pattern Recognition by a Self-Organizing Neural Network. IEEE COMPUTER. (1988) 77-886.
6. Lalanne, C., Lorenceau, J.: Crossmodal Integration for Perception and Action. J. Physiology, Paris. (2004) 265-279
7. Carpenter, G.A., Grossberg, S.: ART 2: self-organization of stable category recognition codes for analog input patterns. APPLIED OPTICS. 26(23) (1987) 4919-4930
8. Breazeal, C. Scassellati, B.: Challenges in Building Robots that Imitate People. Imitation in Animals and Artifacts. (2002) 363–390

An Online Blind Source Separation for Convolutive Acoustic Signals in Frequency-Domain*

Wu Wenyan and Zhang Liming

Department of Electronics Engineering, Fudan University, Shanghai, 200433
{042021032, lmzhang}@fudan.edu.cn

Abstract. In this paper we propose a scheme for online convolutive blind acoustic source separation in frequency domain. A restriction term of DOA (Direction of Arrival) for sources is added to the cost function of mutual information. Minimizing the modified mutual information in each frequency bin, the source order can be adjusted on line. So the problem of permutation ambiguity can be settled on line, and it does not need the location information of receivers like classical DOA method. Simulation results show that the proposed algorithm has better performance than other existing methods.

1 Introduction

Blind source separation (BSS) is a signal processing technique to estimate the source signals only from the observed signals of the receivers without any prior knowledge about the source signals and the mixing environment. Independent Component Analysis (ICA) is an important method for BSS. But in real environment, the convolution of acoustic signals called Convolutive BSS due to the delay and reflection makes the blind separation of acoustic signals very difficult.

The frequency-domain CBSS is to transform the convolutive mixtures in time-domain to instantaneous mixtures in frequency-domain, where we can apply complex-valued ICA for instantaneous mixtures in each frequency bin, such as Minimization of Mutual Information [1] combined with natural gradient [2], FastICA, cumulant-based ICA etc, to solve CBSS. The frequency-domain method has less computation complexity compared to the time-domain methods. However, since ICA is calculated in each frequency bin separately, the permutation and scaling ambiguity becomes a serious problem. The scaling ambiguity can be settled by making the determinant of separating matrix to unity [3] while the permutation problem is more difficult. We need to align the permutation in each frequency bin so that the reconstructed signals in time domain contains frequency components are from the same source signals. Nowadays, there are some proposed methods to solve permutation problem such as the correlation of adjacent frequency bins [4], direction of arrival (DOA) estimation of the source signals [5] and combining DOA and correlation method [6]. The correlation approach is to align the permutation by compare the amplitude correlation

* This research was supported by the grant from the National Natural Science Foundation of China(NSF60571052).

L. Jiao et al. (Eds.): ICNC 2006, Part I, LNCS 4221, pp. 451 – 460, 2006.

between adjacent frequency bins, but it is not robust enough as the misalignment of one frequency bin may cause consecutive misalignment and it can not work on line. As for the DOA approach, it is to analyze the directivity patterns formed by the separation matrix and to estimate the source directions, but it is not able to estimate the direction of sources in some low frequency bins. The method combining DOA and correlation is to use DOA to settle the order on some frequency bins and correlation method on the remaining frequency bins, whose performance is better. However, existing methods mostly need to do batch post-processing after finishing the source separation in each frequency bin. And what is more, the DOA related method needs the location information of the receivers.

In this paper, we propose a scheme by adding a DOA restriction term with the Lagrange multipliers to the cost function of mutual information. This approach can adjust the permutation inconsistency on-line. Meanwhile, using DOA as a restriction can omit the location information of the receivers.

The remaining parts of this paper are organized as follows. In section 2, we introduce the basic approach of frequency-domain CBSS. Section 3 presents our proposed method of on-line frequency-domain CBSS with DOA as a restriction. And the experiment results and conclusion are presented in section 4 and 5.

2 CBSS in Frequency-Domain

A basic CBSS problem with N independent sources $\boldsymbol{S}(t)=[s_1(t),\cdots,s_N(t)]^T$ and M receivers $\boldsymbol{X}(t)=[x_1(t),\cdots,x_M(t)]^T$ can be described as follows. The $x_j(t)$ signal received from the jth microphone is:

$$x_j(t)=\sum_{i=1}^{N}\sum_{p=0}^{P}h_{ji}(p)s_i(t-p) \qquad (j=1,\cdots,M), \tag{1}$$

where h_{ji} is the impulse response from the ith source to the jth receiver, p is the index of delay and P is length of the convolution filter.

The aim of CBSS is to find a separation filter $\boldsymbol{W}$ with length Q so that the recovered signals $\boldsymbol{Y}(t)=[y_1(t),\cdots,y_N(t)]^T$ are mutually independent:

$$\boldsymbol{Y}(t)=\sum_{q=0}^{Q}\boldsymbol{W}(q)\boldsymbol{X}(t-q) \tag{2}$$

Using L-point short-time Fourier transformation for Eq.(2), $\boldsymbol{X}(t)$ in time-domain is converted to a block index-series vector $\boldsymbol{X}(f,k)$ in frequency-domain, where $f=0,f_s/L,\cdots,f_s(L-1)/L$ (f_s is the sampling frequency) and k is the block index.

$$\boldsymbol{X}(f,k)=\boldsymbol{H}(f)\boldsymbol{S}(f,k), \tag{3}$$

where $\boldsymbol{H}(f)$ is the mixing matrix at frequency bin f , $\boldsymbol{S}(f,k)$ and $\boldsymbol{X}(f,k)$ are source and observed signal vector respectively.

Frequency-domain CBSS is to find a separating matrix $W(f)$ in each frequency bin, so that the components of recovered signal $Y(f)$ are mutually independent.

$$\mathbf{Y}(f,k) = \mathbf{W}(f)\mathbf{X}(f,k), \tag{4}$$

where $Y(f,k) = [Y_1(f,k)\cdots Y_N(f,k)]^T$ and $W(f)$ is a $N \times M$ separation matrix.

$W(f)$ can be obtained by minimizing the following mutual information [1]:

$$I(Y,W(f)) = \sum_{i=1}^{N} H_a(y_i(f)) - H_a(Y(f)) \quad f = 0, f_s/L, \cdots, f_s(L-1)/L, \tag{5}$$

where $H_a(Y(f))$ and $H_a(y_i(f))$ are the entropy of recovered vector $\mathbf{Y}(f)$ and its element respectively. In simplicity, the index k is omitted.

3 On-Line Frequency-Domain CBSS

3.1 DOA (Direction of Arrival)

DOA is an approach to decide the order of recovered signals in each frequency bin by estimating the directions of source signals. Fig. 1 shows the M microphones arrayed on a line. Taking the first microphone as the reference, the distance between the reference and other microphones is d_j, j=2,3,...M, $d_1 = 0$ .Assume that the sources are far from the microphones, so the arrival angles from one source to all microphones are the same and the signal propagation is a plane wavefront.

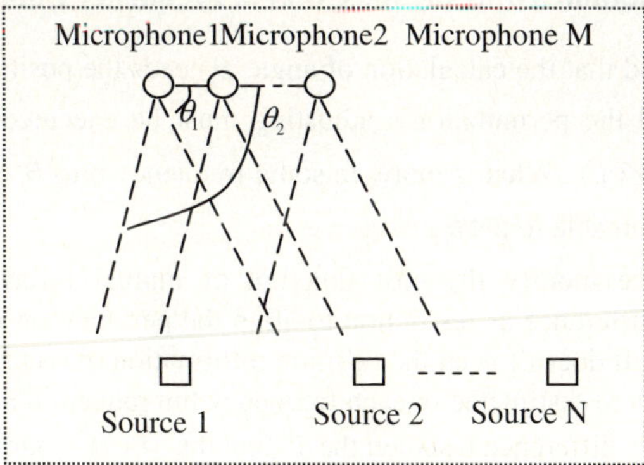

Fig. 1. Sources and microphones layout

Let θ_i be the angle from the ith source to all microphones (the orthogonal direction to the microphones array is 90°). According to beamforming theory, the propagation function from source i to microphone j at frequency bin f H_{ji} can be written as:

$$H_{ji}(f) = A_{ji} e^{j2\pi f c^{-1} d_j \cos\theta_i} (j = 1, \cdots, M, i = 1, \cdots, N), \tag{6}$$

where c is the propagation velocity, f is the frequency and A_{ji} is the amplitude. Let Eq.(6) be the $(j,i)^{th}$ element of matrix $\boldsymbol{H}$ in Eq.(3). The separating matrix $\boldsymbol{W}(f)$ should satisfy $\boldsymbol{\Lambda}(f)\boldsymbol{P}_m(f)\boldsymbol{W}(f)\boldsymbol{H}(f)=\mathbf{I}$, where $\boldsymbol{\Lambda}(f)$ is a diagonal matrix and $\boldsymbol{P}_m(f)$ is the permutation matrix. Then we have $\boldsymbol{H}(f)=\boldsymbol{W}(f)^{-1}\boldsymbol{P}_m(f)^{-1}\boldsymbol{\Lambda}(f)^{-1}$, and the ratio between elements of the ith column H_{ji} and H_{li} is:

$$\frac{H(f)_{ji}}{H(f)_{li}}=\frac{[\boldsymbol{W}(f)^{-1}\boldsymbol{P}_m(f)^{-1}\boldsymbol{\Lambda}(f)^{-1}]_{ji}}{[\boldsymbol{W}(f)^{-1}\boldsymbol{P}_m(f)^{-1}\boldsymbol{\Lambda}(f)^{-1}]_{li}}=\frac{[\boldsymbol{W}(f)^{-1}]_{j\Pi(i)}}{[\boldsymbol{W}(f)^{-1}]_{l\Pi(i)}} \quad for \quad j\neq l \tag{7}$$

For the same column, the coefficient of inverse diagonal matrix is the same. Only permutation matrix will affect the ratio of Eq.(7). Π refers to the permutation. From Eq.(6) and Eq.(7) we can get the angle of the ith source, θ_i [6], as

$$\theta_i=\arccos(B_i) \quad B_i=\frac{\arg\left(\dfrac{[\boldsymbol{W}(f)^{-1}]_{j\Pi(i)}}{[\boldsymbol{W}(f)^{-1}]_{l\Pi(i)}}\right)}{2\pi fc^{-1}(d_j-d_l)} \quad \theta_i\in(0,\pi), \tag{8}$$

where $\arg(\cdot)$ is the argument of complex number. From Eq.(8) we can estimate the direction of the ith source from the ith column of $\boldsymbol{W}(f)^{-1}$. According to the order of N sources in each frequency bin, we can rearrange the order of rows of $\boldsymbol{W}(f)$, so that the permutation is consistent in each frequency bin.

3.2 Online Permutation Adjusting for CBSS in Frequency Domain

From Eq.(8), we find that the calculation of angle θ_i needs the position information of receivers (d_j), and the permutation's adjusting must be executed after we get the separating matrix $\boldsymbol{W}(f)$. What is more, in some frequency bins B_i may go beyond its range ($|B_i|>1$), it is unable to get θ_i.

In this paper, we modify the cost function of mutual information Eq.(5) by inserting an angle difference as restriction to align the permutation in each frequency bin during iteration. It doesn't need the position information of receivers. We hope the directions of the sources estimated in each frequency bin ranked in increasing order.

Define the angle's difference between the ith and the $(i+1)^{th}$ sources as:

$$\theta_{i+1}-\theta_i=\arccos(B_{i+1})-\arccos(B_i) \tag{9}$$

Assume that the ith source's angle is smaller than the $(i+1)^{th}$ source's angle, ($\theta_{i+1}-\theta_i>0$). Considering the "**arccos**" is a monotone decreasing function in $[0\sim\pi]$ and the denominator of B_i for two receivers at a given frequency bin f are the same, we can omit it in Eq.(8) and define DOA_i instead of θ_i as

$$DOA_i = \arg(\frac{[\boldsymbol{W}(f)^{-1}]_{j\Pi(i)}}{[\boldsymbol{W}(f)^{-1}]_{l\Pi(i)}}) = \arg(z) \tag{10}$$

Since the values of DOA_i and z are no limited, the un-solution case like $|B|>1$ in Eq.(8) will not happen and what is more, the location information of receivers is no longer needed. Replace θ_i by DOA_i in Eq.(9) the restriction can be written as:

$$g_i(\boldsymbol{W}(f)) = DOA_{i+1} - DOA_i = \arg(\frac{[\boldsymbol{W}(f)^{-1}]_{j\Pi(i+1)}}{[\boldsymbol{W}(f)^{-1}]_{l\Pi(i+1)}}) - \arg(\frac{[\boldsymbol{W}(f)^{-1}]_{j\Pi(i)}}{[\boldsymbol{W}(f)^{-1}]_{l\Pi(i)}}) < 0 \qquad i = 1,2,\ldots N-1 \tag{11}$$

Combining the cost function of mutual information in Eq.(5) with the restriction of Eq.(11), using augmented Lagrange function [7], the modified cost function of mutual information is written as:

$$P(\boldsymbol{W}(f),\mu) = I(\boldsymbol{W}(f)) + \frac{1}{2\gamma}\sum_{i=1}^{N-1}\{[\max\{0,\mu_i + \gamma g_i(\boldsymbol{W}(f))\}]^2 - \mu_i^2\}, \tag{12}$$

where $\mu_i, i=1,2,..N$, is Lagrange multiplier and γ is a positive constant. The initial value of μ_i is zero, and when the condition of Eq.(11) is satisfied the second term of Eq.(12) will be zero, otherwise μ_i is obtained using the following iteration equation $\mu_i(m+1) = \max\{0, \mu_i(m) + \gamma g_i(m)\}$. Here m is the step in iteration. Using gradient decent algorithm for Eq.(12), combined with the iteration equation of μ_i, we can get the iteration formula for $\boldsymbol{W}(f)$:

$$\vartriangle\boldsymbol{W}_{ij}(f) = -\eta(\frac{\partial I(\boldsymbol{W}(f))}{\partial \boldsymbol{W}_{ij}} + \sum_{i=1}^{N-1}\mu_i\frac{\partial g_i(\boldsymbol{W}(f))}{\partial \boldsymbol{W}_{ij}}), \tag{13}$$

where the first term is to take derivative with respect to Eq.(5), combining natural gradient, the iteration equation for the first term is shown as follows [2] :

$$-\eta\frac{\partial I(\boldsymbol{W}(f))}{\partial \boldsymbol{W}(f)} = \eta[\boldsymbol{I} - \boldsymbol{\Phi}(\boldsymbol{Y}(f))\boldsymbol{Y}^H(f)]\boldsymbol{W}(f), \tag{14}$$

where $\boldsymbol{Y}(f)$ is recovered source vector at frequency f, and $\boldsymbol{\Phi}$ is nonlinear function. Considering $\boldsymbol{Y}$ as a complex vector in frequency domain, $\boldsymbol{\Phi}$ can be written as $\boldsymbol{\Phi}(\boldsymbol{Y}) = \tanh(re(\boldsymbol{Y})) + i\tanh(im(\boldsymbol{Y}))$ [3], where $re(\boldsymbol{Y})$ and $im(\boldsymbol{Y})$ are the real and imaginary part of $\boldsymbol{Y}$ respectively, $\boldsymbol{I}$ is the identity matrix. The second part of Eq.(13) is more complicated but it can still be calculated. Take N=M=2 as an example, here (N-1) =1, the restriction in Eq.(12) has only one item g_1 :

$$
\begin{aligned}
g_1(\boldsymbol{W}(f)) &= \arg\left(\frac{[W(f)^{-1}]_{22}}{[W(f)^{-1}]_{12}}\right) - \arg\left(\frac{[W(f)^{-1}]_{21}}{[W(f)^{-1}]_{11}}\right) \\
&= \arg\left(-\frac{[W(f)]_{21}}{[W(f)]_{22}}\right) - \arg\left(-\frac{[W(f)]_{11}}{[W(f)]_{12}}\right)
\end{aligned} \tag{15}
$$

By using gradient learning rule combining Eq.(13), we can get the iteration formula of $\boldsymbol{W}$ (See appendix):

$$
\begin{aligned}
\boldsymbol{W}_{m+1}(f) &= \boldsymbol{W}_m(f) + \eta[\boldsymbol{I} - \Phi(Y)Y^H]\boldsymbol{W}_m(f) \\
&\quad - \eta\begin{bmatrix} \mu_1(-W_{11}^I + jW_{11}^R)/|W_{11}|^2 & \mu_1(W_{12}^I - jW_{12}^R)/|W_{12}|^2 \\ \mu_1(W_{21}^I - jW_{21}^R)/|W_{21}|^2 & \mu_1(-W_{22}^I + jW_{22}^R)/|W_{22}|^2 \end{bmatrix},
\end{aligned} \tag{16}
$$

$W(f)_{ij}^R$ and $W(f)_{ij}^I$ are the real and imaginary part of the $(i, j)^{th}$ element of $\boldsymbol{W}(f)$.

As for the scaling ambiguity, we normalize the $\boldsymbol{W}(f)$ by its determinant at each iteration [3] $\boldsymbol{W}(f) \leftarrow \boldsymbol{W}(f)/|\boldsymbol{W}(f)|^{1/N}$, where N is source number, so that the scale of all the sources in each frequency bin is consistent.

4 Experimental Results

Several experiments are designed to evaluate our proposed method. In Experiment 1and 2, we use signal-to-interference (SIR) to test the performance of CBBS with some given mixing filters and sources. Exp 1 is to compare the on-line method by minimizing classical mutual information and our modified one. Exp 2 compares the separation performance between our method and the other existing methods to solve permutation ambiguity. Real world acoustic signal separation is shown in Exp 3.

Suppose N acoustic signals are mixed by a known mixing filter $\boldsymbol{H}$. The separation matrix $\boldsymbol{W}(f)$ can be iterated. Signal-to-interference (SIR_i) [8] is defined as:

$$SIR_i = SIR_{Oi} - SIR_{Ii} \quad i = 1, \cdots, N \tag{17}$$

$$SIR_{Oi} = 10\lg \left\langle \frac{\sum_f \|[\boldsymbol{WH}]_{ii}(f)\|^2 \cdot \|\boldsymbol{S}_i(f)\|^2}{\sum_{l\neq i}\sum_f \|[\boldsymbol{WH}]_{il}(f)\|^2 \cdot \|\boldsymbol{S}_l(f)\|^2} \right\rangle \quad i,l = 1, \cdots, N \tag{18}$$

$$SIR_{Ii} = 10\lg \left\langle \frac{\sum_f \|[\boldsymbol{H}]_{ii}(f)\|^2 \cdot \|\boldsymbol{S}_i(f)\|^2}{\sum_{l\neq i}\sum_f \|[\boldsymbol{H}]_{il}(f)\|^2 \cdot \|\boldsymbol{S}_l(f)\|^2} \right\rangle \quad i,l = 1, \cdots, N \tag{19}$$

where $S_i(f)$ and $S_l(f)$ are the ith and l^{th} source signals in frequency bin f and $\langle\cdot\rangle$ denotes the mean of all blocks. SIR is the ratio of the target signal to the interference signal. SIR_{Oi} and SIR_{Ii} are the output and input SIR, respectively. It is obvious that larger SIR_i means better performance.

Experiment 1: Two speech signals sampled by 8KHz are located in the position far from two microphone array at angle 110° and 150° respectively. Their mixing filter $\boldsymbol{H}$ can be calculated by Eq.(6). Two online methods are used in this experiment: classical minimization of mutual information (Eq.(5)) and our modified method(Eq. (12)). The result is shown in Fig.2.

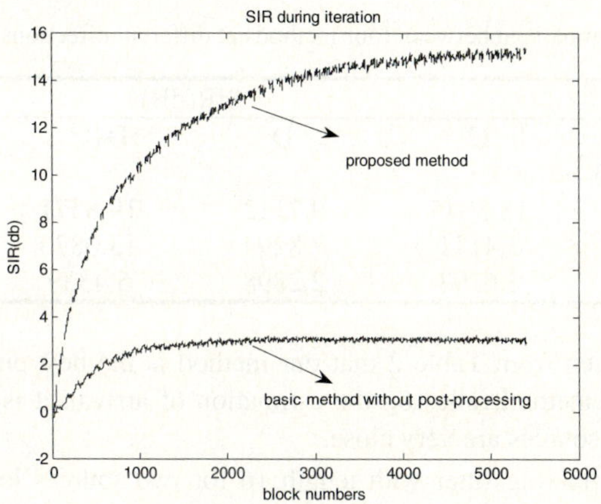

Fig. 2. SIR during iteration

In Fig. 2 We use mixing filter $\boldsymbol{H}$ with length of 256 and the SIR is the average of two channels. We can see from Fig. 2 our proposed method solve the permutation ambiguity during iteration. So the SIR is much higher than the original frequency-domain CBSS method.

Experiment 2: In this experiment, a comparison between our method and other three methods (Correlation of adjacent frequency bins, DOA approach and DOA combined with correlation method) to solve permutation ambiguity is shown in Table. 1.

Table 1. Performance comparison of four methods to solve permutation ambiguity. (Here C: correlation method [4], D: DOA approach [5], D+C: DOA combined with correlation [6]).

	SIR(dB)			
Filter length	C	D	D+C	Proposed method
512	3.4171	7.8294	13.0870	14.1714
256	6.8283	7.5485	15.0221	15.3810
64	7.9986	7.0882	17.6479	17.8585
16	9.0610	12.4189	18.9820	18.9892

We can see from Table 1 that our method is the best one among the four methods. The correlation approach is not robust enough. If there is a frequency bin which is misalignment it may cause consecutive misalignment, so the performance is actually bad (SIR is low). As for DOA approach, it is unable to get the direction in some frequency bins, which limits the performance. Our proposed method is a little better than the DOA combined with correlation approach. Note that our proposed method can align the order during iteration which is suitable for on-line processing and D+C method must be batch processing.

For a given mixing filter with length 512, change the directions of the sources to compare the performance of four methods. The results are shown in Table. 2.

Table 2. Comparison between four methods at different directions of sources

Source Direction		SIR(dB)			
S1 (degree)	S2 (degree)	C	D	D+C	Proposed method
160	100	18.5845	9.7112	19.8571	20.5718
110	150	3.4171	7.8294	13.0870	14.1714
160	150	2.6293	2.2898	5.4339	6.7462

It can also be seen from Table 2 that our method is the best one among the four methods. As DOA method relies on the estimation of arrival, it is not good enough when the angles of sources are very close.

Fig.3 shows the mixing filter with length 16 for two sources located at 160° and 100°. By using our proposed method, the separation matrix is $\boldsymbol{W}$. The second graph shows the global transfer function $\boldsymbol{G}$, combining the mixing and recovered matrix, and its $(i, j)^{th}$ element is defined as

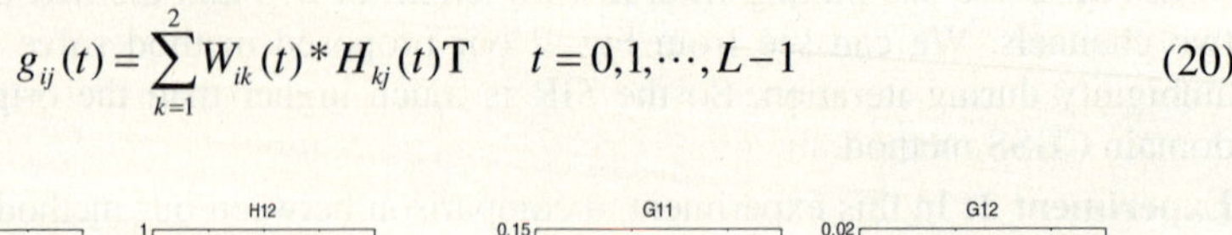

$$g_{ij}(t) = \sum_{k=1}^{2} W_{ik}(t) * H_{kj}(t)\mathrm{T} \quad t = 0,1,\cdots,L-1 \qquad (20)$$

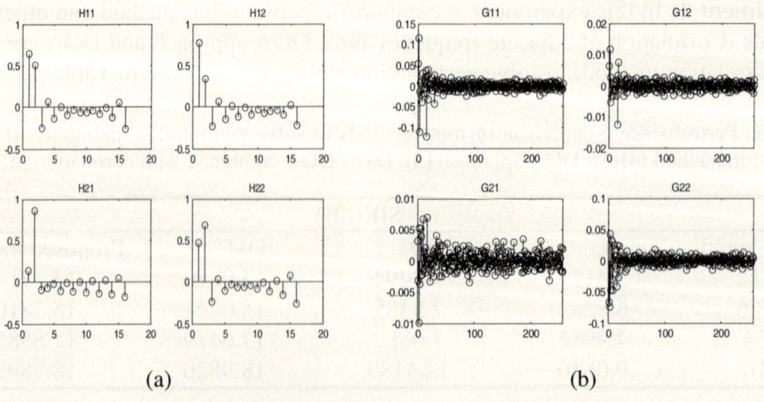

Fig. 3. (a) Mixing matrix b) $\boldsymbol{G}$ after iteration

We can see from Fig.3, the diagonal elements of matrix $\boldsymbol{G}$ are about 10 times to the off-diagonal elements. When listening to the separated signal by ears, it is very clear for each channel and we can hardly hear the audio signal from the other channel.

Experiment 3: We use two speech signals in real world downloaded from website [9] to test our method's performance. This is a convolutive blind signal separation in a room. There is no information about the position of receivers, source signals and the mixing matrix, so it is unable to calculate SIR. Fig. 4 shows the separated signals and mixed signals provided by the website and the separated signals of our proposed method.

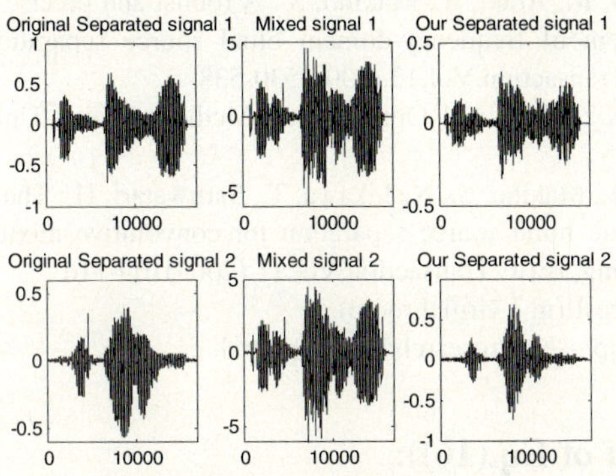

Fig. 4. Real world signals

In Fig.4, the first column is the separated signals using batch processing method proposed by the website, the second column is the mixed signals and the third column is the separated signals using our proposed method. The performances are almost the same. When we listen by human eras, the separated signals are quite good.

It can be seen form Fig.4, although in Eq.(6) we assume the propagation from sources to receivers is plane wavefront and no reverberation, the estimation is still good to decide the permutation in reverberant condition .

5 Conclusions

In this paper, we propose an online convolution blind source separation scheme by modifying the cost function of mutual information with DOA as a restriction. It can solve the permutation ambiguity on line and does not need the information of positions of receivers. It has better performance compared with the correlation between adjacent frequencies bins method, the DOA approach and DOA combined correlation approach. It is meaningful for the real world signal separation. However, our proposed method is also not suitable for separating sources located on the same position like other DOA related approach, which needs to be studied in the future.

References

1. Bell, A., Sejnowski, T.: An information-maximization approach to blind separation and blind deconvolution. Nerualcomputing.Vol.7. (1995)1129-1159
2. Amari, S.: Natural gradient works efficiency in learning. Neuralcomputing. Vol.10.(1998)251-276

3. Smaragdis, P.: Blind separation of convolved mixtures in the frequency domain. Neurocomputing.Vol.22.(1998)21-34
4. Jiang W.D.: Blind source separation of speech signals based on the amplitude correlation of neighbor bins. Journals of Circuits and System.(2005)1-4
5. Ikram, M.Z., Morgan, D.R.: A beamforming approach to permutation alignment for multichannel frequency-domain blind speech separation. Proc. ICASSP.(2002)881-884
6. Sawada, H., Mukai, R., Araki, S., Makino, S.: A robust and precise method for solving the permutation problem of frequency-domain blind source separation. Speech and Audio Processing, IEEE Transaction.Vol.12.(2004)530-538
7. Tang H.W., Qin X.Z.: Practical Optimization Method. Dalian University of Technology Press(2004)202-205
8. Araki, S., Mukai, R., Makino, S., Nishikawa, T., Saruwatari, H.: The fundamental limitation of frequency domain blind source separation for convolutive mixtures of speech. Speech and Audio Processing, IEEE Transaction.Vol.11.(2003)109-116
9. Convolutive mixturesII(in a virtual room)
http://www.ism.ac.jp/~shiro/research/blindsep.html

Appendix (Proof of Eq.(16)):

N=M=2, the restriction is Eq.(15), where $W_{ij} = W_{ij}^R + jW_{ij}^I$. As the value range of **arg** is $(-\pi,\pi)$, while the value range of **arctan** is $(-\pi/2,\pi/2)$, we can get:

$$\begin{aligned}\arg(-[W(f)]_{21}/[W(f)]_{22}) &= \arctan((W_{21}^R W_{22}^I - W_{22}^R W_{21}^I)/(W_{21}^R W_{22}^R + W_{21}^I W_{22}^I)) \\ &+\pi/2\times(sign(W_{21}^R W_{22}^I - W_{22}^R W_{21}^I))(1-sign(W_{21}^R W_{22}^R + W_{21}^I W_{22}^I))\end{aligned} \tag{21}$$

$$\begin{aligned}\arg(-[W(f)]_{11}/[W(f)]_{12}) &= \arctan((W_{11}^R W_{12}^I - W_{12}^R W_{11}^I)/(W_{11}^R W_{12}^R + W_{11}^I W_{12}^I)) \\ &+\pi/2\times(sign(W_{11}^R W_{12}^I - W_{12}^R W_{11}^I))(1-sign(W_{11}^R W_{12}^R + W_{11}^I W_{12}^I))\end{aligned} \tag{22}$$

Taking derivative with respect to $g_1(\boldsymbol{W}(f))$, we can get

$$\frac{\partial g_1(\boldsymbol{W}(f))}{\partial W_{11}} = \frac{-W_{11}^I + jW_{11}^R}{|W_{11}|^2}.$$

It is the same that

$$\frac{\partial g_1(\boldsymbol{W}(f))}{\partial W_{12}} = \frac{W_{12}^I - jW_{12}^R}{|W_{12}|^2}, \quad \frac{\partial g_1(\boldsymbol{W}(f))}{\partial W_{21}} = \frac{W_{21}^I - jW_{21}^R}{|W_{21}|^2}, \quad \frac{\partial g_1(\boldsymbol{W}(f))}{\partial W_{22}} = -\frac{-W_{22}^I + jW_{22}^R}{|W_{22}|^2}$$

So we can get

$$\boldsymbol{W}_{m+1}(f) = \boldsymbol{W}_m(f) + \eta[\boldsymbol{I} - \boldsymbol{\Phi}(Y)Y^H]\boldsymbol{W}_m(f)$$

$$-\eta\begin{bmatrix} \mu_1(-W_{11}^I + jW_{11}^R)/|W_{11}|^2 & \mu_1(W_{12}^I - jW_{12}^R)/|W_{12}|^2 \\ \mu_1(W_{21}^I - jW_{21}^R)/|W_{21}|^2 & \mu_1(-W_{22}^I + jW_{22}^R)/|W_{22}|^2 \end{bmatrix}, \quad \text{the same as Eq.(16).}$$

GPS/INS Navigation Filter Designs Using Neural Network with Optimization Techniques

Dah-Jing Jwo and Jyh-Jeng Chen

Department of Communications and Communications Engineering,
National Taiwan Ocean University, 20224 Keelung, Taiwan, China
djjwo@mail.ntou.edu.tw

Abstract. The Global Positioning System (GPS) and inertial navigation systems (INS) have complementary operational characteristics and the synergy of both systems has been widely explored. Most of the present navigation sensor integration techniques are based on Kalman filtering estimation procedures. For obtaining optimal (minimum mean square error) estimate, the designers are required to have exact knowledge on both dynamic process and measurement models. In this paper, a mechanism called PSO-RBFN, which combines Radial Basis Function (RBF) Network and Particle Swarm Optimization (PSO), for predicting the errors and to filtering the high frequency noise is proposed. As a model nonlinearity identification mechanism, the PSO-RBFN will implement the on-line identification of nonlinear dynamics errors such that the modeling error can be compensated. The PSO-RBFN is applied to the loosely-coupled GPS/INS navigation filter design and has demonstrated substantial performance improvement in comparison with the standard Kalman filtering method.

1 Introduction

Global positioning systems (GPS) are capable of providing accurate position information. Unfortunately, the data is prone to jamming or being lost due to the limitations of electromagnetic waves, which form the fundamental of their operation. The system is not able to work properly in the areas due to signal blockage and attenuation that may deteriorate the overall positioning accuracy [1].

The inertial navigation systems (INS) is a self-contained system that integrates three acceleration components and three angular velocity components with respect to time and transforms them into the navigation frame to deliver position, velocity and attitude components. The three orthogonal linear accelerations are continuously measured through three-axis accelerometers while three gyroscopes sensors monitor the three orthogonal angular rates in an inertial frame of reference. For short time intervals, the integration with respect to time of the linear acceleration and angular velocity monitored by the INS results in an accurate velocity, position and attitude. However, the error in position coordinates increase unboundedly as a function of time.

The GPS/INS integration is the adequate solution to provide a navigation system that has superior performance in comparison with either a GPS or an INS stand-alone

L. Jiao et al. (Eds.): ICNC 2006, Part I, LNCS 4221, pp. 461 – 470, 2006.

system [2]. The GPS/INS integration is typically carried out through Kalman filter (KF). However, the fact that KF highly depends on a predefined dynamics model forms a major drawback [3,4]. To achieve good filtering results, the designers are required to have good knowledge on both the dynamic process and measurement models, in addition to the assumption that both the process and measurement are corrupted by zero-mean Gaussian white sequences. If the input data does not reflect the real model, the KF estimates may not be reliable.

The error model for INS is augmented by some sensor error states such as accelerometer biases and gyroscope drifts. Actually, there are several random errors associated with each inertial sensor [5]. Noise contributions in typical optical gyroscope systems include white noise, correlated random noise, bias instability and angle random walk [6,7]. It is usually difficult to set a certain stochastic model for each inertial sensor that works efficiently at all environments and reflects the long-term behavior of sensor errors. The difficulty of modeling the errors of INS raised the need for a model-less GPS/INS integration technique.

In this paper, a new approach is proposed for improvingr GPS/INS navigation system designs. The method makes use of the radial basis function network (RBFN) and Particle Swarm Optimization (PSO) techniques, resulting in an aiding mechanism called PSO-RBFN. The PSO-RBFN is employed into the navigation systems as a dynamic model corrector, which aids the Kalman filter for real-time identification of nonlinear dynamics errors, especially when the modeling of uncertainty is concerned. By monitoring the measurement residuals, this mechanism will be able to provide real-time error correction for the dynamic modeling errors to prevent divergence of the Kalman filter. The proposed GPS/INS integration architecture is shown in Fig. 1.

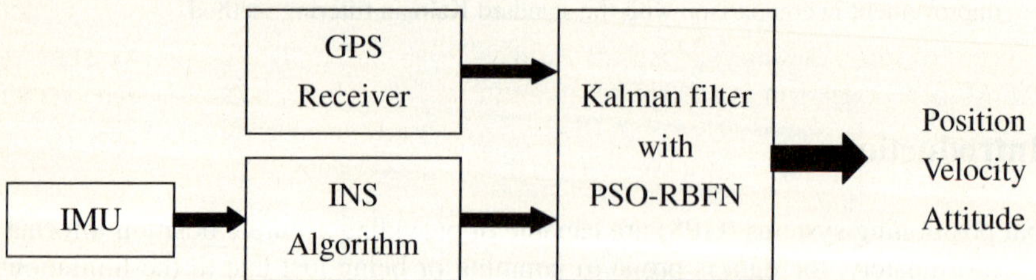

Fig. 1. Proposed GPS/INS integration architecture

2 GPS/INS Navigation Filter

Although it would seem straightforward to model the inertial errors simply as the difference between the estimates and the truth, it turns out the resulting so-called 'perturbation error' equations have a complex coupling which can be avoided if another approach is taken. The alternative uses the so-called 'psi-angle' equations [8].

The nomenclature used in this section is summarized for convenience:

Body frame (b-frame): frame fixed to the vehicle.

Computer frame (c-frame): local level frame at the computed position.

Platform frame (p-frame): frame which the transformed accelerations and angular rates from the accelerometers and gyros are resolved.

Earth frame (e-frame): located at the earth center.
True frame (n-frame): true local level (NED) frame at the true position.
psi-angle (ψ): the angle between c-frame and *p*-frame.
phi-angle (Φ): the angle between n-frame and *p*-frame.
theta-angle (ψ): the angle between n-frame and c-frame.
(The three are then related: $\Phi = \theta + \psi$)
DCM: direction cosine matrix.
C_m^o *: DCM* from m-frame to o-frame.
ω_{kl}^j : angular rate between *k*-frame and *l*-frame resolved in *j*-frame.

2.1 Inertial Error Modeling in State Space

The details are derived in the inertial literature in the list of references but the psi-angle position, velocity and attitude errors are given as follows:
Position Error:

$$\delta\dot{\mathbf{R}} = -\boldsymbol{\omega}_{en}^n \times \delta\mathbf{R}^n + \delta\mathbf{V}^c \tag{1}$$

Velocity Error

$$\delta\dot{\mathbf{V}} = \mathbf{C}_b^p \delta\mathbf{f}^b + \begin{bmatrix} \frac{-g}{r_e} & 0 & 0 \\ 0 & \frac{-g}{r_e} & 0 \\ 0 & 0 & \frac{2g}{r_e+h} \end{bmatrix} \delta\mathbf{R} - \boldsymbol{\Psi} \times (\boldsymbol{\omega}_{ec}^c + 2\boldsymbol{\omega}_{ie}^c) \times \delta\mathbf{V}^c \tag{2}$$

Psi-angle:

$$\dot{\boldsymbol{\Psi}} = -(\boldsymbol{\omega}_{ec}^c + \boldsymbol{\omega}_{ie}^c) \times \boldsymbol{\Psi} + \mathbf{C}_b^p \delta\boldsymbol{\omega}_{ib}^b \tag{3}$$

In order to introduce the concept of Kalman Filter-based INS aiding, we start by assuming there exist some generic positioning system whose noise. A state vector with 15 states is employed. Nine so-called inertial error states (position, velocity and psi-angle), three accelerometer bias states and three gyro bias states:

$$\begin{bmatrix} \delta\dot{\mathbf{R}} \\ \delta\dot{\mathbf{V}} \\ \dot{\boldsymbol{\Psi}} \\ \dot{\boldsymbol{\varepsilon}}_a \\ \dot{\boldsymbol{\varepsilon}}_g \end{bmatrix} = \begin{bmatrix} \mathbf{F}_{11} & \mathbf{F}_{12} & \mathbf{0} & \mathbf{0} & \mathbf{0} \\ \mathbf{F}_{21} & \mathbf{F}_{22} & \mathbf{F}_{23} & \mathbf{C}_b^n & \mathbf{0} \\ \mathbf{0} & \mathbf{0} & \mathbf{F}_{33} & \mathbf{0} & \mathbf{C}_b^n \\ \mathbf{0} & \mathbf{0} & \mathbf{0} & -\frac{1}{\tau_g}\mathbf{I} & \mathbf{0} \\ \mathbf{0} & \mathbf{0} & \mathbf{0} & \mathbf{0} & -\frac{1}{\tau_g}\mathbf{I} \end{bmatrix} \begin{bmatrix} \delta\mathbf{R} \\ \delta\mathbf{V} \\ \boldsymbol{\Psi} \\ \boldsymbol{\varepsilon}_a \\ \boldsymbol{\varepsilon}_g \end{bmatrix} + \begin{bmatrix} \mathbf{0} \\ \mathbf{0} \\ \mathbf{0} \\ \mathbf{u}_a \\ \mathbf{u}_g \end{bmatrix} \tag{4}$$

where the '**0**'s are 3×3 matrices of zeros, and $\mathbf{F}_{11} = [-\boldsymbol{\omega}_{en}^n \times]$; $\mathbf{F}_{12} = \mathbf{I}$;

$$\mathbf{F}_{21} = \begin{bmatrix} \frac{-g}{r_e} & 0 & 0 \\ 0 & \frac{-g}{r_e} & 0 \\ 0 & 0 & \frac{2g}{r_e+h} \end{bmatrix};\ \mathbf{F}_{22} = [-(\boldsymbol{\omega}_{en}^n + 2\boldsymbol{\omega}_{ie}^n)\times];\ \mathbf{F}_{23} = [\mathbf{f}^b \times];\ \mathbf{F}_{33} = [-(\boldsymbol{\omega}_{en}^n + \boldsymbol{\omega}_{in}^n)\times].$$

2.2 Non-linear Filter Implementation

Since the uncertainties in heading can be very large during the initial stage, non-linear state models need to be used. An extended Kalman filter is implemented to fuse the information and estimate the states. The discrete form of the filter is:

$$\dot{\mathbf{x}}_{k+1} = \mathbf{f}_k(\mathbf{x}_k) + \mathbf{G}_k\mathbf{u}_k \quad (5)$$

$$\mathbf{z}_k = \mathbf{H}\mathbf{x}_k + \mathbf{v}_k \quad (6)$$

where $\mathbf{u}_k$ and $\mathbf{v}_k$ are white noise sequences with strength $\mathbf{G}_k$ and $\mathbf{R}_k$.; $\mathbf{z}_k$ is the measurements containing the velocity and position difference between the INS output and the external GPS information. At the estimation time t_k , the system is linearized around the previous state estimate. The Jacobian matrix $\mathbf{F}_k$ of $\mathbf{f}_k$ is evaluated to obtain the conditional covariance matrix $\mathbf{P}_{k,k-1}$:

$$\mathbf{P}_{k,k-1} = \mathbf{F}_k\mathbf{P}_{k-1,k-1}\mathbf{F}_k^T + \mathbf{G}_k\mathbf{Q}_{k-1}\mathbf{G}_k^T \quad (7)$$

As $\mathbf{x}_{k-1}$ is used to correct the INS states, the first 9 states of $X_{k,k-1}$ are all zeros:

$$\mathbf{x}_{k,k-1} = [\mathbf{0}_{9,9};\mathbf{I}_{6,6}]\times\mathbf{f}_k(\mathbf{x}_{k-1}) \quad (8)$$

Finally the Kalman gain, state and covariance update are evaluated:

$$\mathbf{K}_k = \mathbf{P}_{k,k-1}\mathbf{H}_k^T[\mathbf{H}\mathbf{P}_{k,k-1}\mathbf{H}^T + \mathbf{R}_k]^{-1} \quad (9)$$

$$\mathbf{P}_{k,k} = [\mathbf{I} - \mathbf{K}_k\mathbf{H}_k]\mathbf{P}_{k,k-1} \quad (10)$$

$$\mathbf{x}_k = \mathbf{x}_{k,k-1} + \mathbf{K}_k(\mathbf{z}_k - \mathbf{H}\mathbf{x}_{k,k-1}) \quad (11)$$

3 PSO-RBFN Mechanism Design for Function Mapping with Low Pass Filtering

3.1 Particle Swarm Optimization (PSO)

Particle swarm optimization (PSO) is a population based stochastic searching technique developed by Kennedy and Eberhart in 1995 [9]. Similar to Genetic Algorithm (GA), PSO is also a stochastic searching optimization technology. Unlike the drawback of expensive computational cost of GA, PSO has better convergence speed. A *swarm* consists of a set of particles moving around the search space, each representing a potential solution (fitness). Each particle has a position vector (x_i), a velocity vector (v_i), the position at which the best fitness ($pbest_i$) encountered by the particle, and the index of the best particle (*gbest*) in the swarm. In each generation, the velocity of each particle is updated to their best encountered position and the best position encountered by any particle using (1).The position of each particle is updated every generation. This is done by adding the velocity to the position vector

$$v_i = w\times v_i + C_1\times rand()\times(Pbest_i - x_i) + C_2\times rand()\times(Gbest - x_i) \quad (12)$$

$$x_i = x_i + v_i \quad (13)$$

The parameters C_1 and C_2 are set to constant values, which are normally taken as 2 whereas rand() represent uniformly distributed random values, uniformly distributed in [0, 1] and w is called as inertia weight, the inertia weight is employed to control the impact of the previous history of velocities on the current one. Accordingly, the parameter regulates the trade-off between the global and local exploration abilities of the swarm. A large inertia weight facilitates global exploration (searching new areas), while a small one tends to facilitate local exploration. Fig. 2 shows the flow chart for the PSO algorithm.

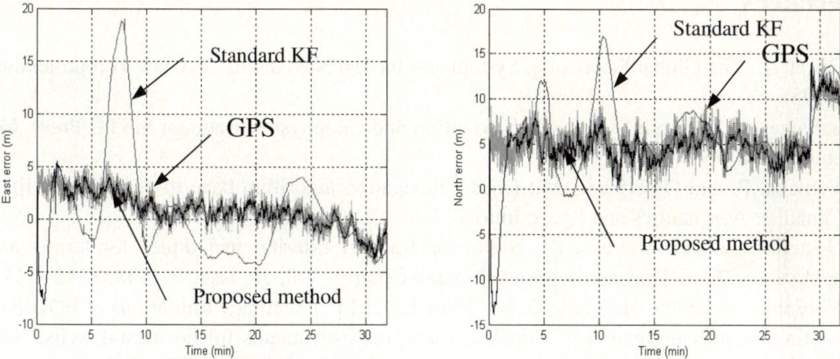

Fig. 6. Positioning errors – without differential correction

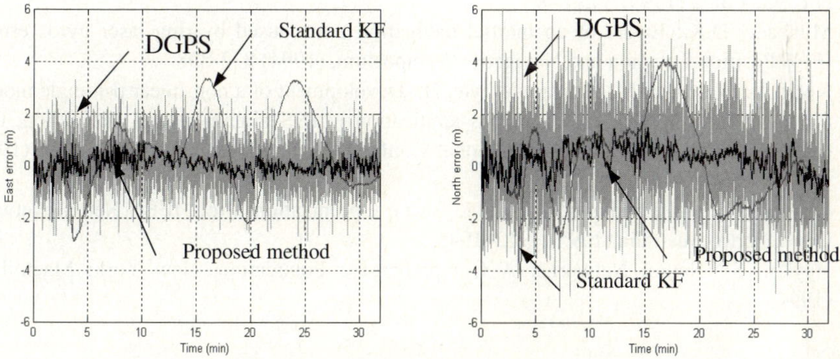

Fig. 7. Positioning errors – DGPS mode

5 Conclusions

Incorporation of PSO-RBFN mechanism into the Kalman filter design has been proposed. The PSO-RBFN is employed as a model nonlinearity identification mechanism for implementing the on-line identification of nonlinear dynamics errors such that the unknown plant is identified and the modeling error can be compensated. The proposed method has been applied to the loosely-coupled GPS/INS navigation filter design. The proposed PSO-RBFN mechanism is able to estimate the Kalman filter parameters and therefore improve the positioning performance. The application for GPS/INS navigation filter designs using the proposed approach has demonstrated significant performance improvement in comparison with the standard Kalman filtering method.

Acknowledgement

This work has been supported in part by the National Science Council of the Republic of China under grant no. NSC 94-2212-E-019-003.

References

1. Farrell, J.: The Global Positioning System and Inertial Navigation, McCraw-Hill professional. (1998)
2. Salychev, O.: Inertial Systems in Navigation and Geophysics, Bauman MSTU Press, Moscow. (1998)
3. Vanicek, P. and Omerbasic, M.: Does a navigation algorithm have to use Kalman filter?, Canadian Aeronautics and Space Journal, Vol. 45, No. 3, (1999)
4. Hosteller, L. and Andreas, R.: Nonlinear Kalman filtering techniques for terrain-aided navigation, IEEE Transactions on Automatic Control, Vol. 28, Issue 3 (1983) 315 -323.
5. Noureldin, A. Irvine-Halliday, D. and Mintchev, M.P.: Accuracy limitations of FOG-based continuous measurement-while-drilling surveying inslruments for horizontal wells," IEEE Transactions on Instlumentation and Measurement, 51(6), (2002) I177 ~ 1191
6. Lobo, J. Lucas, P. Dias, J. and Tram de Almeida, A.: Inertial navigation system for mobile land vehicles," Proceedings of the IEEE International Symposium on industrial electronics N E '95, Vol. 2 (1995) 843-848
7. Mynbaev, D.K.: Errors of an inertial navigation unit caused by ring laser gyros errors, IEEE Position Location and Navigation Symposium, (1994) 833-838
8. Kong, X., Nebot, E. M., Durrant-Whyte, H.: Development of a non-linear psi-angle model-for large misalingment errors and its application in INS alignment and calibration, Proceedings of the 1999 IEEE International Conference on Robotics & Automation, (1999) 1430~1435
9. Kennedy, J., and Eberhart, R.: Particle Swam Optimization, Proc. IEEE Neural Network Conf., Perth, Australia (1995) 1942–1945
10. Haykin, S.: Neural Networks: A Comprehensive Foundation. New York: Macmillan, (1994)

An Adaptive Image Segmentation Method Based on a Modified Pulse Coupled Neural Network

Min Li[1], Wei Cai[1], and Xiao-yan Li[2]

[1] Xi'an Research Inst. of Hi-Tech Hongqing Town, 710025, Shaanxi Province, P.R.C.
limin@mailst.xjtu.edu.cn

[2] Academy of Armored Force Engineering Department of Information Engineering, 100082, Beijing, P.R.C.

Abstract. Pulse-coupled neural network (PCNN) based on Eckhorn's model of the cat visual cortex has great significant advantage in image segmentation. However, the segmented performance depends on the suitable PCNN parameters, which are tuned by trial so far. Focusing on the famous difficult problem of PCNN, this paper establishes a modified PCNN, and proposes adaptive PCNN parameters determination algorithm based on water region area. Experimental results on image segmentation demonstrate its validity and robustness.

1 Introduction

The studies of the applications of PCNN, which was derived from the neuron model of Echorn's, hadn't been carried out until 1990's. Because the algorithm of PCNN model derives directly from the studies of visual properties of the mammal, it is very suitable for some applications, such as image segmentation, image smoothness, and noise reduction [1,2].

Up to now, the parameters are most adjusted manually and it is a difficult task to determine PCNN parameters automatically. During recent years, some work on determining the optimal values of PCNN parameters has been done [3,4,5,6,7]. Based on the research fruits, we establish a modified PCNN model and propose a multi-threshold approach according to water region area in histogram to determine PCNN parameters image segmentation automatically.

2 PCNN Neuron Model

As showed in Fig.1, each PCNN neuron is divided into three compartments with characteristics of the receptive field, the modulation field, and the pulse generator (see Fig. 1). The receptive field is comprised of feeding field and linking field. Let N_{ij} denote the ijth neuron. The feeding field of N_{ij} receives the external stimulus S_{ij} and the pulses Y_{kl} from the neighboring neurons. A signal denoted by F_{ij} is outputted. The linking field receives the pulses from the neighboring neurons and outputs the signal denoted by L_{ij}. In the modulation field, F_{ij} and L_{ij} are inputted and modulated. The modulation result U_{ij} called internal activity signal is sent to the spike generator, in which U_{ij} is compared with the dynamic threshold θ_{ij} to form the pulse output Y_{ij}.

L. Jiao et al. (Eds.): ICNC 2006, Part I, LNCS 4221, pp. 471 – 474, 2006.

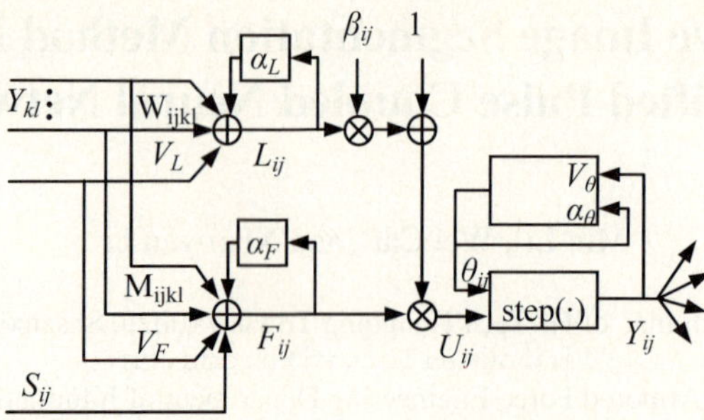

Fig. 1. PCNN neuron model

In the feeding field and linking field, there are six parameters, i.e., three time decay constants (α_F, α_L, α_θ) and three amplification factors (V_F, V_L, V_θ). M and W are the linking matrix, and normally W=M, $\beta_{i,j}$ is the linking coefficient, *step*(•) is the unit step function.

3 Adaptive Parameters Determination Based on Modified PCNN

3.1 Multi-threshold Approach Using Water Region Area Method

We propose a novel auto-detection algorithm of thresholds in image histogram. In accordance with the intuitionistic features of the histogram, the peaks of the histogram are considered as watersheds, each valley including two neighboring peaks and a bottom points. We call the maximal water capacity in each valley as 'water region area'.

Step1.Draw image histogram and smooth it to decrease noise influence if necessary

Step2. Seek all peaks and bottom points in the histogram.

Step3. Calculate the water region area from the left bottom point. Here, define Θ as a lower limitation ranging from 0.01 to 0.03. The smaller the value of Θ is, the more threshold points we will get. When the water region area is larger than Θ, the corresponding bottom point will be kept in threshold array T_m. Meanwhile, the corresponding left side peak point will be kept in peak points array P_m. Otherwise, the valley will be taken as invalid. At this situation, the larger of the two peaks located in the valley's two side will be treated as the new left peak point.

Step4. Iteratively execute step 3 until all bottom points have been processed and then we can get the threshold array T_m (m=1,…M and $T_1<\ldots<T_M$) and the corresponding peak array P_m(m=1,…M+1 and $P_1<\ldots<P_{M+1}$).

Fig.2(c) in Section 4 shows image together with its water regions and corresponding thresholds.

3.2 Modified Pulse Coupled Neural Network

We have established a modified PCNN, which is implemented by applying iteratively the equations

$$L_{ij}[n]=\sum W_{ijkl}Y_{kl}[n-1] \tag{1}$$

$$U_{ij}[n] = S_{ij}(1 + \beta_{i,j}[n]L_{ij}[n]) \tag{2}$$

$$Y_{ij}[n] = \begin{cases} 1, & U_{ij}[n] > T_{ij}[n] \\ 0, & \text{otherwise.} \end{cases} \tag{3}$$

The indexes i and j refer to the pixel location in the image, indexes k and l refer to the dislocation in a symmetric neighborhood around a pixel, and n refers to the time (number of iteration). $L_{ij}[n]$ is linking from a neighborhood of the pixel at location (i,j), $U_{ij}[n]$ is internal activity at location (i,j) , which is dependent on the signal value S_{ij} (pixel value) at (i,j) and linking value. $\beta_{i,j}[n]$ is the PCNN linking parameter, and $Y_{ij}[n]$ is the output value of the PCNN element at (i,j). $T_{ij}[n]$ is a threshold value. We use a set of fixed threshold values, $T_m(m=1,\ldots M)$ determined by water region area method mentioned above.

If $Y_{ij}[n]$ is 1 at location (i,j) at $n=t$, we say that the PCNN element at the location (i,j) fires at t. The firing due to the primary input S_{ij} is called the natural firing. The second type of firing, which occurs mainly due to the neighborhood firing at the previous iteration, we call the excitatory firing, or secondary firing.

Considering those pixels whose intensities are smaller than peak point P_m ought not to be captured at T_m even if they have the largest linking value 1, so in the iteration loop at T_m, the value of β_m is chosen to be

$$\beta_m = \frac{T_m}{P_m} - 1 \tag{4}$$

Because P_1 may be 0, we choose the value of β_1 to be 0.1-0.3 at this situation.

4 Experiments

Extensive experiments has been performed and Fig.2 shows one example. The propose approach to determine multi-thresholds by water region area are shown in Fig.2(c). The segmented images by traditional PCNN and modified PCNN are shown in Fig.2(b) and Fig.2 (d) respectively. Table.1 shows the parameters of traditional PCNN which were tuned by trial in order to get perfect segmentation performance. Comparing with the visual segmented effect and automatically parameters determination method, we can see our proposed method outperforms traditional PCNN and it is very important in expanding the application range of PCNN.

(a) lab image

(b) traditional PCNN segmentation result

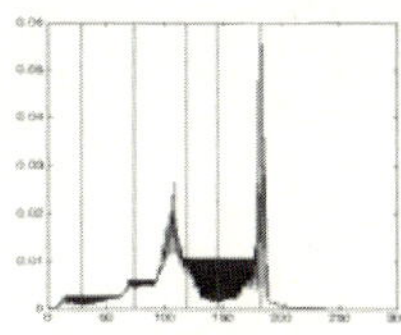

(c) water regions and multi-thresholds

(d) modified PCNN segmentation result

Fig. 2. The image (256-level, size of 640×480) and segmented result

Table 1. Values of parameters in 'lab' image segmentation by traditional PCNN

parameters	β	α_F	α_L	α_θ	V_F	V_L	V_θ	r	N
value	0.3	1	1	2.5	10	10	100	1	2

5 Conclusion

In order to determine PCNN parameters adaptively, this paper brings forward an adaptive segmentation algorithm based on a modified PCNN with the multi-thresholds determined by water region area method. Experimental results show its good performance and robustness. Furthermore, the proposed approach expands the application range of PCNN.

References

1. Eckhorn R., ReitBoeck H.J., et al.: Feature linking via synchronization among distributed assemblies: simulation of results form cat visual cortex. Neural Computation, Vol. 2. (1990) 293–307
2. Kuntimad G., Ranganath H.S.: Perfect image segmentation using pulse coupled neural networks. IEEE Trans. Neural Networks, Vol.10. (1999) 591–598
3. Ma Y.D., Dai R.L., Li L.: Automated image segmentation using pulse coupled neural networks and image's entropy. Journal of China Institute of Communications, Vol.23. (2002) 46–51
4. Liu Q., Ma Y.D., Qian ZH.B.: Automated image segmentation using improved PCNN model based on cross-entropy. Journal of Image and Graphics, Vol.10. (2005) 579–584
5. Gu X.D., Guo Sh.D, Yu D.H.: A new approach for automated image segmentation based on unit-linking PCNN. Proceedings of the first International Conference on Machine learning and Cybernetics, Beijing, China, (2002) 175–178
6. Karvonen J.A.: Baltic sea ice SAR segmentation and classification using modified pulse-coupled neural networks. IEEE Trans. Geoscience and Remote Sensing, Vol.42. (2004) 1566–1574
7. Bi Y.W., Qiu T.SH.: An adaptive image segmentation method based on a simplified PCNN. ACTA ELECTRONICA SINICA, Vol.33. (2005) 647–650

A Edge Feature Matching Algorithm Based on Evolutionary Strategies and Least Trimmed Square Hausdorff Distance

Li JunShan[1], Han XianFeng[1], Li Long[1], Li Kun[2], and Li JianJun[1]

[1] Xi'an Research Inst. Of High-tech Hongqing Town, 710025 Xi'an, China
lijunshan403@163.com

[2] Institute of Intelligent Information Processing, Xidian University, 710000 Xi'an, China

Abstract. Aimed at problems of low orientation precision of traditional gray correlation matching and bad real-time feature based on partial hausdorff distance matching, a edge feature matching algorithm based on evolutionary strategies and least trimmed square hausdorff distance is presented. Experiments show that it has good matching effect.

1 Introduction

Scene matching orientation guidance is an important method of improving long-distance missile hitting precision. Due to the difference of shooting height and time between reference image and real-time image, it can exist gray transformation and scale error. Edge feature is one of important image features. Different images may exist difference in the same region, but edge feature isn't in the influence of gray and has strong stability. Also, edge detecting method is maturate, has low noise sensitivity and high steadiness, and has high extraction speed. So it can get good matching effect.

Evolutionary strategies (ES) model and least trimmed square HD (LTS-HD) measure is introduced in this article. Aimed at real-time image existing rotate transformation, a matching algorithm based on evolutionary strategies and LTS-HD is presented. It retains good matching effect and can fulfill the need of real-time feature.

2 ES Model

ES is a algorithm which simulate nature evolution principle to resolve parameter optimizing problem. ES model[1] can be described as follows:

(1) The problem is defined to seek the n dimension vector $\mathbf{x}$ correlated to function's extremum, $F(\mathbf{x}): R^n \to R$.

(2) Choose the original colony of farther vector randomly from each possible range.

(3) Filial generation is produced by joining a Gaussian random variable of zero mean square deviation and choosing standard deviation of $\mathbf{x}$ in advance into its farther vector $\mathbf{x}_i (i = 1, 2, \ldots, p)$.

(4) Decide to retain which vectors according to sorting errors $F(x_i')(i = 1, ..., p)$. Vectors those owns the least deviation become next new farther generation.

L. Jiao et al. (Eds.): ICNC 2006, Part I, LNCS 4221, pp. 475–478, 2006.

In practice, when searching space is too huge, ES can't enumerate all local optimal answers, and only can choose the best one from several local optimal answers as the approving answer. An improved compositing ES is presented which can choose the individual owning the best adaptive feature as the approving answer.

3 Least Trimmed Square Hausdorff Distance

The HD measure computes the max and min distance value between two sets of points. It describes the resemblance between them. Because the HD measures the distance of the worst matching point between two point sets, it is very sensitive to noise points and leak points far away from the center. Also the sensitivity is unavoidable when extracting feature point set from image. To overcome the shortcoming, Sim et al proposed LTS-HD[4]. The $h_{LTS}(A,B)$ is defined by a linear combination of order statistics

$$h_{LTS}(A,B) = \frac{1}{k}\sum_{i=1}^{k} \min \| a - b \| (i) \tag{1}$$

The measure $h_{LTS}(A,B)$ is minimized by remaining distance values after large distance values are eliminated. In the proposed algorithms, a full search is adopted for finding the optimal matching location, that is, the point that yields a minimum distance is selected because the proposed algorithm are based on a distance measure.

4 Matching Algorithm Based on ES and LTS-HD

Scene matching algorithm based on LTS-HD can retain good matching results, additionally ES is overall optimized searching method, so introducing ES into matching algorithm based on LTS-HD can improve the matching effects obviously. In this paper matching parameters are chosen as follows:

(1) Generation of original colony

ES adopts real number encoding. Original colony consists of μrandomly generated individuals. μindividuals is chosen equally from searching space. Each source individual $X = \{x, \sigma\}$ is composed of two parts: target variableness x and standard deviation σ :

$$X = \{x, \sigma\} = ((x_1, x_2, x_3), (\sigma_1, \sigma_2, \sigma_3)) \tag{2}$$

(2) Evolution operation in ES

In order to accelerate the speed of the algorithmic, the ES here has no crossing operation, instead, it uses individual mutation to generate new filial generation individual. We use the improved mutation function to realize the evolution of the whole colony.

$$\begin{cases} \sigma'(x_i) = \sigma(x_i) \cdot \exp(N(0,1)) \\ X' = X + N(0, \sigma') \end{cases}, \quad 1 \le i \le 3 \tag{3}$$

Here N (0, 1) is a random variable of normal distribution which mean value is 0, and square deviation value is 1.

(3) Calculation of adaptation degree

Adaptation degree is defined as weighing the superiority and inferiority of individuals. Here we use LTS-HD as adaptation degree of edge feature matching algorithm.

(4) Ending conditions of evolution

Ending conditions of evolutionary course usually have two formations: threshold ending and maximized evolutionary algebra ending. In scene matching, threshold τ is different according to different feature and complexity of images.

(5) Choice of matching position

Calculate the reverse LTS-HD of those optimal matching points which are found out by ES. The minimal distance is matching point.

5 Experiments Results and Analysis

Main parameters in experiments are: μ=5, λ=30, matching parameters are $f_1 = f_2 = 0.85$. Experimental results are show as Fig.1, where reference image is optic image (Fig.1 (a), size: 256×256), and real-time image is daytime infrared image existing ten degrees rotating transformation relative to reference image (Fig.1 (b), size: 64×64). Fig.2 shows edge binary feature images corresponding to Fig.1.

(a) Reference image

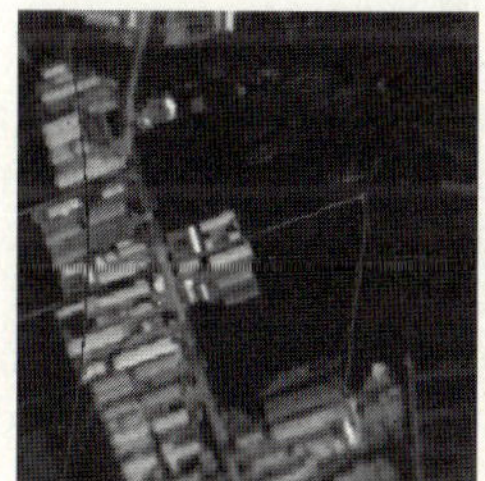

(b) Real time image (infrared)

Fig. 1. Experiment image

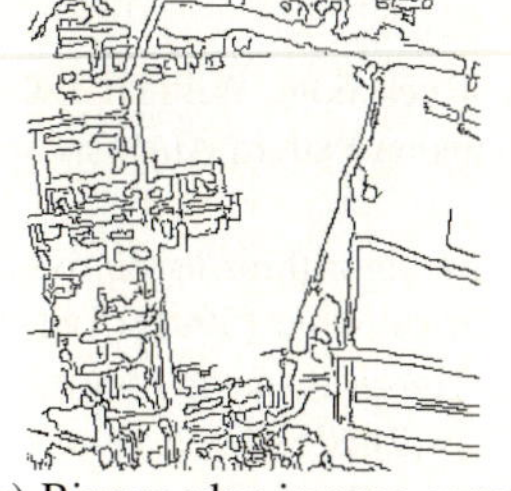

(a) Binary edge images corresponding to reference image

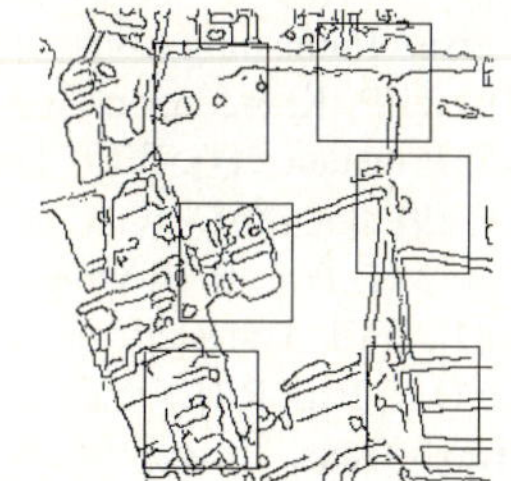

(b) Binary edge images corresponding to real-time image

Fig. 2. Binary edge images corresponding to Fig.1

Table 1. Compare of matching performances

Real-time image No.	Accurate matching position	Partial HD		Algorithm in this paper		
		Matching position	Matching time(/s)	Matching position	Matching time(/s)	Thres hold τ
1	(60,8)	(190,32)	40.21	(61,10)	3.32	4
				(60,8)	4.51	6
2	(63,85)	(61,84)	58.73	(62,87)	4.11	4
				(63,85)	5.62	6
3	(41,161)	(31,72)	52.97	(41,161)	3.56	4
				(41,161)	4.99	6
4	(168,13)	(170,10)	43.36	(168,13)	3.31	4
				(168,13)	4.68	6
5	(169,84)	(167,86)	38.04	(169,84)	3.41	4
				(169,84)	4.78	6

By analyzing the contrast between matching algorithm based on ES and LTS-HD in this paper and that based on partial HD measure, the experimental results are shown in Tab.1. It can be seen that, the real-time feature and precision in this paper are both much better than that of partial HD measure.

6 Conclusion

Aimed at problems of low orientation precision of traditional gray correlation matching and bad real-time feature based on partial hausdorff distance matching, through improving ES and partial HD measure, an edge feature matching method based on ES and LTS-HD is presented. The results of experiment show that orientation precision and real-time feature retained from this paper's algorithm are obviously improved. It's an effective scene matching method.

References

1. Cai Zi-xing, Xu Guang-you. Artificial Intelligence And Its Application [M]. Beijing: TsingHua University Publishing Com., 2004.8.
2. Huttenlocher Daniel P, Klanderman Gregory A, Ruck lidge William J. Comparing images using the hausdorff distance [J]. IEEE Transactions on Pattern Analysis and Machine Intelligence, 1993, 15 (9): 850~863.
3. Sim D G, Kwon O K, Park R H. Object matching algorithms using robust Hausdorff distance measures [J]. IEEE Transactions on Image Processing, 1999, 8(3):425~429.
4. Sim D G, Kwon O K, Park R H. Object matching algorithm using robust Hausdorff distance measures[J]. IEEE Trans. Image Process, 1999, 8(2):425~429.5. Wong R.Y., Hall E.L., Scene Matching with Invariant Moments, CGIP(1978)(8):16~24.
5. H.K.Sardana, M.F. Daemi , M.K.lbrahim, Global Description of Edge Patterns using Moments, Pattern Recognition(1994)27(1).

Least Squares Interacting Multiple Model Algorithm for Passive Multi-sensor Maneuvering Target Tracking

Liping Song and Hongbing Ji

School of Electronic Engineering, Xidian Univ., Xi'an, China
lpsong@lab202.xidian.edu.cn

Abstract. In bearings-only passive target tracking, the state of the target has a nonlinear relation with the bearings measurements. Existing methods are mainly focus on the process of linearization. However, in this process, precision decreasing is obviously unavoidable and even filter divergence will be occur so as to losing the target. Therefore a new algorithm is proposed in the paper. The state of the target is approximately estimated by least squares at first which is taken as pseudo measurements for kalman filter, and then IMM algorithm is employed for maneuvering target tracking.

1 Introduction

In recent years, passive localization and tracking are widely studied. There mainly include passive Direction Of Arrival (DOA) system, passive Time Difference Of Arrival (TDOA) system [5] and bearings-only passive systems. One of the bearings-only passive systems utilizes one passive sensor which maneuvers to track a target [4], the other is constructed by multiple passive sensors which associate the measurements from multi-sensor for target tracking [1, 8].

In maneuvering target tracking, the behavior of a target cannot be characterized at all times by a single model, but a finite number of models can adequately describe its behavior in different regimes, based on which Interacting Multiple Model (IMM) algorithm [6, 7] is proposed and has many successful applications. Existing applications of the IMM algorithm have emphasized the case of radars (active sensors). The case of a passive sensor was first considered in [3] where the azimuth and elevation measured by an infrared search and track (IRST) sensor was fused with the range and azimuth measured by a radar. Tracking a 3D maneuvering target with two infrared sensors is studied in [2] where the bearings measurements are linearized firstly and IMM is applied to tracking a maneuvering target. However, in the process of linearization, precision decreasing is obviously unavoidable. To avoid the computational complexity and the precision decrease from the nonlinear feature in bearings-only target tracking, the state of the target is approximately estimated by least squares algorithm at first and then the IMM is employed in this paper.

2 Least Squares for Passive Multiple Sensors

The elevation and azimuth measured by one passive sensor may determine a line of sight (LOS). Without noise, the lines of sight from multiple sensors would meet with

L. Jiao et al. (Eds.): ICNC 2006, Part I, LNCS 4221, pp. 479 – 482, 2006.

a point which is the position of the target. However, with the noisy measurements, they would not meet with a point. Least squares can estimate the position of the target [8]. It estimates a point which has minimum sum of the distance to these lines of sight as the position of the target.

3 Bearings-Only Multi-sensor Maneuvering Target Tracking

For bearings-only multi-sensor maneuvering target tracking, if kalman filter or IMM algorithm is directly applied by utilizing the bearings measurements from the passive sensors, it need solve the nonlinear problem unavoidable. The process of linearization would result in the computational complexity, the precision decrease and even divergence so as to lose the target. Therefore a new algorithm is proposed here. The state of the target is approximately estimated by least squares at first which is taken as pseudo measurements for kalman filter, and then IMM algorithm is employed for maneuvering target tracking. The block of the proposed algorithm is shown in fig. 1.

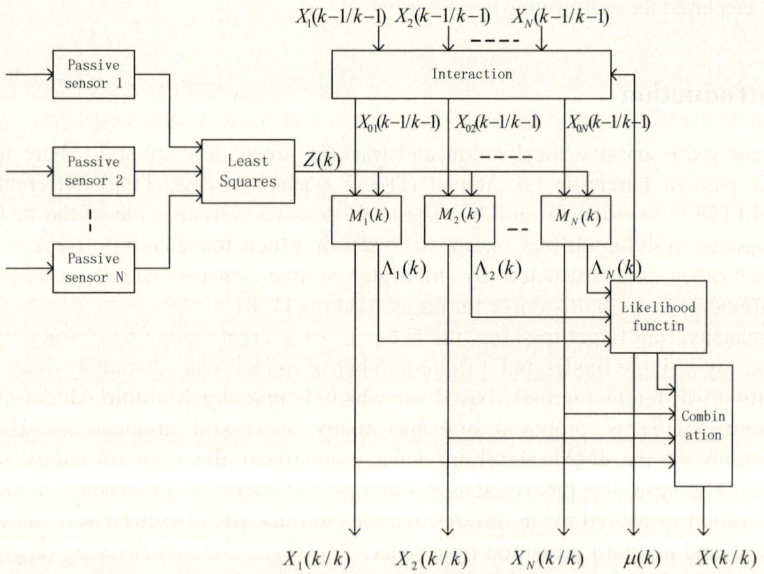

Fig. 1. The block of the least squares interacting multiple model algorithm

Assume a set of models is M . A jump linear fixed-structure hybrid system with mode transition modeled by a semi-Markov process, such as the state-dependent Markov chain, can be described by the equations

$$X(k+1) = \Phi_j(k)X(k) + \Gamma_j(k)v_j(k) \quad \forall j \in M \tag{1}$$

$$Z(k) = H_j(k)X(k) + w_j(k) \quad \forall j \in M \tag{2}$$

Where Γ_j is the process noise gain matrix. v_j and w_j are the mode-dependent process and measurement noise sequences with mean $\overline{v}_j$ and $\overline{w}_j$ and covariance Q_j and R_j.

[6, 7] give a reference to the basic IMM algorithm. In the IMM algorithm, the measurements Z are substituted by the estimate of the least squares and the covariance of measurement noise R is substituted by the variance of the estimate error meanwhile. It is the key of the proposed algorithm.

4 Simulation

Turn models are employed here. Assume a target flight in a plane parallel horizontal and maneuvers in the plane. The initial position and the velocity of the target are [2km, 10km, 3km] and [172m/s, -246m/s, 0]. The interval of samples is T=1s. In the period of first 20 samples it flight by constant velocity of 300m/s. From 21 to 45 samples, it turns left by the turn velocity of 3.74°/s and the acceleration of 2g. From 46 to 60, it flight by constant velocity again, and then turns right by the turn velocity of 3.74°/s and the acceleration of 2g lasting 20 samples and change the acceleration to 0.8g for 10 samples. At last it flight from 91 to 100 samples at constant velocity.

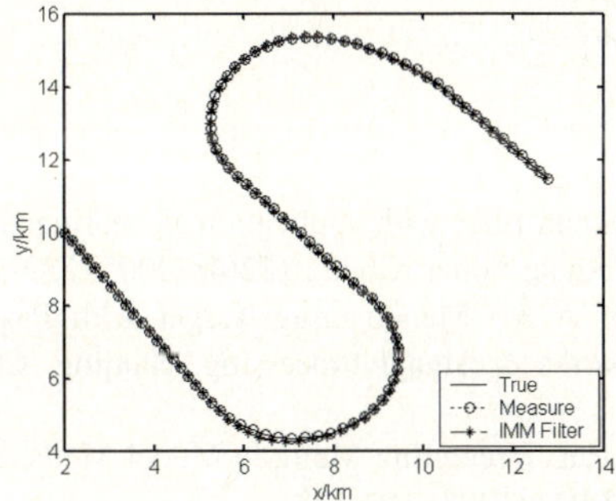

Fig. 2. Tracking result of least squares interacting multiple model algorithm in x-y plane

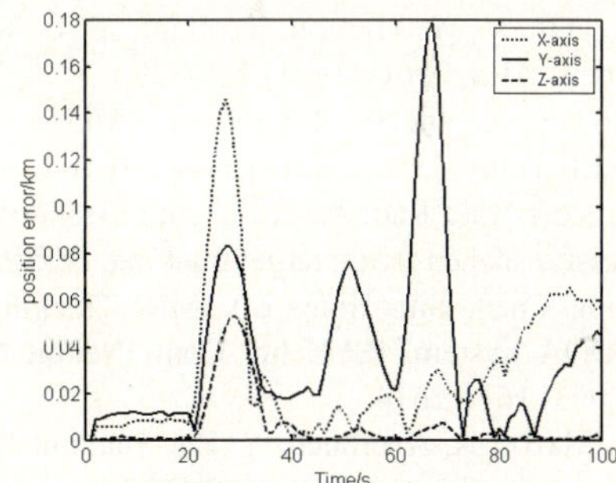

Fig. 3. Position error at three axes with LS-IMM algorithm

Three models are adopted here, corresponding to ω=0, 1g, 2g respectively. Markov transition probability matrix is

$$p = \begin{bmatrix} 0.9 & 0.05 & 0.05 \\ 0.05 & 0.9 & 0.05 \\ 0.05 & 0.05 & 0.9 \end{bmatrix}$$

The initial probability of three models are $\mu_1(0) = 0.9$, $\mu_2(0) = 0.05$ and $\mu_3(0) = 0.05$. The initial covariance is set P(0)= diag($10^{-6} \times$ [400,100,25,400,100,25, 400,100,25]), and the variance of the process noise of different models are 5m/s^2,

$10m/s^2$ and $20m/s^2$. The standard variance of the bearings measurements is $\sigma = 10^{-4} rad$.

The target tracking result of least squares interacting multiple model (LS-IMM) algorithm in *x-y* plane is shown in fig. 2, and the position error at three axes are shown in fig. 3.

5 Conclusion

A new algorithm is proposed in the paper. Firstly the state of the target is estimated from the bearings measurements acquired from the multiple passive sensors by utilizing the least squares algorithm, and then IMM algorithm is employed for tracking a maneuvering target. It avoids the process of linearization, so it can acquire a high precision in maneuvering target tracking.

References

1. Xiu Jian-Juan, He You, Xiu Jianhua: Bearing Measurements Association Algorithm in Passive Location System, Int. Conf. Computational Electromagnetics and Its Applications Proceedings, (2004) 356-359.
2. F. Dufour, M. Mariton: Tracking a 3D Maneuvering Target with Passive Sensors, IEEE Trans. AES, **27(4)** (1991) 725-739.
3. Houles, A., and Bar-Shalom, Y.: Multisensor tracking of a maneuvering target in clutter, IEEE Trans. AES, **25(2)** (1989) 176-188.
4. S. Koteswara Rao: Modified gain extended Kalman filter with application to bearings-only passive maneuvering target tracking, IEE Proc. Radar Sonar Navig., **152(4)** (2005) 239-244.
5. Ling Chen, ShaoHong Li: IMM Tracking Of A 3D Maneuvering Target with Passive TDOA System, IEEE Int. Conf. Neural Networks & Signal Processing, Nanjing, China (2003) 1611-1614.
6. E. Mazor, A. Averbuch, Y. Bar-Shalom, J. Dayan: Interacting Multiple Model Methods in Target Tracking: A Survey, IEEE Trans. AES, **34(1)** (1998) 103-123.
7. Blom H A, Bar-Shalom. Y.: The Interacting Multiple Model Algorithm for Systems with Markovian Switching Coefficient, IEEE Trans. AC, **33(8)** (1988) 780~783.
8. Song Li-ping, Ji Hong-bing, Gao Xin-bo: Least squares adaptive algorithm for bearings-only multi-sensor maneuvering target passive tracking, Journal of Electronics & Information Technology, **27(5)** (2005) 793-796. (In Chinese)

Multiple Classifiers Approach for Computational Efficiency in Multi-scale Search Based Face Detection

Hanjin Ryu, Seung Soo Chun, and Sanghoon Sull

Department of Electronics and Computer Engineering, Korea University, 5-1 Anam-dong, Seongbuk-gu, Seoul, 136-701, Korea
{hanjin, sschun, sull}@mpeg.korea.ac.kr

Abstract. The multi-scale search based face detection is essential to use a window scanning technique where the window is scanned pixel-by-pixel to search for faces in various positions and scales within an image. Therefore, detection of faces requires high computation cost which prevents from being used in real time applications. In this paper, we present face detection approach by using multiple classifiers for reducing the search space and improving detection accuracy. We design three face classifiers which take different feature representation of local image[1]: gradient, texture, and pixel intensity features. The designed three face classifiers are trained by error back propagation algorithm. The computational efficiency is achieved by coarse-to-fine classification approach. A coarse location of a face is first classified by the gradient feature based face classifier where the window is scanned in large moving steps. From the coarse location of a face, the fine classification is performed to identify the local image as a face where the window is finely scanned. In fine classification, the output of each face classifier is combined and then used for a reliable judgment on the existence of face. Experimental results demonstrate that our proposed method can significantly reduce the number of scans compared to the exhaustive full scanning technique and provides the high detection rate.

1 Introduction

The detection of face in an image has been intensively studied and a wide variety of techniques have been proposed so far. Among various face detection methods, image based methods recognize face patterns by classifying a local image within a fixed size window into face and non-face prototype classes using statistic models, such as neural network [1, 2, 3], support vector machine [4, 5] and principal components analysis [6]. In order to find faces in various scales and positions within an image, the fixed size window is scanned at all positions for a pyramid of image that is obtained by subsampling the input image. Therefore, the fixed size window that is the basis unit for classifying a face is scanned for multiple images at various scales. Since the fixed size window is exhaustively scanned to find face in images at various resolutions, this method is often referred to as the multi-scale search technique.

Although multi-scale search based face detection methods can provide high detection accuracy on low quality images, they require high computational cost. Therefore,

[1] For convenience, the image within a scanning window is called a local image.

L. Jiao et al. (Eds.): ICNC 2006, Part I, LNCS 4221, pp. 483 – 492, 2006.

in order to reduce the computational cost accompanied by the window scanning procedure on the whole image, some approaches [7, 8] use skin color or object motion to provide prior information on the estimate location of face. Although these approaches can reduce much the computational cost, they cannot be applied to gray scale and static images.

In order to overcome above limitations, coarse-to-fine search approaches have been proposed. The method in [3] proposed a two-stage scheme to overcome the problem of exhaustive full search. In the first stage, a candidate face classifier was used to quickly discard non-face regions, and in the second stage a more complex classifier was used to perform final classification on the local image that passed from first stage successfully. However, the detection rate is lower than the full search process.

Approaches proposed in [9, 10] made use of grid based search method. In each sub-sampled image, each intersection point of a regular grid was tested by a face classifier. If the output value of the face classifier at the intersection points of a grid was greater than a threshold value, the fine search can be started around those points. The grid based search method heavily relies on the grid step.

In this paper, we present multiple classifiers based face detection approach. The multiple face classifiers, which are taken a different feature such as gradient, texture and pixel intensity, are designed to reduce the computational cost while maintaining the high detection rate.

For computational efficiency, we also use coarse-to-fine classification approach. The coarse-to-fine classification is based on improvement of window scanning process which is achieved by increasing the moving step of scanning window. The sub-optimal moving step of scanning window is empirically determined by the sensitivity analysis of each face classifier. Especially, the translation invariant property of adopted gradient feature contributes to improvement of the scanning process in coarse classification stage. The gradient based face classifier is used to find the coarse location of a face where the window is scanned in large moving step. Then, the local image is identified as a face using multiple face classifiers where the window is finely scanned.

The rest of this paper is organized as follows. Section 2 gives an overview of the proposed system. Section 3 addresses the feature representation for multiple face classifiers. Section 4 presents face detection based on coarse-to-fine classification. In order to demonstrate the effectiveness of proposed approach, the experimental results are provided in Section 5. In Section 6, the concluding remarks are drawn.

2 System Overview

The proposed face detection approach is based on multiple classifiers which are composed of three face classifiers. Each face classifier is trained by error back propagation algorithm and taken a different feature representation such as gradient, texture and pixel intensity.

Fig. 1 illustrates the proposed overall system architecture. A pyramid of multi-resolution of the input image is obtained by a scaling factor 1.2. Before classification, each local image is converted to gray image and then pre-processed to reduce the intensity variation. In pre-processing step, a face mask is applied to remove any piece of the background image. Subsequently, the intensity normalization which consists of a correct lighting [1] and histogram equalization is used to alleviate the variation of lighting condition within local image. After pre-processing, the features of the local image are extracted and then passed to each face classifier. The each face classifier returns a result between 0.0 and 1.0.

For computational efficiency, the coarse-to-fine classification which is based on improvement of window scanning process is utilized. In order to find coarse location of a face, the window is scanned in large moving steps and the local image that might contain a face is examined by the gradient based 1st face classifier. From the coarse location of a face, the other face classifiers identify the local image as a face where the window is finely scanned. As a confidence measure for identifying a face, we apply a weighted sum of the output values from two other face classifiers including 1st face classifier. The identified regions in each scale are mapped back to the input image scale.

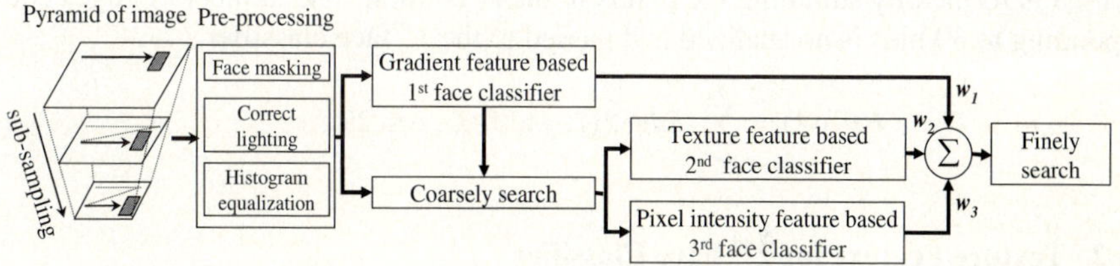

Fig. 1. The overall system architecture based on multiple face classifiers

3 The Feature Representation for Multiple Face Classifiers

Since mixture of various classifiers may give more reliable judgment for a face than using only single classifier, we design the multiple face classifiers which are taken different representations of face patterns. The employed gradient and texture features are represented for global face appearance and the pixel intensity feature is for local face appearance.

3.1 Gradient Feature for 1st Face Classifier

The 1st face classifier is based on gradient feature obtained from the horizontal gradient projection [11]. As shown in Fig. 2, the gradient feature contains the integral information of the pixel distribution, which retains certain invariability among facial features. It is noticeable that the positions of facial features are quite stable even under translating the face center regardless of different amount of gradient strength. This fact provides a clue to improve the window scanning process. That is, if the center of the window falls within permissible bound from the center of face, the 1st face classifier may identify a local image as a face pattern. Therefore, the determination of the permissible bound is a main problem of improving window scanning process, and the solution is described in detail in section 4.

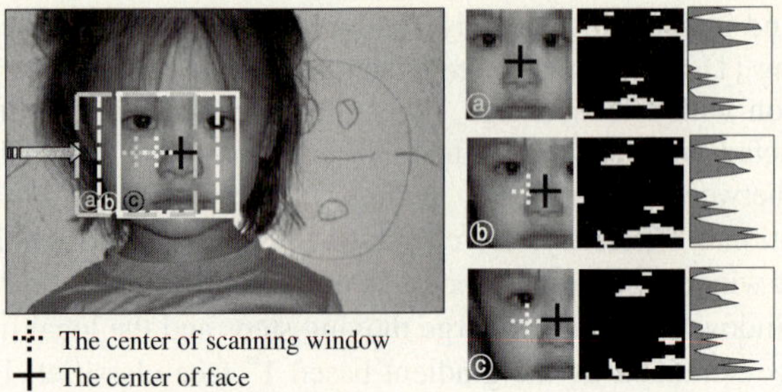

Fig. 2. The gradient feature's characteristic and its translation invariant property

In order to obtain the gradient feature, the horizontal binary edge image (*Edge*(*i, j*)) is generated by applying the Sobel edge operator with horizontal mask. The gradient feature is defined as equation (1). The *HP*(*j*) is the jth entry in the horizontal projection which is formed by summing the pixels in the ith column. The number of edges corresponding to 30 bins is normalized and passed to the 1st face classifier.

$$HP(j) = \sum_{i=0}^{29} Edge(i, j),\ 0 \le j \le 29, \tag{1}$$

3.2 Texture Feature for 2nd Face Classifier

Texture is one of the most important defining characteristics of an image. A face image can be thought of as a regular and symmetric texture pattern. Although a human face has a distinct texture pattern compared to other objects, this property has not been utilized widely in developing face detection. In our system, the texture feature is derived from gray level co-occurrence matrix [12]. The $(i,j)^{th}$ element of the co-occurrence matrix represents the number of times that the pixel with value *i* occur, in adjacent distance (*d*) along a direction (θ), related to a pixel with value *j* in an image.

The texture features are extracted by three measures; correlation, variance and entropy. The correlation is related to the joint probability occurrence of the specified pixel pairs. The variance measures the amount of local variations in an image, whereas the entropy measures the disorder of an image. Feature extraction is processed as follow:

1. The input local image (30×30) is reduced to 10×10 smoothed image by applying the average filter in 3×3 size.
2. The each pixel is quantized into 25 bins for the computational efficiency.
3. Obtain texture features through the following measures;

$$Correlation: \frac{\sum_i \sum_j (i-\mu_x)(j-\mu_y)p(i,j)}{\sqrt{\sigma_x \sigma_y}}, \tag{2}$$

$$Variance: \sum_i \sum_j (i-\mu)^2 p(i,j), \tag{3}$$

$$Entropy := -\sum_i \sum_j p(i,j)\log(p(i,j)), \tag{4}$$

where $p(i, j)$ is the $(i, j)^{th}$ entry of the normalized co-occurrence matrix and $\mu = \mu_x = \mu_y$, because of symmetric matrix.

The extracted 9 texture features (3 measures×3 directions (0°, 45° and 90°)) are passed to the 2nd face classifier.

3.3 Pixel Intensity Feature for 3rd Face Classifier

The pixel intensity feature is the most commonly used input vector for neural network based object detection. The methods that use pixel intensity have yielded promising detection performance so far. Especially, the regions of facial features have been proven valuable clues for classifying faces. In our system, the pixel intensity feature is extracted from eye region, because eye region is more reliable than nose and mouth region for determining face pattern.

To extract the feature, a 10×10 smoothed image is first reconstructed from the local image by sub-sampling the local image with 3×3 average-mask. The normalized pixel intensity values of 40 pixels corresponding to eye region (10×4) are finally obtained and passed to the 3rd face classifier.

4 Coarse-to-Fine Classification for Computational Efficiency

The computational efficiency is achieved by using coarse-to-fine classification approach. In coarse-to-fine classification, the problems that must be solved are related to following two matters. The first matter is how to improve the window scanning process in the coarse classification and the second is how to reliably identify the local image as a face in the fine classification. In order to improve the window scanning process in the coarse classification stage, we use the translation invariant property of the gradient feature that is used in 1st face classifier. That is, if we know the permissible bound of translation, we can easily increase the window moving step to find coarse location of a face. For applying the translation invariant property for the 1st face classifier, we analyze the sensitivity of 1st face classifier with respect to the degrees of shift. That is, we collected a set of 50 images for sensitivity analysis, each of them was cropped around center of face in both x and y directions.

Fig. 3(a) presents the detection rate of the 1st face classifier with respect to shift in both x and y direction when the threshold value was strictly set to 0.8. The detection rate was over 80% when the images were shifted within 10 pixels in x direction and within 4 pixels in y direction. This allows the window moving step for scanning, in the coarse classification stage, to be up to 10 pixels in x direction and 4 pixels in y direction (see Fig. 4(a)). From the coarse location, the fine search is started where the window is shifted by 2 pixels in both x and y directions (see Fig. 4(b)). This is based on observations that the 2nd and the 3rd face classifiers, which are performed in the fine classification stage, have a detection rate of over 80%, when the images were shifted by 2 pixels in both x and y directions as shown in Fig. 3(b) and 3(c), respectively. If the 1st face classifier can not identify the local image as a face, the coarse classification process is applied again.

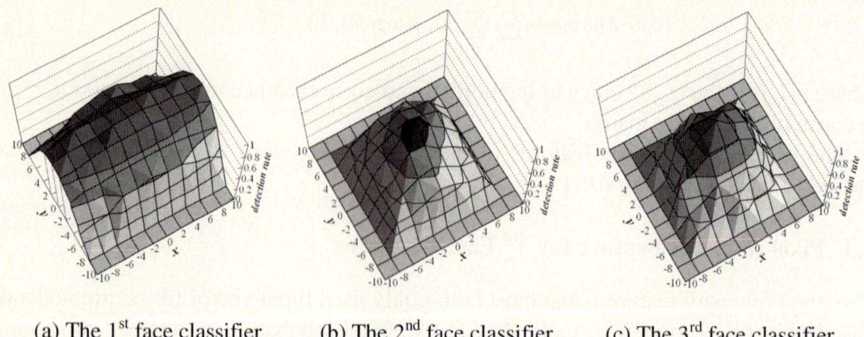

(a) The 1st face classifier (b) The 2nd face classifier (c) The 3rd face classifier

Fig. 3. The results of sensitivity analysis with respect to shift images

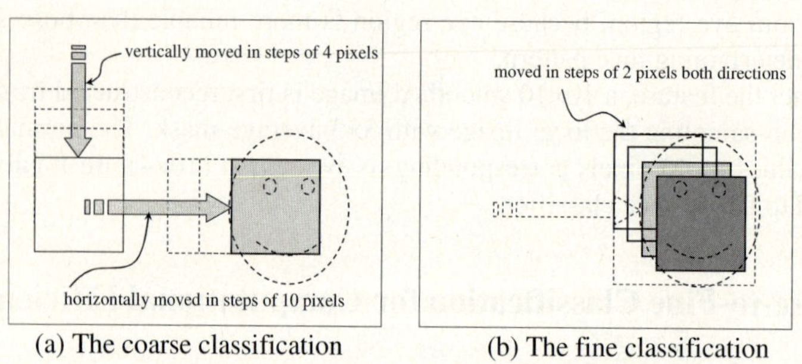

(a) The coarse classification (b) The fine classification

Fig. 4. The moving steps of scanning window

In the fine classification stage, a weighted sum of the results from the multiple face classifiers is utilized to identify the local image as a face. If the weighted sum value is grater than the threshold value (τ), the local image is identified as a face.

5 Experimental Results

5.1 Training Face Classifiers

Each face classifier is trained by error back propagation algorithm and a logistic sigmoid activation function is used in each unit. The numbers of hidden unit of the 1st, 2nd and 3rd face classifier are 10, 4 and 12, respectively.

The 1,056 training images of face pattern came from the benchmark face database (AT&T[2], BioID[3], Stirling[4], Yale [13] dataset) and World Wide Web. The face patterns were manually normalized to 30×30 rectangle including the outer eye corners and upper eyebrows. In addition, we included the mirror-reverse and two rotation angles (5°, -5°) of each image and produced a total of 4,224 examples of faces. The non-face patterns were collected via an iterative bootstrapping procedure [1]. Before training, we used an initial training set of 2,080 non-face patterns from background images. After bootstrapping process, 15,798 non-face patterns were obtained.

---

[2] http://www.uk.research.att.com/facedatabase.html

[3] http://www.bioid.com/downloads/facedb

[4] http://pics.psych.stir.ac.uk/

5.2 Results on Several Databases

To evaluate the performance of our proposed method, we compared to an exhaustive full scanning method with several databases which were not used in the training process. The test database consisted of four different test sets (IMM[5], Caltech[6], AR database [14] and World Wide Web). The face databases were publicly available on the World Wide Web and often used for the benchmarking of face detection algorithm. The images from the IMM (640×480) and the AR database (768×576) which had a uniform background with various poses, expressions and illuminations, while the Caltech database (896×592) varied a lot with respect to background. The images obtained from World Wide Web varied a lot with respect to image resolution and background.

Table 1 shows a tabulated comparison for the proposed method and the exhaustive full scanning method on test databases. Examples of detection results are shown in Fig. 5 and 6. The applied threshold value (τ) and weight factors (w_1, w_2, and w_3) of the proposed method were empirically set to 0.65, 0.25, 0.35, and 0.4 respectively. As shown in Table 1, the proposed method achieved a detection rate between 93.0% and

Table 1. Experimental results

Test DB	Detection results						Reduction rates of # of scans
	Exhaustive full scanning method			Proposed scanning method			
	Detection rate	# of false	# of scans per image	Detection rate	# of false	# of scans per image	
IMM	96.2 %	28	755,418	95.7 %	8	72,273	90.4 %
Caltech	94.5 %	12	1,369,067	93.0 %	10	176,674	87.1 %
AR	95.7 %	22	1,128,541	95.0 %	6	142,136	87.3 %
WWW	80.7 %	46	4,312,203	83.7 %	12	513,152	88.1 %

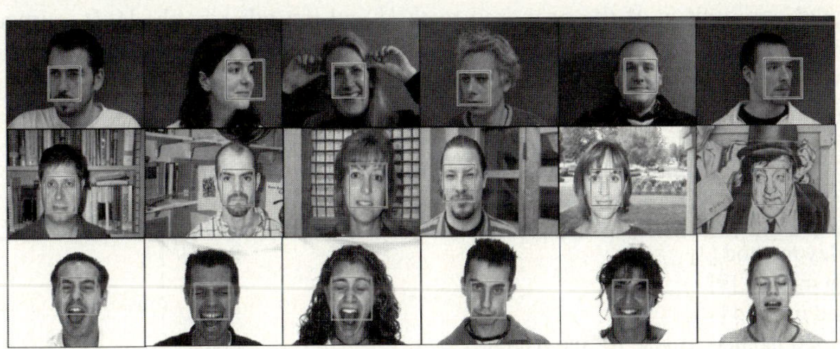

Fig. 5. Detection results. The IMM, Caltech, and AR database (from top to bottom).

95.7% which is almost the same as the detection rate achieved by the full scanning method. In terms of the computational cost, we obtained the total number of scans per image used to detect the faces. It can be seen that the proposed method can reduce the number of scans up to 90.4% compared to the exhaustive full scanning method.

5 http://www2.imm.dtu.dk/~aam/

6 http://vision.caltech.edu/html-files/archive.html

Fig. 6. Examples of detection results obtained from World Wide Web

5.3 Comparison with Other Methods

To compare to the detection rate of our proposed method with other methods, we used CMU database [3] which is widely used for testing face detectors. Table 2 shows a tabulated comparison of the proposed method and the other methods. It is important to observe that the detection rate of proposed method is similar to the other methods even though the search space is reduced by increasing the moving step of scanning window. Some detection results are shown in Fig. 7.

Table 2. The performance comparison with other methods[7]

	Detection rate	# of false
Rowley method [3]	86.2%	23
Froba method [9]	87.8%	120
Feraud method [10]	86.0%	8
Proposed method (coarse-to-fine search)	86.6%	19
Proposed method (full search)	89.1%	32

6 Conclusions

In this paper, we suggest a way to overcome the computational inefficiency of exhaustive full search which is commonly used in multi-scale search based face detection.

[7] Since we trained 30×30 face pattern differently, the images were scaled up by a factor 1.5 before testing.

Fig. 7. Examples of detection results obtained from CMU database

The proposed multiple face classifiers using coarse-to-fine classification is based on the improvement of window scanning process. In order to improve the window scanning process, we empirically determined the sub-optimal moving step of scanning window by analyzing the detection rate of each classifier for various moving step sizes. Furthermore, multiple face classifiers were designed for the reliable judgment on existence of face. Experimental results shows that our proposed method can reduce a significant amount of computational complexity with a negligible change in detection rate compare to exhaustive full search method.

References

1. Sung, K-K., Poggio, T.: Example based learning for view based human face detection. IEEE Trans. on Pattern Analysis and Machine Intelligence, Vol. 20 (1998) 39-51
2. Juell, P., Marsh, R.: A hierarchical neural network for human face detection. Pattern Recognition, Vol. 29 (1996) 781-787
3. Rowley, H., Baluja, S., Kanade, T.: Neural network based face detection, IEEE Trans. on Pattern Analysis and Machine Intelligence, Vol. 20 (1998) 23-38
4. Osuna, E., Freund, R., Girosi, F.: Training support vector machines: an approach to face detection. Pro. Conf. Computer Vision and Pattern Recognition, (1997) 130-136
5. Shih. P., Liu, C.: Face detection using discriminating feature analysis and support vector machine. Pattern Recognition, Vol. 39 (2006), 260-276
6. Turk, M., Pentland, A.: Face recognition using eigenfaces. Pro. Conf. Computer Vision and Pattern Recognition (1991) 586-591

7. Yang, J., Waibel, A.: Tracking human faces in real time. Tech. Report CMU-CS-95-210 (1995)
8. Soriano, M., Martinkauppi, B., Hunvinen, S., Laaksonen, M.: Adaptive skin color modeling using the skin lucus for selecting training pixels. Pattern Recognition, Vol. 3 (2003) 681-690
9. Froba, B., Kublbech, C.: Robust face detection at video frame rate based on edge orientation features. Proc. Conf. Automatic Face and Gesture Recognition, (2002) 327-332
10. Feraud, R., Bernier, O. J., Viallet, J-M., Collobert, M.: A fast and accurate face detector based on neural networks. IEEE Trans. on Pattern Analysis and Machine Intelligence, Vol. 23 (2002) 42-53
11. Bebis, G., Uthiram, S., Georgiopoulos, M.: Face detection and verification using generic search. Artificial Intelligence Tools, Vol. 9 (2000) 225-246
12. Peter, R. A., Strickland, R. N: Image complexity metrics for automatic target recognizers. Automatic Target Recognizer System and Technology Conference (1990)
13. Georghiades, A. S., Belhumeur, P. N., Kriegman, D. J.: From few to many: Illumination cone models for face recognition under variable lighting and pose. IEEE Trans. On Pattern Analysis and Machine Intelligence, Vol. 23 (2001) 643-660
14. Martinez, A. M., Benavente, R.: The AR face database. CVC Tech, Report #24 (1998)

A Blind Watermarking Algorithm Based on HVS and RBF Neural Network for Digital Image

Cheng-Ri Piao, Seunghwa Beack, Dong-Min Woo, and Seung-Soo Han

Department of Information Engineering & NPT Center, Myongji University
Yongin, Kyunggi, 449-728, South Korea
shan@mju.ac.kr

Abstract. This paper proposes a new blind watermarking scheme in which a watermark is embedded into the discrete wavelet transform (DWT) domain. The method uses the HVS model, and radial basis function neural networks (RBF). RBF will be implemented while embedding and extracting watermark. The human visual system (HVS) model is used to determine the watermark insertion strength. The inserted watermark is a random sequence. The secret key determines the beginning position of the image where the watermark is embedded. This process prevents possible pirates from removing the watermark easily. Experimental results show that the proposed method has good imperceptibility and high robustness to common image processing attacks.

1 Introduction

The watermarking technique has recently become a very active area of multimedia security [1]. The non-blind watermarking methods [2-3] proposed above are very effective both in terms of image quality and robustness against various attacks. However the blind watermarking technique, which doesn't require the original image while detecting watermark, is widely used in watermarking area.

In this paper, a new blind watermark embedding/extracting algorithm using the RBF neural networks is introduced. The DWT is used to overcome the blocking phenomenon problems in DCT. First, the original image is 4-level DWT transformed, and decided the watermarking strength according to HVS. When embedding watermark, a secret key is used to determine the watermark embedding beginning location, and after that, embed and extract the watermark by using the trained RBF. The experimental results show that the watermarked image has at least 45dB in peak signal-to-noise ratio (PSNR) and show good imperceptibility and high robustness against common image processing attacks.

2 Related Theories

2.1 Discrete Wavelet Transform (DWT) and HVS

In this paper, the linear-phase 9/7 biorthogonal filters are used for DWT, and the watermark is embedded into LL4, LH4, HL4 and HH4 subband for robustness. We

L. Jiao et al. (Eds.): ICNC 2006, Part I, LNCS 4221, pp. 493 – 496, 2006.

also use HVS [4] to decide the watermarking strength of DWT coefficients. The HVS presented by Watson *et al.* for biorthogonal wavelet basis 9/7, gained the value of quantization matrix [4]. From the quantization matrix, the maximal values of quantization error in LL4, HL4, HH4 and LH4 band is about 7. So the random sequence value should be resisted in the range of $|w_i|<7$ in this paper.

2.2 The Radial Basis Function Neural Networks (RBF) and Training Procedures

In this paper, the data are used to approximate a linear function, and RBF neural network is implemented. Radial basis function networks have a basis function layer and a linear discriminating layer. RBF networks represent the posterior probabilities of the training data by a weighted sum of Gaussian basis functions with diagonal covariance matrices. The information is stored in the centers and the width of the basis functions and the center is set to zero while the width is set to one. RBF will be used to learn the characteristics of image for improving the performance of the proposed watermarking scheme in the section 3. The RBF neural network training procedures are as Fig.1.

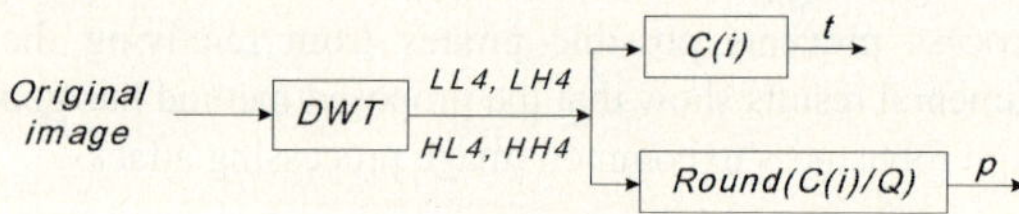

Fig. 1. RBF training procedures

In the Fig.1, *C(i)* is the LL4, LH4, HL4, HH4 band coefficient when DWT transform is performed on original image, *Q* is the Quantization value, *p* is an input value for the RBF, *t* is the desired output value for RBF.

3 Watermark Embedding and Extracting

Fig. 2 shows the block diagram of the proposed watermarking system and the embedding procedures are as follows:

Step1: Transform an original image using the 4-level DWT transform. In Fig. 2, *C(i)* is the LL4, LH4, HL4, and HH4 sub-band coefficient.

Step2: Select the beginning position of watermark embedding coefficient *C(i)* using the secret key.

Step3: Quantize the DWT coefficient *C(i+key)* by *Q*, as the input value of RBF then get the output *RBF(round(C(i+key)/Q))*

Step4: Embed the watermark according to the equation 1 which uses the output value of the RBF neural network (*RBF(round(C(i+key)/Q))*)and the *Q*.

$$C'_{i+key} = RBF\ (Round\ (\frac{C_{i+key}}{Q})) + x_i \qquad (|\ x_i\ | \leq 7) \tag{1}$$

where x_i is the random sequence watermark, Q is a quantization value, and C'_i is the coefficient value when watermark is embedded. Then perform IDWT to get the watermarked image.

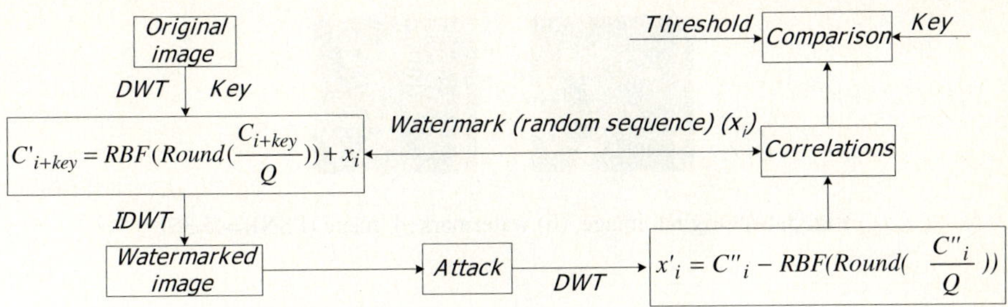

Fig. 2. Block diagram of the proposed watermarking system

The watermark extracting procedures are as follows:

Step1: Transform the watermarked image by the 4-level DWT transform. In Fig. 2, *C''(i)* is the LL4, LH4, HL4, and HH4 subband coefficient.
Step2: Quantize the DWT coefficient *C''(i)* by *Q*, as the input value of RBF then get the output *RBF(round(C''(i)/Q))*.
Step3: Extract the watermark (x'_i) using the equation 2 below, using the output of the RBF neural network (*RBF(round(C''(i)/Q))*) and coefficient *C''(i)*.

$$x'_i = c''(i) - RBF\ (round\ (\frac{c''(i)}{Q})) \tag{2}$$

Step 4. Measure the similarity of the extracted watermark *x'* and the original watermark *x* by equation (3).

$$sim\ (x, x') = \frac{x' \cdot x}{\sqrt{x' \cdot x'}} \tag{3}$$

Step 5. Use *sim*(*x, x'*), threshold, as a key to judge if there is a embedded watermark or not. If *sim*(*x, x'*) is larger than threshold and the location is equal to the key, the watermark can be affirmed . The threshold is determined by standard deviation and false positive error probability that is implemented as 10^{-6}. Based on those values, the threshold value is set to 20.8.

4 Experimental Results and Conclusions

In this paper, 8 bit 512×512 size Lena image is used as original image. Fig.3 shows the original image and watermarked image. The image quality metric is based on the PSNR. The PSNR of the watermarked image is decreasing with increasing *Q*, but the PSNR is bigger than 45dB in any case. This proves that the watermarked image has a good image quality.

Fig.4 shows the test results of the robustness of the proposed method. The peak similarity value in each figures shows that the watermark is successfully detected. This correctly detecting capability proves the robustness of that algorithm.

This paper proposes a new blind watermarking scheme in which a watermark is embedded into the DWT domain. It also utilizes RBF, which learns the characteristic of the image, and then watermark will be embedded and extracted using the trained

Fig. 3. (a) Original image, (b) watermarked image (PSNR=45.35)

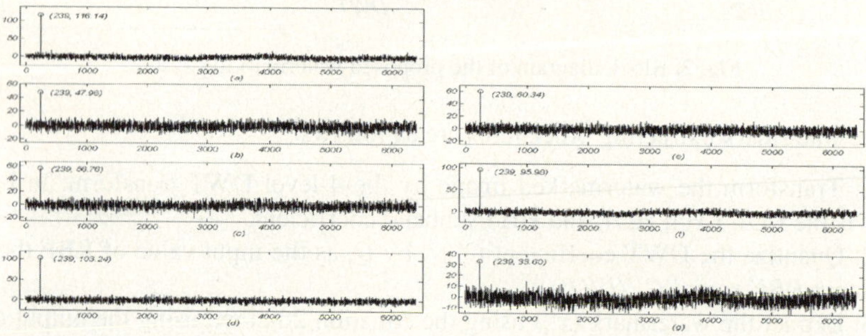

Fig. 4. Correlations for watermark detection (psnr=45.35, key=239). (a) No attack, (b) JPEG compression (quality factor=20), (c) Resize (350*350) (d) Gaussian noise added (SNR=10dB, signal power=25dB), (e) Gaussian low-pass filtering (filter size=3*3, standard deviation=100), (f) Media filtering, (g) SPIHT compression (rate = 0.4bpp).

RBF. The embedding scheme results in good quality of the watermarked image. Due to the learning and adaptive capabilities of the RBF, the embedding/extracting strategy can greatly improve the robustness against various attacks.

Acknowledgement

This work was supported by the ERC program of MOST/KOSEF (Next-generation Power Technology Center).

References

1. Hartung, F., Kutter, M.: Multimedia watermarking techniques. Proceedings of the IEEE, 1999. pp. 1079-1094.
2. Cox, I.J.; Kilian, J.; Leighton, F.T.; Shamoon, T.: Secure spread spectrum watermarking for multimedia. Image Processing, IEEE Transactions on Vol. 6, 1997. pp. 1673 – 1687.
3. Jiwu Huang; Shi, Y.Q.; Yi Shi: Embedding image watermarks in dc components, Circuits and Systems for Video Technology, IEEE Transactions on Vol 10, 2000 pp. 974 - 979
4. A. B. Watson, G. Y. Yang, J. A. Solomon, and J. Villasenor, “Visual thresholds for wavelet quantization error,” in Proc. SPIE Human Vision and Electronic Imaging, 1996, vol. 2657, pp. 381–392.

Multiscale BiLinear Recurrent Neural Network with an Adaptive Learning Algorithm

Byung-Jae Min, Chung Nguyen Tran, and Dong-Chul Park

ICRL, Dept. of Information Engineering, Myong Ji University, Korea
{mbj2000, tnchung, parkd}@mju.ac.kr

Abstract. In this paper, a wavelet-based neural network architecture called the Multiscale BiLinear Recurrent Neural Network with an adaptive learning algorithm (M-BLRNN(AL)) is proposed. The proposed M-BLRNN(AL) is formulated by a combination of several BiLinear Recurrent Neural Network (BLRNN) models in which each model is employed for predicting the signal at a certain level obtained by a wavelet transform. The learning process is further improved by applying an adaptive learning algorithm at each resolution level. The proposed M-BLRNN(AL) is applied to the long-term prediction of MPEG VBR video traffic data. Experiments and results on several MPEG data sets show that the proposed M-BLRNN(AL) outperforms the traditional MultiLayer Perceptron Type Neural Network (MLPNN), the BLRNN, and the original M-BLRNN in terms of the normalized mean square error (NMSE).

1 Introduction

The implementation of various multimedia services over Asynchronous Transfer Mode (ATM) networks has seen rapid growth in recent years. ATM networks are especially designed to efficiently support bursty traffic sources such as interactive data and motion video data. The dynamic nature of these bursty traffic data, however, may cause severe network congestion when a number of bursty sources are involved. Therefore, the demand for dynamic bandwidth allocation to optimally utilize the network resources and satisfy Quality of Service(QoS) requirements should be taken into account.

In order to dynamically adapt for bandwidth allocation, prediction of the future network traffic generated by end-users according to the observed past traffic in the network plays a very important role. By using the predicted traffic volume, the bandwidth can be reallocated dynamically. Various traffic prediction models have been proposed for VBR video traffic prediction. Classical linear models such as the Autogressive (AR) model [1] and adaptive linear model [2] have been widely used in practice. However, these models may be not suitable for predicting traffic over ATM networks due to the bursty characteristics of these networks.

A number of new nonlinear techniques have been proposed for VBR video traffic prediction. Among them, the neural network (NN)-based models have

L. Jiao et al. (Eds.): ICNC 2006, Part I, LNCS 4221, pp. 497–506, 2006.

received significant attention [3]. These recent studies reported that satisfactory traffic prediction accuracy can be achieved for a single-step prediction,i.e., the prediction for only next frame. However, the sing-step prediction may not be suitable in application such as dynamic bandwidth allocation since it is impractical to reallocate the bandwidth frequently for a single frame. Therefore, multi-step prediction of VBR video traffic should be explored.

In this paper, a wavelet-based neural network architecture called the Multiscale BiLinear Recurrent Neural Network with an adaptive learning algorithm (M-BLRNN(AL)) is proposed. The proposed M-BLRNN(AL) is applied to the long-term prediction of VBR video traffic. The proposed M-BLRNN(AL) is formulated by a combination of several individual BLRNN [4] models in which each individual model is employed for predicting the signal at a certain level obtained by the wavelet transform. The learning process of the proposed M-BLRNN(AL) is further improved by application of an adaptive learning algorithm at each resolution level. The proposed learning algorithm is adapted at each level of resolution by an iterative procedure with respect to data using the gradient-descent method. By using an adaptive learning algorithm at each resolution level, the individual BLRNN model at each resolution level can learn the characteristics of data more efficiently.

The remainder of this paper is organized as follows: Section 2 presents a review of multiresolution analysis with the wavelet transform. A brief review of the BLRNN is given in Section 3. The proposed M-BLRNN(AL) and its adaptive learning algorithm are presented in Section 4. Section 5 presents some experiments and results on several MPEG data sets including a performance comparison with the traditional MLPNN and BLRNN models. Concluding remarks provided in Section 6 close the paper.

2 Multiresolution Wavelet Analysis

The aim of a multiresolution analysis is to analyze a signal at different frequencies with different resolutions. It produces a high quality local representation of a signal in both the time domain and the frequency domain. The wavelet transform [5,6], a novel technology developed in the signal processing community, has been proven suitable for the multiresolution analysis of time series data [7].

The *à* trous wavelet transform was first proposed by Shensa [6]. The calculation of the *à* trous wavelet transform can be described as follows: First, a low-pass filter is used to suppress the high frequency components of a signal and allow the low frequency components to pass through. A scaling function associated with the low-pass filter is then used to calculate the average of elements, resulting in a smoother signal.

The smoothed data $c_j(t)$ at a given resolution j can be obtained by performing successive convolutions with the discrete low-pass filter h,

$$c_j(t) = \sum_k h(k) c_{j-1}(t + 2^{j-1}k) \tag{1}$$

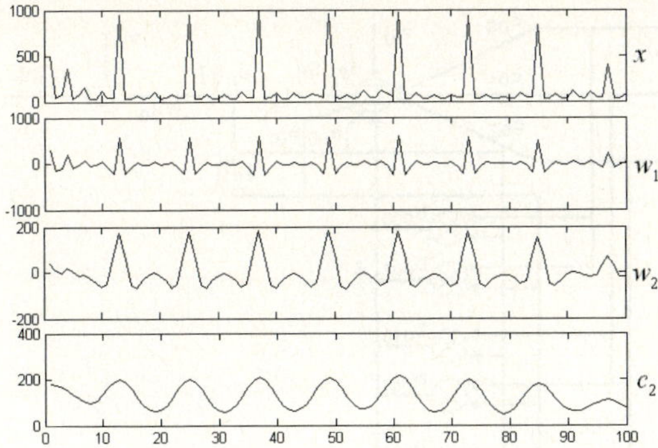

Fig. 1. Example of the wavelet coefficients and the scaling coefficients

where h is a discrete low-pass filter associated with the scaling function and $c_0(t)$ is the original signal. A suitable low-pass filter h is the B_3 spline, defined as $(\frac{1}{16}, \frac{1}{4}, \frac{3}{8}, \frac{1}{4}, \frac{1}{16})$.

From the sequence of the smoothing of the signal, the wavelet coefficients are obtained by calculating the difference between successive smoothed versions:

$$w_j(t) = c_{j-1}(t) - c_j(t) \tag{2}$$

By consequently expanding the original signal from the coarsest resolution level to the finest resolution level, the original signal can be expressed in terms of the wavelet coefficients and the scaling coefficients as follows:

$$c_0(t) = c_J(t) + \sum_{j=1}^{J} w_j(t) \tag{3}$$

where J is the number of resolutions and $c_J(t)$ is the finest version of the signal. Eq.(3) also provides a reconstruction formula for the original signal.

Fig. 1 shows an example of the wavelet coefficients and the scaling coefficients for two levels of resolution over 100 samples from the *"Star Wars"* video trace. From the top to the bottom are the original signal, two levels of the wavelet coefficients, and the finest scaling coefficients, respectively.

3 BiLinear Recurrent Neural Networks

The BLRNN is a simple recurrent neural network, which has a robust ability in modeling dynamically nonlinear systems and is especially suitable for time-series data. The model was initially proposed by Park and Zhu [4]. It has been successfully applied in modeling time-series data [4,8]. Fig. 2 illustrates a simple 3-1-1 BLRNN with 2 feedback taps.

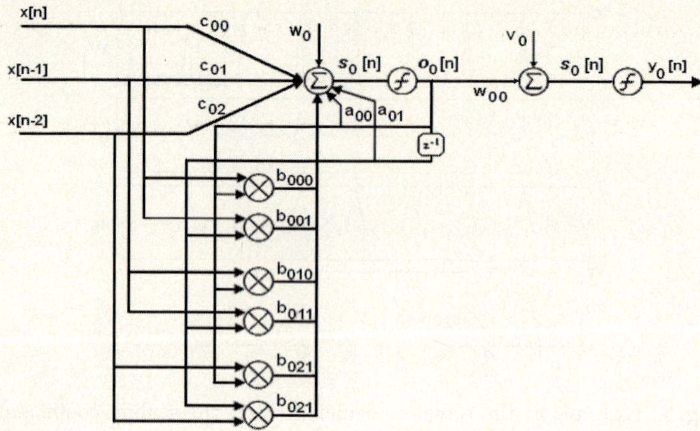

Fig. 2. Simple BLRNN with structure 3-1-1 and 2 recursion lines

For an one-dimensional input/output case, the output value of a bilinear recurrent neuron is computed by the following equation:

$$s[n] = \sum_{i=1}^{N-1} a_i y[n-i] + \sum_{i=0}^{N-1}\sum_{j=1}^{N-1} b_{ij} y[n-j]x[n-i] + \sum_{i=0}^{N-1} c_i x[n-i] \quad (4)$$

where x[i] is the input, y[i] is the output, and N is the order of recursion.

In the following, we explain about a simple BLRNN that has N input neurons, M hidden neurons and where $K = N-1$ degree polynomials is given. The input signal and the nonlinear integration of the input signal to hidden neurons are:

$$\boldsymbol{X}[n] = [x[n], x[n-1], ..., x[n-K]]^T$$
$$\boldsymbol{O}[n] = [o_1[n], o_2[n], ..., o_M[n]]^T$$

where T denotes the transpose of a vector or matrix and the recurrent term is a $M \times K$ matrix

$$\boldsymbol{Z}_p[n] = [o_p[n-1], o_p[n-2], ..., o_p[n-K]]$$

And

$$\begin{aligned} s_p[n] &= w_p + \sum_{k_1=0}^{N-1} a_{pk_1} o_p[n-k_1] \quad (5) \\ &+ \sum_{k_1=0}^{N-1}\sum_{k_2=0}^{N-1} b_{pk_1k_2} o_p[n-k_1]x[n-k_2] \\ &+ \sum_{k_2=0}^{N-1} c_{pk_2} x[n-k_2] \\ &= w_p + \boldsymbol{A}_p^T \boldsymbol{Z}_p^T[n] + \boldsymbol{Z}_p[n]\boldsymbol{B}_p^T\boldsymbol{X}[n] + \boldsymbol{C}_p^T\boldsymbol{X}[n] \end{aligned}$$

where w_p is the weight of bias neuron. $\boldsymbol{A}_p$ is the weight vector for the recurrent portion, $\boldsymbol{B}_p$ is the weight matrix for the bilinear recurrent portion, and $\boldsymbol{C}_p$ is the weight vector for the feedforward portion and $p = 1, 2..., M$. Let ϕ be the activation function of the hidden neuron, the output of pth hidden neuron is then:

$$o_p[n] = \phi(s_p[n]) \tag{6}$$

From the hidden layer to the output layer, it is the same as a traditional feedforward-type neuron network

$$s_l[n] = v_l + \sum_{p=0}^{N_h-1} w_{lp} o_p[n] \tag{7}$$

where v_l is the weight of bias neuron, w_{lp} is the weight between the hidden and the output neurons, and N_h is the number of hidden neurons. The final output is obtained by applying the activation function

$$y_l[n] = \phi(s_l[n]) \tag{8}$$

More detailed information on the BLRNN and its learning algorithm can be found in [4,8].

4 Multiscale BiLinear Recurrent Neural Network with an Adaptive Learning Algorithm

4.1 Multiscale BiLinear Recurrent Neural Network

The M-BLRNN is a combination of several individual BLRNN models where each individual BLRNN model is employed to predict the signal at each resolution level obtained by the wavelet transform. Fig. 3 illustrates an example of the M-BLRNN with three levels of resolution.

The prediction of a time-series based on the M-BLRNN can be separated into three stages. In the first stage, the original signal is decomposed into the wavelet coefficients and the scaling coefficients based on the number of resolution levels. In the second stage, the coefficients at each resolution level are predicted by an individual BLRNN model. It should be noted that the predictions of coefficients at each resolution level are independent and can be done in parallel. In the third stage, all the prediction results from each BLRNN are combined together using the reconstruction formula given in Eq.(3):

$$\hat{x}(t) = \hat{c}_J(t) + \sum_{j=1}^{J} \hat{w}_j(t) \tag{9}$$

where $\hat{c}_J(t)$, $\hat{w}_j(t)$, and $\hat{x}(t)$ represent the predicted values of the finest scaling coefficients, the predicted values of the wavelet coefficients at level j, and the predicted values of the time-series, respectively.

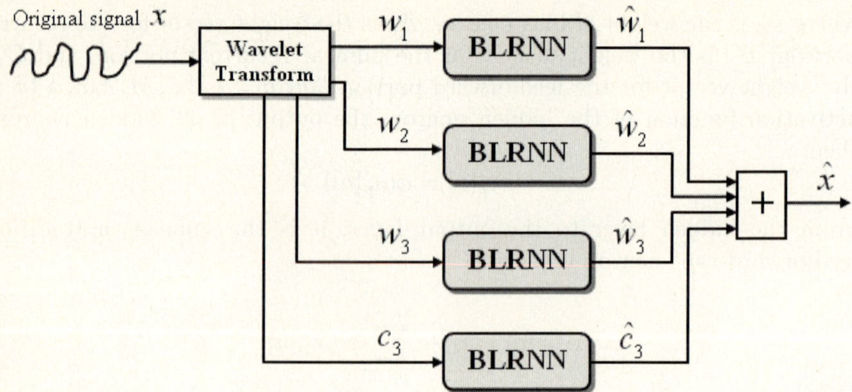

Fig. 3. Example of Multiscale BiLinear Recurrent Neural Network with 3 resolution levels

The improvement of the M-BLRNN over traditional MLPNN and BLRNN models can be explained as follows: In the traditional MLPNN and the BLRNN, the complicated and chaotic signals are used directly for training and predicting. Therefore, the traditional models attain little learning of the underlying relationship between the input and output in the signals. By adopting the wavelet transform to decompose the complicated signal into several simpler sub-signals, the M-BLRNN not only can learn the signals at each level more efficiently but also can learn the correlation structure and internal information hidden in the original signals.

4.2 Multiscale BiLinear Recurrent Neural Network with an Adaptive Learning Algorithm

The proposed M-BLRNN(AL) is based on the M-BLRNN and employs an adaptive learning algorithm for training each individual BLRNN model in the M-BLRNN so as to increase the learning speed and yield a good generalization performance [9]. The adaptive learning algorithm is derived by using an adjustable activation function at each resolution level. Typically, the activation function used in Eq.(6) and Eq.(8) is the following logistic function

$$\phi(x) = \frac{1}{1 + e^{-\lambda x}} \qquad (10)$$

where λ is the slope of the activation function. In the traditional training algorithms, an identical value of λ, $\lambda = 1$, is applied for training all individual BLRNN models at all resolution levels. In contrast, in the adaptive learning algorithm, the slope parameter of the adjustable activation function at each individual BLRNN model is adapted iteratively at each training step using the gradient-descent method.

Assume that the cost function is defined as

$$E = \frac{1}{2}\sum_{l}(t_l - y_l)^2 \tag{11}$$

At the output layer, the slope parameter λ_l at each output neuron l can be iteratively updated by

$$\lambda_l(n+1) = \lambda_l(n) + \mu_\lambda(t_l - y_l)\frac{s_l e^{-\lambda_l s_l}}{(1+e^{-\lambda_l s_l})^2} \tag{12}$$

Similarly, at the hidden layer, the slope parameter λ_p at each hidden neuron p can be iteratively updated by

$$\lambda_p(n+1) = \lambda_p(n) + \left(\sum_{l}(t_l - y_l)\frac{\lambda_l e^{-\lambda_l s_l}}{(1+e^{-\lambda_l s_l})^2} w_{lp}\right)\frac{s_p e^{-\lambda_p s_p}}{(1+e^{-\lambda_p s_p})^2} \tag{13}$$

As can be seen from Fig. 1, the characteristics of coefficients at each resolution level are different. By using the adaptive learning algorithm at each resolution level, the BLRNN model can learn the characteristics of the coefficients more efficiently. This implies that the proposed M-BLRNN(AL) can yield a better generalization performance than the M-BLRNN.

5 Experiments and Results

The experiments were conducted based on several MPEG trace sequences provided by the University of the Wuerzburg, Wuerzburg, Germany. These trace sequences can be downloaded at

http://www3.informatik.uni-wuerzburg.de/MPEG/

We selected 4 typical trace sequences for training and testing: "*Star Wars*", "*Mr. Bean*", "*New Show*", and "*Silence of the Lambs*". From these sequences, the first 1,000 frames of the "*Star Wars*" sequence were used for training while the remaining of "*Star Wars*" and other sequences were saved for testing. All data were subsequently normalized in a range (0,1) to render it suitable for inputs of neural networks.

Table 1. Prediction performance in NMSE for *"Mr. Bean"*

	1	10	20	30	40	50	60	70	80	90	100	Avg.
MLPNN	0.123	0.166	0.330	0.601	0.988	1.413	1.492	1.996	2.511	3.053	3.716	1.489
BLRNN	0.085	0.117	0.253	0.488	0.830	1.237	1.404	2.022	2.681	3.096	3.166	1.398
M-BLRNN	0.042	0.052	0.090	0.141	0.185	0.217	0.217	0.237	0.291	0.356	0.432	0.205
M-BLRNN(AL)	0.033	0.049	0.087	0.132	0.169	0.205	0.216	0.236	0.283	0.354	0.412	0.197

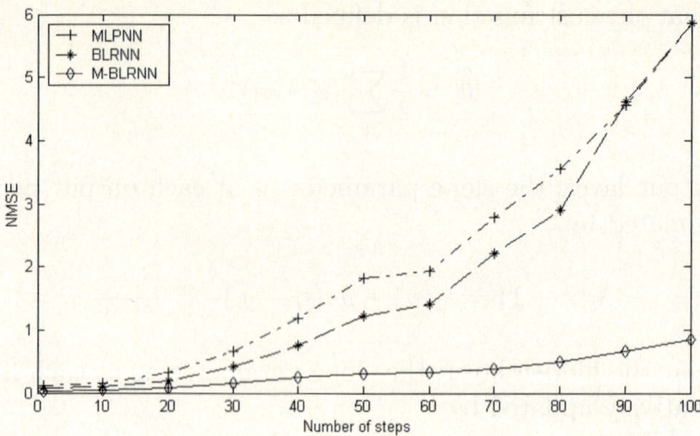

Fig. 4. Prediction performance versus number of steps for the *"Star Wars"* video trace

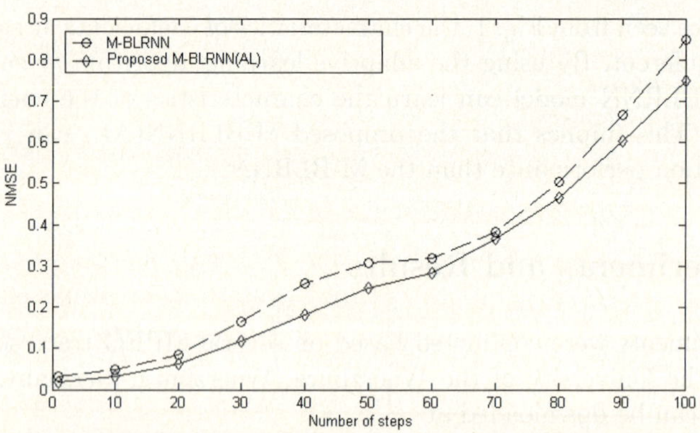

Fig. 5. Prediction performance versus number of steps for the *"Star Wars"* video trace

Table 2. Prediction performance in NMSE for *"Silence of the Lambs"*

	1	10	20	30	40	50	60	70	80	90	100	Avg.
MLPNN	0.198	0.234	0.394	0.723	1.130	1.585	1.631	2.226	2.821	3.663	4.706	1.755
BLRNN	0.170	0.195	0.316	0.539	0.791	1.128	1.252	1.818	2.516	3.508	4.762	1.545
M-BLRNN	0.147	0.135	0.180	0.240	0.309	0.370	0.395	0.455	0.539	0.643	0.734	0.377
M-BLRNN(AL)	0.080	0.125	0.179	0.244	0.308	0.332	0.332	0.388	0.471	0.576	0.684	0.338

Table 3. Prediction performance in NMSE for *"News Show"*

	1	10	20	30	40	50	60	70	80	90	100	Avg.
MLPNN	0.138	0.186	0.294	0.505	0.754	1.014	1.117	1.545	1.886	2.522	3.248	1.200
BLRNN	0.133	0.154	0.243	0.409	0.585	0.799	0.928	1.338	1.747	2.467	3.442	1.113
M-BLRNN	0.109	0.101	0.139	0.201	0.248	0.298	0.339	0.403	0.453	0.536	0.606	0.312
M-BLRNN(AL)	0.061	0.099	0.144	0.194	0.236	0.254	0.288	0.360	0.409	0.481	0.558	0.280

In order to demonstrate the generalization ability of the M-BLRNN(AL) model, a M-BLRNN(AL) model with 3 resolution levels using the adaptive learning algorithm is employed. Based on the statistical analysis of correlations, each individual BLRNN model in the proposed M-BLRNN(AL) model shares a 24-10-1 structure and 3 recursion lines in which the indices denote the number of neurons in the input layer, the hidden layer and the output layer, respectively. To demonstrate the improvements attained through adaptive learning, a M-BLRNN using a traditional learning algorithm is derived for comparison. A traditional BLRNN employing a structure of 24-10-1 with 3 recursion lines and a MLPNN model employing a structure of 24-10-1 are also employed for a performance comparison. The proposed M-BLRNN(AL) was trained with 700 iterations while the others were trained with 1,000 iterations. The iterated multistep prediction [10] was employed to perform the multistep prediction of the real-time VBR video traffic. To measure the performance of the multistep prediction, the normalized mean square error (NMSE) was employed [7].

Fig. 4 shows the prediction performance on the remainder of the *"Star Wars"* sequence. As can be seen from Fig. 4, the M-BLRNN that employs the wavelet-based neural network architecture outperforms both the traditional MLPNN and the BLRNN. In particular, the M-BLRNN can predict up to a hundred steps with a very small degradation of performance whereas the traditional MLPNN and the BLRNN fail to do so. Fig. 5 illustrates the improvement in performance for the *"Star Wars"* sequence attained by the proposed M-BLRNN(AL) relative to the M-BLRNN employing a traditional learning algorithm. As can be seen from Fig. 5, the proposed M-BLRNN(AL) model yields a better performance than the M-BLRNN model while requiring fewer iterations for training.

Tables 1, 2, and 3 summarize the prediction performance for the *"Mr. Bean"*, the *"Silence of the Lambs"*, and the *"New Show"* sequences over the number of steps, respectively. As can be seen from these tables, the M-BLRNN and the proposed M-BLRNN(AL) employing the multiresolution representation always outperform both the traditional MLPNN and the BLRNN. Similar to the results from the *"Star Wars"* sequence, both the M-BLRNN and M-BLRNN(AL) models can predict up to hundreds of frames without suffering significant degradation in performance while the traditional MLPNN and BLRNN suffer severe degradation in performance. These tables also show that the proposed M-BLRNN(AL) model always gives better performance than the M-BLRNN model employing the traditional learning algorithm while requiring fewer iterations for training.

6 Conclusion

A wavelet-based neural network architecture called a Multiscale BiLinear Recurrent Neural Network (M-BLRNN) with an adaptive learning algorithm (M-BLRNN(AL)) is proposed in this paper. The wavelet-based neural network architecture is formulated by a combination of several BLRNN models and a multiresolution representation using the wavelet transform. The adaptive learning algorithm employing an adjustable neural activation function is applied to training the individual BLRNN model at each resolution level. The proposed M-BLRNN(AL) is applied to the long-term prediction of VBR video traffic. Experiments and results on several MPEG data sets show a significant improvement in comparison with the traditional MLPNN and BLRNN as well as an M-BLRNN trained with a traditional learning algorithm. This confirms that the proposed M-BLRNN(AL) is an efficient tool for dynamic bandwidth allocation in ATM networks.

Acknowledgment

This work was supported by the Korea Research Foundation Grant funded by the Korea Government (MOEHRD-KRF-2005-042-D00265).

References

1. Nomura, N., Fujii, T., Ohta, N.: Basic Characteristics of Variable Rate Video Coding in ATM Environment. IEEE J. Select. Areas Commun., Vol. 7 (1989) 752-760.
2. Adas, A.M.: Using Adaptive Linear Prediction to Support Real-time VBR Video under RCBR Network Service Model. IEEE/ACM Trans. Networking 6 (1998) 635-644.
3. Bhattacharya, A., Parlos, A.G., Atiya, A.F.: Prediction of MPEG-coded Video Source Traffic using Recurrent Neural Networks. IEEE Trans. on Acoustics, Speech, and Signal Processing 51 (2003) 2177 - 2190.
4. Park, D.C., Zhu, Y.: Bilinear Recurrent Neural Network. IEEE ICNN, Vol. 3, (1994) 1459-1464.
5. Mallat, S.G.: A Theory for Multiresolution Signal Decomposition: the Wavelet Representation. IEEE Trans. Pattern Anal. Machine Intell. 11 (1989) 674-693.
6. Shensa, M.J.: The Discrete Wavelet Transform: Wedding the À Trous and Mallat Algorithms. IEEE Trans. Signal Proc. 10 (1992) 2463-2482.
7. Liang, Y., Page, E.W.: Multiresolution Learning Paradigm and Signal Prediction. IEEE Trans. Sig. Proc. 45 (1997) 2858-2864.
8. Park, D.C., Jeong, T.K.: Complex Bilinear Recurrent Neural Network for Equalization of a Satellite Channel. IEEE Trans on Neural Network 13 (2002) 711-725.
9. Kruschke, J.K., Movellan, J.R.: Benefits of Gain: Speeded Learning and Minimal Hidden Layers in Back-propagation Networks. IEEE Trans. on Systems, Man and Cybernetics 21(1) (1991) 273-280.
10. Parlos, A.G., Rais, O.T., Atiya, A.F.: Multi-step-ahead Prediction using Dynamic Recurrent Neural Networks. IJCNN '99. Int. Joint Conf. on Neural Networks, Vol. 1, (1999) 349 - 352.

On-Line Signature Verification Based on Dynamic Bayesian Network

Hairong Lv and Wenyuan Wang

Department of Automation, Tsinghua University, Beijing 100084, China
lvhairong98@mails.thu.edu.cn
wwy-dau@tsinghua.edu.cn

Abstract. In this paper, we present a novel approach to automatic on-line signature verification using **d**ynamic **B**ayesian **n**etwork (**DBN**). On-line signatures can be viewed as time-series data. For modeling time-series data, it's natural to use directed graphical models, which can capture the fact that time flows forward. The model is called dynamic Bayesian network and it hasn't been used in online signature verification. Experimental evaluation on a database containing a total of 3500 signatures of 100 individuals shows promising results compared to other classical methods.

1 Introduction

The handwritten signature is a biometric attribute. It is the most popular validation tool for documents or commercial transactions. Signatures can be verified either on-line or off-line. Recently, many methods and models have been developed for automatic on-line signature verification, such as Dynamic Time Warping (DTW) [1], Regional Correlation [1], Tree Matching [1] and Hidden Markov Models [2]. For literature surveys, see [3].

Selecting a good model is the most important step in designing a signature verification system. In this paper, we present a novel approach to automatic on-line signature verification using dynamic Bayesian network (DBN).

As a result of combination of physical differences inherent in the hand and the learned writing habits, signatures of different individuals contain the writer-related information, and this information can be used to discriminate between writers. The advantages of using DBNs in signature verification lie in two aspects: (1) Time series data of a writer's signature can be represented in a unifying statistical framework. (2) Some prior knowledge can be described by DBNs conveniently. Our experimental results also show that DBNs are a promising way for modeling the signature variability.

This paper is organized as the following: as DBNs hasn't been used in on-line signature verification community, we give a brief introduction in section 2. In section 3, we propose details of preprocessing and feature extraction methods. Topology design of DBNs is discussed in section 4. In section 5, database and experimental results are discussed. Finally, we give a conclusion in section 6.

L. Jiao et al. (Eds.): ICNC 2006, Part I, LNCS 4221, pp. 507 – 510, 2006.

2 Dynamic Bayesian Network

A DBN is a specific type of Bayesian network and is almost always assumed to satisfy the following two conditions: (1) it has the same structure at each time slice t and (2) the cross-slice arcs can only be extended from slice t to slice $t+1$. Condition (1) means that DBNs are time-invariant and according to condition (2), DBNs satisfy the Markov assumption. Fig.1. is a simple example of DBNs. Here, $\mathbf{O}$ can be considered as observation, $\mathbf{Q}$ is state variable that drives observations $\mathbf{O}$, and $\mathbf{X}$ is other factor variable.

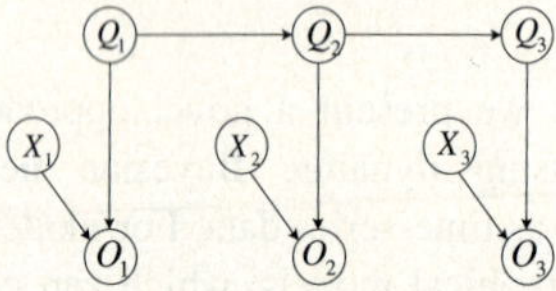

Fig. 1. A simple dynamic Bayesian network

Given some known sequence of observations $\mathbf{O}$, we can learn the structure and the parameters of a DBN using EM algorithm [4]; On the other hand, given the DBN model M, we can calculate the posterior probabilities $P(\mathbf{O} \mid M)$ [4].

3 Feature Extraction

Each sample point (raw data vector) of the signatures consists of values for the horizontal (*x*) position, vertical (*y*) position and pressure (*p*).

Dynamic features such as trajectory tangent angles θ_t and instantaneous velocities v_t, are known to have good discriminative potential. They are computed as:

$$\theta_t = \arctan\frac{\dot{y}_t}{\dot{x}_t}, \quad v_t = \sqrt{\dot{x}_t^2 + \dot{y}_t^2} \tag{1}$$

Here, $\dot{x}_t$, $\dot{y}_t$ indicate first derivatives of x_t and y_t with respect to time. The basic feature vector consists of five values:

$$\tilde{o}_t = [x_t, y_t, p_t, \theta_t, v_t]' \tag{2}$$

Each individual feature $\tilde{o}_{dt}$ ($d = 1,2,3,4,5$) is conformed to a zero-mean, unit variance normal distribution using:

$$o_{dt} = (\tilde{o}_{dt} - \bar{o}_d) / \sigma_{o_d} \tag{3}$$

Fig.2.(a) shows a signature example and Fig.2.(b) shows the pressure (p), velocity (v) and trajectory tangent angle (θ) features (before normalization) of Fig.2.(a).

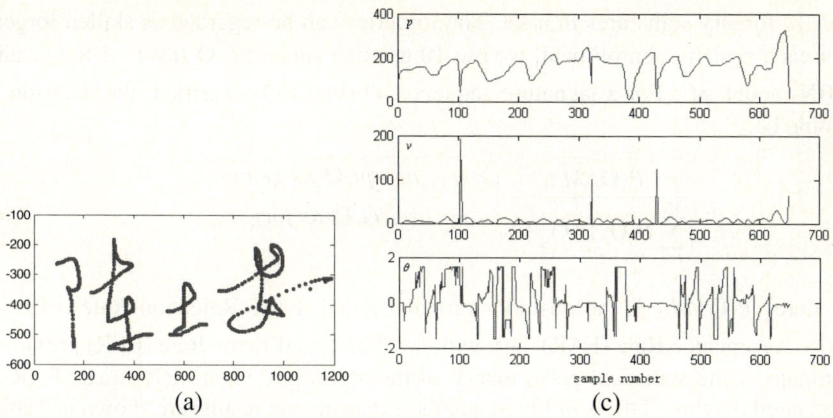

Fig. 2. A signature sample and its basis feature data

4 Topology Design of DBNs

In our experiments, we define the topology of the DBNs as Fig.3. Fig.3.(a) is a simple **H**idden **M**arkov **M**odel (HMM); Fig.3.(b) and (c) are called **C**oupled HMM (CHMM). In Fig.5.(a), O_t represent $(x_t, y_t, p_t, \theta_t, v_t)$; In Fig 5.(b), the first chain O_t^1 represent the static features (x_t, y_t, θ_t) and the second chain O_t^2 represent the dynamic features (p_t, v_t); In Fig.5.(c), O_t^1 represent (x_t, y_t), O_t^2 represent (v_t, θ_t) and the last chain p_t. In our DBNs, $Q_t^i (i = 1,2,3)$ are discrete. For observable node $O_t^i (i = 1,2,3)$ with hidden node Q_t^i, $p(O_t^i \mid Q_t^i = k)$ is a Gaussian distribution. We can use EM algorithm to estimate the parameters of the Gaussian distribution from training sequences [4].

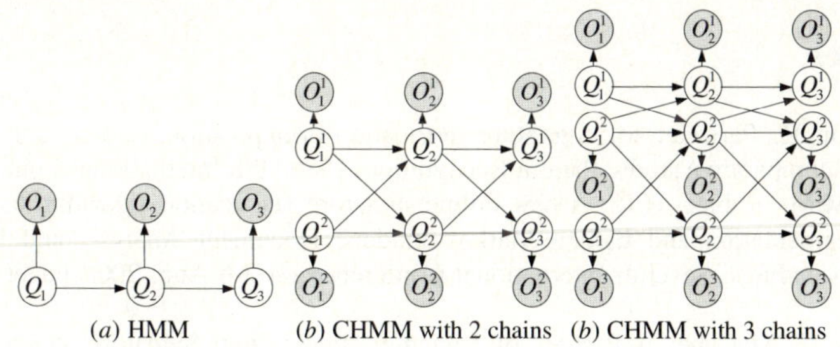

Fig. 3. The DBNs used in our experiments

5 Database and Experiments

The signature database consists of 3500 signature sequences. They are organized into 100 sets, and each set corresponds to one signature enrollment. There are 25 genuine and 10 forgery signatures in a set. The forgeries can be regarded as skilled forgeries. For each signature enrollment, we use 10 genuine signature $\mathbf{O}_i (i = 1 \sim 10)$ to train its DBN model M. For a signature sequence $\mathbf{O}$ that to be verified, the decision rule should be:

$$\frac{P(\mathbf{O} \mid M)}{\sum_{i=1}^{10} P(\mathbf{O}_i \mid M) \Big/ 10} \begin{cases} \geq w & \textit{accept } \mathbf{O} \textit{ as genuine} \\ < w & \textit{reject } \mathbf{O} \textit{ as forgery} \end{cases} \tag{4}$$

Here, $P(\mathbf{O} \mid M)$ is the posterior probability [4]. **F**alse **R**ejection **R**ate (FRR) and **F**alse **A**cceptance **R**ate (FAR) vary with w. The **E**qual **E**rror **R**ate (EER) provides an estimate of the statistical performance of the algorithm when FRR equals FAR. We compared the three DBNs in Fig.3. and the experimental results are shown in Table 1.

Table 1. Experimental results under different DBN models

DBN Models	HMM	CHMM (2 chains)	CHMM (3 chains)
EER (%)	2.9	1.5	1.9

6 Conclusion

This paper presents an approach of using dynamic Bayesian network in online signature verification. We discuss how to extract features of signature sequence, how to design the topology of DBNs and how to perform the verification. The experimental results show that DBNs are a promising way for signature verification. We can see that the CHMM with 2 chains achieves better performance than that with 3 chains. The possible reason maybe that CHMM with 2 chains is more suited for modeling signature procedure.

References

1. Plamondon, R., Parizeau, M.: Signature verification from position, velocity and acceleration signals: a comparative study. Pattern Recognition, 1988., 9th International Conference on
2. Shafiei, M.M., Rabiee, H.R.: A new online signature verification algorithm using variable length segmentation and hidden Markov models. Document Analysis and Recognition, 2003. Proceedings. Seventh International Conference on 3-6 Aug. 2003 Page(s):443 - 446 vol.1
3. G. Gupta, A. McCabe: A Review of Dynamic Handwritten Signature Verification. tech. rep., James Cook University, Computer Science Dept., 1997
4. Murphy, K.: Dynamic Bayesian Networks: Representation, Inference and Learning. PhD. Thesis, U.C. Berkeley, 2002

Multiobjective RBFNNs Designer for Function Approximation: An Application for Mineral Reduction

Alberto Guillén, Ignacio Rojas, Jesús González, Héctor Pomares, L.J. Herrera, and Francisco Fernández

University of Granada

Abstract. Radial Basis Function Neural Networks (RBFNNs) are well known because, among other applications, they present a good performance when approximating functions. The function approximation problem arises in the construction of a control system to optimize the process of the mineral reduction. In order to regulate the temperature of the ovens and other parameters, it is necessary a module to predict the final concentration of mineral that will be obtained from the source materials. This module can be formed by an RBFNN that predicts the output and by the algorithm that designs the RBFNN dynamically as more data is obtained. The design of RBFNNs is a very complex task where many parameters have to be determined, therefore, a genetic algorithm that determines all of them has been developed. This algorithm provides satisfactory results since the networks it generates are able to predict quite precisely the final concentration of mineral.

1 Introduction

Many dynamic optimization problems can be found during the process of the nickel production with the CARON technology. These problems require reaching a balance between the immediate gaining and the optimum behavior of the systems through the time. As an example, it can be considered the process of the mineral reduction. In this process, an expenditure of technologic petroleum is used to establish the thermic profile for the ovens that determine the different chemist reactions for the correct process. This is a complex task that nowadays requires a human operator to take decisions based on his experience and intuition. Therefore, it would be very helpful if a support decision system can be designed and implemented. Figure 1 shows the input, output and control variables that will be used to characterize the model. All these variables are registered by a SCADA system that generates the data used for the experiments.

The problem consists in the optimization of the necessities of the technological petroleum through the analysis of the data dynamically obtained. Since there are several necessities, we are tackling a multiobjective optimization problem. These kind of problems do not have an unique solution since the set of solutions cannot be completed sorted because, for some cases, it is impossible to decide

L. Jiao et al. (Eds.): ICNC 2006, Part I, LNCS 4221, pp. 511–520, 2006.

which solution is better. The set of solutions that cannot be improved is know as the optimal Pareto.

The mineral reduction process can be characterized by the following functions:

- Extractions = f_1 (Input mineral, Oven temperature, Reducing agents)
- Oven temperatures = f_2 (Input mineral, Chamber temperatures)
- Reducing agents = f_3 (Input mineral, Additive petroleum, Petroleum in chambers)

The problem then consists in the learning of the three different functions that relate the input vectors with the corresponding output. This is possible since the data will be measured directly from the source, and once these functions are learned, it will be possible to generate new values not defined in the training sets that will help to take decisions in order to optimize the process.

Figure 2 shows an hybrid process for the dynamic optimization. In the process, there is a module that has to approximate the behavior of the system in order to predict it and to give this information to the Neuro-programming optimization module that will provide the information to take decisions.

Most of the dynamic neuro-programming methods start with an initial sequence of control actions that are used to compute an initial value of the fitness function that will be used. This initial situation can be improved by modifying the the initial control actions. The algorithm proposed in this paper will be used in the function approximation module shown in Figure 2. This module has to include a predictor element, the RBFNN, and an algorithm that designs the network since the number of inputs vectors grows dynamically each 8 hours. This time frame is big enough to be able to use a genetic algorithm that will design a RBFNN that will approximate the input data.

The algorithm presented in this paper is able to design this kind of networks providing excellent results as it will be shown in the experiment section.

2 RBFNN Description

The problem to be tackled consists in designing an RBFNN that approximates a set of given values. The use of this kind of neural networks is a common solution since they are able to approximate any function [11, 10]. Formally, a function approximation problem can be formulated as, given a set of observations $\{(\boldsymbol{x}_k; y_k); k = 1, ..., n\}$ with $y_k = F(\boldsymbol{x}_k) \in \mathbb{R}$ and $\boldsymbol{x}_k \in \mathbb{R}^d$, it is desired to obtain a function $\mathcal{F}$ so $\sum_{k=1}^{n} ||y_k - \mathcal{F}(\boldsymbol{x}_k)||^2$ is minimum. The purpose of the design is to be able to obtain outputs from input vectors that were not specified in the original training data set.

An RBFNN $\mathcal{F}$ with fixed structure to approximate an unknown function F with n entries and one output starting from a set of values $\{(\boldsymbol{x}_k; y_k); k = 1, ..., n\}$ with $y_k = F(\boldsymbol{x}_k) \in \mathbb{R}$ and $\boldsymbol{x}_k \in \mathbb{R}^d$, has a set of parameters that have to be optimized:

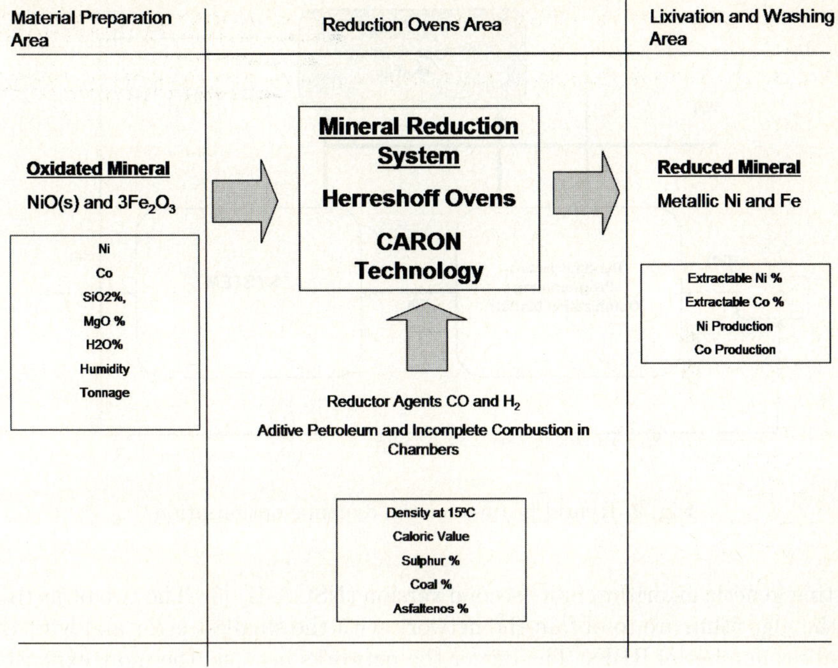

Fig. 1. Mineral reduction process

$$\mathcal{F}(\boldsymbol{x}_k; C, R, \Omega) = \sum_{j=1}^{m} \phi(\boldsymbol{x}_k; \boldsymbol{c}_j, r_j) \cdot \Omega_j \tag{1}$$

where $C = \{\boldsymbol{c}_1, ..., \boldsymbol{c}_m\}$ is the set of RBF centers, $R = \{r_1, ..., r_m\}$ is the set of values for each RBF radius, $\Omega = \{\Omega_1, ..., \Omega_m\}$ is the set of weights and $\phi(\boldsymbol{x}_k; \boldsymbol{c}_j, r_j)$ represents an RBF. The activation function most commonly used for classification and regression problems is the Gaussian function because it is continuous, differentiable, it provides a softer output and improves the interpolation capabilities [2,13].

The procedure to design an RBFNN starts by setting the number of RBFs in the hidden layer, then the RBF centers $\boldsymbol{c}_j$ must be placed and a radius r_j has to be set for each of them. Finally, the weights Ω_j can be calculated optimally by solving a linear equation system [4].

3 Multiobjective Algorithm for Function Approximation: MOFA

This section describes the algorithm that could be used in the prediction and function approximation module within the system described in the previous section. This algorithm is based in the popular multiobjective non-dominated

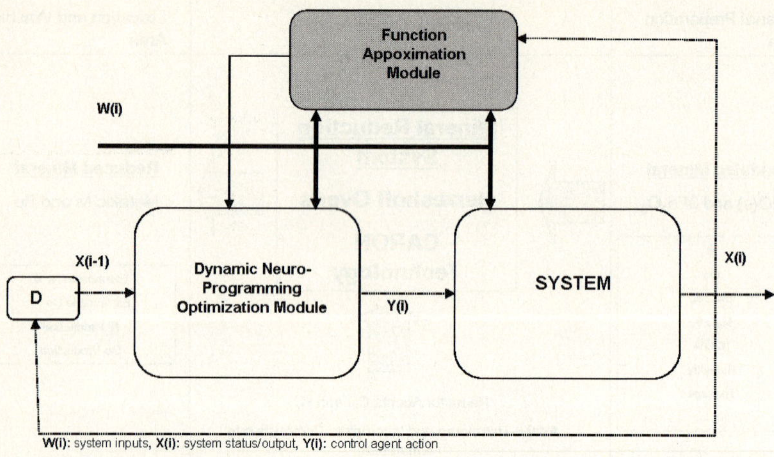

Fig. 2. Hybrid process for the dynamic optimization

sorting genetic algorithm in its second version (NSGA-II) [3]. The two objectives in the algorithm are to obtain the network with the smallest error and with the smallest number of RBFs. The bigger the networks become, the more expensive is its manipulation within the genetic algorithm, making it run very slowly and the network must be retrained each 8 hours. This section will introduce the new elements that have been incorporated to fit the original algorithm to the design of RBFNNs.

3.1 Representing RBFNN in the Individuals

As it was shown in the Introduction, to design an RBFNN it is needed to specify:

1. the number of RBFs
2. the position of the centers of the RBFs
3. the length of the radii
4. the weights for the output layer

The individuals in the population of the algorithm will contain the first three elements in a vector of real numbers. Instead of including the weights, the approximation error is stored in order to save computational effort by the time the individuals will be compared.

In the following subsections the concept of local error of an RBF will be referred. The local error is defined as the sum of the errors between the real output and the output generated by the RBFNN but, instead of considering all the input vectors, only the ones that activate each RBF will be selected. To know if an input vector activates a neuron, its activation function is calculated for each input vector and if it is higher than a determined threshold, the input vector activates the neuron.

3.2 Initial Population

The initial population is generated using clustering algorithms in order to supply good individuals that will make easier and faster to find good solutions. These clustering algorithms are:

- Fuzzy C-means (FCM): This clustering algorithm [1] performs a fuzzy partition of the input data where the same input vector can belong to several clusters at the same time with a membership degree.
- Improved Clustering for Function Approximation (ICFA): this algorithm [5] uses supervised clustering in order to identify the areas where the function is more variable. To do this, it defines the concept of estimated output of a center to assign a value for the center in the output axis.
- Possibilistic Centers Initializer (PCI) and Fuzzy-Possibilistic Clustering for Function approximation (FPCFA): these algorithms [6] modify the way the input vectors are shared between the centers of the clusters. In the ICFA algorithm, a fuzzy partition was defined. In these two algorithms the fuzzy partition is replaced by the ones used in [14] and in [9] respectively.

There are also included individuals generated randomly in order to not to loose diversity in the population.

After this initialization of the population, very few iterations of a local search algorithm ($Levenberg-Marquardt$ [8]) are run and the results are concatenated to the population. This have been proved to improve the quality of the results because the population becomes more diverse since the clustering algorithms are quite robust and could generate individuals that are too similar.

The size of the RBFNNs belonging to the first generation should be small for two reasons:

1. make the initialization as fast as possible
2. allow the genetic algorithm to determine the sizes of the RBFNNs from an incremental point of view, saving the computational effort that would suppose to deal with big networks from the first generations.

The cross operators will have the chance to increment the number of RBFs and with the mutation operators there will be the possibility of removing useless RBFs.

3.3 Crossover Operators

The original crossover operator over a binary or real coded chromosome cannot be performed with the individuals of this algorithm because each groups of genes have different meanings. Two crossover operators were designed for these individuals, and experimentally it was concluded that the application of both operators with the same probability provided better results than applying only one of them.

Crossover Operator 1: Neurons Exchange. This crossover operator, conceptually, would be the most similar one to the original crossover. Since the individuals represent an RBFNN with several neurons, the cross of two individuals will be the result of exchanging one neuron. This is exchange is represented in Figure 3. The advantages of this crossover operator is that it exploits the genetic material of each individual without modifying the structure of the network, the other advantage is its simplicity and efficiency.

Crossover Operator 2: Addition of the Neuron with the Smallest Error. This operator consists in the addition of the neuron with the smallest local error belonging to the other individual and it is represented in Figure 3. If the neuron with the smallest local error is very similar to another in the other network, the neuron with the second smallest error is chosen and so on. This operator will give the opportunity to increase the number of RBFs in one individual, allowing the algorithm to explore more topologies. A refinement step is performed right after the crossing, this refinement consists in the prune of the RBFs which does not influence the output of the RBFNN, to do this, all the weights that connect the processing units to the output layer are calculated and the neurons that do not have a significant weight will be removed.

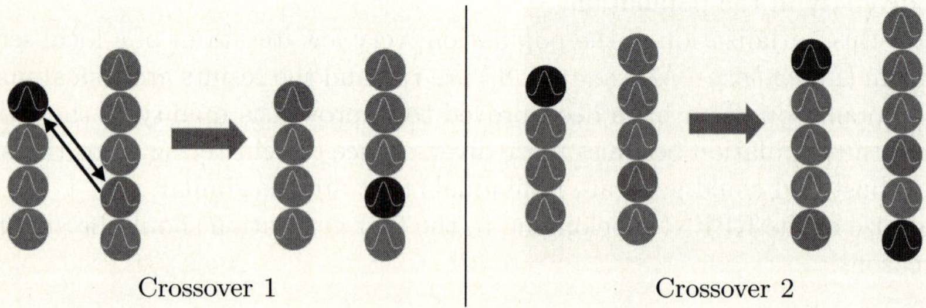

Fig. 3. Crossover operators 1 and 2

3.4 Mutation Operators

The mutation operators proposed for this algorithm can be separated in two categories:

- mutations without any knowledge
- mutations using expert knowledge

The mutation without any knowledge refers to those changes that are performed in a random way, those changes can affect both the structure and the parameters of the RBFNNs. The objective of these operators is to add randomness in the search process to avoid the convergence to local minima. The mutation operators with expert knowledge are mutations that affect also the structure and the parameters of the RBFNNs but using some information in such a way that the changes won't be completely random. As it occurs with the crossover operators, if we divide these subset of mutation operators and perform different runs, the results obtained are worse than if we run the algorithm using both kind of mutation operators.

Mutations without any Knowledge. There are four operators that are completely random:

- The first one is the deletion of an RBF in one random position over the input vectors space setting his radio also with a random value. All the random values are in the interval [0,1] since the input vectors and their output are normalized.
- The second operator is the opposite to the previous one, deleting an existing RBF. This mutation must be constrained and not be applied when the individual has less than two neurons.
- The third one adds to all the coordinates of a center a random distance which value is chosen in the interval [-0.5,0.5].
- The forth one has exactly the same behavior than the third one but changing the value of the radius of the selected RFB.

The two fist operators modify the structure of the network meanwhile the third and the forth modify the parameters of the network. The third and the fourth operators refer to the real coded genetic algorithms as presented in [7].

Mutations with Expert Knowledge. These mutation operators use the information provided by the output of the function to be approximated. As the previous operators, these will take care of the structure of the network, adding and removing RBFs in a RBFNN, and will also modify the value of the parameters of the RBFs. The operators are:

- The first operator inserts one RBF in the position of the input vector with the highest error. To select this position the output of the RBFNN is calculated and then it is compared with the output of the target function, the center will be placed in the position of the point where the difference between the real output and the generated output is greater.
- The second operator removes one RBF from the RBFNN, the RBF to be removed is the one with less local error. This could seem not too logical at first, but it allows to keep more diversity in the population since one of the cross operators adds the neuron of the individual with less local error, so combining this to elements, the genes will remain in the population but avoiding redundant elements and allowing to search for new areas in the input vector space.
- The third operator consists in the application of a local search algorithm (Levenberg-Mardquardt) to tune the positions of the centers and their radii but being aware that with these movements the error will be decreased for sure. This operator must be used carefully and only few iterations should be done, otherwise the population will converge too fast to a local minima.

4 Experiments

The data used in the experiments were obtained by measuring the following parameters from the real system:

– *Inputs*: Ni, Fe, Co, Si, Mg, Ton, Temp Ovens (9), *Output*: Additive petroleum index

The data were obtained measuring each 8 hours the different elements, so the prediction of the next value must be done in 8 hours time, the proposed genetic algorithm is able to provide appropriate results in that time frame. For the experiments, it was used the data obtained in 70 days so the size of the data set has 210 instances of 26 variables each. From this 210 instances, 180 were used for training and the rest for test.

The algorithm proposed in this paper will be compared with other evolutionary strategy presented in [12] where the authors propose a new evolutionary procedure to design optimal RBFNNs. It defines a self-organizing process into a population of RBFs based on the estimation of the fitness for each neuron in the population, which depends on three factors: the weight and width of each RBF, the closeness to other RBFs, and the error inside each RBF width. The algorithm also uses operators that, according to a set of fuzzy rules, transform the RBFs. All these elements allowed the authors to define cooperation, speciation, and niching features in the evolution of the population.

Table 1 shows respectively the approximation errors using the training set and the test set. The two evolutionary algorithms are compared also with the CFA algorithm that was designed specifically to perform the initialization step in the design of RBFNN for function approximation. Once the algorithms were executed a local search algorithm was applied to their results. These tables show how the proposed algorithm overcomes the other approaches when approximating the training data and the test data set. The other two algorithms are able to

Table 1. Mean of the approximation error (NRMSE) for the training and test data

Training				Test			
RBFs	**CFA**	**Rivera**	**MOFA**	RBFs	**CFA**	**Rivera**	**MOFA**
3	0.623	0.428	0.282	3	4.963	4.963	1.995
4	0.566	0.327	0.140	4	1.204	2.235	0.954
5	0.590	0.313	0.170	5	1.644	1.382	1.405
7	0.182	0.169	0.0140	7	1.309	1.410	0.870
8	0.061	0.054	3.333e-5	8	1.495	1.403	0.047
9	0.006	0.007	6.264e-6	9	1.479	1.339	0.047
10	0.003	0.002	4.513e-6	10	1.125	1.129	0.047
11	0.026	0.006	4.385e-6	11	1.128	1.105	0.043
12	0.017	0.002	2.946e-6	12	1.024	0.995	0.045
13	0.004	0.001	2.939e-6	13	0.730	0.755	0.047
14	8.264e-4	4.854e-5	2.896e-6	14	0.830	0.520	0.030
16	6.264e-4	3.644e-5	2.527e-6	16	0.519	0.312	0.029

approximate the training set with a very small error when many RBFs are used, however, when they try to approximate the test data, the error increases significantly. The networks generated by the proposed algorithm have the ability of approximating the training error quite precisely but without loosing generality so the test error is still small, not like when the other algorithms are used.

5 Conclusions

This paper has presented a system that can be used to control and optimize the mineral extraction from source materials. One of the modules that builds the system is in charge of predicting the final amount of extracted mineral from empirical data obtained previously. The module consists in an RBFNN, that is able to predict quite precisely the real output of material, and in a genetic algorithm that trains the network within the time frame required by the system. A multiobjective genetic algorithm that designs the RBFNNs for the prediction module was presented, obtaining a very good performance when it was compared against other techniques for the design of RBFNNs.

Acknowledgements. This work has been partially supported by the Spanish CICYT Project TIN2004-01419 and the European Commission's Research Infrastructures activity of the Structuring European Research Area programme, contract number RII3-CT-2003-506079 (HPC-Europa)

References

1. J. C. Bezdek. *Pattern Recognition with Fuzzy Objective Function Algorithms.* Plenum, New York, 1981.
2. A. G. Bors. Introduction of the Radial Basis Function (RBF) networks. *OnLine Symposium for Electronics Engineers*, 1:1–7, February 2001.
3. Kalyanmoy Deb, Samir Agrawal, Amrit Pratap, and T. Meyarivan. A fast and elitist multiobjective genetic algorithm: NSGA-II. *IEEE Trans. Evolutionary Computation*, 6(2):182–197, 2002.
4. J. González, I. Rojas, J. Ortega, H. Pomares, F.J. Fernández, and A. Díaz. Multiobjective evolutionary optimization of the size, shape, and position parameters of radial basis function networks for function approximation. *IEEE Transactions on Neural Networks*, 14(6):1478–1495, November 2003.
5. A. Guillén, I. Rojas, J. González, H. Pomares, L.J. Herrera, O. Valenzuela, and A. Prieto. Improving Clustering Technique for Functional Approximation Problem Using Fuzzy Logic: ICFA algorithm. *Lecture Notes in Computer Science*, 3512: 272–280, June 2005.
6. A. Guillén, I. Rojas, J. González, H. Pomares, L.J. Herrera, O. Valenzuela, and A. Prieto. A possibilistic approach to rbfn centers initialization. *Lecture Notes in Computer Science*, 3642:174–183, 2005.
7. F. Herrera, M. Lozano, and J. L. Verdegay. Tackling real-coded genetic algorithms: operators and tools for the behavioural analysis . *Artificial Intelligence Reviews*, 12(4):265–319, 1998.

8. D. W. Marquardt. An Algorithm for Least-Squares Estimation of Nonlinear Inequalities. *SIAM J. Appl. Math.*, 11:431–441, 1963.
9. N. R. Pal, K. Pal, and J. C. Bezdek. A Mixed C–Means Clustering Model. In *Proceedings of the 6th IEEE International Conference on Fuzzy Systems (FUZZ-IEEE'97)*, volume 1, pages 11–21, Barcelona, July 1997.
10. J. Park and J. W. Sandberg. Universal approximation using radial basis functions network. *Neural Computation*, 3:246–257, 1991.
11. T. Poggio and F. Girosi. Networks for approximation and learning. In *Proceedings of the IEEE*, volume 78, pages 1481–1497, 1990.
12. A. J. Rivera Rivas, J. Ortega Lopera, I. Rojas Ruiz, and M. J. del Jesus Daz. Coevolutionary Algorithm for RBF by Self-Organizing Population of Neurons. *Lecture Notes in Computer Science*, (2686):470–477, June 2003.
13. I. Rojas, M. Anguita, A. Prieto, and O. Valenzuela. Analysis of the operators involved in the definition of the implication functions and in the fuzzy inference proccess. *International Journal of Approximate Reasoning*, 19:367–389, 1998.
14. J. Zhang and Y. Leung. Improved possibilistic C–means clustering algorithms. *IEEE Transactions on Fuzzy Systems*, 12:209–217, 2004.

A New Time Series Forecasting Approach Based on Bayesian Least Risk Principle

Guangrui Wen and Xining Zhang

State Key Laboratory for Manufacturing System Engineering,
Xi'an Jiaotong University, 710049 Xi'an, P.R. China
{grwen, zhangxining}@mail.xjtu.edu.cn

Abstract. Based on the principle of Bayesian theory-based forecasting, a new forecasting model, called Bayesian Least Risk Forecasting model, is proposed in this paper. Firstly, the principle and modeling idea of Bayesian forecasting are illustrated with the explanation of the meaning of least risk forecasting. Then the advantages and learning algorithm of this model are discussed explicitly. In order to validate the prediction performance of Bayesian Least Risk Forecasting model, a simulated time series and practical data measured from some rotating machinery are used to compare the ability of prediction with classical artificial neural networks model. The results show that the bayesian model can contribute to a good accuracy of prediction.

1 Introduction

The principle, models and methods of Bayesian forecasting have been developed extensively for many years. This development has involved thorough investigation of mathematical and statistical aspects of forecasting models and related techniques [1]. With this has come experience with application in a variety of areas in commercial and industrial, scientific and socio-economic fields [2]. In this paper, we present a new method of least risk forecasting based on Bayesian theory. At the same time, the Principle, Modeling, Knowledge Expression, Learning Algorithm and Application of Bayesian Least Risk Forecasting are also given out.

2 Principle and Model of Bayesian Learning and Forecasting

In fact, some random variables are obtained continuously. For the continuous random variables, Bayesian theorem can be described as [3,4]:

$$\pi(\theta \mid x) = \frac{p(x \mid \theta)\pi(\theta)}{\int_{\theta} p(x \mid \theta)\pi(\theta)d\theta} \tag{1}$$

where $\pi(\theta)$ is priori probability distribution density, $p(x \mid \theta)$ is the sample information, and $\pi(\theta \mid x)$, called posterior probability distribution density [5]. θ 's

L. Jiao et al. (Eds.): ICNC 2006, Part I, LNCS 4221, pp. 521 – 524, 2006.

conditional probability distribution density function, the equation also can be expressed as:

$$\pi(\theta) = \frac{\pi(\theta \mid x) p(x)}{p(x \mid \theta)} \tag{2}$$

Here we take $\theta = x_{n+1}$, and then the above equation can be converted to:

$$\pi(x_{n+1}) = \frac{\pi(x_{n+1} \mid x) p(x)}{p(x \mid x_{n+1})} \tag{3}$$

where the right three items can be obtained by estimation of known information before

Supposed that the time series of forecasting is $\{x_i\}$, $i = 1, 2, 3, \cdots, n$, the model of next time can be expressed as:

$$x_k = x_x \mid \max P(x_x) \tag{4}$$

where $P(x_x) = \int_{x-\delta}^{x+\delta} \frac{\pi(x_{i+1} \mid x') p(x')}{p(x' \mid x_{i+1})} d\delta = \int_{x-\delta}^{x+\delta} \pi(x_{n+1}) d\delta$.

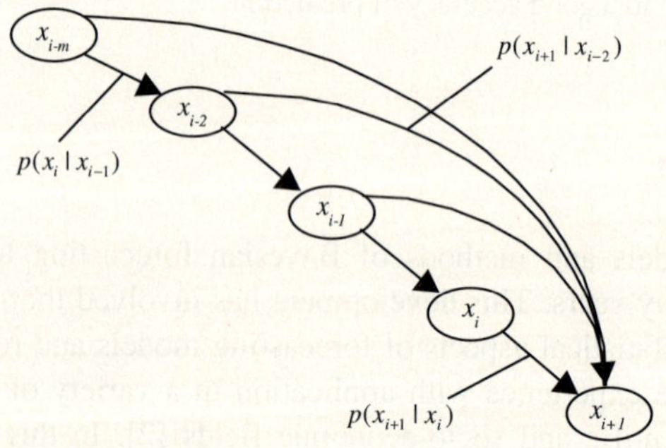

Fig. 1. Bayesian Least Risk Forecasting model

We adopt multiple data-in method to improve $\pi(x_{n+1})$'s calculation accuracy. So, the multiple dimension probability distribution density function such as $P(x), P(x_i \mid x_{i-1}), P(x_i \mid x_{i-2}), P(x_i \mid x_{i-1} x_{i-2}), \cdots$ can be obtained according to known information in advance. Supposed that the data for forecasting is m time interval before the time of n, and the conditional probability is $p(x_i \mid x_{i-1}), \ldots, p(x_i \mid x_{i-m})$. Then the Bayesian Least Risk Forecasting model can be expressed in Fig.1.

Therefore, probability distribution $\pi(x_{n+1})$ at some time $n+1$ can be obtained:

$$\pi(x_{n+1}) = a_0 p(x_i \mid x_{i-1}) + a_1 p(x_i \mid x_{i-2}) + a_2 \, p(x_i \mid x_{i-3}) + a_3 \, p(x_i \mid x_{i-4}) + \cdots \tag{5}$$

where $a_i, i = 1, 2, \cdots, k$ is weight coefficients.

3 Experimental Verification and Engineering Application

The experimental verification was conducted by two time series: a simulated time series and a practical vibration peak-to-peak value time series measured from large rotating machine In order to measure the ability of Bayesian Least Risk Forecasting model, the mean square error, also called prediction error, has been utilized [8].

The simulated time series is given by following equation:

$$x(t) = 5 + 10\sin(150\pi t) + 5 * \sin(94\pi t) + e \quad \text{where } t = (1:1024)*0.0005 \tag{6}$$

Here we initialize parameter t from $t=0$ to $t=1024$. Under the same running condition, Table 1 shows the contractive result in time costs and prediction errors between classical artificial neural network (ANN) model [9] and Bayesian Least Risk Forecasting model over the same training data and predicting data respectively. Fig.2 displays the training & prediction curves by using Bayesian forecasting model.

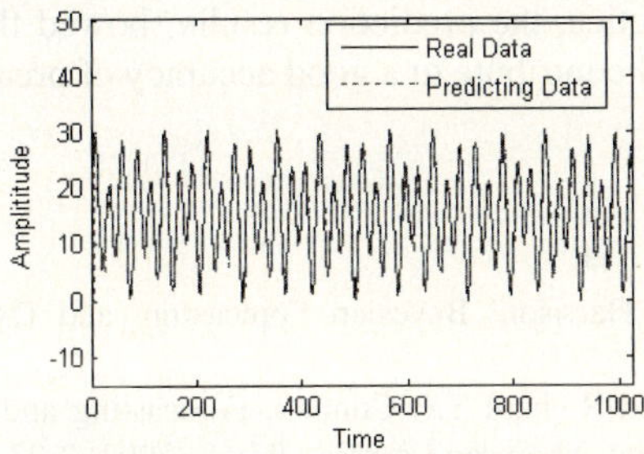

Fig. 2. Full training and prediction curve by using Bayesian forecasting model

Table 1. Comparision of corresponding parameters using two forecasting model

Item	Bayesian Model (512-1024)	ANN Model (512-1024)
Prediction Error	0.8206	6.1033
Training Time (S)	10.961	356.64

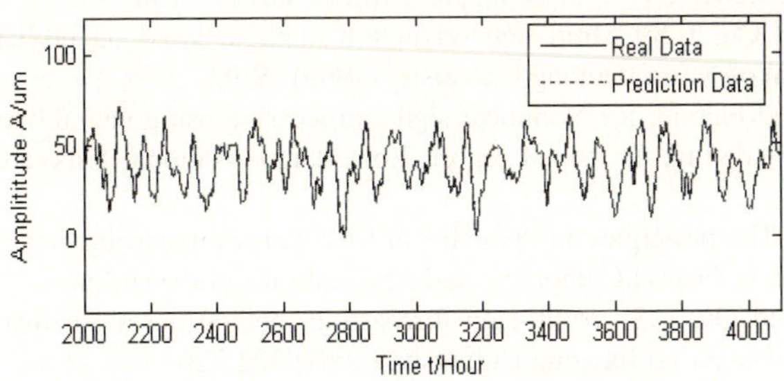

Fig. 3. Practical forecasting results using Bayesian Least Risk Forecasting model

The Trending method is well known in condition monitoring of large rotating machinery [10,11]. Moreover, in monitoring systems the peak-to-peak values of vibration signal is used as the trending object to supervise if there is any change in vibration condition of machine. Accordingly, we use Bayesian Least Risk Forecasting model to monitor the tendency of peak-to-peak value with time. Fig.3 shows the predicting trend of peak-to-peak value of this machine with hour by using this model. The results show that the Bayesian Least Risk Forecasting model possesses a better predictability in practice. It is evident that Bayesian Least Risk Forecasting model's prediction accuracy is much higher than that of classical ANN model.

4 Conclusion

The paper introduces the basic principle of Bayesian Least Risk Forecasting, and then constructs the model of forecasting. The advantages and learning algorithm of this forecasting model is also discussed in detail. By comparison with classical ANN model in prediction practice, the prediction results showed that Bayesian Least Risk Forecasting model could contribute to a good accuracy of prediction.

References

1. Mike West and Jeff Harrison: Bayesian Forecasting and Dynamic Models, Springer-Verlag, (1989) 1-8
2. Bruce L.Bowernan and Richard T.O'Connell: Forecasting and Time Series: An Applied Approach, Third Edition. Thomson Learning Press. (2003) 2-23
3. Rafael A.Calvo and Marwan Jabri: Benchmarking Bayesian neural networks for time series forecasting, http://www-2.cs.cmu.edu/~rafa/docs/acnn97/footnode.html#5, (1997)
4. Wang Jun and Zhou Weida: Research and Process of Bayesian networks, Electronic Science, (1999) 6-7
5. Wang Wei, Chen Enrong and Wang Xufa: Knowledge discovery based on Bayesian Approach. Journal of china university of science and technology, Vol 30(4) (2000) 468-472
6. Len Shanin, Tern Fengzhan and Lu Yuchang: Construction and applications in data mining of Bayesian networks, Journal of Tsinghua Univ (Sci & Tech), Vol 4(1) (2001) 49-52
7. Hu Zhenyu and Lin Shimin:Bayesian Learing of Bayesian Network, Journal of Guangxi Academy of Sciences, Vol.16(4&Supplement), (2000) 145-150
8. Xu, G.H and Qu, L.S.: Multi-step prediction method based on probability neural networks, Journal of Xi'an Jiaotong University (1999) 89-93
9. Laepes, A. and Farben, R.: Nonlinear signal processing using neural networks: prediction and system modeling, Technical report, Los Alamos National Laboratory, Los Alamos, NM (1987)
10. Hoptroff, R.: The principles and practice of time series forecasting and business modeling using neural nets, Neural Computing and Application 1 (1993) 59-66
11. Wen, G.R. and Qu, L.S.: Multi-step forecasting method based on recurrent neural networks, Journal of Xi'an Jiaotong University (2002) 722-726

Feature Reduction Techniques for Power System Security Assessment

Mingoo Kim[1] and Sung-Kwan Joo[2]

[1] Digital Communication Infra Division, Samsung Networks Inc.,
Seoul, 135-798, South Korea
mingoo.kim@samsung.com

[2] School of Electrical Engineering, Korea University,
Seoul, 136-713, South Korea
skjoo@korea.ac.kr

Abstract. Neural Networks (NN) have been applied to the security assessment of power systems and have shown great potential for predicting the security of large power systems. The curse of dimensionality states that the required size of the training set for accurate NN increases exponentially with the size of input dimension. Thus, an effective feature reduction technique is needed to reduce the dimensionality of the operating space and create a high correlation of input data with the decision space. This paper presents a new feature reduction technique for NN-based power system security assessment. The proposed feature reduction technique reduces the computational burden and the NN is rapidly trained to predict the security of power systems. The proposed feature reduction technique was implemented and tested on IEEE 50-generator, 145-bus system. Numerical results are presented to demonstrate the performance of the proposed feature reduction technique.

1 Introduction

Neural networks (NN) have been applied to assess the security of power systems and have shown great potential for achieving fast and accurate evaluation. The first step in applying NN to power system security assessment is the creation of an appropriate training data set. A common approach is to simulate the system in response to various disturbances and then collect a set of pre-disturbance system features along with a corresponding system security index. One of the most important factors in achieving good Neural Network (NN) performance has proven to be the proper selection of system features.

"Feature reduction" refers to the process of reducing the dimensions of the feature vector while preserving the needed information. There are two basic approaches to feature management: feature selection and feature extraction.

Feature selection involves reducing the dimensions of the input vector by selecting only features that are highly correlated with the desired analysis. A commonly used criterion that can be used for feature selection is Fisher discriminant function proposed by R.A. Fisher in 1936 [1]. This function seeks to find the optimal linear separation for two classes of data.

L. Jiao et al. (Eds.): ICNC 2006, Part I, LNCS 4221, pp. 525 – 534, 2006.

On the other hand, feature extraction is a transformation of the original data set into a new space of lower dimensions. Well known feature extraction techniques include the principal-components algorithm and the Karhuren-Loeve expansion [2]-[5].

Feature extraction has advantages over feature-selection methods in that all of the features are used to form a new set of composite features, therefore, in some cases, may preserving more of the original information [3]-[6]. This paper presents a new NN-based feature reduction technique for power system security assessment. By this neural network, the computational burden is reduced and the neural network is rapidly trained to predict the security index of large power systems.

This paper is organized as follows. In Section 2, feature selection techniques are described. In Section 3, the NN-based feature reduction technique is presented. Finally, numerical results are presented to validate the effectiveness of the proposed feature reduction technique in Section 4.

2 Feature Selection

A feature in and of itself could be important, yet when added to other important features form a less effective combination. This is the basic difficulty behind feature selection. Many ideas are applicable to classifying systems such as neural networks and are covered in a number of sources [1]-[10].

2.1 Fisher's Linear Discriminant: Selection Criteria

The Fisher approach is based on the projection of D-dimensional data onto a line [8]. For a two class example, a given training set

$$H = \{x_1, x_2, \cdots, x_n\} = \{H_1 \cup H_2\} \tag{1}$$

is partitioned into an $n_1 \leq n$ training vector in subset H_1, corresponding to class w_1, and an $n_2 \leq n$ training vector in set H_2, corresponding to class w_2, where $n_1 + n_2 = n$.

The task is to find a linear mapping $y = \mathbf{w}^T \mathbf{x}$ such as to maximize

$$F(\mathbf{w}) = \frac{|m_1 - m_2|^2}{\sigma_1^2 + \sigma_2^2}, \tag{2}$$

where m_i is the mean of class H_i and σ_i^2 is the variance of H_i.

The criterion function $\boldsymbol{F}$ can be rewritten as an explicit function of $\mathbf{w}$ as

$$F(\mathbf{w}) = \frac{\mathbf{w}^T S_B \mathbf{w}}{\mathbf{w}^T S_W \mathbf{w}} \tag{3}$$

where S_B is referred to as the "*between-class scatter*" matrix and S_W is the "*within-class scatter*" matrix. The *within-class scatter* S_W is defined as

$$S_W = S_1 + S_2 \tag{4}$$

where S_1 and S_2 are

$$S_i = \sum_{\mathbf{x} \in C_i} (\mathbf{x} - m_i)(\mathbf{x} - m_i)^T \tag{5}$$

and

$$m_i = \frac{1}{n_i} \sum_{\mathbf{x} \in H_i} \mathbf{x}. \tag{6}$$

The *between-class scatter* S_B is

$$S_B = (m_1 - m_2)(m_1 - m_2)^T. \tag{7}$$

Finally, the solution for w to maximize F can be written as

$$\mathbf{w} = S_W^{-1}(m_1 - m_2). \tag{8}$$

2.2 Forward Sequential Search Technique: Search Procedure

The main problem of any feature selection algorithm based on Fisher approach is how to deal with computational complexity. If we have a total of D possible features, then, since each feature can be either present or absent, there are, in all, 2^D possible feature subset to be considered. For a relatively small number of features, we might consider simply searching through all possible combinations and calculate their respective Fisher values. For a large number of features, however, such a method would become prohibitively expensive.

If we have already decided that we want to select d features, the number of is given by

$$N = \frac{D!}{(D-d)!d!}. \tag{9}$$

This number may be significantly smaller than 2^D but still be impracticably large in many applications.

In the paper, a "sequential forward feature-selection" algorithm based on the Fisher linear discriminant function is used for comparative purpose. The sequential forward method begins by considering each of the variables individually and selecting the one which gives the largest Fisher value. Then the entire feature set is built up by adding, one by one, the next feature that works best with those previously chosen. The full algorithm can be found in [4].

3 Neural-Network Feature-Extraction (NNFE)

This section describes an NN-based feature extraction technique, called Neural-Network Feature-Extraction (NNFE). A neural network can be used effectively for

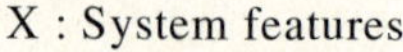

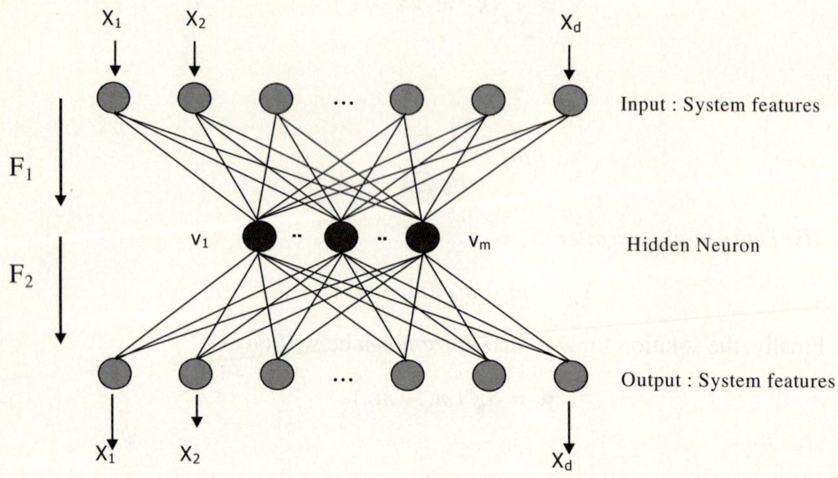

d : Number of features

m : Number of hidden neuron

Fig. 1. Forward selection illustrated for a set of four input features

non-linear dimensionality reduction, thereby overcoming some of the limitations of linear principal-component analysis. Fig. 1 shows a schematic NN structure used to achieve dimensionality reduction.

This neural network has *d* inputs neurons, *d* outputs neurons and *m* hidden neurons, with $m < d$. The input vector constituted by a training-data set, represented by *d*-dimensional vector ***X***, is presented sequentially as the input and output of the neural network. In other words, the neural network is designed and trained to reproduce its own input vector. This is done by means of an attempting to map each input vector onto itself.

The optimization goal is to minimize a sum-of –squares error of the form

$$E = \frac{1}{2}\sum_{n=1}^{N}\sum_{k=1}^{d}\left\{y_k(x^n) - x_k^n\right\}^2 \tag{10}$$

where *d* is the number features and *N* is the number of training data set.

After the neural network is trained, the vector ***v*** of the hidden variables represents the extracted features of the system. This network can be understood as a two-stage successive mapping operation F_1 and F_2. F_1 is a projection from the large *d*-dimensional input onto a small *m*-dimensional sub-space *S*. This is done through the input to hidden layer of the neural network. F_2 is a mapping from that subspace back to the full *d*-dimensional space. This process is represented by the hidden to output layer of the neural network.

Because learning in this network is non-linear, there is the risk of finding a sub-optimal local minimum for the error function. Also, one must take care to choose an appropriate number of m-neurons. Assuming that the two-stage mapping has been successively achieved, the middle layer of this network represents the extracted features and is a valid representation of the original system.

Such a network is said to be an "auto-associator" or "auto-encoder." This feature-reduction system works with any kind of activation function. If the activation functions are set to be linear, this procedure is similar to the conventional principal component approach. On the contrary, if the hidden neurons have non-linear activation functions, such a network effectively performs a non-linear principal component analysis and has the advantage of not being limited to linear transformations. After the NN of Fig. 1 is trained, the outputs of its hidden layer are used as inputs to train the security-assessment network.

By this two-step method, the computational burden is placed on the feature-extraction network, which is used off-line and only once per system. In the second step, the NN is rapidly trained to identify the security index on-line.

4 Performance Comparison of Feature Selection and NNFE

As a feature-selection technique, a forward sequential feature-selection algorithm based on the Fisher linear discriminant function is used for comparison. Even though there are many feature-selection algorithms, the Fisher linear discriminant function has been shown to provide an optimal discriminating function for linear classifiers [1]. In [11], such a technique was applied for power-system security assessments and excellent results were demonstrated.

The case study has been done on the IEEE 50-generator system to illustrate the effectiveness of the feature-extraction technique proposed.

4.1 The Data Set Generation

Before feature selection/extraction and neural-network training can be carried out, a representative data set must be generated. The greater the spread of data in the operating space, the greater effect on the discrimination of the neural network for security assessment. A fully representative training data set is critical for NN accuracy. The training data must contain enough operating conditions so that the neural network can adequately learn to predict the security of the system.

In the paper, Critical Clearing Time (CCT) is used as a security index. CCT is the maximum elapsed time from the initiation of a fault until its isolation such that the power system remains transiently stable. The ETMSP (Extended Transient-Midterm Stability Program) software was used to compute the CCT for various contingency.

4.2 Training/Testing Features

A power system has thousands of measurements, such as line flow, current, and voltage. These are all variables which can be used to assess the security of the system as a whole. A good choice of features is said to be half of the success. Although theories exist as

Table 1. Key features for dynamic security assessment

No.	Features
1	Pbus
2	Qbus
3	Total System Load - P,Q
4	Generator P
5	Generator Q
6	Vbus
7	Bus Angles or Bus Angular Separations
8	Generator Internal Angles
9	Real Flow on a Branch
10	Reactive Flow on a Branch
11	Total System Generation
12	System Exports – P,Q
13	Interface Flows, P
14	Generator Var Reserves – Q
15	System Var Reserve – Q
16	Generator Speed at Fault Clearing (FC)
17	Kinetic Energy at FC
18	Corrected Kinetic Energy at FC
19	Driving Point (Transfer) Impedance
20	Generator Accelerations at FC

guides for feature selection/feature extraction, this is not the case for finding good features to begin with.

Several approaches have been proposed for selecting an appropriate subset of features. Table 1 is a list of key features for dynamic security assessment as suggested by ABB (Asea Brown Boveri) in their ERPI (Electric Power Research Institute) project [12]. They have been tested for dynamic security problems using NN [13].

4.3 IEEE 50-Generator 145-Bus System Case Study

The IEEE 50-generator 145-bus system was used to test the proposed feature extraction technique. This system is derived from a representative model of a realistic power system in North America [14]. Fig. 2 is a one-line diagram of a portion of the high-voltage lines. The IEEE 50-generator system consists of 50 generators, 145 buses, and 453 transmission lines. Among the 50 generators, 44 represented using the classical model and the remaining 6 generators are represented using a two-axis model with a one-gain one-time constant exciter.

In this test, a three-phase fault was simulated at bus #10 and cleared by opening the line between bus #10 and bus #6. 2000 patterns were generated by changing the generation and load levels between 70 % and 140 % of their nominal values. The data set was divided into 1000 training patterns and 1000 test patterns. Fig. 3 is a histogram of CCT that shows the distribution of CCT ranging from 0 sec to 0.5 sec.

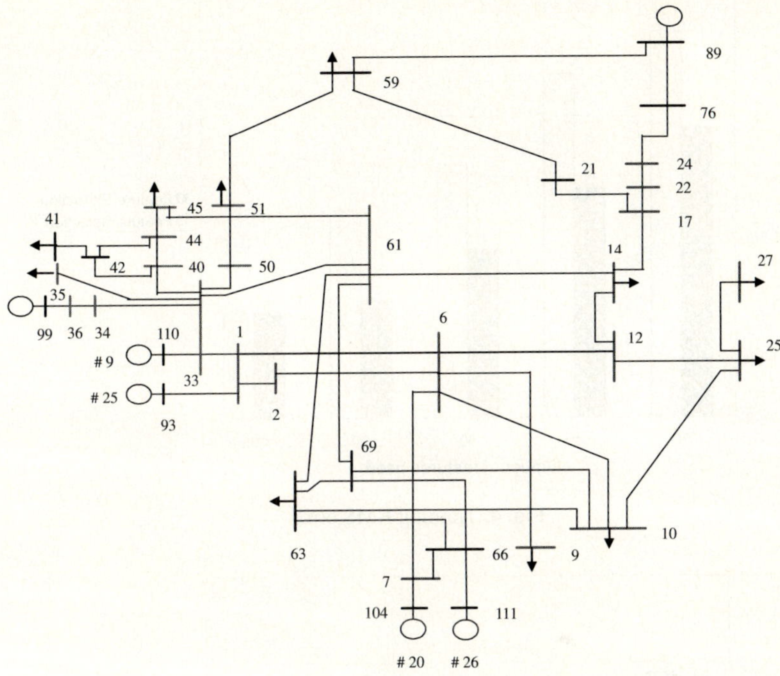

Fig. 2. A portion of the high voltage lines of the IEEE 50-generator 145-bus system

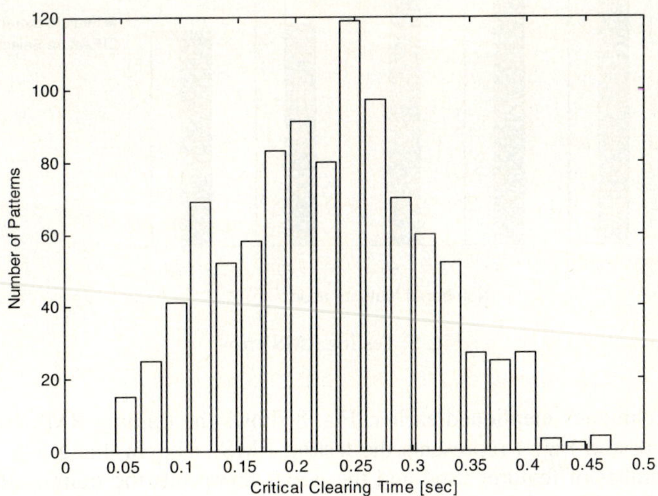

Fig. 3. Distribution of CCT

The total generated data set comprises 102 features, which are the Real and Reactive power output of each of the 50 generators and the corresponding Total System Load. The test cases are carried out for 30, 40, 50 or 60 features respectively, by both

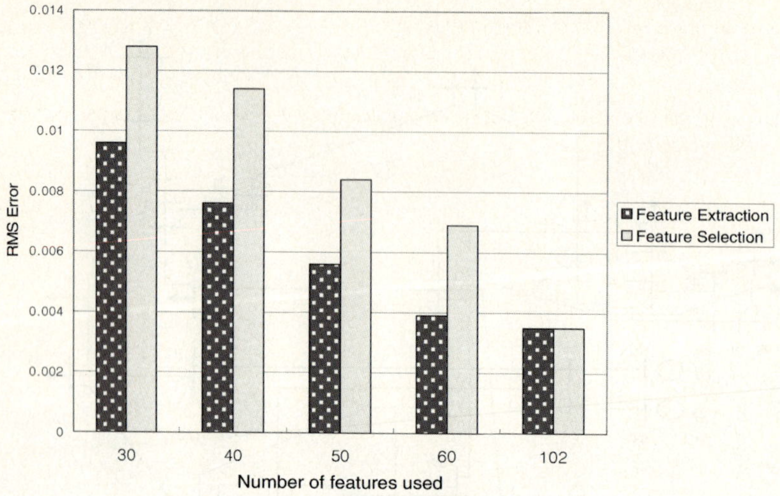

Fig. 4. Training RMS error

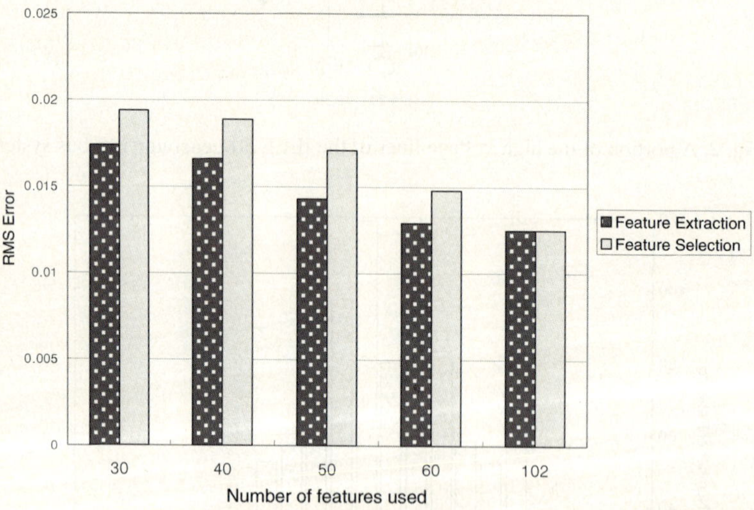

Fig. 5. Testing RMS error

of the two techniques mentioned earlier. Fig. 8 shows the training RMS error for the neural network and Fig. 4 represents the testing RMS error. In this graph, the *x-axis* shows the number of features used and the *y-axis* represents the training RMS error. The left bar of each pair indicates the RMS error when NNFE is applied. The right bar indicates the RMS error when feature selection technique is applied. As can be expected, a steady decline of RMS error rate is shown. As the number of features increases, the RMS error decreases.

As can be seen in this graph, *the training and testing results were notably improved by use of the feature extraction technique.* The extracted features should contain all of the information of the original features even though the resulting set has a lower dimensionality. That is the reason why the test performance of NNFE is superior to the feature-selection technique.

We obtained the smallest RMS value when we use the entire set of features, because most of the raw features have a high correlation with power-system status and redundancy has been minimized. We obtained almost same results, however, using just 60 features, which is 59 % of original number. This means that NNFE is very attractive in the case of larger systems.

5 Conclusion

This paper presents a new NN-based feature reduction technique for power system security assessment. It is evident that neural networks can be used for feature extraction as well as for power system security assessment. It is shown in the case study that the training and testing results can be notably improved by use of the feature extraction technique. When feature-extraction techniques are used, the training and testing times are substantially reduced. This should facilitate the task of on-line assessment of power system security. The next phase of this research is to extend the proposed feature reduction technique to anticipate the security of power systems for cascading events [15] which lead to severe consequences and in loss of service to a large number of customers.

References

1. Fisher, R.A.: The Use of Multiple Measurements in Taxonomic Problems. Annals of Eugenics, Vol. 7. (1936) 179-188
2. Weerasooriya, S. and El-Sharkawi, M.A.: Use of Karhunen-Loe've Expansion in Training Neural Networks for Static Security Assessment. Proceedings of the First International Forum on Applications of Neural Networks to Power Systems. (1991) 59-64
3. Schalkoff, R.J.: Pattern Recognition: Statistical, Structural and Neural Approaches. John Wiley & Sons (1992)
4. Bishop, C.M.: Neural Networks for Pattern Recognition. Oxford University Press. (1995)
5. Ripley, B.D.: Pattern Recognition and Neural Networks. Cambridge University Press. (1996)
6. Zayan, M.B., El-Sharkawi, M.A., and Prasad, N.R.: Comparative Study of feature Extraction Techniques for Neural Network Classifier. Proceedings of Intelligent Systems Applications to Power Systems. (1996) 400-404
7. Rao, C.R.: The Utilization of Multiple Measurements in Problems of Biological Classfication. Journal of the Royal Statistical Society Series (1948) 159-203
8. Duda, R.O., and Hart, P.E.: Pattern Classification and Scene Analysis. John Wiley & Sons. (1973)
9. Kittler, J., Etemadi, A., and Choakjarernwanit, N.: Feature Selection and Extraction in Pattern Recognition. In Pattern Recognition and Image Processing in Physics. Proceedings of the 37th Scottish University Summer School in Physics. (1990)

10. Jensen, C.A.: Application of Computational Intelligence to Power System Security Assessment. Ph.D. Dissertation. University of Washington. (1999)
11. Jensen, C.A., El-Sharkawi, M.A., and Marks, R.J.: Power System Security Assessment Using Neural Networks: Feature Selection Using Fisher Discrimination. IEEE Transactions on Power Systems. (2001) 757 -763
12. EPRI Report: Dynamic Security Analysis – Feasibility Evaluation Report. EPRI. (1994)
13. Mansour, Y., Vaahedi, E., and El-Sharkawi, M.A.: Dynamic Security Contingency Screening and Ranking Using Neural Networks. IEEE Transactions on Neural Networks.Vol. 8. (1997) 942-950
14. IEEE Committee Report: Transient Stability Test Systems for Direct Stability Methods. IEEE Transactions on Power Systems. (1992) 37-44
15. Taylor, C.W., and Erickson, D.C.: Recording and Analyzing the July 2 Cascading Outage. IEEE Computer Applications in Power, Vol. 10. (1997) 26-30

Harmonic Source Model Based on Support Vector Machine

Li Ma, Kaipei Liu, and Xiao Lei

School of Electrical Engineering, Wuhan University, Wuhan, China
mali99532@126.com

Abstract. To analyze the harmonics in power system efficiently, a new harmonic source model is proposed in this paper. And this new model takes advantage of support vector machine (SVM) theory to find the relationship between the harmonic current and all voltage components. Then a comparison between the linear regressive model and nonlinear regressive models with different kernel functions has been made. The computer simulation has revealed that the model implemented by the nonlinear regression with Polynomial kernel is more precise, and is superior to other regressions.

1 Introduction

Conventional ac electric power systems are designed to operate with sinusoidal voltages and currents. However, nonlinear and electronically switched loads will distort steady state ac current and voltage waveforms, which can result in additional heating in power system equipment, unmotivated switching of breakers, blowing of fuses, and interference with communication systems [1], [2].

The IEEE Standard 519-1992 (IEEE Recommended Practices and Requirements for Harmonic Control in Electrical Power Systems) has established a set of limits to the acceptable level of current harmonics in the power system which ensures that each power consumer plays his or her part in keeping the harmonic distortion levels low.

So a brief and accurate model for the source of harmonics is necessary. This model can help to predicate harmonics with the special power supply situation and eliminate those harmonics in time. The simple method for modeling can be implied in figure 1 which gives a series of functional blocks.

There are two main sources of harmonics in conventional power system: devices involving electronic switching and devices involving arc-furnaces or ferromagnetic reactors. There are a number of techniques presently being used for power system harmonic analysis. These techniques vary in terms of data requirements, modeling complexity, problem formulation and solution algorithms. According to different working principle, the former sources can be expressed as a set of differential algebra equations, while the others can be set by the characters between voltages and currents.

In this paper, it firstly presents a simplified model based on support vector machine. It approximately denotes characters of harmonic sources with a zero

L. Jiao et al. (Eds.): ICNC 2006, Part I, LNCS 4221, pp. 535–544, 2006.

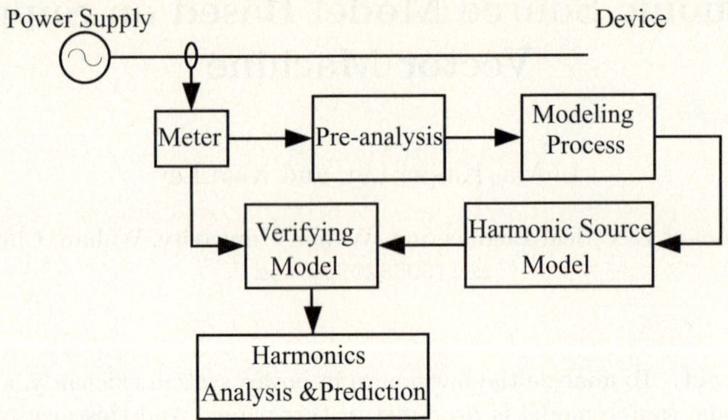

Fig. 1. Harmonic analysis and predication system

fundamental voltage phase angle. And then the brief introduction of algorithm is given. Finally, this paper has compared the accuracy between linear regressive models and nonlinear regressive models with different kernel functions. The simulation has showed that the nonlinear model with polynomial kernel is more precise than linear one.

2 The Simplified Model of Harmonic Sources

The simplified harmonic source model can be built up with the notion that releasing the relationship between the input voltages and harmonic currents. This relationship can be found by some special methods, this paper takes advantages of support vector machine algorithm to find the best solution.

At first, when voltages are known, the sources of harmonics are modeled as a supply voltage-dependent current source [3]:

$$\dot{I}_h = F_h(\dot{V}_1, \dot{V}_2, \ldots, \dot{V}_h, \ldots, C) \qquad h = 2, \ldots, N, \ldots \quad . \tag{1}$$

Where $\dot{I}_h$ is the h*th* vector of harmonic current absorbed by loads. $\dot{V}_1, \dot{V}_2, \cdots$ are fundamental and harmonic vectors of the supply voltages, C is a set of control variables of load characters. Although the configurations and parameters in harmonic sources vary with time randomly, it can be approximately regarded as invariability in a short period. In this condition, harmonic currents in harmonic sources only rely on harmonic voltages of all orders. In the power system, harmonic components of current usually decrease when the harmonic orders increase, so the voltages of high orders can be ignored, and (1) is simplified as:

$$\dot{I}_h = F_h(\dot{V}_1, \dot{V}_2, \ldots, \dot{V}_h, \ldots, \dot{V}_N) \qquad h = 2, \ldots, N \quad . \tag{2}$$

With a zero fundamental voltage phase angle, (2) can be written as (3) with the bias angle between every harmonic voltage component and fundamental voltage:

$$I_h\angle\theta_{Ih} = F_h(V_1, 0\,, \ldots, V_h, \theta_{Vh}, \ldots, V_N, \theta_{VN}) \quad h = 2, \ldots, N \quad . \tag{3}$$

Suppose:

$$\begin{cases} V_h\angle\theta_{Vh} = V_{hr} + j\cdot V_{hi} = V_h\cos\theta_{Vh} + j\cdot V_h\sin\theta_{Vh} \\ I_h\angle\theta_{Ih} \;= I_{hr} + j\cdot I_{hi} = I_h\cos\theta_{Ih} + j\cdot I_h\sin\theta_{Vh} \quad . \end{cases} \tag{4}$$

Separating the real and imaginary part of every vector, (4) can get:

$$\begin{cases} I_{hr} = G_{hr}(V_1, V_{2r}, V_{2i}, \ldots, V_{Nr}, V_{Ni}) \\ I_{hi} = G_{hi}(V_1, V_{2r}, V_{2i}, \ldots, V_{Nr}, V_{Ni}) \quad . \end{cases} \tag{5}$$

This function implies that every harmonic current component has direct relationship with all input voltage components, and following this notion a modeling process based on support vector machine can be proposed.

3 The Algorithm Based on Support Vector Machine

For brief illustration a new vector with real and imaginary part of every voltage component can be supposed:

$$\mathbf{V} = [V_1, V_{2r}, V_{2i}, \ldots, V_{Nr}, V_{Ni}]^T \quad . \tag{6}$$

Based on the theory of support vector machine [4], [11], [12], (5) can be briefly written as:

$$I = \langle\, \varphi(\mathbf{V}) \cdot \omega \,\rangle \quad . \tag{7}$$

Where $\varphi(\mathbf{V})$ is the activation function vector bases on the vector $\mathbf{V}$, ω is weight vector. Generally, $\varphi(\mathbf{V})$ is known, the main purpose is to find the value of ω. And this is the simplified model with a zero fundamental voltage phase angle based on the SVM theory. After getting all value of vector ω, a model of harmonic source can be obtained. For the case of nonzero fundamental voltage phase angle, it can be transferred from the one with a zero fundamental voltage phase angle.

Get the sampling data I_{t_k} ,V_{t_k} $(k = 1, 2, \cdots l)$, and the goal function of support vector machine is:

$$\begin{aligned} &Min\ \ \frac{1}{2}\|\omega\|^2 + C\sum_{k=1}^{l} L(\xi_k^{(*)}) \\ &st.\ I_{t_k} - \langle\, \varphi_k \cdot \omega \,\rangle \le \varepsilon + \xi_k \\ &\quad \langle\, \varphi_k \cdot \omega \,\rangle - I_{t_k} \le \varepsilon + \xi_k^* \\ &\quad \xi_k^{(*)} \ge 0 \quad . \end{aligned} \tag{8}$$

Where $\varphi_k = \varphi(V_{t_k})$, $L(\xi_k^{(*)})$ means this function has two variables ξ_k and ξ_k^*, the optimizing generalization should be obtained by minimizing (8). Then the dual form of (8) is defined as:

$$
\begin{aligned}
Min \quad L_{PD} = &\frac{1}{2}\|\omega\|^2 + \sum_{k=1}^{l} \alpha_k (I_{t_k} - \omega^T \varphi_k - \varepsilon) \\
&+ \sum_{k=1}^{l} \alpha_k^* (-I_{t_k} + \omega^T \varphi_k - \varepsilon) \\
&+ C \sum_{k=1}^{l} L(\xi_k^{(*)}) - \sum_{k=1}^{l} \Big[\beta_k \xi_k + \beta_k^* \xi_k^* + \alpha_k \xi_k + \alpha_k^* \xi_k^* \Big] \ .
\end{aligned} \tag{9}
$$

Where C is the penalization constant of such deviations when regressive precision exceeds permitted value, and C>0. ξ_k and ξ_k^* are the regressive slack variables which give the yardstick for whether exceeding the final target value or not. And $\beta_k, \beta_k^*, \alpha_k, \alpha_k^*$, are Lagrange multipliers. Suppose:

$$
\begin{cases}
e_k = I_{t_k} - \omega^T \varphi_k & e_k^* = -I_{t_k} + \omega^T \varphi_k \\
\lambda_k = \dfrac{2\alpha_k}{I_{t_k} - \omega^T \varphi_k - \varepsilon} & \lambda_k^* = \dfrac{2\alpha_k^*}{-I_{t_k} + \omega^T \varphi_k - \varepsilon}
\end{cases} \ . \tag{10}
$$

If $\Phi = [\varphi_1, \varphi_2, \varphi_3, , \ldots \varphi_l]$, then the corresponding Karush-Kuhn-Tucker complementarity conditions [5] are:

$$
\frac{\partial L_{PD}}{\partial \omega} = \omega - \Phi^T D_\lambda [I - \Phi\omega - \varepsilon] + \Phi^T D_{\lambda^*} [-I + \Phi\omega - \varepsilon] = 0 \ . \tag{11}
$$

$$
\frac{\partial L_{PD}}{\partial \xi_k^{(*)}} = C \frac{dL(\xi_k^{(*)})}{d\xi_k^{(*)}} - \alpha_k^{(*)} - \beta_k^{(*)} = 0 \qquad \forall k = 1, \ldots, l \ . \tag{12}
$$

$$
\beta_k^{(*)} \xi_k^{(*)} = 0 \qquad \forall k = 1, \ldots, l \ . \tag{13}
$$

And I is the vector of sampling data. If $D_{\lambda+\lambda^*}$ denotes the diagonal matrix whose main diagonal kth element is $\lambda_k + \lambda_k^*$; $D_{\lambda-\lambda^*}$ denotes the diagonal matrix whose main diagonal kth element is $\lambda_k - \lambda_k^*$. With simply calculation, the value of vector ω is derived from (11) as follows:

$$
\omega = \Big[1 + \Phi^T D_{\lambda+\lambda^*} \Phi\Big]^{-1} \Big[\Phi^T D_{\lambda+\lambda^*} I - \Phi^T D_{\lambda-\lambda^*} \varepsilon\Big] \ . \tag{14}
$$

And equation (14) can be rewritten after the corresponding supposes that $\omega = \Phi^T \delta$ is introduced:

$$
\delta = \Big[1 + D_{\lambda+\lambda^*} K\Big]^{-1} \Big[D_{\lambda+\lambda^*} I - D_{\lambda-\lambda^*} \varepsilon\Big] \ . \tag{15}
$$

Where Φ is called the feature space which is mapped from the original input space **V**. $K_{ij} = \varphi^T(V_{t_i})\varphi(V_{t_j})$ is the suitable kernel function for the SVM algorithm which definitely satisfy the Mercer's theorem. In practice the approach defines a kernel function directly, gets feature space and works out the inner product in that space. Finally it computes that value of kernel in terms of the original inputs. Briefly, the value of ω with the algorithm above depends on which kernel function you choose.

According to (10), (12) and (13), we get following equation for value of λ_k, λ_k^*:

$$\lambda_k^{(*)} = \begin{cases} 0 & e_k^{(*)} - \varepsilon < 0 \\ \dfrac{2C}{e_k^{(*)} - \varepsilon} \left. \dfrac{dL(\xi^{(*)})}{d\xi^{(*)}} \right|_{\xi^{(*)}} & e_k^{(*)} - \varepsilon \geq 0 \end{cases} . \tag{16}$$

Finally, the detailed process of model foundation with the SVM theory can be concluded as follows:

Step1: On the condition of invariable load characes, change the waveform of supplied voltage, get learning set with different harmonic vectors of voltages and currents;

Step2: Start with an arbitrary δ_0, set $n = 1$;

Step3: Calculate errors: $e_k = I_{t_k} - \omega^T \varphi_k$, $e_k^* = -I_{t_k} + \omega^T \varphi_k$;

Step4: Calculate λ_k, λ_k^* as given in (16);

Step5: Solve (15) to obtain δ_n;

Step6: Come to end if $\|\delta_n - \delta_{n-1}\|$ is smaller than a small value or process times reach the maximum value ; otherwise, set $n = n+1$, then return to step3;

Step7: Solve $\omega = \Phi^T \delta$ to obtain ω.

4 Analysis of Algorithm

For validating the rationality of above simplified harmonic source model, this paper takes three-phase full-bridge controlled rectifier with load of resistance and inductance in series for example in figure 2. The alpha firing angle is 30 degree. The amplitudes of harmonic voltage are set randomly in the range (0, 0.1)pu. The phase of every harmonic voltage component is also set randomly in the range $(-0.5\pi, 0.5\pi)$. As we know, harmonic orders or multiples of the fundamental frequency of this kind of rectifier are $6k \pm 1$, viz. the $5th$, $7th$, $11th$, $13th$ etc. What's more, harmonic components of current usually decrease when the harmonic orders increase. So in this simulation, harmonics are only set as the $5th$, $7th$, $11th$ and $13th$ without considering high order harmonics.

4.1 Linear Regressive Model

At first, define:

$$\varphi(\mathbf{V}) = \mathbf{V} = [V_1, V_{2r}, V_{2i}, \ldots, V_{Nr}, V_{Ni}]^T \quad . \tag{17}$$

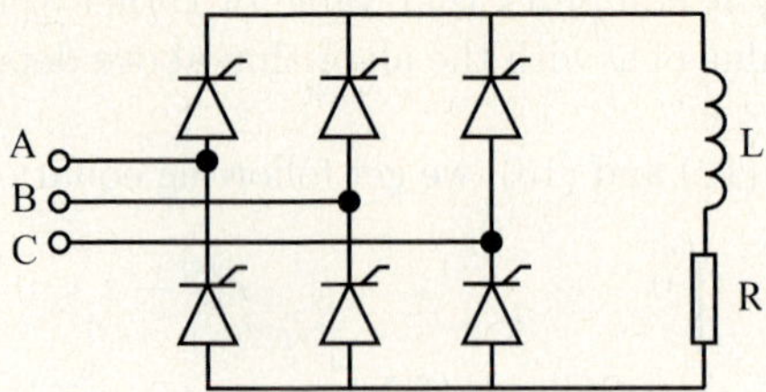

Fig. 2. Three-phase full-bridge controlled rectifier

So the $K_{ij} = \varphi^T(V_{t_i})\varphi(V_{t_j}) = V_i^T V_j$ can be defined. And the cost function is set as Huber function [6]:

$$L(\xi) = \begin{cases} \frac{1}{2\sigma}\xi_k^2 & 0 < \xi_k < \sigma \\ \xi_k - \frac{1}{2}\sigma & \xi_k \geq \sigma \end{cases} . \tag{18}$$

Considering $\xi_k^{(*)} = e_k^{(*)} - \varepsilon$, we can get:

$$\lambda_k = \begin{cases} \frac{2C}{e_k - \varepsilon} & e_k > \sigma + \varepsilon \\ \frac{2C}{\sigma} & \varepsilon \leq e_k \leq \sigma + \varepsilon \\ 0 & elsewhere \end{cases} \quad \lambda_k^* = \begin{cases} \frac{2C}{e_k^* - \varepsilon} & e_k^* > \sigma + \varepsilon \\ \frac{2C}{\sigma} & \varepsilon \leq e_k^* \leq \sigma + \varepsilon \\ 0 & elsewhere \end{cases} . \tag{19}$$

Then the vector ω can be obtained from the SVM algorithm steps, in table 1, we list out the value of vector ω (we only research the $5th$ and $7th$ harmonic wave). And because of the linear regression, the harmonic model is defined as follow:

$$\begin{aligned} I_{5r} &= a_1 V_1 + a_{5r} V_{5r} + a_{5i} V_{5i} + a_{7r} V_{7r} + a_{7i} V_{7i} \\ &+ a_{11r} V_{11r} + a_{11i} V_{11i} + a_{13r} V_{13r} + a_{13i} V_{13i} \end{aligned} . \tag{20}$$

$V_1$is fundamental component, V_{*r} is $*th$ harmonic voltage real components, V_{*i} is $*th$ harmonic voltage imaginary components. I_{5i}, I_{7r}, I_{7i} shares the same model frame as (20).

In the simulation, the SVM harmonic model is constructed by using the data sampled from three-phase full-bridge controlled rectifier circuit with the number of 500. And the regression model is tested with another 1000 sampling data.

Using the SVM model foundation process mentioned above, the value of weight in the linear model can be derived from the loop calculation. The table 1 has the details.

The testing error is evaluated by the average of absolute value from relative error.The testing errors of real part and image part in $5th$ harmonic current are

Table 1. Part coefficients in simplified model of three-phase full-bridge controlled rectifier

Current	a_1	a_{5r}	a_{5i}	a_{7r}	a_{7i}
I_{5r}	0.0068	0.0209	0.0266	-0.0164	0.0355
I_{5i}	0.0084	0.0087	0.0075	-0.0312	-0.0316
I_{7r}	0.0013	-0.0038	-0.0427	0.0173	0.0128
I_{7i}	-0.0038	0.0181	-0.0066	0.0062	0.0149
Current	a_{11r}	a_{11i}	a_{13r}	a_{13i}	
I_{5r}	0.0029	-0.0238	0.0031	-0.0245	
I_{5i}	0.0098	0.0175	-0.0046	0.0159	
I_{7r}	0.0051	0.0266	0.0141	0.0169	
I_{7i}	0.0134	0.0057	0.0058	0.0086	

about 4.6273% and 4.1481% respectively. At the same time, the testing error of real part and image part in *7th* harmonic current are about 15.3301 % and 5.3831% respectively. At the same time, harmonic voltages of all orders can affect the harmonic currents of the same order and other orders, besides the fundamental voltage.

4.2 Polynomial Regressive Model

In general, complex factual applications require more expressive hypothesis spaces than linear functions. In this case, linear learning machine can not exploit more abstract features of the data. Kernel representations offer an alternative solution by projecting the data into a high dimensional feature space to increase the computational power of the linear learning machine [7]. In following simulation, the polynomial function is used as kernel function:

$$K_{ij} = \left(1 + V_i^T V_j\right)^2 \quad . \tag{21}$$

With this kernel function K_{ij}, the structure of $\varphi(\mathbf{V})$ vector can be induced. But with large number elements, the structure of vector $\varphi(\mathbf{V})$ is too complex to list in this paper. This step maps the original 9 dimension input space $\mathbf{V}$ to a 55 dimension feature space $\varphi(\mathbf{V})$, see the figure 3.

This high dimensional feature space has more information about the relationships between currents and voltages, because the linear regression feature space just provides the respective voltage components for calculation. However, nonlinear regression feature space contains the square of voltage components and the across multiplication with two distinct voltage components. Using high dimensional feature space can get more precise result and the following simulation result can prove it. But dimension of vector ω is too big to list in this paper, only the testing error is given.

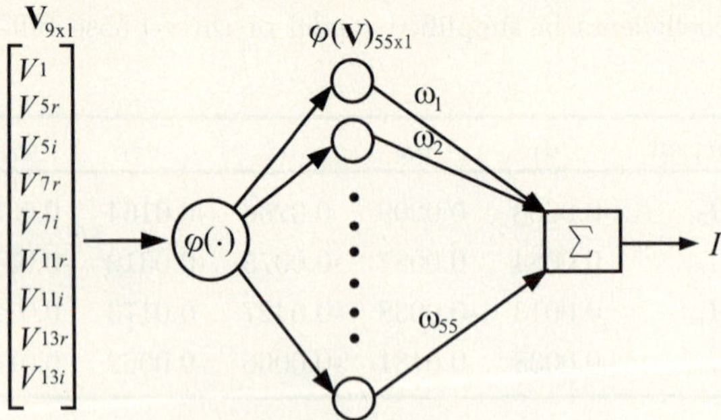

Fig. 3. Harmonic source model with polynomial kernel function

The testing errors of real part and image part in 5*th* harmonic current are about 0.9100% and 2.8967% respectively. At the same time, the testing error of real part and image part in 7*th* harmonic current are about 9.7360 % and 3.3166% respectively.

4.3 Radial Basis-Function [8] Regressive Model

As another non-linear kernel function, radial basis-function (RBF) kernel can be introduced in algorithm of this paper.

$$K_{ij} = exp\left(-\frac{1}{2\sigma^2}\left\|V_i - V_j\right\|^2 \right) . \tag{22}$$

With RBF kernel function, solving the $\varphi(\mathbf{V})$ is difficult, and the result of $\varphi(\mathbf{V})$ is obscure [9]. Without detail structure of $\varphi(\mathbf{V})$, the relationship implied by (7) impossibly exist. But, if the fore suppose $\omega = \varphi^T\delta$ is taken in (7):

$$I = \langle \varphi(\mathbf{V}) \cdot \Phi^T \delta \rangle = \sum_{m=1}^{l} \delta_m \cdot exp\left(-\frac{1}{2\sigma^2}\left\|V - V_m\right\|^2 \right) = \sum_{m=1}^{l} \delta_m K_m \quad . \tag{23}$$

The vector δ can be got by iteration of (15), V_m means the input data for regression, and the l is the number of input training data for the process of regression. The (23) gives a method to use RBF kernel function in this paper's algorithm, the model structure is in figure 4. But this model can not give a real relationship between input voltages and the harmonic currents, because it's influenced by the data provided for training regressive process.

And the regressive performance of RBF is predictable worse than Polynomial kernel [10]. Because the RBF is a limitary function, the polynomial function has not this characteristic. In regressive process the unlimited training is always more accurate. The following simulation results give the evidence.

In the simulation, data for modeling and testing is same as the data used in polynomial regression. And the testing result is also evaluated by the average

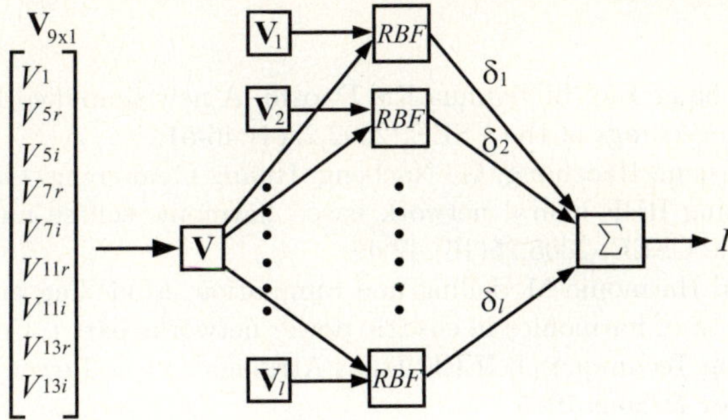

Fig. 4. Harmonic source model with RBF kernel function

Table 2. Accuracy comparison between linear and non-linear regression

Current	Linear (%)	Polynomial (%)	RBF (%)
I_{5r}	4.6273	0.9100	12.5550
I_{5i}	4.1481	2.8967	15.3277
I_{7r}	15.3301	9.7360	23.0995
I_{7i}	5.3831	3.3166	18.2525

of absolute value from relative error. According to (18) and (19), by using the same cost function, the δ can be calculated with $\sigma = 1$. For the complexity of vector δ, it's hard to list in this paper, only testing error is given.

The testing errors of real part and image part in $5th$ harmonic current are 12.5550% and15.3277% respectively. At the same time, the testing errors of real part and image part in $7th$ harmonic current are about 23.0995% and 18.2525% respectively.

Compare the accuracy of regressions with different functions. The result has indicated that the precision of using polynomial kernel in non-linear regression is better than linear regression and the non-linear regression with RBF kernel, The table 2 has the details.

5 Conclusion

This paper proposes a new harmonic source model which takes advantage of support vector machine (SVM) theory. In this modeling process, the different SVM regressive methods are introduced in finding the accurate relationship between the input voltage and the harmonic current. With the simulation research for modeling the 3-phase converter circuit, it proves that this method can define the proper harmonic source model. In future work, this method can be used in the real-time modeling process, and give the useful information for power harmonics elimination.

References

1. Zhao Yong, Zhang Tao, Li Jianhua,Xia Daozhi. A new simplified harmonic source Model [J]. Proceedings of the CSEE, 2002,22(4):46-51.
2. Zhan Yong, Cheng Haozhong, Ge Nacheng, Huang Guangbing. Generalized growing and pruning RBF neural network based harmonic source modeling [J]. Proceedings of the CSEE, 2005,25(16):42-46
3. Task Force on Harmonic Modelling and Simulation .Modelling and simulation of the propagation of harmonics in electric power networks part I: Concepts, Models, and Simulation Techniques[J].IEEE Trans Arrillaga J et al, Power system harmonics. John Wiley & Sons.1985
4. Li Ma; Kaipei Liu; Lanfang Li. "Harmonic and inter-harmonic detecting based on support vector machine," Transmission and Distribution Conference and Exhibition: Asia and Pacific, 2005 IEEE/PES15-18 Aug. 2005 Page(s):1 - 4
5. Fernando Pérez-Cruz, Angel Navia-Vázquez, Aníbal R. Figueiras-Vidal, Antonio Artés-Rodríguez. Empirical Risk Minimization for Support Vector Classifiers. IEEE Trans on neural networks,2003,14(2):296-303.
6. A. J. Smola, B. Schölkopf, and K.-R. Müller, "General cost functions for support vector regression," in Proc. 9th Australian Conf. Neural Networks,Brisbane, Australia, 1998, pp. 79-83.
7. Nello Cristianini, John Shawe-Taylor. An introduction to support vector machines and other kernel-based learning methods[M]. Cambridge University Press, 2000
8. Daniel J. Sebald, James A. Bucklew. Support Vector Machine Techniques for Nonlinear Equalization. IEEE Trans on signal processing,2000, 48(11):3217-3226
9. A. J. Smola, B. Schölkopf. A tutorial on support vector regression. Statistics and Computing 14: 199-222, 2004
10. Simon Haykin. Neural Networks: A Comprehensive Foundation, 2nd Edtion, Inc, publishing as Prentice Hall PTR, 1999.
11. Kecman, V. Learning and Soft Computing: Support Vector machines, Neural Networks and Fuzzy Logic Models. The MIT Press, Cambridge, MA, 2001.
12. Wang, L.P. Support Vector Machines: Theory and Application. Springer, Berlin, Heidelberg, New York, 2005.

Sound Quality Evaluation Based on Artificial Neural Network

Sang-Kwon Lee, Tae-Gue Kim, and Usik Lee

Acoustic Noise Signal Processing Labbatory, Dapartment of Mechanical Engineering, Inha University, 253 Yonhhyun Dong, Nam Gu, Inchon, 402-751, Korea
{sangkwon, taegue, Ulee}@inha.ac.kr
http://ansp.inha.ac.kr

Abstract. Booming index has been developed recently to evaluate the sound characteristics of passenger cars. Previous work maintained that booming sound quality is related to loudness and sharpness--the sound metrics used in psychoacoustics--and that the booming index is developed by using the loudness and sharpness for a signal within whole frequency between 20Hz and 20kHz. In the present paper, the booming sound quality was found to be effectively related to the loudness at frequencies below 200Hz; thus the booming index is updated by using the loudness of the signal filtered by the low pass filter at frequency under 200Hz. The relationship between the booming index and sound metric is identified by an artificial neural network (ANN).

1 Introduction

Recently, noise control technology for passenger cars can reduce the A-weighted noise level in the car compartment to as low as possible. Unfortunately, this technology often brings out sound quality problems inside of cars since A-weighted sound level does not tell the whole story as far as the customer is concerned [1]. Therefore, sound quality is becoming increasingly important as a part of vehicle design and many research papers on booming sound quality in a vehicle have been published [2, 3, 5]. Few of these papers [2, 3], however, are concerned with the objective evaluation of booming sound quality using subjective parameters [6], because almost all of the research papers are interested in the control of A-weighted sound level with a commercial sound level meter. The subjective parameters are based on sound metrics such as loudness, sharpness, roughness and fluctuation used in psychoacoustics theory. Consequently, the objective evaluation of the booming sound quality using these sound metrics is nearer to the subjective evaluation than the A-weighted sound level. In general, the booming index is used for the objective evaluation of the booming sound quality in a vehicle. According to previous research results [2], the booming index is related to the loudness, but recently it was found that the sharpness is also important [7]. This loudness and sharpness for the interior sound of a passenger car is used for the input data of the artificial neural network (ANN), which is a tool to identify the correlation between booming index and the subjective evaluation of interior sound of a passenger car. However, this boom index often has less correlation with

L. Jiao et al. (Eds.): ICNC 2006, Part I, LNCS 4221, pp. 545 – 554, 2006.

subjective evaluation when the interior sound of a car includes high frequency components. In this paper, it is found that the booming sound quality is effectively related to the loudness at a frequency below 200Hz, and thus the booming index is updated by using the loudness of the interior sound filtered by the low pass filter. ANN has also been employed in the development of the booming index of a passenger car.

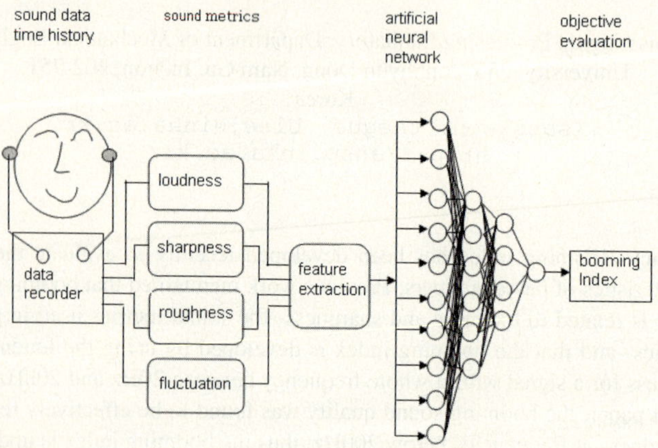

Fig. 1. Artificial neural network-based flowchart for booming index development

The structure of the ANN system used throughout this paper is shown in Fig. 1. The developed booming index is the output of ANN; the input of ANN is the sound metric of the interior sound of the passenger car. The output of ANN is the objective rate of the booming sound quality for the interior sound. If this objective rate has good correlation with the subjective rate of the booming sound quality evaluated by the passenger, the output of ANN is a good booming index. Therefore, before the output of ANN is used as the booming index, the weights of connectors of the neurons in ANN should be optimized throughout the training process.

ANN very loosely simulates a biological neural system (there is an extensive literature on ANN [7,8,9]). The multi-layer feedforward network is used throughout this paper. The training algorithm used with this network is back-propagation [10], which are mostly used in the analysis of mechanics problems. The main goal of back-propagation neural networks is mapping of input, *i.e.*, vector $\mathbf{x} \in R^N$ into output, i.e., vector $\mathbf{y} \in R^M$. This can be written in short

$$x_{N\times 1} \rightarrow y_{M\times 1} \,. \tag{1a}$$

and in general

$$x^{(p)} \rightarrow y^{(p)} \text{, for } p = 1, 2, \ldots, P,. \tag{2b}$$

where p is the number of patterns. The mapping is performed by a network composed of processing units (neurons) and connections between them. In Fig. 2(a) single neuron i is shown. Input signals x_j are accumulated in the neuron summing block Σ and activated by function F to have only output y_i:

$$y_i = F(z_i), z_i = \sum_{j=1}^{N} w_{ij} x_j + b_i \,. \tag{2}$$

where z_i – active potential, $w_{i,j}$ –weights of connection, b_i – threshold parameter. From among various activation functions the sigmoid functions are commonly used:

$$F(z) = \frac{1}{1+e^{-\mu z}} \in (0,1) \quad \text{for } \mu > 0,. \tag{3}$$

In Fig. 2(b), a standard multi-layer feedforward networks are shown. The network is composed of the input, hidden and output layers, respectively.

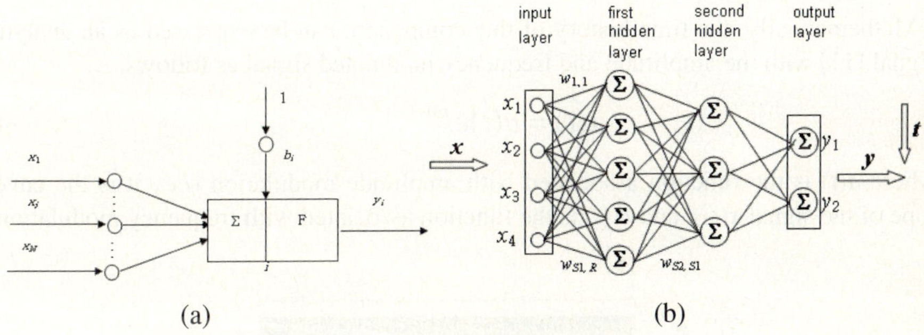

Fig. 2. Structure of artificial neural network for booming index: (a) single neuron *i*, (b) three-layer, back-propagation network

2 Booming Sound in a Passenger Car

Basically, the interior sound of a car consists of very complex frequency spectra since it has many excitation sources, resonance systems and parts of sound radiation [11]. However, it is known that the firing frequency component of the interior sound influences the booming sound quality [2,3,4,5,6]. Other frequency components play roles of background noise. For example, the firing frequency component of the interior sound of a car loaded with an in-line 4 cylinder engine is twice the rotating speed of the crankshaft of the engine. For a V6 engine, the firing frequency is three times the rotating speed of the crankshaft. Therefore, it is inferred that the booming sound quality of a passenger car is related to the amplitude change of the sound at firing frequency. Fig.3 (a) shows a waterfall analysis for the sound pressure inside of a car loaded with an in-line 4-cylinder engine. The speed of the engine increases from 1500 rpm to 5000rpm. In the figure, the horizontal axis designates the frequency and the vertical axis shows the sound pressure level inside of the car. From this figure, we can see that the pressure level of the sound at the firing frequency is dominant and the firing frequency is related to the rotating speed of the crankshaft (*i.e.,* rpm). So if we change the amplitude of this frequency component, the booming sound quality for the interior sound of this car will also be influenced.

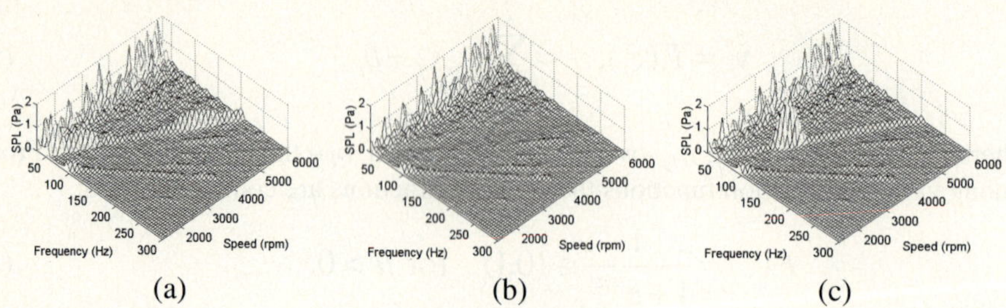

Fig. 3. Waterfall analysis for interior sound of a passenger car loaded with an in-line 4-cylinder engine: (a) base sound, (b) without firing component sound, (c) modified sound

Mathematically, the time history of this component can be expressed as an analytic signal [11] with the amplitude and frequency modulated signal as follows:

$$x(t) = a(t)e^{j\phi(t)} . \tag{4}$$

where $a(t)$ is the function associated with amplitude modulation (*i.e.,* it is the envelope of the signal $x(t)$), and $\phi(t)$ is the function associated with frequency modulation.

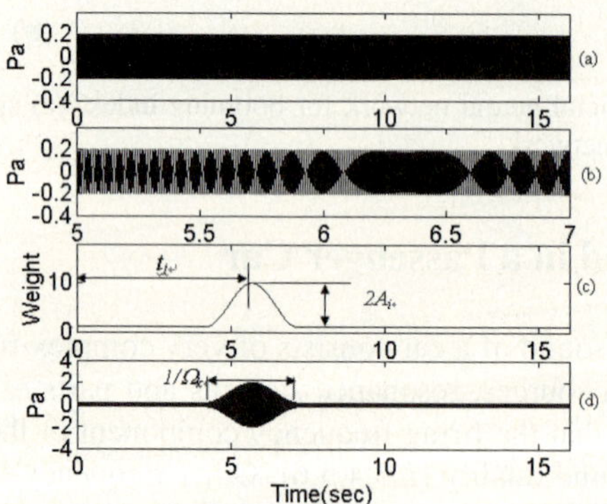

Fig. 4. Modification of firing frequency component sound for production of the 200 interior sounds

Fig. 4(a) represents the time history of the firing frequency component sound. It is obtained by filtering the interior sound as shown in Fig. 3(a) by using a Kalman order adaptive filter [12,13]. Fig. 3(b) shows the waterfall analysis of the interior sound obtained by removing the firing frequency component of the original interior sound. The signal of the sound with only firing frequency component is expressed as a form of the analytic signal [11] explained in equation (4). The envelope and frequency modulation functions are $a(t)$ and $\phi(t)$ respectively. We can produce interior sounds with booming sound qualities of various subjective rates by modifying the envelope of the signal as shown in Fig. 4(a) and adding it to the background noise as shown in

Fig. 3(b) because the background noise influences the booming sound quality. In this paper, the envelope of the analytic signal is modified as follows:

$$\begin{cases} \mathsf{A}\ \ (t) = [A_j \sin \Omega_k (t - t_i) + A_j + 1] \cdot a(t), & t_i - \dfrac{1}{2\Omega_k} \le t \le t_i + \dfrac{1}{2\Omega_k}, \\ i = 1..10,\ j = 1..5, \text{and } k = 1..4 & \\ \mathsf{A}\ \ (t) = 1 \cdot a(t), & \text{otherwise} \end{cases} \quad (5)$$

where $a(t)$ is the envelope of the firing frequency component of the analytic signal; t_i is the ith time where the amplitude modulation takes place; A_j is the jth magnitude for presenting the magnitude of amplitude modulation; and Ω_k represents the kth frequency for determining the duration of amplitude modulations. Fig. 4(c) shows one example of the modified envelopes $\mathsf{A}(t)$ and illustrates the roles of the parameters. Fig. 4(d) displays the analytic signal $x(t)$ modified by using the modified envelopes $\mathsf{A}(t)$. The modified analytic signal is given by

$$x(t) = \mathsf{A}\ \ (t)\exp(j\phi(t)). \quad (6)$$

In order to get the synthetic interior sounds with different booming sound quality, these modified analytic signals with various values for the Ω_k, t_i, and A_j are added to the background noise as shown in Fig. 3(b). Fig. 3(c) shows the waterfall analysis for the synthetic interior sound using the modified analytic signal as shown in Fig. 4. With this method, the 200 synthetic interior sounds with booming sound quality of various subjective rates are completed. The subjective rates of these interior sounds are used for the target of the ANN.

3 Subjective Evaluation

For the target of the ANN, the 200 synthetic interior sounds were subjectively evaluated by 21 passengers (17 males and 4 females). Before subjective evaluation of the 200 synthetic signals, 50 non-booming-like sounds were removed. In addition to the synthetic interior sounds, the interior sounds of 16 mass-produced passenger cars were also used for the subjective evaluation. Ten sounds are the interior sounds of the mass-produced cars loaded with an in-line 4 cylinder engine, four sounds are those with a V6 engine and two sounds are those with a V8 engine. Therefore, the subjective evaluation consists of a total of 166 interior sounds. The playback system and headphone of Head Acoustics Company were used for subjective evaluation. The 166 interior sounds were randomly evaluated; the subjective rate was evaluated for point 4 to point 9. Fig. 5 (a) shows the results of subjective evaluation for the 166 signals. The averaged subjective rates for 150 synthetic interior sounds are plotted from the left side of the graphic from low rate to high rate. The right end part of the graphic is for the interior sounds of 16 mass-produced passenger cars. Fig. 5 (b) illustrates the standard deviation with 95% confidence interval.

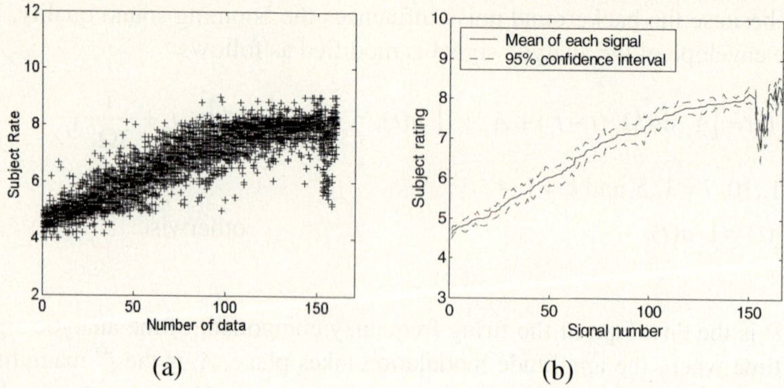

Fig. 5. Subjective rates for the 166 interior sounds of passenger cars: (a) raw subjective rate, (b) averaged subjective rate and standard deviation with 95% confidence interval

4 Sound Metrics

The sound metric for the 166 interior sounds was calculated for the input data of ANN. Four major sound metrics [6] such as loudness, sharpness, roughness and fluctuation strength were used. These sound metrics for 166 interior sounds are shown in Fig. 6 to Fig. 9, and these results are used for the input of the ANN to be used as the booming index. Loudness represents the auditory perception character related to the magnitude of sounds [6]. There are many models [6,14,15] for calculating the loudness. In this paper, the Zwicker model [6] is used for the calculation of the loudness for the 166 interior sounds. Fig. 6 (a) shows the loudness measured for the 166 sounds. From the graphic, it is concluded that the subjective ration is proportional to 1/loudness. However, the loudness for interior sounds of real cars is not smoothly related to the loudness for the synthetic signals. Therefore the loudness for interior sounds of real cars is filtered by a low pass filter at 200Hz frequency since the timbre of booming sound is a lower frequency sound [2,3,5]. Fig. 6 (b) shows the new loudness. The loudness for interior sounds of real cars is now smoothly related to the loudness for the synthetic signals. Sharpness describes auditory perception related to the spectra correlation of a sound. Bismarck [16] and Aures [17] introduce the calculation model of sharpness. In this paper, Bismarck's model is also adopted for the calculation of sharpness. One acum is the sharpness for a pure tone sound with amplitude of 60dB at 1kHz. Fig. 7 shows the sharpness measured for the 166 sounds. According to these results, the maximum sharpness is about 0.9 acum. From the graphic, it is concluded that the sharpness has a nonlinearly increasing relationship with human perception for the booming sound. Roughness is the auditory perception characteristic related to the amplitude modulation and frequency modulation for sound with frequency modulation at middle frequency around 70Hz.

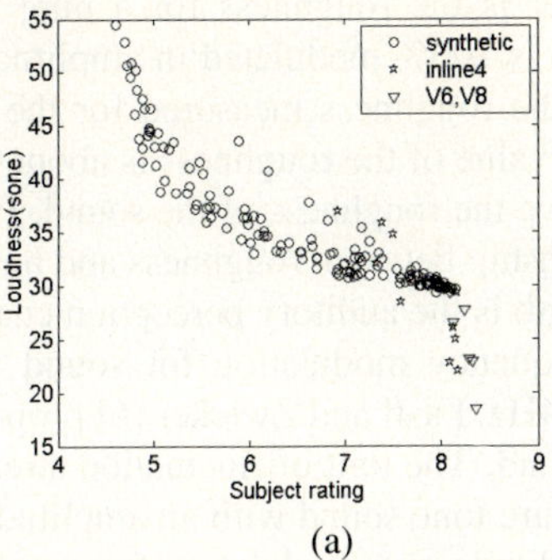

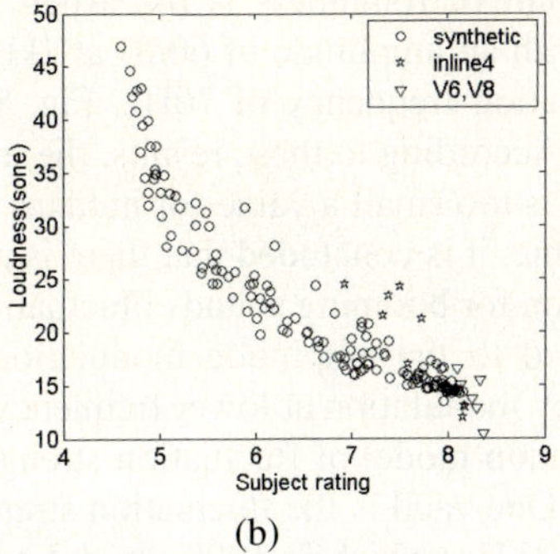

Fig. 6. Loudness for the 166 interior sounds of passenger cars using Zwicker's method: (a) Loudness for the signal without filtering, (b) Loudness for the signal filtered by low pass filter at 200Hz frequency

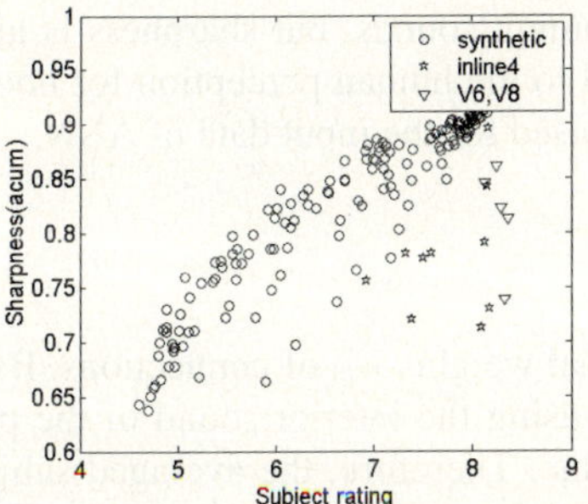

Fig. 7. Sharpness for the 166 interior sounds of passenger cars using Bismarck's model

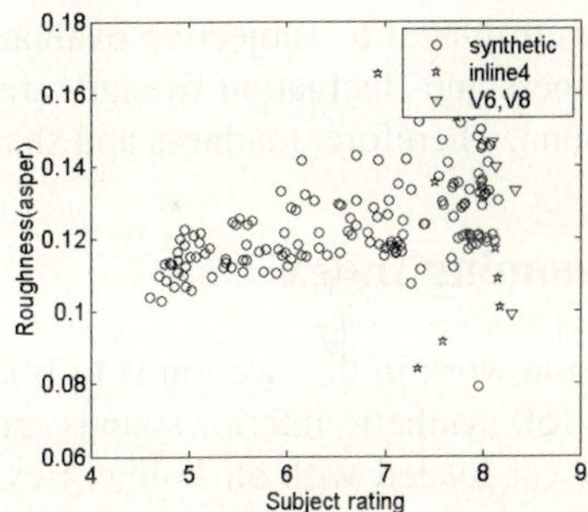

Fig. 8. Roughness for the 166 interior sounds of passenger cars using Aures' model

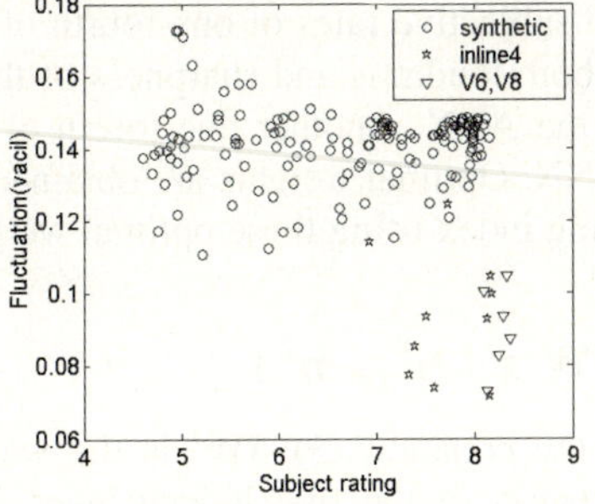

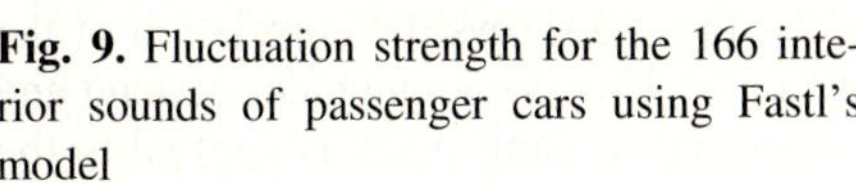

Fig. 9. Fluctuation strength for the 166 interior sounds of passenger cars using Fastl's model

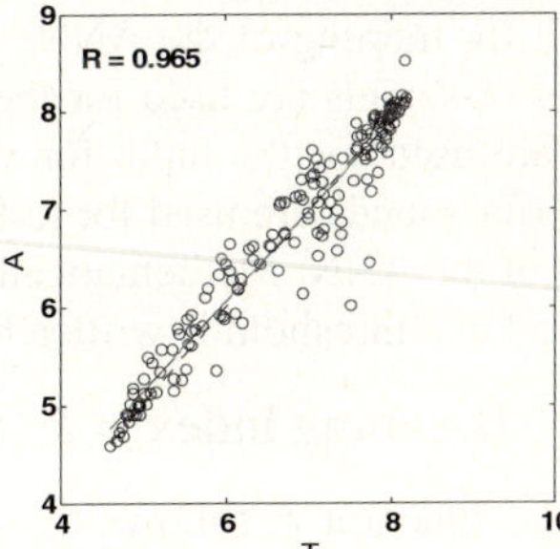

Fig. 10. Correlation between the output of ANN and the averaged subjective rates

The unit of roughness is the asper. One asper is the roughness for a pure tone sound with an amplitude of 60dB at 1kHz, which is 100% modulated in amplitude at a modulation frequency of 70Hz. Fig. 8 shows the roughness measured for the 166 sounds. According to these results, the maximum value of the roughness is about 0.18 asper. It is too small a value for humans to perceive the roughness of the sound. From the graphic, it is concluded that there is no relationship between roughness and human perception for booming sounds.Fluctuation Strength is the auditory perception character related to the amplitude modulation and frequency modulation for sound with frequency modulation at lower frequency around 4Hz. Fastl and Zwicker [6] proposed a calculation model of fluctuation strength for sound. The unit of fluctuation strength is vacil. One vacil is the fluctuation strength for pure tone sound with an amplitude of 60dB at 1kHz, which is 100% modulated in amplitude at a modulation frequency of 4Hz. Fig. 9 shows fluctuation strength measured for the 166 sounds. According to these results, the maximum value of fluctuation strength is about 0.18 vacil. It is also too small a value for humans to perceive. From the graphic, the fluctuation strength has no relationship with human perception for booming sounds.

From the psychoacoustical analysis of the 166 interior sounds, not only is loudness very well related to subjective evaluation of booming sounds, but sharpness is also. Roughness and fluctuation strength are no related to the human perception for booming sound. Therefore, loudness and sharpness are used for the input data of ANN.

5 Booming Index

The main work in this section is to find the optimal weights $w_{i,j}$ of connections. Basically, 150 synthetic interior sounds are made by using the interior sound of the passenger car loaded with an in-line 4-cylinder engine. Therefore, the averaged subjective rates for the 150 synthetic interior sounds and for the 10 interior sounds of the mass-produced passenger cars are used for the target of the ANN being used for the booming index. Both loudness and sharpness of those sounds are used for the input of the ANN. The averaged subjective rates of half of the 160 interior sounds are used for the target of the ANN, and both loudness and sharpness of those sounds are used for the input for training of the ANN. The averaged subjective rates of one-fourth of the 160 interior sounds are used for the target, and both loudness and sharpness of those sounds are used for the input for validation of the ANN. Another one-fourth of the 160 interior sounds are used for testing of the ANN. Optimal weights are obtained by training of the ANN. Mathematically, the booming index using these optimal weights of connect and threshold is written by

$$\text{Booming Index} = F^2(\mathbf{LW}^2 F^1(\mathbf{IW}^1\mathbf{x} + \mathbf{b}^1) + \mathbf{b}^2). \tag{7}$$

where the function F follows the form of the the equation (3), $\mathbf{IW}^1$ is the weight matrix in the input layer, $\mathbf{LW}^2$ is the weight matrix of the first hidden layer. The booming index is the output of the trained ANN. Fig. 10 shows the correlation of the output of the ANN and the averaged subjective rates. They very much correspond and have a good corrrelation of 96.5%. The booming index using ANN is applied to the estimation of the subjective rate for the interior sounds of 10 passenger cars loaded with in-line 4-cylinder engines.These estimated subjective rates are compared with the averaged subjective rates.

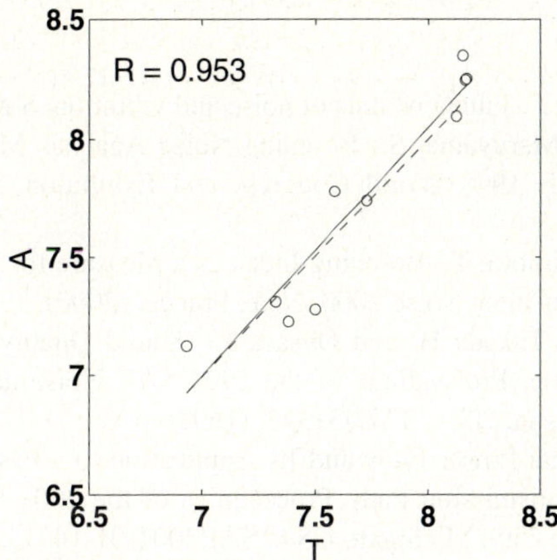

Fig. 11. Correlation between the output of ANN and the averaged subjective rates for the 10 interior sounds of passenger cars with in-line 4-cylinser engines using the loudness for the signals with filtering processing

Fig. 11 shows their correlation, which is 95.3%. From these results, it is concluded that the output of the ANN trained by using the 150 synthetic interior sounds and the 10 interior sounds of passenger cars loaded with in-line 4-cylinder engines can be used as an index for the booming sound quality of passenger cars loaded with in-line 4-cylinder engines based on human sensibility. It can be used to estimate objectively the subjective rates of booming sound qualities of passenger cars without subjective evaluation by a skilled engineer.

6 Conclusions

An artificial neural network has been applied to the development of the sound quality index of the booming sound of passenger cars. The 150 synthetic signals and 10 interior sounds of the real passenger cars with in-line 4 cylinder engines are used for the training of the ANN. It is found that both loudness and sharpness for those sounds have a strong relationship with the averaged subjective rates of those sounds. In particular, the loudness of the signals filtered by a low pass filter is required to get a good correlation. The correlation between neural output and the averaged subjective rate for those sounds is 96.5%. It is concluded that the output of the trained ANN can be used for the booming index for the interior sounds of passenger cars loaded with in-line four cylinder engines. This has been confirmed with the application of the trained ANN to the estimation of the subjective rates for the booming sound qualities of the interior sounds of 10 passenger cars. The output of the trained ANN has 95.3% correlation with the averaged subjective rates evaluated by 21 passengers.

Acknowledgement

"This work was supported by INHA UNIVERSITY Research Grant. (INHA-2006)"

References

1. Fahy, F. and Walker, J.: Fundamentals of noise and vibration. S & FN SPON, (1998)
2. Matsuyama, S. and Maruyama, S.: Booming Noise Analysis Method Based on Acoustic Excitation Test. SAE 1998 World Congress and Exhibition, Detroit, Michigan, USA, SAE980588, (1998).
3. Hatano, S. and Hashimoto, T.: Booming Index as a Measure for Evaluating Booming Sensation. Proceedings of Inter-Noise 2000, Nice, France, (2000).
4. Murata, H., Tanaka, Takada H. and Ohsasa Y.: Sound Quality Evaluation of Passenger Vehicle Interior Noise. Proceedings of the 1993 SAE Noise and Vibration Conference, Traverse City, Michigan, USA, SAE931347, (1993).
5. Lee, S. K.: Vibrational Power Flow and Its Application to a Passenger Car for Identification of Vibration Transmission Path. Proceedings of the 2001 SAE Noise and Vibration Conference, Traverse City, Michigan, USA, SAE2001-01-1451, (2001).
6. Zwicker, E. and Fastl, H.: Psychoacoustics: Facts and Models. Springer-Verlag, Berlin, 2nd Edition, (1999).
7. Lee, S. K. Chae, H. C, Park, D. C. and Jung, S. G.: Sound Quality Index Development for the Booming Noise Of Automotive Sound Using Artificial Neural Network Information Theory. Sound Quality Symposium 2002 Dearborn, Michigan USA, CD N0.5, (2002).
8. Bishop, C. M.: Neural Networks for Pattern Recognition. Oxford University Press, (1995).
9. Matrn H.: Neural Network Design. PWS Publishing Company, (1996).
10. Waszczyszyn, Z. and Ziemianski, L. Neural Networks in Mechanics of Structures and Materials-New Results and Prospects of Applications Computers & Structures (2001) 2261-2276.
11. Lee, S. K.: Adaptive Signal Processing and Higher Order Time Frequency Analysis for Acoustic and Vibration Signatures in Condition Monitoring. Ph.D. Thesis, ISVR, University of Southampton. (1998)
12. Herlufsen, G. H., Hansen, H. K and Vold, H.: Characteristics of the Vold-Kalman Order Tracking Filter. Bruel & Kjaer Technical Review, (1998) 1-50.
13. Lee, S. K. and White, P. R.: The Enhancement of Impulsive Noise and Vibration Signals for Fault Detection in Rotating and Reciprocating Machinery. Journal of Sound and Vibration, (1998) 485-505.
14. Stevens, S. S. : Perceived Level of Noise by MarkII and Decibels. The Journal of the Acoustic Society of America, (1971) 575-601.
15. Moore, B. C. J. and Glasberg, B. R.: A Revision of Zwicker's Loudness Model. Acoustica, (1996) 335-345.
16. Bismarck, V. Sharpness as an Attribute of the Timbre of Steady Sounds. Acoustica, (1974) 159-172.
17. Aures, W.: The Sensory Euphony as a Function of Auditory Sensations. Acoustica, (1985) 282-290.
18. Aures, W. A Procedure for Calculating Auditory Roughness. Acoustica, (1985) 268-281.
19. Laux, P. C. Using Artificial Neural Networks to Model the Human Annoyance to Sound. Ph.D. Thesis School of Mechanical Engineering, Purdue University. (1998)

SOC Dynamic Power Management Using Artificial Neural Network

Huaxiang Lu, Yan Lu, Zhifang Tang, and Shoujue Wang

Neural Network Laboratory, Institute of Semiconductors, Chinese Academy of Sciences, 100083, Beijing, China
luhx@semi.ac.cn

Abstract. Dynamic Power Management (DPM) is a technique to reduce power consumption of electronic system by selectively shutting down idle components. In this article we try to introduce back propagation network and radial basis network into the research of the system-level power management policies. We proposed two PM policies-Back propagation Power Management (BPPM) and Radial Basis Function Power Management (RBFPM) which are based on Artificial Neural Networks (ANN). Our experiments show that the two power management policies greatly lowered the system-level power consumption and have higher performance than traditional Power Management(PM) techniques—BPPM is 1.09-competitive and RBFPM is 1.08-competitive vs. 1.79、1.45、1.18-competitive separately for traditional timeout PM、 adaptive predictive PM and stochastic PM.

1 Introduction

When integrated circuits develop to the Very Deep Submicron (VDSM) level, SOC chips consisted of millions of gates which will work under the clock frequency of several hundreds MHz and the power they needed may amount to tens even hundreds Watts. Power dissipation in a very large scale integration system is a primary design consideration. Thus, greater attention has to be paid to efficient power management techniques. The problem of power management for an embedded system is to reduce system level power dissipation by shutting off parts of the system when they are not being used and turning them back on when requests have to be serviced. The aim is to give high performance with the least power dissipation [1].

The PM approaches can be classified into two major classes. 1) Using heuristically nonlinear regression to predict workload and address the uncertainty in the optimization problem. 2) Using the controlled Markov process to describe optimal PM and applying stochastic control to realize the optimization of PM. Both of these two approaches are based on accurate model of chips and statistical characters of workloads. They attempt to fix on the PM policy in SOC designing stage, so there are drawbacks in their applicability due to the uncertainty of workloads and chip model.

Neural networks have emerged as an attractive technique for modeling various complex and non linear relationships. Neural networks possess the capability to learn arbitrary nonlinear mappings between input and output parameters. Neural network training is a self-organizing process designed to determine an appropriate set of connection weights that allow the activation of simple processing units to achieve a

L. Jiao et al. (Eds.): ICNC 2006, Part I, LNCS 4221, pp. 555 – 564, 2006.

desired state of activation that mimics the relationship between a given set of samples. For the good characteristics of ANN we proposed two PM policies based on ANN's －BPPM and RBFPM.

2 Problem Statement and Definitions

2.1 Modeling of Power Management

System-level power management saves power of subsystems (also called devices) and displays is the most widely adopted system-level power management on PCs [2]. Fig.1 illustrates the concept of power management. When there are requests, the device is busy; otherwise, it is idle. Here, the device is idle between T_1 and T_4. When the device is idle, it can be shut down to enter a low-power sleeping state. In this illustration, the device is shut down at T_2 and woken up at T_4, when requests arrive again. Changing power states will take time; T_{sd} and T_{wu} are the shutdown and wake-up delays. Furthermore, wake up a sleeping device may take extra energy. In other words, power states transfer has power consumption. If there were no power consumption, power management would be trivial: Just shut down a device whenever it is idle. Unfortunately, there is delay and/or energy consumption. Consequently, a device should sleep only if the saved energy justifies the consumption [3]. The rules to determine whether to shut down a device are called policies. For simplicity, this article addresses only single device and single stream of requests. Because power management does not change requests, we are only concerned about the power consumption when a device is idle in either the working or sleeping state. We don't consider with the power consumption required to serve requests.

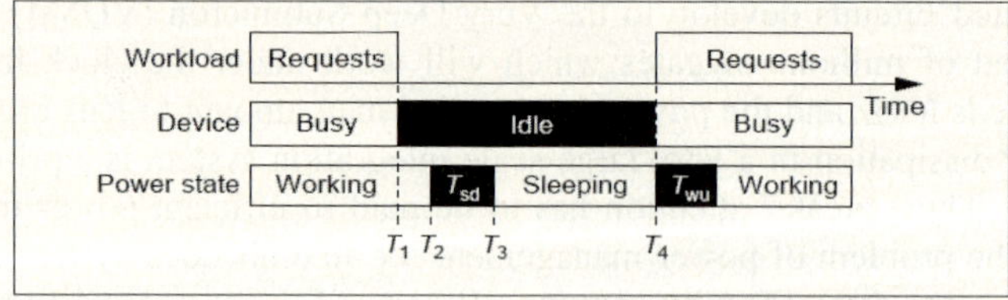

Fig. 1. Power management from the workload, device, and power state points of view

In most practical instances, we can model a Power Manageable Components (PMC) by a finite-state representation called Power State Machine (PSM). States are the various modes of operation that span the tradeoff between performance and power. State transitions have a power and delay cost. In general, low-power states have lower performance and larger transition latency than states with higher power. This simple abstract model holds for many single-chip components like processors and memories as well as for devices such as disk drives, wireless network interfaces and displays, which are more heterogeneous and complex than a single chip. The PSM model of StrongARM SA-1100 is shown in Fig.2. States are marked with power dissipation and performance values, edges are marked with transition times. In our experiment, we adopt such a PSM.

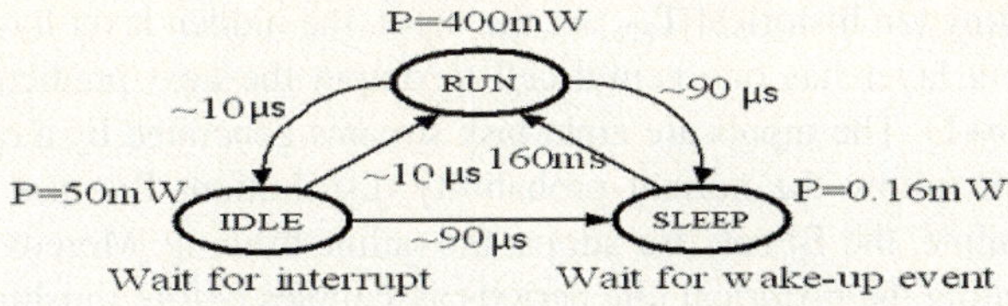

Fig. 2. Power state machine for the StrongARM SA-1100

2.2 Traditional PM Methods

Three classes of power management policies have been proposed in the past [4] [5]: timeout, predictive, and stochastic policies. The fixed timeout policy shuts down the system after a fixed amount of idle time [6]. Adaptive timeout policies are more efficient because they change the timeout according to the previous history. In contrast with timeout policies, predictive techniques[7][8] do not wait for time to expire, but shut down the system as soon as it becomes idle if they predict that the idle time will be long enough to amortize the cost of shutting down. Some predictive techniques are based on extensive off-line analysis of usage traces. Adaptive prediction policies overcome this limitation by adopting an exponential average prediction scheme. Both timeout and predictive policies have been applied only to systems with a single sleep state. A stochastic approach was proposed in [9] where general systems and user requests were modeled as Markov chains. This approach provides a polynomial-time exact solution for the search of optimal power management policies under performance constraints. The main drawback of this approach is the assumption that the Markov model of the workload is stationary and known. This limitation is addressed in [10], where adaptive Markov policies are investigated.

2.3 Competitive Analysis

In this article we adopt the Oracle Power Manager (OPM) as the standard to evaluate the other PM policies. OPM adopts competitive analysis [11]. It is usually used as a standard to evaluate the other PM policies. For example, for the same trace of requirements μ, $P_{opm}(\sigma)$ represents the minimum consumption when using OPM and $P_A(\sigma)$ represents the minimum consumption of PM policy A. If for all the input, $P_A(\sigma) \leq r \cdot P_{opm}(\sigma)$,then the competitive rate of policy A is r. In this article if some policy's competitive rate is r, it means that the consumption of policy A is r times of the OPM's under the worst condition.

3 PM Approaches Based on Neural Network

3.1 BPnet in Power Management (BPPM)

In our experiment, BPPM policy adopts the single-step prediction way: the network inputs m historic T_{idle} and outputs T_{idle} (n+m+1) value. Here, m=10, k=1, and n is the present idle periods. The net has three layers and the output layer has ten neural cells.

The corresponding ten historical T_{idle} are the input, the hidden layer has twenty neural cells. The output layer has one neural cell to output the next predictive idle period value T_{idle} (n+m+1). The inputs are eight task streams generated by a random number generator which follow the normal probability distribution. Because the inputs are generated on online, the BPnet also adopts the online training. Moreover the network needs accumulating the historical idle periods as samples before finishing the training, the BPnet should work with the adaptive predictive algorithm [8].

For the first one hundred idle periods, apply the adaptive predictive algorithm to do the power management and at the same time use the same first one hundred idle periods to train the BPnet.

The idle period $T_{idle}(101)$、$T_{idle}(102)$... are the first one hundred idle periods predicted by the BP network . If the predictive idle period $T_{idle}>T_{be}$ [1], then shut down the system at the beginning of the idle periods; if $T_{idle}<T_{be}$, keep the idle state of the devices waiting the next requirement. Evaluate the BPPM policy after every 500 idle periods, compare the consumption between BPPM and OPM. If the consumption of BPPM reach the class A of the OPM (the consumption of BPPM is not more than 1.15 times of the OPM's) or the BP network reach the upper limit of the training times, then the BP network is fixed.

Next, use the 101~600 idle periods to predict the idle period, the results using the trained BPnet are shown in the Fig.3. The blue broken lines represent the idle period's real value and the red real lines represent the idle period's predictive result. The average error of idle periods is 1.3796s and correct shutdowns rate is 44.98%. We can find the predictive idle value is smaller than the real value.

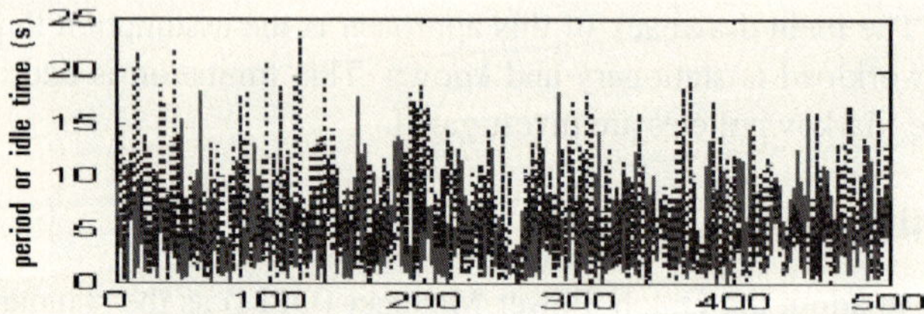

Fig. 3. The predictive value and the real value of idle time by BPnet after the first training

Then use the 401~600 idle periods to train the network again, the predictive result of 601~1100 idle periods after the second training is shown in the Fig.4. Compared with the first prediction, the second predictive idle periods average error is 1.0544s and the relative correct shutdowns rate is 41.58%.

Continue to train the BPnet with the 1001~1100 idle periods. The 1101~1600 idle periods' predictive result by the BPnet after the third training is shown in the Fig.5. Comparing with the first and second prediction, the idle period average error declines to 0.6878s and the correct shutdowns rate has improved to 43.26%.

---

[1] T_{be} is short for Break-Even Time and it is the threshold value in Timeout PM policy. T_{be} is the minimum time that makes the system consumption saved in sleep state equal to the consumption restarting the system.

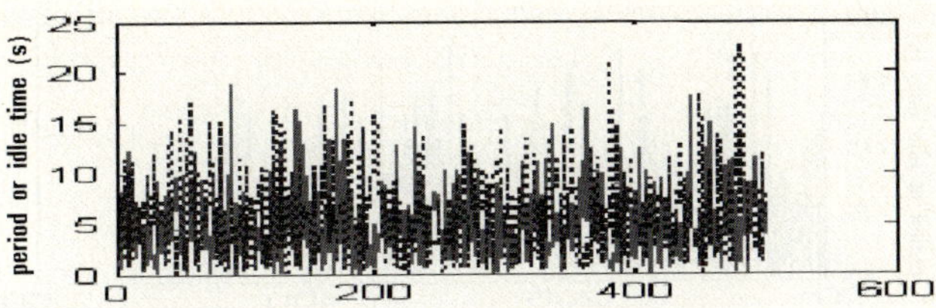

Fig. 4. The predictive value and the real value of idle time by BPnet after the second training

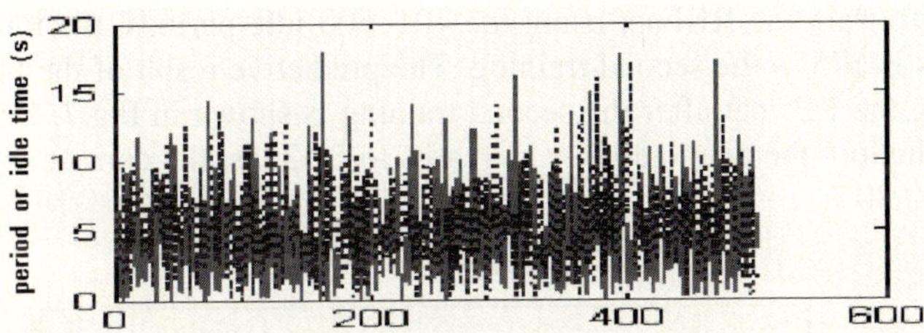

Fig. 5. The predictive value and the real value of the idle time by BPnet after the third training

Apply the three times predictive results into the power management, we got the evaluations of the power consumption are all in A class (they are 1.0355、1.0600、1.0312 times of OPM's separately), the BPnet can be fixed.

3.2 RBF Algorithm in Power Management (RBFPM)

RBF [12] is a neural network learning algorithm which extends the input vectors or pretreats vectors into the high dimension space to solve some problems. Our lab explores the general artificial neural network computer CASSANN-II. It adopts the two weighted cell [13] model and can easily simulate the RBFnet and other artificial neural networks [14].

In the RBFPM experiment we use a two layers RBFnet, the RBFnet's input connects to the first layer cells, the net's output connects to the second layer linear cells by the weight. Construct the RBFnet by increasing a RBF cell each time in order to ensure that the sum of the error squares is less than the initial target error until the cells reaches the maximum number. In such a way we can avoid fixing on the hidden layer cells' number by experience. Repeat the process and make the input vector of which the error decreases the most to generate the RBF cell.

Use the single-step prediction method to predict the next idle period: input m historic T_{idle} (n), T_{idle} (n+1),……,T_{idle} (n+m), output T_{idle} (n+m+1) and m=10, k=1, n is the present idle period. Same with BPPM, RBFPM adopts the online training and use the same requests traces: At first, use the first 100 idle periods to predict by the traditional prediction policies. At the same time, apply the 100 periods to train the RBFnet as samples. The total training process of RBFnet is the same with BPnet. The total number of the RBFnet cells is 75 after the first training. The RBFnet predictive results using the 101~600 idle periods are shown in Fig.6. The blue broken lines represent the idle period's real value and the red real lines represent the predictive value, the idle period average error is 0.2992s, the relative correct shutdowns rate is 47.10%.

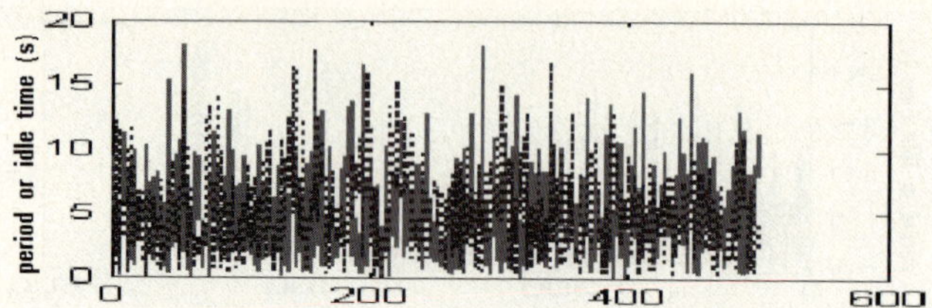

Fig. 6. The predictive value and the real value of the idle time by RBFnet after the first training

Continue to train the RBFnet using the 401~600 idle periods, the total number of RBFnet cells is 175 in the second training. The predictive result of the 601~1100 idle periods using the RBFnet after the second training is shown in Fig.7. Compared with the first prediction, the average error increases to 0.3295s, the correct shutdowns rate declines to 40.80%.

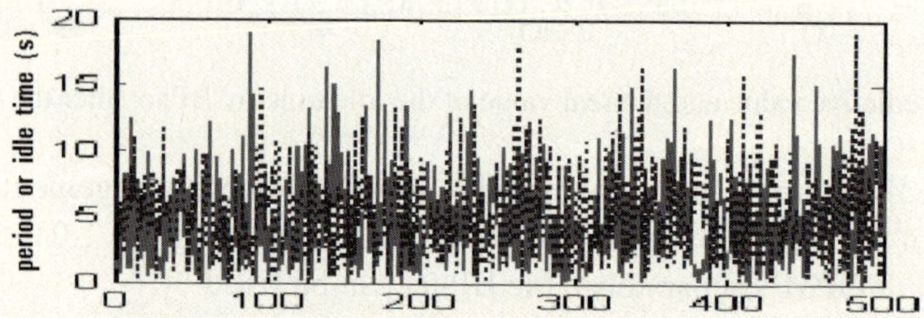

Fig. 7. The predictive value and real value using RBFnet after the second training

Train the RBFnet again using the 1001~1100 idle periods, the total RBFnet cells number decreases to 75. The predictive result using the 1101~1600 idle periods is shown in Fig.8. Compared with the first two predictions the third prediction accuracy decreases. The average error is 0.4285s, but the correct shutdowns rate increases to 45.67%.

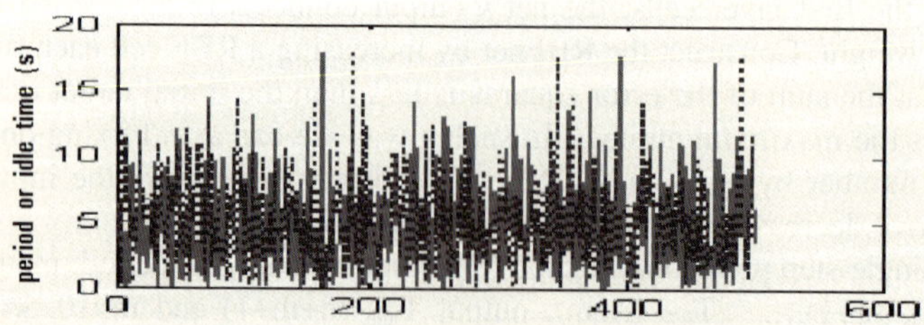

Fig. 8. The predictive value and real value using RBFnet after the third training

After three times network training, apply such a network into power management, we get the consumption evaluation of class A. (They are 1.0611, 1.0536, 1.0302 times of the consumption of OPM policy), so the RBF network can be fixed.

Use the fixed RBFnet to predict the idle periods of any one requests trace, the average error is 0.3308s, the correct shutdowns rate is 44.24%, and the consumption is 1.0830 times of OPM. The Fig.9 shows us the comparison of prediction (They are separately the result of using first 100 samples without RBFnet, the first、second、third trained RBFnet and the fixed RBFnet) in details of power, number of shutdowns and correct shutdowns.

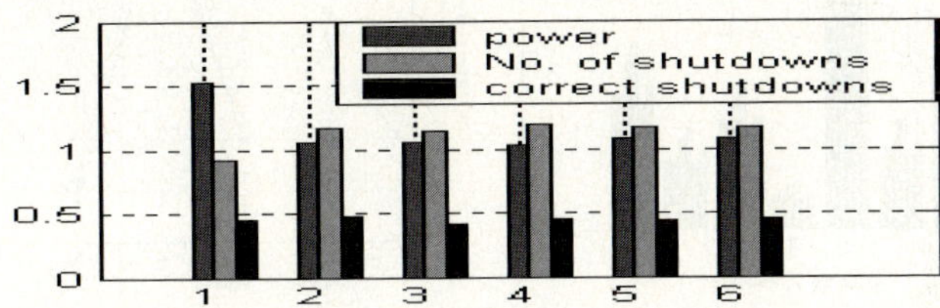

Fig. 9. Comparison for each prediction result in our RBFnet experiment

For the same 8 traces of requests, RBFnet's total average prediction error is 0.3308s which is less than BPPM's 0.5035s. For the single trace requests, the RBFPM's average prediction error is 0.4285s and the BPnet's is 1.0524s. The training time o f RBFnet is far less than the BPnet.

4 Experiment Analysis

Based on the PM experiment with BPnet and RBFnet, We take an IBM mobile disk system of which the consumption can be controlled to compare the traditional PM methods and the PM methods based on the artificial neural network.
All the PM policies compared are as follows:

1) Oracle Power Manager (OPM): Assume this policy know the traces of the tasks and the relating responding time. If the next idle time is more than $T_{be,}$ OPM will shut off the system, or keep the system idle contrarily. So OPM make the best decision, other PM policies will take it as the standard;
2) Timeout Policy;
3) Predictive Policy;
4) Stochastic Policy based on Markov(short for Markov);
5) BPnet:
6) RBnet.

Under the same environment and same eight request traces, the Fig.10 shows the total consumption and shutdowns comparison of the six PM policies. And the Table 1 shows us the system consumption ratio of Timeout、Markov、Predictive、BPnet and RBFnet to OPM's for the same eight traces of requests.

From the Fig.10, Table 1and Table 2 we can find that because the OPM policy knows requests' all information in advance, it can make the best decision and the consumption is the least. The timeout policy are more accurate in predicting whether the idle periods are longer than T_{be} than other policies and the shutdowns are the least. However, timeout policy's consumption is also the most due to the consumption

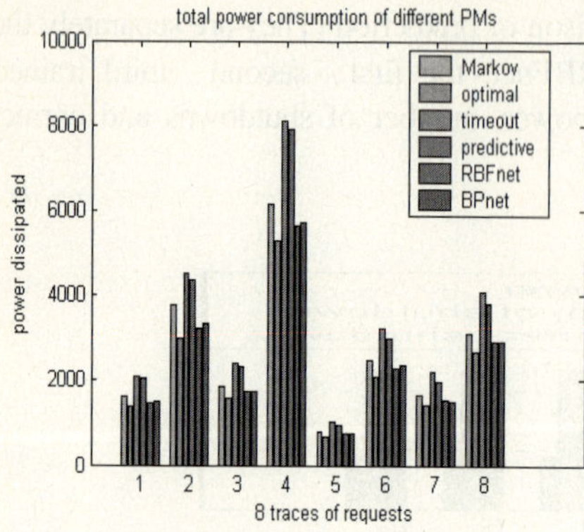

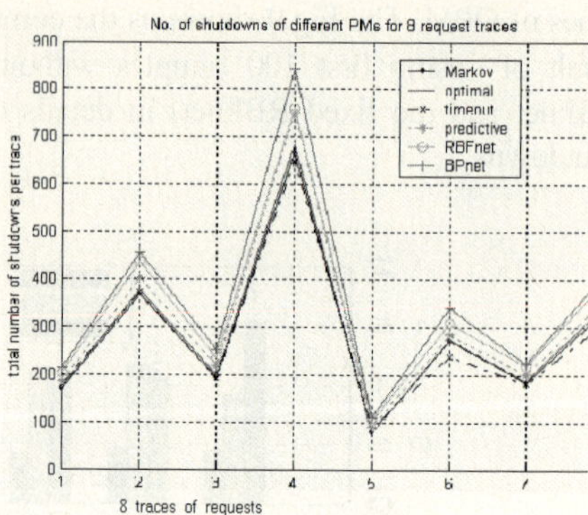

Fig. 10. Left: Total consumption comparison of the six different PM policies. Right: Total NO. of shutdowns comparison of the six different PM policies.

Table 1. The ratio of system consumptions to OPM's for each request trace

	Timeout	Markov	Predictive	Bonnet	RBFnet
Trace 1	1.8747	1.1721	1.4804	1.0501	1.0699
Trace 2	1.7650	1.2577	1.4600	1.0802	1.1104
Trace 3	1.7328	1.1668	1.4734	1.0938	1.1024
Trace 4	1.7708	1.1619	1.4911	1.0644	1.0782
Trace 5	1.6717	1.1515	1.4210	1.0954	1.0943
Trace 6	1.9688	1.1754	1.4230	1.0881	1.1177
Trace 7	1.7313	1.1644	1.3988	1.1000	1.0672
Trace 8	1.84801	1.1611	1.5239	1.0868	1.0908
average	1.7954	1.1764	1.4471	1.0941	1.0824

Table 2. Some statistical datum for all the 8 tace of requests. Here, ATC = Average Total Consumption Ratio to OPM's, AST = Average Shutdowns Times Ratio to OPM's, ACSR=Average Correct Shutdowns Rate.

	Timeout	Markov	Predictive	BPnet	RBFnet
ATC	1.7954	1.1635	1.3520	1.0915	1.0788
AST	1.0229	0.9740	1.0999	1.1672	1.1725
ACSR%	97.76	43.18	43.40	43.80	45.29

waiting for reaching the threshold. The Markov Policy has less consumption and shutdowns even than OPM, but in our experiment Markov policy is based on the accurate modeling of the system. The Morkov model concludes Service Requester state set(SR), Service Provider state set(SV),Power Manager function(PM) and Cost Manager(CM).SP and SR are only approximate estimation by experience, thereby the Morkov policy are not the very accurate method. Despite that the shutdowns of BPPM and RBFPM are a little more than the traditional PM policies, but the difference is so small and the correct shutdowns ratio of BPPM and RBFPM is improved a lot, the relative expense and the consumption are both decreased greatly, even less than Markov's. As a result the system capability is improved very much.

BPPM and RBFPM policies we proposed obviously have better performance than the traditional PM policies.

However, how to determine the BPnet and RBFnet's structure is important. In RBF network if the hidden layer cells are inadequate , the accuracy can not be ensured; if there are too many hidden layer cells, the network may be weakened ,and the network is easy to reach the partial minimum point.

Additionally, the system relay in the PM should be considered. After shutting off the system, it needs time to initialize again and then responds to the requirements, then the intervals between the idle periods change. The system delay is an important parameter. In this article we do not consider the system delay and we will research in it after.

5 Conclusions

In all, the traditional PM policies always rely on the exact modeling for the system model and the workload's statistics features. In most applications, neither the workloads nor the system inner structure can be modeled accurately. But the BPPM and RBFPM policies do not need any hypothesis of the workload requirements and they study from the system historic idle periods and are attractive techniques for modeling various complex nonlinear relationships in the power management. Our experiments show that the policies based on artificial neural networks is so effective that their performance exceeds the traditional ones.

Acknowledgment

This paper is supported by the National Nature Science Foundation of China (NO.60576033, NO 90207008).

References

1. Shrirang Yardi, Channakeshava, K., Hsiao, M.S., Martin, T.L., Ha, D.S.: A formal framework for modeling and analysis of system-level dynamic power management. Computer Design, 2005. Proceedings. 2005 International Conference, 2-5 Oct. 2005 Page(s):119 – 126.
2. Ren, Z., Krogh, B.H., Marculescu, R.: Hierarchical adaptive dynamic power management. Computers, IEEE Transactions, Volume 54, Issue 4, April 2005 Page(s):409 - 420.
3. Mihic, K., Simunic, T., De Micheli, G.: Reliability and power management of integrated systems, Digital System Design, 2004. DSD 2004. Euromicro Symposium, 31 Aug.-3 Sept. 2004 Page(s):5 – 11.
4. Yung-Hsiang Lu, De Micheli, G.: Comparing system level power management policies. Design & Test of Computers, IEEE, Volume 18, Issue 2, March-April 2001 Page(s):10 – 19.
5. D.Ramanathan, S.Irani, R.K.Gupta: An analysis of system level power management algorithms and their effects on latency, IEEE Trans. Computer-Aided Design, vol.21.No.3,Mar 2002.

6. Rong Zheng, Hou, J.C., Lui Sha: On time-out driven power management policies in wireless networks, Global Telecommunications Conference, 2004. GLOBECOM '04. IEEE, Volume 6, 29 Nov.-3 Dec. 2004 Page(s):4097 - 4103.
7. M. Srivastava, A. Chandrakasan. R.Brodersen : Predictive system shutdown and other architectural techniques for energy efficient programmable computation, IEEE Transactions on VUL Systems, vol. 4, no. 1, pp. 42-55,March 1996.
8. C.-H. Hwang, A. Wu: A predictive system shutdown method for energy saving of event-Driven computation. Proceedings of the Int.1 Conference on Computer Aided Design, pp. 28-32, 1997.
9. Norman, G., Parker, D., Kwiatkowska, M., Shukla, S.K., Gupta, R.K.: Formal analysis and validation of continuous-time Markov chain based system level power management strategies. High-Level Design Validation and Test Workshop, 2002. Seventh IEEE International,27-29 Oct. 2002 Page(s):45 - 50.
10. Qu, Q., Pedram, M.: Stochastic modeling of a power-managed system-construction and optimization. Computer-Aided Design of Integrated Circuits and Systems, IEEE Transactions, Volume 20, Issue 10, Oct. 2001 Page(s):1200 – 1217.
11. Irani, S., Shukla, S., Gupta, R.: Competitive analysis of dynamic power management strategies for systems with multiple power saving states. Design, Automation and Test in Europe Conference and Exhibition, 2002. Proceedings,4-8 March 2002 Page(s):117 – 123.
12. Guang-Bin Huang, Saratchandran, P., Sundararajan, N.: A generalized growing and pruning RBF (GGAP-RBF) neural network for function approximation. Neural Networks, IEEE Transactions, Volume 16, Issue 1, Jan. 2005 Page(s):57 – 67.
13. Wang Shoujue, Shi Jingpu, Chen Chuan, Li Yujian: Direction-basis-function neural networks. Neural Networks, 1999, International Joint Conference, Volume 2, 10-16 July 1999 Page(s):1251 - 1254.
14. Wang Shoujue, etc.: Priority ordered neural networks with better similarity to human knowledge representation. Chinese Journal of Electronics, Vol.8, No.1, 1999 Page(s):1-4.

Effects of Feature Selection on the Identification of Students with Learning Disabilities Using ANN

Tung-Kuang Wu[1,*], Shian-Chang Huang[2], and Ying-Ru Meng[3]

[1] Dept. of Information Management, National Changhua University of Education
tkwu@mail.tkwu.net

[2] Dept. of Business Administration, National Changhua University of Education
shhuang@cc.ncue.edu.tw

[3] Dept. of Special Education, National HsinChu University of Education
myr321@mail.nhcue.edu.tw

Abstract. Due to the implicit characteristics of learning disabilities (LD), the identification and diagnosis of students with learning disabilities has long been a difficult issue. Identification of LD usually involves interpreting some standard tests or checklist scores and comparing them to norms that are derived from statistical method. In our previous study, we made a first attempt in adopting two well-known artificial intelligence techniques, namely, artificial neural network (ANN) and support vector machine (SVM), to the LD identification problem. The preliminary results are quite satisfactory, and indicate that we may be going in the right direction. In this paper, we go one step further by combining various feature selection algorithms and the ANN model. The outcomes show that the correct identification rate has improved quite a lot over what we achieved previously. The combined selected features and the ANN classifier can be used as a strong indicator in the LD identification process and improve the accuracy of diagnosis.

1 Introduction

The term "learning disabilities" (LD) was first used in 1963 ([1]). However, so far experts in this field have not yet completely reach an agreement on the definition of LD and its exact meaning ([2]). As a result, the identification of students with LD has long been a difficult and prolonged process. In fact, there is little consensus about what is the best procedure to distinguish LD from non-LD ([2]). Instead, the procedures are based on empirical findings from scholarly research. In the United States, the so called "Discrepancy Model" ([3]), which states that a severe discrepancy between intellectual ability and academic achievement has to exist in one or more of these academic areas: (1) oral expression, (2) listening comprehension, (3) written expression, (4) basic reading skills, (5) reading comprehension, and (6) mathematics calculation, is commonly adopted to evaluate if a student is eligible for special education services.

In Taiwan, the identification procedure pretty much follows the "Discrepancy Model" and is roughly separated into 4 steps: (1) application for screening of potential

[*] Corresponding author.

L. Jiao et al. (Eds.): ICNC 2006, Part I, LNCS 4221, pp. 565 – 574, 2006.

LD students by parents, general education teachers and/or junior-level evaluation personnel, (2) identification of potential LD students by junior-level evaluation personnel, (3) diagnosis of possible LD students by senior-level evaluation personnel, and (4) final confirmation by special education specialists (usually college or university professors with LD major). The sources of input parameters required in such prolonged process include parents, general education teachers, students' academic performance and a number of standard achievement and IQ tests. To guarantee collection of required information regarding to students suspected with LD, usually checklists of some kind are developed to assist parents and general education teachers. The Learning Characteristics Checklists (LCC), a Taiwan locally developed LD screening checklist, is the most commonly used one. Among the standard tests, the Wechsler Intelligence Scale for Children, Third Edition (WISC-III) plays the most important role in current LD diagnosis model. The WISC-III is composed of 13 sub tests. The sub-test scores are used to derive 3 IQs, which include Full scale IQ (FIQ), Verbal IQ (VIQ) and Performance IQ (PIQ), and 4 indexes, which include Verbal Comprehension Index (VCI), Perceptual Organization Index (POI), Freedom from Distractibility Index (FDI) and Processing Speed Index (PSI). There are also a number of locally developed standard achievement tests. Identification of LD then involves mainly interpreting the checklist, WISC-III and achievement test scores and comparing them to the norms that are derived from statistical method. As an example, in case the difference between VIQ and PIQ is greater than 20, representing significant discrepancy between a student's cultural knowledge, verbal ability, etc and his/her ability in recognizing familiar items, interpreting action as depicted by pictures, etc, is a strong indicator in differentiating between students with or without LD. A number of similar indicators together with the students' academic records and descriptive data (if there is any) are then used as the basis for the final decision (by senior evaluation personnel and special education specialists). Confirmed possible LD students are then evaluated for a full year before admitting to special education.

The first issue with the above procedure is the extensive manpower and resources that are required. In current Taiwan's school system, there is no psychologist or counselor (as in the U.S.) to assist the identification or diagnosis related activities. The evaluation personnel in step 2 and 3 are in fact selected special education teachers, who received days (junior level) or weeks (senior level) of extensive training. Accordingly, some of the already overloaded special education teachers have to leave their students and class for a period of time and later make up overtime (with only limited pay) for the absence. Furthermore, the interpretation of detailed student records and assessment results require a strong background in both psychology and statistics. Unfortunately, those were not included in special education teachers' training at their college level. And days or weeks of on-job training certainly does not make them qualified psychologists. Accordingly, the quality of interpretation varied and the pressure is primarily on the special education specialists in the final stage.

The second issue with the LD identification procedure is the lack of nationally regulated standard. As a result, the procedure and its outcomes varied from county to county. In most cases, the difference can be quite significant. To be more specific, the percentage of LD students with respect to the total number of students with disabilities of each county/city ranges between 4% and 34%, with an average of 19% ([4]). It does not take an expert to realize that there must be something wrong behind these numbers. In addition, among approximately three millions students enrolls between grade 1 and 12 level in Taiwan, only 0.5% of them are diagnosed as having a learning disability ([4]). As a comparison, the 2004 National Health Interview Survey ([5]) in the United States shows that almost 8% of children age between 3-17 years had a learning disability. It is a reasonable guess that a lot more potential LD students in Taiwan may be out there un-identified. And chances are they will remain un-identified throughout their life unless some actions are taken.

In order to relieve special education teachers' workload so as to ensure better quality in special education and to include students that are entitled to special education, the above issues have to be resolved. In our previous study, we made our first attempt in adopting two well-known artificial intelligence techniques, namely, artificial neural network (ANN) and support vector machine (SVM) to the LD identification problem. The preliminary results are quite satisfactory, and indicate that we may be going in the right direction. In this paper, we go one step further by combining various feature selecting algorithms and the ANN model. The outcomes show that the correct identification rate has improved quite a lot over what we achieved in previous study.

This rest of the paper is organized as follows. Section 2 first describes the history of applying artificial intelligence techniques to special education related applications, and then provides a brief description of the multilayer perceptron ANN model as a classification technique. Section 3 presents the experiment setting, corresponding results and comments. Finally, Section 4 gives a summary of the paper and lists some issues that deserve further investigation.

2 Related Works

Artificial intelligence techniques have long been applied to special education. However, most attempts occurred in more than a decade ago and mainly used expert system to assist special education in various ways.

In 1985, Moore and a panel of technology and special education experts ([6]) had identified expert system could help in screening, diagnosis, and placement issues in special education. Hofmeister and Ferrara ([7]) developed a series of expert system prototypes, one of which was used to address the classification of students as learning disabilities. They concluded that a need for similar technology exists in special education and that it was possible to develop practical expert system with the tools and research and development resources available at that time. A preliminary finding by Hofmeister et al ([8]) also indicated that expert systems can perform as well as humans in specific areas.

Geiman and Nolte ([9]) proposed an expert system for the diagnosis of students with disabilities, which revealed quite positive results. Baer et al ([10]) designed an expert system to assist educators to reduce bias in the process of referring students with suspected disabilities. However, field test of such system showed that it had no statistically significant impact on the percentage of students referred for special education, or the ratio of students placed to students assessed. In addition, variables predicting referral for special education assessment were not consistent. On the other hand, experiments conducted by Vinsonhaler et al. ([11]) showed a more positive outcome. The decisions of their developed expert system were found to agree with those of a majority of teachers in 90% of the cases.

In addition to expert systems, numerous classification techniques have been developed and widely used in various applications ([12]). For a classification problem, it is necessary to first try to estimate a function $f : R^N \rightarrow \{\pm 1\}$ using training data, which are l N-dimensional patterns $\mathbf{x}_i$ and class labels y_i, where

$$(\mathbf{x}_1, y_1), \ldots, (\mathbf{x}_l, y_l) \in R^N \times \{\pm 1\} \tag{1}$$

such that f will classify new samples $(\mathbf{x}, y)$ correctly.

Among all the classification techniques, artificial neural network (ANN) has received lots of attentions due to their demonstrated performance and has gained widely acceptance beginning from the 1990s ([13]). An artificial neural network is a mathematical representation that is inspired by the way the brain process information. Many types of ANN models have been suggested in the literature, with the most popular one for classification being the multilayer perceptron (MLP) with back propagation ([14]). This type of ANN is known as a supervised network because it requires a desired output in order to learn. The goal of this type of network is to create a model that correctly maps the input to the output using historical data so that the model can then be used to produce the output when the desired output is unknown. Multilayer feed-forward ANN is typically composed of an input layer, one or more hidden layers and an output layer, each consisting of several neurons. Each neuron processes its inputs and generates one output value that is transmitted to the neurons in the subsequent layer.

Although artificial neural network has been used successfully in assisting medical diagnosis and decision ([15]), none has been reported in the diagnosis of students with learning disabilities. It did, however, previously used as an arithmetic training tool for the children with learning disabilities ([16]), and recently used for prediction of successful or unsuccessful completion of special education programming for students diagnosed with SED (Serious Emotional Disturbance) with up to 64% of accuracy ([17]).

In our previous study ([4]), we adopted ANN and SVM classifier and applied to the LD identification problem. Our findings and experiences learned from the study are: (1) SVM technique achieves in average better accuracy than the ANN model, (2) ANN performs better in terms of lower false positive rate, which makes it a preferred model for identification purpose, (3) with ANN model, the number of hidden layers required to achieve the highest correct identification rate never exceeds two, and (4) it seems that some of the input data sets (AT, LCC, and WISC-III) contribute to the correct decision-making at one time, while at the other time pull it away from the correct direction.

The last finding leads us toward the thought that may be not all the input features should be included. In addition, a number of researches also indicate that application of data dimensionality reduction pre-processing step prior to the classification procedure does improve the overall classification performance ([18], [19], [20]). As a result, our goal in this continuing research is to explore the potential of combining various feature selecting algorithms and the ANN model in identifying and diagnosing students with learning disabilities.

3 Experimental Settings and Results

The data set we used in this study is the same as used in ([4]), which consists of 125 cases of possible LD students. Among them, 31 (24.8%) are later identified as students with LD. Each case contains three groups of standard test scores, include the Wechsler Intelligence Scale for Children-III (WISC-III, with 7 features), the Learning Characteristics Checklists (LCC, with six features), and achievement test scores (AT, with three features). Table 1 lists the sixteen features associated with the three data sets. The tools we use for feature selection and ANN classification are YALE (Yet Another Learning Environment) ([21]) and XLMiner 3.0 ([22]), respectively.

Table 1. Set of 16 features from different standard tests and checklist

Standard Test	Features	Feature Size
WISC-III	FIQ, VIQ, PIQ, VCI, POI, FDI, PSI	7
LCC	LCC full scale index, LCC-A, LCC-B, LCC-C, LCC-D, and LCC-E	6
AT	Chinese recognition, reading, and Math	3

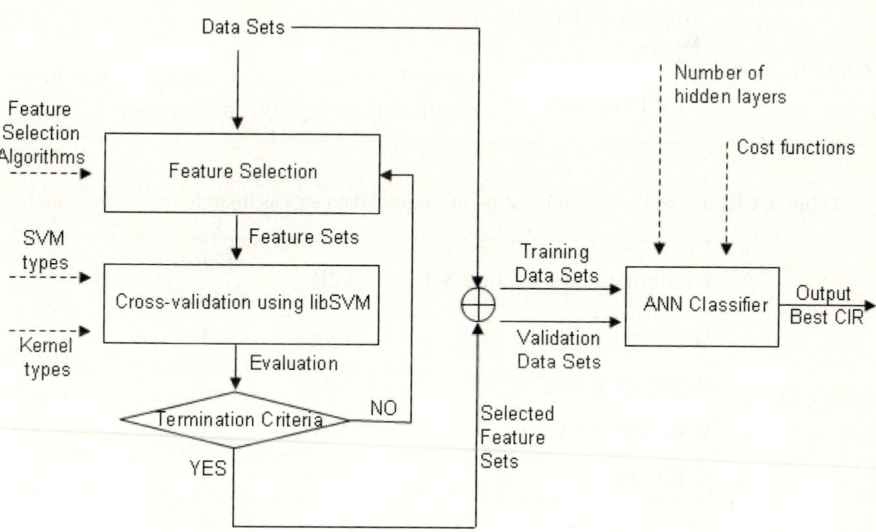

Fig. 1. Illustration of procedures employs in the experiment

We have experimented with four feature selection algorithms that are built in the YALE package, includes brute-force (BF) algorithm, greedy algorithms with forward selection (GF), greedy algorithms with backward elimination (GB), and the genetic algorithm (GA). Within each feature selection application, we provide the whole data sets and ten-fold cross-validate the performance of the selected features with libSVM learner using different SVM and kernel types against some pre-defined evaluation criteria (accuracy in this case). We then randomly divide the data set into two halves, each serves as the training (contain 60% of the samples) and validation (the rest of 40% samples) data for use in ANN classification. With each selected feature set from the previous phase, we apply the training data set to train the multilayer feedforward ANN model and then validate the model using the validation data set repeatedly with various ANN settings until we get the best correct identification rate (CIR). The feature selection and classification procedures are illustrated in Figure 1. Available parameters in each sub-stage of the procedures are listed in Table 2.

Table 2. Parameters used in the procedure as illustrated in Figure 1

Stage	Parameters Type	Available Parameters
Feature Selection	Feature Selection Algorithms	Brute-force (BF), Greedy with forward selection (GF), Greedy with backward elimination (GB) and Genetic algorithm (GA)
	SVM Types	C_SVC, nu_SVC, one class, epsilon_SVR, nu_SVR
	Kernel Types	linear, poly, rbf, sigmoid
Classification	Number of hidden layers	1, 2
	Cost Functions	squared error, cross entropy, maximum likelihood, perceptron convergence

Table 3. CIR derived with combination of original data sets as input to the ANN model

Combination of Feature Sets	CIR	False Positive
WISC-III + LCC + AT	72%	3
WISC-III + LCC	74%	2
WISC-III + AT	74%	3
WISC-III	78%	3

As a reference to evaluate how the selected feature sets perform, we first measure the CIR using combinations of the original data sets with ANN classification algorithm. The results, include the features used, their corresponding correct identification rate and number of false positive cases, are shown in Table 3.

The experimental results are organized and presented in Table 4. Note that some combinations of parameters (like SVM types or kernel types for SVM learner and evaluator) are not supported by YALE and are not listed. In addition, for each combination of parameters with genetic algorithm, it may result in multiple solutions. In that case, we list produced feature sets that achieve more than 80% in evaluation accuracy.

Table 4. Selected Feature Sets and their corresponding best CIR

F. S. Methods	Parameters (SVM type) (Kernel type)	Selected Features	CIR with Selected Features	False Positive
BF	C-SVC / rbf	PIQ, PSI	86%	1
GF	C-SVC / nu-SVR, rbf	POI, PSI	86%	1
GF	epsilon-SVR / rbf	PIQ, POI, PSI	86%	2
GF	C-SVC / linear	LCC-A, PIQ, POI, FDI	82%	5
GB	nu-SVR / rbf	LCC-B, LCC-E, FIQ, PIQ, POI, FDI	84%	2
GB	epsilon-SVR / linear	LCC-E, POI, PSI	86%	3
GA	C-SVC / epsilon-SVR, rbf	POI, PSI	86%	1
GA	C-SVC / epsilon-SVR, rbf	PIQ, PSI	86%	1
GA	C-SVC / epsilon-SVR, linear	LCC-D, VIQ, VCI, POI	84%	3
GA	C-SVC / epsilon-SVR, rbf	LCC-D, FIQ, VCI, POI	82%	3
GA	C-SVC / epsilon-SVR, rbf	LCC-D, FIQ, VIQ, PIQ, VCI, POI	82%	2

A number of observations can be drawn base on the results of Table 4. The most important one being that the use of feature selection as the pre-processing step to the original data set in general achieves a much better performance both in terms of CIR and number of false positive cases. Secondly, the combined best result (higher CIR and lower number of false positive) can be achieved by as less as two features (PIQ+PSI or POI+PSI) in our experiment, which are supported by many different feature selection algorithms. Note that although features within the achievement test (AT) are not shown in Table 4, the Chinese reading does have been selected by a few applications of the genetic algorithm. However, the resulted CIR never exceeds 78% if Chinese reading is included in the ANN classifier. This seems to indicate that AT tests do not contribute to the diagnosis decision (at least not with the data set we use in this experiment), which is somewhat in contraction to the so called discrepancy model. However, the result still needs further verification before any major conclusion is made. Overall, the implication behind these findings is that some tests or checklists may not be needed and students can be relieved from taking unnecessary tests. Accordingly, the loading of teachers (both general and special education teachers) and evaluation personnel can also be reduced as they do not have to prepare or take into account a large number of features. Most important of all, in case the outcomes here can be further verified with more samples and experiments, the combined features and the ANN classifier can be used as a strong indicator in the LD identification process and improve the overall accuracy of LD diagnosis.

To minimize the possibility of error due to imperfect data partition, we also experiment with five-fold validation using the four feature sets that achieve higher CIR (86% and 84%). We first divide the data set into five subsets with similar sizes and evenly distributing classes (LD and Non-LD). The tests are then performed five times, each with one of the dataset as the validation set and the union of the others as the training sets. The results, as presented in Table 5, are quite consistent with those shown in Table 4.

Table 5. Results of the fivefold cross validation experiment

	Using all features	PIQ, PSI	POI, PSI	PIQ, POI, PSI	LCC-D, VIQ, VCI, POI
CIR (Dataset 1)	76%	84%	80%	84%	88%
CIR (Dataset 2)	64%	76%	76%	72%	68%
CIR (Dataset 3)	64%	80%	76%	72%	76%
CIR (Dataset 4)	64%	88%	88%	80%	80%
CIR (Dataset 5)	80%	92%	92%	92%	84%
Average CIR	70%	84%	82%	80%	79%

To explore the correlation among the feature sets, we look into the classification results and derive the statistics of how the four selected feature sets (as shown in Table 4) contribute to the correct decision making. The results are listed as follow: (1) in 87 (out of 125) cases, all four feature sets can result in correct diagnosis, (2) in 12 (out of 125) cases, none of the feature sets could result in correct diagnosis, (3) in the rest of cases, some feature set(s) contributes to correct decision, while the other (s) does/do not. These numbers indicate that it may not be possible to correctly distinguish a number of cases with the ANN classifier we used and the available combination of features. It is thus important to either choose (through more experiments) other classification algorithms or find out features that are missing that could potentially assist our ANN classifier in making correct decision. On the other hand, for those correct identification cases, it is also imperative that we identify the essential features and rules that can assist special education community in refining their diagnosis procedure and our AI classification model. As an example, when we feed a selected feature set containing POI and PSI to a J48 learner, which is a YALE class that generates a C4.5 decision tree, we derive the following rule:

"(POI > 96) and (PSI <= 90)" → "Learning Disabilities = YES"

The rule has 16 supporting samples and 100% in accuracy. When we apply this rule to our newly collected data set containing 159 samples, we correctly identify 18 out of the 76 students with LD without any false positive case. This is quite a remarkable performance if it can be validated with further research and generalized to more samples. Currently this finding and more similar rules are under careful interpretation by our special education colleague.

4 Conclusion and Future Work

The identification and diagnosis of students with learning disabilities, which requires a lot of man power, resources and expertise, have never been an easy job. This is particularly true in Taiwan. Artificial intelligence (like artificial neural network) has been applied successfully to solve classification problems in numerous fields, include assisting medical personnel in making decisions. However, there seem few activities happening in special education, which in some way is similar (but less critical) to making medical decisions. In fact, if proven feasible, computer-aided diagnosis can not only save time and manpower, but also has the advantage of eliminating possible human bias.

In this study, we continue our previous work by combining feature selection algorithm with the well-known artificial intelligence techniques (ANN) and apply to the LD identification and diagnosis problem. The results are quite encouraging, not just in terms of classification accuracy, but in demonstrating the potential of AI techniques in uncovering little-known information, e.g., the essentiality of certain features. And, in our opinion, the AI techniques can be a strong indicator to be included in the procedure of identifying students with learning disabilities.

Our future work would be seeking possible ways in combining our classifier with some other algorithms that can produce meaningful explanations or rules for the prediction. This is essential as most special education teachers or professionals we talked to tend to be skeptical to this kind of black-box predictor. In addition, application of data mining techniques may also be useful in uncovering more information or answering questions that may be currently under controversy in learning disabilities community.

References

1. Kirk, S. A. Behavioral diagnosis and remediation of learning disabilities. *Proceedings of the Conference on the Exploration into the Problems of the Perceptually Handicapped Child,* Evanston, IL (1963).
2. Fletcher, J. M., W. A.Coulter, D. J. Reschly and S. Vaughn, Alternative Approach to the Definition and Identification of Learning Disabilities: Some Questions and Answers, *Annals of Dyslexia*, Vol. 54, No. 2, pp. 304-331 (2004).
3. Schrag, J., Discrepancy approaches for identifying learning disabilities, Alexandria, VA: National Association of State Directors of Special Education (2000).
4. Wu, Tung-Kuang, Shian-Chang Hwang and Ying-Ru Meng, Identifying and Diagnosing Students with Learning Disabilities using ANN and SVM, *Proceedings of the 2006 IEEE International Joint Conference on Neural Networks*, July, 16-21, 2006, Vancouver, BC, Canada.
5. National Health Interview Survey, 2006. http://www.cdc.gov/nchs/data/series/sr10/sr10_227.pdf, extracted on Jan. 16th, 2006.
6. Moore, G. B., Robotics, Artificial Intelligence, Computer Simulation: Future Applications in Special Education, COSMOS Corp., Washington, DC. (1985)
7. Hofmeister, Alan. M. and J. M., Ferrara, Artificial Intelligence Applications in Special Education: How Feasible? Final Report, Logan, Utah State U. (1986).

8. Hofmeister, Alan M. and Margaret M. Lubke, Expert Systems: Implications for the Diagnosis and Treatment of Learning Disabilities, *Learning Disability Quarterly*, Vol. 11, No. 3 pp. 287-91 (Summer, 1988).
9. Geiman, R. M. and W. L. Nolte, An Expert System for Learning Disability Diagnosis, *Proceedings of the IEEE Conference on System Engineering*, pp. 363-366 (Aug. 1990).
10. Baer, R., Referral Consultant: An Expert System for Guiding Teachers in Referring Students for Special Education Placement, Logan, Utah State U., Center for Persons with Disabilities (1991).
11. Vinsonhaler, J., Using an Experts System to Study the Classroom Placement Decision for Students Who Are Deaf or Hard-of-Hearing, *Journal of Special Education Technology*, Vol. 12, No. 2, pp. 135-148 (Fall, 1993).
12. Baesens, B, T Van Gestel, S Viaene, M Stepanova, J Suykens and J Vanthienen, Benchmarking state-of-the-art classification algorithms for credit scoring, *Journal of the Operational Research Society* (2003) 54, 627–635.
13. Razi, M. A. and K. Athappilly, A comparative analysis of neural network (NNs), nonlinear regression and classification and regression tree (CART) models, *Expert Systems with Application*, 29, pp. 65-74 (2005).
14. NeuroDimension Incorporated, What is a Neural Network?, http://www.nd.com/neurosolutions/products/ns/whatisNN.html
15. Sordo, M., Introduction to Neural Networks in Healthcare. *OpenClinical.:Knowledge Management for Medical Care*. Harvard (2002).
16. Mead, W. C., Development of Neural Network Tools and Techniques for Arithmetic Training of Children with Learning Disabilities: Phase I - Theory and Simulations, *AERA '97*, March 24-28, Chicago, IL (1997).
17. Linstrom, K. R. and A. J. Boye, Neural Network Prediction Model for a Psychiatric Application, *Proceedings of the Sixth International Conference on Computational Intelligence and Multimedia Applications* (2005).
18. Fu, X.J., Wang, L.P.: Data dimensionality reduction with application to simplifying RBF network structure and improving classification performance. *IEEE Trans. System, Man, Cybernetics, Part B*, 33 (2003) 399-409
19. Fu, X.J., Wang, L.P.: A GA-Based Novel RBF Classifier with Class-Dependent Features. *Proc. 2002 IEEE Congress on Evolutionary Computation (CEC 2002)*, vol.2 (2002) 1890 ¨C 1894
20. Raymer, M.L., Punch, W.F., Goodman, E.D., Kuhn, L.A., Jain, A.K.: Dimensionality reduction using genetic algorithms. *IEEE Trans. Evolutionary Computation*, 4 (2000) 164 – 171.
21. Rittho, Oliver, Ralf Klinkenberg, Simon Fischer, Ingo Mierswa and Sven Felske, Yale: Yet Another Learning Environment, *LLWA 01 - Tagungsband der GI-Workshop-Woche Lernen - Lehren - Wissen - Adaptivität*, No. 763, pages 84--92, Dortmund, Germany, 2001.
22. XLMiner, http://www.xlminer.net/, extracted on Jan. 29th, 2006.

A Comparison of Competitive Neural Network with Other AI Techniques in Manufacturing Cell Formation

Gurkan Ozturk[1], Zehra Kamisli Ozturk[2], and A. Attila Islier[1]

[1] Eskisehir Osmangazi University, Industrial Engineering Department, Bademlik 26030, Eskisehir, Turkey
{gurkano, aislier}@ogu.edu.tr

[2] Anadolu University, Open Education Faculty No:312, 26470, Eskisehir, Turkey
zkamisli@anadolu.edu.tr

Abstract. The cell formation (CF) problem aims to transform the incidence matrix into block diagonal form. Numerous techniques are developed for this purpose ranging from mathematical programming to heuristic and AI techniques. In this study a simple but effective competitive neural network algorithm (CNN) is applied and compared with genetic algorithms, tabu search, simulated annealing and ant systems by making use of some well known data sets from literature. As a result at 14 out of 15 cases, better results are obtained by CNN.

Keywords: Cell formation, cellular manufacturing, competitive neural network, AI techniques.

1 Introduction

Wemmerlov and Hyer [1] assert that cellular manufacturing is an application of group technology in which all or a portion of firm's manufacturing system has been converted into cells. A manufacturing cell is a cluster of machines or processes located in close proximity and dedicated to the manufacture of a family of parts. The parts are similar in their processing requirements, such as operations, tolerances, and machine tool capacities.

Cell formation (CF) problem is a type of NP-complete problems [2]. There are lots of algorithms for solving the CF problems ranging from heuristics to mathematical programming in the literature. The heuristic methods developed by Ferreira and Pradin [3], Lee and Diaz [4], Purcheck [5] and Boe and Cheng [6] are based on a penalty function, network flow, hospitality and close neighbor concepts. Another common concept used in several techniques is the similarity coefficient. Shafer and Rogers [7] discuss the improvement of various coefficients by classifying them. Chu and Hayya [8] use fuzzy logic to cope with alternative routings.

As artificial techniques, Lee et al. [9], Goncalves et al. [10], Nsakanda et al. [11], Venugopal and Narendran [12] and Islier [13] use genetic algorithms, Sofianopoulou [14], Jayaswal and Adil [15], Tam [16] and Boktor [17] apply simulated

L. Jiao et al. (Eds.): ICNC 2006, Part I, LNCS 4221, pp. 575–583, 2006.

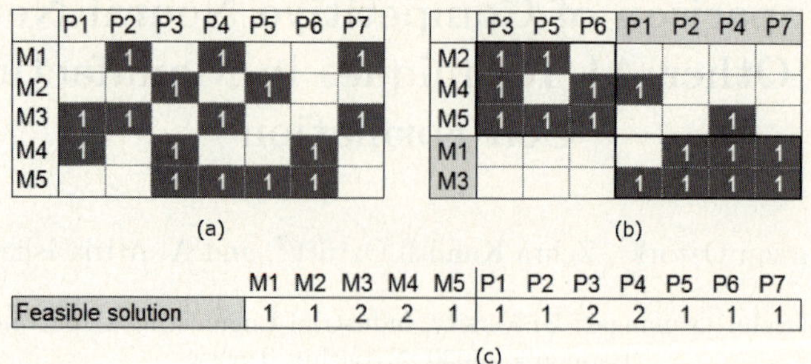

Fig. 1. An example for incidence matrix and its diagonalized form

annealing, Islier [18] employs Ant Systems and Chu [19], [20], Dobado et al. [21], Soleymanopuor et al. [22], Guerro et al. [23], Ozturk and Ozturk [24] and Venugopal [25] apply neural networks.

All of these techniques usually use part-machine incidence matrix $\{A_{ij}\}$ in which the columns and rows represent the parts and machines, to get a block diagonalized structure of the matrix. The (ij)th element of $\{a_{ij}\}$ is 1 if the jth part visits the ith machine; otherwise it is 0. An incidence matrix, its diagonalized structure and representation of the solution as a string are given in Figure 1-(a-b-c) respectively.

Following section briefly introduces the AI techniques by emphasizing their common structure. Numerical experience on section 3 is followed by conclusion.

2 AI Techniques in Cell Formation Problem

Artificial Intelligence (AI) techniques are appropriate tools to solve the complicated real-life problems within a reasonable time by generating high-quality solutions. Consequently, exhaustive research and development is carried out in this field and numerous alternative techniques are generated. Abundance of available techniques, in turn, causes confusion deciding on the technique to be used for a particular problem. So an analysis is necessary to compare these techniques. Since, the performance of AI techniques is problem specific, this analysis should be made for distinct application fields separately. So this study is restricted with application of AI techniques to Cell Formation Problem [18].

The pseudo codes of the most fundamental forms of genetic algorithms (GA), simulated annealing (SA), tabu search (TS) and ant systems (AS) are presented in Figure 2. Interested reader may refer to references for further information.

All this techniques and competitive neural network (CNN) developed by Ozturk and Ozturk [24] use the same string representation for feasible solutions as given in 1-(c). This string corresponds to a chromosome in GA, a crystal structure in SA, a grouping in TS, a tour in AS and the maximum weights on sum of weighted inputs in CNN.

Competitive neural network structure presented by [24] consists of three different components. First and second components represent input pattern that

```
Genetic Algorithms (GA)
Initialize;
repeat
        evaluate the individuals;
        repeat
              select parents;
              generate offspring;
              mutate if enough solutions are generated;
        until population number is reached;
        copy the best fitted individuals into population as they were;
until required number of generations are generated.
```

```
Simulated Annealing (SA)
initialize;
repeat
        generate a candidate solution;
        evaluate the candidate;
        determine the current solution;
        reduce the temperature;
until termination condition is met;
```

```
Tabu Search (TS)
initialize;
repeat
        generate all of the permissible neighborhood solutions;
        evaluate the generated solutions;
        nominate the best of them as the candidate solution
        if no candidate is found then choose the best of forbidden solutions as the candidate;
        update the tabu list; move to candidate solution;
        if sufficient number of solutions are generated, diversify;
until termination condition is met;
```

```
Ant Systems (AS)
Initialize;
i← 0;
repeat
        generate a feasible solution;
        evaluate goodness η;
        if η > η_max then begin update elite list; shift the bounds end;
        if i mod α = 0 then alter the solution;
        if i mod β = 0 then intensify elite pheromone traces;
        update pheromone trails;
        i← i + 1;
until i=σ;
```

Fig. 2. Pseudo code for GA, SA, TS and AS

correspond to machines and parts respectively and finally the third component is output pattern which represent cells. Connection weights between input and output neurons are denoted as w_{ij}, v_{pj} . Indices $i = 1, \ldots, m$; $p = 1, \ldots, n$ and $j = 1, \ldots, k$ are used respectively for machines, parts and cells.

Figure 3 depicts the proposed network topology.

Pseudo code of competitive neural network algorithm for CF problem is presented in Figure 4. This algorithm is based on the algorithm given by Chu [19]. As mentioned above, the difference between CNN [24] approach and Chu's is in determination of input pattern. In this way, parts and machines are simultaneously grouped into cells.

At the initialization step, parameters of CNN algorithm (learning rate l, number of epoch n_{epoch}) are given and weights are randomly generated then all weights in the network are normalized. Each machine and each part in the incidence matrix has its own unique pattern. Hence exactly $m + n$ distinct patterns are

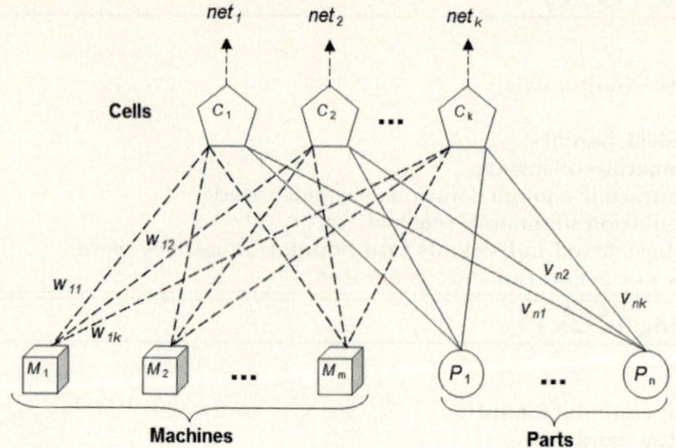

Fig. 3. Topology of the Competitive Neural Network

```
Initialize;
s ← 0 ;
Repeat
          s ← s +1;
          q ← 0;
          create a random permutation π_{m+n}
          Repeat
                    q ← q +1;
                    set all the elements of the pattern to zero
                    generate π[q]^th input pattern
                    Calculate the value of net_j for all output neurons (cells)
                    Determine the winner output neuron
                    Update the weights
          Until q=m+n ;
Until s =n_epoch;
```

Fig. 4. Pseudo code for competitive neural network

formed. The algorithm mainly incorporates two loops: outer and inner. The outer loop generates a random permutation π length of $m + n$ at each iteration. The inner loop uses this permutation to generate the distinct patterns one at a time. This generation is made in accordance with $\pi[q]$.

If $\pi[q] \leq m$ then $\pi[q]^{th}$ element of the pattern set to '1'. Then $\pi[q]^{th}$ row of incidence matrix is copied onto last n elements of pattern. Otherwise transpose of $(\pi[q] - m)^{th}$ column of incidence matrix is copied onto first m elements of the pattern. Then the $\pi[q]^{th}$ element of the pattern is set to '1'. For example, by considering incidence matrix in Figure 1, input patterns for 2nd machine and 4th part are generated as [01000|0010100] and [10101|0001000] respectively.

In the following step, the net inputs of neurons in output layer are calculated by using Equation (1). The winner neuron is determined by maximum net input value of output neurons. Weights are updated using Equations (2)-(3) according to learning rate, l. At the last step of the inner loop, weights of the connections between input and output neurons are normalized. To illustrate the algorithm a step by step example is given in the following sub-section.

Illustrative Steps for CNN. Illustrative steps of the algorithm is given on Figure 5. Initializations, loops and results are shown there. First, termination parameter n_{epoch}, learning rate l and initial weights are set. The permutation produced by outer loop and the all corresponding patterns follow the initializations. In the inner loop of the Figure 5, the sixth pattern which is the first

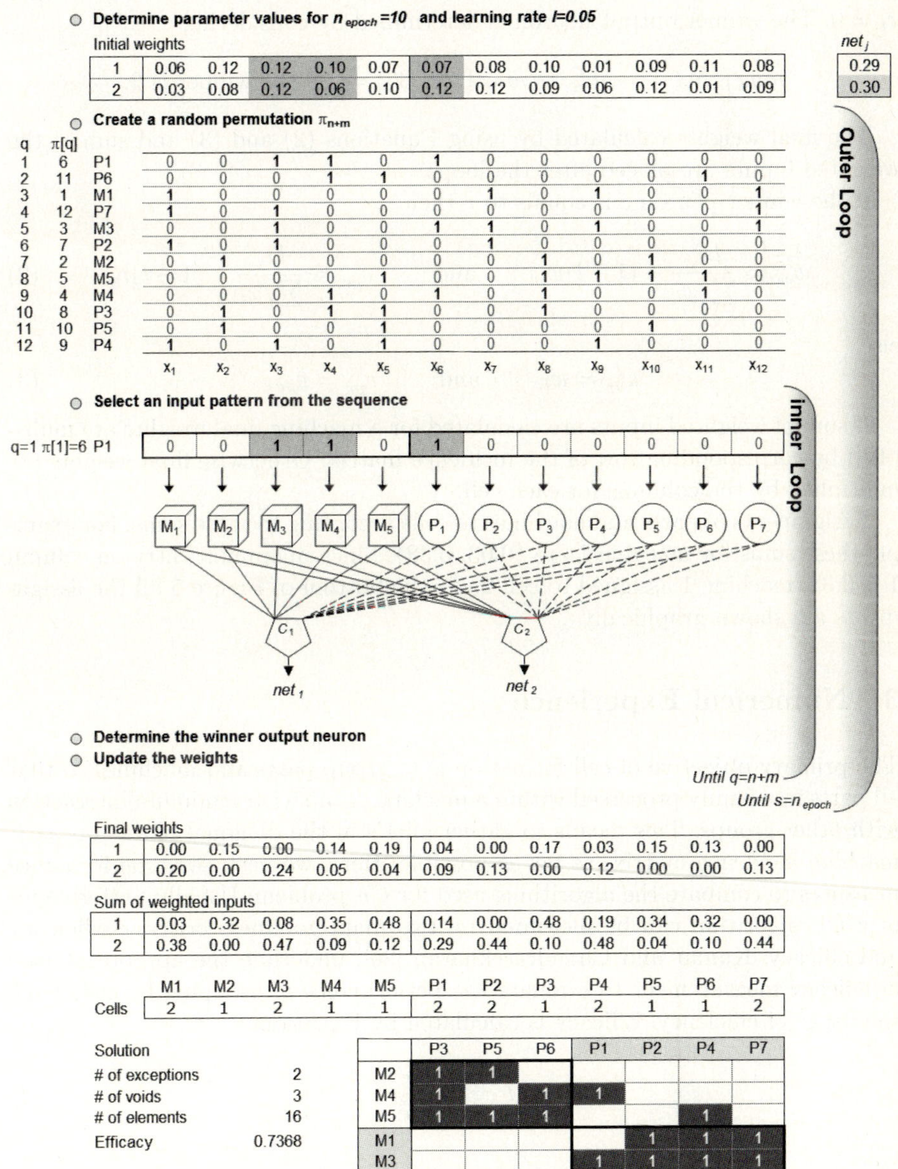

Fig. 5. The illustrative s teps of the CNN algorithm

element of the permutation is given. The net_j values calculated by using this pattern and making use of Equation (1) are given at top right of the figure.

$$net_j = \sum_{i=1}^{m}\sum_{t=1}^{m} w_{ij}x_t + \sum_{j=1}^{n}\sum_{t=m+1}^{m+n} v_{pj}x_t \qquad \forall j \tag{1}$$

where input pattern is x_t , $(t = 1, \ldots, m+n)$, if neuron is active $x_t = 1$ else $x_t = 0$. The winner output neuron is determined by considering

$$max_j\{net_j\} = net_c \rightarrow max\{.29, .30\} = .30 \text{ the winner is 2.}$$

The final weights calculated by using Equations (2) and (3) and sum of the weighted inputs are placed after the loops.

If the winner neuron c is equals to r then

$$w'_{ir} = \frac{lx_t}{\sum_t x_t} + (1-l)w_{ir} \qquad \text{and} \qquad v'_{pr} = \frac{lx_t}{\sum_t x_t} + (1-l)v_{pr} \tag{2}$$

else

$$w'_{ir} = w_{ir} \qquad \text{and} \qquad v'_{pr} = v_{pr}. \tag{3}$$

If sum of weighted inputs are calculated for a machine final weights are multiplied by corresponding row of the incidence matrix. Otherwise final weights are multiplied by the columns for each cell.

Assignments of parts and machines to cells according to these sums. For example these sums for machine 1 are {0.03, 0.38}. Since maximum entry on column 1 is 0.38, machine 1 assigned to cell 2. At the bottom of Figure 5 all the assignments are shown graphically.

3 Numerical Experience

The primary objective of cell formation is to group parts and machines so that all parts in a family processed within a machine group with minimum interaction with other groups. This means to gather all 1's in the diagonal blocks on part-machine incidence matrix, as far as possible. Here, we need some performance measures to compare the algorithms used for CF problem. Usually performance of a CF algorithm can be measured by two main goodness criteria: efficiency and efficacy. Kumar and Chandrasekharan [26], underline the appropriateness of efficacy measurement to evaluate the performance of grouping by criticizing the usage of efficiency. Efficacy is calculated by Equation 4

$$\Gamma = \frac{1-\Psi}{1+\Phi} \tag{4}$$

where

$$\Psi = \frac{\text{\# of exceptional elements}}{\text{total \# of operations}}, \qquad \Phi = \frac{\text{\# of voids in the diagonal blocks}}{\text{total \# of operations}}$$

Table 1. Efficacy values of AI techniques

Source	Set	m	n	k	Ultimate efficacy	CNN	AS	SA	TS	GA
Data sets 1 to 7 of	A	24	40	7	1.00	1.00	.7047	.1485	.1826	.2252
Chandrasekharan and	B	24	40	7	.851	.851	.6149	.1367	.1826	.2524
Rajagopalan -1989	C	24	40	7	.735	.735	.4667	.1423	.1903	.2336
	D	24	40	7	.707	.755	.4971	.1682	.2000	.2217
	E	24	40	7	.368	.419	.3575	.1446	.1739	.2284
	F	24	40	7	.322	.397	.3208	.1383	.1940	.2342
	G	24	40	7	.295	.385	.3100	.1494	.1937	.2182
Carrie -1973	H	20	35	4	.751	.757	.6940	.1992	.2678	.3839
Askin and Standrige -1993	J	6	8	3	.889	.621	.7391	.6000	.4444	.7391
Askin and Standrige -1993	J	6	8	3	.889	.889	.8889	.6190	.5217	.8889
Burbidge -1977	K	16	43	5	.438	.511	.3925	.1757	.2110	.2925
Chandrasekharan and	L	40	100	10	.823	.840	.3956	.0853	.1177	.1320
Chan and Milner -1982	M	10	15	3	.818	.818	.8182	.3784	.3846	.7857
Waghodekar and Sahu -1984	N	5	7	2	.737	.737	.7368	.7368	.5455	.7368
deWitte -1980	O	12	12	3	.531	.570	.5500	.3571	.3455	.5288

Hence we use efficacy as the performance measure to compare the AI techniques in this study .

The data sets and their characteristics used in this study are given in Table 1. This table gives also the goodness (efficacy) of the best known solutions in the literature.

As seen from Table 1 CNN reached all the best known values (except case I). Even it outperformed the ultimate solutions given in the literature, in problems D,E,F,G,H,J,K,L,N and O . All of the solutions obtained by CNN are given in Figure-6. According to results we can also say that closest competitor of CNN is AS.

4 Conclusions

In this study, we presented a competitive neural network procedure to group parts and machines into cells simultaneously. To test the success of this CNN in CF problems, its performance is compared with those of other AI techniques. The results of comparisons are summarized in Table 1. As can be seen from the results CNN outperformed the other AI techniques for all efficacy values except one. Only AS attained the performance of CNN in problem I.

The performance of proposed method can be explained by its success to use the knowledge on the incidence matrix as a whole at each iteration.

As future research, the good aspects of the other AI techniques can be merged to CNN to form hybrid structures. CNN can also be applied to different kinds of intricate cell formation problems like problems with alternative routes of parts and capacitated cell formation problems.

Acknowledgement. This study is partially supported by TUBITAK - The Scientific and Technological Research Council of Turkey.

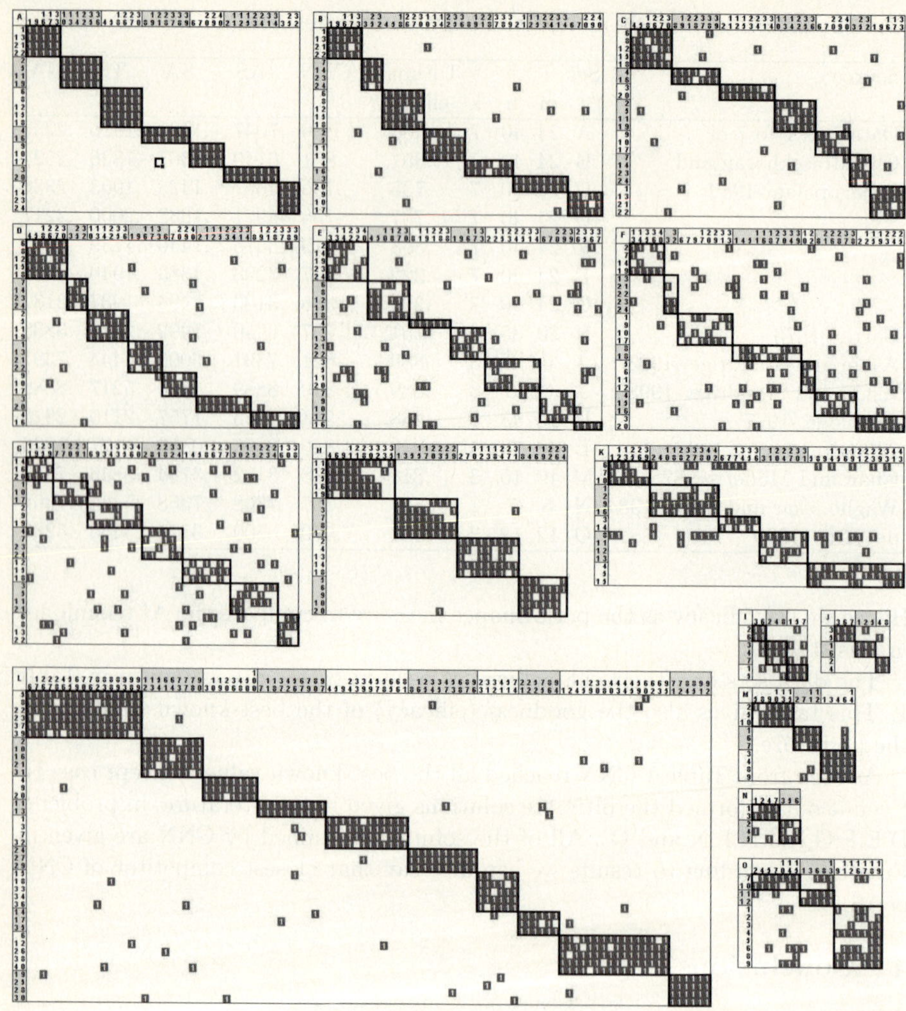

Fig. 6. Solutions of test problems

References

1. Wemmerlov, U., Hyer, N.L.: Research issues in cellular manufacturing. International Journal of Production Research **25** (1987)
2. King, J.R., Nakornchai, V.: Machine–component group formation in group technology: review and extension. International Journal of Production Research **20** (1982)
3. Ferreira, J.F., Pradin, B.: A methodology for cellular manufacturing design. International Journal of Production Research **31** (1993) 235–250
4. Lee, H., Diaz, G.: A network flow approach to solve clustering problems in gt. International Journal of Production Research **31** (1993)
5. Purcheck, G.: Machine-component group formation: an heuristic method for flexible production cells and fms. International Journal of Production Research **23** (1985) 911–943

6. Boe, W.J., Cheng, C.H.: A close neighbor algorithm for designing cellular manufacturing systems. International Journal of Production Research **29** (1991) 2097–2116
7. Shafer, S.M., Rogers, D.F.: Similarity and distance measures for cellular manufacturing part i. a survey. International Journal of Production Research **31** (1993)
8. Chu, C.H., Hayya, J.C.: A fuzzy clustering approach to manufacturing cell formation. International Journal of Production Research **29** (1991) 1475–1487
9. Lee, M.K., Luong, H.S., Ahbary, K.: A genetic algorithm based cell design considering alternative routing. Computer Integrated Manufacturing Systems **10** (1997)
10. Goncalves, J.F., Resende, M.G.C.: An evolutionary algorithm for manufacturing cell formation. Computers & Industrial Engineering **47** (2004) 247–273
11. Nsakanda, A.L., Diaby, M., Price, W.L.: Hybrid genetic approach for solving large–scale capacitated cell formation problems with multiple routings. European Journal of Operational Research **171** (2006) 1051–1070
12. Venugopal, V., Narendran, T.T.: A genetic algorithm approach to the machine component grouping problem. Computers & Industrial Engineering **22** (1992) 469–480
13. Islier, A.A.: Forming manufacturing cells by using genetic algorithm. Anadolu University Journal of Science and Technology **2** (2001) 137–157
14. Sofianopoulou, S.: Application of simulated annealing to a linear model for the formulation of machine cells in group technology. International Journal of Production Research **35** (1997)
15. Jayaswal, S., Adil, G.K.: Efficient algorithm for cell formation with sequence data, machine replications and alternative process routings. International Journal of Production Research **42** (2004)
16. Tam, K.Y.: A simulated annealing algorithm for allocating space to manufacturing cells. International Journal of Production Research **30** (1992)
17. Boktor, F.: A linear formulation of the machinepart cell formation problem. International Journal of Production Research **29** (1991) 343–356
18. Islier, A.A.: Group technology by ants. International Journal of Production Research **43** (2005)
19. Chu, C.H.: Manufacturing cell formation by competitive learning. International Journal of Production Research **31** (1989) 829–843
20. Chu, C.H.: An improved neural network for manufacturing cell formation. Decision Support Systems **20** (1997) 279–295
21. Dobado, D., Lozano, S., Bueno, J., Larraneta, J.: Cell formation using a fuzzy min–max neural network. International Journal of Production Research **40** (2002)
22. Soleymanpour, M., Vrat, P., Shankar, R.: A transiently chaotic neural network approach to design of cellular manufacturing. International Journal of Production Research **40** (2002)
23. Guerrero, F., Lozano, S., Smith, K.A., Canca, D., Kwok, T.: Manufacturing cell formation using a self–organizing neural network. Computers & Industrial Engineering **42** (2002) 377–382
24. Ozturk, G., Ozturk, Z.K.: A competitive neural network approach to manufacturing cell formation. In: Proceedings of the 35th International Conference on Computers and Industrial Engineering, Istanbul, Turkey (2005) 1549–1554
25. Venugopal, V., Narendran, T.T.: Machine–cell formation through neural network models. International Journal of Production Research **32** (1994)
26. Kumar, C.S., Chandrasekharan, M.P.: Grouping efficacy: a quantitative criterion for goodness of block diagonal forms of binary matrices in group technology. International Journal of Production Research **28** (1990)

Intelligent Natural Language Processing

Wojciech Kacalak[1], Keith Douglas Stuart[2], and Maciej Majewski[1]

[1] Technical University of Koszalin, Department of Mechanical Engineering
Raclawicka 15-17, 75-620 Koszalin, Poland
{wojciech.kacalak, maciej.majewski}@tu.koszalin.pl

[2] Polytechnic University of Valencia, Department of Applied Linguistics
Camino de Vera, s/n, 46022 Valencia, Spain
kstuart@idm.upv.es

Abstract. In this paper, a natural language interface is presented which consists of the intelligent mechanisms of human identification, speech recognition, word and command recognition, command syntax and result analysis, command safety assessment, technological process supervision as well as human reaction assessment. In this paper, a review is carried out of selected issues with regards to recognition of speech commands in natural language given by the operator of the technological device. A view is offered of the complexity of the recognition process of the operator's words and commands using neural networks made up of a few layers of neurons. The paper presents research results of speech recognition and automatic recognition of commands in natural language using artificial neural networks.

1 Intelligent Two-Way Speech Communication

In the developed intelligent system for natural language processing [1,4,5,6,7,8], if the operator is identified and authorized by the natural language interface, a command produced in continuous speech is recognized by the speech recognition module and processed in to a text format. Then the recognised text is analysed by the syntax analysis subsystem. The processed command is sent to the word and command recognition modules using artificial neural networks to recognise the command, which is sent to the effect analysis subsystem for analysing the status corresponding to the hypothetical command execution, consecutively assessing the command correctness, estimating the process state and the technical safety, and also possibly signalling the error caused by the operator. The command is also sent to the safety assessment subsystem for assessing the grade of affiliation of the command to the correct command category and making corrections. The command execution subsystem signals commands accepted for executing, assessing reactions of the operator, defining new parameters of the process and run directives [2]. The subsystem for voice communication produces voice commands to the operator [9].

L. Jiao et al. (Eds.): ICNC 2006, Part I, LNCS 4221, pp. 584–587, 2006.

2 Recognition of Commands in Natural Language

In the automatic command recognition system shown in Fig. 1a, the speech signal is processed to text and numeric values with the module for processing voice commands to text format. The speech recognition engine is a continuous density mixture Gaussian Hidden Markov Model system which uses vector quantization for speeding up the Euclidean distance calculation for probability estimation [3]. The separated words of the text are the input signals of the neural network for recognizing words. The network has a training file containing word patterns. The network recognizes words as the operator's command components, which

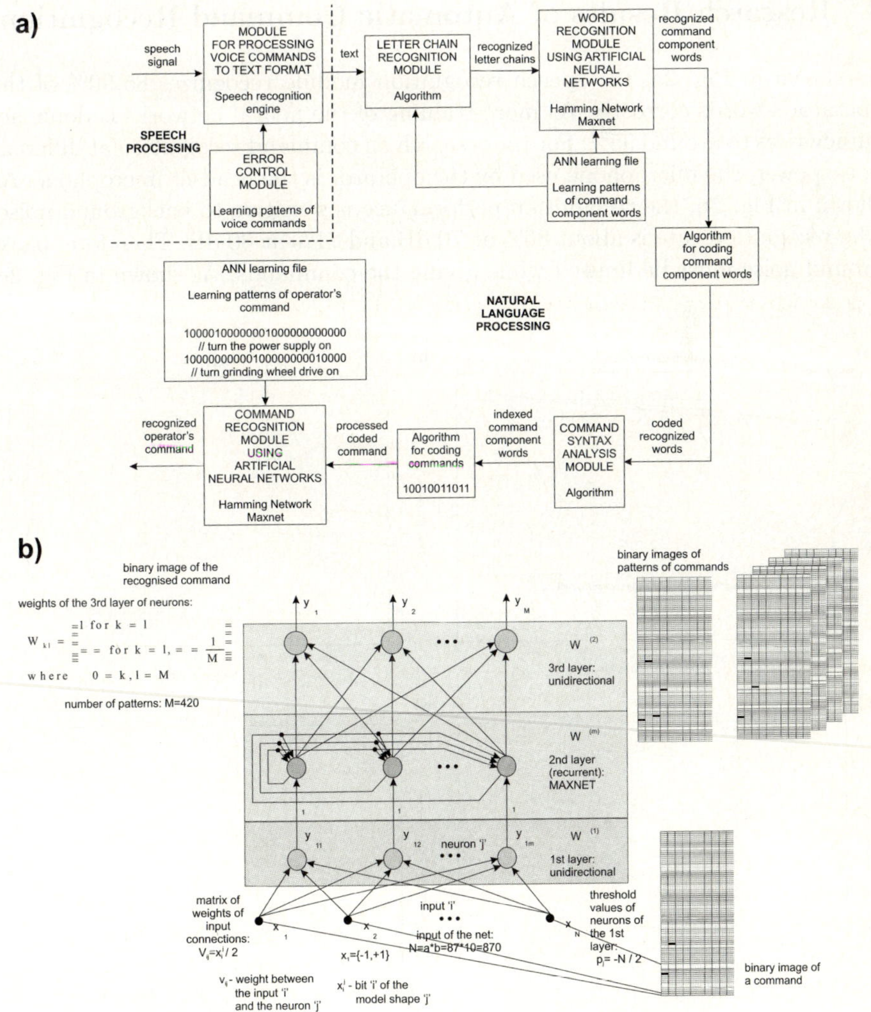

Fig. 1. Scheme of the automatic command recognition system

are represented by its neurons. The recognized words are sent to the algorithm for coding words. Then, the coded words are transferred to the command syntax analysis module. It is equipped with the algorithm for analysing and indexing words. The module indexes words properly and then they are sent to the algorithm for coding commands. The commands are coded as vectors and they are input signals of the command recognition module using a neural network. The module uses the 3-layer Hamming neural network in Fig. 1b, either to recognize the operator's command or to produce the information that the command is not recognized. The neural network is equipped with a training file containing patterns of possible operator commands.

3 Research Results of Automatic Command Recognition

As shown in Fig. 2a, the speech recognition module recognizes 85-90% of the operator's words correctly. As more training of the neural networks is done, accuracy rises to around 95%. For the research on command recognition at different noise power, the microphone used by the operator is the headset microphone. As shown in Fig. 2b, the recognition performance is sensitive to background noise. The recognition rate is about 86% at 70 dB and 71% at 80 dB. Therefore, background noise must be limited while giving the commands. As shown in Fig. 2c,

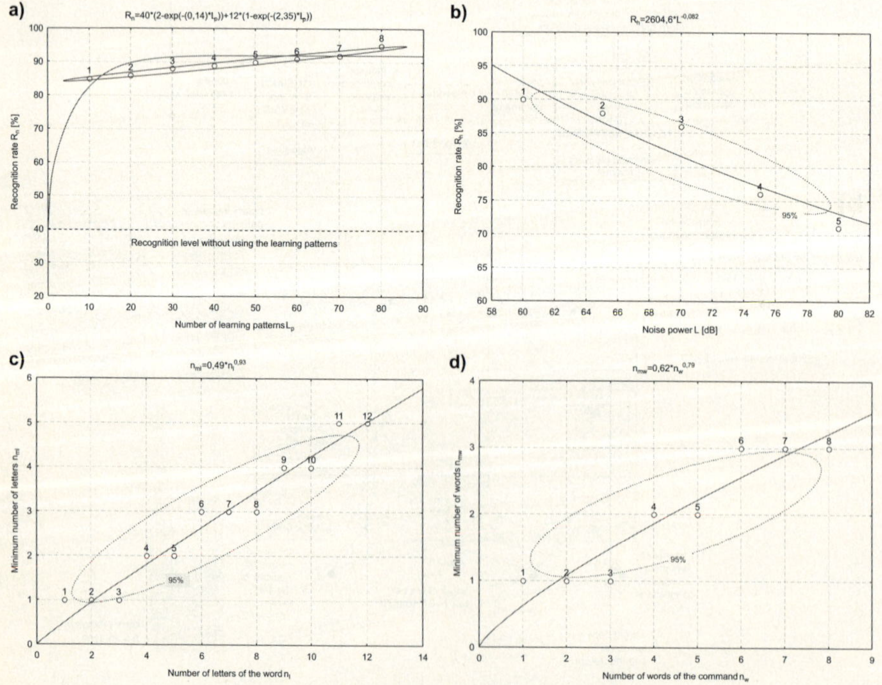

Fig. 2. Speech and command recognition rate

the ability of the neural network to recognise the word depends on the number of letters. The neural network requires the minimal number of letters of the word being recognized as its input signals. As shown in Fig. 2d, the ability of the neural network to recognise the command depends on the number of command component words. Depending on the number of component words of the command, the neural network requires the minimal number of words of the given command as its input signals.

4 Conclusions and Perspectives

The condition of the effectiveness of the presented system is to equip it with mechanisms of command verification and correctness. The aim of this research to develop an intelligent layer of two-way speech communication is difficult, but the prognosis of the technology development and its first use shows a significant efficiency in supervision and production humanization.

References

1. Majewski, M., Kacalak, W., Douglas Stuart, K.: Selected Problems of Intelligent Natural Language Processing. 7th International Conference on Computing CORE2006, Mexico. International Journal of Research on Computing Science (2006)
2. Majewski, M., Kacalak, W., Douglas Stuart, K., Cuesta, D.: Selected Problems of Recognition and Evaluation of Natural Language Commands. 7th Int. Conf. CORE2006, Mexico. Int. Journal of Research on Computing Science (2006)
3. Majewski, M., Kacalak, W.: Automatic Recognition and Evaluation of Natural Language Commands. Int. Symposium on Neural Networks ISNN2006, Chengdu, China. LNCS 3973, Advances in Neural Networks, Springer-Verlag (2006) 1155-1160
4. Majewski, M., Kacalak, W.: Natural Language Human-Machine Interface using Artificial Neural Networks. Int. Symposium on Neural Networks ISNN2006, Chengdu, China. LNCS 3973, Advances in Neural Networks, Springer-Verlag (2006) 1161–1166
5. Majewski, M., Kacalak, W.: Intelligent Two-Way Speech Communication System between the Technological Device and the Operator. Artificial Neural Networks in Engineering ANNIE2005, St. Louis. ASME Press, New York (2005) 841-850
6. Majewski, M., Kacalak, W.: Intelligent Human-Machine Voice Communication System. Engineering Mechanics International Journal 12(3) (2005) 193-200
7. Majewski, M., Kacalak, W.: Intelligent Layer of Human-Machine Voice Communication. 9th International Conference KES2005. Melbourne, Australia. Lecture Notes in Computer Science 3683. Springer-Verlag (2005) 930-936
8. Majewski, M., Kacalak, W.: Intelligent Human-Machine Speech Communication System. International Conference on Intelligent Computing ICIC2005, Hefei Anhui, China (2005) 3441-3450
9. O'Shaughnessy, D.: Speech Communications: Human and Machine. IEEE Press, New York (2000)

Optimal Clustering-Based ART1 Classification in Bioinformatics: G-Protein Coupled Receptors Classification*

Kyu Cheol Cho, Da Hye Park, Yong Beom Ma, and Jong Sik Lee

School of Computer Science and Engineering
Inha University
#253, YongHyun-Dong, Nam-Ku,
Incheon 402-751, South Korea
{landswell, audrey57}@empal.com, myb112@hanmail.net,
jslee@inha.ac.kr

Abstract. Protein sequence data have been revealed in current genome research and have been noticed in demand of classifier for new protein classification. This paper proposes the optimal clustering-based ART1 classifier for the GPCR data classification and processes the GPCR data classification. We focuses on a demand of optimal classifier system for protein sequence data classification. The optimal clustering-based ART1 classifier reduces processing cost for classification effectively. We compare classification success rate to those of Backpropagation Neural Network and SVM. In experimental result of the optimal clustering-based ART1 classifier, classification success rate of ClassA group is 99.7% and that of the others group is 96.6%. This result demonstrates that the optimal clustering-based ART1 classifier is useful to the GPCR data classification. The classification processing time of the optimal clustering-based ART1 classifier is the 27% less than that of the Backpropagation Neural Network and is the 39% less than that of the SVM in an optimal clustering rate which is 15%. And the classification processing time of the optimal clustering-based ART1 classifier is the 39% less than that of the optimal clustering-based ART1 classifier in a prediction success rate which is 96%. This result demonstrates that the optimal clustering-based ART1 classifier provides the high performance classification and the low processing cost in the GPCR data classification.

1 Introduction

Prediction classification technology, which extracts from pattern information about various data, has been researched for a long time. Prediction classification transacts and analyzes perceptional information through a sense organ of human. Currently, two methods on prediction classification have been noticed; one method previously decides kind of pattern. Its input pattern determines a pattern of a certain model. The

* This research was supported by the MIC(Ministry of Information and Communication), Korea, under the ITRC(Information Technology Research Center) support program supervised by the IITA(Institute of Information Technology Assessment).

L. Jiao et al. (Eds.): ICNC 2006, Part I, LNCS 4221, pp. 588 – 597, 2006.

other method determines a pattern of at least two classification patterns which automatically study examples. These predictions classification are used in bioinformatics researches. Signal transduction is basically biological phenomenon in physiological function and directly influences growth, specialization and extinction of cells or organization. This signal transduction controls various miniatures. The GPCR (G-Protein Coupled Receptor) [1], which takes an important part at recent biology and industry, is one of the receptor-ligand interactions in signal transduction. The GPCR discusses various automatic classification methods. Currently pharmaceutical research focuses on the GPCR because the GPCR plays an important part in many diseases. Also, it is increasing demand of classifier which can be automatic classification prediction of new proteins. We induct to classification method with optimal ART1 (Adaptive Resonance Theory 1) [2] clustering and implement the optimal clustering-based ART1 classifier for examining its efficiency.

This paper is organized as follows: Section 2 describes bioinformatics applications with neural network, SVM for GPCR Classification and the ART1 classification network. Section 3 describes ART1 Optimal-clustering for GPCR Classification. Section 4 demonstrates processing time reduction and evaluates the performance of optimal clustering-based ART1 classifier. The conclusion is in Section 5

2 Related Work

2.1 Bioinformatics Applications with Neural Network

ART1 is a nerve network model as one of the Neural Network. Neural Network [3] solves a problem by prediction and decreases error by employing repetitive training. The ANN (Artificial Neural Nctwork) [4] is a simple operating model, which is practiced in the biological nerve system. The ANN is used in various medicine diagnosis fields and is compared to doctor diagnosis and the existing consultation fractionation [5]. The ANN training network repeats output and obtains accurate results. First, with the false-positives detection tool [6], ANN applies digital chest radiation, when searching for a lung cancer. Then, the ANN is widely used to minimize chest tumor false-positive detection [7] from digital chest radiation. The GANN (Genetic Algorithm Neural Network) [8] applying genetic algorithm is the most known recognizer and is used to searching in clinically variable set. In case of train, classification processing time increases, but GANN gets accurate result because gradient descent optimization does not fall a minimum [9]. And, The FNN (Fuzzy Neural Network) [10] is used to fast gene differentiation and is relatively advanced ANN model.

2.2 SVM for GPCR Classification

SVM [11] solving pattern classification prediction problem with statistical study and is known to an optimal solution of binary classification. Trained SVM uses a kernel function tool for accurate prediction of classification qualifications in the GPCR classification and measures similarity between two classifications, using small subfamily classification prediction. It does FSV (Fisher score vectors) [12] mapping and makes the SVM library for protein classification and classifies result about all library. The result of the SVM is compared with nearest-neighbor method which is based on the

SVM approximation method. Lately, it applies machine learning community and various learning problems for subfamily distinction of the GPCR class A and C to the SVM, and using an extended binary SVM [13] with a 1-to-N training method.

3 Optimal Clustering-Based ART1 Classification

3.1 ART1-Based GPCR Classification

The ART1 makes learned clusters with new learning data and selects new cluster for new classification category learning and keeps that the existing contents is deleted by new excess input. And, the ART1 automatically integrates new learning knowledge into all knowledge base. The ART1 ascertains structure and function of protein, and plays an important part in constituting new protein family. The ART1 is advantage which can identify data without trained data pattern group in online state. Also the ART1 is advantage which can divide the GPCR classes of two or above.

The ART1-based classifier is a module processing classifier with the ART1 clustering algorithm for GPCR classification. This classifier is available when user knows an optimized threshold value for high success rate of classification. Classification success rate and classification processing time of the ART1-based classifier are set by threshold value. And, threshold value for proper performance is known by user continuously training. But, disadvantage of this classifier is that does not assure expands of data.

3.2 Optimal Clustering-Based ART1 Classification Algorithm

The optimal clustering-based ART1 classifier guarantees the minimum classification success rate and searches the minimum classification processing time. The optimal clustering-based ART1 classifier adjusts cluster information, classification processing time, and classification success rate through data-set training, and controls threshold. It is useful to process having a small amount of data. Also, the optimal clustering-based ART1 classifier applies in different data and transformed data.

In fig. 1, it is a flowchart which guarantees the minimum classification success rate (95%) and uses binary pattern that half value of a previous value for getting the minimum threshold. Algorithm within the optimal clustering-based ART1 classifier chooses the first input as representative pattern of the first cluster. After, if an input comes into it, the input compares with the first representative pattern. If distance of the first representative cluster is shorter than threshold, the input is classified into first cluster. If not so, the input generates new cluster. This process is applied to all input.

Therefore, cluster number is getting more and more increase in process of time. And the optimal clustering-based ART1 classifier algorithm generates different result according to distance measurement method between input and representative pattern of cluster. Vigilance threshold decides discord tolerance between input pattern and stored pattern. If vigilance threshold is large, the optimal clustering-based ART1 classification algorithm identifies a little difference between input pattern and expectation pattern and classifies new category. But, if is small, this algorithm is a great difference between input pattern and expectation pattern and roughly divides input patterns. This processing applies threshold to data training machine and progresses same processing.

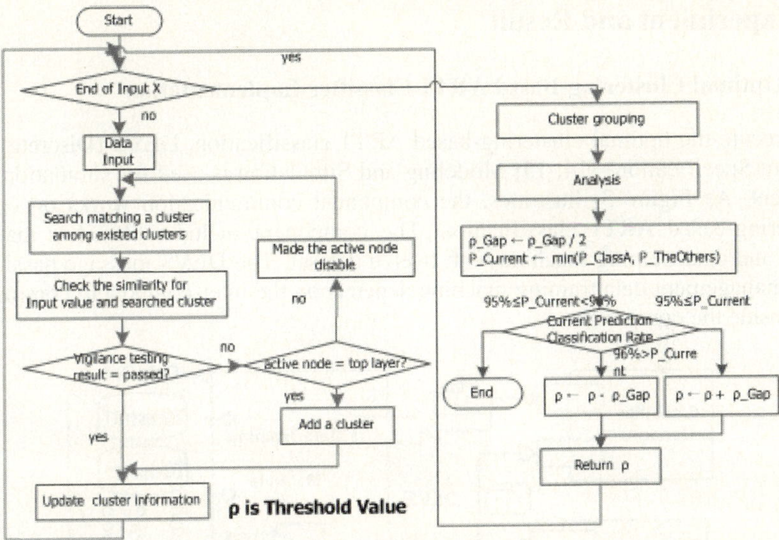

Fig. 1. Flowchart of optimal clustering-based ART1 classification

3.3 Optimal Clustering-Based ART1 Classifier

Fig.2 shows component progress in the optimal clustering-based ART1 classifier. Clustering manager component searches for appropriate threshold through classification success rate and processing time. Searching for appropriate threshold reduces cost and classification processing time, because number of cluster is different according to data. Also, clustering manager component guarantees classification success rate. Clustering manager component can find out appropriate threshold using binary pattern that half value of a previous value. Data training machine component reads data for the ART1 clustering, Data training machine component generates cluster about data and progresses training. Next time, if this component is connected, data training machine progresses training based on the existing data and sends cluster information, classification processing time, and classification success rate to tester component. Tester component progresses test through cluster information, classification processing time, and classification success rate, and sends results to clustering manager component.

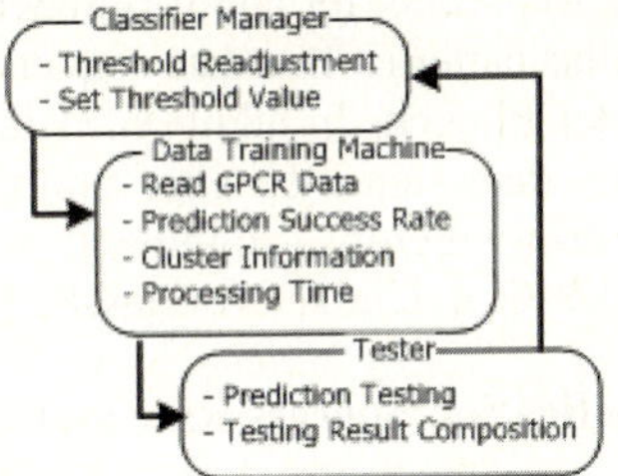

Fig. 2. Components flow of the optimal clustering-based ART1 classifier

4 Experiment and Result

4.1 Optimal Clustering-Based ART1 Classifier Implementation

To execute the optimal clustering-based ART1 classification, DEVS (Discrete Event System Specification) [14, 15] Modeling and Simulation is used for simulation environment. As Figure 3 illustrates, the component communication works on optimal clustering-based ART1 classification. The component includes classifier manager, tester and data training machine with twelve dataset. The DEVS message passing for data management data training machine depends on the inter-component communication inside the component.

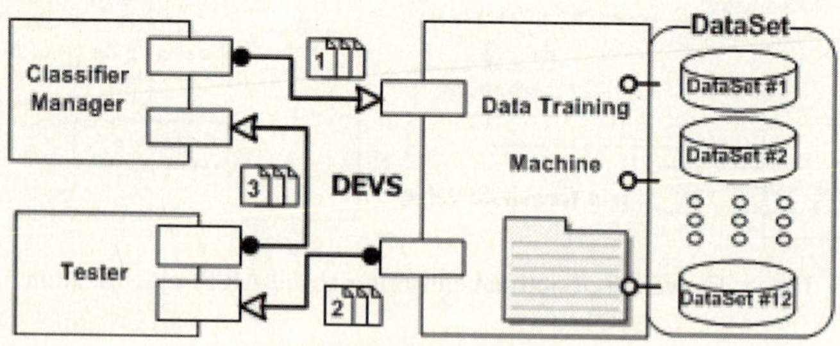

Fig. 3. Simulation Architecture of the optimal clustering-based ART1 classification

4.2 Experiment and Result

4.2.1 GPCR Sequence Data Preprocessing

According to the information system of GPCRDB, superfamily is divided into five main GPCR classes (ClassA, ClassB, ClassC, ClassD, ClassE) and decoy. The basic data using in the experiment is composed [16] of 692 ClassA, 56 ClassB, 16 ClassC, 11 ClassD, 3 ClassE and 99 decoy. The sequence data comparing the input layer requires a fixed-length layer and requires data preprocessing for real-number conversion. At first, it is the process of converting 877 sequence data into fixed-length feature vectors. ClustalX [17] was used for multiple alignments. ClustalX is a multiple-alignment program which is developed by Julie D. Thompson in 1994 and is used to search biological meaning through global multiple alignment of DNA or protein. This program compares similarity among target ranks, and sorts it, and calculates equality and similarity among selected ranks. That is, this program provides function which can visually see equality, similarity and difference. This program inputs character (-) after searching similar patterns. If a data sequence is long, the data sequence after multiple-alignment is lengthened. In addition, if there are many differences of length among sequence data, data sequence after multiple-alignment is lengthened. Therefore, this paper performs multiple-alignment of 766 total data sequences (ClassA: 634, ClassB: 41, ClassC: 0, ClassD: 8, ClassE: 3).

4.2.2 Experiment 1: Classification Success Rate and Classification Processing Time

This experiment demonstrates that generated clusters in the optimal clustering-based ART1 classifier are resolved by use of a threshold. And, it detects optimum classification environments through correlation between a cluster number and threshold. If the optimal clustering-based ART1 classifier consists of many clusters, it lengthens total classification processing time and has high classification success rate. If the optimal clustering-based ART1 classifier has few clusters, it reduces classification processing time through the adjustment of the cluster number, because it has low classification success rate.

Fig. 4-(a) demonstrates measurement each classification success rate through adjusting the border threshold and cluster generation rate. ClassA records 93% classification success rate when the clustering rate is 3%, and records 95% classification success rate when the clustering rate is 5%. The ClassA group almost records a 100% classification success rate when the clustering rate is 10%. The others group records 87% classification success rate when the clustering rate is 3% and records 95% classification success rate when the clustering rate is 15%. These results demonstrate that data classification improves in quality through increasing the clustering rate. In particular, these results demonstrate that it provides distinguished efficiency on the GPCR sequence data classification. This paper compares the ART1 classifier's classification processing time and classification success rate to those of the Backpropagation Neural Network and the SVM. The SVM uses the SVM classification method [18] and the Backpropagation Neural Network uses the ANN classifier applying delta learning algorithm [19].

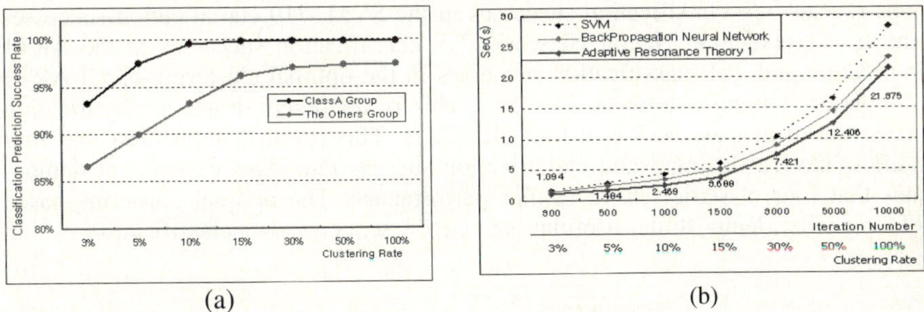

(a) (b)

Fig. 4. (a) Classification success rate with the optimal clustering-based classifier, (b) Classification processing time

Fig. 4-(b) shows variations of classification processing time through adjusting iteration number and the clustering rate. Classification processing time is learning time and test time of the GPCR sequence data. Generated cluster numbers vary according to data quantity, sequence size and the threshold. However, this experiment measures the clustering rate according to the threshold and inquires into the optimum clustering rate. The classification processing time is increases when the clustering rate is increases. The ART1 classifier's classification processing time records 1.094 seconds when the clustering rate is 3%. It records 21.375 seconds when the clustering rate is 100%. Processing cost is increasing when classification processing time is increasing. The clustering rate demands at least 15% and classification processing time needs 3.688 seconds when the clustering rate is 95%. In the backpropation neural network's classification processing time is records as 5.012 seconds and the clustering rate is 27% when the iteration number is 1500. In the classification of SVM classification processing time records 6.032 seconds and the clustering rate is 39% when iteration number is 1500. These results demonstrate that the ART1 provides reliable efficiency over two classifiers.

4.2.3 Experiment 2: Classification Success Rate (SVM vs. Backpropagation Neural Network vs. ART1 Clustering Classifier vs. Optimal Clustering-Based ART1 Classifier)

This experiment measures ClassA and the others in the Backpropagation nural ntwork, the SVM, the ART1 clustering classifier and the optimal clustering-based ART1 classifier. This experiment compares the optimal clustering-based ART1 classifier to the SVM, the bckpropagation neural network and the ART1 clustering classifier. ART1 demonstrates usefulness in GPCR sequence data classification. It chooses 12 data sequence sets at random from 744 data sequences. This experiment divides a 12 data sequence set into a training data set and test data set, and training sequences in ClassA and other classes are 572, and the test sequences are 62. 744 test results are obtained through 12 data sequence sets. Fig. 5 presents the GPCR data classification success rate using these three classifiers.

Result of ClassA demonstrates 742 classification successes in the SVM, 741 classification successes in the backprogagation neural network, 742 classification successes in the ART1 classification and 742 classification successes in the optimal clustering-based ART1 classification, using a total of 744 data. This result reveals a classification success rate of at least 99.5% in the three classifiers. The result of the other groups reveals 720 classification successes in the SVM, 719 classification successes in the Backpropagation Neural Network, 719 classification successes in the ART1 classification and 721 classification successes in the optimal clustering-based ART1 classification, among a total of 744 data. This result demonstrates a classification success rate of at least 96.5% in four classifiers. This result demonstrates that four classifiers have a distinguished classification success rate. This experiment demonstrates that four classifiers have similar performance. The optimal clustering-based ART1 classifier demonstrates usefulness in GPCR sequence data classification.

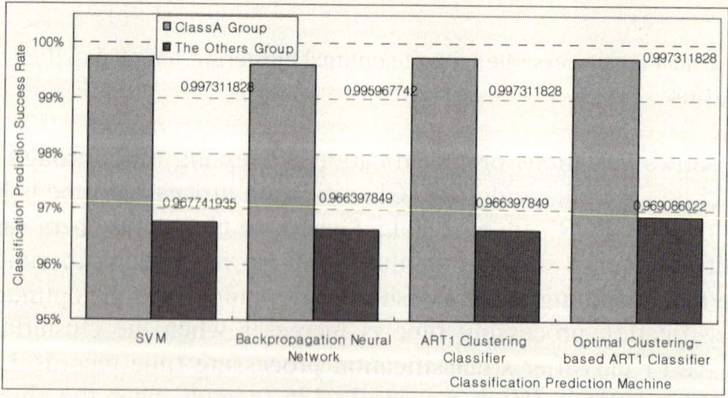

Fig. 5. Classification success rate

4.2.4 Experiment 3: Classification Processing Time (ART1 Clustering Classifier vs. Optimal Clustering-Based ART1 Classifier)

This experiment 3 compares processing time of the optimal clustering-based ART1 classifier to the ART1 clustering classifier. This experiment adjusts data-set number of two classifiers and assures the optimal classification rate (95%). This experiment measures processing time until threshold value uses the optimal classification processing time. We demonstrate that any classifier is useful through this experiment, because it enables to measure a possible data-set number and can estimate processing cost. We compose GPCR data into 12 data-set through measure processing time of classifier.

Fig. 4-(b) shows variations of classification processing time through adjusting iteration number and the clustering rate and evaluates performance of SVM, Backpropagation Neural Network and the ART1 clustering classifier. Fig. 6 evaluates classification processing time performance of the ART1 clustering classifier and the optimal clustering-based ART1 classifier.

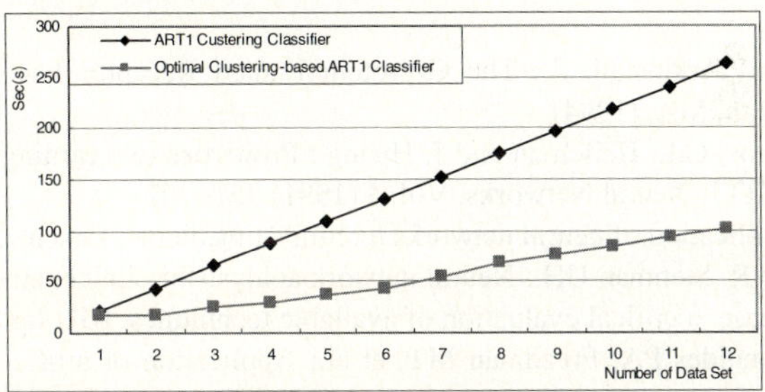

Fig. 6. Classification processing time (ART1 clustering classifier vs. optimal clustering-based ART1 classifier)

Processing time by data-set number is on the increase. If the optimal clustering-based ART1 classifier uses 12 data-set, processing time records about 102 seconds. And, if the ART1 clustering classifier uses 12 data-set, processing time records about 262 seconds. This result shows that processing time of the optimal clustering-based ART1 classifier is the 39% less than that of the ART1 clustering classifier. This result demonstrates that the optimal clustering-based ART1 classifier reduces cost and classification processing time than the ART1 clustering classifier.

5 Conclusion

There is increasing importance of research about protein sequence and have been noticed in demand for a classifier for new protein classification. This paper focuses on a demand of classification system for protein sequence data classification. We propose the optimal clustering-based ART1 classifier for the GPCR data classification and processes the GPCR sequence data classification. We divide GPCR data into ClassA and the others. And, classification success rate of two groups are measured by the Backpropagation Neural Network, the SVM and the ART1. Measured results using the ART1 classifier are that ClassA is 99.7% classification success rate and the others group is 96.6%.

Also, we compare classification processing time of the ART1 classifier is compared to those of backpropgation neural network and SVM. The ART1 classifier's classification processing time is the 27% less than that of the Backpropagation Neural Network and is the 39% that of the SVM when the classification success rate is 95% and the clustering rate is at least 15%. And we design the optimal clustering-based ART1 classifier and compare classification processing time of the optimal clustering-based ART1 classifier to those of the ART1 clustering classifier. The optimal clustering-based ART1 classifier reduces processing cost for classification effectively. The classification processing time of the optimal clustering-based ART1 classifier is the 39% less than that of the ART1 clustering classifier in a success rate which is 96%. This result demonstrates that the optimal clustering-based ART1 classifier provides the high performance classification and the low processing cost in the GPCR data classification and is useful in GPCR data classification.

References

1. Watson, S. and Arkinstall, S.: The G-protein Linked Receptor Facts Book. Academic Press, Burlington, MA. (1994)
2. M. Georgiopulos, G.L. Heileman and J. Huang.: Properties of Learning Related to Pattern Diversity in ART1. Neural Networks, Vol. 4 (1991) 751~757
3. Baxt WG.: Application of neural networks to clinical medicine. Lancet, 346 (1995) 1135-8
4. Finne P, Finne R, Stenman UH.: Neural network analysis of clinicopathological factorss in urological disease: a critical evaluation of available techniques. BJU Int 88 (2001) 825-831
5. Lin JS, Ligomenides PA, Freedman MT, et al.: Application of artificial neural networks for reduction of false-positive detections in digital chest radiographs. Proc Annu Symp Comput Appl Med Care (1993) 434-438
6. Wu YC, Doi K, Giger ML, et al.: Reduction of false positives in computerized detection of lung ndodules in chest radiographs using artificial neural networks, discriminant analysis, and a rule-based scheme. J Digit Imaging 7 (1994) 196-207
7. Biganzoli E, Boracchi P, Mariani L, et al.: Feed forward neural networks for the analysis of censored survival data: a partial logistic regression approach. Stat Med 17 (1998)1169-1186
8. Goldberg DE.: Genetic algorithms in search, optimization and machine learning. 1st edition. Reading, MA: Addison-Welsey Publishing Co. (1989)
9. Jefferson MF, Narayanan MN, Lucas SB.: A neural network computer method to model the INR response of individual patients anticoagulated with warfarin. Br J Haematol 89(1):29, (1995)
10. Noguch H, Hanai T, Honda H, Harrison LCm Kobayashi T.: Fuzzy neural network-based prediction of the motif for MHC class II binding peptides. J Biosci Bioeng 92 (2001)227-31
11. Vapnik V.: The Nature of Statistical Learning Theory. Springer-Verlag, NewYork (1995)
12. Jaakkola, and D. Haussler.: Exploiting generative models in discriminative classifiers. In Advances in Neural Information Processing Systems 11, Morgan Kauffmann, San mateo, Ca, (1998)
13. N.Cristianini and J.Shawe-Taylor.: An Introduction to Support Vector Machines. Cambridge University Press, New York (2000)
14. Zeigler B.P., et al.: The DEVS Environment for High-Performance Modeling and Simulation. IEEE C S & E, Vol. 4, No3 (1997) 61-71
15. Zeigler B.P., et al.: DEVS Framework for Modeling, Simulation, Analysis and Design of Hybrid Systems in Hybrid II. Leture Notes in CS. Springer-Verlag, Berlin (1996) 529-551
16. Horn F., Weare J., Beukers M.W., Horsch S., Bairoch A., Chen W., Edvardsen O., Campagne F., Vriend G., Gpcrdb G.: an information system for g protein-coupled receptors, Nucleic Acids Res 26 (1998) 277-281

17. J. Thompson D., D. Higgins G., Gibson T.J., CLUSTAL W:improving the sensitivity of progressive multiple sequence alignment through sequence weighting, positions-specific gap penalties and weight matrix choice, Nucleic Acids Res 22 (1994) 4673-4680.
18. Jaakkola T., Diekhans M., Haussler D., A discriminative framework for detecting remote protein homologies, J. Comput. Biol. 7 (2000).
19. Poggio T., Girosi F., Networks for Approximation and Learning, Proc. IEEE 78 (1990) 1481-1497.

Trawling Pattern Analysis with Neural Classifier

Ying Tang[1], Xinsheng Yu[2], and Ni Wang[1]

[1] College of Information Science and Technology
[2] College of Marine Geo-Science
Ocean University of China, 5 Yu Shan Road, Qingdao, P.R. China, 266003
xsyu@ouc.edu.cn

Abstract. It has been noticed that bottom trawling not only caused the decline of major fish stocks, but also damaged the biomass of non-target species and habitats as well. This paper proposes a method for identification of trawling marks from video images. The proposed method adopts a pattern recognition approach based on the extraction and the analysis of pattern shape of seabed images. At first, an approach of stationary wavelet transform based edge detection and line segment trace algorithm is developed for line detection. Second, based on the extracted line segments, shape features are computed and classified with a neural network classifier. Experiments on a variety of real seabed images are presented.

1 Introduction

Fishing is the most widespread anthropogenic activity in the marine environment, so during the past few decades, the environmental effects of bottom trawling have aroused a growing degree of concern [1]. Bottom trawling can be expected to cause a number of direct and indirect damages in the ecosystem. Intensive and repeated trawl in the same area may lead to long-term changes in both benthic habitat and communities [2]. Optical image offers the competitive opportunity to provide topographical information with high resolution and is suitable for discriminating detailed structure of sea floor [3]. Analyzing optical images is a tedious and high time-consuming task for the human operator. Therefore, an image processing system is required to automatically detect the trawling marks from video data.

There are several difficulties in trawling marks detection for optical image, such as changed illumination condition, vehicles' movement, high variability due to the different type of seabed. Thus, the commonly use feature set based on the grey level or textural measures are not robust. Instead of using textural information of seabed image, a method using neural network to detect shape of trawling marks is proposed. It begins with a wavelet transform to detect the edge and group the detected edges into segments, and then a neural classifier is applied to those features (see Fig. 2). The experiment results demonstrate that the proposed method can be used to solve specific real world problems, and it is shown a new opportunity for applying computer vision and pattern recognition methods with fishery management applications.

L. Jiao et al. (Eds.): ICNC 2006, Part I, LNCS 4221, pp. 598–601, 2006.

2 Feature Extraction Method

Trawling marks appeared on seabed images mainly have three class of characteristics, namely geometric feature, grey level and orientation. In some cases, because of seabed current and sediment's movement, long trawling marks are broken into several segments and some parts formed as curve-lines. In order to detect these features, a method of multi-scale line tracking scheme and feature extraction approach was developed.

2.1 Line Segment Detection with Stationary Wavelet Transform

The main strength of stationary wavelet transform (SWT) is its properties of locality, time-invariance and multi-scale analysis, which are useful in noise image processing [4]. In this study, each time the image is decomposed, one approximation image and three high-frequency detail sub-images are produced. A shrunk algorithm is applied to the decomposition coefficients to reduce the influence of noises [5], and then the edge detection based on contextual filter method is adopted for gradient images [6]. Each point of gradient images is checked according to four directions (horizontal, vertical, top-left to bottom-right, and top-right to bottom-left) in a square window to see weather it has a greater relative maximum than its neighbors. If it is a local maximum, this point is noted as candidate of edge and the counterpoint of candidate edge image will be set to 1, otherwise it is set to 0.

According to the multiresolution and persistence properties of wavelet, we can use the edge information in the candidate edge images and fused them to form a new edge image. A tracking algorithm is then applied to trace and connect edge points into complete line segments [7]. Some edge segments which length is less than a threshold will be considered as noise edge and will be deleted.

2.2 Feature Expression for Trawling Marks

When trawling marks boundary have been found, a sequence of feature components was calculated according to trawling marks' characteristics (see Fig. 1). The feature set used in this study is defined as following:

Aspect ***(A)*****:** It reflects the straightness of a line segment and is measured by ratio between height *(H)* and width *(W)* of the enclosure rectangle containing the segment.
Curliness ***(C)*****:** This feature is based on the ratio of the length *(L)* of the line segment and the Euclidian distance *(Ed)* of two endpoint of line. It describes the deviation from the straight lines.
Extension ***(E)*****:** It is obtained by the width *(W)* of the enclosure rectangle normalized with respect to the length of the line segment.
Grey Level ***(G)*****:** It represents the grey level (gi, i=1,2) difference between two sides of line mark. For a trawling mark, the gray level difference is often close to zero.
Parallelism ***(P)*****:** It represents the parallel degree of line segments according to their orientations in an image. It indicates weather the line segment in this image exhibit some sort of privileged direction. Here, *Np* is the number of parallel segments of lines, N is the total of segments in the image.

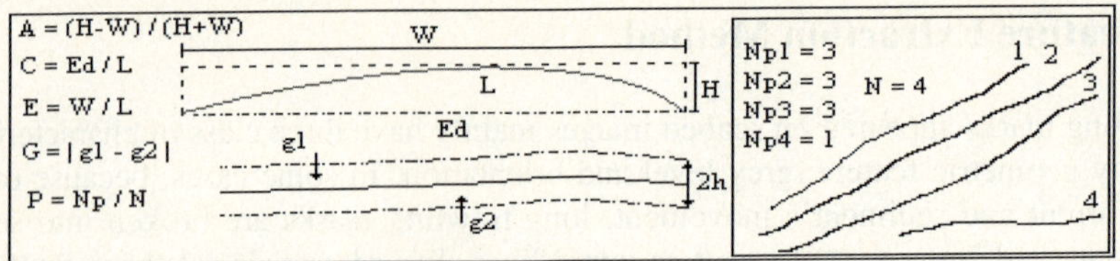

Fig. 1. Schematic overview of feature expression

3 Neural Classification

Nowadays, back-propagation neural network has been widely recognized as a powerful tool for learning input-output mapping in many real world applications. Thus, a three layered back-propagation neural network was used to classify extracted features into trawling category. The Levenberg-Marguardt optimizing algorithm was applied for network training [8]. A total of 16 seabed images which were taken from different type of seabed were used for the study. As the image dataset was relatively small, the neural classifier was evaluated with "leave-one-out" method. True Positive Rate (TPR) and False Positive Rate (FPR) are used as performance measurement. Fig. 3 is performance results of 16 real world seabed images. The results showed that the average recognition accuracy is 83%.

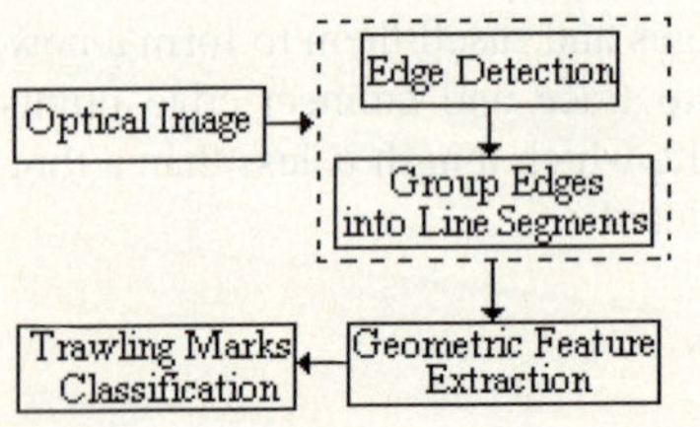

Fig. 2. Schematic of trawling marks detection

Fig. 3. FPR, FPR and RR results of 16 images

The classification result of a real world image is shown in Fig.4 and the recognized trawling marks are highlighted in Fig. 4 (c). It has been seen that there are some classification errors on left side of image, although most of line segments have been classified successfully. This left an open area for the future research.

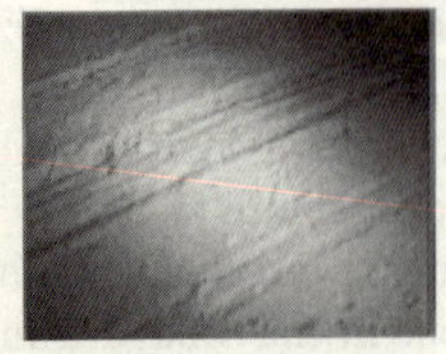

(a) Original image.

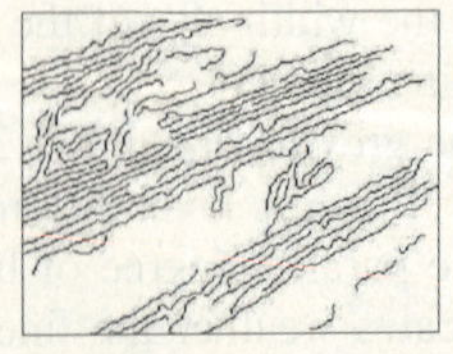

(b) SWT Line segments.

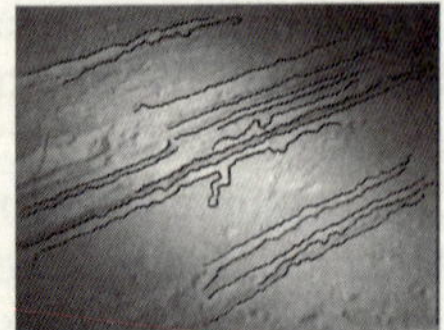

(c) Identification results.

Fig. 4. Real image classification results

4 Conclusions

An assessment of the natural processes in relation to the human impacts of the seabed is required to balance the natural resource management and protection policy, as well as the exploitation of the sea. However, as far as known, works on the recognition of trawling marks with computer remains rare. In this paper, an approach to address automated recognition of trawling marks based on SWT analysis and neural network was presented. Experimental results demonstrated the acceptable performance of overall 83% recognition accuracy was achieved. The method can be helpful to deal with trawling marks detection for real time applications.

References

1. Jennings, S., Kaiser, M.J. Kaiser, M.J.: The effects of fishing on marine ecosystems. Adv. Mar. Biol. 34 (1998) 202-304
2. Linnane, A., Ball, B., Munday, B. et al.: A Review of Potential Techniques to Reduce the Environmental Impact of Demersal Trwals, Irish Fisheries Investigations (New Series) 7 (2000) 1- 43
3. Roberts, J.M., Harvey, S.M., Lamont, P.A., Gage, J.D., Humphery, J.D.: Seabed photography, environmental assessment and evidence for deep-water trawling on the continental margin west of the Hebrides. Hydrobiologia, 441 (2000) 172-183
4. Pesquet, J.C., Kim, H., Carfantan, H.: Time-invariant orthonormal wavelet representations. IEEE Trans. Signal Processing. 44 (1996) 1964-1970
5. Shin, M.Y., Tseng, D.C.: A wavelet based multiresolution edge detection and tracking, Image and Vision Computing. 23 (2005) 441-451
6. Donoho, D.L., Johnstone, I.M., Adapting to unknown smoothness via wavelet shrinkage. J. Amer. Statist. Assoc. 90 (1995) 1200-1224
7. Ville, K., Heikki, K.: Combination of local and global line extraction. Real Time Imaging, 6 (2000) 79-91
8. Zhao, G., Si, J.: Advanced neural network training algorithm with reduced complexity based on Jacobian deficiency, IEEE Transactions on Neural Networks. 9 (1998) 448-453

Model Optimization of Artificial Neural Networks for Performance Predicting in Spot Welding of the Body Galvanized DP Steel Sheets*,**

Xin Zhao, Yansong Zhang, and Guanlong Chen

School of Mechanical Engineering, Shanghai Jiao Tong University,
Shanghai 200030, China
zhxin@sjtu.edu.cn

Abstract. This paper focused on the performance predicting problems in the spot welding of the body galvanized DP steel sheets. Artificial neural networks (ANN) were used to describe the mapping relationship between welding parameters and welding quality. After analyzing the limitation existed in standard BP networks, the original model was optimized based on lots of experiments. Lots of experimental data about welding parameters and corresponding spot weld quality were provided to the ANN for study. The results showed that the improved BP model can predict the influence of welding currents on nugget diameters, weld indentation and the shear loads ratio of spot welds. The forecasting precision was so high that can satisfy the practical need of engineering and have some application value.

1 Introduction

Resistance spot welding (RSW) was crucial to the automotive industry and it took 90% of the whole assembly workload. So it is very important to have a good monitoring or control with the quality of the spot welds. Because of the complicated condition during RSW process and lots of interferential factors, especially the short-time property of the process and the non-visibility of nugget formation, it's very difficult to build an accurate mathematic model for the welding process.

Artificial neural networks (ANN) had been used in spot-welding for many applications, such as welding process-control, welding defect inspection and so on [1]-[3]. It had a strong ability of non-linear mapping and was applicable for multi-signal coupling through training. It could extract key characters of the studied data automatically by the generalization ability of the non-linear mapping. All of these characters described the mapping relationship between spot-welding parameters space and welding joint quality space. So ANN could predict the quality of joints [1]-[2]. Recently some correlative researches all over the world mainly focused on the welding of different materials and simulated with different welding types. But the precision of

* The subject is sponsored by the National Natural Science Foundation of China. (NSFC, NO 50575140)

** The subject is also sponsored by the Specialized Research Fund for the Doctoral Program of Higher Education. (SRFDP, NO 20050248028)

L. Jiao et al. (Eds.): ICNC 2006, Part I, LNCS 4221, pp. 602 – 605, 2006.

these predictions wasn't satisfied enough and it needed more improvements on the model optimization and prediction effects [3].

This paper focused on the spot-welding of galvanized DP steel sheets which were used in the body manufacturing widely. Their welding performance differed from other low-carbonic steels because of the influence of the zinc-layer [4]. Through lots of experimental data about welding parameters and corresponding spot weld quality were provided to the model for training, we could optimize the ANN model with the studying results and the predicting precision was improved greatly.

2 System Modeling and Optimization

The research methods were that: the nugget diameter (d) of the welding spot was measured by the metallographic test, under different effective welding lobe of two different steel (GMW2 and DP600 which were marked as A and B respectively); we measured the shear loads (F_t) of the welding spot by the all-purpose material test machine; we got the weld indentation values by the SRTC-model servo-gun which was installed in a FANUC robot. If we let the shear loads ratio n= F_t/F and F is the shear loads of the raw materials, then three parameters (d, s, n) could be considered as the quality estimation parameters of the spot welds.

In the research, the back propagation network algorithm (BP) was adopted first because of its uncomplicated calculations and strong parallel property [5]. Its net inputs were the efficient current values in every welding cycle and the electrode voltage, while its net outputs were nugget diameter, weld indentation or the shear loads ratio of spot welds. Because of the immaturity property of the traditional BP algorithm such as easy to "immerse" in the partial least point and so on [5], we considered five improved BP algorithms (GDX, RP…) and optimized the ANN model [6]. At last, a synthetically optimized model was formed. We did 10000 times training for our 20-10-3 models and the result was shown as figure 1 (*left*). It could be found that the dynamic root-mean-square error (RMSE) is 0.01052 compared with the results using original BP algorithms (*right*) and RMSE=0.01969. Thus it can be seen that the optimized algorithm is much better than the originals in the training efficiency (*convergence velocity*), the learning stability, the storage and computation etc. Limited by the paper length, the details of algorithm analyzing were omitted.

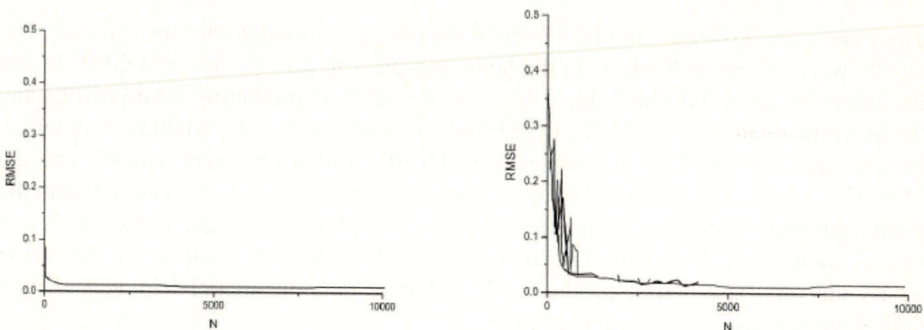

Fig. 1. Network training process of the generally optimized BP model and the original BP model

3 Performance Predicting of Spot Welding and Results Analyzing

In the research, 77 training samples and 16 verifying samples of model were determined after the data processing. After training, the model could finish the prediction of the nugget diameters, the weld indentation and the shear loads under the same welding conditions. The illustrations of them were gained too and only one was showed below because of the paper length limit.

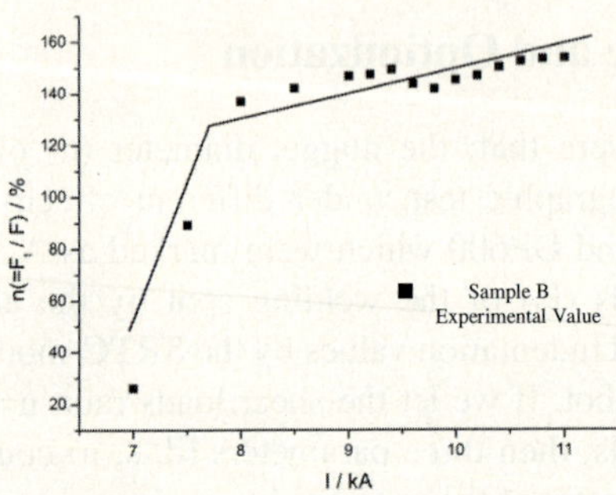

Fig. 2. Relationship between welding currents and shear loads ratio predicted by ANN model

From figure 2 above, it could be seen that the predicting curves I-n had the approximately step-up trends which accorded with the actual situation. The bias between experimental and predicting value was obvious because of the "splash" phenomenon. It would accelerate the deterioration of welding performance of steels especially the DP steels because of their thick zinc-layers. The needed current of welding the high-strength galvanized steel sheets (*such as DP600*) should be 8 kA at least comparing with the common galvanized steel (*such as GMW2*). If not, the strength of formed nugget wouldn't be enough. So the optimized model was efficient and it could accurately predict the influence law of the welding current to the galvanized steel sheets' spot welding performance.

4 Analyzing the Precision of Prediction

There were 16 verifying samples to verify the optimized model after the training. The results were shown as Table 1. In the table, sample 1 to 8 is the material GMW2, and the others are DP600. From Table 1, the average efficient predicting error (AEPE) and the maximal relative error (MRE) of the three parameters (d, s, F_t) could be concluded. For example, the AEPE of the nugget diameter (d) wouldn't be more than 4% and the MRE was 9.87%. The abnormal samples with much bigger error were resulted from the input error and some noises which induced these samples anamorphic. (Like NO 2, 3, 9 and 13 samples in the table). The disturbance didn't affect the outputs of the whole model and no fault-output produced. It reflected that the model had very strong fault-tolerance and classification abilities.

Table 1. Sample verifying results of DP600 and GMW2 galvanized steel sheets

NO.	Mean-value of the input Para.		Actual value (mm)			Predicting value (mm)			Relative error (%)		
	I(kA)	V(v)	d	s	F_t	d	s	F_t	d %	s %	F_t %
1	8.024	0.859	4.1	1.41	3382	4.178	1.429	3450	1.90	1.35	2.01
2	8.198	0.865	5.3	1.36	3695	4.823	1.411	4150	***9.00***	3.75	***12.31***
3	8.421	0.871	5.6	1.29	3903	5.712	1.391	4100	2.00	***7.83***	5.05
4	8.619	0.865	5.9	1.34	4034	6.002	1.385	4208	1.73	3.36	4.31
5	8.817	0.862	6.0	1.39	4211	6.111	1.372	4450	1.85	1.29	5.68
6	9.013	0.871	5.9	1.34	4044	6.119	1.360	4270	3.41	1.49	5.59
7	9.207	0.884	6.3	1.30	4210	6.397	1.344	4507	1.54	3.38	7.05
8	9.452	0.899	6.6	1.28	4425	6.725	1.328	4620	1.89	3.75	4.41
9	9.058	0.892	6.1	1.36	8407	5.498	1.315	8605	***9.87***	3.31	2.36
10	9.461	0.903	5.5	1.32	8539	5.591	1.303	8971	1.65	1.29	5.06
11	9.615	0.918	5.7	1.26	8627	5.821	1.278	9024	2.12	1.43	4.60
12	9.805	0.933	5.9	1.19	8715	6.052	1.231	9185	2.58	3.45	5.39
13	10.011	0.947	6.1	1.13	8819	6.324	1.210	9210	3.67	***7.07***	4.43
14	10.196	0.961	6.3	1.14	8912	6.412	1.181	9351	1.78	3.60	4.93
15	10.622	0.979	6.6	1.13	8929	6.725	1.174	9488	1.89	3.89	6.26
16	11.072	0.988	6.1	1.15	8911	6.338	1.188	9575	3.90	3.31	7.45

5 Conclusions

After synthetical optimization, the model stability and convergence rate improved greatly, meanwhile training time shortened efficiently. It provided a useful reference of ANN's in-depth application in the welding industry. After training, the optimized model can appropriately reflect the basic principle of the spot welding accurately. The forecasting precision of the model are so high that can satisfy the practical need of engineering and have some application value. It can be an important basis of accurately on-line monitoring to the RSW quality and the nondestructive inspection.

References

1. Ivezic, N., Alien, J.D.: Neural network-based resistance spot welding control and quality prediction. Intelligent Processing and Manufacturing of Materials, (1999)
2. Zhongdian, Zhang; Yan, Li; Xingping, He: Artificial neural network estimating of spot welding mechanical property. Hanjie Xuebao, (1997)
3. Yongjoon, Cho, Rhee, S.: Quality estimation of resistance spot welding by using pattern recognition with neural networks. Instrumentation and Measurement, IEEE Transactions on Vol. 53, April (2004) 330 -334
4. M. Marya, X. Q. Gayden: Development of requirements for resistance spot welding dual-phase (DP600) steels. Welding Journal, (2005)
5. Zhongdian, Zhang etc. Selection of the learning rate when modeling ANN using BP algorithm. Welding, (2004)
6. Peixian, Jin: The study and application of different improved BP algorithms. Nanjing Hangkonghangtian Daxue Xuebao, (1994)

Robust Clustering Algorithms Based on Finite Mixtures of Multivariate *t* Distribution

Chengwen Yu, Qianjin Zhang, and Lei Guo

College of Automatic Control, Northwestern Polytechnical University, Xi'an 710072, China
cwyu@mail.nwpu.edu.cn

Abstract. Providing protection against outlier in clustering data is a difficult problem. We proposed two robust clustering algorithms which integrate two modified versions of EM algorithm for mixtures *t* model with a model selection criterion respectively. The proposed methods can select the number of clusters component automatically by a combined component annihilation strategy and can also avoid the drawbacks of traditional mixture-based clustering algorithms - highly dependent on initialization and may converge to the boundary of the parameter space [7]. Experiment results show the contrast among different algorithms and demonstrate the effectiveness of our algorithms.

1 Introduction

Finite mixtures model treats clustering issues like the selection of the number of clusters or the assessment of the validity of a given model in a formal way. For the reason of computational convenience, attention has focused on the use of multivariate mixtures normal model for clustering [1]. Mixtures *t* model provides a more robust approach to the fitting of multi-component gauss data with background noise for its relationship with M-estimation [2]. For solving mixtures of multivariate *t* distribution, ML estimates of its parameters were obtained by the standard EM algorithm in [3] and an application of Expectation Conditional Maximization (ECM) algorithm was also given in [4].

In this paper, we proposed two robust clustering algorithms by integrating two modified versions of EM algorithm for mixtures *t* model with Minimum Message Length (MML) criterion respectively. The rest of this paper is organized as follows: In section 2, we describe the new versions of EM algorithm, and the corresponding robust clustering algorithms using a component annihilation strategy. Section 3 gives the results of experiments and the conclusion.

2 Robust Clustering Algorithms

Learning of mixtures *t* distribution is typically a missing data problem for which the EM algorithm appears to be useful. Denoting the complete parameter set of mixtures *t* distribution as $\boldsymbol{\Psi} = (\pi_1,\cdots,\pi_m,\nu_1,\ldots,\nu_m,\boldsymbol{\theta}_1,\ldots,\boldsymbol{\theta}_m)$, where $\pi_1,\cdots,\pi_m$ are mixing proportions, $\nu_1,\ldots,\nu_m$ are degrees of freedom, $\boldsymbol{\theta}_i$ consist of mean $\boldsymbol{\mu}_i$ and covariance $\boldsymbol{\Sigma}_i$. The EM algorithm produces a sequence of estimates of

L. Jiao et al. (Eds.): ICNC 2006, Part I, LNCS 4221, pp. 606–609, 2006.

$\{\hat{\boldsymbol{\Psi}}^{(k)}, k = 0,1,2,\cdots\}$ by iteratively applying E-step (to update $\tau_{ij}^{(k)}$ - prior $\pi_i^{(k)}$'s posterior probability and $u_{ij}^{(k)}$ - a robustness tuning parameter according to current fit $\hat{\boldsymbol{\Psi}}^{(k)}$) and M-step (to update $\pi_i^{(k+1)}$, $\boldsymbol{\mu}_i^{(k+1)}$, $\boldsymbol{\Sigma}_i^{(k+1)}$ and $\nu_i^{(k+1)}$) [1] [9].

2.1 The Proposed Modified Versions of EM

- **Algorithm 1.** Component-wise Multi-cycle version of ECM(CMECM)

We combine the idea of Component-wise EM algorithm for Mixtures (CEM^2) [5] and the Multi-cycle version of ECM [6], and obtain a new version of EM for fitting mixtures *t* model, CMECM. We partition the parameter $\boldsymbol{\Psi}$ as $(\boldsymbol{\Psi}_1, \boldsymbol{\Psi}_2)$, where $\boldsymbol{\Psi}_1 = (\pi_1,\cdots,\pi_m,\boldsymbol{\theta}_1,\cdots\boldsymbol{\theta}_m)$ and $\boldsymbol{\Psi}_2 = (\nu_1,\cdots,\nu_m)$. In our presentation, the components are updated successively, starting from $i = 1,\cdots m$ and repeating this after *m* iterations. Therefore the component updated at iteration *k* is given by $i = k - \lfloor k/m \rfloor m + 1$, where $\lfloor \bullet \rfloor$ denoting the integer part. The *k*th iteration of the CMECM is as follows.

E-step: for $j = 1,\cdots,n$, compute $\tau_{ij}^{(k)}$ and $u_{ij}^{(k)}$.

CM-step 1: Calculate $\boldsymbol{\Psi}_1^{(k+1)}$ by maximizing $Q(\boldsymbol{\Psi};\hat{\boldsymbol{\Psi}}^{(k)})$ with $\boldsymbol{\Psi}_2$ fixed at $\boldsymbol{\Psi}_2^{(k)}$.That is to update $\pi_i^{(k+1)}$, $\mu_i^{(k+1)}$ and $\Sigma_i^{(k+1)}$ when ν_i fixed at $\nu_i^{(k)}$, and for $\ell \neq i$, $\pi_l^{(k+1)} = \pi_l^{(k)}, \boldsymbol{\theta}_l^{(k+1)} = \boldsymbol{\theta}_l^{(k)}$.

Additional E-step: Instead $(\boldsymbol{\Psi}_1^{(k)}, \boldsymbol{\Psi}_2^{(k)})$ with $(\boldsymbol{\Psi}_1^{(k+1)}, \boldsymbol{\Psi}_2^{(k)})$, for $j = 1,\cdots,n$, compute $u_{ij}^{(k)}$.

CM-step 2: Calculate $\boldsymbol{\Psi}_2^{(k+1)}$ by maximizing $Q(\boldsymbol{\Psi};\hat{\boldsymbol{\Psi}}^{(k)})$ with $\boldsymbol{\Psi}_1$ fixed at $\boldsymbol{\Psi}_1^{(k+1)}$. That is to update $\nu_i^{(k+1)}$.

- **Algorithm 2.** Extension of CMECM (ECM^2)

We have notice that it is most time consuming for solving $\nu_i^{(k+1)}$, and the additional E-step in CMECM is an incomplete E-step, using only the information of $u_{ij}^{(k)}$.So we use the CEM^2 only for $\boldsymbol{\Psi}_1$ and get the newest information of $\tau_{ij}^{(k)}$ after *m* iterations, then, update $\nu_i^{(k)} (i = 1,\cdots,m)$ together with $u_{ij}^{(k)}$.Then we obtain ECM^2.

2.2 MML Criterion and Component Annihilation

Figueiredo and Jian [7] derived a model selection criterion-MML and applied it into mixtures normal model. We adapt it for the situation of mixtures *p*-variate *t* Distribution and get $\hat{\boldsymbol{\Psi}} = \arg\min_{\boldsymbol{\Psi}}\{\mathcal{L}(\boldsymbol{\Psi}, \boldsymbol{Y})\}$ with

$$\mathcal{L}(\boldsymbol{\Psi}, \boldsymbol{Y}) = (N/2)\sum_{i=1}^{m} \log(\pi_i) - \log p(\boldsymbol{Y} \mid \boldsymbol{\Psi}) + [m(N+1)/2][1 + \log(n/12)] \qquad (1)$$

where $N = 1 + p + p(p+1)/2$ is the number of parameters specifying each component.

The minimization of (1) is realized by EM algorithm, the M-step of EM algorithm for minimization of (1) has following form [8]:

$$\pi_i^{(k+1)} = \max\left\{0, (\sum_{j=1}^{n} \tau_{ij}^{(k)}) - N/2\right\} / \sum_{i=1}^{m} \max\left\{\{0, (\sum_{j=1}^{n} \tau_{ij}^{(k)}) - N/2\right\} \quad (2)$$

Other parameters update as usually when $\pi_i^{(k+1)} > 0$. Thus, MML criterion and EM are integrated together: for fixed number of component (when $\pi_i^{(k+1)} = 0$, the ith component is killed off), the model fitting procedure of clustering data is represented by EM and the best model presentation of clustering data is selected by MML.

We use a linear Deterministic Annealing (DA) method to search in a limited range $i_{\min} \sim i_{\max}$ including the true number of component i_{true} .We notice that the component may be killed off by (2) when searching from $i_{\max}$ to $i_{\min}$, so we introduce a new concept- *the combined Component Annihilation (CA) strategy*, which consist of a *strong rule CA* and a *compelling CA*. Strong rule CA is controlled by $N/2$ in (2); Compelling CA is controlled by the convergence threshold value γ of MML criterion.

2.3 The Complete Algorithms

According the description of section 2.1 and 2.2, we get two robust clustering algorithms (RCA) that corresponding to two new versions of EM: CMECM and ECM^2. In RCAs, the updating of $\pi_i^{(k+1)}$ in proposed EM algorithms is substituted by (2), and Component Annihilation is controlled by strong rule CA before the convergence of the proposed EM algorithms (for $\mathcal{L}(\boldsymbol{\Psi}, \boldsymbol{Y})$, $\left|\mathcal{L}^{(k-1)} - \mathcal{L}^{(k)}\right| < \gamma \mathcal{L}^{(k-1)}$), and CA is controlled by the compelling CA after the convergence of the proposed EM algorithms, which kill off the most least probable component (with smallest $\pi_i^{(k+1)}$), then return the proposed EM, the procedure is repeated until the number of component is reach $i_{\min}$ .The number of component and model parameters that corresponding to the minimum of $\mathcal{L}(\boldsymbol{\Psi}, \boldsymbol{Y})$ were chosen for the best presentation of clustering data.

3 Experimental Results and Conclusion

Our test is run in Matlab environment at PC with P4 1.73G CPU and 512M RAM. The test data set is consisting of 3 random normal distributions mixtures and background noise. The parameters of the 3 random normal distributions mixtures are: mixing proportion, $\pi_1 = \pi_2 = \pi_3 = 1/3$; mean, $\mu_1 = [0,0]^T$, $\mu_2 = [0,4]^T$, $\mu_3 = [0,-4]^T$; covariance matrix, $\Sigma_1 = \Sigma_2 = \Sigma_3 = [3, 0.2; 0.2, 0.4]$ .Background noise is a uniform distribution over the range -8 to +8 on each variant. There are totally 1000 sample points for the test data, and the ratio of noise samples is from 0 to 20%.

We test the algorithms for robust effectiveness to noise and the convergence velocity. So we compared the results of different algorithms: Clustering algorithm 1(CA1) based mixtures normal model (CEM^2); RCA1 based mixtures t model (standard EM); RCA2 based mixtures t model (CEM^2); RCA3 based mixtures t model (proposed CMECM); RCA4 based mixtures t model (proposed ECM^2). RCA1 uses

Bayesian inference criterion (BIC) for selection of number of component while the others using the same combined Component Annihilation strategy.We run the above algorithms with lots of experiments, and show the statistic results in Fig.1,

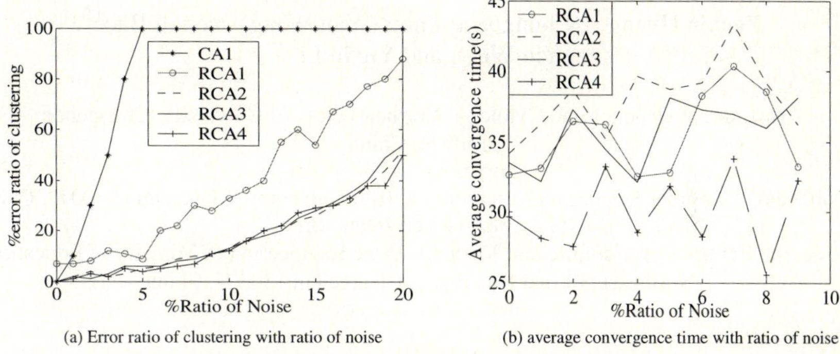

(a) Error ratio of clustering with ratio of noise (b) average convergence time with ratio of noise

Fig. 1. Performance comparison of algorithms

Experiment results show that our robust clustering algorithms(RCA3 and RCA4) absorb the benefit of mixtures of multivariate t model, multi-cycle version of ECM, CEM^2, MML and DA, have much more robust effectiveness to outlier than mixtures normal model, and have more robust effectiveness to outlier while converging faster than conventional EM/ECM for solving mixtures t model.

References

[1] A.K.Jain and R.Dubes, Algorithms for Clustering Data. Englewood Cliffs, N.J.:Prentice Hall,1988.

[2] G. McLachlan and D. Peel, Finite Mixture Models. New York: John Wiley & Sons, 2000.

[3] G. McLachlan and D. Peel, Robust Cluster Analysis via Mixtures of Multivariate t Distribution. in LNCS .vol.1451.Berlin:Springer-Verlag,pp, 658-666 .1998

[4] D. Peel, G. McLachlan J. Robust Mixture Modeling using the t Distribution, Statistics and Computing, vol.10 , pp. 339–348. 2000

[5] G. Celeux, S. Chretien, F. Forbes, A. Mkhadri. A Component Wise EM Algorithm for Mixtures. J. of Computational and Graphical Statistics, vol.16, no.10, pp. 697-712, 2001

[6] Liu C, Rubin D B. ML Estimation of the t Distribution using EM and its Extensions, ECM and ECME, Statistica Sinica , no.5, pp. 19–39. 1995

[7] MAT. Figueiredo, A..K. Jain. Unsupervised Learning of Finite Mixture Models. IEEE Trans. on Pattern Analysis and Machine Intelligence, vol.24, no.3, pp.381-396. Mar. 2002.

[8] J.Bernardo and A.Smith. Bayesian Theory. Chichester, UK: J.Wiley&Sons, 1994.

A Hybrid Algorithm for Solving Generalized Class Cover Problem

Yanxin Huang[1,2], Chunguang Zhou[2], Yan Wang[2], Yongli Bao[1,3], Yin Wu[1,3], and Yuxin Li[1,3]

[1] Institute of Genetics and Cytology, Northeast Normal University, Changchun 130024, China
huangyx@jlu.edu.cn

[2] College of Computer Science and Technology, Jilin University, Changchun 130012, China
cgzhou@jlu.edu.cn

[3] Research Center of Agriculture and Medicine Gene Engineering of Ministry of Education, Northeast Normal University, Changchun 130024, China
nenucc@sina.com

Abstract. A generalized class cover problem is presented in this article, and then reduced to a constrained multi-objective optimization problem. Solving this problem is significantly important to construct a robust classification system. Therefore, three algorithms for solving the generalized class cover problem are proposed, which are greedy algorithm, binary particle swarm optimization algorithm, and their hybrid algorithm. Comparison results of these three methods show that the hybrid algorithm can get better solutions in less runtime.

1 Introduction

C. Adam *et al* presented the class cover problem (CCP) and proved it is a NP-hard problem firstly [1]. Most recent works on solving CCP are mainly based on greedy algorithms [2][3][4]. A generalized class cover problem (GCCP) based on CCP are proposed in this paper, which can be applied in many terrains such as pattern recognition, machine learning, data mining, and so on. Meanwhile, three solution algorithms for GCCP, which are greedy algorithm, binary particle swarm optimization (bPSO) algorithm, and their hybrid algorithm, are proposed. The comparison experimental results show that the proposed algorithms are feasible and effective. Moreover, the hybrid algorithm, which combines with greedy algorithm and bPSO, can get better solutions in less runtime especially.

2 Generalized Class Cover Problem

C. Adam *et al* defined the CCP as follows: Let B be the set of points in class one, and R be the set of points in class two, with |R|+|B|=n. Then the CCP is described as:

Minimize K

$$s.t.\ \max_{v\in B}\{d(v,S)\} < \min_{w\in R}\{d(w,S)\}$$

L. Jiao et al. (Eds.): ICNC 2006, Part I, LNCS 4221, pp. 610 – 619, 2006.

where $S \subseteq B$, $|S|=K$, $d(.,.)$ denotes the distance between two points or a point to a set.

In order to design robust classification systems (e.g. fuzzy neural system architecture in [5]), we propose the generalized class cover problem (GCCP) as follows:

Minimize K, Var

$$s.t.\begin{cases} r_\beta(v,R) \le \gamma MaxD_\beta, \forall v \in S \\ |\{w \mid w \in B, and\ \exists v \in S, d(v,w) \le r_\beta(v,R)\}| \ge \alpha |B| \end{cases}$$

where $S \subseteq B$, $|S|=K$, $Var = \sqrt{\frac{1}{|S|-1}\sum_{v\in S}(r_\beta(v,R)-\bar{r})^2}$, $\bar{r} = \frac{1}{|S|}\sum_{v\in S} r_\beta(v,R)$, $\gamma,\alpha,\beta \in [0,1]$, $r_\beta(v,R)$ denotes the cover radius centered on the point $v \in S$, and $MaxD_\beta$ denotes the most cover radius in the points in B. The parameter γ is used to control producing the class covers with relatively even radii, and the parameters α and β are used to make the system more noise-resistant, in which, α indicates that at least $\alpha|B|$ points in B must be covered by the class covers produced from S, and β indicates that a class cover centered on a point in B is permitted to cover at most $\beta|R|$ points in R. In other words, GCCP is a multi-objective problem, which to find a minimum cardinality set of covering balls with center points in S and relatively radii. The union contains at least $\alpha|B|$ points in B and each covering ball contains at most $\beta|R|$ points in R. The CCP can be regarded as a special GCCP with $\gamma=1, \alpha=1, \beta=0$.

3 Improved Greedy Algorithm for GCCP

Towards GCCP, we propose an improved greedy algorithm as follows.

The Improved Greedy Algorithm:

Let $S=\phi$, and $C=B$,

(1) $\forall x \in B$, calculate $d_\beta(x,R)$, which equals to the $\beta|R|+1$th smallest distance between x and the points in R, let $MaxD_\beta = \max_{x\in B}\{d_\beta(x,R)\}$, then $\forall x \in B$ calculate $r_\beta(x,R)$ as:
$$\begin{cases} r_\beta(x,R) = \gamma MaxD_\beta, if\ d_\beta(x,R) > \gamma MaxD_\beta \\ r_\beta(x,R) = d_\beta(x,R), if\ d_\beta(x,R) \le \gamma MaxD_\beta \end{cases};$$

(2) Produce digraph $G=(B,E)$ as follows: $\forall x \in B$, for all $y \in B$, if $d(x,y) \le r_\beta(x,R)$, then produce a edge from x to y , namely, $(x,y) \in E$;

(3) $\forall x \in C$, calculate $cover(x) = \{y \mid y \in C, and\ (x,y) \in E\}$;

(4) Take $z \in C$, and $cover(z) = \max_{x\in C}\{cover(x)\}$. If $|C| < (1-\alpha)|B|$, then output S and terminate the procedure, else set $S = S \cup \{z\}$, $C = C - \{x \mid x \in C, (z,x) \in E\}$ and go to (3).

Let $|B|=N$,$|R|=M$, and $|S|=\rho(S)$, then the algorithm has the time complexity of $O(NM+N(N-1)+N\rho(S))$. Suppose $N\approx M$, and $\rho(S)<<N$, then the algorithm has the time complexity of $O(N^2)$ approximately.

4 Binary Particle Swarm Algorithm for GCCP

J. Kennedy *et al.* invented the real-valued particle swarm optimization (PSO) model in 1995[6]. Now, the PSO demonstrates good performance in many optimization problems, and a great effort has been put into validating that the PSO is a competitive optimization method[7]. In 1997, J. Kennedy *et al.* presented the binary particle swarm optimization (bPSO) model for solving discrete optimization problem[8], but the bPSO model still needs further research till now[7]. In order to obtain better solutions than that of the greedy algorithm, we proposed a new bPSO algorithm for GCCP in this section.

4.1 Data Pretreatment

In order to meet the requirement of our following bPSO steps, the data are pretreated firstly. The data points in R are stored in the array of $Array\_R$. And the data points in B are stored in the array of $Array\_B$ with size of N by the below algorithm, which the nearer data points in the Euclid space in B are stored as nearer as possible in $Array\_B$.

(1) Let $k=0, C=B$;

(2) If $C=\phi$, then terminate, else go to (3);

(3) Let $k=k+1$, if $k=1$, take a point y in B randomly, let $Array\_B[k]=y$, $C=C-\{y\}$, go to (2), else take a point y in B subject to $d(Array\_B(k-1),y)$ $=\min_{x\in B}\{d(Array\_B(k-1),x)\}$. Let $Array\_B[k]=y, C=C-\{y\}$, go to (2).

4.2 Computation of Class Cover Relational Matrix

After the data pretreatment procedure, the class cover relational matrix of B relative to R is constructed as follows.

(1) Produce a digraph $G=(B,E)$ as same as the improved greedy algorithm, and store the cover radii of the points in $Array\_B$ into an array with size of N accordingly.

(2) Based on the digraph $G=(B,E)$, the 0-1 class cover relational matrix M_G of B relative to R with size of $N\times N$ is calculated, where $M_G(i,j)=1$ or 0 indicates there is a edge from i to j or not, $i,j\in[1,2,...,N]$.

Obviously, M_G is not a symmetrical matrix because of the digraph $G=(B,E)$.

4.3 Binary Encoding of Particles

Let L be the population size of the particle swarm, then a particle I_k can be encoded as $I_k = b_1^{(k)} b_2^{(k)} \ldots b_N^{(k)}$, where $b_i^{(k)} = 1$ or 0 indicates the point $Array\_B[i]$ is included in S or not, $i \in [1,2,\ldots,N]$, $k \in [1,2,\ldots,L]$.

4.4 Fitness Function of Particles

According to the definition of GCCP, the fitness of the particle I_k mainly depends on three factors: (1) the cardinality $C(I_k)$ of S corresponding to the particle I_k, where $C(I_k) = b_1^{(k)} + b_2^{(k)} + \ldots + b_N^{(k)}$, And the smaller $C(I_k)$ is, the bigger fitness value of particle I_k tends to be; (2) the number $R(I_k)$, which indicates the number of points in B which are covered by the class covers corresponding to particle I_k. And the bigger $R(I_k)$ is, the bigger the fitness value of particle I_k tends to be. Especially, when $R(I_k) \geq \alpha |B|$, which satisfies the constrained conditions of GCCP, a reward should be given to particle I_k; (3) the sample standard deviation $Var(I_k)$ of the class cover radii corresponding to particle I_k. And the smaller $Var(I_k)$ is, the bigger the fitness value of particle I_k tends to be.

According to these factors mentioned above, we define the fitness of the particle I_k as:

$$F(I_k) = \begin{cases} \lambda_1 \times \dfrac{N - C(I_k)}{N} + \lambda_2 \times \dfrac{R(I_k)}{N} + \lambda_3 \times \dfrac{Q - Var(I_k)}{Q}, if \;\; R(I_k) < \alpha N \\ \lambda_1 \times \dfrac{N - C(I_k)}{N} + \lambda_2 \times \dfrac{R(I_k)}{N} + \lambda_3 \times \dfrac{Q - Var(I_k)}{Q} + 0.2, if \;\; R(I_k) \geq \alpha N \end{cases} \quad (1)$$

where λ_1, λ_2, and λ_3 are weight coefficients, which denote the importance of the three factors on scaling the fitness of a particle. Q is the least upper bound of the function $Var(I_k)$ which is calculated by the following theorem 4.1.

Theorem 4.1

Let $a_1, a_2, \ldots, a_n$ be a sequence of real numbers, and $a_{i_1}, a_{i_2}, \ldots, a_{i_k}$ be its any subsequence, $i_1, i_2, \ldots, i_k \in \{1,2,\ldots,n\}$, $2 \leq k \leq n$, then $Var(a_{i_1}, a_{i_2}, \ldots, a_{i_k}) \leq Var(a_{\min}, a_{\max})$, where $Var(a_{i_1}, a_{i_2}, \ldots, a_{i_k}) = \sqrt{\frac{1}{k-1} \sum_{j=1}^{k} (a_{i_j} - \bar{a}_k)^2}$,

$\bar{a}_k = \frac{1}{k} \sum_{j=1}^{k} a_{i_j}$, $a_{\min} = \min\{a_1, a_2, \ldots, a_n\}$, $a_{\max} = \max\{a_1, a_2, \ldots, a_n\}$.

Proof

Let $f(a_{i_1}, a_{i_2}, \ldots, a_{i_k}) = Var(a_{i_1}, a_{i_2}, \ldots, a_{i_k})^2 = \frac{1}{k-1} \sum_{j=1}^{k} (a_{i_j} - \bar{a}_k)^2$, then calculate $\frac{\partial f}{\partial a_{i_j}} = 0$, where $a_{\min} \leq a_{i_j} \leq a_{\max}$, $j = 1,2,\ldots,k$, we know that the minimum of function f is zero at

$a_{i_1}=a_{i_2}=...=a_{i_k}$ within the closed region: $a_{\min}\le a_{i_j}\le a_{\max}$, $j=1,2,...,k$, $2\le k\le n$. Next we calculate the function value of f on the boundary of the closed region. Without losing generality, suppose the subsequence $a_{i_1},a_{i_2},...,a_{i_k}$ has p a_{max} and k-p $a_{\min}$ $p\in Z$, $1\le p\le k-1$, then $f=\frac{1}{k-1}\frac{1}{k^2}.(a_{\max}-a_{\min})^2\ .((k-p)^2p+p^2(k-p))$, regarding p as an independent variable, we know that the maximum of function f is gotten in $p=\frac{k}{2}$, and the function f is a unimodal function. Because p is an integer, when k is an even number, we obtain $f=\frac{1}{k-1}\frac{1}{k^2}.(a_{\max}-a_{\min})^2.((k-\frac{k}{2})^2\frac{k}{2}+(\frac{k}{2})^2(k-\frac{k}{2}))\ =\frac{1}{4(k-1)}(a_{\max}-a_{\min})^2$, and because $k\ge 2$, we get $f\le\frac{1}{2}(a_{\max}-a_{\min})^2\ =Var(a_{\max}-a_{\min})^2$. When k is an odd number, we obtain $f=\ \frac{1}{k-1}\frac{1}{k^2}.(a_{\max}-a_{\min})^2.((k-\frac{k-1}{2})^2\frac{k-1}{2}\ +(\frac{k-1}{2})^2(k-\frac{k-1}{2}))$ $=\frac{k+1}{4k}(a_{\max}-a_{\min})^2$, and because $k\ge 2$, we get $f\le\frac{1}{2}(a_{\max}-a_{\min})^2$ $=Var(a_{\max}-a_{\min})^2$. Sum up, we come to the conclusion that $Var(a_{i_1},a_{i_2},...,a_{i_k})\le Var(a_{\min},a_{\max})$. The proof is completed.

According to theorem 4.1, we can set $Q=Var(R_{\min},R_{\max})$, such that $0\le\frac{Q-Var(I_k)}{Q}\le 1$, where $R_{\min}$ and $R_{\max}$ denote the least radius and the most radius in the class covers obtained from S respectively.

Let $M_G(i,1:N)$ and $M_G(1:N,j)$ be the ith row vector and the jth column vector of M_G respectively, $i,j\in\{1,2,..,N\}$, then $R(I_k)$ is calculated as follows:

(1) Let $p=1$, the particle $I_k=b_1^{(k)}b_2^{(k)}...b_N^{(k)}$, and $T=[0,0,...,0]^T$ presents a zero vector;

(2) If $b_p^{(k)}=1$, then set $T=T\vee M_G(1:N,p)$, where operator "$\vee$" denotes "or" operation, and go to (3); else go to (3) directly;

(3) If $p<N$, set $p=p+1$, go to (2), else set $R(I_k)=\sum_{i=1}^{N}T(i)$, where $T(i)$ presents the ith element of vector T, and terminate the procedure.

4.5 Recursive Equations of bPSO

Let L be the population size of particle swarm (generally set L=20[7]), then each particle presents a candidate solution in the search space. Each particle has four state variables: $\vec{v}(t)$, $\vec{x}(t)$, $\vec{x}^{(p)}(t)$ and $\vec{x}^{(g)}(t)$, which present its current velocity, current position, previous best position and the best position of all the particles, respectively. The velocity and position of the ith particle are updated with the following formulae:[8]

$$v_{ij}(t+1) = wv_{ij}(t) + c_1 r_{1j}(t)(x_{ij}^{(p)}(t) - x_{ij}(t)) + c_2 r_{2j}(t)(x_j^{(g)}(t) - x_{ij}(t)) \tag{2}$$

$$x_{ij}(t+1) = \begin{cases} 0, if & \rho \geq Sig(v_{ij}(t+1)) \\ 1, if & \rho < Sig(v_{ij}(t+1)) \end{cases} \tag{3}$$

where $i = 1,2,...,L$; $j \in \{1,2,...,N\}$, j represents the jth element of N-dimensional vector, $r_{1j}(t) \sim U[0,1], r_{2j}(t) \sim U[0,1]$, w, c_1 and c_2 are *acceleration coefficients*, $\rho \sim U[0,1]$, and $Sig(x) = \frac{1}{1+\exp(-x)}$ [8].

5 Hybrid Algorithm for GCCP

Compared with the bPSO, the improved greedy algorithm runs at a faster speed, which is its strongest character. However, it is hard to get best optimal solutions. On the contrary, the bPSO can get better optimal solutions but needs much more running epochs. Therefore, a hybrid algorithm is proposed to take all the advantages of these two algorithms, which mainly involves two steps: (1) initialize a particle with the solution which got by the improved greedy algorithm, generally which is the best particle of all particles at the beginning; (2) choose parameters of bPSO carefully to enhance local search ability near the best position of all the particles. Meanwhile, it should avoid particles convergence to be premature. So the hybrid algorithm can get better optimal solutions than the improved greedy algorithm and spend fewer running epochs than the bPSO.

Obviously, two particles with binary encoding which are near in Hamming distance does not mean that their corresponding data points in B are near in Euclid distance as well. That will lead to invalidation local search in the hybrid algorithm. So we deal with this problem as follows. Firstly, the data points in B are stored same as section 4.1, which makes the near data points in B are as near as possible in $Array\_B$ in Euclid space. Next, the recursive equation (2) is revised as below:

$$v_{ij}(t+1) = wv_{ij}(t) + c_1 r_{1j}(t)(x_{ij}^{(p)}(t) - x_{ij}(t)) + c_2 r_{2j}(t)(x_j^{(g')}(t) - x_{ij}(t)) \tag{2'}$$

where $x^{(g')}(t)$ presents a potential solution of the problem near the known best position $x^{(g)}(t)$. By updating $x^{(g')}(t)$ step by step, the hybrid algorithm enhances local search near the best position of all particles. $x^{(g')}(t)$ is calculated as follows:

(1) Set $k_1 = -1, k_2 = 0, k_3 = 0, Current\_Pos = 0, B\_String = x^{(g)}(t)$;

(2) Set $k_2 = k_2 + 1$, if $k_2 > N$, set $k_2 = k_2 - 1$, go to (5), else go to (3);

(3) If $B\_String(k_2) = 0$, then go to (2), else go to (4);

(4) If $k_1 = -1$, set $k_1 = k_2 - (Current\_Pos + 1), Current\_Pos = k_2$, go to (2), else go to (5);

(5) Set $k_3 = k_2 - (Current\_Pos + 1)$, go to (6);

(6) Let $S = \{Current\_Pos - \left\lceil \frac{k_1}{2} \right\rceil, Current\_Pos - \left\lceil \frac{k_1}{2} \right\rceil + 1, ..., Current\_Pos - 1,$ $Current\_Pos, Current\_Pos + 1, ..., Current\_Pos + \left\lfloor \frac{k_3}{2} \right\rfloor\}$, where $\lceil x \rceil$ indicates the least integer greater than or equal to *x*; $\lfloor x \rfloor$ indicates the biggest integer smaller than or equal to *x*, and then a position number *i*, $i \in S$, is picked by Roulette Wheel Selection[9], such that $B\_String(i) = 1$, and $\forall j \in S - \{i\}, set\ B\_String(j) = 0$. $\forall x \in S$, the probability that *x* is picked out is calculated by the bilateral Gauss function as Eq.(4):

$$f(x) = \begin{cases} \exp(-(x-c)^2/\sigma_1^2), x \le c \\ \exp(-(x-c)^2/\sigma_2^2), x > c \end{cases} \tag{4}$$

where $\sigma_1 = k_1/3$; $\sigma_2 = k_3/3$; $c = Current\_Pos$. If $k_2 = N$, set $x^{(g')}(t) = B\_String$, terminate the algorithm, else set $k_1 = k_3$, $Current\_Pos = k_2$, go to (2).

In order to enhance local search near the best position of all the particles, the parameters of the recursive equation of particles is determined as follows: (1) set the biggest velocity of particles $v_{\max} = 5 \times (t/T)^{\kappa}$, where *t* is the current recursive epoch number, *T* is total recursive epoch number set beforehand, and κ is the curvature parameter (set κ=1.2 in this paper); (2) set $w = 1, c_1 + c_2 = 2, c_2 = 2/(1 - \exp(-\theta\tau))$, where τ is a recursive parameter, which increases 1 when the current recursive epoch number t increases 1, and it will be cleared when $x^{(g')}(t)$ is updated, θ is the curvature parameter (set θ=0.4 in this paper).

The Hybrid Algorithm:

(1) Initialize the position of one particle by the solution gotten from the improved greedy algorithm, and initialize all the other particles randomly, which means a bit of binary code of a particle takes 1 with 0.2 probability, and 0 with 0.8 probability. The initial velocities of the particles are set by random numbers uniformly distributed on [-0.4, +0.4]. And set *T*=300;

(2) Update the positions and the velocity of all particles based on Eq.(2)′ and Eq.(3);

(3) Update the positions of the best particle of all the particles and the previous best position of every particle.

(4) If the position of the best particle of all the particles unchanged in consecutive *Y* epochs (set *Y*=15 in this paper), then re-calculate $x^{(g')}(t)$, and clear τ;

(5) Reinitialize those inactive particles randomly to avoid premature convergence. If the position of a particle unchanged after consecutive *E* epochs, it is regarded as an inactive particle (set *E*=7 in this paper).

(6) Set *t*=*t*+1, if $t \le T$, go to (2); else output *S*, and then terminate the algorithm.

6 Experimental Results

The taste signals of 11 kinds of mineral waters[5] are used as experimental data in this paper. The experimental data consist of 1100 points with 100 points for each taste signal is shown in Fig. 1.

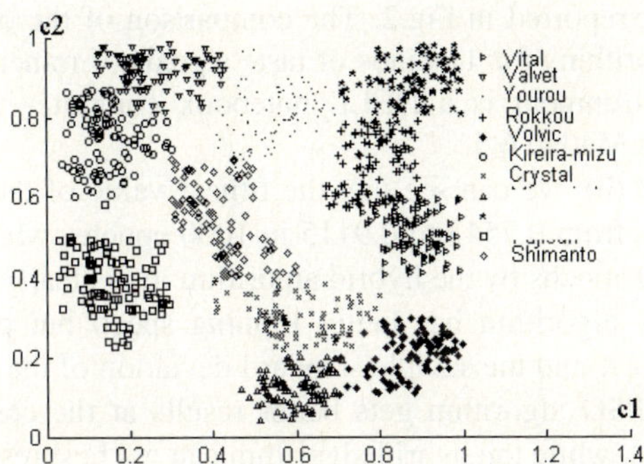

Fig. 1. 11 kinds of taste signals of mineral waters

Table 1. Comparison of the obtained class covers from three algorithms for 11 kinds of taste signals of mineral waters

	Greedy Algorithm		bPSO Algorithm		Hybrid Algorithm	
	K	*Var*	*k*	*Var*	*k*	*Var*
Vital	13	0.0009874	10	0.002677	10	0.001947
Valvet	9	0.0009513	5	0.000470	5	0.000309
Yourou	11	0.0005199	8	0.000319	9	0.000051
Rokkou	20	0.0003759	28	0.000624	23	0.000343
Volvic	5	0.0024951	2	0.000006	2	0.000014
Kireira-mizu	9	0.0005067	7	0.000609	9	0.000073
Crystal	11	0.0000523	16	0.000490	10	0.000786
Minmi-Alps	5	0.0011321	2	0.000519	3	0.000194
Pierval	10	0.0004432	16	0.000574	14	0.000308
Fujisan	4	0.0032171	2	0.000009	2	0.000002
Shimanto	10	0.0006482	11	0.000458	11	0.000243
Running Time	8.503sec.		347.139sec. (1000 epochs)		124.137sec. (300 epochs)	

When calculate the class covers of a taste signal, let the taste signal be the set of B and all the others be the set of R. The class covers of the taste signals are calculated by the improved greedy algorithm (parameter: $\gamma = 0.4$), the bPSO algorithm (parameter set: $v_{max}=4$; w=c_1=c_2=1; $\lambda_1=0.55$, $\lambda_2=0.1$, $\lambda_3=0.15$; $\alpha=0.96$, $\beta=0.01$, $\gamma=0.4$), and the hybrid algorithm (parameter set: $v_{max}=5\times(t/T)^{12}$; w=1, c_1+c_2=2, $c_2=2/(1-\exp(-0.4\tau))$; $\lambda_1=0.55$, $\lambda_2=0.1$, $\lambda_3=0.15$; $\alpha=0.96$, $\beta=0.01$, $\gamma=0.4$), respectively. The varying curves of the fitness of the best particle of all particles in the bPSO algorithm and the hybrid algorithm for the Shimanto signals are reported in Fig.2. The comparison of the obtained class covers from these three algorithms for 11 kinds of taste signals of mineral waters is listed in Tab.1. The test environment is a DELL notebook computer with P4 2G, 512M, Windows XP OS and Matlab 6.5.

From Fig.2(a) and (b), we can see that the fitness value of the best particle of the bPSO algorithm rises from 0.7543 to 0.9115 in 1000 epochs, while rises from 0.9029 to 0.9265 only in 300 epochs by the hybrid algorithm. It is clearly shown in Tab.1 that the improved greedy algorithm has better running speed but poorer results in the number of class cover *K* and the sample standard deviation of the obtained class cover radii *Var*, and the bPSO algorithm gets better results at the cost of most recursive epochs (running time), while the hybrid algorithm can get best results in less recursive epochs (running time).

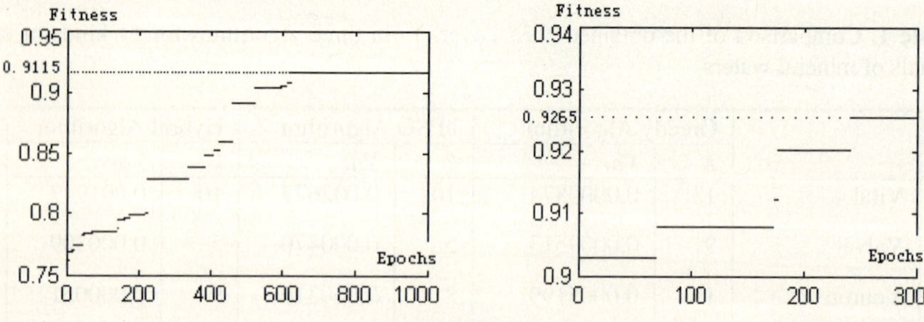

Fig. 2. (a) The varying curves of the fitness of the best particle of all particles in the bPSO algorithm for the Shimanto signals; (b) The varying curves of the fitness of the best particle of all particles in the hybrid algorithm for the Shimanto signals

7 Conclusions

A generalized class cover problem is proposed in this paper and three algorithms are adopted in solving it, which are the improved greedy algorithm, the bPSO algorithm, and their hybrid algorithm. The comparative results of these three algorithms show that the hybrid algorithm can get better solutions in less runtime.

Acknowledgement

This work was supported by the National Natural Science Foundation of China under Grant No. 60433020, and a key grant(No. 20020502) of Jilin Science & Technology Committee.

References

1. Cannon, A., Cowen, L.: Approximation Algorithms for the Class Cover Problem. Annals of Mathematics and Artificial Intelligence, .40(3) (2004) 215-223
2. David, J.M., Carey E.P.: Characterizing the Scale Dimension of a High Dimensional Classification Problem. Pattern Recognition, 36(1) (2003) 45-60
3. Priebe CE, Marchette DJ, DeVinney J, Socolinsky D.: Classification using Class Cover Catch Digraphs. Journal of Classification, 20(1) (2003) 3-23
4. Abhay, K.P.: Analysis of a Greedy Heuristic for Finding Small Dominating Sets in Graphs. Information Processing Letters, 39(5) (1991) 237-240
5. Yanxin, H., Chunguang, Z.: Recognizing Taste Signals Using A Clustering-based Fuzzy Neural Network. Chinese Journal of Electronics, 14(1) (2005) 21-25
6. Kennedy, J., Eberhart, R.C.: Particle Swarm Optimization. In Proceeding of IEEE International Conference on Neural Networks, Volume IV, Perth, Australia: IEEE Press. (1995) 1942-1948
7. Van, Den, Bergh, F.: An Analysis of Particle Swarm Optimizers [PH.D thesis]. Pretoria: Natural and Agricultural Science Department, University of Pretoria, 2001.
8. Kennedy, J., Eberhart R.C.: A Discrete Binary Version of the Particle Swarm Algorithm. Proceedings of the 1997 Conference on Systems, Man, and Cybernetics. Piscataway, NJ, IEEE Press (1997) 4104-4109
9. Zhengzhi, W., Tao, B.: Evolutionary Computation. Chang Sha, China, National University of Defense Technology Press (2000) 26-37

Cooperative Co-evolutionary Approach Applied in Reactive Power Optimization of Power System

Jianxue Wang[1], Weichao Wang[2], Xifan Wang[1], Haoyong Chen[1], and Xiuli Wang[1]

[1] School of Electrical Engineering, Xi'an Jiaotong University
Xi'an, 710049, P.R. China
JXWang@mailst.xjtu.edu.cn
[2] Xi'an Electric Power Institute
Xi'an, 710032, P.R. China

Abstract. Cooperative Co-evolutionary Approach (CCA) is a new architecture of evolutionary computation. Based on CCA, the paper proposes a new method for reactive power optimization problem in power system, which is non-convex, non-linear, discrete, and usually with a large number of control variables. According to the decomposition-coordination principle, the reactive power optimization problem is decomposed into a number of sub-problems, which is optimized by a single evolutionary algorithm population. The populations interact with each other through a common system model and co-evolve and result in the continuous evolution of the whole system. The reactive power optimization problem is solved when the co-evolutionary process ends. Simulation results show that compared with conventional Genetic Algorithm (GA), CCA not only can obtain better optimal results, but also has better convergence property. CCA reduce the over-long computational time of GA and is more suitable for solving large-scale optimization problems.

1 Introduction

Cooperative Co-evolutionary Approach (CCA) simulates the co-evolution mechanism in nature and adopts the notion of ecosystem. The early application of co-evolution concept can be traced back to the Hillis host model [1] and the multiple co-evolution model in workshop dispatch of Husbands [2]. Evolutionary Computation Laboratory of George Mason University directed by professor De Jong has significant progress in the research of cooperative co-evolutionary computation [3]. Compared with GA, CCA can reasonably divide the control variables into different populations and jump out of the local minimum and find better optimization solution to the larger scale system. There have been some examples successfully using CCA in power system. Chen applies the cooperative co-evolutionary model to power system unit commitment [4] and power market [5], and gets quite encouraging results.

The reactive power optimization can improve the voltage level and decrease system loss. It takes switchable shunt capacitor, transformer tap and generator output as control means, and it belongs to a nonlinear combinatorial optimization problem. Conventional optimization methods have been used in this problem for many years.

L. Jiao et al. (Eds.): ICNC 2006, Part I, LNCS 4221, pp. 620 – 628, 2006.

However, they always need some hypothesis and often get a local minimum. In recent years, for overcoming the defect of conventional optimization methods, genetic algorithm, simulated annealing, tabu searching and interior point method have been widely used in the reactive power optimization.

But for reactive power optimization is rather complex and has a great deal of control variables, the CCA is the better choice to solve the problem. And through setting different populations sort methods, CCA can conveniently realize reactive power optimization according voltage level or district. This paper applies CCA to reactive power optimization, and compares the calculation results with simple GA and improved GA with annealing selection. The results show that proposed method can find optimization solution in greater range, and CCA is suitable and effective to reactive power optimization.

2 Reactive Power Optimization Model

To minimize the active power loss and the voltage deviation under some operational constraints, reactive power optimization in power system need find a optimal solution, containing the reactive power output of generators (or voltage of generators V_G), the reactive power compensation capacity(including the capacity of shunt capacitors Q_c and reactance Q_L) and transformer tap-settings (T). It has a significant influence on security and economic operation of power systems.

The augmented objective function is to minimize the active power loss ΔP and the voltage deviation ΔV_i:

$$F = Min\left[\Delta P + \lambda \sum_{i=1}^{n} \Delta V_i^2\right] \tag{1}$$

Where n is the node number and λ is the equivalent coefficient of nodal voltage deviation ΔV_i. Supposing $V_{i\max}$ is the maximum voltage of node i, $V_{i\min}$ is the minimum voltage of node i, the definition of ΔV_i is given as below:

$$\Delta V_i = \begin{cases} V_i - V_{i\max} & (V_i > V_{i\max}) \\ 0 & (V_{i\min} \le V_i \le V_{i\max}) \\ V_{i\min} - V_i & (V_i < V_{i\min}) \end{cases} \tag{2}$$

Equality constraints contain active and reactive power balance constrains in the following.

$$\begin{aligned} \Delta P_i &= P_{Gi} - P_{Li} - V_i \sum_{j\in i} V_j \left(G_{ij}\cos\theta_{ij} + B_{ij}\sin\theta_{ij}\right) = 0 \\ \Delta Q_i &= Q_{Gi} - Q_{Li} + N_{ci}\Delta Q_{ci} - N_{Li}\Delta Q_{Li} - V_i \sum_{j\in i} V_j \left(G_{ij}\sin\theta_{ij} - B_{ij}\cos\theta_{ij}\right) = 0 \end{aligned} \tag{3}$$

Where P_{Gi} and Q_{Gi} are active power and reactive power supply of bus i, P_{Li} and Q_{Li} are active and reactive load of bus i, V_i and V_j is the voltage module of bus i and j, θ_{ij} is the difference angle between bus i and j. ΔQ_{ci} and ΔQ_{Li} are the unit capacity of capacitor and reactance of bus i. N_{ci} and N_{Li} are working span of capacitor and reactance of bus i.

Inequality constraints contain:

$$V_{Gi\min} \le V_{Gi} \le V_{Gi\max} \tag{4}$$

$$N_{ci\min} \le N_{ci} \le N_{ci\max} \tag{5}$$

$$N_{Li\min} \le N_{Li} \le N_{Li\max} \tag{6}$$

$$T_{ij\min} \le T_{ij} \le T_{ij\max} \tag{7}$$

$$Q_{Gi\min} \le Q_{Gi} \le Q_{Gi\max} \tag{8}$$

$$S_{ij} \le S_{ij\max} \tag{9}$$

Where V_{Gi}, N_{ci}, N_{Li}, T_{ij} are voltage of generator, working span of capacitor, working span of reactance and transformer tap. All of them are control variables. Q_{Gi} and S_{ij} are reactive power output of generator i and power flow on branch ij, and they are state variables.

3 CCA in Reactive Power Optimization

3.1 Basis Thought of Cooperative Co-evolutionary Approach

The main conventional methods, such as linear planning and nonlinear planning, need to suppose that the control variables are continuous and require that the objective function is differentiable. Because of better convergence and adaptability, GA is used to solve some nonlinear combinatorial optimization problem. But when optimization problem has numerous control variables and great question scale, GA has long computation time and slow convergence speed. Furthermore GA has a strong tendency to converge prematurely. While the new evolution algorithm- CCA can improve these problems. CCA introduces ecosystem into conventional algorithm [3]. Conventional GA often regard the whole problem as single population evolution, but CCA regards this problem as ecosystem in which the individuals affect each other and solve the optimization problem by ecosystem evolution. To large-scale problem, CCA adopts the thought of decomposition-coordination, i.e. complex problem is decomposed into interactive and simple sub-problem, and each sub-problem is regarded as one population in ecosystem which can use genetic approach to evolve separately. The populations interact with each other in a common model and have cooperative relationship. So the CCA can simplify the research spaces, improve the ability of finding global optimization solution and avoid local optimization solution. It is obvious that CCA is a highly abstract algorithm framework. Fig.1 and Fig.2 show the calculation process of the conventional GA and CCA. The normal operation in GA and CCA is similar, such as of selecting, crossing and mutating in the following figures. The main difference is that CCA need to divide the large-scale problem into many sub-problems, so each sub-problem need coding and initializing respectively, and then cooperatively evolves. In addition, CCA is a parallel algorithm essentially and can directly adopt parallel computation. So it can largely increase the calculation speed and has wide potential application in engineering calculation.

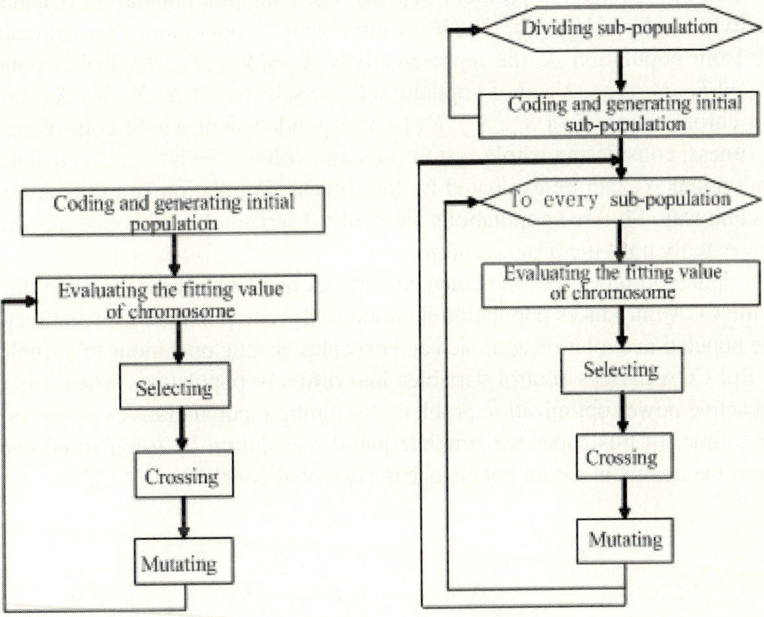

Fig. 1. The process of the conventional GA **Fig. 2.** The process of CCA

The core of CCA is how to divide numerous variables into different populations. For reactive power optimization, we can define some sort method from practical operation. For instance, the power system should keep reactive power to balance in local neighborhood and avoid the remote transmission of reactive power. So we can divide control variables in same supplied area into one population. Sometimes it is necessary to control reactive power devices in different voltage grades in order to adjust voltage by different grades and avoid disordering voltage adjustment. So we can divide control variables under same voltage grade into one population. If each control equipment must be operated one by one, we can divide the same kind of control methods into one population. We can also take several requirements into account when sorting populations.

3.2 CCA in Reactive Power Optimization

Here we offer a sample system to explain how to use CCA in reactive power optimization. This system is made up of 3 power supply areas (A, B and C), as shown in Fig.3, and each area is a reactive power optimization sub-problem represented by one population in CCA, as shown in Fig.4.

In Fig.4, we can see that each sub-problem can be evolved in its own population independently. At the same time, in order to construct the complete problem, the other sub-problems select their typical representative to participate in the cooperation evolution when the special sub-problem evolves. For example, population A generates a new individual $X_1 = [V_{G1}, V_{G2}, T_1, N_{C1}]$ after genetic operations. To evaluate individuals from population A, the representatives $X_2 = [V_{G3}, T_2, N_{C2}]$ of population B and $X_3 = [T_3, T_4, N_{C3}, N_{C4}]$ of population C are selected, then X_1 X_2 X_3 can make up of a chromosome $X = [X_1, X_2, X_3]$ corresponded with whole control variables. So the fitness, considering whole system loss and voltage deviation, is calculated. The new individual X_1 can be evaluated by this fitness. Population B and C are evolved in the same way. So three populations are evolved separately and cooperate with each other repeatedly until the program stops.

The calculation process of the conventional GA and CCA show the main difference is that the CCA introduces population iteration to the computation. Conventional GA is a single population evolution approach and executes genetic operations to a single population. But CCA divides control variables into different populations when it is used to solve reactive power optimization problem. So multiple populations exist and evolve at the same time. In this paper we simulate parallel evolution by using serial evolution. However, the treatment would not change the essential conclusions of CCA.

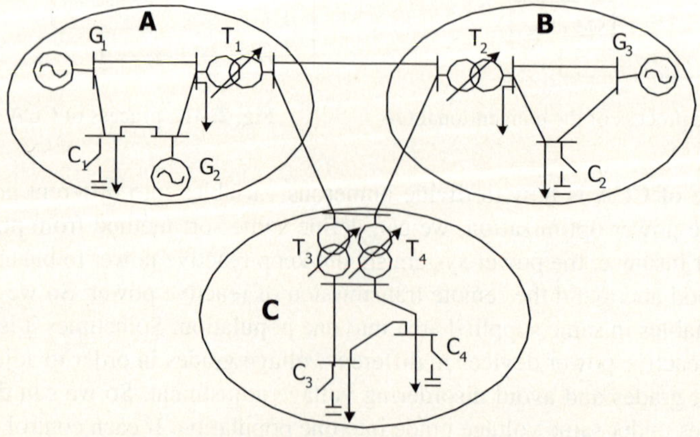

Fig. 3. The simple power system

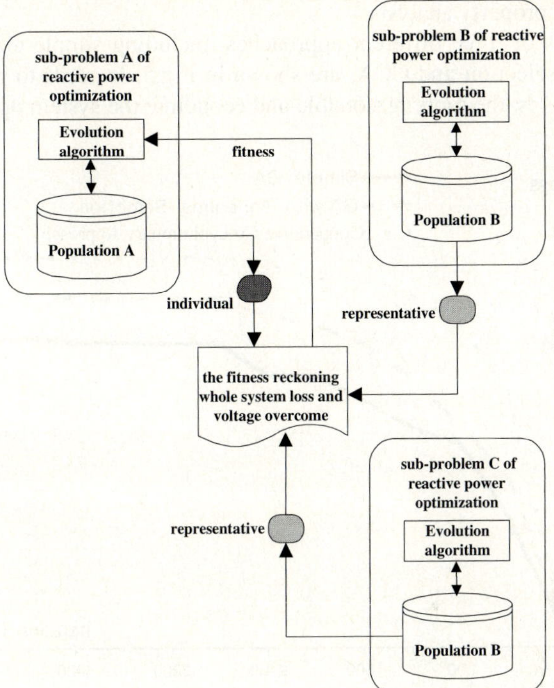

Fig. 4. Cooperative Co-evolutionary Approach diagram

4 Simulation Results

To test the suggested method, we calculated the reactive power optimization of a real power system. This system has 120 nodes, 142 branches, 2 generators, 41 transformers and 35 reactive power compensation devices. Voltages of nodes can vary from 1.0 to 1.10. Here the populations are divided according to power supply areas.

1) Basic results using CCA

The basic results are given in table 1. From table 1 we can see that the voltage level has been improved, all bus voltage are in their permissible limits, system loss decreasing rate is 4.83%. Operation of the system is more reasonable and economic.

Table 1. The results of real system for Var optimization

Cases	Loss (MW)	Bus voltage (p.u.)			Number of over-standard bus
		Maximum	Minimum	Average	
Before optimizing	6.42	1.102	0.943	1.0334	3
After optimizing	6.11	1.088	1.004	1.0563	0

2) Convergence property analysis

Convergences of three different approaches, including simple GA, improved GA with annealing selection and CCA, are shown in Fig.5. It needs to point out that the bigger the fitness is, the more reasonable and economic the system operation is.

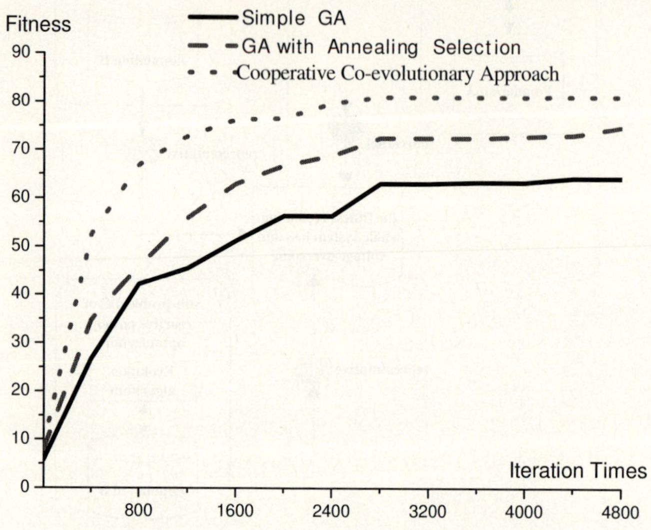

Fig. 5. Convergence curve of different algorithms

Compared with simple GA and the GA with annealing selection, CCA can appropriately sort control variables, get away from local optimal solution and get a more satisfying solution to large-scale system. Its convergence is better than simple GA and GA with annealing selection.

3) Quality of optimization solutions

To compare the solution quality, the three approaches are executed 100 times. Fig.6 shows the results. Where X-axis stands for system loss, Y-axis stands for happening times. From Fig.6, it can be seen that the simple GA's results is not satisfying because they are more dispersive and the good solution is hard to find. GA with annealing selection is improved in these parts, but the improvement is not obvious. At the same time, CCA improves the results obviously. CCA offers the better solutions that other approaches are difficult to find. By using CCA the optimal solutions are better and happen in a larger probability. CCA is a more available approach.

4) CPU time Analysis of CCA

In this part we counted the CPU time of IEEE14, 30, 57-bus system and a practical system in a PC computer with 2.80 GHz. Table 2 shows the results. We can see that system scale has a great influence on the CPU time of CCA. Because of simulating parallel evolution by using serial evolution in this paper, along with the larger scale, CPU time increases obviously. But it is possible that CPU time will be reduced if parallel methods are used.

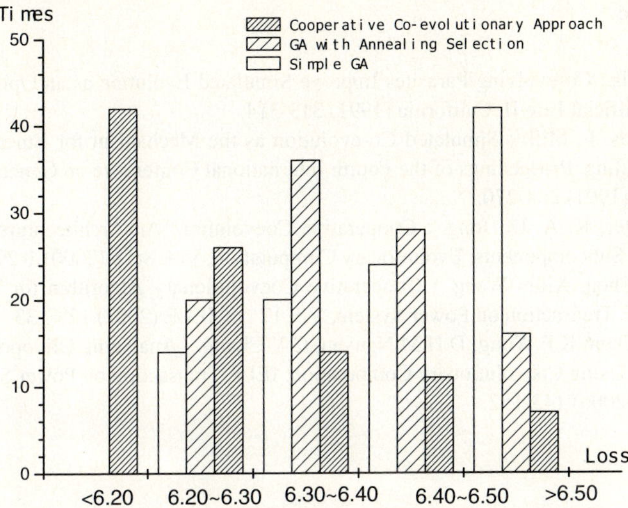

Fig. 6. Solution distribution of practical system optimization

Table 2. The computation time of various systems

Systems	Bus number	CPU time(s)
IEEE14	14	8
IEEE30	30	22
IEEE57	57	52
Practical system	120	81

5 Conclusions

Reactive power optimization is a complex non-linear problem. Based on the thought of decomposition-coordination, CCA divides this complex problem into multiple interactive sub-problems. These sub-problems interact and co-evolve with each other by a common system model. The algorithm is evaluated on IEEE test systems and a practical system. The simulations show that, compared with simple GA and the GA with Annealing Selection, CCA can get global optimal solution in larger probability. Its convergence and solution quality are both better. Furthermore, CCA can adopt parallel calculation method, which has wide potential application for reactive power analysis on line or other engineering computation.

Acknowledgment

This work is supported by Special Fund of the National Basic Research Program of China(2004CB217905) and Teaching Scholarship of XJ Group Corporation.

References

1. D. W. Hillis.: Co-evolving Parasites Improve Simulated Evolution as an Optimization Procedure. Artificial Life II. California (1991) 313-314.
2. P. Husbands, F. Mill. : Simulated Co-evolution as the Mechanism for Emergent Planning and Scheduling. Proceedings of the Fourth International Conference on Genetic Algorithms. California (1991) 264-270.
3. M. A. Potter, K. A. De Jong. : Cooperative Coevolution: An Architechture for Evolving Coadapted Subcomponents. Evolutionary Computation, Vol. 8(1). (2000) 1-29.
4. Haoyong Chen, Xifan Wang. : Cooperative Coevolutionary Algorithm for Unit Commitment. IEEE Transaction on Power System, Vol.17 (1). IEEE (2002) 128-133.
5. Haoyong Chen, K.P.Wong, D.H.M.Nguyen, C.Y.Chung. : Analyzing Oligopolistic Electricity Market Using Coevolutionary Computation. IEEE Transaction on Power System, Vol.21 (1). IEEE (2006) 143-152.

Evolutionary Algorithms for Group/Non-group Decision in Periodic Boundary CA*

Byung-Heon Kang, Jun-Cheol Jeon, and Kee-Young Yoo**

Dept. of Computer Engineering, Kyungpook National University,
Daegu, Korea, 702-701
{bhkang, jcjeon33}@infosec.knu.ac.kr, yook@knu.ac.kr

Abstract. Cellular automata (CA) have been used for modeling the behavior of complex systems such as the VLSI technology and parallel processing system. Recently, Das et al. have reported the characterization of reachable/non-reachable CA states. Their scheme has only offered characterization under a null boundary CA (NBCA). In the simplifications of hardware implementation, however, a periodic boundary CA (PBCA) is suitable for constructing cost-effective schemes such as a linear feedback shift register (LFSR) structure because of its circular properties. Thus, this paper presents two evolutionary algorithms for the identification of the reachable/non-reachable state and the group/non-group CA based on periodic boundary conditions.

1 Decision Algorithm of Reachable or Non-reachable State

The characterizations of reachable/non-reachable CA states and group/non-group CA based on a null boundary CA were reported by Das et al. [1]. We present that a state of CA is either reachable or non-reachable based on a PBCA. A state having n cells is presented by $<S_0\ S_1 \cdots S_{n\text{-}2}\ S_{n\text{-}1}>$. First, we assign each possible neighborhood configuration with a rule and present each rule as *rules*[n]. Second, we classify four cases according to the value of the first and last cell of a state. The classified four cases are (i) $<0\ S_1 \cdots S_{n\text{-}2}\ 0>$, (ii) $<0\ S_1 \cdots S_{n\text{-}2}\ 1>$, (iii) $<1\ S_1 \cdots S_{n\text{-}2}\ 0>$ and (iv) $<1\ S_1 \cdots S_{n\text{-}2}\ 1>$. The values of the leftmost cell and the rightmost cell of NBCA are 0. Therefore, if those values are 0, then the next values of cell 1 and cell 4 are represented as d (*don't care*).

The following Program 1 classifies state transition arrays according to the value of the first and last cell of a state based on a formal program code.

```
Program 1. FindStateTransitionArrayOnPBCA
void FindStateTransitionArray_PBCA(rules[n],state[n],n)
{
   var remain;
```

* This research was supported by the MIC (Ministry of Information and Communication), Korea, under the ITRC (Information Technology Research Center) support program supervised by the IITA (Institute of Information Technology Assessment). This work was also supported by the Brain Korea Project 21 in 2006.

** Corresponding author.

L. Jiao et al. (Eds.): ICNC 2006, Part I, LNCS 4221, pp. 629 – 632, 2006.

```
array Rule[n][8]={0,};
for(int i=0; i<n; i++) {
     for(int j=0; rules[i]>1; j++) {
        remain=rules[i]%2;
        rule[i]=rules[i]/2;
        Rule[i][j]=remain;
     }
     Rule[i][j]=rules[i];
  }
 if(state[0]=0 && state[n-1]=0) {
    for(i=4; i<=n; i++)
      Rule[0][i]='d';
    for(i=1; i<=n; i=i+2)
      Rule[3][i]='d';
 }
 else if(state[0]=0 && state[n-1]=1) {
    for(i=1; i<=n; i=i+2)
      Rule[3][i]='d';
 }
 else if(state[0]=1 && state[n-1]=0) {
    for(i=1; i<=n; i++)
       Rule[0][i]='d';
 }
 else(state[0]=1 && state[n-1]=1) ;
}
```

To determine whether a state is reachable or non-reachable, we define the following two properties and these properties are presented by a two-dimensional array.

(i) Two-dimensional array $p[n][]$ is $(i+1)$th decimal states transferred from ith decimal states. If the ith decimal state is m, the $(i+1)$th decimal states are changed to $(2m)$ *mod* 8 and $(2m+1)$ *mod* 8. For instance, if the ith decimal state m is 2, the $(i+1)$th decimal states are changed to {4, 5}.

(ii) Two-dimensional array $q[n][]$ is the value that the $(i+1)$th bit of a state is equal to the present state in the $(i+1)$th cell.

The following Program 2 determines whether a state is reachable or non-reachable. It returns 1 if the state is non-reachable. Otherwise, it returns 0.

Program 2. FindReachableOrNonreachableOnPBCA

```
int FindReachableOrNonreachable_PBCA(Rule[n][8],state[n],n)
{
   var m;
   array s[n][8], p[n][8], q[n][8];

   for(int i=0, j=0; i<8; i++) {
      Find s[0] where Rule[0][i]==state[0].
   }
   for(int i=0; i<8; i++) {
      If s[0] has 4,5,6 and 7, replaced by 0,1,2 and 3 respectively.
   }

   if(s[0]==ϕ)
      Return 1 as the state is non-reachable.
   for(int i=1; i<n; i++) {
      for(int j=0, k=0; j<8; j++) {
         Find p[i] such that if m ∈ s[i-1],
          ((2m) mod 8) and ((2m+1) mod 8) are in p[i].
      }
      for(int j=0, k=0; j<8; j++) {
         Find q[i], where q[i]=={j} and Rule[i][j]==state[j].
```

```
        }
        for(int j=0, k=0; j<8; j++) {
            Find s[i]=(p[i][j]∩q[i][j]).
        }
        for(int j=0; j<8; j++) {
            If s[i] has 4,5,6 and 7, replaced by 0,1,2 and 3 respectively.
        }
        if(s[i]==φ)
           Return 1 as the state is non-reachable.
    }
    Return 0 as the state is reachable.
}
```

2 Decision Algorithm of Group or Non-group CA

We classify that a PBCA is a group or non-group in this section. Each cell in a PBCA has only two states with three-neighborhood interconnections. The decision algorithm is as follows:

(i) First, we find that *s*[*i*][] has 0 or 1 with the first rule respectively, where *i* has 0 and 1. Then, we check the numbers between *s*[0][] and *s*[1][]. If the numbers between *s*[0][] and *s*[1][] are not the same, the PBCA is a non-group.

(ii) And, we determine the next decimal states for *q*[*i*][] such that that if $m \in s[i]$, $((2m) \bmod 8)$ and $((2m + 1) \bmod 8)$, where *i* has 0 and 1. Then, we distribute these decimal states into *p*[*j*][] and *p*[*j*+1][], such that *Rules*[*i*][8] contains 0 and 1, respectively. If the numbers between *p*[*j*][] and *p*[*j*+1][] are not the same, the PBCA is a non-group. We remove duplicate sets from *p*[*j*][] and *p*[*j*+1][]. Then, we assign the sets of *p*[*j*][] and *p*[*j*+1][] to *s*[*i*][] for the next step, where *i* has 0 and 1.

(iii) Finally, if the numbers between *s*[0][] and *s*[1][] are not the same, the PBCA is a non-group. Otherwise, we can determine that the PBCA is a group.

Program 3 determines whether the PBCA is a group or non-group. It returns 1 if the state is a non-group. Otherwise, it returns 0.

Program 3. CheckGroupOrNongroupOnPBCA

```
int CheckGroupOrNongroup_PBCA (Rule[n][8],n)
{
    array s[2][8], p[n][8], q[2][8];
    for(int i=0; i<2; i++) {
        for(int j=0, k=0; j<8; j++) {
            Find s[i]={j}, where Rule[0][j]==(i-1).
        }
        for(int m=0; m<8; m++) {
            If s[i] has 4,5,6 and 7, replaced by 0,1,2 and 3 respectively.
        }
    }
   If |s[0]|!=|s[1]|, return 1 as the state is a non-group.


   for(int i=1; i<n; i++) {
       for(int j=0; j<2; j++) {
           for(int k=0; k<8; k++) {
               Determine the next decimal states q[j] such that
                  if m ∈ s[j], ((2m) mod 8) and ((2m+1) mod 8).
           }
           for(int k=0; k<8; k++) {
```

```
            Distribute these decimal states into p[j] and p[j+1], such that
              p[j] and p[j+1] contain 0 and 1 for Rule[i], respectively.
          }
          If |p[j]|!=|p[j+1]|, return 1 as the state is a non-group.
          for(int k=0; k<8; k++) {
            If s[k] has 4,5,6 and 7, replaced by 0,1,2 and 3 respectively.
            If s[k+1] has 4,5,6 and 7, replaced by 0,1,2 and 3 respectively.
          }
          If (|p[j]| mod 2)==1 or (|p[j+1]| mod 2)==1,
            return 1 as the state is a non-group.
      }
    }

    If |s[0]|!=|s[1]|, return 1 as the state is a non-group.
    Otherwise, return 0 as the state is a group.
}
```

3 Discussion and Conclusion

We have presented two decision algorithms for the identification of reachable/non-reachable states and a group/non-group on a PBCA. The proposed schemes have the same time complexity as Das et al.'s scheme which has only offered the characterization under null boundary conditions, while our schemes are based on periodic boundary conditions. Moreover, we have reduced the number of loops in comparison with Algorithm 2 of Das et al.'s algorithms. We have removed one loop in Program 3 by proceeding simultaneously with step 3 and 4 of Algorithm 2 in Das et al.'s. Therefore, our scheme reduced the program processing time. The execution time of our algorithms depends on n. Hence, the complexity of all proposed algorithms is $O(n)$.

For hardware implementation, a PBCA is suitable for the construction of cost-effective schemes such as a LFSR structure because of its circular properties. Thus, we expect that our classification can be used in applications regarding the above-mentioned criteria and VLSI technology.

Reference

1. S. Das, B. K. Sikdar and P. P. Chaudhuri: Characterization of Reachable/Nonreachable Cellular Automata States. Lecture Notes in Computer Science, Vol. 3305. Springer-Verlag, Berlin Heidelberg (2004) 813-822

of the GAs prior to the Selection Operator. This hybrid approach was firstly proposed by [7] as a modification of the roulette selection method, to regulate the high selective pressure it imposes, giving birth to the Hawk-Dove Roulette selection method – where Hawk-Dove is an evolutionary game presented in [10], and Roulette is the traditional implementation of this method.

In [1] this approach was followed and the Hawk-Dove Tournament method was proposed. Based on genetic studies of behavior, it added a new chromosome to individuals, representing the behavior (strategy) they adopt in the games. This originates what is known as Genetic Coding and Transmission of Behavior. The new chromosome also could suffer the crossover and mutation operations, causing the strategies to proliferate during the GA's generations.

This approach is formalized in [2] as a new macro-step of the GA, making it independent from any selection method that is adopted, opening the doors to countless combinations of games vs. selection methods. The inclusion of the intelligent agent to regulate the GA's parameters presented in this work is made upon this approach, minding to better the already good results obtained in the works formerly mentioned.

3 Selection of Parameters for Genetic Algorithms and Dynamic Parameterization

The aforementioned choice of parameters for GAs, such as crossover and mutation rates and size of population is a rather hard task, due to the enormous possibilities of variations in the modeling of the problem and the cost (fitness) function. Traditional GAs base themselves upon the generation of several random factors in the creation of the initial population, crossover and mutation. Therefore, two executions with the same initial parameters of execution can produce significantly different results [6].

According to [8], this matter should lead to the conclusion that there are several methods of crossover and mutation, as much as hybrid and generic possibilities. Comparing all these possibilities and generating an optimal parameter set is an extremely hard task, as the results may well be intrinsically linked to the problem in question. The works of [6] and [8] try to find this optimal set of parameters, or at least point to sets of problems on which the best results may be obtained.

One of the first to study this area was [13]. He was able, through the definition of a wide set of tests to a few simple problems, to establish some parameters considered as optimal. These parameters began to be largely used by the community, such as: population size between 50 and 100 individuals, crossover rate higher than 60% and mutation rate equal to 0.1% for each gene of the population.

Despite the high degree of adoption of these parameters in the literature, there is no evidence that they are the best and ought to be applied to each and every problem, yet there are other works that ratify their use [8]. Grefenstette (1986) in [6] proposes a genetic meta-algorithm that should be used to find the set of parameters for the problem. In this definition, each execution with different parameters represents an individual of a "macro-GA", whose evaluation function is inferred on the performance of a whole execution. Despite the high computational cost involved in this approach, it is good in the sense of a meta-optimization, and also avoids the excess of heuristics in the selection of parameters.

Both the mentioned approaches, as well as most of the others, take into consideration finding a set of parameters that will be defined before the start of execution and persist until the end of it. However, this set does not necessarily reflect the real needs for values of the parameters according to the current state of the population in a given generation.

This work also consists in a dynamic approach of parametric variation, allowing a detailed adjustment, according to the state of the population. Nevertheless, the approach utilized is of an external fuzzy intelligent regulator to the GA, which because of its similarity in perceiving the environment and taking action is called "agent".

This ability to externally control the environment inspired the selection of the name "GAIA" to be the agent's name. In the Greek Mythology, Gaia is the goddess of fertility, representing the personification of the Earth as a goddess.

Taking this as a base, this entity would intercede in a population aiming to guide its evolution, ensuring an acceptable degree of genetic variability and regulating aspects that were kept random up to then. This control is possible through Fuzzy reasoning, and the respective project to implement its idea is described in the next section.

4 Project and Implementation of the GA Fuzzy Regulator

The prime intention of the dynamic parameter regulator for genetic algorithms was to build an "intelligent" agent that, with the aid of Fuzzy Logic, would be able to periodically evaluate certain variables, considered indicators of evolution (good coverage of the search space). Then, based on these indicators, it would act in order to regulate the values of some parameters of execution, as well as operate directly on the population, altering its individuals or replacing them if necessary.

Fig. 1 shows the structure of the GA, highlighting the moment of the execution in which the fuzzy agent performs the tasks of supervision and intervention in the population.

The first step to build the agent consisted of an analysis of the candidate output variables, representing what the agent should control. The following candidate variables were evaluated: crossover rate, mutation rate, percentage of surviving individuals (the ones that go through to the next generation), and the number of generations of the execution (halt controlled by the agent).

Due to the high subjectivity related to the fuzzification of the latter, the following were adopted as output variables: crossover rate, mutation rate, percentage of surviving individuals.

After defining the output variables, the candidate input variables were named, meaning the ones which have a great significance for the chosen parameters of control: mean value of the population, standard deviation of the population, optimal value/best fitness of the generation, frequency of the best value in the last generations, percentage of duplicate individuals in the current generation, distance from a supposed optimal value.

So, after considerations the applicability and capability of fuzzification of each candidate variable, taking into account the main objectives of optimization and evolution and minding to maintain the variability, the following variables remained: optimal value in the current generation, frequency of best value in the past n generations, and the rate of duplicate individuals in the current generation.

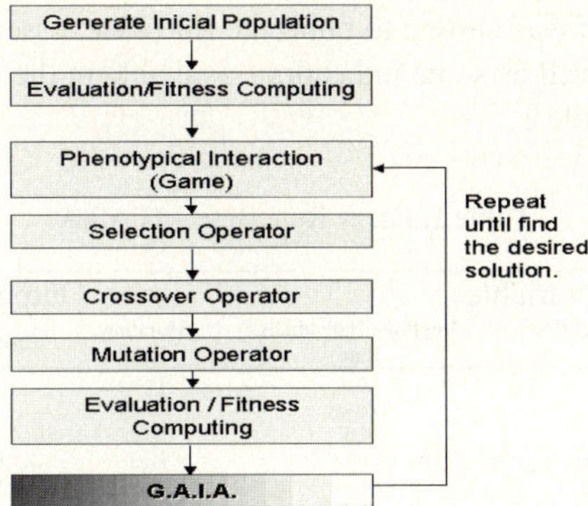

Fig. 1. Proposed structure for the GA after the insertion of the step when the fuzzy agent intervenes, at the end of each iteration

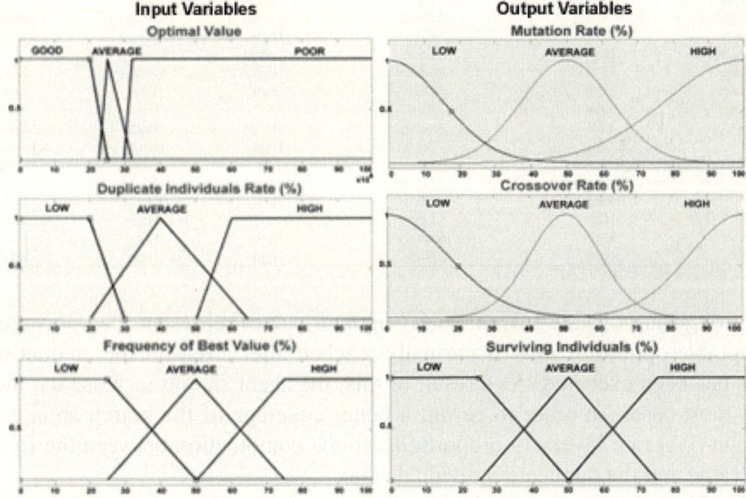

Fig. 2. Mapping and membership functions for the G.A.I.A. input and output variables

For each variable selected, either input or output, its domain was defined and partitioned in three fuzzy linguistic values, defined as "low", "average" and "high" for most of the variables.

The trapezoid and triangular membership functions were used to model the values of the input variables. The output variable for surviving individuals also used the trapezoid and triangular functions. For their turn, the variables for mutation and crossover rates were modeled using Gaussian membership functions, as these permit a smoother transition [9]. Fig. 2 displays graphically the mapping of both the input and output variables.

The singleton fuzzifier was used, along with the weighted average defuzzifier, and minimum inference operator and Mamdani composition (MaxMin) [12].

A heuristic approach was chosen to build the rule base, essentially founded on empirical knowledge, as well as some indications available in the literature. The full rule base is displayed in Table 1.

Table 1. Fuzzy Rule Base – G.A.I.A

Input Variables			Output Variables		
Optimal Value	**Frequency of Best Value**	**Duplicate Individuals Rate**	**Mutation Rate**	**Crossover Rate**	**Surviving Individuals Rate**
Good	Low	Low	Low	High	High
Good	Low	Average	Average	High	High
Good	Low	High	High	High	Average
Good	Average	Low	High	Average	High
Good	Average	Average	High	Average	Average
Good	Average	High	High	Average	Low
Good	High	Low	High	Average	Average
Good	High	Average	High	Average	Low
Good	High	High	High	Low	Low
Average	Low	Low	Low	High	High
Average	Low	Average	Average	High	High
Average	Low	High	High	High	Average
Average	Average	Low	Average	High	High
Average	Average	Average	Average	Average	Average
Average	Average	High	High	Average	Average
Average	High	Low	Average	Average	Average
Average	High	Average	High	Average	Low
Average	High	High	High	Low	Low
Poor	Low	Low	Low	High	High
Poor	Low	Average	Average	High	High
Poor	Low	High	High	High	Average
Poor	Average	Low	Low	High	Average
Poor	Average	Average	Average	High	Average
Poor	Average	High	High	High	Low
Poor	High	Low	Average	High	Low
Poor	High	Average	High	High	Low
Poor	High	High	High	High	Low

The first group of nine rules ("good" optimal value) represent the convergence to an acceptable optimum, what is normally reached after a reasonable amount of generations has been executed. As a result of this, the agent should increase the mutation rate for most cases, in order to permit a better coverage of the search area, and promote a survival rate inversely proportional to the combination between the frequency of best value and the duplicate individuals rate.

As for the second group of rules ("average" optimal value), typically they indicate cases of "half-life" of an execution, meaning the GA is exploring the search surface aiming to get the best values. The rules state that the mutation rate varies to a certain extent proportionally to the duplicate individuals and the amount of times the best value appears, thus trying to conduct the search away from possible local optima, and try to explore other areas where the global optimum could be.

Finally, the third group ("poor" optimal value) refers typically to rules activated at the beginning of an execution. Thereby, the crossover rate should be kept high, the mutation rate follows the proportion of duplicate individuals, and the surviving individuals rate decreases as the frequency of the best value increases. Needless to say, near the beginning of an execution, the expected optimal value is poor.

Generally, the agent (G.A.I.A.) receives the population as an input, after the crossover and mutation operations, and calculates the numeric values for the parameters. These are then converted in fuzzy values, and activate one of the rules in the base. The fuzzy values for the output variables are then defuzzified and these converted numeric values are used to alter the population, eventually generating new individuals according to the rate for surviving individuals. Also, the crossover and mutation rates may vary, and the new values are applied in the next generation.

The test platform was built upon the Java language, with the aid of the FuzzyJ API [5] for the implementation of the intelligent agent's fuzzy inference engine. The tools supplied by the API guaranteed a significant gain in the programming productivity.

The use of the regulator has its own function modes that can be set by the user at the beginning of each execution. These are related to the frequency it should execute: at each generation ("permanent"), or at fixed intervals ("periodic"), or establishing a probability that the agent could be executed at each generation ("stochastic"). These options are important in the sense of evaluating and comparing the real need of dynamic control of the parameters at a fixed period (periodic/permanent), or exposing the environment to a "simulation of natural disasters" (stochastic).

5 Simulation and Evaluation of the Purpose

In order to evaluate the purposes properly, a platform for running the tests had to be built. It would provide a mean of comparing the performance of the different configurations of parameters that ought to be tested, for the same initial population.

The symmetric traveling salesman problem was selected for implementation, applied to the road distances among 26 cities, representing the Brazilian state capitals [3], formulated in such a way similar to [7], for comparison purposes.

Table 2. Comparative Evaluation of the Main Results

Execution Type	Best (km)	Generation Found (1st)
GAIA Periodic-100, HDR(20,30)	20586	1364
GAIA Stochastic-25%, HDR(20,30)	21069	833
GAIA Periodic-10, HDT(30,20)	21181	723
GAIA Stochastic-50%, HDT(30,20)	21254	357
GAIA Stochastic-25%, HDR(20,30)	21324	4641
GAIA Periodic-100, HDR(20,30)	21528	1212
GAIA Stochastic-75%, HDR(20,30)	21946	1883
GAIA Periodic-10, HDR(20,30)	21957	2843
GAIA Periodic-20, Tournament	22061	4793
GAIA Stochastic-25%, HDR(20,30)	22185	1577
GAIA Periodic-50, HDR(20,30)	22242	2534
GAIA Periodic-20, HDR(20,30)	23764	312
GAIA Stochastic-50, HDR(20,30)	24079	4394
GAIA Permanent, HDR(20,30)	24260	2256
GAIA Periodic-10, Roulette	24487	790
GAIA Permanent, HDR(20,30)	24487	3960
GAIA Periodic-10, HDR(20,30)	24782	4680
GAIA Permanent, HDR(20,30)	25440	1776

The G.A.I.A. was used in addition to following selection methods: Roulette; Tournament; Hawk-Dove Roulette C=20,V=30 and C=30,V=20; and Hawk-Dove Tournament C=20,V=30 and C=30,V=20. Therefore, several tests were run (the most significant are presented in Table 2), considering the same parameters at the start of the execution:

- Same initial population;
- Number of individuals: 100 for all generations;
- Number of generations: 5000;
- Initial crossover rate: 100%;
- Initial mutation rate: 0%;
- Number of matches (for the methods involving hybrid evolutionary games [2]): 100 per generation.

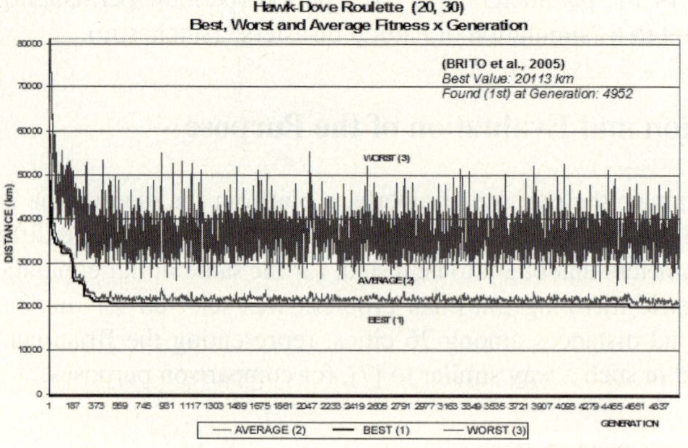

Fig. 3. Execution Hawk-Dove Roulette, C=20, V=30

If one analyses in detail the results that were obtained, especially if placed under the light of the desired objectives, there is noticeable evidence that the agent leads to an increase of the variability of the population, as it forces the elimination of some individuals, given the degree of duplication in the population. It also avoids the stagnation of the optimal value, regulating it accordingly through the tuning of the crossover and mutation parameters.

Fig. 3, extracted from [2], shows an execution with the same parameters applied in the G.A.I.A. tests, except the crossover and mutation rates, which are static throughout the whole execution. It reveals a quick evolution and a somewhat premature convergence, as in a generation close to number 500 a "best fitness value" is found and persists virtually until the end of the execution (generation 5000), when a mutation leads to the best optimum.

When G.A.I.A. is applied, as illustrated in Fig. 4, the evolution is considerably different. This particular graphic is from the execution that produced the best overall result, when the agent intervened each 100 generations. It is noticeable that there are significant changes (leaps) each time the agent is activated, trying to escape from stagnation, whether for the best value or the average. The population evolves to a supposed local optimum and starts to stagnate until a new intervention from the agent renews the parameters and replaces some individuals for new ones, possibly reintroducing lost routes in the population, and enabling it to produce individuals with a better fitness value.

Fig. 4 also shows that the execution could well be run without a limit for its maximum number of generations. This threshold for ending the execution, e.g. a target acceptable optimal value, could be defined by the user, so as this value is reached the execution would be halted.

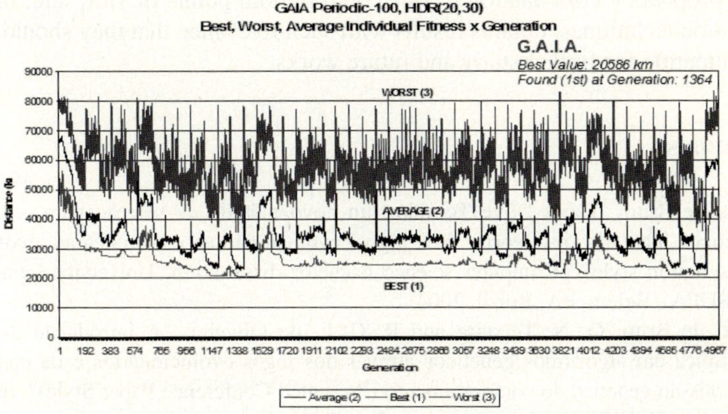

Fig. 4. Execution Hawk-Dove Roulette, C=20, V=30, G.A.I.A. Periodic-100

6 Final Remarks

The results obtained were mostly very satisfactory, as the main purpose was to obtain an optimal value, minding to preserve the variability of the population.

In the tests run using the problem of [7] as a model, the best values found were always close to the best ever found (20093 km in [11]). Besides that, from a qualitative point of view, the majority of results were superior from other works on this same problem, as it was guaranteed a broader and more diversified exploration of the search surface, noticeable by the analysis of the variability of the individuals throughout the executions.

The main difficulties in modeling and building the agent had to do with the relative subjectivity of mapping real values to fuzzy values, especially because these are not firmly supported by the literature. Instead, the construction of the rule base had to sit essentially upon empirical knowledge.

The introduction of the role of an "agent" that regulates the evolution of a population according to its own rules opens new possibilities of search in the area. New values for the parameters on the case of study can be tried, including new techniques to control the rate of surviving individuals. There is also room for definition of new input and output variables, as presented in section 4.

It was noticed that different adjustments in the configurations, such as changes in the numeric values correspondent to the fuzzy variables, can favor one or another mode of execution of the regulating agent, periodic or stochastic.

Among the modes of G.A.I.A., the permanent one tended to present the worst results, which may lead to the conclusion that its influence in the evolutionary process may be too forceful, which somehow affects the process, covering it up.

The success achieved in the implementation of an intelligent agent controlling the evolutionary process is somewhat similar to the controversial approach of the Intelligent Design Theory [14], which is defended by many scientists as an answer to several aspects that are not well explained by the Neodarwinist Theory.

The present work is not assuming as truth either of the aforementioned theories. Instead, it proposes a combination of aspects from both points of view and, in terms of optimization techniques, obtains results with such relevance that they should receive a special attention for further study and future works.

References

1. F. H. de Brito, "Hawk-Dove Torneio: um novo método de seleção para os algoritmos genéticos baseado na teoria dos jogos evolucionários e estratégias evolucionárias (Dissertation style)," Computer science bachelor dissertation, University Center of Pará (CESUPA), Belém, PA, Brazil, 2004.
2. F. H. de Brito, O. N. Teixeira and R. C. L. de Oliveira, "A introdução da interação fenotípica em algoritmos genéticos através dos jogos evolucionários e da codificação e transmissão genética do comportamento (Presented Conference Paper Style)", presented at VII SBAI/II IEEE LARS: Simpósio Brasileiro de Automação Inteligente / II Latin-American IEEE Robotics Symposium, São Luís, MA, September 18-23 2005.
3. DNER. Departamento Nacional de Estradas e Rodagens. November 01 2004. Online. Internet. Available WWW: http://www.dnit.gov.br/rodovias/distancias/distancias.asp.
4. R. Eberhart, P. Simpson and R. Dobbins, Computational intelligence PC tools: an indispensable resource for the latest in fuzzy logic, neural network and evolutionary computing (Book Style). American Press Inc., 1996.
5. FuzzyJ Toolkit and FuzzyJess. October 14 2005. Online. Internet. Available WWW: http://www.iit.nrc.ca/IR_public/fuzzy/fuzzyJToolkit2.html.
6. R. L Haupt and S. E. Haupt, Practical genetic algorithms (Book Style). 2.ed. John Wiley & Sons, Inc., 2004.
7. C. Lehrer. "Operador de seleção para algoritmos genéticos baseado no jogo hawk-dove (Dissertation Style)," M.S. Dissertation, Federal University of Santa Catarina, Florianópolis, SC, Brazil, 2000.
8. M. Mitchell, An introduction to genetic algorithms (Book Style). MIT Press, 1999.
9. I. S. Shaw and M. G. Simões, Controle e modelagem fuzzy (Book Style). São Paulo: Edgard Blücher, 1999.
10. J. M. Smith, Evolution and the theory of game (Book Style). Cambridge University, 1982.
11. O. N. Teixeira, "Computação evolucionária: dos aspectos filosóficos à implementação dos algoritmos genéticos na solução do problema do caixeiro viajante simétrico (Dissertation Style)," Computer science bachelor dissertation, Dept. Inf., Federal University of Pará, Belém, PA, Brazil, 2004.
12. L-X. Wang, A course in fuzzy system and control (Book Style). Prentice-Hall International, 1997.
13. De Jong, K. A. “An Analysis of the Behavior of a Class of Genetic Adaptive Systems (Ph.D. Thesis Style)”. University of Michigan, Ann Arbor, 1975.
14. Intelligent Desing and Evolution Awareness Center. January 20 2006. Online. Internet. Available WWW: http://www.ideacenter.org/.

An Interactive Preference-Weight Genetic Algorithm for Multi-criterion Satisficing Optimization

Ye Tao[1,2], Hong-Zhong Huang[3], and Bo Yang[3]

[1] Key Lab. for Precision and Non-traditional Machining Technol. of Ministry of Education, Dalian University of Technology, Dalian 116023, P.R. China
[2] School of Information Engineering, Dalian Fisheries University, Dalian 116023, P.R. China
taoye18@163.com
[3] School of Mechatronics Eng., University of Electronic Science and Technology of China, Chengdu, Sichuan 610054, P.R. China
hzhuang@uestc.edu.cn

Abstract. After review of several weight approaches, this paper proposes an interactive preference-weight method for Genetic Algorithm (GA). Decision makers (DMs) can pre-select a few feasible solutions, arrange these sample points based on their binary relations, and then design satisfaction degree ratio as the adaptive feasible regions. Through minimizing the weighted L_p-norm of the most satisfactory and unsatisfactory points, DMs can obtain inaccurate weight information for multi-criterion satisficing optimization in current population, and use it to formulate evaluation function as the preferred optimization direction for Pareto GA. Finally, DMs can acquire the corresponding optimal satisficing solution. The optimization of two-bar plane truss is used as an example to illustrate the proposed method.

1 Introduction

Most practical design problems are often formulated as a multi-objective optimization (MOP) problem, where multiple and conflicting objectives are concerned. In MOP, the complete optimal solution is difficult, and sometimes impossible to obtain. The ultimate goal of satisficing optimization is to select the satisfactory solutions instead of the traditional optimal ones. Within the satisficing design stage, the satisfaction degree of each objective will be developed, which is called multi-criterion satisficing problem. Just same as MOP, in satisficing optimization, it is important for decision makers (DMs) to find the proper tradeoff to evaluate and select the best satisfactory alternatives.

There are two major kinds of approaches to evaluate multi-criterion satisficing solution, namely, generating approaches and preference-based approaches. The first is to find the satisficing set of Pareto solutions [1]. However, when the problem involves large number of objectives, it is very difficult for DMs to handle the extremely excessive dimension of tradeoffs between the satisfaction degrees of any two objectives. The other is to transform multi-objective to single-objective problem through the weighted-sum approach [2]. This method requires the same precondition to assign

L. Jiao et al. (Eds.): ICNC 2006, Part I, LNCS 4221, pp. 643 – 652, 2006.

specified weights to satisfaction degree functions of each objective. In this case, the solution is Pareto-optimal, and DM can adjust weights until more satisfactory solution is obtained. Considering simplicity and comparability in practice, the weighted-sum approach is widely applied for multi-criterion satisficing problem.

The weight coefficients represent the preferences of object satisfaction degree. However, the selection of appropriate and exact weight values is very complicated and difficult, mainly depending on the experiment and knowledge of DM. Recently, in MOP, several weight methods are proposed. For example, with conventional method, which is also called fixed-weight approach, weight coefficients are fixed in the whole evolution of Genetic Algorithm (GA). The major disadvantage of this method is that DM must preset proper weight values. Murata *et al.* [3] proposed the random-weight approach for MOP in GA. This approach assigns each individual with the equally selected opportunity to reach the Pareto frontier, and ignores the useful preference information in each generation. Strictly speaking, random-weight approach belongs to generating method in conception. Zheng *et al.* [4] presented an adaptive weight approach. In this method, weight coefficients may utilize some information in the population to adjust themselves adaptively, and acquire the direction to ideal point. However, aforementioned several methods cannot completely reflect the preference of DM at each generation of GA. In many situations, it is difficult for DM to determine the right preference relation among different objectives at the beginning of optimization. It means that DM possibly needs to analyze and compare the optimal results after each or several generations in GA, and interactively modifies the weights, which can effectively guarantee GA to evolve toward the direction preferred by DM. In order to realize the above idea, Wang *et al.* [5] proposed an interactive GA method. Unfortunately, the binary relation of preference structure is lack of integrity. The proper definition for $\prec\!\prec$ and $\succ\!\succ$ are not presented. In addition, if there is no threshold between any two different preference relations, it is difficult for DM to compare and classify the sample points using the results of the population in advance for practical issues.

This paper proposes a classification model to evaluate two points in GA, and used it to construct the feasible region for nonlinear optimization. The rest of the paper is organized as follows. Section 2 gives the definition of satisfcing theory and the formulation of multi-criterion satisficing optimization. Section 3 designs binary relations between the sample points, and the L_p-norm model to obtain the optimal weights. Section 4 presents the procedure of interactive preference-weight GA. Section 5 gives an application example to illustrate the proposed method. Conclusions are given in Section 6.

2 Satisficing Theory

In 1947, Simon, awarded Nobel Prize, first introduced satisficing criterion and proposed the satisfactory solutions in place of the traditional optimal ones in some situations, which provides a new approach to solve optimization problems [6]. Later,

Takatsu presented the basic mathematical theory and characteristics [7]. Goodrich studied the theory of satisficing control, and applied it to some classical control problems [8]. Jin gave the satisfaction solution theory of neural computing [9]. Satisficing theory presents powerful potential in engineering fields.

Definition 1 [10]. Given a feasible solution set *X*, $X \subseteq \mathbf{R}^n$, a multivariate function can be defined as follows:

$$\left.\begin{array}{l} f: X \to Q \\ q = f(x) \in Q, x \in X \end{array}\right\}. \tag{1}$$

The function *f* is called quality criterion function. *Q* is the quality set, $Q \in \mathbf{R}^m$, and *q* is used to describe the quality of solution *x*, $x \in X$.

Definition 2 [11]. Given a feasible solution set *X*, $X \subseteq \mathbf{R}^n$, and quality set *Q*, a mapping function *h*(.) can be defined as follows:

$$\left.\begin{array}{l} h: Q \to [0,1]^m \\ s_{\tilde{F}}(x) = h(q) = h(f(x)) \\ \forall x \in X, q \in Q, s_{\tilde{F}}(x) \in [0,1]^m \end{array}\right\}, \tag{2}$$

where $\tilde{F}$ means satisfactory solution set of *X* for *f*; $s_{\tilde{F}}(.)$ is the satisfactory function of $\tilde{F}$, and $s_{\tilde{F}}(x)$ is called the satisfaction degree of *x*, $x \in X$. For simplicity, the satisfactory solution set $\tilde{F}$ is often denoted by {*X*, *f*, *h*}. *f* and *h* represent quality criterion mapping and satisfactory mapping, respectively. In application, *X*, *f*, and *h* are often called three fundamental elements in satisfactory solution set. If there exists multiple quality criterion mappings or multiple satisfactory mappings, $\tilde{F}$ may be denoted as {*X*, *F*, *h*} or {*X*, *F*, *H*}.

2.1 Multi-criterion Satisficing Optimization

Let us assume that the vector of decision variables (design parameters) is as follows:

$$X = \{(x_1, x_2, ..., x_n) \mid x_i \in \mathbf{R}, i = 1,2,...,n\}, x \in X \subseteq \mathbf{R}^n. \tag{3}$$

The quality criterion set is denoted by:

$$Q = \{q_1, q_2, ..., q_m\}, Q \in \mathbf{R}^m, q \in Q, \tag{4}$$

where $q_k = f_k(x), k = 1,2,...,m$. Given a satisfaction degree function $h_k : \mathbf{R} \to [0,1]$, $s_k = h_k(q_k), k = 1,2,...,m$, the vector of satisfaction degree function is depicted by :

$$S = (s_1, s_2, ..., s_m) = (h_1(q_1), h_2(q_2), ..., h_m(q_m)), S \subseteq [0,1]^m. \tag{5}$$

Define the overall satisfaction degree function $f_{\mathrm{op}} : [0,1]^m \to [0,1]$. In the practical design, f_{op} may be formulated as several styles depending on the problem's properties.

In this paper, a linear weighted method is adopted to describe the overall satisfaction degree as follows:

$$sw = f_{\mathrm{op}}(s) = \sum_{i=1}^{m} \omega_i s_i \ . \ \omega_i \in [0,1], \sum_{i=1}^{m} \omega_i = 1 \,. \tag{6}$$

Therefore, the multi-criterion satisficing is formulated by the constraint model Eq. 7 [12]. We will focus on Eq. 7, i.e. the overall satisfaction degree and its weights for multi-criterion optimization in GA in this paper.

$$\left.\begin{array}{ll} \text{Max} & sw = f_{\mathrm{op}}(s) \\ & s = h(q) \\ & q = f(x) \\ & x \in X \subseteq \mathbf{R}^n, q \in Q \subseteq \mathbf{R}^m \\ \text{s.t.} & g_i(x) \le 0 \quad i = 1,2,\ldots,k \end{array}\right\} . \tag{7}$$

3 Preference-Weight Genetic Approach for Multi-criterion Satisficing Optimization

3.1 Individual Binary Relation

In the population of GA, due to the different preference relations among individuals, DM pays a specific attention to them. Therefore, this paper uses the symbols shown in Table 1 to describe the binary relation between two points.

Table 1. Binary relations and symbols

Symbol	Meanings	The threshold condition
$\approx$	Equal	$\mid d \mid < \varepsilon$
$\succ$	Superior	$\varepsilon \le d < \alpha$
$\succ\succ$	More superior	$\alpha \le d < \beta$
$\prec$	Inferior	$-\alpha < d \le \varepsilon$
$\prec\prec$	More inferior	$-\beta < d \le -\alpha$

In Table 1, d represents the difference of overall satisfaction degrees between two sample points chosen by DM. If taking the sample points of x_k , x_{k-1} , $x_k, x_{k-1} \in X$ as example, the meaning of d_k is depicted by:

$$d_k = sw_k - sw_{k-1} = \sum_{i=1}^{m} \omega_i s_{ki} - \sum_{i=1}^{m} \omega_i s_{(k-1)i} \,. \tag{8}$$

d_k reflects the preference degree between two sample points. If d_k is a positive number, it means the solution x_k is superior to x_{k-1}. Reverse, it means the solution x_k is inferior to x_{k-1}. If d_k is close to zero, it means that x_k is equal to x_{k-1}. Considering the

the symmetric relation between positive and negative ones, we only study the design problem for $d_k \geqslant 0$ in this paper.

Note also that the 3rd column in Table 1 lists every symbol and its relative threshold condition, which can be defined as the rules to determine the preference degree of two points. ε is a very small positive number, such as, 10^{-5}, which is used as the upper limit of $|d|$ for the relation of $\approx$. $\alpha, \beta\,(0 < \varepsilon < \alpha < \beta < 1)$ are also defined as the upper limits for the relation of $\succ$ and $\succ\succ$, respectively. In fact, they are measures of the preference intensity about different sample points. Figure 1 gives a 3D plot using three sample points and two kinds of binary relation, $\succ$ and $\succ\succ$.

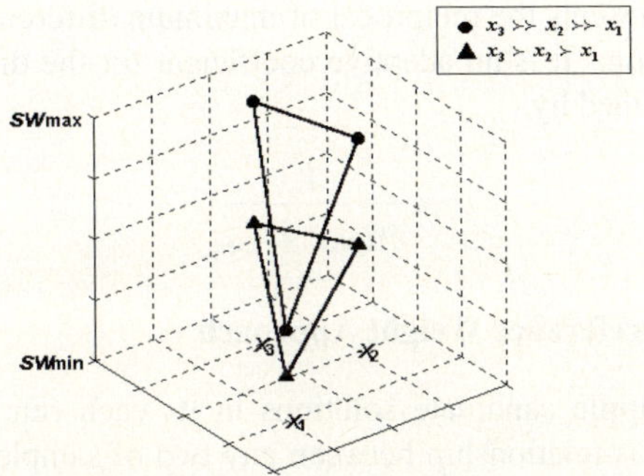

Fig. 1. The comparison between two kinds of relations about three sample points

If we observe the tendency of overall satisfaction degree of three sample points, obviously, under the binary relation $\succ\succ$, the larger preferred tendency can be obtained. In practical application, DM can use α, β to control preference degree, and they may be initiated through DM's knowledge or expert's experience in advance.

3.2 Weight Optimization Using Preference Feasible Region

According to the features of aforementioned weighted optimization, GA needs to get the proper weights to guide each generation towards the preferred evolution direction. Consequently, it is necessary to transform the preference of some limit sample points into the preference of weight coefficients for overall satisfaction degree. In order to cope with such situation, the following optimization method is introduced.

Assume that L is the sample points set chosen by DM and l is the set length, i.e.,

$$L = \{x_k, x_{k-1}, \ldots, x_{k-l-1} \mid x_i \in \mathbf{R}^n, i = k, k-1, \ldots, k-l-1\}\,, \tag{9}$$

where the preference relation of these sample set L is :

$$x_k \approx x_{k-1} \succ\succ x_{k-2} \succ \ldots \succ x_{k-l-1}\,. \tag{10}$$

Regarding Eq.10 as the preference feasible region, and using the weighed L_p-norm of the satisfaction degree for x_k, x_{k-l-1} as the objective, the nonlinear programming model is defined as:

$$\left.\begin{array}{ll} \text{Find} & (\omega_1, \omega_2, \ldots, \omega_m) \\ \text{Min} & r(s_k; p, \omega) = \| s_k - s_{k-l-1} \|_{p,\omega} = (\sum_{i=1}^{m} \omega_i^p \mid s_{ki} - s_{(k-l-1),i} \mid^p)^{1/p} \\ \text{s.t.} & g_1: \ \mid cd_k \mid < \varepsilon \\ & g_2: \ \alpha \le cd_{k-1} < \beta \\ & g_3: \ \varepsilon \le cd_{k-2} < \alpha \\ & \quad \vdots \\ & g_l: \ \varepsilon \le cd_{k-l} < \alpha \end{array}\right\} . \tag{11}$$

In Eq. 11, c represents the reciprocal of maximum difference among the population at current generation. It is an adaptive coefficient for the difference of overall satisfaction degree, defined by:

$$c = \frac{1}{sw_{\max} - sw_{\min}} . \tag{12}$$

3.3 Interactive Preference Weight Approach

GA can offer multiple candidate solutions in its each run, so variable satisfaction information, such as relationship between any two of sample points can be obtained. DM chooses some distinct sample points to form the sample set L, and according to their binary relations based on the information of DM's preference, the preference feasible region may be constructed correspondingly. After solving Eq.11, the nonlinear programming problem, the DM can find the optimal solution used as the weight coefficients of every satisfaction degree objective, and design the new evaluation function. This same procedure is repeated in GA's iterations. By comparing the series of sample points in population, DM interactively modifies the weight coefficients and evaluation value, until he or she reaches a satisficing solution. Therefore, the proposed approach consists of two main stages: solving nonlinear programming and evaluation value reconstruction. The individuals in GA, with most preferred information by DM, can be preferentially selected and crossed over, which effectively avoids the disadvantages resulted from fixed and random weighed-methods.

In the first run, DM may create the initial weight coefficients roughly. After several generation and interactive optimization, the population is gradually added with the DM's inaccurate preference information as the search direction of GA, and the weights can be refined furthermore, until GA evolves and produces satisficing individuals.

4 The Search by GA

In order to search the Pareto-optimal solutions by GA, it is necessary to find the whole Pareto solution set for the optimization problem, and distribute them on the Pareto frontier as widely as possible, which can ensure that DM performs the tradeoff among the solutions on the frontier. Goldberg [13] has presented the well-known Pareto-based ranking method and fitness assignment, named Pareto GA method, and made it more possible to find the non-dominated solutions from the population.

The proposed method in this paper, which combines Pareto GA with preference weight (direction) method is extremely appealing to DM. Firstly, through Pareto GA, Pareto-optimal solution set can be easily obtained. Secondly, the overall satisfaction degree of individuals using the optimal weight coefficients tends to move towards the preferred direction, as shown in Fig.2. When the optimization ends, DM can directly acquire preferred Pareto solutions.

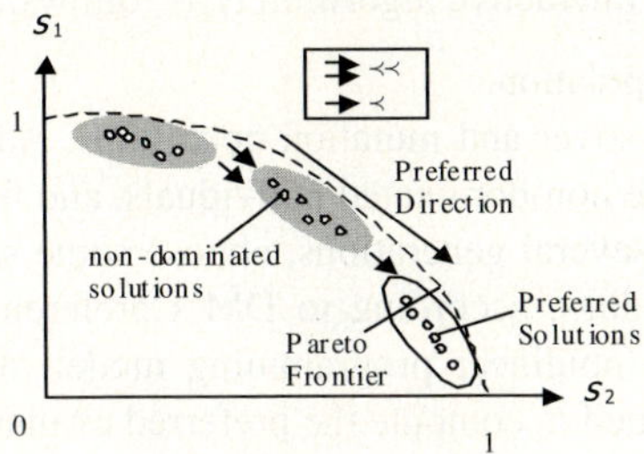

Fig. 2. The direction of movement determined by DM

We employ the selection and sharing method proposed by Kiyota *et al.* [14] in this paper. This method is composed of two kinds of selections: Pareto selection and tournament selection. Pareto selection's main job is to select all the non-dominated in a population at each generation. The latter can compare two individuals randomly chosen from population and find the winner according to a certain rule. Define the parameter R_{sw} for two individuals *i* and *j* as follows:

$$R_{sw} = \frac{1}{1+\exp(b(c-z))}, \tag{13}$$

where

$$z = \frac{1+sw_i}{1+sw_j} \quad (sw_i \le sw_j). \tag{14}$$

b and *c* are the shape coefficients of R_{sw}. For the diversity of individuals in the population, the niche technique is applied to the tournament selection. The following power law function is commonly used by the sharing function [13].

$$\mathrm{Sh}(d_{ij}) = \max\left\{0, 1-\left(\frac{d_{ij}}{\sigma_{\text{share}}}\right)^{\alpha}\right\}, \tag{15}$$

where σ_{share} is the niche radius. d_{ij} represents the distance between two individuals *i* and *j*, shown as Eq.16. The sharing function Sh(.) is maintained as a measure of d_{ij}.

$$d_{ij} = d(s(x_i), s(x_j)). \tag{16}$$

The niche count m_i of the *i*th individual is the sum of sharing function values of all individuals in the population, that is:

$$m_i = \sum_{j=1}^{\text{pop\_size}} \mathrm{Sh}(d_{ij}). \tag{17}$$

Then, the strategy of tournament selection is described as follows [14]:

1. Generate a uniform random number $R(0 \le R \le 1)$, and choose randomly two individuals.
2. If $R_{sw} \le R$, then select the individual with the larger sw_i as the winner.
3. If $R_{sw} > R$, then select the individual with the smaller m_i as the winner.

The algorithm of the interactive algorithm is as follows:

Step 1. Initiate the population.
Step 2. Apply the crossover and mutation operations. Adopt Pareto-optimal ranking method to acquire the non-dominated individuals, and fill them into the population.
Step 3. After each or several generations, choose some sample points with different satisficing fitness values, according to DM's preference, design the binary relations, and solve the nonlinear programming model of Eq.11. Then, the optimal weights can be obtained to compute the preferred evaluation function *sw*.
Step 4. Apply the tournament selection and sharing approach to select new individuals into population, and produce the next generation. If DM cannot find the different preferred individuals, then algorithm ends; else go to step 2.

5 A Case Study

In this paper, the proposed method is applied to a practical example, multi-criterion satisficing optimization for two-bar plane truss problem [15].

Two bars with completely same dimensions locate in the symmetrical position about axis *y*. Design variables are $x_1 = d/h, x_2 = A/A_0$, where *A* is cross section area of each bar. The constraint conditions are stress constraints, $\sigma_i \le [\sigma_0]$, *i*=1,2, and the design variables need to satisfy the relative border conditions. There are two satisficing criterions, namely, the mass of bar and displacement at the 3rd joint point. The corresponding formulation is denoted as:

$$\left.\begin{aligned} &\text{Min } f_1(x) = ph\sqrt{1+x_1^2+x_2^4+x_1^6}\,/(\sqrt{2}Ex_1x_2A_0) \\ &\text{Min } f_2(x) = 2\rho hx_2A_0\sqrt{1+x_1^2} \\ &\text{s.t.}\quad p(1+x_1)\sqrt{(1+x_1^2)}\,/(2\sqrt{2}x_1x_2A_0\rho_0) \le 1 \\ &\qquad p(x_1-1)\sqrt{(1+x_1^2)}\,/(2\sqrt{2}x_1x_2A_0\rho_0) \le 1 \\ &\qquad 0.01 \le x_1 \le 2.5 \quad x_2 \in \{x \mid x = 0.5+0.1k, k=0,1,\dots,20\} \end{aligned}\right\}, \tag{18}$$

where x_1 is a continuous variable and x_2 is a discrete variable. E=2.1×10^{5}MPa, ρ=7.8×10^{-5}N/mm^3, P=10^{4}N, A_0=100mm^2, σ_0=235MPa. The satisfaction degree function of f_1 and f_2 is:

$$s_i = (f_{i\text{Max}} - f_i)/(f_{i\text{Max}} - f_{i\text{Min}}), i = 1,2\,. \tag{19}$$

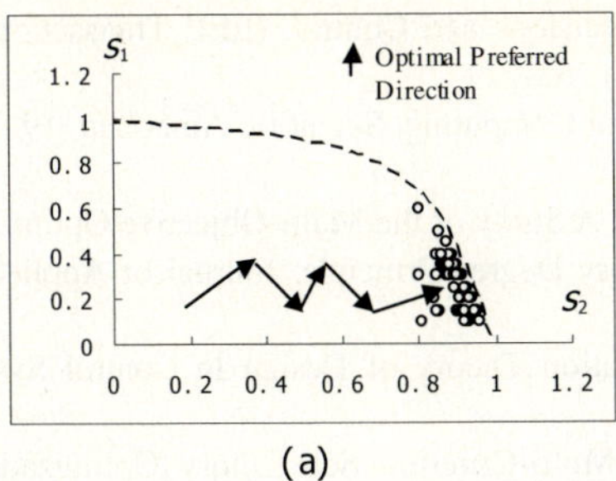

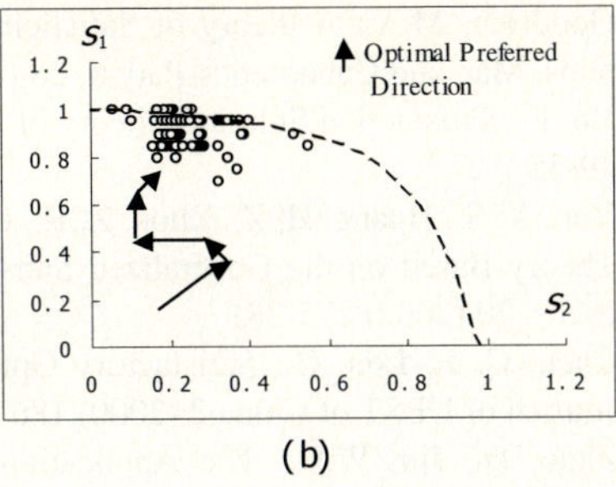

Fig. 3. Distribution of satisficing solutions after using five interactive preferred directions

Suppose that there are two kinds of preference directions, DM chooses three sample points to compute the optimal weights ω_1, ω_2 every ten generations to rebuild the evaluation function *sw*. After fifty generations, distribution of every individual's satisfaction degree s_1, s_2 in a population are shown in Fig.3. Note that in two situations, GA can reach the Pareto frontier closely, and at the same time, the movement direction of individuals in final population is consistent with the preferred by DM, which can ensure DM performs tradeoff more conveniently and effectively.

6 Conclusions

This paper introduces an interactive preference-weight approach for GA and applies it to satisficing optimization. According to arrangement of several sample points in the population at generations, DM may control and adjust the evolution direction, which ensures that the final satisficing solution is of non-dominated and preferred characteristics simultaneously. It also can effectively reduce the difficulties for tradeoff by DM. A design example was presented to illustrate the effectiveness of the proposed method.

References

1. Stadler, W.: A Survey of Multi Criteria Optimization or the Vector Maximization Problem: I .1776-1960. Journal of Optimization Theory and Applications.69 (1979) 1-52
2. Zadeh, L.: Optimality and Non-scalar-valued Performance Criteria. IEEE Transactions on Automatic Control. 8 (1963) 59
3. Murata, T., Ishibuchi, H., Tanaka, H.: Multi-objective Genetic Algorithm and its Application to Flow Shop Scheduling. Computers and Industrial Engineering. 30 (1996) 957-968
4. Zheng,. D., Gen, M., Cheng, R.: Multi objective Optimization Using Genetic Algorithms. Engineering Valuation and Cost Analysis. 2 (1999) 303-310
5. Wang, X. P., Cao, L. M.: Genetic Algorthm. Xi'an Jiaotong University Press, Xi'an (2002)
6. Simon, H A.: The New Science of Management Decision. China Social Sciences Press, Beijing (1998)
7. Takatsu, S.: Latent satisficing decision criterion. Information Science. 25 (1981) 145-152

8. Goodrich, M.A.: A theory of Satisficing Decisions and Control. IEEE Transaction Systems, Man and Cybernetics-Part A. 28 (1990) 763-779
9. Jin, F.: Satisfaction Solution Theory of Neural Computing. Scientific American. 19 (1992) 49-55
10. Yao, X. S., Huang, H. Z., Zhou, Z. R., et al.: A Study of the Multi-Objective Optimization Theory Based on the Generalized Satisfactory Degree Principle. Journal of Applied Sciences. 20 (2002) 275-281
11. Chen, C. J., Lou, G.: Satisfactory Optimization Theory of Design in Control Systems. Journal of UEST of China.2 (2000) 186-189
12. Zhao, D., Jin, W.D.: The Application of Multi-Criterion Satisfactory Optimization in Fuzzy Controller Design. Proceedings of the EPSA.5 (2003) 64-69
13. Goldberg, D.E.: Genetic Algorithms in Search. Optimization and Machine Learning. Addison Wesley, Reading, Ma, USA, 1999
14. Kiyota, T., Tsuji, Y., Kondo, E.: An Interactive Fuzzy Satisficing Approach Using Genetic Algorithm for Multi-objective Problems. The proceeding of Joint 9th IFSA World Congress and 20th NAFIPS International Conference. Vancouver, BC, Canada. 2(2001): 757-762
15. Zhu, X. J., Pan, D., Wang, N. L., et al.: Multi objective Optimization Design with Mixed-Discrete Variables in Mechanical Engineering via Pareto Genetic Algorithm. Journal of Shanghai Jiaotong University, 3 (2000) 411-414

A Uniform-Design Based Multi-objective Adaptive Genetic Algorithm and Its Application to Automated Design of Electronic Circuits

Shuguang Zhao, Xinquan Lai, and Mingying Zhao

School of Electronic Engineering, Xidian University
Xi'an 710071, China

Abstract. The uniform design technique was integrated into an adaptive genetic algorithm for the sake of optimal design of circuits. The approach proposed features a dynamic evaluation mechanism of multi-objectives, an efficient encoding-decoding scheme based on preferred values, and a classified adaptation strategy of genetic parameters. It was validated by experiments.

1 Introduction

Circuit design is a typical problem of multi-objective optimization. A fitness function in the form of *weighted sum of the objective functions* was frequently used in solving such a problem with an evolutionary algorithm (EA) [1-4]. However, it confines a host EA to just one search direction implied by the weight vector and at most one Pareto-optimal solution per run, preventing the EA from obtaining uniformly distributed solutions. Treating less important objectives as constraints leads to similar faults [2-4]. Considering that multiple Pareto-optimal solutions could be obtained by using specific weight-vectors which suggest multiple search directions uniformly directed at the Pareto-frontier, we have developed a multi-objective adaptive genetic algorithm (UMOAGA) based on the Uniform Design Technique (UDT) [5] and our previous works [1]. The UMOAGA was introduced and verified hereafter.

2 UDT Based Fitness Functions and Crossover Operators

The UDT is a useful technique in experiment design, which can select q representative combinations from q^n combinations resulted from n factors and q levels. As the q combinations are generally related with q points uniformly scattered in the combination space, the components of a weight vector (i.e., weight coefficients) derived from them may have expected proportion relationships with each other in the sense of uniformity and integrality [3]. Thus, to compose m fitness functions for k objective functions, we can form a *uniform design matrix*, $\boldsymbol{U}(k,m)=[\boldsymbol{U}_{i,j}]_{m\times k}$, by looking up the UDT tables [5] and use it in the following way,

$$w_{i,j} = U_{i,j} / \sum_{j=1}^{k} U_{i,j} \quad i=1,\cdots,m, \quad j=1,\cdots,k \tag{1}$$

L. Jiao et al. (Eds.): ICNC 2006, Part I, LNCS 4221, pp. 653 – 656, 2006.

Then, m normalized fitness functions can be composed as follows, each of which will be used to update $1/m$ of the population so as to let the UMOAGA search in m uniformly scattered directions towards the Pareto frontier,

$$fit_i = \sum_{j=1}^{k} w_{i,j} \cdot f_j(X), \quad i = 1, \cdots, m \tag{2}$$

The UDT was also applied to the crossover operation. In contrast with traditional ones imitating zoogamy, the UDT based crossover operators we designed choose and mate multiple individuals to make multiple offspring. For a binary-coded GA whose individuals are encoded as $P_i = b_{i,1} \& b_{i,2} \ldots \& b_{i,L}$, $b_{i,j} \in \{0,1\}$, the multi-parent crossover operation can be briefly explained as follows: (1) After determining the number of parents, m, choose a natural number, $n<m$. Look up the UDT tables to obtain a *uniform design matrix*, $\boldsymbol{U}(n,m) = [U_{i,j}]_{m\times n}$. (2) Randomly divide the set of individuals′ subscript into n subsets, S_i, which satisfies $S_i \cap S_j = \phi$ for $i \neq j$ and $\cup_i S_i = S$, $i=1\ldots n$. (3) Randomly choose m parents in a proper manner. Sort them in a degressive order of fitness. Relate each of them with its sequence number ranges from 1 to m. (4) Iteratively produce m offspring with a exclusive row of $\boldsymbol{U}(n,m)$ each time. For the ith offspring, cope the jth partition of the parent indicated by the element $\boldsymbol{U}_{i,j}$ as its jth partition, referring to subset S_j that prompts the subscripts involved.

As to a real-coded or integer-coded GA, the multi-parent crossover operation is similar to that mentioned above. But each of m offspring is produced by computing a weighted sum of n parents on the basis of $\boldsymbol{U}(n,m)$, as shown in Equation (3).

$$X_i' = \sum_{j=1}^{n} (U_{i,j} \bullet X_j) / \sum_{j=1}^{n} U_{i,j} \quad i = 1, \cdots, m \tag{3}$$

3 Representing Scheme and Parameters Adaptation Strategy

Based on a few restrictions on circuit construction other than a fixed topological structure, an individual in the UMOAGA is encoded in a format like a net-list, e.g.,

$$C_i = [type_i, node1_i, node2_i, walue_i] \tag{4}$$

Where $type_i$, $node1_i$, $node2_i$ and $value_i$ denote the type, the (two) nodes connected and the value of the ith component, respectively. The components′ values are deliberately encoded with *preferred values* (e.g., a 1% precision series using just 96 numbers to span a value interval of one decade) commonly used with discrete components, in order to get a shorter chromosome (e.g., 9 bits other than 18 bits for a component value between 1 and 10^5) and accordingly less computation amount and more effective results. Moreover, the crossover probability P_c and the mutation probability P_m are allowed to vary with the individual diversity, which is estimated as

$$f_d(t) = fit_{avg}(t) \;/\; [\,\varepsilon + fit_{max}(t) - fit_{min}(t)] \tag{5}$$

Where $fit_{avg}(t)$, $fit_{max}(t)$ and $fit_{min}(t)$ are the average, maximum and minimum fitness of the individuals, respectively. The P_c is allowed to adapt in the following way

$$P_c(t) = P_{c0} \cdot \exp(-b_1 \cdot t / t_m) \;/\; f_d(t) \tag{6}$$

Because genes of different categories usually have distinct effects on the circuit performances, two distinct adaptation rules are designed for the P_m as follows

$$P_{ms}(t)=P_{ms0}\cdot \exp(-b_2\cdot t/t_m)\cdot f_d(t) \tag{7}$$

$$P_{mv}(t)=\begin{cases}0 & t<t_0\\ P_{mv0}\cdot[1-e^{-b_3\cdot(t-t_0)/t_{\max}}]\cdot f_d(t) & t_0\le t<t_1\\ P_{mv0}\cdot[e^{-b_3\cdot(t-t_1)/t_{\max}}-e^{-b_3\cdot(t-t_0)/t_{\max}}]\cdot f_d(t) & t_1\le t<t_m\end{cases} \tag{8}$$

Where $P_{mv}(t)$ is assigned to those related to component values (i.e., $value_i$), $P_{ms}(t)$ is assigned to others, t_m is the allowed generation number, and the other terms are positive constants. With the above multi-stage adaptation strategy, the UMOAGA can be expected to generate and optimize the circuits automatically.

4 Experimental Results and Conclusions

With the UMOAGA, some experiments on active filters were completed successfully, using Pspice simulation based fitness evaluation [1]. For example, an experiment on low-pass active filters was performed with the following restriction to the circuit scale: one operational amplifier, no more than 8 nodes (as shown in Fig. 1) and no more than 15 resistors or capacitors. The four objective functions to be minimized are: relative error of transition frequency, $f_1=|f_c - f_{co}|/f_{co}$, where f_{co} and f_c are the

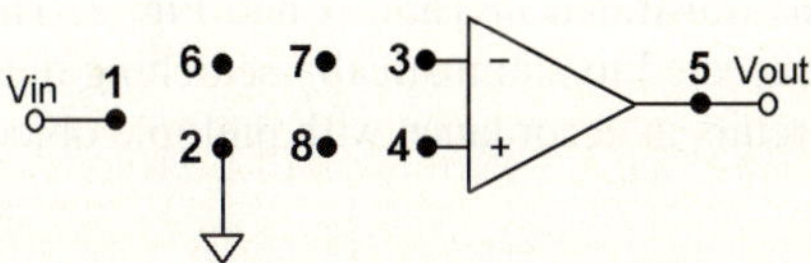

Fig. 1. The circuit framework used in the experiments on active filters

Table 1. Some experimental results on a low-pass active filter (f_{co}=2kHz, G_0=20dB, G_s=-60dB)

No.	Simplified chromosomes of the evolved circuits	Objective values
1	[R,1,6,2.20K], [C,6,2,51.0nF], [R,3,6,22.0K], [C,5,3,200pF], [R,5,6,22.0K]	$f_1=0.13, f_2=10.5$ $f_3=0.05$, $\mathbf{f_4=6}$
2	[R,1,6,1.20K], [C,2,6,62.0nF], [R,6,3,5.10K], [C,3,5,22.0nF], [R,5,6,5.10K], [R,4,2,5.10K], [R,4,5,51.0K]	$\mathbf{f_1=0.01}$, $f_2=8.2$ $f_3=0.11, f_4=8$
3	[R,1,6,330], [C,2,6,1.0uF], [R,6,7,750], [C,7,2,43.0nF], [C,3,5,1.0nF], [R,3,7,5.10K], [R,7,5,5.10K], [R,4,5,47.0K], [R,4,2,4.70K]	$f_1=0.07, f_2=3.7$ $\mathbf{f_3=0.02}$, $f_4=10$
4	[R,1,7,2.20K], [C,7,2,360nF], [R,7,6,22.0K], [R,6,4,20.0K], [C,4,2,9.10nF], [C,5,6,2.00nF], [R,2,3,22.0K], [R,3,5,200K]	$f_1=0.12$, $\mathbf{f_2=1.1}$ $f_3=0.35, f_4=9$

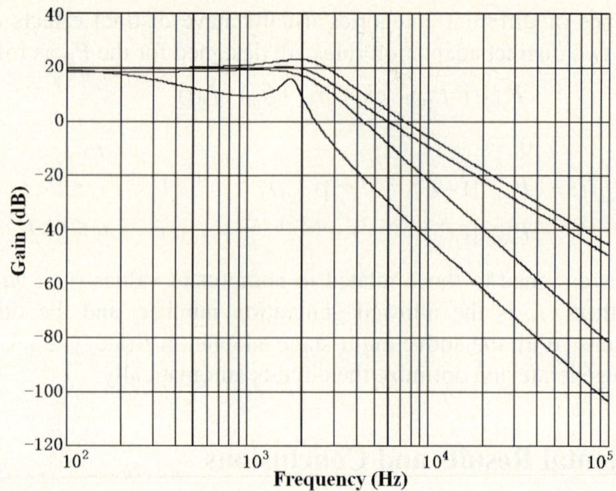

Fig. 2. Amplitude-frequency response curves of the filters listed in Table 1

expected and actual transition frequency, respectively; pass-band undulation, $f_2 = |G_{max} - G_{min}|/G_0$, where G_{max}, G_{min} and G_0 are maximum, minimum and the expected *Gain* in the pass-band, respectively; relative bandwidth of the transition-band, $f_3 = |f_s - f_c|/f_c$, where f_s is the first frequency with the expected attenuation, G_s; circuit complexity, f_4 = *number of the components*. With a uniform design array, ***U***(4,5), a set of effective circuits were evolved, as illustrated in Table 1 and Fig. 2. These results suggest that the UMOAGA can be expected to automatically searching out a set of Pareto-optimal solutions or effective circuits in accordance with multiple objectives.

References

1. Zhao, S.: Study of the evolutionary design methods of electronic circuits. PhD dissertation (in Chinese). Xidian University, Xi′an (2003)
2. Fonseca C. M., Fleming P. J.: An overview of evolutionary algorithms in multi-objective optimization. Evolutionary Computation 1 (1995) 1–16
3. Leung Y.W., Wang Y. P.: Multiobjective programming using uniform design and genetic algorithm. IEEE Trans. on System Man and Cybernetics–Part C 3 (2000) 293–304
4. Coello Coello C. A.: An Updated Survey of GA-Based Multiobjective Optimization Techniques. ACM Computing Surveys 2 (2000) 109–143
5. Fang K. T., Ma C. X.: Orthogonal and Uniform Experiment Design (in Chinese). Science Press, Beijing (2001)

The Research on the Optimal Control Strategy of a Serial Supply Chain Based on GA

Min Huang[1], Jianqin Ding[1], W.H. Ip[2], K.L. Yung[2], Zhonghua Liu[1], and Xingwei Wang[1]

[1] Faculty of Information Science and Engineering, Northeastern University, P.R. China
mhuang@mail.neu.edu.cn

[2] Department of Industrial and Systems Engineering, The Hong Kong Polytechnic University, Hong Kong
mfwhip@polyu.edu.hk

Abstract. Determination of optimal control strategy is one of the key factors for a successful supply chain management. This paper focuses on the research of an inventory control strategy of a serial supply chain. First, it proposes an optimization model of an inventory control that is based on the combination of a nonlinear integer programming model and the generally used push/pull control model. Then, the optimal control strategy is ratified by the fusion of genetic algorithm and simulation analysis. Case studies have demonstrated the effectiveness of the method.

Keywords: Supply chain management (SCM), inventory control, genetic algorithm, push control, pull control.

1 Introduction

Under globalization and the rapid development of computers and information technology, all enterprises face new chances as well as new challenges. This breeds the concept of a serial supply chain, a value-added chain that is composed of a series of enterprises: raw material suppliers, parts suppliers, producers, distributors, retailers, and transportation enterprisers. Clients finally get their products, which are manufactured and handled systematically by the enterprises of the chain, started from either the raw material suppliers or the parts suppliers. This series of activities are the total activities of a complete serial supply chain, that is, from the supplier's suppliers to the clients' clients [1,2].

Supply chain management (SCM) aims at lowering the system cost, enhancing the product quality, and improving service level by collaborating and controlling the conduct of each entities of the supply chain. The goal is to upgrade the overall competitive ability of the whole system. Hence, the inventory management of the serial supply chain is important. The inventory control strategy of an enterprise affects the cost directly and the revenue indirectly. Therefore, the target of an optimal inventory is both to actualize the degree of clients' satisfaction and to minimize the overall cost [3].

L. Jiao et al. (Eds.): ICNC 2006, Part I, LNCS 4221, pp. 657 – 665, 2006.

Inventory makes the supplying, producing, and selling systems of an enterprise become discrete sections. Each section operates independently. This helps to enlighten the blow that comes from the deviation of demand forecast, and makes good use of resources when turmoil happened due to demand changes and market changes. On the other hand, capital is needed for setting up an inventory. The cost includes the capital used for inventory and products in process, the space used for inventory, the expenses on management, maintenance, and discarding of defected products. Inappropriate inventory management even affects the operation efficacy of the enterprise.

The characteristic of indetermination of a supply chain increases the overall inventory of the whole chain system. It also brings unnecessary cost to the node enterprises of the supply chain. In order to avoid the "bullwhip effect" caused by the indetermination of demand and supply, the traditional inventory strategy has to be revised. Inventory standard of a supply chain can be boosted by strategies like shared technology, contract system, and integrated enterprises. Thus, the competitive ability of a supply chain is strengthened.

Generally speaking, there are two kinds of production/inventory systems: the push and the pull system. The current worldly popular production/inventory control systems of Manufacturing Resource Planning (MRPII) and Just-in-time (JIT) belong to the system, respectively. The push production control system adopts a central control method and organizes production by forecasting the future demand. Therefore, production period is estimated in advance. The pull production control system adopts a dispersed control method and production is organized according to the real demand[4]. Each method has its own advantages [5, 6]. People try to fuse the two methods to attain better performances [7-9], CONWIP (CONstant Work in Process), proposed by Spearman et al. in 1990, is an example of combined push-pull control method[10]. In 2003, Oscar has suggested the model of CONWIP control system for a serial supply chain and also shows the corresponding simulation analysis [11]. But, up till this moment, all researches on control system of supply chain only deal with single specific control strategy like the push system, the pull system, the classic combined push-pull CONWIP control system[10,12-13], or the simple combination of push-pull system. A generally used model of push-pull control strategy and its research is still absent. The aim of the paper is to establish a generally used method of the inventory system of a serial supply chain to replace those traditional classic control systems.

In the systems of Kanban and CONWIP, system performances rely on the billboard quantities. Similarly, in the serial push-pull system, distribution of circulating cards (the number of circulating cards at different stages) determines the control model, guides the production time and production quantity of the generally used system. Therefore, determination of circulating cards becomes a key factor affecting the operating efficacy of a generally used inventory system of a serial supply chain.

This paper proposes an optimal control model that tackles a series of multi-stages of inventory control system of a supply chain. The model is based on the fusion of nonlinear integer programming and the generally used push-pull system of the inventory of a serial supply chain. It determines the distribution of circulating cards by integrating the genetic algorithm and simulation analysis. Both the research and case studies have proved that the results from genetic algorithms are logical and effective.

2 Problem Description

Fig 1 shows that there are N nodes on the whole serial supply chain. Each node represents upstream/downstream node enterprises like raw material suppliers, manufacturers, distributors, retailers, and clients. Since the final target of a supply chain is to satisfy the clients' demands, each enterprise operates its production and sales under the generally used push-pull inventory system control. That is, production of each node enterprise is affected by the raw material supply of the upstream enterprise and the demand of the downstream enterprise. One important goal of the supply chain is to reach an all-win status, to maximize the profits of the whole chain instead of any individual enterprise. In order to control the quantity of products in process of the chain, there is a fixed product standard on the feedback from the demand on upstream enterprise i from downstream enterprise j and marks it as circulating card number K_{ij}^{u}. Only when the real quantity of products in process of node enterprise i is less than the forecast product quantity of each downstream enterprise, then it is allowed to proceed with the manufacture. Once the product of a unit is allowed to be manufactured by node enterprise i, the node's blank circulating card is attached to its manufactured vessel (container), then the product has finished processing it is sent to the next node enterprise i+1. The attached circulating card is detached and returns to node enterprise i as a blank circulating card and authorizes further manufactures of other products. When the value of K_{ij}^{u} is ∞, it means that there is no feedback control on upstream node i from downstream node j. If there is no feedback control on node i from all downstream nodes, then node i is under the push control.

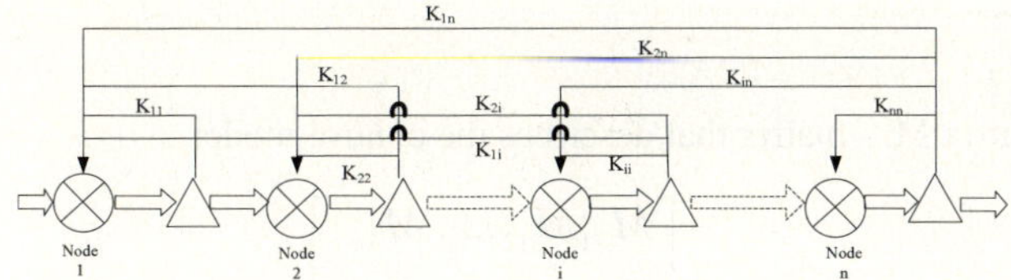

Fig. 1. The push-pull inventory control strategy of a serial supply chain

The commonly used push-pull system has the following properties and assumptions:

1) Clients' demands fall within the upper and lower limits of a normal distribution.
2) The supply chain manufactures only one kind of product.
3) All upstream enterprises can obtain the return circulating cards from any manufacturing node enterprises without delay (no return delay).
4) There is ample raw material supply from the initial node enterprise of the supply chain.

The main problem is how to choose an appropriate control strategy so that the overall cost of the chain is minimum, and at the same time, fulfills the degree of clients' satisfaction as well as falls within the standard deviation of invested orders.

3 Simulation of the Control Strategy

Two definitions are needed for description of the strategy.

Definition 1:

Matrix K of the circulating cards: element K_{ij}^{u} of matrix K represents the circulating card number in the control cycle sent by downstream node j to upstream node i. $K_{ij}^{u} = \infty$ implies that there is no pull control from node j to node i. Since all nodes are under the control of their downstream nodes, so the lower half of matrix K is meaningless. Thus, values of these elements are fixed as –1. That is:

$$K=\begin{bmatrix} K_{11}K_{12}.........K_{1n} \\ \quad K_{22}.....K_{2n} \\ . \\ . \quad .. \quad ... \quad K_{nn} \end{bmatrix} \tag{1}$$

$$K_{ij} = \begin{cases} -1 & (i > j) \\ 1 \sim K_{ij}^{u} \ +1 & (i \le j) \end{cases} \tag{2}$$

where K_{ij}^{u} represents the upper limit of the circulating card number K_{ij}^{u} + 1 represents +∞.

Definition 2:

Control matrix M: - matrix that describes the control model.

$$M=\begin{bmatrix} M_{11}M_{12}......M_{1n} \\ M_{21}M_{22}......M_{2n} \\ . \\ . \\ M_{n1}M_{n2}......M_{nn} \end{bmatrix} \tag{3}$$

$$M_{ij} = \begin{cases} 1 & (K_{ij} \ne \infty) \\ 0 & (K_{ij} = \infty) \end{cases} \quad i \le j \tag{4}$$

Matrix M shows what kind of control strategy each node of the supply chain has. With 0-1 dimension, when element M_{ij} of matrix M equals to 1 means there is pull control on upstream node i from downstream node j. When M_{ij} equals to 0 means that there is no pull control on upstream i from downstream node j.

Property 1: When the sum of all elements of row i of matrix M equals to 0, there is no pull control on node i from all its downstream nodes. Actually, at that moment, node i is under the push control from its upstream node.

Descriptions of a two-tertiary model are presented to cope with the circumstances. The first-tertiary model is a nonlinear integer programming model. It optimizes the distribution of circulating cards and guarantees that the average overall cost of the supply chain of the system is the smallest, under the constraints of the degree of clients' satisfaction F_{r0} and the range of the standard deviation SDO_0 of invested orders. In another word, solve all elements K_{ij} of matrix K.

Target function: $\min(ACR(K))$ (5)

$$st: \begin{cases} F_r(K) \geq F_{r0} & (6) \\ SDO(K) \leq SDO_0 & (7) \end{cases}$$

K_{ij} is the integer between 1 and $K_{ij}^u + 1$ where $i \leq j$ (8)

where K_{ij} represents the upper limit of the circulating card number and $K_{ij} + 1$ represents $+\infty$.The second-tertiary model shows simulation of the inventory system of the whole serial supply chain.

The parameters delivered from the first-tertiary model to the second-tertiary model are circulating card distributions of each node. The parameters delivered from the second-tertiary model to the other model are clients' degree of satisfaction, standard deviation of the invested orders, and average overall cost of the supply chain.

Solutions of the two-tertiary model are handled separately.

4 The Algorithm of the Model

This section shows the solutions of the above model.

4.1 The First-Tertiary Model

The first-tertiary model is a mixture of combination problem and integer programming. If there are n node enterprises, the circulating card number that needed to be determined will be $n(n+1)/2$. When node enterprise i is under the control of node enterprise j in the supply chain and the greatest possible circulating card number is K_{ij}^u, there are $\prod_{i=1}^{n}\prod_{j=i}^{n}(K_{ij}^u + 1)$ states of searching spaces. To process every example of each state is possible only if both the scale of the supply chain and the circulating cards number are small. Heuristic algorithms are needed to solve practical problems.

Genetic algorithm is more suitable for solving optimization problem as it has the characteristics of parallel search, which decrease the probability of fall into a local solution [14]. Therefore, it is selected to solve the first-tertiary model.

4.1.1 Genetic Coding

Integer coding is adopted to fit into the characteristic of the problem. Each individual unit represents the element of the upper triangular part of matrix K accordingly, and there are $n(n+1)/2$ units in total. Fig 2 shows the actual coding procedure.

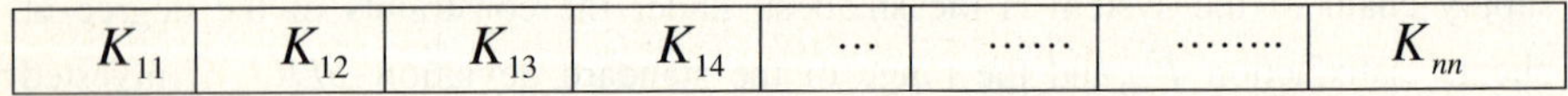

Fig. 2. Coding procedures

The range of each unit's value is from 1 to $k_{ij}^{u}+1$ where the value of K_{ij}^{u} is determined by the ability of each production node. Making $k_{ij}^{u}+1$ of a certain node as an infinite integer means that its downstream nodes have no limit of circulating card number on it. For instance, if the production ability of the first node of a supply chain is 30, then $K_{ij}^{u}\geq 30$. If the production ability of the second node is 20, then K_{ij}^{u} will contain the control cycles of the two nodes, that is, $k_{12}^{u}\geq(30+20)$.

Property 2: Following this coding rule, there are different upper and lower limits of card number when solving each individual that is having different number of nodes.

4.1.2 Fitness Function

Due to the minimizing property of the target function, fitness function has to be introduced for solving the problem. It is obtained by subtracting the target function from a bigger number $f_{\max}$.

The procedures are:

$$F(K)=f_{\max}-f(K) \tag{9}$$

$$f(K)=ACR(K)+\alpha_1*[SDO(K)-SDO_0]^{+}+\alpha_2*[F_{r0}-F_r(K)]^{+} \tag{10}$$

where $\alpha_1*[SDO(K)-SDO_0]+\alpha_2*[F_{r0}-F_r(K)]$ are the penalties for not satisfying the constraints of (6) and (7), α_1 and α_2 are penalty coefficients, and $f_{\max}$ is a given colossal number to guarantee that the overall fitness value is non-negative, and

$$[y]^{+}=\begin{cases} y & y>0 \\ 0 & otherwise \end{cases}$$

4.1.3 Operators Design

Each initial solution is obtained by creating an integer within the range of 1 to $K_{ij}^{u}+1$ randomly for each bit.

Two points crossover is adopted here. Two intersecting points are chosen from the chromosomes. Crossover is taken in the space between the two intersecting points and the rest is still inheriting the parent genes.

Mutation is also applied. First, creating a number between 0-1 randomly. When the number created is smaller than the mutation probability, mutation will happen by creating an integer that lies within the limits of the circulating cards randomly.

According to the basic theory of GA, the commonly used roulette mechanism is chosen as the choice strategy [14]. The biggest iteration number is chosen as the criterion for algorithm termination.

4.1.4 The Elitist Mechanism

In order to maintain the best chromosome for the next generation, an elitist mechanism is implemented in the choice process. If the best chromosome of the last generation is not duplicated in the next generation, then the next generation will randomly delete a chromosome so that the best chromosome of the last generation will be duplicated directly.

4.2 The Second-Tertiary Model

A simulation is given in the second-tertiary model to determine all kinds of performance indices of the supply chain system under a given parameter K_{ij} of circulating card distribution. These performance indices are clients' degree of satisfaction, standard deviation of invested orders, and overall cost of the supply chain.

5 Simulation Analysis

In order to test the efficiency of genetic algorithm and the generally used inventory control model, many simulations are done on different inventory systems of the supply chains. To analyze the performance the algorithm runs for 100 times and obtains the best solution (The optimal solution).

In order to analyze the performance of the algorithm, many examples were simulated, due to the limit of the space, only the problems with 4-nodes (10^{18}), 6 nodes (10^{40}) are analyzed here, taking suitable parameters obtained by simulation. The result is shown in Table 1, where the optimal value is the best value obtained during the running of the algorithm.

Table 1 showed that, after the scale of the 6-node problem has expanded by 10^{22} times as compared with a 4-node problem, the CPU time is increased by 18S and the optimization percentage is decreased by 6%. Though these are all caused by the expanded complexity of the problem, the algorithm still possesses a better optimization percentage. The increase in the CPU time is within an acceptable range. Moreover, time increase is mainly caused by the influences of the expansion of scale of the problem on simulation. Thus, conclusively speaking, the optimal performance of the algorithm is not greatly affected and it is still a fairly stable algorithm.

Table 1. The simulation result of the algorithm

Scale of the problem	NP	NG	PC	PM	$\Delta f / f^{*}$ (%)	The optimal value	The optimization percentage	T(s)
10^{18} (4- node)	200	150	1.0	0.3	0.00000%	892611.49023	0.92	21
10^{40} (6-ode)	200	150	1.0	0.3	0.00000%	890379.74511	0.86	39
NP: number of populations PC: probability of crossover					NG: number of generations PM: probability of mutation			

6 Conclusion

Determination of the optimal control strategy is a key factor for a successful supply chain. This paper has made researches on the inventory control strategy of a serial supply chain and presents the description of a two-tertiary model. The first-tertiary model is a nonlinear integer programming model. Its main purpose is to determine the control strategy which gives the least overall cost of a supply chain under some constraints. These constraints include the degree of clients' satisfaction not lower than the demand standard and the standard deviation of manufacture less than a given value. When the inventory control strategy is determined, the second-tertiary model is to obtain all kinds of economic indices of the supply chain of the first-tertiary model. The first model reaches optimization through genetic algorithm. The second model actualizes the generally used push-pull model of inventory of a serial supply chain by making use of simulation with C procedures. The main characteristic of the second-tertiary model is that, the choice of control, push or pull, of a node is determined by whether it is under the feedback control of its downstream nodes. The two-tertiary model makes use of a fused method of genetic algorithm and simulation analysis to determine the control strategy. That is, it solves the fixed number of circulating cards and becomes the optimal control of inventory system of a serial supply chain. Further more, it becomes an effective mathematic tool to determine a better optimal control strategy.

Base on the characteristics of the problem, a combined method of genetic algorithm and simulation analysis is proposed. It offers a way to determine the fixed number of circulating cards. Instances of different scales are compared. Instance simulations have ratified the efficacy and the solving efficiency of the method. The cost of the whole supply chain is minimized after satisfying the clients' demands and decreasing the "bullwhip effect". It collaborates production rhythm and shared benefits of each node enterprise of a supply chain and gives a quantitative support of rational organization of purchase, production, transportation, and sales. In the future, we will try other intelligent algorithms for this problem in order to increase the optimization percentage for the large-scale problem.

Acknowledgments

This work was supported by Program for New Century Excellent Talents in University (Project no. NCET-05-0295, NCET-05-0289), the NSFC under Grant No. 70101006, 60003006, 70431003 and 60473089, the NSF of Liaoning under Grant No. 20032019 and 20032018, Shenyang City Natural Science Foundation of China (Project no. 1041006-1-03-03), Modern Distance Education Engineering Project of MoE, P.R.China, Open foundation from key laboratory of process industry automation by Ministry of Education and Liaoning province of China, the project "Hybrid Intelligent Decision Support System for Process Design and Optimization of Product Assembly" (Project A/C Code A-PE49) by the Hong Kong Polytechnic University.

References

1. Lan B.X., Zheng X N., Xu X. Supply Chain Management in E-commerce era. Chinese Journal of Management Science, vol.(3) 17 (2000).
2. Zhao Lindu. Theory and Practice of Electronic Commerce, Beijing: People's Posts & Telecommunications Publishing House , (2001) 1-10.
3. Zhao Lindu. Theory and Practice of Supply Chain and Logistics Management. Mechanical Industry Publishing House , (2003)34-36.
4. Gstettner S., Kuhn H., Analysis of Production Control Systems Kanban and CONWIP. International Journal of Production Research,vol. 34(11).(1996)3253-3273.
5. Deleersnyder J. L., Hodgson T. J., Mueller(-Malek) Hetal, Kanban Control Led Pull System: Analysis Approach. Management Science, vol. 35(9).(1989)1079-1091.
6. Sarker B. R., Fitzsimmons J. A., The Performance of Push and Pull Systems: A Simulation and Comparative Study. International Journal of Production Research, vol. 27(11). (1989) 1715-1732.
7. Flapper S. D. P., Miltenburg G. J.,Wijugard J., Embedding JIT into MRP. International Journal of Production Research, vol.29(2). (1991)329-341
8. Larsen N. E. Alting L. Criteria for selecting production control philosophy. Production Planning & Control, vol.4(1).(1993)54-68
9. Villa A., Watanabe T. Production management: Beyond the Dichotomy Between "Push" and "Pull". Computer Integrated Manufacturing Systems, vol.6(1).(1993)53-63
10. Spearman, M. L., Woodruff, D. L., and Hopp, W. J. CONWIP: a Pull Alternative to Kanban. International Journal of Production Research, VOL. 28(5). (1990) 879-894.
11. Oscar R., Adolfo C. M., Exploring the Utilization of a CONWIP System for Supply Chain Management. International Journal of Production Economics,2003,83:195-215
12. Herer Y. T., Masin M. Mathematical Programming Formulation of CONW IP Based Production Lines and Relationships to MRP. International Journal of Production Research, vol.35 (4). (1997) 1067-1076.
13. Gerald J. .Bose .Implementing JIT with MRP Creates Hybrid Manufacturing Environment. IE. (1988) 14-18.
14. Michalewicz Z. Genetic algorithms + data structure = evolution programs. Seconded. New York: Springer-Verlag, (1994).

A Nested Genetic Algorithm for Optimal Container Pick-Up Operation Scheduling on Container Yards

Jianfeng Shen, Chun Jin, and Peng Gao

Institute of System Engineering, Dalian University of Technology, 116024, Dalian, P.R. China
jinchun@dlut.edu.cn

Abstract. For the optimization problem on container pick-up operation scheduling, a multi-stage mathematical programming model is established to minimize the total operation cost. This model is comprised of two parts: rehandling operation scheduling and the shortest path search. A nested genetic algorithm with two layers is proposed, where the inner layer algorithm is embedded in the outer one. The outer algorithm is responsible for optimizing rehandling operation scheduling, and the inner is for searching the shortest path of rehandling operation. Two reformative operations are introduced, including the parent population involving selection and optimal individual maintaining. With an actual example, the algorithm is testified, the result shows that the algorithm is effective to solve this problem and has a high efficiency and speed of convergence.

1 Introduction

The import container handling operation on the container yard is to pick up a container being stored on a container block and to carry it out of the gate of the container yard. In the process of pick-up operation, container rehandling operation occurs frequently that causes higher material handling cost and lower operation efficiency. So, it is quite important to optimize the schedule of pick-up operation to meet the operation requirement for fast increase of container logistics.

To the problem on the import container pick-up operation, Kim [1] proposed an algorithm to determine the expected number of rehandling operation. Kozan [2] pointed out that the container pick-up problem was NP-hard and genetic algorithm (GA) could be used to reduce rehandling time. Preston [3] proposed an objective function to determine the optimal storage strategy for container-handling schedules.

This paper considers that the container pick-up operation is carried out within several bays, and the objective is to determine a container-handling schedule minimizing the total operation cost. This problem refers to path search and slot selection in the process of rehandling operation, and so on. A nested algorithm based on GA is proposed to solve the optimal pick-up operation scheduling problem.

2 Formulation of the Optimization Model

2.1 Description of the Pick-Up Operation

The configuration of the container stacking is shown in Fig. 1. The location of a slot in a stack is specified by a triple (bay, row, tier). For any container to handle, the

L. Jiao et al. (Eds.): ICNC 2006, Part I, LNCS 4221, pp. 666–675, 2006.

rehandle operation has to be carried out if the container could not be picked up immediately. Taking the container located at position (5, 3, 1) shown in Fig. 1 as an example, there are 3 rehandles: containers (5, 3, 2), (5, 3, 3) and (5, 3, 4). The container pick-up operation scheduling focuses on how to move the rehandles to slots following yard operation rules, which are called destination positions of rehandles in this paper. Optimization on the pick-up operation aims to minimize the total handling operation cost. It is assumed as follows:

1) The analysis scope of pick-up operation is restricted to a single container block.

2) For picking up one container, only one kind of material handling machine is used to handle or rehandle containers in the whole operation process.

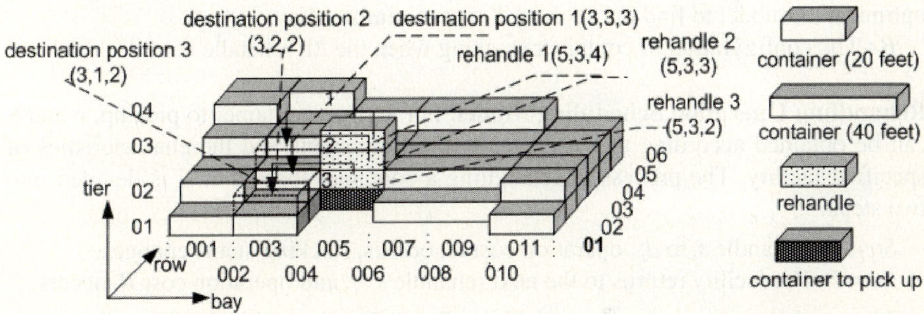

Fig. 1. Configuration of the container stacking and pick-up operation

The constraints are defined according to the operation rules. The main operation rules are presented as follows:

- Rule 1. Even for the same container stacking status, different facilities operate in different modes. So, the destination position and rehandling route for a given pick-up operation must be fit for the operation character of specified facility.
- Rule 2. Containers with size of 20 feet and those of 40 feet cannot be located at the same bay.
- Rule 3. It is forbidden to locate one container upon another that can't be pressed.
- Rule 4. All the containers to pick up are called booked containers. It is not allowed to locate containers upon a booked container.
- Rule 5. The destination position of a rehandle should be limited within the range of specified number of bays.

2.2 Description of the Optimization Model

The container pick-up operation optimization model consists of two parts: rehandling operation scheduling model and the shortest path search model; the later model is embedded into the former one to provide the minimum values of the path cost.

The definitions of notations in the model are listed as follows:

b: the number of bays of the block;
r: the number of rows of the block;
t: the number of tiers of the block;
n: the total number of rehandles;

s_i: the original position of slot for the ith rehandle in the block, $s_i=(sb_i,sr_i,st_i)$;

$\boldsymbol{S}$: the set of s_i for all rehandles, $\boldsymbol{S}=\{s_1, \ldots, s_n\}$, where, i is the rehandle's sequence number in the rehandling operation;

d_i: the destination position of the ith rehandle in the block, $d_i=(db_i,dr_i,dt_i)$;

$\boldsymbol{D}_i$: the set of destination positions for the ith rehandle, $\boldsymbol{D}_i=\{d_{i1},d_{i2}\ldots d_{ij}\ldots d_{im}\}$, where, m is the number of destination positions that should meet the operation rules;

$\boldsymbol{H}_i$: the minimum operation costs of conveying the ith rehandle from s_i to d_i;

$\boldsymbol{R}_i$: the minimum return cost for machine to move from d_i to s_{i+1} without load;

$\boldsymbol{C}$: the total cost of the whole rehandling operation, its value is equal to $\sum \boldsymbol{H}_i + \boldsymbol{R}_i$;

$\boldsymbol{X}$: candidate solution of the problem, $\boldsymbol{X}=\{d_{1\,j1}\ldots d_{i\,ji}\ldots d_{n\,jn}\}$, $d_{i\,ji}$, which is selected from $\boldsymbol{D}_i$, is the destination position of the ith rehandle (s_i). Thus, it is the object of optimization model to find out such an $\boldsymbol{X}$ that minimizes $\boldsymbol{C}$;

$\boldsymbol{B}_i$: The configuration of container stacking when the ith rehandle is to deal with.

Rehandling Operation Scheduling Model. For a given container to pick up, n and S can be obtained according to the container stacking status and the characteristics of specified facility. The process of rehandling a container located at si is devided into two steps:

Step 1: Rehandle s_i to d_{ij}, operation cost $\boldsymbol{H}_i$ occurs, stacking status changes.

Step 2: The facility returns to the next rehandle s_{i+1}, and operation cost $\boldsymbol{R}_i$ occurs.

The stacking status of container block $\boldsymbol{B}_i$ is changing with the rehandles being moved to their destination positions one by one. Supposing that the current status of the operation system is $\boldsymbol{Q}_i = [\boldsymbol{B}_i, s_i, \boldsymbol{D}_i]^{\mathrm{T}}$, then, status transition function is:

$$\boldsymbol{B}_{i+1}=\boldsymbol{T}_i(\boldsymbol{D}_i, d_{ij}) \tag{1}$$

Where, $\boldsymbol{T}_i\,(\boldsymbol{D}_i, d_{ij})$ means selecting d_{ij} from $\boldsymbol{D}_i$ as the destination position for the ith rehandle located at s_i, accordingly, the container stacking status is updated to $\boldsymbol{B}_{i+1}$. Thus, the destination position set of the $(i+1)$ th rehandle is obtained only just the ith rehandle's slot is specified. Therefore, the rehandling operation process is complicated, combinatorial and dynamic; optimization on container rehandling operation can be attributed to a multi-stage decision-making problem, where one stage means a process comprised of two steps mentioned above. The operation cost $(\boldsymbol{H}_i+\boldsymbol{R}_i)$ in ith stage occurs and is added cumulatively to the total operation cost $\boldsymbol{C}$, and the selection of d_{ij} from $\boldsymbol{D}_i$ to generate $\boldsymbol{X}$ that minimizes $\boldsymbol{C}$ is the object of the model.

The mathematical programming model is formulated as equation (2). Where, $f_k\,(d_{ij})$ are the constraint functions according to the operation rules described in section 2.1, equation $f_k\,(d_{ij})=1$ means the destination position d_{ij} obeys the kth constraints. h and r are cost functions that compute $\boldsymbol{H}_i$ and $\boldsymbol{R}_i$ respectively, they are important parameters provided by the inner model, which will be introduced in detail in the next part.

$$\begin{aligned}
&\min \quad C = \sum_{i=1}^{n} \left(H_i + R_i\right) \\
&H_i = h\left(s_i, d_{ij}\right) = h\left(sb_i, sr_i, st_i, db_i, dr_i, dt_i\right) \\
&R_i = r\left(s_{i+1}, d_{ij}\right) = r\left(db_i, dr_i, dt_i, sb_{i+1}, sr_{i+1}, st_{i+1}\right) \\
&s.t. \qquad f_k\left(d_{ij}\right) = 1 \qquad \left(k = 1,2,3,4,5\right)
\end{aligned} \tag{2}$$

The Shortest Path Search Model. The unit operation cost is various with the facility's moving in different directions with a container loaded or not. There are also many choices of routes to move a container from original position to the destination one, so it is significant to determine the shortest path to minimize H_i and R_i.

Under specified stacking status, the values of $\boldsymbol{H_i}$ and $\boldsymbol{R_i}$ are determined by the start point $\boldsymbol{S}$ and destination $\boldsymbol{T}$ of the path, supposed that there are t bays between $\boldsymbol{S}$ and $\boldsymbol{T}$, the set of positions of a bay container can pass is $\boldsymbol{P_i}\{p_{i1}, p_{i2} \ldots p_{i\,ki}\}$ (k_i is the total number of members in $\boldsymbol{P_i}$, 1<=i<=t). When container moves between bays, it starts from $\boldsymbol{S}$, pass though several p_{ij} and arrives at $\boldsymbol{T}$, so the path can be denoted by $\boldsymbol{U}\{\boldsymbol{S}, p_{1\,g1}, p_{2\,g2}, \ldots p_{tgt}, \boldsymbol{T}\}$ ($p_{i\,gi}$ is the gith member of $\boldsymbol{P_i}$).

It is supposed that $\boldsymbol{C_1}$, $\boldsymbol{C_2}$, $\boldsymbol{C_3}$, $\boldsymbol{C_4}$ are the unit operation cost weights when the facility moves in the direction of bay, row, tier up, and tier down, respectively; $\boldsymbol{M_{bi}}$, $\boldsymbol{M_{ri}}$, $\boldsymbol{M_{ui}}$, $\boldsymbol{M_{di}}$ are the total distances in four directions mentioned above for route i in the whole operation process, respectively. The mathematical programming model of the shortest path search is denoted by formula (3) and (4). Function v computes the cost of container's moving between two adjacent bays.

$$\begin{aligned}
&\min H_i = h(s_i, d_{ij}) = v(s_i, p_{1k}) + \sum_{e=1}^{t} v(p_{eq}, p_{e+1k}) + v(p_{tk}, d_{ij}) \\
&\min R_i = r(s_{i+1}, d_{ij}) = v(s_{i+1}, p_{1q}) + \sum_{e=1}^{t} v(p_{eq}, p_{e+1k}) + v(p_{tk}, d_{ij}) \\
&1 \le q, k \le k_i, 1 \le e < t, p_{ij} \in P_i
\end{aligned} \tag{3}$$

$$\begin{aligned}
&v(S,T) = C_1 M_{bi} + C_2 M_{ri} + C_3 M_{ui} + C_4 M_{di} \\
&M_{bi} = |S_{bi} - T_{bi}| \\
&M_{ri} = |S_{ri} - T_{ri}| \\
&M_{ui} = \begin{cases} T_{ti} - S_{ti}, & T_{ti} \ge S_{ti} \\ 0, & T_{ti} < S_{ti} \end{cases} \\
&M_{di} = \begin{cases} S_{ti} - T_{ti}, & T_{ti} < S_{ti} \\ 0, & T_{ti} \ge S_{ti} \end{cases}
\end{aligned} \tag{4}$$

3 Algorithm Design

Based on the analysis of the optimization model, it is known that the outer model is for combinatorial optimization problem and the inner model is for the shortest path search problem. GA is chosen due to the good results that have been reported about solving these two kinds of problems. Li [4] proposed a Partheno-Genetic Algorithm for combinatorial optimization and analyzed its schema theorem and global convergence. Rocha [5] studied an order based representation genetic and evolutionary algorithms in combinatorial optimization problems. Gomez-Albarran [6] proposed a routing strategy based on GA. Davies [7] utilized GA to find the shortest path in a dynamic network, where the weights changed as known functions of time.

3.1 Design of Nested Genetic Algorithm

This paper proposed a nested genetic algorithm (NGA) as shown in Fig. 2, where the inner GA is embedded into the outer GA. The outer GA is used to optimize rehandling operation schedule, and the inner GA is to search the shortest path between two slots, namely minimal value of H_i/R_i, which are provided to the outer GA. Both of them call a common genetic algorithm procedure but have different fitness functions.

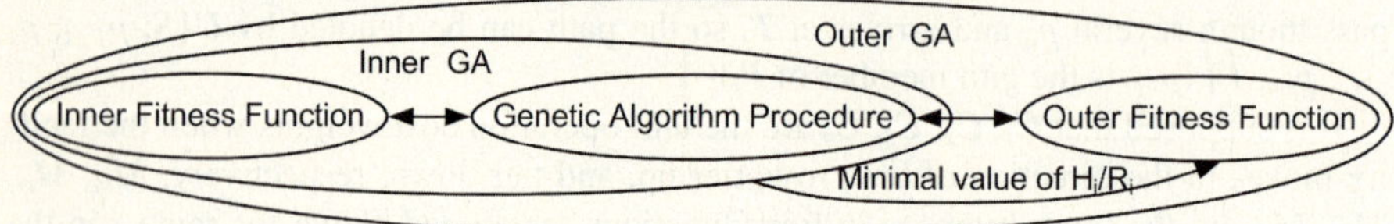

Fig. 2. The structure of nested genetic algorithm

For the container pick-up problem, the key points of GA are the convergence speed and global convergence to its optimal solution; so two methods are adopted in this algorithm. One is the parent population involving selection, which is used to quicken the speed of convergence; another is optimal individual maintaining to ensure the convergence to its optimal solution. The procedure is described as follows:

Step 1: Initialization. The GA parameters, including crossover probability P_c, mutation probability P_m and population size *popsize*, are initialized; then, the initial population X_0 is generated randomly, and the counter of generation is set as t=0.

Step 2: Crossover. The crossover operation is implemented independently upon the population X_t, and a new population Y_t is generated.

Step 3: Mutation. The mutation operation is implemented independently upon the population Y_t, and a new population Z_t is generated.

Step 4: Evaluation. The fitness of each chromosome in population X_t+Z_t is calculated, and the chromosome with the best fitness ($BEST_t$) is selected [8]. The inner GA, whose procedure is similar to the outer one, is invoked here. The fitness function is used in this fitness evaluation process. According to Formula (2) in section 2, the inner GA provides the values of H_i and R_i to the outer GA as parameters.

Step 5: Optimal individual maintaining. the $BEST_t$ is compared with global optimal individual *opt*, and the one with better fitness value will be kept as *opt*. Thus, the individual with the best fitness value among all generations is kept at last, and this method is called optimal individual maintaining.

Step 6: Selection. The method of roulette is adopted for the selection operation upon population X_t+Z_t, and the new population X_{t+1} is generated. X_t is the parent population, so this method is called parent population involving selection.

Step 7: Terminal rule checking. If the terminal condition is met, the procedure will stop and go to next step; otherwise it will set t=t+1 and return to step 2.

Step 8: Optimal cost obtaining. The optimal container pick up operation cost C is gained according to *opt*.

3.2 Design of Encoding and Decoding

The integral encoding scheme is adopted, each individual, $\boldsymbol{G}$ $\{g_1, g_2 \ldots g_n\}$, is composed of n integers, and g_i, as the ith member of $\boldsymbol{G}$, denotes an order number. For the outer GA, it means the g_ith member of $\boldsymbol{D}_i$, and for the inner GA, it means the g_ith member of $\boldsymbol{P}_i$. So this encoding scheme can be considered as an indirect representation, with the genes keeping the order of d_{ij} in $\boldsymbol{D}_i$, or p_{ij} in $\boldsymbol{P}_i$, but not their value.

The decoding process means decoding $\boldsymbol{G}$ into candidate solution. For the outer model, the candidate solution is $\boldsymbol{X}$, as shown in Fig. 3. And g_i denotes that the g_ith member of $\boldsymbol{D}_i$ (expressed as $d_{i\,gi}$) is selected as the destination position of s_i and added into $\boldsymbol{X}$ as its ith member.

Similarly, for the inner model, the candidate solution is $\boldsymbol{U}$, as shown in Fig. 4. g_i is a number denoting the order of $\boldsymbol{P}_i$*'s* member in $\boldsymbol{P}_i$, and it indicates that $p_{i\,gi}$ is selected as a node of the path connecting original slots to the destination, and $\boldsymbol{U}$ records all nodes in the shortest path.

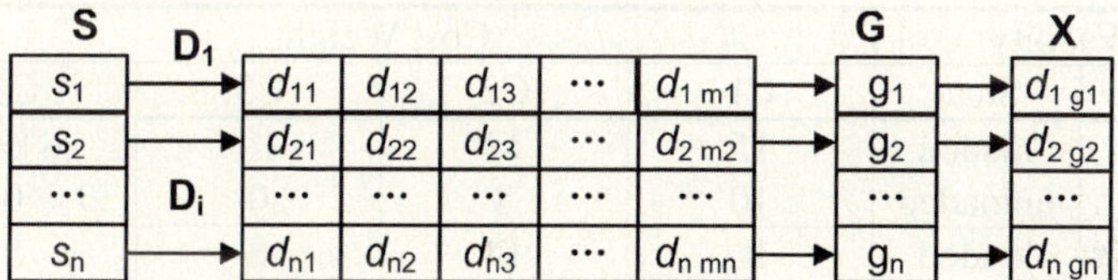

Fig. 3. Outer GA: Decoding G into candidate solution X

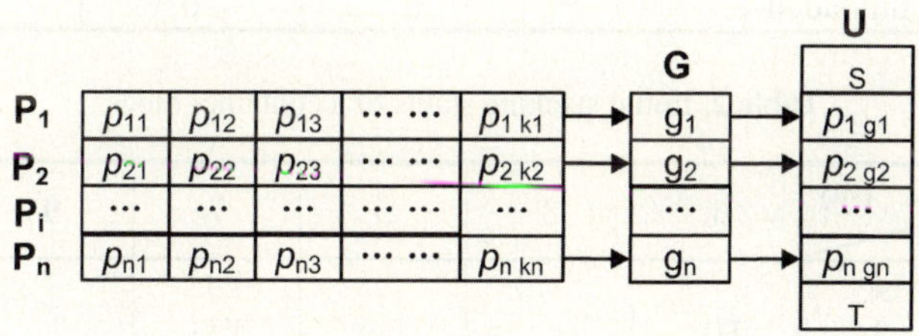

Fig. 4. Inner GA: Decoding G into candidate solution U

3.3 Design of Fitness Function

The fitness function of genetic algorithm has close relationship to the objective function in the optimization model, so the fitness functions are deduced from Formula (2) and (3). An estimated value is specified because the object function is to find the minimum value.Therefore, the fitness function of outer GA and inner GA are:

$$Fit_1\,(\boldsymbol{C}) = \boldsymbol{C}_{max} - \boldsymbol{C},\ \boldsymbol{C} <= \boldsymbol{C}_{max}\,. \tag{5}$$

$$Fit_2\,(\boldsymbol{v}) = \boldsymbol{V}_{max} - \boldsymbol{v},\ \boldsymbol{v} <= \boldsymbol{V}_{max}\,. \tag{6}$$

Where $\boldsymbol{C}_{max}$ is the maximum estimated value of the operation cost $\boldsymbol{C}$, and $\boldsymbol{V}_{max}$ is the maximum estimated value of $\boldsymbol{H}_i$ or $\boldsymbol{R}_i$.

4 Case Study and Result Analysis

An actual example is shown to illustrate the validity and efficiency of this proposed algorithm. The operation cost weights of three kinds of facilities: *front handling mobile crane* (**FMC**), *forklift* (**FL**) and *transfer crane* (**TC**) are shown in Table 1.

The initial stacking status of a container block with 6 bays, 4 rows and 4 tiers is shown in Table 2, where ***Txx*** denotes container code, the blanks without ***Txx*** mean empty positions, ***T30*** (7,1,2) in the broad-brush blank is the container to be picked up. Expected execution time of the algorithm is not longer than 2 seconds.

Table 3 shows the optimization results with NGA for picking up container T30 (7, 1, 2) under the condition shown in Table 1 and Table 2. The rehandling range includes all empty slots of bay 1,3,5,7 and 9. Results according to NGA, standard genetic algorithm (SGA) are list respectively.

Table 1. Cost weights of different facilities

Facility		Cost Weight			
Type	State	C1	C2	C3	C4
FMC	loaded	15	13	10	6
	unloaded	10	8	0	0
FL	loaded	16	14	8	5
	unloaded	10	8	0	0
TC	loaded	20	11	5	3
	unloaded	15	6	0	0

Table 2. Initial stacking status of a container block

Row \ Tier \ Bay		1	3	5	7	9	12
1	4				T32		T45
	3	T03			T31		T44
	2	T02	T12	T21	**T30**	T38	T43
	1	T01	T11	T20	T29	T37	T42
2	4						
	3	T06					T48
	2	T05	T14	T23	T34		T47
	1	T04	T13	T22	T33		T46
3	4						
	3						
	2	T08		T25		T40	T50
	1	T07	T15	T24	T35	T39	T49
4	4		T19				
	3		T18	T28			
	2	T10	T17	T27			T52
	1	T9	T16	T26	T36	T41	T51

Table 3. Optimization results

Facility	FMC	FL	TC
Value of n, m	n=3, m=5	n=6, m=7	n=2, m=14
P_c/P_m	0.3/0.02	0.3/0.02	0.3/0.02
popsize (outer/inner)	20/10	20/10	20/10
Rehandles and their destination positions	***T36***:(9,4,2) ***T32***:(7,3,2) ***T31***:(5,2,3)	***T36***:(9,4,2) ***T35***:(9,4,3) ***T34***:(5,4,4) ***T33***:(9,4,4) ***T32***:(1,4,3) ***T31***:(1,4,4)	***T32***:(7,2,3) ***T31***:(7,2,4)
NGA cost	370	842	244
SGA cost	378	853	256

From Table 3, it is known that the optimization results determined by NGA are better than SGA; accordingly, the optimal rehandling scheduling and the shortest path can be obtained by NGA, but they might not be found by SGA. So, the validity of NGA to this problem can be verified.

The convergence processes of NGA and that of SGA are compared by simulation run in Fig. 5(1), and the curves indicate their trend of minimal operation cost for **TC**. It can be seen that the convergence speed of NGA is faster than that of SGA, mature convergence of NGA exists at simulation run=11 (position *a*), but SGA is immaturely convergent to suboptimal cost at simulation run=36 (position *b*). There is a phenomenon of degeneration exists in SGA during simulation run (in interval of simulation run from 15 to 20, and 30 to 35), but NGA does not. The main reason is that NGA uses the method of parent population involving selection and optimal individual maintaining, but SGA does not.

Therefore, there are two advantages of NGA: one is that the global optimal solution can be definitely obtained by NGA but not be definitely obtained by SGA; the other is that the convergence speed of NGA is much faster than that of SGA.

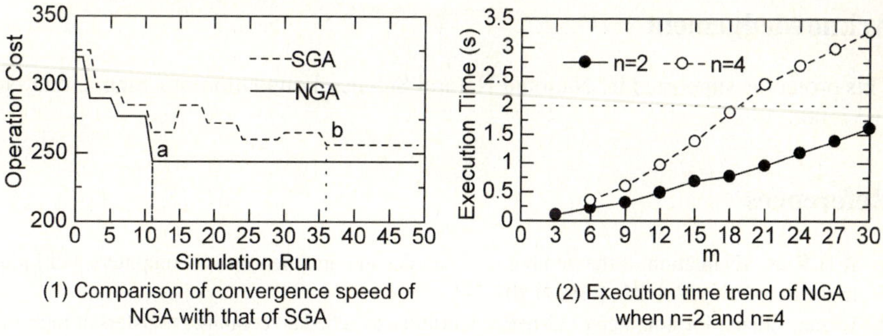

Fig. 5. Convergence speed and execution time of NGA

To analyze the execution efficiency, the execution time of the algorithm is shown in Fig. 5(2) which indicates the relationship between execution time and parameter m (the number of destination positions for rehandles). The curves show the time change when m = 3, 6, … , 30 under the condition of n=2 (2 rehandles) and n=4 (4 rehandles) with TC as the operation machine. From Fig. 5(2), it is found that when n=2, the execution time of NGA is shorter than 2 seconds for all m; but when n=4, the execution time exceeds 2 seconds after m=18. It is also found that the slope of curve of n=4 is much larger than that of n=2 when the value of m is assigned equally.

By a further analysis, it is known that, the search space of the problem is up to m^n, which has an exponential relation to n, but the times of outer GA's calling inner GA is $2n \times popsize \times generation$, it is a linear function of n approximately. Therefore, the search efficiency of NGA is a synthetic result of exponential function and linear function. *generation* is determined by the convergence speed, so, the NGA will have a higher efficiency other than search all nodes in the state space if the control variables of GA are set properly.

5 Conclusions

In this paper, the optimization of container pick-up operation scheduling is considered to be a multi-stage decision problem including optimization on rehandling operation and the shortest path search. In the first place, the mathematical programming model is established to minimize the total operation cost. Then, a nested genetic algorithm (NGA) with two layers is proposed; the outer GA is responsible for optimizing rehandling schedule and the inner GA for searching the shortest rehandling operation path. Two reformative operations are introduced: parent population involving selection is used to quicken the speed of convergence, and optimal individual maintaining ensure the convergence to its global optimal solution. Thirdly, the validity and efficiency of the NGA is testified with an actual example, the results from NGA have shown that it is convergent to global optimal solution and the convergence speed is faster than that of SGA. Therefore, NGA possesses better result and higher efficiency for the optimal operation scheduling.

Acknowledgement

This project is supported by National Natural Science Foundation of China, Grant No. 70571008.

References

1. K H Kim.: Evaluation of the number of rehandles in container yards. Computers and Industrial Engineering,Vol. 32. 4(1997) 701-711
2. Kozan, Erhan., Preston, Peter.: Genetic algorithms to schedule container transfers at multimodal terminals. International Transactions in Operational Research, Vol.6.3(1999)311-329
3. Preston, Peter., Kozan, Erhan.: An approach to determine storage locations of containers at seaport terminals. Computers and Operations Research, Vol. 28.10(2001) 983-995

4. Maojun Li., Shaosheng Fan., An Luo.: A Partheno-genetic Algorithm for Combinatorial Optimization. Lecture Notes in Computer Science, Vol. 3316. Springer-Verlag, Berlin Heidelberg New York (2004)224-229
5. Miguel Rocha., Carla Vilela., José Neves.: A Study of Order Based Genetic and Evolutionary Algorithms in Combinatorial Optimization Problems. Lecture Notes in Computer Science, Vol. 1821. Springer-Verlag, Berlin Heidelberg New York (2000)601-611
6. Gomez-Albarran, Ma de las Mercedes., Fernandez Pampillon Cesteros, Ana Ma., Sanchez-Perez, Juan Manuel.: A routing strategy based on genetic algorithms. Microelectronics Journal (Incorporating Journal of Semicustom ICs). Vol. 28. 6(1997)641-656
7. Davies, Cedric., Lingras, Pawan.: Genetic algorithms for rerouting shortest paths in dynamic and stochastic networks. European Journal of Operational Research. Vol. 144.1(2003)27-38
8. Si-Ho Yoo., Sung-Bae Cho.: Partially Evaluated Genetic Algorithm Based on Fuzzy c-Means Algorithm. Lecture Notes in Computer Science. Vol. 3242. Springer-Verlag, Berlin Hei-delberg New York (2004)440-449

A Genetic Algorithm for Scale-Based Product Platform Planning*

Zhen Lu and Jiang Zuhua

School of Mechanical Engineering, Shanghai Jiao Tong University, Shanghai, P.R.C.
{lzhen, zhjiang}@sjtu.edu.cn

Abstract. Product platform planning paves the way for DFMC (Design for Mass Customization). In this paper, we mainly focus on the development of a method for scale-based product platform planning using genetic algorithm (GA) to satisfy a set of customer requirements. Different from many usual methods for platform planning that need users to specify product platform prior to optimize it, the new GA-based method proposed by this paper focuses on searching a proper balance between the commonality of product platform and the performance of product family derived from the platform. The method improves as more commonality of the product platform as possible, within the satisfactory of the diverse customer needs, and then determines the variable product attributes and their variation ranges, and the common parameters of product platform and their optimal values. This method of product platform planning is validated by a case study of the small-size induction motor design. At last we compare the results from our GA-based method with the benchmark products that are individually designed to the optimal performance.

1 Introduction

Product variety strategy has been recognized as an effective means to implement mass customization, and the rationality of developing product platforms as a means to achieve product variety while maintaining economy of scale has been well recognized in both academia and industry [1]. According to Meyer and Lehnerd (1997) [2], a product platform is the set of parts, subsystem, interfaces that shared among a set of products, and allow the development of derivative products with cost and time saving. A product family can derived from the product platform by variegating designs to satisfy individual customer requirements. The key to a successful product family is the product platform from which it is derived either by adding, removing, or substituting one or more modules to the platform, or by scaling the platform in one or more dimensions to target specific market niches. According to the different ways by which product family is derived from the platform, there are mainly two kinds of product platforms [3]: module-based product platforms and scale-based product platforms. This paper mainly concerns the latter.

* This research work obtained financial support from the 973/National Basic Research Program of China (2003CB317005) and Shuguang Program of Shanghai Educational Committee (No. 05SG15).

L. Jiao et al. (Eds.): ICNC 2006, Part I, LNCS 4221, pp. 676 – 685, 2006.

Scale-based product platform has received less attention in the academia than the former one. However, many companies have been successful with it. Boeing designs many of its aircraft around scalable platforms that are stretched to carry more passengers or vary the flight range [4]. Black and Decker designed a universal motor platform that could be scaled along its stack length to generate a wide variety of power outputs while significantly increasing economies of scale and reducing labor costs [2]. In the academia, Simpson and his co-workers have done a lot of work on the scale-based product platforms; they have brought out some methodologies that involve the robust concept exploration method, product platform concept exploration method and variation-based platform design method [5, 6, 7].

A key feature of a product platform is commonality. By introducing higher levels of commonality within a product platform, a manufacturing is able to reduce part in inventory and lower procurement costs. The only snag being that increased commonality has an associated tradeoff with performance. The challenge, then, is to find an acceptable tradeoff between commonality and performance during product platform planning. Genetic algorithms appear well suited for solving combinatorial problems typical in product platform planning. Unlike usual method, that need designers predetermine the common parameters and scaling variables, the new method proposed by this paper focuses on improving the commonality of the product family within the satisfactory of the diverse customer needs, and then determine the variable product attributes and their variation range, and the common parameters of product platform and their optimal values.

2 Evolutionary Product Platform Planning

An effective product platform planning method should focus on improving the commonality of the product family within the satisfactory of the diverse customer needs, rather than determining the common and variable parameters beforehand. In the process of planning the scale-based product platform, the key target is to identify which design variables to make common to the platform and which to make unique for each product in the family; and also determine the optimal value for each common parameter in the platform, and variation range for each variable attribute. Toward that end, we have developed a genetic algorithm based method for product platform planning to determine the variable product attributes and their variation range, as well as the common parameters of product platform and their optimal values.

(1) Analyze market requirements

In order to take advantage of commonality in product platform planning, the customer demands should be analyzed and clustered according to the similarity among them [8].In general, the customer demands can be described as a set of attributes, $CD=\{cd_1, cd_2, ..., cd_m\}$. Suppose all customers comprise a set, $C=\{c_1, c_2, ..., c_n\}$, the demand vector of customer c_i can be depicted as follow:

$$CD(c_i) = [cd_k(c_i) \mid k = 1,...,m]$$

Where $cd_k(c_i)$ denotes the kth customer demand attribute of customer c_i. The distance between the c_i and c_j can be calculated using follow formula:

$$d(c_i,c_j)=\sqrt{\frac{1}{m}\sum_{k=1}^{m}\left(\frac{cd_k(c_i)-cd_k(c_j)}{\max\limits_{\forall f\in C} cd_k(c_f)-\min\limits_{\forall f\in C} cd_k(c_f)}\right)^2}$$

After calculating all the distance between every two customers, we can get a distance metric. The set of n customers could be clustered into s customer groups with the tool of clustering analysis [9]. The customer demand metric is s×m: s columns stand for s customer groups; m rows stand for m customer demand attributes

(2) Build mathematical product model

Before the product platform planning, we need to construct a mathematical model for the product, which connects the design variables and performance parameters [10]. The mathematical model is just like a bridge between the customer demand domain and the engineer design domain. With the model and customer requirements, we can get several sets of individually design variables that could meet every customer's demands. n_i design vectors are obtained from a set of demand variables (cd_1, cd_2, …, cd_m) through mathematical model. On the basis of the n_i design vectors, n_i possible products (P_1, P_2, …, P_{ni}) will be able to satisfy the customers' demands. Sometimes, the n_i will be 1, that means only one set of design variables could make the product's performance attributes satisfy the customers' demand. In the same way, the n_i will also be 0, that means there is no products could be produced to satisfy the demands. The product mathematical model should be inexpensive to run. If the analysis relating design variables to performance parameters are too computationally expensive to warrant modeling, this step can be skipped [11].

(3) Compute design variable metrics

On the basis of s customer groups' requirements and mathematical model mentioned in the previous two steps, s sets of design variables could be gained through the model. Each set may contain more than one design variable vectors; because there may be several different design variable vectors could satisfy a customer group's demand. As to the several vectors, we could choose the best one according to some rules; however, the integration of partially best one may not be the best as a whole. When designing a product platform, the focus is more on the product family as a whole rather than the determining what works best for the individual products. So, with satisfying the customer demands, several design variable vectors that are kept in process will pave the way for product platform's evolution to an optimal one.

(4) Construct design variable chromosome and initial population

With the s product design variable metrics obtained in previous step, we can construct design variable chromosome and initial population for following evolution. We choose one design variable vector randomly from every design variable metric to form a chromosome, which contains s units. Each unit is a set of design variables relating to one customer group. Because of the randomness, the probable number of different chromosomes will be a huge number. On account of the genetic algorithm's good behavior in optimization, we use it as a tool to search for an optimal solution for product platform.

(5) Perform evolution of design variable population

In the process of evolution, we mainly use three kinds of operators that are reproduction, crossover and mutation operators [12]. As to N_{size} chromosomes in initial population, we compute each chromosome's fitness at first. In section 3.3, how to calculate each chromosome's fitness will be mentioned in detail. Then, according to their fitness, one operator is chosen for them. In the process of assigning different operators to every chromosome, our principle is that reproduction operator is assigned for chromosomes with higher fitness, crossover operator for ones with middle fitness, and mutation operator for ones with lower fitness. We use P_r, P_c and P_m to denote the probability of reproduction, crossover and mutation operators.

After one generation, we gain a new population, in which the best chromosome is sought and compared with previous generations' best one, and then update the best chromosome in the whole process of evolution. After a certain number of loops, that is the generation size; the evolution stops and outputs the best chromosome.

(6) Develop product platform

The best design variable chromosome is a metric, in which s columns denote s customer groups and n rows denote n design variables.

As to every row of the metric, we analyze number of different design variables. If all the design variables in a row are the same, the number of different design variables is one. Then, the design variable related to that row should be common parameter in product platform; and other rows with more than one different design variables should be variable design attributes. If there is no row in which all the design variables are the same, we choose the row with the least different design variables; for example, the least is two, so we can construct two product platforms, with that two variables as their common parameters respectively.

3 Analysis on Some Technology Bases

3.1 Design Variables Clustering

It is the premises for calculating the chromosome's fitness to determine the number of different design variables. With that purpose, we adopt clustering analysis as a tool. For example, a certain row design variables are: {1.203, 1.211, 1.193, 1.523, 1.492}. After clustering, we obtain two different variables: 1.2 and 1.5, so the number of different variables is 2 rather than 5. As to a chromosome, Fig.1 illustrates a detailed instance for the usage of design variable clustering.

3.2 Fitness Function

A chromosome's fitness is related to the quality of the product platform with the s design vectors in the chromosome as the platform parameters. With the satisfactory of

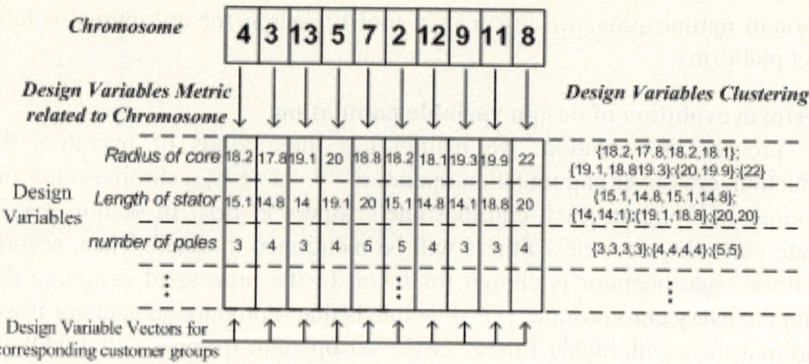

Fig. 1. Design variables clustering instance

diverse customer demands, the higher the level of commonality is, the better the product platform is and the lower the manufacturing costs become. Therefore, the higher the level of commonality is, the larger the chromosome's fitness is, vice versa.

Every chromosome is a metric with *s* columns and *n* rows; *s* columns denote *s* customer groups and *n* rows denote *n* design variables. We calculate the chromosome's fitness according to the commonality of each row, which denotes a design variable. We use N_i to denote the number of different variables:

$$N_i \in \{1,2,\cdots,s\} \qquad (i=1,2,\ldots,n)$$

If all the variables in a row are the same, the N_i is 1 and the row's fitness should be maximum; whereas, if all the variables in a row are the different from each other, the N_i is s and the row's fitness should be minimum. We define that when N_i=1, the fitness is *s* and when N_i=s, the fitness is 1. All the rows' fitness could be calculated as follow formula:

$$a_i = s + 1 - N_i$$

The chromosome's fitness is the weighted sum of all rows' fitness.

$$Fit(I) = \sum_{i=1}^{n} a_i \times \sigma_i = \sum_{i=1}^{n} (s + 1 - N_i) \times \sigma_i$$

Where σ_i is weight of the *ith* row. The higher the weight is, the design variable is more influential to the product cost; vice versa.

4 Case Study

We take the small-size induction motor with power less than 30kw as an example to validate the feasibility of our method. In this paper, we only take these six performance parameters into consideration: power, efficiency, power factor, maximum torque, starting current and starting torque. In addition, as to various design variables of motor, we only consider five representative design variables: number of poles, number of rotor slots per phase·pole, number of stator slots, radius of core, and effective length of the stator.

(1) Market requirements clustering

Before the product platform planning, customer groups should be clustered on the basis of market investigation. With method of clustering analysis, 10 customer groups are obtained, shown as table.1. In this table, each column is one customer group (in roman numerals) and each row denotes one performance parameter cared by customers. Among those parameters, as to efficiency, power factor, maximum torque and starting torque, the higher they are, the better is; while, as to starting current, the lower it is, the better is.

Table 1. The customer demand metric

Customer group	I	II	III	IV	V	VI	VII	VIII	IX	X
Power（kw）	8	10	11	13	15	18	20	22	25	30
Volt（V）	380	380	380	380	380	380	380	380	380	380
Efficiency	86%	87%	87%	88%	89%	89%	90%	91%	91%	91%
Power factor	0.80	0.80	0.81	0.82	0.83	0.84	0.85	0.86	0.87	0.86
Max torque (Nm)	220	250	290	320	340	380	420	500	550	620
starting current (A)	90	110	120	150	180	200	220	250	280	300
starting torque (Nm)	100	130	180	200	220	250	280	300	320	350

(2) Build product mathematical model

A frequently quoted example of a scale-based product platform in academia is Simpson's universal electric motor. The detail information about the mathematic model for that universal motor is mentioned in the paper "Product platform design: method and application" [3]. In this paper, we use a mathematical model developed by ourselves for the small-size induction motor that is different from Simpson's model for the universal motor. According to customers' requirements, we could gain several sets of design variables that could make the product's performance parameters satisfy customers' requirements. As to the input parameters of this model, that is the customers' requirements, we only take these six performance parameters into consideration: power, efficiency, power factor, maximum torque, starting current and starting torque. While as to the output parameters of this model, that is the design variables, we only consider five representative design variables: number of poles, number of rotor slots per phase·pole, number of stator slots, radius of core, and effective length of the stator.

(3) Compute product design variable metrics

On the basis of the customer demand metric, we calculate the design variables for each customer group through the mathematical model for the three-phase induction motor. Among various design variables of motor, we only take five representative design variables into the demonstration for our method: number of poles, number of rotor slots per phase·pole, number of stator slots, radius of core, and effective length of the stator.

As to each customer group, we could obtain several different design variable vectors that satisfy the customer group's demands. For example, 5 design variable vectors are gained for customer group I (shown in Table.2). 6 design variable vectors are gained for customer group II (shown in Table.3). In the same way, from group III to X, we

Table 2. The design variable vectors for customer group I

	Vector1	Vector 2	Vector 3	Vector 4	Vector 5
number of poles	3	3	3	4	4
number of rotor slots	3	4	4	3	4
number of stator slots	66	58	88	88	112
radius of core (m)	0.18	0.18	0.19	0.19	0.20
effective length of stator (m)	0.15	0.15	0.14	0.14	0.16

Table 3. The design variable vectors for customer group II

	Vector1	Vector 2	Vector 3	Vector 4	Vector 5	Vector 6
number of poles	2	3	3	3	3	3
number of slots	4	3	3	4	4	4
number of stator slots	62	66	68	58	82	88
radius of core (m)	0.17	0.19	0.19	0.19	0.20	0.20
effective length of stator (m)	0.13	0.17	0.18	0.18	0.17	0.17

obtain 8, 5, 7, 6, 5, 3, 3, 4 design variable vectors respectively. Due to the limited space, we only show the 5 design variable vectors for customer group I , the 6 design variable vectors for customer group II, and other groups are omitted.

(4) Construct chromosome and initial population
According to the coding rule for chromosomes mentioned in the Section 3.1, we code the chromosomes on the basis of the 10 design variables metrics for 10 customer groups obtained in previous step. There are 10 customer groups, so the length of chromosome is 10. Since 5 design variable vectors are gained for customer group I (shown in Table.2), a random integer from 1 to 5 is assigned to the first unit of chromosome. 6 design variable vectors are gained for customer group II (shown in Table.3), a random integer from 1 to 6 is assigned to the second unit of chromosome. In the same way, other units of chromosome are assigned a random integer according the total number of design variable vectors gained for respective customer groups.

Because of the randomness, a lot of different chromosomes will be obtained in that way. According to a thumb rules mentioned in Section 2.5, population size = 10×the number of units in a chromosome, we determine the population size as 100. Then 100 different chromosomes form the initial population for the later evolution.

(5) Perform evolution to find the optimal chromosome
Each customer group is related to several different design variable vectors. One vector is chosen from every customer group, then there 10 vectors are obtained, which construct one possible product platform for our target market. According to the probability theory, there are 9072000 kinds of possible combinations (5×6×8×5×7×6×5×3×3×4). By means of genetic algorithm, we search for one or more optimal combination. With the consideration of different importance of each design variable, we give them with different weight. The higher the weight is, the relating design variable has more influence on the cost of the product.

We write a specialized program for our product planning method using C++ language. Using that program, we calculate for the problem mentioned above. In the evolution process, we set the population size as 100, crossover probability as 0.80, mutation probability as 0.05. The program has run for 2000 generations, and finds one optimal solution in the 1576*th* generation, shown as Table.5:

Table 4. The optimal design variables

Customer group	I	II	III	IV	V	VI	VII	VIII	IX	X
number of poles	3	3	3	3	3	2	2	2	2	3
number of rotor slots	4	4	4	4	4	4	4	4	4	4
number of stator slots	88	58	56	88	88	58	56	58	56	56
radius of core (m)	0.19	0.19	0.19	0.22	0.22	0.20	0.20	0.20	0.22	0.27
effective length of stator (m	0.14	0.18	0.17	0.19	0.19	0.18	0.19	0.20	0.20	0.23

(6) Product platform strategy

Based on the above optimal solution, we planning the product platform. The number of rotor slots in the table is the same, all of them are 4; so we choose that design variable ("number of rotor slots per phase·pole") as the common parameter with its optimal value as 4. Then, other 4 design variables become the variable attributes, where the domain of "number of poles" is {3, 2}, "number of stator tooth" is {56, 58, 88}, "radius of core" is {0.19, 0.20, 0.22, 0.27} and "effective length of stator" is {0.14, 0.17, 0.18, 0.19, 0.20, 0.23}.

5 Comparison with Benchmark Results

In this section, a group of 10 individual designed motor is created as a set of benchmark motors for the comparison with the result gained in the previous step. In the set of 10 benchmark motors, the performance parameters are achieved to the best. Here, we choose one of the performance parameters, the efficiency, as the objective in evaluating the optimal performance of motors. Therefore, the 10 benchmark motors are the ones with highest efficiency for the 10 customer groups respectively.

On the basis of the motor mathematical model, we calculate design variables for the 10 benchmark motors with the highest efficiency within satisfying respective customer group's demand. The 10 benchmark motors' design variables and corresponding efficiency are shown in the right part of Table.6. As to the 10 motors designed by the method of product platform planning proposed in this paper, their design variables and efficiencies are also shown in the left part of the Table.6. The notes for the meaning of design variables 1 to 5 are list at the bottom of the table.

The differences of efficiency between the benchmark motors and a family of motors based on the platform are list in the last column of Table.6. The average of the differences of efficiency is -1.1%, which means that the family of motors based on the platform loses 1.1% in efficiency more than the benchmark motors.

However, with the loss in performance, the family of motors based on the platform has better structure for design variables than the benchmark motors, because the numbers of different values as to 5 design variables in the former than the latter. For example, as to the number of poles (design variable 1), the former is 2 while the latter is 3. In the same way, as to other 4 design variables, the former is 1, 3, 4, and 6; while the latter is 2, 6, 6, and 6. It means that the family of motors based on the product platform will have the lower cost than the benchmark motors in developing products.

Table 5. The comparison between benchmark motors and a family of motors based on the platform

Customer Groups	Product family motors						Benchmark motors						Diff. of η
	Design variables					η(%)	Design variables					η(%)	
	1	*2*	*3*	*4*	*5*		*1*	*2*	*3*	*4*	*5*		
I	3	4	88	0.19	0.14	89.2	3	4	88	0.19	0.14	89.2	0
II	3	4	58	0.19	0.18	90.3	3	3	66	0.19	0.17	91.2	-0.9
III	3	4	56	0.19	0.17	88.9	4	3	86	0.22	0.18	90.8	-1.9
IV	3	4	88	0.22	0.19	90.6	3	3	56	0.21	0.17	91.4	-0.8
V	3	4	88	0.22	0.19	90.2	2	3	56	0.18	0.19	91.8	-1.6
VI	2	4	58	0.20	0.18	91.5	2	4	62	0.20	0.17	92.1	-0.6
VII	2	4	56	0.20	0.19	92.0	2	4	64	0.20	0.18	92.5	-0.5
VIII	2	4	58	0.20	0.20	91.6	2	4	64	0.21	0.19	92.7	-1.1
IX	2	4	56	0.22	0.20	92.6	2	4	62	0.22	0.18	93.0	-0.4
X	3	4	56	0.27	0.23	91.9	3	3	64	0.24	0.21	93.3	-1.4
Num of diff. variables	**2**	**1**	**3**	**4**	**6**		**3**	**2**	**6**	**6**	**6**		Avg: **-1.1**

Note: Design variable 1: number of poles; Design variable 2: number of rotor slots per phase·pole; Design variable 3: number of stator slots; Design variable 4: radius of core (m); Design variable 5: effective length of stator (m);

6 Conclusions

The strategy of product platform is one effective way to mass customization, and research on the platform planning is also the hotspot in academia. Thanks to that flurry of activity in the engineering design community, this nascent field of research has matured rapidly in the past decade. There has been considerable progress in planning, designing and optimizing product platforms and families of products derived from them. In this paper, we propose a product platform planning method based on the genetic algorithm, which focuses on improving the commonality of the product platform within the satisfactory of the diverse customer needs, and then determine the variable product attributes and their variation ranges, and the common parameters of product platform and their optimal values. We use a specialized program to realize the method and also validate its feasibility using an example of induction motor. At last we compare the results with the benchmark products that are individually designed to the optimal performance.

The core of the product platform planning is the search for a proper tradeoff between the commonality within the product platform and performance for the family of products derived from the platform. Genetic algorithms are well suited for solving combinational problems owning to their stochastic nature. Therefore, the proposed method will become more practical and powerful if the number of design variables, and the number of customer groups are relatively large. However, this algorithm is just suitable to scale-based product platform, and the parameters in product's mathematical model should be explicit and definable. Therefore, as to the module-based product platform, or some products that is difficult for building mathematical model, this algorithm will not work. As to those limitations, we will make further research in future.

References

[1] Jiao Jianxin. Design for Mass Customization by Developing Product Family Architecture. A dissertation of PhD of the Hong Kong university of science and technology, 1998

[2] Meyer M, Lehnerd A. The Power of Product Platforms. New York: The Free Press, 1997

[3] Simpson T W, Maier J R A, Mistree F A. Product platform design: method and application. Research in Engineering Design, 2001, 13(1): 2-22

[4] Sabbagh K. Twenty-first century jet: the market and marketing of Boeing 777. Scribner, New York, 1996

[5] Hernandez G, Allen J K, Simpson T W, et al. Robust design of family for products with production modeling and revaluation [J]. ASME Journal of Mechanical Design. 2001, 6: 183-190

[6] Nayak R U, Chen W, Simpson T W. A variation-based method for product family design. Engineering Optimization, 2002, 34(1): 65-81

[7] Simpson T W, Maier J R A, Mistree F A. Product platform concept exploration method for product family design. ASME Design Theory and Methodology, 1999(9): 1-219

[8] Tseng M M, Jiao Jianxin. Design for mass customization. Annuals of CIRP, 1996, 45(1): 153-156

[9] Han J, Kamber M. Data mining. Concepts and techniques. San Francisco: Morgan Kaufmann Publishers; 2001

[10] Javier P. Gonzalez-Zugasti, Kevin N. Otto, et al. A method for architecting product platforms. Research in Engineering Design, 2000, 12(1): 61-72

[11] Messac A, Martinez M P, Simpson T W. Effective product family design using physical programming. Engineering Optimization, 2002, 34(3): 245-261

[12] Muhlenbein H. Evolutionary algorithm: theory and applications. GMD Birlinhoven, 1995

possible. As mentioned in some basic researches [12], premature convergence is one of the major problems in GA. This problem relates to the uniformity of the population's genotypic content and the inability of the evolutionary process to produce new genetic material when it reaches local optima. This problem is due to the fact that after producing some good chromosomes, they try to distribute their genotypic content using selection and recombination operators. So, other chromosomes have been deleted from the population after some iteration. In the traditional GAs, the selection operator uses the chromosome's fitness as the only criterion to select chromosomes for the next generation. However, in *SSO*, each chromosome is evaluated using two parameters: fitness and a new criterion based on the current chromosome's dissimilarity with the best chromosome in the population. We called this measure the Genetic Advantage Criterion (*GAC*). The main motivation to introduce *GAC* is to preserve the genotypic content of the population. In the other word, it was assumed that if one chromosome has different genotypic content than the other ones in the population, then it is one of the valuable individuals. In the SSO, we compare the genotypic structure of that chromosome with the genotypic structure of the best individual in the population using Hamming Distance as the measuring criterion. So, *GAC* is calculated using *HD(C, Best)*, where *C* is the current chromosome and *Best* is the best chromosome of the population with respect to its fitness. After calculating *GAC*, Survival Probability (*SP*) of each chromosome is calculated using equation 3.

$$SP = w_1 GAC + w_2 FIT \tag{3}$$

It is clear that *SP* is an averaged value between genotypic (*GAC*) and phenotypic (*FITness*) evaluation of the current chromosome.

5 XCSF with FAPGA

In this section, we describe the overall architecture of our proposed extension to XCSF. As mentioned in [7], the Widrow-Hoff update rule, encounters some difficulties in certain range of input variables. In [7], some solutions to overcome these issues are proposed. In this paper, we propose a new method to approximate the payoff functions with FAPGA. This new extension is called XCSF-PG. the overall architecture of XCSF-PG is as same as XCSF with some modifications. The main difference between XCSF-PG and XCSF is in classifier's structure. Classifier's structure in XCSF-PG is modified as shown in Figure 3 to support FAPGA. In this type of classifier, prediction is calculated using a secondary population of chromosomes which is allocated for each rule separately.

To calculate the prediction value for each rule, we follow this procedure: At first, we choose the best chromosome of its secondary population. Then, we calculate the prediction value by applying the evolved polynomial by the best chromosome of the population on the environmental input for the current state. We call this value *G(x)*.

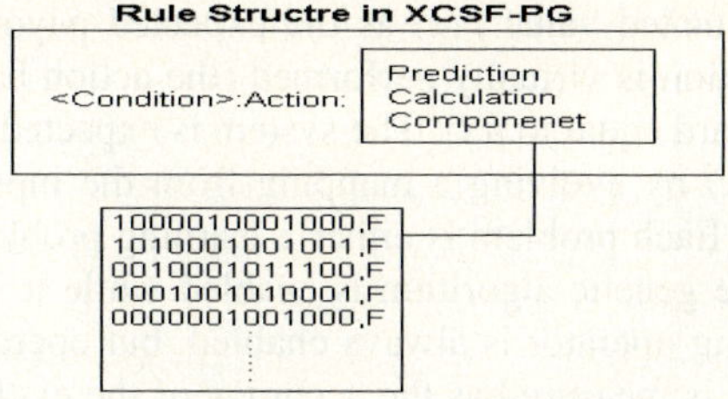

Fig. 3. The Structure of the Classifiers in XCSF-PG

5.1 Secondary Population's Architecture

As we mentioned in previous section, to calculate the prediction value for each rule, we must choose the best chromosome of its secondary population. So calculating the fitness of the secondary population is one of the most important procedures in XCSF-PG. In the fitness calculation routine, the most important goal is to select an individual with lowest error rate in the previous prediction estimation epochs. Therefore, in XCSF-PG fitness calculation must be done using rule's prediction error. To do so, in each iteration, a tuple is made of the environmental input and reward. We call this tuple an Estimation Twin (*ET*). *ETs* are listed and memorized for each fired rule which was involved in the reward gathering procedure. When the number of the stored ET's of a specified rule reaches a predefined threshold n_{ET}, then the fitness of the chromosome in the secondary populations is calculated as follows: For each chromosome, the estimated output for each input are calculated using previously described method separately and the overall error is calculated using equation 4.

$$E_j = \frac{\sum_{i=1}^{n} \left[G_j(x_i) - P(x_i)\right]^2}{n_{ET}} \tag{4}$$

Where $G_j(x_i)$ is the estimated value of the jth chromosome for x_i and $P(x_i)$ is the environmental reward for the same input (from *ET*s). Then fitness of the jth chromosome is calculated using equation 5.

$$F_j = e^{-E_j} \tag{5}$$

When fitness of the secondary population is calculated, the mating pool is constructed using SSO. Then the crossover procedure occurs as described before and the new generation of the secondary population is constructed.

6 Design of the Experiments

All of the experiments discussed in this paper are single step problems and chosen due to the standard design that is used in the literature [7]. In each experiment the system has to learn to approximate a target function *f(x)*; each experiment consists of a finite number of problems that system must solve. For each problem, an example <*x,f(x)*> of the target function *f(x)* is randomly selected; *x* is the input for the system who computes the approximated value *f'(x)* as the expected payoff of the only available dummy action; the action is virtually performed (the action has no actual effect), and system receives a reward equal to *f(x)*. The system is expected to learn to approximate the target function *f(x)* by evolving a mapping from the inputs to the payoff of the only available action. Each problem is either a learning problem or a test problem. In learning problems, the genetic algorithm is enabled while it is turned off during test problems. The covering operator is always enabled, but operates only if needed. The system's performance is measured as the accuracy of the evolved approximation *f'(x)* with respect to the target function *f(x)*. To evaluate the evolved approximation, we measure the Mean Absolute Error (*MAE*) defined as:

$$MAE = \frac{1}{n}\sum_{x}\left| f(x) - \hat{f}(x) \right| \tag{6}$$

Where *n* is number of the points for which *f(x)* is defined. In particular we use the average *MAE* over the performed experiments. Our benchmark problems are some random generated polynomials that are produced using a random sequence of integers between -8 and 8 as the order and coefficient of the relevant terms. For example if our random number generator produces the following sequence: 2, -4, 5, 3, 8 ... then, the resulted polynomial has 2 terms and will be in the form of: $-4x^5 + 3x^8$. The input range of our benchmark problems is [-1, 1]. Table 1 represents these benchmark problems.

Table 1. Benchmark Problem Definition

	Definition
Problem1	$2x^5$
Problem 2	$6x^4 - 9x^3 + 2x$
Problem 3	$3x^2 + 5x^7 - 4x^6$
Problem 4	$-7x^2 - 3x^3 + 4x^8 - 2x^4$

The approximated polynomials are drawn as follows:

1. The system is trained using *<x,y>* tuples that are drawn from the problem definition.
2. After training phase (50000 iterations), input variables are generated using a uniform distribution between -1 and 1 (the input range of the problems) with sample rate of 0.01.
3. Input variables are fed to the system and the obtained results are stored as *f'(x)*.
4. The above procedure is repeated 150 times and *f'(x)* is averaged over the results of these 150 independent runs.

In our experiments four different systems are compared with each other using the above procedure. These systems are XCSF, XCSFH, XCSFGH and XCSF-PG. XCSF and XCSF-PG are described in details in this paper but XCSFH and XCSFGH are described briefly in the following. XCSFH [9] is our nick name for a newly introduced version of XCSF which is able to approximate higher order polynomials and utilize a corrected method of linear approximation to overcome the input range dependency issue in XCSF. Also XCSFGH is introduced in [6] and is an extension to XCSFG [5] which is able to approximate higher order polynomials. XCSFGH is very similar to XCSF-PG but uses simple GA operators instead of the pattern-based ones which are described earlier

6.1 The Experimental Results

To start our experiments, the XCSF's Parameters are initiated as in [1] and XCSFH parameters are also initiated as [9]. Also size of the secondary populations in XCSF-PG and XCSFGH are set to 50, and the other parameters are initiated as follows: Len_{order}= 3, Len_{coeff} = 8, $SSO\text{-}w_1$ =0.7, $SSO\text{-}w_2$ =0.3, $U=2^{len-1}-1$ and $L=-(2^{len-1}-1)$ for coefficient calculation, $U=2^{len-1}$ and $L=0$ for order calculation, P_{th}=.5 and P_{add}=.3.

Table 2. MAE for XCSF(H, GH, -PG) in the benchmark problems for the input rage of [-1,1] and the input range if [999,1001] (Averaged between 150 runs)

[-1, 1]	PR1	PR2	PR3	PR4
XCSF	0.142	0.254	0.351	0.295
XCSFH	0.004	0.198	0.299	0.287
XCSFGH	0.000	0.246	0.248	0.341
XCSF-PG	0.003	0.124	0.148	0.152
[999, 1001]	PR1	PR2	PR3	PR4
XCSF	0.698	1.872	1.652	1.243
XCSFH	0.02	0.228	0.220	0.301
XCSFGH	0.05	0.251	0.224	0.295
XCSF-PG	0.02	0.115	0.137	0.179

Also in Table 3, the number of the final classifiers for XCSF (H, GH, -PG) is represented. The number of macro classifiers is extracted using the following procedure: in the end of each experiment, all of the macro-classifiers in the system's population are sorted with respect to their numerosity in descending order. Then the top ones in the queue are selected till all of the problem space was covered by the selection set. Then, the average size of the selection set over 150 independent runs is inserted in the Table 3. Also we define another performance criterion called Optimal Reach Point (*ORP*) which is defined as follows: *ORP* for a given experiment is the first point which the systems *MAE* in its previous 500 iterations was less than a predefined threshold. This threshold is set to 0.3 in our experiments.

To quantify these results and to test whether it is statistically significant we apply these experiments 150 times with different random generators independently and then

Table 3. Number of the obtained Macro Classifiers in XCSF(H, GH, -PG)'s rule BASE FOR the benchmark problems with the input in [-1,1] interval. (Averaged between 150 runs)

	PR1	PR2	PR3	PR4
XCSF	88	70	75	81
XCSFH	16	11	13	21
XCSFGH	10	15	19	17
XCSF-PG	12	14	20	15

Table 4. ORP for XCSF(H, GH, -PG) in the benchmark problems for the input rage of [-1,1] and the input range if [999,1001] (Averaged between 150 runs). If an algorithm does not reach the desired error threshold (0.3) the coresponding cell will be filled by NA.

[-1, 1]	PR1	PR2	PR3	PR4
XCSF	32550	36552	46231	39984
XCSFH	25669	36552	47005	40895
XCSFGH	26951	38765	40251	49872
XCSF-PG	20512	31225	34952	36552
[999,1001]	PR1	PR2	PR3	PR4
XCSF	NA	NA	NA	NA
XCSFH	27685	36652	37596	49932
XCSFGH	26558	34991	39885	49865
XCSF-PG	22542	29887	32568	40259

apply one-tailed Wilcoxon signed rank test [13] on the resulted values demonstrated in Table 2, 3 and 4. The test results are represented in Table 5. With respect to these *P-values*, it can be concluded and statistically verified that XCSF-PG can reach the optimal value faster than both XCSFH and XCSFGH and also can produce more accurate approximation for the given problems.

Table 5. The Resulted P-Values of WILCOXON Test (The cell with significance level above 99.5% are indicated with gray background)

	PR1	**PR2**	**PR3**	**PR4**
XCSF-PG vs. XCSF (MAE) [-1,1]	.000	.000	.000	.000
XCSF-PG vs. XCSFH (MAE) [-1,1]	.125	.048	.035	.047
XCSF-PG vs. XCSFGH (MAE) [-1,1]	.143	.038	.045	.027
XCSF-PG vs. XCSF (MAE) [999,1001]	.000	.000	.000	.000
XCSF-PG vs. XCSFH (MAE) [999,1001]	.458	.039	0.025	0.031
XCSF-PG vs. XCSFGH (MAE) [999,1001]	.264	.031	0.029	0.026
XCSF-PG vs. XCSF (Number of the macro classifiers)	.000	.000	.000	.000
XCSF-PG vs. XCSFH (Number of the macro classifiers)	.425	.493	.399	.445
XCSF-PG vs. XCSFGH (Number of the macro classifiers)	.436	.358	.492	.417
XCSF-PG vs. XCSF (ORP) [-1,1]	.000	.000	.000	.012
XCSF-PG vs. XCSFH (ORP) [-1,1]	.001	.003	.015	.009
XCSF-PG vs. XCSFGH (ORP) [-1,1]	.007	.011	.013	.008
XCSF-PG vs. XCSF (ORP) [999,1001]	.000	.000	.000	.000
XCSF-PG vs. XCSF (ORP) [999,1001]	.020	.006	.007	.005
XCSF-PG vs. XCSFH (ORP) [999,1001]	.015	.010	.016	.002

References

[1] Wilson, S. W. 2002. Classifiers that Approximate Functions. Journal of Natural Computating 1, 211-234.

[2] Wilson, S. W. 1995. Classifier Fitness Based on Accuracy. Evolutionary Computation 3 (2), 149-175.

[3] Lanzi, P. L., Loiacono D., Wilson, S.W., and Goldberg, D.E. 2005. XCS with Computable Prediction for the Learning of Boolean Functions. Technical Report 2005007, Illinois Genetic Algorithms Laboratory.

[4] Wilson, S. W. 2001. Function Approximation with a Classifier System. In proceedings of the GECCO-2001, San Francisco, California, USA, pp. 974-981.

[5] Hamzeh, A., Rahmani, A. 2005. An Evolutionary Function Approximation Approach to Compute Prediction in XCSF. In Proceedings of 16th European Conference of Machine Learning. Porto, Portugal.

[6] Hamzeh, A., Rahmani, A. 2006. Extending XCSFG beyond Linear Approximation. To be appeared in the proceedings of the IEEE world congress on Evolutionary Computation (CEC 2006), Vancouver, Canada.

[7] Lanzi, P. L., Loiacono D., Wilson S.W., and Goldberg D.E. 2005. Generalization in the XCSF Classifier System: Analysis, Improvement, and Extension. Technical Report 2005012, Illinois Genetic Algorithms Laboratory.

[8] Lanzi, P. L., Loiacono D., Wilson S.W., and Goldberg D.E. 2005. XCS with Computable Prediction in Multistep Environments. Technical Report 2005008, Illinois Genetic Algorithms Laboratory.

[9] Lanzi, P. L., Loiacono D., Wilson S.W., and Goldberg D.E. 2005. Extending XCSF beyond Linear Approximation. Illinois Genetic Algorithms Laboratory University of Illinois at Urbana-Champaign.

[10] Hamzeh, A., Rahmani, A. 2005, A New Selection Method for Genetic Algorithms based on Genotypic Information of the Population. In Proceedings of 10th Annual Conference of Computer Society of Iran, Tehran.

[11] Hamzeh, A. Rahmani, A. 2004. An Adaptive Pattern-Based Uniform Crossover for Genetic Algorithms. In Proceedings of 9th Iranian Computer Society, Computer Conference, Tehran, Iran.

[12] Holland, J. H. 1975. Adaptation in Natural and Artificial Systems. Ann Arbor: University of Michigan Press. Republished by the MIT press, 1992.

[13] Kanji, G. 1994. 100 Statistical Tests, SAGE Publications.

Genetic Algorithm Based on the Orthogonal Design for Multidimensional Knapsack Problems*

Hong Li[1,2], Yong-Chang Jiao[1], Li Zhang[2], and Ze-Wei Gu[2]

[1] National Laboratory of Antennas and Microwave Technology
Xidian University, Xi'an, Shaanxi 710071, China
lihong@mail.xidian.edu.cn, ychjiao@xidian.edu.cn
[2] School of Science, Xidian University, Xi'an, Shaanxi 710071, China

Abstract. In this paper, a genetic algorithm based on the orthogonal design for solving the multidimensional knapsack problems is proposed. The orthogonal design with the factor analysis, an experimental design method, is applied to the genetic algorithm, to make the algorithm be more robust, statistically sound and quickly convergent. A crossover operator formed by the orthogonal array and the factor analysis is presented. First, this crossover operator can generate a small, but representative sample of points as offspring. After all of the better genes of these offspring are selected, an optimal offspring better than its parents is then generated in the end. Moreover, a check-and-repair operator is adopted to make the infeasible chromosomes generated by the crossover and mutation operators feasible, and make the feasible chromosomes better. The simulation results show that the proposed algorithm can find optimal or close-to-optimal solutions with less computation burden.

1 Introduction

The multidimensional knapsack problem (MKP) is a combinatorial optimization problem, which is also a NP-hard problem. The general MKP can be expressed as

$$\begin{cases} \text{maximize} & f(x) = \sum_{i=1}^{n} c_i x_i \\ \text{subject to} & \sum_{i=1}^{n} a_{ij} x_i \le b_j, \quad j = 1, \cdots, m \\ & a_{ij} \ge 0, \quad c_i \ge 0 \\ & x_i \in \{0,1\}, \quad i = 1, \cdots, n, \end{cases} \tag{1}$$

where $x = (x_1, x_2, \cdots, x_n)$ is regarded as the item set in the knapsacks, x_i is set as follows:

$$x_i = \begin{cases} 1, & i\text{th item in every knapsack}, \\ 0, & \text{otherwise}, \end{cases}$$

* This work was supported by the National Natural Science Foundation of China under grants 60171045 and 60374063.

L. Jiao et al. (Eds.): ICNC 2006, Part I, LNCS 4221, pp. 696–705, 2006.

n is the number of items, and m is the number of knapsack constraints with capacities b_j $(j = 1, 2, \cdots, m)$. Each item x_i requires a_{ij} units of resource consumption of jth knapsack $(j = 1, 2, \cdots, m)$ and yields c_i units of profit upon inclusion. The goal is to find a subset of items that yields maximum profit without exceeding the resource capacities. The much simpler case with single constraint $(m = 1)$ is known as the single knapsack problem, which is not strongly NP-hard, and effective approximate algorithms have been developed for obtaining its near-optimal solutions. The general case corresponding to $m \geq 2$ is known as the multidimensional knapsack problem (MKP), which is strongly NP-hard. Many practical problems can be formulated as the MKP, for example, the capital budgeting problem, allocating processors and databases in a distributed computer system, and the project selection and cargo loading. Many algorithms, such as the dynamic programming method, the enumerative method, the branch-and-bound method and the heuristic algorithms, are proposed for solving the MKP.

In this paper we apply the orthogonal experimental design method with factor analysis to the genetic algorithm, so that the resulting algorithm can be more robust, statistically sound and quickly convergent. In practice, the experimental design methods have been used to solve a few optimization problems with discrete or continuous variables. Zhang and Leung [1] proposed an orthogonal genetic algorithm for multimedia multicast routing problems, in which the orthogonal design, an experimental design method, is used to design a new crossover operator. Subsequently, Leung and Wang [2] also proposed an orthogonal genetic algorithm with quantization for global numerical optimization with continuous variables, but their algorithm has heavy computational burden. Tsai et.al. [3] proposed a hybrid Taguchi-genetic algorithm, which combined the orthogonal array with factor analysis, for global numerical optimization with continuous variables, and got the better results with low computational cost. Ho et.al. [4] proposed two intelligent evolutionary algorithms, in which the recombination operator is based on the orthogonal experimental design, the factor analysis as well as the divide-and-conquer approach for large parameter optimization problems with continuous variables. Both algorithms also have better performances. Inspired by these existing algorithms, we present a genetic algorithm based on the orthogonal design (OGA), in which the orthogonal design with factor analysis is incorporated into the genetic algorithm, for solving the multidimensional knapsack problems (MKPs). Our algorithm is similar to but different from the algorithms in [1] [2] [3] [4]. A crossover operator formed by the orthogonal array and the factor analysis can generate a small, but representative sample of points, and can exploit an optimal offspring inheriting all of the better genes of parents. Moreover, a check-and-repair operator based on the greedy approach is adopted to make infeasible chromosomes generated by the crossover and mutation operators feasible, and make feasible chromosomes better. The simulation results show that the proposed algorithm can find optimal or close-to-optimal solutions with less computation burden.

This paper is organized as follows. The orthogonal experimental design method with factor analysis is introduced in Section 2, and a genetic algorithm based on the orthogonal design is then proposed in Section 3. Experimental results and comparison are also presented in Section 4. We finally conclude our paper in Section 5.

2 Orthogonal Experimental Design Method with Factor Analysis

The statistical experiment design is the process of planning experiments so that appropriate data will be collected, a minimum number of experiments will be performed to acquire necessary technical information, and suitable statistic methods will be used to analyze the collected data.

An efficient way to study the effect of several factors simultaneously is to use the orthogonal experimental design with both the orthogonal array and the factor analysis. The factors are the variables, which affect response variables, and a setting of a factor is regarded as a level of the factor. A "complete factorial" experiment would make measurements at each of all possible level combinations. However, the number of level combinations is often so large that this is impractical, and a subset of level combinations must be judiciously selected to be used, resulting in a"fractional factorial" experiment [1], [2], [3]. The orthogonal experimental design utilizes properties of fractional factorial experiments to efficiently determine the best combination of factor levels to use in design problems. The detailed description of the orthogonal experimental design can be found in [1], [2], [3].

Two-level orthogonal arrays are used in this paper. The general symbol for two-level standard orthogonal arrays is $L_N(2^{N-1})$, where $N = 2^k$ is the number of experimental runs, k is a positive integer such that $k > 1$, $N - 1$ denotes the number of columns in the orthogonal array. The letter "L" comes from "Latin", and the idea that orthogonal arrays are used for the experimental design have been associated with Latin square designs from the outset. The two-level standard orthogonal arrays used in this paper include $L_8(2^7)$, $L_{12}(2^{11})$, $L_{16}(2^{15})$, $L_{24}(2^{23})$, $L_{32}(2^{31})$ and $L_{64}(2^{63})$.

After evaluation of the N combinations, the summarized data are analyzed using the factor analysis. Factor analysis can evaluate the effects of individual factors on the objective (or fitness) function, rank the most effective factors, and determine the better level for each factor. In this paper, the factor analysis in [3] is used in our algorithm. Since the objective function is to be maximized (larger-the-better), we choose $\eta_i = y_i^2$. Let y_i denote the function evaluation value of experiment i, $(i = 1, 2, \cdots, N)$, where N is the total number of experiments. The effects of the various factors can be given as follows:

$$E_{fl} = \text{sum of } \eta_i \text{ for factor } f \text{ at level } l \tag{2}$$

where i is the experiment number related to level l.

Two chromosomes from the initial population are randomly chosen to execute the matrix experiments of the orthogonal array. The primary goal in conducting this matrix experiment is to determine the best or the optimal level for each factor. The optimal level for a factor is the level that gives the highest value of E_{fl} in the experimental region. For a two-level problem, if $E_{f1} > E_{f2}$, the optimal level is level 1 for factor f. Otherwise, level 2 is the optimal one. After the optimal levels for each factor are selected, one also obtains an optimal chromosome. Therefore, the new offspring has the best or nearly the best function value among those of 2^{N-1} combinations of factor level, where 2^{N-1} is the total number of experiments needed for all combinations of factor levels.

3 A Genetic Algorithm Based on the Orthogonal Design

We adopt binary code due to the nature of the problem. A string $x = (x_1 \cdots x_n)$ is used to express a chromosome, where $x_i \in \{0, 1\}$, $i = 1, 2, \cdots, n$, which represents a trial solution of problem (1).

3.1 Generation of the Initial Population

We generate the *pop* chromosomes by the following `Algorithm 1`, where *pop* denotes the population size.

`Algorithm 1:`

Step1. Generate n-dimensional random vector $\gamma = (\gamma_1, \cdots, \gamma_i, \cdots, \gamma_n)$, where $\gamma_i \in (0, 1)$.

Step2. For $i \in \{1, \cdots, i, \cdots, n\}$, if $\gamma_i < 0.5$, let $x_i = 0$; Otherwise, $x_i = 1$. Thus, produce a chromosome $(x_1 \cdots x_n)$.

Step3. If $\sum_{i=1}^{n} a_{ij} x_i \leq b_j, \ \ j = 1, \cdots, m$, then this chromosome $(x_1 \cdots x_n)$ is feasible; Otherwise, go to Step1.

Step4. Repeat the above steps *pop* times and produce *pop* initial feasible chromosomes.

3.2 Fitness Function

The objective function is directly chosen as the fitness function. Since the objective function is to be maximized, the larger the fitness function, the better the chromosome is. Before the fitness value of a chromosome is evaluated, its feasibility must be checked in advance. For infeasible chromosome, we adopt repair operator discussed in Subsection 3.3 to change it into a feasible one.

3.3 Check-and-Repair Operator

Obviously, the chromosomes generated by the crossover and mutation operators may be infeasible. In order to guarantee their feasibility, a check-and-repair operator based on the greedy algorithm is applied, as described in [5].

We first calculate the profit density $\delta_{ij} = b_j \cdot c_i/a_{ij}$ of every item in every knapsack, and only consider the lowest value $\delta_i = min\{b_j \cdot c_i/a_{ij}\}$ for every item. Second, sort and relabel items by increasing or decreasing values of δ_i. Then, items are successively taken and included in the knapsack if they fit in it.

For the infeasible chromosome, we change it into feasible chromosome by the following `Algorithm 2`.

`Algorithm 2:`

Step1. Calculate the profit density $\delta_{ij} = b_j \cdot c_i/a_{ij}$ for every item in every knapsack.

Step2. Compute the lowest value of the profit density $\delta_i = min\{b_j \cdot c_i/a_{ij}\}$ for every item.

Step3. Sort and relabel items according to the ascending order of δ_i.

Step4. Remove the corresponding item with lowest values of δ_i from the item set (i.e. change corresponding gene 1 into gene 0 in the chromosome).

Step5. Repeat Step4 until a feasible chromosome is achieved.

In addition, in order to get a better chromosome in the mutation operator, we also improve the feasible chromosome by using the check-and-repair operator. A better chromosome than the original chromosome is then generated by the following `Algorithm 3`.

`Algorithm 3:`

Step1. Calculate the profit density $\delta_{ij} = b_j \cdot c_i/a_{ij}$ of every item out of the knapsack.

Step2. Compute the lowest value of the profit density $\delta_i = min\{b_j \cdot c_i/a_{ij}\}$ for every item.

Step3. Sort and relabel items according to the descending order of δ_i.

Step4. Add the corresponding item with highest values of δ_i into the item set (i.e. change corresponding gene 0 into gene 1 in the chromosome).

Step5. If one of knapsack constraints is not satisfied, then stop, and output the resulting chromosome. Otherwise, return to Step4.

3.4 Crossover Operator Based on the Orthogonal Experimental Design Method with Factor Analysis

Crossover offspring is generated by using the orthogonal experimental design method with factor analysis. There are n genes in a chromosome, which are regarded as factors, and each variable has two values 0 or 1, that is to say, each factor has two levels. Thus a two-level orthogonal array is used here. Let $L_N(2^{N-1})$ represent $N-1$ columns and N individual experiments corresponding to N rows, and $n \leq N-1$. If $n = N-1$, we directly adopt the orthogonal array $L_N(2^{N-1})$. If $n < N-1$, only the first n columns are used. We ignore the other $N-1-n$ columns, just like the method used in [3]. For problems with high dimensions, no larger orthogonal array can be used directly. So we must divide each parent chromosome into $N-1$ parts (or gene segments) in order to adopt appropriate orthogonal array, as similar to the method in [1], [4].

Let p_c be the crossover probability. Randomly select two parent chromosomes with probability p_c: $x^1 = (x_1^1 \cdots x_n^1)$ and $x^2 = (x_1^2 \cdots x_n^2)$. According to dimension of the problem, we choose appropriate two-level orthogonal array $L_N(2^{N-1})$ to execute the matrix experiments, and generate N offspring chromosomes. The purpose for the use of the orthogonal experimental design method is to produce a better chromosome from two randomly generated chromosomes. By the factor analysis, we obtain a potentially best combination of better levels of factors and then generate an optimal child chromosome, which inherit good genes of parents. If this optimal chromosome is infeasible, we make it feasible by using the check-and-repair operator given in Subsection 3.3 and evaluate its fitness value.

Now we illustrate the application of the orthogonal experimental design method and factor analysis in the crossover operator. For example, we choose the objective function as $f(x) = \sum_{i=1}^{7} x_i$ and two constraints as $\sum_{i=1}^{7} ix_i \leq 21$ and $4x_1 + 2x_2 + 5x_3 + 7x_4 + x_5 + 4x_6 + 5x_7 \leq 16$. The objective function would be maximized.

First, two chromosomes x^1 and x^2, each consists of seven factors corresponding to x_i of $f(x)$, are randomly chosen to execute various matrix experiments of an orthogonal array. Regarded as level 1, the values of seven factors in chromosome x^1 are 1, 1, 1, 1, 0, 0, and 0; those in chromosome x^2 are 0, 0, 0, 0, 1, 1, and 1, which are regarded as level 2. Obviously, x^1 and x^2 are all feasible. Therefore, $f(x^1) = 4$ and $f(x^2) = 3$. The orthogonal array $L_8(2^7)$ has been chosen, because each chromosome has seven genes (factors). Then, the seven genes in chromosomes x^1 and x^2 correspond to the factors A, B, C, D, E, F, and G, respectively, as defined in Table 1.

Next, as shown in Table 2, the values of level 1 and level 2 are assigned to the level cells in the orthogonal array $L_8(2^7)$. The function values of each row (offspring chromosome) are then calculated. The η_i for each experiment number i is also calculated.

Obviously, an optimal chromosome $\hat{x}$ generated by the orthogonal design and factor analysis with $f(\hat{x}) = 7$ is infeasible. By the repair operator, we obtain a feasible optimal chromosome $x^\star$ with $f(x^\star) = 5$, which is much better than its parent chromosomes x^1 and x^2 with $f(x^1) = 4$ and $f(x^2) = 3$. It is obvious that, instead of executing all combinations of factor levels, the orthogonal design method with factor analysis can offer an efficient approach toward finding the optimal chromosome by only executing eight experiments.

Table 1. Corresponding factors of x^1 and x^2

Factors	A	B	C	D	E	F	G
Level 1 (x^1)	1	1	1	1	0	0	0
Level 2 (x^2)	0	0	0	0	1	1	1

Table 2. Generating a better child chromosome from two parent chromosomes by using the orthogonal experimental design with factor analysis and the repair operator

Experiment number i	Factors							Function value	η_i
	A	B	C	D	E	F	G		
1	1	1	1	1	0	0	0	4	16
2	1	1	1	0	1	1	1	6	36
3	1	0	0	1	0	1	1	4	16
4	1	0	0	0	1	0	0	2	4
5	0	1	0	1	1	0	1	4	16
6	0	1	0	0	0	1	0	2	4
7	0	0	1	1	1	1	0	4	16
8	0	0	1	0	0	0	1	2	4
E_{f1}	72	72	72	64	40	40	40		
E_{f2}	40	40	40	48	72	72	72		
Optimal level	1	1	1	1	2	2	2		
Optimal chromosome $\hat{x}$ (infeasible)	1	1	1	1	1	1	1	7	
Feasible optimal chromosome $x^\star$ (after repair)	1	1	1	0	1	1	0	5	

3.5 Mutation Operator

Let p_m be the mutation probability. Randomly select a chromosome in the crossover offspring with probability p_m. A mutation procedure is performed, which mutates some randomly selected genes in the selected child chromosome from 0 to 1 or vice versa.

By using check-and-repair operator, we transform the infeasible chromosome into a feasible chromosome and convert the feasible chromosome into a better chromosome.

3.6 Selection Operator

We first compare the fitness values of all the chromosomes, including those in the current population and all the offspring generated by the crossover and mutation operators, sort the fitness values in the descending order. The first *pop* chromosomes constitute the next population. In addition, we retain the best chromosome of every generation.

3.7 Stopping Criterion

If the algorithm is executed to the maximal number of generations M, then stop. The best chromosome found in the last population is then taken as the approximate global optimal solution.

Now we present the proposed algorithm for solving problem (1) as follows.

`Algorithm 4:`

Data. Choose population size *pop*, crossover probability p_c, mutation probability p_m, maximal number of generations M.
Step 0. Generate the initial population set by Algorithm 1, and compute the fitness values.
Step 1. Crossover operation.
Step 1.1. Select a suitable two-level orthogonal array for matrix experiments.
Step 1.2. Choose randomly two chromosomes with possibility p_c to execute the matrix experiment.
Step 1.3. Calculate the function values of chromosomes generated by the matrix experiment.
Step 1.4. Calculate the effects of the various factors (E_{f1}, E_{f2}), generate an optimal chromosome and check its feasibility.
Step 1.5. If this optimal chromosome is infeasible, we make it feasible by `Algorithm 2`, and calculate its fitness value.
Step 1.6. Repeat Steps 1.2-1.5 *pop* times, produce pop_c offspring chromosomes, which form a temporary population.
Step 2. Mutation operation.
Step 2.1. Choose randomly a chromosome in all crossover offspring with possibility p_m, and choose randomly two gene bits in this chromosome.
Step 2.2. Change these two genes from 0 to 1 or vice versa.
Step 2.3. Repeat Steps 2.1-2.2 pop_c times, produce pop_m offspring chromosomes.
Step 3. Check-and-repair operation. For each mutation offspring, we check its feasibility. If it is feasible, we make it better by Algorithm 3; Otherwise, we make it feasible by Algorithm 2.
Step 4. Selection. Sort the fitness values in the descending order among parents and offspring populations. Select the better *pop* chromosomes as parents of the next generation, and retain the best chromosome.
Step 5. If the stopping criterion is met, then stop, and record the best chromosome as the approximate global optimal solution of problem (1). Otherwise, set $k = k + 1$, and go to Step 1.

4 Experimental Results and Comparison

To validate its efficiency, we carry out the proposed algorithm (OGA) on a PC with Pentium IV 2.4GHz CPU. The algorithm is implemented in VC++. We test 7 benchmark problems, which can be found in WWW address http:// people.brunel.ac.uk/ mastjjb/jeb/orlib/files/mknap1.txt.

Parameters of OGA for these problems are set to: $pop = 100$, $M = 50$, $p_c = 0.8$, $p_m = 0.3$. Two-level orthogonal arrays $L_8(2^7)$, $L_{12}(2^{11})$, $L_{16}(2^{15})$, $L_{24}(2^{23})$, $L_{32}(2^{31})$ and $L_{64}(2^{63})$ are used here. We performed 30 independent runs for each benchmark problem. The statistical results of OGA are summarized in Table 3. We compare OGA against the Partheno-Genetic Algorithm (PGA) in [6], as shown in Table 4.

From Tables 3 and 4, we find that optimal solutions or near-optimal solutions can be found by OGA for all the benchmark problems. Furthermore, OGA can find the optimal solutions or near-optimal solutions with the higher speed than PGA.

Table 3. Statistical results obtained by OGA for 7 benchmark problems over 30 independent runs. "MNFE" stands for the mean number of fitness evaluations and "MNG" for the mean number of generations when the best solution is found.

	Problem size (n/m)	Optimal value	Best value	Best solution	CPU time (second)	MNFE	MNG
T1	6/10	3800	3800	011001	0.001	357	1
T2	10/10	8706.1	8706.1	0101100101	0.001	488	1
T3	15/10	4015	4015	110101101100011	0.047	4645	8
T4	20/10	6120	6120	10000000010001111111	0.031	7971	8
T5	28/10	12400	12390	11100000100001101001 11111111	0.047	10059	9
T6	39/5	10618	10618	11010101101010111111 0010101110110111111	0.094	17100	9
T7	50/5	16537	16524	00010101101111111010 00100111111101111111 1111001111	0.172	22659	12

Table 4. Comparison of OGA with PGA for 7 benchmark problems

	Problem size (n/m)	Optimal value	Best value		CPU time (second)	
			OGA	PGA	OGA	PGA
T1	6/10	3800	3800	3800	0.001	0.3
T2	10/10	8706.1	8706.1	8706	0.001	0.3
T3	15/10	4015	4015	4015	0.047	0.3
T4	20/10	6120	6120	6120	0.031	0.3
T5	28/10	12400	12390	12400	0.047	0.4
T6	39/5	10618	10618	10618	0.094	0.6
T7	50/5	16537	16524	16537	0.172	0.5

5 Conclusion

In this paper, a genetic algorithm based on the orthogonal design (OGA) for solving multidimensional knapsack problems is proposed. The orthogonal experimental design with factor analysis is incorporated into the genetic algorithm, so that the resulting algorithm can be more robust, statistically sound and quickly convergent. OGA has also been compared with one existing algorithm, PGA, by solving 7 benchmark problems. Numerical results show that OGA is efficient and effective.

References

1. Qingfu Zhang, Yiu-Wing Leung: An Orthogonal Genetic Algorithm for Multimedia Multicast Routing. IEEE Trans. on Evolutionary Computation **3** (1999) 53-62
2. Yiu-Wing Leung, Yuping Wang: An Orthogonal Genetic Algorithm with Quantization for Global Numerical Optimization. IEEE Trans. on Evolutionary Computation **5** (2001) 41-53
3. Jinn-Tsong Tsai, Tung-Kuan Liu, and Jyh-Horng Chou: Hybrid Taguchi-Genetic Algorithm for Global Numerical Optimization. IEEE Trans. on Evolutionary Computation **8** (2004) 365-377
4. Shinn-Ying Ho, Li-Sun Shu, and Jian-Hung Chen: Intelligent Evolutionary Algorithms for Large Parameter Optimization Problems. IEEE Trans. on Evolutionary Computation **8** (2004) 522-541
5. Carlos Cotta, Jose Ma Troya: A Hybrid Genetic Algorithm for the 0-1 Multiple Knapsack Problem. Artificial Neural Nets and Genetic Algorithms 3, New York (1998) 250-254
6. Jian-cong Bai,Hui-you Chang,Yang Yi: An Partheno-Genetic Algorithm for Multidimensional Knapsack Problem. Proceedings of the Fourth International Conference on Machine Learning and Cybernetics, Guangzhou (2005) 2962-2965

A Markov Random Field Based Hybrid Algorithm with Simulated Annealing and Genetic Algorithm for Image Segmentation*

Xinyu Du, Yongjie Li, Wufan Chen, Yi Zhang, and Dezhong Yao

Center of NeuroInformatics
School of Life Science & Technology
University of Electronic Science and Technology of China
Chengdu, 610054 PR China
duxinyu126@126.com
{liyj, zhangyi, dyao}@uestc.edu.cn

Abstract. In this paper, a simulated algorithm-genetic (SA-GA) hybrid algorithm based on a Markov Random Field (MRF) model (MRF-SA-GA) is introduced for image de-noising and segmentation. In this algorithm, a population of potential solutions is maintained at every generation, and for each solution a fitness value is calculated with a fitness function, which is constructed based on the MRF potential function according to Metropolis algorithm and Bayesian rule. Two experiments are selected to verify the performance of the hybrid algorithm, and the preliminary results show that MRF-SA-GA outperforms SA and GA alone.

1 Introduction

In the computer vision field there has been increasing interest in use of statistical techniques for modeling and processing image data [1][2]. Most of these works has been directed toward application of Markov Random Field (MRF) models to problems in image segmentation and de-noising.

In an image, the feature of a pixel is highly depended on the features of pixels around it. The dependence can be described precisely and quantitatively by MRF (Markov Random Field) model [3][4]. In 1984 Geman emphasized the equivalence between MRF and Gibbs distributing, so that MRF could be defined by Gibbs distributing and be titled as GRF (Gibbs Random Field) [5]. Because of the flexible cliques and effective prior models, MRF is used in lots of image processing areas, such as medicine, remote sensing, radar and aviation.

Image de-noising and segmentation denote a process by which a raw input image is partitioned into smoothing and non-overlapping regions such that each region is homogeneous, connected and smoothing. It can be seen as an application of classification for knowledge discovery [6]. So that we can use MRF to characterize an image and also can use it for segmentation and de-noising.

* Supported by the grants from the 973 Project (#2003CB716100), NSFC (#90208003, #30525030, # 30500140).

L. Jiao et al. (Eds.): ICNC 2006, Part I, LNCS 4221, pp. 706 – 715, 2006.

Using MRF, we aim to find an optimal instantiation of the random label field given the image pixel gray value. There are already a lot of literatures about how to embed aptitude computation algorithms into MRF models. In the work of Xiao Wang et al [7], an evolutionary algorithm was used with Markov Random Field prior in segmentation. Their results showed that the algorithm worked well in texture segmentation and simple segmentation.

Din-Chang Tseng adopted a GA for MRF-based segmentation [8]. In their work, GA is only used as an initialization for the traditional partial optimization- iterated conditional modes (ICM) algorithms. Suchendra M. used SA-GA for image segmentation, but the algorithm is not based on MRF [9]. It is based on edge detection. Salima Ouadfel et al adopted ant colony system algorithm (AA) to segment MRF-based images [10]. Comparing GA, AA needs more time to reach the aim.

In this paper, we propose a novel SA-GA algorithm to de-noise and segment images. SA has strong capability to climb hills but slow convergence speed. GA can search larger space but has weaker climbing ability. Spontaneously we can combine SA and GA to segment and de-noise images. We stretch fitness in GA, so that premature can be avoided using roulette wheel selection in early time, and contrast can be increased in later time. At the same time the crossover rate and mutation rate are also declined exponentially with the evolution number increasing. In the coding stage, we use 2 dimension chromosome coding scheme [11]. As for the population initialization, we shake the threshold to generate different individual. In the selection stage, the strategy of roulette wheel selection is adopted. In the recombination and mutation stage, we combine MRF model 2 rank clique with 2 dimension windows in an individual. As we use Bayesian rule and MRF model in Metropolis and Annealing Algorithm, the stretched potential function of MRF can be used as fitness function of GA.

Consequently and in order to be self-contained, the rest of the paper is organized as follows. Section 2 presents a brief review on image modeling using MRF. Section 3 describes the SA-GA algorithm and investigates the application of SA-GA for MRF-based image de-noising and segmentation. In section 4 we present the experimental results, and finally a conclusion is drawn in section 5.

2 MRF Model for Image De-noising and Segmentation

MRF presents that one pixel is only depended on its neighbors effectively [12]. According to Hamersley-Clifford Theory, MRF can be equivalently depicted as GRF [3]:

$$P(x) = \exp(-U(x)) / Z \tag{1}$$

Where $Z = \sum_{x \in X} \exp(-U(x))$, $U(x) = \sum_{c \in C} \phi_c(x)$, X is a random field in label space and C is the clique space.

In the following sub-sections, two different characterized model to embody MRF model.

2.1 Multi-level Logistic Model

According to MLL (Multi-Level Logistic Model), we can get two point potential functions on the kind of clique c [13]:

$$\phi(X_s, X_t) = -\beta \quad X_s = X_t$$
$$\phi(X_s, X_t) = +\beta \quad X_s \neq X_t \tag{2}$$

Where $X = (X_s \in S)$ is a random field in gray space and $(X_s, X_t \in [0,1]^2), t \in V_s$ is the neighborhood of s. β is the Gibbs parameter and has a positive value. The meaning of β is the penalty of different pixels in the neighborhood of the wanted pixel. β is also the measurement of potential between the center and its neighbors. This model can be used to simulate district distribution.

2.2 Binomial Model

In an image *L*, the fact that the pixel *s'* value is x_s can be seen as event A, and the fact that the pixel *s'* neighborhood is $x_t, t \in \eta_s$,(η_s is the neighborhood of *s*) can be seen as event *B* and *m* is the times that the pixel *s'* value is x_s. It is a binomial distribution. So we can get [14]:

$$P(x_s = m / x_t, t \in \eta_s) = C_n^m p^m q^{n-m} \qquad m = 0, 1, \cdots\cdots, n \tag{3}$$

Where, *p* can be computed accordance to (1). This model can be used to simulate the texture in the district.

2.3 Image Segmentation Based on Maximum A Posteriori (MAP) and MRF

By referring to the theorems and formula in [15], we can explain the principles of MAP-MRF below.

Assuming *Y* is the original image and *X* is the distinct distribution of *Y*. The aim of segmentation is to find *X* from *Y*. Using MAP rule, that is:

$$P(X^* / Y) = \max_x P(X / Y) \tag{4}$$

As $P(X/Y) = P(Y/X)P(X)/P(Y)$ and $P(Y)$ is not correlative to *X*, we can get:

$$\max_X P(X / Y) \propto \max_X P(Y / X) P(X) \tag{5}$$

Where *P (x)* and *P (X/Y)* can be seen as Gibbs distribution and be calculated by (1), as:

$$P(Y / X)P(X) = \frac{1}{Z_{Y(X)}} \exp\{-U_Y(Y / X)\} \times \frac{1}{Z_X} \exp\{-U_X(X)\} \tag{6}$$

To find MAP segmentation, $P(Y/X)P(X)$ must be maximized in all *X*. According theorems in [15], we can get:

$$\max_X P(X/Y) \propto \max_X \{\frac{P(Y/\hat{X})P(\hat{X})}{P(Y/X)P(X)}\} \tag{7}$$

$\hat{X}$ is the same as X except for the pixel x_k, which is:

$\hat{X} = \{x_1, x_2, \cdots\cdots, x_{k-1}, \hat{x}_k, x_{k+1}, \cdots\cdots, x_M\}$

So we will find

$$\frac{P(Y/\hat{X})}{P(Y/X)} = \frac{p(y_k / y_1, \cdots, y_{k-1}, y_{k+1}, \cdots, y_M, x_1, \cdots, \hat{x}_k, \cdots, x_M)}{p(y_k / y_1, \cdots, y_{k-1}, y_{k+1}, \cdots, y_M, x_1, \cdots, x_k, \cdots, x_M)} \tag{8}$$

$$\frac{P(\hat{X})}{P(X)} = \frac{p(\hat{x}_k / x_1, \cdots, x_{k-1}, x_{k+1} \cdots, x_M)}{p(x_k / x_1, \cdots, x_{k-1}, x_{k+1} \cdots, x_M)} \tag{9}$$

$$\frac{P(Y/\hat{X})P(\hat{X})}{P(Y/X)P(X)} = \frac{p(y_k / \eta_k, \hat{x}_k)}{p(y_k / \eta_k, x_k)} \times \frac{p(\hat{x}_k / \delta_k)}{p(x_k / \delta_k)} \tag{10}$$

Where η_k and δ_k are neighborhood of y_k and x_k.

Furthermore, (7) can be written as below:

$$\max_X P(X/Y) \propto \max_X \{\frac{p(y_k / \eta_k, \hat{x}_k)}{p(y_k / \eta_k, x_k)} \times \frac{p(\hat{x}_k / \delta_k)}{p(x_k / \delta_k)}\} \tag{11}$$

That is the essential of Metropolis Algorithm.

In Eq. (11), $p(y_k / \eta_k, \hat{x}_k)$ can be calculated by (3) using binomial model, and $p(y_k / \eta_k, x_k)$, $p(\hat{x}_k / \delta_k)$, $p(x_k / \delta_k)$ can be calculated by (2) and (1) using MLL.

3 SA-GA for MAP-MRF

As described in section 1, SA has strong capability to climb hills but slow convergence speed. GA can search larger space but has weaker climbing ability. Spontaneously we can combine SA and GA to segment and de-noise images.

Summarizing, six improvements are proposed as below:

First, we stretch fitness function, so that premature can be avoided using roulette wheel selection in earlier stage of evolution, and contrast can be increased in later stage. Following expression is adopted to realize the stretchment:

$$f_i = \frac{e^{f_i/T}}{\sum_{i=1}^{M} e^{f_i/T}} \tag{12}$$

f_i is the fitness value of the *i-th* chromosome.

Second, the crossover rate and mutation rate are declined exponentially with the evolution number increasing, that is:

$$\begin{aligned} Pcross &= e^{(-(n-1)/(Gen-1))} Pcross \\ Pmute &= e^{(-(n-1)/(Gen-1))} Pmute \end{aligned} \tag{13}$$

The reason is that when crossover rate and mutation rate are larger in the first several generations, SA-GA can get larger solution space so that SA-GA can get quicker convergence. When crossover rate and mutation rate are smaller in the last several generations, SA-GA can get better climbing ability and can segment and de-noise images more precisely.

Third, in the coding stage, we use a two-dimension of chromosome coding.

Fourth, during the initialization of population, the threshold is disturbed in order to generate different individuals.

$$T_i(i,j) = T_i(i,j) - rand \times T_i(i,j) \times k \tag{14}$$

Where, k is a small swing value and *rand* is a random value.

Fifth, as for the operation of recombination and mutation, the MRF model 2 rank clique and 2 dimension windows are combined in one individual. As 2 rank MRF model is used, 8 points neighborhood as a 2 dimension window are adopted in an individual.

Sixth, the temperature table used in SA-GA is:

$$T = \frac{c_0}{\ln(c_1 + n)} \tag{15}$$

Where, c_0 and c_1 are constants, and n is the current generation.

Seventh, the fitness function is similar as the MRF potential function used in (1):

$$f_i = U(x) = \sum_{c \in C} \phi_c(x) \tag{16}$$

Now, the SA-GA for MAP-MRF can be summarized as follows:

I. Initializing parameters, such as population number, crossover rate, mutation rate, temperature table parameters, β of MLL, etc.
II. Initializing population using (14)
III. For every individual, scanning the image pixel by pixel, and generating a new pixel randomly.
IV. Computing (11), based on (1)~(10).

V. If (11)>1, the new pixel replaces the original one, else generating a random number between 0 and 1. If the random number $>X^*$ (determined by eq.(11)), go to the next step, else the replacement is done.
VI. Computing the fitness function using (16) and (12)
VII. Conducting the selection based on the roulette wheel.
VIII. Computing the crossover and mutation rate using (13)
IX. Recombining with 2-dimension window.
X. Implementing mutation using 2-dimension window.
XI. Declining temperature using (15)
XII. If the temperature is not low enough, return to III, else terminate.

The time complexity of the algorithm is:

$$T(n) = SAn \cdot Ps \cdot Cn \cdot Nn \cdot k \cdot n = O(n) \tag{17}$$

Where, *SAn*, *Ps*, *Cn*, *Nn* and *k* are constant parameters representing the iterative number, population size, classification number, the number of neighborhood pixels and foundational instruments independent of *n*. The number of pixels to process in a image is denoted by *n*.

4 Experimental Results

In this section, two experiments are employed to verify the performance of the proposed algorithm. One is for de-noising and the other is for de-noising and segmentation. They are all based on MAP-MRF. In the first one, the results are compared between SA described in [15] and the SA-GA described in this paper. The results in the second one are compared between GA described in [7] and the SA-GA described in this paper.

The following parameters are experimentally selected: Population Size: 20, Evolution Generations: 50, Crossover Rate: 0.1, Mutation Rate: 0.01, C_0=1.0, C_1=2.0, The End of Temperature for SA [15]: T_e=0.18, and the MLL parameter, β =-1.

4.1 Experiment I

Fig. 1(a) is a binary image selected from Matlab. Fig. 1(b) is the image added with Gaussian noises. Fig 1(c) is the de-noised image using Median Filtering (MF). Fig 1(d) is the de-noised image using Adaptive Wiener Filtering (AWF). Fig 1(d) is the de-noised image using SA described by [15]. Fig 1 (e) is the de-noised image using MRF-SA-GA designed by this paper.

The classification error rates are listed in Table 1:

Table 1. Error Rate of Experiment 1

Algorithms	MF	AWF	SA	SA-GA
Error Rate	10.19%	6.96%	2.85%	1.83%

4.2 Experiment II

To de-noise and segment an image in multi-part (>2) simultaneously is an area full of challenge in image analysis. The original image (Fig 2(a)) is a brain dissection image from IBSR [16]. Fig 2(b) is the image added with Gaussian noises. Fig 2(c) is the de-noising and 4-segmented image using GA described by [7], and Fig 2(d) is the de-noising and 4-segmented image using SA-GA of this paper.

Obviously, Fig 2(d) is much better than Fig 2 (c). Compared to Fig.2(c), more details about tissue edges are reserved in Fig.2 (d).

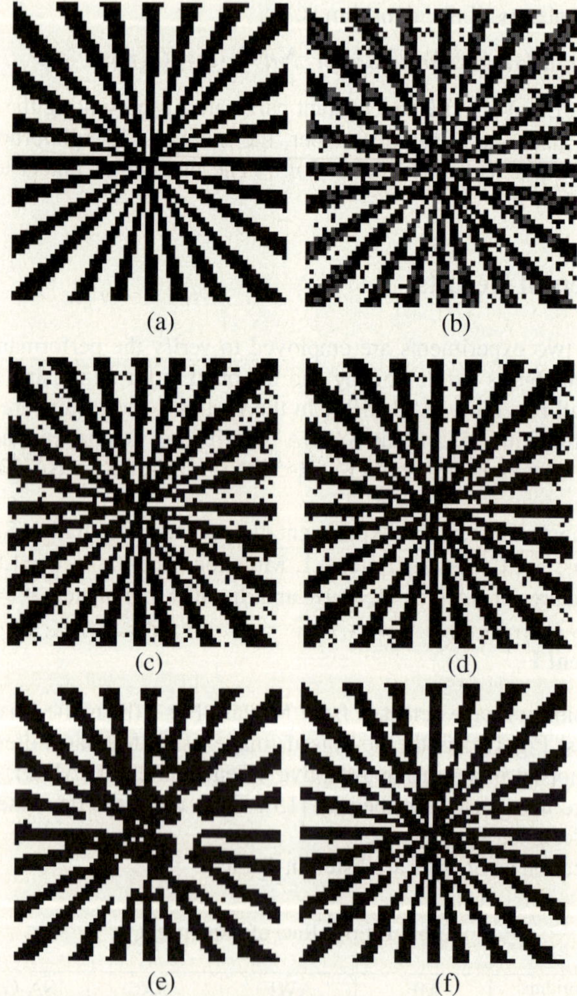

Fig. 1. The results of experiment 1: (a) original image; (b) original image added with noise(c) de-noised by MF; (d) de-noised by AWF; (e) de-noised by SA; (e) de-noised by SA-GA

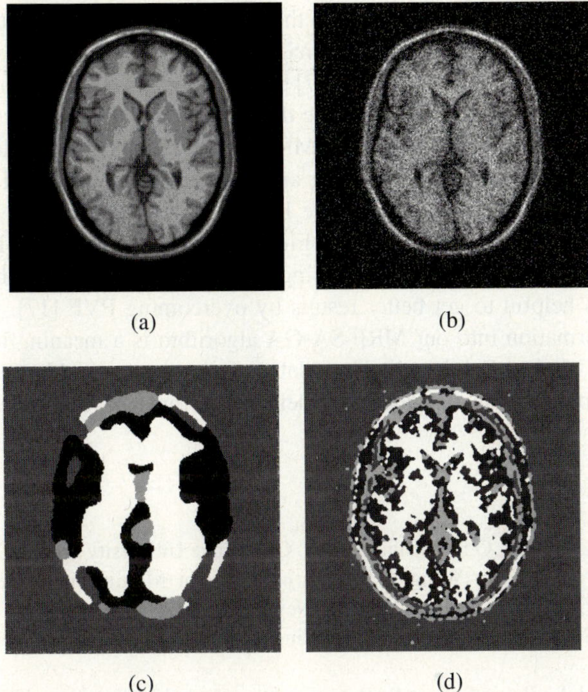

Fig. 2. The results of experiment 2: (a) the original image; (b) the original image added with noise; (c) segmented by GA; (d) segmented by SA-GA

5 Discussions and Conclusion

In this paper, a hybrid algorithm named MRF-SA-GA was proposed for image denoising and segmentation based on the MRF. From experiment 1, we can find that SA-GA is better than SA. Although SA can find the global best value theoretically, there is a conflict between the quality of solution and the computation efficiency. It is difficult to judge whether SA gets the balance point at every temperature, as the length of Markov chain cannot be controlled easily. SA has good capability to climb hills, but the result is not satisfying by using SA alone. On the contrary, GA can search larger space but its climbing ability is weaker. Selection shows the approach to the best value of all, recombination generates the new solution randomly and mutation overlays the best value of all. So we can combine SA and GA to generate the powerful algorithm, SA-GA.

From experiment 2, we can also find that SA-GA is much better than GA used in [7]. The algorithm in [7] divides an image into many 4-point square clique and one point is processed by evolutionary algorithm and other three points is processed by Gibbs sampling. Although Gibbs Sampling can process image more precisely in three points, evolutionary algorithm will break the approaching steady state in the one point, which causes a larger search space in one point with the cost of slower convergence speed in the other three points. So the asynchronism can degrade the segmentation and de-noising results. In medical image processing area, precise segmentation is an important issue. So the algorithm in [7] cannot be used well in this area.

From the experimental results and the discussions mentioned above, we could restrainedly conclude that the proposed MRF-SA-GA algorithm has the potential to improve the results of image de-noising and segmentation, compared to the SA and GA alone. However, the results here are limited and preliminary, several directions will be conducted in our future works. First, Partial Volume Effect (PVE) is an important factor in medical image processing. It has been demonstrated that the fuzzy information is helpful to get better results by overcoming PVE [17]. How to embed the fuzzy information into our MRF-SA-GA algorithm is a meaningful subject. Second, different Markov Random Field models have been proposed by several researchers. It is important to select a suitable model for different images segmentation.

References

1. Wesley, S., Hairong, Q.: Machine Vision. Cambridge University Press (2004)
2. Tan, Y.P., Yap, K.H., Wang, L.P. (Eds.): Intelligent Multimedia Processing with Soft Computing. Springer, Berlin Heidelberg New York (2004)
3. Stan, Li.: Markov Random Field Modeling in Image Analysis. Springer, Tokyo, Japan (2001)
4. Karvonen, J.A.: Baltic Sea ice SAR segmentation and classification using modified pulse-coupled neural networks. IEEE Trans. Geosciences and Remote Sensing, 42 (2004) 1566-1574
5. Stuart, G., Donald, G.: Stochastic Relaxation, Gibbs Distributions, and the Bayesian Restoration of Images. IEEE Trans.on Pattern Analysis and Machine Intelligence, Vol. 6, No. 6 (1984) 721-741
6. Halgamuge, S., Wang, L.P. (eds.), Classification and Clustering for Knowledge Discovery, Springer, Berlin (2005)
7. Xiao, W., Han, W.: Evolutionary Optimization with Markov Random Field Prior. IEEE Trans.on Evolutionary Computation, Vol. 8, No 6 (2004) 567-579
8. Dinchang, T., Chihching, L.: A Genetic Algorithm for MRF-based Segmentation of Multi-spectral Textured Images. Patter Recognition Letters 20 (1999)1499-1510
9. Suchendra, M, B., Hui, Z.: Image Segmentation Using Evolutionary Computation. IEEE Tran.on Evolutionary Computation Vol. 3, No. 1(1999)1-21
10. Salima, O., Mohamed, B,: MRF-based Images Segmentation Using Ant Colony System. Electronic Letters on Computer Vision and Image Analysis 2(2) (2003) 12-24
11. Jinhao, X,. Yujin, Z., Xinggang, L.: Dynamic Image Segmentation Using 2-D Genetic Algorithms. ACTA AUTOMATICA SINICA Vol.26, No.5 (2000) 685-689
12. Yanqiu, F., Wufan, C.: A New Algorithm for Image Segmentation Based on Gibbs Random Field and Fuzzy C-Means clustering. ACTA ELECTRONICA SINICA. Vol32, No.4 (2004) 645-647
13. Lahlou, A., Wojciech, P.: Hierarchical Markov Fields and Fuzzy Image Segmentation. Second IEEE International Conference on Intelligent Processing Systems (ICIPC'98). Gold Coast , Australia,August,4-7, 1998
14. Hu, R., Fahmy, M, M.: Texture Segmentation based on a hierachical MRF model. Signal processing. 26 (1992) 285-305

15. Zhaobao, Z.: Markov Random Field Method for Image Analysis. Wuhan Measurement Technology University Press (2000)
16. http://www.cma.mgh.harvard.edu/ibsr/
17. Dembele, D., Kastner, P.: Fuzzy C-Means Method for Clustering Microarry Data. Bioinformatics, Vol.19, No.8 (2003)973-980

Genetic Algorithm Based Fine-Grain Sleep Transistor Insertion Technique for Leakage Optimization

Yu Wang, Yongpan Liu, Rong Luo, and Huazhong Yang

Department of Electronics Engineering, Tsinghua University,
Beijing, 100084, P.R. China
wangyuu99@mails.tsinghua.edu.cn

Abstract. Fine-grain Sleep Transistor Insertion (FGSTI) is an effective leakage reduction method in VLSI design optimization. In this paper, a novel Genetic Algorithm (GA) based FGSTI technique is presented to decide where to put the sleep transistors (ST) when the circuit slowdown is not enough to assign sleep transistors everywhere in the combinational circuits. Penalty based fitness function with a built-in circuit delay calculator is used to meet the performance constraint. Although optimal FGSTI problem is proved to be NP-hard, our method can steadily give a flexible trade-off between runtime and accuracy. Furthermore a Successive Chromosome Initialization method is proposed to reduce the computation complexity when the circuit slowdown is 3% and 5%. Our experimental results show that the GA based FGSTI technique can achieve about 75%, 94% and 97% leakage current saving when the circuit slowdown is 0%, 3% and 5% respectively.

1 Introduction

With the development of the fabrication technology, leakage power dissipation has become comparable to switching power dissipation [1]. It is known that leakage power may make up 42% of total power at the 90nm technology node [2]. Thus various techniques are proposed to reduce the leakage power from system level down to physical level. Among these, Multi-Threshold CMOS (MTCMOS) technique is the most effective one, in which sleep transistors (ST) are placed between the gates and the power/ground (P/G) net in order to put the circuit into sleep mode when it is standby.

MTCMOS technique can be mainly categorized into two approaches: block based sleep transistor insertion (BBSTI) technique [3-6] and fine-grain sleep transistor insertion (FGSTI) technique [7-9]. In BBSTI, all the gates in the circuits are clustered into sizable blocks and then these blocks are gated using large ST; all the gates are assumed to have a fixed slowdown. On the other hand, FGSTI technique assigns ST with appropriate size to individual gates in the circuit while the circuit performance constraints are still satisfied as shown in Fig. 1. It is easier to guarantee circuit functionality in a FGSTI technique [8], since ST sizes are not determined by the worst case current of large circuit blocks which is quite difficult to determine without comprehensive simulation [3]. In addition, FGSTI technique leads to a smaller simultaneous switching current when the circuit changes between standby mode and active

L. Jiao et al. (Eds.): ICNC 2006, Part I, LNCS 4221, pp. 716–725, 2006.

mode comparing to BBSTI technique. Furthermore, better circuit slack utilization can be achieved as the slowdown of each gate is not fixed, and then leads to a further reduction of leakage and area [7] [9].

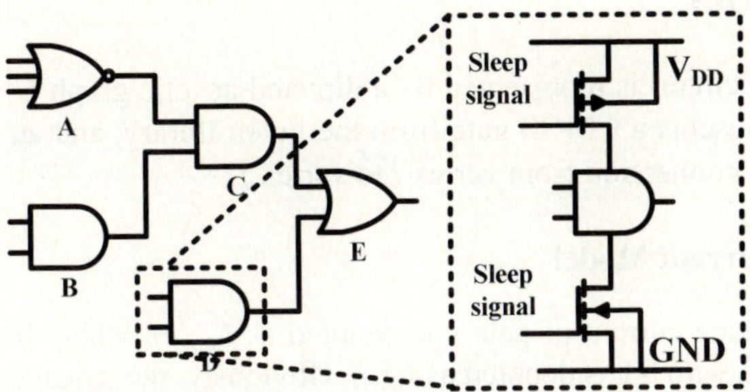

Fig. 1. Fine-grain sleep transistor insertion technique

The most different thing is that FGSTI technique can be performed when the circuit speed is not influenced, while BBSTI technique will definitely induce certain circuit slowdown, about 5% or more for most combinational circuits [9]. Thus the FGSTI technique is changed into a slack distribution problem to determine which gate can be assigned with ST. Recently, [9] use a one-shot heuristic algorithm to determine where to put ST in a FGSTI design, but how to perform FGSTI technique isn't addressed when the circuit slowdown is 0% and the one-shot heuristic algorithm may easily fall into a local optimal result. Our previous work [7] presents a mixed integer programming (MLP) model for FGSTI technique. Since the MLP problem is proved to be NP-hard [10], the MLP problem for large size circuit may take unbearable time to converge.

Ever since the genetic algorithm was introduced by Holland [11], lots of empirical evidences have indicated that GA can find good solutions to some complex problems. In this paper, a novel GA based FGSTI technique for leakage optimization is proposed. Our contributions include:

1. To our best knowledge, this is the first work to use GA based techniques to decide where to put ST in an FGSTI problem which is NP-hard. Penalty based fitness function is adopted to perform genetic search from both feasible and unfeasible solution space. Furthermore, our method can give a flexible trade-off between runtime and leakage saving which is becoming more important while the problem size is growing up. The computation complexity of our method turns out to be quite stable.
2. A Successive Chromosome Initialization (SCI) method is invented based on the successive attribute of the chromosome to further reduce the computation time. Our experiments show that this method could reduce the computation time significantly.

The rest of our paper is organized as follows. The preliminaries are given in Section 2. The detailed genetic algorithm optimization framework is illustrated in

Section 3. Section 4 is devoted to our SCI method. The implementation and experimental results are shown and analyzed in Section 5. Finally, we draw the conclusions.

2 Preliminaries

A combinational circuit is represented by a directed acyclic graph $G = (V, E)$ where a vertex $v \in V$ represents a CMOS gate from the given library, and an edge $(i, j) \in E$, $i, j \in V$ represents a connection from vertex i to vertex j.

2.1 Leakage Current Model

The original leakage current of gate v is denoted as $I_{w/o}(v)$, while the leakage current of gate v assigned with ST is denoted as $I_w(v)$. Obviously, the leakage current of gate v with ST depends on the ST's size. We choose the largest ST size $(W/L)v = 16$ for simplicity, which leads to the minimum delay overhead as shown below. Due to the stacking effect, $I_{w/o}(v)$ is about two orders of magnitude larger than $I_w(v)$. Thus if more gates in the circuit are assigned with ST, more leakage saving is achieved.

Extensive HSPICE simulations are used to create two leakage current look up tables for all the gates in the circuits to represent these two values: $I_{w/o}(v)$ and $I_w(v)$.

2.2 Delay Model

As shown in [12], the gate delay is influenced by the ST insertion. The load dependent delay $d_{w/o}(v)$ of gate v without ST is given by:

$$d_{w/o}(v) = \frac{KC_L V_{DD}}{(V_{DD} - V_{THlow})^{\alpha}} \tag{1}$$

where C_L, V_{THlow}, α, K are the load capacitance at the gate output, the low threshold voltage, the velocity saturation index and the proportionality constant respectively. The propagation delay $d_w(v)$ of gate v with ST can be expressed as:

$$d_w(v) = \frac{KC_L V_{DD}}{(V_{DD} - 2V_x - V_{THlow})^{\alpha}} \tag{2}$$

where V_x is the drain to source voltage of the ST. Suppose that $I_{ON}(v)$ is the current flowing through ST during the active mode, it can be expressed as given by [9]:

$$I_{ON}(v) = \mu_n C_{ox}(W/L)_v((V_{DD} - V_{THhigh})V_x - \frac{V_x^2}{2}) = \mu_n C_{ox}(W/L)_v(V_{DD} - V_{THhigh})V_x \tag{3}$$

where μ_n is the N-mobility, C_{ox} is the oxide capacitance, V_{THhigh} is the high threshold voltage, $(W/L)_v$ represents the size of the ST inserted to gate v. The voltage drop V_x in gate v due to ST insertion can be expressed as:

$$V_x = \frac{I_{ON}(v)}{\mu_n C_{ox}(V_{DD} - V_{THhigh})} \times \frac{1}{(W/L)_v} \tag{4}$$

Combining equation (1), (2) and (4), the propagation delay $d_w(v)$ of gate v with ST can be rewrite as:

$$d_w(v) = d_{w/o}(v) + \left(\left(1 - \frac{2 \frac{I_{ON}(v)}{\mu_n C_{ox}(V_{DD} - V_{THhigh})} \times \frac{1}{(W/L)_v}}{V_{DD} - V_{THlow}} \right)^{-\alpha} - 1 \right) d_{w/o}(v) = d_{w/o}(v) + \varphi((W/L)_v) d_{w/o}(v) \quad (5)$$

where $d_{w/o}(v)$ is constant which can be extracted from the technology library. Referring to equation (5), a larger $(W/L)_v$ leads to a smaller delay overhead. Here the largest ST size $(W/L)_v = 16$ is still chosen which makes $\varphi((W/L)_v)$ a constant.

3 Genetic Algorithm Based FGSTI Technique

The FGSTI problem is first formulated as a mathematical model:

$$\begin{cases} \min \quad I_{leak} = \sum_{v \in V} \left(I_{w/o}(v) \times (1 - ST(v)) + I_w(v) \times ST(v) \right) \\ \text{Subject to:} \\ \quad D_{circuit} \le T_{req} \\ \quad ST(v) \in \{0,1\}, \ \forall v \in V \end{cases} \quad (6)$$

where I_{leak} is the total leakage current in the circuit; $ST(v)$ is used to represent sleep transistor state of gate v: $ST(v) = 1$ means that gate v is assigned with sleep transistors, $ST(v) = 0$ means that gate v is not modified; $D_{circuit}$ is the longest path delay of the modified circuit; T_{req} represents the circuit performance constraint. $D_{circuit}$ is derived by a built-in longest path calculator using the delay models in Section 2.

Genetic Algorithm is widely applicable for complex problems and makes few assumptions from the problem domain; and it is not biased towards local minimums. A genetic algorithm based FGSTI technique is developed to solve above problem. The representation structure, chromosome initialization fitness function and genetic operator of our genetic algorithm are shown as follows.

3.1 Encoding and Chromosome Initialization

A binary vector $B = (ST(v_1), ST(v_2), \ldots, ST(v_N))$ is used as a chromosome to represent where to assign ST in a combinational circuit, and N refers to the total gate number. Apparently, if every gate in the circuit is not modified, the performance constraints are satisfied. Hence a chromosome (0, 0, …, 0) is a surely feasible chromosome. Suppose the population size is M, the other M-1 chromosomes are randomly chosen in order to gain a better capability to search the whole state space.

3.2 Fitness Function

Penalty terms are used in our fitness functions in order to perform genetic search from both feasible and infeasible parts in the search space towards the global optimal results. Therefore, two terms are included in our fitness function: the total leakage current for feasible solution and the penalty term for infeasible solution. Assuming B^k is the kth chromosome in the current population, N is the length of the chromosome, and

$ST^k(v_i)$ is the binary variable for each gate. The total leakage current of the circuit for the kth chromosome is directly derived by equation (6):

$$f_k(B^k) = \sum_{i=1}^{N} \left(I_{w/o}(v_i) \times (1 - ST^k(v_i)) + I_w(v_i) \times ST^k(v_i) \right) \tag{7}$$

For some chromosomes, the modification to the original circuit leads to a large total circuit delay variance which may violate the circuit performance constraints in equation (6). Those chromosomes project to infeasible solutions. The penalty coefficient P_k is proportional to the difference between the modified circuit delay and the performance required.

$$P_k = \begin{cases} 0, & \text{if constraint in (6) is satisfied} \\ \alpha_0 (D_{circuit} - T_{req}), & \text{else} \end{cases} \tag{8}$$

where α_0 is a large user-specified positive penalty value.

Hence our fitness function can be given as:

$$eval(B^k) = \frac{1}{f_k(B^k) + P_k}, \; k = 1, 2, \ldots, M \tag{9}$$

where M is the population size.

Referring to our previous work on Static Timing Analysis (STA) [13], an extended Breadth-first search is used to calculate the circuit delay of modified circuits. The computation complexity of BFS in a DAG $G = (V, E)$ is $O(V+E)$; thus our algorithm runs in time $O(N)$, where N is the total gate number, and also the length of the chromosome in our genetic algorithm.

3.3 Elitist Selection Strategy

The chromosomes are selected by ranking of the chromosomes according to their fitness from initial to final stage of genetic search. This mechanism can maintain the diversity of the species in the beginning of genetic search, as the fitnesses are scaled down so that the influence of high fitness is diminished; while at the later stage of the genetic search, when most of the chromosomes have similar high fitnesses, fitness ranking can address the effect of higher fitness and thus facilitate selection of the best chromosome for faster convergence [14].

The tournament selection is adopted to preserve the best chromosome from the current generation to the next generation.

3.4 Chromosome Crossover and Mutation

A "Scattered" crossover function is used; it first creates a random binary vector with the same length as the chromosome and then selects the genes from the first parent where the vector is a 1, and the genes from the second parent where the vector is a 0. With further research in the future, genes in the chromosome may be grouped to several highly dependent parts due to the circuit attributes. Hence the scattered crossover function can be easily adapted to crossover the fixed groups of genes.

An adaptive mutation method is used to avoid disrupting a good chromosome from based on non-uniform mutation method [15]. The key idea is that mutations are probabilistically performed more towards the weak chromosomes in order to explore different regions, meanwhile the best few chromosomes are disrupted with much less probability than those with weak fitnesses in order to find the optimum solutions. This strategy is especially important at the later stage of generations.

As shown above, the computation complexity of each generation is $O(N*N_C)$, where N is the chromosome length and N_C is the population size. Suppose the genetic algorithm takes M generation to converge to an acceptable results, the total complexity of our GA based FGSTI is $O(M*N*N_C)$.

4 Successive Chromosome Initialization

Generally speaking, the solution space of a GA problem consists of two parts: feasible solution region and infeasible solution region. Referring to our specialized GA problem, T_{req} which corresponds to the performance constraint, can be chosen from the original circuit delay to 1.05 times the original circuit delay. That is, the modified circuit delay may vary in the range of 5% from the original circuit delay. Suppose when the circuit performance is not influenced by adding sleep transistors to various gates in the circuits, the feasible solution region is Region A in Fig. 2.

Obviously, every solution in Region A must be a feasible solution when the performance constraint T_{req} changes to 1.03 and 1.05 times the original circuit delay. Furthermore every feasible solution of 3% circuit slowdown must be a feasible solution when the circuit slowdown is 5%. Thus the feasible solution regions of 3% and 5% are represented using Region B and C respectively in Fig. 2.

It should be clear that the solution spaces need not to be convex or continuous as shown in Fig. 2. Suppose solution a and b are the best individual in Region A and Region B respectively, which means a is the best individual when the circuit slowdown is 0% and b is the best individual when the circuit slowdown is 3%. When we solve the problem under the 3% circuit slowdown constraint, a is definitely a feasible solution. As the solution space is 2^N, where N is the chromosome length, the possibility of which almost all the initial chromosomes are in the infeasible solution region is very large. It may take a long time for genetic search to find the feasible solution region. Naturally, it is reasonable that the best solution of 0% circuit slowdown is used as one of the initial chromosome when the circuit slowdown is 0%, and this may reduce the computation complexity because a may be very near to b the best solution of Region B. Furthermore, we can use the last generation of 0% circuit slowdown as the initial generation of 3% circuit slowdown. It is the same case that the best solution b or the last generation of 3% circuit slowdown can be used for initial point of 5% circuit slowdown. This chromosome initialization mechanism is called "Successive Chromosome Initialization" in our specialized GA problem.

From Fig. 2 we can further confirm why using a penalty based fitness function. Suppose the best solution of 5% circuit slowdown is c_2, as shown in Fig. 2 the distance

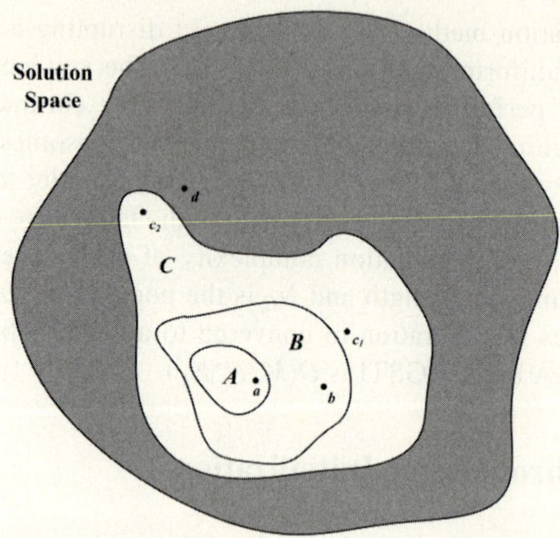

Fig. 2. Solution space: feasible solution region for different constraints and infeasible solution region in grey

between c_2 and infeasible solution d is much less than the distance between c_2 and feasible solutions a, b, c_1. Since we do not know where the best solution is in the solution space, the genetic search should be performed from both directions: feasible solution region and the infeasible solution region.

5 Implementation and Experimental Results

ISCAS85 benchmark circuit is used to verify our GA based FGSTI technique, all the netlists are synthesized using Synopsys Design Compiler and a TSMC 0.18μm standard cell library. The two leakage current look up tables for all the standard cells with and without sleep transistors are generated using HSPICE. The values of various transistor parameters have been taken from the TSMC 0.18μm process library, i.e. V_{DD}=1.8V, V_{THhigh}=500mV, V_{THlow}= 300mV, and I_{ON}= 200μA for all the gates in the circuit. The genetic algorithm are implemented using MATLAB.

We assume $(W/L)_v$= 16, corresponding to a delay variance of 6% if we assign sleep transistors to all the gates in the circuit [7]. Thus when the circuit slowdown varies in the range of 6% circuit original delay, we can not assign sleep transistors to every gate in the circuit.

The gate number N of the circuit is also the chromosome length as shown in Table 1, the search space of the problem is very large. When N is smaller than 1000, we set the population size to 200 and the max generation number to 1500; when N is larger than 1000, we set the population size to 500 and the max generation number to 4000. Table 1 shows the leakage saving using our GA based FGSTI technique; these results are the best ones of five runs.

Comparing to the leakage savings with MLP method [7]: 79.8%, 94%, 95% for 0%, 3%, 5% circuit slowdown respectively, the GA based FGSTI method is comparable. The MLP model for a certain circuit consists of about 7N variables and

Table 1. Leakage reduction using GA based FGSTI technique

ISCAS85 Benchmark Circuits	Gate Number *N*	Original leakage current (pA)	0% circuit slowdown (pA)	3% circuit slowdown (pA)	5% circuit slowdown (pA)
C432	169	4609	1764	479	211
C499	204	21375	14530	1080	109
C880	383	9261	684	318	179
C1355	548	11874	6495	1666	806
C1908	911	23418	3065	1027	316
C2670	1279	35191	2081	564	372
C3540	1699	40370	3470	1154	284
C5315	2329	56292	2938	1008	634
Leakage saving	N/A	N/A	74.8%	94.6%	97.7%

the corresponding constraints for each variable, hence the problem size is becoming extremely large when the gate number increases. It is unstable since it does not converge well for some circuits in our experiments. However, our GA based FGSTI technique can give a better solution for the circuits that MLP model can not converge well; this leads to a larger leakage saving when the ciruit slowdown is 3% and 5%.

In our GA based FGSTI technique, the population size and the max generation number are controllable; meanwhile the genetic algorithm can perform a linear scale down of the object function. Therefore, it is more flexible to solve this problem using GA compare to MLP method which is unstable and with less controllable.

Table 2 shows the computation complexity using our Successive Chromosome Initialization method. There are two different Successive Chromosome strategies: 1. using the best individual as one of the initial point; 2. using the last generation including the best individual as the initial pool. They are represented as Successive Chromosome Initialization with best individual (SCI_BI), Successive Chromosome Initialization with last generation (SCI_LG). The population size is set to 200, the generation number when the best results are achieved are compared shown in Table 2. These results are average of five runs. As we can see, the SCI_BI strategy and the SCI_LG strategy is efficient compared to the original one.

Table 2. Computation complexity comparison using generation number (Original Chromosome Initialization, SCI_BI, SCI_LG)

	C432 0% Slow down	C432 3% slow down	C432 5% slow down	C499 0% slow down	C499 3% slow down	C499 5% slow down
Original	107	132	167	110	156	176
SCI_BI	N/A	37	24	N/A	117	59
SCI_LG	N/A	30	22	N/A	102	45

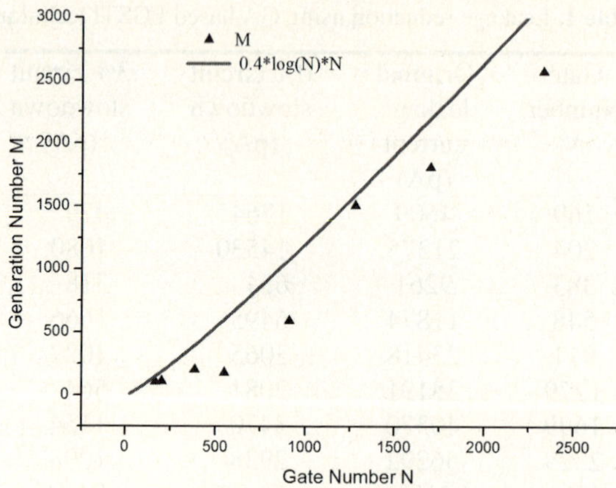

Fig. 3. Complexity analysis (Gate Number N vs Generation Number M)

Furthermore, we look into the complexity of our GA based FGSTI technique. In this case, all the population size is set to 200; the search is stopped when the leakage saving is within 5% compared to the results shown in Table 1. Fig. 3 shows the relationship between the gate number N (chromosome length) and the final generation number M.

From Fig. 3, it is shown that our genetic algorithm is stable, since all the generation number M is below the red line which corresponds to function 0.4*log(N)*N. From our previous complexity analysis, the total complexity of our GA based FGSTI is $O(M*N*N_C)$, where N is the chromosome length and N_C is the population size, M is the generation number. As N_C is assumed to be a constant, the computation complexity of our GA based FGSTI technique is $O(N*\log(N)*N)$ based on our experimental results.

6 Conclusions

In this paper, a novel genetic algorithm based FGSTI technique is proposed, which can assign sleep transistors to appropriate gates in order to achieve the max leakage saving when the circuit is standby. Penalty based fitness function with a built-in circuit delay calculator is used to meet the performance constraint. The GA based FGSTI technique can achieve about 75%, 94% and 97% leakage current saving when the circuit slowdown is 0%, 3% and 5% respectively. Furthermore, the experimental results show that our SCI mechanism leads to further reduction of the computation time when the circuit slowdown is 3% and 5%. Our genetic algorithm is stable based on our complexity analysis, which is superior to the MLP approaches.

Acknowledgement

The authors would like to thank the grants from the National 863 project of China (No. 2005AA1Z1230), National Natural Science Foundation of China (No. 60506010).

Thanks to Qian Ding for helpful suggestions during coding and using MATLAB. We would like to thank all the anonymous reviewers.

References

1. G. Moore, "No exponential is forever: But forever can be delayed," in IEEE ISSCC Dig. Tech. Papers, 2003, pp. 20 - 23
2. J. Kao, S. Narendra, A. Chandrakasan, "Subthreshold Leakage modeling and reduction techniques", in Procs. of ICCAD, 2002, pp 141 – 149
3. J. Kao, S. Narendra, A. Chandrakasan, "MTCMOS hierarchical sizing based on mutual exclusive discharge patterns," in Procs. of DAC, 1998, pp. 495–500
4. M. Anis, S. Areibi, and M. Elmasry, "Dynamic and leakage power reduction in MTCMOS circuits using an automated efficient gate clustering technique," in Procs of DAC, 2002, pp. 480–485
5. W. Wang, M. Anis, S. Areibi, "Fast techniques for standby leakage reduction in MTCMOS circuits" in Procs. of IEEE SOC, 12-15 Sept. 2004 pp 21 – 24
6. C. Long, L. He; "Distributed sleep transistors network for power reduction" in Procs. of DAC, 2-6 June 2003 pp. 181 – 186
7. Y. Wang, H. Lin, HZ. Yang, R. Luo, H. Wang, "Simultaneous Fine-grain Sleep Transistor Placement and Sizing for Leakage Optimization", in Procs. of International Symposium on Quality Electronic Design (ISQED)'06, March 2006, pp. 723-728
8. B. H. Calhoun, F. A. Honoré, and A. P. Chandrakasan, "A Leakage Reduction Methodology for Distributed MTCMOS," IEEE JSSC Vol. 39, No. 5, May 2004, pp. 818 - 826
9. V. Khandelwal, A. Srivastava; "Leakage Control Through Fine-Grained Placement and Sizing of Sleep Transistors ," in Procs. of ICCAD 2004, pp 533 - 536
10. Gerard Sierksma, Linear and Integer Programming: theory and practice, Marcel Deccckker, 2002
11. J. H. Holland, Adaptation in Natural and Artificial Systems (Univ. of Michigan Press, Ann Arbor, MI, 1975; reprinted by MIT Press, Cambridge, MA, 1992).
12. S. Mutoh et al. "1-V Power Supply High Speed Digital Circuit Technology with Multithreshold Voltage CMOS," in IEEE JSSC, Vol. 30, No. 8 August 1995
13. Y. Wang, HZ. Yang, H. Wang, "Signal-path Level Dual-Vt Assignment for Leakage Power Reduction", in Journal of Circuits, System and Computers Vol. 15, No. 2 (2006)
14. David E. Goldberg, Genetic Algorithms in Search, Optimization, and Machine Learning. Addison-Wesley, 1989
15. T. C. Fogarty, "Varying the probability of mutation in genetic algorithms," in Proceedings of Third International Conference on Genetic Algorithms, pp. 104-109, 1987

Self-adaptive Length Genetic Algorithm for Urban Rerouting Problem*

Li Cao[1,3], Zhongke Shi[2], and Paul Bao[3]

[1] Collage of Civil Aviation, Nanjing University of Aeronautics & Astronautics, 210016, China
caoli@nuaa.edu.cn

[2] Department of Automatic Control, Northwestern Polytechnical University, 710072, China
zkeshi@nwpu.edu.cn

[3] University of South Florida, FL 33803, USA
Pbao@usf.edu

Abstract. Vehicle rerouting can reduce the travel cost in high-volume traffic network by real-time information. However, it is a computational challenge. A self-adaptive string length evolution strategy was presented so as to meet the inconstant intersection number in different potential routes. String length would be adjusted according to the intersection number. Simulation results showed it could work well in the urban rerouting problem.

1 Introduction

As the intelligent transportation system (ITS) developing, drives can receive the updated directions based on the time-varying and real-time traffic information as they approach an intersection [1]. Thus, drives can reroute the ways once they are in congested networks. To implement the schema, some researchers were studying in different ways. Oscar Franzese *etc.* [1] optimized routes by traffic simulation models with the depth information of the network. Seongmoon Kim *etc.* [2,3] presented a systematic approach for non-stationary stochastic shortest path problem with real-time information. Waller S T *etc.* [4] proposed an online method on the shortest path problem with only local arc and time dependencies. Fu L. [5] discussed implementations of real-time vehicle routing based on estimation of mean and variance travel times.

However, as the road number increasing, the route selecting becomes a computationally challenge [2]. Bielli M *etc.* [6] used a GA in bus network optimization to improve computational performance. However, the intersection number of different potential ways for the same origin-destination (O-D) problem is inconstant in urban traffic networks so that the current GAs have not ways to treat with it. The goal of this paper is to deal with the rerouting problem with real-time traffic information when the intersection number was changeable.

---

* The work is supported by National Natural Science Foundation of China (60134010) and Talent Recruitment Foundation of NUAA (S0398-071).

L. Jiao et al. (Eds.): ICNC 2006, Part I, LNCS 4221, pp. 726 – 729, 2006.

2 Self-adaptive String Length GA

For an urban traffic network, there are many ways for an O-D problem. Fig. 1 shows a network with 6 nodes. There are 16 paths from 1st node to 6th node. Some ways contain unequal number of node. Obviously, the current GAs have no ways to solve it. To deal with this problem, the following evolution strategy was presented.

Encoding and initializing: Here, we will use decimal string to represent the potential route. Each bit represents an intersection. The string length depends on the intersection number that its relevant route has. For example, P_1 and P_2 contain 3 and 4 intersections in Fig. 1.

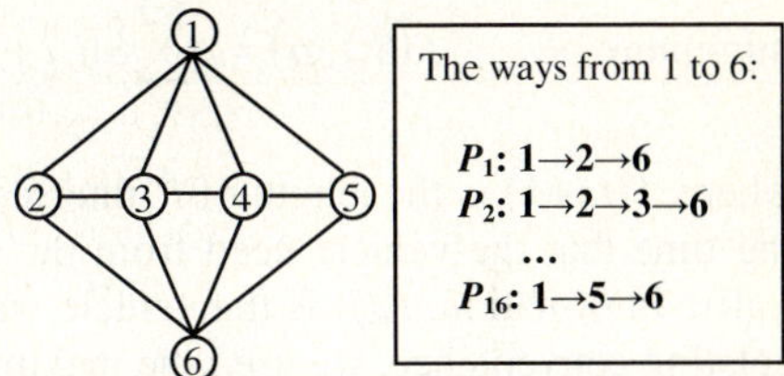

Fig. 1. The illustration of a traffic network

To insure the initial population being logical and high quality, we will link the origin with the destination by a series of curves in the map (traffic network). Thus, we can get a set of interim intersections around each curve. Each intersection's set will represent a potential route.

Evolution strategy: To implement the uncertainty length GA, we had to introduce a notion, **gene segment**, firstly. It is a section of genome that represents the genes between two points of a string. In the dynamic routing problem, gene segment is actually a sub-path between two intersections. In our strategy, we regard the gene segment as the bit in GAs. Selection is done as the current GAs. Crossover and mutation are treated as the following:

1. **Crossover:** Once two strings have been chosen for crossover, we will merge their genetic information in order to create a new string. Fig. 2 illustrates our **gene segment crossover** process. Matching two parents, two gene segments are gotten that both have the same begin and end bits (6 and 27). Exchanging the gene segments will get offspring.
2. **Mutation:** For the validity of the route, we use the **gene segment mutation** to replace the traditional mutation. Fig. 3 shows its working. N_{gs} represents a new gene segment. Two cutting positions (6 and 27) are chosen randomly in P. Substituting N_{gs} for the corresponding section of P, we will get a new string O.

P_A: 1→3→**6**→12→13→15→21→**27**→30

P_B: 1→4→**6**→9→18→26→**27**→30

O_A: 1→3→**6**→9→18→26→**27**→30

O_B: 1→4→**6**→12→13→15→21→**27**→30

Fig. 2. Gene segment crossover working

P: 1→3→**6**→12→13→15→21→**27**→30

N_{gs}: **6**→9→18→26→**27**

O: 1→3→**6**→9→18→26→**27**→30

Fig. 3. Gene segment mutation working

3 Dynamic Route Optimizing by the Proposed Method

Problem statement: The new real-time traffic information can be updated slightly before vehicle reaches the next intersection. Here, we study the single-vehicle dynamic routing problem by the minimum travel cost. It contains the distance and travel time from the current intersection to the destination. So we described it as follows:

$$\text{minimum:} \qquad f(S,t,\rho)=\rho\sum_{i=1}^{m-1}S(i,i+1)+(1-\rho)\cdot\alpha\cdot\left(t_{est}+\sum_{j=1}^{m-2}t_w(j)\right), \tag{1}$$

where, $S(i,i+1)$ is the length of ith intersection to $(i+1)^{th}$ intersection. t_{est} is the estimating time that the vehicle need from the current location to its destination by current traffic information. $t_w(j)$ is the vehicle waiting time at the jth intersection. For the calculating convenience, we used the maximum waiting time. Here, we set the maximum waiting time as a constant, 20. $\rho \le 1$ is the weight coefficient and α is a constant coefficient adjusting the power of time cost.

Simulation testing: Fig. 4 was a simulative urban traffic network. It contains 110 links and 60 nodes. Each node represented an intersection and each link expressed a path. Some information had shown in this figure. The dashed were used to initialize. The same way was to be used to produce the gene segment of mutation. Our testing was run on a 2.4 GHz and 512MB RAM PC. All the simulations were implemented by Matlab 6.0. The size of the population was 10. Each result was calculated in 20 times.

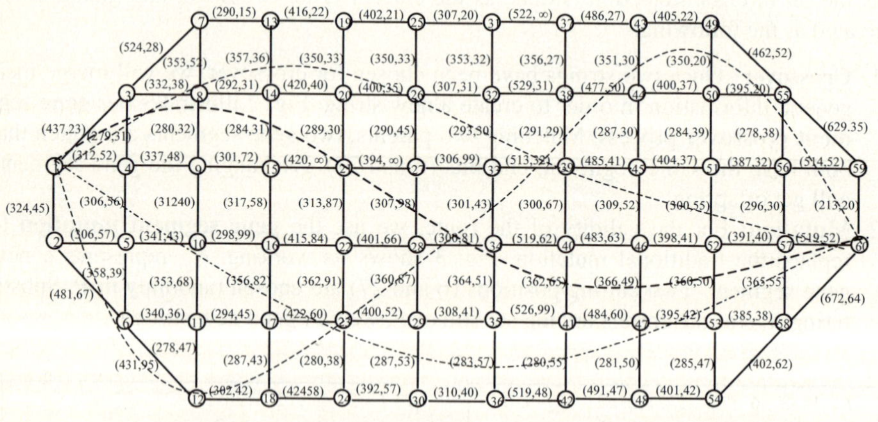

Fig. 4. The simulative urban traffic network (including 60 nodes and 110 links)

Table 1 showed the calculating results by the proposed method. From the table, we could not find that as ρ changed from 1 to 0, the route length would increase and the traveling time would decreased. The same conclusion could get from equation (1). When ρ=1, our question is actually a shortest path problem. And ρ=0, it is a minimum traveling time problem. On the other situations, the routes were the optimal routes with different ρ. The less ρ, the more traveling time cost we considered. Here, we set α=8.

Table 1. The optimal routes and the convergence by different ρ

	ρ	Route	Route length	Traveling time	Avg. steps	Max. steps	Min. steps
1	1	1→4→9→15→21→27→33→39→45 →51→56→59→60	4597	∞	31	137	12
2	0.75	1→4→9→15→16→22→28→34→40 →46→52→57→60	4684	829	35	202	17
3	0.5	1→2→6→11→17→23→29→35→34 →40→46→52→57→60	5243	775	34	183	14
4	0.25	1→2→6→11→17→23→29→35→34 →33→39→45→51→56→57→60	5838	767	35	215	16
5	0	1→3→7→13→19→25→31→32→38 →37→43→49→50→55→59→60	6088	503	40	288	15

The convergence of the proposed method was also shown in table 1. We listed the average, maximum and minimum iteration steps in different situations. There were great ranges between the max and min steps that the algorithm needed. It showed that quality of initialization population could greatly influence the convergence of the algorithm. The iteration steps would be saved more when the initialization population was good enough. Otherwise, it would need more iteration steps. On the other hand, 90% optimizing could finish within 30 steps so the average steps were not big during the experimenting. It showed our initialization method were much efficient.

4 Conclusions

In this paper, we proposed a changeable string length GA strategy for dynamic routing optimization with real time information. The initialization method of the algorithm could help to reduce CPU time obviously. Simulation experiments showed that the proposed method could fit for rerouting in urban traffic networks.

References

1. Franzese Oscar, Joshi Shirish: Traffic simulation application to plan real-time distribution route. *Winter Simulation Conference Proceedings*, vol. 2, 2002, 1214-1218
2. Seongmoon Kim, Lewis M. E. and White C. C. III.: Optimal vehicle routing with real-time traffic information. *IEEE Trans. Intell. Transp. Syst.*, vol. 6, no. 2, 2005, 178-188
3. Seongmoon Kim, Lewis M. E., White, C. C. III: State space reduction for nonstationary stochastic shortest path problems with real-time traffic information. *IEEE Trans. Intell. Transp. Syst.*, vol. 6, no. 3, 2005, 273-84
4. Waller S. T., Ziliaskopoulos A. K.: On the online shortest path problem with limited arc cost dependencies. *Networks*, vol. 40, no. 4, 2002, 216-227
5. Fu L.: An adaptive routing algorithm for in vehicle route guidance systems with real-time information. *Transp. Res.*, vol. 35, no. 8, pt. B, 2001, 749-765
6. Bielli, M., Caramia, M. and Carotenuto, P.: Genetic algorithms in bus network optimization. *Transp. Res.*, vol. 10, no. 1, pt. C, 2002, 19-34

A Global Archive Sub-Population Genetic Algorithm with Adaptive Strategy in Multi-objective Parallel-Machine Scheduling Problem

Pei-Chann Chang[1], Shih-Hsin Chen[1], and Jih-Chang Hsieh[2]

[1] Department of Industrial Engineering and Management, Yuan-Ze University, Ne-Li, Tao-Yuan, Taiwan, China
iepchang@saturn.yzu.edu.tw
[2] Department of Finance, Vanung University, Chung-Li 32061, Tao-Yuan, Taiwan, China

Abstract. This research extends the sub-population genetic algorithm and combines it with a global archive and an adaptive strategy to solve the multi-objective parallel scheduling problems. In this approach, the global archive is applied within each subpopulation and once a better Pareto solution is identified, other subpopulations are able to employ this Pareto solution to further guide the searching direction. In addition, the crossover and mutation rates are continuously adapted according to the performance of the current generation. As a result, the convergence and diversity of the evolutionary processes can be maintained in a very efficient manner. Intensive experimental results indicate that the sub-population genetic algorithm combing the global archive and the adaptive strategy outperforms NSGA II and SPEA II approaches.

1 Introduction

Parallel-machine production systems are commonly used in practical manufacturing activities. Parallel machines are able to make the workstations free from being bottlenecks. Regardless of the popularity, the scheduling for parallel machines is still complicated. Garey and Johnson (1979) have shown that two identical parallel machines scheduling with minimizing makespan is NP-hard. Brucker (1998) further indicated that parallel machine scheduling is even strong NP-hard as long as the number of machines is greater than two. The previous works reflect that scheduling for parallel machines is still a great challenge. Because of the NP-hard property, optimality becomes neither effective nor efficient. Many heuristic algorithms were ever proposed for parallel machine scheduling problem such as Hsieh et al. (2003). Among these heuristic algorithms, the performance of genetic algorithms (GA) is convincing and approved by many successful applications such as Neppali et al. (1996), and Sridhar and Rajendran (1996).

Total tardiness time and makespan are considered in this parallel-machine scheduling problem. Total tardiness time reflects if the production meets due-dates and makespan indicates the utilization of the shop floor. Several genetic algorithms

L. Jiao et al. (Eds.): ICNC 2006, Part I, LNCS 4221, pp. 730 – 739, 2006.

have ever been derived for bi-objective or multi-objective optimization problems. Schaffer (1985) proposed vector evaluated genetic algorithm (VEGA), which was the first idea to extend the simple genetic algorithm for multi-objective optimization. Murata and Ishibuchi (1996) proposed multi-objective genetic algorithm (MOGA). MOGA assigns each objective a weight and the weight changes along with the evolving process. Through the weight-changing, MOGA can search Pareto optimal solutions toward different directions. Murata et al. (1996) addressed that MOGA outperforms VEGA on multi-objective flowshop scheduling problem. Zitzler et al. (2002) modified SPEA as SPEA II for multiobjective optimization. Deb et al. (2000) proposed non-dominated sorting genetic algorithm II (NSGA II) by accommodating elitism strategy and crowding distance. Hsieh (2005) proposed grid-partitioned objective space approach based on genetic algorithm with considering multiple objectives. More and more sophisticated genetic algorithms are expected to be developed for solving optimization problems effectively and efficiently.

There are some researchers who propose their subpopulation-like approaches Cochran et al. (2003) proposed a multi-population genetic algorithm (MPGA). Chang et al. (2005) have proposed a two-phase sub population genetic algorithm (TPSPGA) for parallel machine scheduling problems. TPSPGA outperforms NSGA II and MOGA in the numerical experiments. There are still other approaches, such as Segregative Genetic Algorithms (Affenzeller, 2001), Multisexual Genetic Algorithm (Lis and Eiben, 1997), and MO Particle Swam Optimization (Coello et. al, 2004; Mostaghim and Teich, 2004).

Chang et al. (2006) proposed an adaptive multi-objective genetic algorithm for drilling operations scheduling problem in printed circuit board industry. The result indicated that adaptive strategy could be able to improve the solution quality.

Inspired by these pioneer works as discussed above, the SPGA proposed by Chang et al. (2005) is modified by using a global archive Pareto solution and embedded adaptive strategies in this research. In SPGA, the subpopulation works independently; however, according to previous research of these subpopulation algorithms, which create a chance for these subpopulations to be able to exchange information, it may improve the solution quality. Therefore, the modified SPGA considers how to make these subpopulations able to interchange information.

The rest of the research is organized as follows: Section 2 introduces the modified SPGA algorithm, including the global archive technique and adaptive strategies. Because the better parameter settings are not available, the research Design of Experiment is able to obtain better configuration. Then the experimental results of global archive and adaptive strategies are given in Section 3. In addition, the solution of the modified SPGA is compared with the SPGA, NSGA II, and SPEA II. Finally, the conclusion is discussed and the performance of the algorithm is evaluated.

2 Methodology

In this research, using a global archive first modifies SPGA and then an adaptive strategy is embedded in the modified SPGA. The description of SPGA and global archive technique can be found in section 2.1 and the detail procedure is shown in section 2.2. Finally, the adaptive strategies and performance metric are illustrated in section 2.3 and 2.4 respectively.

2.1 The Concept of SPGA and Global Pareto Archive

In order to prevent the searching procedures from being trapped into local optimality, the research applies and modifies the SPGA proposed by Chang et al. (2005). There are two main characteristics of the subpopulation-like method: (1) numerous small sub-populations are designed to explore the solution space; and (2) the multiple objectives are scalarized into a single objective for each sub-population.

In SPGA, each subpopulation works independently and cannot communicate with each other. However, from previous research works of Affenzeller (2001), and Lis and Eiben (1997), the subpopulation should communicate with each other so that it may bring better convergence and diversity. Therefore, this research uses a global archive to exchange information of better solutions among sub-populations while they are exploring the different solution space together. The framework of the global archive is shown as in Fig 1.

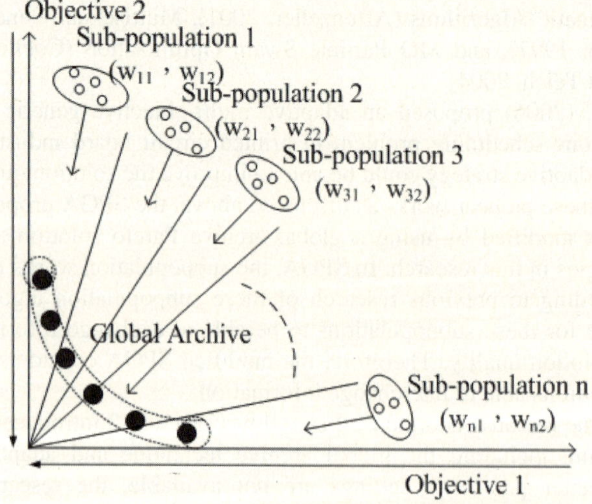

Fig. 1. The framework of the global archive for SPGA

2.2 Procedures of Modified Sub-Population Genetic Algorithm

Because SPGA does not share the Pareto archives with each other, it might lose the chance to obtain a better solution; the other sub-populations cannot apply it to improve the solution quality. Consequently, the main idea of the modified SPGA is to collect Pareto optimal solutions from sub-populations as a global Pareto archive. The global Pareto archive is expected to improve the solution quality and maintain the diversity.

The algorithmic procedure of the modified SPGA is explained in the following:

```
Algorithm: The modified SPGA()
Initialize()
DividePopulation()
AssignWeightToEachObjectives()
counter ← 0
while counter < Iteration do
    for i = 1 to ns do
        Selection and Elitism(i)
        Crossover(i)
        Mutation (i)
        EvaluateSolutions(i)
        Fitness(i)
        UpdateglobalParetoarchive()
        Replacement(i)
    end for
    counter ← counter + 1
end while
exit
```

The procedure *initialize* is used to generate the chromosomes of a population, whose size is determined by user. The procedure *DividePopulation* is to divide the original population into *ns* sub-populations.

At the procedure *AssignWeightToEachObjectives*, each sub-population is assigned different weight values and the individuals in the same sub-population share the same weight value. Because the research focuses on bi-criteria problem, the vector size is two. The equation of combination of weight value below:

$$(W_{n1}, W_{n2}) = \left(\frac{1}{N_s+1} \cdot n\,,\ 1 - \frac{1}{N_s+1} \cdot n\right) \tag{1}$$

where n is the nth sub-population.

After the weight value assignment, the corresponding scalarized objective value of the two objectives in sub-population can be written as equation 2.

$$f(x) = W_{n1} \cdot Z_{TT}(x) + W_{n2} \cdot Z_{TC}(x) \tag{2}$$

where Z_{TT} and Z_{TC} denote total tardiness time and makespan for each solution x.

Because the scales of the two objectives are different, the objective values are normalized in a unit interval. The *Elitism* strategy at the first stage randomly selects a number of individuals from non-dominated set into mating pool, so that individuals can be selected while the *crossover* procedure. The *Elitism* strategy for global archive SPGA is to collect best non-dominated solution from all subpopulations and it copies a proportional elites into the selection procedure. The binary tournament selection is employed in the *selection* operation. The smaller objective value has better chance to be selected. Besides, it also employs some elites from the global archive. Finally, the replacement strategy is the total replacement one, which means the offspring substitutes the parent solution entirely.

2.3 Procedure of the Adaptive Sub-Population Genetic Algorithm

The procedure of the adaptive SPGA includes measure diversity of the population and the adaptive crossover and adaptive mutation operator will apply the result into their own operations.

Two adaptive strategies are embedded into the modified SPGA. The first one and the second one were proposed by Srinivas and Patnaik (1994), and the Zhu and Liu (2004) respectively.

The method of Srinivas and Patnaik (1994) has been widely applied, which depends on the fitness judgment and the fitness normalization. The goodness of a solution is judged by the average fitness value. If the smaller fitness value means a better solution, the definition of better solution here is the solution whose fitness value is lower than the average fitness. Thus, these solutions apply smaller crossover rate and mutation rate. The scale of the probability is based on the normalization ratio among solutions. On the other hand, the worsen solution is mated or mutated in a higher probability. The adaptive crossover and mutation operators are as equation (3) and (4) respectively:

$$p_c = \begin{cases} k_1(f_i - f_{\min})/(f_{\max} - \overline{f}) & \text{if} \quad f_i \le \overline{f} \\ k_3 & \text{otherwise} \end{cases} \tag{3}$$

$$p_m = \begin{cases} k_2(f_i - f_{\min})/(f_{\max} - \overline{f}) & \text{if} \quad f_i \le \overline{f} \\ k_4 & \text{otherwise} \end{cases} \tag{4}$$

Another adaptive strategy proposed by Zhu and Liu (2003) consists of three steps the distance measure, diversity measure, and diversity control, which can be expressed by the following equation (5), (6), and (7):

$$\text{Step 1: } Ham(\text{x, y}) = \sum_i |\,\text{sgn}(x[i] - y[i])\,| \tag{5}$$

$$\text{Step 2: } gD(P) = \frac{1}{2}\sum_{i \ne j} Ham(P[i], p[j]) \tag{6}$$

$$\text{Step 3: } p' = \max(p_{min}, \min(p_{max}, p\left[1 + \frac{\xi(gD_t - gD)}{gD}\right])) \tag{7}$$

where gD_t: The target population diversity.

In equation (5), the distance measure applies the hamming distance between two solutions. The diversity measure evaluates the diversity of all solutions in equation (6). It detects how the "health level" of the population. Finally, the diversity control is to modify the rate according to the target population diversity. If the population diversity is higher than the target diversity, the rate is decreased. Otherwise, the rate is increased.

2.4 Evaluation Metric

The research uses $D1_R$ to evaluate the solution quality. Knowles and Corne (2002) indicated that $D1_R$ considers the convergence and diversity at the same time. After a run, an algorithm obtains a set of Pareto solutions, which is compared with a reference set. Thus, the $D1_R$ value is obtained. The lower $D1_R$ value, the better the solution quality. Therefore, the $D1_R$ provides a basis for comparing the performance among different algorithms in the study. The equation of $D1_R$ is represented as equation (8) and (9).

$$D1_R(A_j) = \frac{1}{\left|Z^*\right|} \sum_{y \in Z^*} \min\{d_{xy} \mid x \in A_j\} \tag{8}$$

$$dxy = \sqrt{\sum_{i=1}^{n} (f_i^*(y) - f_i(y))^2} \tag{9}$$

where A_j: A set of Pareto solution obtained by an algorithm

Z^*: The reference solution or true Pareto solution

$\left|Z^*\right|$: The number of reference solution

3 Numerical Experiments

The data collected from a printed circuit board factory are applied to be the test instances[1]. Three job/machines combinations are considered, i.e., 35/10, 50/15, 65/18.

3.1 Experiment Design for Parameters Settings in SPGA

The subsection tries to determine the optimal parameter setting in the algorithms. Then, the next experiment applies the result of these parameter settings and compares it with NSGA II and SPEA II.

There are several parameters that may influence the performance of the algorithm. For example, the larger population size may find better solution quality but cost higher computational expense. When the number of sub-populations is larger, it may have better diversity. However, it may also be a trade-off that to reduce the number of generations. Moreover, the secondary crossover and mutation operator are also considered because it may provide better solution quality. The crossover rate and mutation rate are set to 0.9 and 0.1 respectively. The factors and treatments of these factors are as shown in Table 1. The detail ANOVA result can be obtained at our website and the suggested parameter settings is presented in Table 2.

---

[1] The data can be assessed at our website: http://ppc.iem.yzu.edu.tw/download.html

Table 1. The default parameter setting and the treatments of different factors

Factor	Treatment
Number of job (A)	35/10, 50/15, 65/18 (jobs/ machines)
Number of sub-population (B)	10, 20, 30, 40
Population Size (C)	100, 155, 210
Secondary Crossover Operator (D)	Apply multiple crossover (1), not using it (0)
Secondary Mutation Operator (E)	Apply multiple mutation (1), not using it (0)

Table 2. The suggested parameter for the modified SPGA

Factor	Treatment
Crossover Rate	0.9
Mutation Rate	0.1
Population Size	210
Number of sub-population	40

3.2 Comparisons for Adaptive Strategies

The experiment compares different adaptive strategies, including the method of Srinivas and Patnaik (1994) and the adaptive strategy of Zhu and Liu (2004). They are coded as 0, 1, and 2 in the experiment. The ANOVA table is available on our website and it represents the interaction between the instance and method which causes significant difference. Therefore, the interaction plot of the two factors is depicted in Fig 2. It shows that the adaptive strategies do not perform better in small size instances, while they outperform in large size instances. Duncan grouping method is

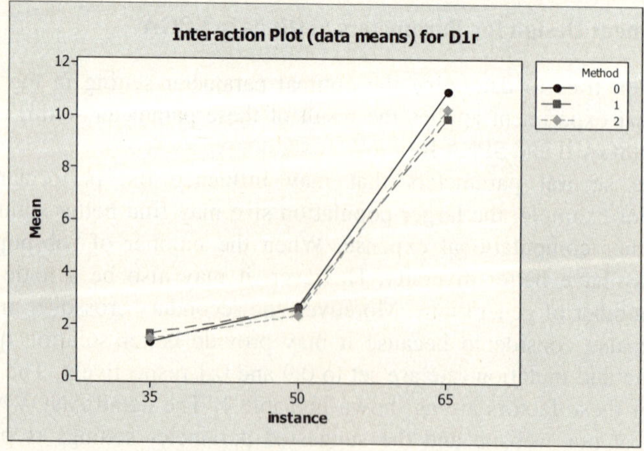

Fig. 2. The interaction plot between the instance and methods

Duncan Grouping	Mean	N	Method
A	6.8502	60	0
B	6.3144	60	2
B	6.2674	60	1

Fig. 3. The Duncan grouping method

applied to distinguish the group for the adaptive strategies. The grouping result is shown in figure 3. It shows that the modified SPGA with the adaptive strategies are better than the modified SPGA without adaptation. Since the time-complexity of Zhu and Liu (2004) is higher than Srinivas and Patnaik (1994), the study suggests using the later one when researchers would like to apply the adaptive strategy.

3.3 Numerical Results

After the study obtains the result of adaptive strategy for SPGA, the section compares the result with Modified SPGA, NSGA II and SPEA II, by three testing instances. Table 3 shows the statistics result of instances of 35 jobs and 10 machines, 50 jobs and 15 machines, and 65 jobs and 18 machines.

From the three instances, the modified SPGA is superior to the SPGA, NSGA II, and SPEA II in minimum, average, and maximum value. There is only one exception that the maximum value of modified SPGA is not better than SPGA and SPEA II in the instance of 65 jobs and 18. Then, the adaptive SPGA is better than the Modified SPGA 5.07% through the three instances.

Table 3. The min, average, and max value of different algorithms of the three instances

Instance	Algorithm	Min	Avg.	Max
35/10	Adaptive SPGA	0.494	1.667	3.391
	Modified SPGA’	0.56	1.4722	2.5147
	NSGA II	5.16	11.82	22.22
	SPEA II	4.8	10.39	22.48
50/15	Adaptive SPGA	1.418	2.609	3.554
	Modified SPGA’	1.72	2.876	3.901
	NSGA II	9.68	11.74	13.79
	SPEA II	7.65	10.27	12.89
65/18	Adaptive SPGA	5.092	9.925	13.192
	Modified SPGA’	7.537	10.611	13.941
	NSGA II	20.97	23.08	25.43
	SPEA II	7.7	10.3	12.9

4 Conclusion and Future Works

A modified SPGA and an adaptive SPGA were proposed for solving parallel machine scheduling problem with minimizing total tardiness time and makespan. Production data collected from a printed circuit board factory were applied as test instances. The numerical result indicated that SPGA with adaptive strategy perform better in large size test instances than SPGA without adaptation. Two genetic algorithms for multi-objective optimization, NSGA II and SPEA II, were compared with the proposed methods in this research. Extensive studies were conducted and the result reported that the adaptive SPGA and modified SPGA proposed in this research tend to outperform NSGA II and SPEA II especially in the large size problems. This also means the adaptive SPGA and modified SPGA are potential in the future works.

In the future research, although the algorithm is attractive to implement MO problem, we can still consider to combine SPGA with local search algorithms that may bring better solution quality.

References

1. Affenzeller, M. (2001). New Generic Hybrids Based Upon Genetic Algorithms. *Institute of Systems, Science Systems Theory and Information Technology*, Johannes Kepler University.
2. Brucker, P. (1998). *Scheduling Algorithm*. Berlin: Springer.
3. Chang, P.C., Chen, S.H., Lin, K.L. (2005). Two-phase sub population genetic algorithm for parallel machine-scheduling problem. *Expert Systems with Applications*, 29 (3), 705-712.
4. Chang, P.C. Hsieh, J.C., Wang, C.Y. (2006). Adaptive multi-objective genetic algorithms for scheduling of drilling operation in printed circuit board industry. To appear in *Applied Soft Computing*.
5. Cochran, J.K., Horng, S., Fowler, J.W. (2003). A multi-objective genetic algorithm to solve multi-objective scheduling problems for parallel machines. *Computers and Operations Research*, 30, 1087-1102.
6. Coello, C.A.C., G.T. Pulido, M.S. Lechuga (2004), "Handling multiple objectives with particle swarm optimization," IEEE Transactions on Evolutionary Computation, 8(3), 256- 279.
7. Deb, K., Amrit Pratap, S.A., Meyarivan, T. (2000). A fast and elitist multi-objective genetic algorithm-NSGA II. *Proceedings of the Parallel Problem Solving from Nature VI Conference*, 849-858.
8. Garey, M.R., Johnson, D.S. (1979). *Computers and Intractability: A guide to the theory of NP-completeness*. San Francisco, CA: New York Freeman.
9. Hsieh, J.C. (2005). Development of grid-partitioned objective space algorithm for flowshop scheduling with multiple objectives. Proceedings of the the 6th Asia Pacific Industrial Engineering and Management Systems Conference 2005.
10. Hsieh, J.C., Chang, P.C., Hsu, L.C. (2003). Scheduling of drilling operations in printed circuit board factory. *Computers and Industrial Engineering*, 44(3), 461-473.
11. Knowles, J.D. and Corne, D.W. (2002). On metrics for comparing non dominated sets. *Proceedings of the 2002 congress on evolutionary computation conference* (*CEC02*), 711-716. New York: IEEE Press.

12. Lis, J. and Eiben, A.E. (1997). A multisexual Genetic Algorithm for multicriteria optimization. *In Proceedings of the 4th IEEE Conference on Evolutionary Computation*, 59-64.
13. Murata, T., Ishibuchi, H. (1996). MOGA: Multi-objective genetic algorithm. *Proceedings of the Second IEEE International Conference on Evolutionary Computation*, 170-175.
14. Murata, T., Ishibuchi, H., Tanaka, H. (1996). Genetic algorithm for flowshop scheduling problem. *Computers and Industrial Engineering*, 30, 1061-1071.
15. Mostaghim , S., Teich, J. (2004), "Covering Pareto-optimal fronts by subswarms in multi-objective particle swarm optimization," *Evolutionary Computation*, 2, 1404 - 1411.
16. Neppali, V.R., Chen, C.L., Gupta, J.N.D. (1996) Genetic algorithms for the two-stage bicriteria flowshop problem. *European Journal of Operational Research*, 95, 356-373.
17. Schaffer, J.D. (1985) Multiple objective optimization with vector evaluated genetic algorithms. *Proceedings of First International Conference on Genetic Algorithms*, 93-100.
18. Sridhar, J., Rajendran, C. (1996). Scheduling in flowshop and cellular manufacturing systems with multiple objectives – a genetic algorithm approach. *Production Planning and Control*, 7, 374-382.
19. Srinivas, M. and Patnaik, L. M. (1994). Adaptive Probabilities of Crossover and Mutation in Genetic Algorithms, *IEEE Transactions on Systems, Man and Cybernetics*, 24 (4), pp. 656-667.
20. Zhu, K.Q., Liu, Z. (2004). Population diversity in permutation-based genetic algorithm. *Proceedings of the 15th European Conference on Machine Learning*, ECML 2004, Pisa, Italy. Springer. 537--547.
21. Zitzler, E., Laumanns, M., Thiele, L. (2002). SPEA 2: improving the strength Pareto evolutionary algorithm for multiobjective optimization, *Evolutionary Methods for Design, Optimisation and Control*, 1-6, Giannakoglou, K., Tsahalis, D., Periaux, J., Papailiou, K., and Fogarty, T. eds. CIMNE, Barcelona, Spain.

A Penalty-Based Evolutionary Algorithm for Constrained Optimization*

Yuping Wang[1] and Wei Ma[2]

[1] School of Computer Science and Technology
Xidian University, Xi'an 710071, China
ywang@xidian.edu.cn
[2] School of Science
Xidian University, Xi'an 710071, China
kunpengzhanchi_999@sina.com

Abstract. Evolutionary algorithm based on penalty function is a new kind of efficient method for constrained optimization problems, however, the penalty parameters are usually difficult to control, and thus the constraints can not be handled effectively. To handle the constraints effectively, a new constraint handling scheme based on a continuous penalty function with only one control parameter is proposed. Moreover, it does not add extra local optimal solutions to the primary problem. In order to decrease the amount of computation and enhance the speed of the algorithm, the uniform design is combined into the proposed algorithm to design a new crossover operator. This operator can generate a diversity of points and exploit the search space efficiently. Based on these, a novel evolutionary algorithm is proposed and the simulation on 5 benchmark problems is made. The results show the efficiency of the proposed algorithm with less computation, higher convergent speed for all test problems.

1 Introduction

The general form of the nonlinear constraint optimization problems is as follows

$$\begin{cases} \min f(X) \\ s.t.\ \ g_i(X) \le 0,\ i = 1 \sim m_1 \\ \qquad h_j(X) = 0,\ j = 1 \sim m_2 \\ \qquad BX \le d \\ \qquad AX = b \\ \qquad X \in [L, U] \end{cases} \tag{1}$$

where $X = (x_1, x_2, \cdots, x_n)^T \in R^n$ is an n dimensional column vector, $f(X)$, $g_i(X)$ and $h_j(X)$ are nonlinear functions of X, $A \in R^{m_4 \times n}$ is an $m_4 \times n$ matrix, $b \in R_4^m$ is an m_4 dimensional column vector, $B \in R^{m_3 \times n}$ is an $m_3 \times n$ matrix, $d \in R_3^m$ is an m_3 dimensional column vector, $[L, U] = \{X | l_i \le x_i \le u_i, 1 \le i \le$

* This work was supported by the National Natural Science Foundation of China (60374063).

L. Jiao et al. (Eds.): ICNC 2006, Part I, LNCS 4221, pp. 740–748, 2006.

$n\}$ with given vectors $L = (l_1, l_2, \cdots, l_n)^T$ and $U = (u_1, u_2, \cdots, u_n)^T$. This problem frequently arises in engineering, management and operations research, etc. When $f(X)$, $g_i(X)$ and $h_j(X)$ are non-convex, non-differentiable, even or non-continuous on $[L, U]$, it is very hard to get global optimal solution by traditional optimization methods.

Evolutionary algorithms are one of the most efficient algorithms for this problem. In the last decade, evolutionary algorithms for solving constrained optimization problems have received much attention. Many new algorithms have been proposed (e.g., [1] $\sim$ [8]). Among them penalty function methods are the most common methods. In these methods, a penalty term is added to the objective function, and the penalty will increase with the degree of constraint violation (static penalty) or the degree of constraint violation and generation number (dynamic penalty). In general the weakness of penalty methods is that they often require multiple parameters (to adjust the relative weights of each constraint in the penalty, and the weight of the penalty against the objective function), and the penalty terms may be non-differentiable even non-continuous, and thus may add extra local optimal solutions. However, due to their simplicity and ease of implementation they are still the most common methods used in solving real world problems.

In this paper a novel fitness function is first introduced. This fitness function can easily handle the constraints, and uses fewer control parameters (only one control parameter) than the common used ones. Moreover, it can not add extra local optimal solutions to the primary problem. In order to decrease the amount of computation and enhance the speed of the algorithm, the uniform design is combined into the proposed algorithm to design a new evolutionary operator. Based on these, a novel evolutionary algorithm is proposed. At last, we execute the proposed algorithm to solve some benchmark problems. The results also show the efficiency of the proposed algorithm with less computation, higher convergent speed for all test problems.

The remainder of this paper is organized as follows. Section 2 presents new techniques for constraint handling. Section 3 briefly introduces the uniform design and the new evolutionary operator. Section 4 proposes the new evolutionary algorithm. Section 5 describes the simulation results of the proposed algorithm. Section 6 presents the conclusions.

2 Constraint Handling Technique

First, we handle the equality constraints. Suppose that A has full row rank and $AX = b$ is consistent, i.e., all rows of A are linearly independent and $AX = b$ has solutions. In fact, if rows of A are not linearly independent, one can use Gaussian elimination method to eliminate some equation(s) from $AX = b$ such that the coefficient matrix in the resulted equations has full row rank and the resulted equations are consistent.

Since $AX = b$ is consistent and A has full row rank, $AX = b$ can be written as $A_1X^1 + A_2X^2 = b$, where A_2 is an invertible matrix and X consists of X^1 and X^2. Similarly, $BX \leq d$ can be written as $B_1X^1 + B_2X^2 \leq d$. Thus X^2 can expressed by X^1 by formula $X^2 = A_2^{-1}(b - A_1X^1)$. Deleting linear equations $AX = b$ from

(1) and inserting X^2 into (1) we get a lower dimensional constrained optimization problem with variables X^1 as follows

$$\begin{cases} \min f(X^1, A_2^{-1}(b - A_1X^1)) \\ s.t.\ \ g_i(X^1, A_2^{-1}(b - A_1X^1)) \leq 0,\ i = 1 \sim m_1 \\ \quad\ \ h_j(X^1, A_2^{-1}(b - A_1X^1)) = 0,\ j = 1 \sim m_2 \\ \quad\ \ B_1X^1 + B_2A_2^{-1}(b - A_1X^1) - d \leq 0 \\ \quad\ \ X^1 \in [L_1, U_1] \end{cases} \tag{2}$$

where L_1 and U_1 are lower and upper bounds of X^1 corresponding to $[L, U]$. For a given small tolerance $\delta > 0$, convert $h_j(X) = 0$ into $g_{m_1+j}(X) = |h_j(X)| - \delta \leq 0$ for $j = 1 \sim m_2$. Then problem (2) is transformed into a problem only with inequality constraints. For notation convenience, we also use X to denote X^1, use $f(X)$ to denote $f(X^1,, A_2^{-1}(b - A_1X^1))$ and use $g_i(X)$ to denote all functions in inequality constraints, respectively, and use $[L, U]$ to denote $[L_1, U_1]$. Thus we can simply write problem (2) as

$$\begin{cases} \min f(X) \\ s.t.\ \ g_i(X) \leq 0,\ i = 1 \sim m \\ \quad\ \ X \in [L, U] \end{cases} \tag{3}$$

where we also denote the dimension of the problem by n, i.e., $X \in R^n$.

Second, we handle the inequality constraints. Note that problem (3) can be transformed to the following problem with one constraint

$$\begin{cases} \min f(X) \\ s.t.\ \ \max\limits_{1 \leq i \leq m} \{g_i(X)\} \leq 0 \\ \quad\ \ X \in [L, U] \end{cases} \tag{4}$$

Since the constraint function $\max\limits_{1 \leq i \leq m} \{g_i(X)\}$ may not be continuous even if each $g_i(X)$ is differentiable. Thus if the constraint in problem (4) is handled by using traditional penalty function or traditional augmented Lagrangian function directly, a non-continuous fitness function caused by a non-continuous penalty term will be produced. As a result, this fitness function may have extra local optimal points which may not be the local or global optimal points of the problem (1). To avoid this problem, a continuous function

$$g(X, p) = \frac{1}{p} \ln \sum_{i=1}^{m} \exp(p g_i(X)) \tag{5}$$

is used to approximate the constraint function $\max\{g_i(X) | i = 1 \sim m\}$ of problem (4), and to be as unique penalty term in the fitness function. Of course, handling the constraint in this way will result in a fitness function with only one penalty parameter p. Obviously, this fitness function is continuous when each $g_i(X)$ is continuous, and has the following property.

Lemma 2.1 For any $X \in R^n$ and $p > 0$, $g(X, p)$ satisfies the following inequalities

$$g(X, p) - (\ln m)/p \leq \max_{1 \leq i \leq m} \{g_i(X)\} \leq g(X, p)$$

Based on function $g(X,p)$, a novel fitness function $F(X,p)$ can be proposed as follows

$$F(X,p) = \begin{cases} f(X), & \text{if } g(X) \leq 0 \\ \sup_{X \in \Omega} \{f(X)\} + g(X,p), & \text{otherwise} \end{cases} \quad (6)$$

It can be seen from formula (6) that fitness function $F(X,p)$ has the following properties

- Any feasible solution is better than any infeasible solution.
- In the feasible region, $F(X,p) = f(X)$. Therefore, the fitness function defined by formula (6) can not add extra local optimal solutions to the primary problem. In the infeasible region, $F(X,p)$ is continuous if each $g_i(X)$ is continuous. Furthermore, $F(X,p)$ has only one control parameter p.
- For feasible solutions X, the smaller $f(X)$ is, the smaller the fitness is, and thus the better the solution is. For infeasible solutions, the less the violation of constraints is, the smaller the fitness is, and the better the corresponding infeasible solution can be considered.

3 Uniform Design and Genetic Operators

3.1 Crossover Operator Using Uniform Design

Uniform design is an efficient method to generate a set of uniform distributed points on a region. In the following we briefly introduce the uniform design and the method to use it to construct a new crossover operator. For details of uniform design please refer to [9].

Let $U(q,b) = (a_{ij})_{q\times(q-1)}$ be a $q \times (q-1)$ matrix generated by a uniform design method in the following form

$$U(q,b) = (a_{ij})_{q\times(q-1)} = \begin{bmatrix} 1 \; b & \cdots \; b^{q-2} \\ 2 \; 2b & \cdots \; 2b^{q-2} \\ \cdots\cdots & \\ q \; qb & \cdots \; qb^{q-2} \end{bmatrix} \pmod q \quad (7)$$

where q is a chosen prime with $n \leq q-1$, b is a proper positive integer which is determined by q. For different q, it will correspond to different values of b (see [9]). For example, when $q = 5$, it can be looked for from [9] that $b = 2$. In the simulations in section 5 we adopt $q = 5$ and $b = 2$.

Now we use matrix $U(q,b)$ to design a crossover operator. The detail is as follows. For two parents

$$X = (x_1, x_2, \cdots, x_n)^T \text{ and } Y = (y_1, y_2, \cdots, y_n)^T,$$

using each row of matrix $U(q,b)$ can define an offspring of X and Y as follows. Define

$$W = (w_1, w_2, \cdots, w_n)^T \text{ and } Z = (z_1, z_2, \cdots, z_n)^T, \quad (8)$$

where $w_i = \min\{x_i, y_i\}$ and $z_i = \max\{x_i, y_i\}$ for $i = 1 \sim n$. When $n > q-1$, denote

$$W = (w^1, w^2, \cdots, w^{q-1})^T \text{ and } Z = (z^1, z^2, \cdots, z^{q-1})^T, \quad (9)$$

where $w^j = (w_{r_{j-1}+1}, \cdots, w_{r_j})$ and $z^j = (z_{r_{j-1}+1}, \cdots, z_{r_j})$ are $(r_j - r_{j-1})$ dimensional sub-vectors for $j = 1 \sim q-1$, and $r_1, r_2, \cdots, r_{q-2}$ are randomly generated integers on $[1, n)$ satisfying $0 = r_0 < r_1 < \cdots < r_{q-1} = n$. Then the i-th offspring O^i can be defined by

$$O^i = \begin{cases} (o_1^i, o_2^i, \cdots, o_n^i)^T, & \text{if } n < q \\ (v_1^i, v_2^i, \cdots, v_{q-1}^i)^T, & \text{else} \end{cases} \tag{10}$$

for $i = 1 \sim q$, where $o_j^i = w_j + \frac{2a_{ij}+1}{2q}(z_j - w_j)$ for $j = 1 \sim n$, and $v_j^i = w^j + \frac{2a_{ij}+1}{2q}(z^j - w^j)$ for $j = 1 \sim q-1$.

3.2 Mutation Operator

Suppose that $O = (o_1, o_2, \cdots, o_n)$ is chosen to undergo the mutation. It generates an offspring $\tilde{O}$ by Gaussian Mutation, i.e., it generates offspring $\tilde{O}$ by

$$\tilde{O} = O + \Delta O \tag{11}$$

where the i-th component ΔO_i of ΔO was generated by a random number generator which obeys the Gaussian distribution with mean 0 and deviation σ_i^2, denoted by $\Delta O_i \sim N(0, \sigma_i)$ for $i = 1, 2, \cdots, n$ or

$$\Delta O \sim N(0, \sigma) = (N(0, \sigma_1), N(0, \sigma_2), \cdots, N(0, \sigma_n)),$$

$N(0, \sigma_i)$ represents the Gaussian distribution with mean 0 and deviation σ_i^2, and n components ΔO_1, $\Delta O_2, \cdots$, ΔO_n of ΔO are independent. In simulation, $\sigma_i = 1$, $i = 1 \cdots, n$.

4 The Proposed Evolutionary Algorithm

Algorithm 1

1. (Initialization) Given crossover probability p_c and population size N. Let $k = 1$. Generate initial population $POP(k) = \{X^1, X^2, \cdots, X^N\}$.
2. (Crossover) In generation k, generate N random numbers $r_1^k, r_2^k, \cdots, r_N^k \in [0, 1]$. If $r_i^k < p_c$, then $X^i \in POP(k)$ is chosen for crossover for $i = 1 \sim N$. Suppose that τ_k pairs of points are chosen. For each pair of chosen points, generate q offspring using the crossover operator in section 3.1. All these offspring generated by crossover constitute a set denoted as $OFF(k)$.
3. (Mutation) In generation k, each offspring O^i in $OFF(k)$ undergoes the mutation to generate an offspring $\tilde{O}^i$ by mutation operator in section 3.2. All these offspring form a new set denoted by $OFF(\tilde{O})$.
4. (Selection) Select the best N points (with the smallest fitness function values) from $POP(k) \cup OFF(k) \cup OFF(\tilde{O})$ using the new fitness function defined in section 2 as the $(k+1)$-th generation population $POP(k+1)$.
5. (Stop criteria) If stop criteria are satisfied, stop; otherwise, let $k = k+1$, go to Step 2.

5 Simulation Results and Applications

5.1 Test Problems

Five commonly-used test problems are chosen as the benchmark problems from the literatures ([1]~[5]).

$$
\begin{aligned}
G4: \min f(X) &= 5.3578547x_3^2 + 0.8356891x_1x_5 + 87.293239x_1 - 40792.141\\
s.t.\ g_1(X) &= 85.334407 + 0.005658x_2x_5 + 0.0006262x_1x_4 - 0.0022053x_2x_5 \geq 0\\
g_2(X) &= 85.334407 + 0.005658x_2x_5 + 0.0006262x_1x_4 - 0.0022053 \leq 92\\
g_3(X) &= 80.51249 + 0.0071317x_2x_5 + 0.00029955x_1x_2 + 0.0021813x_3^2 \geq 90\\
g_4(X) &= 80.51249 + 0.0071317x_2x_5 + 0.00029955x_1x_2 + 0.0021813x_3^2 \leq 110\\
g_5(X) &= 0.300961 + 0.0047026x_3x_5 + 0.0012547x_1x_3 + 0.00019085x_3x_4 \geq 20\\
g_6(X) &= 0.300961 + 0.0047026x_3x_5 + 0.0012547x_1x_3 + 0.00019085x_3x_4 \leq 25\\
& 78 \leq x_1 \leq 102,\ 33 \leq x_2 \leq 45,\ 27 \leq x_i \leq 45,\ i = 3, 4, 5
\end{aligned}
$$

$$
\begin{aligned}
G7: \min f(X) &= x_1^2 + x_2^2 + x_1x_2 - 14x_1 - 16x_2 + (x_3 - 10)^2 + 4(x_4 - 5)^2 + (x_5 - 3)^2\\
&\quad + (x_6 - 1)^2 + 5x_7^2 + 7(x_8 - 11)^2 + 2(x_9 - 10)^2 + (x_{10} - 7)^2 + 45\\
s.t.\ g_1(X) &= 105 - 4x_1 - 5x_2 + 3x_7 - 9x_8 \geq 0\\
g_2(X) &= -10x_1 + 8x_2 + 7x_7 - 2x_8 \geq 0\\
g_3(X) &= 8x_1 - 2x_2 - 5x_9 + 2x_{10} + 2 \geq 0\\
g_4(X) &= -3(x_1 - 2)^2 = 4(x_2 - 3)^2 - 2x_362 + 7x_4 + 20 \geq 0\\
g_5(X) &= -5x_1^2 - 8x_2 - (x_3 - 6)^2 + 2x_4 + 40 \geq 0\\
g_6(X) &= -x_1^2 - 2(x_2 - 4)^2 + 2x_1x_2 - 14x_5 + 6x_6 \geq 0\\
g_7(X) &= -0.5(x_1 - 8)^2 - 2(x_2 - 4)^2 + 3x_5^2 + x_6 + 30 \geq 0\\
g_8(X) &= 3x_1 - 6x_2 - 12(x_9 - 8)^2 + 7x_{10} \geq 0\\
& -10 \leq x_i \leq 10,\ i = 1 \sim 10
\end{aligned}
$$

$$
\begin{aligned}
G8: \min f(X) &= \frac{\sin^3(2\pi x_1)\sin(2\pi x_2)}{x_1^3(x_1 + x_2)}\\
s.t.\ g_1(X) &= x_1^2 - x_2 + 1 \leq 0\\
g_2(X) &= 1 - x_1 + (x_2 - 4)^2 \leq 0\\
& 0 \leq x_i \leq 10,\ i = 1, 2
\end{aligned}
$$

$$
\begin{aligned}
G9: \min f(X) &= (x_1 - 10)^2 + 5(x_2 - 12)^2 + x_3^4 + 3(x_4 - 11)^2\\
&\quad + 10x_5^6 + 7x_6^2 + x_7^4 - 4x_6x_7 - 10x_6 - 8x_7\\
s.t.\ g_1(X) &= 127 - 2x_1^2 - 3x_2^4 - x_3 - 4x_4^2 - 5x_5 \geq 0\\
g_2(X) &= 282 - 7x_1 - 3x_2 - 10x_3^2 - x_4 - 5x_5 \geq 0\\
g_3(X) &= 196 - 23x_1 - x_2^2 - 6x_6^2 + 8x_7 \geq 0\\
g_4(X) &= -4x_1^2 - x_2^2 + 8x_1x_2 - 2x_3^2 - 5x_6 + 11x_7 \geq 0\\
& -10 \leq x_i \leq 10,\ i = 1 \sim 7
\end{aligned}
$$

$$
\begin{aligned}
G10: \min f(X) &= x_1 + x_2 + x_3\\
s.t.\ g_1(X) &= 1 - 0.0025(x_4 + x_6) \geq 0\\
g_2(X) &= 1 - 0.0025(x_5 + x_7 - x_4) \geq 0\\
g_3(X) &= 1 - 0.01(x_8 - x_5) \geq 0\\
g_4(X) &= x_1x_6 - 833.33252x_4 - 100x_1 + 83333.333 \geq 0\\
g_5(X) &= x_2x_7 - 1250x_5 - x_2x_4 + 250x_4 \geq 0\\
g_6(X) &= x_3x_8 - x_3x_5 + 2500x_5 - 1250000 \geq 0\\
& 100 \leq x_1 \leq 10000,\ 1000 \leq x_2, x_3 \leq 10000\\
& 10 \leq x_i \leq 1000,\ i = 4 \sim 8
\end{aligned}
$$

5.2 Results and Comparison

In the execution of the proposed algorithm we adopt the following parameter values: $\delta = 0.0001$, $q = 5$, $b = 2$, $p = 1000$, $p_c = 0.2$, $N = 100$, and the maximum number of the generations was set to 1000. The parameter values taken in this manner cnsure the proposed algorithm uses fewer number of function evaluations than the compared algorithms. For example, the proposed algorithm used average 200×1000 function evaluations for each of these problems, while algorithms in [1] and [2] used 70×20000 and 200×1750 function evaluations, respectively.

For all test problems the proposed algorithm is carried out 50 independent runs. We record the objective function value of the best solution (denoted as Best for short), the objective function value of the worst solution (denoted as Worst for short), the mean of the objective function values of 50 found best solutions (denoted as Mean for short) and the standard deviation of the objective function values of 50 found solutions (denoted as Stdv for short) in 50 runs.

Tables 1 to 4 summarize the results obtained by the proposed algorithm, the algorithms in [1], [2], and [4], respectively.

Table 1. Results obtained by proposed algorithm for $G4$, $G7$ to $G10$

Func.	Opt.	Best	Mean	Worst	Stdv.
$G4$	-30665.5	-30669.3	-30665.9	-30659.3	123.3
$G7$	24.306	24.358	24.365	24.913	5.19E-1
$G8$	-0.095825	-0.095825	-0.095825	-0.095825	0
$G9$	680.63	680.63	680.631	680.633	8.3E-3
$G10$	7049.33	7049.33	7049.81	7066.87	0.512

Table 2. Results obtained by Farmani and Wright in [1] for $G4$, $G7$ to $G10$

Func.	Opt.	Best	Mean	Worst	Stdv.
$G4$	-30665.5	-30665.5	-30665.2	-30663.3	4.85E-1
$G7$	24.306	24.48	26.58	28.40	1.14E+0
$G8$	-0.095825	-0.095825	-0.095825	-0.095825	0
$G9$	680.63	680.64	680.72	680.87	5.92E-2
$G10$	7049.33	7061.34	7627.89	8288.79	3.73E+2

It can be seen from Tables 1 to 4 that the proposed algorithm found the better or optimal solutions for all test problems. It should also be noted that the standard deviations for these test problems are all very small except for problem $G4$. This indicates the proposed algorithm is stable. Although the standard deviation for problem $G4$ is relatively large, the best solution -30669.3 found

Table 3. Results obtained by Runarsson and Yao in [2] for $G4$, $G7$ to $G10$

Func.	Opt.	Best	Mean	Worst	Stdv.
$G4$	-30665.5	-30665.539	-30665.539	-30665.539	2.0E-5
$G7$	24.306	24.307	24.374	24.642	6.6E-2
$G8$	-0.095825	-0.095825	-0.095825	-0.095825	2.6E-17
$G9$	680.63	680.63	680.656	680.763	3.4E-2
$G10$	7049.33	7054.316	7559.192	78835.655	5.3E+2

Table 4. Results obtained by Koziel and Michalewicz in [4] for $G4$, $G7$ to $G10$

Func.	Optimal.	Best	Mean	Worst	Stdv.
$G4$	-30665.5	-30664.5	-30655.3	-30645.9	NA
$G7$	24.306	24.620	24.826	25.069	NA
$G8$	-0.095825	-0.095825	-0.0891568	-0.0291438	NA
$G9$	680.63	680.91	681.16	683.18	NA
$G10$	7049.33	7147.9	8163.6	9659.3	NA

by the proposed algorithm is much better than the known "optimal" solution -30665.5. Moreover, even the mean solution found by the proposed algorithm is better than the known "optimal" solution.

6 Conclusion

This paper introduces a continuous penalty evolutionary algorithm for constrained optimization. A new constraint handling scheme is first introduced based on a novel fitness function. This fitness function does not add extra local optimal solutions to the primary problem. Then new genetic operators are designed. Based on these, a novel evolutionary algorithm is proposed. At last, the proposed algorithm is tested on 5 benchmark problems. The results indicate the efficiency of the proposed algorithm.

References

1. R. Farmabi, J. A. Wright, "Self-adaptive fitness formulation for constrained optimization," *IEEE Trans. On Evolutionary Computation*, vol.7, no.5, pp.445-455, 2003.
2. T. P. Runarsson, X. Yao, "Stochastic ranking for constrained evolutionary optimization," *IEEE Trans. On Evolutionary Computation*, vol.4, no.4, pp.284-294, 2000.
3. K. Deb, "An efficient constraint handling method for genetic algorithms," *Comput. Meth. Appl. Mech. Eng.*, vol.186, pp.311-338, 2000.

4. S. Koziel, Z Michalewicz, "Evolutionary algorithms, homomorphous mapping, and constrained parameter optimization," *Evolutionary Computation*, vol.7, no.1, pp.19-44, 1999.
5. S. Ben-Hamida, M. Schoenauer, "An adaptive algorithm for constrained optimization problems," in *Proc. Parallel Problem Solving form Nature*, vol.VI, 2000, pp.529-538.
6. Z Michalewicz, *Genetic algorithms +data structures = evolution programs.* 3rd Edition. Berlin: Springer-Verlag, 1999.
7. Min-Jea Tahk, Byung-Chan Sun, "Coevolutionary augmented Lagrangian methods for constrained optimization," *IEEE Trans. On Evolutionary Computation*, vol.4, no.1, pp.114-124, 2000.
8. Carlos A, Clello Coello, "Treating Constraints as objectives for single-objective evolutionary optimization," *Engineering Opimization*, vol.32, no.3, pp.275-308, 2000.
9. Fang, K. T. and Wang, Y., *Number-Theoretic Method in Statistics*, Chapman and Hall, London, 1994.

Parallel Hybrid PSO-GA Algorithm and Its Application to Layout Design

Guangqiang Li[1], Fengqiang Zhao[2], Chen Guo[1], and Hongfei Teng[3]

[1] College of Automation and Electrical Engineering, Dalian Maritime University, Dalian 116026, People's Republic of China
gqlimail@163.com
[2] College of Electromechanical & Information Engineering, Dalian Nationalities University, Dalian 116600, People's Republic of China
[3] School of Mechanical Engineering, Dalian University of Technology, Dalian 116024, People's Republic of China

Abstract. Packing and layout problems belong to NP-Complete problems theoretically and have found a wide utilization in practice. Parallel genetic algorithms (PGA) are relatively effective to solve these problems. But there still exist some defects of them, e.g. premature convergence and slow convergence rate. To overcome them, a parallel hybrid PSO-GA algorithm (PHPSO-GA) is proposed based on PGA. In PHPSO-GA, subpopulations are classified as several classes according to probability values of improved adaptive crossover and mutation operators. And in accordance with characteristics of different classes of subpopulations, different modes of PSO update operators are introduced. It aims at making full use of the fast convergence property of particle swarm optimization. Adjustable arithmetic-progression rank-based selection is introduced into this algorithm as well. It not only can prevent the algorithm from premature in the early stage of evolution but also can accelerate convergence rate in the late stage of evolution. To be hybridized with simplex method can improve local search performance. An example of layout design problem shows that PHPSO-GA is feasible and effective.

1 Introduction

Packing and layout problems [1] are to study how to put objects into limited space reasonably under constraints (for example, no interference, increasing space utilization ratio). In complex problems (e.g. the layout design of spacecraft modules), some extra behavioral constraints should be taken into consideration, such as the requirements for equilibrium, connectivity and adjacent states. This kind of problems is often faced in many engineering fields and they are of great importance. They usually directly affect some performance indices of design, as reliability and economy. Since these problems belong to NP-Complete problems, it is quite difficult to solve them satisfactorily.

Ref. [1] and [2] summarized the common methods for solving packing and layout problems. According to the algorithm trend and solution quality, it shows that the robust algorithms based on physical computation (especially genetic algorithms) are of

L. Jiao et al. (Eds.): ICNC 2006, Part I, LNCS 4221, pp. 749 – 758, 2006.

advantage. They are particularly fit to solve medium or large-scale problems. But there still exist some defects with regard to genetic algorithms themselves, such as premature convergence and slow convergence rate [3]. To overcome them, some measures are taken and an improved parallel hybrid PSO-GA algorithm (PHPSO-GA) is proposed based on PGA. It aims at solving packing and layout problems more effectively.

2 Parallel Hybrid PSO-GA Algorithm

2.1 Improved Adaptive Crossover and Mutation

To prevent genetic algorithms from premature effectively as well as protect superior individuals from untimely destruction, Srinivas and Patnaik [4] proposed the concept of adaptive crossover and mutation. But according to these operators, crossover and mutation rate of the best individual among a population are both zero. It may lead to rather slow evolution in the early stage. To avoid its occurrence, it's better to let the individuals possess due crossover and mutation rates, whose fitness values are equal or approximate to the maximal fitness. Therefore, based on Ref. [4], improved adaptive crossover rate p_c and mutation rate p_m are presented, see formula (1) and (2).

$$p_c = \begin{cases} k_1 \exp[(F_{max} - F') \cdot (\ln k_3 - \ln k_1)/(F_{max} - F_{avg})], & F' \geq F_{avg} \\ k_3, & F' < F_{avg} \end{cases} \tag{1}$$

$$p_m = \begin{cases} k_2 \exp[(F_{max} - F) \cdot (\ln k_4 - \ln k_2)/(F_{max} - F_{avg})], & F \geq F_{avg} \\ k_4, & F < F_{avg} \end{cases} \tag{2}$$

where F_{max} and F_{avg} denote the maximal and average fitness of current population. F' denotes the greater fitness of the two individuals that take part in crossover operation. F denotes the fitness of the individual that take part in mutation operation. k_1, k_2, k_3, k_4 are constants. And there exist $0<k_1, k_2, k_3, k_4 \leq 1.0$, $k_1<k_3$, $k_2<k_4$.

2.2 Adjustable Arithmetic-Progression Rank-Based Selection

There is one independent parameter in this operator, dominance coefficient λ. It denotes the ratio of the maximal individual selection probability p_{max} to the minimal one p_{min} within a generation, $p_{max}=\lambda \cdot p_{min}$. It numerically shows the superiority that the better individuals are reproduced into the next generation during selection and it is changeable along with evolution. In the early stage, lesser λ can maintain population diversity and prevent algorithm from premature; in the late stage, greater λ can benefit accelerating convergence. Let $\lambda=f(K)$, K and f denote the generation number and a increasing function respectively. We adopt linear increasing function. Let λ_{max} and λ_{min} denote the maximum and minimum of dominance coefficient respectively, then

$$\lambda = \frac{(K-1)(\lambda_{max} - \lambda_{min})}{K_{max} - 1} + \lambda_{min} \, . \tag{3}$$

where K_{max} is the maximal generation number set in algorithm. Our experiments show that λ_{max} and λ_{min} may be chosen in the interval [6, 15] and [1.5, 5] respectively.

To calculate the selection probability of every individual, first of all, we should arrange all the individuals within a population in descending order based on their fitness values. Let Ind_i represent the ith individual within a population as well as F_i and p_i represent its fitness and selection probability respectively. There exist Ind_i (i=1,2, …, M) and $F_i > F_{i+1}$(i=1,2, , …,M-1). M is the population size. Suppose that the selection probability values of all the individuals form an arithmetic progression. We set its first term and last term are $p_1=\lambda a$ and $p_M=a$ respectively. Obviously, the sum of all the individual selection probability is 1, i.e. subtotal of arithmetic progression

$$S_M=[(p_1+p_M)\,M]\,/\,2=[(\lambda a+a)\,M]\,/\,2=1\ . \tag{4}$$

We get

$$p_M=a=2\,/\,[M(1+\lambda)]\ . \tag{5}$$

Therefore we obtain the common difference of the arithmetic progression

$$\Delta=(p_1-p_M)\,/\,(M-1)=[a\,(\lambda-1)]\,/\,(M-1)=[2(\lambda-1)]\,/\,[M\,(1+\lambda)(M-1)]\ . \tag{6}$$

And there exists

$$p_i= p_1-(i-1)\Delta=\lambda a-(i-1)\Delta\ . \tag{7}$$

Substituting formula (5) and (6) into formula (7), it is easy to find that

$$p_i=\frac{2\lambda(M-i)+2(i-1)}{M\,(M-1)(1+\lambda)}\ ,\ i=1,2,\ \ldots,\ M\ . \tag{8}$$

In the process of selection, we firstly reproduce the best individual of current generation and put its copy into next generation directly based on elitist model. And then figure out selection probability of all individuals according to formula (8). Finally generate the remaining M-1 individuals of next generation by fitness proportional model. Compared with traditional rank-based selection, the advantage of proposed selection operator is that it can conveniently change the selection probability of individuals by changing dominance coefficient and is more adaptive to algorithm run.

2.3 Multi-subpopulation Evolution

We classify all the subpopulations of proposed algorithm into four classes (named class α, β, γ and δ) according to their crossover and mutation rates (P_c and P_m). Suppose that there is only one subpopulation within every class, named class α, β, γ and δ subpopulation respectively. Their parametric features are shown in Table 1.

Table 1. Parametric features of four classes of subpopulations

Subpopulation	Class α	Class β	Class γ	Class δ
Crossover rate	k_1=0.8; k_3=1.0	k_1=0.5; k_3=0.8	k_1=0.2; k_3=0.5	k_1=0.1; k_3=0.2
Mutation rate	k_2=0.3; k_4=0.4	k_2=0.2; k_4=0.3	k_2=0.1; k_4=0.2	k_2=0.05; k_4=0.1
Initial fitness	Minimal	Medium	Greater	Maximal

According to their properties of initial fitness and crossover & mutation rates, we can see that it is easier for class α subpopulation to explore new parts of solution space and guard against premature. Class γ subpopulation is mainly to consolidate local search. Class β subpopulation is a transitional subpopulation. And the function of class δ subpopulation is to keep stability and diversity of superior individuals. After random initialization, PHPSO-GA arranges all the generated individuals according to their fitness values. The initial individuals with the maximal fitness are allocated to class δ subpopulation; the initial individuals with relatively greater fitness are allocated to class γ subpopulation; the initial individuals with the minimal fitness are allocated to class α subpopulation; the rest of initial individuals are allocated to class β subpopulation.

At intervals of migration cycle, PHPSO-GA copies the best individuals in class α, β and γ subpopulation and saves them into class δ subpopulation, then update class δ subpopulation. Meanwhile, it selects some individuals from class δ subpopulation and makes them migrate to class α, β and γ subpopulation respectively. The migration individuals will replace inferior individuals in above subpopulations respectively as well. This migration strategy can accelerate convergence. In addition, we set control parameter K_m. When generation number K is multiples of K_m, PHPSO-GA merges all the subpopulations together and arrange all individuals according to their fitness. Then it reallocates individuals to every subpopulation respectively according to their fitness values.

2.4 PSO Update Operators

Basic Theory of Particle Swarm Optimization. Kennedy and Eberhart [5, 6] presented the idea of particle swarm optimization (PSO). In PSO, each particle as an individual in genetic algorithms represents a potential solution. There are mainly two forms of PSO at present, i.e. global version and local version.

With regard to global version of PSO, in the n-dimensional search space, M particles are assumed to consist of a population. The position and velocity vector of the ith particle are denoted by $\boldsymbol{X}_i=(x_{i1}, x_{i2}, \ldots, x_{in})^{\mathrm{T}}$ and $\boldsymbol{V}_i=(v_{i1}, v_{i2}, \ldots, v_{in})^{\mathrm{T}}$ respectively. Then its velocity and position are updated according to the following formulas.

$$v_{id}^{k+1} = w \cdot v_{id}^{k} + c_1 \cdot rand() \cdot \left(p_{id}^{k} - x_{id}^{k}\right) + c_2 \cdot rand() \cdot \left(p_{gd}^{k} - x_{id}^{k}\right). \quad (9)$$

$$x_{id}^{k+1} = x_{id}^{k} + v_{id}^{k+1}. \quad (10)$$

where i=1,2,…, M; d=1,2,…, n; k and k+1 are iterative numbers. $\boldsymbol{p}_i=(p_{i1}, p_{i2},\ldots, p_{in})^{\mathrm{T}}$ is the best previous position that ith particle searched so far and $\boldsymbol{p}_g=(p_{g1}, p_{g2}, \ldots, p_{gn})^{\mathrm{T}}$ is the best previous position for whole particle swarm. *rand*() denotes a uniform random number between 0 and 1. Acceleration coefficients c_1 and c_2 are positive constants (usually $c_1=c_2=2.0$). w is inertia weight and it showed that w decreases gradually along with iteration can enhance entire algorithm performance effectively [7].

It is usually set limitation to a particle velocity. Without loss of generality, assume that relevant following intervals are symmetrical. There exists $v^{k}_{id} \in [-v_{d,\max}, +v_{d,\max}]$. $v_{d,\max}$ (d=1,2,…, n) determine the resolution with which regions between present position and target position are searched. If $v_{d,\max}$ is too high, particles may fly past good solutions. While, if it is too small, the algorithm may be stuck to local optima. Suppose that the range of definition for the dth dimension of a design variable is $[-x_{d,\max}, +x_{d,\max}]$, i.e. $x^k_{id} \in [-x_{d,\max}, +x_{d,\max}]$. Usually let $\pm v_{d,\max} = \pm k x_{d,\max}$, $0.1 \le k \le 1$.

In local version of PSO, particle i keeps track of not only the best previous position of itself, but also the best position $\boldsymbol{p}_{li}=(p_{li,1}, p_{li,2},\ldots, p_{li,n})^T$ attained by its local neighbor particles rather than that of the whole particle swarm. Typically, the circle-topology neighborhood model is adopted [6]. Its velocity update formula is

$$v_{id}^{k+1} = w \cdot v_{id}^{k} + c_1 \cdot rand() \cdot \left(p_{id}^{k} - x_{id}^{k}\right) + c_2 \cdot rand() \cdot \left(p_{li,d}^{k} - x_{id}^{k}\right). \tag{11}$$

And its position update formula is same as that of the global version of PSO. Compared with global version of PSO, local version of PSO has a relatively slower convergence rate but it is not easy to be stuck to local optima.

PSO has been applied to many fields and results are satisfactory [8]. It is easy to be implemented and has quite fast convergence rate among evolutionary algorithms. But it also has the limitations such as low precision and premature. Noticed that genetic algorithms and PSO are both based on swarm intelligence and can match each other fairly well. To make full use of fast convergence property of PSO and global convergence ability of genetic algorithms, we propose this hybrid algorithm. Specifically, let velocity and position update formulas together serve as a new operator (PSO update). After conventional genetic operation, individuals go on with PSO update operation. It hopes to make hybrid algorithm possess more superior global performance.

Operators Realization. In PHPSO-GA, different mode PSO update operators are introduced into different subpopulations. Global mode PSO update operator is introduced into class δ subpopulation in order to accelerate the convergence of its individuals to global optima. Average mode PSO update operator is introduced into class γ subpopulation so as to help its individuals to consolidate local search around discovered superior solutions. Random mode and synthesis mode PSO update operator are introduced into class α and β subpopulation respectively. The former matches the function of exploring solution space of class α subpopulation and helps to prevent algorithm from premature. The latter matches the function of class β subpopulation and gives consideration to the balance of exploration and exploitation in solution space.

Every mode PSO update operator has the same position update formula, see formula (10). But their velocity update formulas are different. Global mode update operator is on the basis of global version PSO completely. Synthesis mode update operator integrates global version PSO with local version PSO together. In its individual velocity update formula, see formula (12), three best positions are chased, i.e. the best position an individual visited so far, the best position attained by its local neighbor particles and the best position obtained so far by the whole population.

$$\begin{aligned} v_{id}^{k+1} &= w \cdot v_{id}^{k} + c_1 \cdot rand() \cdot \left(p_{id}^{k} - x_{id}^{k}\right) + c_2 \cdot rand() \cdot \left(p_{gd}^{k} - x_{id}^{k}\right) \\ &\quad + c_3 \cdot rand() \cdot \left(p_{li,d}^{k} - x_{id}^{k}\right). \end{aligned} \tag{12}$$

where c_3 is the acceleration coefficient of the newly added item. According to Ref. [9], we set $c_1 = c_2 = 1.5$ and $c_3 = 1.1$ in synthesis mode update operator.

In random mode update operator, the neighborhood N_i of particle i is composed of s particles. Apart from particle i itself, the other s-1 particles are randomly selected from whole population. In this mode update operator, particle i keeps track of the best previous position of itself and the best position attained within its random neighborhood N_i. In the broad sense, random mode PSO can be regarded as a special kind of local version PSO. Merely its topology structure of neighborhood is dynamic and stochastic. Therefore it helps to explore solution space thoroughly and prevent from premature. We usually set $s=int(0.1\sim0.15M)$ and $int(\cdot)$ denotes round-off function.

As for average mode PSO update operator, we first arrange the best position of every particle $\boldsymbol{p}_i$ (i=1,2,…,M) in descending order according to their corresponding fitness. Then select the front u best positions, here denoted by $\boldsymbol{p}_{gj}=(p_{gj,1}, p_{gj,2},\ldots, p_{gj,n})^{\mathrm{T}}$, j=1, 2,…, u. And change velocity of particle i based on its own best previous position $\boldsymbol{p}_i$ and average of $\boldsymbol{p}_{gj}$ (j=1, 2,…, u), i.e. $\overline{\boldsymbol{p}}_g=(\overline{p}_{g1}, \overline{p}_{g2},\cdots, \overline{p}_{gn})^{\mathrm{T}}$, as follows.

$$v_{id}^{k+1} = w \cdot v_{id}^{k} + c_1 \cdot rand\,() \cdot \left(p_{id}^{k} - x_{id}^{k}\right) + c_2 \cdot rand\,() \cdot \left(\overline{p}_{gd}^{k} - x_{id}^{k}\right). \tag{13}$$

$$\overline{p}_{gd}^{k} = \frac{1}{u} \cdot \sum_{j=1}^{u} p_{gj,d}\,. \tag{14}$$

Usually $1\le u\le int(0.15M)$ and this mode PSO will reduce to global version PSO if u=1.

We lay emphasis on two parameters in PSO update operators, i.e. inertia weight w and maximal velocity $\boldsymbol{V}_{\max}$. Usually there exist $w\in[0.3, 1.5]$, $\pm v_{d,\max}=\pm kx_{d,\max}$ ($0.1\le k\le 1.0$). If they select greater values, the update operator is more likely to find out new parts of solution space. Otherwise the update operator is good at local search. According to characteristics of different subpopulations, we set the range of w and $\boldsymbol{V}_{\max}$ of every update operator in Table 2. Based on adaptation idea [7], we let w and k (coefficient of maximal velocity) decrease linearly along with evolution from their maximal values to the minimal values.

Table 2. Relevant settings of PSO update operators of all classes of subpopulations

Subpopulation	Class α	Class β	Class γ	Class δ
Update operators	Random mode	Synthesis mode	Average mode	Global mode
Inertia weight w	$w_{\max}$=1.5; $w_{\min}$=1.0	$w_{\max}$=1.1; $w_{\min}$=0.6	$w_{\max}$=0.7; $w_{\min}$=0.4	$w_{\max}$=0.6; $w_{\min}$=0.3
Coefficient k	$k_{\max}$=1.0; $k_{\min}$=0.7	$k_{\max}$=0.7; $k_{\min}$=0.4	$k_{\max}$=0.5; $k_{\min}$=0.2	$k_{\max}$=0.3; $k_{\min}$=0.1

2.5 Hybrid Strategy

To further improve local search ability of the algorithm, it is necessary to apply hybrid strategy. Taking the matching problem into consideration, we hybridize simplex method [10] with proposed algorithm. Simplex method possesses relatively fast local convergence rate and doesn't involve derivative information. Allowing for the problem of computational efficiency, the hybrid algorithm should give full play to the global search ability of genetic algorithm in the early stage, while to the local search ability of simplex method in the late stage. Therefore in PHPSO-GA, we set parameter K_s, randomly select N_s individuals to form initial simplex and search C_s

turns by simplex method at intervals of K_s generations. To enhance the local search ability of PHPSO-GA in the late stage and accelerate convergence rate, N_s and C_s are set in direct proportion to generation number in the proposed algorithm.

2.6 The Procedures of Algorithm

Flow chart of the proposed parallel hybrid PSO-GA algorithm is shown in Fig. 1.

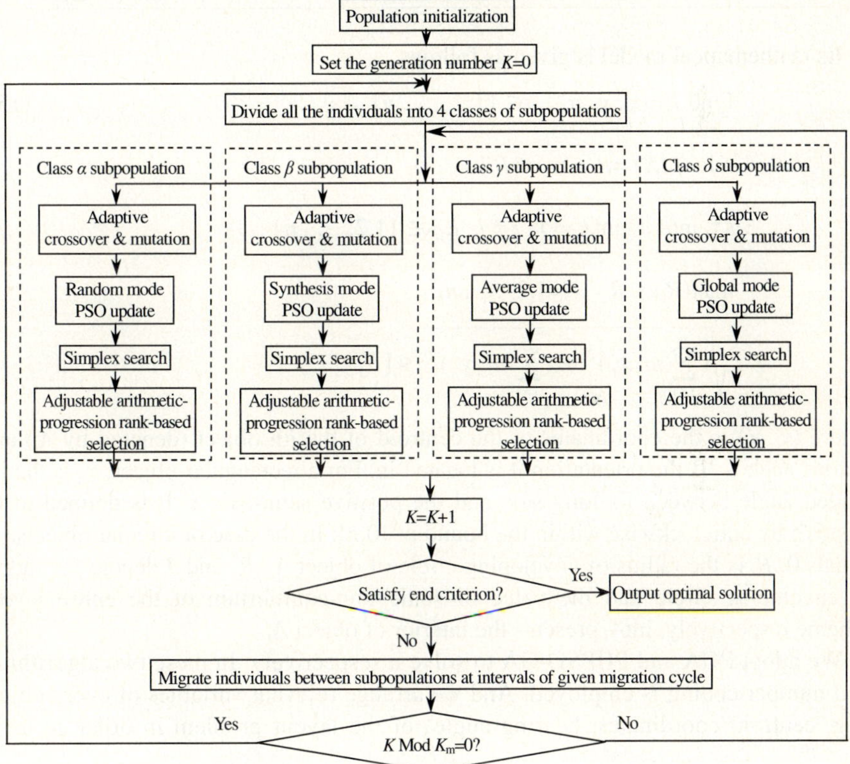

Fig. 1. Flow chart of the proposed parallel hybrid PSO-GA algorithm (PHPSO-GA)

3 Numerical Example

Quoted from Ref. [11], 5 rectangular objects and 5 circular objects will be located on a bearing plate with radius R=1750mm. Relevant data are given in Table 3, in which a_i and b_i are the longer and shorter edge of a rectangular object while r_i is the radius of a circular object, and m_i denotes the mass of every object. The allowable value of static non-equilibrium is $[\delta_J]$ = 1.5Kg·mm. Try to locate each object such that these objects highly concentrate on the center of the container as well as the constraints of no interference and static equilibrium behavior are satisfied. All computation is performed on PC with CPU at 667MHz and RAM size of 256MB.

Table 3. Dimensions and mass of objects of the example

No.	a_i /mm	b_i /mm	r_i /mm	m_i / Kg	No.	a_i /mm	b_i /mm	r_i /mm	m_i / Kg
1	730	730		4.0	6			450	4.0
2	850	600		4.5	7			400	2.0
3	500	450		3.6	8			340	2.6
4	760	560		3.9	9			130	1.5
5	660	280		2.0	10			400	2.5

Its mathematical model is given as follows.

$$\text{Find } \boldsymbol{X}=(x_i, y_i, \alpha_i)^{\mathrm{T}},\ i\in\{1,2,\ldots,n\}\ . \tag{15}$$

$$\min f(\boldsymbol{X}) = R^* = \max\{R_i\}\ .$$

$$\text{s.t. int}A_i \cap \text{int}A_j = \varnothing \quad i \neq j,\ i,j \in \{1,2,\ldots,n\}\ .$$

$$\max(R_i) \leq R \qquad i=1,2,\ldots,n\ .$$

$$\sqrt{(\sum_{i=1}^{n} m_i x_i)^2 + (\sum_{i=1}^{n} m_i y_i)^2} \leq [\delta_J]\ .$$

where (x_i, y_i) is the coordinates of the centroid of the ith object (denoted by A_i) and bearing angle α_i is the orientation it is located in. For a rectangular object, α_i is the included angle between its long side and the positive semi-axis x. It is defined to be positive by anticlockwise within the bound of $[0,\pi]$. In the case of circular objects, α_i equals 0. R_i is the radius of enveloping circle of object A_i. R^* and J denote the radius of enveloping circle and the value of static non-equilibrium of the entire layout scheme respectively. intA_i presents the interior of object A_i.

We adopt PGA and PHPSO-GA to solve it respectively. In these two algorithms, real number coding is employed. And we arrange relevant variables of every object (e.g. centroid coordinates, bearing angle) of the layout problem in order so as to

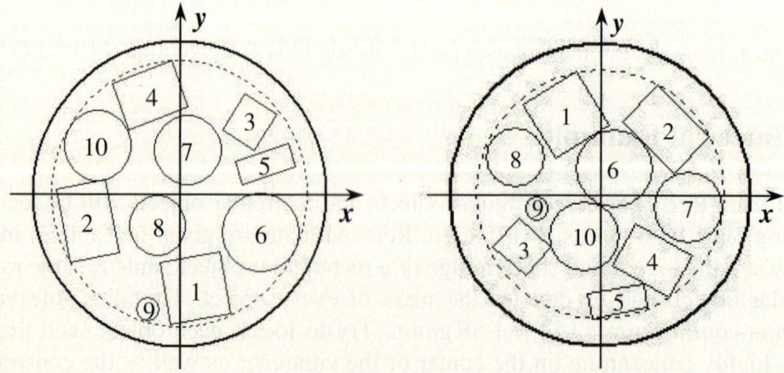

Fig. 2. Obtained best layout patterns of the example by PGA (*left*) and PHPSO-GA (*right*)

form a digital string (an individual). Both algorithms are calculated 20 times. The best layout patterns by them are shown in Fig. 2. Both satisfy the noninterference constraint. For the best layout by PGA, R^*, J and computation time t are 1524.529mm, 0.082Kg·mm and 226.69s; for the best layout by PHPSO-GA, R^*, J and t are 1443.305mm, 0.017Kg·mm and 217.31s. When obtained $R^* \leq 1524.529$mm, $J \leq 0.082$Kg·mm by PHPSO-GA, it takes 182.01s. So in the sense of best results, to reach the same precision, PHPSO-GA reduces the cost of time by 19.71% compared with PGA.

Table 4 lists relevant average values of obtained twenty optimal results of the example by two algorithms. In this table, ΔS represents the interference area and K represents elapsed generation number for an optimal result.

Table 4. Comparison of average values of optimal results by two algorithms of the example

Algorithms	R^*/ mm	J/ Kg·mm	ΔS/ mm^2	K
PGA	1585.783	0.702	0.000	1120
PHPSO-GA	1459.518	0.266	0.000	752

Table 4 shows that compared with PGA, on an average, PHPSO-GA reduces enveloping circle radius R^*, the value of static non-equilibrium J and elapsed generation number K by 7.96%, 62.11% and 32.86%, i.e. from 1585.783mm to 1459.518mm, from 0.702Kg·mm to 0.266Kg·mm and from 1120 to 752 respectively.

4 Concluding Remarks

Taking layout design as background, to overcome two main defects of PGA, we take several measures on it and propose PHPSO-GA. These measures involve introducing adjustable arithmetic-progression rank-based selection, PSO update operators, hybrid strategy and multi-subpopulation evolution based on adaptive crossover and mutation. An example of layout design problem shows that PHPSO-GA is feasible and effective. Because proposed PHPSO-GA is a universal algorithm, it also can be adopted to solve other complex optimization problems. In fact, to completely test its performance, we have done lots of numerical computation on various examples [12]. Results show that it is really superior to PGA in accuracy and convergence rate.

Acknowledgements

We would like to express our gratitude to the National Natural Science Foundation of China (No. 50275019, No. 50575031), the Specialized Research Foundation for the Doctoral Programs of Ministry of Education of P.R.C (No. 20040151007) and the Application foundation Research Project of Ministry of Communications of P.R.C. (No. 200432922504) for financial support of our work.

References

1. Dowsland, K. A., Dowsland, W. B.: Packing Problems. European Journal of Operational Research. 56 (1992) 2-14
2. Qian, Z. Q., Teng, H. F.: Algorithms of Complex Layout Design Problems. China Mechanical Engineering. 13 (2002) 696-699
3. Davis, L. D.: Handbook of Genetic Algorithms. Van Nostrand Reinhold, New York (1991)
4. Srinivas, M., Patnaik, L. M.: Adaptive Probabilities of Crossover and Mutation in Genetic Algorithm. IEEE Transaction on System Man and Cybernetics. 24 (1994) 656-667
5. Kennedy, J., Eberhart, R.: Particle Swarm Optimization. Proceedings of the IEEE International Conference on Neural Networks. Perth, Australia (1995) 1942-1948
6. Eberhart, R., Kennedy, J.: A New Optimizer Using Particle Swarm Theory. Proceedings of the Sixth International Symposium on Micro Machine and Human Science. Nagoya, Japan (1995) 39-43
7. Shi, Y., Eberhart, R.: A Modified Particle Swarm Optimizer. Proceedings of the IEEE Conference on Evolutionary Computation. Anchorage, AK, USA (1998) 69-73
8. Eberhart, R., Shi, Y.: Particle Swarm Optimization: Developments, Applications and Resources. Proceedings of the 2001 Congress on Evolutionary Computation. Seoul Korea (2001) 81-86
9. Clerc, M.: The Swarm and the Queen: Towards a Deterministic and Adaptive Particle Swarm Optimization. Proceedings of the 1999 Congress on Evolutionary Computation. Washington, USA. 3 (1999) 1951-1957
10. Shamir, R.: Efficiency of the Simplex Method: a Survey. Management Science. 33 (1987) 301-334
11. Teng, H. F., Sun, S. L., Liu, D. Q.: Layout Optimization for the Objects Located within a Rotating Vessel—a Three-dimensional Packing Problem with Behavioral Constraints. Computer & Operations Research. 28 (2001) 521-535
12. Li, G. Q.: Evolutionary Algorithms and their Application to Engineering Layout Design. Postdoctoral Research Report, Tongji University, Shanghai, China (2005)

Knowledge-Inducing Interactive Genetic Algorithms Based on Multi-agent

Yi-nan Guo, Jian Cheng, Dun-wei Gong, and Ding-quan Yang

School of Information and Electronic Engineering,
China University of Mining and Technology,
221008 Xuzhou, China
nanfly@126.com

Abstract. Interactive genetic algorithms lack a common model to effectively integrate different assistant evolution strategies including knowledge-based methods and fitness assignment strategies.Aiming at the problems,knowledge-based interactive genetic algorithm based on multi-agent is put forward in the paper combined with the flexibility of multi-agent systems.Five kinds of agents are abstracted based on decomposed-integral strategy of MAS.A novel implicit knowledge model and corresponding inducing strategy are proposed and realized by knowledge-inducing agent.A novel substitution strategy for evaluating fitness by an online model instead of human is proposed and implemented in fitness-estimation agent.State-switch conditions of above agents are given using agent-oriented programming. Taking fashion design system as a testing platform, the rationality of the model and the effective of assistant evolution strategies proposed in the paper are validated. Simulation results indicate this algorithm can effectively alleviate users' fatigue and improve the speed of convergence.

1 Introduction

Interactive genetic algorithms (IGAs) are a kind of genetic algorithms in which fitness of individuals is evaluated by human subjectively. Now they have been applied to optimization problems whose objectives can not be expressed by explicit fitness functions, such as music composition, production design and so on [1].However there exits users' fatigue in evaluation, which limits population size and the number of evolution generations. Moreover, human subjective fitness mainly reflects users' cognition and preference which is related to domain knowledge. So IGAs need more knowledge than other genetic algorithms for explicit optimization functions. Though many knowledge-based methods have been proposed, few of them specialize in IGAs. And there is a lack of a common model to effectively integrate different kinds of strategies together which limits its application.

Domain knowledge is classified into priori knowledge and implicit knowledge by Giraldez [2].In IGAs, the latter one which reflects users' psychological requirement is of interest. The key problems about it are how to extract and utilize it so as to alleviate users' fatigue and improve the speed of convergence. Up to now, many knowledge-based methods have been proposed. According to the operations induced by knowledge

L. Jiao et al. (Eds.): ICNC 2006, Part I, LNCS 4221, pp. 759 – 768, 2006.

in an evolution, they are divided into fitness-estimation strategies and knowledge-inducing strategies. In the former, human is replaced by fitness-estimation models to evaluate fitness of individuals so as to reduce the number of individuals evaluated by users. Fitness-estimation models based on artificial neural networks and a phase evaluation strategy were given in[3][4].Support vector machines, sparse expression and hybrid methods were adopted in evaluation process to reduce users' burden[5]-[7].In above methods, implicit knowledge only acts on evaluation process. Aiming at knowledge embodied in an evolution, many knowledge-inducing strategies were proposed, in which knowledge direct evolution operators, such as mutation. Knowledge models based on artificial neural networks were adopted to update fitness of bad individuals so as to drive them to better solutions[8][9].Rules-based knowledge models are proposed to improve the generalization of solutions[10][11].Based on dual structure, co-evolutionary GA and cultural algorithms were proposed, in which schema was extracted and utilized to induce evolution[12][13].But few of them specialized in IGAs. So a novel knowledge-inducing strategy is proposed aiming at IGAs in this paper.

Above two kinds of knowledge-based methods are all needed in an evolution. But there is a lack of a common model to rationally synthesize and effectively manage them which limit the research and application of knowledge-inducing IGAs(KIGAs). Multi-agent systems(MAS) consist of multiple autonomous agents by some cooperation mechanism which has powerful integration capability and management functions. Agent can autonomously interact with environments and cooperate with other agents which correspond to the requirements of IGAs. Now MAS have been used in GAs. Parmee proposed an interactive evolutionary design station based on agents [14], in which agents act as communication controller and interface like in[15]. Agent-based GAs in fine granularity were put forward by Li and Zhong, in which each agent denoted an individual [16]-[18].These indicate MAS provide a execution mechanism for GAs. But how to combine MAS with IGAs in coarse granularity so as to build a common model for KIGAs is still an open problem. Aiming at above problems, a knowledge-inducing interactive genetic algorithm based on multi-agent (KIGA-MAS) is put forward in this paper. It makes the best of autonomy of agent and cooperation capability among agents to realize function groupware of KIGAs which alleviate users' fatigue and make the algorithms more intelligent.

In the rest of the paper, the structure of KIGA-MAS and five kinds of agent are given in Section2.Function descriptions and state-switch conditions of knowledge-inducing agent and fitness-estimation agent are illustrated in following sections. To validate rationality of the algorithm, experiments based on fashion design system and testing results are analyzed in Section5.At last, future work planned to extend the algorithm to multi-user is included.

2 Structure of KIGA-MAS

In KIGAs, implicit knowledge is extracted and utilized to estimate fitness of individuals and direct the evolution so as to alleviate users' fatigue. In order to effectively integrate and autonomously manage above algorithm, decomposed-integral strategy of MAS is adopted. Different function groupware are abstracted and their semantic descriptions and derivative relationships are given using agent-oriented programming. The structure of KIGA-MAS is shown in Fig.1.

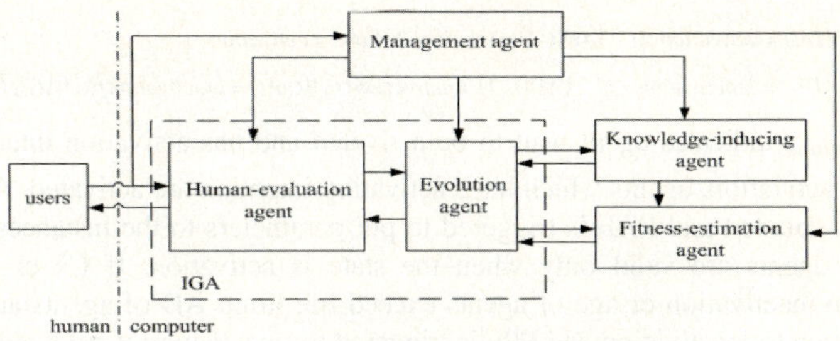

Fig. 1. The structure of KIGA-MAS

Users evaluate individuals according to the phenotype of individuals via human-machine interface in human-evaluation agent. Evolution agent implements elemental genetic operations, such as crossover and mutation, according to the genotype of individuals and corresponding fitness evaluated by users or fitness-estimation model. Above two agents implement the functions of canonical IGAs. Implicit knowledge which reflects users' cognition and preference embodied in the evolution is extracted and utilized in knowledge-inducing agent. Fitness-estimation model is trained to simulate users' preference and adopted to estimate fitness of individuals in fitness-estimation agent. Above agents are all supervised and scheduled by management agent. The latter three agents realize knowledge-based assistant evolution strategies. In this paper, knowledge-inducing agent and fitness-estimation agent are of interest.

Agent is the basic of MAS. Different agents have different functions. According to the analysis of different agents' functions in the model, the normal kernel structure of agent is abstracted using agent-oriented programming as follows [19].

<Cagent>::=<resource RS>< function FD>
<RS>::=<unit description CD><environment description ED>
<CD>::=<symbol ><state AD><age AG><life limit AL>
<FD>::=<universal function CF><extended function SF>
<CF>::=<interface management IM><state control function SCF>
<SCF>::=<state transfer function A><parameter activation function PAF> <parameter preservation function PPF>

According to the requirement of knowledge-based interactive genetic algorithm, five kinds of derived classes are inheritance from Cagent so as to supervise and manage an evolution. They are COagent, CEagent, CMagent, CFagent and CKagent which realize corresponding functions of human-evaluation agent, evolution agent, management agent, fitness-estimation agent, knowledge-inducing agent respectively.

Agents have three kinds of inner states, which can be switched each other, denoted by AD={activation,waiting,inactivation}. Inner states are controlled by SCF: $AD(t) = A(AD(t-1), VR(t), Ca(t), AG(t))$ .Ca={activation,inactivation} is outside collaboration signal [20]. Aiming at the instances of COagent, CEagent,CFagent and CKagent, Ca is offered by management agent. But Ca of CMagent is given by human. When an application solved by KIGA-MAS startup, Ca of CMagent is activation. It is obvious that inner state at next time is decided by self-information and outside signal. Different derived classes of agent have different state transfer function. But the basic rules for state-switch are same, shown as follows:

Rule 1: $AD(t) = activation \quad if \quad (AD(t-1) = waiting) \wedge (Ca(t) = activation)$

Rule 2: $AD(t) = inactivation \quad if \quad (AD(t-1) = activation) \wedge ((Ca(t) = inactivation) \vee (AG(t) > AL))$

$AD(t) = waiting$ indicates agent wait to be activated and has activation intention. Only when Ca is activation, agents which have activation intention are activated. At the same time, age is counted and PAF is triggered to put parameters to the instances of agents. Outputs of agents are valid only when the state is activation. If Ca changes from activation to inactivation or age of agents exceed life limit, AD of agents are switched from activation to inactivation and PPF is triggered to save values of parameters.

3 Function Description of Knowledge-Inducing Agent

In knowledge-inducing agent, genotype of individuals from evolution agent and corresponding fitness from human-evaluation agent make up of samples. Implicit knowledge is extracted from samples and utilized to direct the evolution.

3.1 Implicit Knowledge Model

In IGAs, implicit knowledge reflects users' cognition and preference, which can not be expressed by explicit functions. Based on the definition of gene-meaning-units (GM-units)[21],a novel implicit knowledge model is defined as

$$C = [W \quad P \quad Rv]^T \tag{1}$$

$W = [w_1 \quad w_2 \quad \cdots \quad w_n]$ where w_i expresses the distribution of attributes which reflect users' cognition tendency to i^{th} GM-unit. $P = [p_1 \quad p_2 \quad \cdots \quad p_n]$ where p_i expresses the importance of GM-units which reflect the degree of users' preference. A GM-unit is described by $V_i = \{v_{i1}, v_{i2}, \cdots, v_{il_i}\}$ where v_{ij} denotes j^{th} attribute of this unit. Assuming n is the number of GM-units and l_i is the number of attributes included in the unit.

The reliability of knowledge is defined as $Rv^t = e^{-Ns^t \bullet Sd^t \bullet t_w}$.Letting $Ns^t = \sum_{i=1} N \bullet \rho_e^t$ denotes the total number of evaluated individuals until t^{th} generation where N is population size and ρ_e^t is the percentage of individuals evaluated by users. Letting $Sd^t = \frac{N(N-1)}{2} \Big/ \sum_{i=1}^{N-1} \sum_{j=i+1}^{N} d(x_i, x_j)$ denotes the similarity of population where $d(x_i, x_j)$ is the Hamming distance between x_i and x_j. t_w is waiting time during evaluation.When users feel tired during the evolution, fitness evaluated by users can not reflect users' true preference any more. This lowers the reliability of extracted knowledge. So the degree of users' fatigue *Fa* is in inverse proportion to *Rv*: $Fa^t = 1/Rv^t$.

3.2 Extraction and Utilization of Implicit Knowledge

1) Extraction about knowledge

Implicit knowledge is extracted from sample-population $\overline{S}$ by statistical learning method.Because individuals with higher fitness contain more information about good solution,they are chose as samples from ranked population by fitness in descending order.

In a GM-unit,attributes which users most like have more probability to exist in offspring.Assuming $\rho_{v_{ij}}$ is the survival probability of jth attribute in i^{th} GM-unit.

$$\rho_{v_{ij}} = \frac{1}{K}\sum_{k=1}^{K} f\left(d\left(s_i^k, v_{ij}\right)\right) \tag{2}$$

$$f\left(d\left(s_i^k, v_{ij}\right)\right) = \begin{cases} 1 & d\left(s_i^k, v_{ij}\right) = 0 \\ 0 & d\left(s_i^k, v_{ij}\right) \neq 0 \end{cases} \tag{3}$$

where s_i^k is the value of ith GM-unit of kth sample. K is sample-population size. Defined average survival probability of ith GM-unit as $\eta_{v_i} = 1/l_i$ .Then users' cognition tendency of ith GM-unit is extracted by following rules.

Rule A: $w_i = \rho_{v_{ij}} \quad if \quad (\rho_{v_{ij}} > \eta_{v_i}) \wedge (\rho_{v_{ij}} > \rho_{v_{il}}) \quad (\forall v_{ij}, v_{il} \in V_i, j \neq l, j, l \leq l_i)$

Rule B: $w_i = 0 \quad if \quad \rho_{v_{ij}} \leq \eta_{v_i} \quad (\forall v_{ij} \in V_i, j \leq l_i)$

Rule C: $w_i = 0 \quad if \quad (\rho_{v_{ij}} = \rho_{v_{il}}) \wedge (\rho_{v_{ij}} > \eta_{v_i}) \wedge (\rho_{v_{il}} > \eta_{v_i}) \quad (\forall v_{ij}, v_{il} \in V_i, j \neq l, j, l \leq l_i)$

Rule1 indicates attribute with maximum survival probability is extracted as schema which will direct the evolution to good solutions.But if users' cognition tendency is fuzzy or one more attributes have maximum survival probability,it indicates there is no valuable information embodied in this GM-unit.

Based on W, characteristic-individual $Vo = [vo_1 \ vo_2 \ \cdots \ vo_n]$ is extracted in which $vo_i = \begin{cases} v_{ij} & w_i \neq 0 \\ (*)^{m_i} & w_i = 0 \end{cases}$.

The degree of users' preference is defined as $p_i = w_i / \sum_{i=1}^{n} w_i$ which reflects how deep users prefer this unit.

2) Update about knowledge

Users' cognition varies and becomes clearer during the evolution.So characteristic-vector needs to be updated dynamically so as to exactly track the change of users' cognition in time.

How to update samples is a key issue,which influences the speed of track. Individuals in evaluated population are added to $\bar{s}$ if they provide more information and contribute to the evolution.By comparison of fitness between individuals and samples,a novel update strategy is shown as follows.

Step1:Rank population according to $F^t(x_i)$ which is fitness of x_i in t^{th} generation.

Step2:Individual with maximum fitness is added to $\bar{s}$ directly.

Step3:Other individuals which satisfy $\mu_{x_i} > \alpha$ are added to $\bar{s}$ where μ_{x_i} is a probability generated randomly and is addition probability.

Step4:Rank new sample-population according to fitness.

Step5:Individuals with minimum fitness are move from new sample-population. The number of moved individuals is equal to the number of added individuals.

How many generations between two neighboring updates is a key problem, which influences the efficiency of computation. More samples are similar, the diversity of $\bar{s}$ is worse and redundant information embodied in $\bar{s}$ is more. So the period of extracting knowledge is defined as $\tau_u = e^{d(\bar{s})}$ where $d(\bar{s}) = \frac{K(K-1)}{2} \Big/ \sum_{i=1}^{K-1}\sum_{j=i+1}^{K} d(\bar{s}_i, \bar{s}_j)$ denotes the similarity of $\bar{s}$.

3) Direction about knowledge

Each genetic operation can be influenced by implicit knowledge.Here,a novel knowledge-inducing strategy for replacing individuals is proposed.

Vo is adopted as the constraint to limit the number of similar individuals in new population of next generation.Considering users' preference to each GM-unit,the weighted Hamming distance is defined as $d_s = Rv\sum_{i=1}^{n} p_i f(d(x_i, Vo_i))$.The detail steps of the strategy are shown as follows.

Step1:Calculate d_s between *Vo* and each individual in offspring population.

Step2:Move individuals which satisfy $d_s = 0$.Go back to Step1 with a new randomly generated individual.

Step3:Individuals which satisfy $d_s \neq 0$ are placed in new population.

3.3 State-Switch Condition of Knowledge-Inducing Agent

According to above analysis and function description, the state of knowledge-inducing agent is switched from inactivation to waiting when u is satisfied.Assuming tlast is last generation in which knowledge is updated and tnow is current generation.The condition for activation intention is $\{AD_{CKagent} = waiting | Q_{pre}^{CKagent} := [t_{now} = t_{last} + \tau_u] \cap [t_{now} \leqslant AL]\}$. When Ca is activated,the state of this agent is changed from waiting to activation: $AD_{CKagent}(t) = activation \quad if \quad (Ca_{CMagent}(t) = activation) \wedge (AD_{CKagent}(t-1) = waiting) | AD_{CMagent} = activation$.

4 Function Description of Fitness-Estimation Agent

Users are easy to feel tired if they evaluate all individuals in each generation.In order to solve this problem,fitness-estimation models are adopted to evaluate fitness of individuals instead of users.How to obtain the model and utilize it to evaluate fitness are of interest. Above functions are realized in fitness-estimation agent.

1) Online modeling method based on artificial neural networks

Users' cognition to a special objective is memorized in fitness-estimation model. And the model needs to sense and track the change of users' cognition.Now most of modeling methods about IGAs are offline.So the models can not be updated in time. Here,a novel online modeling method is put forward based on BP networks.

Letting $fu'(x_i)$ and $fm'(x_i)$ denote fitness of x_i evaluated by model and users respectively. Assuming $ef'(x_i) = \sqrt{(fu'(x_i) - fm'(x_i))^2}$ is evaluation error of a individual. Then evaluation error of population is define as $ef'(X) = \frac{1}{N}\sum_{i=1}^{N} ef'(x_i)$ which reflects generalization of the model. Large $ef'(x_i)$ shows estimation precision is bad and the model needs to be correct. Letting Ψ denotes the threshold for evaluation error.The detail steps about the online modeling method are shown as follows.

Step1:Evaluate individuals and calculate evaluation error of population.

Step2:When $ef'(x_i) < \Psi$ is satisfied,go to setp1 or else go to step3.

Step3:Individuals which satisfy $ef'(x_i) > \Psi$ are added to training set as training data.

Step4:Train BP networks using above training data.

2) Evaluation strategy based on fitness-estimation model

Now few of researches concern when to startup the model and how many individuals are evaluated by the model in each generation.Here,these problems are of interest.

When *Fa* is more then the threshold ,it indicates users feel tired and may not evaluate individuals according to their real preference.So fitness-estimation model is startup to evaluate individuals instead of users.Define the number of individuals evaluated by the model as $N_M^t = N\rho_m^t$ where $\rho_m^t = e^{\overline{ef}'(X)\cdot(Fa^t-\varepsilon)}$ is substitution probability.Letting $\overline{ef}'(X)=\frac{1}{N}\sum_{i=1}^{W}\left(ef'(x_i)/fu'(x_i)\right)$ denotes normalized evaluation error of population.It is obvious that substitution probability is increasing while users feel more tired or the model is more close to users' cognition.

According to *Fa* and $\overline{ef}'(x_i)$,fitness-estimation model is utilized to evaluate fitness of individuals instead of human according to following rules.

Rule A: $N_M^t = 0 \quad if \quad (F_a^t \leqslant \varepsilon) \wedge (ef'(X) > \Psi)$

Rule B: $N_M^t = N\rho_m^t \quad if \quad (F_a^t \leqslant \varepsilon) \wedge (ef'(X) < \Psi)$

Rule C: $N_M^t = N \quad if \quad (F_a^t > \varepsilon)$

When the degree of users' fatigue is lower,the model is not startup if estimation precision and generalization of it is bad.But if evaluation error is little,it indicates the model can reflect users' cognition accurately.So the model is startup and the number of individuals evaluated by it is in direct proportion to substitution probability.When users feel tired,fitness evaluated by users may not reflect users' real cognition any more and even can not be used as samples.So fitness of all individuals is evaluated by this model,namely ρ_m^t=1.

3) State-switch condition of fitness-estimation agent

According to above function description,the state of fitness-estimation agent is changed from inactivation to waiting when startup conditions are satisfied: $\{AD_{CFagent} = waiting \mid Q_{pre}^{CFagent} := [F_a < \varepsilon] \cup [ef'(X) < \Psi] \cap [t_{now} \leqslant AL]\}$.The activation rules are as same as CKagent.

5 Simulations and Analysis

In this paper, fashion design system is adopted as a typical background to validate the rationality of KIGA-MAS.Visual Basic 6.0 as programming tool for human-machine interface and Access as database are utilized.Matlab 6.5 is adopted as modeling tool. The main window for user interface is shown in Fig. 2.

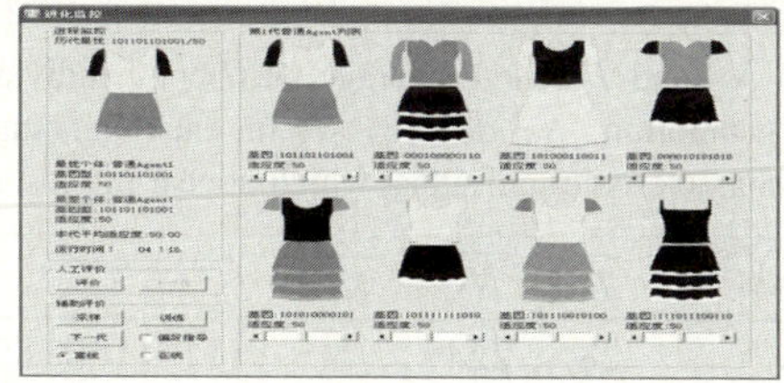

Fig. 2. Human-machine interface in fashion deign system based on KIGA-MAS

In fashion design system, each dress is composed of collar, skirt and sleeve. Each part has two factors including pattern and color which described by two bits.So each dress is expressed by 12 bits,which act as 6 GM-unit as shown in Table.1[19].

Table 1. The attributions of gene-meaning-units

name	code of attributions							
	1		2		3		4	
	name	code	name	code	name	code	name	code
collar's pattern	medium collar	00	high collar	01	wide collar	10	gallus	11
sleeve's pattern	long sleeve	00	medium sleeve	01	short sleeve	10	non-sleeve	11
skirt's pattern	long skirt	00	formal skirt	01	medium skirt	10	short skirt	11
color	pink	00	blue	01	black	10	white	11

In order to validate the rationality of KIGA-MAS,two groups of experiments are designed.They have different desired objectives which reflect different psychological requirements of human.Desired objective in Group1 is to find a favorite dress fitting for summer without the limit of color.Desired objective in Group2 is to find a favorite dress fitting for summer and the color is pink.In experiments,when the instance of each derived classes of agent is activated,corresponding parameters are initialized by PAF:Pc=0.5,Pm=0.01,N=8,K=20,T=30,ε=0.3,Ps=0.6, ψ=0.1,α=0.5.

1) 21 persons are divided into three groups to do experiments adopting IGA,KIGA, KIGA-MAS respectively.Testing results in Table.2 show KIGA-MAS converges fastest than others averagely.

Compared with KIGA and IGA $\overline{Ns}$ averagely reduces 62.1%.These indicate implicit knowledge model is beneficial to alleviating users' fatigue. $\overline{Ns}$ adopting KIGA-MAS averagely reduces 82.1% and 82.9% than others and $\overline{Ns}'$ adopting KIGA-MAS is less than others.These show MAS provide effective management for KIGAs and online fitness-estimation model can reduces users' burden for evaluation.

Vo in different experiments are different,but reflect key characters of desired objectives accurately. This shows *Vo* which extracted through implicit knowledge model can track users' preference exactly.Moreover,comparison of results between two groups show while the constraint of desired objectives are more,the speed of convergence is slower.This matches the rule of users' cognition.

Table 2. Testing results($\overline{Ns}$ and $\overline{T}$ denote average number of individuals evaluated by users and average evolution generation. $\overline{Ns}'$ denotes $\overline{Ns}$ in each generation. x^* and *Vo* denote optimum individual and characteristic-individual which appear in most of results.)

No.	methods	$\overline{T}$	$\overline{Ns}$	$\overline{Ns}'$	x^*	*Vo*
Group1	IGA	28	224	8	111011001111	-
	KIGA	10	80	8	111011001111	****11**11**
	KIGA-MAS	7	31	4	111111001111	****11**11**
Group2	IGA	40	240	8	111011011110	-
	KIGA	16	128	8	110011001111	****110011**
	KIGA-MAS	11	49	4	100011001111	****110011**

2) The degree of users' fatigue *Fa* during the evolutions using different algorithms are plotted in Fig. 3. It is obvious that user feel more tired along with the evolution.

Based on the same algorithms,*Fa* under desired objective in Group1 is lower than it in Group2 on each generation.In Group1,desired objective have less constraint than it in Group2.This means users are easy to feel tired when they concern more unit.This matches the physiological rules of human.

Under same desired objectives,*Fa* adopting KIGA-MAS is less than them using other methods which indicates KIGA-MAS can effectively alleviate users' fatigue.

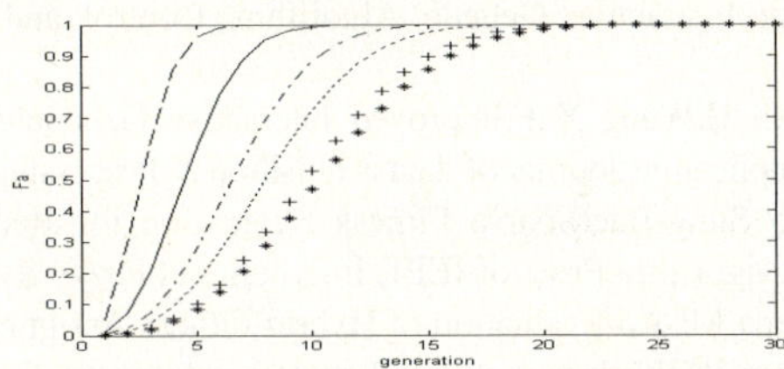

Fig. 3. Comparison of *Fa* by same algorithms with different desired objectives(Under desired objectives of Group1 and Group2,solid line and dashed,dotted line and dashdotted line,plus line and star line denote the change of *Fa* using IGA,KIGA,KIGA-MAS respectively)

6 Conclusions

In order to effectively integrate knowledge-based methods and fitness assignment strategies aiming at alleviate user fatigue, a knowledge-based interactive genetic algorithm based on multi-agent is put forward combined with the flexibility of multi-agent systems, which integrate knowledge-based methods and fitness assignment strategies with IGAs. Five kinds of agents are abstracted to implement the operations in KIGA-MAS. In knowledge-inducing agent, a novel implicit knowledge model and corresponding operators are proposed. A novel substitution strategy for evaluating fitness based on online modeling method is put forward and implemented in fitness-estimation agent. Semantic descriptions and state-switch conditions of above agents are given using agent-oriented programming. Taking fashion design system as a testing platform, the rationality of the model and the effective of assistant evolution strategies are validated by comparison with canonical IGAs aiming at different psychological requirements of human. Simulation results indicate KIGA-MAS can effectively alleviate users' fatigue and improve the speed of convergence, which makes users absorbed in more creative design work. The multi-population KIGA-MAS for multi-user is the future research.

Acknowledgements

This work was supported by the National Postdoctoral Science Foundation of China under grant 2005037225, the Postdoctoral Science Foundation of Jiangsu under grant 2004300, the Youth Science Foundation of CUMT under grant OC 4465.

References

1. H.Takagi: Interactive Evolutionary Computation:System Optimization Based on Human Subjective Evolution.Proc.of IEEE Conference on Intelligent Engineering System.(1998)1-6
2. R.Giraldez,J.S.Aguilar-ruiz,J.C.Riquelme:Knowledge-based Fast Evaluation for Evolutionary Learning.IEEE Trans. on SMC-Part C:Application and Review.Vol.35.(2005) 254-261
3. Biles J A,Anderson P G,Loggi L W: Neural Network Fitness Functions for A Musical IGA. Proc.of the Symposium on Intelligent Industrial Automation&Soft Computing,(1996)39-44.
4. Zhou yong,Gong Dun-wei,Hao Guo-sheng,et al: Neural Network Based Phase Estimation of Individual Fitness in Interactive Genetic Algorithm. Control and decision,Vol.20.(2005) 234-236
5. Wang Sh F,Wang Sh H,Wang X F:Improved Interactive Genetic Algorithm Incorporating with SVM and Its Application.Journal of data acquisition & Processing,Vol.18.(2003)429-433
6. Lee Joo-Young,Cho Sung-Bae:Sparse Fitness Evaluation for Reducing User Burden in Interactive Genetic Algorithm.Proc. of IEEE International Fuzzy Systems,(1999)998-1003
7. Sugimoto F,Yoneyama M: An Evaluation of Hybrid Fitness Assignment Strategy in Interactive Genetic Algorithm.5th Workshop on Intelligent&Evolutionary Systems,(2001) 62-69
8. Furuya H: Genetic Algorithm and Multilayer Neural Network.Proc.of Calculation and Con-trol,(1998)497-500
9. Gu Hui,Gong Yu-en,Zhao Zhen-xi:A Knowledge Model Based Genetic Algorithm. Computer engineering,Vol.26.(2000)19-21
10. Sebag M,Ravise C,Schoenauer M: Controlling Evolution by Means of Machine Learning. Evolutionary Programming, (1996)57-66
11. Fan Lei,Ruan huai-zhong,Jiao Yu,et al: Conduct Evolution Using Induction Learning. Journal of University of Science and Technology of China, Vol.31.(2001)565-634
12. H.Handa,T.Horiuchi,O.Katai,et al:Co-evolutionary GA with Schema Extraction by Machine Learning Techniques and Its Application to Knapsack Problem. IEEE Conference of Evolutionary Computation, (2001)1213-1219
13. Reynolds,G.R: An Introduction to Cultural Algorithms.Proc. of the 3rd Annual Conference on Evolutionary Programming,(1994)131-139
14. I.C.Parmee,D.Cvetkovic,A.H.Watson: Multi-objective Satisfaction within An Interactive Evolutionary Design Environment. Evolutionary Computation,(2000)197-222
15. Jiang Shan-shan,Cao xian-bing,Wang xi-fa: User's Agent Model and Design Using IGA. Pattern recognition and Artificial Intelligence,Vol.17.(2004)244-249
16. Li Zhi-hua, Chen De-zhao, Hu Shang-xu: The Construction and Realization of Intelligent Multi-agent System to Implement Genetic Algorithm. Computer Engineering and Application,Vol.11.(2002)41-43
17. Zhong Weicai, Liu Jing, Xue Mingzhi, Jiao Licheng:A Multiagent Genetic Algorithm for Global Numerical Optimization. IEEE Trans. on System, Man, and Cybernetics-Part B, Vol.34.(2004) 1128-1141
18. J.Liu, W.Zhong, L. Jiao: A Multiagent Evolutionary Algorithm for Constraint Satisfaction Problems. IEEE Trans. on System, Man, and Cybernetics-Part B, Vol.36.(2006) 54-73
19. Guo Yi'nan,Gong Dunwei,Zhou Yong:Cooperative Interactive Evolutionary Computation Model Based on Multi-agent System.Journal of System Simulation,Vol.17.(2005)1548-1552
20. Albert J.N.van Breemen: Agent-Based Multi-Controller Systems,Ph.D. thesis.Twente University,Netherlands (2001)
21. Hao G S,Gong D W,Shi Y Q: Interactive Genetic Algorithm Based on Landscape of Satisfaction and Taboos.Journal of China University of Mining&Technology,Vol.34.(2005) 204-208

Concurrent Design of Heterogeneous Object Based on Method of Feasible Direction and Genetic Algorithm

Li Ren, Rui Yang*, Dongming Guo, and Dahai Mi

Key Laboratory for Precision and Non-traditional Machining Technology of Ministry of Education, Dalian University of Technology, Dalian 116024, P.R. China
renlihappy@sohu.com, rayy@student.dlut.edu.cn

Abstact. In this paper, we propose a new approach combining the method of feasible direction (MFD) with the genetic algorithm (GA) to the concurrent design of geometry and material for the heterogeneous object. For a component made of heterogeneous materials, its functional requirements can be satisfied by changing the component's configuration or materials' distribution. Thus, its geometric feature and material feature are coupled with each other. For the reason of the coupling of geometry with material and the non-linearity of design problem, the conventional gradient-based algorithm is not very competent. To address this issue, the combining algorithm is used in such a way that, the increments of geometric variables can be calculated by MFD while that of material variables are obtained through GA, which implements their simultaneous optimization and solves their coupling. An isothermal heat utensil made of ZrO_2 and Ni is designed and the optimization result shows that the method proposed is of good engineering applicability.

1 Introduction

Heterogeneous object composed of two or multi-primary materials, both keeping the properties of each constituent material and developing some better properties than primary materials, is an emerging area with wide-spread application in manufacturing, bio-medicine, pharmacology, geology and physics. Traditional design and fabrication systems have focused on developing homogeneous models of physical object to capture their geometry. Due to the development of structural optimization and layered manufacturing, the research in engineering has paid much more attention to the design and fabrication of heterogeneous materials.

Recently, Regarding design optimization, only a few researches [1-4] have been reported. Qian and Dutta [1] tied B-spline representation of a turbine blade to a physics process. Huang and Fadel [2] mentioned the optimizer DOT, genetic and annealing algorithms for solving the heterogeneous flywheel optimization. However, for a heterogeneous object, its geometry feature and material feature coupled with each other are all the complex functions of spatial positions, the sensitivity analysis is difficult and thus the conventional gradient-based algorithm is not attractable especially for largescale problem while the GA is time consuming heavily. Chen and

---

* Corresponding author.

L. Jiao et al. (Eds.): ICNC 2006, Part I, LNCS 4221, pp. 769 – 778, 2006.

Feng [3] classified the heterogeneous materials design into geometric design and material design based on axiomatic design principles. The geometric design was done first and followed by material design. Huang and Fadel [4] also presented a two-step methodology for the design of heterogeneous injection mold cooling systems. They all did not consider the coupling relation between geometric design and material design. In this paper, the heterogeneous object design is regarded as a concurrent design for geometry with material and expressed as a constrained nonlinear optimization problem including both geometric variables and material variables. The simultaneous optimization is implemented by combining the MFD with the GAs at each optimization iteration. The effectiveness and applicability are verified by an example for design of an isothermal heat utensil.

The remainder of the paper is organized as follows. In Section 2 the heterogeneous object is introduced. The concurrent optimization design of heterogeneous object based on MFD and GAs is presented in Section 3. Section 4 provides a numeral example and the conclusion is drawn in Section 5.

2 Description of Heterogeneous Object

Consider a heterogeneous object consisting of n primary materials. It can be abstracted to a vector space $T = R^3 \times R^n$. R^3 is the geometry space and R^n is the material space with each dimension representing a primary material.

The space of volume fraction V can be defined as:

$$V = f(x, y, z) = \{\bar{v} = (v_1, v_2, \cdots, v_n) \in R^n \ \ \| \bar{v} \|_1 \equiv \sum_{i=1}^{n} v_i = 1 \ \ v_i \geq 0\} \cdot \quad (1)$$

where f is the mapping from geometry space to material space, v_i represents the volume fraction of material i .

Based on the linear rule of mixtures, the overall properties Π^P at any point $P \in \Omega$ inside the heterogeneous object are determined by the linear combination of the volume fractions and material properties of the constituent materials:

$$\Pi^P = \sum_{i=1}^{n} v_i \cdot \pi_{M_i} \cdot \quad (2)$$

Where π_{M_i} is the properties of material M_i at point P .

3 Concurrent Optimization Design of Heterogeneous Object

Based on the functional requirements (FRs) of consumer and boundary conditions, the design of heterogeneous object can be transformed into a multi-objective optimization problem with constraints. Here, its geometric feature of is expressed by the coordinates of a set of key points and material feature is denoted by volume fractions of fictitious nodes discretized in the design domain of the component. The design of heterogeneous object is implemented by the perturbation of position and volume fraction of each fictitious node, in which the fictitious node moves caused by the changing of key points' coordinates.

3.1 Formulation of Optimization Model

Assume that the heterogeneous object is composed of two primary materials, the concurrent optimization of heterogeneous object, including geometric optimization and material optimization, can be regarded as a constrained non-linear optimization problem defined in the following general form:

$$\begin{aligned} \min \quad & OBJ = \sum_{i=1}^{L} w_i S_{obj}^{(i)}(h, X) \\ s.t. \quad & g_j(h, X) - b_j \leq 0, \quad j = 1,2,\cdots,\mathrm{m} \end{aligned} \quad (3)$$

where $S_{obj}^{(i)}(h, X)$ and w_i are the objective function and the weight factor, respectively, $i = 1,2,\cdots,L$, $g_j(h, X) - b_j$ is the inequality constraints, $j = 1,2,\cdots,\mathrm{m}$, h is the geometric design variable, X is the material design variable.

3.1.1 Geometric Design Variable

In the optimization design of heterogeneous object, the geometric design variables are the variables which represent the profiles of the optimized boundaries. Here, the geometric design variable $h = [h_0, h_1 \cdots, h_K]$ is defined as the coordinates of a set of key points on the changeable boundaries. The Bezier curves are constructed to represent the optimized boundaries through these key points. For 3D problems, $h_i = (x_i, y_i, z_i)$ $i = 1,2,\cdots,K$, any point in the Bezier curve can be expressed as:

$$h(t) = \sum_{i=0}^{K} \binom{K}{i} t^i (1-t)^{K-i} h_i \quad 0 \leq t \leq 1 . \quad (4)$$

where $\binom{K}{i} = \dfrac{K!}{i!(K-i)!}$.

3.1.2 Material Design Variable

In the optimization design of heterogeneous object, the material design variables are the variables which represent the space of volume fraction. Here, the material design variable $X = [v_1, v_2, \cdots, v_n]$ is the volume fraction of one primary material in each fictitious node, $0 \leq v_i \leq 1$, n is the number of fictitious nodes distributed in the design domain of the component.

For heterogeneous object, the material properties are complex functions of spatial position. So, the finite element method (FEM) is utilized to perform numerical analysis and the remeshing process must be carried out since the geometric boundary usually changes during the optimization process. The volume fraction in the finite element is interpolated by the volume fractions of fictitious nodes and shape function based on the moving least squares (MLS) approximation [5]. For example, the volume fraction $v(x)$ of the space coordinate x is expressed as:

$$v(x)=\sum_{I=1}^{m}\phi_I(x)v_I \ . \tag{5}$$

where m is the number of fictitious nodes in the component, $\phi_I(x)$ denotes the local shape function associated with the fictitious node I of the space coordinate x.

Then, based on the volume fraction at any point of the component obtained from Eq. (5), the material properties can be calculated approximately from Eq. (1) and (2).

3.1.3 Constraint Conditions

For a heterogeneous component, the material distributions are complex functions of spatial position, the local stress concentration will occur caused by an abrupt change in constituent composition. For reducing the local stress and prolong the working life, the von Mises stresses of the component must not exceed the limit value σ_{adm} :

$$\max(\sigma_{von,i})-\sigma_{adm}\le 0 \ . \tag{6}$$

where based on the fourth stress strength theory, the von Mises stress of the i-th element $\sigma_{von,i}$ is expressed as follows:

$$\sigma_{von,i}=(\sigma_{x,i}^2+\sigma_{y,i}^2-\sigma_{x,i}\sigma_{y,i}+3\tau_{xy,i}^2)^{1/2} \ . \tag{7}$$

where $\sigma_{x,i}$, $\sigma_{y,i}$ and $\tau_{xy,i}$ are stress components of the i-th element of the heterogeneous object.

At the same time, the side constraints should be imposed to prevent topology changes in configurations of geometric model, which may yield non-real geometries.

3.2 An Approach Combining the Method of Feasible Direction (MFD) and the Genetic Algorithm (GA)

For heterogeneous object, its geometry variable and material variable are all the complex functions of spatial positions, the sensitivity analysis is difficult and thus the conventional gradient-based algorithm is not attractable especially for largescale problem while the GA is time consuming heavily. So, as the number of geometric design variable is small and the sensitivities can be calculated by the finite difference method, the increment of geometric variable is gained by MFD [6] at each optimization iteration. At the same time, the number of material variable is large and its increment is obtained through GA [7-10]. Thus implements the simultaneous optimization of geometry and material and saves the computer time.

3.2.1 The Method of Feasible Direction (MFD)

According to the finite difference, the sensitivities of the objective function and constraint conditions (Eq.(3)) with respect to geometric design variable at the k-th optimization iteration can be calculated by:

$$\frac{\partial OBJ^k}{\partial h}=\frac{OBJ(h^k+\delta h,X^k)-OBJ(h^k,X^k)}{\delta h}$$
$$\frac{\partial g_i^{\,k}}{\partial h}=\frac{g_i(h^k+\delta h,X^k)-g_i(h^k,X^k)}{\delta h} \qquad i=1,2,\cdots,m \tag{8}$$

The basic steps in MFD involve solving a nonlinear subproblem to find the direction vector and then finding the step-length along this direction. After updating the current point, the above steps are repeated until the termination criterion is satisfied. The general algorithm of MFD can be described in **Algorithm 1** as follows:

- Find or give an initial feasible vector $h^0=[h_0^0,h_1^0,\cdots,h_K^0]$. Set $k=0$.
- Repeat
  1. Direction Finding Subproblem (DFS): Given a feasible iteration h^k, based on the values of $\frac{\partial OBJ^k}{\partial h}$ and $\frac{\partial g_i^k}{\partial h}$, set up the DFS to find S^k, a feasible direction of descent at h^k.
  2. One Dimensional Search: Find a^k such that
     (i) $h^k+\alpha^k S^k$ is feasible;
     (ii) $OBJ(h^k+\alpha^k S^k)<OBJ(h^k)$.
  3. Set $h^{k+1}=h^k+a^k S^k, k=k+1$.
- Until terminating condition is satisfied

Algorithm 1. Optimization procedure of general algorithm of MFD

3.2.2 Real Coded Genetic Algorithm

Genetic Algorithms are non-deterministic stochastic search/optimization methods that utilize the theories of evolution and natural selection to solve a problem within a complex solution space, firstly stated by Holland in 1975. With various improvements in decades past, GAs have demonstrated remarkable merits over gradient-based optimization methods in widespread engineering areas. Generally, it is known that the real coded genetic algorithms save the process of coding and decoding and perform better than the binary coded genetic algorithms for high precision optimization problems.

3.2.2.1 Individual Representation. For the material optimization of heterogeneous object (Eq.(3)), the volume fraction of one primary material in each fictitious node v is expressed as a gene. The chromosome corresponds to material design variable and can be represented as $[v_1, v_2, \cdots, v_n]$, $v_i \in [0,1]$, n is the number of fictitious nodes.

3.2.2.2 Genetic Operators. In this paper, we adopt the Arithmetic crossover operators [7] and non-uniform mutation operator [7]. The process of crossover and mutation is shown as:

Crossover. Assume that X_1 and X_2 are two chromosomes that have been selected to apply the crossover operator, then two offspring chromosomes X_1' and X_2' are expressed as:

$$\begin{matrix} X_1' = \lambda \cdot X_1 + (1-\lambda) \cdot X_2 \\ X_2' = \lambda \cdot X_2 + (1-\lambda) \cdot X_1 \end{matrix}, \lambda \in (0,1). \tag{9}$$

Mutation. Suppose $X = [v_1, \cdots, v_k, \cdots, v_n]$ is a parent chromosome, $v_k \in [0,1]$ is a gene to be mutated. $X' = [v_1, \cdots, v_k', \cdots, v_n]$ is the offspring chromosomes. The gene v_k' is given by:

$$v_k' = \begin{cases} v_k + \Delta(t, 1 - v_k) & \text{if a random digit >0.5} \\ v_k - \Delta(t, v_k) & \text{if a random digit <=0.5} \end{cases} \tag{10}$$

where t is the number of generations performed and the function $\Delta(t, y) = y \cdot r \cdot (1 - \frac{t}{T})^b$.

3.2.2.3 Search Process. The objective function is represented by fitness and the rejection of infeasible individual method [9] is used to handle constraints. After an initial population of individuals is generated, these individuals are evaluated according to Eq.(3). The fitter individuals are chosen to undergo reproduction, crossover, and mutation operations in order to produce a population of children for the next generation. This procedure is continued until either convergence is achieved, or a sufficiently fit individual has been discovered. The practical operation steps of genetic algorithms are summarized in **Algorithm 2** as follows:

- Choose randomly initial population of $nind$ individuals in the range [0,1]
- Evaluate each individual's fitness function and constraint functions
- Repeat
  1. Select individuals to reproduce based on the fitness and constraints
  2. Mate pairs at random
  3. Apply crossover operator based on probability of crossover p_c
  4. Apply mutation operator based on probability of mutation p_m

5. Evaluate each individual's fitness function and constraint functions in the offspring's population
6. Create new population by comparing each individual's fitness of parents and offspring

- Until terminating condition is satisfied

Algorithm 2. Optimization procedure of Genetic algorithms

3.2.3 The Concurrent Optimization Process of Heterogeneous Object

Considering the coupling of material variable with geometric variable, the optimization process of heterogeneous object follows:

1. Based on the functional requirements (FRs) of consumer and boundary conditions, formulate the optimization model (Eq.(3)).
2. Select two materials based on material properties from material database.
3. Give an initial feasible vector $h^0 = [h_0^0, h_1^0, \cdots, h_K^0]$ and generate randomly initial population of $nind$ individuals in the range [0,1]. Set $k = 0$.
4. Repeat
   4.1. Select one or several generations, genetic algorithms are performed. The search algorithm is shown in **Algorithm. 2**. Then, the best individual is obtained $X^k = [v_1^k, v_2^k, \cdots, v_n^k]$.
   4.2. Perform an optimization iteration of geometric variable by **Algorithm 1**. and the update geometric variable is gained $h^k = [h_0^k, h_1^k, \cdots, h_K^k]$.
   4.3. Set $k = k + 1$.
5. Until terminating condition is satisfied

4 Numerical Example

A steady-state isothermal heat utensil made of ZrO_2 and Ni is designed. The 2D initial design domain is shown in Fig.1. The dimension is in mm. The aim is to find the optimal shape and material distribution that ensure the uniformity of temperature distribution (temperature difference less than 1 $°C$) on the boundary GH under stress constrains and side constraints. The convection occurs between the ambient air and the heat utensil during heating process. A prescribed heat flux $q = 5000kW/m^2$ on Boundary AB. The convection properties $T_a = 15°C, h = 5W/m^2 \cdot °C$. The inherent allowable stress of the utensil σ_0 =7.165 GPa.

By symmetry, a left half of the utensil is considered. The geometric and material design variables are shown in Fig.2. The coordinates of selected points on the boundaries CE are the geometric variables $h = [h_1, h_2, h_3, h_4]^T, -2 \le h_j \le 10$ $j = 1,2,3,4$. The points C and E remain fixed during design process. The volume fraction of ZrO_2 in each fictitious node of the design domain CEIJ is material design variable $X = [v_1, v_2, \cdots, v_n]^T$, $v_i \in [0,1]$, $n = 121$.

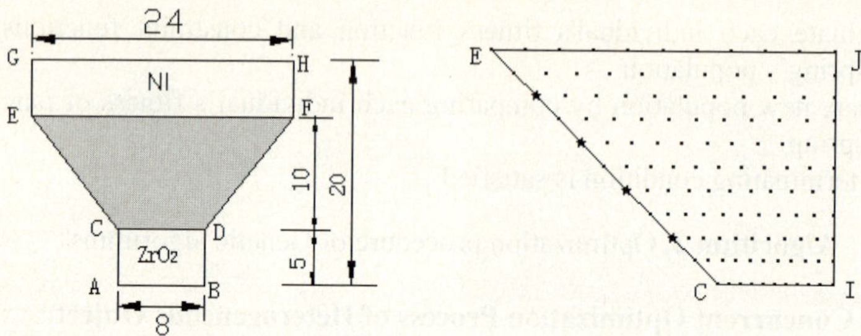

Fig. 1. The initial design domain **Fig. 2.** Geometric and material design variables

The method of feasible direction (MFD) and the real coded genetic algorithm (GA) are used together for solving the design of heterogeneous utensil. The genetic operators and the parameters used for this genetic algorithm are taken to be: 25.3% of all individuals undergo crossover (p_c =0.253), 18% of all individuals are mutated (p_m =0.18) and 100 individuals is initialized randomly ($nind$ =100).

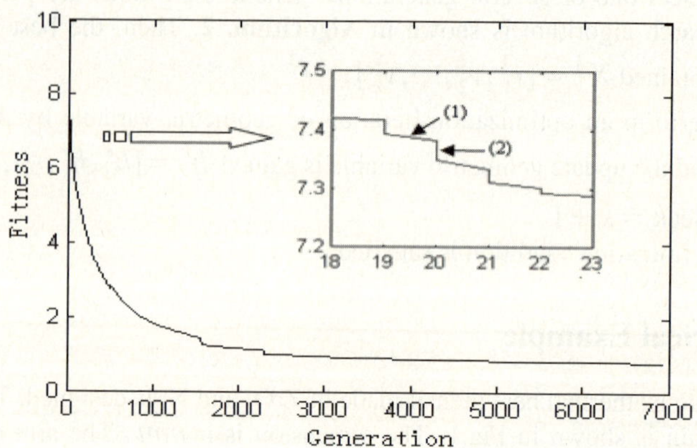

Fig. 3. Evolution history and the local evolution process

Fig.3 presents the evolution history of the fittest individual in the population. According to Fig.3, convergence has been achieved and the optimal geometry boundary and material volume-fraction distribution are obtained. For displaying the coupled optimization of geometric variables and material variables, the local evolution process is zoomed out. Fig.3 (1) indicates the drop of fitness caused by material variables and Fig.3 (2) indicates the drop caused by geometric variables, which implements the simultaneous optimization of them.

The coordinates of selected points on the boundaries CE for pre-and post-optim are listed in Table 1. Material distribution and geometry shape of the utensil are shown in Fig.4. We can see that much fraction of material ZrO_2 with low thermal conductivity is distributed in the middle region while little one in the ambient region and the lower region on the changeable boundaries is concave while the upper region is convex, which achieves the uniformity of temperature distribution on the boundary GH.

Table 1. The coordinates of selected points on the boundaries CE for pre-and post-optimization

The selected points	h_1	h_2	h_3	h_4
Pre-optim (mm)	(6.40,7.00)	(4.80,9.00)	(3.20,11.00)	(1.60,13.00)
Post-optim (mm)	(10.02,7.00)	(5.73,9.00)	(-2.02,11.00)	(-2.02,13.00)

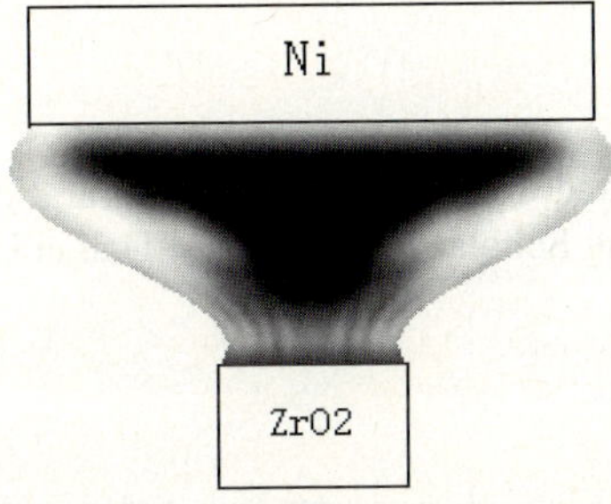

Fig. 4. Material distribution and geometry shape of the utensil

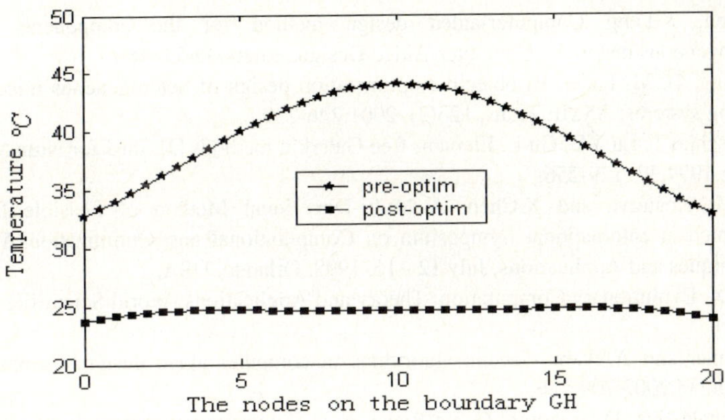

Fig. 5. The temperature of the nodes on the boundary GH for pre-and post-optimization

In order to get a distinct contrast, assume 50% ZrO_2 distributed in the initial design domain for pre-optimization, the temperature of the nodes on the boundary GH for pre- and post-optimization are plotted in Fig.5. The maximum of temperature difference of pre-optimization is $11.212\,°C$; while that of post-optimization is only $0.986\,°C$, which is greatly lower than the former and satisfies the design requirement.

5 Conclusion

In this paper, the concurrent design of geometry and material for the heterogeneous object is solved by combining the method of feasible direction (MFD) with the genetic algorithm (GA). The combined algorithm not only saves the computer time for GA's time consuming heavily but also implements the simultaneous optimization of geometry and material and solves the coupling of them. An isothermal heat utensil made of ZrO_2 and Ni is designed and the optimal results verify the effectiveness of the proposed method.

Acknowledgement

This project is supported by the National Natural Science Foundation of China No.50275018 and Young Scholar Foundation of Dalian University of Technology.

References

1. X.Qian, D.Dutta. Design of heterogeneous turbine blade. Computer-Aided Design, 35,2003:319- 329.
2. J.Huang, G. M. Fadel. Heterogeneous flywheel modeling and optimization. Materials and Design, 21, 2000:111-125.
3. K.Chen, X.Feng Computer-aided design method for the components made of heterogeneous materials. Computer-Aided Design, 35,2003:453- 466.
4. J.Huang, G. M. Fadel. Bi-objective optimization design of heterogeneous injection mold cooling systems. ASME Trans, 123(2), 2001:226-239.
5. Belytschko T, Lu YY, Gu L. Element-free Galerkin methods [J]. Int J for Num Methods in Engrg,1994,37: 229-256.
6. M. M. Kostreva and X.Chen. A Multi-Directional Method of Feasible Directions, presented at International Symposium on Computational and Optimization Algorithms, Techniques and Applications, July 12 - 16, 1998. Orlando, USA.
7. Yao, X. Evolutionary Computation: Theory and Applications. World Scientific, Singapore (1999).
8. G.Renner and A.Ekart. Genetic algorithms in computer aided design. Computer-Aided Design, 35,2003:709-726.
9. Z.Michalewicz, D.Dasgupta, G. Le Riche, et al. Evolutionary algorithms for constrained engineering problems. Computers ind. Engng, 30(4),1996: 851-870.
10. Tan, K.C., Lim, M.H., Yao,X., Wang L.P. (Eds.): Recent Advances in Simulated Evolution And Learning. World Scientific, Singapore (2004).

Genetic Algorithm-Based Text Clustering Technique

Wei Song and Soon Cheol Park

Division of Electronics and Information Engineering
Chonbuk National University, Korea
songwei9988@yahoo.com.cn, scpark@moak.chonbuk.ac.kr

Abstract. A modified variable string length genetic algorithm, called MVGA, is proposed for text clustering in this paper. Our algorithm has been exploited for automatically evolving the optimal number of clusters as well as providing proper data set clustering. The chromosome is encoded by special indices to indicate the location of each gene. More effective version of evolutional steps can automatically adjust the influence between the diversity of the population and selective pressure during generations. The superiority of the MVGA over conventional variable string length genetic algorithm (VGA) is demonstrated by providing proper text clustering.

1 Introduction

Clustering is an unsupervised pattern classification technique which is defined as group *n* objects into *m* clusters without any prior knowledge. In most real life situations the number of clusters in a data set is not known in advance. The real challenge is to be able to automatically evolve a proper number of clusters. Genetic algorithms (GAs) [1] based on the principle of evolution and heredity, are able to search in complex, large and multimodal landscapes. We can apply the searching capability of GAs to evolving proper number of clusters and providing appropriate clustering. A variable string length genetic algorithm (VGA) provided in [2, 3] employs simple chromosomes to evolve the number of clusters. However, the compact chromosome encoding causes reduction of chances of getting the optimal centers combination. We propose a modified variable length GA (MVGA) using gene indices to encode the chromosomes which have more chance to obtain the special centers combination and find the optimal number of clusters. We also propose dynamic evolutional steps to evolve. The details of the algorithm are described in section 3. Experiment results are given in section 4. Discussion and conclusions are given in section 5.

2 Genetic Algorithms for Clustering

GAs are able to search in complex, large and multimodal landscapes. The parameters in the search space are represented in the form of chromosome. A collection of such chromosomes is called population. An objective and fitness function is associated

L. Jiao et al. (Eds.): ICNC 2006, Part I, LNCS 4221, pp. 779 – 782, 2006.

with each chromosome that represents the degree of fitness. Biologically inspired operators like selection, crossover and mutation are applied to yield new child chromosomes. These operators continue several generations till the termination criterion is satisfied. The fittest chromosome seen up to the last generation provided the best solution to the clustering problem. We modified VGA chromosome using gene index to indicate the location of each gene. Such method is described in the next section.

3 MVGA for Texts Clustering

3.1 Encoding Chromosome

In MVGA clustering, each chromosome C_i in the population initially encodes a number of K_i centers which are chosen randomly from the data set. These points are distributed in the random positions of the chromosome. We use gene index to denote the relative position of each point. Let us consider the following example.

Example: Let the random number K_i and K_j be 5 and 4 for chromosome C_i and chromosome C_j respectively. The VGA chromosomes encoding are shown as follows, C_i [Z_{i1}, Z_{i2}, Z_{i3}, Z_{i4}, Z_{i5}] and C_j [Z_{j1}, Z_{j2}, Z_{j3}, Z_{j4}]. We know that a random exchange point crossover happened in VGA makes their offspring hold the same length as their parents. For example, if the crossover point is 3 their offspring generated are as follows, C_i' [Z_{i1}, Z_{i2}, Z_{i3}, Z_{j4}] and C_j' [Z_{j1}, Z_{j2}, Z_{j3}, Z_{i4}, Z_{i5}]. MVGA uses gene index to indicate the position of each gene. We generate a string of random numbers for gene locations. When random numbers are [0, 2, 6, 7, 9] and [0, 3, 4, 8], the C_i and C_j are shown as:

C_i: [Z_{i1}, Z_{i2}, Z_{i3}, Z_{i4}, Z_{i5}] gene index: [0, 2, 6, 7, 9]
C_j: [Z_{j1}, Z_{j2}, Z_{j3}, Z_{j4}] gene index: [0, 3, 4, 8]

If the crossover point is 5, we only need to exchange the genes whose gene index is greater than 5. The offspring generated are shown as:

C_i': [Z_{i1}, Z_{i2}, Z_{j4}] gene index: [0, 2, 8]
C_j': [Z_{j1}, Z_{j2}, Z_{j3}, Z_{i3}, Z_{i4}, Z_{i5}] gene index: [0, 3, 4, 6, 7, 9]

We can find that the lengths of offspring are different from their parents. So MVGA chromosome provides more chance to evolve proper number of clusters.

3.2 Evolution Principle of MVGA

We chose dynamic evolutional steps to evolve, which is shown in Fig1. The termination criterion is iteration of best fitness value without improvement exceeding consecutive N_{max} iterations. The criterion I happens when there have been no improvements in best fitness value for consecutive n_{max} iterations. The fittest concept selection, classical single-point crossover and Gaussian Mutation [4] are adopted in this paper. Fitness function is defined to be $1/DB$, where DB is Davies-Bouldin index [2, 5].

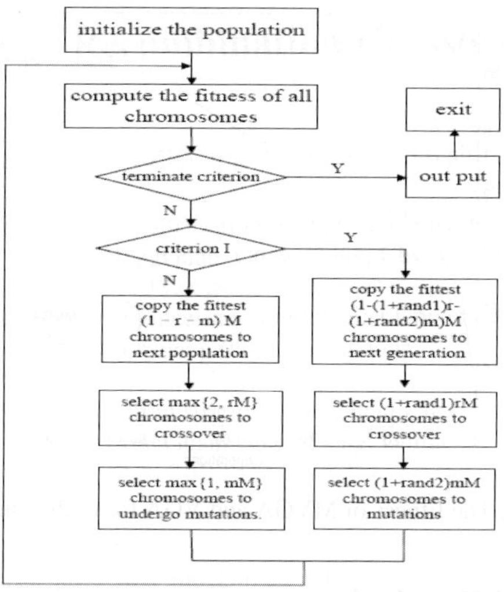

Fig. 1. Dynamic evolutional steps of MVGA. *M* is the size of population. *r* and *m* are the proportion of chromosomes to crossover and mutation respectively.

4 Experiments

We chose 500 texts from the Reuter-21578 collection as data set which includes five topics (acq, crude, earn, interest, trade). After being processed by word extraction, stop words removal, and stemming, there are 6,233 terms in the vocabulary. MVGA is implemented with the following parameters: $r = 0.4$, $m = 0.15$, $N_{max} = 20$, $n_{max} = 15$ and the population size is 100. We then applied MVGA and VGA to clustering respectively. MVGA obtained 5 clusters, while VGA only got 3 clusters in the end. We used precision, recall and F-measure to evaluate the clustering results. Precision of cluster C_i is defined to be the number of texts correctly assigned divided by the total number of texts in cluster C_i. We define Recall of cluster C_i to be the number of texts correctly assigned divided by the total number of texts that should be assigned. Let P be average precision. Let R be average recall. F-measure is defined to be $2RP/(R+P)$. Table1 shows the evaluation of MVGA and VGA.

Table 1. The Precision, Recall and F-measure of MVGA and VGA

Name	Precision (%)	Recall (%)	F-measure (%)
MVGA	75.5	73.1	74.3
VGA	52.7	71.8	60.8

To clustering problem, the real evaluation criterion is the fitness function $1/DB$ (Davies-Bouldin index). We choose the fitness of the best chromosome in each generation to represent generation fitness. In Fig2 the final fitness of 1.86 provided by MVGA is superior to that of VGA. Moreover, MVGA ends with fewer generations.

greedy manner[11]. Although ERX has been brought forward and improved for many years, most research work for it is applied to the symmetrical TSP. Little literature about the ERX for the asymmetrical TSP (ATSP) is found. Based on the experience of the pioneers, this paper presents a new crossover operator for ATSP which is called Directed Edge recombination crossover (DERX). The operator was evaluated on a number of well-known benchmarks in the TSPLIB. Comparison between the DERX and the conventional ERX shows the improvement of DERX in the ATSP.

2 The Conventional ERX for the TSP

2.1 Traveling Salesman Problem and the GA

Traveling salesman problem can be described as: Given a set $\{c_1, c_2, \ldots, c_n\}$ cities and distance $d(c_i, c_j)$ for each pair $(c_i,\ c_j)$ of distinct cities [4], find a roundtrip of minimal total length visiting each city exactly once. The objective of the TSP is to find an order π (Hamiltonian cycle) of the cities that minimizes the quantity:

$$\sum_{i=1}^{n-1} d(c_{\pi(i)}, c_{\pi(i+1)}) + d(c_{\pi(n)}, c_{\pi(1)}) \quad (1)$$

According to relation between the distance and the trip direction of each pair of distinct cities, the TSP can be divided into two categories: the symmetrical TSP and the asymmetrical TSP (ATSP). In the symmetric TSP, the distance satisfies the condition $d(c_i, c_j) = d(c_j, c_i)$ for $1 \le i,\ j \le$ n, and has nothing to do with the direction of the tour route. On the contrary, in the ATSP, the distance from city i to city j and the distance from city j to city i may be different.

The TSP is NP-hard, and the GA is an effective approach for it. In the GA for the TSP, the order π is encoded to a chromosome, and the gene of the chromosome consists of city names. The finite length of the chromosome corresponds to the size of a given TSP instance. The fitness is computed by the sum of all the distances between cities on the Hamiltonian cycle, which is denoted by expression (1).

In the past decade, many researchers proposed hopeful genetic operators and analyzed their effectiveness applied to the TSP. Crossover is a principal operator of genetic algorithm (GA). There are many hopeful Crossovers applied to the TSP, such as partially matched crossover (PMX), cycle crossover (CX) and so on. Among those crossovers, Edge recombination Crossover (ERX) shows more excellent performance.

However, up to now, most of the ERX are developed for the symmetric TSP, and assumes the condition: $d(c_i, c_j) = d(c_j, c_i)$. But in the ATSP, the condition is not valid. Based on the research work of the pioneers, this paper presents Directed Edge recombination for ATSP. In order to better understand how DERX work, the family of Edge recombination operators for the symmetric TSP is explained in detail below.

2.2 Edge Recombination Crossover (ERX) for the Symmetric TSP

Edge recombination is a kind of crossover that takes two chromosomes as parents and produces one child. The aim of the ERX is to inherit as much edges as possible from parent to child. Figure 1 is an example of crossover for 10 cities. The parents is showing as (a) and (b, while (c) is a possible offspring [6].

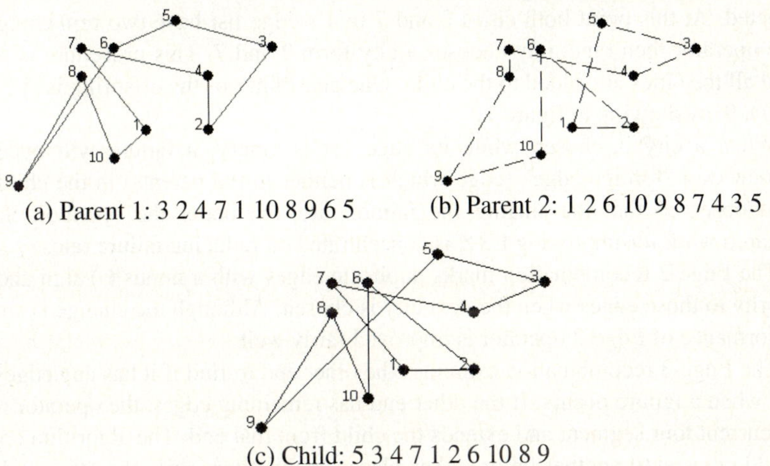

(a) Parent 1: 3 2 4 7 1 10 8 9 6 5 (b) Parent 2: 1 2 6 10 9 8 7 4 3 5

(c) Child: 5 3 4 7 1 2 6 10 8 9

Fig. 1. Example of the origin Edge Recombination operator

Unlike other crossover, Edge recombination operator emphasizes adjacency information of the cities in the tour. The original Edge operator starts by collecting the edge information contained in both parents in an edge table. As an example, the edge table of the mentioned example above is showing in table 1. For every city, there is an edge list that consists of all the cities connected to it in two parent tours. To construct an offspring, Edge recombination first chooses a random starting city and removes all its edges in the edge table. The operator then chooses the next city by finding the city connected to the current city with the least number of remaining edges. If there are some connecting cities that have the same number of remaining edges, the next city is chosen randomly among them. If at any time, a city is chosen such that its edge list is empty, a failure occurs. In such case, the original edge operator does not attempt to correct the failure, but instead chooses a new city randomly, adds it to the child tour,

Table 1. The edge table of the two parents in the example

City	Edge list	City	Edge list
1	7,10,5,2	6	9,5,2,10
2	3,4,1,6	7	4,1,8
3	5,2,4	8	10,9,7
4	2,7,3	9	8,6,10
5	6,3,1	10	1,8,6,9

and continues the method described above. This introduces a "foreign edge" (edge which is neither in the parents) in the child. The operator stops when all the cities have been included in the offspring.

Suppose that city 5 is randomly selected as a staring city. ERX operator first removes all edges linked to 5 in the edge table. Then the edge list of 5 is examined to search for the city that has smallest number of remaining edges. City 3 is chosen next. Then ERX removes the edges list of 5. The operator continues this way until city 4 is selected. At this point both cities 2 and 7 in 4's edge list have two remaining edges. The operator then randomly chooses a city form 2 and 7. This procedure is repeated until all the cities are added to the child. One possibility of the offspring is: 5 3 4 7 1 2 6 10 8 9, as showing in figure 1.

When a city is chosen while its edge list is empty, a failure will occur. This introduces a "foreign edge" (edge which is neither in the parents) in the child. It has been observed that the smaller the failure rate, the better the operator. So, most research work for improving ERX is concentrated on reducing failure rate.

The Edge-2 recombination marks duplicate edges with a minus (-) sign and give a priority to those edges when the next city is chosen. Although the change is small, the performance of Edge-2 operator is improved fairly well.

The Edge-3 recombination examines the other end to find if it has any edges left or not, when a failure occurs. If the other end has remaining edges, the operator reverses the current tour segment and extends the child from that end. The algorithm continues to add cities until another failure takes place. At this time, since the other end cannot extended any more, the next city will be chosen at random.

The Edge-4 recombination tries to decrease the failure that the both end of partially built tour can not be extended. When two end cities have no edge to continue, it must be the case that all of their edges are already contained in the partially built tour.

The Edge-5 and the Edge-6 is derived from the Edge-4, and introduced some new criterion which make better performance [6].

3 Directed Edge Recombination Crossover (DERX)

3.1 The Adjacent Relation

Edge recombination Crossover emphasizes inheritance of the adjacent relation of the cities in the tour, and the objective of the ERX is to inherit as much edges as possible from parent to child. For the convenience of analysis, we give some definition about adjacent relation the in the following.

Definition 1. adjacent relation: two cities are contiguous to each other in a chromosome.

Definition 2. right adjacent relation: Given a certain city c_i, if a city c_j is contiguous to c_i at the right, we call c_j right adjacent to c_i.

Definition 3. left adjacent relation: Given a certain city c_i, if a city c_j is contiguous to c_i at the left, we call c_j left adjacent to c_i.

If exists adjacent relation between the city c_i and c_j in a chromosome, there must be an edge that links the two cities. Suppose c_j is right adjacent to c_i, c_i has a right adjacent edge which is denoted by $c_i \rightarrow c_j$. Analogously, if c_i is left adjacent to c_j, c_j has a left adjacent edge which is denoted by $c_i \leftarrow c_j$.

In the symmetric TSP, whether c_j is adjacent to c_i left or right, the distance between the two cities satisfies the condition $d(c_i, c_j) = d(c_j, c_i)$. It can be considered that left adjacent edge $c_i \rightarrow c_j$ and right adjacent edge $c_i \leftarrow c_j$ are the same edge. In other word, if $c_i \rightarrow c_j$ is an edge of parent chromosomes, $c_i \leftarrow c_j$ is also an edge of them without all doubt.

While in the ATSP, the condition is not valid, hence the two adjacent edges $c_i \rightarrow c_j$ and $c_i \leftarrow c_j$ are different. Obviously, the edge $c_i \rightarrow c_j$ is in the edge table can not assure or certify that the edge $c_i \leftarrow c_j$ is also in the table. Accordingly, we can not treat the adjacent relation in the ATSP as in the symmetric TSP.

All the ERX described in section above are regardless of the direction of the tour. When they are applied to the symmetric TSP, they can make excellent performance. While they are employed to the ATSP, the left and the right adjacent relation of a city are mapped to different adjacent edge. Consequently, some edges listed in the above edge table will become "foreign edge" at this time.

3.2 Directed Edge Recombination Crossover

According to the analysis above, Directed Edge Recombination crossover is proposed for the asymmetric TSP, which are developed on the base of the Edge-3 recombination. Because the Edge-3 recombination is regardless of the direction of the tour, some amendment must be applied before it adapts to the ATSP.

The edge table showing in table 1 includes all the adjacent relation of the parents and can not distinguish the right and the left adjacent relation. It is unsure that the adjacent edge should be place on the right or the left of a chosen city. Therefore, the DERX divide the edge table into two parts: the right adjacent edge table and the left adjacent table, which deals with the right and the left adjacent relation respectively.

In the example for the origin Edge operator, under the environment of the ATSP, the tour route of the two parent chromosomes can be expressed as: 3→2→4→7→1→10→8→9→6→5→3and 1→2→6→10→9→8→7→4→3→5→1. The edges linked to city 1 are 7, 10, 5, 2. Among those edges, city 7 and city 5 which link city 1 at the left belong to the left adjacent edge table, while city 10 and city 2 at the right belong to the right adjacent edge table. Obviously, the edges 7→1, 5→1, 1→10, 1→2 are the edges of the parent chromosomes, but the edges 7←1, 5←1, 1←10, 1←2 are not (the direction is reversed), and are "foreign edges". The right and the left adjacent table of the example are showing in table 2 and table 3 respectively.

For every city, there are a right and a left adjacent edge list which are recorded in the right and the left adjacent table respectively. The right adjacent edge list consists of the cities connected to it right in two parent tours. During the procedure of producing offspring, the DERX can but connect these cities to the tour at the right. Oppositely, the cities in the left adjacent edge list should be placed at the left.

Table 2. The right adjacent edge table

City	Edge list	City	Edge list
1	7,5	6	9,2
2	3,1	7	4,8
3	5,4	8	10,9
4	2,7	9	8,10
5	6,3	10	1,6

Table 3. The left adjacent edge table

City	Edge list	City	Edge list
1	7,5	6	9,2
2	3,1	7	4,8
3	5,4	8	10,9
4	2,7	9	8,10
5	6,3	10	1,6

If there exists a duplicate edge $c_i \rightarrow c_j$ in the parents, only city c_j will list in the right adjacent edge list of c_i, while only city c_i will list in the left adjacent edge list of c_j. Therefore, it is not necessary to mark a duplicate edge with a minus in the DERX.

As the Edge-3 recombination does, when a failure occurs, the DERX examines the other end to find if it has any edges left or not. If the other end has remaining edges, the operator extends the offspring from that end. The algorithm uses a direction variant D to control the searching direction. If failure takes place both ends and the offspring can not be extended any more, the searching direction will be randomly set, and then the next city will be chosen at random at that direction. After collecting the edge information contained in both parents into the two edge table, the detailed procedure of the DERX is given in the following.

step 1: (initialization)
(1) Choose a random city as current city;
(2) add current city into offspring
(3) set the direction variant D=1,
where D=1 means operator select edge from the right adjacent edge list;
D= -1 means operator select edge from the left adjacent edge list;
D=0 means both the right and the left adjacent edge list have not remaining edge.

step 2: (select edge from the right adjacent edge list)
If D=1 then do the following
(1) set the current city be the city at the right end of the offspring
(2) if the right adjacent edge list of current city is empty,
set D= -1;
go to step 3;
(3) Choose the next city by finding the city connected to current city with the least number of remaining edges in the right adjacent edge list;
(4) remove all edges of current site in both the right and the left adjacent edge table;

remove both the right and the left adjacent edge list of current site;
(5) add chosen city to the offspring at the right
(6) go to step 5.

step 3: (select edge from the left adjacent edge list)
If D= -1 then do the following
(1) set the current city be the city at the left end of the offspring
(2) if the left adjacent edge list of current city is empty,
set D= 0;
go to step 4;
(3) Choose the next city by finding the city connected to current city with the least number of remaining edges in the left adjacent edge list;
(4) remove all edges of current site in both the right and the left adjacent edge table;
remove both the right and the left adjacent edge list of current site;
(5) add chosen city to the offspring at the left
(6) go to step 5.

step 4: (random select searching direction and chosen city)
IF D=0 then do the following
(1) randomly set D=-1 or D=1;
(2) if D=1 then do the following
I. randomly choose a city with the least number of remaining edges in the right adjacent edge table, add the chosen city at the right of the offspring
II. go to step 5
(3) if D=-1 the do the following
I. randomly choose a city with the least number of remaining edges in the left adjacent edge table, add the chosen city at the left of the offspring
II. go to step 5

step 5: If all the cities have been added into the offspring then stop, else go to step 2.

4 Experiments

To investigate the efficiency of the DERX, we conduct experiments on a common PC computer. The algorithm is implemented in the MATLAB environment. We use benchmark problem instances taken from the TSPLIB [10]. Three benchmarks p43.atsp, ftv170.atsp, rbg443.atsp are chosen. Initial individuals are generated by the nearest neighbor heuristic [14] in advance. The algorithm uses the 'NormGorm' select operator which is a ranking selection function based on the normalized geometric distribution, and the shift mutation operator. The following fixed common parameters are selected for the algorithm: population size 200, Probability of crossover 1.0, Probability of mutation 0.1. In order to make a comparison, we also select some other crossovers apply to the ATSP benchmarks problem, including: the Edge-3 and the Edge-4 recombination crossover. For every crossover, a different genetic algorithm is given with the same setting as the DERX. The computations are terminated at 10,000 generations, and are performed on the AMD Anthon™ XP 1700 MHz computer. The result of the experiments is list in tables and figure below.

Table 4 shows the best searched by every crossover. It is clearly that the DERX can get better optimal result than the others. The edge-4 is the worst operator here, because reversing the tour section will introduce "foreign edges". Oppositely, owing to its lowest failure rate, the DERX is the best.

Table 4. The best tour length searched by every crossover

	Edge-3	Edge-4	DERX
p43.atsp	5620	5620	5620
ftv170.atsp	2800.8	2840.8	2760.6
rbg443.atsp	2750.5	2850.5	2738.9

Table 5 gives the time consume of computation for each GA. Without question, the DERX is most timesaving, since it chooses city at an edge list which contains two cities at most While the other two operators select one in four cities at most.

Table 5. Time consume of the computation

	Edge-3	Edge-4	DERX
p43.atsp	380.3	440.3	332.1
ftv170.atsp	830	923.2	779
rbg443.atsp	2983.2	3283.2	2490.3

Figure 2 gives a comparison of convergence among the three crossovers. Wherein, (a) shows the comparative convergence for the p43.atsp benchmark; (b) for the ftv170.atsp; (c) for the rgb443.atsp. The Optimum lines in the figure give the known optimal of each benchmark. Obviously, the DERX is superior to the other operators. It can be observed here that the operator with smaller failure has better convergence characteristic. It is also proved that the edges in the ATSP are directional.

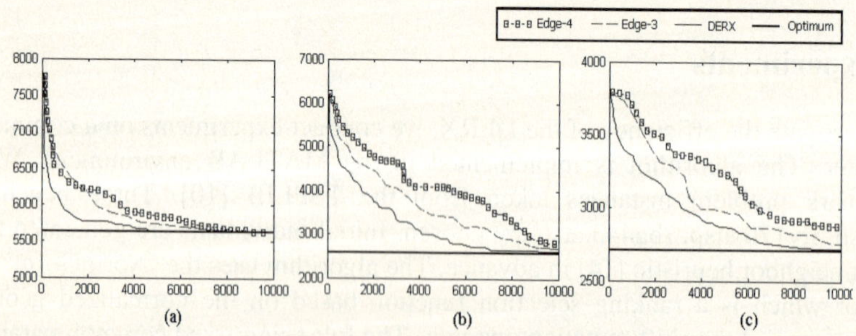

Fig. 2. Comparative convergence of different crossovers

5 Conclusions

The edge recombination crossover emphasizes transition highly fit sub-tours from parents to offspring; hence it shows better performance than other crossovers. To transit highly fit sub-tours of the parents, it must be assured that the edge of the parents can be inherited rightly by the children. In the symmetric TSP, the distance satisfies the condition $d(c_i,c_j)=d(c_j,c_i)$, It can be considered that left adjacent edge $c_i \rightarrow c_j$ and right adjacent edge $c_i \leftarrow c_j$ are the same edge. Therefore the edge in the symmetric TSP has nothing to do with the direction. On the contrary, the edge in the ATSP is directional, and treating the adjacent relation in the ATSP as in the symmetric TSP is not suitable.

This paper proposes the DERX crossover operator, which was applied to the asymmetric TSP. Unlike the ERX proposed before, the DERX divides the edge table into two parts: the right adjacent edge table and the left adjacent table, which record the right and the left edges respectively. The operator extends the offspring tour at both ends. The right and left adjacent edges can only link to the right and the left end of the offspring respectively.

Experiments show it is better than the conventional ERX and some other crossovers clearly, especially for the large scale ATSP. It can be concluded from the Experiments that the DERX reduce the failure in the ATSP, and then lower failure rate leads to its better performance. The DERX make much improvement for the former ERX.

References

1. D.E. Goldberg, Genetic Algorithms in Search, Optimization, and Machine Learning, Addison-Wesley, (1989).
2. Kengo Katayama, Analysis of Crossovers and Selections in a Coarse-Grained Parallel Genetic Algorithm Mathematical and Computer Modelling 38 (2003) 1275-1282
3. Seppala, Dwain Alan, Genetic algorithms for the traveling salesman problem using edge assembly crossovers. Master degree dissertation, University of Nevada, Las Vegas. 2003
4. K. Katayama, H.Hirabayashi and H.Narihisa, Performance analysis of a new genetic crossover for the traveling salesman problem, IEICE tins. Fundamentals E81-A, 738-750, (1998).
5. D. Whitley, T.Starkweather, and D.Shaner, "Traveling Salesman and Sequence Scheduling: Quality Solutions Using Genetic Edge Recombination", In Handbook of Genetic Algorithms, Van Nostrand 1990
6. Nguyen H.D., I.Yoshihara, M Yasunaga, Modified edge recombination operators of genetic algorithms for the traveling salesman problem Industrial Electronics Society, 2000. IECON 2000. 26th Annual Conference of the IEEE , Volume: 4 , 22-28 Oct. 2000 Pages:2815 - 2820 vol.4
7. Nagata Y, Criteria for designing crossovers for TSP Evolutionary Computation, 2004. CEC2004. Congress on , Volume: 2 , 19-23 June 2004 Pages:1465 - 1472
8. M. Srinivas, L.M.Patnaik,, Adaptive probabilities of crossover and mutation in genetic algorithms Systems, Man and Cybernetics, IEEE Transactions on , Volume: 24 , Issue: 4 , April 1994 Pages:656 – 667
9. Jun Zhang, H.S.H. Chung, B.J Hu, Adaptive probabilities of crossover and mutation in genetic algorithms based on clustering technique Evolutionary Computation, 2004. CEC2004. Congress on , Volume: 2 , 19-23 June 2004 Pages:2280 - 2287
10. Gerhard Reinelt's homepage http://www.iwr.uni-heidelberg.de/groups/comopt/software/TSPLIB95/
11. Chuan-Kang Ting. Improving Edge Recombination through Alternate Inheritance and Greedy Manner, Lecture Notes in Computer Science, Volume 3004/2004, Page: 210 – 219

Research on the Convergence of Fuzzy Genetic Algorithm Based on Rough Classification

Fachao Li[1,2], Panxiang Yue[2], and Lianqing Su[2]

[1] College of Economics and Management, Hebei University of Science and Technology, Shijiazhuang, Hebei 050018, P.R. China
lifachao@tsinghua.org.cn

[2] College of Science, Hebei University of Science and Technology, Shijiazhuang, Hebei 050018, P.R. China
yuepanxiang@163.com

Abstract. Genetic algorithm, as a global optimization method, has attracted considerable attention and been applied successfully in many domains. Using rough classification method, propose a kind of fuzzy genetic algorithm to deal with the problem with fuzzy fitness value and point out how to implement it as well. Then consider its convergence using Markov chain theory and analyze its performance through simulation. All these indicate that this kind of algorithm is of feasibility, operability and could be widely used in many problems.

1 Introduction

In 1960s, three kinds of algorithms, genetic algorithm (GA), evolution strategy (ES) and evolution programming (EP) had been proposed by Holland [1], Rechenberg [2], Fogel and Owens [3] respectively, which could simulate the evolution principle of nature from different aspects. Today, the three have become the main component of evolutionary computation which has been greatly used in many domains. In theory, most researches are mainly focused on Holland's Simple Genetic Algorithm, concerning some improvements on the computation accuracy, the convergence or the premature. What needs to pay much attention is that the results are all obtained only for problems with certainty. Speaking concretely, in the process of evaluating the fitness of individuals to the environment, only the fitness function is considered as the measure way, that is to say, the reproduction process is carried out only according to the fitness of individuals, which affect directly the next two processes of crossover and mutation. For problems with uncertainty, the fitness of an individual cannot be expressed as a real number, so it is difficult to compare accurately the fitness of individuals, and then the selection and reproduction process cannot be done only according to the conventional proportion way, which restricts greatly the application of GA to uncertainty optimization problems. Considering this, we give a kind of fuzzy genetic algorithm based on rough classification, saying BRC-FGA for short, by using the compound quantification strategy for uncertain information and the classification model for rough data [4], and then present the concrete implementation steps and consider its convergence using Markov chain theory.

L. Jiao et al. (Eds.): ICNC 2006, Part I, LNCS 4221, pp. 792–795, 2006.

2 Structure of BRC-FGA

Since possessing the properties of both numbers and sets, the theory of fuzzy numbers [5, 6] has become a most common used tool for describing uncertain information. In this paper, we assume that E^1 is the class of all fuzzy numbers. For a fuzzy number A, let $A_\lambda = \{x \mid A(x) \geq \lambda\}$ $(0 < \lambda \leq 1)$ be the λ-cuts of A, A_0 the closure of the support set $\text{supp}A = \{x \mid A(x) > 0\}$ of A, and specially $A = (a, b, c)$ a triangular fuzzy number.

Let S denote the individual space, S^N the population space, $\vec{X}(n) = \{X_1, X_2, \cdots, X_N\}$ the nth generation of population, f the fuzzy value fitness function (a mapping from S to $E^1(+) = \{A \in E^1 \mid A_0 \subset [0, +\infty)\}$), T_s the selection operator (a mapping from S^N to S). Using the strategy of compound quantification and the Max-Min classification model proposed in [4], we may define the selection operator T_s according to following steps.

Step 1: Determine the centralized quantification values $a_1, a_2, \cdots, a_N$ of fuzzy fitness values $f(X_1), f(X_2), \cdots, f(X_N)$;

Step 2: Determine the subsidiary quantification indexes $(a_{k1}, a_{k2}, \cdots, a_{ks})$ of a_k;

Step 3: Select an effect synthesizing function $\theta(x_1, x_2, \cdots, x_s)$ for the subsidiary indexes, and acting $\theta(a_{k1}, a_{k2}, \cdots, a_{ks})$ onto a_k will obtain $a_1^*, a_2^*, \cdots, a_N^*$;

Step 4: Set $P(T_s(X(n)) = X_k) = J(a_k^*)\left[\sum_{i=1}^{N} J(a_i^*)\right]^{-1}$, $k = 1, 2, \cdots, N$.

Here $J(a_k^*) = \text{int}\{1 + [(\delta + a_k^* - \min_{1\leq i\leq N} a_i^*)/(\delta + \max_{1\leq i\leq N} a_i^* - \min_{1\leq i\leq N} a_i^*)] \times (M-1)\}$, M is an appropriate large natural number and δ is an appropriate small positive number. We view a_1^*, $a_2^*, \cdots, a_N^*$ as the rough data reflecting the size character of $f(X_1), f(X_2), \cdots, f(X_N)$, M as the classification parameter, δ as the classification adjustment parameter, and $J(a_k^*)$ as the class of a_k^* with respect to $a_1^*, a_2^*, \cdots, a_N^*$.

If we take the selection operator as above, and others coincide with those proposed by Holland, then we obtain a new kind of GA that can process the case with fuzzy fitness, and written as BRC-FGA.

3 Convergence of BRC-FGA

Definition 1. [7] *Let* $\vec{X}(n) = \{X_1(n), X_2(n), \cdots, X_N(n)\}$ *be the* n^{th} *population of GA,* $Z_n = \max\{f(X_i(n)) \mid i = 1, 2, \cdots, N\}$, $f^* = \max\{f(X) \mid X \in S\}$. *If* $P\{Z_n = f^*\} \to 1\ (n \to \infty)$, *then we say the genetic sequence* $\{\vec{X}(n)\}_{n=1}^{\infty}$ *convergent.*

According to the definition above, when analyzing the convergence of BRC-FGA, we need to consider the problems of the ranking for fuzzy numbers and the difference metric between fuzzy numbers. In this paper, we will use the integral value ranking method proposed by Liou's in [5] and the L_p-metric D_p proposed by Diamond in [6]. Thus we may describe the equivalent relationship $A = B$ as $D_p(A, B) = 0$, and $P\{Z_n = f^*\} \to 1\ (n \to \infty)$ as $P\{D_p(Z_n, f^*) = 0\} \to 1\ (n \to \infty)$.

From the definition of Markov chain, we can know that the genetic sequence $\{\vec{X}(n)\}_{n=1}^{\infty}$ of BRC-FGA is a Markov chain. And with further analysis we can obtain the following conclusions.

Lemma 1. *Genetic sequence* $\{\vec{X}(t)\}_{t=1}^{\infty}$ *of BRC-FGA is a Markov chain which is homogenous and mutually attainable.*

Lemma 2. *The genetic sequence* $\{\vec{X}(t)\}_{t=1}^{\infty}$ *of BRC-FGA is an ergodic Markov chain.*

Using the Lemmas above and Markov chain theory we can deduce that:

Theorem 1. *Genetic sequence* $\{\vec{X}(t)\}_{t=1}^{\infty}$ *of BRC-FGA is not convergent to the global optimal solution.*

Theorem 2. *The genetic sequence* $\{\vec{X}(t)\}_{t=1}^{\infty}$ *of BRC-FGA including the strategy of reserving the optimal individual is convergent to the global optimal solution.*

4 Application Examples

Example 1. Consider the following fuzzy optimization question:

$$\max f(x_1, x_2) = A(x_1^2 - x_2)^2 - B(1 - x_1)^2 ,\ \text{s.t.}\ -2.048 \le x_1 \le 2.048 , -2.048 \le x_2 \le 2.048 ,$$

where $A = (99, 100, 101)$ and $B = (0.8, 1, 1.2)$ are all triangular fuzzy number.

By using the BRC-FGA, we got the maximum value shown on Fig. 1 after 100 times of iterations. The optimal solutions are $x_1 = -2.048$ and $x_2 = -2.048$, and the centralized quantification value of the fuzzy maximum value is 3.8873×10^3 . When *A* and *B* degenerate to real numbers *A*=100 and *B*=1, the real optimal solutions are $x_1 = -2.048$, $x_2 = -2.048$, and $\max f(x_1, x_2) = 3.9059\times10^3$.

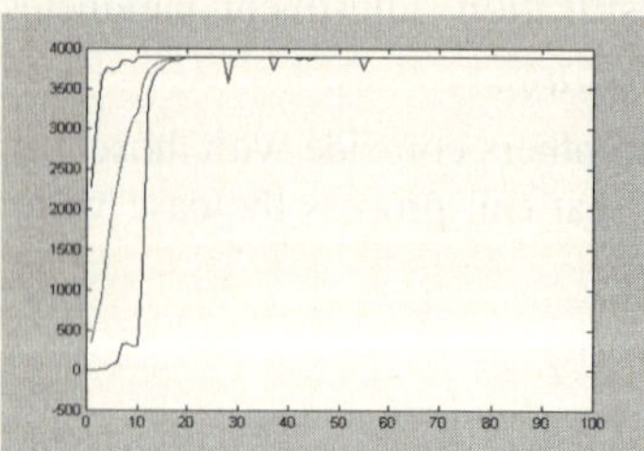

Fig. 1. The results of 100 times iterations of Example 1

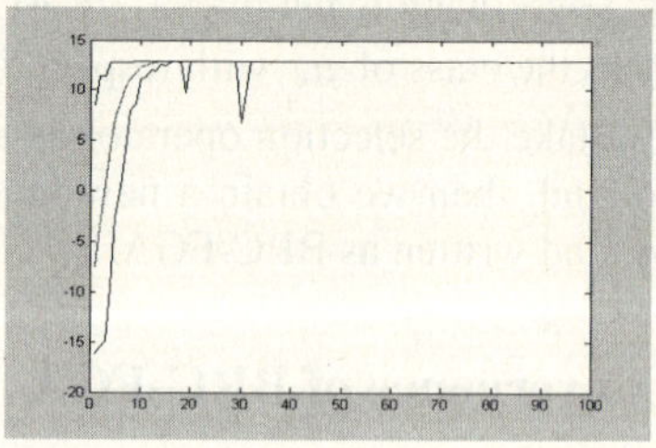

Fig. 2. The results of 100 times iterations of Example 2

Example 2. Consider the following nonlinear fuzzy programming

$$\min f(x_1, x_2) = (0.8, 1.0 ,1.2)x_1^2 + (3.6, 4.0, 4.4)x_2^2 ,$$

$$\text{s.t.}\quad (2.7, 3.0, 3.3)x_1 + (3.5, 4.0, 4.5)x_2 \gtrsim (12.6 ,13, 13.4) ,\ \text{and}\ x_1, x_2 \in E^1(+) .$$

Using the effect synthesizing quantification value of fuzzy numbers to describe the fuzzy inequality, with BRC-FGA, we got the minimum value shown on Fig. 2 after 100 times of iterations. The optimal solutions are $x_1 = (1.4826, 3.0000, 4.5174)$ and

$x_2 = (0.8986, 1.0000, 1.7610)$, and the centralized quantification value of the fuzzy minimum value is 12.9845. And for the fixed form of this fuzzy programming, that is $\min f(x_1, x_2) = x_1^2 + 4x_2^2$, s.t. $3x_1 + 4x_2 \geq 13$, the optimal solutions are $\min f(x_1, x_2) = 13$, $x_1 = 3$ and $x_2 = 1$.

In the Fig.1 and Fig.2, horizontal ordinate represents the times of iteration, and the vertical ordinate the centralized quantification value of minimum value with the form of fuzzy number.

In the examples above, the algorithm is carried on by population size of 80, and for the fuzzy number A, selects $f^*(A) = a/(1+\beta m(A_0))^\alpha$ as its effect synthesizing value, where, a represents the centralized quantification value of A, $m(A_0)$ the length of A_0, and α, β all some kind of decision consciousness. Besides, the classification parameter M=40 in Example 1 and M=50 in Example 2, and the classification adjustment parameter δ is a random number smaller than 10 in Example 1 and 5 in Example 2. In order to test the speed of BRC-FGA, we took 50 times of independent optimizations as one experiment in computer simulation, and took the mean of evolutionary generations as the standard of weighing the algorithm speed. If it is still not convergent to the optimization threshold value after the algorithm has evolved 100 generations, then it is thought that this computation did not converge. Take the number of times at which there was no convergence in 50 experiments as the standard weighing the stability of this algorithm. The test result demonstrates that the average evolutionary generation of this algorithm is 20.8 for Example 1 and 28.2 for Example 2, and the number of times not convergent is 0.

The discussion above indicates that the BRC-FGA has a better performance not only in the aspect of convergence but also in the aspect of stability, it can effectively solve the uncertain optimization problems.

Acknowledgement

This work is supported by the Natural Science Fund of Hebei Province (F2006000346) and the Ph. D. Fund of Hebei Province (05547004D-2, B2004509).

References

1. Holland, J.H.: Genetic Algorithms and the Optimal Allocations of Trials. SIAM J of Computing **2** (1973) 88-105
2. Rechenberg, I.: Evolutions Strategies: Optimerung Technischer Systemenach Prinzipien der Biologischen Evolution. Stuttgart: From mann-Holzboog (1973)
3. Fogel, D.B., Owens, A.G., Walsh, M.J.: Artificial Intelligence through Simulated Evolution. John Wiley New York (1966)
4. Li F., Liu M., Wu C.: Max-Min Classification for Rough Data. Journal of Tsinghua University (Science and Technology) **4** (2004) 114-117
5. Liou T.-S., Wang M.-J.J.: Ranking Fuzzy Numbers with Integral Value. Fuzzy Sets and Systems **50** (1992) 247-255
6. Diamond, P., Kloeden, P.: Metric Space of Fuzzy Set: Theory and Applications. Word Scientific Singapore (1994)
7. Zhang W., Leung Y.: Mathematical Foundation of Genetic Algorithms. Xi'an Jiaotong University Xi'an (2003)

Continuous Optimization by Evolving Probability Density Functions with a Two-Island Model

Alicia D. Benítez and Jorge Casillas

Dept. Computer Science and Artificial Intelligence
University of Granada. Spain, E-18071
casillas@decsai.ugr.es

Abstract. The work presents a new evolutionary algorithm designed for continuous optimization. The algorithm is based on evolution of probability density functions, which focus on the most promising zones of the domain of each variable. Several mechanisms are included to self-adapt the algorithm to the feature of the problem. By means of an experimental study, we have observed that our algorithm obtains good results of precision, mainly in multimodal problems, in comparison with some state-of-the-art evolutionary methods.

1 Introduction

Continuous (real-parameter, global) optimization implies an important problem for engineering, because, it is hard to find a method able to obtain solutions constituting global optimum for the problem we are dealing with. Metaheuristics, such as Genetic Algorithms (GAs), Evolutionary Strategies (ESs), and Memetic Algorithms (GAs hybrided with local search techniques), are the ones being most currently applied to continuous optimization.

There are algorithms specifically designed for this type of optimization, Estimation of Distribution Algorithms (EDAs) [3], which do not use crossover operators, or mutation operators, unlike GAs and ESs. In order to evolve population evolution they use a mechanism consisting of individuals sampling from a given density of probability. In the same way of this algorithm, we find continuous PBIL [5], with a density of probability being normal for each variable, from which a population updating the normal ones is generated. Mechanisms to adapt standard deviation are also included.

In many cases, a number of populations working in a parallel way [1] can give rise to a good behavior of the algorithm. This kind of algorithms are known as Multideme Parallel Evolutionary Algorithms (PEAs) or island models. They consist of a number of subpopulations evolving independently which, occasionally, exchange information among them.

In this work, a new algorithm is proposed, as a result from taking essential ideas of several metaheuristics: GAs, ESs, Memetic Algorithms, EDAs, and

L. Jiao et al. (Eds.): ICNC 2006, Part I, LNCS 4221, pp. 796–805, 2006.

PEAs. The basic algorithm is based on evolving a mixture of normal distributions by keeping the best obtained solutions. It is embedded on an algorithm that evolves several set of them to avoid premature convergence and that considers two islands of different behaviors to self-adapt the algorithm to the problem.

According to that, this contribution is organized as follows: Section 2 describes the algorithm; Section 3 shows an empirical study of the algorithm compared with other proposals; and finally, Section 4 shows some conclusions and future works.

2 Description of the EvolPDF-2 Algorithm

For a better understanding, we have distinguished between what we call the *basic algorithm* and the final proposed algorithm (called EvolPDF-2). The basic algorithm, that can be considered the search process' core, is described in subsections 2.1 and 2.2. This basic algorithm would not work properly as such due to its high risk of falling in local optima and its high dependence with respect to the value parameters set. To face this, we have designed a more complex algorithm based on the basic one that is described in subsections 2.3 and 2.4.

2.1 Basic Components: Representation, Initialization, Sampling, and Replacement

The basic algorithm consists on four main components: *representation*, that describes how solutions are coded; *initialization*, that creates a set of initial solutions; *sampling*, that generates new solutions with the probability density functions (PDFs) estimated from the current best solutions; and *replacement*, that update the set of solutions considered to generate new ones. The mentioned components are designed as follows:

- *Representation*: The algorithm maintains a set of k solutions during the search process. Each solution is represented as follows: $S_i = (\mu_i^1, ..., \mu_i^v, ..., \mu_i^n)$, with n being the number of continuous variables of the problem.
- *Initialization*: The algorithm begins generating randomly k solutions according to a uniform distribution.
- *Sampling*: In each iteration, a number of solutions is generated from the current set. To do that, we consider that the current set of solutions constitutes a PDF (mixture of normal distributions) for each variable:

$$PDF_v = \sum_{i=1}^{k} \omega_i \cdot N(\mu_i^v, \sigma), \sum_{i=1}^{k} \omega_i = 1 \quad (1)$$

where

$$N(\mu, \sigma) = \frac{1}{\sigma\sqrt{2\pi}} e^{-\frac{(x-\mu)^2}{2\sigma^2}} \quad (2)$$

It should be noticed that the standard deviation is common to all the normals.

Table 3. Ranking and total mean errors obtained for multimodals functions (functions 7 to 14) for dimension 30

Algorithm	*Ranking*					*Total*
	1st	*2nd*	*3rd*	*4th*	*5th*	*mean error*
DMS-L-PSO	**3**	0	2	0	1	2.83
G-CMA-ES	2	2	0	2	1	**2.30**
L-SaDE	2	1	2	2	0	2.72
K-PCX	2	1	1	0	0	3.58
L-CMA-ES	2	1	0	1	0	4.61
BLX-GL50	1	1	1	0	2	3.03
EDA	1	0	0	0	0	4.95
DE	0	1	0	1	1	3.31
SPC-PNX	0	1	0	0	0	2.87
BLX-MA	0	0	1	1	2	3.23
CoEVO	0	0	0	0	0	6.30
EvolPDF-2	0	0	1	1	1	4.38

Table 4. Ranking and total mean errors obtained for highly multimodals functions (functions 15 to 25) for dimension 30

Algorithm	*Ranking*					*Total*
	1st	*2nd*	*3rd*	*4th*	*5th*	*mean error*
SPC-PNX	1	1	2	0	3	5.34
DE	1	1	1	0	0	6.34
BLX-MA	1	0	5	0	1	6.13
BLX-GL50	1	0	2	3	2	5.87
G-CMA-ES	0	5	1	0	1	5.69
K-PCX	0	4	0	2	2	6.99
L-CMA-ES	0	1	3	4	0	7.31
DMS-L-PSO	0	0	0	1	1	n/a
CoEVO	0	0	0	1	0	9.11
L-SaDE	0	0	0	0	1	n/a
EDA	0	0	0	0	0	n/a
EvolPDF-2	**9**	0	1	0	0	**1.79**

For hybridations of multimodals problems (Table 4), our algorithm is the most appeared at the first position with 9 times and it is followed of SPC-PNX, DE, BLX-MA and BLX-GL50 with 1 first position. From a global view of the ranking obtained by the algorithms according to positions, we can say that the three algorithms that have the best values are: our algorithm with 10 times and BLX-GL50, K-PCX and L-CMA-ES with 8. With respect to the total mean error, our algorithm obtains the lowest error.

To sum up, we can say that our algorithm obtains very good results for problems not solved by any other analyzed algorithm. We can also observe that where we obtain the lowest error are between functions number 15 and 25 (all of them being highly multimodal problems). These results suggest that our algorithm, thanks to the mechanisms included to avoid local optima, works better in multimodal problems.

4 Conclusions and Future Works

We have proposed a simple algorithm for continuous optimization. It involves using several PDF to model the interest region of each variable and evolving them to focus the search region. Some mechanisms to avoid premature convergence and local optima are included. The algorithm does not belong to any specific metaheuristic paradigm but it takes ideas from some of them such as GA and EDA. The obtained empirical results lead us to think that the algorithm has a high performance, obtaining the best results in multimodal problems. As further works, we suggest to consider self-adaptation of more parameters (e.g., σ_0 and σ_f) as well as to do a deeper characterization of the problems where the algorithm has a competitive advantage.

Acknowledgment

The authors would like to thank to Carlos García and Daniel Molina (both from the University of Granada, Spain) for providing us with the tools used in the empirical comparison. This work was supported in part by the Spanish Ministry of Science and Technology under grant no. TIC2003-00877 and by ERDF.

References

1. E. Alba, J.M. Troya. A survey of parallel distributed genetic algorithms. Complexity, 4:303-346, 1999.
2. K. Deb, D. Corne, Z. Michalewicz, Special Session on Real-parameter Optimization, IEEE Congress on Evolutionary Computation 2005, Edimburgh, UK, September 2005.
3. P. Larrañaga, J.A. Lozano (Eds.) (2001). Estimation of distribution algorithms. A new tool for evolutionary computation. Kluwer Academic Publishers.
4. Numerical Recipes in C. The Art of Scientific Computing. Second Edition. William H. Press, Saul A. Teukolsky, William T. Vetterling, Brian P. Flannery. Cambridge University Press.
5. M. Sebag, A. Ducoulombier (1998). Extending population-based incremental learning to continuous search spaces. Lecture Notes in Computer Science 1498: 418-427.

Make Fast Evolutionary Programming Robust by Search Step Control

Yong Liu[1] and Xin Yao[2]

[1] School of Computer Science
China University of Geosciences
Wuhan, 430074, P.R. China
yliu@u-aizu.ac.jp

[2] School of Computer Science
The University of Birmingham
Edgbaston, Birmingham, U.K.
X.Yao@cs.bham.ac.uk

Abstract. It has been found that both evolutionary programming (EP) and fast EP (FEP) could get stuck in local optima on some test functions. Although a number of methods have been developed to solve this problem, nearly all have focused on how to adjust search step sizes. This paper shows that it is not enough to change the step sizes alone. Besides step control, the shape of search space should be changed so that the search could be driven to other unexplored regions without getting stuck in the local optima. A two-level FEP with deletion is proposed in this paper to make FEP robust on finding better solutions in function optimisation. A coarse-grained search in the upper level could lead FEP to generate a diverse population, while a fine-grained search in the lower level would help FEP quickly find a local optimum in a region. After FEP could not make any progress after falling in a local optimum, deletion would be applied to change the search space so that FEP could start a new fine-grained search from the points generated by the coarse-grained search.

1 Introduction

Evolutionary programming (EP) can be defined as a population-based variant of generate-and-test algorithms [1]. EP uses search operators, such as mutation, to generate new solutions and a selection scheme to test which of the newly generated solutions should survive to the next generation. The generate-and-test formulation of EP indicates that mutation is a key search operator which generates new solutions from the current ones.

Classical EP (CEP) uses Gaussian mutation while fast EP (FEP) applies Cauchy mutation [2]. The difference between CEP and FEP is the length of their search steps. Generally speaking, Cauchy mutation generates larger steps than Gaussian does. It has been discovered that FEP performed much better than CEP on most of the functions among a suite of 23 functions [2]. The larger

L. Jiao et al. (Eds.): ICNC 2006, Part I, LNCS 4221, pp. 806–815, 2006.

search step size in FEP was attributed to FEP's success, while the smaller step size in CEP led to the CEP's worse performance.

Although FEP has displayed their fast search speed, it has also been discovered that FEP could turn to be slow when the search points are close to a local optimum. Meanwhile, CEP might overtake FEP after the search points are near to a local optimum. As pointed out in a previous study [2,3,4], there is an optimal search step size for a given problem. A step size which is too large or too small will have a negative impact on the performance of search. The optimal search step depends on the distance between the current search point and the global optimum. A large step size is beneficial (i.e., increases the probability of finding a near-optimal solution) only when the distance between the current search point and the global optimum is larger than the step size, or else a large step size may be detrimental to finding a near-optimal solution. Since the global optimum is usually unknown for real-world problems, it is impossible to know a priori what search step sizes should be used in EP.

A number of methods have been developed to determine a good search step size in EP. The search step in EP is the product of two terms in which one term is a random number, and the other term is the parameter of standard deviations. Depending on whether the random numbers or the parameters are adjusted, these methods can belong to one between the two approaches discussed as follows.

The first approach is to mix different types of random numbers, such as Gaussian and Cauchy random numbers. In general, different mutations based on different types of random numbers have different search step sizes, and are appropriate for different problems as well as different evolutionary search stages for a single problem. Rather than using either Gaussian or Cauchy mutations, both could be used at the same time. There are two ways to use both. One way is to generate two candidate offsprings by Gaussian and Cauchy mutations, respectively. The better one will survive as the single child. The other way is to take the simple average or the weighted average of the two mutations. The idea of mixing can be generalised to Lévy mutation. Lévy probability distribution can be turned to generate any distribution between the Gaussian and Cauchy probability distributions.

The second approach is to adaptively adjust the standard deviation parameters. In order to prevent the standard deviations from being too small, a fixed lower bound or a dynamic lower bounb could be set up for the standard deviations.

To determine an optimal search step is only the half way to the success of finding the global optimum. It should be noticed that the optimal search step only has the higher probability to reach a point that is close to the global optimum. A point near to the global optimum might not have a better object function value than the current local optimum found. If a point from the optimal search step could not survive in the selection, the optimal search step would have no effect on moving the search points away from the local optimum where they are located.

In order to let the points generated from the optimal search step to survive in the evolutionary process, it is essential to punish the regions that have been searched so that the search could go on in other unexplored regions that might contain better solutions. Fitness sharing is one way to punish the points that are close to each other so that the population could be diverse. However, certain number of points near to a local optimum found should be kept in the population. Otherwise, search might come back to the region where the local optimum is. In another word, the search might come back to the attraction basin of the local optimum. In order to prevent the search from going back to the attraction basins of local optima that have been found, certain number of points in each basin should be kept. A fixed population cannot cope with the increased number of basins when the number of local minima becomes too large.

Instead of using fitness sharing, minimum deletion could deal with the situation when the number of local optima is large. The idea of minimum deletion is to use $f(x) + p(x, m)$ to replace $f(x)$ when a local minimum m is found, where $f(x)$ is the original object function, and $p(x, m)$ is a punishment to "cancel out" the local optimum at m without disturbing $f(x)$ too much. If another local optimum would be found after the search escapes from the local optimum m, another punishment term should be added to the modified objection function. By minimum deletion, different local optima are able to be found.

Minimum deletion is different to the fitness sharing that could only re-shape the search space temporarily. For example, once a lot of points fall in a small region **A**, fitness sharing would punish the points in the region **A** so that points in other regions would have more chances to survive. However, when the points in the region **A** are reduced, the punishment on the points in the region **A** would be lessened or released so that the region **A** might become a possible region again. In contrast, minimum deletion would re-shape the search space permanently.

A two-level FEP with minimum deletion is proposed in this paper to make FEP robust on finding better solutions in function optimisation. A coarse-grained search in the upper level could lead FEP to generate a diverse population, while a fine-grained search in the lower level would help FEP quickly find a local optimum in a region. After FEP could not make any progress after falling in a local optimum, deletion would be applied to change the search space so that FEP could start a new fine-grained search from the points generated by the coarse-grained search.

The rest of this paper is organized as follows: Section 2 describes the global minimization problem considered in this paper and FEP used to solve it. Section 3 presents the two-level FEP with deletion, and discusses the experimental results. Finally, Section 4 concludes with some remarks and future research directions.

2 Function Optimization by Fast Evolutionary Programming

A global minimization problem can be formalized as a pair (S, f), where $S \subseteq R^n$ is a bounded set on R^n and $f : S \mapsto R$ is an n-dimensional real-valued function.

The problem is to find a point $\mathbf{x}_{min} \in S$ such that $f(\mathbf{x}_{min})$ is a global minimum on S. More specifically, it is required to find an $\mathbf{x}_{min} \in S$ such that

$$\forall \mathbf{x} \in S : f(\mathbf{x}_{min}) \leq f(\mathbf{x}),$$

where f does not need to be continuous but it must be bounded. This paper only considers unconstrained function optimization.

The FEP applied to function optimization can be described as follows [2]:

1. Generate the initial population of μ individuals, and set $k = 1$. Each individual is taken as a pair of real-valued vectors, $(\mathbf{x}_i, \eta_i)$, $\forall i \in \{1, \cdots, \mu\}$, where $\mathbf{x}_i$'s are objective variables and η_i's are standard deviations for Cauchy mutations (also known as strategy parameters in self-adaptive evolutionary algorithms).
2. Evaluate the fitness score for each individual $(\mathbf{x}_i, \eta_i)$, $\forall i \in \{1, \cdots, \mu\}$, of the population based on the objective function, $f(\mathbf{x}_i)$.
3. Each parent $(\mathbf{x}_i, \eta_i)$, $i = 1, \cdots, \mu$, creates a single offspring $(\mathbf{x}_i{}', \eta_i{}')$ by: for $j = 1, \cdots, n$,

$$\eta_i{}'(j) = \eta_i(j) \exp(\tau' N(0,1) + \tau N_j(0,1)), \quad (1)$$
$$x_i{}'(j) = x_i(j) + \eta_i{}'(j)\delta_j \quad (2)$$

   where $\eta_i(j)$, $\eta_i{}'(j)$, $x_i(j)$, and $x_i{}'(j)$ denote the j-th component of the vectors η_i, $\eta_i{}'$, $\mathbf{x}_i$ and $\mathbf{x}_i{}'$, respectively. The factors τ and τ' are commonly set to $\left(\sqrt{2\sqrt{n}}\right)^{-1}$ and $\left(\sqrt{2n}\right)^{-1}$ [5]. $N(0,1)$ denotes a normally distributed one-dimensional random number with mean 0 and standard deviation 1. $N_j(0,1)$ indicates that the random number is generated anew for each value of j. δ_j is a Cauchy random variable with the scale parameter $t = 1$, and is generated anew for each value of j. The one-dimensional Cauchy density function centered at the origin is defined by:

$$f_{C(t)}(x) = \frac{1}{\pi} \frac{t}{t^2 + x^2}, \qquad -\infty < x < \infty, \quad (3)$$

   where $t > 0$ is a scale parameter [6](pp.51). The shape of $f_t(x)$ resembles that of the Gaussian density function but approaches the axis so slowly that an expectation does not exist. As a result, the variance of the Cauchy distribution is infinite.
4. Calculate the fitness of each offspring $(\mathbf{x}_i{}', \eta_i{}')$, $\forall i \in \{1, \cdots, \mu\}$.
5. Conduct pairwise comparison over the union of parents $(\mathbf{x}_i, \eta_i)$ and offspring $(\mathbf{x}_i{}', \eta_i{}')$, $\forall i \in \{1, \cdots, \mu\}$. For each individual, q opponents are chosen uniformly at random from all the parents and offspring. For each comparison, if the individual's fitness is no smaller than the opponent's, it receives a "win."
6. Select the μ individuals out of $(\mathbf{x}_i, \eta_i)$ and $(\mathbf{x}_i{}', \eta_i{}')$, $\forall i \in \{1, \cdots, \mu\}$, that have the most wins to be parents of the next generation.
7. Stop if the halting criterion is satisfied; otherwise, $k = k+1$ and go to Step 3.

3 Experimental Analyses

This section presents experimental analyses of FEP's behaviour in the evolution process. The experiment results had shown that it was hard for both FEP and EP to find the global minimum for the generalised Griewank function [2]:

$$f_{11}(x) = \frac{1}{4000}\sum_{i=1}^{n} x_i^2 - \prod_{i=1}^{n} \cos\left(\frac{x_i}{\sqrt{i}}\right) + 1 \tag{4}$$

where x is in $[-600, 600]^n$, and n is 30. The purpose of experiments designed in this section is to find out why FEP had failed on finding the global optimum of this function.

3.1 Why FEP Fails

Two experiments are conducted in this section to show the behaviour of FEP in details by observing how the following 11 values change in the evolutionary process:

1. Best of FEP. The best function value among the population.
2. Average of FEP. The average objective function value of the population.
3. Distance to the global minimum. The average distance from the population to the global minimum.
4. Average distance among the population. The average distance between two individuals in the population.
5. Survival rate. The percentage of the offsprings surviving to the next generation.
6. Average winning step size. The average search step size made by the offsprings surviving to the next generation.
7. Average losing step size. The average search step size made by the offsprings not selected to the next generation.
8. Average value of the remaining parents. The average objective function value of the parents surviving to the next generation.
9. Average value of the replaced parents. The average objective function value of the parents not selected to the next generation.
10. Average value of the winning offsprings. The average objective function value of the offsprings surviving to the next generation.
11. Average value of the losing offsprings. The average objective function value of the offsprings not selected to the next generation.

The lower bound set for the standard deviation parameter in Eq.(1) is set to 0.001 in Experiment 1, and 1.0 in Experiment 2, respectively. The smaller lower bound could lead FEP search to become fine-grained, while the larger lower bound only allows FEP to have the coarse-grained search. The evolutionary process of 11 values in 2000 generations given by the fine-grained FEP in Experiment 1 and the coarse-grained FEP in Experiment 2 are shown in the left side and the right side of Figs. 1 and 2., respectively.

From the best function value and average function value of the population in Fig. 1(a), it can be found that the fine-grained FEP could find the better solutions than the coarse-grained FEP. The best function value and average function value of the population become nearly the same at the later evolutionary stage in the fine-grained FEP while the two values can be different through the evolution process in the coarse-grained FEP. It suggests that the fine-grained FEP could lead the whole population to converge to the same point. Meanwhile, the coarse-grained FEP is able to keep the population diverse.

The distance to the global minimum in Fig. 1(b) clearly shows that the fine-grained FEP got stuck in a local minimum without being able to make progress. The coarse-grained FEP could not get closer to the global minimum either. Such results tell that it is hard for both the fine-grained FEP and the coarse-grained FEP to find the global minimum. It suggests that two searches should be combined in the evolutionary process.

The average distance among the population in Fig. 1(c) became smaller and smaller in the fine-grained FEP. At the same time, the average distance among the population could be kept at a certain value through the evolution process in the coarse-grained FEP. It suggests that the diversity among the population can be kept in the coarse-grained FEP but disappear in the fine-grained FEP.

The survival rate in Fig.2(a) was little larger at the early evolution stage compared to the value at the later stage in both the fine-grained FEP and the coarse-grained FEP. There is little difference in the survival rate between the the fine-grained FEP and the coarse-grained FEP after the average function values of the population have no significant change. It suggests that the survival rate could be at certain level when the population converge.

The average winning step size in Fig.2(b) reduced quickly before the population converged to a local minimum in the fine-grained FEP. After the population converged to a local minimum, the average winning step size changed very litter. The similar behaviour had been found in the coarse-grained FEP. The only difference is that the population in the coarse-grained FEP did not converge to a local minimum but became stable at a certain value. Compared to the winning step sizes, the losing step sizes were rather large.

Fig.2(c) shows that the average values of the remaining parents, the replaced parents, and the winning offsprings are rather closer from beginning to the end in the fine-grained FEP. In contrast, in the coarse-grained FEP, only the average values of the replaced parents and the winning offsprings are similar. The average values of the remaining parents in the coarse-grained FEP can be much smaller than the average of the winning offsprings. It suggests that some rather poor offspring with large function values could survive in the coarse-grained FEP but have little chances to be selected in the fine-grained FEP.

In summary, the fine-grained FEP might likely get stuck in a local minimum because the small search step size would make the search points get closer and closer to the local minimum. By contrast, the coarse-grained FEP would hardly fall in a local minimum because the large search step would keep the search points at a certain distance to the local minimum.

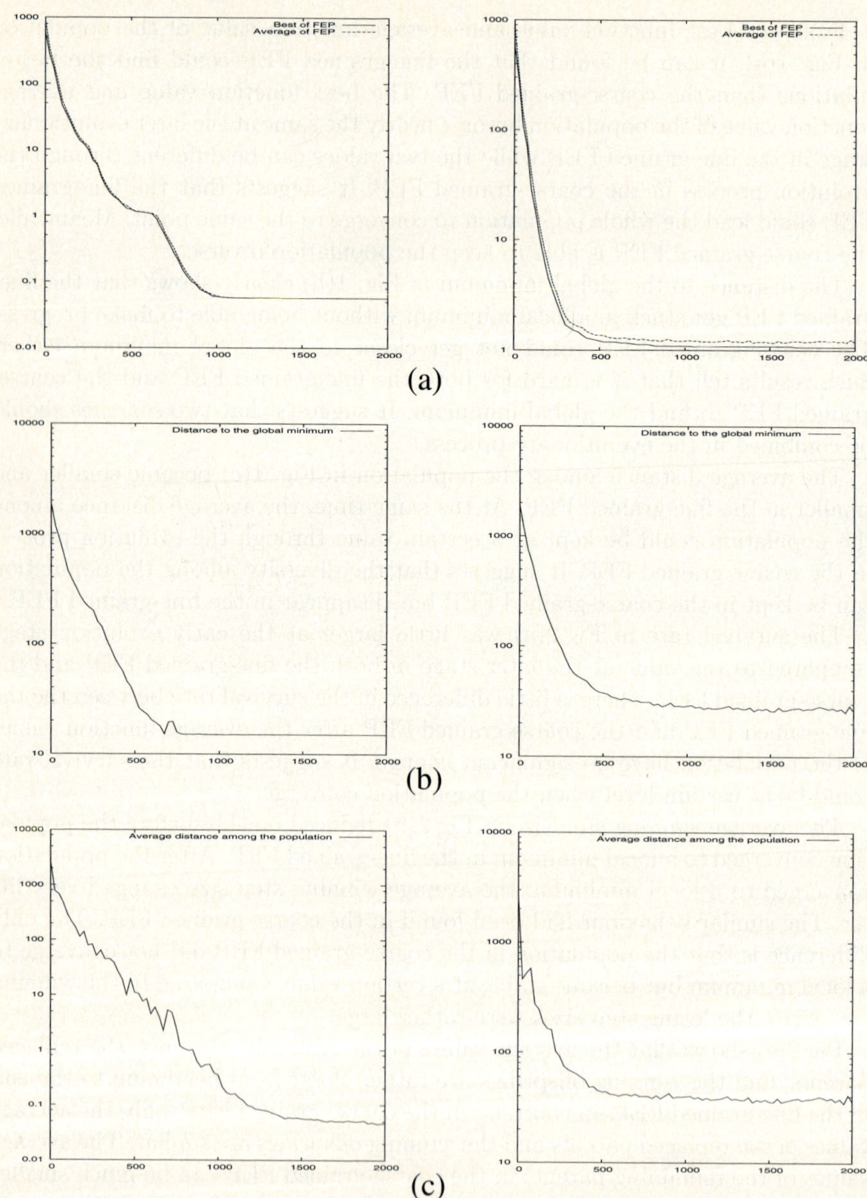

Fig. 1. Comparison between the fine-grained FEP (left) and the coarse-grained FEP (right). (a) The best function value among the population and the average function value of the population. (b) The average distance from the population to the global minimum. (c) The average distance among the population.

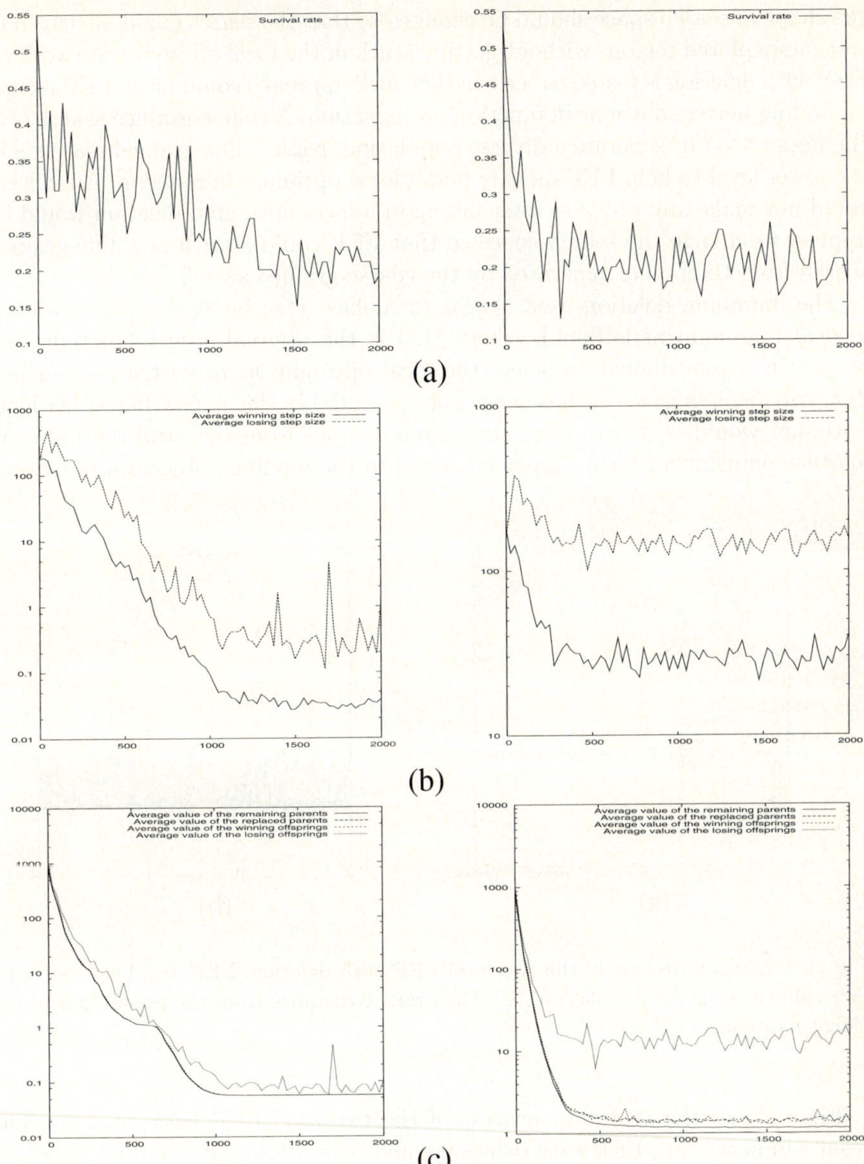

Fig. 2. Comparison between the fine-grained FEP (left) and the coarse-grained FEP (right). (a) The survival rate of offsprings. (b) The average step sizes of the winning offsprings and losing offsprings. (c) The average function values of the remaining parents, the replaced parents, the winning offsprings, and losing offsprings.

3.2 Two-Level FEP with Deletion

From the experimental analyses given to Experiment 1 and Experiment 2, it suggests that it is not enough to change the step sizes alone. Besides step control, the shape of search space should be changed so that the search could be driven to some unexplored regions without getting stuck in the local optimum. A two-level FEP with deletion is tested to see whether such approach could make FEP robust on finding better solutions in function optimisation. A coarse-grained search is in the upper level to generate a diverse population, while a fine-grained search is in the lower level to help FEP quickly find a local optimum in a region. After FEP could not make any progress after falling in a local optimum, deletion would be applied to change the search space so that FEP could start a new fine-grained search from the points generated by the coarse-grained search.

The minimum deletion used here is to replace $f(x)$ by $f(x) + p(x, m)$ when a local minimum m is found, where $f(x)$ is the original object function, and $p(x, m)$ is a punishment to delete the local optimum at m without disturbing $f(x)$ too much. $p(x, m)$ is chosen as $exp(-|x-m|^2)$ in this paper. If another local optimum would be found after the search escapes from the local optimum m, another punishment term should be added to the modified objection function.

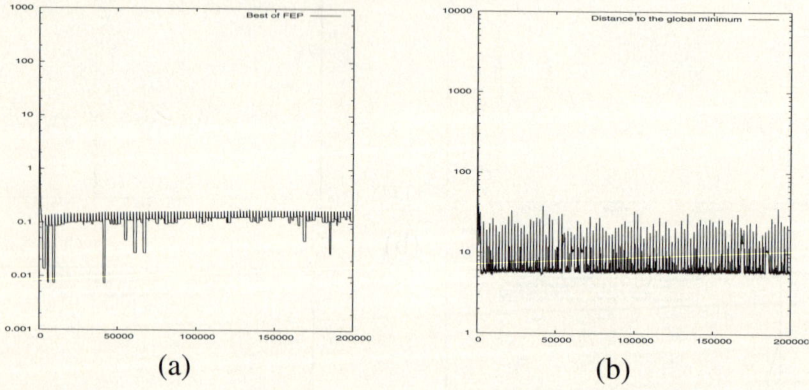

Fig. 3. Evolution process of the two-level FEP with deletion. FEP. (a) The best function value among the population. (b) The average distance from the population to the global minimum.

Fig.3 shows the evolution process of the two-level FEP with deletion. The results indicate that FEP with deletion could search from one local minimum to the next. 100 different local minima had been found in 200,000 generations. The minimal distance between 2 local minima among 100 local minima is 1.99691, and the average distance between 2 local minima is 7.21691. Although none of local minima found is the global minimum, it at least shows the best local optimum among the 100 local minima could be chosen as the final solution. If the time would allow, the two-level FEP with deletion could continue to find new local minima. After all, the global minimum is just the best local minimum so that the two-level FEP with deletion is able to find it.

4 Conclusions

The experimental analyses show that the coars-grained search in the upper level is good at finding a diverse population while the fine-grained search could locate a local optimum effectively. By introducing deletion into FEP, the search space can be re-shaped so that the evolutionary search could go on in other unexplored regions. The experimental results have clearly shown that the two-level FEP with deletion is so robust that it could move away from one local optimum easily and lead the search into other possible regions that might contain the global optimum.

Such two-level search with deletion can also be applied to other evolutionary algorithms, such as evolution strategies, to design faster and more robust optimization algorithms.

Acknowledgement

This work was supported by the National Natural Science Foundation of China (No.60473081).

References

1. Fogel, D. B. (1991) *System Identification Through Simulated Evolution: A Machine Learning Approach to Modeling.* Needham Heights, MA 02194: Ginn Press.
2. Yao, X., Liu, Y., and Lin, G. (1999) "Evolutionary programming made faster," *IEEE Transactions on Evolutionary Computation*, vol. 3, no. 2, pp.82–102.
3. Yao, X., and Liu, Y. (1998) "Scaling up evolutionary programming algorithms," *Evolutionary Programming VII: Proc. of the Seventh Annual Conference on Evolutionary Programming (EP98), Lecture Notes in Computer Science*, vol. 1447, Springer-Verlag, Berlin, pp.103–112.
4. Yao, X., Lin, G., and Liu, Y. (1997) "An analysis of evolutionary algorithms based on neighbourhood and step sizes," *Evolutionary Programming VI: Proc. of the Sixth Annual Conference on Evolutionary Programming*, P. J. Angeline, R. G. Reynolds, J. R. McDonnell, and R. Eberhart, eds., Lecture Notes in Computer Science, vol. 1213, pp. 297–307, Springer-Verlag.
5. Bäck, T., and Schwefel, H.-P. (1993) "An overview of evolutionary algorithms for parameter optimization," *Evolutionary Computation*, vol. 1, no. 1, pp. 1–23.
6. Feller, W. (1971) *An Introduction to Probability Theory and Its Applications*, John Wiley & Sons, Inc., 2nd ed.
7. Hunt, R. A. (1986) *Calculus with Analytic Geometry*, Harper & Row Publ., Inc, New York, NY 10022-5299.
8. Potter, M. A., and De Jong, K. A. (1994) "A cooperative co-evolutionary approach to function optimization," in *Proc. of the Third International Conference on Parallel Problem Solving from Nature*, pp. 249–257, Springer-Verlag.

Improved Approach of Genetic Programming and Applications for Data Mining

Yongqiang Zhang and Huashan Chen

School of information and electrical engineering, Hebei University of Engineering, Handan, 056038, P.R. China

yqzhang@hebeu.edu.cn, chs811013@163.com

Abstract. Genetic Programming (GP for short) is applied to a benchmark of the data fitting and forecasting problems. However, the increasing size of the trees may block the speed of problems reaching best solution and affect the fitness of best solutions. In this view, this paper adopts the dynamic maximum tree depth to constraining the complexity of programs, which can be useful to avoid the typical undesirable growth of program size. For more precise data fitting and forecasting, the arithmetic operator of ordinary differential equations has been made use of. To testify what and how they work, an example of service life data series about electron parts is taken. The results indicate the feasibility and availability of improved GP, which can be applied successfully for data fitting and forecasting problems to some extent.

Keywords: GP; dynamic tree depth; ordinary differential equation; data mining.

1 Introduction

Genetic Programming needs a good random tree-creation algorithm to create trees that form the initial population, and then trees in the following generations can be created by GP operators. Its search space is potentially unlimited and programs may grow in size during the evolutionary process. Unfortunately, it also increases the size of programs without improving their fitness. For data fitting and forecasting, the common GP algorithm usually can't get the satisfied solution or even so poor [2]-[3].

In this view, dynamic tree depth is adopted, which is closely related to the fitness of the solution in one generation. Meanwhile, the ordinary differential equation is used too, which can improve the efficiency of GP. In order to testify them, this paper applies them into the practical data mining, which can get better model for data fitting and forecasting. The results indicate that this improved approach of GP can be applied successfully for data fitting and forecasting.

2 Improved Approach of GP Algorithm

In GP, a piece of redundant code can be improved from an individual without affecting its fitness. Therefore, it adopts an approach, which is designated as dynamic tree depth

L. Jiao et al. (Eds.): ICNC 2006, Part I, LNCS 4221, pp. 816 – 819, 2006.

limit: it is initially set with a low value, it is increased when needed to accommodate an individual found during the run. If the new individuals are deeper than the maximum depth, reject it and consider one of their parents for the new individuals instead. Otherwise, consider it acceptable to participate in the new generation. If the new individuals are deeper than the dynamic depth limit (but no deeper than the maximum depth) measure its fitness and which includes as follow:

- If the individuals have better fitness than the current best individuals of the run, increase the dynamic limit to match the depth of the individuals and consider it acceptable to the new generation.
- Otherwise, leave the dynamic level unchanged and reject the individuals, considering one of their parents for the new generation instead [1].

As we know, we can do some works by GP algorithm to meet the need of data fitting and forecasting for a time series. However, the common GP isn't always able to obtain satisfied solutions of the given problems. Therefore, this paper makes use of the ordinary differential equation during GP operation, which can improve the dynamic description capability and reform the complexity of the model.

During the course of fitness calculating, it is necessary to do some multistage integral transformation by ordinary differential equation used. Assume the dynamic system is described by n number of functions that $x_1(t), x_2(t), \cdots, x_n(t)$. Given the real observation data at time $t_j (j = 0,1,2,\cdots,n)$, then X defined as $X(t)=(x_1(t), x_2(t), \cdots, x_n(t))$

While, $t_j = t_0 + j \times \Delta t$ (Δt means step length between two time-intervals), so the ordinary differential equation is defined by $dx_i/dt = f_i(x_1, x_2, \cdots x_n, t) \quad (i = 1,2 \cdots ;n) \quad x_i^0 = x_i(t_0)$. Following by this expression, the solving process is impelled by the points' order step by step. And can be deduced by $y_{n-1}, y_{n-2}, \cdots y_1$. It defines the last function for solutions of this system as $f(t,X)=[f_1(t,X), f_2(t,X), \cdots, f_n(t,X)]$, whose sub-function is related both the current time t and the observation data series $x_j(t)$, which is given as: $f_j(t,X) = f_j(t, x_1(t), x_2(t), \cdots, x_n(t))$. Finally, a vector will be obtained, which is described as $X^*(t) = f(t, x^*(t), x^*(t))$ in order to make the difference $\min(\|X^* - X\|)$, and $\|X^* - X\| = \sqrt{\sum_{i=0}^{n} [x^*(t-i) - x(t-i)]^2}$. The expression $f(t, x^*)$ to meet the need of $dx^* / dt = f(t, x^*)$.

The differential quotient is instead by difference coefficient, which is shown (1).

$$\frac{dy}{dx}\Big|_{(t_n, y_n)} = \frac{y(t_n) - y(t_{n-1})}{t_n - t_{n-1}} = f(t_n, y(t_n)) \tag{1}$$

Assuming that $y_n \approx y(t_n)$ and $y_{n+1} \approx y(t_{n+1})$, it can obtain $y_{n+1} = y_n + \Delta t \cdot f(t_n, y_n)$.

3 Applications for Data Mining

Based on the improved GP algorithm above, the programming for a time series of service life about electron parts have been run, under the circumstances of MATLAB6.5.1, which is shown in Tab.1 whose simulation result is given at Fig.1(a).

Table 1. Failure time series of electron parts

t	1	2	3	4	5	6	7	8	9	10
f(t)	2.01	4.35	6.92	9.61	12.20	15.49	18.74	22.31	26.06	30.03

The new GP model created making use of cynamic maximum tree depth is (2):

$$f(t) = \left(t + \sqrt{t}\right) \cdot \ln t \tag{2}$$

Comparing with this GP model, the model of ordinary differential equations is also have been created, which are shown as equations (3) its simulation given in Fig.1(b).

$$\begin{aligned} dy &= 0.7 \cdot u(n-1) \cdot u(n-1) - 0.6 \cdot y(n-1) - 0.3 \cdot y(n-2) - 0.1 \\ du &= 0.700147 \cdot u(n-1) \cdot u(n-1) - 0.2997 \cdot y(n-2) - 0.600474 \cdot y(n-1) - 0.147109 \\ y(n) &= y(n-1) + dy \end{aligned} \tag{3}$$

In order to testify their characteristics, four relative criteria are given below. The differences of the two models comparing with these criteria shown in the Tab.2.

Table 2. Diffrences of the two models for Data Fitting and Forecasting

Criteria	Ordinary differential GP	GP model
Covariance analysis	106.8935	109.9401
Least square error	13.2514	14.7720
Mean values	14.0629	14.2672
Correlation coefficient	0.9996	0.9992

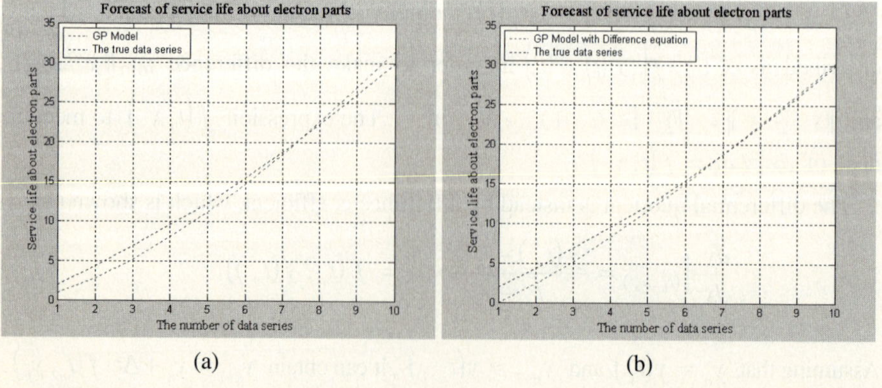

(a) (b)

Fig. 1. Data fitting simulation on the left and data forecasting simulation on the right

4 Conclusions

This paper shows improved approach for GP modeling, including dynamic maximum tree depth, which is useful to aviod the typical undesirable growth of program size, and the ordinary differential GP operations, which can efficiently improve the precision of data fitting and forecasting. The results of example testifies the feasibility and availability of them, both of which should be study further and deeper.

Acknowledgements

Project 60573088 supported by National Natural Science Foundation of China (NSFC); Project 603407 supported by Natual Science Foundation of Hebei Province.

References

1. Sara Silva1 and Jonas Almeida. Dynamic Maximum Tree Depth A Simple Technique for avoiding Bloat in Tree-Based GP.Biomathematics Group, Instituto de Tecnologia Qu'ımica e Biol'ogica Universidad Nova de Lisboa, PO Box 127, 2780-156 Oeiras, Portugal.2002.
2. Kilian Stoffel, Lee Spector. High-Performance, Parallel, Stack-Based Genetic Programming. Proceeding of the First Annual Conference.1996. 224-229.
3. Sean Luke, Liviu Panait. Fighting Bloat with Nonparametric Parsimony Pressure. Proceeding of the First Annual Conference.2000.
4. John R.Koza. Genetic Programming. Encyclopedia of Computer Science and Technology.1997 (8.18) 2-4.

Niching Clonal Selection Algorithm for Multimodal Function Optimization

Lin Hao[1], Maoguo Gong[1], Yifei Sun[1], and Jin Pan[2]

[1] Institute of Intelligent Information Processing, P.O. Box 224, Xidian University 710071, Xi'an, P.R. China

[2] Lab of Network Security and Countermeasure, Xi'an Communications Institute, Xi'an, Shaanxi 710106, China

haolin0901@163.com, maoguo_gong@hotmail.com

Abstract. Interest in multimodal function optimization problems is expanding rapidly since real-world optimization problems often require the location of multiple optima in the search space. In this paper, a new niching strategy, deterministic replacement policy, is put forward in the proposed algorithm named Niching Clonal Selection Algorithm (NCSA). In order to improve the algorithm's performance, it advances a new selection method- meme selection, based on the knowledge of meme. The numerical experiment on four typical multimodal function optimization problems attests to the proposed algorithm's validity. Finally, the study compares NCSA with some niching evolution algorithms. From the experimental results, we can see that the proposed algorithm is superior to them on all of the tested functions.

1 Introduction

Clonal Selection Algorithm [1]~[3] is an optimization technique which based on theoretical immunology. **A** standard Clonal selection algorithm will eventually converge to a single peak, even in domains characterized by multiple peaks of equivalent fitness. Nowadays, more and more optimization problems require not only one solution, but a set of good solutions. In order to generate more than one solution, niching strategies have been developed. In the nature, life-form is apt to lives in an environment where other life-form is similar with itself. It copulates and produces offspring with these individuals. The given environment is called a niche. In a multimodal domain, each peak can be regarded as a niche. The analogy from nature is that within the environment there are different subspaces (niches) that can support different types of life (species, or organisms). Many researches have been done to design diverse niching methods including fitness sharing [4], sequential niche technique [5] and some other niche techniques [6], [7] to extend genetic algorithms to domains that require the location and maintenance of multiple solutions. Such domains include classification and machine learning, multimodal function optimization, multiobjective function optimization, and simulation of complex and adaptive systems [8]~[11]. But all the researches are about niching evolution algorithms. We know that, compared to evolution algorithms, such as genetic algorithm, Clonal Selection Algorithm has many predominant performances.

L. Jiao et al. (Eds.): ICNC 2006, Part I, LNCS 4221, pp. 820 – 827, 2006.

In the paper, a new algorithm named Niching Clonal Selection Algorithm (NCSA) is put forward for multimodal optimization problems. In the algorithm a new niching strategy, deterministic replacement policy, is proposed. In order to improve the algorithm's performance, a new selection method, meme selection is put forward. Meme, relative to gene, was put forward by Richard Dawkins [12] in "The selfish Gene" in 1976 firstly. Similar to gene, it is able to copy, mutate and take some message. It absorbs information from surroundings and spread it, so it is the radical unit in culture evolution. Based on this knowledge we introduced meme selection.

The rest of the paper is organized as follows: Section 2 gives some the background knowledge of standard clonal selection algorithm; Section 3 describes the niching clonal selection algorithm; Section 4 gives the main loop of NCSA; Section 5 validates the new algorithm based on four typical multimodal function optimization problems; Finally, concluding remarks are presented in section 6.

2 Standard Clonal Selection Algorithm (SCSA)

The clonal selection algorithm only adopted monoclonal operator is called standard clonal selection algorithm (SCSA).

We suppose that $f: R^m \to R$ is the optimized object function, without loss of generality, we consider the maximum of affinity function. The affinity function $\Phi: I^n \to R$ where n is population size; m is the number of the optimized variable, viz. $a = \{x_1, x_2, \ldots x_m\}$..The details of the overall algorithms are as Follows.

Step1 k=0, randomly generate the original antibody population $A(0) = \{a_1(0), a_2(0), \ldots a_n(0)\} \in I^n$. Set the initial conditions.

Step2 Do clonal operation according to clonal size: $A'(k) = \{A(k), T_c^c(A(k))\}$.

Step3 Do mutational operation to $A'(k): A''(k) = T_m^c(A'(k))$.

Step4 Calculate the affinity: $A''(k): \{\Phi(A''(k))\}$.

Step5 Do clonal selection operation: $A(k+1) = T_s^c(A''(k))$; then update the antibody population *A(k)*.

Step6 *k=k+1*.The algorithm will be halted when meet the restricted iterative number, the Termination Criterion or the two blending. Where Termination Criterion can defined as follows: $\left|f^* - f^{best}\right| < \varepsilon$ .Or else, return to Step2.

3 Description of Niching Clonal Selection Algorithm

We assume the antibody space is I and antibody population is $A = \{a_1, a_2, \ldots a_n\}$.

(1) Clonal proliferation T_c^p. In immunology, clone means asexual propagation so that a group of identical cells can be descended from a single common ancestor. In artificial immune response, the clonal proliferation T_c^p on population $A = \{a_1, a_2, \ldots a_n\}$ is defined as:

$$T_c^p(A) = T_c^p(a_1 + a_2 + ... + a_n) = T_c^p(a_1) + T_c^p(a_2) + ... + T_c^p(a_n)$$
$$= \{a_1^1 + a_1^2 + ... + a_1^{nc}\} + \{a_2^1 + a_2^2 + ... + a_2^{nc}\} + ... + \{a_n^1 + a_n^2 + ... + a_n^{nc}\} \quad (1)$$

Where $T_c^p(a_i) = \{a_i^1 + a_i^2 + ... + a_i^{nc}\}$, $a_i^j = a_i$, $j = 1, 2...nc$, nc is clonal scale. The antibody population after T_c^p can be denoted as:

$$A^{(1)} = T_c^p(A) = \{a_1^1, a_1^2, ...a_1^{nc}, a_2^1, a_2^2, ...a_2^{nc}, a_n^1, a_n^2, a_n^{nc}\} \quad (2)$$

(2) Clonal mutation T_c^m . Mutation is a way to achieve affinity maturation. The clonal mutation T_c^m worked on the antibody population $A^{(1)}$ after clonal proliferation is defined as:

$$T_c^m(A^{(1)}) = T_c^m(\{a_1^{1'} + a_1^2 + ... + a_1^{nc}\} + \{a_2^1 + a_2^2 + ... + a_2^{nc}\} + \{a_n^1 + a_n^2 + ... + a_n^{nc}\})$$
$$= \{T_c^m(a_1^1) + T_c^m(a_1^2) + ... + T_c^m(a_1^{nc})\} + ... + \{T_c^m(a_n^1) + T_c^m(a_n^2) + ... + T_c^m(a_n^{nc})\} \quad (3)$$
$$= \{a_1^{1'} + a_1^{2'} + ... + a_1^{nc'}\} + \{a_2^{1'} + a_2^{2'} + ... + a_2^{nc'}\} + ... + \{a_n^{1'} + a_n^{2'} + ... + a_n^{nc'}\}$$

Where $T_c^m(a_i^j) = a_i^{j'}$, $i = 1, 2, ...n$, $j = 1, 2, ...nc$. After clonal mutation operation, the antibody population can be denoted as:

$$T_c^m(\{a_1 + a_1^{1'} + a_1^{2'} + ... + a_1^{nc'}\} + \{a_2 + a_2^{1'} + a_2^{2'} + ... + a_2^{nc'}\} + ... + \{a_n + a_n^{1'} + a_n^{2'} + ... + a_n^{nc'}\})$$
$$= T_c^m(\{a_1 + a_1^{1'} + a_1^{2'} + ... + a_1^{nc'}\}) + T_c^m(\{a_2 + a_2^{1'} + a_2^{2'} + ... + a_2^{nc'}\}) + ... + T_c^m(\{a_n + a_n^{1'} + a_n^{2'} + ... + a_n^{nc'}\}) \quad (4)$$
$$= a_1^{'} + a_2^{'} + ... + a_n^{'}$$

(3) Clonal selection T_c^s . Different from the selection operation of evolutionary algorithms, clonal selection T_c^s selects an excellent individual from the sub-population generated by clonal proliferation. For antibody

$$A^{(3)} = A + A^{(2)} = \{a_1, a_1^{1'}, a_1^{2'}, ...a_1^{nc'}, a_2, a_2^{1'}, a_2^{2'}, ...a_2^{nc'}, a_n, a_n^{1'}, a_n^{2'} ...a_n^{nc'}\} \quad (5)$$

clonal selection T_c^s is defined as:

$$T_c^s(\{a_1 + a_1^{1'} + a_1^{2'} + ... + a_1^{nc'}\} + \{a_2 + a_2^{1'} + a_2^{2'} + ... + a_2^{nc'}\} + ... + \{a_n + a_n^{1'} + a_n^{2'} + ... + a_n^{nc'}\})$$
$$= T_c^s(\{a_1 + a_1^{1'} + a_1^{2'} + ... + a_1^{nc'}\}) + T_c^s(\{a_2 + a_2^{1'} + a_2^{2'} + ... + a_2^{nc'}\}) + ... + T_c^s(\{a_n + a_n^{1'} + a_n^{2'} + ... + a_n^{nc'}\}) \quad (6)$$
$$= a_1^{'} + a_2^{'} + ... + a_n^{'}$$

Where $T_c^s(\{a_1 + a_i^{1'} + a_i^{2'} + ... + a_i^{nc'}\}) = a_i^{'}$. The antibody $a_i^{'}$ has the highest affinity in sub-population $A_i^{'} = \{a_1, a_i^{1'}, a_i^{2'}, ..., a_i^{nc'}\}$. After clonal selection thc population is denoted as $A^{(4)} = \{a_1^{'}, a_2^{'}, ...a_n^{'}\}$.

(4) Deterministic replacement policy T^r . In the antibody population, there is one niche at least. Realign the individuals in population $A^{(4)}$ according to the affinity. Then the individual $a_1^{'}$ has the highest affinity in population $A^{(4)}$, we suppose that individual $a_1^{'}$ is a member of niches, i.e. $a_1^{'} \in A^{(5)}$. $A^{(5)}$ denotes the niches population.

Then the other individuals in population $A^{(4)}$ are judged in turn by the following rules. For any antibody $a_j^{'}$ ($2 \le j \le n$), if all the distances with $a_j^{'}$ and $a_k^{'}$ ($1 \le k \le j-1$) are far than the niche radius σ, $a_j^{'}$ is considered to be a different niche, then add $a_j^{'}$ to $A^{(5)}$. Otherwise the individual $a_j^{'}$ is considered to be a non niche member and replaced it by a new individual. And then examine it according to the above criterion. Repeating this process until all the individuals in $A^{(4)}$ have been added to $A^{(5)}$. The deterministic replacement policy T^r is defined as:

$$T^r(A^{(4)}) = T^r(\{a_1^{'} + a_2^{'} + ... + a_n^{'}\}) = \{a_1^{''} + a_2^{''} + ... + a_n^{''}\} \tag{7}$$

After deterministic replacement the population is denoted as $A^{(5)} = \{a_1^{''}, a_2^{''}, ... a_n^{''}\}$, which is deemed to be the niches set needed to be optimized.

After the above operations, the evolution of the first generation is finished and the original niches are formed viz. $A(k+1) = A^{(5)}$. Then do clonal proliferation T_c^P and clonal mutation T_c^m in turn on the new population A. The population after clonal mutation operation is denoted as:

$$A^{(7)} = \{a_1^{1'''}, a_1^{2'''}, ... a_1^{nc'''}, a_2^{1'''}, a_2^{2'''}, ... a_2^{nc'''}, a_n^{1'''}, a_n^{2'''}, ... a_n^{nc'''}\} \tag{8}$$

(5) Meme selection T_m^s. Similar to clonal selection, meme selection selects an excellent individual from the sub-population generated by clonal proliferation. The difference is that in meme selection operation each sub-population is divided into two parts denoted niche-in individuals and niche-out individuals respectively by dint of meme. The selected excellent individual must belong to the niche-in part. Here meme denotes the niche radius σ. For population $A^{(8)} = A + A^{(7)}$, meme selection T_m^s is defined as:

$$\begin{aligned} &T_m^s(\{a_1 + a_1^{1'''} + a_1^{2'''} + ... + a_1^{nc'''}\} + \{a_2 + a_2^{1'''} + a_2^{2'''} + ... + a_2^{nc'''}\} + ... + \{a_n + a_n^{1'''} + a_n^{2'''} + ... + a_n^{nc'''}\}) \\ &= T_m^s(\{a_1 + a_1^{1'''} + a_1^{2'''} + ... + a_1^{nc'''}\}) + T_m^s(\{a_2 + a_2^{1'''} + a_2^{2'''} + ... + a_2^{nc'''}\}) + ... + T_m^s(\{a_n + a_n^{1'''} + a_n^{2'''} + ... + a_n^{nc'''}\}) \\ &= a_1^{'''} + a_2^{'''} + ... + a_n^{'''} \end{aligned} \tag{9}$$

The population after meme selection is denoted as $A^{(8)} = \{a_1^{'''}, a_2^{'''}, ... a_n^{'''}\}$.

If the halting criterion is not meet, perform deterministic replacement operation again. Then repeat the above operations. And else end the evolution.

4 NCSA Main Loop

In order to solve the multimodal function optimization problem, the algorithm is designed as follows.

Step 1. Generate the initial antibody population $A = \{a_1, a_2, ... a_n\}$. Set the termination criterion and other parameters. $k = 1$.

Step 2. Perform clonal proliferation operation on $A(k)$: $A^{(1)}(k) = T_c^P(A(k))$.

Step 3. Perform clonal mutation operation on $A^{(1)}(k)$: $A^{(2)}(k) = T_c^m(A^{(1)}(k))$.

Step 4. if $k = 1$,go to step 5, else go to step 7.

Step 5. Perform clonal selection operation on $A^{(2)}(k)$: $A^{(3)}(k) = T_c^s(A(k) + A^{(2)}(k))$.

Step 6. Perform deterministic replacement operation T^r on $A^{(3)}(k)$: $A(k+1) = T^r(A^{(3)}(k))$. $k = k+1$. Go to step 2.

Step 7. Perform meme selection T_m^s on $A^{(2)}(k)$: $A^{(3)}(k) = T_m^s(A(k) + A^{(2)}(k))$

Step 8. If the termination criterion is satisfied, stop the algorithm. Otherwise, perform deterministic replacement operation T^r on $A^{(3)}(k)$: $A(k+1) = T^r(A^{(3)}(k))$. $k = k+1$. Go to step 2.

Additionally, there is something to be explained:

(1) The algorithm adopts real-number encoding.

(2) We design the clonal mutation as a bit mutation method viz. for the new antibody $a_i^{(1)}(k)$ in the antibody population $A^{(1)}(k)$, replacing its certain numbers by a random integer between 0 and 9 with probability p_m.

(3) The termination criterion is defined as the maximum number of iterations.

5 Numerical Experiment

In order to compare with the results presented in reference [8], the very same test functions are used, and the halting condition is also uniform.

5.1 Test Function

Experiments are conducted in four well known one-dimensional multimodal functions [8], see figure 1. F1 is a function with five peaks of the same height and width. These peaks are uniformly distributed in the [0, 1] interval (0.1, 0.3, 0.5, 0.7, 0.9). F2 is a function with five peaks of the same width hut different height. These peaks are uniformly distributed in the [0, 1] interval (0.1, 0.3, 0.5, 0.7, 0.9). F3 is a function with five peaks of the same height hut different width. These peaks are non-uniformly distributed in the [0, 1] interval (0.08, 0.25, 0.45, 0.68, 0.93). F4 is a function with five peaks of different height and width. These peaks are non-uniformly distributed in the [0, 1] interval (0.08, 0.25, 0.45, 0.68, 0.93).

5.2 Parameter Settings

In the paper, we use real-valued representation to carry out the proposed algorithm. It is supposed that when the iterative generations ngen=100, the algorithm will be halted. For each test function, a population size of 5 is used and other parameters are set as follows, niche radius (σ) is 0.1, clonal scale (n_c) is 15 and mutation probability (p_m) is 0.8.

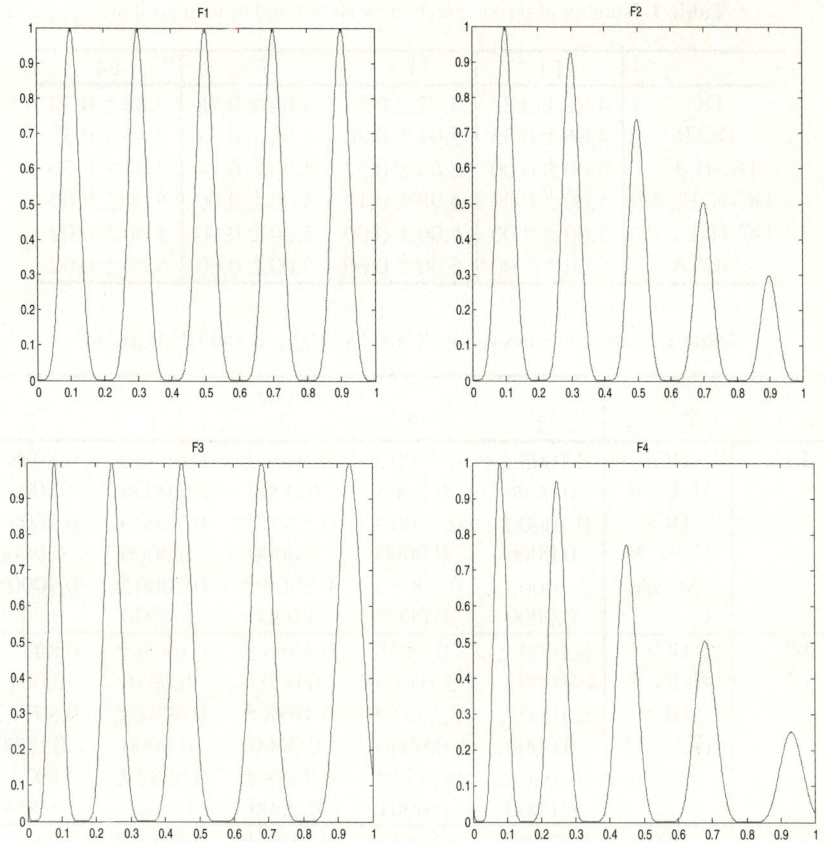

Fig. 1. Multimodal Functions

5.3 Results

Table 1 shows the number of peaks (niches) searched by NCSA and some other algorithms given in reference [8] on the tested functions. Where DC is a niching strategy namely deterministic crowding. DC-R means DC with a maximum distance of 0.1 at the phenotype level for accepting the replacement. H_AE_A means Hybrid Adaptive Evolutionary Algorithm. M means mating restriction scheme.

A value $a \pm b$ indicates that an average of a peaks was discovered, with standard deviation of b .

From the compared results given in table 1, we can see that in 50 independent runs, NCSA and DC-H_AE_A-R found 5 peaks for each function in all the 50 runs while DC-H_AE_A-M found 5 peaks in 49 different runs for F2. And the performance of other algorithms was inferior to the above three algorithms.

Table 1. Number of peaks searched by NCSA and other algorithms

	F1	F2	F3	F4
DC	4.92 ± 0.27	3.02 ± 0.14	4.64 ± 0.48	3.88 ± 0.32
DC-R	4.86 ± 0.35	3.04 ± 0.20	4.58 ± 0.57	3.86 ± 0.35
DC-H_AE_A	5.00 ± 0.00	2.64 ± 0.52	4.98 ± 0.14	3.04 ± 0.20
DC-H_AE_A-M	5.00 ± 0.00	4.98 ± 0.14	5.00 ± 0.00	5.00 ± 0.00
DC-H_AE_A-R	5.00 ± 0.00	5.00 ± 0.00	5.00 ± 0.00	5.00 ± 0.00
NCSA	5.00 ± 0.00	5.00 ± 0.00	5.00 ± 0.00	5.00 ± 0.00

Table 2. Searched niches using NCSA, DC-H_AE_A-R and DC-H_AE_A-M

		Niches				
		1	2	3	4	5
F1	DC-H_AE_A-R	0.1003 ± 0.0008	0.3000 ± 0.0005	0.5001 ± 0.0006	0.6998 ± 0.0006	0.8998 ± 0.0008
	DC-H_AE_A-M	0.1000 ± 0.0000	0.3000 ± 0.0000	0.5000 ± 0.0000	0.7000 ± 0.0000	0.9000 ± 0.0000
	NCSA	0.1000 ± 0.0000	0.3000 ± 0.0000	0.5000 ± 0.0000	0.7000 ± 0.0000	0.9000 ± 0.0000
F2	DC-H_AE_A-R	0.1000 ± 0.0001	0.2993 ± 0.0004	0.4986 ± 0.0005	0.6979 ± 0.0016	0.8977 ± 0.0012
	DC-H_AE_A-M	0.1000 ± 0.0000	0.2994 ± 0.0000	0.4988 ± 0.0000	0.6982 ± 0.0000	0.8977 ± 0.0000
	NCSA	0.1000 ± 0.0000	0.2994 ± 0.0001	0.4988 ± 0.0000	0.6982 ± 0.0000	0.8977 ± 0.0000
F3	DC-H_AE_A-R	0.0798 ± 0.0012	0.2468 ± 0.0006	0.4505 ± 0.0006	0.6814 ± 0.0009	0.9336 ± 0.0010
	DC-H_AE_A-M	0.0797 ± 0.0000	0.2466 ± 0.0000	0.4506 ± 0.0000	0.6814 ± 0.0000	0.9339 ± 0.0000
	NCSA	0.0797 ± 0.0000	0.2467 ± 0.0000	0.4506 ± 0.0000	0.6814 ± 0.0000	0.9339 ± 0.0000
F4	DC-H_AE_A-R	0.0797 ± 0.0001	0.2463 ± 0.0000	0.4494 ± 0.0004	0.6789 ± 0.0009	0.9298 ± 0.0014
	DC-H_AE_A-M	0.0797 ± 0.0000	0.2463 ± 0.0000	0.4495 ± 0.0000	0.6792 ± 0.0000	0.9301 ± 0.0000
	NCSA	0.0797 ± 0.0000	0.2463 ± 0.0000	0.4495 ± 0.0000	0.6792 ± 0.0000	0.9302 ± 0.0000

Table 2 shows the average best individual per niche after 100 generations using NCSA, DC-H_AE_A-R and DC-H_AE_A-M, respectively. In general, NCSA is able to find an individual very close to each local optimum. The distance between the searched average solution and the local optimum is less than 0.001 in every case. The performance of NCSA is similar to DC-H_AE_A-M and much better than DC-H_AE_A-R.

6 Conclusion

In this paper, a new algorithm named NCSA was put forward and some heuristic rules including clonal proliferation, clonal selection, meme selection and deterministic replacement policy were introduced. The effectiveness of the algorithm is demonstrated via experiments on a collection of multimodal function optimization problems. Compared with the results in reference [8], we can see that the performance of NCSA is significantly superior to DC, DC-R, DC-H_AE_A which were introduced in reference [8]. There are limitations in both DC-H_AE_A-R and DC-H_AE_A-M while NCSA can outperform them successfully.

The new work is characterized as follows:

(1) It presented a new niching strategy viz. deterministic replacement policy and a new selection method namely meme selection. Based on this condition, a new algorithm called NCSA is proposed.

(2) It simulated the new algorithm on some multimodal function optimization problem and its validity has been proved.

References

1. Du, H. F., Jiao, L. C., Liu, R. C.: Adaptive Polyclonal Programming Algorithm with Applications. Proceedings of the Fifth International Conference on Computational Intelligence and Multimedia Applications (ICCIMA'03), IEEE. (2003) 350-355
2. Castro, D., Zuben, J. V.: Learning and Optimization Using the Clonal Selection Principle. IEEE Transactions On Evolutionary Computation, vol. 6, no.3, June. (2002) 239-251
3. Du H.F., Jiao L.C., Gong M.G., Liu R.C.: Adaptive Dynamic Clone Selection Algorithms. In: Zdzislaw, P., Lotfi, Z. (eds.): Proceedings of the Fourth International Conference on Rough Sets and Current Trends in Computing. Uppsala, Sweden (2004) 768-773
4. Goldberg, D. E., Richardson, J.: Genetic algorithms with sharing for multimodal function optimization. Proc. 2nd ICGA. (1987) 41-49
5. Beasley, D., Bull, D. R., Martin, R. R.: A sequential niche technique for multimodal function optimization Evolutionary Computation, vol.1, no. 2. (1993) 101-125
6. Mahfoud, S. W.: Crowding and preselection revisited. In R. Manner & B. Manderick (Eds.). Parallel problem solving from nature, Vol. 2. (1992) 27-36. Elsevier
7. Lee, C. G., Cho, D. H., Jung, H. K.: Niching Genetic Algorithm with Restricted Competition Selection for Multimodal Function Optimization. IEEE Transactions On Magnetics, vol. 35, Issue. 3, May. (1999) 1722-1725
8. Gomez, J.: Self Adaptation of Operator Rates for Multimodal Optimization. CEC2004 Congress on Evolutionary Computation, vol. 2, June (2004) 1720-1726
9. Horn, J., Nafpliotis, N., Goldberg, D. E.: A Niched Pareto Genetic Algorithm for Multiobjective Optimization. Evolutionary Computation, 1994. IEEE World Congress on Computational Intelligence. Proceedings of the First IEEE Conference, vol.1, June. (1994) 82-87
10. Miller, B. L., Shaw, M. J.: Genetic Algorithms with Dynamic Niche Sharing for Multimodal Function Optimization. Proceedings of IEEE International Conference on Evolutionary Computation, May. (1996) 786-791
11. Laplante, P., Flaxman, H.: The Convergence of Technology and Creativity in the Corporate Environment. IEEE Transactions on Professional Communication, vol. 38, no. 1. (1995) 20-23
12. Richard Dawkins. The Selfish Gene. 1976

A New Macroevolutionary Algorithm for Constrained Optimization Problems*

Jihui Zhang[1] and Junqin Xu[2]

[1] School of Automation Engineering, Qingdao University, Qingdao 266071, China
[2] School of Mathematics Sciences, Qingdao University, Qingdao 266071, China

Abstract. Macroevolutionary algorithm (MA) is a new approach to optimization problems based on extinction patterns in macroevolution. It is different from the traditional population-level evolutionary algorithms such as genetic algorithms. In this paper, a new macroevolutionary algorithm based on uniform design is presented for solving nonlinear constrained optimization problems. Constraints are handled by embodying them in an augmented Lagrangian function, where the penalty parameters and multipliers are adapted as the execution of the algorithm proceeds. The efficiency of the proposed methodology is illustrated by solving numerous constrained optimization problems that can be found in the literature.

1 Introduction

Many problems in several scienti=c areas are formulated as nonlinear programming problems. In most cases they consist not only of a nonlinear objective function that has to be optimized, but of a number of linear and/or nonlinear constraints as well that must be satisfied by the solution. Due to the complex nature of many of these problems, conventional optimization algorithms are often unable to provide even a feasible solution. For example, gradient optimization techniques have only been able to tackle special formulations, where continuity and convexity must be imposed. Obviously, the development of efficient algorithms for handling complex nonlinear optimization problems is of great importance. In this work we present a new framework for solving such problems that belongs to the family of stochastic search algorithms, known as evolutionary algorithms.

Evolutionary algorithms have many advantages compared to the traditional nonlinear programming techniques, among which the following three are the most important:

(i) They can be applied to problems that consist of discontinuous, non-differentiable and non-convex objective functions and/or constraints.
(ii) They do not require the computation of the gradients of the cost function and the constraints.
(iii) They can easily escape from local optima.

* Supported by SRF for ROCS, SEM; Taishan Scholarship program of Shandong Province.

L. Jiao et al. (Eds.): ICNC 2006, Part I, LNCS 4221, pp. 828–837, 2006.

Nonetheless, until very recently evolutionary algorithms have not been widely accepted, due to their poor performance in handling constraints. Evolutionary algorithms for constrained optimization problems can be classified into the following four categories:

(i) Methods based on preserving feasibility of solutions.
(ii) Methods based on penalty functions.
(iii) Methods based on a search for feasible solutions.
(iv) Other hybrid methods.

Among the evolutionary algorithms the methods based on penalty functions have proven to be the most popular. These methods augment the cost function, so that it includes the squared or absolute values of the constraint violations multiplied by penalty coefficients. However, penalty function methods are also characterized by serious drawbacks, since small values of the penalty coefficients drive the search outside the feasible region and often produce infeasible solutions, while imposing very severe penalties makes it difficult to drive the population to the optimum. The above observations drove the research towards methods that are able to adapt the penalty parameters as the algorithm proceeds.

This paper mainly considers the following form of constrained optimization problems:

$$\textbf{(P)} \quad \min_x f(x), \qquad x \in R^n \tag{1}$$

subject to

$$g_i(x) \leq 0, i = 1, \cdots, m \tag{2}$$

$$h_j(x) = 0, \ j = 1, \cdots, \ell \tag{3}$$

$$L_i \leq x_i \leq U_i, \ i = 1, \cdots, n \tag{4}$$

As an evolutionary algorithm, MAs have some of their own shortcomings, too. For example, premature convergence is one common problem in almost all the evolutionary algorithms. Another problem much cared is the convergence speed of the stochastic search methods. Many solution methods are presented to improve the performance of this kind of algorithms. The key problem is to keep a proper balance between "exploration" and "exploitation". "Exploration" is concerned with the ability to search new region and find good solutions, while "exploitation" is concerned with the convergence speed. In fact, much effort is needed to implement this kind of difficult balance in practice.

The rest of the paper is organized as follows: section 2 gives a brief introduction to macroevolutionary algorithm. The main idea of uniform design is followed in section 3. Our method for constrained optimization problem is presented in section 4 in detail. Section 5 presents some numerical examples and experimental analysis. Finally are the conclusions and further research directions.

2 Macroevolutionary Algorithm

Macroevolutionary algorithm (MA) is first proposed by Jesús and Ricard [2]. The biological model of macroevolution (MM) is a network ecosystem where the dynamics are based only on the relation between species. The links between units/species are essential to determine the new state (alive or extinct) of each species at each generation. The state of species i at generation t is defined as

$$S_i(t) = \begin{cases} 1, \text{ if state is "alive"} \\ 0, \text{ if state is "extinct".} \end{cases} \quad (5)$$

In this model, time is discretized in "generations" and that each generation constitutes a set of P species where P is constant. The relationship between species is represented by a connectivity matrix W, where each entry $w_{i,j}(t)(i, j \in \{1, 2, \cdots, P\})$ of the matrix W measures the influence of species j on species i at t with a continuous value within the interval $[-1, 1]$ (in ecology, this influence is interpreted as the trophic relation between species). At the end of each generation, all extinct species are replaced by the existing species. Briefly, each generation in the biological model consists of a set of steps (the rules) that will be translated to the MA model.

1) *Random variation*: For each species i, a connection $w_{i,j}(t)$ is chosen randomly, and a new random value between -1 and 1 is assigned.
2) *Extinction*: The relation of each species to the rest of the population determines its survival coefficient h defined as

$$h_i(t) = \sum_{j=1}^{P} w_{i,j}(t) \quad (6)$$

where t is the generation number. The species state in the next generation is updated synchronously as

$$S_i(t+1) = \begin{cases} 1, \text{ (alive) if } h_i(t) \geq 0 \\ 0, \text{ (extinct) otherwise.} \end{cases} \quad (7)$$

This step allows for the selection and extinction of species.
3) *Diversification*: Vacant sites freed by extinct species are colonized with surviving species. Specifically, a colonizer c will be randomly chosen from the set of survivors. For all vacant sites (i.e., those such that $s_k(t) = 0$) the new connections will be updated as

$$\begin{aligned} w_{k,j} &= w_{c,j} + \eta_{k,j}, \\ w_{j,k} &= w_{j,c} + \eta_{j,k}. \end{aligned} \quad (8)$$

where is a small random variation and $s_k(t+1) = 1$.

The main idea of MA is that the system will choose, through network interactions, which are the individuals to be eliminated so as to guarantee exploration by new individuals and exploitation of better solutions by further generations. To this purpose, it is essential to correctly establish a relationship between individuals. This is described by the following criteria.

c1) Each individual gathers information about the rest of the population through the strength and sign of its couplings $w_{i,j}$. Individuals with higher inputs h_i will be favored. Additionally, they must be able to out-compete other less-fit solutions.

c2) Some information concerning how close two solutions are in Ω is required (although this particular aspect is not strictly necessary). Close neighbors will typically share similar f-values and will cooperate. In this context, the connection $w_{i,j}$ is defined as

$$w_{i,j} = \frac{f(X_i) - f(X_j)}{\| X_i - X_j \|} \tag{9}$$

where $X_i = (x_i^1, x_i^2, \cdots, x_i^n)$ is the ith individual.

The main ingredients of MA include:

1) *Selection operator*: It allows calculating the surviving individuals through their relations, i.e., as a sum of penalties and benefits. The state of a given individual S_i will be given by

$$S_i(t+1) = \begin{cases} 1, \text{ if } \sum_{j=1}^{P} w_{i,j}(t) \leq 0 \\ 0, \text{ otherwise} \end{cases} \tag{10}$$

where t is generation number and $w_{i,j} = w(X_i, X_j)$ is calculated according to (9).

2) *Colonization operator*: It allows filling vacant sites that are freed by extinct individuals (that is, those such that $S_i = 0$). This operator is applied to each extinct individual in two ways. With a probability τ a totally new solution $X' \in \Omega$ will be generated. Otherwise exploitation of surviving solutions takes place through colonization. For a given extinct solution X_i, one of the surviving solutions, say X_b. Now the extinct solution will be "attracted" toward X_b. A possible (but not unique) choice for this colonization of extinct solutions can be expressed as

$$X_i(t+1) = \begin{cases} X_b(t) + \rho\lambda(X_b(t) - X_i(t)), & \text{if } \xi > \tau \\ X_i', & \text{if } \xi \leq \tau \end{cases} \tag{11}$$

where $\xi \in [0, 1]$ is a random number, $\lambda \in [-1, 1]$ (both with uniform distribution) and ρ and τ are given constants of the algorithm. It can be seen that ρ describes a maximum radius around surviving solutions and τ acts as a "temperature". Parameter τ can be set as that in simulated annealing. For example, τ can take the following forms

$$\tau(t, G) = 1 - \frac{t}{G} \tag{12}$$

$$\tau(t) \propto \exp(-\Gamma t). \tag{13}$$

Comparison [2] between the performance of MA's and that of genetic algorithm with tournament selection shows that MA is a good alternative to standard GA's, showing a fast monotonous search over the solution space even for very small population sizes. A mean field theoretical analysis also shows that symmetry-breaking (i.e., the choice among one of the two equal peaks) typically occurs because small fluctuations and the presence of random solutions eventually shifts the system toward one of the two peaks.

3 Uniform Design

Uniform design is one important experimental design technique which was proposed by K.T. Fang and Y. Wang [8,10] in 1981. Its main objective is to sample a small set of points from a given set of points such that the sampled points are uniformly scattered. Suppose there are n factors and q levels per factor. When n and q are given, the uniform design selects q combinations out of q^n possible combinations, such that these q combinations are scattered uniformly over the space of all possible combinations. The selected q combinations are expressed in terms of a *uniform array* $U(n,q) = (U_{i,j})_{q\times n}$, where $U_{i,j}$ is the level of the jth factor in the ith combination. If q is a prime and $q > n$, it was shown that $U_{i,j}$ is given by

$$U_{i,j} = (i\sigma^{j-1} \mod q) + 1, \tag{14}$$

where σ is a parameter which is different for different experiment [5,9].

4 The New Macroevolutionary Algorithm

First, unconstrained optimization problem is considered.

4.1 Generation of the Initial Population

The initial population is selected using uniform design technique such that the individuals are evenly distributed on the whole search space. It is necessary to divide the search space into smaller subspaces if the search space is very large. After this quantization, there are Q_0^n points for sampling in this subspace. In the same manner, Q_0 points are selected from each of the other subspaces, then totaly sQ_0 points are selected. From each one of s groups, the best $\lfloor N/s \rfloor$ candidates are selected to compose the initial population. If $s\lfloor N/s \rfloor < N$, other $N - s\lfloor N/s \rfloor$ candidates can be produced randomly in the search space.

4.2 Selection and Extinction Process

The equation (7) is changed to

$$S_i(t+1) = \begin{cases} 1, \text{(alive) if } h_i(t) \geq 0 \\ 1, \text{(alive) if } h_i(t) \geq 0 \ \& \ \eta' > \tau' \\ 0, \text{(extinct) otherwise.} \end{cases} \tag{15}$$

Where, the meaning of η' and τ' is the same with that in equation (20). After selection process, $N-\ell$ new species are generated, so the remaining $N-\ell$ species will be generated by the following colonization process (here N is the population size, while ℓ is the extinct specie number).

Colonization Process. This process allows filling vacant sites that are freed by extinct individuals. The remaining $N-\ell$ individuals will be generated by two ways: with a probability $\tau \in (0,1)$, a totally new solution is generated. Otherwise, it is generated based on the current search knowledge. It can be expressed by the following formulae:

$$x_i^j(t+1) = \begin{cases} x_b^j(t) + \eta \cdot \Delta x_i^j(t) + \alpha \cdot sx_i^j(t) & \text{if } \xi > \tau \\ x_{new}^j & \text{otherwise} \end{cases} \tag{16}$$

for $j = 1, 2, \cdots, n$. $x_i^j(t)$ is the jth entry of X_i at generation t. t is the current generation number. ξ is a random variable distributed uniformly on $[0,1]$. $X_b(t)$ is the best individuals in generation t. $sx_i^j(t)$ and $\Delta X_i^j(t)$ have the following forms:

$$\Delta x_i^j(t) = (x_b^j(t) - x_i^j(t)) \cdot |\mathcal{N}(0,1)| \tag{17}$$

$$sx_i^j(t+1) = \eta \cdot acc^i(t) \cdot \Delta x_i^j(t) + \alpha \cdot sx_i^j(t) \tag{18}$$

where $x_i^j(k)$ is the jth variable of an ith individual at the tth generation. η and α are *learning rate* and *momentum rate* respectively. $\mathcal{N}(0,1)$ is standard Gaussian random variable. $\Delta x_i^j(t)$ is the amount of change in an individual, which is proportional to the temporal error, and it drives the individual to evolve close to the best individual at the next generation. It can be viewed as a tendency of the other individuals to take after or emulate the best individual in the current generation. $sx_i^j(t)$ is the evolution tendency or momentum of previous evolution. It accumulates evolution information and tends to accelerate convergence when the evolution trajectory is moving in a consistent direction. $acc^i(t)$ is defined as follows:

$$acc^i(t) = \begin{cases} 1, \text{ if the current update has improved cost,} \\ 0, \text{ otherwise.} \end{cases} \tag{19}$$

τ in equation (16) is defined as follows

$$\tau(t, G) = 1 - \frac{t}{G} \tag{20}$$

where G is the recycling generations. New individuals X_{new} in equation (16) is generated using uniform design technique. The implementation details are similar to that in section 4.1, but each time the sampling subspace is different such that different sampling points will be trailed.

4.3 Local Search

The idea is to line up the solutions in current generation in a descending order according to the corresponding values of the objective function. The crossover and mutation operations are then applied to the solutions as follows: During the crossover operation, for each adjacent pair the difference vector is computed, weighted by a random number between 0 and 1 and added on the first vector of the pair. The produced solution replaces the first vector, if it produces an objective function value that is lower than the fitness value of the second vector. At the end of the crossover operation, the solutions are lined up again. Then the mutation operation is applied, taking into account that the worse members of the population should be altered substantially, while only small changes should be made to the best solutions. In order to achieve this, a different probability of mutation is calculated for each solution in the list, which is reduced from the top to the bottom. This probability defines the number of variables in each solution that will undergo the mutation operation. The nonuniform mutation is utilized, since it adds to the algorithm more local search capabilities. Using this approach, in the first iterations the variables which are mutated can be located anywhere in the input space. As the algorithm proceeds, more conservative moves are preferred and thus, search is located on a more local level. Let *maxiter* be the maximum number of iterations and *iter* be the number of current iterations. Its details can be described as the flowing algorithm.

Algorithm 1. Local search

(1) Compute the objective function value corresponding to each solution $f(X_i)$, $i = 1, \cdots, N$.
(2) Arrange the solutions so that they formulate a line in a descending order: $X_1, \cdots, X_N$ where X_i precedes X_j if $f(X_i) > f(X_j)$, $i, j = 1, \cdots, N$.
(3) Apply the crossover operator:
FOR $i = 1, \cdots, N-1$
$X_{i,\text{new}} = X_i + r \cdot (X_{i+1} - X_i)$, where r is a random number between 0 and 1, if $f(X_{i,\text{New}}) < f(X_{i+1})$ then $X_i = X_{i,\text{new}}$
END
(4) Arrange the solutions so that they formulate a line in a descending order: $X_1, \cdots, X_N$ where X_i precedes X_j if $f(X_i) > f(X_j)$, $i, j = 1, \cdots, N$.
(5) Apply the nonuniform mutation operation:
FOR $i = 1, \cdots, N$
$p_{m,i} = \frac{N-i+1}{N}$
FOR $j = 1, \cdots, N$
Generate a random number r between 0 and 1
IF $b = 0$ **THEN** $X_{i,\text{new}}(j) = X_i(j) + (U(j) - X_i(j)) \cdot r \cdot \exp(-\frac{2*iter}{maxiter}) \cdot \text{sig}(\frac{maxiter}{2} - iter)$
IF $b = 1$ **THEN** $X_{i,\text{new}}(j) = X_i(j) - (X_i(j) - L(j)) \cdot r \cdot \exp(-\frac{2*iter}{maxiter}) \cdot \text{sig}(\frac{maxiter}{2} - iter)$
END
END
IF $f(x_i, \text{new}) < f(X_i)$ **THEN** $X_i = X_{i,\text{new}}$
END
Where sig( ) is the Sigmoid function which defined as $\text{sig}(x) = \frac{1}{1+\exp(-x)}$.

4.4 Termination Condition

In this paper, a fixed generation number is used as the termination condition of the procedure.

Now we return back to constrained optimization problem 1. Constraints are handled by the following augmented Lagrange function

$$F(X,a,b,p,q) = f(X) + \sum_{j=1}^{\ell} a_j[(h_j(X)+p_j)^2 - p_j^2] + \sum_{i=1}^{m} b_i[\langle g_i(X)+q_i\rangle_+^2 - q_i^2], \tag{21}$$

where $\langle g_i(X)\rangle_+ = max\{0, g_k(X)\}$. Our method for constrained optimization problems can be described by the following algorithm.

Algorithm 2. (New Macroevolutionary algorithm for constrained optimization problems)

Initialization process

(1) Select the maximum number of iterations $maxjter$.
(2) Assign values greater than 1 to the scalar factors w_1 and w_2.
(3) Assign a very large positive number to the maximum constraint violation cv.
(4) Initialize the penalty parameters by the following procedure:
  (4.1) assign very small values to the penalty parameters;
  (4.2) set the Lagrange multipliers equal to 0;
  (4.3) solve the unconstrained optimization problem defined by (21);
  (4.4) assign the corresponding value of $\boldsymbol{a}$ and $\boldsymbol{b}$ of minimum of (21) to the penalty parameters.
(5) Set the number of current iterations $jter = 0$.

Iterative Procedure

(1) Increase the number of iterations by 1, $jter = jter + 1$.
(2) Use Algorithm 1 to solve the formulated unconstrained optimization problem.
(3)Determine the maximum constraint violation as follows

$$cv_{\max} = \max\{\max_j |h_j|, \max_i \langle g_i\rangle\} \tag{22}$$

and identify the equality and inequality constrains, for which violation is not improved by the factor w_1:

$$S_E = \{j : |h_j| > cv/w_1\} \text{ and } S_I = \{i : \langle g_i\rangle_+ > cv/w_1\}.$$

(4) If $cv_{\max} \geq cv$ then

$$\text{Let } a_j = a_j w_2, \quad p_j = p_j/w_2 \text{ for all } j \in S_E, \tag{23}$$

$$\text{Let } b_i = b_i w_2, \quad q_i = q_i/w_2 \text{ for all } i \in S_I. \tag{24}$$

(5) If $cv_{\max} < cv$ then

Set $cv = cv_{\max}$, $p_j = h_j(X_b) + p_j$, $q_i = \langle g_i(X_b) + q_i \rangle_+$, for all $j \in S_E, i \in S_I$, (25)

where x_b is the vector that has produced the best fitness function value so far.

(6) If $cv_{\max} \leq cv/w_1$,

$$\text{Let } a_j = a_j/w_2, \quad p_j = p_j w_2 \text{ for } j = 1, \cdots, \ell. \tag{26}$$

$$\text{Let } b_i = b_i/w_2, \quad q_i = q_i w_2 \text{ for } i = 1, \cdots, m. \tag{27}$$

(7) If $jter = maxjter$ stop the algorithm. Otherwise turn the algorithm to step 1.

5 Numerical Experiments

In this section, the benchmark problems, G1, G4-G7, G9-G10, are used to test the search performance of our algorithm. All the benchmark examples are taken from [11].

For each benchmark problem, simulation is repeated for 20 times. The best solution and the constraint violation in each simulation are recorded. Their average performance and deviation in the 20 simulations are computed and used to evaluate the performance of our algorithm.

For all test problems, our algorithm has a good performance in terms of both search speed and solution accuracy. For problem G1, a very good solution is obtained after 200 generation using our algorithm, but in [12], it is obtained after about 1500 generations. For other test problems, the solution accuracy is improved significantly, too. The detailed statistics are given in Tables 1. Moreover, our algorithm has a very stable performance, this is meaningful for practical use.

Table 1. Experimental results over 20 runs

Problem	*G1*	*G4*	*G5*	*G6*	*G7*	*G9*	*G10*
optimal	−15.0000	−30616.1046	5113.5460	−6766.0042	24.7819	681.7033	7034.5683
average	−14.9602	−30671.9115	5147.0216	−6984.7735	25.2267	684.3867	7066.6624
Deviation	0.0004	8.3671	11.8431	12.6608	1.3714	5.1912	9.7067
Convergence	52	65	55	68	54	96	73

6 Conclusions

Macroevolutionary algorithm is a new approach to optimization problems based on extinction patterns in macroevolution. It is different from the traditional population-level evolutionary algorithms. In this paper, a new version of MA is proposed to solve complicated constrained optimization problems. Numerical simulation results show the power of this new algorithm. Solving multi-objective optimization problems using MA should be our further research topics.

References

1. Reynolds, R.G. and Zhu, S.: Knowledge-based function optimization using fuzzy cultural algorithms with evolutionary programming. IEEE Trans. System, Man and Cybernetics. **31**(2001) 1–18
2. Marín, J. and Solé, R.V.: Macroevolutionary algorithms: a new optimization method on fitness landscapes. IEEE Trans. on Evolutionary Computation. **3**(1999) 272–286
3. Montgomery, D.C.: Design and Analysis of Experiments. 3rd Edition. Wiley, New York (1991)
4. Hicks, C.R.: Fundamental Concepts in the Design of Experiments. 4th Edition. Sauders, New York (1993)
5. Winker, P. and Fang, K.: Application of threshold accepting to the evaluation of the discrepancy of a set of points. SIAM J. Numer. Anal. **34** (1998) 2038–2042
6. Zhang, J. and Xu, X.: An efficient evolutionary algorithm. Computers & Operations Research **26**(1999) 645-663
7. Wang, Y. and Fang, K.: A note on uniform distribution and experimental design. KEXUE TONGBAO **26** (1981) 485–489
8. Fang, K. and Li, J.: Some New Uniform Designs, Hong Kong Baptist Univ., Hong Kong, Tech. Rep. Math-042 (1994)
9. Fang, K. and Wang, Y.: Number-Theoretic Methods in Statistics. Chapman & Hall, London, U.K. (1994)
10. Fang, K.: Uniform Design and Design Tables. Science, Beijing, China (1994)
11. Michalewicz, Z., G. Nazhiyath, and M. Michalewicz (1996). "A note on usefulness of geometrical crossover for numerical optimization problems". In P. J. Angeline and T. Bäck (Eds.), *Proceedings of the 5th Annual Conference on Evolutionary Programming*.
12. M. Tahk and B. Sun. "Coevolutionary augmented Lagrangian methods for constrained optimization". *IEEE Trans. Evolutionary Computation*, 2000, 4(2), 114–124.

Clonal Selection Algorithm with Search Space Expansion Scheme for Global Function Optimization

Yifei Sun, Maoguo Gong, Lin Hao, and Licheng Jiao

Institute of Intelligent Information Processing, Xidian University,
P.O. Box 224, Xi'an, 710071, P.R. china
yifeis@hotmail.com, maoguo_gong@hotmail.com

Abstract. Unlike evaluation strategy (ES) and evaluation programming (EP), clonal selection algorithm (CSA) strongly depends on the given search space for the optimal solution problem. The interval of existing optimal solution is unknown in most practical problem, then the suitable search space can not be given and the performance of CSA are influence greatly. In this study, a self-adaptive search space expansion scheme and the clonal selection algorithm are integrated to form a new algorithm, Self Adaptive Clonal Selection Algorithm, termed as SACSA. It is proved that SACSA converges to global optimum with probability 1.Qualitative analyzes and experiments show that, compared with the standard genetic algorithm using the same search space expansion scheme, SACSA has a better performance in many aspects including the convergence speed, the solution precision and the stability. Then, we study more about the new algorithm on optimizing the time-variable function. SACSA has been confirmed that it is competent for solving global function optimization problems which the initial search space is unknown.

1 Introduction

Clonal Selection Algorithms (CSAs) are stochastic global optimization method inspired by the biological mechanisms of evolution and heredity. It is related to Darwin's theory of natural selection but applied to the cell populations within the immune system. CSAs have been put forward by De Castro [12], Jiao [7] et al. respectively, and have been verified that they have a great deal of useful mechanisms from the viewpoint of programming, controlling, information processing and so on [8-12].

This paper aims at solving the global optimization problems that do not know the suitable bound of variables. Although CSAs have many merits and big potential, they still strongly depend on the given search space for the optimal solution problem. The interval of existing optimal solution is unknown in most practical problem, then the suitable search space can not be given and the performance of CSAs are influence greatly. Therefore, many scientists think the Evolutionary Strategy (ES) and the Evolutionary Programming (EP) are better choices when there is no relevant information about the search space in that ES and EP can self adaptive adjusts the search interval with the parameter. But these ways are very complex [3].

In order to solve this problem, based on algorithm named Search Space Expansion Scheme (S^2ES) in reference [2], Zhong et al. [1] proposed another new algorithm,

L. Jiao et al. (Eds.): ICNC 2006, Part I, LNCS 4221, pp. 838 – 847, 2006.

named Self-Adaptive Coarse-to-Fine Search Space Expansion Scheme (SA-S^{2}ES). Inspired by these, a novel algorithm, SACSA is proposed by integrating the new scheme and clonal selection algorithm. It is proved that SACSA converges to global optimum with probability1.Qualitative analyzes and experiments show that, compared with the corresponding genetic algorithm, SACSA has a better performance in many aspects and it has been confirmed SACSA is competent for solving global numerical optimization.

2 Self-Adaptive Coarse-to-Fine Search Space Expansion Scheme

SA-S^{2}ES can adjust the search space adaptively. The initial search space is specified as $\left[l^t, u^t\right] = \left\{\left[l_k^t, u_k^t\right], k = 1, 2, \cdots n\right\}$, t is the search interval changed generation. $\left[l^t, u^t\right]$ is updated through Coarse Step and Fine Step. First the initial search space must be satisfied l_k^0<0 and u_k^0>0, then the method that suit to any initial search space is proposed.

Coarse Step: Check every parameter of the optimal antibody x^* in current population. If $u_k^t > x_k^* > \frac{u_k^t}{2} > 0$, let $u_k^{t-1} = u_k^t$ and $u_k^t = 2x_k^*$. On the contrary, if $x^* < 0$, update the lower bound by the similar way. A coarser search interval including overall situation optimal solution can be found quickly. Once the search space of same parameter change, generate N_p antibodies from the new space and calculate their affinities. Then select the better antibodies whose affinity is larger than the others from the new N_p individuals and the olds. If the search intervals of all parameters don't expand, go to the fine step.

Fine step: Check x^* in current population. If $x^* > 0$, renew the upper bound as follows:

$$u_k^t = u_k^{t-1}, u_k^{t-1} = 0.5u_k^{t-1} \quad \text{where } x_k^* < u_k^{t-1} \tag{1}$$

$$u_k^t = 0.5(u_k^{t-1} + u_k^t) \quad \text{where } u_k^{t-1} \le x_k^* < 0.75u_k^{t-1} + 0.25u_k^t \tag{2}$$

$$u_k^{t-1} = 0.5(u_k^{t-1} + u_k^t), u_k^t = 1.2x_k^* \quad \text{where } x_k^* > 0.25u_k^{t-1} + 0.75u_k^t \tag{3}$$

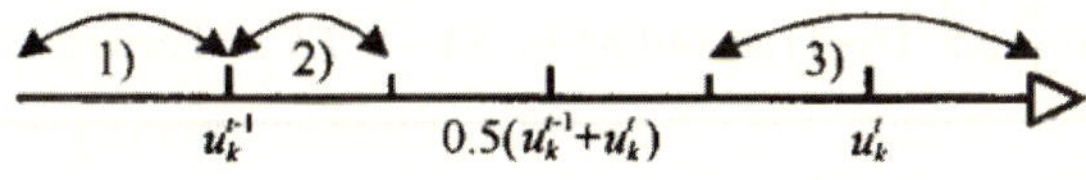

Fig. 1. The three cases of the upper bound that need to update

On above, the precision of SACSA is obvious better than SASGA's for Fn1 and Fn6; and the standard deviation is much smaller than it. This shows that the stability of SACSA is superior to SASGA. Meanwhile, the abilities of the global search and the characteristic of avoiding prematurity have been validated. For Fn2, the probability of trapping into the local ones is much larger which show that SASGA is easier trapped than SACSA. For Fn3, Fn4, Fn5, Fn7, SACSA's performance is better than SASGA's in both the mean number of function evaluations and the average solution precision.

The above results are more clearly illustrated by the Figure 3.The real line is represented the performance of SACSA, and the other is meant SASGA.

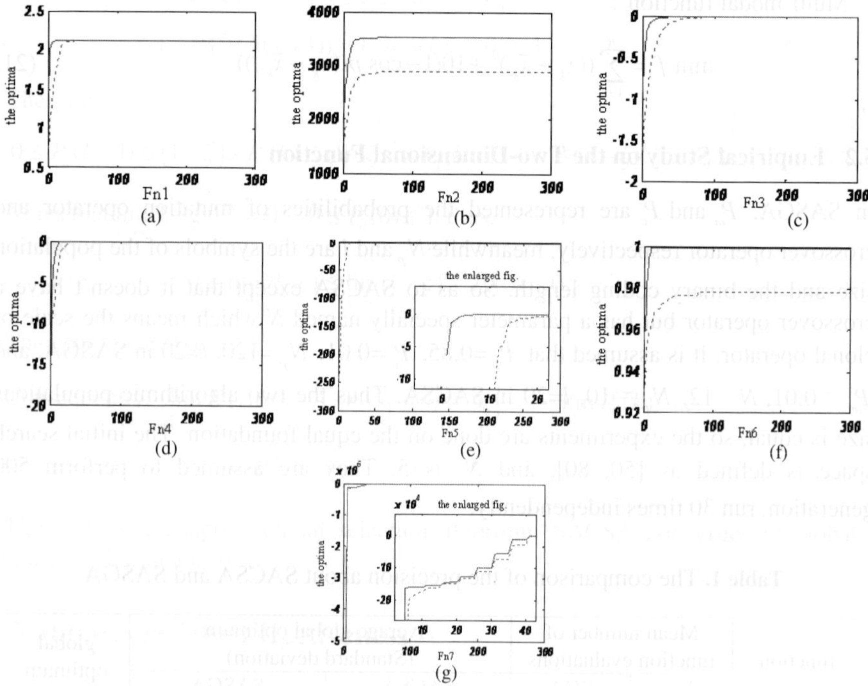

Fig. 3. The average evolution curve of each functions (from Fn1 to Fn7) optima are shown respectively in (a), (b), (c) till (g). The horizontal axis is the evolution generation, and the vertical axis is the average optimum of the evaluation function out of 30 runs.

In order to test the new algorithm adequately, we specify the precision to test that how many generations they need respectively. It is assumed the max generation is 5000. The main parameters are same as the above. Especially, ε is the symbol of solution precision.

As shown above, SACSA's performance is much better than that of SASGA. And SACSA has a strong ability of avoiding the local optima; meanwhile the capability of evolving to the global optima is superior to SASGA. On the other side, the stabilization of SACSA is also excelled to SASGA as the standard deviation indicated distinctly.

Table 2. The test of the convergence rate

function	ε	The generation of the search need				The times of getting optimum	
		SACSA		SASGA		SACSA	SASGA
		mean	std	mean	std		
Fn1	5^{-1}	113	153.8	1032	1687.61	30	26
Fn2	5^{-1}	523	140.4	4310	1186.57	30	14
Fn3	10^{-1}	21	8.375	1126	1778.15	30	27
Fn4	10^{-1}	45	31.99	1288	1777.89	30	29
Fn5	10^{-1}	27	33.28	4102	1679.83	30	10
Fn6	10^{-1}	28	11.99	1802	2302.48	30	21
Fn7	10^{-1}	20	26.89	1540	2303.60	30	21

5.3 The Test of the Time-Variable Function in SACSA

Objective function changing with time varying is challenge of the optimization problems. Here we use this kind of function to test the performance of SACSA.

In practical engineering optimization the global function is few, and Fn3 is not global function, we rebuild the function to a global one. Let x changed with the optima's change. In Fn3, $\overline{x}$ changed every 500 generations and the value is [0, 0], [4, 4], [8, 8] in turn.

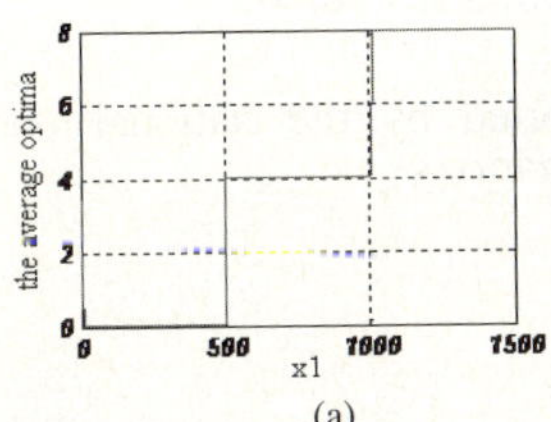

(a)

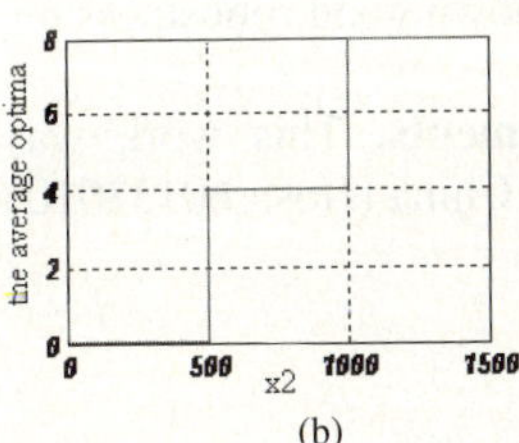

(b)

Fig. 4. The evolution processes of the variable x_1 and x_2 in the time-variable Fn3 are shown respectively in (a) and (b). The horizontal axis is the generation of the evolution, and the vertical axis is the average optimum of the evaluation function out of 30 runs.

In the same way, Fn7 is also an objective time-variable function. Due to Fn7 is a global function, the rebuild operator is not needed. Let $\overline{x}$ changed every 200 generation, and the optima is [800, 1000], [1600, 1800], [1000, 1200] in turn. The part enlarged diagrams are made to be observed better.

The figure below illustrates that the SA-S^2*ES* could adapt the dynamic condition commendably. If the optimum $\overline{x}$ is changed, there is a very short iterative approach to confirm the appropriate search space. From the trial upper we know that no matter $\overline{x}$ is expanded or shrunk, the SA-S^2*ES* could find the optima rapidly and accurately. All these are verified the self adaptability of the SACSA.

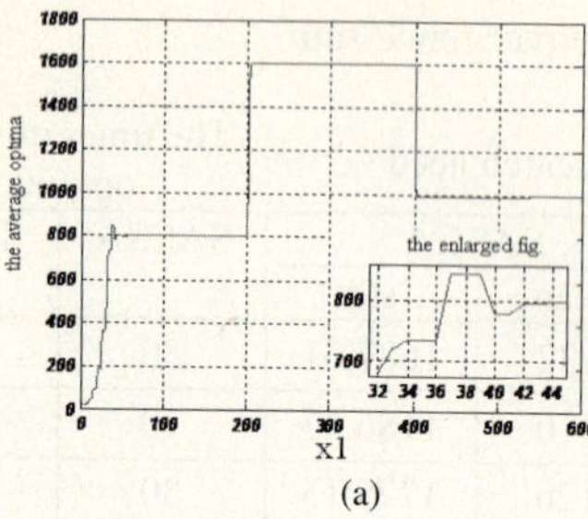

(a)

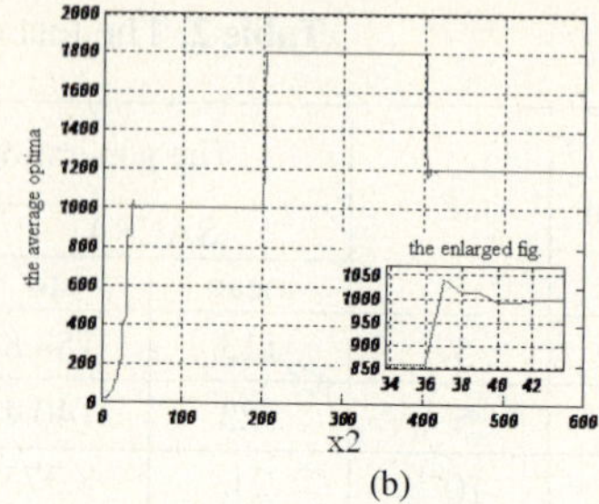

(b)

Fig. 5. The evolution processes of the variable x_1 and x_2 in the time-variable Fn7 are shown respectively in (a) and (b). The horizontal axis is the generation of the evolution, and the vertical axis is the average optimum of the evaluation function out of 30 runs.

6 Concluding Remarks

In this paper, a novel algorithm, Self-Adaptive Clonal Selection Algorithm (SACSA), is proposed. The aim was to apply the search space expansion scheme to enhance the clonal selection algorithm, so that it could be more practicably and statistically sound. In particular, we adopted the Self-Adaptive method to complement the search space expansion scheme, and proved SACSA converges to global optimum with probability1. When compared with SASGA, SACSA is more effective for global numerical optimization problems, time-variable function problem and so on. Besides these, it has shown good robustness on the problems studied.

Acknowledgements. This work was supported by the national Natural Science Foundation of China (Nos.: 60133010 and 60372045).

References

1. Zhong W.C, Liu J., Xue M.Z., and Jiao L.C.: A Multiagent Genetic Algorithm for Global Numerical Optimization [J].IEEE Trans. on Syst., Man, Cybern. A, Vol.34, No.2 (2004) 1136
2. Cheng S., Hwang C.: Optimal Approximation of Linear Systems by a Differential Evolution Algorithm [J].IEEE Trans. on Syst., Man, Cybern. A, Vol.31, No.6 (2001) 698-707
3. Deb K., Beyer H.G.: Self-Adaptive Genetic Algorithm with Simulated Binary Crossover [J]. Evolutionary Computation, Vol.9, No.2 (2001) 137-221
4. Cooper K.D., Hall M.W., Kennedy K.: Procedure cloning[C]. Proceedings of the 1992 International Conference on Computer Languages (1992) 96-105
5. Zhang Z.K., Chen H.C.: Stochastic Process (in Chinese). Xi'an: Xidian University Press (2003) 140-193
6. Swinburne R.: Bayes's Theorem, Oxford: Oxford University Press (2002)
7. Du H.F., Jiao L.C., Liu R.C.: Adaptive Polyclonal Programming Algorithm with Applications. Proceedings of the Fifth International Conference on Computational Intelligence and Multimedia Applications (ICCIMA'03), IEEE (2003)

8. Du H.F., Jiao L.C., Gong M.G., Liu R.C.: Adaptive Dynamic Clone Selection Algorithms. In: Zdzislaw, P., Lotfi, Z. (eds.): Proceedings of the Fourth International Conference on Rough Sets and Current Trends in Computing. Uppsala, Sweden (2004) 768-773
9. Lydyard P. M., Whelan A., Fanger M.W.: Instant notes in immunology [MI. Beijing: BIOS Scientific Publishers Limited (2000)
10. Balazinska M., Merlo E., Dagenais M., et al: Advanced clone-analysis to support object-oriented system refactoring [C]. Proceedings Seventh Working Conference on Reverse Engineering (2000) 98-107
11. Smaili N.E, Sammut C., Shirazi G.M.: Behavioural cloning in control of a dynamic system[C]. IEEE International Conference on Systems, Man and Cybernetics Intelligent Systems for the 21st Century. March (1995) 2904-2909
12. Hybinette M., Fujimoto R.: Cloning: A Novel Method for Interactive Parallel Simulation[C]. Proceedings of the 1997 Winter Simulation Conference (1997) 444- 451

Network Evolution Modeling and Simulation Based on SPD

Chen Yang, Zhao Yong, Xie Hongsheng, and Wu Chuncheng

Institute of Systems Engineering, HuaZhong University of Science and Technology, 430074 Wuhan, China
cchensunny@163.com, zhiwei98530@sohu.com,
hshx_hust@126.com, wuchuncheng@sohu.com

Abstract. The researches on evolution of complex network focused on "how to form", but ignored "why form like this". From the view of cooperation evolution, based on spatial game theory, the individuals are classified into two types, and the micro-mechanism of individual choice is analyzed. At last, the model for social network structure evolving is built. The simulation is implemented by the multi-agent system simulation tool——Repast. By the criterion of degree distribution, clustering coefficient and average shortest path, the result is given. The conclusion shows that the evolving network has obvious small world property, and the cooperation evolution can explain why the real network forms like this to some extent. Another revelatory conclusion is that high social total profit will be gained through the improvement of the network structure even if the cooperators are few.

1 Introduction

The research for complex network grow quickly recently. More and more network model advanced, from regular network to small world network, scale-free network and the improved scale-free network [1]. The researchers pay much attention to the problem how the real network forms in social and information areas. These network models exhibit the characters of real network, but become more and more complicated. But the important thing is that we must not only research the "how to form", but also the problem "why to form like this".

The cooperation phenomena happen around us, it's the foundation of the civilization [2]. People cooperate to work and study in the society, similarly, animals cooperate to prey and live in the biology. The cooperation is an important for the individuals to live and develop, and individuals can benefit from the cooperation. The phenomena give an lively explanation to the problem why the individuals in society or biology contact each other. So in the paper, form the view of cooperation, the reason for the evolution of real network is studied. Nowak is influential in the research of evolution of cooperation; he suggested the famous spatial prisoner's dilemma model (SPD). The SPD model has become a paradigm to study the cooperation. Many researchers studied the evolution of cooperation on various network topologies, for example, regular network and small world network [3,4,5,6,7,8]. These works focus on how the network topologies influence the evolution of cooperation. In above models, individuals

L. Jiao et al. (Eds.): ICNC 2006, Part I, LNCS 4221, pp. 848 – 855, 2006.

just can adjust strategies to improve their profit. Our model gives up the above assumption; the individuals can't changes their strategies, but only can change their ego-networks to improve the profit instead. That is, the network topology can influence the cooperation, constantly; the cooperation can influence the network topology indeed.

2 Model

2.1 Spatial Prisoner's Dilemma

Prisoner's dilemma game (PD) is the basic model in the research of cooperation, and it clearly illustrates the difficulty of maintaining cooperation between individuals. Considering only memoryless strategies: each individual chooses just cooperation (denoted by C) or defection (denoted by D) in each round. When an individual chooses C, it receives payoff R (reward) or S (sucker) as the opponent chooses C or D, respectively. An individual that chooses D receives T (temptation) or P (punishment) as the opponent chooses C or D, respectively. Given T >R>P >S, an individual is always tempted to defect no matter whether the opponent takes C or D. The combination of D and D, with which both of the individuals get unsatisfactory payoff P, is the unique Nash equilibrium in a single game.

A payoff matrix for the PD is subject to two restrictions:

1) $T > R > P > S$

The first restriction lists a preference ranking for the payoffs, ensuring that mutual cooperation is more beneficial than mutual defection.

2) $R > (S+T)/2$

The second restriction dictates that mutual cooperation should be more beneficial than mutual exploitation (defecting when the other player cooperates and vice versa).

Table 1. General form of the PD payoff matrix

payoff		**Individual B**	
		cooper-ate	defect
Indi-vidualA	cooper-ate	R， R	S， T
	defect	T， S	P， P

Spatial prisoner's dilemma: The SPD extend the PD to spatial version. A network $G=(V,E)$ denotes the space that individuals locate. V is the vertex set, the individuals is denoted by a vertex. E is the edges set, that is a sub-set of the set V×V, denoting the connection between the individuals. The number of individuals is $n=|V|$. Assuming k is the average degree of the network, then the number of the edges is $e=|E|=n\times k$.

Individual can only adopt pure strategy, namely cooperation (denoted by C) or defection (denoted by D). The strategy set is denoted by S. each round, every individual can game with every neighbor, and calculates the sum of the payoffs, denoted by p_i.

2.2 Individual Classifications

After a round ends, individuals can decide how to act according as the sum-payoff p_i in the round. The aim of the individuals is to improve its payoff, so individual decided to change its ego-network when it is unpleased with its payoff, likewise, the pleased individual will not change when it is pleased with its payoff. Three rules are given to classify the individual to the "pleased" and "unpleased".

1) If $\exists v_j \in V, (v_i, v_j) \in E$, make $p_i \geq p_j$, then the individual v_i is pleased.

After a round, individual finds that the sum payoff it gets isn't the lowest, it is pleased, and namely, the individual is pleased when it considers that it isn't the worst one.

2) If $\forall v_j \in V, (v_i, v_j) \in E$, make $s_j = C$, then the individual v_i is pleased

After a round, the individual finds all its neighbors cooperator with it, hen the individual is pleased, even if its sum payoff is the lowest compared to its neighbors.

3) If $\forall v_j \in V, (v_i, v_j) \in E$, make $p_i < p_j$, and $\exists v_j \in V, (v_i, v_j) \in E$, $s_j = D$, then the individual v_i is unpleased.

After a round, the individual finds its sum payoff is lowest compared to its neighbors and some neighbors are defectors, then this individual is unpleased.

2.3 Adaptation Rules

To the pleased individuals, they have no desire to change. But to unpleased individuals, they would like to change his connection to other individuals. On contrast to the past research, these unpleased individuals don't change their strategies, but change its connections to other individuals.

In PD, there are three connects between the individuals, that is C-C, C-D, D-D. The C-D connect will make defectors l exploit the cooperator's benefit, to the unpleased cooperators; they would not like to be exploited by defectors. To the unpleased defector, they also would not like to keep connection because they hope to connect with other cooperator to exploit. So the connection "C-D" and "D-D" is not steady when a pleased individual exists.

The adaptation rule is: the unpleased will randomly cut a connection with a defector in its neighbors, and select another individual to build connect randomly. Along with the time gone, the individuals in the network will all be pleased individuals, the network will arrive at the steady state.

2.4 Network Evolution Criterion

The network topology characters don't change any more when the system evolves to the steady state. In the study of complex network, there are three most common geometrical means to describe the network statistical characters: degree distribution, clustering coefficient (denoted by CC) and average path length(denoted by APL). The three attributes are used to evaluate the network property in this paper. Degree distribution is used to describe the factions of vertices in the network that have various degrees. Average path length describes how to extend the network is connected. In some cases, there exist vertex pairs that have no connecting path. To avoid this problem, an approach is to define average path length to be the "harmonic mean" geodesic distance between all pairs. This is suggested by Newman.

$$l^{-1} = \frac{2}{n(n+1)} \sum_{i \geq j}^{n} d_{ij}^{-1} \tag{1}$$

Where *dij* is the geodesic distance from vertex *i* to vertex *j*.

Clustering coefficient is defined as follow:

$$CC = \frac{3 \times number\ \ of\ \ trangles\ \ in\ \ the\ \ network}{number\ \ of\ \ connected\ \ triples\ \ of\ \ vertices} \tag{2}$$

Besides, the total payoff of all the individuals is an important criterion to evaluate how the whole system evolves. The total payoff is defined as:

$$p_{sum,t} = \sum_{i=1}^{n} p_{i,t} \tag{3}$$

Note that $p_{sum,t}$ is the total payoff, and $p_{i,t}$ is the payoff of individual v_i after round *t*.

2.5 Network Evolution Algorithm

The following algorithm briefly outlines the steps involved in the experiment:

1) Initial population of individuals (agents) and the random network.
2) Repeat to steady state:
   A) For each individual in the population:
      i) Play against all it's neighbors.
         a) Record he actions of it's neighbors
         b) Determine payoff using payoff matrix.
   B) For each individual in the population:
      i) determine the type: pleased or unpleased
      ii) the unpleased individuals change it's ego-network according to the adaptation rules
   C) Update the neighbor list for each individual.
   D) Calculate the sum payoffs of the whole population.
3) Determine the degree distribution, CC and APL at steady state.

3 Simulations and Results

The character of the regular network is high clustering coefficient and short average path length, the random network has low clustering coefficient and long average path length. To show the characters changes from initial state to steady state, the ER random network is selected to be the initial network, and the average degree is 8. The simulation tool is Repast, designed for complex adaptive system and developed by Chicago University. The values of the payoff matrix are set follow Axelrod: R=3, S=0, T=5, P=1.

3.1 Network Topology Graph

Fig. 1 shows the initial network topology and the steady network topology respectively. The proportion of the cooperators is 4%. The yellow blocks represent defector, and the blue blocks represent cooperators. The figures show that most individuals tend to connect with the cooperators at the steady state. The degree of the cooperator is much higher than the defectors. The reason is the most individuals is defectors in the network, the most of these defectors are unpleased because their benefit are low and can't exploit the cooperators, so they change the connections to build connection with the cooperators. This is why the cooperators have very high degree.

3.2 Curves of CC and APL

Fig. 2 shows the CC and APL at the steady state under various proportions of cooperators. Curve 1 and 2 represent respectively the ratio of CC and APL for regular 2-dimension lattice network to that for random network. Curve 3 and 4 represent the ratio of CC and APL for evolving steady network to that for random network. The curves show that the steady network has the approximate APL to the random network when the proportion of the cooperators is less than 30%, the ratio lies between 0.85 and 0.96. But the CC is much higher than the random network close to regular network and is close to the CC of regular network, the ratio lies between 2.33 and 9.78.

The evolving network shows the character of high CC and low APL, the character is just the "small world" property. The conclusion is that the cooperation is an important reason for why the real network shows the "small world" effect.

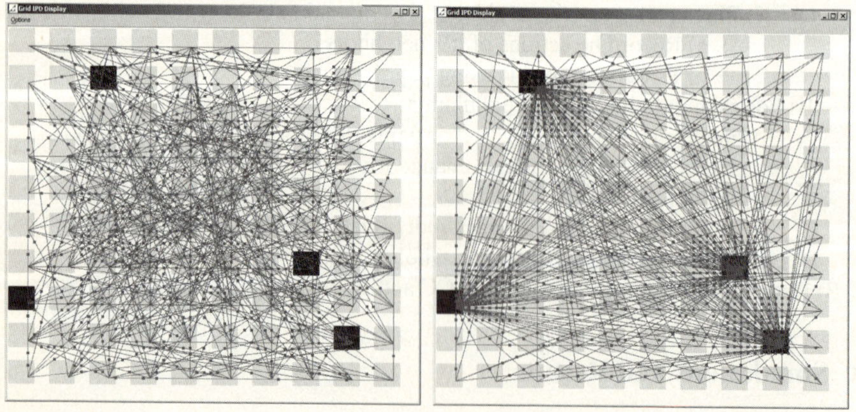

Fig. 1. The initial random network and the steady network (the proportion of the cooperators is 4%)

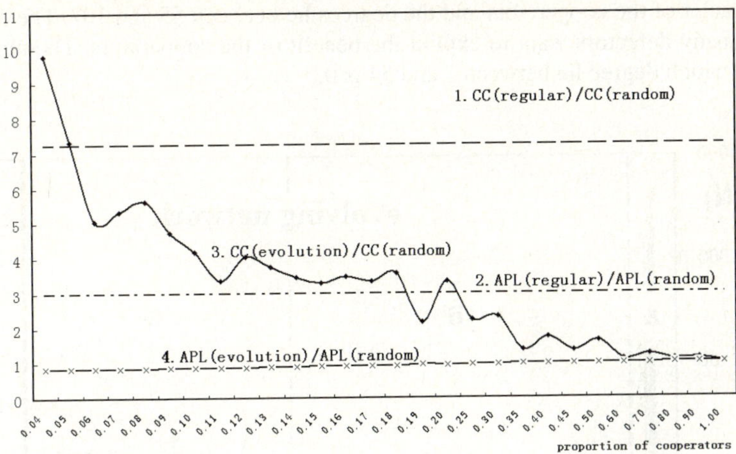

Fig. 2. The CC and APL under various proportions of cooperators

3.3 The Degree Distribution of the Evolving Network

To describe the degree distribution of the evolving network, a network having 6400 individuals is studied. Fig. 3 shows the degree distribution of initial random network. Just like the conclusion before, it is a Poisson distribution just like a "bell".

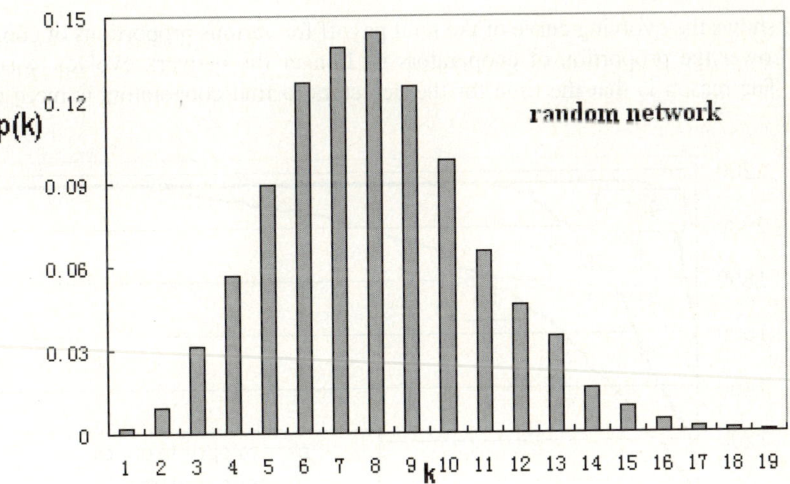

Fig. 3. The degree distribution of random network

Fig. 4 shows the degree distribution of the steady network when 5% cooperators exists. To describe distinctly, the y-axis adopt the number of the vertices and use the logarithmic coordinate. The degree distribution is obvious separate. The A area shows the character of the defectors and the degrees lie between 3 and 6. The C area shows the character of the cooperators and the degrees lie between 55 and 107. The reason is that so many defectors want to exploit the benefit of the cooperators. The number of vertices which degree lie between 7 and 54 is 0.

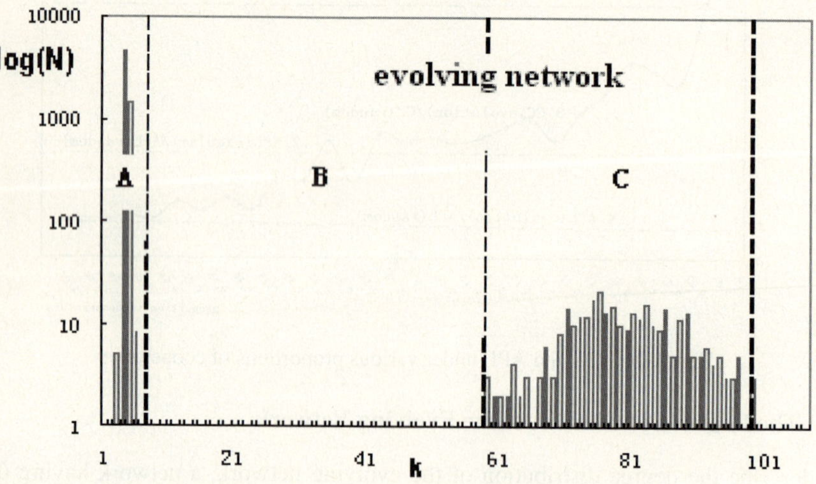

Fig. 4. The degree distribution of evolving network

3.4 The Total Payoff Curve

Fig. 5 shows the evolving curve of the total payoff for various proportions of cooperators. Lower the proportion of cooperators is, Longer the network evolves to steady state. The reason is that the time for the defectors to find cooperators is more when

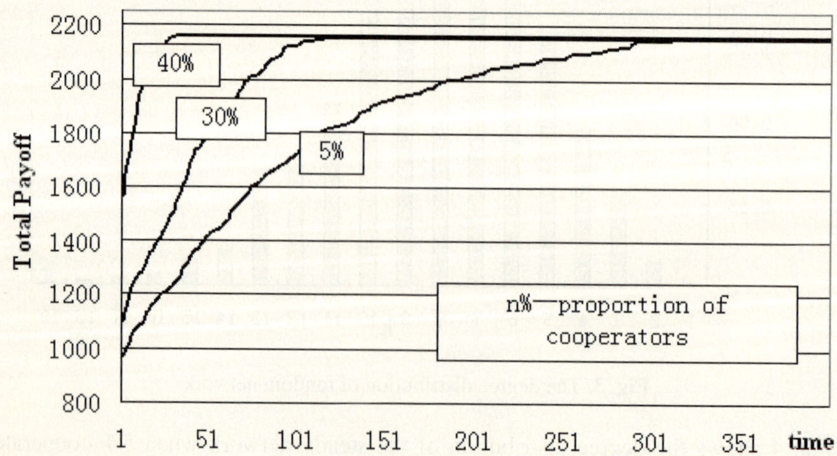

Fig. 5. The total payoff curve for various proportion of cooperators

the cooperators are. This conclusion is accordant with our cognition. But the amazing result is that the total payoffs are tend to be same for various proportion of cooperators. We conclude that the more cooperators can do litter contribution to improve the social wealth level, but the improvement of the network topology can do a lot to improve the social total wealth.

4 Conclusion

The evolving network model based on SPD is implemented to study the reason for network formation. The conclusions indicate that the cooperation is an important factor for the real network to exhibit the small world effect, and suggest that for the aim to improve the social wealth we must increase the number of the cooperators and the more important is to improve the connection between the individuals even though the cooperators are few.

References

1. Newman M E J.: The structure and function of complex networks. SIAM Review. 45(2003) 167-256
2. Axelrod, R.: The Evolution of Cooperation, Basic Books, New York (1984)
3. Nowak M A , May R M.: Evolutionary Games and Spatial Chaos. Nature. 359(1992) 826-829
4. D.J. Watts, S.H. Strogatz.: Collective Dynamics of 'small-world' Networks. Nature. 393(1998) 440-442
5. Guillermo Abramson, Marcelo Kuperman.: Social games in a Social Network. Physical Review E. 63(2001) 030901(R)
6. Naoki Masuda, Kazuyuki Aihara.: Spatial Prisoner's Dilemma Optimally Played in Small-World Networks. Physics Letters A. 313(2003) 55-61
7. Yao, X.: Evolutionary Computation: Theory and Applications. World Scientific, Singapore (1999)
8. Tan, K.C., Lim, M.H., Yao, X., Wang L.P. (Eds.): Recent Advances in Simulated Evolution And Learning. World Scientific, Singapore (2004)

Intelligent Optimization Algorithm Approach to Image Reconstruction in Electrical Impedance Tomography

Ho-Chan Kim and Chang-Jin Boo

Dept. of Electrical Eng., Cheju National Univ., Cheju, 690-756, Korea
{hckim, boo1004}@cheju.ac.kr

Abstract. In electrical impedance tomography (EIT), various image reconstruction algorithms have been used in order to compute the internal resistivity distribution of the unknown object with its electric potential data at the boundary. Mathematically the EIT image reconstruction algorithm is a nonlinear ill-posed inverse problem. This paper presents two intelligent optimization algorithm techniques such as genetic algorithm (GA) and simulated annealing (SA) for the solution of the static EIT inverse problem. We summarize the simulation results for the modified Newton-Raphson, GA, and SA algorithms.

1 Introduction

EIT is an imaging technique used to reconstruct the resistivity distribution within an domain from boundary electrode voltage measurements [1]. The physical relationship between inner resistivity and boundary surface voltage is governed by the nonlinear Laplace equation with appropriate boundary conditions. In general, the internal resistivity distribution of the unknown object is computed using the boundary voltage data based on various reconstruction algorithms. Yorkey et al. [2] developed a modified Newton-Raphson (mNR) algorithm and compared it with other existing algorithms such as backprojection, perturbation and double constraints methods. However, in real situations, the mNR method is often failed to obtain satisfactory images from physical data due to large modeling error and ill-posed characteristics.

In this paper, we will discuss the EIT image reconstruction based on intelligent optimization approaches via two-step approach. In the first step, each mesh is classified into three mesh groups: object, background, and temporary groups. In the second step, the values of these resistivities are determined using GA [3] and SA [4] algorithms.

2 Proposed Method

The forward problem in EIT calculates boundary potentials with the given electrical resistivity distribution, and the inverse problem, known as the image reconstruction, takes potential measurements at the boundary to update the resistivity distribution.

When electrical currents are injected into the domain $\Omega \in R^2$ through electrodes attached on the boundary $\partial\Omega$ and the resistivity distribution ρ is known over Ω, the corresponding induced electrical potential u can be determined uniquely from the following nonlinear Laplace equation [1]

L. Jiao et al. (Eds.): ICNC 2006, Part I, LNCS 4221, pp. 856 – 859, 2006.

$$\nabla \cdot (\rho^{-1} \nabla u) = 0 \text{ in } \Omega \tag{1}$$

The numerical solution for the forward problem can be obtained using the finite element method (FEM). In the FEM, the potential at each node is calculated by discretizing (1) into $YU_c = I_c$, where U_c (I_c) are the vector of boundary potential(injected current patterns) and the matrix Y is a functions of the unknown resistivities.

The inverse problem consists in reconstructing the resistivity distribution ρ from potential differences measured on the boundary of the domain. The methods used for solving the EIT problem search for a resistivity distribution minimizing some sort of residual involving the measured and calculated potential values.

In two-component visualization systems, we may assume that there are only two different representative resistivity values (object and background). We will discuss the EIT image reconstruction using GA and SA algorithms via two-step approach. In the first step, after a few initial mNR iterations performed without any grouping, we classify each mesh into one of three mesh groups: background(BGroup), object(OGroup), and temporary(TGroup) group. All meshes in BGroup and in OGroup are forced to have the same but unknown resistivity value (ρ_{back} and ρ_{tar}), respectively. However, all meshes in TGroup can have different resistivity vaules.

GA Algorithm. In the simplest implementation of GA in EIT, a set (population) of EIT images is generated, usually at random. The EIT chromosome is a sequence of n resistivities. After mesh grouping, in this paper, we will determine the values of these resistivities using two GA's. The first GA searches for the optimal range of resistivities by generating and evolving a population of individuals whose chromosome consists of two real genes (ρ_{back} **and** ρ_{tar}). The second GA solves the EIT problem, searching for the resistivity distribution minimizing the reconstruction error. The computed resistivities($\rho_{temp,i}$, $i = 1, \cdots, n-2$) in TGroup are constrained between the ρ_{back} and ρ_{tar} obtained in the first GA.

We will iteratively reconstruct an image that fits best the measured voltages U_l at the l-th electrode. To do so, we will calculate at each iteration the pseudo voltages $V_l(\rho)$ that correspond to the present state of the reconstructed image. We assume that the reconstructed image will converge towards the sought-after original image.

The fitness function is the reciprocal of the reconstruction error, the relative difference between the computed and measured potentials on the domain boundary

$$f_c = M \left[\sum_{i=1}^{M} \left| \frac{U_i - V_i(\rho)}{U_i} \right| \right]^{-1} \tag{2}$$

SA Algorithm. The SA reconstruction algorithm for EIT can be formulated as follows. We choose as cost function a following function of the relative difference between the computed and measured potentials on the domain boundary

$$E(\rho) = \frac{1}{M} \sum_{i=1}^{M} \left(\frac{U_i - V_i(\rho)}{U_i} \right)^2 \tag{3}$$

SA starts with an initial schedule ρ_k and generates another schedule $\rho_{new} = \rho_k + \Delta\rho$ within a neighborhood. In a minimization problem if $\Delta E = E(\rho_{new}) - E(\rho_k) < 0$ (i.e., ρ_{new} is better), then our new move(schedule) is accepted; otherwise, it is accepted with a probability of $P_r(k) = \exp(-\Delta E / T_k)$, where T_k is the temperature at the kth iteration. Note that as T_k decrease, the lower the probability to accept worse schedules. T_k is controlled by the cooling schedule function $T_k = \alpha T_{k-1}$, where $\alpha \in (0,1)$ is the temperature decay rate.

3 Computer Simulation

The proposed algorithm has been tested by comparing its results for numerical simulations with those obtained by mNR method. For the current injection the trigonometric current patterns were used. For the forward calculations, the domain Ω was a unit disc and the mesh of 3104 triangular elements with 1681 nodes and 32 channels was used as shown in Fig. 1(a). A different inverse mesh system with 776 elements and 453 nodes was adopted as shown in Fig. 1(b). In this paper, under the assumption that the resistivity varies only in the radial direction within a cylindrical coordinate system [5], the results of the three inverse problem methods can be easily compared. The resistivity profile is divided into 9 radial elements ($\rho_1, \cdots, \rho_9$) in Fig. 1(b).

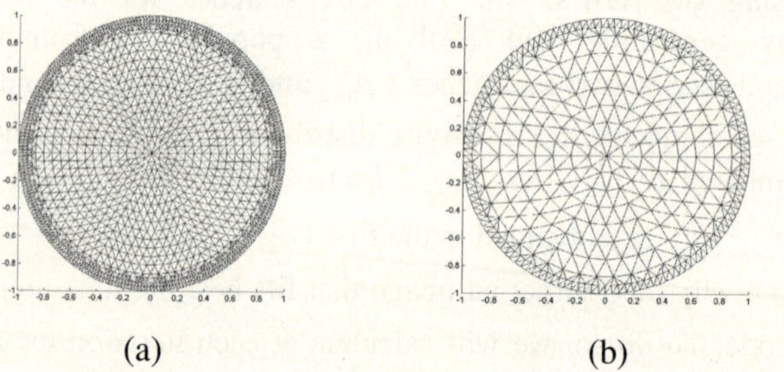

(a) (b)

Fig. 1. Finite element mesh (a) mesh for forward solver, (b) mesh for inverse solver

Synthetic boundary potentials were computed for idealized resistivity distributions using the finite element method described earlier. The resistivity profile appearing in Fig. 2 contains two large discontinuities in the original resistivity distribution. The present example is a severe test in EIT problems because there are large step changes at r/R=0.56 and 0.81 preventing electric currents from going into the center region.

We started the mNR iteration without any mesh grouping with a homogeneous initial guess. In Fig. 2, we see that the mNR algorithm may roughly estimate the given true resistivities. Since the mNR have a large error at the boundary of object and background in Fig. 2, we can not obtain reconstructed images of high spatial resolution. The inverted profile using GA and SA matches the original profile very well near the wall at r/R=1.0 as well as the center at r/R=0.0. Furthermore, the SA reconstruction is practically perfect for the jump of resistivty at r/R=0.56 and 0.81.

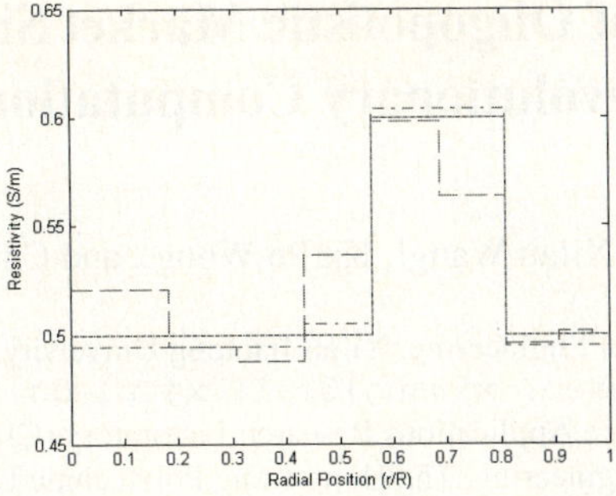

Fig. 2. True resistivities(solid line) and coumputed resistivities using mNR(dashed line), GA (dashdot line), and SA(dotted line)

4 Conclusion

In this paper, EIT image reconstruction methods based on intelligent optimization algorithms via two-step approach were presented to improve the spatial resolution. A technique based on intelligent optimization algorithm with the knowledge of mNR was developed for the solution of the EIT inverse problem. Although intelligent optimization algorithms such as GA and SA are expensive in terms of computing time and resources, which is a weakness of the method that renders it presently unsuitable for real-time tomographic applications, the exploitation of a priori knowledge will produce very good reconstructions.

References

1. Webster, J.G.: Electrical Impedance Tomography, Adam Hilger (1990)
2. Yorkey, T.J., Webster, J.G., Tompkins, W.J.: Comparing Reconstruction Algorithms for Electrical Impedance Tomography. IEEE Trans. on Biomedical Eng.. 34 (1987) 843-852.
3. Goldberg, D.E.: Genetic Algorithms in Search, Optimization and Machine Learning, Addison Wesley (1989)
4. Kirkpatrick, S., Gelatt, C.D., Vecchi, M.P.: Optimization by Simulated Annealing. Science. 220 (1983) 671-680
5. Kim, M.C., Kim, S., Kim, K.Y., Lee, J.H., Lee, Y.J.: Reconstruction of Particle Concentration Distribution in Annular Couette Flow Using Electrical Impedance Tomography. J. Ind. Eng. Chem. 7 (2001) 341-347

A Framework of Oligopolistic Market Simulation with Coevolutionary Computation*

Haoyong Chen[1], Xifan Wang[1], Kit Po Wong[2], and Chi-yung Chung[2]

[1] Department of Electrical Engineering, Xi'an Jiaotong University, Xi'an 710049, China
{hychen, xfwang}@mail.xjtu.edu.cn
[2] Computational Intelligence Applications Research Laboratory (CIARLab), Department of Electrical Engineering, The Hong Kong Polytechnic University, Hung Hom, Kowloon, Hong Kong
{eekpwong, eecychun}@polyu.edu.hk

Abstract. The paper presents a new framework of oligopolistic market simulation based on coevolutionary computation. The coevolutionary computation architecture can be regarded as a special model of the agent-based computational economics (ACE), which is a computational study of economies modeled as dynamic systems of interacting agents. The supply function equilibrium (SFE) model of an oligopolistic market is used in simulation. The piece-wise affine and continuous supply functions which have a large number of pieces are used to numerically estimate the equilibrium supply functions of any shapes. An example based on the cost data from the real-world electricity industry is used to validate the approach presented in this paper. Simulation results show that the coevolutionary approach robustly converges to SFE in different cases. The approach is robust and flexible and has the potential to be used to solve the complicated equilibrium problems in real-world oligopolistic markets.

1 Introduction

In recent years different equilibrium models have been used in analysis of strategic interaction between participants in oligopolistic markets, such as electricity markets. Among them the Supply Function Equilibrium (SFE) and Cournot models are the most extensively used models [1], [2].

The Cournot oligopoly model assumes that strategic firms employ quantity strategies: each strategic firm decides its quantity to produce, treating the output level of its competitors as a constant [2].

The SFE model offers a more realistic view. The general SFE model was introduced by Klemperer and Meyer [3] and first applied to oligopolistic market analysis by Green and Newbery [4], in which each firm chooses its strategy as a "supply function" relating its quantity to its price.

Baldick and Hogan [5] consider a supply function model of an oligopolistic market where strategic firms have capacity constraints. To find stable equilibria, they numerically solve for the equilibrium by iterating in the function space of allowable

* This work was supported by National Natural Science Foundation of China (No. 50207007) and Special Fund of the National Priority Basic Research of China (2004CB217905).

L. Jiao et al. (Eds.): ICNC 2006, Part I, LNCS 4221, pp. 860–869, 2006.

supply functions. Baldick et al. [6] consider a SFE model of interaction in an oligopolistic market, assuming a linear demand function.

A recent fast-developing area rests on the application of coevolutionary computation [7], [8] to oligopolistic market analysis. Coevolutionary computation is an extension of conventional evolutionary algorithms (EAs). It models an ecosystem consisting of two or more species. Multiple species in the ecosystem coevolve and interact with each other and result in the continuous evolution of the ecosystem [7], [8]. Coevolutionary computation models have been gradually obtaining extensive applications in different areas [9].

The coevolutionary computation model can be regarded as a special form of the agent-based computational economics (ACE) model [10], in which each participant in the market under investigation is represented by an agent. The ability of ACE to capture the independent decision-making behavior and interactions of individual agents provides a very good platform for the modeling and simulation of oligopolistic markets. The coevolutionary computation model has many advantages compared to other agent-based models in that it uses the well-developed GA (or other EAs) for agent "strategic learning". GAs and other EAs are found to be effective analogues of economic learning [11], [12].

The coevolutionary computation models for oligopolistic market equilibrium analysis are presented in [2]. Coevolutionary computation model is found to be an effective and powerful approach for oligopolistic market simulation. This paper focuses on the analysis of SFE models based on coevolutionary computation and shows its high computational efficiency and great potentials in practical application.

The rest of the paper is organized as follows: Market models used for analysis are presented in Section 2. The framework of coevolutionary computation model is proposed in Section 3 and then validated by an example based on the cost data from the real-world electricity industry in Section 4. Finally, Section 5 concludes the paper.

2 Market Model Formulation

The paper uses the similar general model as in [6]. Suppose there are I firms in the market, each firm has a strictly convex quadratic cost function:

$$C_i(q_i) = \frac{1}{2} c_i q_i^2 + a_i q_i \qquad i = 1, \ldots, I \,, \tag{1}$$

where q_i is the quantity generated by firm i; c_i, a_i are the coefficients of the firm's cost function with $c_i > 0$ and $a_i \geq 0$.

Assume that the supply function of each firm is $S_i(p)$ $i = 1, \ldots, I$, where p is the market price, if the market-clearing condition is satisfied, the aggregate demand $D(p,t)$ at pricing-period t will be equal to the total output of all the firms in the market as shown below:

$$D(p,t) = \sum_{i=1}^{I} S_i(p) \,. \tag{2}$$

We use the form of the demand curve as follows:

$$D(p,t) = D_0(t) - \gamma p \ . \tag{3}$$

The underlying time-continuous demand is specified by $D_0(t)$ and for each period t, the demand is linear in p with slope $dD/dp = -\gamma$. The coefficient γ is assumed to be positive.

An equilibrium consists of a set of supply functions, one for each firm, such that each firm is maximizing its profits, given the supply functions of the other firms. Suppose firm i produces the residual demand of the other firms (that is, total demand less the other firms' supply) to meet the market-clearing condition, its profit π_i at each period can be written as:

$$\pi_i(p,t) = p\left(D(p,t) - \sum_{j \neq i} S_j(p)\right) - C_i\left(D(p,t) - \sum_{j \neq i} S_j(p)\right) . \tag{4}$$

If p^* is an equilibrium price at pricing-period t, it must satisfy

$$\pi_i(p^*, t) = \max_p \pi_i(p,t) \quad i = 1, \ldots, I \ . \tag{5}$$

The first-order condition for profit maximization is:

$$S_i(p) = \left(p - C_i'(S_i(p))\right)\left(-\frac{dD}{dp} + \sum_{j \neq i} \frac{dS_j(p)}{dp}\right) . \tag{6}$$

Any solution to the coupled differential equations (6) for $i = 1, \ldots, I$ such that $S_i(p)$ is non-decreasing is a SFE. These equations do not involve the underlying time-continuous demand $D_0(t)$ but depend on the demand slope dD/dp.

To consider complicated conditions in real oligopolistic markets, the paper numerically estimates the equilibrium supply functions of any shapes with piece-wise affine and continuous functions having a large number of pieces.

The supply function is defined as a piece-wise affine non-decreasing function with break-points evenly spaced between $(\underline{p} + 0.1)$ and $(\overline{p} - 0.1)$, where $\underline{p}$ and $\overline{p}$ are the lower and upper limits of the market price respectively. No capacity constraint is considered. An illustration of firm i's piece-wise affine supply function is given in Fig. 1.

The kth $(1 \le k \le K)$ linear section of firm i's supply function existing in the price range $p_{i,k-1}$ to $p_{i,k}$ and the capacity range $q_{i,k-1}$ to $q_{i,k}$ can be defined as:

$$q = q_{i,k-1} + \frac{1}{\beta_{i,k}}(p - p_{i,k-1}) = q_{i,k-1} + \frac{q_{i,k} - q_{i,k-1}}{p_{i,k} - p_{i,k-1}}(p - p_{i,k-1}) , \tag{7}$$

where $\beta_{i,k}$ is the slope of the kth linear section.

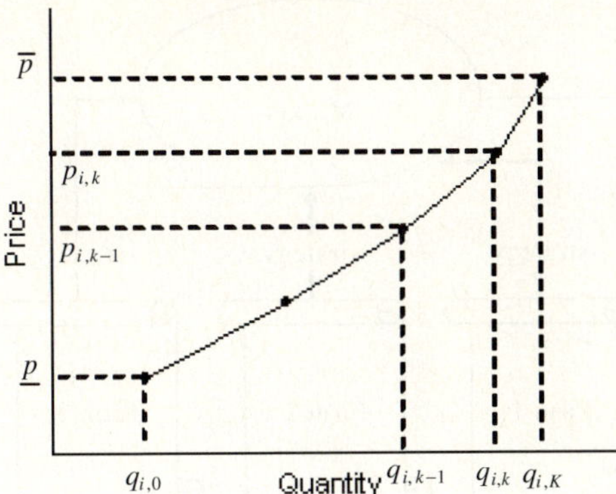

Fig. 1. Firm i's piece-wise affine supply function

3 Oligopolistic Market Simulation Based on Coevolutionary Computation

This section introduces the framework of oligopolistic market simulation based on coevolutionary computation [2].

The above market models can be illustrated by Fig. 2. Each firm submits its optimal trading strategy to the market. For the SFE model, the trading strategy is the supply function relating its quantity to its price. Then the market price is calculated using (2) and (3) according to the demand characteristics and market rules. Each firm can calculate its profit with the market price and its trading strategy according to (4). Fig. 2 suggests an agent-based simulation method for oligopolistic market, in which each participant in the market under investigation is represented by an agent in the model. Each agent makes its decisions based on the knowledge of itself and the environment. The agents interact with each other through the system model. Here the system model consists of the market rules and demand function.

This paper presents an agent based simulation approach of oligopolistic market based on coevolutionary computation, in which the agents simulate the species in an ecosystem. The species are genetically isolated and interact with one another within a shared domain model. Coevolutionary computation is a relatively new area in the research of evolutionary computation and are still under rapid developing [7], [8], [9].

The basic coevolutionary computation model is shown in Fig. 3 [7], which is an analogue of Fig. 2. Each firm in Fig. 2 is represented by a species in the ecosystem. Each species evolves a bundle of individuals, which represent the candidate trading strategies of the corresponding firm. The market is modeled with the domain model. Each species is evolved through the repeated application of a conventional EA. Fig. 2 shows the fitness evaluation phase of the EA from the perspective of species 1. To evaluate an individual (trading strategy) from species 1, collaboration is formed with representatives (representative trading strategies) from each of the other species. The domain model solves for the system variable (market price). Then species 1 can use

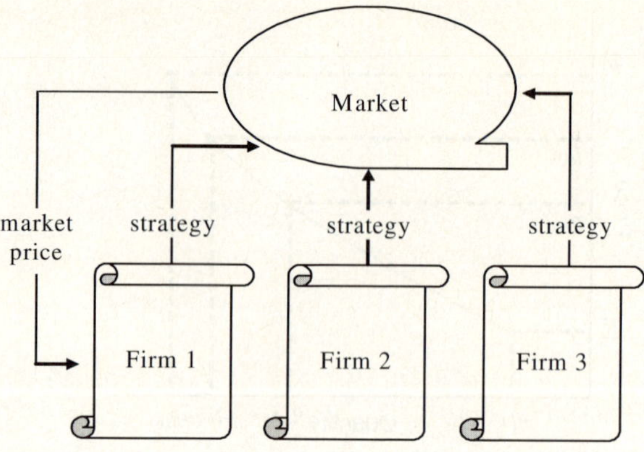

Fig. 2. Illustration of oligopolistic market models

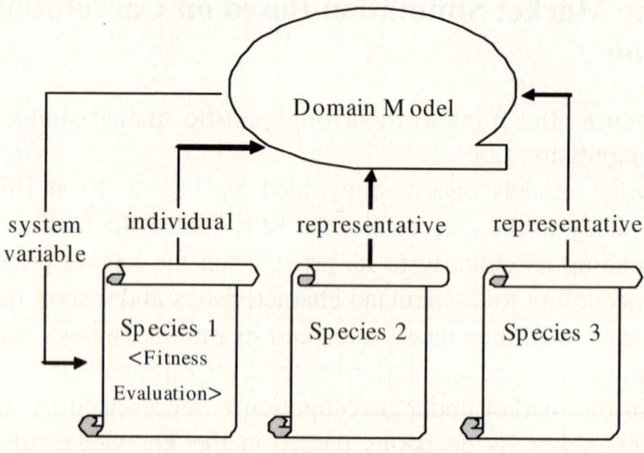

Fig. 3. Framework of coevolutionary computation model

the system variable to evaluate the fitness of its individual. Here the fitness is the profit of the corresponding firm. The5re are many possible methods for choosing representatives with which to collaborate. An obvious one is to simply let the current best individual from each species be the representative [8].

The pseudo-code of a coevolutionary genetic algorithm (CGA) is given as follows, in which the evolution of each species is handled by a standard GA.

```
k = 0
for each species i do
  begin
    initialize the species population Pop[i][0]
    evaluate fitness of each individual Ind[i][0][n] in
      Pop[i][0]
```

```
    choose a representative Rep[i][0] from Pop[i][0]
  end
while termination condition = false do
  begin
    for each species i do
      begin
        reproduction from Pop[i][k] to get Mate[i][k]
        crossover and mutation from Mate[i][k] to get
          Pop[i][k+1]
        evaluate fitness of each individual
          Ind[i][k+1][n] in Pop[i][k+1]
        choose a representative Rep[i][k+1] from
          Pop[i][k+1]
      end
    k = k + 1
  end.
```

From Section 2 we know that to fully describe a piece-wise affine supply function, we should keep the quantity and price values of each break-point. Since the price break-point values are specified as in Section 2, we only need to keep the quantity break-point values. We keep the quantity value at the price $\underline{p}$ and the slope reciprocal values of each linear section of the piece-wise affine supply function instead of the quantity values of each break-point. That is, the coding structure of a chromosome is shown as Fig. 4. The chromosome consists of $K+1$ binary genes, each of which encodes a corresponding variable value.

1	2	3		K+1
$q_{i,0}$	$1/\beta_{i,1}$	$1/\beta_{i,2}$	...	$1/\beta_{i,K}$

Fig. 4. Coding structure of chromosome for piece-wise affine supply function model

4 Case Study

We use the 5-firm example system in [5] to validate the coevolutionary approach described in this paper. The example system is based on the cost data for the five strategic firm industry of electricity in England and Wales subsequent to the 1999 divestiture. The firms' cost data are reproduced in Table 1 below.

Table 1. Firms' cost data

Firm's Cost Function	Firm No. i	c_i [£/(MW)^{2}h]	a_i (£/MWh)
$C_i(q_i)=\frac{1}{2}c_i q_i^2 + a_i q_i$	1	2.687	12
	2	4.615	12
	3	1.789	8
	4	1.930	8
	5	4.615	12

A demand slope of $\gamma = 0.1$ GW/(£/MWh) and underlying time-continuous demand characteristic of:

$$D_0(t) = 10 + 40(1-t),\ 0 \le t \le 1\ , \tag{8}$$

with quantities measured in GW are used in our simulation, where the time argument t is normalized to 0 to 1 and $t = 0$ corresponds to peak demand conditions. We suppose that one trading day equals 48 pricing-periods, and viz. one pricing-periods spans $1/48$ time interval.

As in [5], the piece-wise affine supply function has 40 break-points evenly spaced between $(\underline{p} + 0.1)$ and $(\overline{p} - 0.1)$. A price cap $\overline{p} = 40$ £/MWh and a price minimum of $\underline{p} = 12$ £/MWh are used in our simulation. The evolutionary process needs to start from an estimate of the equilibrium supply function in simulation. We should define a set of starting functions as the initial estimates of the supply functions. As in [5], three different starting functions are used:

1. "competitive", where the supply functions are the inverses of the marginal cost functions.
2. "affine SFE", where the supply functions are assumed to be affine and satisfy the simultaneous differential equations (6).
3. "Cournot", where quantities and prices under Cournot competition are calculated for each $t \in [0,1]$, and a supply function is drawn through the resulting price-quantity pairs.

An 8-bit binary code chromosome is used to encode a variable in each GA population. The evolutionary process starts from the representatives encoding the starting functions and runs for a maximum 100 generations. A two-point crossover and a bit-flip mutation genetic operators are used. The elitist strategy preserving a single copy of the best individual is adopted. The crossover and mutation probability

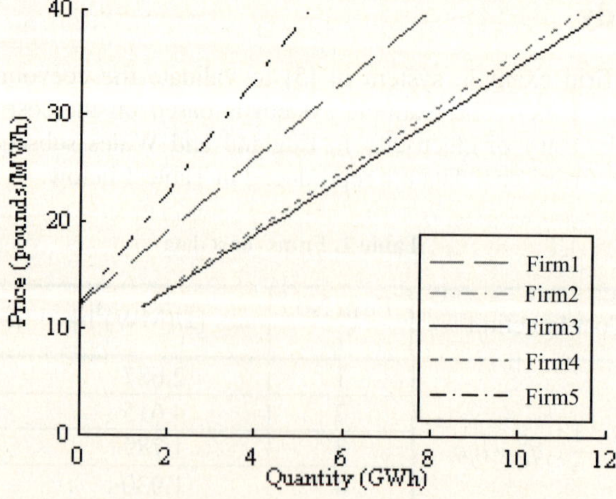

Fig. 5. Affine supply functions

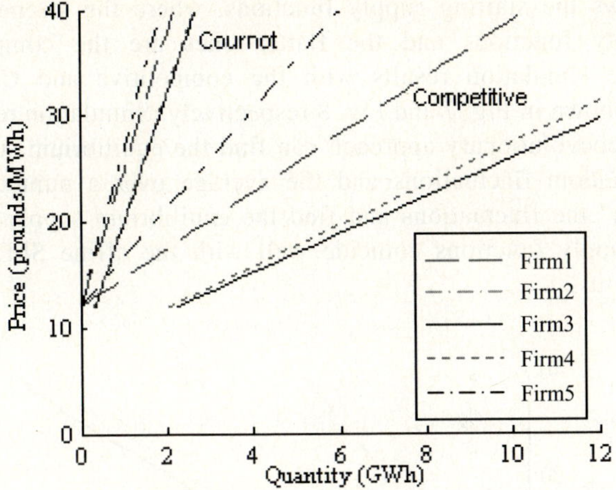

Fig. 6. Starting Supply functions

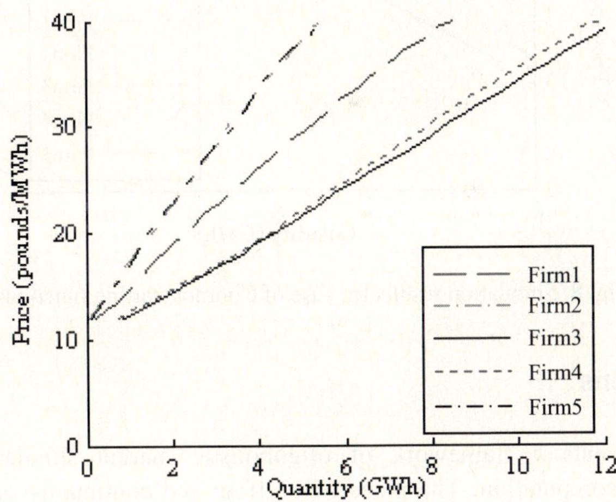

Fig. 7. Simulation results for case of competitive starting functions

are set to 0.9 and 0.06 respectively. The simulation results are shown in Fig. 5, Fig. 7, and Fig. 8. Since the results of the approach are stochastic, we run the algorithm for 20 times in each simulation and calculate the average results. The average computational time of each time run is 20s on a PC Pentium 4 with 2.8GHz CPU.

The affine supply functions are shown in Fig. 5. In this figure and all subsequent figures, firm 1 is shown as a dashed line, firms 2 and 5 have identical costs and are shown superimposed as a dash-dot line, firm 3 is shown as a solid line and firm 4 is shown as a dotted line. The evolutionary process starting from the affine supply function sticks to the starting functions. That means the affine SFE is "stable".

Fig. 6 shows the starting supply functions, where the steeper lines are the Cournot supply functions and the flatter ones are the competitive supply functions. The simulation results with the competitive and Cournot starting functions are shown in Fig. 7 and Fig. 8 respectively. Simulation results show that generally the coevolutionary approach can find the equilibrium supply functions with minor random fluctuations and the average over a number of runs will “smooth down” the fluctuations and find the equilibrium supply functions. The equilibrium supply functions coincide well with the affine SFE, which is the theoretical result.

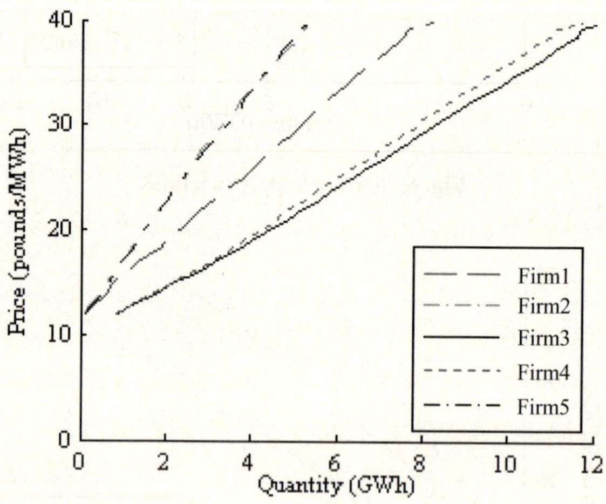

Fig. 8. Simulation results for case of Cournot starting functions

5 Conclusions

The paper presents a framework of oligopolistic market simulation based on coevolutionary computation. The piece-wise affine and continuous supply functions which have a large number of pieces are used to numerically estimate the equilibrium supply functions of any shapes. Simulation results show that the coevolutionary approach is effective and robustly converges to SFE in all simulation cases.

The paper makes several key points concerning simulation of SFE based on the coevolutionary computation model: simulation of strategic best-response behavior of the market participants allows convergence to the solution of the differential equation describing the equilibrium, and the simulation obtains the similar results but takes much less computation time than the numerical framework in [5].

Since the coevolutionary approach utilizes the general genetic operators in strategic learning, it is robust and flexible and has the potential to be used to solve the equilibrium problems in real oligopolistic markets.

References

1. Baldick R.: Electricity Market Equilibrium Models: The Effect of Parametrization. IEEE Trans. Power Syst. 4 (2002) 1170-1176
2. Chen H., Wong K. P., Nguyen D. H. M., Chung C. Y.: Analyzing Oligopolistic Electricity Market Using Coevolutionary Computation. IEEE Trans. Power Syst. 1 (2006) 143-152
3. Klemperer P. D., Meyer M. A.: Supply Function Equilibria in Oligopoly under Uncertainty. Econometrica 6 (1989) 1243-1277
4. Green R. J., Newbery D. M.: Competition in the British Electricity Spot Market. J. Political Econ. 5 (1992) 929-953
5. Baldick R., Hogan W.: Capacity Constrained Supply Function Equilibrium Models of Electricity Markets: Stability, Non-Decreasing Constraints, and Function Space Iterations. Univ. Calif. Energy Inst. [Online]. Available: http://www.ucei.berkeley.edu/PDF/pwp089.pdf (2001)
6. Baldick R., Grant R., Kahn E. P.: Theory and Application of Linear Supply Function Equilibrium in Electricity Market. J. Regulatory Econ. 2 (2004) 143-167
7. Potter M. A., De Jong K. A.: Cooperative Coevolution: An Architecture for Evolving Coadapted Subcomponents. Evol. Comp. 8 (2000) 1-29
8. Wiegand R. P.: An Analysis of Cooperative Coevolutionary Algorithms. Ph.D. Thesis. George Mason University (2003)
9. Tan, K.C., Lim, M.H., Yao, X., Wang L.P. (eds.): Recent Advances in Simulated Evolution and Learning. World Scientific, Singapore (2004)
10. Tesfatsion L.: Agent-Based Computational Economics Growing Economies from the Bottom Up. [Online]. Available: http://www.econ.iastate.edu/tesfatsi/ace.htm
11. Riechmann T.: Genetic Algorithm Learning and Evolutionary Games. J. Econ. Dynamics Contr. 6-7 (2001) 1019-1037
12. Yao, X. (ed.): Evolutionary Computation: Theory and Applications. World Scientific, Singapore (1999)

Immune Clonal MO Algorithm for 0/1 Knapsack Problems

Ronghua Shang, Wenping Ma, and Wei Zhang

Institute of Intelligent Information Processing, P.O. Box 224, Xidian University, Xi'an, 710071, P.R. China
shangronghua1980@163.com

Abstract. In this paper, we introduce a new multiobjective optimization (MO) algorithm to solve 0/1 knapsack problems using the immune clonal principle. This algorithm is termed Immune Clonal MO Algorithm (ICMOA). In ICMOA, the antibody population is split into the population of the nondominated antibodies and that of the dominated antibodied. Meanwhile, the nondominated antibodies are allowed to survive and to clone. A metric of Coverage of Two Sets is adopted for the problems. This quantitative metric is used for testing the convergence to the Pareto-optimal front. Simulation results on the 0/1 knapsack problems show that ICMOA, in most problems, is able to find much better spread of solutions and better convergence near the true Pareto-optimal front compared with SPEA, NSGA, NPGA and VEGA.

1 Introduction

Many real-world optimization problems involve optimization of several criteria. Since multiobjective optimization (MO) searches for an optimal vector, not just a single value, one solution often cannot be said to be better than another and there exists not only a single optimal solution, but a set of optimal solutions, called Pareto front. In recent years, many algorithms for MO problems have been introduced. Schaffer put forward VEGA [1]. Horn et al's NPGA and Srinivas et al's NSGA attracted more attentions [2]. In recent years, a lot of newly improved algorithms were proposed, such as Deb et al's NSGA and NSGA-II [3], Zitzler's SPEA [4] and SPEA2 [5]. Just like evolutionary algorithms, artificial immune system (AIS) constructs new intelligent algorithms with immunology terms and fundamental [6, 7, 8].

The immune system is one of the most important biological mechanisms humans possess since our life depends on it. In recent years, several researchers have developed computational models of the immune system that attempt to capture some of their most remarkable features such as its self-organizing capability[9,10,11]. The immune system establishes the idea that the cells are selected when they recognize the antigens and proliferate. When exposed to antigens, the immune cells which may recognize and eliminate the antigens can be selected in the body and mount an effective response against them. Its main ideas lie in that the antigen can selectively react to the antibody, which are native production and spread on the cell surface in the form of peptides. The reaction leads to cell proliferate clonally and the colony has the same antibody. Some clonal cell divide into antibody producing cells, and others become

L. Jiao et al. (Eds.): ICNC 2006, Part I, LNCS 4221, pp. 870–878, 2006.

immune memory cells to boost the second immune response. From the point view of the Artificial Intelligence, some biologic characters such as learning, memory and antibody diversity can be used to artificial immune system [6].

Based on the immune clonal theory, a new MO algorithm-Immune Clonal MO Algorithm (ICMOA) is introduced and ICMOA algorithm is used to solve 0/1 knapsack problems. The simulation results on the **test problems** show that ICMOA outperforms SPEA, NSGA, NPGA and VEGA in terms of finding a diverse set of solutions and in converging near the true Pareto-optimal set.

2 The Multiobjective 0/1 Knapsack Problems

The knapsack problem is an abstract of restricted resource problems with some purposes, which has broad projective background. A 0/1 knapsack problem is a typical combinatorial optimization problem and is difficult to solve (NP-hard). Multiobjective 0/1 Knapsack Problems with k knapsacks (i.e. k objectives and k constraints) and m items in Zitzler and Thiele [4] can be written as follows:

$$\text{Maximize} \quad \mathbf{f}(x) = (f_1(\mathbf{x}), f_2(\mathbf{x}), \ldots, f_n(\mathbf{x})) \tag{1}$$

$$\text{subject to} \quad \sum_{j=1}^{m} w_{ij} x_j \le c_i \quad i = 1, 2, \cdots, k \tag{2}$$

$$\text{where} \quad f_i(x) = \sum_{j=1}^{m} p_{ij} x_j \quad i = 1, 2, \cdots, k \tag{3}$$

In this formulation,

$x_j = 1 \, (j = 1, \cdots, m)$ iff item j is selected,

$\mathbf{x} = (x_1, x_2, \ldots, x_m) \in \{0,1\}^m$ is an m-dimensional binary vector,

p_{ij} = the profit of item j according to knapsack i,

w_{ij} = the weight of item j according to knapsack i,

c_i = the capacity of knapsack i.

Each solution $\mathbf{x}$ is handled as a binary string of length m in evolutionary multiobjective optimization (EMO) algorithms.

Zitzler and Thiele [4] examined the performance of several EMO algorithms using nine test problems where both the number of knapsacks and the number of items were varied. These test problems with two, three and four objectives and 250, 500 and 750 items. In this paper, in order to validate our algorithm, we use the nine test problems.

3 The Artificial Immune System (AIS)

The main goal of the immune system is to protect the human body from the attack of foreign organisms. The immune system is capable of distinguishing between the normal components of our organism and the foreign material that can cause harm. These foreign organisms are called *antigens*. In AIS, antigen usually means the problem and its constraints. Especially, for the multiobjective 0/1 knapsack problems, we have

$$\begin{aligned} &max. && y = f(x) = (f_1(x), f_2(x), \ldots, f_n(x)) \\ &subject\ to && \sum_{j=1}^{m} w_{i,j} \bullet x_j \le c_i, x = (x_1, x_2, \ldots, x_m) \in \{0,1\}^m \end{aligned} \tag{4}$$

The molecules called *antibodies* play the main role on the immune system response. The immune system response is specific to a certain foreign organism (antigen). When an antigen is detected, those antibodies that best recognize an antigen will proliferate by cloning. This process is called *clonal selection principle* [9]. In the traditional AIS, the new cloned cells undergo high rate mutations or hypermutation in order to increase their receptor population (called repertoire). Antibodies represent candidates of the problem. For the multiobjective 0/1 knapsack problem, every candidate adopts binary coding with the length m, each binary bit represents the value of variable $x_i \mid i \in [1, m]$.

4 ICMOA Algorithm for 0/1 knapsack Problems

Our Algorithm in this paper is based on the AIS previously described. The ICMOA algorithm for 0/1 knapsack problems is the following:

Step1: Give the population size N, *the* clonal size R, the termination generation G_{max}, Set the mutation rate MR, Generate randomly the initial antibody population $A(0) = \{a_1(0), a_2(0), \cdots a_N(0)\}$, $it := 1$.

Step2: Determine for each antibody in the antibody population $A(it)$, whether it is Pareto dominated or not. Delete the dominated antibodies in $A(it)$.

Step3: Modify $A(it)$ with the greedy repair method as reported in reference [4], obtain the antibody population $A(it)$ which satisfies the constrained conditions.

Step4: Create a number R of copies of all of the non-dominated antibodies in $A(it)$ and we get $B(it)$.

Step5: Assign a mutation rate (MR) to each clone and apply mutation rate MR to each clone in $B(it)$.

Step6: Determine for each antibody in the new antibody population $B(it)$, whether it is Pareto dominated or not.

Step7: Delete all of the dominated antibodies in the antibody population $B(it)$ and the corresponding individuals in the Pareto optimal set.

Step8: If the size of the antibody population $B(it)$ is lager than N, delete the antibodies in the antibody population based on the crowding distance [1] until the size is N.

Step9: $A(it+1) := B(it)$, $it := it+1$ and go back to step3 until $it > G_{max}$.

5 Simulation Results

5.1 Performance Measures

Unlike in single-objective optimization, there are two goals in a multiobjective optimization: 1) convergence to the Pareto-optimal set and 2) maintenance of diversity in solutions of the Pareto-optimal set. In this paper, we adopt the performance metric (proposed in [13]) that is a quantitative metric for testing the convergent extent to the Pareto-optimal front. For a multiobjective optimization problem, the goal is to find as many different Pareto-optimal or near Pareto-optimal solutions as possible. The Coverage of Two Sets used in this paper is defined as follows:

Definition (Coverage of Two Sets). Let $A', A'' \subseteq X$ be two sets of decision vectors. The function ς maps the ordered pair (A', A'') to the interval [0, 1]:

$$\varsigma(A', A'') \triangleq \frac{\left|\{a'' \in A''; \exists a' \in A' : a' \rhd a''\}\right|}{|A''|} \tag{5}$$

Where $\rhd$ means Pareto dominate or equal. If all points in A' dominate or are equal to all points in A'', then by definition $\varsigma(A', A'') = 1$. $\varsigma(A', A'') = 0$ implies the opposite situation. In general, $\varsigma(A', A'')$ and $\varsigma(A'', A')$ have to be considered because $\varsigma(A', A'')$ is not necessarily equal to $1 - \varsigma(A'', A')$. The advantage of this metric is that it is ease to calculate and provides a relative comparison based upon dominance numbers between generations or algorithms.

5.2 Simulation Results

In order to validate the algorithm, we compare the algorithm with another four algorithms in solving the nine multiobjective 0/1 knapsack problems on Zitzler's homepages (http://www.tik.ee.ethz.ch/~zitzler/testdata.html/). They are Zitzler's Strength Pareto Evolutionary Algorithm (SPEA) [4], Schaffer's Vector Evaluated Genetic Algorithm (VEGA) [7], the niched Pareto genetic algorithm (NPGA) [10] and Srinivas' and Deb's Nondominated Sorting Genetic Algorithm (NSGA) [11]. The test data sets are available from Zitzler's homepages, where two, three, and four knapsacks with 250, 500, 750 items are taken under consideration. A greedy repair method adopted is also applied in ICMOA that the repair method removes items from the infeasible solutions step by step until all capacity constraints are fulfilled. The order in which the items are deleted is determined by the maximum profit/weight ratio per item. 30 independent runs of IDCMA are performed per test problem. The reported results of SPEA, NPGA, NSGA and VEGA are directly gleaned from Zitzler's website.

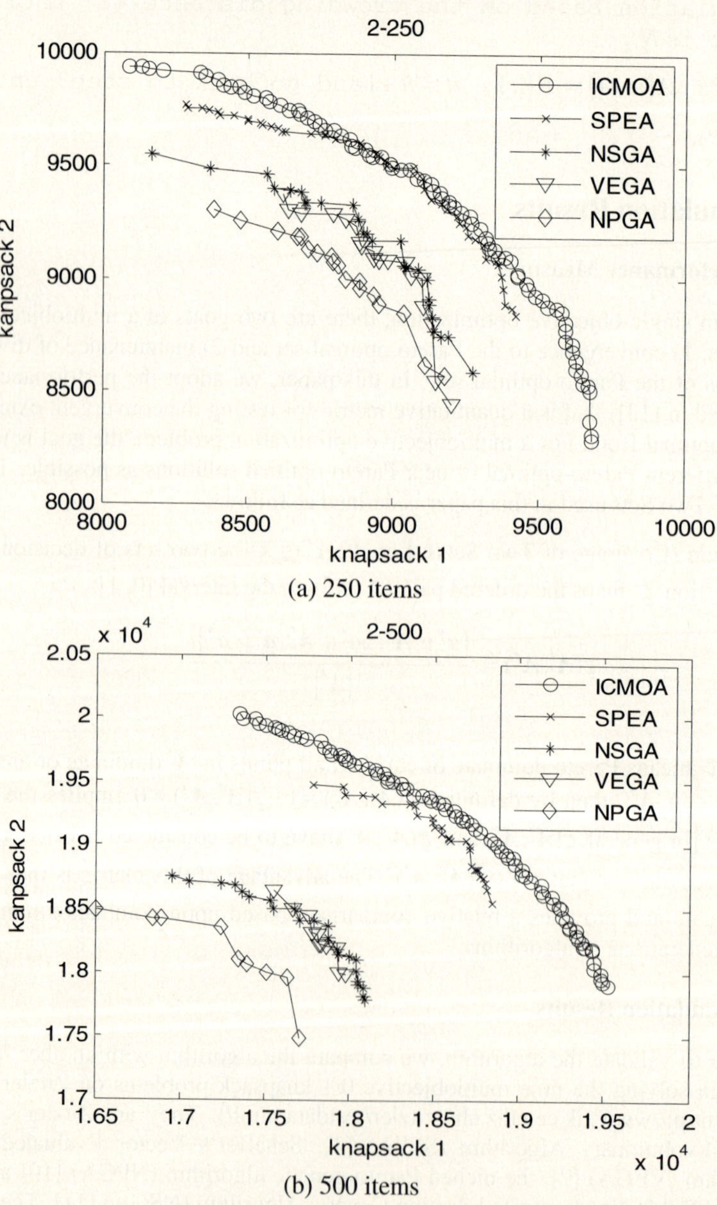

Fig. 1. Trade-off Fronts for 2 knapsack

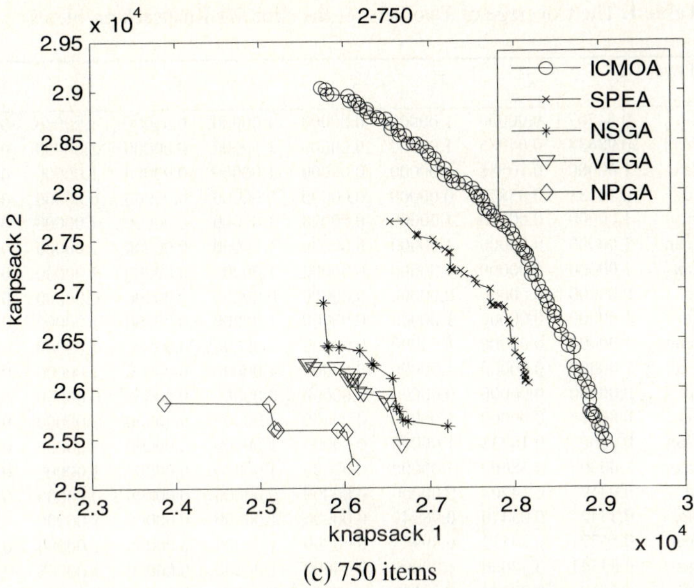

(c) 750 items

Fig. 1. *(continued)*

For a typical example, the Pareto fronts of the problems with two knapsacks and 250, 500 and 750 items for the five algorithms are shown in Figure 1. Here, the non-dominated solutions regarding the first 5 runs are plotted. For better visualization, the points achieved by a particular method are connected by real lines.

The simulation results of two knapsacks with 250, 500 and 750 items are shown in figure 1. For the two knapsacks with 250 items ICMOA can find the solutions widely and obtain a better distribution as shown in (a). The convergence to the true Pareto-optimal front of SPEA is similar to that of ICMOA. But NSGA, VEGA and NPGA have much poorer performance for the problem. (b) and (c) show the better performance of ICMOA than SPEA, NSGA, VEGA and NPGA in both the convergence and diversity of obtained solutions on two knapsacks with 500 items and 750 items.

The direct comparison of the different algorithms based on the ς measure is shown in table 1.

Table 1 shows the min, the mean, the max and the standard deviation of the metric Coverage of Two Sets obtained by ICMOA, SPEA, NPGA, NSGA and VEGA for the nine 0/1 knapsack test problems. In the table, ς_S^I means $\varsigma(X^I, X^S)$ and ς_I^S indicates $\varsigma(X^S, X^I)$ where X^I denotes the solutions solved by ICMOA and X^S denotes the solutions solved by SPEA. ς_{Ns}^I , ς_I^{Ns} , ς_V^I , ς_I^V , ς_{Np}^I and ς_I^{Np} indicate $\varsigma(X^I, X^{Ns})$, $\varsigma(X^{Ns}, X^I)$, $\varsigma(X^I, X^V)$, $\varsigma(X^V, X^I)$, $\varsigma(X^I, X^{Np})$ and $\varsigma(X^{Np}, X^I)$ respectively. X^{Ns} , X^{Np} and X^V denote the solutions obtained by NSGA, NPGA and VEGA respectively.

Table 1. The Coverage of Two Sets for the Nine 0/1 Knapsack problems

Coverage of two set		ς_S^I	ς_I^S	ς_{Ns}^I	ς_I^{Ns}	ς_V^I	ς_I^V	ς_{Np}^I	ς_I^{Np}
	Min	0.64707	0.00000	1.00000	0.00000	1.00000	0.00000	1.00000	0.00000
2-250	Mean	0.98634	0.04965	1.00000	0.00000	1.00000	0.00000	1.00000	0.00000
	Max	1.00000	0.10588	1.00000	0.00000	1.00000	0.00000	1.00000	0.00000
	Std	0.10181	0.05871	0.00000	0.00000	0.00000	0.00000	0.00000	0.00000
	Min	1.00000	0.00000	1.00000	0.00000	1.00000	0.00000	1.00000	0.00000
2-500	Mean	1.00000	0.00000	1.00000	0.00000	1.00000	0.00000	1.00000	0.00000
	Max	1.00000	0.00000	1.00000	0.00000	1.00000	0.00000	1.00000	0.00000
	Std	0.00000	0.00000	0.00000	0.00000	0.00000	0.00000	0.00000	0.00000
	Min	1.00000	0.00000	1.00000	0.00000	1.00000	0.00000	1.00000	0.00000
2-750	Mean	1.00000	0.00000	1.00000	0.00000	1.00000	0.00000	1.00000	0.00000
	Max	1.00000	0.00000	1.00000	0.00000	1.00000	0.00000	1.00000	0.00000
	Std	0.00000	0.00000	0.00000	0.00000	0.00000	0.00000	0.00000	0.00000
	Min	0.8611	0.00000	1.00000	0.00000	1.00000	0.00000	1.00000	0.00000
3-250	Mean	0.9866	0.10733	1.00000	0.00000	1.00000	0.00000	1.00000	0.00000
	Max	1.0000	0.58000	1.00000	0.00000	1.00000	0.00000	1.00000	0.00000
	Std	0.0319	0.14704	0.00000	0.00000	0.00000	0.00000	0.00000	0.00000
	Min	0.3342	0.03470	0.98947	0.00000	1.00000	0.00000	1.00000	0.00000
3-500	Mean	0.56271	0.24133	0.99965	0.00000	1.00000	0.00000	1.00000	0.00000
	Max	0.83541	0.32000	1.00000	0.00000	1.00000	0.00000	1.00000	0.00000
	Std	0.1118	0.00434	0.00192	0.00000	0.00000	0.00000	0.00000	0.00000
	Min	0.57674	0.00000	0.69795	0.00000	0.89703	0.00000	0.95070	0.00000
3-750	Mean	0.78607	0.10124	0.84577	0.00145	0.97861	0.01596	0.98741	0.00124
	Max	0.90785	0.24158	0.96831	0.08000	0.99682	0.07250	0.99527	0.00278
	Std.	0.09058	0.08974	0.06546	0.01006	0.05679	0.02576	0.08425	0.02785
	Min	0.1101	0.13333	0.66515	0.00000	0.74903	0.00000	0.80731	0.00000
4-250	Mean	0.32948	0.50808	0.85047	0.02866	0.93261	0.00033	0.92765	0.02000
	Max	0.55200	0.69697	0.97831	0.28000	0.99682	0.01000	0.99701	0.03120
	Std.	0.11436	0.07236	0.08335	0.01756	0.06369	0.00182	0.05180	0.04068
	Min	0.25324	0.12698	0.94851	0.00000	0.97849	0.00000	0.98893	0.00000
4-500	Mean	0.59467	0.29629	0.98606	0.00000	0.99871	0.00000	0.99939	0.00000
	Max	0.80176	0.49206	1.00000	0.00000	1.00000	0.00000	1.00000	0.00000
	Std	0.12219	0.10415	0.01726	0.00000	0.00455	0.00000	0.00217	0.00000
	Min	0.58624	0.03500	0.99435	0.00000	1.00000	0.00000	0.99725	0.00000
4-750	Mean	0.78526	0.12866	0.99965	0.00000	1.00000	0.00000	0.99991	0.00000
	Max	0.92958	0.38000	1.00000	0.00000	1.00000	0.00000	1.00000	0.00000
	Std	0.09928	0.01756	0.00117	0.00000	0.00000	0.00000	0.00050	0.00000

It can be seen from table 1 that for 2 knapsacks with 250 items, the results of Coverage of Two Sets tell us that the solutions produced by ICMOA in a certain extent dominates or equals those generated by SPEA and clearly dominates or equals to those obtained by NSGA, NPGA and VEGA, which indicates that the behavior of ICMOA is better than that of the other four algorithm for this test problem. For 2 knapsacks with 500 items and with 750 items, the values of ς_S^I, ς_{Ns}^I, ς_V^I and ς_{Np}^I are all 1 in 30 independent runs, which indicates that the solutions produced by ICMOA clearly dominates or equals those generated by the other four algorithms. For 3 knapsacks with 250 items, the measure shows that solutions produced by ICMOA in a certain extent dominates or equals those generated by SPEA and clearly dominates or equals to those obtained by NSGA, NPGA and VEGA. For 3 knapsacks with 500 items and 750 items, the measure shows that solutions produced by ICMOA in a certain extent dominates or equals those generated by SPEA and NSGA and dominates or equals to those obtained by NPGA and VEGA. For 4 knapsacks with 250 items, the minimum, the mean and the maximum of ς_S^I are all smaller than that of ς_I^S, which indicates a very similar behavior from ICMOA and SPEA but SPEA performs better than ICMOA for this test problems. For 4 knapsacks with 500 items and 750 items, the measure shows that solutions produced by ICMOA also in a certain extent dominates or equals those generated by SPEA NSGA and dominates or equals to those obtained by NPGA and VEGA.

So for these nine problems, the behavior of ICMOA is better than that of SPEA, NSGA, NPGA and VEGA in most of the problems and much better performance in convergence of obtained solutions of ICMOA is observed. So Among the five algorithms, ICMOA seems to provide the best performance in the convergence to the Pareto-optimal front.

Figure 1 and table 1 reveal the superiority of ICMOA over the compared five algorithms in terms of the quality of the nondominated solutions in solving the nine multiobjective 0/1 knapsack problems.

6 Conclusion

In this paper, we have introduced a new multiobjective optimization algorithm- Immune Clonal MO Algorithm (ICMOA) to solve 0/1 knapsack problems. ICMOA is compared with SPEA, HLGA, NPGA, NSGA and VEGA. From the numerical results of the metrics, Coverage of Two Sets, we can see that the solutions obtained from ICMOA dominate those obtained from SPEA, NSGA, NPGA, and VEGA obviously. The Pareto fronts of the two knapsacks with 250, 500 and 750 items for the five algorithms show that ICMOA, in most problems, is able to find much better spread of solutions and better convergence near the true Pareto-optimal front compared with SPEA, NSGA, NPGA and VEGA.

References

1. Schaffer, J.D.: Multiple objective optimization with vector ecaluated genetic algorithms. PhD thesis, Vanderbilt University (1984)
2. Abido, M.A.: Environmental economic power dispatch using multiobjective evolutionary algorithms. IEEE Trans. Power Systems, Vol.18, No. 4, November (2003)
3. Deb, K., Pratap, A., Agarwal, S., Meyarivan, T.: A fast and elitist multiobjective genetic algorithm: NSGA-II. IEEE Transactions on Evolutionary Computation 6(2002) 182-197.
4. Zitzler, E., Thiele, L.: Multiobjective Evolutionary Algorithms: A Comparative Case Study and the Strength Pareto Approach. IEEE Trans. Evolutionary Computation. Vol. 3, No. 4, November (1999)
5. Zitzler, E., Laumanns , M., Thiele. L.:SPEA2: Improving the Strength Pareto Evolutionary Algorithm. Technical Report 103, Computer Engineering and Networks Laboratory (TIK), Swiss Federal Institute of Technology (ETH) Zurich, Gloriastrasse 35, CH-8092 Zurich, Switzerland (2001)
6. Jiao, L.C., Du, H.F.: Artificial immune system: progress and prospect (in Chinese). ACTA ELECTRONICA SINICA, 2003, Vol.31, No.10: 1540-1548

7. Jiao, L.C., Gong, M.G., Shang, R.H., Du, H.F., Lu, B.: Clonal Selection with Immune Dominance and Anergy Based Multiobjective Optimization. Proceedings of the Third International Conference on Evolutionary Multi-Criterion Optimization, EMO 2005, Guanajuato, Mexico, March 9-11, 2005. Springer-Verlag, LNCS 3410, (2005) 474 – 489
8. Shang, R.H., Jiao, L.C., Gong, M.G., Lu, B..: Clonal Selection Algorithm for Dynamic Multiobjective Optimization. Proceedings of the International Conference on Computational Intelligence and Security, CIS 2005. Springer-Verlag, LNAI 3801, (2005) 846 – 851
9. Jiao, L.C., Wang, L.: A novel genetic algorithm based on immunity. IEEE Transactions on Systems, Man and Cybernetics, Part A. Vol.30, No.5, Sept. 2000
10. de Castro, L. N., Von Zuben, F. J.: Learning and Optimization Using the Clonal Selection Principle. IEEE Transactions on Evolutionary Computation, Vol.6, No. 3, 2002: 239-251
11. Coello, Coello, C., A. and Nareli, C. C.: An Approach to Solve Multiobjective Optimization Problems Based on an Artificial Immune System. In: Jonathan Timmis and Peter J. Bentley (editors). Proceedings of the First International Conference on Artificial Immune Systems (2002) 212-221
12. Xue, F., Sanderson, A.C., Graves, R.J.: Pareto-based multi-objective differential evolution. In: Proceedings of the 2003 Congress on Evolutionary Computation (CEC'2003). Volume 2, Canberra, Australia, IEEE Press (2003) 862-869
13. Zitzler, E.: Evolutionary Algorithms for Multiobjective Optimization: Methods and Applications. A dissertation submitted to the Swiss Federal Institute of Technology Zurich for the degree of Doctor of Technical Sciences. Diss. Eth No. 13398. 1999

Training Neural Networks Using Multiobjective Particle Swarm Optimization

John Paul T. Yusiong[1] and Prospero C. Naval Jr.[2]

[1] Division of Natural Sciences and Mathematics,
University of the Philippines-Visayas, Tacloban City, Leyte, Philippines
jtyusiong@up.edu.ph

[2] Department of Computer Science, University of the Philippines-Diliman,
Diliman, Quezon City, Philippines
pcnaval@up.edu.ph

Abstract. This paper suggests an approach to neural network training through the simultaneous optimization of architectures and weights with a Particle Swarm Optimization (PSO)-based multiobjective algorithm. Most evolutionary computation-based training methods formulate the problem in a single objective manner by taking a weighted sum of the objectives from which a single neural network model is generated. Our goal is to determine whether Multiobjective Particle Swarm Optimization can train neural networks involving two objectives: accuracy and complexity. We propose rules for automatic deletion of unnecessary nodes from the network based on the following idea: a connection is pruned if its weight is less than the value of the smallest bias of the entire network. Experiments performed on benchmark datasets obtained from the UCI machine learning repository show that this approach provides an effective means for training neural networks that is competitive with other evolutionary computation-based methods.

1 Introduction

Neural networks are computational models capable of learning through adjustments of internal weight parameters according to a training algorithm in response to some training examples. Yao [20] describes three common approaches to neural network training and these are:

1. for a neural network with a fixed architecture find a near-optimal set of connection weights;
2. find a near-optimal neural network architecture;
3. simultaneously find both a near-optimal set of connection weights and neural network architecture.

Multiobjective optimization (MOO) deals with simultaneous optimization of several possibly conflicting objectives and generates the Pareto set. Each solution in the Pareto set represents a "trade-off" among the various parameters that optimize the given objectives.

L. Jiao et al. (Eds.): ICNC 2006, Part I, LNCS 4221, pp. 879–888, 2006.

In supervised learning, model selection involves finding a good trade-off between at least two objectives: accuracy and complexity. The usual approach is to formulate the problem in a single objective manner by taking a weighted sum of the objectives but Abbass [1] presented several reasons as to why this method is inefficient. Thus, the MOO approach is suitable since the architecture and connection weights can be determined concurrently and a Pareto set can be obtained in a single computer simulation from which a final solution is chosen.

Several multiobjective optimization algorithms [7] are based on Particle Swarm Optimization (PSO) [12] which was originally designed to solve single objective optimization problems. Among the multiobjective PSO algorithms are Multiobjective Particle Swarm Optimization (MOPSO) [5] and Multiobjective Particle Swarm Optimization-Crowding Distance (MOPSO-CD) [16]. MOPSO extends the PSO algorithm to handle multiobjective optimization problems by using an external repository and a mutation operator. MOPSO-CD is a variant of MOPSO that incorporates the mechanism of crowding distance and together with the mutation operator maintains the diversity in the external archive.

Some researches used PSO to train neural networks [18], [19] while others [2]-[3], [9], [23] compared PSO-based algorithms with the backpropagation algorithm and results showed that PSO-based algorithms are faster and get better results in most cases. Likewise, researches that deal with finding a near-optimal set of connection weights and network architecture simultaneously have been done [11], [13], [15], [20]-[22].

The remainder of the paper is organized as follows. Section 2 presents the proposed approach to train neural networks involving a PSO-based multiobjective algorithm. The implementation, experiments done and results are discussed in Section 3. The paper closes with the Summary and Conclusion in Section 4.

2 Neural Network Training Using Multiobjective PSO

The proposed algorithm called NN-MOPSOCD is a multiobjective optimization approach to neural network training with MOPSO-CD [16] as the multiobjective optimizer. The algorithm will simultaneously determine the set of connection weights and its corresponding architecture by treating this problem as a multiobjective minimization problem. In this study, a particle represents a one-hidden layer neural network and the swarm consists of a population of one-hidden layer neural networks.

2.1 Parameters and Structure Representation

The neural network shown in Figure 1 is represented as a vector with dimension D containing the connection weights as illustrated in Figure 2. The dimension of a particle is:

$$D = (I + 1) * H + (H + 1) * O \tag{1}$$

where I, H and O respectively refer to the number of input, hidden and output neurons. The connection weights of a neural network are initialized with random values from a uniform distribution in the range of

$$\left[\frac{-1}{\sqrt{fan-in}}, \frac{1}{\sqrt{fan-in}}\right] \tag{2}$$

where the value of *fan-in* is the number connection weights (inputs) to a neuron.

The number of input and output neurons are problem-specific and there is no exact way of knowing the best number of hidden neurons. However, there are rules-of-thumb [17] to obtain this value. We set the number of hidden neurons to 10. This will be sufficient for the datasets used in the experiments.

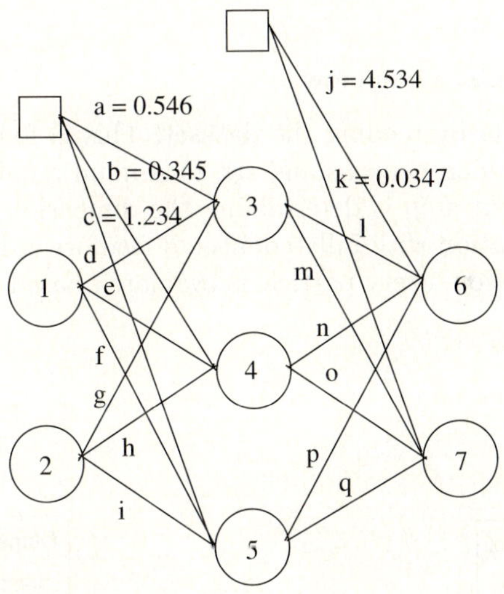

Fig. 1. A fully-connected feedforward neural network

Node	3			4			5			6				7			
Link	a	d	g	b	e	h	c	f	i	j	l	n	p	k	m	o	q
Index	0	1	2	3	4	5	6	7	8	9	10	11	12	13	14	15	16

Fig. 2. The representation for a fully-connected feedfoward neural network

Node Deletion Rules. One advantage of this representation is its simultaneous support for structure (hidden neuron) optimization and weight adaptation. Node deletion is accomplished automatically based on the following idea: *A connection (except bias) is pruned if its value is less than the value of the smallest bias of the network. Thus, a neuron is considered deleted if all incoming connection weights are pruned or if all outgoing connection weights are pruned.*

A neuron is deleted if any of these conditions are met:

1. incoming connections have weights less than the smallest bias of the network,
2. outgoing connections have weights less than the smallest bias of the network.

In Figure 1, the neural network has five biases and these are a, b, c, j, and k, with corresponding values of *0.546*, *0.345*, *1.234*, *4.534* and *0.0347*. Among these biases, k is the smallest bias of the network, thus (1) if the weights of incoming connections e and h are smaller than k then neuron 4 is deleted or (2) if the weights of outgoing connections n and o are smaller than k then neuron 4 is deleted. In short, *if a neuron has incoming connections or outgoing connections whose weights are smaller than the smallest bias of the network then the neuron is automatically removed.* Thus, this simple representation of the NN enables NN-MOPSOCD to dynamically change the structure and connection weights of the network.

2.2 NN-MOPSOCD Overview

NN-MOPSOCD starts by reading the dataset. This is followed by setting the desired number of hidden neurons and the maximum number of generation for MOPSO-CD. The next step is determining the dimension of the particles and initializing the population with fully-connected feedforward neural network particles. In each generation, every particle is evaluated based on the two objective

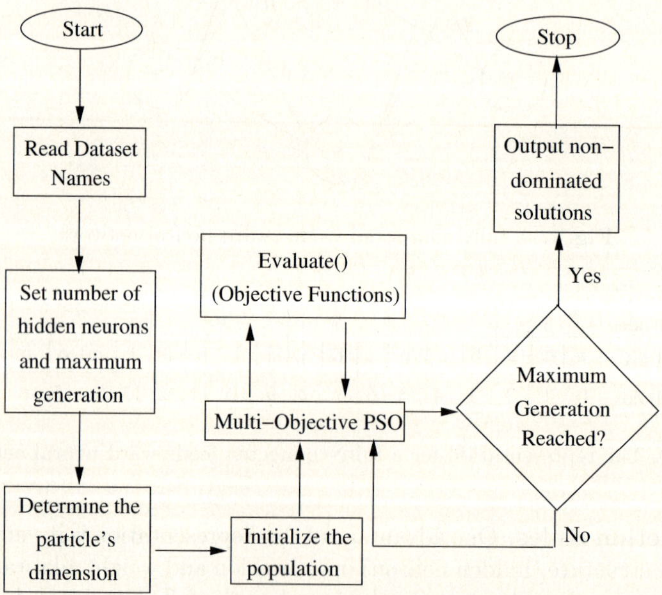

Fig. 3. The procedure for training neural networks with NN-MOPSOCD

functions and after the maximum generation is reached the algorithm outputs a set of non-dominated neural networks. Figure 3 illustrates this procedure.

2.3 Objective Functions

Two objective functions are used to evaluate the neural network particle's performance. The two objective functions are:

1. Mean-squared error (MSE) on the training set
2. Complexity based on the Minimum Description Length (MDL) principle

Mean-squared Error. This is the first objective function. It determines the neural network's training error.

Minimum Description Length. NN-MOPSOCD uses MDL as its second objective function to minimize the neural network's complexity. It follows the MDL principle described in [4], [8]. As stated in [10], the MDL principle asserts that the best model of some data is the one that minimizes the combined cost of describing the model and describing the misfit between the model and the data. MDL is one of the best methods to minimize complexity of the neural network structure since it minimizes the number of active connections by considering the neural network's performance on the training set.

The total description length of the data misfits and the network complexity can be minimized by minimizing the sum of two terms:

$$MDL = Error + Complexity \tag{3}$$

$$Error = 1.0 - \frac{num\ of\ CorrectClassification}{numofExamples} \tag{4}$$

$$Complexity = \frac{num\ of\ ActiveConnections}{TotalPossibleConnections} \tag{5}$$

where: *Error* is the error in classification while *Complexity* is the network complexity measured in terms of the ratio between active connections and the total number of possible connections.

3 Experiments and Results

The goal is to use a PSO-based multiobjective algorithm to simultaneously optimize architectures and connection weights of neural networks. To test its effectiveness, NN-MOPSOCD had been applied to the six datasets stated in Table 1. For each dataset, the experiments were repeated thirty (30) times to minimize the influence of random effects. Each experiment uses a different randomly generated initial population.

Table 1 shows the number of training and test cases, classes, continuous and discrete features, input and output neurons as well as the maximum generation count used for training. These datasets are taken from the UCI machine learning repository [14]. For the breast cancer dataset, 16 of the training examples were deleted due to some missing attribute values thus reducing the total number of training examples from 699 to 683. Table 2 shows the parameter settings used in the experiments.

Table 1. Description of the datasets used in the experiments

Domains	Train Set	Test Set	Class	Cont.	Disc.	Input	Output	MaxGen
Monks-1	124	432	2	0	6	6	1	500
Vote	300	135	2	0	16	16	1	300
Breast	457	226	2	9	0	9	1	300
Iris	100	50	3	4	0	4	3	500
Heart	180	90	2	6	7	13	1	500
Thyroid	3772	3428	3	6	15	21	3	200

Table 2. Parameters and their corresponding values of NN-MOPSOCD

Parameters	NN-MOPSOCD
Optimization Type	Minimization
Population Size	100
Archive Size	100
Objective Functions	2
Constraints	0
Lower Limit of Variable	-100.0
Upper Limit of Variable	100.0
Probability of Mutation (pM)	0.5

3.1 Results and Discussion

For each dataset, the results from each of the 30 independent runs of the NN-MOPSOCD algorithm were recorded and analyzed. Tables 3-6 show the results obtained from the experiments. The results show that the average MSE on the training and test sets are small and the misclassification rate are slightly higher.

Least-Error versus Least-Complex. Among the neural networks in the Pareto set, two neural networks are considered, the least-error and least-complex neural networks. Least-error neural networks are those that have the smallest value on the first objective function in the Pareto sets while least-complex neural networks are those that have the smallest value on the second objective function in the Pareto sets. Tables 7-9 compare the least-error and least-complex neural networks. It can be seen that least-error neural networks involve a higher number of connections than least-complex neural networks but perform significantly better than the least-complex neural networks.

Comparison of Performance and Complexity. The purpose of the comparison is to show that the performance of this PSO-based multiobjective approach is comparable to that of the existing algorithms which optimize architectures and connections weights concurrently in a single objective manner. When presented with a completely new set of data, the capability to generalize is one of the most significant criteria to determine the effectiveness of neural network learning. Table 10 compares the results obtained with that of MGNN [15] while Table 11 compares the results with EPNet [21] in terms of the error on the test set and the number of connections used.

Table 3. Average number of neural networks generated

Datasets	Average	Median	Std. Dev.
Monks-1	22.500	21.500	9.391
Vote	31.633	30.000	15.566
Breast	28.700	28.000	6.444
Iris	20.300	20.000	7.145
Heart	21.600	20.500	10.180
Thyroid	22.167	23.000	5.025

Table 4. Average mean-squared error on the training and test set

Datasets	Training Set			Test Set		
	Average	Median	Std. Dev.	Average	Median	Std. Dev.
Monks-1	3.38%	3.40%	0.027	4.59%	4.80%	0.033
Vote	2.06%	1.81%	0.009	2.28%	2.04%	0.008
Breast	1.32%	1.20%	0.003	1.68%	1.61%	0.003
Iris	2.39%	1.46%	0.020	4.58%	4.39%	0.019
Heart	7.37%	7.40%	0.007	7.62%	7.44%	0.012
Thyroid	5.02%	4.97%	0.007	5.21%	5.14%	0.006

Table 5. Average percentage of misclassification on the training and test set

Datasets	Training Set			Test Set		
	Average	Median	Std. Dev.	Average	Median	Std. Dev.
Monks-1	8.79%	8.28%	0.075	12.40%	11.49%	0.094
Vote	4.53%	3.79%	0.021	5.34%	4.83%	0.021
Breast	2.94%	2.76%	0.007	3.93%	3.92%	0.007
Iris	2.97%	1.92%	0.027	6.24%	5.96%	0.026
Heart	18.19%	18.36%	0.024	19.34%	18.82%	0.041
Thyroid	7.11%	6.91%	0.008	7.42%	7.22%	0.006

Table 6. Average number of connections used

Datasets	Average	Median	Std. Dev.
Monks-1	32.72	32.02	10.32
Vote	67.34	63.65	28.34
Breast	48.13	48.46	13.18
Iris	66.02	65.50	7.16
Heart	65.35	67.05	26.04
Thyroid	149.27	145.41	30.91

Table 7. Average mean-squared error on the training and test set

Datasets	Training Set		Test Set	
	Least-Error	Least-Complex	Least-Error	Least-Complex
Monks-1	2.16%	5.70%	3.74%	6.66%
Vote	1.56%	2.82%	2.07%	2.66%
Breast	0.87%	2.57%	1.45%	2.66%
Iris	1.42%	4.77%	3.81%	6.84%
Heart	6.30%	9.04%	6.87%	8.99%
Thyroid	4.12%	7.02%	4.43%	6.98%

Table 8. Average percentage of misclassification on the training and test set

Datasets	Training Set		Test Set	
	Least-Error	Least-Complex	Least-Error	Least-Complex
Monks-1	5.54%	13.71%	10.08%	16.75%
Vote	3.56%	5.48%	5.11%	5.46%
Breast	1.94%	4.72%	3.39%	5.32%
Iris	2.20%	5.40%	5.20%	9.20%
Heart	16.04%	21.09%	17.70%	21.59%
Thyroid	7.00%	7.27%	7.57%	7.23%

Table 9. Average number of connections used

Datasets	Least-Error	Least-Complex
Monks-1	49.00	17.63
Vote	113.70	35.47
Breast	89.80	17.87
Iris	77.37	55.23
Heart	108.17	28.60
Thyroid	195.77	117.17

Table 10. Performance comparison between MGNN and NN-MOPSOCD

Algorithms	**MSE on Test set**		**Number of Connections**	
	Breast	**Iris**	**Breast**	**Iris**
MGNN-ep	3.28%	6.17%	80.87	56.38
MGNN-rank	3.33%	7.28%	68.46	**47.06**
MGNN-roul	3.05%	8.43%	76.40	55.13
NN-MOPSOCD	**1.68%**	**4.58%**	**48.13**	66.02

Table 11. Performance comparison between EPNet and NN-MOPSOCD

Algorithms	**MSE on Test set**			**Number of Connections**		
	Breast	**Heart**	**Thyroid**	**Breast**	**Heart**	**Thyroid**
EPNet	**1.42%**	12.27%	**1.13%**	**41.00**	92.60	208.70
NN-MOPSOCD	1.68%	**7.62%**	5.21%	48.13	**65.35**	**149.27**

4 Conclusion

This work dealt with the neural network training problem through a multi-objective optimization approach using a PSO-based algorithm to concurrently optimize the architectures and connection weights. Also, the proposed algorithm dynamically generates a set of near-optimal feedforward neural networks with their corresponding connection weights. Our approach outperforms most of the evolutionary computation-based methods found in existing literature to date.

References

1. Abbass, H.: An Evolutionary Artificial Neural Networks Approach to Breast Cancer Diagnosis. Artificial Intelligence in Medicine, 25(3) 2002 265-281
2. Alfassio Grimaldi, E., Grimaccia, F., Mussetta, M. and Zich, R.: PSO as an Effective Learning Algorithm for Neural Network Applications. Proceedings of the International Conference on Computational Electromagnetics and its Applications, Beijing - China (2004) 557-560
3. Al-kazemi, B. and Mohan, C.: Training Feedforward Neural Networks using Multi-phase Particle Swarm Optimization. Proceedings of the 9th International Conference on Neural Information Processing (ICONIP 2002), Singapore (2002)
4. Barron, A., Rissanen, J. and Yu, B.: The Minimum Description Length Principle in Coding and Modeling. IEEE Trans. Inform. Theory, 44 (1998) 2743-2760
5. Coello, C. and Lechuga, M.: MOPSO: A Proposal for Multiple Objective Particle Swarm Optimization. Proceedings of the IEEE Congress on Evolutionary Computation (CEC 2002). Honolulu, Hawaii USA (2002)
6. Deb K., Pratap A., Agarwal S. and Meyarivan, T.: A Fast and Elitist Multiobjective Genetic Algorithm: NSGA-II. IEEE Transactions on Evolutionary Computation, vol. 6(2) (2002) 182-197
7. Fieldsend, J.: Multi-Objective Particle Swarm Optimisation Methods. Technical Report # 419, Department of Computer Science, University of Exeter (2004)

8. Grunwald, P.: A Tutorial Introduction to the Minimum Description Length Principle. Advances in Minimum Description Length: Theory and Applications. MIT Press (2004)
9. Gudise V. and Venayagamoorthy G.: Comparison of Particle Swarm Optimization and Backpropagation as Training Algorithms for Neural Networks. IEEE Swarm Intelligence Symposium, Indianapolis, IN, USA (2003) 110-117
10. Hinton, G. and van Camp, D.: Keeping Neural Networks Simple by Minimizing the Description Length of the Weights. Proceedings of COLT-93 (1993)
11. Jin, Y., Sendhoff, B. and Körner, E.: Evolutionary Multi-objective Optimization for Simultaneous Generation of Signal-type and Symbol-type Representations. Third International Conference on Evolutionary Multi-Criterion Optimization. LNCS 3410, Springer, Guanajuato, Mexico (2005) 752-766
12. Kennedy, J. and Eberhart, R.: Particle Swarm Optimization. Proceedings of the 1995 IEEE International Conference on Neural Networks, Perth, Australia, vol. 4 (1995) 1942-1948
13. Liu, Y. and Yao, X.: A Population-Based Learning Algorithm Which Learns Both Architectures and Weights of Neural Networks. Chinese J. Advanced Software Res., vol. 3, no. 1 (1996) 54-65
14. Newman, D., Hettich, S., Blake, C. and Merz, C.: UCI Repository of machine learning databases. Irvine, CA: University of California, Department of Information and Computer Science (1998)
15. Palmes, P., Hayasaka, T. and Usui, S.: Mutation-based Genetic Neural Network. IEEE Transactions on Neural Networks, vol. 16, no 3 (2005) 587-600
16. Raquel, C. and Naval, P.: An Effective Use of Crowding Distance in Multiobjective Particle Swarm Optimization. Proceedings of the 2005 Conference on Genetic and Evolutionary Computation (Washington DC, USA, June 25-29, 2005). H. Beyer, Ed. GECCO '05. ACM Press, New York, NY (2005) 257-264
17. Shahin, M., Jaksa, M. and Maier, H.: Application of Neural Networks in Foundation Engineering. Theme paper to the International e-Conference on Modern Trends in Foundation Engineering: Geotechnical Challenges and Solutions, Theme No. 5: Numerical Modelling and Analysis, Chennai, India (2004)
18. Sugisaka, M. and Fan, X.: An Effective Search Method for Neural Network Based Face Detection Using Particle Swarm Optimization. IEICE Transactions 88-D(2) (2005) 214-222
19. van den Bergh, F.: Particle Swarm Weight Initialization in Multi-layer Perceptron Artificial Neural Networks. Development and Practice of Artificial Intelligence Techniques (Durban, South Africa) (1999) 41-45
20. Yao, X.: Evolving Artificial Neural Networks. Proceedings of the IEEE, vol. 87 (1999) 1423-1447
21. Yao, X. and Liu, Y.: Evolving Artificial Neural Networks through Evolutionary Programming, Presented at the Fifth Annual Conference on Evolutionary Programming, 29 February-2 March 1996, San Diego, CA, USA MIT Press (1996) 257-266
22. Yao, X. and Liu, Y.: Towards Designing Artificial Neural Networks by Evolution. Applied Mathematics and Computation. vol. 91 no. 1 (1998) 83-90
23. Zhao, F., Ren, Z., Yu, D., and Yang Y.: Application of An Improved Particle Swarm Optimization Algorithm for Neural Network Training. Proceedings of the 2005 International Conference on Neural Networks and Brain. Beijing, China, Vol. 3 (2005) 1693-1698

New Evolutionary Algorithm for Dynamic Multiobjective Optimization Problems*

Chun-an Liu[1,2] and Yuping Wang[1]

[1] School of Computer Science and Engineering, Xidian University, Xi'an 710071, China
ywang@xidian.edu.cn

[2] Faculty of Mathematics, Baoji College of Arts and Science, Baoji 721013, China
Liu2006@126.com

Abstract. In this paper, a new evolutionary algorithm for dynamic multi-objective optimization problems(DMOPs) is proposed. first, the time period is divided into several equal subperiods. In each subperiod, the DMOPs is approximated by a static multi-objective optimization problem(SMOP). Second, for each SMOP, the static rank variance and the static density variance of the population are defined. By using the two static variance of the population, each SMOP is transformed into a static bi-objective optimization problem. Third, a new evolutionary algorithm is proposed based on a new mutation operator which can automatically check out the environment variation. The simulation results indicate the proposed algorithm is effective.

1 Introduction

Many real-world problems naturally fall within the purview of optimization problems, in which the objective function is not only decided by the decision variables but also varies with the time (or termed as environment). Thus, the optimum solution also varies with the environment changing. This kind of optimization problem is called dynamic optimization problems(DOPs). DOPs include dynamic simple-objective optimization problems(DSOPs) and dynamic multi-objective optimization problems(DMOPs). When DSOPs is considered, several studies are available in the literature [1], [2], [3]. However, when DMOPs is concerned, very few studies are available in the literature [4], [5].

In this paper, a new evolutionary algorithm for DMOPs is proposed. we divide the time period into several smallest equal subperiods. In each subperiod, the DMOPs is approximated by a static multi-objective optimization problem(SMOP). Thus, the DMOPs is approximately transformed into several SMOPs. This paper also develops a new changing feedback operator which can automatically check out the environment variation. The simulation is made and the results demonstrate the effectiveness of the proposed algorithm.

* This work was supported by the National Natural Science Foundation of China (60374063).

L. Jiao et al. (Eds.): ICNC 2006, Part I, LNCS 4221, pp. 889–892, 2006.

2 Transformation of DMOPs

We consider the following dynamic multi-objective optimization problems

$$\begin{cases} \min\limits_{x\in[L,U]_t} f(x,t) = (f_1(x,t), f_2(x,t), \cdots, f_M(x,t)) \\ s.t. \quad g_i(x,t) \leq 0 \quad i = 1,2,\cdots,p \end{cases} \tag{1}$$

Where $t \in [t_0, t_s] \subset R$, $x \in R^n$ is called decision vector, $g_i(x,t)$ is called constraint conditions depending on time variable t. $\Omega(t) = \{x|g_i(x,t) \leq 0, i = 1,2,\cdots,p\}$ is called decision vector space, $[L,U]_t = \{x = (x_1, x_2, \cdots, x_n)|l_i(t) \leq x_i \leq u_i(t), i = 1,2,\cdots,n\}$ is called search space.

2.1 Continuous Time Variable Discretization

For DMOPs, the time period $[t_0, t_s]$ is divided into several equal and disintersection subperiods $[t_{i-1}, t_i], i = 1 \sim s$, for $\forall i \in \{1,2,\cdots,s\}$, denote $\Delta t_i = t_i - t_{i-1}$, when $\Delta t_i \longrightarrow 0$, each period Δt_i is regard as a fixed environment t_i; thus, the Pareto solutions of the DMOPs can be approximated by superimposition of the Pareto solutions of limits SMOPs in different fixed environment t.

2.2 Static Rank Variance

Under fixed environment t, suppose that the $k-$th generation population $p^k(t)$ is made of individuals $x_1^k(t), x_2^k(t), \cdots, x_N^k(t)$. $n_i^k(t)$ is the number of individual $x_i^k(t)$ dominated by all individuals in $p^k(t)$, then $1 + n_i^k(t)$ is called the rank of individual $x_i^k(t)$, and usually it can be written in the form of $r_i^k(t) = 1 + n_i^k(t)$. Let $\overline{r}(t) = \frac{1}{N}\sum\limits_{i=1}^{N} r_i^k(t)$ denote the mean of rank of all individual in the population $p^k(t)$, then $r(t) = \frac{1}{N}\sum\limits_{i=1}^{N}(\overline{r}(t) - r_i^k(t))^2$ can be defined as the static rank variance of the $k-$th generation population $p^k(t)$.

2.3 Static Density Variance

Under fixed environment t, suppose that the $k-$th generation population $p^k(t)$ is made of individuals $x_1^k(t), x_2^k(t), \cdots, x_N^k(t)$. $U_1^k(t), U_2^k(t), \cdots, U_N^k(t)$ are the solutions corresponding to the individuals $x_1^k(t), x_2^k(t), \cdots, x_N^k(t)$ in objective space, respectively. Suppose that $D_i^k(t) = \min\{dist(U_i^k(t), U_j^k(t))|j \neq i, j = 1 \sim N\}$ is the distance between $U_i^k(t)$ and $U_j^k(t)$. Denote $\bar{D}_i^k(t) = \frac{1}{N}\sum\limits_{i=1}^{N} D_i^k(t)$, then, for each of solution $U_i^k(t)(i = 1,2,\cdots,N)$, $|m_i^k(t)|$ is called the density of individual $x_i^k(t)$ and it was usually written in the form of $\rho_i^k(t) = |m_i^k(t)|$, where $m_i^k(t) = \{j|\|U_j^k(t) - U_i^k(t)\|_p \leq \bar{D}_i^k(t), j = 1,2,\cdots,i-1,i+1,\cdots,N\}$. So the density variance of the $k-$th generation population in objective space is defined as $\rho(t) = \frac{1}{N}\sum\limits_{i=1}^{N}(\bar{\rho}(t) - \rho_i^k(t))^2$, where $\bar{\rho}(t) = \frac{1}{N}\sum\limits_{i=1}^{N}\rho_i^k(t)$.

2.4 New Model of Static Multi-objective Optimization

It follows that if the above variance are regarded as the objective function to be optimized, then, under fixed environment t, the problem (1) can be approximated by problem (2) as following. As a result, utilizing formula (2), the DMOPs can be approximately transformed into several SMOPs.

$$\min\{r(t), \rho(t)\} \tag{2}$$

3 New Evolutionary Algorithm for DMOPs

3.1 Environment Changing Feedback Operator

In order to make the algorithm adapt and recover from the change of environment, a environment changing feedback operator is proposed as follows.

$$\varepsilon(t) = \frac{\sum_{i=1}^{N} \|f(x_i, t) - f(x_i, t-1)\|_p}{N\|R(t) - U(t)\|_p} \tag{3}$$

where $R(t), U(t)$ represent the worst solution and the best solution of problem in objective space under fixed environment t. $x_i(i = 1, 2, \cdots, N)$ are the individuals used to test the change of environment. N is the population size.

3.2 New Dynamic Multi-objective Evolutionary Algorithm(DMEA)

1. Divide $[t_0, t_s]$ into several smallest subperiods according to 2.1. Suppose that different environment $t_0, t_1, \cdots, t_s$ are obtained, let $t = t_0$.
2. For fixed environment t_i, generate initial population $P^0(t_i)$ in $[L, U]_t$ and let the number of generation $k = 0$.
3. Select a pair of parents $(x_i^k(t_i), x_j^k(t_i))$ from $p^k(t_i)$, adopt the arithmetic crossover operator to generate two offsprings. All the offsprings are kept in the set $c^k(t_i)$.
4. Select parent from $c^k(t_i)$, and utilize the mutation operator ([4]) to generate an offspring. The set of all these offsprings is denoted as $\bar{p}^k(t_i)$.
5. Utilize the fitness function $F_i^k(t_i) = \frac{1}{1+r_i^k(t_i)}$(where $r_i^k(t_i)$ represents the rank of individual $x_i^k(t_i)$)to select N individuals from $p^k(t_i) \bigcup \bar{p}^k(t_i) \bigcup c^k(t_i)$ and constitute next population $p^{k+1}(t_i)$.
6. Computer $\varepsilon(t_i)$, if $\varepsilon(t_i) > \eta(\eta$ is a user-defined parameter), let $t_i = t_{i+1}$, go to step2; otherwise, let $k = k + 1$, go to step3.

4 Simulation Results

To evaluate the efficiency of DMEA, we choose two dynamic multiobjective functions. $G1$ is the function structured by the author and $G2$ is borrowed from [5].

$$G1 : \min f(x,t) = (t \cdot x^2, (1-t) \cdot (x-2)^2), s.t.\ x \in [0,2], t \in [0,1]$$

$$G2 : \begin{cases} \min f(x,t) = (f_1(x,t), f_2(x,t)) \\ s.t.\ \ f_1(x,t) = t(x_1^2 + (x_2-1)^2) + (1-t)(x_1^2 + (x_2+1)^2 + 1) \\ \qquad f_2(x,t) = t(x_1^2 + (x_2-1)^2) + (1-t)((x_1-1)^2 + (x_2^2+2)) \\ \qquad -2 \le x_1, x_2 \le 2, t \in [0,1] \end{cases}$$

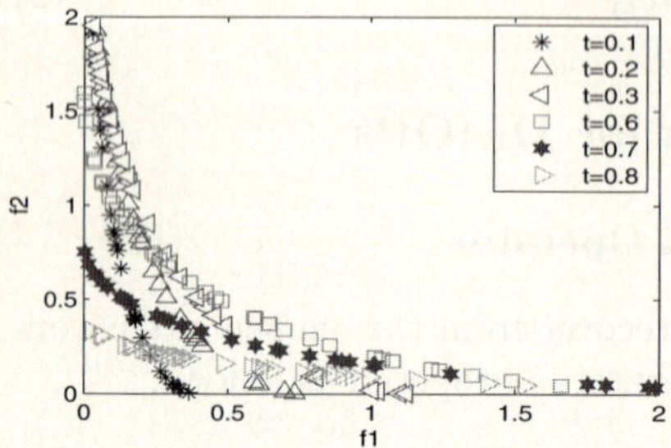

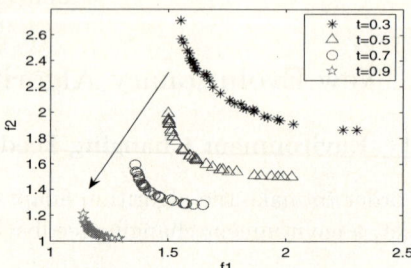

Fig. 1. The Pareto front obtained by DMEA for test problem $G1$

Fig. 2. The Pareto front obtained by DMEA for test problem $G2$

In the simulation, the population size $N = 100$, crossover probability $p_c = 0.9$, mutation probability $p_m = 0.05$. $\eta = 0.01$. For $G1$, we divide the time period $[0,1]$ into several subperiod and obtained different environment(seeing in Fig.1). For $G2$, we adopted the same value for t as the reference.

In Figure 1 and 2, Pareto fronts achieved by DMEA under different environment t are visualized. From a careful observation of these two figures show that the DMEA can find the true Pareto front in different environment for DMOPs and the solutions in the Pareto front can uniformly distributed.

References

1. C. Ronnewinkel, C. O. Wilke, T. Martinetz.: Genetic Algorithms in Time-Dependent Environments. In Theoretical Aspects of Evolutionary Computing, L. Kallel, B. Naudts, and A. Rogers, Eds. Berlin, Germany: Springer-Verlag (2000) 263-288
2. K. E. Parsopoulos, M.n. Vrahatis.: Unified Particle Swarm Optimization in Dynamic Environments. In Proc. Evo Workshops 2005, LNCS 3449, F. Rothlauf et al. Eds. Berlin, Germany: Springer-Verlag (2005) 590-599
3. Blanckwell, T., Branke, J.: Multi-Swarm Optimization in Dynamic Environment. In: Lecture Notes in Computer Science. Volume 3005. Springer-Verlag (2004) 489-500
4. Z. Bingul, A. Sekmen, S. Zein-Sabatto.: Adaptive Genetic Algorithms Applied to Dynamic Multiobjective Programs. In Proc. Artificial Neural Networks Engineering Conf., C.H. Dagli, A. L. Buczak, J. Ghosh, M. Embrechts, O. Ersoy, and S. Kercel, Eds. New York: Springer-Verlag (2000) 273-278
5. Yaochu Jin, Bernhard Sendhoff.: Constructing Dynamic Optimization Test Problems Using the Multiobjective Optimization Concept. In Proc. Evo Workshops 2004, LNCS 3005, G. R. Raidl et al. Eds. Berlin, Germany: Springer-Verlag (2004) 525-536

Simulation for Interactive Markov Chains*

Xiying Zhao[1], Lian Li[1], and Jinzhao Wu[1,2]

[1] The Information Engineering College of Lanzhou University,
Lanzhou 730000, China
zhaoty03@st.lzu.edu.cn

[2] Chengdu Institute of Computer Applications
Chinese Academy of Sciences, Chengdu 610041, China
hiwujz@web.de

Abstract. Interactive Markov chains (IMCs) are compositional performance evaluation models which can be used to powerfully model concurrent systems. Simulations that are abstracted from internal computation have been proven to be useful for verification of compositely defined transition system. In the literature of stochastic extension of this transition system, computing simulation preorders are rare. In this paper strong(weak) simulation is introduced for IMCs. The main result of the paper is that we give algorithms to decide strong(weak) simulation preorder for IMCs with a polynomial-time complexity in the number of states of the transition system.

1 Introduction

Process algebra is a prominent abstract specification language used to describe concurrent systems. It views systems as composition of many small subsystems and provides powerful operators to support such compositional description. On the other hand, continuous-time Markov chains (CTMCs) are one of the most important classes of stochastic processes. They are widely used in performance analysis model and offer numerous means to solve the state-based probabilities available. Interactive Markov chains (IMCs) are one of models, which combine interactive processes and CTMCs orthogonally. IMCs inherit both merits of the process algebra and CTMCs and form an elegant framework for compositional performance evaluation.

Labelled transition systems (LTSs) together with a variety of widely accepted equivalence relations (e.g. bisimulation, trace or failure equivalence) and preorders (e.g. simulation or testing preorders) have proved to be very useful for modeling and analyzing concurrent processes. Typically, the equivalences or preorders are defined as relations on the state space of a LTS but they can be extended for the comparison of two processes. Many equivalences and preorders relations have been defined on probabilistic distributed transition systems[1,19,24,16]. Baier[2] gives the definition of strong (weak) (Bi)simulation relations on CTMCs, and [23] difined the bisimulation equivalences and simulation preorders for IMCs. Many deciding algrithms for bisimulations and simulation for probabilistic transition systems and CTMCs have been presented in

* This work is supported by National Natural Science Foundation of China (90612016) and supported by the "Hundreds-Talent Program" of Chinese Academy of Sciences.

L. Jiao et al. (Eds.): ICNC 2006, Part I, LNCS 4221, pp. 893–902, 2006.

the literature, e.g. bisimulation can be decided in time $O(m \cdot \log n)$[22], weak bisimulation in time $O(n^3)$ [8,18] and weak simulation in time $O(n^4 \cdot m)$ [9], where n is the number of states and m the number of transitions of underlying transition system. But, algorithms for testing simulations for IMCs are missing now.

The main contributions of this paper are that we present algorithms for testing strong and weak simulations for IMCs in the sense of [23,4]. They appear to be rather natural extensions of the corresponding testing algorithms in probabilistic and non-probabilistic cases. The main idea of our algorithms is to reduce the question of whether a state s of an IMC simulates a state s' to a two-phased refinement technique. One for action transition simulation, another for CTMC transition simulation. Before a CTMC simulation test, we have to eliminate internal action transitions by the maximal process assumption of IMCs. Furthermore, nondeterminism is considered in action transitions of our algorithms. When an IMC reduces to a CTMC, our algorithms just reduce to the original one. The coincidence benefits from the orthogonal construction of IMCs. As a result, we get two polynomial time algorithms for testing strong (weak) simulation on IMCs.

The rest of this paper is organized as following. Section 2 introduces our system model, interactive Markov chains. In Section 3, we give the definitions of strong and weak simulation on IMCs. The testing algorithms for simulation are given in Section 4 and Section 5, and Section 6 concludes the paper.

2 Interactive Markov Chains

Interactive Markov chains are proposed by H. Hermanns [12], which combine interactive processes and CTMCs together and aim at compositional performance evaluation. In this section, we give a brief introduction to IMCs, serving as our underlying model.

Let AP be a fixed finite set of atomic propositions, and $\mathbb{R}_{\geqslant 0}$ denote the set of non-negative reals.

Definition 1. A (labelled) *CTMC* is a tuple $C = (S, \mathbf{R}, L)$ where:

- S is a countable set of states,
- $\mathbf{R} : S \times S \rightarrow \mathbb{R}_{\geqslant 0}$ is a *rate matrix*, and
- $L : S \rightarrow 2^{AP}$ is a labelling function which assigns to each state $s \in S$ the set $L(s)$ of atomic propositions that are valid in s.

Intuitively, $\mathbf{R}(s, s') > 0$ iff there is a transition from s to s', and the probability of this transition taking place within t time units is $1 - e^{-\mathbf{R}(s,s')\cdot t}$, an exponential distribution with rate $\mathbf{R}(s, s')$. If $\mathbf{R}(s, s') > 0$ for more than one state s', a *race* between the outgoing transition from s exists, and the probability of which the state s' wins the race is given by $\mathbf{P}(s, s') = \mathbf{R}(s, s')/E(s)$ where $E(s) = \sum_{s' \in S} \mathbf{R}(s, s')$, denoting the *total rate* at which any transition outgoing from state s is taken. If $E(s) = 0$, we call s an absorbing state, and define $\mathbf{P}(s, s') = 0$. Consequently, when there exists race condition, the probability to move from s to s' within t time units is given by $\mathbf{P}_t(s, s') = \mathbf{P}(s, s') \cdot (1 - e^{-E(s)\cdot t})$. For $C \subseteq S$, let $\mathbf{R}(s, C) = \sum_{s' \in C} \mathbf{R}(s, s')$ and $\mathbf{P}(s, C) = \sum_{s' \in C} \mathbf{P}(s, s')$ for simplicity.

Definition 2. Let *Act* denote the universal set of actions, ranged over by $a, b, \cdots$. An *interactive Markov chain (IMC)* is a quadruple $M = (S, \mathcal{A}, \longrightarrow, \dashrightarrow)$, where

- S is a nonempty set of states,
- $\mathcal{A} \subseteq Act$ is a countable set of actions,
- $\longrightarrow \subset S \times \mathcal{A} \times S$ is a set of *interactive transitions*, and
- $\dashrightarrow \subset S \times \mathbb{R}_{\geqslant 0} \times S$ is a set of *Markovian transitions*.

Here, we restrict the Markovian transition to further satisfaction that for each pair of states (s_1, s_2) there is at most one Markovian transition between them. We can see that IMCs combine the interactive processes and CTMCs as orthogonal to each other as possible except that the labelling function of CTMCs is absent. In fact, each interactive process is isomorphic to an interactive Markov chain, and each CTMC is isomorphic to an interactive Markov chain as well [12].

3 Strong and Weak Simulation Relations

Since there are two different kinds of transitions in the IMCs, it is important how these two different transitions perform when they are both present. As to internal action, we assume that its occurrence can not be delayed. This assumption is called the *maximal progress assumption* [12]. Therefore, when both an internal interactive transition and a Markovian transition emanate from one state, the internal interactive transition is performed and the Markovian transition is ignored. For observational action, its occurrence depends on the sojourn time of the state where this interactive transition emanates.

Simulation of two states is defined in terms of simulation of their successor states [20,16,4]. In IMCs action transitions and Markovian transitions coexist. Meaningful preorder should reflect their coexistence. We use [23] as the definition of simulation.

Let τ ($\tau \in Act$) represent the internal action and $s \stackrel{\tau}{\not\longrightarrow}$ to denote the absence of such internal transitions. Let $M = (S, \mathcal{A}, \longrightarrow, \dashrightarrow)$ be an IMC. We now define strong and weak Simulation on IMCs as follows:

Definition 3. Let $\mu, \mu' \in Dist(S), \mathcal{R} \subseteq S \times S$. A *weight function* for μ and μ' with respect to $\mathcal{R}$ is a function $\Delta : S \times S \to [0, 1]$ such that:

1. $\Delta(s, s') > 0$ implies $s\mathcal{R}s'$,
2. $\mu(s) = K_1 \cdot \sum_{s' \in S} \Delta(s, s')$ for any $s \in S$,
3. $\mu'(s') = K_2 \sum_{s \in S} \Delta(s, s')$ for any $s' \in S$
   where $K_1 = \sum_{s \in S} \mu(x)$ and $K_2 = \sum_{s' \in S} \mu'(s')$.

We write $\mu \sqsubseteq_{\mathcal{R}} \mu'$ iff there exists a weight function for μ and μ' w.r.t. $\mathcal{R}$.

Definition 4. A preorder relation $\mathcal{R}$ on S is a *strong simulation* iff for all $s_1\mathcal{R}s_2$:

1. $s_1 \stackrel{a}{\longrightarrow} s_1', a \in \mathcal{A} \Rightarrow \exists s_2' \in S, s_2 \stackrel{a}{\longrightarrow} s_2'$ and $s_1'\mathcal{R}s_2'$,
2. $s_1 \stackrel{\tau}{\not\longrightarrow}$ and $s_2 \stackrel{\tau}{\not\longrightarrow} \Rightarrow \mathbf{P}(s_1, \cdot) \sqsubseteq_{\mathcal{R}} \mathbf{P}(s_2, \cdot)$ and $E(s_1) \leqslant E(s_2)$.

s_2 strongly simulates s_1, denoted $s_1 \precsim s_2$, iff there exists a strong simulation relation $\mathcal{R}$ on S with $s_1\mathcal{R}s_2$.

Example 1. Figure 1 illustrates a strong simulation relation on IMC. We have $s_1 \precsim s_2$. The simulation for interactive transition is obvious. For Markovian transitions, we note that $E(s_1) = 2 < E(s_2) = 3$, $\mathbf{P}(s_1, u_1) = \mathbf{P}(s_1, u_2) = \frac{1}{2} = \frac{3}{6}$, $\mathbf{P}(s_2, v_1) = \frac{1}{3} = \frac{2}{6}$, $\mathbf{P}(s_2, v_2) = \frac{2}{3} = \frac{4}{6}$, and the weight function is defined by $\Delta(u_1, v_1) = \frac{2}{6}$, $\Delta(u_1, v_2) = \frac{1}{6}$, $\Delta(u_2, v_2) = \frac{3}{6}$. By simple computation we can check that it satisfies the definition of strong simulation.

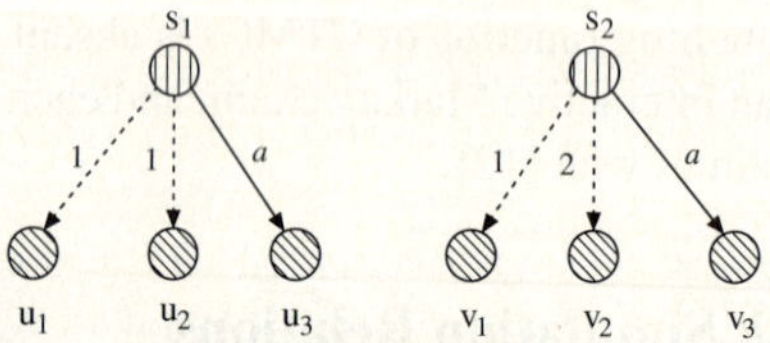

Fig. 1. Strong simulation preorder on IMCs

Let $\stackrel{\tau}{\Longrightarrow}$ represent the reflexive and transitive closure of $\stackrel{\tau}{\longrightarrow}$, and $\stackrel{a}{\Longrightarrow}$ denote $\stackrel{\tau}{\Longrightarrow}\stackrel{a}{\longrightarrow}\stackrel{\tau}{\Longrightarrow}$. Note that $\stackrel{\tau}{\Longrightarrow}$ is possible without actually performing an internal action, but $\stackrel{a}{\Longrightarrow}$ must perform exactly one transition $\stackrel{a}{\longrightarrow}$ proceeded and followed by arbitrary (possibly empty) internal actions.

Definition 5. A preorder relation $\widetilde{\mathcal{R}}$ on S is a *weak simulation* iff for all $s_1 \widetilde{\mathcal{R}} s_2$:

1. $s_1 \stackrel{a}{\Longrightarrow} s_1', a \in \mathcal{A} \backslash \{\tau\} \Rightarrow \exists s_2' \in S, s_2 \stackrel{a}{\Longrightarrow} s_2'$ and $s_1' \widetilde{\mathcal{R}} s_2'$,
2. $s_1 \stackrel{\tau}{\Longrightarrow} s_1', s_1' \stackrel{\tau}{\not\longrightarrow} \Rightarrow \exists s_2' \in S, s_2 \stackrel{\tau}{\Longrightarrow} s_2', s_2' \stackrel{\tau}{\not\longrightarrow}$ and there exist functions $\Delta : S \times S \to [0,1], \delta_i : S \to [0,1]$ and sets $U_i, V_i \subseteq S\,(i = 1,2)$ with

$$U_i = \{u_i \in S \mid \mathbf{R}(s_i', u_i) > 0 \wedge \delta_i(u_i) > 0\} \text{ and}$$
$$V_i = \{v_i \in S \mid \mathbf{R}(s_i', v_i) > 0 \wedge \delta_i(v_i) < 1\}$$

such that:
   - $v_1 \widetilde{\mathcal{R}} s_2'$ for any $v_1 \in V_1$ and $s_1' \widetilde{\mathcal{R}} v_2$ for any $v_2 \in V_2$,
   - $\Delta(u_1, u_2) > 0$ implies $u_1 \in U_1, u_2 \in U_2$ and $u_1 \widetilde{\mathcal{R}} u_2$,
   - $K_1 \cdot \sum_{u_2 \in U_2} \Delta(w, u_2) = \delta_1(w) \cdot \mathbf{P}(s_1', w)$ and $K_2 \cdot \sum_{u_1 \in U_1} \Delta(w, u_1) = \delta_2(w) \cdot \mathbf{P}(s_2', w)$ for all $w \in S$,
   - $K_1 \cdot E(s_1') \leqslant K_2 \cdot E(s_2')$.

   where $K_i = \sum_{u_i \in U_i} \delta_i(u_i) \cdot \mathbf{P}(s_i', u_i)$ for $i = 1, 2$.

s_2 weakly simulates s_1, denoted $s_1 \precapprox s_2$, iff there exists a weak simulation $\widetilde{\mathcal{R}}$ on S with $s_1 \widetilde{\mathcal{R}} s_2$.

Figure 2 gives an intuitive imagination of weak simulation relation. The successor states of weakly similar states are grouped into two subsets according to function δ_i, denoted by U_i and V_i respectively. The transitions to the V_i-states are viewed as "internal" moves and such transitions are taken totally with probability $1 - K_i$. Accordingly, the transitions to the U_i-states are considered as "observable" moves and such transitions

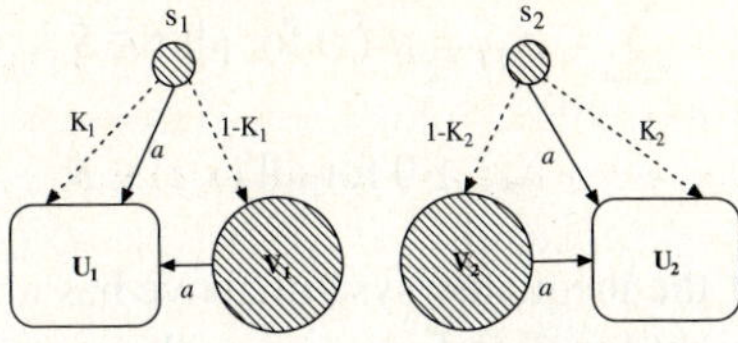

Fig. 2. Scenario of weak simulation relation on IMCs

are taken totally with probability K_i. The U_1-state and U_2-state are related by a weight function Δ. It is a weight function for the probability distributions $\delta_i(\cdot) \cdot \mathbf{P}(s_i, \cdot)/K_i$.

Example 3. Figure 3 illustrates a weak simulation on IMCs. Let $\delta_1(s_3) = \frac{1}{2}$ and $U_1 = \{s_2, s_3\}$, $V_1 = \{s_3\}$ then with $\delta_1(s_2) = \delta_2(s_2^{'}) = \delta_2(s_3^{'}) = 1$, $U_2 = \{s_2^{'}, s_3^{'}\}$, $V_2 = \emptyset$ and $R = \{(s_1, s_1^{'}), (s_2, s_2^{'}), (s_3, s_1^{'}), (s_4, s_4^{'}), (s_3, s_3^{'}), (s_2, s_4^{'})\}$ the conditional probabilities for the U_i-states are related via weight function. There $K_1 = \frac{3}{4}$, $K_2 = 1$, $\Delta(s_2, s_2^{'}) = \frac{2}{3}$ and $\Delta(s_3, s_3^{'}) = \frac{1}{3}$

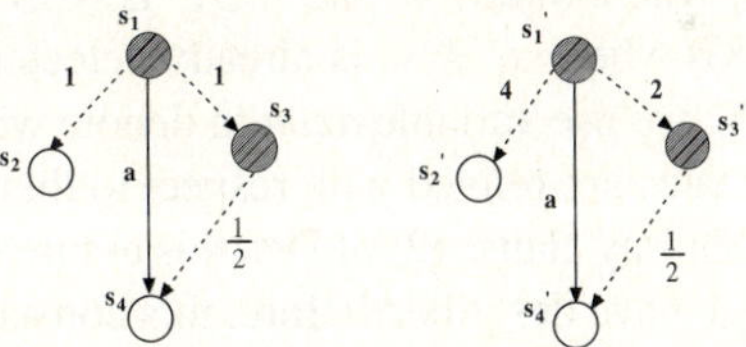

Fig. 3. Weak simulation preorder on IMCs

4 Computing Strong Simulation Preorder

In this section, we present an algorithm that computes the strong simulation preorder of an IMC. The key idea of our algorithm is as in process algebra case [15]: we start with the trivial preorder $R = S \times S$ and then successively remove those pairs (s_1, s_2) from R where $s_1 \not\lesssim_R s_2$. In [12] Hermman gave the partition refinement algorithms for rong and weak bisimulation on IMCs. Similarly, based on Def. 4, a two-phase algorithm will be used to compute the strong simulation preorder. One for interactive transitions, and another for CTMCs transitions. For the interactive transitions refinement, we know that the strong simulation deciding method [15] can be used. For clause (2) of Def. 4, it is reduced to decide whether a weight function exist.

4.1 Deciding Whether $\mu \sqsubseteq_R \mu'$

We show that the question whether two distributions are related via a weight function, i.e., whether $\mu \sqsubseteq_R \mu'$, can be reduced to consider a linear inequality system.

Let $R \subseteq S \times S$ be a binary relation, for every $(s, t) \in R$, given variable $x_{s,t}$. From the definition of weight function, we have:

$$\sum_{t \in S, (s,t) \in R} x_{s,t} = \mu(s) \text{ for all } s \in S$$

$$\sum_{s \in S,(s,t) \in R} x_{s,t} = \mu'(s) \text{ for all } t \in S$$

$$x_{s,t} \geq 0 \text{ for all } (s,t) \in R$$

It is true that $\mu \sqsubseteq \mu'$ iff the inequality system above has a solution. In this case, the solution $x_{s,t \in R}$ yields a weight function for (μ, μ') with respect to R. The above system has $|R| = O(n^2)$ variables and $|R| + 2|S| = O(n^2)$ equations.

Lemma 1. *The following are equivalent:*

1. *There exists a weight function Δ for (μ, μ') w.r.t. R.*
2. *The inequality system above has a solution.*

Proof. It is a straightforward of Def.3.

4.2 Deciding Strong Simulation

Our algorithm for strong simulation on an IMC starts with relation $R_{init} := \{(s_1, s_2) \in S \times S : act(s_1) \subseteq act(s_2)\}$ where $act(s) = \{a | a \in \mathcal{A}, \exists s' \in S \text{ such that } s \xrightarrow{a} s'\}$. We will remove the pairs (s_1, s_2) where $s_1 \not\precsim s_2$ is already detected. To avoid unnecessary tests for $\mathbf{P}(s_1, \cdot) \sqsubseteq_R \mathbf{P}(s_2, \cdot)$, we use variable *asim* to denote whether clause (1) of Def. 4 is satisfied. Only stable states are refined with respect to their weight function, using method offered above. In this way clause (2) of Def. 4 is assured. This is a crucial difference w.r.t. the algorithms we have described before, this consistency with the *maximal progress assumption* of an IMC.

Algorithm 1. *Computing Strong Simulation of IMCs*

Initialization:
 S_Part:=$(s' \in S | s' \overset{\tau}{\not\rightarrow}) - \emptyset$; $R := \{(s_1, s_2) \in S \times S : act(s_1) \subseteq act(s_2)\}$;
Repeat
 $R_{old} := R$; $R := \emptyset$
 For all $(s_1, s_2) \in R_{old}$ do
  $*$ asim:=*true*;
  $*$ for all $a \in act(s_1)$ do
   • If $\forall a \in act(s_2)$ such that $s_1 \xrightarrow{a} s_1'$, $s_2 \xrightarrow{a} s_2'$ and $(s_1', s_2') \notin R$ then
    $asim := false$
  $*$ if asim then do
   • if $s_1, s_2 \in S\_Part$ and $\mathbf{P}(s_1, \cdot) \not\sqsubseteq_R \mathbf{P}(s_2, \cdot)$ or $E(s_1) > E(s_2)$ then
    $asim := false$;
  $*$ if *asim* then $R := R \cup \{(s_1, s_2)\}$
Until $R_{old} = R$
Return R

Theorem 2. *The algorithm 1 computes strong simulation on S. It can be halts after at most n^2 iterations and returns the simulation preorder.*

Proof. Let $R_0 = R_{init}$ and R_j the relation R after the $j-th$ iteration. Then , $R_0, R_1, R_2, \cdots$ is a decreasing sequence of subsets of $S \times S$. Thus, there is some $J \leq |S \times S| = n^2$ with $R_0 \supset R_1 \supset \cdots \supset R_{J-1} = R_J$. After the $J-th$ iteration, the algorithm returns R_J. From the Def 4, it is easy known that R_J is the strong simulation preorder.

5 Computing Weak Simulation Preorder

When computing the weak simulation preorder of a finite state IMC, this computation procedure explicitly relies on a test whether s_2 weakly simulates (under a fixed R) s_1 on a CTMC. According to Def. 5, we will show that this problem can be reduced to a *linear programming (LP) problem.*

5.1 Deciding Weak Simulation on Markovian Transitions

Given $M = (S, \mathcal{A}, \longrightarrow, \dashrightarrow)$ and a binary relation R on S. Let $(s_1, s_2) \in R$, and $\mathbf{Post}(s) = \{s'|R(s, s') > 0\}$. The notation $s \downarrow_R = \{s' \in S|(s', s) \in R\}$ is the downward closure of s with respect to R, and similarly $s \uparrow_R = \{s' \in S|(s, s') \in R\}$ is the upward closure of s with respect to R. Note that for case $\mathbf{Post}(s_1) \subseteq s_2 \downarrow_R$ and $E(s_2) = 0$, the checks whether $s_1 \lessapprox s_2$ can be done in polynomial time [6]. We only consider the case: s_2 is non-absorbing and s_1 has at least one successor state $u_1 \in \mathbf{Post}(s_1)$ such that $u_1 \notin s_2 \downarrow_R$. As $u_1 \not\lessapprox s_2$, and all states in V_1 have to be simulated by s_2 , state $u_1 \in U_1$.Thus, $K_1 > 0$, and, by of Def. 5, $K_2 > 0$. Consider the following variables:

- x and y which stand for the values $x = \frac{1}{K_1}$ and $y = \frac{1}{K_2}$, respectively
- x_u for $u \in S$ with $(u, s_2) \in R$ which stands for the value $x_u = \frac{\delta_1(u)}{K_1}$
- y_u for $u \in S$ with $(s_1, u) \in R$ which stands for the value $y_u = \frac{\delta_2(u)}{K_2}$
- z_{u_1,u_2} for each pair of states $(u_1, u_2) \in R$.

We write $\Delta(u_1, u_2)$ instead of z_{u_1,u_2}, and put: $x_u = x$ if $u \in S \setminus s_2 \downarrow_R$, $y_u = y$ if $u \in S \setminus s_1 \uparrow_R$. For each state u in $\mathbf{Post}(s_1) \setminus s_2 \downarrow_R$ has to be put completely in U_1. Thus, $\delta_1(u) = 1$, and hence:

$$x_u = x = \frac{1}{K_1} = \frac{\delta_1(u)}{K_1}$$

By a symmetric argument, we put $y_u = y$ if $u \notin s_1 \uparrow_R$.
The linear program now consists of the following equations and inequalities:

$$\sum_{u_1 \in u_2 \downarrow_R} \Delta(u_1, u_2) = \mathbf{P}(s_2, u_2) \cdot y_{u_2} \text{ for } u_2 \in S$$

$$\sum_{u_2 \in u_2 \uparrow_R} \Delta(u_1, u_2) = \mathbf{P}(s_1, u_1) \cdot x_{u_1} \text{ for } u_1 \in S$$

$$\sum_{u_1 \in S} x_{u_1} \cdot \mathbf{R}(s_1, u_1) = E(s_1)$$

$$\sum_{u_2 \in S} y_{u_2} \cdot \mathbf{R}(s_2, u_2) = E(s)$$

$$x \geqslant 1, \qquad y \geqslant 1$$

$$x \geqslant x_u \geqslant 0 \text{ if } u \in s_2 \downarrow_R$$

$$y \geqslant y_u \geqslant 0 \text{ if } u \in s_1 \uparrow_R$$

$$y \cdot E(s_1) \leqslant x \cdot E(s_2)$$

This **LP** problem has $O(|S|^2)$ variables and $4 \cdot |S| + 5$, i.e., $O(|S|)$ equations. It is clear to see that any solutions to the **LP** problem mentioned above inducing components in Def. 5 are fulfilled. Vice versa, components δ_i, U_i, V_i, K_i and Δ in Def.5 induce a solution of the above linear program. By solving **LP** problems, the test whether a state weakly simulates another one can be performed in polynomial time.

5.2 Deciding Weak Simulation on IMCs

In the sequel, we use $s \overset{\tau}{\Longrightarrow}$ to indicate that s may internally evolve to a stable state s', and s is called time-convergent state. In our computing of weak simulation, we first compute the weak action transitions relation $\Longrightarrow$ from $\longrightarrow$, then, we will substitute "$\longrightarrow$" in algorithm 1 with "$\Longrightarrow$", and strong Markovian simulation with weak Markovian simulation. Weak simulation algrithm on IMCs can be obtained. Let $wact(s) = \{a | a \in \mathcal{A} \setminus \tau, \exists s' \in S$ such that $s \overset{a}{\Longrightarrow} s'\}$. We have:

Algorithm 2. Computing Weak Simulation of IMCs

Initialization:
  Compute weak transition $\Longrightarrow$ from $\longrightarrow$;
  Let TC_Part:=$\{s' \in S | s' \overset{\tau}{\Longrightarrow}\} - \emptyset$; $R := \{(s, s') \in S \times S : wact(s) \subseteq wact(s')\}$;
 Repeat
  $R_{old} := R$; $R := \emptyset$
  For all $(s_1, s_2) \in R_{old}$ do
   ∗ asim:=*true*;
   ∗ for all $a \in wact(s_1)$ do
    • If $\forall a \in act(s_2)$ such that $s_1 \overset{a}{\Longrightarrow} s_1', s_2 \overset{a}{\Longrightarrow} s_2'$ and $(s_1', s_2') \notin R$ then
      $asim := false$
   ∗ if asim then do
    • if $s_1, s_2 \in$ *TC_Part* and s_2 not Markovian weak simulaton s_1 then
      $asim := false$;
   ∗ if *asim* then $R := R \cup \{(s_1, s_2)\}$
 Until $R_{old} = R$
Return R

Complexity analysis. Algorithm 1 and algorithm 2 can be performed in polynomial time by using the efficient method for solving LP problems. the number of iterations of LP is bounded $|S^2|$, Thus, one has to solve at most $|S|^2$ LP problems, each being linear in $|S|$.

6 Conclusions

We have presented two polynomial-time algorithms for computing the strong and weak simulation preorders of a finite-state Interactive Markov chain. Indeed, we have continued the work of [12,23,7], and studied how to lift simulation concepts from classical labelled transition systems and purify stochastic settings to IMCs. Due to the combination characteristic of IMCs, our algorithms can easily be reduced to decide simulations preorder of LTSs and CTMs . The crux of our algorithms is to consider the check whether a state strongly or weakly simulates another one as a linear programming problem.

As the maximal progress assumption of IMCs, our algorithms have to consider the internal action transitions of states before the Markov simulation is tested as well as nondeterministic of action transitions. These are difficult points of our algorithms. But, we can use powerful mathematical tool to solve the LP problems efficiently, so our work is valuable .

In this paper we only work out the algorithms to decide simulation of IMCs, but how to improve our basic algorithms is not considered here, it should be interesting to find if a better method exists, it is also our future work. We expect that techniques from, e.g.,[10,25], can be employed to speed up the algorithm.

References

1. C. Baier, B. Engelen, and M. Majster-Cederbaum. Deciding bisimilarity and similarity for probabilistic process. *J. of Comp. and System Sc.*, 60(1): 187-231, 2000.
2. C. Baier, H. Hermanns, J.-P. Katoen and V. Wolf. Comparative branching-time semantics for Markov chains. In R. de Simone and D. Lugiez(eds), *Concurrency Theory*, LNCS, 2003.
3. C. Baier and H. Hermanns. Weak bisimulation for fully probabilistic system. In O. Grumberg(ed), *Computer-Aided verification*, LNCS 1256, pp. 119-130, 1997.
4. C. Baier, J.-P. Katoen, H. Hermanns and B. Haverkort. Simulation for continuous-time Markov chains. In L. Brim *et al.(eds), Concurrency Theory*, LNCS 2421, pp. 338-354, 2002.
5. C. Baier, Boudewijn R. Haverkort, Holger Hermanns, and Joost-Pieter Katoen. Model-checking algorithms for continuous-time Markov chains. *IEEE Trans. Software Eng.*, 29(6): 524-541, 2003.
6. C. Baier, Boudewijn R. Haverkort, Holger Hermanns, and Joost-Pieter Katoen. Model checking meets performance evaluation. *SIGMETRICS Performance Evaluation Review*, 32(4): 10-15, 2005.
7. M. Bravetti. Revisting interactive Markov chains. In W. Vogler and K. G. Larsen(eds), *Models for Time-Critical Systems*, BRICS Notes Series NS-02-3, pp. 60-80, 2002.
8. T. Bolognesi, S. Smolka: Fundamental Results for the Verification of Observational Equivalence: a Survey, Protocl Specification, Testing and Verification, Elsevier Science Publishers, IFIP, pp 165-179, 1987.

9. R. Cleaveland, J. Parrow, B. Steffen: A Semantics-Based Verification Tool for Finite State Systems, Protocl Specification, Testing and Verification IX, Elsevier Science Publishers, IFIP, pp 287-302, 1990.
10. R. Gentilini, C. Piazza and A. Policriti. Simulation as coarsest partition problem. In Joost-Pieter Katoen and P. Stevens (eds), *Tools and Algorithms for the Construction and Analysis of Systems*, LNCS 2280, pp. 415-430, 2002.
11. H. Hermanns, J. Meyer-Kayser, and M. Siegle. Multi-terminal binary decision diagrams to represent and analyse continuous-time Markov chains, 1999.
12. H. Hermanns. Interactive Markov chains. PhD thesis, Universität Erlangen-Nürnberg, 1998.
13. H. Hermanns and Joost-Pieter Katoen. Performance evaluation : =(process algebra + model checking) Markov chains. In *CONCUR*, pages 59-81, 2001.
14. H. Hermanns, Joost-Pieter Katoen, Joachim Meyer-Kayser, and Markus Siegle. Towards model checking stochastic process algebra. In *IFM* pages 420-239, 2000.
15. M. Henzinger, T. Henzinger, P. Kopke: Computing Simulations on Finite and Infinite Graphs, *in* Proc. FOCS'95, PP 43-462, 1995.
16. B. Jonsson. Simulations between specifications of distributed systems. In J.C.M. Baeten and J.F.Groote(eds), *Concurrency Theory*, LNCS 527, pp. 346-360, 1991.
17. A. Jensen. Markov chains as an aid in the study of Markov process. *Skand. Aktuarietidskrift* **3**: 87-91, 1953.
18. P. Kannelakis, S. Smolka: CCS Expressions, Finite State Processes and Three Problems of Equivalenec, Proc. 2nd ACM Symposium on the Prinicples of Distributed Computating, pp 228-240, 1983.
19. K. G. Larsen and A. Skou. Bisimulation through probabilistic testing. *Inf. and Comp.*, **94**(1): 1-28, 1992.
20. R. Milner. Communication and Concurrency, *Prentice Hall*, 1989.
21. A. Philippou, I. Lee and O. Sokolsky. Weak bisimulation for probabilistic systems. In C. Palamidessi (ed), *Concurrency Theory*, LNCS 1877, pp. 334 - 349, 2000.
22. R. Paige, R. Tarjan: Three Partition Refinement Algorithms,*SIAM Journal of Computing*, Vol. 16, No. 6, pp. 973-989, 1987.
23. Guangping Qin and Jinzhao Wu. Branching time equivlences for interactive Markov chains. In *FORTE Workshops*, volume 3232 of *Lecture Notes in Computer Science 3236*, pages 156-169. Springer, 2004.
24. R. Segala. N. Lynch: Probabilistic Simulations for Probabilistic Processes, Proc. *CONCUR* 94, Theories of Concurrency: Unification and Extension, LNCS 836, Springer-Verlag, pp. 492-493, 1994.
25. L. Tan and R. Cleaveland. Simulation revisited. In T. Margaria and W. Yi (eds), *Tools and Algorithms for the Construction and Analysis of Systems*, LNCS 2031, pp. 480-495, 2002.

On Parallel Immune Quantum Evolutionary Algorithm Based on Learning Mechanism and Its Convergence

Xiaoming You[1,2], Sheng Liu[1,2], and Dianxun Shuai[1]

[1] Dept of Computer Science and Technology, East China University of Science and Technology, Shanghai 200237, China
yxm6301@163.com

[2] College of Electronic and Electrical Engineering, Shanghai University of Engineering Science, Shanghai 200065, China

Abstract. A novel Multi-universe Parallel Immune Quantum Evolutionary Algorithm based on Learning Mechanism (MPMQEA) is proposed, in the algorithm, all individuals are divided into some independent sub-colonies, called universes. Their topological structure is defined, each universe evolving independently uses the immune quantum evolutionary algorithm, and information among the universes is exchanged by adopting emigration based on the learning mechanism and quantum interaction simulating entanglement of quantum. It not only can maintain quite nicely the population diversity, but also can help to accelerate the convergence speed and converge to the global optimal solution rapidly. The convergence of the MPMQEA is proved and its superiority is shown by some simulation experiments in this paper.

1 Introduction

Research on merging evolutionary algorithms with quantum computing has been developed since the end of the 90's, this research can be divided in two different groups: one that focus on developing new quantum algorithms; and another which focus on developing quantum–inspired evolutionary algorithms which can be executed on classical computers.

QGA (Quantum-Inspired Genetic Algorithm), proposed by Ajit Narayanam, introduces the theory of many universes in quantum mechanics into the implementation of genetic algorithm [1]. It proves the efficiency of the strategy that uses multiple colonies to search in parallel, and uses a joint crossover operator to enable the information exchange among colonies.

Although quantum evolutionary algorithms are considered powerful in terms of global optimization, they still have several drawbacks regarding local search: (i) lack of local search ability, and (ii) premature convergence.

In recent years, the study on immune evolutionary algorithms has become an active research field. A number of researchers have experimented with biological immunity-based optimization approaches to overcome these particular drawbacks implicit in evolutionary algorithms. Author has proposed a quantum evolutionary algorithm based on the adaptive immune operator (MQEA) in [2], the convergence of the MQEA is proved and its superiority is shown by some simulation experiments. When

L. Jiao et al. (Eds.): ICNC 2006, Part I, LNCS 4221, pp. 903 – 912, 2006.

such problems are too hard, they require a very long time to reach the solution. This has led to different efforts to parallelize these programs in order to accelerate the process.

In this paper, A novel Multi-universe Parallel Immune Quantum Evolutionary Algorithm based on Learning Mechanism (MPMQEA) is proposed, in the algorithm, all individuals are divided into some independent sub-colonies, called universes, each universe evolving independently uses the immune quantum evolutionary algorithm, information among the universes is exchanged by adopting emigration based on the learning mechanism and quantum interaction simulating entanglement of quantum. In order to evaluate MPMQEA, a set of standard test functions were used and its performance is compared with that of MQEA in [2]. Specifically, Section 2 presents the Quantum Evolutionary Algorithm based on Immune operator. Section 3 proposes a Multi-universe Parallel Immune Quantum Evolutionary Algorithm based on Learning Mechanism (MPMQEA). Section 4 then proves that the novel algorithm is convergence with probability one by using Markov chain model. In Section 5 the performance of MPMQEA is evaluated by some well-known test functions.

2 Immune Quantum Evolutionary Algorithm and Related Work

In quantum computing, quantum bit (Q-bit) is the smallest unit of information stored in a two-state quantum computer. The characteristic of the Q-bit representation is that any linear superposition of solutions can be represented. This superposition can be expressed as follows: $\psi = \alpha|0\rangle + \beta|1\rangle$, (α, β) is a pair of complex invariables, called probability amplitude of Q-bit, which satisfies $|\alpha|^2 + |\beta|^2 = 1$. Evolutionary computing with Q-bit representation has a better characteristic of population diversity than other representation, since it can represent linear superposition of state's probabilities.

Quantum Evolutionary Algorithm (QEA) [3], [8] is efficacious, in which the probability amplitude of Q-bit was used for the first time to encode the chromosome and the quantum rotation gate was used to implement the evolving of population. Quantum evolutionary algorithm has the advantage of using a small population size and the relatively smaller iterations number to have acceptable solution, but they still have several drawbacks such as premature convergence.

Immune Algorithms (IA) are evolutionary algorithms [4], [5] based on physiological immune systems. Physiological immune systems have affinity maturation mechanism and the immune selection mechanism. Affinity maturation conduces to immune system self-regulates the production of antibodies and diverse antibodies, these higher-affinity matured cells are then selected to enter the pool of memory cells.

The merits of IA are as follows:

1. IA has the pool of memory cells, which guarantees fast convergence toward the global optimum.
2. IA has an affinity maturation routine, which guarantees the diversity of the immune system.
3. The immune response can enhance or suppress the production of antibodies.

2.1 Quantum Evolutionary Algorithm Based on Immune Operator

The flowchart of Quantum Evolutionary Algorithm based on Immune operator (MQEA)[2] is as follows:

```
MQEA ()
{  t=0;
   initialize Q(0) ;
   make P(0) by observing the states of Q(0) ;
   evaluate P(0) ;
   store the optimal solutions among P(0) ;
 while (not termination-condition) do
  {t=t+1;
   update Q(t-1) using Q-gate U(t-1);
   make P(t) by observing the states of Q(t);
   evaluate P(t) ;
   store the optimal solutions among P(t) ;
   implement the immune operator for Q(t),P(t):
 {Clonal proliferation;
  Self-adaptively cross-mutate each cell;
  suppress similar antibody;
  select antibody with higher affinity as memory cells;
    }
  } }
end.
```

In the step "make P(0)", generates binary solutions in P(0) by observing the states of Q(0), where P(0) = $\{x_0 1, x_0 2, \cdots, x_0 n\}$, Q(0)= $\{q_0 1, q_0 2, \cdots, q_0 n\}$. One binary solution, $x_0 j$, is a binary string of length m, which is formed by selecting either 0 or 1 for each bit by using the probability, either $|\alpha_{0j} i|^2$ or $|\beta_{0j} i|^2$ of $q_0 j$ in equation (1), respectively. Each binary solution $x_0 j$ is evaluated to give a level of its fitness. The initial best solutions are then selected among the binary solutions P(0). Quantum gate U (t) is a variable operator of MQEA, it can be chosen according to the problem.

2.2 Immune Operator

The clonal selection and affinity maturation principles [6] are used to explain how the immune system improves its capability of recognizing and eliminating pathogens. Clonal selection states that antigen can selectively react to the antibodies, if the antibody matches the antigen sufficiently well; its B cell becomes stimulated and can produce related clones. The cells with higher affinity to the invading pathogen differentiate into memory cells. This whole process of mutation plus selection is known as affinity maturation. Inspired by the above clonal selection and affinity maturation principles, the cross-mutation operator could be viewed as a self-adaptive mutation operator. Self-adaptive mutation plays the key role in MQEA; generally, cells with low affinity are mutated at a higher rate, whereas cells with high affinity will have a lower mutation rate. This mechanism offers the ability to escape from local optima on an affinity landscape.

Cross-mutation operator can act as follows: give a randomly position i of the chromosome q_{0j}

$$q_{0j} = \begin{pmatrix} \alpha_{0j}1 & \alpha_{0j}2 & \ldots\ldots & \alpha_{0j}m \\ \beta_{0j}1 & \beta_{0j}2 & \ldots\ldots & \beta_{0j}m \end{pmatrix} \tag{1}$$

If $|\beta_{0j}i|^2 < p$ (p is mutation rate) then let $(\alpha_{0j}i, \beta_{0j}i)$ become $(\beta_{0j}i, \alpha_{0j}i)$. In order to avoid trapping in a local optimal solution and to ensure the searching capability of near global optimal solution, the high affinity antibody suffers a smaller mutation rate *pl*, whereas the low affinity one undergoes a larger mutation rate *ph*, $pl<ph$.

3 Multi-universe Parallel Immune Quantum Evolutionary Algorithm Based on Learning Mechanism

This article proposed the Multi-universe Parallel Immune Quantum Evolutionary Algorithm based on Learning Mechanism (MPMQEA). MPMQEA uses the multi-universe parallel structure, each universe evolving independently uses the immune quantum evolutionary algorithm, and information among the universes is exchanged by adopting emigration based on the learning mechanism and quantum interaction simulating entanglement of quantum. It can help to accelerate the convergence speed and converge to the global optimal solution rapidly.

3.1 Multi-universe Parallel Topology

Parallel Evolutionary Algorithms (PEA) can be classified into three different models: (1) Master-slaves PEA. (2) Fine-grained PEA. (3) Coarse-grained PEA [7].In coarse-grained model, the population is divided into several subpopulations; each subpopulation applies the classical EA process independently with its own parameters. The design of the parallel model is intended to reduce the communications, to this purpose, the following options have been chosen:

Migration policy: To reduce communications, migrations do not take place from one subpopulation to any other, but only to the next one (Fig.1. (b)). The N cooperative subpopulations thus form a ring.

Migration Criteria: Populations can send and receive individuals to and from the rest of the populations. They are randomly chosen with a probability proportional to their fitness.

To reduce communications, we improve the coarse-grained model, in which information among universes is exchanged by adopting emigration based on the learning mechanism and interaction simulating entanglement of quantum, and propose an easily expandable architecture (Fig.1.(a)). The proposed topology for parallel architecture is basically organized with coarse-grained algorithm and each coarse-grained PEA is connected to other coarse-grained PEA group like ring topology.

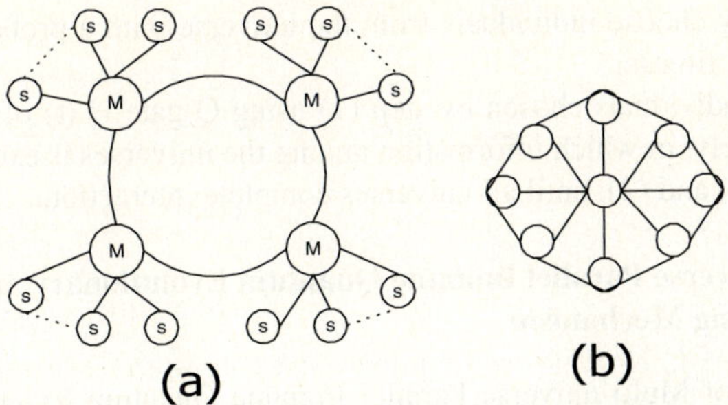

Fig. 1. Improved coarse-grained model & coarse-grained model

3.2 Migration Operator and Learning Operator Between the Universes

Parallel Evolutionary Algorithms (PEA) performance is based on the interchange of individuals among different sub-populations. PEA will be able to send and receive individuals from and to any other sub-population. Plans best known for individuals interchange are migration and diffusion. Here we adopt modified migration mechanism based on learning mechanism, namely, exchanging information between slave and master population, as show in Fig.2.

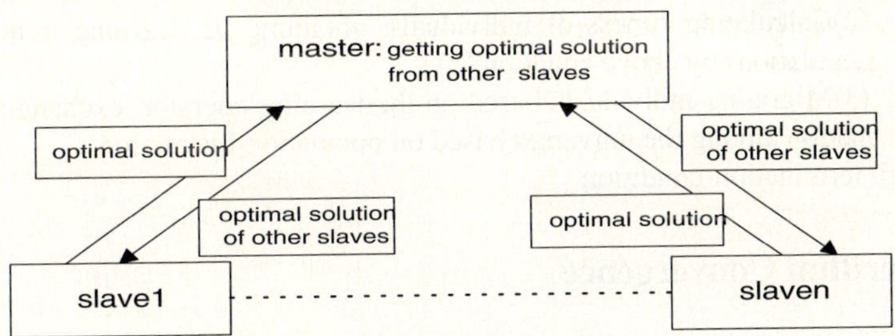

Fig. 2. Migration policy based on learning mechanism

Learning operator between slave and master population is as follows:

1. ps is learning frequency of slave population , calculated by equation (2).

$$ps = \mu \frac{maxfitness - avefitness}{maxfitness - minfitness} \tag{2}$$

(μ is 1.0 in general)

2. Generating random $R \in [0, 1]$;
3. If $R<ps$, getting optimal solution from master on admissible condition;
4. If $R \geq ps$, sending optimal solution of slave population to master.

3.3 Population Interaction

We proposed that information among the universes is exchanged by quantum interaction simulating entanglement of quantum, it is as follows:

(1)Randomly choose individuals from the universe with a probability proportional to their average fitness;

(2)Update individuals chosen by step (1) using Q-gate U (t) of neighborhood universe respectively, in which information among the universes is exchanged;

(3)Repeat (1) and (2), until all universes complete interaction.

3.4 Multi-universe Parallel Immune Quantum Evolutionary Algorithm Based on Learning Mechanism

The flowchart of Multi-universe Parallel Immune Quantum Evolutionary Algorithm based on Learning Mechanism (MPMQEA):

1. Initialize all sub-colonies of universes.
2. A master population was selected as the virtual host server; other colonies are the slave populations.
3. Begin loop

   (1)Each sub-population evolves independently certain generations:

   Update Q (t-1) using Q-gate U (t-1);
   Make P (t) by observing the states of Q (t);
   Implement the immune operator for Q (t), P (t):
   Evaluate P (t);
   Store the optimal solutions among P (t);

   (2)Calculating fitness of individuals, obtaining *ps* (learning frequency of population) by above equation (2)

   (3)Migrating individuals based on the learning operator; exchanging information among the universes based on population Interaction.

   Until termination-condition

4 Algorithm Convergence

Theorem 1: Population sequence of Multi-universes Parallel Immune Quantum Evolutionary Algorithm based on Learning Mechanism (MPMQEA) {A(n), n>= 0} are finite stochastic Markov chain.

Proof. Population sequence shift may express as the following stochastic states:

$$A(K) \xrightarrow{mutation} A1(K) \xrightarrow{measure} A2(K) \xrightarrow{immune} A3(K)$$

$$\xrightarrow{select} A4(K) \xrightarrow{cross} A5(K) \xrightarrow{emigration} A(K+1)$$

MPMQEA uses Q-bit chromosome q_0j; q_0j is then represented by above equation (1). $\alpha_{0j}i$ and $\beta_{0j}i$ have a limited precision, assume that the precision is e(e is 10^{-5} or 10^{-6}), so the dimension of $\alpha_{0j}i$ in Q-bit chromosome q_0j is $(Qh-Ql)/e$, where Qh is the upper bound of $\alpha_{0j}i$, Ql is the lower bound of $\alpha_{0j}i$. For Q-bit, Qh=1, Ql=-1, let $V=(Qh-Ql)/e$, therefore $V=2/e$. Assume that the length of chromosome is M, the size of population is N, the number of universes is L, thus the size of population states space is V^{M*N*L}, therefore the population sequence is finite.

$A(K+1)=T(A(K))= Te\circ Tc\circ Ts\circ Ti\circ Tm\circ Tu\ (A(K))$, Te, Tc, Ts ,Ti, Tm and Tu are transition matrices of stochastic states, denoting migration operator, interaction operator, selection operator, immune operator, measurement operator, quantum rotation gate mutation operator respectively, they have nothing to do with the number of iterations K, thus A(K+1) only have to do with A (K). Therefore $\{A(n),n>= 0\}$ are finite stochastic Markov chain, which completes the proof.

We assume that S is the feasible solutions space and f* is the optimal solutions of S, let $A^*= \{A|\ \max (f (A)) = f^*,\ \forall A\in S\}$

Definition 1: $\{A(n), n>= 0\}$ are stochastic states, $\forall\ S_0\in S$, S_0 is the initial solution , If $\lim_{k\to\infty} P\{A(k)\in A^*|A(0)= S_0\}=1$, Then the stochastic states $\{A(n), n>= 0\}$ is called convergence with probability one[9].

Let P_k denote $P\{A(k) \in A^*|A(0)=S_0\}$, then $P_k=\sum_{i\in A^*} P\{A(k)= i\ |A(0)=S_0\}$.

Let Pi (k) denote $P\ \{A (k) = i\ |A (0) =S_0\}$, then

$$P_k=\sum_{i\in A^*} Pi\ (k) \tag{3}$$

Let $Pij\ (k) = P\ \{A\ (k) =j\ |A\ (0) =i\}$.

Under elitist approach (the best individual survives with probability one), we have two special equations [9]:

$$\text{When } i\in A^*, j\notin A^*,\quad Pij\ (k)=0 \tag{4}$$

$$\text{When } i\in A^*, j\in A^*,\quad Pij\ (k)=1 \tag{5}$$

Theorem 2: MPMQEA is convergence with probability one.

Proof. From the above equation (3) $P_k=\sum_{i\in A^*} Pi\ (k)$

From $\sum_{j\notin A^*} Pij(k) + \sum_{j\in A^*} Pij(k)=1$ Thus $P_k=\sum_{i\in A^*} Pi(k)$

$$=\sum_{i\in A^*} Pi(k)\Big(\sum_{j\notin A^*} Pij(k) + \sum_{j\in A^*} Pij(k)\Big)= \sum_{i\in A^*}\sum_{j\in A^*} Pi(k)\ Pij\ (1) + \sum_{i\in A^*}\sum_{j\notin A^*} Pi(k)\ Pij(1)$$

From above equation (4) $\sum_{i\in A^*}\sum_{j\notin A^*} Pi\ (k)\ Pij\ (k) =0$, so $P_k = \sum_{i\in A^*}\sum_{j\in A^*} Pi(k)Pij(1)$

$\{A(n), n>= 0\}$ of MPMQEA is finite stochastic Markov chain (By Theorem 1).

Thus $P_{k+1}= \sum_{i\in A^*}\sum_{j\in A^*} Pi\ (k)\ Pij\ (1) + \sum_{i\notin A^*}\sum_{j\in A^*} Pi\ (k)\ Pij\ (1)$

So $P_{k+1}=P_k + \sum_{i\notin A^*}\sum_{j\in A^*} Pi(k)\ Pij(1)>P_k$, So that, $1 \geq P_{k+1}>P_k>P_{k-1}>P_{k-2}\ldots\ldots \geq 0$,

therefore $\lim_{k\to\infty} P_k=1$. By definition 1, MPMQEA is convergence with probability one.

5 Simulation Results

In this section, MPMQEA is applied to the optimization of well-known test functions and its performance is compared with that of MQEA algorithm. The test examples used in this study are listed below:

$$f1(x_1,x_2)=4x_1^2-2.1x_1^4+x_1^6/3+x_1x_2-4x_2^2+4x_2^4,\quad -5\le x_i\le 5 \tag{6}$$

$$f2(x_1,x_2)=100(x_1^2-x_2)^2+(1-x_1)^2,\quad -2\le x_i\le 2 \tag{7}$$

f2(Rosenbrock function): Rosenbrock's valley is a classic optimization function. The results for the case of Rosenbrock's function with three variables, averaged over 20 trials, are shown in Fig.3. and Fig.4., where solid line denotes MPMQEA and dot line denotes MQEA. Comparison of the results indicates that MPMQEA offers a significant improvement in the results compared to the conventional MQEA. In Fig.3., for 90 populations, the optimization value was obtained with MQEA after 50 generations, whereas the parallel method MPMQEA was able to obtain after 20 generations. In Fig.4. , for 30 populations, the optimization value was obtained with MQEA after 120 generations, whereas the parallel method MPMQEA was able to obtain after 35 generations. This certainty of convergence of the MPMQEA may be attributed to its ability to maintain the diversity of its population. As a result, fresh feasible antibodies are constantly introduced, yielding a broader exploration of the search space. In fact, the superior performance of parallelism may be due to their very good diversity and multi-solution capabilities.

f1 (Camelback function): It has six local minima, two of them are global minimum, f min= f (-0.0898, 0.7126) = f (0.0898, -0.7126) = -1.0316285.It was used to evaluate the diversity of MPMQEA influencing by variable number of universes.

The result for the case of Camelback function, averaged over 20 trials, are shown in Fig.5., solid line denotes MPMQEA-1(8 universes), and dot line denotes MPMQEA-2(3 universes).Comparison of the result indicates that the number of universes can influence the diversity and the convergence speed, for 24 populations, the optimization value was obtained with MPMQEA-1(8 universes) after 20 generations, whereas MPMQEA-2(3 universes) was able to obtain after 40 generations.

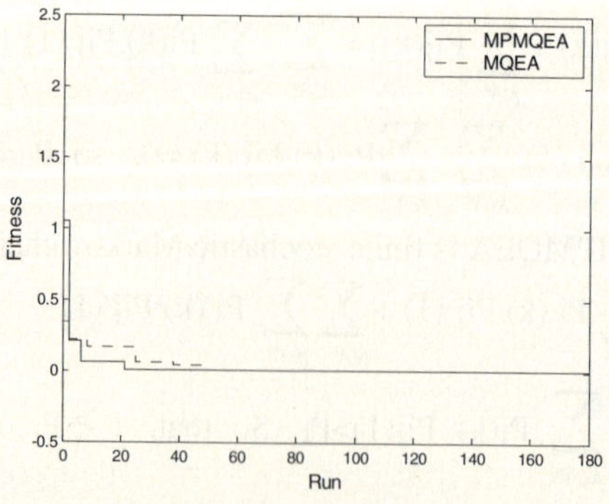

Fig. 3. MPMQEA: 3 universes, 30 populations; MQEA: 90 populations, 180 generations, average 20 trials

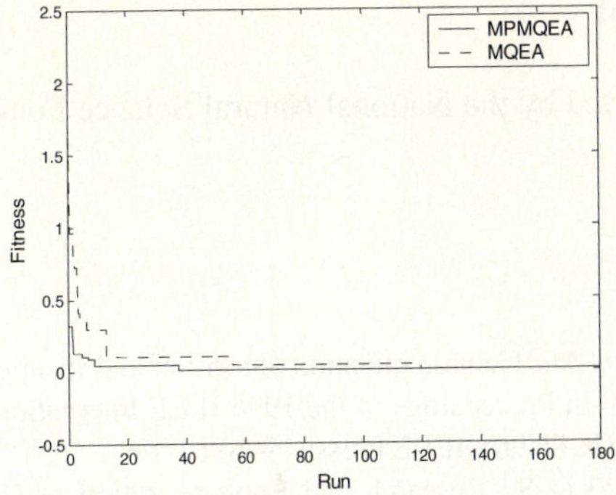

Fig. 4. MPMQEA: 3 universes, 10 populations; MQEA: 30 populations, 180 generations, average 20 trials

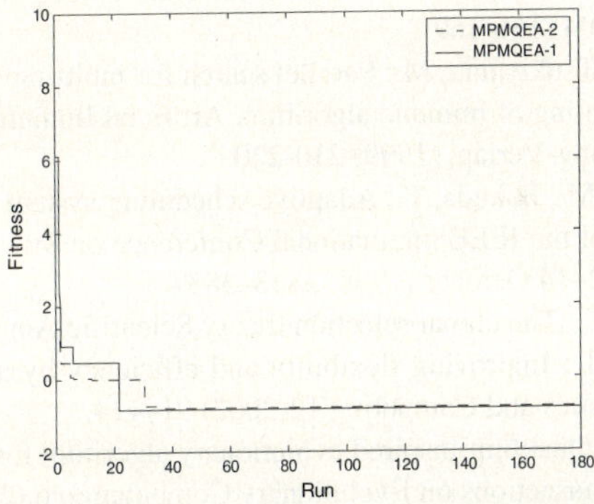

Fig. 5. Comparison of the diversity of MPMQEA having variable number of universes, 24 populations, 180 generations, average 20 trials

6 Conclusions

In this study, we proposed a novel Multi-universe Parallel Immune Quantum Evolutionary Algorithm based on Learning Mechanism (MPMQEA).It is efficacious, helping in alleviating premature convergence problems and in speeding up the convergence process. We also prove that the algorithm is convergence with probability one by using Markov chain model.

The efficiency of the approach has been illustrated by application to a number of test cases. The results show that integration of the parallelism in the quantum evolutionary algorithm procedure can yield significant improvements in both the convergence rate and solution quality. The further work is exploiting more reasonable parameters of MPMQEA, such as the number of populations, the number of individuals per population, the number of migrating individuals, the frequency of migration and so on.

Acknowledgement

This work was supported by the National Natural Science Foundation of China under Grant No. 60575040

References

1. Narayanan, A., Moore, M.: Genetic quantum algorithm and its application to combinatorial optimization problem. In Proceedings of the 1996 IEEE International Conference on Evolutionary Computation (ICEC96), IEEE Press (1996) 61–66
2. You XM, Shuai DX, Liu S.: Research and Implementation of Quantum Evolution Algorithm Based on Immune Theory. In Proceedings of the 6th World Congress on Intelligent Control and Automation (WCICA06), Da Lian,China, (2006) Accepted for publication.
3. Han, K.H., Kim, J.H.: Quantum-Inspired Evolutionary Algorithms with a New Termination Criterion, Hε Gate, and Two-Phase Scheme. IEEE Transactions on Evolutionary Computation, IEEE Press, 8(2004) 156-169
4. Fukuda, T., Mori, K., Tsukiyama, M.: Parallel search for multi-modal function optimization with diversity and learning of immune algorithm. Artificial Immune Systems and Their Applications. Berlin: Spring-Verlag, (1999) 210-220.
5. Mori, K., Tsukiyama, M., Fukuda, T.: Adaptive scheduling system inspired by immune systems. In Proceedings of the IEEE International Conference on Systems, Man, and Cybernetics, San Diego, CA, 12–14 October (1998) 3833–3837
6. Ada,G.L.,Nossal,G.J.V.: The clonal selection theory,Scientific American 257(1987)50–57.
7. Enrique, A., Jose, M.T.: Improving flexibility and efficiency by adding parallelism to genetic algorithms, Statistics and Computing 12(2002) 91–114.
8. Han, K.H., Kirn, J.H.: Quantum-inspired evolutionary algorithm for a class of combinatorial optimization. IEEE Transactions on Evolutionary Computation 6 (2002) 580–593
9. Pan ZJ, Kang LS, Chen YP.: Evolutionary Computation [M]. Bei-jing:Tsinghua University Press,(1998)

Self-Organization Particle Swarm Optimization Based on Information Feedback

Jing Jie[1,2], Jianchao Zeng[2], and Chongzhao Han[1]

[1] School of Electronic and Information Engineering,
Xi'an Jiaotong University
710049, Xi'an City, China
Jjing277@sohu.com

[2] Division of System Simulation & Computer Application,
Taiyuan University of Science & Technology
030024, Taiyuan City, China
zengjianchao@263.net

Abstract. The paper develops a self-organization particle swarm optimization (SOPSO) with the aim to alleviate the premature convergence. SOPSO emphasizes the information interactions between the particle-lever and the swarm-lever, and introduce feedback to simulate the function. Through the feedback information, the particles can perceive the swarm-lever state and adopt favorable behavior model to modify their behavior, which not only can modify the exploitation and the exploration of the algorithm adaptively, but also can vary the diversity of the swarm and contribute to a global optimum output in the swarm. Relative experiments have been done; the results show SOPSO performs very well on benchmark problems, and outperforms the basic PSO in search ability.

1 Introduction

Particle swarm optimization (PSO) is a swarm-intelligence-based algorithm and inspired originally by the social behavior lying in the bird flocking. The algorithm initialized with a population of random solutions, called particles. During a search process, each particle has a tendency to fly towards better search areas with a velocity, which is dynamically adjusted according to its own and its companion's historical behavior. Since the original version introduced in 1995 by Kennedy and Eberhart [1], PSO has attracted lots of attention from researchers around the world and lots of research results have been reported in the literatures. As a stochastic algorithm, PSO own some attractive features such as simple concept, few parameters, and easy implementation. Now, PSO has been applied successfully in many areas[2-6], including evolving neural networks, solving multidimensional complex problems, solving multiobjective optimizations, and dealing with the reactive power and voltage control, etc.. However, PSO also suffers from the premature convergence problem as others stochastic algorithms, especially in the large scale and complex problems.

As far as the premature convergence is concerned, a main reason is that a fast information interaction among particles in PSO leads to the swarm-diversity declining

L. Jiao et al. (Eds.): ICNC 2006, Part I, LNCS 4221, pp. 913 – 922, 2006.

and the particles clustering quickly. In order to avoid the premature convergence, a feasible idea is to maintain an appropriate swarm-diversity during a search. Some techniques have been proposed following the idea. Suganthan[7] introduced the spatial neighborhood into PSO; Li[8] used species to choose the neighbourhood adaptively. Xie[9] developed a self-organizing PSO based on the dissipative struction; Riget[10] proposed an attractive and repulsive PSO(ARPSO) to adjust the diversity, etc.. According to the relative literatures, those methods can improve the global convergence ability of PSO in certain degree but fewer can solve the premature convergence problem truly.

In order to overcome the premature problem, the present paper develops a self-organization particle swarm optimization (SOPSO). Different from the previous improved methods, SOPSO emphasizes the interactions between the particle-lever and the swarm-lever, and introduce feedback to simulate the functions. Through the feedback information, the particles can apperceive the swarm-lever state and adopt favorable behavior model to modify their behavior, which not only can modify the diversity of the swarm adaptively, but also can adjust the exploitation and the exploration of the algorithm, and contributes to a global optimum output in the swarm. Relative experiments have been done; the results show SOPSO performs very well on benchmark optimization problems, and outperforms the basic PSO in search ability.

The remaining of the paper is organized as the follows. Section 2 provides some relative work on PSO. Section 3 describes the SOPSO in details. Section 4 presents the experiments and the relative results. Finally, Section 5 concludes with some remarks.

2 Some Previous Work on PSO

2.1 The Basic PSO Model

PSO adopts a swarm of particles to represent the potential solutions of an optimization problem. In order to search an optimum, each particle is supposed to flying through the search space. During the flying, each individual can adjust its trajectory according to its and other particle's flying experience.

In basic PSO model[11], the flying of a particle is described by its velocity and position, the update equations is defined as the following:

$$v_{id}(t+1)=Wv_{id}(t)+c_1r_1\big(P_{id}(t)-x_{id}(t)\big)+c_2r_2\big(P_{gd}(t)-x_{id}(t)\big) \tag{1}$$

$$x_{id}(t+1)=x_{id}(t)+v_{id}(t+1) \tag{2}$$

Here, let the particles fly through a d-dimensional space, X_i and V_i represent the position vector and velocity vector of the ith individual respectively, while P_i and P_g represent the personal best position and the global best position for the ith individual; the inertia weight ω is a scaling factor controlling the influence of the old velocity; c_1 and c_2 are constants known as "cognitive" and "social" coefficients which determine the weight of P_i and P_g respectively; r_1 and r_2 are two random numbers generated by uniformly distribution in the range [0,1] separated.

2.2 The Lbest and Gbest Models

There are two models for the basic PSO, one is the Gbest model and the other is the Lbest model. The Gbest model offers a faster convergent rate at the expense of robustness. In this model, only a single "best solution" is regarded as the global best particle referred by all particles in the swarm. Due to the model with a star topology, there is a fast information flow between the global best particle and other particle, so the Gbest model converges quickly but tends to converge prematurely. In order to overcome the premature convergence, the Lbest model adopts different neighborhood topologies [12], such as the "ring" or the "Wheel" topology, which can help the Lbest model to maintain a high diversity and multiple attractors in the swarm. Though the convergent speed of the Lbest model is lower than the one of the Gbest model, it can prevent the swarm trapped in an inferior local optimum validly.

2.3 The Improved BPSO Models

In order to keep the balance between "exploration" and "exploitation", shi and Eberhart introduced the inertia weight (ω) to modify the original PSO [11]. They analyzed the effect of the inertia weight by a series of experiments and found that a larger ω can bring in a better global search while a smaller ω can lead to a better local search. They also manipulated PSO with a decreased linearly ω over time[13], the method performs well in unimodal functions, but often failure in multimodal functions. Later, Shi and Eberhart introduced a fuzzy controller to modify the inertia weight dynamically [14]. The adaptive fuzzy controller is a promising technique for optimizing the parameter, but difficult for implementation.

The literatures above embody the concept of control in developing PSO, but they couldn't regarded the PSO system as a closed dynamic system and manipulate the model only based on the particle lever. Almost improved methods disregard the particles would percept the swarm-lever state information and be influenced by the global information during a search for an optimum, so the control or modification to the algorithm are blind or experiential at some extent, which baffles PSO developing into a global optimization.

Recently, R.Ursem has developed a model called the Diversity-Guided Evolutionary Algorithm(DGEA)[15], the main contribution of the model lies in adopting the diversity measure to guide the algorithm alternating between exploring and exploiting behavior. Riget has followed Ursem's idea and developed an attractive and repulsive PSO(ARPSO) model[10]. In the model, the "attraction" and the "repulsion" are adopted to modify the velocity as the diversity-increasing operator and the diversity-decreasing operator respectively; the switch between the two operators is controlled by the diversity of the swarm. The relative results show the diversity-guided technique has a potential in overcoming the premature convergence.

Another literature is deserved to be referred that is "Adaptive Particle Swarm Optimization via Velocity feedback", an improved model proposed by Yasuda[16]. In the model, the average velocity is adopted as the swarm's information to control the parameters shifting between two kinds of parameter sets. According to the analysis of Yasuda, some parameter sets can lead to particle's velocity converging, while others will lead to the velocity diverging in a run, so the transition of parameters sets between a divergent one and a convergent one can modify the search to alter between exploitation and exploration adaptively, and improve the success rate for a goal optimum.

3 The Self-Organization PSO Model

3.1 Analysis of PSO Based on Self-Organization Theory

Considering the difficulties in developing PSO into a global optimization, according to our notion, we should go back to analyze its simulation mechanisms. How to do and what can prompt PSO to be a successful and robust optimization? Maybe, The answer just lies in its headstream----Swarm Intelligence. In generally, "Swarm Intelligence" can be defined as "the emergent complex intelligence inspired by the collective behavior of social insect colonies and other animal societies". Here, the "Swarm" is consisted of some individuals with simple behavior. In swarms, no one is in charge, no one give out direct orders; but every individual can perceive and receive amount of information from different directions. Owing to series of interactions among individuals, the swarm's collective complex behaviors can be produced. So, the Swarm can be regarded as a self-organization system. According to Bonabeau[17], a SO system is characterized with four basic components: positive feedback, negative feedback, random fluctuations and multiple interactions, which just can explain why the complex collective production can emerge in a system through the interactions among the low-lever individuals.

According to the literatures, it's easy to know the previous study of PSO just simulate two characters of SO: random fluctuations and multiple interactions occurring only in the individual lever, but regardless of the information feedback between the individual lever and the swarm lever, what just play an important role to prompt a high quality emergence or stabilize a system.

Based on the concept of Self-organization, the PSO system should be considered as a dynamic closed system, and its dynamic behavior should be understood from two aspects: its low-lever ---the individual lever and its high-lever----the swarm lever, an emergence of a high-lever quality in such a system is mainly due to the internal interactions between the two levers. One aspect, the low-lever particles in the swarm continually interact the local information each other and try to get a collective output cooperatively; another aspect, each particle can apperceive the swarm-lever state and make some decisions to modify its behavior intelligently. In other words, the swarm-lever outputs have an important influence on the behavior of the particles, which just provide a guiding direction and prompt a higher-lever quality to emerge in the swarm.

According to the analysis above, we introduce information feedback into BPSO and develop a self-organization particle swarm optimization (SOPSO).

3.2 The Construction of SOPSO

Fig.1 Presents the construction of SOPSO. Obviously, SOPSO consists of four components, including a decision controller, a swarm of particles, a statistic analyzer and a swarm state set. The function of each component is discussed in details in the follows:

(1) Swarm State Set

Here, Swarm State Set is defined as $\{D(t), V(t), P_g(t)\}$ to record the swarm state information. $D(t)$ means the diversity of the swarm, $V(t)$ is the average velocity and $P_g(t)$ is the best solution of the swarm in *t*th iteration. According to our analysis, the three values can describe the dynamic state of the swarm in different sides; we can choose a single one or a combination of them to guide particles' flying adaptively. Considering the premature convergence, $D(t)$ is the most important one in the three values.

(2) A swarm of particles

During a search, the particles of the swarm fly through the search space following the computation model of the basic PSO. At the same time, they not only interact local information each other, but also apperceive the swarm's state and decide how to modify their behavior adaptively.

(3) Decision Controller

Obviously, the decision controller provides a feedback bridge for the information interaction between the swarm-lever and the particle-lever in the system. Its input comes from the swarm state set and its output gives out a decision command that indicates the particles how to modify their behavior adaptively.

(4) Statistic Analyzer

Here, Statistic Analyzer is a tool used to get the swarm-lever state information based on the information of the particle-lever.

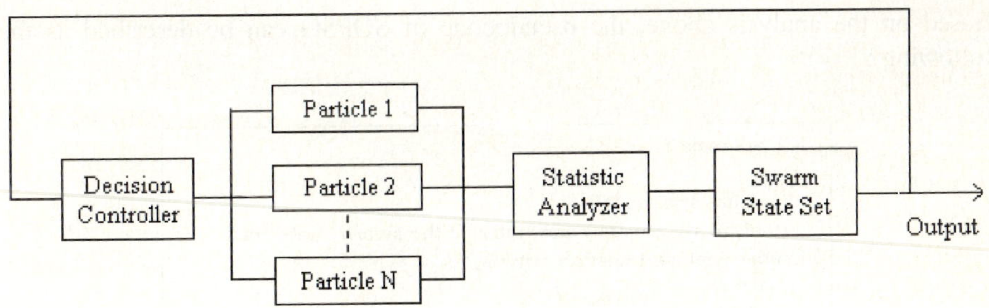

Fig. 1. Construction of SOPSO

3.3 The Design of Decision Controller

Obviously, the decision controller is a key component in SOPSO; its design mainly depends on the task and the controlled objective. Here, we try to modify the behavior of every particle under the guide of the diversity. Considering the premature problem lying in basic PSO, the main reason is that the fast information flow among swarms leads to particles clustering and the diversity declining quickly. So, the diversity is an important factor to describe the dynamic swarm-lever state in SOPSO. During a search, there are three behaviors models for every particle to chose and follow, including the Gbest model, the Lbest model and the gauss mutation model. The Gbest model has a quick convergent speed and can conduct the exploitation around the best solution; the Lbest model can keep the diversity of a swarm in a higher lever, The Gauss mutation take advantage of the $P_g(t)$ or the average point $\overline{P}$ of the swarm to calculate the mutation direction of each particle, which can manipulate the exploration among the search space. Based on the analysis above, the relative rules for the decision controller are designed as the following:

Rule 1: If $D(t)>D_{high}$, Then each particle complies with the Gbest Model to update its velocity and position.

Rule 2: If $D_{low}<D(t)<D_{high}$, Then each particle follows the Lbest model to update its velocity and position.

Rule 3: If $D(t)<D_{low}$, Then each particle adopts the gauss mutation to update its velocity and position.

Here, D_{low} and D_{high} are two thresholds of the diversity used to manipulate the particle shifting among the three behavior models above. $D(t)$ is the diversity of the swarm and computed according to the following diversity–measure[15]:

$$D(t) = \frac{1}{|S|\cdot|L|}\cdot\sum_{i=1}^{|s|}\sqrt{\sum_{j=1}^{N}(P_{ij}-\overline{P}_j)^2} \tag{3}$$

Where S is the swarm, $|S|$ means the size of the swarm, $|L|$ is the lenth of the longest diagonal in the search space, N is the dimension of the objective problem, P_{ij} is the jth value of the ith particle and $\overline{P}_j$ is the jth value of the average point $\overline{P}$.

3.4 The Pseudocode of SOPSO

Based on the analysis above, the pseudocode of SOPSO can be described as the following :

```
SOPSO main {
   t=0
   Initialize swarm S(0)
   Analyze swarm S(0) and Initialize the swarm state sets
   while (not (termination criteria)) {
      t=t+1
      if (D(t)>Dhigh)
          Particle behavior model= "Gbest model"
      if (Dlow<D(t)<Dhigh)
          Particle behavior model= "Lbest model"
      elseif (D(t)<Dlow)
          Particle behavior model= "the Gauss Mutation"
      Update S(t)
      Analyze swarm S(t) and Update the swarm state sets
   }
}
```

Fig. 2. Pseudocode for SOPSO

4 Experiments and Results

4.1 The Benchmark Functions

The following benchmark functions have been used to test the performances of SOPSO in the experiments.

$$f_1 = \sum_{i=1}^{N} x_i^2 \tag{4}$$

$$f_2 = \sum_{i=1}^{N} [100(x_{i+1}^2 - x_i)^2 + (1 - x_i)^2] \tag{5}$$

$$f_3 = \sum_{i=1}^{N} (x_i^2 - 10COS(2\pi x_i) + 10) \tag{6}$$

$$f_4 = \frac{1}{4000} \sum_{i=1}^{N} x_i^2 - \prod_{i=1}^{N} \cos(\frac{x_i}{\sqrt{i}}) + 1 \tag{7}$$

$$f_5 = 20 + e - 20\exp(-0.2\sqrt{\frac{1}{n}\Sigma_{i=1}^{n} x_i^2}) - \exp(\frac{1}{n}\Sigma_{i=1}^{n}\cos 2\pi x_i) \tag{8}$$

$$f_6 = \frac{1}{N} \sum_{i=1}^{N} (x_i^4 - 16x_i^2 + 5x_i) \tag{9}$$

The test functions above can be divided into two categories according to their complexities. f_1 ~f_2 are unimodal functions with different unique features, which are relatively easy to optimize, but the difficulty increases with the dimension growing. f_3 ~f_6 are multimodal functions with many local optima and they are very difficult to solve for many optimization algorithms. The basic information of the test functions in experiments is listed in Table.1.

Table 1. Basic information about the test functions

Test Func	Solution space	Global minimum
f_1-sphere	$[-100,100]^N$	0
f_2-Rosenbrock	$[-100,100]^N$	0
f_3-Rastrigrin	$[-5.12,5.12]^N$	0
f_4-Griewank	$[-600,600]^N$	0
f_5-Ackley	$[-32,32]^N$	0
f_6-2^n minima	$[-5,5]^N$	-78.3323

4.2 Experiments and Results

In the following experiments, both BPSO and SOPSO adopted a popular set of parameters that is ω=0.7298 and c_1=c_2=1.4962. Problems dependent parameters such as the *swarmsize* and v_{max} were tuned in each test case. The diversity thresholds were set to D_{low}=0.5×10^{-4} and D_{high}=0.10 respectively.

A series of experiments have been done to make a comparison on the convergent performance between SOPSO and others optimization algorithms. The relative results are showed in the following tables.

Table 2. Comparison Between SOPSO and BPSO based on the mean best function values. All results were averaged over 50 runs, where the dimension of the functions is 50 and the max iteration is 5000. The swarm size adopted by every algorithm is 50.

Function	BPSO with Lbest	BPSO with Gbest	SOPSO
f_1--sphere	7.45×10^{-10}	4.52×10^{-11}	3.67×10^{-10}
f_2- Rosenbrock	35.2992	30.5438	24.7632
f_3-Rastrigrin	1.76644	3.76215	0.76192
f_4-Griewank	0.078162	0.096582	0.005439
f_5-Ackley	1.45×10^{-3}	6.64×10^{-3}	7.67×10^{-5}
f_6-2^n minima	-72.7665	-70.4527	-77.6326

In Table 2., SOPSO is compared to BPSO with Gbest model and Lbest model respectively. It's apparent that SOPSO outperforms the two models of BPSO for the multimodal functions. Though SOPSO has a worse result for the unimodal function f_1 compared to BPSO with Gbest model, that just is due to SOPSO keep a higher diversity, which makes its convergent speed slower during the search. As far as the multimodal optimization problems is concerned, BPSO seems easy to be trapped in local minima while SOPSO is easy to escape from the local minima and converge to the global minimum with a high precision.

Table 3. The mean best values found by the different algorithms. All results were averaged over 50 runs, where "D" stands for the dimension of the functions and the max iteration=50 • D. The swarm size adopted by every algorithm is 400. The results of ARPSO and DGEA come from the references [10] and [15] respectively.

Function	SOPSO	ARPSO	DGEA
f_2 (20D)	13.6705	2.34	96.007
f_2 (50D)	20.4359	10.43	315.395
f_2 (100D)	85.4035	103.46	1161.55
f_3 (20D)	3.72×10^{-3}	0	2.21E-5
f_3 (50D)	0.46582	0.02	0.01664
f_3 (100D)	8.68941	0.438	0.15665
f_4 (20D)	0.027623	2.5×10^{-2}	7.02E-4
f_4 (50D)	0.008981	3.05×10^{-2}	4.40E-3
f_4 (100D)	0.003913	9.84×10^{-2}	0.01238
f_5(20D)	3.85×10^{-9}	3.3×10^{-8}	8.05E-4
f_5(50D)	7.68×10^{-5}	0.027	4.61E-3
f_5(100D)	0.005629	0.218	0.01329

Table 3. shows the comparisons between SOPSO and other two improved methods guided by the diversity, including ARPSO and DEGA. For f_2 and f_4, SOPSO performs worse in lower dimension space but outperforms ARPSO and DEGA in higher dimension space. Though SOPSO performs worse a little than ARPSO and DEGA for f_3, it outperforms BPSO a lot. According to Table.3, The best performance of SOPSO occurs in function f_5, which shows SOPSO is a robust optimizitor for the complex optimization problems.

5 Conclusions

The paper has developed a self-organization particle swarm optimization (SOPSO) and evalutes its performance on a number of benchmark problems. The relative experimental results show that SOPSO outperforms BPSO for the complex optimization problems, especially for the multimodal functions in a high dimension space. Through the information feedback , SOPSO is susceptible to escape from the local optima and convenge to the global optimum successfully. Though SOPSO shows a slower convergent speed when applied to the unimodal functions or the simple optimizations in a low dimention space, SOPSO still is a rubust optimizer for the complex optimization problems compared to BPSO.

Acknowledgments. This work is supported by the key science and technology fund project by the Ministry of Education of P.R.China under the grant No.204018, and the science and technology project by the Ministry of Education of Shanxi under the grant No.20051310.

References

1. Kennedy, J. and Eberhart, R.C.: Particle swarm optimization. In: Proc. IEEE Conference on Neural Networks, Perth, Australia: IEEE Service Center. 11(1995) 1942–1948
2. Bergh, F.V.D. and Engelbrecht, A.: Particle swarm weight initialization in multi-layer perception artificial neural networks. In: Development and Practice of Artificial Intelligence Techniques, Durban, South Africa. (1999)41–45
3. Bergh, F.V.D., Engelbrecht, A.P.: Cooperative Learning in Neural Networks using Particle Swarm Optimizers. In: South African Computer Journal, vol.26.No.11.(2000) 84–90
4. Clerc, M., and Kennedy, J.: The Particle Swarm--Explosion, Stability, and Convergence in a Multidimensional Complex Space. In: IEEE Transactions on Evolutionary Computation, Vol. 6. No.1. (2002) 58–73
5. Fukuyama, Y., Yoshida, H.:A Particle Swarm Optimization for Reactive Power and Voltage Control in Electric Power Systems. In: Proc. Congress on Evolutionary Computation, Seoul, Korea. Piscataway, NJ: IEEE Service Center. (2001) 87–93
6. Zeng, J.C., Jie, J., Cui, Z. H.: Particle Swarm Optimization. Beijing, Science Press, (2004)
7. Suganthan, P.N.:Particle Swarm Optimizer with Neighborhood Operator. In: Proc. Congress on Evolutionary Computation, Washington D.C, USA, July, Piscataway, NJ: IEEE Service Center, (1999) 1958–1961
8. Li, X.D.: Adaptively Choosing Neighborhood Using Species in a Particle Swarm Optimizer for Multimodal Function Optimization. In: Proceedings of the Genetic and Evolutionary Computation Conference. (2004)105–116

9. Xie, X.F., Z, W.J. and B, D.C.: Optimizing Semiconductor Devices by Self-organizing Particle Swarm, Congress on Evolutionary Computaion. Oregon,USA: (2004)2017-2022
10. Jacques, R., Jakob, S. V.: A Diversity-Guided Particle Swarm Optimizer –the ARPSO. http://citeseer.nj.nec.com/riget02diversityguided.html
11. Shi, Y. and Eberhart, R.C.: A modified particle swarm optimizer In: Proc. Conference on Evolutionary Computation: IEEE Press. Piscataway. (1998)69–73
12. Kennedy, J. Small Worlds and Mega-Minds: Effects of Neighborhood Topology on Particle Swarm Performance, Proceedings of the Congress of Evolutionary Computation, vol. 3, IEEE Press. 1999, pp:1931-1938
13. Shi, Y. and Eberhart, R. C.: Empirical study of particle swarm optimization. In: Proc. Congress on Evolutionary Computation, Piscataway, NJ: IEEE Service Center,(1999) 1945-1950
14. Shi, Y. and Eberhart, R. C.: Fuzzy Adaptive Particle Swarm Optimization. In: Proc. Congress on Evolutionary Computation, Seoul, Korea: IEEE service Center.(2001)101–106
15. Ursem, R. K.: Diversity-Guided Evolutionary Algorithms. In: Proc. the 7th international conference on parallel problem solving from Nature, Lecture Notes in Computer Science, Vol. 2439. (2002)462-474
16. Iwasaki, N. and Yasuda, K.: Adaptive Particle Swarm Optimization via Velocity Feedback, The 36th ISCIE International symposium on Stochastic Systems Theory and Its Applications, B7-5 (2004)116-117
17. Bonabeau, E., Dorigo, M. and Theraulaz, G.: Swarm Intelligence From Natural to Artificial Systems. Oxford Univergity Press Inc. ISBN 0-19-513158-4; ISBN 0-19-513159-2,(1999) 1–22

An Evolving Wavelet-Based De-noising Method for the Weigh-In-Motion System

Xie Chao[1,2], Huang Jie[2], Wei Chengjian[1], and Xu Jun[1,2]

[1] College of Information Science and Engineering, Nanjing University of Technology, Nanjing, Jiangsu, 210009, P.R. China
xiejacob@gmail.com

[2] Research Center of Information Security, Dept. of Radio Engineering, Southeast University, Nanjing, Jiangsu, 210096, P.R. China

Abstract. In order to enhance the roads maintenance, a number of weigh-in-motion (WIM) systems are used in the high way to collect the weight of every vehicle pass by, but normally the outcome is rough. The one important reason of this problem is that lots of complex low-frequency noises exist in the weighing signal, which can not be filtered by the Fourier series analysis. In this paper[1], firstly we apply the de-noising method via thresholding of wavelet coefficients to purify the signal, and then we propose an evolving wavelet thresholding method for the WIM system using evolutionary programming (EP), further we evolves the best threshold by this method and improve the weighing signal processing. Finally, we use our thoughts to model a simulated WIM signal denosing system and experimental results shows that this novel system provides more accurate data compared to other existing analyzing measures for weighing signal.

1 Introduction

With the rapid development of the economy in the developing country, the transportation has made great progress, and Weigh-In-Motion (WIM) system plays a key role in the pavement research for over a few years [1]. Although the WIM system provides many reasonable means of weighing signal processing for the weight enforcement, the accuracy is still by far not satisfactory. Based on many experiences by scale testing [2], the error of WIM scale is mainly caused by two facts, first one is high and low frequency signal interference in output, and another is the scale mechanical structure and physical status of pavements. In this paper, we only focus on the improvement of scale signal processing rather than other facts.

In fact, interferences exist in the low frequency is a primary part of the whole error. In order to clear this kind of noise, people normally use Fourier analysis means to filter and analyze the sampling data in the time-frequency. The Fourier series analysis

[1] This research is supported by National High Technology Research and Development Program of China under Grant No.2005AA147040 and Natural Science Foundation of China under Grant No.60133010. Special thanks Mettler-Toledo (Switzerland) International Scale & System Ltd., Co for collecting the weigh station data of this research.

L. Jiao et al. (Eds.): ICNC 2006, Part I, LNCS 4221, pp. 923 – 926, 2006.

is perfectly applied to convert the time signal to frequency time, but losing the important time information. Meanwhile the wavelet transformation can give us an overview landscape in both time and frequency level, so we use wavelet transform to decompose weighing signal in specified levels and run the signal de-nosing via wavelet thresholding to eradicate the low noise by virtue of the WIM scale noise character.

Furthermore, we consider the wavelet thresholding as a critical factor affecting the whole de-nosing quality, but the global best threshold is difficult to find out by experience, and it is a nonlinear and NP problem. Because of the advantage of the evolutionary intelligence [3], we propose a new evolving wavelet-based de-nosing method using the EP and model an appropriate fitness function, which can be used by the EP to evolve the ideal threshold. Based on the above consideration, we finally construct a simulated weighing signal de-nosing system using the real data from Mettler-Toledo Company to testify our new de-nosing approach successfully.

2 Evolutionary Wavelet-Based De-noising Method

Wavelets are mathematical functions that cut up data into different frequency components, and then study each component with a resolution matched to its scale. So we naturally analyze a hierarchy of signal frequency distribution using the discrete wavelet transform. After executing the multi-level wavelet transform, we can obtain both high and low frequency in respective level without losing time space information. Consequently, we use specified de-noising method via thresholding of wavelet coefficients in the low frequency and rebuild better signal for computing the weight.

Recent works on signal de-noising via wavelet thresholding of Donoho [4] have shown that various wavelet thresholding schemes for de-noising have an advantage to de-noise the non-stationary signal and also achieve effective results. So, we apply this kind of de-noising method to the WIM scale. Suppose the output signal $f(t)$ when a truck passes the scale is the addition of original signal and noise, we sample the $f(t)$, and get the wavelet coefficient $Wf(j,k)$. Then we use the recursive method to compute wavelet coefficient by the double scale equation for the reason of simple computation.

With above knowledge, we note the $Wf(j,k)$ as $w_{j,k}$ and $Ws(j,k)$ as $u_{j,k}$ for short and describe the general flow of wavelet thresholding de-noising as below:

1) Analyzing original signal $f(k)$ with wavelet transform and obtaining a group of wavelet coefficients $w_{j,k}$.
2) Thresholding the $w_{j,k}$, estimating the wavelet coefficients $\hat{w}_{j,k}$ and minimize the $\left\|\hat{w}_{j,k} - u_{j,k}\right\|$.
3) Reconstructing the signal with $\hat{w}_{j,k}$, and getting the final result signal $\hat{f}(k)$.

And then we choose the combinatory method of soft-thresholding and hard-thresholding to estimate the wavelet coefficient.

So we can see that the threshold estimation is a very important step for getting a better de-noising effect, if the threshold is good, thus the wavelet de-noising can purify the signal well. At the same time many sources [3] testify that finding the good threshold is a non-linear and NP problem, so this leads us to use evolutionary method to solve this problem. Evolutionary Programming (EP) is a useful bio-inspired method of optimization when combinatory and real-valued function optimization in which the optimization surface or fitness landscape is "rugged", possessing many locally optimal solutions [4]. Thus, we use EP to find the best threshold in the wavelet-based de-noising.

We choose the weight error as the fitness function for efficient computation and use this fitness function to judge the evolved threshold λ in every recursion. In the initial step, we set the $\lambda_{init} = \sigma\sqrt{2\log(N)}$ according to the Donoho's thoughts [5] in every decomposition level. Then we continue our method using the EP execution flow as below:

1) Choose an initial POPULATION of trial solutions at random.
2) Each solution is replicated into a new population.
3) Computing its fitness function assesses each offspring solution.

With above smart thoughts, we recursively apply evolutionary programming to search the best threshold λ for wavelet-based de-noising, and gain the better signal output results. All experiments will be described in the next section.

3 Experiments and Results

Based on the above consideration, we use MATLAB environment to construct a simulated weighing signal system to testify our new de-nosing approach. The size of testing weighing data is 120 groups collected by Mettler-Toledo (Switzerland) International Scale & System Ltd., Co and these data are categorized into 5km/h and 10km/h two-speed level in 8 meter and 2.2 meter two types of scale respectively. Five sample rates, 240bps, 480bps, 960bps, 1920bps and 3840bps, are applied in the system which according to the industry sampling standard.

Firstly, we use the Fourier series analysis method mentioned above to de-nosing the output signal. The number of composition level is 5 and the type of the wavelet is “db3”. So we can get that the average weight errors are around 3%, which equals to the performances of current WIM scales. Next, we use the evolving wavelet thresholding method to de-noising the original output signal. We set mutation rate $\alpha = 0.01$, the number of evolving steps is 2000 and $POPULATION = 100$. So we can gain the magical result as figure 1.

In the above experiment, due to the relationship between the signal's variance and the threshold, we can not get the global best value of λ at the first step of our evolving method, but five best values (280,448,887,500,800) are evolved in different level as the threshold during the course of method execution. So we can select threshold among five good values as a global value by computing original signal's variance, and filter the unwanted wavelet coefficient more accurately in deeper level, then continue the operation of de-noising. Consequently, we can see that white line which represents the de-nosing signal error (0.9617% in 0.8m and 0.56% in 2.2m) almost is lower than the black line which represents the original signal error (2.198% in 0.8m and 1.215% in 2.2m) in both type of WIM scales' testing. Thus, we get the excellent average error that is mainly lower than 1.0%, which is much better than Fourier-based method and other WIM scales' performances.

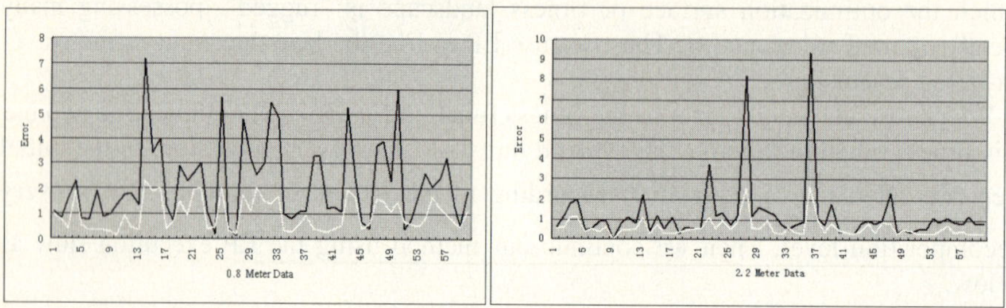

Fig. 1. 0.8 and 2.2 Meter weight error results. Black line stands for original results and white line stands for results of our method.

Finally, we indeed get the effective de-noising method with the evolutionary implementation. The result of the threshold leads to the best result among the current WIM scales. We also get the ideas that wavelet-based thresholding de-noising method is better than normal Fourier-based method because of the multi-level feature leads to a more information of the low frequency which can be investigated during the wavelet de-nosing.

However, considering the fixed interference produced by the different truck and scale, we will conceive a self-learning de-noising system to alter the threshold for dealing with various kinds of truck weight. Further on, we will design the good wavelet base with evolutionary method and hope to identify every kind of noise. We hold the faith that our method would have broad practicability in the WIM system.

References

1. Bernard Jacob, Bill Newton, "COST 232 project final report," Proceedings of the Second European Conference on Weigh-in-Motion of Road Vehicles, ISBN 92-828-6786-2, 1999.
2. States' Successful Practices Weigh-in-Motion Handbook. Federal Highway Administration, U.S. Department of Transportation, 1997.
3. Rajan C.C.A., Mohan M.R., "An evolutionary programming method for solving the unit commitment problem," IEEE Power Engineering Society General Meeting, pp.11149, 2004.
4. David L. Donoho, Martin Vetterli, R. A. DeVore, and Ingrid Daubechies, "Data compression and harmonic analysis," IEEE Trans. Inform. Theory, 44(6):2435--2476, 1998.
5. D.L.Donoho. "De-noising by soft-thresholding," IEEE Trans. Inform. Theory, vol.41, pp.613-627, May 1995.

SAR Image Classification Based on Clonal Selection Algorithm

Wenping Ma and Ronghua Shang

Institute of Intelligent Information Processing, P.O. Box 224, Xidian University, Xi'an, 710071, P.R. China
wpma@mail.xidian.edu.cn

Abstract. This paper presents a new classification method based on the Clonal Selection Principle, named Clonal Selection Algorithm (CSA). The new algorithm can carry out the global search and the local search in many directions rather than one direction around the same antibody simultaneously, and obtain the global optimum quickly. The implementation of new algorithm composes of three main processes: firstly, selecting training samples and choosing clustering centers randomly. Secondly, training samples using CSA, and obtaining optimal clustering center based on three main clonal operations: cloning, clonal mutation and clonal selection. Finally, output the classification results according to clustering center obtained. To show the usefulness of this approach, experiment with simulated SAR image was considered. The classification results are evaluated by comparing with three well-known algorithms, UAIC, K-means, and fuzzy K-means. Accroding to the overall accuracy and Kappa coefficient, CSA has high classification precision and can be used in SAR images classification.

1 Introduction

Immune system is an important information processing function of the organism. Clonal Selection Principle[1] is very important for the immunology. Clone means reproducing or propagating asexually. A group of genetically identical cells are descended from a single common ancestor, such as a bacterial colony whose members arose from a single original cell as a result of binary fission. The idea attracts such great attentions that some new algorithms based on clonal selection theory are proposed successively[2~4].

Synthetic Aperture Radar (SAR) is a system that possesses its own illumination and produces images with a high capacity for discriminating objects. Data SAR possesses a random behavior that is usually explained by a multiplicative model[5]. Even though AIS have been successfully utilized in several fields[3,6,7], few applications have been reported in SAR images classification . This may be due to the fact that it is difficult to apply current AIS algorithms to SAR images classification owing to the huge data volumes associated with SAR images.

The Clonal Selection Principle incorporates the global search with local search, and obtain the global optimum quickly. And the clustering algorithm based on it is feasible for the large data sets with mixed numeric and categorical values. We attempt

L. Jiao et al. (Eds.): ICNC 2006, Part I, LNCS 4221, pp. 927 – 934, 2006.

to solve SAR images classification using clonal selection algorithm. In order to show the new algorithm's advantages, the performances of the CSA is evaluated via computer simulations and compared with UAIC[8] and conventional classification algorithms: K-means[9], fuzzy K-means[10].

2 SAR Image Classification Based on CSA

2.1 Clonal Selection Theory

The immune system's ability to adapt its B-cells to new types of antigen is powered by processes known as clonal selection and affinity maturation by hypermutation. The first immune optimization algorithm may be Fukuda, and Mori's algorithm. But the clonal selection algorithm for optimization has been popularized mainly by de Castro and Von Zuben's CLONALG[13].

The clonal selection theory (F. M. Burnet, 1959) is used by the immune system to describe the basic features of an immune response to an antigenic stimulus[11]. The references [2] to [4] simulate this clonal selection mechanism above from different viewpoints and put forward the Clonal Selection Algorithms one after another. Just as the same as the Evolutionary Algorithms (EAs) , the Artificial Immune System Algorithms depend on the encoding of the parameter set rather than the parameter set itself. When exposed to antigens, the immune cells which may recognize and eliminate the antigens can be selected in the body and mount an effective response against them. Its main ideas lie in that the antigen can selectively react to the antibody, which is a native production and spreads on the cell surface in the form of peptide. Some clonal cells divide into antibody producing cells, and others become immune memory cells to boost the second immune response.

The clonal selection is a dynamic process of the immune system stimulated by the self-adapting antigen. From the viewpoint of the Artificial Intelligence, some biologic features such as learning, memory and antibody diversity can be used in artificial immune system[12].

2.2 Basic Definition

The new algorithm composes of three main processes: firstly, selecting training samples for every region in the SAR image. Secondly, training these samples using CSA, and obtain optimal clustering center of every region. Finally, output the classification results of SAR image according to clustering center obtained.

Let us establish the major elements of clonal selection algorithm:

Antigen: In AIS, antigen usually means the problem and its constraints. Especially to classification problems, antigens are simulated as feature vectors based on image gray-level co-occurrence matrix which are presented to the system during training and testing.

Antibody: Antibodies represent candidates of the problem. Especially, for the classification problem, the antibodies as candidate clustering centers experience a form of clonal process after being presented with an input image data, every candidate adopts binary coding with the length l, namely, l = n*m, in which n represents the dimension of feature vectors and m represents the length of binary bit for every feature.

Antibody-Antigen Affinity: The reflection of the total combination power locates between antigen and antibodies. For classification problem, given the antibody population $A=\{a_1,a_2,\cdots,a_n\}$, clustering center is $C=\{c_1,c_2,\cdots,c_n\}$, which is the mean value of all antibodies. Antibody-Antigen Affinity is defined as (1):

$$d_j=\sum_{i=1}^{n}|a_i - c_i| \quad j=1,2,\cdots,k$$
$$affinity_j=-d_j \tag{1}$$

where k is the sample number of certain region. So that the affinity between antigens and antibodies is between the range [0, 1].

Cloning T_c^C : Define

$$Y(k)=T_c^C(A(k))=[T_c^C(a_1(k)) \quad T_c^C(a_2(k)), \quad \cdots \quad ,T_c^C(a_n(k))]^T \tag{2}$$

where $T_c^C(\boldsymbol{a}_i(k))=\boldsymbol{I}_i\times \boldsymbol{a}_i(k)$ $i=1,2\cdots n$, I_i is q_i dimension row vectors which is called as q_i clone of antibody a_i, $i=1,2,\cdots,n$.

$$q_i(k)=g(n_c, affinity(a_i(k))) \tag{3}$$

n_c is a given value relating to the clone scale. After clone, the population becomes:

$$Y(k)=\{Y_1(K),Y_2(K),\cdots,Y_n(k)\} \tag{4}$$

where $Y_i(K)=\{Y_{ij}(k)\}=\{Y_{i1}(k),Y_{i2}(k),\cdots,Y_{iq_i}(k)\}$ and $Y_{ij}(k)=a_i(k), j=1,2,\cdots,q_i$.

Clonal Mutation T_m^C : According to the mutation probability p_m, the cloned antibody populations are mutated as follows:

$$\boldsymbol{Z}(k)=T_m^C(\boldsymbol{Y}(k))=(-1)^{random\le p_m}\boldsymbol{Y}(k) \tag{5}$$

Clonal Selection_ T_s^C : $\forall i=1,2,\cdots n$, if there are mutated antibodies

$a_i'(k)=\max\{\boldsymbol{Z}_i(k)\}=\{\boldsymbol{Z}_{ij}(k)\mid \max affinity(\boldsymbol{Z}_{ij}) \quad j=1,2,\cdots,q_i\}$, the probability of $a_i'(k)$ taking place of $a_i(k)\in A(k)$ is:

$$T_s^k\left(a_i(k)=a_i'(k)\right)=\begin{cases}1 & \text{when } affinity(a_i(k))<affinity(a_i'(k))\\ 0 & \text{when } affinity(a_i(k))\ge affinity(a_i'(k))\end{cases} \tag{6}$$

After the clonal selection, the new population is:

$$A(k+1)=\{a_1(k+1),a_2(k+1),\cdots,a_i'(k+1),\cdots,a_n(k+1)\} \tag{7}$$

where $a_i'(k+1)=a_j(k+1)\in A(k+1)\ i\ne j$ and

$affinity(a_i'(k+1))=affinity(a_j(k+1))$, in which $a_j(k+1)$ is one of the best antibodies in $A(k+1)$.

2.3 Description of the Algorithm

The Clonal Selection Principle was proposed as the basic features of an immune response to an antigenic stimulus[11]. Inspired by the Clonal Selection Principle, we proposed a novel clonal selection algorithm for classification of SAR imagery which can be implemented as following:

Step 1: Selecting training samples for every region in the image according to the number of classes. Assign every antigen to one of classes.

Step 2: Generate the original antibody population $A(0)=\{a_1(0),a_2(0),\cdots a_{n_b}(0)\}\in I^{n_b}$, by coding the antigens. Give the size of antibody population n_b, the clonal scale n_c. Set the mutation probability p_m, and coding length l. k:=0.

Step 3: Calculate the antibody-antigen affinities of all the antibodies in $A(k)$.

Step 4: Compute the clonal proportion $q_i(k)$ of each antibody a_i in $A(k)$ according to antibody-antigen affinity and the clonal scale n_c.

Step 5: Implement the Cloning T_c^C at $A(k)$ and get the antibody population $Y(k)$ after clonal operation.

Step 6: Implement the Clonal Mutation T_m^C at $Y(k)$ with the probability p_m and get the antibody population $Z(k)$, $Z(k)=T_m^C(Y(k))$.

Step 7: Implement the Clonal Selection T_s^C at $Z(k)$ and get the antibody population $A'(k)$, $A'(k)=T_s^C(Z(k))$.

Step 8: If stop condition is satisfied , export $A'(k)$ as the output of the algorithm, Stop. Otherwise, $A(k+1) = A'(k)$, k:=k+1, go to Step 3. Here the stop condition is to set a fixed threshood T, if the total affinity difference between fore-and-aft generations is in the range of T, the algorithm stops.

Step 9: Get the optimal clustering center, which is the center of last population obtained.

Step 10: Repeat Step 2 to Step 9 for every class, get the optimal clustering center for every class.

Step 11: After traing all the samples, calculate the distance between all pixels and clustering centers, assign every pixel to one class with the least distance.

3 Experimental Results

We tested the new algorithm using the SAR image (256×256 pixels), which is shown in Fig.1. The observed image was expected to fall into three classed: river, vegetation and crop. For a convenient comparison, we selected 100 labeled samples for each class.

In order to show the new algorithm's advantages, the performances of the CSA is evaluated via computer simulations and compared with UAIC and conventional algorithms: K-means, fuzzy K-means. The parameters setting of CSA is as follows: the size of antibody population n_b = 100, the clonal scale n_c=5, coding length $l = n*m$, in our experiment, we extracted 3-dimension texture features from gray-level co-occurrence matrices. The mutation probability $p_m=2/l$, the threshold T = 0.05. The parameter settings of UAIC are the same as those in [8].

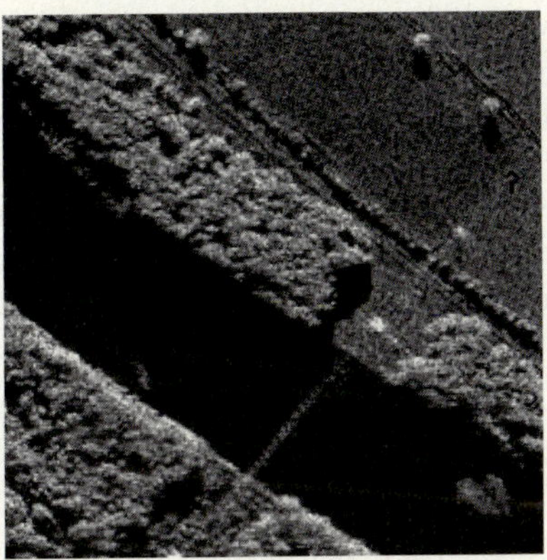

Fig. 1. Simulated SAR image

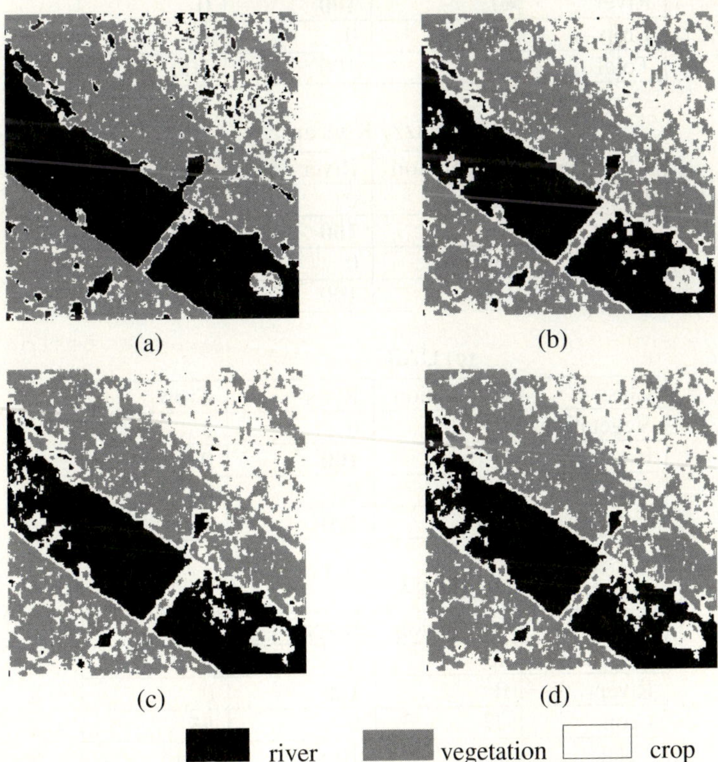

Fig. 2. Classification images for Fig.1. (a) CSA. (b) UAIC. (c) Fuzzy K-means. (d) K-means.

The visual comparisons of the four classifications in Fig.1. It can be found from the classification images (Fig. 2) that four classifiers have similar classification results in the river class. But in Fig.2.(b), UAIC misclassifies many vegetation pixels to the crop class. K-means and fuzzy K-means create similar classification maps, but in Fig.2.(c) and Fig.2.(d), many vegetation pixels and river pixels are misclassified to the crop class. By contrast, our new algorithm CSA achieves the best visual accuracy in the river class than other classifiers, and also performs satisfactorily to the vegetation and crop classes. As a result, those using CSA have better results for three classes.

For a more detailed verification of the results, we compared truth data with the classified images and assess the accuracy of each classifier quantitatively using both the overall accuracy measure and the Kappa coefficient[14]. Table 1-2 list the results of comparisons between the truth data and classified images obtained by four algorithms:

Table 1. Comparison of four methods of classification in confusion matrix

(a) K-means

Class	Vegetation	River	Crop
Vegetation	**54**	0	5
River	0	**100**	0
Crop	46	0	**95**
Total	100	100	100

(b) Fuzzy K-means.

Class	Vegetation	River	Crop
Vegetation	**64**	0	5
River	0	**100**	0
Crop	36	0	**95**
Total	100	100	100

(c) UAIC

Class	Vegetation	River	Crop
Vegetation	**71**	0	5
River	0	**100**	0
Crop	29	0	**95**
Total	100	100	100

(d) CSA

Class	Vegetation	River	Crop
Vegetation	**73**	0	5
River	0	**100**	0
Crop	27	0	**95**
Total	100	100	100

Table 2. Comparison of four methods of classification in overall accuracy and Kappa coefficient

Accuracy	K-means	Fuzzy K-means	UAIC	CSA
Overall accuracy	83.0%	86.3%	88.7%	89.3%
Kappa coefficient	0.7450	0.7950	0.8300	0.8400

From Tables 1 and Table 2 it is apparent that the CSA produces better classification results than other algorithms. The details are as follows: CSA exhibits the best overall classification accuracy, i.e., the best percentage of correctly classified among all the testing pixels considered, with a gain of 6.3%, 3.00%, 0.60% over the K-means, Fuzzy K-means, and UAIC. Respectively, CSA improves the Kappa coefficient from 0.7450 to 0.8400, an improvement by 0.0950. This is due to that the conventional algorithms require ideal conditions, however, the ideal conditions are not often met in real classification. As a result, these conventional classification methods have a low precision. And our new algorithm has a faster global convergence speed, which is the reason of better than UAIC.

4 Concluding Remarks

A novel algorithm based on the Clonal Selection Principle CSA, was designed for SAR images classification in this paper. The new algorithm can carry out the global search and the local search in many directions rather than one direction around the same antibody simultaneously. The clonal selection algorithm can be characterized as a cooperative and competitive approach, where individual antibodies are competing but the whole population will cooperate as an ensemble of individuals to present the final solution, and has a faster global convergence speed. The experimental results consistently show that the proposed algorithm has high classification precision. When compared with other three classifiers, K-means , fuzzy K-means, and UAIC, the average performance of CSA is better than them.

References

1. F. M. Burnet. The Clonal Selection Theory of Acquired Immunity. Cambridge University Press, 1959.15. F.M. Burnet. Clonal selection and after. Theoretical Immunology. New York: Marcel Dekker, 1978: 63-85
2. L. N.De Castro, F. J.Von Zuben.: The Clonal Selection Algorithm with Engineering Applications. Proc. of GECCO'00, Workshop on Artificial Immune Systems and Their Applications, 2000:36-37
3. J. Kim and P. Bentley, "Toward an artificial immune system for network intrusion detection: An investigation of clonal selection with a negative selection operator," in Proc. Congress on Evolutionary Computation (CEC), Vol. 2, Seoul, Korea, (2001) 1244–1252

4. Haifeng DU, Licheng JIAO, Sun'an Wang. Clonal Operator and Antibody Clone Algorithms. Proceedings of the First International Conference on Machine Learning and Cybernetics, Beijing, (2002) 506-510
5. Frery, A.C. Mueler, H.J. Yanasse, C.C.F. and Sant'ana, S.J.S.: A model for extremely heterogeneous clutter. IEEE Transactions on Geoscience and Remote Sensing, Vol.35, No. 3, (1997) 648–659
6. S. A. Hofmeyr and S. Forrest, "Immunity by design: An artificial immune system," in Proc. Genetic and Evolutionary Computation Conf. (GECCO), (1999) 1289–1296
7. J. Timmis, M. Neal, and J. E. Hunt, "An artificial immune system for data analysis," Biosystem, vol. 55, no. 1/3, pp. 143–150, 2000
8. Yanfei Zhong, Liangpei Zhang, Bo Huang, and Pingxiang Li: An Unsupervised Artificial Immune Classifier for Multi/Hyperspectral Remote Sensing Imagery. IEEE Transactions on Geoscience and Remote Sensing, Vol. 44, No. 2. (2006) 420-431
9. D. Hall and G. Ball, "Isodata: A novel method of data analysis and pattern classification," Stanford Res. Inst., Stanford, CA, Tech. Rep., 1965.
10. P. Thitimajshima, "A new modified fuzzy c-means algorithm for multispectral satellite images segmentation," in Proc. IGARSS, Vol. 4, (2000) 1684–1686.
11. Zhou Guangyan. Principles of Immunology. Shang Hai Technology Literature Publishing Company, 2000. (In Chinese)
12. Dasgupta D, Forrest S Artificial immune systems in industrial applications. In: Proceedings of the Second International Conference on Intelligent Processing and Manufacturing of Materials. IEEE press, (1999) 257~267
13. Leandro N. de Castro: Larning and Optimization Using the Clonal Slection Principle.IEEE Transactions on Evolutionary Computation, Vol. 6, No. 3. (2002) 239-251
14. Giles M. Food. Status of Land Cover Classification Accuracy Assessment[A]. Remote Sensing of Environment, 2002, 80: 185-201
15. Licheng Jiao, Lei Wang. A novel genetic algorithm based on immunity. IEEE Transactions on Systems, Man and Cybernetics, Part A. Vol.30, No.5, Sept. 2000

Crossed Particle Swarm Optimization Algorithm*

Teng-Bo Chen, Yin-Li Dong, Yong-Chang Jiao, and Fu-Shun Zhang

National Laboratory of Antennas and Microwave Technology
Xidian University, Xi'an, Shanxi 710071, P.R. China
c t b210@sina.com, ychjiao@xidian.edu.cn

Abstract. The particle swarm optimization (PSO) algorithm presents a new way for finding optimal solutions of complex optimization problems. In this paper a modified particle swarm optimization algorithm is presented. We modify the PSO algorithm in some aspects. Firstly, a contractive factor is introduced to the position update equation, and the particles are limited in search region. A new strategy for updating velocity is then adopted, in which the velocity is weakened linearly. Thirdly, using an idea of intersecting two modified PSO algorithms. Finally, adding an item of integral control in the modified algorithm can improve its global search ability. Based on these strategies, we proposed a new PSO algorithm named crossed PSO algorithm. Simulation results show that the crossed PSO is superior to the original PSO algorithm and can get overall promising performance over a wide range of problems.

1 Introduction

Several evolutionary algorithms, such as genetic algorithm, can be used to solve complex problems, and obtain ideal optimal results. However these methods have heavy computational burden and hence are time-consuming. Particle Swarm Optimization (PSO) was originally introduced by J. Kennedy et. al. in 1995, which is an evolutionary algorithm based on the swarm intelligence [1] [2], and motivated from the simulation of social behavior. Shi and Eberhart found a significant improvement in the performance of the PSO with a linearly varying inertia weight(LPSO)[3] over the generation. Reference [4][5] introduces an modification of the PSO intended to solve the problems of premature convergence observed in many applications of PSO, which is accomplished by using the ratio of the relative fitness and the distance of other particles to determine the direction in which each component of the particle's position needs to be changed. A new PSO algorithm named the crossed PSO algorithm is introduced and adopted to optimize a series of functions. The simulation results show that the crossed PSO algorithm is more effective than the original PSO algorithm and the other modified PSO algorithms.

* This work was supported by the National Natural Science Foundations of China (60171045, 60374063 and 60133010).

L. Jiao et al. (Eds.): ICNC 2006, Part I, LNCS 4221, pp. 935–938, 2006.

2 A Modified PSO Algorithm

In PSO algorithms, the particles have initial positions and velocities, where the positions and velocities are iterated. At each generation, two "best positions" are chased to update the particle. The first is an optimal solution found by the particle, which called personal best position. The other, called global best position, is an optimal solution founded by the entire population. The ith particle in a D-dimensional solution space is determined by a fitness function value. Denote the position of ith particle by $x_i = (x_{i1}, \cdots, x_{id}, \ldots, x_{iD})$, and the velocity of the ith particle by $v_i = (v_{i1}, \cdots, v_{id}, \ldots, v_{iD})$. Let$p_i = (p_{i1}, \cdots, p_{id}, \ldots, p_{iD})$ be the position vector for an individual particle's best fitness, which is the personal best position, and $p_g = (p_{g1}, \cdots, p_{gd}, \ldots, p_{gD})$ be the global best position among all the particles.

The original PSO algorithm has a high convergence speed, but it is easy to fall into local optima. In order to overcome this weakness, some improvements are applied to the origin PSO algorithm. Based on the information intersecting between two modified PSO algorithms, the velocities and positions are updated by the crossed PSO algorithm. Here the first update formulae are

$$v_{1id}^t = w^t * v_{id}^t + c_1 * rand * (p_{id}^t - x_{id}^t) + c_2 * rand * (p_{gd}^t - x_{id}^t), \quad (1)$$

$$x_{1id}^t = x_{id}^t + m_c * v_{1id}^t + \tau(I - f_i),, \quad (2)$$

Where w is the inertia weight, c_1 and c_2 are the learn factors, and $rand$ is a random number in the range [0,1], m_c is the contractive factor. This algorithm adds the third dimension to the calculation. Each particle also learn the experience from the neighboring particles that have the better fitness value than itself. This calculation changes the velocity update equation, but the position update equation remain unchanged. The simple and robust strategy is chosen to increase the benefit of all the associated computations, the dth dimension of the ith particle's velocity is updated using a particle called p_n, with prior best position p_j, chosen to

$$\max_j \frac{Fitness(p_j) - Fitness(x_i)}{0.00001 + |p_{jd} - x_{id}|}, \quad (3)$$

$|p_{jd} - x_{id}|$ denote the absolute value. It is presumed that the fitness function is to be maximized. Then the second update formulae are

$$\begin{aligned} v_{id}^{t+1} = w^t * v_{1id}^t + c_1 * rand * (p_{id}^t - x_{1id}^t) + c_2 * rand * (p_{gd}^t - x_{1id}^t) \\ + c_3 * rand * (p_{nd}^t - x_{1id}^t), \end{aligned} \quad (4)$$

$$x_{id}^{t+1} = x_{1id}^t + m_c * v_{id}^{t+1} + \tau(I - f_i), \quad (5)$$

Where

$$w^t = w_{max} - \frac{w_{max} - w_{min}}{T} * t,$$

$$c_1^t = c_{1max} - \frac{c_{1max} - c_{1min}}{T} * t, \; c_2^t = c_{2max} - \frac{c_{2max} - c_{2min}}{T} * t,$$

Here, one effective strategy is adding an item of integral control into the position update equation. where $\tau \geq 0$ is the integral factor, I is a constant such that $I \leq f_{min}$, f_{min} stands for the minimum fitness value. The item of integral control can improve the performance of the original PSO algorithm. When $f_i = I$, the item is zero. In other cases $I - f_i < 0$, and the item of integral control can make x_{id} become smaller and escape from the local minimum. If τ is a fixed value, the item does not work well. Thus at each generation the value of τ is determined according to

$$\tau = \begin{cases} \tau_0 / \max\limits_i(|f_i|), & \text{if } \min\limits_i(|f_i|) > c \\ \tau_0, & \text{otherwise} \end{cases}$$

Here τ_0 and c both are positive. We suggest to choose $c = 1$. For Benchmark functions, we choose $I = 0$. This algorithm is found to be more successful than variations such as selecting a single particle in whole direction all velocity components are updated.

3 Experiments and Results

To throughly investigate the performance of the crossed PSO algorithm, five benchmark test functions(Sphere,Rosenbrock, Rastrigrin, Griewank, Schwefel) and a multimodal function f_6 are used here.

Table 1. Minimum fitness values founded by three algorithms

Function	Algorithm	Minima Achieved	success %
Sphere	PSO	6.3e-20	85
	LPSO	5.8e-43	67
	CPSO	0	86
Rosenbrock	PSO	0.061	32
	LPSO	0.042	28
	CPSO	8.1e-7	76
Rastrigrin	PSO	0.996	52
	LPSO	0.994	39
	CPSO	1.6289e-012	81
Griewank	PSO	0	53
	LPSO	0	65
	CPSO	0	93
Schwefel	PSO	136.432	32
	LPSO	118.167	36
	CPSO	1.2728e-004	56
f_6	PSO	17.5693	34
	LPSO	16.9953	47
	CPSO	15.8319	68

$$f_6 = \sum_{i=1}^{n}[(5\sin x_i)^2 + (x_i + 5)^2] \quad (6)$$

The inertial weight (w)is varied linearly from 0.9 to 0.4, c_{1max} and c_{2max} are both equal to 2.0 , c_{1min} and c_{2min} are both equal to 1.2, c_3 is equal to 1, contractive factor m_c is equal to 0.38. The population size is set as 20, and D is equal to 10. Each algorithm is terminated after 2000 generations, in each simulation. The simulation result of the functions with the same settings were shown in Table 1. 100 tests are carried out for each function to evaluate the performance in terms of the rate of convergence and optima solution of the crossed PSO in comparison with the LPSO and the original PSO. From Table 1, it can be easily seen that the crossed PSO algorithm is able to reach most of global optima of the six test functions, whereas the other two algorithms may not be able to reach them. Especially for the schwefel function, the LPSO method and the original PSO algorithm can not converge to its global optimum, however the crossed PSO algorithm could reach it.

4 Conclusion

In this paper, the performance of the crossed PSO algorithm has been extensively investigated by experimental studies of a series of benchmark functions. The optimization result illuminate that the new algorithm, for most cases, can quickly search towards the global optimum. Such algorithm is feasible for solving complex global optimization problems. The cross strategy with an item of integral control can enables the particles to develop more area in search space, meanwhile the chance of premature convergence can also be decreased. Such algorithm can be used to solve other kinds of optimization problems.

References

1. Clerc, M., Kennedy, J.: The Particle Swarm—Explosion, Stability, and Convergence in a Multidimensional Complex Space. IEEE Transactions on Evolutionary Computation **6** (2002) 58-73
2. Parsopoulos, K.E., Vrahatis, M.N.: On the Computation of All Global Minimizers Through Particle Swarm Optimization. IEEE Transactions on Evolutionary Computation **8** (2004) 211-224
3. Shi, Y.H., Eberhat, R.C.,: A Modified Particle Swarm Optimization. Proceedings of IEEE International Congress on Evolutionary Computation(1998) 69-73
4. Thanmaya, P., Kalyan, V.,Chilukuri, K.M.: Fitness-Distance-Ratio Based Particle Swarm Optimization. Proceedings of IEEE Swarm Intelligence Symposium(2003) 174-181
5. Kennedy, J.: Small Worlds and Megaminds: Effects of Neighbourhood Topology on Particle Swarm Performance. IEEE Press. Proceedings of the 1999 Congress Evolutionary Computation **3** (1999) 1931-1938

A Dynamic Convexized Function with the Same Global Minimizers for Global Optimization

Wenxing Zhu

Department of Computer Science and Technology, Fuzhou University,
Fuzhou 350002, China
wxzhu@fzu.edu.cn

Abstract. We consider the box constrained continuous global minimization problem. We present an auxiliary function $T(x,k,p)$, which has the same global minimizers as the problem if p is large enough. The minimization of $T(x,k,p)$ can escape successfully from a previously converged local minimizer by taking the value of k increasingly. We propose an algorithm to find a global minimizer of the box constrained continuous global minimization problem by minimizing $T(x,k,p)$ dynamically. Numerical experiments on two sets of standard testing problems show that the algorithm is effective, and is competent with some well known global minimization methods.

Keywords: Box constrained global minimization problem, auxiliary function, local minimizer.

1 Introduction

Many methods have been developed for continuous global optimization problems. Roughly, these methods can be classified into two categories [7]. The first category of methods have exhaustive characteristics, which include the Branch and Bound method and the cutting plane method [7,8]. These methods are applicable to problems with some analytical properties.

The second category of methods include methods that use heuristic search or stochastic search, and generally have two phases: global search phase and local search phase. An important class of methods in this category emphasize on global search phase, which evaluate the objective function at a suitably chosen sample of points randomly or deterministically, and then in the local search phase manipulate these sample points to find good local (and hopefully global) minimizers [1].

Another important class of methods in the second category emphasize on local search phase. They use a local search method to minimize the objective function firstly, and then switch to global search phase. When the objective function has a large number of local minimizers, local optimization techniques are likely to get stuck before the global minimum is reached. Various mechanisms have been developed to escape from local minima, e.g., the simulated annealing [11], the

L. Jiao et al. (Eds.): ICNC 2006, Part I, LNCS 4221, pp. 939–948, 2006.

taboo search [3], and the methods that minimize some auxiliary functions to descend from one local minimizer to another better one of the objective function. The methods using auxiliary functions include the diffusion equation method [9], the terminal repeller unconstrained sub-energy tunnelling method (TRUST) [2], the filled function method [5,6], and the tunnelling method [10,14]. These methods rely heavily on the successful construction of auxiliary functions to bypass the previously converged local minima.

The diffusion equation method [9] transforms the original objective function into a family of smoothed functions via integration of the objective function. Such integrations are too expensive to compute at run time. The terminal repeller unconstrained sub-energy tunnelling method (TRUST) [2] modifies the objective function $f(x)$ as a sub-energy tunnelling function, which has an advantage that it keeps all local minimizers of $f(x)$ in the region lower than the current local minimizer. However, to minimize the sub-energy tunnelling function, this method has to integrate a dynamical system rather than use a local optimization method as a subroutine. The filled function method [5,6] modifies the original objective function as a filled function, and then use a local optimization method to minimize the filled function to find a local minimizer of the original objective function with a lower function value. The filled functions constructed up to now have a disadvantage that they have some parameters which are difficult to adjust. And it is not clear at present whether the constructed filled functions can preserve the descent directions of $f(x)$ or not in the region where $f(x) \geq f(x_1^*)$.

In this paper, we consider the box constrained continuous global minimization problem. We give in section 2 an auxiliary function $T(x, k, p)$, which has the same local minimizers and global minimizers as the problem in the region where $f(x) < f(x_1^*) - \frac{1}{\sqrt{p}}$, and keeps some descent directions of $f(x)$ in the region where $f(x) \geq f(x_1^*)$ if the value of parameter k is chosen suitably. Moreover, the minimization of $T(x, k, p)$ can bypass a previously converged local minimizer if the value of k increases. In Section 3, we design an algorithm basing on the auxiliary function. In Section 4, we test the algorithm on two sets of standard testing problems. Numerical experiments show that this algorithm is competent with some well known global optimization methods.

2 An Auxiliary Function and Its Properties

Consider the following global minimization problem

$$(P) \quad \begin{cases} \min f(x) \\ s.t. \;\; x \in X, \end{cases}$$

where X is a bounded closed box in R^n, i.e., $X = \{x \in R^n : a \leq x \leq b\}$, and $f(x)$ is a continuously differentiable function on X.

Suppose that x_1^* is the current local minimizer of problem (P), which can be found by any local minimization method. Moreover, suppose that f^* is the global minimal value of problem (P), and $x_0 \in X$ is a local minimizer of problem (P) such that $f(x_0) \geq f(x_1^*)$.

Construct the following auxiliary function

$$T(x,k,p) = f(x) + kG(\|x - x_0\|)u(1 - p(\max\{0, f(x_1^*) - f(x)\})^2), \quad (1)$$

where k and p are nonnegative parameters, $G(t)$ is a continuously differentiable univariate function which satisfies that

$$G(0) = 0, \quad G'(t) \geq a > 0 \ \text{ for all } \ t \geq 0, \quad (2)$$

and $u(t)$ is a continuously differentiable univariate function which satisfies that

$$u(t) = 0 \text{ for all } t \leq 0, \text{ and } u(t) > 0 \text{ for all } t \in (0,1]. \quad (3)$$

Construct the following auxiliary global minimization problem

$$(AP) \quad \begin{cases} \min T(x,k,p) \\ s.t. \ x \in X. \end{cases}$$

The objective of our method in this paper is by solving problem (AP) to find a local minimizer of problem (P) lower than its current one x_1^*. Firstly, we analyze properties of the function $T(x,k,p)$ on X.

2.1 Local Minimizer x_0

Theorem 1. *If x_0 is a local minimizer of problem (P) with $f(x_0) \geq f(x_1^*)$, then x_0 is a local minimizer of problem (AP).*

Theorem 1 does not consider the case that x_0 is not a local minimizer of $f(x)$. In fact, during solution of problem (P), if x_0 is not a local minimizer of problem (P), then we minimize $f(x)$ on X starting from x_0 to get a local minimizer x', if $f(x') \geq f(x_1^*)$, then we set $x_0 = x'$; otherwise if $f(x') < f(x_1^*)$, then we set $x_1^* = x'$ and $x_0 = x'$.

2.2 Properties of $T(x,k,p)$ Dependent on p

In this subsection, we analyze properties of $T(x,k,p)$ relative to parameter p. We have the following lemmas.

Lemma 2. *$\forall x \in \{x \in X : f(x) \geq f(x_1^*)\}$, $T(x,k,p) = f(x) + ku(1)G(\|x - x_0\|)$; and $\forall x \in S = \{x \in X : f(x) \leq f(x_1^*) - \frac{1}{\sqrt{p}}\}$, $T(x,k,p) = f(x)$.*

Lemma 3. *$\forall x \in S = \{x \in X : f(x) \leq f(x_1^*) - \frac{1}{\sqrt{p}}\}$, and $\forall y \in X - S = \{x \in X : f(x) > f(x_1^*) - \frac{1}{\sqrt{p}}\}$, it holds that $T(x,k,p) < T(y,k,p)$.*

By Lemma 3, it is obvious that the following corollary holds.

Corollary 4. *If x_1^* is not a global minimizer of problem (P), and p is large enough such that*

$$p > \frac{1}{(f(x_1^*) - f^*)^2}, \tag{4}$$

where f^* is the global minimal value of problem (P), then $\{x \in X : f(x) \leq f(x_1^*) - \frac{1}{\sqrt{p}}\} \neq \emptyset$, and all global minimizers of problem (AP) are in the set $\{x \in X : f(x) < f(x_1^*)\}$.

Using the above lemmas, we can prove the following theorem.

Theorem 5. *Suppose that x_1^* is not a global minimizer of problem (P), and p is large enough such that inequality (4) holds. For $y \in \{x \in X : f(x) \leq f(x_1^*) - \frac{1}{\sqrt{p}}\}$, if y is a local minimizer of problem (AP), then y is a local minimizer of problem (P), and vice versa.*

Generally speaking, we do not know the global minimal value of problem (P). So it seems hard to take a value of p such that inequality (4) holds. But for practical considerations, for a given sufficiently small number $\epsilon > 0$, problem (P) might be considered solved if we can find a point $x \in X$ such that $f(x) < f^* + \epsilon$. Thus if $f(x_1^*) < f^* + \epsilon$, then x_1^* is an approximate global minimizer of problem (P); otherwise if $f(x_1^*) \geq f^* + \epsilon$, then take p such that

$$p > \frac{1}{\epsilon^2}, \tag{5}$$

and in this case $f(x_1^*) - \frac{1}{\sqrt{p}} \geq f(x_1^*) - \epsilon \geq f^*$.

2.3 Properties of $T(x, k, p)$ Dependent on k

Theorem 6. *Let L be a constant such that $\|\nabla f(x)\| \leq L$ for all $x \in X$. For the function $T(x, k, p)$, we have:*

1. For all $x \in G_k = \{x \in X : f(x) \geq f(x_1^*), \|\nabla f(x)\| < kau(1)\}$, $x \neq x_0$, $x_0 - x$ is a descent direction of $T(x, k, p)$ at x.

2. Especially, Especially, if

$$k > \frac{L}{au(1)}, \tag{6}$$

then for all $x \in S_1 = \{x \in X : f(x) \geq f(x_1^*)\}$, $x \neq x_0$, $x_0 - x$ is a descent direction of $T(x, k, p)$ at x.

By Theorem 6, if we take an initial point in $G_k = \{x \in X : f(x) \geq f(x_1^*), \|\nabla f(x)\| < kau(1)\}$ to minimize $T(x, k, p)$, then the minimization sequence will converge to x_0, or to a point in the set $\{x \in X : f(x) \geq f(x_1^*), \|\nabla f(x)\| \geq kau(1)\}$, or to a point in the set $\{x \in X : f(x) < f(x_1^*)\}$. Especially, if k satisfies inequality (6), then while minimizing $T(x, k, p)$ from any initial point in X, the minimization sequence will converge to the prefixed minimizer x_0, or converge to a point in the set $\{x \in X : f(x) < f(x_1^*)\}$.

Note that in the attraction region of a local minimizer of $f(x)$ lower than x_1^*, starting from an initial point x with $f(x) \geq f(x_1^*)$ to minimize $f(x)$ on X will converge to the local minimizer lower than x_1^*. So we have one question that whether or not the minimization of $T(x,k,p)$ on X from the initial point could converge to the local minimizer. The essence of such a question is whether or not we can make $T(x,k,p)$ keep the descent directions of $f(x)$ at the points in the region $\{x \in X : f(x) \geq f(x_1^*)\}$. In fact, we have the following result.

Theorem 7. *Suppose that d is a descent direction of $f(x)$ at $x \in X$ such that $f(x) \geq f(x_1^*)$, i.e., $d^T \nabla f(x) < 0$. Then d is a descent direction of $T(x,k,p)$ if and only if one of the following conditions holds.*

1. $k = 0$.

2. $k > 0$ and $d^T(x - x_0) \leq 0$.

3. $k > 0$, $d^T(x - x_0) > 0$, and $k < -\frac{d^T \nabla f(x) \|x - x_0\|}{u(1) G'(\|x - x_0\|) d^T \|x - x_0\|}$.

Theorem 7 means that $T(x,k,p)$ could not keep the descent directions of $f(x)$ at points in the region $\{x \in X : f(x) \geq f(x_1^*)\}$ if k is too large. So for the sake of finding a local minimizer of $f(x)$ lower than x_1^*, while minimizing $T(x,k,p)$ on X from an initial point in the attraction region of a local minimizer of $f(x)$ lower than x_1^*, k should not be too large.

But by Theorem 6, to bypass the previously converged local minimizers while minimizing $T(x,k,p)$ on X, k should be large enough. This contradicts the above conclusion of Theorem 7. So in the algorithm presented in the next section, we take $k = 0$ initially, and increase the value of k sequentially.

3 Algorithm

Now we present an algorithm to solve problem (P) by solving problem (AP). We suppose that the value of p satisfies inequality (4) or (5). The basic idea of the algorithm is as follows.

We take $k = 0$ initially, and take randomly a starting point in X to minimize $T(x,k,p)$ on X. If the minimization sequence converges to a point $x' \neq x_0$ and $f(x') \geq f(x_1^*)$, then increase the value of k, and minimize $T(x,k,p)$ on X from x'. If at this time the minimization sequence converges to a point $x'' \neq x_0$ and $f(x'') \geq f(x_1^*)$, then by Theorem 6, the value of k is too small, and we increase the value of k and minimize $T(x,k,p)$ on X from x'' again, till the minimization sequence converges to x_0 or to a point in $\{x \in X : f(x) < f(x_1^*)\}$.

If the minimization sequence converges to x_0, then we repeat the above process. If the minimization sequence converges to a point in $\{x \in X : f(x) < f(x_1^*)\}$, then we have found a point lower than x_1^*. And we start from the point to minimize $f(x)$ on X to find a better local minimizer of $f(x)$ on X, and repeat the above process.

The Algorithm

Step 1. Select randomly a point $x \in X$, and start from which to minimize $f(x)$ on X to get a local minimizer x_1^* of problem (P). Let p be a large positive number, N_L be a sufficiently large integer, and let δ_k be a positive number. Set $N = 0$.

Step 2. Select a point $x_0 \in X$ such that x_0 is a local minimizer of problem (P) and $f(x_0) \geq f(x_1^*)$. Construct a function $T(x, k, p)$ with k, p, x_1^* and x_0.

Step 3. Set $k = 0$ and $N = N + 1$. If $N \geq N_L$, then go to Step 6; otherwise draw randomly an initial point y in the bounded closed box X and go to Step 4

Step 4. Minimize $T(x, k, p)$ on X from y using any local minimization method. Suppose that x' is an obtained local minimizer.

If $x' \neq x_0$ and $f(x') \geq f(x_1^*)$, then set $k = k + \delta_k$, $y = x'$, and repeat Step 4.

If $x' = x_0$, then go to Step 3.

If $f(x') \leq f(x_1^*) - \frac{1}{\sqrt{p}}$, then let $x_1^* = x'$, and go to Step 2.

If $f(x_1^*) - \frac{1}{\sqrt{p}} < f(x') < f(x_1^*)$, then go to Step 5.

Step 5. Minimize $f(x)$ on X using x' as an initial point, and obtain a local minimizer x_2^* of $f(x)$ on X. Let $x_1^* = x_2^*$ and go to Step 2.

Step 6. Stop the algorithm, output x_1^* and $f(x_1^*)$ as an approximate global minimal solution and global minimal value of problem (P) respectively.

It should be remarked that we have only four cases in Step 4. If the minimization of $T(x, k, p)$ on X converges to $x' \neq x_0$ with $f(x') \geq f(x_1^*)$, then by Theorem 6 the algorithm has not found a point lower than x_1^*. So we increase the value of k by δ_k, and minimize $T(x, k, p)$ on X from x' again.

If the minimization of $T(x, k, p)$ on X converges to x_0, then by Theorem 6, the value of parameter k is large enough. So we reset the value of k by taking $k = 0$ and repeat Steps 3 and 4.

If the minimization of $T(x, k, p)$ on X converges to x' such that $f(x') \leq f(x_1^*) - \frac{1}{\sqrt{p}}$, then by Theorem 5, x' is a local minimizer of problem (P) lower than x_1^*, we let $x_1^* = x'$, and go to Step 2.

Furthermore, if the minimization of $T(x, k, p)$ on X converges to x' such that $f(x_1^*) - \frac{1}{\sqrt{p}} < f(x') < f(x_1^*)$, then we have to minimize $f(x)$ on X from x' to obtain a local minimizer of problem (P).

4 Numerical Experiments

In this section, we test our algorithm on two sets of standard testing problems. The testing problems are from Dixon and Szegö [4], and the test problems which are considered hard and challenging in the first contest on evolutionary optimization (ICEO) at the ICEO'96 conference. These problems are summarized in Tables 1 and 2.

In $T(x, k, p)$, we choose $G(t) = t$, $u(t) = \max^2\{0, t\}$, and $p = 1000$. We use the BFGS local search method as the local minimization method in the algorithm.

To compare our numerical results with those of some well known methods [1,8,9,12,13]. The stopping criterion of our algorithm is taken as: If our algorithm finds a solution x such that

$$\frac{f(x) - f^*}{f^*} < 10^{-4} \text{ for } f^* \neq 0, \text{ and } \frac{f(x) - f^*}{f^* + 1} < 10^{-4} \text{ for } f^* = 0, \qquad (7)$$

then stop the algorithm. In the other respect, during practical implementations, if the stopping criterion (7) has not been met after 12,000 function calls for problems 1-9, and after 100,000 function calls for the other 10 problems, then we stop the algorithm.

Table 1. Dixon and Szegö's functions [4]

Problem Label	Test Function	Dimension	Bounded Box
S_5	Shekel 5	4	$[0, 10]^4$
S_7	Shekel 7	4	$[0, 10]^4$
S_{10}	Shekel 10	4	$[0, 10]^4$
H_3	Hartman 3	3	$[0, 1]^3$
H_6	Hartman 6	6	$[0, 1]^6$
GP	Goldstein-Price	2	$[-2, 2]^2$
BR	Branin	2	$[-5, 10] \times [0, 15]$
C6	Six-hump camel	2	$[-3, 3] \times [-2, 2]$
SHU	Shubert	2	$[-10, 10]^2$

Table 2. ICEO test functions

Problem Label	Test Function	Dimension	Bounded Box
Sph_5	Sphere model	5	$[-5, 5]^5$
Sph_{10}	Sphere model	10	$[-5, 5]^{10}$
Gri_5	Griewank's function	5	$[-600, 600]^5$
Gri_{10}	Griewank's function	10	$[-600, 600]^{10}$
She_5	Shekel's foxholes	5	$[0, 10]^5$
She_{10}	Shekel's foxholes	10	$[0, 10]^{10}$
Mic_5	Michalewicz's function	5	$[0, \pi]^5$
Mic_{10}	Michalewicz's function	10	$[0, \pi]^{10}$
Lan_5	Langerman's function	5	$[0, 10]^5$
Lan_{10}	Langerman's function	10	$[0, 10]^{10}$

We run our algorithm 25 times by taking $\delta_k = 3$ on every testing problem.We record the number of function calls to reach a global minimizer. The test results are put in Table 3.

In Table 3, every number in the column 'min' is the minimal number of function calls to reach a global minimizer among 25 runs of our algorithm; every number in the column 'max' is the maximal number of function calls to reach a global minimizer among successful runs of our algorithm; every number in the column 'med' is the average number of function calls of successful runs of our algorithm; and every number in the column 'fail' is the number of runs that the stopping criterion has been met but the optimum has not been reached.

It can be seen from Table 3 that our algorithm works very well for most of the testing problems, except for the Griewank's function with dimension 5, the Michaelwitz's function with dimension 10, and the Shekel's foxholes function with dimension 10. However, if we stop our algorithm after 600,000 function calls, then it succeeds in finding a global minimizer of the Griewank's function with dimension 5 in all 25 runs, and the average number of function calls is 331319. Moreover, for the Shekel's foxhole function with dimension 10, if we stop our algorithm after 200,000 function calls, then it succeeds in finding a global minimizer of the function in all 25 runs, and the average number of function calls is 39335.

Table 3. Performance of our algorithm

Problem	min	max	med	fail
S_5	63	333	159	0
S_7	27	751	237	0
S_{10}	68	794	299	0
H_3	17	108	35	0
H_6	44	110	69	0
GP	32	1572	298	0
BR	20	49	28	0
C6	23	88	43	0
SHU	131	8268	1938	0
Sph_5	35	121	81	0
Sph_{10}	87	356	127	0
Gri_5	31253	77470	53007	21
Gri_{10}	4318	37937	17391	0
She_5	2358	16846	9069	0
She_{10}	12161	93415	33437	2
Mic_5	54	3546	1489	0
Mic_{10}	20026	81416	45650	3
Lan_5	3253	41553	17734	0
Lan_{10}	2148	37552	15662	0

To understand the competence of our algorithm with $\delta_k = 3$, we compare in Tables 4 and 5 the numerical results of our algorithm with those of some well known methods [1,8,9,12,13]. The comparison criterion is the average number of function calls among all successful runs.

Table 4. Comparison of our algorithm with some algorithms

Method	S_5	S_7	S_{10}	H_3	H_6	GP	BR	C6	SHU
Boender et al. [1982]	567	624	755	235	462	398	235		
Snyman-Fatti [1987]	845	799	920	365	517	474		178	
Kostrowiki-Piela [1991]	*	*	*	200	200	120		120	
Jones et al. [1993]	155	145	145	199	571	191	195	285	2967
Storn-Price [1997]	6400	6194	6251	476	7220	1018	1190	416	1371
Huyer-Neumaier [1999]	83	129	103	79	111	81	41	42	69
Our algorithm	159	237	299	35	69	298	28	43	1938

An asterisk means that a global minimizer was not found within 12000 function calls.

All except the last rows of Tables 4 and 5 are taken from the papers cited. From Table 4, it can be seen that on the 9 testing functions, our algorithm outperforms the method by Boender et al. [1], the method by Snyman and Fatti [12], the method by Kostrowiki and Piela [9], and the method by Storn and Price [13]. Moreover, our algorithm is competent with the method by Jones et al. [8] on the 9 testing functions.

Table 5. Comparison of our algorithm with DE

Method	Sph_5	Sph_{10}	Gri_5	Gri_{10}	She_5	She_{10}	Mic_5	Mic_{10}	Lan_5	Lan_{10}
DE1	736	1892	5765	13508	76210	-[1]	1877	10083	5308	44733
DE2	463	1187	5157	16228	67380	-[2]	2551	18158	4814	-[3]
Our algorithm	81	127	53007	17391	9069	33437	1489	45650	17734	15662

[1]744250, [2]203350, [3]174006 function calls.

A dash means that a global minimizer was not found within 100000 function calls.

From Table 5, it can be seen that our algorithm outperforms the differential evolution method [13] on the Sph_5, Sph_{10}, She_5, She_{10}, Mic_5, and Lan_{10} functions. But the differential evolution method [13] outperforms our algorithm on the Gri_5, Gri_{10}, Mic_{10}, and Lan_5 functions.

Next, we study how the value of parameter k affects the performance of our algorithm. We take $\delta_k = 0, 1, 2, 3, 5, 10, 15$ respectively, and use our algorithm to minimize the Michaelwitz's function. During practical implementations, if the stopping criterion (7) has not been met after 1,000,000 function calls, then we stop the algorithm. The test results are put in Table 6.

It can be seen from Table 6 that the Michaelwitz's function with dimension n=5 is easy for our algorithm, since our algorithm can find a global minimizer of the function successfully using relatively small number of function calls, even for our algorithm with $\delta_k = 0$.

Table 6. Performance of our algorithm for different δ_k

δ_k	dimension n=5				dimension n=10			
	min	max	med	fail	min	max	med	fail
$\delta_k = 0$	504	5287	2073	0	305934	774745	580344	20
$\delta_k = 1$	54	2823	1408	0	1180	115236	38235	0
$\delta_k = 2$	54	8237	1995	0	7280	198749	59221	0
$\delta_k = 3$	54	3546	1489	0	20026	81416	45650	3
$\delta_k = 5$	54	6130	1833	0	16175	165267	67396	5
$\delta_k = 10$	54	3513	1283	0	289959	843893	581157	8
$\delta_k = 15$	54	12250	2518	0	305213	875520	585367	19

But the Michaelwitz's function with dimension n=10 seems difficult for our algorithm. Roughly, in Table 6 the number of function calls increases when the value of δ_k increases (except for $\delta_k = 0$), not only in the 'min' column, but also in the 'max' and 'med' columns. Moreover, by the 'fail' column, if the value of δ_k increases (except for $\delta_k = 0$), then the probability of failing to find a global minimizer of the Michaelwitz's function with dimension n=10 increases. So during practical implementations, the value of δ_k could not be taken too large.

Furthermore, our algorithm with $\delta_k = 0$ performs worse than with $\delta_k = 1, 2, 3, 5, 10$. So it implies that the value of k should not be too small. In fact, intuitively a small value of δ_k would make our algorithm take more effort to escape from a local minimizer.

5 Conclusions

In this paper, we have proposed an auxiliary function $T(x, k, p)$ for problem (P), which has the same local minimizers and the same global minimizers as problem (P) in the region $\{x \in X : f(x) \leq f(x_1^*) - \frac{1}{\sqrt{p}}\}$. By taking the value of parameter k increasingly, the minimization of $T(x, k, p)$ can escape from a previously converged local minimizer successfully. An algorithm has been designed to minimize $T(x, k, p)$ on X to find a global minimizer of problem (P). Numerical experiments show that the algorithm is competent with some well known global minimization methods.

Acknowledgements. Research supported by the National Science Foundation of China under Grants 10301009 and 10431020.

References

1. Boender, C. and Rinnooy Kan, A., Stougie, L. and Timmer, G.: A stochastic method for global optimization. Mathematical Programming 22(1982)125-140
2. Cetin, B., Barhne, J. and Burdick, J.: Terminal repeller unconstrained subenergy tunneling (TRUST) for fast global optimization. Journal of Optimization Theory and Applications 77(1)(1993)97-126
3. Cvijović, D. and Klinowski, J.: Taboo search: an approach to the multiple minima probelm. Science 267(1995)664-666
4. Dixon, L. and Szegö, G.:The Global Optimization Problem: an Introduction. In Dixon, L.C.W. and Szegö, G.P. (eds.), *Towards Global Optimization 2*, North-Holland, Amsterdam, 1-15 (1978)
5. Ge, R.: A filled function method for finding a global minimizer of a function of several variables. Mathematical Programming 46(1990)191-204
6. Ge, R. and Qin, Y.: A class of filled functions for finding global minimizers of a function of several variables. Journal of Optimization Theory and Applications 52(1987)240-252
7. Horst, R., Pardalos, P. and Thoai, N.: Introduction to Global Optimization (2nd edition), Kluwer Academic Publishers, Dordrechet, The Netherlands, 2000
8. Jones, D., Perttunen, C. and Stuckman, B.: Lipschizian optimization without Lipstchitz constant. Journal of Optimization Theory and Appliations 79(1993)157-181
9. Kostrowicki, J., Piela, L.: Diffusion equation method of global minimization: performance for standard test functions. Journal of Optimization Theory and Applications 69 (2)(1991)97-126
10. Levy, A. and Montalvo, A.: The tunneling algorithm for the global minimization of functions. SIAM Journal on Scientific and Statistical Computing 6(1985)15-29
11. Locatelli, M.: Simulated Annealing Algorithms For Continuous Global Optimization. in Handbook of Global Optimization II, Kluwer Academic Publishers, 179-230(2002)
12. Snyman, J. and Fatti, L.: A multi-start global minimization algorithm with dynamic search trajectories. Journal of Optimization Theory and Applications 54(1987)121-141
13. Storn, R. and Price, K.: Differntial evolution - a simple and efficient heuristic for global minimization over continuous spaces. Journal of Gloabal Optimization 11(1997)341-359
14. Yao, Y.: Dynamic tunneling algorithm for global optimization. IEEE Transactions on Systems, Man, and Cybernetics 19(1989)1222-1230

Clonal Selection Algorithm with Dynamic Population Size for Bimodal Search Spaces

V. Cutello[1], D. Lee[2], S. Leone[1], G. Nicosia[1], and M. Pavone[1,2]

[1] Department of Mathematics and Computer Science
University of Catania
Viale A. Doria 6, 95125 Catania, Italy
{vctl, nicosia, mpavone}@dmi.unict.it
[2] *IBM-KAIST* Bio-Computing Research Center
Department of BioSystems, KAIST
373-1, Guseong-dong, Yuseong-gu, Daejeon, Republic of Korea
{dhlee, mario}@biosoft.kaist.ac.kr

Abstract. In this article an Immune Algorithm (IA) with dynamic population size is presented. Unlike previous IAs and Evolutionary Algorithms (EAs), in which the population dimension is constant during the evolutionary process, the population size is computed adaptively according to a cloning threshold. This not only enhances convergence speed but also gives more chance to escape from local minima. Extensive simulations are performed on trap functions and their performances are compared both quantitatively and statistically with other immune and evolutionary optmization methods.

Keywords: Immune algorithm, dynamic population size, aging operator, trap functions.

1 The Clonal Selection Algorithm

Size of the population is one of the most important aspects in Evolutionary Algorithms (EAs). To choose the population size too small, can drive EAs towards a fast convergence; whilst if it is too large, EAs need of wide computational resources. Then setting a correct population size can be a critical task in many applications. Several researchers have been interesting to investigate the setting of the optimal population size, for EAs. In this research paper is presented a modified version of the CSA proposed in [1] characterized by dynamic population size, and it is called DYN-IMMALG (*dynamic immune algorithm*). To evaluate the search ability of DYN-IMMALG, and study its evolutionary behavior was used a well-known test-bed: the trap functions, which are complex toy problems often used in evolutionary computation to assess the search ability of a given EA.

In the designed IA, the initial population of candidate solutions is randomly generated using uniform distribution in the relative domains; the algorithm chooses randomly the values 0 or 1 in a binary representation. DYN-IMMALG

L. Jiao et al. (Eds.): ICNC 2006, Part I, LNCS 4221, pp. 949–958, 2006.

creates the initial population with a fixed number of B cells receptor, by parameter d. At each time step $t > 0$, DYN-IMMALG presents a population $P^{(t)}$ of dynamic size, like occurs in nature: this is the main feature of the designed immune algorithm.

Static cloning operator clones each B cell *dup* times producing an intermediate population $P^{(clo)}$, assigning each clone the same *age* of its parent. Parameter *Age* of B cells, determines their life span into the population: when a B cell will have maximum age allowed, then it will die, i.e. will be eliminated from the population. After the hypermutation phase, a cloned B cell which undergoes a successfully mutation, i.e. improving the fitness value, called *constructive mutation*, will be considered to have age equal to 0. Such a scheme, as proposed in [10, 9, 4], intends to give an equal opportunity to each new B cell to effectively explore the landscape. Static cloning operator represents a good strategy working on constant population size, as showed in [5, 1], rather than on variable size. In this last case is more easy to have an exponential growth of the population size just in the first generations ($|P^{(t)}| > 10000$). This feature is called *explosion of the population* (or *Malthusian Catastrophe*). To avoid the explosion phenomenon, a new cloning operator, *dynamic clonal operator*, was designeded, which is based on a cloning threshold θ that determines the percentage of clones created by each B cell: only $\theta \times |P^{(t)}|$ best B cells in the current population will be clones. Nevertheless, using this cloning operator is possible to have *the extinction of the current population.* With respect to the explosion phase, choosing a correct balance between θ, *dup* and age's parameter were obtained better performances by DYN-IMMALG.

The hypermutation operator acts on the B cell receptor of $P^{(clo)}$. The proposed DYN-IMMALG uses *inversely proportional hypermutation* operator, which try to mutate each B cell receptor M times without the explicit usage of a mutation probability. The main feature of the inversely proportional hypermutation operator is that the number of mutations is inversely proportional to the fitness value: as the fitness function value of the current B cell increases, the number of mutations performed by DYN-IMMALG decreases. This operator performs at most $M_i(f(\boldsymbol{x})) = (c \times \ell)\frac{E^*}{f(\boldsymbol{x})}$ mutations, where E^* is the optimum of the given trap function and c is mutation rate. If $M_i(f(\boldsymbol{x})) \geq \ell$, then we have performed ℓ mutations. Moreover, DYN-IMMALG use the *stop at the First Constructive Mutation* (FCM) strategy [9]: if a constructive mutation occurs, the mutation procedure will move on to the next B cell. We adopted such a mechanism to slow down (premature) convergence, exploring more accurately the search space.

The aging operator, used by the algorithm, eliminates old B cells in the populations $P^{(t)}$, and $P^{(hyp)}$, maintaining high diversity in the current population, in order to avoid premature convergence. The maximum number of generations the B cells are allowed to remain in the population is determined by the τ_B parameter: when a B cell is $\tau_B + 1$ old it is erased from the current population, independently what is its fitness value.

After the application of the immune operators the surviving B cells from the populations $P_a^{(t)}$ and $P_a^{(hyp)}$, will constitute the population of the next genera-

Table 1. Pseudo-code of DYN-IMMALG

```
DYN-IMMALG(d, dup, θ, τ_B, c)
t := 0;
P^(t) := Initial_Population(d);
Evaluate(P^(t));
while ( ¬ Termination_Condition() ) do
    Increase_Age(P^(t));
    P^(clo) := Dynamic_Cloning (P^(t), dup, θ);
    P^(hyp) := Hypermutation (P^(clo), c);
    Evaluate(P^(hyp));
    (P_a^(t), P_a^(hyp)):= Aging(P^(t), P^(hyp), τ_B);
    P^(t+1) := (P_a^(t) ∪ P_a^(hyp));
    t := t + 1;
end_while
```

tion. Tackling the trap functions, to maintain high diversity into the population, no redundancy is allowed: each B cell receptor is unique.

The function *Evaluate($P^{(*)}$)* computes the fitness function value of each B cell $\boldsymbol{x} \in P^{(*)}$. Moreover, function *Termination_Condition()* is a boolean function, which returns true if a maximum number of fitness function evaluations (T_{max}) is reached, or the optima solution is found.

In table 1 is showed the pseudo-code of DYN-IMMALG.

Trap Functions. In this research work are used the traps functions, complex toy problem, to assess if DYN-IMMALG is able to reach an optimal solution starting from a randomly initialized population. Our study is restricted on *unitation functions*, which depend entirely upon the number of ones in a bit string but not on their position:

$$f(\boldsymbol{x}) = \hat{f}(u(\boldsymbol{x})) = \hat{f}\left(\sum_{k=1}^{\ell} x_k\right)$$

where ℓ is the length of the bitstring, such that $\boldsymbol{x} \in \mathbb{B}^\ell$, and $\hat{f} : \mathbb{B}^\ell \to \mathbb{R}$ is the *fitness function* that maps the bit string to a real number. To evaluate, mainly the search ability and, then goodness of the results obtained by DYN-IMMALG, were used two different trap functions: basic trap function and complex trap function [2]. The basic (or simple) trap function is characterized by a global optimum, obtained when the input is a string of all 0's, and a local optimum, obtained when the input is a string of ones. The formal definition is:

$$\hat{f}(u) = \begin{cases} \frac{a}{z}(z-u), & \text{if } u \leq z \\ \frac{b}{\ell - z}(u-z), & \text{otherwise.} \end{cases} \tag{1}$$

where a, b and z are parameters (see [2]). The complex trap function is more difficult to investigate, having two different slopes that lead to the local optimum.

It also avoids that bit-flipping operators, which change/reverse the values of all the bits in a string, may find the global optimum easily. This may happen in the simple trap presented above, where the two optima are bit-wise complements of each other. The formal definition is:

$$\hat{f}(u) = \begin{cases} \frac{a}{z_1}(z_1 - u), & if \quad u \leq z_1 \\ \frac{b}{\ell - z_1}(u - z_1), & if \quad z_1 < u \leq z_2 \\ \frac{b(z_2 - z_1)}{\ell - z_1}\left(1 - \frac{1}{\ell - z_2}(u - z_2)\right) & otherwise. \end{cases} \tag{2}$$

The values of the parameters chosen for both functions are the same as are used in [2]. This choice simplifies a comparison with works published previously. Using these parameters, are obtained several different functions, for each kind of trap function. Therefore, was used the following syntax: $S(type)$ and $C(type)$, where S and C mean respectively simple and complex trap function and *type* varying with respect the parameter values used (see table 2).

2 The Algorithm Dynamics

The common population explosion phenomenon is showed in left plot of figure 1, where is possible to see how population size increases very quickly in the first 8 generations. In this plot, one can see how the constructive mutations number is higher than destructive ones, i.e. when a mutation worsen the fitness value. Therefore, the major of the hypermutated cells not will die (each constructive mutation will be assign age 0), staying into the current population and increasing its size. For this reason, DYN-IMMALG reaches the maximum number of fitness function evaluations allowed (T_{max}), in the first 8 generations on each independently run.

Using the threshold ($\theta = 50\%$), instead, the destructive mutations increase more than the constructive, as showed in the right plot of figure 1. Of course, the population increase proportionally to an higher mean life value, like a *"natural"* selection process.

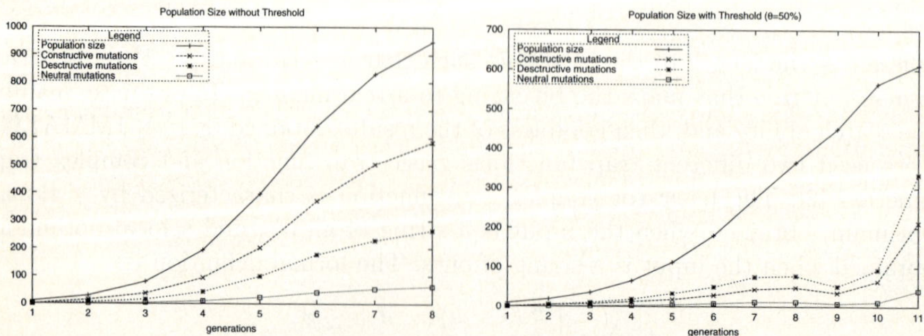

Fig. 1. Explosion of the population (left plot) without cloning threshold, $c = 0.1$, and $\tau_B = 10$. No explosion of the population (right plot) with cloning threshold $\theta = 50\%$, $c = 0.1$, and $\tau_B = 10$.

A set of experiments is performed on the simple and complex trap functions in order to analyse the population dynamics and performances of dynamic cloning with the threshold. All results from the set of experiments presented in this article are obtained with a population size of $d = 10$ and averaged over 1000 independent runs.

Figure 2 reports four snapshoots of Success Rate (SR) obtained by DYN-IMMALG for $\tau_b = (1, 2, 10, 50)$, tackling the simple trap function $S(II)$. From an overall view of such figure, one can see how the best performances, in term of SR, move towards low threshold values with the increase of τ_B parameter (from $SR = 2\%$ in (a) plot to $SR = 96\%$ in (d) plot). In fact, just (a) plot shows worst performances of the IA. Worst performances showed in (a) and (b) plots are obtained because for low values of τ_B : increasing the τ_B parameter value increases the success rate for low value of cloning threshold θ and high mutation rates c.

As written above, using a not correct setting of the parameters, is more easy to obtain either the explosion of the population or its extinction. However, were obtained several *special cases*, where DYN-IMMALG obtained together either success (optima solution), extinction, and failures (i.e. not able to find optima solution), or success and extinction, on 1000 independent runs. These special cases are showed in the figures 3, 4 and 5, using static and dynamic cloning operator, respectively.

Figure 3 shows the interesting case where on 1000 independent runs DYN-IMMALG was able to converge to optima solution, in some runs, and in others no, either because it reached T_{max}, or occurred the population's extinction. This feature is reported for static cloning (left plot) and dynamic cloning with the threshold θ (right plot). Left plot shows a peak in correspondence of the $8th$ generation, where the population size reach the maximum number of B cells (about $|P^{(t)}| = 1000$): around of this point is crucial for the performance of the proposed algorithm, because DYN-IMMALG either take the correct path towards global optima solution (success), or its opposite path towards local optima (failures), reaching, in this case, T_{max} fitness function evaluations (stop criterion). If DYN-IMMALG overtakes this around, then population size decreases, until to extinction ($13th$ generation). Using the threshold (right plot), DYN-IMMALG presents a steady population size, until to an unexpected increase. During this increase, the algorithm converge towards global optima solution.

In figure 4 is showed the population size of the described algorithm versus the generations; both plots present the special case where DYN-IMMALG either obtains the global optima (success) or the population size decrease until $|P^{(t)}| = 0$ (extinction). Never DYN-IMMALG have consumed all T_{max} fitness function evaluations. Left plot shows the behavior obtained not using the threshold: in the first generations, in correspondence of the peak, DYN-IMMALG reaches the optimal solution, increasing the SR value. After the $6th$ generation, i.e. after the peak, when DYN-IMMALG not follows the right path, the population size have a quickly decrease until the extinction. Even in this plot, the around of the peak represents a bound for the performances of the proposed algorithm.

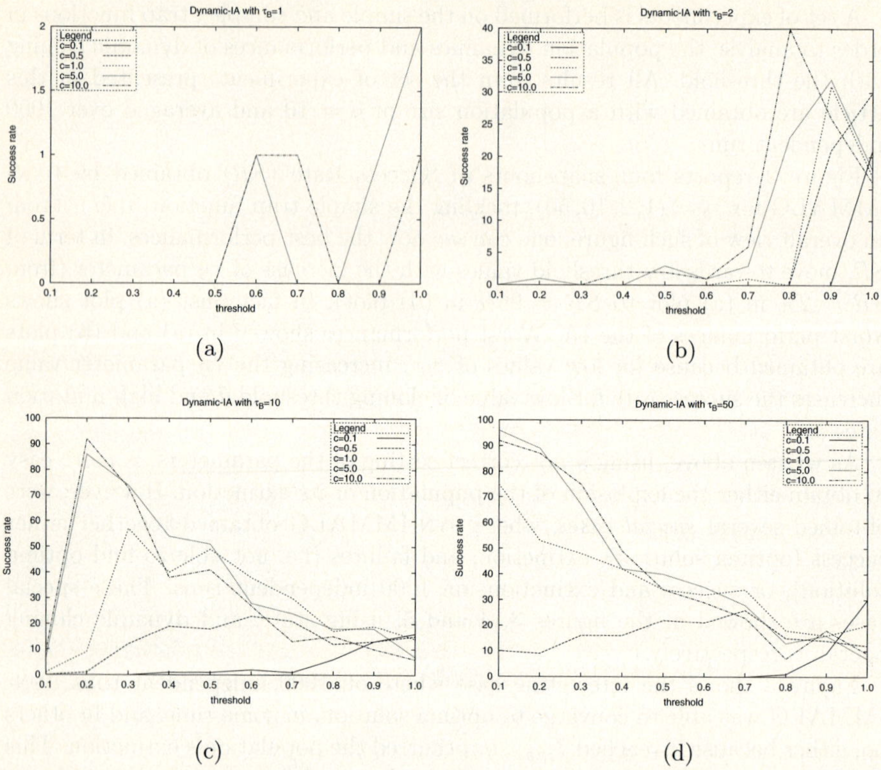

Dynamic-IA with τ_B=1

(a)

Dynamic-IA with τ_B=2

(b)

Dynamic-IA with τ_B=10

(c)

Dynamic-IA with τ_B=50

(d)

Fig. 2. Success Rate (SR) versus cloning threshold θ, and various mutation rates c for $\tau_b = 1, 2, 10, 50$. The plots concern the behaviour of DYN-IMMALG tackling the simple trap function $S(II)$.

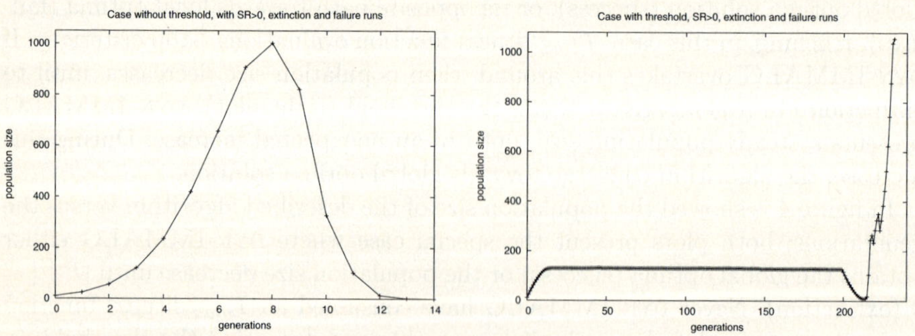

Fig. 3. Special case where DYN-IMMALG obtained together success (i.e. find optima solution), extinction and failures (i.e. is not able to reach the optima solution). These cases are showed both, without (left plot) and with (right plot) threshold θ.

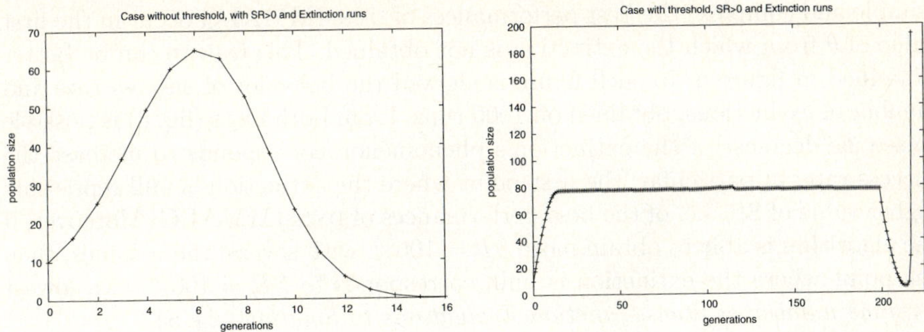

Fig. 4. Special case where DYN-IMMALG obtained both success (i.e. find optima solution) and population's extinction, on 1000 independent runs. This feature is showed using (right plot) or not (left plot) the threshold θ.

Instead, the right plot, shows the population size dynamics using the threshold θ. Like in figure 3 (right plot), the population size stays in a steady state, until the generation number is less than τ_B (in this experiment τ_B was fixed to 200). After $(\tau_B + 1)$ generations, the plot shows two different steps: in the first step, a decrease of the population size, which corresponds to extinction process; in the second step, a quickly increase, where the algorithm reaches the optimal solution.

Figure 5 shows the extinction process, which was obtained in all 1000 independent runs, with and without threshold. In particular, right plot shows a steady state of the population size, for the first τ_B generations ($\tau_B = 200$). Once the generations number is equal to τ_B, the population decreases until the extinction. This extinction phenomenon is obtained to low values of $dup = 1$ and $\theta = 1\%$ ($\theta = 0.01$). It is important to highlight, how using the threshold θ helps DYN-IMMALG to converge towards the right solution in more generations, handling better the population size, and to increase the SR value.

Analyzing deeply the obtained results, using dynamic clonal operator with threshold θ emerge an interesting feature, that is for each kind of trap functions

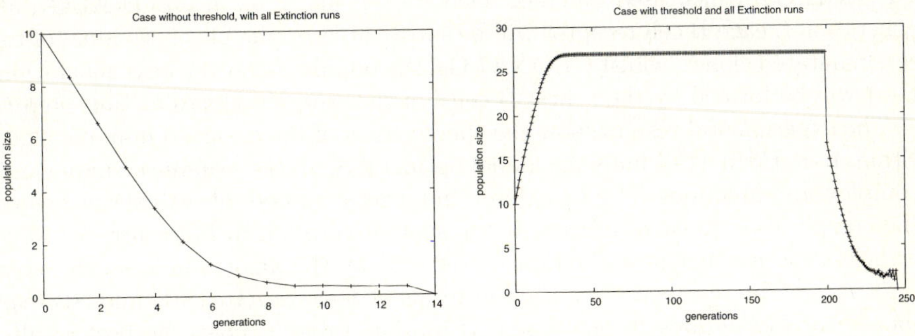

Fig. 5. Extinction. Special case where DYN-IMMALG obtained only extinction on all 1000 independent runs. This case is showed both, without (left plot) and with (right plot) threshold θ.

(simple and complex) the best performances of DYN-IMMALG occur in the first value of θ from which the extinction is not obtained. This feature can be better examined in figure 6. In such figure is showed the behavior of success rate and number of extinctions, obtained on 1000 runs. From both plots (fig. 6) is possible to see, as decrease of the extinction's phenomenon corresponds to increase the success rate: in particular, the first point where the extinction is null represents higher value of SR, i.e. of the best performances of DYN-IMMALG. Moreover, if the algorithm is able to obtain more $SR = 100\%$, with several theta values, then the point where the extinction is null, corresponds to $SR = 100\%$ with lowest *Average number of fitness function Evaluations to Solution* (AES).

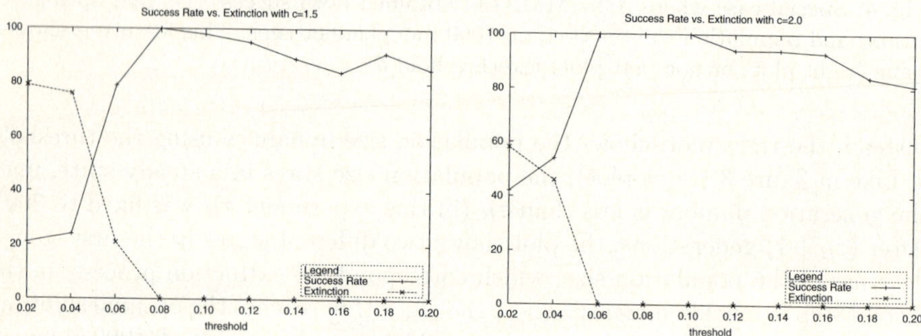

Fig. 6. Success Rate versus extinction, with $c = \{1.5, 2\}$. The plots concern the behaviour of DYN-IMMALG tackling the complex trap function $C(II)$.

3 Experimental Results

This section reports the comparisons between the proposed algorithm and others well-known immune algorithms. Were used two well-known clonal selection algorithms OPT-IA [5] and CLONALG [7]. For each instance, the experimental study was conducted using two versions of CLONALG, CLONALG_1 and CLONALG_2, which used different selection scheme [7, 5]: in CLONALG_1, at generation t, each B cell receptor will be substituted by the best individual of its set of mutated clones, whilst CLONALG_2, the population at the next generation $t+1$ will be formed by the n best B cells' of the mutated clones at time step t.

The experimental results show the effectiveness of the designed immune algorithm. DYN-IMMALG finds the global optimum of all the examined simple and complex trap functions. The T_{max} limit, in terms of number of evaluations before the computation must terminate, is the same used often in literature [5]. Nevertheless the number of evaluations performed by the algorithm is much lower than T_{max}. The results are reported in terms of *Success Rate* (SR) and *Average number of Evaluations to Solutions* (AES). The table 2 shows the best results obtained by DYN-IMMALG on each *trap function* averaged on 1000.

In the simple trap functions, although, DYN-IMMALG and OPT-IA are competitive in terms of SR, the proposed algorithm is more able to find the optima

Table 2. DYN-IMMALG versus OPT-IA, and the two versions of CLONALG

Trap	*SR*	*AES*	*SR*	*AES*	*SR*	*AES*	*SR*	*AES*	*SR*	*AES*	*SR*	*AES*
	DYN-IMMALG		OPT-IA		CLONALG$_1$				CLONALG$_2$			
					$\left(\frac{1}{\rho}\right) e^{(-f)}$		$e^{(-\rho * f)}$		$\left(\frac{1}{\rho}\right) e^{(-f)}$		$e^{(-\rho * f)}$	
S(I)	100	**70**	100	477.04	100	1100.4	100	479.7	100	725.3	100	539.2
S(II)	100	**2291.57**	100	35312.29	100	27939.2	100	174563.4	30	173679.8	31	172191.2
S(III)	100	**4871.04**	100	20045.81	0	-	0	-	0	-	0	-
S(IV)	100	**15089.97**	100	42089	0	-	0	-	0	-	0	-
S(V)	100	**29507.47**	100	80789.94	0	-	0	-	0	-	0	-
C(I)	100	**63.2**	100	388.42	100	272.5	100	251.3	100	254.0	100	218.4
C(II)	100	**894.25**	100	29271.68	100	17526.3	10	191852.7	29	173992.6	24	172434.2
C(III)	**100**	18270.55	24	149006.5	0	-	0	-	0	-	0	-
C(IV)	**72.8**	828.51	2	154925	0	-	0	-	0	-	0	-
C(V)	**65.2**	2021.76	0	-	0	-	0	-	0	-	0	-

solution with lower values of fitness function evaluations. In the complex trap functions, DYN-IMMALG takes the correct path more times than OPT-IA and CLONALG, in particular for $C(III)$, $C(IV)$ and $C(V)$ complex trap functions.

4 Conclusion

In this paper we have introduced a new immune algorithm using dynamic population size to face bimobal search spaces. Unlike most previous IAs and EAs, in which the population dimension is constant during the evolutionary process, the population size is computed adaptively according to a cloning threshold θ. The validity of the proposed algorithm was tested using a widely used test bed. Experimental results show that the IA with dynamic population size, DYN-IMMALG, overcomes premature convergence, escape many local minima and finally finds more suitable solutions than OPT-IA and CLONALG. Future work includes DYN-IMMALG application to parameter extraction of circuit design problems, numerical optimization of dynamic functions (consist of a number of peaks, changing over time in height, width and location), and study on the convergence of the proposed immune algorithm.

Acknowledgments

This work was supported by the National Research Laboratory Grant (2005-01450) from the Ministry of Science and Technology. D.L. and M.P. would like to thank CHUNG Moon Soul Center for BioInformation and BioElectronics for providing research and computing facilities.

References

1. Cutello V., Nicosia G., Pavone M., Narzisi G.; "*Real Coded Clonal Selection Algorithm for Unconstrained Global Numerical Optimization using a Hybrid Inversely Proportional Hypermutation Operator*," in 21st Annual ACM Symposium on Applied Computing (SAC), vol. 2, pp. 950–954 (2006).

2. Nijssen S., Bäck T.; "*An Analysis of the Behavior of Simplified Evolutionary Algorithms on Trap Functions*," IEEE Transaction on Evolutionary Computation, vol. 7, no. 1, pp. 11–22 (2003).
3. Cutello V., Nicosia G.; "*The Clonal Selection Principle for in silico and in vitro Computing*," in de Castro, L. N., and von Zuben, F. J. V., eds., Recent Developments in Biologically Inspired Computing, pp. 104–146, USA: Idea Group Publishing (2004).
4. Cutello V., Morelli G., Nicosia G., Pavone M.; "*Immune Algorithms with Aging Operators for the String Folding Problem and the Protein Folding Problem*," in 5th European Conference on Computation in Combinatorial Optimization (EvoCOP), pp. 80–90 (2005).
5. Cutello V., Narzisi G., Nicosia G., Pavone M.; "*Clonal selection algorithms: A comparative case study using effective mutation potentials*," in 4th International Conference on Artificial Immune Systems (ICARIS), pp. 13–28 (2005).
6. de Castro L. N., Timmis, J.; "*Artificial Immune Systems: A New Computational Intelligence Paradigm*," Berlin, Germany: Springer-Verlag (2002).
7. de Castro L. N., von Zuben F. J. V.; "*Learning and optimization using the clonal selection principle*," IEEE Transaction on Evolutionary Computation, vol. 6, no. 3, pp. 239–251 (2002).
8. Nicosia G., Cutello V., Bentley P., Timmis J.; "*Proceedings of the Third International Conference on Artificial Immune Systems*," Berlin, Germany: Springer-Verlag (2004).
9. Cutello V., Nicosia G., Pavone M.; "*Exploring the capability of immune algorithms: A characterization of hypermutation operators*," in 3rd International Conference on Artificial Immune Systems (ICARIS), pp. 263–276 (2004).
10. Cutello V., Nicosia G., Pavone, M.; "*An immune algorithm with hyper-macromutations for the 2d hydrophilic-hydrophobic model*," in Congress on Evolutionary Computing (CEC), pp. 1074–1080 (2004).
11. Nicosia G.; "*Immune Algorithms for Optimization and Protein Structure Prediction*," Ph.D. Dissertation, Department of Mathematics and Computer Science, University of Catania, Italy (2004).

Quantum-Behaved Particle Swarm Optimization with Adaptive Mutation Operator

Jing Liu, Jun Sun, and Wenbo Xu

Center of Intelligent and High Performance Computing,
School of Information Technology, Southern Yangtze University
No. 1800, Lihudadao Road, Wuxi,
214122 Jiangsu, China
{liujing_novem, sunjun_wx, xwb_sytu}@hotmail.com

Abstract. In this paper, the mutation mechanism is introduced into Quantum-behaved Particle Swarm Optimization (QPSO) to increase the diversity of the swarm and then effectively escape from local minima to increase its global search ability. Based on the characteristic of QPSO algorithm, the two variables, global best position (gbest) and mean best position (mbest), are mutated with Cauchy distribution respectively. Moreover, the amend strategy based on annealing is adopted by the scale parameter of mutation operator to increase the self-adaptive capability of the improved algorithm. The experimental results on test functions showed that QPSO with gbest and mbest mutation both performs better than PSO and QPSO without mutation.

1 Introduction

Particle swarm optimization (PSO), introduced by Kennedy and Eberhart in 1995 [1], is a population-based evolutionary optimization inspired by the collective behaviors of birds. It has already been shown that PSO is comparable in performance with other evolutionary algorithms such as Simulated Annealing (SA) and Genetic Algorithm (GA). [2][3][4]

A lot of revised versions based on basic PSO have emerged since the PSO algorithm was proposed. On one hand, the emphases mainly concentrate on combing PSO with the concepts of evolutionary computation, such as selection, recombination, breeding and so on. As articles written by Angeline in 1999[5], the operators of selection and cross in Evolutionary computation are introduced into PSO algorithm to guarantee the convergence. And in view of breeding and cross, Lovbjerg etc. further apply evolutionary mechanism into PSO algorithm to present a concrete form and the experimental results based on benchmark functions show the validity of algorithm [6]. On the other hand, to improve the global convergence, the diversity maintenance is vital to PSO algorithm. In 2004, from the point of view of probability statistics, Kennedy [7] eliminated the velocity formula and sample from the Gaussian distribution, using a random number generator and experimental results show that some remarkable degree of success can be achieved through simple collaborative probabilistic

L. Jiao et al. (Eds.): ICNC 2006, Part I, LNCS 4221, pp. 959 – 967, 2006.

search within regions defined by particles' success. And here Kennedy questioned whether we could find a simpler algorithm that performs better.

In this paper, following these previous studies, Quantum-behaved particle swarm optimization (QPSO) with adaptive mutation operator, which is originated from Evolutionary computation, is proposed to increase diversity in the latter period of the search to escape from a local minima. The proposed QPSO algorithm in 2004, kept to the philosophy of PSO, is based on Delta potential well. QPSO algorithm is depicted only with the position vector without velocity vector [8]. QPSO algorithm is simpler as well as less parameter to control than PSO algorithm. And the results show that QPSO performs better than standard PSO on several benchmark test functions and is a promising algorithm due to its global convergence guaranteed characteristic.

The rest of this paper is organized as follows. Section 2 briefly describes PSO and Quantum-behaved PSO algorithm. Then the mutation mechanism is introduced into QPSO algorithm to construct the adaptive mutation operator in Section 3. Section 4 shows the experimental settings and the comparative results of QPSO with adaptive mutation operator on test functions. Finally, conclusions are made.

2 PSO and QPSO

2.1 Dynamics of Classical PSO

In a classical PSO system proposed by Kennedy and Eberhart, each particle flies in a D-dimensional space S according to its own historical experience and others. The velocity and location for the i th particle is represented as $\vec{v}_i = (v_{i1},..., v_{id},..., v_{iD})$ and $\vec{x}_i = (x_{i1},..., x_{id},..., x_{iD})$, respectively. The particles are manipulated according to the following equation:

$$v_{id} = w \cdot v_{id} + c_1 \cdot rand() \cdot (p_{id} - x_{id}) + c_2 \cdot rand() \cdot (p_{gd} - x_{id}) \quad (1)$$

$$x_{id} = x_{id} + v_{id} \quad (2)$$

where c1 and c2 are acceleration constants, rand( ) are random values between 0 and 1. In Eq(1),the vector p_i is the best position (the position giving the best fitness value) of the particle i, vector p_g is the position of the best particle among all the particles in the population. Parameter w is the inertia weight to balance the global and local search [9], which does not appear in the original version of PSO [1]. In [10], M.Clerc and J. Kennedy analyze the trajectory and prove that, whichever model is employed in the PSO algorithm, each particle in the PSO system converges to its local point P, whose coordinates are $p_d = (\varphi_{1d} p_{id} + \varphi_{2d} p_{gd}) / (\varphi_{1d} + \varphi_{2d})$ so that the best previous position of all particles will converge to an exclusive global position with $t \to \infty$.

2.2 Dynamics of Quantum PSO

In the quantum model of a PSO, the state of a particle is depicted by wavefunction $\Psi(\vec{x},t)$, instead of position and velocity. The dynamic behavior of the particle is widely divergent from that of the particle in traditional PSO systems in that the exact values of x and v cannot be determined simultaneously. We can only learn the probability of the particle's appearing in position x from probability density function $|\psi(X,t)|^2$, the form of which depends on the potential field the particle lies in.

The particles move according to the following iterative equation[8][11]:

$$x(t+1) = p \pm \beta * |mbest - x(t)| * \ln(1/u) \tag{3}$$

where

$$mbest = \frac{1}{M}\sum_{i=1}^{M} P_i = \left(\frac{1}{M}\sum_{i=1}^{M} P_{i1},\ \frac{1}{M}\sum_{i=1}^{M} P_{i2},\ \ldots,\ \frac{1}{M}\sum_{i=1}^{M} P_{id}\right) \tag{4}$$

$$p_{id} = \varphi * p_{id} + (1-\varphi) * p_{gd}, \varphi = rand() \tag{5}$$

$mbest$ (Mean Best Position) is defined as the mean value of all particles' pbests, φ and u are a random umber distributed uniformly on [0,1] respectively; β, called Contraction-Expansion Coefficient, is the only parameter in QPSO algorithm.

3 Loss of Diversity and Mutation Operator

In QPSO algorithm, although the search space of an individual particle at each iteration is the whole feasible solution space of the problem, the loss of diversity in the population is also inevitable due to the collectiveness, as PSO and other population-based evolutionary algorithms. During the latter search period, the particles are investigated to cluster together gradually and its search area is so limited that the whole swarm is very possible to be trapped in a local minima. So the mechanism of mutating is proposed to increase its search space so as to escape local optima and increase its diversity of the swarm.

In Evolutionary Programming, each individual is mutated by adding a random function either Gaussian or Cauchy function. Through adjusting the variance of the random function or step size, evolutionary programming employed ideally has the ability to fine tune the search area around the optima. While in QPSO algorithm, some of the component of the final global best position obtained is not accordant with the theoretical value. This suggested the possibility of mutating a single component only of a solution vector. Because the global best particle *gbest* attracts all particles among the swarm, and also the intermediate variable *mbest*, mean best position among the particles, is provided with its correlation among each particle's best position. So it is possible to lead the swarm away from a current position by mutating a single individual if the mutated individual becomes the new global best position or mean best position. In this paper, to this end a mutation operator is introduced which mutates the global best position *gbest* or mean best position *mbest*.

3.1 Adaptive Mutation Operators

The mutation is performed independently on each vector element by adding a Gaussian or Cauchy distributed random value with expectation zero and standard deviation.

$$x' = x + \phi D(\cdot) \tag{6}$$

Where x' is mutated from x, $D(\cdot)$ is a random variable of a probability distribution. In this paper, Cauchy distribution is chosen because when Gaussian and Cauchy distribution have the same step sizes, Cauchy distribution are able to make larger perturbation [12]. This implies that Cauchy distribution have higher probability to escape from local optima than Gaussian mutation does. The Cauchy distribution probability definition function is $f(x) = \frac{a}{\pi * (x^2 + a^2)}$, where variable a is called the scale parameter. The scale parameter of Cauchy function is always much related with the objective function and then its ideal optimum is hard to select. However, from the Cauchy distribution probability function, we can see that bigger the scale parameter is, the more probability the algorithm have the great step sizes, which is benefit to coarse search in wide range and overcome the local minima. On the other hand, smaller the scale parameter is, the more probability the algorithm make local smaller perturbation, which is benefit to fine search in local areas. Therefore, to increase the self-adaptable capacity and performance of the algorithm, the amend strategy based on annealing is adopted by the scale parameter a of Cauchy distribution function. In this paper, the scale parameter a is according to the annealing function $T = T_0 * (CR)^k$. The annealing process of scale parameter a can adaptively balance global exploration and local fine exploitation with the generations when the particles are mutated by Cauchy probability function.

3.2 QPSO-Mutation Algorithm Pseudocode

The Quantum-Behaved PSO algorithm with the adaptive operator mutation is described as following:

```
Initialize the population
Do
   β linearly decreases with generations;
   find out the mbest of the swarm;
   mbest=Cauchy(mbest);
   for i=1 to population size M
      if f(pi)<f(xi) then pi=xi;
      pg=argmin(pi);
      update the position of particle according to
       Equation(3);
   end
Until the termination criterion is met
```

The code of QPSO algorithm with *gbest* mutation is the same as above, just Cauchy mutating gbest instead of mbest. The annealing operation of the scale parameter of Cauchy mutation function is according to $T = T_0 * (CR)^k$, where T_0=2, cooling rate *CR* is set to 0.99 and *k* is generation of evolution. The termination criterion is always set to its maximum number of generations of the objective function.

4 Experimental Setting and Results

4.1 Experimental Setting

In our QPSO algorithm with mutation, the only parameter Contraction-Expansion Coefficient β decreases linearly from 1.0 to 0.5 with the generations.

To test the performance of QPSO with *gbest* mutation and *mbest* mutation, three representative benchmark functions are used here for comparison with SPSO and QPSO. Table 1 gives the test function mathematics expression, its initialization range and the corresponding limits to the search space and its function value.

Table 1. Test function and parameter configuration

F	Mathematic expression	Initialization range	fmin	Xmax
Rosen-brock	$\sum_{i=1}^{n}(100(x_{i+1}-x_i^2)^2+(x_i-1)^2)$	(15,30)	0	100
Rastrigrin	$f(x)_3=\sum_{i=1}^{n}(x_i^2-10\cos(2\pi x_i)+10)$	(2.56, 5.12)	0	10
Griewank	$f(x)_4=\frac{1}{4000}\sum_{i=1}^{n}(x_i-100)^2-\prod_{i=1}^{n}\cos(\frac{(x_i-100)}{\sqrt{i}})+1$	(300,600)	0	600

As in [2], for each function, three different dimension sizes, 10,20 and 30 are tested. The corresponding maximum generations are 1000, 1500 and 2000 respectively. And the population size is set to 20, 40 and 80. A total of 50 runs for each experimental setting are conducted.

4.2 Experimental Results

The mean values and standard deviations for 50 runs of each test function are recorded in Table2 to Table 4. The numerical results show that the QPSO with gbest and mbest mutation operator based on annealing operation works better on Rastrigrin benchmark functions and has comparable performance on Rosenbrock and Griewank functions with QPSO and SPSO.

The convergence of QPSO algorithm and the revised algorithm are shown in Figure1 to Figure 3. As shown in figures, due to the increase of diversity and improvement of search space, the convergence of the revised algorithm is relatively slower than original QPSO algorithm during the first of search period; however, during the latter of search period the convergence of the revised algorithm is much more precise.

Table 2. Rosenbrock function

P	D	G	SPSO	QPSO	QPSO-mutation	
					gbest	mbest
			Mean (St.Dev)	Mean (St.Dev)	Mean (St.Dev)	Mean (St.Dev)
20	10	1000	94.1276 (194.3648)	59.4764 (153.0842)	21.2081 (60.0583)	15.3939 (35.1079)
	20	1500	204.336 (293.4544)	110.664 (149.5483)	61.9268 (92.9440)	67.6978 (110.2616)
	30	2000	313.734 (547.2635)	147.609 (210.3262)	86.1195 (127.6446)	76.1894 (115.3902)
40	10	1000	71.0239 (174.1108)	10.4238 (14.4799)	8.1828 (8.3604)	9.5005 (9.5505)
	20	1500	179.291 (377.4305)	46.5957 (39.5360)	40.0749 (68.4074)	55.4853 (63.3355)
	30	2000	289.593 (478.6273)	59.0291 (63.4940)	65.2891 (79.4420)	68.0551 (54.6795)
80	10	1000	37.3747 (57.4734)	8.63638 (16.6746)	7.3686 (8.4972)	6.4841 (5.8200)
	20	1500	83.6931 (137.2637)	35.8947 (36.4702)	30.1607 (33.2090)	38.3067 (38.8141)
	30	2000	202.672 (289.9728)	51.5479 (40.8490)	38.3036 (27.4658)	52.4678 (39.2240)

Table 3. Rastrigrin function

P	D	G	SPSO	QPSO	QPSO-mutation	
					gbest	mbest
			Mean (St.Dev)	Mean (St.Dev)	Mean (St.Dev)	Mean (St.Dev)
20	10	1000	5.5382 (3.0477)	5.2543 (2.8952)	4.2976 (2.5325)	4.7332 (2.6337)
	20	1500	23.1544 (10.4739)	16.2673 (5.9771)	14.1678 (4.9272)	13.6202 (5.592)
	30	2000	47.4168 (17.1595)	31.4576 (7.6882)	25.6415 (6.6575)	27.7975 (7.2808)
40	10	1000	3.5778 (2.1384)	3.5685 (2.0678)	3.2046 (3.0587)	2.8160 (1.8854)
	20	1500	16.4337 (5.4811)	11.1351 (3.6046)	9.5793 (2.8107)	9.9143 (3.2174)
	30	2000	37.2796 (14.2838)	22.9594 (7.2455)	20.5479 (5.0191)	19.8991 (4.5271)
80	10	1000	2.5646 (1.5728)	2.1245 (1.1772)	1.7166 (1.3067)	1.8923 (1.6015)
	20	1500	13.3826 (8.5137)	10.2759 (6.6244)	7.2041 (2.4822)	7.8625 (2.8771)
	30	2000	28.6293 (10.3431)	16.7768 (4.4858)	15.0393 (4.1800)	15.4082 (4.5364)

Table 4. Griewank function

P	D	G	SPSO	QPSO	QPSO-mutation	
					gbest	mbest
			Mean (St.Dev)	Mean (St.Dev)	Mean (St.Dev)	Mean (St.Dev)
20	10	1000	0.09217 (0.08330)	0.08331 (0.06805)	0.0780 (0.0612)	0.0932 (0.0862)
	20	1500	0.03002 (0.03225)	0.02033 (0.02257)	0.0235 (0.0214)	0.0193 (0.0147)
	30	2000	0.01811 (0.02477)	0.01119 (0.01462)	0.0099 (0.0107)	0.0114 (0.0146)
40	10	1000	0.08496 (0.07260)	0.06912 (0.05093)	0.0641 (0.0423)	0.0560 (0.0445)
	20	1500	0.02719 (0.02517)	0.01666 (0.01755)	0.0191 (0.0162)	0.0171 (0.0177)
	30	2000	0.01267 (0.01479)	0.01161 (0.01246)	0.0098 (0.0135)	0.0092 (0.0118)
80	10	1000	0.07484 (0.07107)	0.03508 (0.02086)	0.0460 (0.0209)	0.0554 (0.0500)
	20	1500	0.02854 (0.02680)	0.01460 (0.01279)	0.0186 (0.0164)	0.0123 (0.0131)
	30	2000	0.01258 (0.01396)	0.01136 (0.01139)	0.0069 (0.0094)	0.0111 (0.0133)

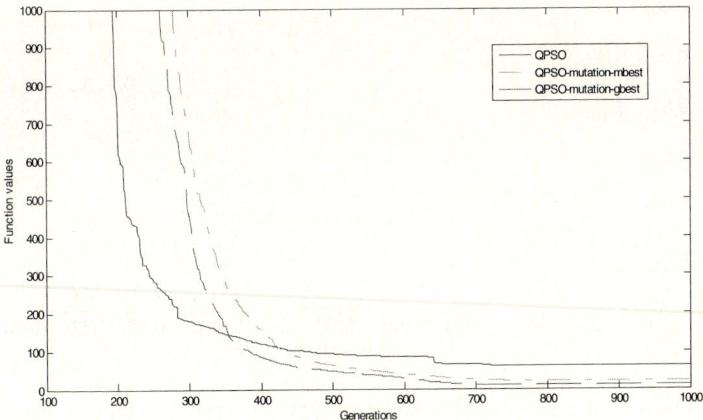

Fig. 1. Convergence of Rosenbrock Function (10 popsize, 10 dimension)

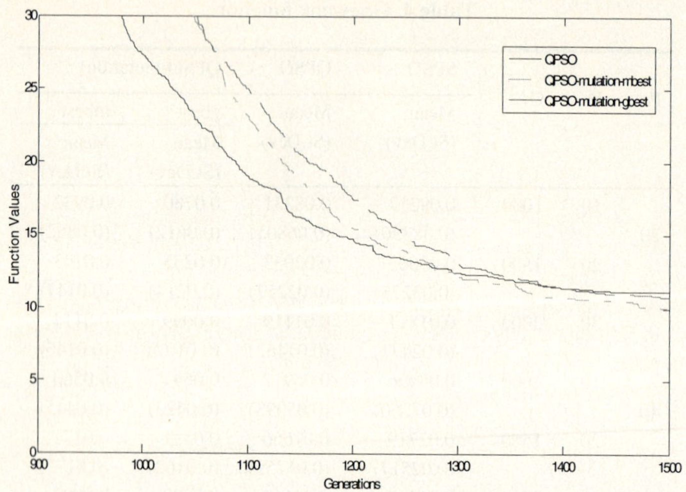

Fig. 2. Convergence of Rastrigrin Function (40 popsize, 20 dimension)

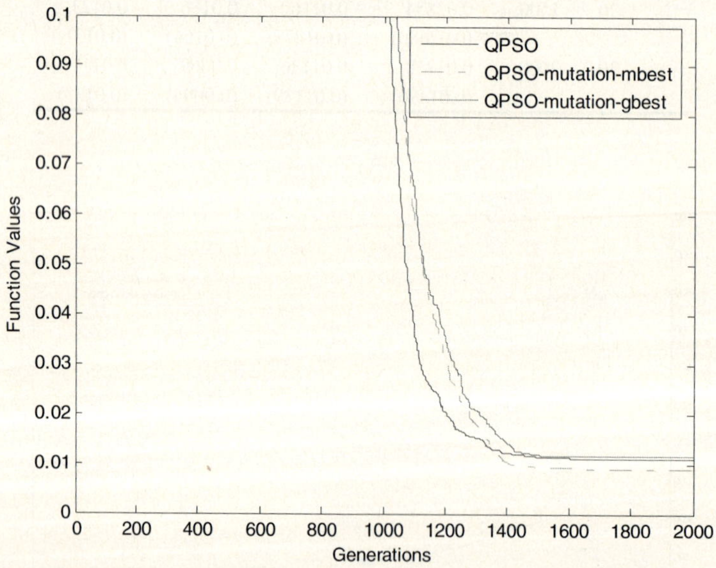

Fig. 3. Convergence of Griewank Function (80 popsize, 30 dimension)

4.3 Conclusions

In this paper, the mutation mechanism was introduced to improve the behavior of Quantum-behaved particle swarm optimization. According to the characteristic of QPSO algorithm, the parameter *gbest* and *mbest* are mutated by Cauchy distributed function respectively. And the scale parameter of Cauchy function is amended by the annealing operation to further improve the self-adaptive capability of the algorithm. Comparison with the original PSO and QPSO without mutation, the experimental results show that the adaptive mutation mechanism provides an extend improvement on global search ability of QPSO algorithm.

References

1. J. Kennedy and R. Eberhart: Particle Swarm Optimization. Proceedings of IEEE Int. Conf. on Neural Network (1995) 1942-1948
2. P. J. Angeline: Evolutionary Optimizaiton Versus Particle Swarm Opimization: Philosophyand Performance Differences. Evolutionary Programming VIII, Springer, Lecture Notes in Computer Science 1477, (1998) 601-610
3. R. C. Eberhart and Y. Shi: Comparison between Genetic Algorithm and Particle Swarm Optimization. Evolutionary Programming VII, Springer, Lecture Notes in Computer Science 1447, (1998) 611-616
4. T. Krink, J. Vesterstrom and J. Riget: Particle Swarm Optimization with Spatial Particle Extension. IEEE Proceedings of the Congress on Evolutionary Computation 2002
5. P. J. Angeline: Using Selection to Improve Particle Swarm Optimization. Proceedings of the IEEE Conference on Evolutionary Computation, ICEC (1998) 84-89
6. M.Lovbjerg, T K Rasussen and T Krink: Hybrid Particle Swarm Optimiser with Breeding and Subpopulations. Proc.of the third Genetic and Evolutionary Computation Conferences, 2001.
7. Kennedy J: Bare Bones Particle Swarms. IEEE Swarm Intelligence Symposium. (2003) 80-87
8. J. Sun, B. Feng and W. Xu: Particle Swarm Optimization with Particles Having Quantum Behavior. IEEE Proc. of Congress on Evolutionary Computation (2004) 325-331
9. Y. Shi and R. Eberhart: Empirical Study of Particle Swarm Optimization. Proc. Congress on Evolutionary Computation (1999) 1945-1950
10. M. Clerc and J. Kennedy: The Particle Swarm: Explosion, Stability and Convergence in a Multi-Dimensional Complex Space. IEEE Transaction on Evolutionary Computation (2002) 58-73
11. J. Sun *et al*: A Global Search Strategy of Quantum-behaved Particle Swarm Optimization. Proceedings of IEEE conference on Cybernetics and Intelligent Systems (2004) 111-116
12. Yao X, Liu Y: Fast Evolutionary Strategies. Proc 6th Conf Evolutionary Programming (1997) 151-161

An Improved Ordered Subsets Expectation Maximization Reconstruction

Xu Lei[1], Huafu Chen[1,2], Dezhong Yao[1], and Guanhua Luo[1]

[1] School of Life Science and Technology, University of Electronic Science and Technology of China, 610054, China

[2] School of Applied Mathematics, University of Electronic Science and Technology of China, 610054, China

ray_sure@163.com, {chenhf, dyao}@uestc.edu.cn

Abstract. Positron emission computerized tomography (PET) based on Ordered Subsets Expectation Maximization (OS-EM) is usually used to computerized tomography, which imposes the radionuclide to emission positron. But it also has some disadvantage as costing long time and bad reconstruction quality. Filtered Back Projection (FBP) that has many advantages such as simple structure and short reconstruction time is firstly introduced to the initialization stage to accelerate reconstruction. Then, the smoothness method is introduced to improve the reconstruction quality after OS-EM algorithm. The results show that this method has the advantages of fast reconstruction speed and good quality.

1 Introduction

Positron emission tomography (PET) [1] is a powerful biomedical-imaging tool because of its ability to image and quantify metabolic and physiological functions. There exist two types of reconstruction that are available on PET scanners: filtered back-projection (FBP) [2] and iterative techniques. The FBP method is good for transmission tomography systems such as computed tomography (CT), since its structure is simple and the computational time is relatively short. But as the noise cannot be suppressed effectively using the FBP method, the performance of FBP is limited. Iterative techniques are often based on the maximum likelihood expectation-maximization (ML-EM) [3], which has played an important role in emission tomography since it incorporates the Poisson nature of raw data. Ordered subsets expectation maximization (OS-EM) algorithm, due to the fact that while retaining the advantages of ML-EM such as accurate modeling of any type of system, provides an order-of-magnitude acceleration over ML-EM. But as it has the inherent instability problem of ML-EM, the noise artifact is magnified after a critical number of iterations.

In this paper, we alleviate the instability of OS-EM by applying the incorporation of suitable preprocessing. FBP is introduced to initialize the iterative dataset to avoid the slow convergence problem in the traditional OS-EM algorithm. Then the smoothness method is implemented to improve the reconstruction quality. Finally, the improved OS-EM algorithm is tested by the simulation data.

L. Jiao et al. (Eds.): ICNC 2006, Part I, LNCS 4221, pp. 968 – 971, 2006.

2 Method

2.1 Maximum Likelihood Expectation Maximization (ML-EM)

The OS-EM algorithm is one of the iterative image reconstruction methods, and it is based on the ML-EM algorithm. The theoretic basis of the ML-EM algorithm is the Poisson process of electronic emission. For image reconstruction model:

$$\mathrm{Y} = \mathrm{AX} + \mathrm{e} \tag{1}$$

Where $Y = \{y_1, y_2, \cdots, y_M\}$ is electron projection data; $X = \{x_1, x_2, \cdots, x_M\}$ is the image intensity data; $e = \{e_1, e_2, \cdots, e_M\}$ is the noise; A denotes the projection matrix with M×N. Under the Poisson assumption, ML-EM algorithm updates the value of pixel x_j at the iteration n according to the following multiplication scheme:

$$x_j^{n+1} = (x_j^n \sum_{i=1}^{M} \frac{y_i p(i \mid j)}{\sum_{j=1}^{N} x_j^n p(i \mid j)}) / (\sum_{j=1}^{N} p(i \mid j)) \quad \text{(j=1,2,...,L)} \tag{2}$$

Where x_j^n, x_j^{n+1} are value of pixel x_j after iteration n and n+1. $y_i = \sum_{j \in T} P(i \mid j) x_j$, i=1, 2, …, N. T denotes the set of pixels, from which gamma-rays finally hit detector i. $P(i \mid j)$ denotes conditional mathematical expectation of j pixel's rays hit detector i.

2.2 OS-EM and Improved OS-EM Reconstruction

Differently from ML-EM using all of the projecting data to updating image data, the OS-EM groups projection data into an ordered sequence of subsets (or blocks) [4]. Every subset updates the whole image intensity data and the image intensity data is updated k times when all the projecting data used (assume there are k subsets). In our study, FBP is used as an initiative scheme. It is acceptable because FBP spends as much time as one iteration step of OS-EM, but the reconstruction quality, compared with the OS-EM first iteration, is greatly improved.

And after the OS-EM algorithm, the smoothness method is proposed to improve the reconstruction speed and quality. Because the emitting caused by infected organs, such as fat and muscle, occupies the main part of the reconstruction image, and the distribution of them are usually smooth. Contrastively, the intensity of radionuclide contained in different organ is changed greatly, especially the focus of infection that usually expresses as singular point when compared with the normal organization. In order to recover the background information and reduce the influence on the infection focus, the smooth operator is defined as follows:

$$x = \frac{1}{w_1 + w_2 \sum_{i=1}^{4} f_i + w_3 \sum_{i=5}^{8} f_i} (w_1 x + w_2 \sum_{i=1}^{4} f_i x_i + w_3 \sum_{i=5}^{8} f_i x_i) \tag{3}$$

Where x_i (i = 1, …, 8) denotes the eight nearest point to x, w_i, i=1, 2, 3 refer to the smooth weight, which determine the influence of the 8 points to the center pixel. f_i equals 1 when $|x - x_i| < t$, and 0 otherwise, where t denotes the threshold. t is usually estimate according to the circumstance of reconstruction object.

2.3 OS-EM Reconstruction Algorithm

S_1, S_2, …, S_L is L sorted subsets. In our study, we adopt the orthogonal partition, so the number of projection data (detectors) in every subset is $J = \frac{M}{L}$. X^0 denotes the initialized image used the FBP algorithm. X^i denotes the image after iteration i. The concrete step of developed OS-EM algorithm is:

1. Set i = 0, initialize the image X= X^0 with FBP algorithm;
2. Iterate the following steps until image X^i satisfies the convergence request:
a) Set X= X^i, i = i+1;
b) For each subset: S_l, (l = 1, 2, …, L) updates image with equation (2).

At this step, we only use part of projecting data, the subset S_l, but update the whole image X in equation (2).

c) If $l \equiv$L+1, then $X^i = X^L$, repeat step a) and b), restart iteration.
3. Finally, operator (3) is introduced to improve the reconstruction quality.

3 Simulation and Result

The phantoms in 20 different slices of head are selected as test model. Figure 1 show the data flow chart of reconstruction process. Receiving the source image (A), PET simulator generates projection data Y (B). Then our algorithm initializes the image with FBP as (C) show. After 2 iteration steps, the image quality is improved (D). At final stage, use the smooth operator to make the further tune to the image (E).

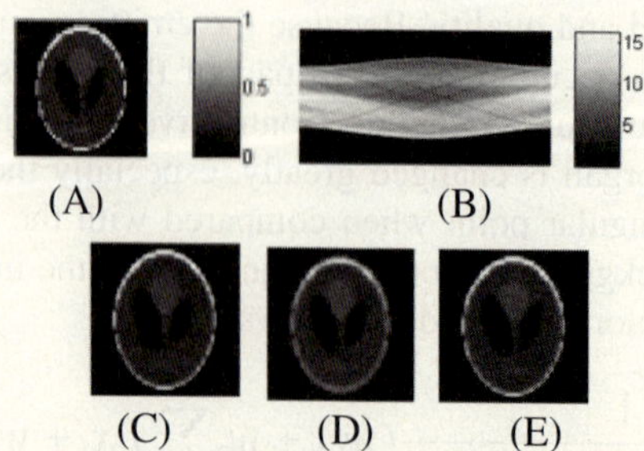

Fig. 1. Data flow chart of Improved OS-EM: (A) The source image, (B) Projecting data, (C) Initialized image (D) 2 iteration steps(E) After smooth process

In order to compare two reconstruction algorithms, average of correlation coefficient (CC) of 20 slices at the same iteration step is used to measure the similarity between the source images and reconstructed one. After 7 iteration steps, average of CCs between source images and reconstruction images of traditional OS-EM is 0.9187, and that of improved OS-EM is increased to 0.9548.

4 Conclusion

In this paper, we introduced an improved OS-EM Reconstruction to improve the reconstruction speed and quality. The simulation results show that our method has the advantages of fast reconstruction speed and good quality.

Acknowledgment

Supported by the 973 Project No. 2003CB716106, NSFC 90208003, 30570507, Fok Ying Tong Education Foundation (91041), New Century Excellent Talents in University.

References

1. Cherry S, Phelps, M.: Imaging brain function with positron emission tomography. In: Toga AW, Mazziotta JC, editors. Brain mapping: the methods, London: Academic Press, (1996) 191-221.
2. Brooks, R.A., Di Chiro, G.: Principles of computer assisted tomography (CAT) in radiographic and radioisotopic imaging. Phys. Med. Biol. (1976) 21, 689–732.
3. Vardi, Y., Shepp, L.A, Kaufman, L.: A statistical model for positron emission tomography. J. Am. Stat. Assoc. (1985) 80, 8-37.
4. Hudson, H.M., Larkin, R.S.: Accelerated image reconstruction using ordered subsets of projection data. IEEE trans. Med. Imaging. (1994) 13: 601-609.

Self-Adaptive Chaos Differential Evolution

Guo Zhenyu, Cheng Bo, Ye Min, and Cao Binggang

Research & Development Center of Electric Vehicle
Xi'an Jiaotong University, Xi'an 710049, China
zhenyu_guo@126.com

Abstract. Differential evolution (DE) is a recently invented global optimization algorithm. In this paper, a new differential evolution algorithm, self-adaptive chaos differential evolution (SACDE) with chaos mutation factor and dynamically changing weighting factor and crossover factor is presented. The evolution speed factor and aggregation degree factor of the population are introduced in this new algorithm. In each iteration process, the weighting factor is changed dynamically based on the current aggregation degree factor, and the crossover factor is changed dynamically based on the current evolution speed factor. The chaos mutation factor is introduced to avoid falling into the local optimum. DE and SACDE are tested with two well-known benchmark functions. The experiments show that the convergence speed of SACDE is significantly superior to DE, the convergence accuracy is also increased.

1 Introduction

Differential evolution (DE) is a kind of global optimization techniques, developed by Rainer Storn and Kenneth Price (1996)[1]. DE is a branch of Evolutionary Algorithms (EA). It has been proven to be a promising candidate to solve real value optimization problems. In this paper, we introduce a new version of DE, Self-adaptive Chaos Differential Evolution (SACDE), where the weighting factor is changed dynamically based on the current aggregation degree factor, and the crossover factor is changed dynamically based on the current speed factor. The chaos mutation factor is introduced to avoid falling into the local optimum.

2 New Strategies

2.1 Aggregation Degree Factor

The ratio of the best individual's fitness to the average fitness in every generation is used to represent population aggregation degree. $F(g)_{best}$ is the fitness of the best individual, and $\overline{F(g)}$ is the average fitness, C is aggregation degree factor[2].

$$C = \frac{\min(F(g)_{best}, \overline{F(g)})}{\max(F(g)_{best}, \overline{F(g)})} . \tag{1}$$

L. Jiao et al. (Eds.): ICNC 2006, Part I, LNCS 4221, pp. 972 – 975, 2006.

C is aggregation degree factor, and its range is [0, 1]. The bigger value of C represents higher aggregation degree, and smaller diversity of the population.

2.2 Evolution Speed Factor

The ratio of the best fitness of the new generation to the best fitness of the previous generation is used to represent the evolution speed factor a [2].

$$a = \frac{\min(F(g)_{best}, F(g-1)_{best})}{\max(F(g)_{best}, F(g-1)_{best})} . \tag{2}$$

$F(g-1)_{best}$ is the best fitness of the previous generation.

2.3 Chaos Mutation Factor

Introducing a sort of chaos mutation factor, this operation can avoid falling into local optimum. In this paper, the chaos mutation factor is Logistic mapping[3]:

$$p_{i,n} = 4p_{i-1,n}(1 - p_{i-1,n}) . \tag{3}$$

Randomness and space ergodicity of chaos mapping is use to increase the range of mutation, and avoid falling into local optimum.

3 Self-Adaptive Chaos Differential Evolution

Mutation operation amends individual value by difference vector. While the aggregation degree of population is high, the convergence speed declines sharply. Wherefore, the weighting factor is changed dynamically based on the current aggregation degree factor. So the weighting factor is expressed as

$$F = C \cdot F_s . \tag{4}$$

Constant F_s is constant.

High values of crossover factor give faster convergence, but fall into local optimum easily. Low values of crossover factor are robust enough for some problem, but give slower convergence. So the crossover factor changes with evolution speed, and expressed as

$$CR = a \cdot CRs . \tag{5}$$

CRs is constant.

The chaos mutation operation is implemented after selection operation, and the individual generated by chaos mutation operation replaces the worst one directly.

4 Test and Analysis of Result

In order to compare DE with SACDE, two benchmark functions, Rosenbrock and Rastrigin are chosen. The tow kinds of algorithms are compared in convergence speed, convergence precision and failure times. The maximum iteration times are 2000.

The average optimum evolution curve of calculation 50 times for the two functions shows in Fig1 and Fig2.

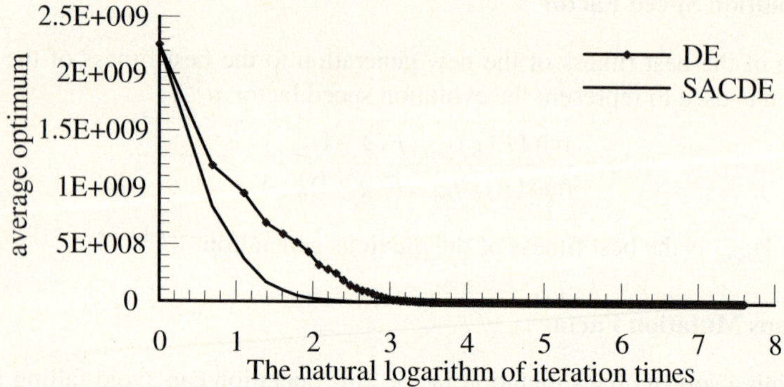

Fig. 1. Function Rosenbrock's average optimum evolution curve

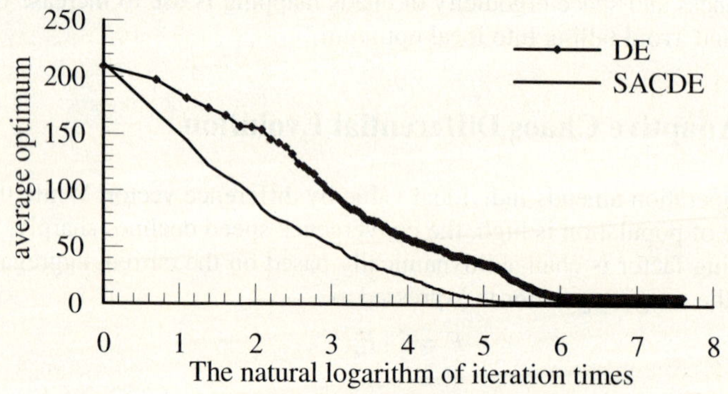

Fig. 2. Function Rastrigrin's average optimum evolution curve

The abscissa is natural logarithm of iteration times, so that the difference of DE and SACDE can be shown clearly. It is obvious that the convergence speed of SACDE is faster than DE in initial stage, but with the iteration times increasing, the difference becomes small.

The average optimums of DE and SACDE at step 2000 of calculation 50 times show in table 1. From the data, the conclusion that SACDE is more precise than DE can be drew easily.

Table 1. Average optimum at step 2000

Method	Rosenbrock	Rastrigrin
DE	0.840212694	6.964711
SACDE	0.806328037	2.686389

The convergence tolerances of the two functions, Rosenbrock and Rastrigrin, are 1 and 10. If the optimum at step 2000 is greater than convergence tolerance, then this calculation is failure. DE and SACDE have been calculated 50 times respectively, the failure times show in table 2.

Table 2. Failure times of DE and SACDE

Method	Rosenbrock	Rastrigrin
DE	13	9
SACDE	3	2

The data of table 2 show that the failure times of SACDE are much less than DE.

5 Conclusions

In this paper, a new differential evolution, self-adaptive chaos differential evolution is presented. In SACDE, the weighting factor is changed dynamically based on the current aggregation degree factor, and the crossover factor is changed dynamically based on the current evolution speed factor. The results of calculation show that this strategy can improve convergence speed and convergence precision of DE. In order to avoid falling into the trap of local minimum, the chaos mutation factor is introduced into SACDE. The failure times of SACDE and DE show that chaos mutation factor is effective.

References

1. DE Homepage. Http: //www.icsi.berkeley. edu/ ~storn/ code. Html
2. Zhang Xuanping, Du Yuping: Adaptive Particle Swarm Algorithm with Dynamically Changing Intertia Weight. Journal of Xi’an Jiaotong University, 2005,39(10),1039-1042
3. Liu Zifa: Optimal Reactive Power Dispatch Using Chaotic Particle Swarm Optimization Algorithm. Automation of Electric Power Systems, 2005, 29(7),53-57

Using the Ring Neighborhood Topology with Self-adaptive Differential Evolution

Mahamed G.H. Omran[1], Andries P Engelbrecht[2], and Ayed Salman[3]

[1] Faculty of Computing & IT, Arab Open University, Kuwait
mjomran@gmail.com
[2] Department of Computer Science, University of Pretoria, Pretoria, South Africa
engel@cs.up.ac.za
[3] Computer Engineering Department, Kuwait University, Kuwait
ayed@eng.kuniv.edu.kw

Abstract. This paper investigates the performance of Self-adaptive Differential Evolution (SDE) using a ring neighborhood topology, and compares the results with other well-known DE approaches. The experiments conducted show that using the ring topology with SDE generally improves the performance of SDE in the benchmark functions.

1 Introduction

Differential Evolution (DE) (Storn and Price 1995) has been successfully applied to solve a wide range of optimization problems. DE is now generally considered as a reliable, accurate, robust and fast optimization technique. However, the user has to find the best values for the problem-dependent control parameters used in DE. Finding the best values for the control parameters is a time consuming task. A new version of DE was proposed by Omran *et al.* (2005a; 2005c) where the control parameters are self-adaptive. The new version was called *Self-adaptive Differential Evolution* (SDE). The experiments conducted show that SDE generally outperform other DE algorithms in all the benchmark functions.

On the other hand, the effect of neighborhoods on DE has been investigated by Omran *et al.* (2005b, 2005d). The results show that the performance of DE can be improved using the ring neighborhood topology.

The purpose of this paper is to investigate the performance of SDE using the ring neighborhood topology. The results of the experiments conducted are shown and compared with other well-known approaches.

2 Self-adaptive Differential Evolution (SDE)

SDE (Omran *et al.* 2005a; Omran *et al.* 2005c) works as follows: For each parent, $\boldsymbol{x}_i(t)$, of generation t, an offspring, $\boldsymbol{x}'_i(t)$, is created by randomly selecting three

L. Jiao et al. (Eds.): ICNC 2006, Part I, LNCS 4221, pp. 976–979, 2006.

individuals from the current population, namely $\boldsymbol{x}_{i_1}(t)$, $\boldsymbol{x}_{i_2}(t)$ and $\boldsymbol{x}_{i_3}(t)$, with $i_1 \neq i_2 \neq i_3 \neq i$ and $i_1, i_2, i_3 \sim U(1,\ldots, s)$, where s is the population size. A random number, r, is then selected with $r \sim U(1,\ldots, N_d)$, where N_d is the number of genes (parameters) of a single chromosome. Then, for all parameters $j = 1,\ldots, N_d$,

$$x'_{i,j}(t) = \begin{cases} x_{i_3,j}(t) + F_i(t)(x_{i_1,j}(t) - x_{i_2,j}(t)) & \text{if } U(0,1) < N(0.5,0.15) \text{, or if } j = r \\ x_{i,j}(t) & \text{otherwise} \end{cases} \tag{1}$$

where

$$F_i(t) = F_{i_4}(t) + N(0,1) \times (F_{i_5}(t) - F_{i_6}(t)) \tag{2}$$

with $i_4 \neq i_5 \neq i_6$ and $i_4, i_5, i_6 \sim U(1,\ldots, s)$.

Thus, each individual i has its own scaling factor F_i which is a function of the scaling factor of randomly selected individuals. The parameter F_i is first initialized for each individual in the population from a normal distribution, $N(0.5,0.15)$, generating values which fits well within the range (0,1].

According to Omran *et al.* (2005a, 2005c), SDE generally outperformed other well-known versions of DE (including other adaptive versions) in all the benchmark functions.

Recently, different neighborhood topologies have been investigated with respect to DE (Omran *et al.* 2005b; Omran *et al.* 2005d). One common neighborhood topology is the *ring* (or *circle*) topology where particles are arranged in a ring. Each particle has some number of particles to its right and left as its neighborhood. According to Omran *et al.* (2005b, 2005d), using DE with the ring topology (known as DE/lbest/1) generally outperformed other well-known versions of DE in all the benchmark functions. Hence, this paper will focus on the ring topology.

2.1 Using SDE with the Ring Topology (SDE/lbest/1)

For SDE/lbest/1, equation (1) is modified as follows:

$$x'_{i,j}(t) = \begin{cases} \hat{y}_{i,j}(t) + F_i(t)(x_{i_1,j}(t) - x_{i_2,j}(t)) & \text{if } U(0,1) < N(0.5,0.15)\text{, or if } j = r \\ x_{i,j}(t) & \text{otherwise} \end{cases} \tag{3}$$

where $\hat{\boldsymbol{y}}_i(t)$ is the best position found so far in the ring neighborhood of the i-th individual (i.e. the two adjacent left and right neighbors of i).

3 Experimental Results

This section compares SDE/lbest/1 with other DE strategies proposed by Price and Storn (2006), DE/lbest/1 and SDE.

For all the algorithms used in this section (except for SDE and SDE/lbest/1), F = 0.5, P_r = 0.9 (these values were suggested by Price and Storn (2006)) and s = 50. All functions were implemented in 30 dimensions. The details of the benchmark functions can be found in Yao *et al.* (1999). Unless otherwise specified, these values were used as defaults for all experiments.

The results reported in this section are averages and standard deviations over 30 simulations. Each simulation was allowed to run for 50 000 evaluations of the objective function. These values are used unless otherwise specified.

3.1 Unimodal Functions

Table 1 summarizes the results obtained by applying the different methods to the unimodal problems. The results show that DE/lbest/1 performed better than (or at least equal to) the other strategies in all the test functions except the Rosenbrock function where DE/lbest/1 performed better than the other strategies. Furthermore, the results show that SDE/lbest/1 performed significantly better than SDE for the Rosenbrock function. Hence, using the ring topology with SDE generally improves the performance of SDE for the unimodal benchmark functions.

Table 1. Mean and standard deviation (±SD) of the uni-modal function optimization results

	SPHERE	SCHWEFEL PROBLEM 2.22	ROSENBROCK	STEP
DE/rand/1	0±0	0±0	32.28566± 19.876539	0±0
DE/best/1	2.096891± 0.829576	23.241731± 14.595706	2697508.244715± 1973397.373999	5918.766667± 2115.501527
DE/lbest/1	0±0	0±0	25.038083± 19.574514	0±0
SDE	0±0	0±0	49.914711± 28.570990	0±0
SDE/lbest/1	0±0	0±0	29.357924± 19.158455	0±0

3.2 Multimodal Functions

Table 2 summarizes the results obtained by applying the different methods to the multimodal problems. The results show that SDE performed consistently better than (or at least equal to) the other strategies in all the functions. The improvement is even more significant for the Rastrigin function. The Rastrigin function has many good local optima, which may trap optimization algorithms. It is therefore a very difficult function to optimize. Furthermore, examining the standard deviations of all the algorithms, SDE achieved the smallest standard deviation, illustrating that SDE is more stable and thus more robust than the other versions of DE. However, it is important to note that although SDE/lbest/1 has not improved on SDE, there is also no significant difference in their results. In other words, even for Rastrigin where SDE/lbest/1's average is worse than SDE, it is not significantly worse.

Table 2. Mean and standard deviation (±sd) of the multi-modal function optimization results

	SCHWEFEL PROBLEM 2.26	RASTRIGIN	ACKLEY	GRIEWANK
DE/rand/1	-10945.633647± 1549.624715	149.262978± 35.487699	-0.000002 ±0	0.004839± 0.008546
DE/best/1	-6910.956747± 711.555594	108.462203± 23.936989	13.704112± 1.443857	44.928188± 14.849933
DE/lbest/1	-18679.249346± 4020.662879	117.288499± 48.678269	-0.000002 ±0	0.001972± 0.004045
SDE	-13330.430005± 399.429446	5.079414± 2.305105	-0.000002 ±0	0±0
SDE/lbest/1	-14976.484062± 1209.077994	8.176275± 3.165998	-0.000002 ±0	0±0

References

1. M. Omran, A. Engelbrecht and A. Salman. Empirical Analysis of Self-adaptive Differential Evolution. Submitted, 2005a.
2. M. Omran, A. Engelbrecht and A. Salman. Empirical Analysis of Using Neighborhood Topologies with Differential Evolution. Submitted, 2005b.
3. M. Omran, A. Salman and A. Engelbrecht. Self-Adaptive Differential Evolution. In *the 2005 International Conference on Computational Intelligence and Security*, December, Xi'an, China, Springer's Lecture Notes in Artificial Intelligence, vol. 3801, pp. 192-199, 2005c.
4. M. Omran, A. Engelbrecht and A. Salman. Using Neighborhood Topologies with Differential Evolution. To appear in the *Workshop Proceedings of the 2005 International Conference on Computational Intelligence and Security*, December, Xi'an, China, 2005d.
5. K. Price and R. Storn. DE Web site, http://www.ICSI.Berkeley.edu/~storn/code.html (visited 2 June 2006), 2006.
6. R. Storn and K. Price. Differential Evolution – A Simple and Efficient Adaptive Scheme for Global Optimization over Continuous Spaces. Technical Report TR-95-012, International Computer Science Institute, Berkeley, CA, 1995.
7. X. Yao, Y. Liu and G. Lin. Evolutionary Programming Made Faster. IEEE Transactions on Evolutionary Computation, vol. 3, no. 2, pp. 82-102, 1999.

Liquid State Machine by Spatially Coupled Oscillators

Andreas Herzog[1], Karsten Kube[1], Bernd Michaelis[1],
Ana D. de Lima[2], and Thomas Voigt[2]

[1] Institute of Electronics, Signal Processing and Communications,
[2] Institute of Physiology, Otto-von-Guericke University Magdeburg,
P.O. Box 4120, 39114 Magdeburg, Germany
andreas.herzog@et.uni-magdeburg.de

Abstract. Liquid State Machines [1] are a new strategie for real-time information processing in recurrent networks. In the present work we show that spatially coupled oscillators can be used as a usable liquid. If inputs stream are synchronized to oscillator phase its temporal dynamics can be be transformed into a high dimensional spatial pattern of oscillator activity. A memory less readout function can extract information about recent inputs. The fading memory is considered as the resynchronisation of oscillator field and can be adjusted by the parameter of small world connection mechanisms.

1 Introduction

Liquid State Machines (LSM) are developed by Natschlaeger et al [1] and Maass et al [2] as a new conceptual framework for real-time neural computation. The information processing in the liquid based on statistical learning theory. The liquid transform a temporal input stream into a high dimensional pattern of internal states. Useful informations about recent inputs are coded in the actual internal state. The liquid itself needs no learning. To decode the information only a linear readout function is need. The liquid assumes the computational role of a kernel for support vector machines automatically [2]. A sufficed liquid is define as an expectable, spatially distributed system, which can receive and integrate inputs at anytime and have a fade out memory. Fade out memory means that input information are lost over time. It is shown that generic neural microcircuits can work as a liquid [2]. The balance of integration input, fade out memory and spatial coupling inside the liquid is a problem which limits practically approaches. The stable state without any input over long time is often define as silence or as only spontaneously uncorrelated activity. But in biological neuronal networks different oscillations occur without external stimulation. Already during early neocortical development spontaneous large-scale wave-like activity typical observed in cell cultures [3,4]. The source of synchronized bursts changes on every episode and distant neurons burst after a certain time delay [5]. This behavior is coherent with predominantly local connected network architectures like the 'Small world' networks [6] which contain a majority of local connections and a few long distance connections. In [7] we investigated conditions and parameters for the emergence of oscillatory network activity in intrinsically driven networks.

Tabak et al. [8] described a global model of synchronized episodic events (bursts) in random connected networks driven by spontaneous activity. Based on this work, we

L. Jiao et al. (Eds.): ICNC 2006, Part I, LNCS 4221, pp. 980–983, 2006.

expand the model to a spatial distribution and demonstrate, that such oscillator field can generate a rich dynamic behavior, depend of the connection properties [9]. In this paper we will show, that it can transform external input signals in patterns of hight dimensional internal states and in this way the field works as a liquid of coupled oscillators.

2 Methods

In order to get a spatial distribution of activity we expand the population model (s-model) of Tabak et al. [8] to a two dimensional distribution of activity. The three differential equations:

$$\tau_a \dot{a}(x,y) = a_\infty (I_{eff}) - a(x,y), \tag{1}$$

$$\tau_d \dot{d}(x,y) = d_\infty (a(x,y)) - d(x,y), \tag{2}$$

$$\tau_s \dot{s}(x,y) = s_\infty (a(x,y)) - s(x,y). \tag{3}$$

describe the oscillators on discrete points (x, y). Where a is the population activity, d the fraction of synapses, which is not effected by the fast synaptic depression, s the fraction of synapses which are not effected by the slow synaptic depression and the sigmoidal functions: $a_\infty (I_{eff}) = 1/\left(1 + e^{-(I_{eff}-\theta_d)/k_a}\right)$, $d_\infty (a) = 1/\left(1 + e^{(a-\theta_d)/k_d}\right)$, $s_\infty (a) = 1/\left(1 + e^{(a-\theta_s)/k_s}\right)$.

An external source I_{ext} is added to the effective input current $I_{eff} = s \cdot d \cdot a + I_{ext}$ of each oscillator. All parameters (thresholds: $\theta_a = 0.18$, $\theta_d = 0.5$, $\theta_s = 0.14$, time constants: $\tau_a = 1.0$, $\tau_d = 0.5$,$\tau_s = 500.0$, and steepness $k_a = 0.05$, $k_d = 0.2$, $k_s = 0.02$) are taken from original paper [8]. Time constants are defined as dimensionless time units. We use time steps of 0.1 time units for calculation (less than $1/10$ of smallest time constant, Runge-Kutter fourth order integration method). To fit physiological data a time unit is approximately $50ms$ ($10s$ period see [4]).

The external input current is the weighted (w_i) sum of i connected oscillators (spatial coupling) and additional external sources I_{in} (inputs):

$$I_{ext}(x,y) = \sum_i (w_i \cdot a_i \cdot d_i \cdot s_i) + I_{in}. \tag{4}$$

In order to simulate a more realistic network behavior we connect spatial distributed oscillators according to strategies used in single cell simulations. As coupling function we use an two dimensional adaptation of the 'small world' topology by Watts and Strogatz [6]. This topology allows a sliding transition from locally coupled networks to purely randomly coupled networks by adjusting only one free parameter ρ, which is the relation of long range connections to local connections [6,10].

3 Oscillator Field as Liquid

An external input signal is generated by locally applying an external current I_{in} in a region of interest (roi, see Fig. 1 left). Each roi overlap several oscillators. So input signals spread over local and global connections into the liquid. The input current I_{in}

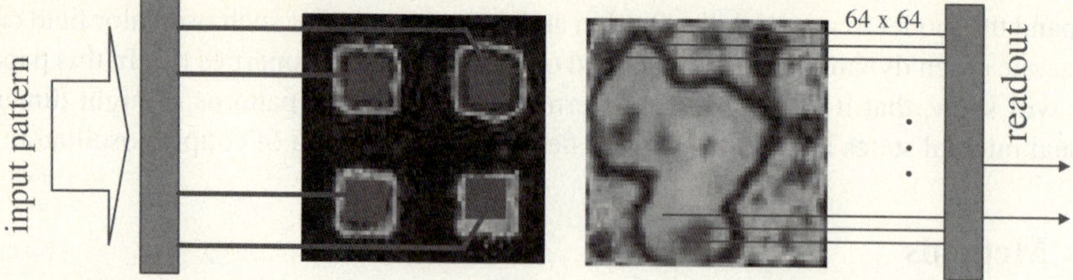

Fig. 1. left: External input to four regions of interests. **right:** Readout by a linear separator.

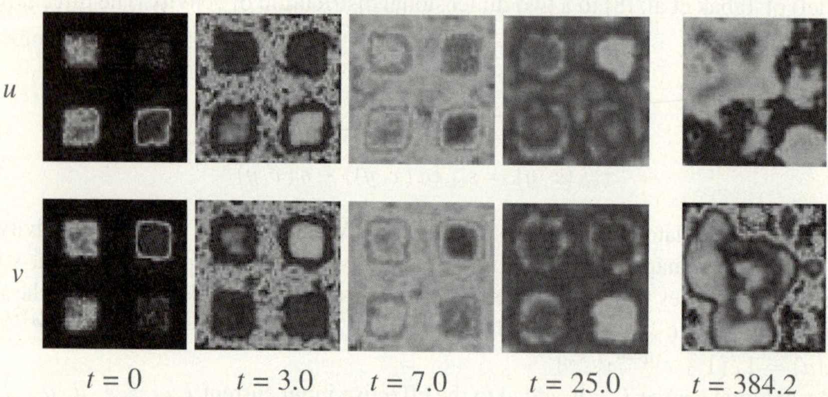

Fig. 2. Activity of the system after stimulation **above:** stream $u(t)$ clockwise **below:** stream $v(t)$ anti-clockwise. t is in dimensionless time units. Every time unit has 10 time steps for calculation.

can varied in position (roi), value and time. In our experiment we use a field of 64x64 oscillators, each locally coupled by Gaussian kernel ($\gamma = 0.7$) and 5 long connections. The internal state at time t of the liquid is the activity $a(x, y)$. The readout function is approximate by a single neuron with 4096 inputs. To test the computational power we train the readout function to distinguish the two different input streams $u(t)$ and $v(t)$. $u(t)$ stimulate the four roi clockwise and $v(t)$ stimulate it anti-clockwise ($length = 50$, $I_{in} = 0.01$). Stimulations occur only in a time window outside the refraction period in second half of silent interval (100 time units after last burst). The different inputs can easily distinguish by significant different states after the stimulation (see Fig. 2). Because the stimulation is mirrored the patterns of activity is also mirrored. Over several periods of oscillations the signals of input streams is transformed into a complex pattern of activity (see Fig. 2 most right). The mirrored activity can be see here too. The (fast) oscillator pase is shifted because noise but the linear classificator is able to separate the patterns stable. Fading out time can adjust by connections weights, part of long distance connections and the level of additional noise.

4 Discussion and Conclusion

We show that spatial coupled oscillators have the capacity to be used as a liquid for real time computation. The used oscillator model from Tabak et. all [8] is primarily designed as a population model for wave-like activity during early development of the neocortex. It can produce oscillations in two different time scale. The fast oscillation of the $a - d$ system during a burst and the slow oscillation of the s system for start and end the bursts. In a small world connected field of such oscillators there are rich temporal interactions on different time scales. The system can serve as source of information about present and past stimuli by a simple linear readout mechanism. A fading memory effect is realize as a resynchronisation of all oscillators after a period without external stimuli. Because the oscillators have a refraction period and input effects depend strongly on time in oscillator phase the real-time computation behaviour is limited. An idea to solve this problem is the using of an input gate with a variable delay line to compress and synchronize input stream to oscillator phase. In this way we can also analyse temporal patterns longer than oscillation period. It is possible that the readout function of an oscillator liquid needs more training patterns as a liquid of asynchronously spiking neurons. But it seems it is easier to set-up parameters and get a stable system.

Supported by the federal state of Saxony-Anhalt FKZ XN3590C/0305M.

References

1. Th. Natschläger, W. Maass, and H. Markram. The liquid computer: A nover strategy for real-time computing on time series. *Special Issue on Foundations of Information Processing of TELEMATIK*, pages 32–36, 2002.
2. W. Maass and H. Markram. On the computational power of recurrent circuits of spiking neurons. *J of Computer and System Sciences*, 69(4):593–616, 2004.
3. Th. Voigt, T. Opitz, and A. D. de Lima. Synchronous oscillatory activity in immature cortical network is driven by gabaergic preplate neurons. *J Neurosc*, 21(22):895–905, Nov. 15 2001.
4. T. Opitz, A. D. De Lima, and Th. Voigt. Spontaneous development of synchronous oscillatory activity during maturation of cortical networks in vitro. *J Neurophys.*, 88:2196–2206, 2002.
5. E. Maeda, H. Robinson, and A. Kawana. The mechanisms of generation and propagation of synchronized bursting in developing networks of cortical neurons. *J Neurosc*, 10(15):6834–6845, 1995.
6. D. Watts and S. Strogatz. Collective dynamics of 'small-world' networks. *Nature*, 393:440–442, 1998.
7. K. Kube, A. Herzog, V. Spravedlyvyy, B. Michaelis, T. Opitz, A. de Lima, and T. Voigt. Modelling of biologically plausible excitatory networks: Emergence and modulation of neural synchrony. In *ESANN 2004*, pages 379–384, 2004.
8. J. Tabak, W. Senn, M. J. ODonovan, and J. Rinzel. Modeling of spontaneous activity in developing spinal cord using activity-dependent depression in an excitatory network. *J Neurosc*, pages 3041–3056, 2000.
9. A. Herzog, K. Kube, B. Michaelis, AD. de Lima, and T. Voigt. Simulation of young neocortical networks by sparcely couplet oscillators. In *IJCNN 2006*, in press 2006.
10. A. Herzog, K. Kube, B. Michaelis, AD. de Lima, and T. Voigt. Connection strategies in neocortical networks. In *Proceedings of the ESANN'2006 European Symposium on Artificial Neural Networks*, pages 215–220, 2006.

Author Index

Lecture Notes in Computer Science

For information about Vols. 1–4133

please contact your bookseller or Springer

Vol. 4238: Y.-T. Kim, M. Takano (Eds.), Management of Convergence Networks and Services. XVIII, 604 pages. 2006.

Vol. 4228: D.E. Lightfoot, C.A. Szyperski (Eds.), Modular Programming Languages. X, 415 pages. 2006.

Vol. 4227: W. Nejdl, K. Tochtermann (Eds.), Innovative Approaches for Learning and Knowledge Sharing. XVII, 721 pages. 2006.

Vol. 4224: E. Corchado, H. Yin, V. Botti, C. Fyfe (Eds.), Intelligent Data Engineering and Automated Learning – IDEAL 2006. XXVII, 1447 pages. 2006.

Vol. 4222: L. Jiao, L. Wang, X. Gao, J. Liu, F. Wu (Eds.), Advances in Natural Computation, Part II. XLII, 998 pages. 2006.

Vol. 4221: L. Jiao, L. Wang, X. Gao, J. Liu, F. Wu (Eds.), Advances In Natural Computation, Part I. XLI, 992 pages. 2006.

Vol. 4219: D. Zamboni, C. Kruegel (Eds.), Recent Advances in Intrusion Detection. XII, 331 pages. 2006.

Vol. 4217: P. Cuenca, L. Orozco-Barbosa (Eds.), Personal Wireless Communications. XV, 532 pages. 2006.

Vol. 4216: M.R. Berthold, R. Glen, I. Fischer (Eds.), Computational Life Sciences. XIII, 269 pages. 2006. (Sublibrary LNBI).

Vol. 4213: J. Fürnkranz, T. Scheffer, M. Spiliopoulou (Eds.), Knowledge Discovery in Databases: PKDD 2006. XXII, 660 pages. 2006. (Sublibrary LNAI).

Vol. 4212: J. Fürnkranz, T. Scheffer, M. Spiliopoulou (Eds.), Machine Learning: ECML 2006. XXIII, 851 pages. 2006. (Sublibrary LNAI).

Vol. 4211: P. Vogt, Y. Sugita, E. Tuci, C. Nehaniv (Eds.), Symbol Grounding and Beyond. VIII, 237 pages. 2006. (Sublibrary LNAI).

Vol. 4209: F. Crestani, P. Ferragina, M. Sanderson (Eds.), String Processing and Information Retrieval. XIV, 367 pages. 2006.

Vol. 4208: M. Gerndt, D. Kranzlmüller (Eds.), High Performance Computing and Communications. XXII, 938 pages. 2006.

Vol. 4207: Z. Ésik (Ed.), Computer Science Logic. XII, 627 pages. 2006.

Vol. 4206: P. Dourish, A. Friday (Eds.), UbiComp 2006: Ubiquitous Computing. XIX, 526 pages. 2006.

Vol. 4205: G. Bourque, N. El-Mabrouk (Eds.), Comparative Genomics. X, 231 pages. 2006. (Sublibrary LNBI).

Vol. 4203: F. Esposito, Z.W. Ras, D. Malerba, G. Semeraro (Eds.), Foundations of Intelligent Systems. XVIII, 767 pages. 2006. (Sublibrary LNAI).

Vol. 4202: E. Asarin, P. Bouyer (Eds.), Formal Modeling and Analysis of Timed Systems. XI, 369 pages. 2006.

Vol. 4201: Y. Sakakibara, S. Kobayashi, K. Sato, T. Nishino, E. Tomita (Eds.), Grammatical Inference: Algorithms and Applications. XII, 359 pages. 2006. (Sublibrary LNAI).

Vol. 4199: O. Nierstrasz, J. Whittle, D. Harel, G. Reggio (Eds.), Model Driven Engineering Languages and Systems. XVI, 798 pages. 2006. (Sublibrary LNBI).

Vol. 4197: M. Raubal, H.J. Miller, A.U. Frank, M.F. Goodchild (Eds.), Geographic, Information Science. XIII, 419 pages. 2006.

Vol. 4196: K. Fischer, I.J. Timm, E. André, N. Zhong (Eds.), Multiagent System Technologies. X, 185 pages. 2006. (Sublibrary LNAI).

Vol. 4195: D. Gaiti, G. Pujolle, E. Al-Shaer, K. Calvert, S. Dobson, G. Leduc, O. Martikainen (Eds.), Autonomic Networking. IX, 316 pages. 2006.

Vol. 4194: V.G. Ganzha, E.W. Mayr, E.V. Vorozhtsov (Eds.), Computer Algebra in Scientific Computing. XI, 313 pages. 2006.

Vol. 4193: T.P. Runarsson, H.-G. Beyer, E. Burke, J.J. Merelo-Guervós, L. D. Whitley, X. Yao (Eds.), Parallel Problem Solving from Nature - PPSN IX. XIX, 1061 pages. 2006.

Vol. 4192: B. Mohr, J.L. Träff, J. Worringen, J. Dongarra (Eds.), Recent Advances in Parallel Virtual Machine and Message Passing Interface. XVI, 414 pages. 2006.

Vol. 4191: R. Larsen, M. Nielsen, J. Sporring (Eds.), Medical Image Computing and Computer-Assisted Intervention – MICCAI 2006, Part II. XXXVIII, 981 pages. 2006.

Vol. 4190: R. Larsen, M. Nielsen, J. Sporring (Eds.), Medical Image Computing and Computer-Assisted Intervention – MICCAI 2006, Part I. XXXVVIII, 949 pages. 2006.

Vol. 4189: D. Gollmann, J. Meier, A. Sabelfeld (Eds.), Computer Security – ESORICS 2006. XI, 548 pages. 2006.

Vol. 4188: P. Sojka, I. Kopeček, K. Pala (Eds.), Text, Speech and Dialogue. XIV, 721 pages. 2006. (Sublibrary LNAI).

Vol. 4187: J.J. Alferes, J. Bailey, W. May, U. Schwertel (Eds.), Principles and Practice of Semantic Web Reasoning. XI, 277 pages. 2006.

Vol. 4186: C. Jesshope, C. Egan (Eds.), Advances in Computer Systems Architecture. XIV, 605 pages. 2006.

Vol. 4185: R. Mizoguchi, Z. Shi, F. Giunchiglia (Eds.), The Semantic Web – ASWC 2006. XX, 778 pages. 2006.

Vol. 4184: M. Bravetti, M. Núñez, G. Zavattaro (Eds.), Web Services and Formal Methods. X, 289 pages. 2006.

Vol. 4183: J. Euzenat, J. Domingue (Eds.), Artificial Intelligence: Methodology, Systems, and Applications. XIII, 291 pages. 2006. (Sublibrary LNAI).

Vol. 4182: H.T. Ng, M.-K. Leong, M.-Y. Kan, D. Ji (Eds.), Information Retrieval Technology. XVI, 684 pages. 2006.

Vol. 4180: M. Kohlhase, OMDoc – An Open Markup Format for Mathematical Documents [version 1.2]. XIX, 428 pages. 2006. (Sublibrary LNAI).

Vol. 4179: J. Blanc-Talon, W. Philips, D. Popescu, P. Scheunders (Eds.), Advanced Concepts for Intelligent Vision Systems. XXIV, 1224 pages. 2006.

Vol. 4178: A. Corradini, H. Ehrig, U. Montanari, L. Ribeiro, G. Rozenberg (Eds.), Graph Transformations. XII, 473 pages. 2006.

Vol. 4176: S.K. Katsikas, J. Lopez, M. Backes, S. Gritzalis, B. Preneel (Eds.), Information Security. XIV, 548 pages. 2006.

Vol. 4175: P. Bücher, B.M.E. Moret (Eds.), Algorithms in Bioinformatics. XII, 402 pages. 2006. (Sublibrary LNBI).

Vol. 4174: K. Franke, K.-R. Müller, B. Nickolay, R. Schäfer (Eds.), Pattern Recognition. XX, 773 pages. 2006.

Vol. 4173: S. El Yacoubi, B. Chopard, S. Bandini (Eds.), Cellular Automata. XV, 734 pages. 2006.

Vol. 4172: J. Gonzalo, C. Thanos, M. F. Verdejo, R.C. Carrasco (Eds.), Research and Advanced Technology for Digital Libraries. XVII, 569 pages. 2006.

Vol. 4169: H.L. Bodlaender, M.A. Langston (Eds.), Parameterized and Exact Computation. XI, 279 pages. 2006.

Vol. 4168: Y. Azar, T. Erlebach (Eds.), Algorithms – ESA 2006. XVIII, 843 pages. 2006.

Vol. 4167: S. Dolev (Ed.), Distributed Computing. XV, 576 pages. 2006.

Vol. 4166: J. Górski (Ed.), Computer Safety, Reliability, and Security. XIV, 440 pages. 2006.

Vol. 4165: W. Jonker, M. Petković (Eds.), Secure, Data Management. X, 185 pages. 2006.

Vol. 4163: H. Bersini, J. Carneiro (Eds.), Artificial Immune Systems. XII, 460 pages. 2006.

Vol. 4162: R. Královič, P. Urzyczyn (Eds.), Mathematical Foundations of Computer Science 2006. XV, 814 pages. 2006.

Vol. 4161: R. Harper, M. Rauterberg, M. Combetto (Eds.), Entertainment Computing - ICEC 2006. XXVII, 417 pages. 2006.

Vol. 4160: M. Fisher, W.v.d. Hoek, B. Konev, A. Lisitsa (Eds.), Logics in Artificial Intelligence. XII, 516 pages. 2006. (Sublibrary LNAI).

Vol. 4159: J. Ma, H. Jin, L.T. Yang, J.J.-P. Tsai (Eds.), Ubiquitous Intelligence and Computing. XXII, 1190 pages. 2006.

Vol. 4158: L.T. Yang, H. Jin, J. Ma, T. Ungerer (Eds.), Autonomic and Trusted Computing. XIV, 613 pages. 2006.

Vol. 4156: S. Amer-Yahia, Z. Bellahsène, E. Hunt, R. Unland, J.X. Yu (Eds.), Database and XML Technologies. IX, 123 pages. 2006.

Vol. 4155: O. Stock, M. Schaerf (Eds.), Reasoning, Action and Interaction in AI Theories and Systems. XVIII, 343 pages. 2006. (Sublibrary LNAI).

Vol. 4154: Y.A. Dimitriadis, I. Zigurs, E. Gómez-Sánchez (Eds.), Groupware: Design, Implementation, and Use. XIV, 438 pages. 2006.

Vol. 4153: N. Zheng, X. Jiang, X. Lan (Eds.), Advances in Machine Vision, Image Processing, and Pattern Analysis. XIII, 506 pages. 2006.

Vol. 4152: Y. Manolopoulos, J. Pokorný, T. Sellis (Eds.), Advances in Databases and Information Systems. XV, 448 pages. 2006.

Vol. 4151: A. Iglesias, N. Takayama (Eds.), Mathematical Software - ICMS 2006. XVII, 452 pages. 2006.

Vol. 4150: M. Dorigo, L.M. Gambardella, M. Birattari, A. Martinoli, R. Poli, T. Stützle (Eds.), Ant Colony Optimization and Swarm Intelligence. XVI, 526 pages. 2006.

Vol. 4149: M. Klusch, M. Rovatsos, T.R. Payne (Eds.), Cooperative Information Agents X. XII, 477 pages. 2006. (Sublibrary LNAI).

Vol. 4148: J. Vounckx, N. Azemard, P. Maurine (Eds.), Integrated Circuit and System Design. XVI, 677 pages. 2006.

Vol. 4147: M. Broy, I.H. Krüger, M. Meisinger (Eds.), Automotive Software – Connected Services in Mobile Networks. XIV, 155 pages. 2006.

Vol. 4146: J.C. Rajapakse, L. Wong, R. Acharya (Eds.), Pattern Recognition in Bioinformatics. XIV, 186 pages. 2006. (Sublibrary LNBI).

Vol. 4144: T. Ball, R.B. Jones (Eds.), Computer Aided Verification. XV, 564 pages. 2006.

Vol. 4143: R. Lämmel, J. Saraiva, J. Visser (Eds.), Generative and Transformational Techniques in Software Engineering. X, 471 pages. 2006.

Vol. 4142: A. Campilho, M. Kamel (Eds.), Image Analysis and Recognition, Part II. XXVII, 923 pages. 2006.

Vol. 4141: A. Campilho, M. Kamel (Eds.), Image Analysis and Recognition, Part I. XXVIII, 939 pages. 2006.

Vol. 4139: T. Salakoski, F. Ginter, S. Pyysalo, T. Pahikkala, Advances in Natural Language Processing. XVI, 771 pages. 2006. (Sublibrary LNAI).

Vol. 4138: X. Cheng, W. Li, T. Znati (Eds.), Wireless Algorithms, Systems, and Applications. XVI, 709 pages. 2006.

Vol. 4137: C. Baier, H. Hermanns (Eds.), CONCUR 2006 – Concurrency Theory. XIII, 525 pages. 2006.

Vol. 4136: R.A. Schmidt (Ed.), Relations and Kleene Algebra in Computer Science. XI, 433 pages. 2006.

Vol. 4135: C.S. Calude, M.J. Dinneen, G. Păun, G. Rozenberg, S. Stepney (Eds.), Unconventional Computation. X, 267 pages. 2006.

Vol. 4134: K. Yi (Ed.), Static Analysis. XIII, 443 pages. 2006.